# 2013 IEEE Custom Integrated Circuits Conference

# (CICC 2013)

San Jose, California, USA
22-25 September 2013

**Pages 1-480**

| IEEE Catalog Number: | CFP13CIC-POD |
|---|---|
| ISBN: | 978-1-4673-6145-3 |

**Copyright © 2013 by the Institute of Electrical and Electronic Engineers, Inc
All Rights Reserved**

*Copyright and Reprint Permissions*: Abstracting is permitted with credit to the source. Libraries are permitted to photocopy beyond the limit of U.S. copyright law for private use of patrons those articles in this volume that carry a code at the bottom of the first page, provided the per-copy fee indicated in the code is paid through Copyright Clearance Center, 222 Rosewood Drive, Danvers, MA 01923.

For other copying, reprint or republication permission, write to IEEE Copyrights Manager, IEEE Service Center, 445 Hoes Lane, Piscataway, NJ 08854. All rights reserved.

***This publication is a representation of what appears in the IEEE Digital Libraries. Some format issues inherent in the e-media version may also appear in this print version.**

| | |
|---|---|
| IEEE Catalog Number: | CFP13CIC-POD |
| ISBN 13: | 978-1-4673-6145-3 |
| ISSN: | 0886-5930 |

**Additional Copies of This Publication Are Available From:**

Curran Associates, Inc
57 Morehouse Lane
Red Hook, NY 12571 USA
Phone:      (845) 758-0400
Fax:         (845) 758-2633
E-mail:     curran@proceedings.com
Web:       www.proceedings.com

# 2013 IEEE Custom Integrated Circuits Conference (CICC 2013)

San Jose, California, USA
22-25 September 2013

**IEEE Catalog Number:** CFP13CIC-POD
**ISBN:** 978-1-46736-145-3

# Table of Contents

## Technical Sessions
*Monday, September 23 – Wednesday, September 25*

## Educational Sessions
*Tuesday, September 24 and Wednesday, September 25*

## Session 1 – Plenary Session

Monday, 9/23/2013, 8:15 am
Oak Ballroom
8:15 am
Welcome and Opening Remarks
Awards Presentations
Keynote Speaker Introduction
Aurangzeb Khan, Altia Systems

**Keynote Presentation**
**Digital Analog Design**
**Mark Horowitz, Stanford University**

The past 30 years have seen an enormous growth in the power and sophistication of digital design tools, while progress in analog tools has been much more modest. Digital tools use many abstractions to allow them to validate implementations match the functional models, and the composition of cells matches the composition of the functional models. While there are many reasons why this is more difficult for analog circuits, it can be done. To prove this point, this talk presents how to leverage the fact that the result surface of analog designs are smooth to create ways to formally validate analog models to instances, define analog fault models, and even efficiently explore the effect of process variations.

## Session 2 -- Microsystems for Biomedical and Sensing Applications

Monday, 9/23, 10:00 am
Oak Ballroom
**Session Chair:** Christophe Antoine, Analog Devices
**Session Co-Chair:** Stephen O'Driscoll, University of California, Davis

10:00 am     **Introduction**

Biomedical and sensing applications present new challenges for IC and system designers. In this session, five papers representing advances in these areas are presented.

10:05 am     **A Broadband Biosensor Interface IC for Miniaturized Dielectric Spectroscopy from**
2-1     **MHz to GHz,** *M. Bakhshiani, M. A. Suster, and P. Mohseni, Case Western Reserve University*    **1**

This paper describes a broadband biosensor interface IC as part of a miniaturized measurement platform for MHz-to-GHz dielectric spectroscopy. Developed in 0.35μm 2P/4M RF CMOS, the IC measures the frequency-dependent S21 magnitude and phase of a microfabricated microfluidic dielectric sensor, when the sensor is loaded with a solution-under-test (SUT).

**10:30 am**
**2-2**

**Capacitive Proximity Communication with Distributed Alignment Sensing for Origami Biomedical Implants,** *M. Loh, A. Emami-Neyestanak, California Institute of Technology*     **5**

Origami implant design is a 3D integration technique which addresses size and cost constraints in biomedical implants. A capacitive proximity interconnect that enables this technique is presented. It embeds an alignment sensor that measures link quality directly and simplifies adaptation to alignment. The sensor and transceiver share functional blocks, saving power and area. Data rates 10-60 Mbps are achieved over 4-12μm parylene-C, with efficiencies up to0.180 pJ/bit.

**10:55 am**
**2-3**

**An Active Rectifier/Regulator Combo Circuit for Powering Biomedical Implants,** *E. Lee, Alfred Mann Foundation*     **9**

A circuit that combines an active rectifier and a linear regulator, which will be referred to as a rectulator, is proposed in this paper. The main transistor in the rectulator is used for both the rectification of the AC input (VAC) and the regulation of the DC output (VO) to reduce the overall voltage drop between VAC and VO. The rectulator has a high power efficiency (nP) and very good amplitude modulation (am) rejection on VAC against AC field strength variations. A full-wave rectulator implemented in a 0.18μm CMOS process showed an AM rejection > 45dB for an AM frequency < 2kHz. A nP of 90.7% was measured for an input frequency of 5MHz, VO = 2.5V and an output power of 10.5mW.

**11:20 am**
**2-4**

**65nW CMOS Temperature Sensor for Ultra-Low Power Microsystems,** *Seokhyeon Jeong, Jae-yoon Sim\*, David Blaauw, Dennis Sylvester, University of Michigan, \*Pohang University of Science and Technology*     **13**

A temperature sensor using a novel process-invariant temperature sensing element and voltage to current converter is proposed for battery-operated ultra-low power micro systems. Measurements from a 180nm CMOS test chip show power consumption of 65nW with an inaccuracy of +1.3°C/-1.4°C over a temperature range of 0°C to 100°C after 2-point calibration.

**11:45 am**
**2-5**

**Design and Characterization of Electronic Sensing System for a 13 x 13 Biomechanical Ground Reaction Sensor Array,** *Q. Guo, M. A. Suster\*, R. Surapaneni, C. H. Mastrangelo and D. J. Young, The University of Utah, \*Case Western Reserve University*     **17**

This paper presents the design and characterization of an electronic sensing system interfaced with a high-density flexible biomechanical ground reaction sensor array. The prototype system can measure real-time ground force, shear strain and sole deformation associated with a human bipedal locomotion, thus providing zero-velocity correction to an inertial measurement unit to improve navigation accuracy.

## Session 3 -- High-Speed Wireline Timing Recovery and PLLs

Monday, 9/23, 10:00 am
Fir Ballroom
**Session Chair:** Samuel Palermo, Texas A&M University
**Session Co-Chair:** Kimo Tam, Analog Devices

**10:00 am**    **Introduction**

This session presents various design techniques for ADC-based and burst-made timing recovery systems and PLLs with peaking-free transfer functions and low-area utilization.

**10:05 am**
**3-1**

**Design Metrics for Blind ADC-Based Wireline Receivers (Invited),** *Ali Sheikholeslami and Hirotaka Tamura\*, University of Toronto and \*Fujitsu Labs*    **21**

This paper compares blind ADC-based receivers against binary and phase-tracking ADC-based receivers in terms of their design complexity and cost, and derives equations that relate the required ADC resolution (ENOB) to channel loss and to the characteristics of the FFE/DFE that follow the ADC.

**10:55 am**
**3-2**

**A 10Gbps, 1.24pJ/bit, Burst-Mode Clock and Data Recovery with Jitter Suppression,** *Ming-Chiuan Su, Wei-Zen Chen, Pei-Si Wu\*, Yu-Hsian Chen\*, Chao-Cheng Lee\*, Shyh-Jye Jou, National Chiao Tung University, \*Realtek Corp.*    **29**

A 10Gbps, 1/5 rate burst mode CDR is reconfigurable between data gating and phase tracking modes to achieve instantaneous phase-locking with low jitter suppression. Incorporated with 1:5 demultiplexer, it achieves a high energy efficiency of 1.24pJ/bit. The prototype chip is fabricated in UMC 55nm CMOS technology.

**11:20 am**
**3-3**

**A 9.2-GHz Digital Phase-Locked Loop with Peaking-Free Transfer Function,** *Sigang Ryu, Hwanseok Yeo, Yoontaek Lee, Seuk Son, Jaeha Kim, Seoul National University*    **33**

This paper describes a digital phase-locked loop (PLL) that realizes a peaking-free jitter transfer. That is, the PLL's second-order transfer function does not have a closed-loop zero. Such a PLL does not exhibit overshoots in the phase step response and achieves fast settling. Unlike the previously-reported peaking-free PLLs, the proposed PLL implements the peaking-free loop filter directly in digital domain without requiring additional components. A time-to-digital converter (TDC) is implemented as, a set of three binary phase-frequency detectors that over sample the timing error with time-varying offsets, achieving a linear TDC gain and PLL bandwidth insensitive to the jitter condition. And a 9.2-GHz digitally-controlled LC oscillator (DCO) with transformer-based tuning realizes a predictable DCO gain set by a ratio between two digitally-controlled currents. The prototype 9.2-GHz-output digital PLL fabricated in a 65nm CMOS demonstrates a fast settling time of 1.58-µs with 700-kHz bandwidth. The PLL has a 3.477-psrms divided clock jitter and -120dBc/Hz phase noise at 10-MHz offset while dissipating 63.9-mW at a 1.2-V supply.

**11:45 am**
**3-4**

**A Sub-200 fs RMS Jitter Capacitor Multiplier Loop Filter-Based PLL in 28 nm CMOS for High-Speed Serial Communication Applications,** *Burak Çatl, Ali Nazemi, Tamer Ali, Siavash Fallahi, Yang Liu, Jaehyup Kim, Mohammed Abdul–Latif, Mahmoud Reza Ahmadi, Hassan Maarefi, Afshin Momtaz, and Namik Kocaman, Broadcom Corporation*    **37**

An 8.0 GHz to 12.2 GHz PLL with a capacitor multiplier-based active loop filter is designed in a 28 nm digital CMOS process. A passive loop filter-based version of the PLL is also implemented for comparison. While the PLL area is comparable to that of digital PLLs, the PLL performance is as good as that of an analog PLL that employs a passive loop filter. The capacitor multiplier-based active loop filter PLL has a jitter performance of 198 fs(rms), while its passive loop filter-based counterpart shows a jitter performance of 195 fs (rms). The PLL occupies 0.093 mm$^2$ and consumes 15.5 mA at 1.0V.

# Session 4 -- RF Building Blocks

Monday, 9/23, 10:00 am
Pine Ballroom
**Session Chair:** Andrea Mazzanti, University of Padova
**Session Co-Chair:** Earl McCune

**10:00 am**　　**Introduction**

Moving beyond traditional technologies this session presents digital PLL nonlinearity cancellation, integrated tunable duplexer, smartphone T/R switch, and a programmable broadband integrated phase shifter.

**10:05 am**　　**Nonlinearity Cancellation in Digital PLLs (Invited),** *S. Levantino, C. Samori, Politecnico*
**4-1**　　　　　*di Milano*　　**41**

The spur level in digital fractional-N PLLs is often bounded by TDC resolution and linearity. Methods for mitigating TDC nonlinearity tend to increase phase-noise level. By contrast, PLL architectures based on digital-to-time converters enable nonlinearity cancellation and spur reduction with no added noise at lower design complexity and power consumption.

**10:55 am**　　**Hybrid Transformer-Based Tunable Integrated Duplexer with Antenna Impedance**
**4-2**　　　　　**Tracking Loop,** *S. Abdelhalem, P. Gudem\*, L. Larson\*\*, University of California at San*
*Diego, \*Qualcomm Inc., \*\*Brown University*　　**49**

Electrical balance between the antenna and the balance network impedances is crucial for achieving high isolation in a hybrid transformer duplexer. In this paper, an auto calibration loop for tuning a novel integrated balance network to track the antenna impedance variations is introduced. It achieves an isolation of more than 50 dB in the transmit and receive bands, with an antenna VSWR within 2:1, and between 1.7 and 2.2 GHz. The duplexer, along with a cascaded direct-conversion receiver, achieves a noise figure of 5.3 dB, a conversion gain of 45 dB and consumes 34 mA. The insertion loss in the transmit path was less than 3.8 dB. Implemented in a 65-nm CMOS process, the chip occupies an active area of 2.2 mm$^2$.

**11:20 am**　　**A Smartphone SP10T T/R Switch in 180nm SOI CMOS with 8kV ESD Protection by**　　**53**
**4-3**　　　　　**Co-Design,** *X.S. Wang, X. Wang\*, F. Lu\*\*, L. Wang\*\*, R. Ma\*\*, Z. Dong\*\*, L. Sun, A.*
*Wang\*\*, C.P. Yue, D. Wang\*\*\*, A. Joseph\*\*\*, University of California, Santa Barbara.,*
*\*OmniVision Technologies., \*\*University of California, Riverside., \*\*\*IBM Microelectronics*

This paper reports the first 8kV+ ESD-protected SP10T transmit/receive (T/R)antenna switch for quad-band (0.85/0.9/1.8/1.9GHz) GSM and multiple WCDMA smartphones fabricated in an 180nm SOI CMOS. A novel physics-based switch-ESD co-design methodology is applied to ensure full-chip optimization for a SP10T test chip and its ESD protection circuit simultaneously.

**11:45 am**　　**A Lumped Component Programmable Delay Element for Ultra-Wideband**　　**57**
**4-4**　　　　　**Beamforming,** *Naga Rajesh, Shanthi Pavan, Indian Institute of Technology Madras*

We introduce a ladder filter based programmable time delay element for beamforming in Ultra-Wideband (UWB) systems. When compared to conventional methods based on the tapped delay line architecture, our technique achieves lower power dissipation, better area efficiency, and finer delay and gain resolution more efficiently. The proposed architecture is more scalable, has better parasitic absorption capability and highly programmable with delay and gain resolution dependent only on transconductor resolution. A prototype delay line designed for the 3.1-10.6 GHz UWB range achieves a delay range of 80 ps with 0.5 ps resolution and a gain range of -30 dB to +10 dB with 0.15 dB step. Fabricated in a 0.25 μm SiGe BiCMOS process, the delay element occupies an active area of 1 mm$^2$ and consumes 47 mW from a 2.5 V supply. A four antenna beamforming system using the delay element can achieve scanning range of +/-51deg with resolution of 0.86 deg for antenna spacing of 10 mm.

## Session 5 -- Beyond 14nm Technology Circuit Interaction

Monday, 9/23, 10:00 am
Cedar Ballroom
**Session Chair**: Rajiv Joshi, IBM TJ Watson Research Center
**Session Co-Chair:** Ramnath Venkatraman, LSI Corp.

10:00 am **Introduction**

This session covers technology-circuit interaction needs beyond 14nm including advanced device and lithographic considerations. Multiple patterning and FinFETs add significant complexity. This session further covers efficient methodology and solutions needed for product metrics applied to a myriad of design-technology choices.

10:05 am **The Past, Present, and Future of Design-Technology Co-Optimization (Invited)**, *G.*
5-1   *Yeric, B. Cline, S. Sinha, D. Pietromonaco, V. Chandra, and R. Aitken, ARM*   **61**

Design-Technology Co-Optimization (DTCO) has evolved into a multi-faceted, multi-lateral co-optimization below 20nm, where double patterning and FinFETs create significant complexities. Effective DTCO now involves end product metrics applied to a myriad of technology choices. A future of even more complex lithography, devices, and reliability will drive continued evolution in DTCO.

10:55 am **From 2D-Planar to 3D-Non_Planar Device Architecture: A Scalable Path Forward?**
5-2   **(Invited)**, *G. Shahidi, IBM T.J. Watson Research Center*   **69**

11:45 am **Foundations for Scaling to 7nm and Beyond (Invited)**, *R. Schenker, V. Singh, Intel*
5-3   *Corporation*   **77**

## Session 6 – Forum Session:  20 nm Design Challenges

Monday, 9/23, 1:30 pm
Oak Ballroom
**Organizers** – Pavan Hanumolu, Kimo Tam, Manoj Sachdev
**Moderator** – Manoj Sachdev

1:30 **Introduction**

1:35 pm **Analog/wireline design in an increasingly digital process**, *Matt Straayar, Maxim*
6-1

Analog design in mostly digital process is a challenge. What are issues of embedding sensitive analog circuits surrounded by noisy digital transistors in 20 nm?   **N/A**

2:00 pm **Tools/rules driving designers vs. designers driving tools?**, *Ravi Subramanian, Berkeley*

6-2    *Design Automation*    **N/A**

Design automation is the key to enhance designer productivity. Are we relying too heavily on tools in scaled geometries? Are design rules & tools are too restrictive in 20 nm?

2:25 pm    **DFM Issues for 20 nm Analog,** *Stacy Ho, Mediatek*    **N/A**
6-3

The emergence of finFETs is an important evolutionary step for transistor scaling? How one can exploit its benefits for analog circuits while ensuring yield and reliability? What for the Design For Manufacturing (DFM) challenges that we should be mindful of?

2:50 pm    **Analog design in 20 nm - putting it all together,** *Madhukar Reddy, Maxilinear*    **N/A**
6-4

Successful implementation of analog circuit in Systems on Chip (SoC) requires experience, expertise.  How one should execute design, exploit tools while ensuring good yield and reliability in 20 nm?

# Session 7 -- Power Management

Monday, 9/23, 1:30 pm
Fir Ballroom
**Session Chair:** Christoph Sandner, Infineon
**Session Co-Chair:** Raj Amirtharajah, UC Davis

1:30 pm    **Introduction**

Effective power management requires innovative techniques to minimize system cost while maximizing efficiency.  This session covers a wide array of advances in power management spanning battery interfaces, high voltage converters, LED lighting, digital control, and energy harvesting.

1:35 pm    **BIF–Battery Interface Standard for Mobile Devices (Invited),** *W. Furtner, S.*
7-1    *Schaecher, M. Littow\*, L. Cimaz\*, and P. Leinonen\*\*, Infineon Technologies AG,*
*\*ST-Ericsson and \*\*Nokia*    **81**

The MIPI® Alliance Battery Interface (BIF) is the first comprehensive battery communication interface standard for mobile devices. MIPI BIF is a robust, scalable and cost-effective single-wire communication interface between the mobile terminal and smart or low cost batteries. It is suited for removable batteries as well as for embedded batteries.

2:00 pm    **A 40V 10W 93%-Efficiency Current-Accuracy-Enhanced Dimmable LED Driver with**
7-2    **Adaptive Timing Difference Compensation for Solid-State Lighting Applications,** *D. Park and H. Lee, University of Texas at Dallas*    **89**

This paper describes a floating buck dimmable LED driver for solid-state lighting applications. Adaptive timing difference compensation is proposed to enable the driver to achieve high accuracy of the average LED current, fast settling time, and high-frequency operation over a wide range of input voltages and number of LED loads. The power efficiency of the proposed LED driver is benefited from the capabilities of using synchronous rectification and having no sensing resistor in the power stage. The synchronous rectification under high input supply voltage is enabled by a proposed high-speed and low-power gate driver with pseudo-digital level shifters. Implemented in a 0.35µm 50V CMOS process, the proposed 40V LED driver can operate at 1MHz and achieve 93% peak power efficiency when driving up to 10 series-connected LEDs. It has only 2.8% current error from the average LED current of 345mA and settles within 8.5μs under different line and load variations. The performances of the proposed driver significantly outperform all state-of-the-art counterparts.

**2:25 pm**
**7-3**

**A Stackable Switched-Capacitor DC/DC Converter IC for LED Drivers with 90% Efficiency,** *Chengrui Le, Mitchell Kline\*, Daniel L. Gerber\*, Seth R. Sanders\*, Peter R. Kinget, Columbia University, \*University of California, Berkeley* **93**

A stackable switched-capacitor DC-DC converter IC for a hybrid-SC-resonant LED driver is presented. The IC can handle a range of input voltages through chip-stacking in the voltage domain. The tested driver delivers 17.6W with 90% peak efficiency and maintains >85% efficiency over a rectified voltage range from 160VDC to 180VDC.

**2:50 pm**
**7-4**

**A 100V Gate Driver with Sub-Nanosecond-Delay Capacitive-Coupled Level Shifting and Dynamic Timing Control for ZVS-Based Synchronous Power Converters,** *Z. Liu and H. Lee, The University of Texas at Dallas* **97**

A high-voltage high-speed gate driver to enable synchronous rectifiers with zero-voltage-switching (ZVS) operation is presented in this paper. A capacitive-coupled level-shifter (CCLS) is developed to achieve negligible propagation delay and static current consumption. With only 1 off-chip capacitor, the proposed gate driver possesses strong driving capability and requires no external floating supply for the high-side driving. A dynamic timing control is also proposed not only to enable ZVS operation in the converter for minimizing the capacitive switching loss, but also to eliminate the converter short-circuit power loss. Implemented in a 0.5μm HV CMOS process, the proposed CCLS of the gate driver can shift up a 5V signal to the 100V DC rail with sub-nanosecond delay, improving the FoM by at least 29 times compared with that of state-of-the-art counterparts. The dynamic dead-time control properly enables ZVS operation in a synchronous buck converter under different input voltages (30V to 100V). The power losses of the high-voltage buck converter are thus greatly reduced under different load currents, achieving a maximum power efficiency improvement of 11.5%.

**3:15 pm**

**BREAK**

**3:30 pm**
**7-5**

**A Compact 120-MHz 1.8V/1.2V Dual-Output DC-DC Converter With Digital Control,** *S. Arora, D.K. Su, B. A. Wooley, Stanford University* **101**

A dual-output cascaded dc-dc converter for embedded applications uses a programmable switching frequency up to 120 MHz, output stage segmentation, and cascoding to achieve high power efficiency with small output inductors. A fast digital constant-off-time controller provides the suppression of cross-regulation among multiple output voltages.

**3:55 pm**
**7-6**

**A Monolithic Digitally Controlled Ripple-Based DC-DC Converter with Digital Inductor Current Sensor,** *Man Pun Chan, Philip K.T. Mok, The Hong Kong University of Science and Technology* **105**

A ripple-based digital controller is presented which has incorporated a digital inductor current sensor that does not require extra ADCs and occupies a small chip. Both the digital sensor and controller are fully synthesizable with the total chip area of 0.048 mm$^2$ in 0.13-μm digital CMOS process.

**4:20 pm**

**A Fully Integrated Battery-Connected Switched-Capacitor 4:1 Voltage Regulator with 70% Peak Efficiency Using Bottom-Plate Charge Recycling,** *T. Tong, X. Zhang,*

7-7      *W. Kim, D. Brooks, G.-Y. Wei, Harvard University*     **109**

This work presents a switched-capacitor (SC) DC-DC voltage regulator that converts a 3.7V battery voltage down to ~0.8V in order to power the 'brain' SoC of a flapping-wing microrobotic bee. A cascade of two 2:1 SC converters offers high efficiency for a 4:1 conversion ratio. A charge recycling technique reduces the flying capacitor's bottom-plate parasitic loss by 50% and overall conversion efficiency reaches 70%. The output droop is less than 10% of the nominal output voltage for a worst-case 47mA load step.

4:45 pm
7-8

**A 110nA Synchronous Boost Regulator With Autonomous Bias Gating for Energy Harvesting,** *Khondker Z. Ahmed and Saibal Mukhopadhyay, Georgia Institute of Technology* **113**

An autonomously bias gated synchronous boost regulator consuming 110nA at 1V is demonstrated in 130nm CMOS. The IC generates regulated 1V output from 30mV input, starts up autonomously (battery-less) at 265mV, and regulates output ranging from 0.78V-3.3V. The peak efficiency is 83% with 10µA and 85% with 10mA load.

5:10 pm
7-9

**A Power Sensor with 80ns Response Time for Power Management in Microprocessors,** *S. Bhagavatula, B. Jung, Purdue University*     **117**

A real-time, on-chip power sensor that estimates load currents and on-chip temperatures concurrently is presented. It occupies an area of 0.11mm X 0.09mm in 130n mCMOS technology. With a simplified 1-point calibration and a response time of 80ns, it shows improvements in input dynamic range by 10 X, response time by 6 X and sensitivity by 3 X over previous such sensors. A current reference with a measured temperature coefficient 91ppm/C (-20Cto 120C) is presented. This reference is used for online calibration of the power sensor to enable greater tolerance to PVT variations and aging effects.

# Session 8 -- AMS Verification in Advanced Technologies

Monday, 9/23, 1:30 pm
Pine Ballroom
**Session Chair:** Hidetoshi Onodera, Kyoto University
**Session Co-Chair:** Yu (Kevin) Cao, Arizona State University

1:30 pm

**Introduction**

Verification of AMS circuits is increasingly challenging. This session presents novel AMS simulation and emulation techniques, and advanced reliability and performance issues with technology scaling.

1:35 pm
8-1

**Discretization and Discrimination Methods for Design, Verification, and Testing of Analog/Mixed-Signal Circuits (Invited),** *J. Kim, J. Lee, D.-G. Song, T. Kim, K.-H. Kim, S. Jung, S. Youn, Seoul National University*     **121**

This paper describes how the difficult problems of designing, verifying, and testing analog circuits in presence of variability can be converted to easier ones by discretizing the search spaces or discriminating one case from another. For instance, discretizing the continuous design space of analog circuits enables the use of an efficient, predictive global circuit optimizer. Also, discretizing the initial condition space of a circuit enables one to establish its global convergence property over the entire space by exploring only a small number of samples. Lastly, discriminating the test responses of the circuit with and without a fault in consideration of the underlying statistical distribution provides a formal guide on how to quantify the fault coverage of analog/mixed-signal circuit tests. It is noteworthy that it is variability that introduces the cross-correlations in the performance metrics, convergence behaviors, and test responses between two nearby candidates in consideration and therefore enables the use of discretization and discrimination methods listed in this paper. The proposed methods are demonstrated on the practical examples of sizing an operational amplifier, verifying the correct start-up of a coupled ring oscillator, and composing a test suite for screening faults in a digitally-controlled phase interpolator circuit.

2:25 pm
8-2

**Indirect Performance Sensing for On-Chip Analog Self-Healing via Bayesian Model Fusion,** *S. Sun, F. Wang, S. Yaldiz, X. Li, L. Pileggi, A. Natarajan\*, M. Ferriss\*, J. Plouchart\*, B. Sadhu\*, B. Parker\*, A. Valdes-Garcia\*, M. Sanduleanu\*, J. Tierno\*, D. Friedman\*, Carnegie Mellon University, \*IBM T.J. Watson Research Center* **129**

On-chip analog self-healing requires low-cost sensors to accurately measure performance metrics. In this paper, we propose a novel approach of indirect performance sensing based upon Bayesian model fusion to facilitate inexpensive-yet-accurate on-chip measurement. A 25GHz differential Colpitts VCO is used to validate the proposed indirect performance sensing and self-healing methodology.

2:50 pm
8-3

**Fast FPGA Emulation of Background-Calibrated SAR ADC with Internal Redundancy Dithering,** *G. Wang, Y. Chiu, University of Texas at Dallas* **133**

A custom FPGA emulation platform for the verification of a slowly adapted, background calibration technique for successive-approximation-register (SAR) analog-to-digital converter (ADC) is demonstrated in an Altera DE4 board. The internal redundancy of a sub-binary SAR is exploited for the identification of ten leading bit weights in a 14.5-bit SAR ADC using pseudorandom bit sequence (PRBS) injection with background correlation. Experimental results reveal that the FPGA emulation achieves a 3000× speedup for the same simulation executed on a general-purpose microprocessor.

3:15 pm

**BREAK**

3:30 pm
8-4

**Circuit Reliability Simulation Using TMI$^2$ (Invited),** *Min-Chie Jeng, Cheng Hsiao, Ke-Wei Su, Chung-Kai Lin, Taiwan Semiconductor Manufacturing Company* **137**

This paper reviews existing circuit aging simulation approaches with focus on TMI$^2$.

4:20 pm
8-5

**Scalable Behavior Modeling for 3D Field-Programmable ESD Protection Structures,** *L. Wang, X. Wang\*, Z. T. Shi\*\*, R. Ma, C. Zhang, Z. Dong, F. Lu, H. Zhao\*\* and A. Wang, University of California, Riverside, \*Omnivision Technologies, \*\*Marvell Semiconductor* **144**

This paper reports new accurate and scalable behavioral modeling for novel 3D field-programmable ESD protection circuits, which enables post-Si on-chip ESD protection design simulation. New field-programmable ESD protection devices were fabricated in CMOS-compatible processes. The behavior models were developed from ESD testing results and verified in SPICE circuit simulation.

4:45 pm
8-6

**Quasi-3D method: Time-efficient TCAD and Mixed-Mode Simulations on finFET Technologies,** *G. Hellings, S-H Chen, D. Linten, M. Scholz, G. Groeseneken, imec,* **148**

The Quasi-3D allows to drastically speed up TCAD and mixed-mode simulations of finFET technologies, by solving on well-chosen 2D finFET cross sections. The method accurately reproduces important transistor metrics requiring only 1/20th of the simulation time.

| | |
|---|---|
| 5:10 pm<br>8-7 | **Gate Stack Resistance and Limits to CMOS Logic Performance,** *R. A. Wachnik, S. Lee, L. H. Pan, N. Lu, H. Li, R. Bingert \*\*, M. Randall, S. Springer, C. Putnam, IBM Corporation, \*\*ST Microelectronics* **152** |

Measured data from five generations of CMOS technology including polysilicon and High-K metal gate stacks shows a trend of increasing gate resistance. The data are analyzed to determine horizontal and vertical components in terms of scalable model parameters. Gate resistance affects performance of a 20nmreplacement gate technology.

## Session 9 -- Wireless Transceivers

Monday, 9/23, 1:30 pm
Cedar Ballroom
**Session Chair:** Julian Tham, Broadcom
**Session Co-Chair:** Jonathan Borremans, imec

| | |
|---|---|
| 1:30 pm | **Introduction** |

This session presents papers on advances in wide-band receiver designs and a calibrated software-defined radio. It also presents WLAN, GPS and ultra low-power receivers and transceivers.

| | |
|---|---|
| 1:35 pm<br>9-1 | **IIP2 and HR Calibration for an 8-Phase Harmonic Recombination Receiver in 28nm,** *B. van Liempd, J. Borremans, S. Cha\*, E. Martens, H. Suys, J. Craninckx, imec vzw, \*Renesas Electronics Corporation* **156** |

Fully integrated CMOS receivers achieve high linearity and low noise due to harmonic recombination, but suffer from limited IIP2 and harmonic rejection due to mismatch and inaccuracies. This paper presents an 8-phase harmonic recombination receiver with independent IIP2, HR3 and HR5 calibration techniques. Calibrated >80dBm IIP2, >70dB HR3 and >75dB HR5 are measured.

| | |
|---|---|
| 2:00 pm<br>9-2 | **Advances in the Design of Wideband Receivers (Invited),** *D. Murphy, M. Mikhemar, A. Mirzaei, H. Darabi, Broadcom Corporation* **160** |

To be practical, wideband receivers must tolerate large out-of-band blockers, which can desensitize the receiver through gain compression or reciprocal mixing with LO phase noise. This paper reviews how a new noise-cancelling receiver architecture – that utilizes 3 important circuit innovations – mitigates gain compression without compromising noise figure. While the architecture is still susceptible to reciprocal mixing, it is shown how a recently proposed reciprocal mixing cancelling technique (if incorporated into the receiver) can eliminate the need for a dramatic rise in LOGEN current.

| 2:50 pm | **An Asymmetric Dual-Channel Reconfigurable Receiver for GNSS in 180nm CMOS,** |
| 9-3 | *Nan Qi, Baoyong Chi, Yang Xu, Zhou Chen, Jun Xie, Yang Xu, Zheng Song, Zhihua Wang, Tsinghua University* **168** |

A fully integrated dual-channel reconfigurable receiver supporting all the GNSS signals is presented. The two channels share the frequency synthesizer and RF front-end circuits, but employ separate asymmetric IF strips to support simultaneous dual-constellation reception. The 2nd IF strip can be configured to a dual-conversion mode to lower the sampling rate.

| 3:15 pm | **BREAK** |

| 3:30 pm | **A 5-GHz 11.6-mW CMOS Receiver for IEEE 802.11a Applications,** *A. Homayoun, B.* |
| 9-4 | *Razavi, University of California, Los Angeles* **172** |

A direct-conversion receiver employs a 1-to-6 transformer as a low-noise amplifier along with passive mixers and noninvasive baseband filters. Realized in 65-nm CMOS technology, the receiver provides an average noise figure of 5.3dB and a sensitivity of -70 dBm at a data rate of 54 Mb/s. The prototype draws 11.6 mW from a 1-V supply and occupies an active area of 0.18 mm$^2$.

| 3:55 pm | **An Adaptive Predistorter for Wireless LAN RFSoC with embedded PA and T/R switch** |
| 9-5 | **in 55nm CMOS,** *K. Muhammad, M.-C. Chen, K.-H. Wang, K.-P. Ma, Y.-L. Hiseh, W.-S. Hsu, Y.-Y. Fu, M.-C. Lee, S.-Y. Hsiao, C.-M. Hung, MStar Semiconductor Inc.* **176** |

We present an adaptive predistortion system for a WLAN transceiver in 55nm CMOS. The forward DSP path utilizes complex gain predistortion while the APD module in the feedback path computes AMAM and AMPM coefficients by comparing ideal transmit signal with the distorted signal from the receiver. This module operates with various calibration signals generated on-chip in addition to TX data. Measurement results show improvement of EVM by 1dB with the proposed approach. Improvement of P1dB of more than 3dB was obtained using fully automatic processing. The total solution utilizes 120k gates.

| 4:20 pm | **A 116nW Multi-Band Wake-Up Receiver with 31-bit Correlator and Interference** |
| 9-6 | **Rejection,** *S. Oh, N. Roberts*, D. Wentzloff*, Samsung, *University of Michigan* **180** |

This paper presents a 116nW wake-up radio complete with crystal reference, interference compensation, and baseband processing, such that a selectable 31-bit code is required to toggle a wake-up signal. The baseband processor detects interferers and dynamically adjusts the receiver's sensitivity, mitigating the jamming problem to previous energy-detection wake-up radios.

| 4:45 pm | **A 11µW Sub-pJ/bit Reconfigurable Transceiver for mm-Sized Wireless Implants,** *A.* |
| 9-7 | *Yakovlev, J. Jang, D. Pivonka, A. Poon, Stanford University* **184** |

A wirelessly powered 11µW transceiver for mm-sized wireless implants supporting TDMA has been designed and demonstrated through 35mm of porcine heart. The communication links have configurability for operation in diverse biological environments. The forward link achieves 4-20Mbps at 0.3pJ/bit, and the reverse link achieves 0.7-2Mbps at 0.7pJ/bit.

## Session 10 – Panel Session

"Can biomedical electronic startups make money??"

Monday, 9/23, 3:30 pm
Oak Ballroom
**Session Chair:** Pedram Mohseni, Case Western Reserve University
**Moderator:** John McNeill, Worcester Polytechnic Institute

-Straight semiconductor startups are marginally viable
-Green/solar startups took the money and ran
-Will bioelectronics fare any better?

Panelists    Arjang Hassibi, UT Austin & Insilixa

Patrick Chiang, Fudan University

Chris Raanes, Viewray

## Monday Poster Session

Monday, 9/23, 5:00 pm – 7:00 pm
Donner/ Siskiyou/ Cascade Ballrooms

M-1    **An All-Digital Time Difference Hold-and-Replication Circuit utilizing a Dual Pulse Ring Oscillator,** *Tetsuya Iizuka, Teruki Someya, Toru Nakura, Kunihiro Asada, University of Tokyo*    **188**

This paper presents a time-domain analog signal hold-and-replication circuit which holds an input time interval of two signal transitions and replicates it any number of times.65nm CMOS implementation accepts 100ps to 1.2ns time interval while occupying40x60µm$^2$ area. A TDC resolution enhancement application is also demonstrated in this paper.

M-2    **A 15-Bit Binary-Weighted Current-Steering DAC with Ordered Element Matching,** *T. Zeng, K. Townsend, J. Duan\*, D. Chen, Iowa State University, \*Broadcom Corporation*    **192**

This paper introduces a 15-bit binary-weighted current-steering DAC in a 130nm CMOS technology. The core area is less than 0.42mm$^2$, among which the MSB area is well within 0.021mm$^2$. Measurement results have shown that the DNL and INL can be reduced from 9.85LSB and 17.41LSB to 0.34LSB and 0.77LSB, respectively.

M-3    **A 500 MS/s 76dB SNDR Continuous Time Delta Sigma Modulator with 10MHz Signal Bandwidth in 0.18µm CMOS,** *Rune Kaald, Bjørnar Hernes\*, Christian Holdø\*, Frode Telstø\*, Ivar Løkken\*, Norwegian University of Science and Technology, \*Hittite Microwave Norway*    **196**

A 5th order continuous-time delta sigma modulator is designed in 0.18um CMOS. At a sampling rate of 500MHz it achieves 76dB SNDR over a 10MHz bandwidth consuming 58mW. 5th order noise shaping is realized with 4 op amp based RC integrators and a VCO realizing an integrator and a 4 bit quantizer. A THD of-82.3dBc is achieved without calibration of feedback DACs. We address two problems related to VCO quantizers which have local feedback to also work as integrators with a high-speed excess loop delay compensation using capacitive summation and a method for reducing the switching activity of the output codes.

M-4    **A 0.1-3GHz Cell-Based Fractional-N All Digital Phase-Locked Loop Using ΔΣ Noise-Shaped Phase Detector,** *Yao-Chia Liu, Wei-Zen Chen, , Mao-Hsuan Chou\*, Tsung-Hsien Tsai\*, Yen-Wei Lee, Min-Shueh Yuan, National Chiao Tung University and \*TSMC*    **200**

A 0.1-3 GHz, cell-based, fractional-N ADPLL with ΔΣ noise-shaped phase detector is presented. By dithering the reference phase and quantization phase error through an additional feedback path, linear phase detection and zero stabilization are accomplished without resort to sophisticated time to digital converter (TDC). The measured rms jitter from a 3GHz carrier is 1.9 ps with a multiplication factor of 60. Implemented in TSMC 40nm general purpose superb CMOS technology, the chip size is 280μm x 240μm. Keywords: TDC, ΔΣ phase detector, fractional-N ADPLL

M-5    **A Direct-Battery Hookup, Fully Integrated Stereo Headphone Module with 82 mW Output Power and 110 dB PSRR,** *Khaled Abdelfattah, Sherif Galal, Iuri Mehr, Alex Jianzhong Chen, Ahmet Tekin\*, Xicheng Jiang, Todd L. Brooks, Broadcom Corp.,\*Semtech*    **204**

A complete stereo ground-referenced headphone module that supports direct battery hookup is integrated on a 40 nm mobile baseband SoC. Several techniques were employed to guarantee the reliability of the module circuitry under high output swing and limited safe operating regions in this low-voltage technology. Additional techniques to reduce area and enable low-cost integration were also employed. The module delivers 3.24 Vpp to a 16Ω load (82 mW) and achieves 100 dB dynamic range (DR), 110 dB PSRR, and 84 dB THD+N with an area of 0.675mm$^2$ on the SoC.

M-6    **A Fast-Locking Digital DLL with a High Resolution Time-to-Digital Converter,** *Dandan Zhang, Hai-gang Yang, Zhujia Chen, Wei Li, Zhihong Huang, Lijiang Gao, Wenrui Zhu, Institute of Electronics, Chinese Academy of Sciences*    **208**

A fast-locking DLL is presented in this paper. By adopting a novel high resolution TDC, the total locking time is reduced to 8 clock cycles and shortened by 80% to 94.6% compared to previous closed-loop architectures .The measured RMS and p-p jitters are 2.3ps and 10ps respectively.

M-7    **A Stochastic Sampling Time-to-Digital Converter with Tunable 180-770fs Resolution, INL less than 0.6LSB, and Selectable Dynamic Range Offset,** *J. Tandon, T. Yamagichi, S. Komatsu, K. Asada, VDEC-D2T, University of Tokyo*    **212**

We introduce a stochastic time-to-digital converter (TDC) that has 180-770fs tunable resolution, less than 0.6LSB INL, and selectable dynamic range offset .Previous arbiter-based TDCs have fine resolution but small dynamic range which is difficult to calibrate. Our approach uses comparators as decision elements to precisely control dynamic range offset.

M-8    **A 50 μW/Ch Artifacts-Insensitive Neural Recorder Using Frequency-Shaping Technique,** *J. Xu, Z. Yang, National University of Singapore*    **216**

This paper presents a frequency-shaping (FS) neural recording interface that can inherently reject electrode offset, 5-10 times increase input impedance,4.5-bit extend system dynamic range, and provide much more tolerance to motion artifacts and 50/60 Hz power noise interferences. It is supposed to be more suitable for long-term brain-machine-interface (BMI) experiments.

M-9    **A Bipolar >40-V Driver in 45-nm SOI CMOS Technology,** *Yousr Ismail, Chang-Jin Kim, Chih-Kong Ken Yang, University of California, Los Angeles*    **220**

A novel, switched-capacitor output driver combining both voltage-conversion and pulse-drive is introduced. The driver is implemented in 45-nm SOI CMOS technology and uses only process-compliant devices. It achieves a maximum output drive of 44 V and has a 36 KΩ output resistance while consuming 28 mA from a 1.5-V supply.

**M-10**
**High-Sensitivity Photodetection Sensor Front-End, Detecting Organophosphourous Compounds for Food Safety,** *L. Wan, Y. Qin, P. Chiang\*, G. Chen, R. Liu, Z. Hong, Fudan University and \*Oregon State University* **224**

A high-sensitivity, high-dynamic range photo detection sensor front-end is presented, suitable for low-cost hand-held food safety systems. This sensor-on-a-chip for detecting organophosphorus compounds incorporates a non-chip deep N-well photo detector, pulse width modulation, and a folded reference. Measurement results show an input optical power dynamic range of 71dB, a sensitivity of $3.6nW/cm^2$.

**M-11**
**A 16-Channel, 359 μW, Parallel Neural Recording System Using Walsh-Hadamard Coding,** *Vahid Majidzadeh, Alexandre Schmid and Yusuf Leblebici, Swiss Institute of Technology (EPFL)* **228**

Application of an algebraic coding to a multichannel parallel neural recording system is presented. The Walsh-Hadamard coding enables back-end hardware sharing between recording channels, using a linear and orthogonal superposition of the analog inputs. Moreover, this technique preserves the temporal information of the channels in contrast to the conventional architectures which use time-multiplexed ADC. In the proposed architecture a single ADC operates on a superposed signal, thereby the dynamic range of the ADC is effectively shared between channels benefiting from the sparsity characteristics of the channels. A 16-channel parallel recording system is implemented as a proof of concept. The system is implemented in a 0.18 μm CMOS technology and occupies $1.99$ mm$^2$ of silicon area. The input-referred noise of a single channel integrated from 10 Hz to 100 kHz equals to $4.1$ μV$_{rms}$, and the effective power consumption of each channel is measured at 22.4 μW from a 1.2 V power supply, which results in a system level NEF of 5.6.

**M-12**
**Analysis of Deviation from Pelgrom Scaling Law in V$_{th}$ Variability of Pocket-implanted MOSFET,** *K. Sakakibara, Y. Miura, T. Kumamoto, S. Tanimoto, Renesas Electronics Corporation* **232**

Deviation from Pelgrom scaling law in threshold voltage variability of pocket-implanted MOSFET is attributed to an increasing behavior of offset-voltage variability in weak and moderate inversion regions. This increasing behavior can be completely removed by using both-side ring gate structure. This means that the deviation is caused by subthreshold hump.

**M-13**
**Low Power ARM® Cortex™-M0 CPU and SRAM Using Deeply Depleted Channel (DDC) Transistors with VDD Scaling and Body Bias,** *V. Agrawal, N. Kepler, D. Kidd, G. Krishnan, S. Leshner, T. Bakishev, D. Zhao, P. Ranade, R. Roy, M. Wojko, L. Clark, R. Rogenmoser, M. Hori\*, T. Ema\*, S. Moriwaki\*, T. Tsuruta\*, T. Yamada\*, J. Mitani\*, and S. Wakayama\*, SuVolta Inc. and \*Fujitsu Semiconductor Ltd.* **236**

130-D Knowles Drive, Deeply Depleted Channel™ (DDC) technology demonstrates more than 50% power reduction for ARM® Cortex™-M0 CPU cores and SRAMs at matched performance via VDD scaling and body biasing. DDC technology also demonstrates 35% speed improvement at matched power, improved SNM, 150mV 8Mb SRAM VDDmin improvement, and 5x SRAM retention leakage reduction.

**M-14**   **Highly Efficient CMOS Rectifier Assisted by Symmetric and Voltage-Boost PV-Cell Structures for Synergistic Ambient Energy Harvesting,** *K. Kotani, Tohoku University*   **240**

A highly efficient CMOS RF rectifier assisted by symmetric PV cells was developed as an example of the synergistic ambient energy harvesting concept. Output-voltage-boosted PV cell structures were also developed to improve the efficiency of this rectifier. Under typical indoor lighting conditions, 4x PCE than a conventional rectifier was achieved.

**M-15**   **A Slew-Rate Based Process Monitor and Bi-directional Body Bias Circuit for Adaptive Body Biasing in SoC Applications,** *S. Lee, E. Boling, A. Kuo, R. Rogenmoser, SuVolta, Inc.*   **244**

A process monitor based on slew-rate measurement has been applied to a body bias control system to detect the process corners and adjust the body bias voltage to meet the power and performance requirements for SoCs. A 55nm testchip includes a new pulse extender and a bi-directional body bias circuit.

**M-16**   **Comparison of Modeling Approaches through Hierarchical Behavioral Modeling of a GNSS Receiver Front-end,** *Z. Chen, Y.Wang, J. Driesen\*, F. Garzia\*\*, S. Koehler\*\*, F. Henkel\*, R. Wunderlich, S. Heinen, IAS RWTH, \*IMST, \*\*IIS Fraunhofer*   **248**

This paper analyzes and compares the mixed-signal modeling approaches (conservative, timed data flow, and event-driven, as well as the base band modeling approach) through the hierarchical behavioral modeling of a GNSS (Global Navigation Satellite System) receiver front-end. Based on the result of the comparison, one hierarchical modeling flow is finally derived comprising multi-modeling approaches with reduced manual modeling effort.

**M-17**   **Pulse Amplification Based Dynamic Synchronizers with Metastability Measurement using Capacitance De-rating,** *B. Giridhar, M. Fojtik, D. Fick, D. Sylvester, D. Blaauw, University of Michigan*   **252**

We present dynamic buffer based synchronizers where only pulses (rather than stable intermediate voltages) cause metastability. This unique feature is exploited by amplifying such pulses to improve MTBF by $>10^6$x over jamb latches and double flip-flops at 2GHz in 65nm CMOS. A new on-chip metastability measurement method is also proposed.

**M-18**   **FireBird: PowerPC e200 Based SoC for High Temperature Operation,** *Radisav Cojbasic, Omer Cogal, Pascal Meinerzhagen, Christian Senning, Conor Slater, Thomas Maeder, Andreas Burg, Yusuf Leblebici, Ecole Polytechnique Federale de Lausanne (EPFL)*   **256**

This work presents FireBird, the first PowerPC based SoC for reliable operation beyond 200C. This paper proposes to customize a PowerPC e200 based SoC by using a dynamically reconfigurable clock frequency, exhaustive clock gating, and electromigration-resistant power supply rings. The custom testing procedure showed the expected maximum operating frequency reduction from 38MHz at room-temperature to 30MHz at 200C. The maximum power dissipation at 3.3V supply voltage was 1.2W and the idle state static leakage current was 3.4mA. Silicon measurements proved that this design outperforms PowerPC based SoCs available in the high-temperature microcontrollers market which are not operational at temperatures above 125C.

**M-19**   **A 1/10000 Lower Error Rate Achievable SSD Controller with Message-Passing Error Correcting Code Architecture and Parity Area Combined Scheme,** *K. Li, M. Ito, A. Esumi, Siglead, Inc.*   **260**

A new Error Correcting Code (ECC) solution to improve the reliability of NAND is proposed. Implemented in SSD controller IC, it is confirmed that more than1/10000 lower error rate, and 1.7x longer endurance of SSD can be achieved. This solution consists of a Message-Passing ECC architecture and a Parity Area Combined ECC scheme.

M-20      **45pW ESD Clamp Circuit for Ultra-Low Power Applications,** *Yen-Po Chen, Yoonmyung Lee, Jae-Yoon Sim\*, Massimo Alioto\*\*, David Blaauw, and Dennis Sylvester, University of Michigan, \*Pohang University of Science and Technology, \*\*University of Siena* **263**

Novel ultralow-leakage ESD power clamp designs are proposed and implemented in 0.18μm CMOS. Limiting both subthreshold leakage and GIDL, the proposed designs consume 43pW at 25°C and 119nW at 125°C with 4500V HBM level and 400V MM level protection, marking an 18-139× leakage reduction over conventional ESD clamps.

M-21      **A 1.14mW 750kb/s FM-UWB Transmitter with 8-FSK Subcarrier Modulation,** *F. Chen, Y. Li, D. Lin, H. Zhuo, W. Rhee, J. Kim\*, D. Kim\*, and Z. Wang, Tsinghua University and \*Samsung Advanced Institute of Technology* **267**

A noninvasive energy-efficient FM-UWB transmitter is implemented in 65nm CMOS for stereo hearing aid. 8-FSK subcarrier modulation is employed to triple data rate by a fast-settling PLL. The FM-UWB signal is generated by an FLL-assisted ring VCO and a class AB PA. The 3.5-4GHz 750kb/s transmitter consumes 1.14mW,achieving 1.5nJ/bit.

M-22      **A 2.4 GHz Energy-Efficient 18-Mbps FSK Transmitter in 0.18 μm CMOS,** *Jingjing Chen, Weiyang Liu, Peng Feng, Haiyong Wang, and Nanjian Wu, Institute of Semiconductors, Chinese Academy of Sciences* **271**

This paper presents a 2.4 GHz energy-efficient phase locked loop (PLL)-based transmitter (TX) integrated in 0.18-μm CMOS technology. By using Twin-VCO transmission scheme, the data rate of the transmitter is free of loop bandwidth of PLL with stable carrier frequency. Measured results show that The TX achieves an energy efficiency less than 0.64 nJ/bit at a data rate of 18 Mbps.

M-23      **A 60GHz, Linear, Direct Down-Conversion Mixer with mm-Wave Tunability in 32nm CMOS SOI,** *M.A.T. Sanduleanu, A. Valdes-Garcia, Y. Liu, B. Parker, S. Shlafman\*, B. Sheinman\*, D. Elad\*, S. Reynolds, D. Friedman, IBM T.J. Watson Research Center and \*IBM Haifa R&D* **275**

The gain/linearity trade-off is exploited to achieve the best linearity performance of a mm-Wave down-conversion system. The achieved linearity (IIP3) for the whole down-conversion chain is better than 11.06dBm for 5.8dB gain at 60GHz. The down-converter occupies 1.38mm$^2$ in 32nm CMOS SOI and consumes 19.2mW from 1V supply.

M-24      **A Fully Integrated Highly Linear Receiver with Automatic IP$_2$ Calibration Schemes for Multi-Standard Applications,** *A. Borna, Y. Wang\*, C. Hull\*, H. Wang\*\*, A. Niknejad, UC Berkeley, \*Intel Corp., \*\* Georgia Institute of Technology* **279**

This paper presents an entire receiver chain with fully integrated self-calibration circuitries for suppressing the 2nd-order intermodulation distortions in Homodyne receivers for multi-standard applications. All the potential sources for IM2 generation are identified and tackled independently by architectural and calibration techniques, which results in a robust IP$_2$

enhancement.

# Session 11 -- Power and Heterogeneous Technology Circuit

Tuesday, 9/24, 9:00 am
Oak Ballroom
**Session Chair**: Takamaro Kikkawa, Hiroshima University
**Session Co-Chair**: Philippe Jansen, Texas Instruments

9:00 am    **Introduction**

This session focuses on technology circuit interactions for heterogeneous 3D stacked-silicon FPGA, advanced high-voltage GaN electronics and future generation nanowire transistors.

9:05 am    **40V MESFETs Fabricated on 32nm SOI CMOS,** *W. Lepkowski, S. Wilk, J. Kam, T.*
11-1     *Thornton, Arizona State University*   **283**

40V N-channel MESFETs fabricated at a commercial 32nm SOI CMOS foundry without changing any of the process flow or including additional mask steps. Current drives of 110mA/mm with peak cut-off frequency of 30.5GHz and maximum oscillation frequency of 34.5GHz were observed.

9:30 am    **Recent Advances in GaN Power Electronics (Invited),** *Karim Boutros, Rongming Chu,*
11-2     *Brian Hughes,HRL Laboratories, LLC*   **287**

Gallium Nitride power devices are poised to replace silicon-based MOSFETs in power switching applications having weight and volume constraints, while simultaneously needing a high overall efficiency.  With its projected 100x performance advantage over silicon, GaN is a game changing technology for energy-efficient power electronics. This paper reviews the advantages of GaN material and devices, the performance of these devices in power circuits, and the potential applications for this technology.

9:55 am    **Prospective for Nanowire Transistors (Invited),** *J.P. Colinge, S. Dhong, Taiwan*
11-3     *Semiconductor Manufacturing Company*   **291**

The multigate nanowire FET architecture allows for ultimate short-channel control and push Moore's law down to sub-5nm gate lengths. This paper reviews nanowire transistor device physics as well as circuit prospects in the fields of CMOS logic, memory, analog, RF and integrated sensor applications.

10:45 am   **BREAK**

11:05 am   **Interconnect and Package design of a Heterogeneous Stacked-Silicon FPGA**
11-4     **(Invited),** *E. Wu, K. Abugharbieh, B. Banijamali, S. Ramalingam, P. Wu, C. Wyland, Xilinx,*
     *Inc.*   **299**

# Session 12 -- Wireline Transmitter and Receiver Design Techniques

Tuesday, 9/24, 9:00 am
Fir Ballroom
**Session Chair**: Dennis Fischette, AMD
**Session Co-Chair**: Jaeha Kim, Seoul National University

**9:00am**   **Introduction**

This session presents state of the art wireline transmitter and receiver circuits, including advanced low-power equalization and impedance-matching techniques.

**9:05 am**   **A 5Gb/s 3.2mW/Gb/s 28dB Loss-compensating Pulse-Width Modulated**
**12-1**   **Voltage-Mode Transmitter,** *S. Saxena, R. K. Nandwana, and P. K. Hanumolu, Oregon State University*   **307**

A voltage mode transmitter employs pulse width modulation (PWM) based equalization of NRZ input data at 5Gb/s and compensates 28dB channel loss at 2.5GHz. Fabricated in a 90nm CMOS process, the proposed transmitter achieves a horizontal eye opening of 0.3UI with BER$<10^{-12}$ and consumes only 16mW power of which 2.5mW is consumed by the digital PLL.

**9:30 am**   **Current-Steering Pre-Emphasis Transmitter with Continuously Tuned Line**
**12-2**   **Terminations for Optimum Impedance Match and Maximum Signal Drive Range,**
*Gerrit W. den Besten, Harold G. Hanson, Ranjeet K. Gupta, NXP Semiconductors*   **311**

A configurable 24-segment current-steering transmitter with linear continuously-tuned active line terminations is presented. A linearized resistor-MOSFET termination topology is proposed for accurate output levels ($\sigma$=1%) and good impedance matching ($\sigma$=2%) enabling larger drive levels by better supply utilization. The concept is implemented in 0.16 µm CMOS for 1-6Gbps.

**9:55 am**   **Design Techniques for CMOS Backplane Transceivers Approaching 30-Gb/s Data**
**12-3**   **Rates (Invited),** *J. Bulzacchelli, IBM T.J. Watson Research Center*   **315**

This paper highlights design techniques for extending backplane transceiver data rates by describing a 28-Gb/s prototype implemented in 32-nm SOI CMOS and featuring a source-series terminated driver with 4-tap FFE, a two-stage peaking amplifier with active feedback, and a 15-tap DFE. Equalization is demonstrated over a 35-dB loss channel.

**10:45 am**   **BREAK**

**11:05 am**   **Design Considerations for Low-Power Receiver Front-End in High-Speed Data Links**
**12-4**   **(Invited),** *S. Shekhar, J. E. Jaussi, F. O'Mahony, M. Mansuri, B. Casper, Intel Corporation*   **323**

This paper presents different design considerations for the receiver front-end (RXFE) in low-power, high-speed data links. Specifications for the RXFE are defined and explained in detail, including their impact on the overall link performance. Based on these specifications, low-power RXFE topologies are then analyzed to illustrate the design and performance tradeoffs. Techniques to properly characterize and measure the RXFE specifications are also provided, supplemented with measurement results from three different low-power links operating at 10Gb/s, 16Gb/s and 20Gb/s.

**11:55 am**   **An 8mW Frequency Detector for 10Gb/s Half-Rate CDR using Clock Phase Selection,**
**12-5**   *M.S. Jalali, R. Shivnaraine, A. Sheikholeslami, M. Kibune\*, H. Tamura\*, University of Toronto, \*Fujitsu Laboratories Limited*   **331**

A half-rate single-loop CDR with a new frequency detection scheme is introduced. The proposed frequency detector selects between the clock phases (I and Q) to reduce cycle slipping, hence improving lock time and capture range. This frequency detector, implemented within a 10Gb/s CDR in Fujitsu 65nm CMOS, consumes only 8mW, but improves the capture range by up to 3.6 . The measured capture range with the FD is from 8.675Gb/s to 11Gb/s.

## Session 13 -- mm-Wave Circuits and Systems

Tuesday, 9/24, 9:00 am
Pine Ballroom
**Session Chair**: John Rogers, Carleton University
**Session Co-Chair:** Howard Luong, Hong Kong University of Science & Tech.

9:00 am          **Introduction**

This session will present the latest advances in mm-Wave (>30 GHz) circuits and systems, including transceivers, power amplifiers, synthesizers, VCOs and dividers.

9:05 am          **A 60 GHz Linear Wideband Power Amplifier using Cascode Neutralization in 28 nm**
13-1              **CMOS,** *Siva V Thyagarajan, Ali M Niknejad, Christopher D Hull\*, University of California Berkeley, \*Intel Corporation*          **335**

This paper presents the design of a 60GHz linear wideband power amplifier (PA) in deeply scaled 28nm CMOS technology. The PA utilizes cascode drain-source neutralization to improve stability and low-k transformer techniques to achieve high bandwidth. The PA delivers a saturated output power of 16.5dBm with a peak PAE of 12.6% and achieves a bandwidth of 11GHz with a peak gain of 24.4dB.

9:30 am          **Compact High-Power 60 GHz Power Amplifier in 65 nm CMOS,** *Payam M. Farahabadi,*
13-2              *Kambiz Moez, University of Alberta*          **339**

This paper presents a compact 60 GHz power amplifier utilizing a novel 4-waymulti-conductor power combiner. Fabricated in 65 nm CMOS process, the measured gain of the 0.19 mm$^2$ power amplifier is 18.8 dB at 60 GHz.  A maximum saturated output power of 18.3 dBm is measured with the 15.9% peak power added efficiency.

9:55 am          **CMOS Low-Power Transceivers for 60GHz Multi Gbit/s Communications (Invited),** *V.*
13-3              *Vidojkovic, V. Szortyka, K. Khalaf,G. Mangraviti, B. Parvais, K. Vaesen, S. Brebels, A. Spagnolo, M. Libois, J. Long\*, K. Raczkowski, P. Raghavan, A. Bourdoux, L. Min, C. Soens, V. Giannini, P. Wambacq, imec, \*Delft University of Technology*          **343**

The availability of 9GHz bandwidth around 60GHz in combination with simple modulations schemes, low-cost radio ICs and small antenna size allows for multi Gbit/s wireless communications. In this article the potential of 60GHz wireless communications is evaluated from system, application and user point of view. Further, design challenges for 60GHz CMOS transceivers are identified. State-of-the-art designs show that short-range high-data rate radio links based on CMOS ICs can be made, potentially helped with beamforming.

10:45 am          **BREAK**

11:05 am          **A CMOS 21-48GHz Fractional-N Synthesizer Employing Ultra-Wideband Injection-**
13-4              **Locked Frequency Multipliers,** *A. Li, S. Zheng, J. Yin, X. Luo\*, H. Luong, The Hong Kong University of Science and Technology, \*Huawei Technologies Co. Ltd.*          **351**

Higher-order LC tanks are proposed to widen the locking range of mm-Wave injection-locked frequency multipliers. Employing a chain of such multipliers, a wideband fractional-N frequency synthesizer is demonstrated in 65nm CMOS. An output tuning range from 20.6-48.2GHz (80.2%) is measured with phase noise<-107dBc/Hz at 1MHz offset while consuming 148mW.

| | |
|---|---|
| 11:30 am<br>13-5 | **A 75.7GHz to 102GHz Rotary-traveling-wave VCO by Tunable Composite Right /Left Hand T-line,** *Shunli Ma, Wei Fei\*, Hao Yu\*, Junyan Ren, Fudan University, \*Nanyang Technological University* **355** |

With the use of tunable composite-right/left-hand transmission line, this paper provides a wide frequency-tuning-range mechanism for Mobius-ring rotary-traveling-wave VCO in millimeter-wave region. Measurement results show 29.5% tuning range with center frequency at 89.3GHz, and phase noise from-100.08dBc/Hz to -98.7dBc/Hz with 10MHz offset, demonstrating state-of-art FOMT of -177.78dBc/Hz.

| | |
|---|---|
| 11:55 am<br>13-6 | **Transformer-Based Dual-Band VCO and ILFD for Wide-Band mm-Wave LO Generation,** *Yue Chao, Howard C. Luong, Hong Kong University of Science and Technology* **359** |

This paper presents wide-band transformer-based mm-wave dual-band VCO and ILFD. Based on two novel design techniques, the circuit is designed and fabricated in TSMC 65nm CMOS process and measures a state-of-art performance.


# Educational Session 1

Tuesday, 9/24, 9:00 am
Cedar Ballroom
**Session Chair**: Foster Dai, Auburn University
**Session Co-Chair:** Earl McCune, RF Communications Consulting

| | |
|---|---|
| E-1<br>9:00 am –<br>10:30 am | **Concurrent Design of ESD Protection and Integrated Circuits for Optimization and Prediction,** *Albert Wang, University of California, Riverside* **363** |

As semiconductor technologies continuously advance into nano nodes, while integrated circuits (IC) become faster and more complex, on-chip ESD protection design quickly emerges as a huge IC design barrier nowadays. Major ESD design challenges include the followings: First, how to conduct simulation-based quantitative design to achieve ESD protection design optimization and prediction? Second, how to minimize ESD-induced parasitic effects that affect IC performance. Third, how to perform co-design of ESD protection and IC to achieve ESD protection and core circuit design optimization simultaneously. This lecture discusses critical aspects and techniques in practical ESD protection designs, including a mixed-mode ESD simulation-design method for design prediction, accurate RF ESD design characterization, complex ESD-IC interactions, ESD+IC co-design for whole-chip design optimization, etc. Real-world design examples will be presented.

10:30 am    **BREAK**


# Educational Session 2

Tuesday, 9/24, 11:00 am
Cedar Ballroom

**Session Chair**: Eric Naviasky, Cadence
**Session Co-Chair:** Gerrit den Besten, NXP Semiconductors

| | |
|---|---|
| E-2<br>11:00 am –<br>12:30 pm | **Characterization of Matching, Variability, and Low-Frequency Noise for Mixed-Signal Technologies,** *Hans Tuinhout, NXP*    **397** |

Parametric mismatches and low frequency noise are major performance limiters as well as notorious causes for redesigns of high performance mixed-signal (HPMS) circuits and systems. Consequently, it is extremely important to measure, analyze, interpret, model and document these effects for mixed-signal technologies.

Part one of this educational lecture discusses parametric mismatch benchmarks and variability characterization challenges for active and passive devices. Part two focuses on low frequency noise, in particular on the emerging challenge in this field, namely variability of 1/f noise and associated Random Telegraph Noise.

These topics will be exemplified with results from (Bi)CMOS technologies, ranging from the current HPMS cash cow technologies (140 to 250 nm minimum dimensions), up to more advanced 40 nm node devices which can be seen as the stepping stone to some of the ultimate challenges of sub 10 nm devices that will mark the end of the CMOS shrink roadmap.

# Luncheon Keynote

Tuesday, Sept. 24, 12:20 – 1:50 pm
Sierra Ballroom
Tickets for the luncheon are for sale at the Registration Desk

**Connecting with the Emerging Nervous System of Ubiquitous Sensing,** *presented by Joseph Paradiso, MIT Media Laboratory*    **N/A**

Embedded sensors are touching every phase of our lives as they diffuse into the objects and environments around us. We'll exhibit a "phase change" within a few years, however, once this sensor information becomes networked and available to applications running outside of each device's domain that will be at least as profound as the web was to computers. Accordingly, this talk will overview the broad theme of interfacing humans to the ubiquitous electronic "nervous system" that sensor networks will soon extend across things, places, and people. I'll illustrate this through two avenues of research - one looking at a new kind of digital "omniscience" (e.g., building different kinds of browsers for sensor network data) and the other looking at buildings & tools as "prosthetic" extensions of humans (e.g., making HVAC systems an extension of your sense of comfort), drawing from many projects that are running in my group at the MIT Media Lab.

# Session 14 – Panel Session

"Do You Need to Plug In to Get Your Fill of Bits?"
Tuesday, 9/24, 2:00 pm
Oak Ballroom
**Moderator:** Sam Palermo, Texas A&M University

Applications such as video streaming and data sharing/backup are driving demand for increased device-to-device data transfer bandwidth for in/adjacent-room communication on the order of 10-20m. While the demand for higher data rates is growing, consumers are also becoming accustomed to and beginning to expect broadband wireless connectivity for their devices. To replace electrical cable-based links, wireless links will need to demonstrate competitive or superior data rates, energy efficiency, reliability, cost, and in some applications, latency. On the other hand, optical interconnects offer the flexibility of traditional electrical cable-based systems at potentially higher data rates and lower power and latency. However, the reliability and cost of optical cable systems are open issues. This panel aims to answer the question: "Do You Need to Plug In to Get Your Fill of Bits?" In other words, can future wireless systems support the 10+Gb/s data rates that future systems will demand? If not, what is best approach? Traditional electrical or emerging optical cable solutions?

Panelists:    Elad Alon, University of California - Berkeley
              Marc Loinaz, Broadcom
              Payam Heydari, University of California - Irvine
              Tirdad Sowlati, Broadcom
              Drew Alduino, Intel

# Session 15 - Forum Session: Electrical and Photonics I/O Test and Debug

Tuesday, 9/24, 2:00 pm
Fir Ballroom
**Session Chair:** Mike Li, Altera
**Session Co-Chair:** Takahiro Yamaguchi, Advantest Laboratories

2:00 pm      **Introduction**

             This session addresses test and debug challenges associated with multi-Gbits/s to more than 30 Gbits/s I/Os built with electrical and photonic ICs.

2:05 pm      **The Future of High Speed Electrical and Photonics IO Testing: Facing Complex**
15-1         **Challenges,** *Salem Abdennadher, Intel Corporation*     **N/A**

             Where is High Speed IO manufacturing Test heading? As technology continues to scale to increase system bandwidth, decrease power dissipation, die area and system cost, the challenges associated with test seem to expand exponentially. There has been a rise in defect occurrences as Serial Electrical IO interfaces instances and test complexity rise. In addition, introduction of optical IO's in main stream products is introducing unprecedented challenges. Technology process variation and process uncertainty is also affecting the performance of these circuits.

             Intel Test community wonder if the current technologies and strategies are adequate in the short term and what they should focus on now to deal with issues that are surfacing in the 2013-2015 timeframe. Current 22nm analog test coverage issues will persist to be an issue with 14nm process and beyond or even get worse. With Signal Headroom becoming too small to design analog circuit with sufficient signal integrity, HVM tests need to screen not only for manufacturing defects but also for design marginality and process uncertainty.

             In this talk will present new HVM test techniques to meet the ever increasing test complexity challenges. Such as developing methodologies to test optical interfaces depending on their level of integration (Hybrid or Full). Providing new innovative approaches in areas such as: Complete No Touch Testing (NTT) methodology and systematic defect capture in Electrical High Speed IO circuits.

2:40 pm      **Testability Improvement for 12.8 GB/s Wide IO DRAM Controller by Small Area**
15-2         **Pre-bonding TSV Tests and a 1 GHz Sampled Fully Digital Noise Monitor,** *T. Nomura, R. Mori, M. Ito\*, K. Takayanagi, T. Ochiai, K. Fukuoka, K. Otsuga, K. Nii, S. Morita, T.*

Hashimoto, T. Kida, J. Yamada, H. Tanaka, Renesas Electronics Corporation, *Renesas Micro Systems Corporation    **453**

A Wide IO DRAM controller chip was designed, and fabricated with Through Silicon Via (TSV) technology. The memory interface consists of 1200 TSVs including 512 bit data signals, which introduces new challenges in testability. To address these challenges, testing schemes by dedicated circuitry are proposed. TSV test circuitry is implemented in the micro-IOs placed in between the fine pitch TSV arrays, which can detect and reject TSV defects prior to stacking process. Another circuitry is for monitoring power noise, where we are aware of 512 bit Data simultaneously switching noise. We also introduced a impedance optimization scheme associated with the noise monitoring circuitry, where Vmin was improved for 30mV by appropriate optimization. We achieved 12.8GB/s operation, while IO power was reduced by 89% compared to LPDDR3.

3:15 pm
15-3

**Design Verification and Testing of High Speed Silicon Photonics Links,** *Brian Welch, Luxtera, T. Nomura, R. Mori, M. Ito\*, K. Takayanagi, T. Ochiai, K. Fukuoka, K. Otsuga, K. Nii, S. Morita, T. Hashimoto, T. Kida, J. Yamada, H. Tanaka, Renesas Electronics Corporation, *Renesas Micro Systems Corporation*    **N/A**

This paper looks at the verification and test techniques that can be deployed in silicon photonics solutions, and how it mirrors those that are used in conventional CMOS design.

3:50 pm    **BREAK**

4:05 pm
15-4

**CMOS Photonics: Product Test and Debug Challenges,** *David Piede, Cisco Systems*    **N/A**

CMOS photonics is a new and exciting technology that offers potential improvements in cost, power, integration, and size over current photonic and electronic technologies. As with any new technology, there are new challenges associated with productization. We will focus on the test and debug challenges associated with known-good-die (KGD), and with the heterous integration of electronics and CMOS photonics in a package format.

4:40 pm    **Panel  Discussion and Q&A**


# Session 16 -- Nyquist Rate A/D Converters

Tuesday, 9/24, 2:00 pm
Pine Ballroom
**Session Chair**: Mohammad Ranjbar, Cirrus Logic
**Session Co-Chair**: John McNeill, Worcester Polytechnic Insittute

2:00 pm    **Introduction**

A/D converters are key building blocks in many electronic systems. Their applications range from low-speed, low-power sensors to high-speed wireless or wireline communication systems.  Papers in this session cover a range of speeds form 250 kSps to 12.8 GSps, and resolutions from 5 to 12 bits.

2:05 pm
16-1

**A 12.8GS/s Time-Interleaved SAR ADC with 25GHz 3dB ERBW and 4.6b ENOB,** *Y. Duan, E. Alon, UC Berkeley*    **457**

This paper presents a 12.8GS/s 32-way hierarchically time-interleaved SAR ADC with 4.6-bit

ENOB in 65nm CMOS. The prototype utilizes multi-stage sampling and a cascode sampler circuit to enable greater than 25GHz 3dB effective resolution bandwidth (ERBW). We further employ a pseudo-differential SAR ADC to save power and area. The core circuit occupies only 0.23mm$^2$and consumes a total of 162mW from dual 1.2V/1.1V supplies.

**2:30 pm**
**16-2**

**A 10GS/s 6b Time-Interleaved ADC with Partially Active Flash sub-ADCs,** *Xiaochen Yang, Robert Payne\*, Jin Liu, University of Texas at Dallas, \*Texas Instruments*  **461**

A 10GS/s 6b time-interleaved ADC in 65nm CMOS is presented. A partially-active flash sub-ADC is proposed to improve the power efficiency, a source-follower based boot-strap T&H circuit reduces input kickback and improve ADC bandwidth, and timing skew is corrected with duty-cycle calibration. The measurement shows a FOM of 197fJ/conv-step.

**2:55 pm**
**16-3**

**An 8-Bit 4-GS/s 120-mW CMOS ADC,** *Hegong Wei, Peng Zhang\*, Bibhu Datta Sahoo\*\*, and Behzad Razavi, University of California-Los Angeles, \*Tsinghua University, \*\*Amrita University*  **465**

A four-channel time-interleaved pipelined ADC employs a new timing calibration technique to suppress mismatch-induced spurs and achieve a Nyquist-rate SNDR of 44.4 dB. Designed in 65-nm CMOS technology, the ADC draws 120 mW, providing an FOM of 219 fJ per conversion step.

**3:20 pm**
**16-4**

**A 7.1-mW 1-GS/s ADC with 48-dB SNDR at Nyquist Rate,** *S. Hashemi, B. Razavi, University of California, Los Angeles*  **469**

A two-stage pipelined ADC employs a double-sampling residue amplifier, two interleaved precharged DACs, and a new calibration scheme to correct for residue gain error, offset, and nonlinearity. Realized in 65-nm CMOS technology and sampling at 1 GHz, the prototype exhibits an FOM of 25 fJ/conversion-step while drawing 7.1 mW from a 1-V supply.

**3:45 pm**    **BREAK**

**4:00 pm**
**16-5**

**A 0.55 V 7-bit 160 MS/s Interpolated Pipeline ADC Using Dynamic Amplifiers,** *J. Lin, D. Paik, S. Lee, M. Miyahara, A. Matsuzawa, Tokyo Institute of Technology*  **473**

This paper presents a 0.55 V, 7-bit, 160 MS/s pipeline ADC using dynamic amplifiers. In this ADC, dynamic amplifiers with a common-mode detection technique are used as residual amplifiers to increase its robustness against supply voltage lowering and to remove the unnecessary static power consumption achieving clock-scalability in power performance.

**4:25 pm**
**16-6**

**A 95-MS/s 11-bit 1.36-mW Asynchronous SAR ADC with Embedded Passive Gain in 65nm CMOS,** *Jae-Won Nam, David Chiong, Mike Shuo-Wei Chen, University of Southern California*  **477**

An asynchronous SAR ADC with embedded passive gain is fabricated in 65nm CMOS. The prototype ADC demonstrates a peak SNDR of 63.1dB and SFDR of 75.2dB at 95MS/s. It dissipates 1.36mW at 1.1V supply and achieves the lowest FoM among the recently published ADCs with similar specification (>10 ENOB, >10MS/s).

**4:50 pm**
**16-7**

**A 24µW 12b 1MS/s 68.3dB SNDR SAR ADC with Two-Step Decision DAC Switching,** *Yung-Hui Chung, Meng-Hsuan Wu, Hung-Sung Li, MediaTek, Inc.*  **481**

A 12-bit SAR ADC employs a new DAC switching technique for improving the ADC linearity and tolerating the DAC settling errors. At 1MS/s, it consumes 24uW from a 0.9V supply. At the Nyquist-rate input, measured SNDR and SFDR are 68.3dB and 82dB, respectively. It achieves a FoM of 11.7fJ/conversion-step.

5:15 pm
16-8

**A 3.3fJ/conversion-step 250kS/s 10b SAR ADC Using Optimized Vote Allocation,** *M. Ahmadi, W. Namgoong, University of Texas at Dallas* **485**

A 10b SAR ADC that supports a flexible differential input swing from 0.4V to 1V is presented. The proposed ADC employs a non-binary architecture along with a majority vote comparison using optimized vote allocation. The prototype achieves ENOB ranging from 7.1b to 9.1b and FOM from 3.3 to 6.8fJ/conversion step.

# Educational Session 3

Tuesday, 9/24, 2:00 pm
Cedar Ballroom
**Session Chair**: Christoph Sandner, Infineon Technologies Austria AG
**Session Co-Chair**: Hoi Lee, University of Texas at Dallas

E-3
2:00 pm –
3:30 pm

**Single-Inductor-Multiple-Output DC-DC Converter Design,** *Philip Mok, Hong Kong University of Science and Technology* **489**

Multiple well-regulated power supplies are essential for reducing power consumption and isolating the coupling noise between different functional blocks in VLSI design. With the increasing number of functional blocks in SoC applications, the need for a cost and efficiency effective solution of multiple power supplies is growing. Single-Inductor-Multiple-Output (SIMO) switching regulator, which provides several output voltages with only one inductor, is one of promising solutions and becomes a hot topic in DC-DC converter design due to the cost and volume reduction. However, with one inductor shared by all the outputs to accumulate and transfer energy from the input, cross regulation easily appears at outputs when a change in the inductive energy is induced by a load transient at one output. These unwanted voltage variations affect the performance or even the function of the loading devices. This talk will discuss the operation principle of SIMO switching converters and their design issues. To minimize cross regulation of a SIMO regulator, several control techniques will be presented and their pros and cons will be discussed.

3:30 pm    **BREAK**

# Educational Session 4

Tuesday, 9/24, 4:00 pm
Cedar Ballroom
**Session Chair**: Foster Dai, Auburn University
**Session Co-Chair**: Jonathan Borremans, IMEC

E-4
4:00 pm –
5:30 pm

**Advanced Digital Phase-Locked Loops,** *Salvatore Levantino, Politecnico di Milano* **524**

This tutorial will introduce the fundamentals of digital phased-locked loops for wireless applications. After reviewing the basic architectures, the tutorial will analyze the mechanisms of generation of limit cycles, which manifest themselves as spurious tones in the output spectrum even when synthesizing integer-N channels. Then, loop-parameter settings and design strategies for spur elimination and phase-noise optimization will be derived. Next, we will move to the fractional-N case, in which quantization and nonlinearity add new sources of spur tones, and we will review the different design techniques which helps mitigate such impairments. Finally, examples of state-of-the-art implementations of frequency synthesizers and direct-FM modulators based on digital PLLs will be discussed.

## Tuesday Poster Session

Tuesday, 9/24, 5:00 pm – 7:00 pm

T-1      **All-Digital 90° Phase-Shift DLL with a Dithering Jitter Suppression Scheme,** *D. H. Jung, K. Ryu, J. H. Park, W. Lee\*, S. O. Jung, Yonsei University, \*Samsung Electronics*    **572**

We propose a 90° phase-shift digital delay-locked loop (DLL) with a new dithering jitter suppression scheme. Delay-line control code dithering is effectively suppressed by comparing the distribution of the input and the output clock jitter. And the phase shift and duty cycle correction accuracy are enhanced by MDLL based phase-shift structure.

T-2      **A 1Gb/s Reconfigurable Pulse Compression Radar Signal Processor in 90nm CMOS,** *Jun Li, Hirohito Mukai\*, Mehmet Parlak, Michiaki Matsuo\*, James F. Buckwalter, University of California San Diego and \*Panasonic Corporation*    **576**

This paper presents a reconfigurable analog signal processing circuit for pulse compression radar. Adapting bandwidth for the range of the target is proposed for radar systems. The baseband signal processor includes a high-speed correlator/integrator, a 4-bit flash analog-to-digital converter (ADC) and a multi-range delay lock loop (DLL). The DLL generates multi-phase clock to align the template signal with the received signal. The circuit is fabricated in90-nm CMOS and can be configured to work from 50Mb/s to 1Gb/s with Barker codes. An SNR of 8.5dB is demonstrated for 1Gb/s. The total power consumption is 33mW at 1Gb/s.

T-3      **A 5GS/s 4-bit Time-Based Single-Channel CMOS ADC for Radio Astronomy,** *A. Macpherson, J. Haslett, L. Belostotski, University of Calgary*    **580**

A 4-bit 5GS/s 65nm time-based analog-to-digital converter (ADC) targeting the next-generation Square Kilometre Array (SKA) is presented. This ADC is composed of an analog voltage-to-time converter (VTC) front end and a digital time-to-digital converter (TDC) back end. The two components can be physically separated to minimize the impact of digital noise from the ADC on high-gain, high-sensitivity receiver chains common in radio telescopes.

T-4      **A 6b 800MS/s 3.62mW Nyquist AC-coupled VCO-Based ADC in 65nm CMOS,** *P. K. Sharma, M. S-W. Chen, University of Southern California*    **584**

A 6-bit 800MS/s Nyquist VCO-Based ADC is proposed. The ADC utilizes an analog differentiator, replacing the conventional digital differentiator to avoid quantization noise shaping and achieve Nyquist operation with embedded DC rejection, first order anti-aliasing filtering and improved VCO linearity without calibration. The ADC achieves peak SNDR of 34dB with over 400MHz input bandwidth and occupies an active area of 0.015mm2 while consuming 3.65mW.

T-5      **A Fully-Digital Beat-Frequency Based ADC Achieving 39dB SNDR for a 1.6mV$_{pp}$**

**Input Signal,** *Bongjin Kim, Weichao Xu, Chris H. Kim, University of Minnesota* **588**

A fully-digital VCO-based ADC employing a beat frequency detection scheme is demonstrated in 65nm. The proposed design is highly effective in measuring extremely small changes in the VCO frequency within a short sampling time. Direct A-to-D conversion of a 1.6mVpp differential input signal with 39dB SNDR was experimentally verified.

T-6 **A 4–15-GHz Ring Oscillator based Injection-Locked Frequency Multiplier with Built-in Harmonic Generation,** *J. Xu, J. Hu, B. Ciftcioglu, H. Wu, University of Rochester* **592**

This paper presents a new inductorless injection-locked frequency multiplier(ILFM) designed to achieve wide locking range and low power dissipation. The ILFM integrates harmonic generation in each stage and realizes multiphase injection simultaneously. A multiply-by-2 ILFM prototype is implemented and demonstrated the wide locking range of the proposed ILFM.

T-7 **WITHDRAWN**

T-8 **Power Management Circuits for a 15-μA, Implantable Pressure Sensor,** *Steve Majerus\* and Steven L. Garverick, Case Western Reserve University, \*APT Rehabilitation Research and Development Center* **596**

An ASIC for wireless bladder pressure sensing incorporates power-management circuitry, limiting active time of instrumentation circuitry and minimizing telemetry rate. Measured results with prerecorded bladder signals indicate that5% of acquired samples merit transmission, resulting in an average telemetry rate of 1.5 Hz and total IC current of 12.8 μA.

T-9 **A Novel Voltage-Programmed Pixel Circuit with $V_T$-Shift Compensation for AMOLED Displays,** *M. Yang, N. Papadopoulos, C-H. Lee, W.S. Wong, M. Sachdev, University of Waterloo* **600**

A novel voltage-programmed pixel circuit using hydrogenated amorphous silicon(a-Si:H) thin-film transistors (TFTs) for active-matrix organic light-emitting diode (AMOLED) displays is proposed. The threshold voltage shift (ΔVT) of the drive TFT due to electrical stress is compensated by the change of gate-to-source voltage (ΔVGS) generated by the ΔVT-dependent charge transfer from the drive TFT to a TFT-based Metal-Insulator-Semiconductor (MIS)capacitor. Another MIS capacitor is used to improve OLED drive currents. Measurement results verify the effectiveness and speed of the proposed pixel circuit.

T-10 **Design for Manufacturing Layout Analyses Correlate Layout to Physico-Chemical Yield Loss Mechanisms,** *C.P. Tan, C. Zhou, Y. Tian, C. Liu, H.-M. Lam, J. Zhang, M. Lu, GLOBALFOUNDRIES Singapore Pte. Ltd.* **604**

We introduce a case-based learning workflow in the foundry for managing layout weakpoints and implementing layout analyses checks. In this work, we describe case studies that demonstrate how layout analyses can be used to detect layout weakpoints and correlate them to actual physico-chemical mechanisms behind defects observed on silicon.

T-11 **A Split-Foundry Asynchronous FPGA,** *B. Hill, R. Karmazin, C. Ortega, J. Tse, R. Manohar, Cornell University* **608**

We present the first published measurements of a complex digital integrated circuit fabricated in both standard and split-foundry processes. Our1.3-million-transistor asynchronous FPGA operates at over 300MHz in 130nm. We discuss the challenges inherent in split design and our automated layout tools that address them.

T-12   **A 40-nm 8T SRAM with Selective Source Line Control of Read Bitlines and Address Preset Structure,** *S. Yoshimoto, S. Miyano\*, M. Takamiya\*\*, H. Shinohara\*, H. Kawaguchi\*, and M. Yoshimoto, Kobe University, \*Semiconductor Technology Academic Research Center, and \*\*University of Tokyo*   **612**

This paper presents a 40-nm 8T SRAM in which bit lines are partially discharged by a selective source line control (SSLC) for low-power operation. The proposed SSLC scheme reduces a read bit line voltage swing in an unselected column with af loating source line (SL) of dedicated read ports.

T-13   **AOT-Controlled Dual-Mode AVP Buck Reglator with AEAF Mechanism,** *Hsin-Lun Li, Chia-Cheng Pao, Bo-Ming Chen, Chien-Hung Tsai, National Cheng Kung University*   **616**

A novel adaptive voltage positioning (AVP) buck regulator using adaptive on-time (AOT) control targeted for applications with low-ESR output capacitors is proposed. In this work, AOT control is adapted to keep the system's switching frequency quasi-fixed or independent of the input supply voltage and the AVP mechanism is realized without the need to use conventional error amplifier compensator or extra current-sensing circuit. For ensuring the system's switching frequency not entering the range of acoustic frequency at light load, an AEAF (avoid entering acoustic frequency) circuit is also proposed. For comparison purpose, the implemented buck regulator can be set to operate under AVP or non-AVP mode. This work has been fabricated and verified with a standard 0.18µm CMOS technology. Experimental results show excellent transient recovery time of 4µs (under AVP mode), ±0.11% switching frequency variation (for the specified input voltage range), and 91% peak conversion efficiency.

T-14   **Switched-Capacitor Filter based Type-III Compensation for switched-mode Buck Converters,** *G. Bawa, A.Q. Huang, North Carolina State University*   **620**

A switched-capacitor filter based Type-III compensation for regulation of Buck converters is presented. Compared to the all-analog filter, the proposed compensator can be fully-integrated onto the die, resulting in reduced footprint and cost. The filter time-constants also scale linearly with the converter's switching time-period, resulting in increased programmability and ease-of-use.

T-15   **Estimation of Passive Mixer Output Bandwidth Using Switched-Capacitor Techniques,** *Essam S. Atalla, Frank Zhang\*, Abdellatif Bellaouar\*, Poras T, Balsara, The University of Texas at Dallas and \*NVIDIA Corp.*   **624**

Passive mixers have become an essential component of SAW-less receivers. It is well known that the passive mixer behaves as a switched-capacitor circuit (SC)but to the authors knowledge, there is no reported analysis of the mixer impedance that truly accounts for the SC behavior. In this paper, we present for the first time a closed form of the passive mixer output impedance based on SC techniques. We prove that the fundamental lower limit of the mixer impedance is proportional to the well-known switched capacitor resistor$1/\left(f_{LO}C\right)$ and different from the previously reported mixer switch ON resistance. We also explain that the equation is useful in estimating output bandwidth of passive mixer based front-ends with general LNA load impedance. We finally show that our bandwidth estimation matches measured results of two receiver front-ends.

T-16     **How to Reduce Power in 3D IC Designs: A Case Study with OpenSPARC T2 Core,** *Moongon Jung, Taigon Song, Yang Wan, Young-Joon Lee, Debabrata Mohapatra\*, Hong Wang\*, Greg Taylor\*, Devang Jariwala\*, Vijay Pitchumani\*, Patrick Morrow\*, Clair Webb\*, Paul Fischer\*, and Sung Kyu Lim, Georgia Institute of Technology, \*Intel Corporation* **628**

The power benefit of 3D ICs is demonstrated with an OpenSPARC T2 core. Four design techniques are explored to optimize power in 3D ICs: 3D floor planning, intra-block metal layer usage control, dual-Vth, and FUB folding. With aforementioned methods, the total power saving of 21.2% is achieved against the2D counterpart.

T-17     **A General-purpose Vision Processor with 160x80 Pixel-Parallel SIMD Processor Array,** *A. Lopich, P. Dudek, University of Manchester* **632**

In this paper we present a vision processor, which incorporates a 160×80 SIMD array of pixel-processors. The processor operates with a 100MHz clock and 1.8Vsupply. The device provides 640 GOPS (binary) and 23 GOPS (greyscale) consuming0.5 W. The chip occupies 50mm^2 and is fabricated in a standard 0.18 µm CMOS process. The I/O interface supports 200 M Pixels/s (greyscale), 1.6 G Pixels/s(binary) and 40 M Pixels/s (address-event readout) data rate, and PE-parallel image sensing mode for embedded high-speed vision applications. Experimental results indicate that the performance of the presented chip approaches the efficiency of recently reported application-specific vision processors, while providing full programmability and thus being adjustable to a wide range of applications.

T-18     **A Programmable Analog Frequency-Locked Loop for VCO Characterization and Test with 8 ppm Resolution,** *S. Aouini, J.F. Bousquet, N. Ben-Hamida, L. Jakober, J. Wolczanski, C. Kurowski, Ciena Corporation* **636**

We present a digitally controlled analog frequency-locked loop for VCO characterization and test. The scheme allows a frequency tuning better than 8ppm. The AFLL comprises a 17-bit frequency counter, a sigma-delta modulator used for dithering the correction signal, a charge-pump and capacitance used as integrator and a VCO.

T-19     **Detection of Early-Life Failures in High-K Metal-Gate Transistors and Ultra Low-K Inter-Metal Dielectrics,** *Y.M. Kim, J. Seomun\*, H.-O. Kim\*, K.-T. Do\*, J.Y. Choi\*, K.S. Kim\*, M. Sauer\*\*, B. Becker\*\*, S. Mitra, Stanford University, \*Samsung Electronics, \*\*University of Freiburg* **640**

We derive signatures for early-life failures (ELF) in 28nm high-K/metal-gate transistors and ultra low-K inter-metal dielectrics. We also demonstrate that the derived ELF signatures can be successfully detected using a clock control technique, activated during periodic on-line self-test and diagnostics in robust systems, without requiring expensive concurrent error detection.

T-20     **A Fully Differential Ultra-Compact Broadband Transformer Based Quadrature Generation Scheme,** *Jong Seok Park, Shouhei Kousai\*, and Hua Wang, Georgia Institute of Technology, \*Toshiba Corporation* **644**

This paper presents a fully differential ultra-compact broadband transformer-based quadrature generation scheme implemented within only one inductor-footprint. A 5GHz design in a 65nm CMOS only occupies 260µm-by-260µmarea and achieves 0.82dB insertion loss (5GHz) with 3.8° maximum phase error and ±0.5dB amplitude mismatch within 13% bandwidth (4.75GHz to 5.41GHz).

T-21    **A -173 dBc/Hz @ 1 MHz offset Colpitts Oscillator using AlN Contour-Mode MEMS Resonator,** *Jabeom Koo, Augusto Tazzoli*, Jeronimo Segovai-Fernandez*, Gianluca Piazza*, Brian Otis, University of Washington, *Carnegie Mellon University*    **648**

A differential Colpitts oscillator using AlN MEMS CMR designed in 0.13 µm CMOS is presented in this work. The oscillator operates at 1.16 GHz, with a total power consumption of 4.2 mW at 1 V supply. It achieves a phase noise of -143.6dBc/Hz, -173.3 dBc/Hz at 100 kHz and 1 MHz offset frequency respectively with a figure of merit (FOM) of 228.3 dB. Current-based temperature compensation was employed to reduce oscillator drift across temperature.

T-22    **An Ultra-Broadband Compact Mm-Wave Butler Matrix in CMOS for Array-Based MIMO Systems,** *J. Park, T. Chi, H. Wang, Georgia Institute of Technology*    **652**

This paper presents an ultra-broadband compact mm-wave Butler Matrix utilizing new transformer-based swapped-port couplers. It is implemented as a 4×4 Butler Matrix at 63GHz in a 65nm CMOS process with 0.335×0.215mm2, and it achieves9.8GHz bandwidth, 2.77dB insertion loss, and better-than 17dB array peak-to-null-ratio over 57GHz and 67GHz.

T-23    **A 1.2 pJ/b 6.4 Gb/s 8+1-Lane Forwarded-Clock Receiver with PVT-Variation-Tolerant All-Digital Clock and Data Recovery in 28nm CMOS,** *Shuai Chen, Hao Li*, Liqiong Yang, Zongren Yang, Weiwu Hu and Patrick Yin Chiang*, Chinese Academy of Science, *Oregon State University*    **656**

This paper presents an energy/area-efficient forwarded-clock receiver fabricated in 28 nm CMOS process. The receiver consists of 8 data lanes plus one forwarded clock lane, and adopts a novel all-digital clock and data recovery (CDR) based on delay-locked loop (DLL). The proposed all-digital DLL-based CDR uses the calibration and the update techniques to achieves a robust PVT-variation tolerance as well as a low power/area consumption. The measurement results show that our receiver can work at a data rate of 6.4 Gb/s with BER<10e-12 and consume only 7.5 mW per lane under 0.85 V power supply. The receiver core merely occupies an area of 0.02 mm*mm per lane.

T-24    **A True 4-Cycle Lock Reference-Less All-Digital Burst-Mode CDR Utilizing Coarse-Fine Phase Generator with Embedded TDC,** *Tetsuya Iizuka, Satoshi Miura*, Yohei Ishizone*, Yoshimichi Murakami*, Kunihiro Asada, University of Tokyo, *THine Electronics, Inc.*    **660**

This paper presents a reference-less all-digital burst-mode CDR using a coarse-fine phase generator with embedded TDC. It achieves true 4-cycle lock without warm-ups, and eliminates dynamic power consumption in a stand-by state. Fabricated in 65nm CMOS, this CDR operates from 1.40 to 2.06Gb/s and consumes9.6mW at 2.06Gb/s with 80x80µm$^2$.

# Session 17 -- Variation and Analog Modeling

Wednesday, 9/25, 9:00 am
Oak Ballroom
**Session Chair:** Trent McConaghy, Solido Design
**Session Co-Chair:** Brian Chen, Agilent

**9:00 am**     **Introduction**

This session discusses modeling process variation in analog and SRAM circuits, as well as analog thermal noise and distortion modeling.

**9:05 am**
**17-1**

**Thermal Noise Modeling of Nano-scale MOSFETs for Mixed-signal and RF Applications (Invited),** *Chih-Hung Chen, David Chen, Ryan Lee, Peiming Lei, and Daniel Wan, United Microelectronics Corporation*     **664**

This paper presents the thermal noise in nano-scale MOSFETs – from measurement, characterization, modeling, and potential technology enhancement for future low power, mixed-signal, and radio-frequency (RF) applications. Experimental data from five CMOS technology nodes, namely 180 nm, 130 nm, 90nm, 65 nm, and 40 nm nodes are presented and discussed.

**9:55 am**
**17-2**

**A Model-Agnostic Technique for Simulating Per-Element Distortion Contributions,** *Nagendra Krishnapura, Rakshitdatta K. S., Indian Institute of Technology*     **672**

The nonlinearity of an element can be altered while maintaining the operating point and first-order terms by appropriately combining two instances of the nonlinear element with complementary scaling factors for incremental voltages above the operating points. Per-element distortion contributions in a circuit can then be determined by altering the nonlinear terms by known factors and simulating the output distortion in each case. This technique can be used in a standard circuit simulator with the appropriate nonlinear device models butr equires no knowledge of the device model details on the part of the circuit designer. The technique is demonstrated by applying it to a common source amplifier with a nonlinear load and a two stage fully differential opamp.

**10:20 am**
**17-3**

**Corner Models: Inaccurate at Best and it Only Gets Worst ...,** *Colin C. McAndrew, Ik-Sung Lim, Brandt Braswell, and Doug Garrity, Freescale Semiconductor*     **676**

Corner (best- and worst-case) models have been a mainstay of integrated circuit design for decades. Obviously they can be effective, especially for digital CMOS design. However, there are significant inaccuracies that arise when digital CMOS corner models are used for analog circuits, or any types of circuits or measures of circuit performance they were not targeted for. This paper details what corner models can and cannot do, and shows their inadequacies for analog CMOS circuits.

**10:45 am**     **BREAK**

**11:05 am**
**17-4**

**Energy Centric Model of SRAM Write Operation for Improved Energy and Error Rates,** *Swaroop Ghosh, University of South Florida*     **680**

We propose an energy centric model of SRAM write operation. The model provides useful insight about energy and write error rate. We employ the proposed model for evaluating write assist mechanisms and their potential in reducing intrinsic memory error rates. We also employ it for optimizing energy of memories.

**11:30 am**
**17-5**

**SRAM Read Current Variability and its Dependence on Transistor Statistics,** *Sriramkumar Venugopalan, Vivek Joshi\*, Luis Zamudio\*, Matthias Goldbach\*, Gert Burbach\*, Ralf VanBentum\*, Sriram Balasubramanian\*, University of California, Berkeley, \*GLOBALFOUNDRIES*     **684**

Our study breaks down the dependence of SRAM read current (Iread) variability($\sigma$Iread) into constituting pass-gate (PG) and pull down (PD) NMOS transistor variability. We report a bottoms-up model for $\sigma$Iread including feedback in stacked transistors and discuss its implications on SRAM performance.

**11:55 am**
**17-6**

**Mismatch Characterization of Small Metal Fringe Capacitors,** *V. Tripathi, B. Murmann, Stanford University* **688**

This paper describes a test structure and measurements results pertaining to the characterization of single-layer, lateral-field, 0.45-fF and 1.2-fF unit metal capacitors in a 32-nm SOI CMOS process. The measurement-inferred average standard deviations for these capacitances are 1.2% and 0.8%, respectively, confirming variance scaling according to Pelgrom's matching law.

# Session 18 -- Energy Efficient SoC Design

Wednesday, 9/25, 9:00 am
Fir Ballroom
**Session Chair:** Visvesh Sathe, Advanced Micro Devices
**Session Co-Chair:** Arif Rahman, Altera Corporation

**9:00 am**

**Introduction**

Processors, accelerators and on-chip clocking implementations for energy-efficient SoC design.

**9:05 am**
**18-1**

**A 1GHz Hardware Loop-Accelerator with Razor-based Dynamic Adaptation for Energy-Efficient Operation,** *S. Das, G. Dasika, K. Shivashankar and D. Bull, ARM Ltd.* **692**

We describe the implementation and silicon measurement results from a Razor-based hardware loop-accelerator (RZLA), implementing the Sobeledge-detection algorithm. We demonstrate robust operation with a large Dynamic Voltage Scaling (DVS) range achieved using 50% of the clock-period for timing-speculation. At 1GHz operating frequency, Razor DVS enables34% energy saving on a per-device basis and 33% overall on the entire batch of devices.

**9:30 am**
**18-2**

**Energy Efficient Recognition and Mining Processor using Scalable Effort Design,** *Vinay Chippa, Hrishikesh Jayakumar, Debabrata Mohapatra, Kaushik Roy, Anand Raghunathan, Purdue University,* **696**

A Recognition and Mining(RM) processor, that exploits the inherent application resilience using scalable effort design is implemented in TSMC-65nm technology. Measurements demonstrate energy savings of 1.2-2.3X with no quality-loss, and2X-20X with modest quality reduction due to cross-layer optimization of algorithm, architecture and circuit level scaling mechanisms.

**9:55 am**
**18-3**

**An Energy-Efficient Coarse-Grained Dynamically Reconfigurable Fabric for Multiple-Standard Video Decoding Applications,** *L. Liu, C. Deng, D. Wang, M. Zhu, S. Yin, P. Cao\*, S. Wei,Tsinghua University,\*Southeast University* **700**

We introduce a coarse-grained dynamically reconfigurable fabric consisting of16x16 Processing Elements. Line-Switched Mesh Connect routing and Hierarchical Configuration Context organization scheme are proposed. Measured results show that the fabric has great advantage in energy efficiency compared with the state-of-art designs when processing video decoding and some other computation-intensive applications.

**10:20 am**
**18-4**
**SURFEX: A 57fps 1080P resolution 220mW Silicon Implementation for Simplified Speeded-Up Robust Feature with 65nm Process,** *L. Liu, W. Zhang, C. Deng, S. Yin, S. Cai, S. Wei, Tsinghua University* **704**

Speeded-Up Robust Feature algorithm is optimized for silicon implementation and a 57fps 1080P 220mW ASIC is presented. Methods including orientation assignment& descriptor extraction reorganization, memory accesses improvement and etc. are introduced. Experimental results show proposed architecture has great advantages in performance and power consumption compared with the state-of-art designs.

**10:45 am**  **BREAK**

**11:05 am**
**18-5**
**Supply-Noise Resilient Adaptive Clocking for Battery-Powered Aerial Microrobotic System-on-Chip in 40nm CMOS,** *Xuan Zhang, Tao Tong, David Brooks, Gu-Yeon Wei, Harvard University* **708**

A battery-powered aerial microrobotic System-on-Chip (SoC) has stringent weight and power budgets, which requires fully-integrated solutions for both clock generation and voltage regulation. Supply-noise resilience is important yet challenging for such SoC systems due to a non-constant battery discharge profile and load current variability. This paper proposes an adaptive-frequency clocking scheme that can tolerate supply noise and improve performance when implemented with an integrated voltage regulator (IVR). Measurements from a `brain' SoC, implemented in 40nm CMOS, demonstrate 2x performance improvement with adaptive-frequency clocking over conventional fixed-frequency clocking. Combining adaptive-frequency clocking with open-loop IVR extends error-free operation to a wider battery voltage range (2.8 to 3.8V) with higher average performance.

**11:30 am**
**18-6**
**Distributed clock generator for synchronous SoC using ADPLL network,** *E. Zianbetov, D. Galayko, F. Anceau, M. Javidan, C. Shan, O. Billoint\*\*, A. Korniienko\*, E. Colinet\*\*, G. Scorletti\*, J. M. Akre, J. Juillard\*\*\*, UPMC LIP6 Lab, \*Ampere lab, \*\*CEA-LETI, \*\*\*Supelec* **712**

This paper presents a novel architecture of on-chip clock generation employing a network of oscillators synchronized by the distributed all-digital PLLs(ADPLLs). The implemented prototype has 16 clocking domains operating synchronously in a frequency range of 1.1-2.4 GHz. The synchronization error between the neighboring clock domains is less than 60 ps. The fully digital architecture of the generation offers flexibility and efficient synchronization control suitable for use in synchronous SoCs.

**11:55 am**
**18-7**
**A 920MHz Quad-core Cryptography Processor Accelerating Parallel Task Processing of Public-key Algorithms,** *Shuai Wang, Jun Han, Yang Li, Yifan Bo, Xiaoyang Zeng, Fudan University* **716**

The wireless access point (AP) devices of the next generation requires to implement the high-complexity public-key ciphers efficiently on programmable processors. Therefore, this paper presents a quad-core processor that accelerates public-key computations by enabling high-speed parallel task processing.

# Session 19 -- Analog Techniques I

Wednesday, 9/25, 9:00 am
Pine Ballroom
**Session Chair:** Don Thelen, ON Semiconductor
**Session Co-Chair:** Jerry (Xicheng) Jiang, Broadcom

9:00 am **Introduction**

This session showcases a collection of analog techniques that enable high-performance analog design.

9:05 am **Parallel Gain Enhancement Technique for Switched-Capacitor Circuits,** *Hariprasath*
19-1 *Venkatram, Benjamin Hershberg, Taehwan Oh, Manideep Gande, Kazuki Sobue*, Koichi*
*Hamashita*, Un-Ku Moon, Oregon State University, *Asahi Kasei Microdevices* **720**

This paper presents a unified classification model for gain enhancement techniques used in the design of high performance amplifiers. A parallel gain enhancement technique is proposed for switched capacitor circuits which combine the best features of the existing gain enhancement techniques found in continuous-time and discrete-time amplifiers. This technique utilizes two dependent closed loop amplifiers to enhance the open loop DC gain of the main amplifier. This replicated parallel gain enhancement (RPGE) technique enables a very high DC gain amplifier with an improved harmonic distortion performance. A proof of concept pipeline ADC in a 0.18 µm CMOS process using RPGE technique achieves 75 dB SNDR, 91 dB SFDR, -87 dB THD at 20 MS/s. The measured 13 bit DNL and INL is +0.75/-0.36 and +0.88/-0.92 LSB respectively. The ADC operates from a supply voltage of 1.3 V, consumes 5.9 mW, occupies 3.06 sq. mm and achieves a figure of merit of 65 fJ/CS.

9:30 am **Sampling Circuits That Break the kT/C Thermal Noise Limit (Invited),** *R. Kapusta, H.*
19-2 *Zhu, C. Lyden, Analog Devices, Inc.* **724**

This paper presents techniques that prove the kT/C limit of sampled thermal noise is, in fact, not a limit at all. A first feedback technique is demonstrated to reduce thermal noise power by nearly 50%, and a second active noise cancellation technique achieves better than 70% noise power reduction.

10:20 am **Blind Background Calibration of Harmonic Distortion Based on Selective Sampling,**
19-3 *Manideep Gande, Ho-Young Lee, Hariprasath Venkatram, Jon Guerber, Un-Ku Moon, Oregon*
*State University* **730**

This paper proposes a blind calibration algorithm for suppressing harmonic distortion in analog to digital converters (ADCs). The proposed algorithm does not need any external calibration signal and is first of its kind. The proposed algorithm relies on the properties of downsampling and orthogonality of sinusoidal signals to estimate the harmonic distortion coefficients. The algorithm can be operated in both foreground and background modes to remove even and odd harmonics simultaneously. The algorithm is demonstrated on a first-order ring oscillator based delta sigma ADC, whose performance is harmonic distortion limited. Built in 0.13 µm, the algorithm improves the SNDR of the ADC by 39dB while improving SFDR by 45 dB.

10:45 am **BREAK**

11:05 am **CMOS Millimeter Wave Phase Shifter Based on Tunable Transmission Lines,** *Wayne*
19-4 *H. Woods, Alberto Valdes-Garcia*, Hanyi Ding, Jay Rascoe, IBM Semiconductor Research and*

*Development, \*IBM T.J. Watson Research Center*   **734**

Design and measurements are presented of a new type of phase shifter, fabricated in a 32 nm SOI technology and operating at 60 GHz, which consists of novel tunable t-line sections that use FET switches to control L and C separately to minimize Z0 variation while changing delay.

11:30 am   **Charge Steering: A Low-Power Design Paradigm (Invited),** *Behzad Razavi, University*
19-5       *of California, Los Angeles*   **738**

Discrete-time charge-steering circuits consume less power than their continuous-time current-steering counterparts even at high speeds. This advantage can be exploited in the design of semi-analog circuits such as latches, demultiplexers, and CDR circuits as well as mixed-mode systems such as ADCs. Employing charge steering in 65-nm CMOS technology, a 25-Gb/sCDR/deserializer consumes 5 mW and a 10-bit 800-MHz pipelined ADC draws 19 mW.

# Educational Session 5

Wednesday, 9/25, 9:00 am
Cedar Ballroom
**Session Chair**: Gordon Roberts, McGill University
**Session Co-Chair:** Takamaro Kikkawa, Hiroshima University

E-5        **Design for Nanoscale Patterning,** *Puneet Gupta, UCLA*   **746**
9:00 am –
10:30 am   This tutorial explains how layout and circuit design interact with lithography choices. Lithography technology is rapidly evolving and has started to impose unusual restrictions on design layout. The tutorial will give a brief introduction to current and upcoming lithography technologies. We especially focus on multi-patterning technologies such as LELE double patterning and SADP. We will discuss design enablement of multi-patterning technologies, especially in context of cell-based digital designs. Models for electrical impact of lithography imperfections such as polysilicon/active rounding and overlay errors will be outlined. We will also briefly explore role of design in lithography technology development.

10:30 am   **BREAK**

# Educational Session 6

Wednesday, 9/25, 11:00 am
Cedar Ballroom
**Session Chair**: Howard Luong, Hong Kong University of Science & Technology
**Session Co-Chair:** Earl McCune, RF Communications Consulting

E-6        **Low Power Chip & System Design for Biomedical Applications,** *Brian Otis, Google, Inc.*
11:00 am –                                                                                              **772**
12:30 pm   Advances in chip and system design will help define the next generation of wireless sensors for biomedical applications. This talk will investigate system and circuit design techniques for body-worn systems, implantable chips, and wireless sensors. These areas present unique challenges at the interface between the IC and the body that cannot be solved by technology scaling alone. Traditional circuit blocks, architectures, and even assembly techniques need to be questioned. Several future applications will demand thin-film realization and biocompatibility of complex systems.   RFID-like techniques are highly useful for many of

these emerging biomedical applications. High-Q RF resonators are useful for minimizing power and size of low power radios. We'll discuss a few examples of the above.

## Session 20 -- Advanced Memory Topics

Wednesday, 9/25, 1:30 pm
Oak Ballroom
**Session Chair**: Koji Nii, Renesas
**Session Co-Chair:** Toshiaki Kirihata, IBM

**1:30 pm**     **Introduction**

This session covers scaling challenges, latest advances, and future trends on spin-torque, MRAM, NAND, and logic-compatible flash, TCAM, and 6T/8T SRAM.

**1:30 pm**     **Introduction**

This session covers scaling challenges, latest advances, and future trends on spin-torque, MRAM, NAND, and logic-compatible flash, TCAM, and 6T/8T SRAM.

**1:35 pm**
**20-1**     **ST-MRAM Fundamentals, Challenges, and Applications (Invited),** *T. Andre, S.M. Alam, D. Gogl, C.K. Subramanian, H. Lin, W. Meadows, X. Zhang, N.D. Rizzo, J. Janesky, D. Houssameddine, J.M. Slaughter, Everspin Technologies*    **799**

MRAM technology emerged from research and development into volume production within the last decade in the form of Toggle MRAM. Spin-Torque MRAM has reached the level of customer sampling, offering higher density and bandwidth. This paper describes the devices, fundamental circuit challenges, and applications of this evolving MTJ based memory.

**2:25 pm**
**20-2**     **Scaling Challenges of NAND Flash Memory and Hybrid Memory System with Storage Class Memory & NAND flash memory (Invited),** *Ken Takeuchi, Chuo University*    **807**

SSDs and emerging storage class non-volatile memories such as PCRAM, ReRAM and MRAM have enabled innovations in various nano-scale VLSI memory systems for personal computers, smart phones, tablets and enterprise servers. This paper discusses the scaling challenges of 2D and 3D NAND flash memory and then provides a state-of-the-art hybrid memory solution with storage class memory and NAND flash memory for the big data solid-state storage system.

**3:15 pm**     **BREAK**

**3:30 pm**
**20-3**     **A 28nm High Density 1R/1W 8T-SRAM Macro with Screening Circuitry against Read Disturb Failure,** *M. Yabuuchi, H. Fujiwara, Y. Tsukamoto. M. Tanaka, K. Nii, Renesas Electronics Corporation*    **813**

We developed a high density 1R/1W SRAM macro based on 8T-SRAM with a novel scheme for Design for Testing. To achieve a smaller Macro area, a differential sense amplifier is introduced to read the data, where the reference voltage for reading 0/1 data is generated by unselected cell array. In addition, we proposed a screening test circuit for read disturb operation.

**3:55 pm**     **A HKMG 28nm 1GHz Fully-Pipelined Tile-able 1MB Embedded SRAM IP with**

20-4     **1.39mm$^2$ per MB,** *M. Z. Kuo, O. Takahashi, P. L. Yang, C. C. Lin, M.J. Wang, P.W. Wang, S. H. Dhong, Taiwan Semiconductor Manufacturing Company*    **817**

A fully-pipelined tile-able 1MB SRAM IP with a 0.127µm2 cell in a HKMG 28nmbulk technology has an area of 1.39mm2/MB with 79.2% array efficiency. It operates with 2-cycle latency up to 1GHz. The no-repair hardware has a circuit limited yield of 99.92 and 53% at 100 and 850MHz, respectively with 0.75V VDD.A Data Retention Voltage of 0.42V has been measured.

4:20 pm    **A Bit-by-Bit Re-Writable Eflash in a Generic Logic Process for Moderate-Density**
20-5      **Embedded Non-Volatile Memory Applications,** *Seung-Hwan Song, Ki Chul Chun, Chris H. Kim, University of Minnesota*    **821**

A bit-by-bit re-writable embedded flash memory is demonstrated in a generic65nm logic process for moderate-density embedded non-volatile memory applications. The proposed 6T embedded flash memory cell improves the overall cell endurance by eliminating redundant program/erase cycles without disturbing cells in the unselected wordlines. A multi-story high voltage switch utilizes four boosted supply levels generated by a compact voltage doubler based on-chip negative charge pump.

4:45 pm    **Tail-Bit Tracking Circuit with Degraded VGS Bit-Cell Mimic Array for a 50%**
20-6      **Search-Time and 200mV Vmin Improvement in a Ternary Content Addressable**
         **Memory,** *Igor Arsovski, Travis Hebig, John Goss, Paul Grzymkowski, Josh Patch , IBM*    **825**

A memory sense-timing circuit uses VGS degradation to emulate the behavior of weak memory tail-bits, improving Tail-Bit Tracking (TBT) across process, voltage and temperature. The TBT circuit generates timing for a TCAM search operation reducing Vmin by 200mV, and improving sense-time by 50%. Implemented in 32nm HKMG SOI process the 2Kx640b TCAM achieves 0.60V and 1G search/sec.

# Session 21 -- Oversampled ADC's

Wednesday, 9/25, 1:30 pm
Fir Ballroom
**Session Chair**: Eric Naviasky, Cadence
**Session Co-Chair**: Hasnain Lakdawala, Qualcomm

1:30 pm    **Introduction**

This session has 4 papers on noise shaping ADC's. The first two papers take advantage of the noise shaping 1~ inherent in a VCO. The last two offer novel solutions to feedback DAC imperfections.

1:35 pm    **A 1.8mW 2MHz-BW 66.5dB-SNDR Delta-Sigma ADC Using VCO-Based Integrators**
21-1      **with Intrinsic CLA,** *Kyoungtae Lee, Yeonam Yoon, Nan Sun, The University of Texas at Austin*    **829**

This paper presents a scaling-friendly continuous-time closed-loop VCO-based Delta-Sigma ADC. It uses the VCO as both quantizer and integrator, and thus, obviates the need for power-hungry scaling-unfriendly OTAs and precision comparators. It arranges two VCOs in a pseudo-differential manner, which cancels out even-order distortions. More importantly, it brings an intrinsic CLA capability that automatically addresses DAC mismatches. The prototype ADC in 130nm CMOS occupies a small area of 0.03mm^2 and achieves 66.5dB

SNDR over2MHz BW while sampling at 300MHz and consuming 1.8mW under a 1.2V power supply. It can also operate with a low analog supply of 0.7V and achieves 65.8dB SNDR while consuming 1.1mW. The corresponding FOMs for the two cases are 0.25pJ/step and 0.17pJ/step, respectively.

**2:00 pm**
**21-2**

**A 50MHz bandwidth, 10-b ENOB, 8.2mW VCO-based ADC enabled by filtered-dithering based linearization,** *Abhishek Ghosh, Sudhakar Pamarti, University of California, Los Angeles* **833**

A dithering technique for linearization of VCO-based ADCs is proposed. The proposed technique conditions the signal to the VCO input to appear as whitenoise thereby eliminating spurious signal content arising out of the VCO nonlinearity. The technique, thus obviates the need for power-hungry digital calibration techniques or expensive front-end loop-filters. A prototype implementation (in 65nm CMOS) based on the technique achieves 10-b ENOB in digitizing signals with 50MHz bandwidth consuming 8:2mW at an FoM of 90fJ/conv.step.

**2:25 pm**
**21-3**

**A Reconfigurable Delta-Sigma Modulator With Up To 100 MHz Bandwidth Using Flash Reference Shuffling,** *T. Caldwell, D. Alldred, Z. Li, Analog Devices, Inc* **837**

A reconfigurable 65 nm continuous-time low-pass delta-sigma modulator operates with a sampling frequency from 491 MHz to 1536 MHz, a signal bandwidth from 10MHz to 100 MHz, and a dynamic range of 75.4 dB to 62.8 dB, respectively. Reference shuffling in the flash ADC is used to improve the linearity of the flash and DAC, while also increasing the highest sampling rate and bandwidth of the modulator.

**2:50 pm**
**21-4**

**A 10-MHz Bandwidth 70-dB SNDR 640MS/s Continuous-Time Sigma-Delta ADC Using Gm-C Filter with Nonlinear Feedback DAC Calibration,** *J. Huang, S. Yang, J. Yuan, Hong Kong University of Science and Technology* **841**

Traditionally, wide-band (>10MHz) continuous-time sigma-delta ADCs with Gm-C filters have poor linearity. This paper introduces a novel on-chip calibration scheme to compensate the Gm-cell's nonlinearity. Measurements of the 640MS/s CTSD modulator show the best SNDR and power efficiency among Gm-C-based modulators. The FOM is also comparable to active-RC-based modulators.

**3:15 pm**     **BREAK**

## Session 22 - AMS System Simulation Techniques

Wednesday, 9/25, 1:30 pm
Pine Ballroom
**Session Chair:** Larry Nagel, Omega Enterprises
**Session Co-Chair:** Colin McAndrew, Freescale

**1:30 pm**     **Introduction**

This session presents new, innovative, and efficient techniques that extend the state of the art for analog mixed-signal (AMS) system-level simulations.

**1:35 pm**
**22-1**

**Algorithmic Nonlinear Macromodeling: Challenges, Solutions and Applications in Analog/Mixed-Signal Validation (Invited),** *C. Gu, Intel Corporation* **845**

Analog/Mixed-Signal validation at the system level is becoming increasingly important as more electrical bugs are caused by the interaction among various circuit blocks. While hand-crafted behavioral models and linear models are still most widely used among designers, there is an increasing need for automatic behavioral modeling tools which capture low-level nonlinear behaviors in the circuit. This paper discusses challenges and difficulties of algorithmic nonlinear macromodeling, and reviews a series of recently developed techniques. In particular, we study the behavioral modeling problem from the perspective of projection in the state space defined by voltages and currents. We review a few nonlinear macromodeling techniques from the projection perspective, and demonstrate the model accuracy and computational efficiency compared to transistor-level models and linear models.

**2:25 pm**
**22-2**

**Event-Driven Simulation of Volterra Series Models in System Verilog,** *J-E. Jang, S-J. Yang, J. Kim, Seoul National University* **853**

This paper presents a true event-driven methodology to simulate weakly-nonlinear analog circuits in System Verilog. We express a continuous-time signal as a linear combination of basis functions and reformulate a Volterra series model into a set of linear differential equations with an explicit notion on initial conditions. Two circuit examples showed 300~1000× speed-up compared to SPICE with the same accuracy.

**2:50 pm**
**22-3**

**A Verilog Piecewise-Linear Analog Behavior Model for Mixed-Signal Validation,** *S. Liao, M. Horowitz, Stanford University* **857**

Mixed-signal validation requires simulating the entire chip through a large number of test vectors, which makes pin-accurate and fast Verilog functional models of analog circuits with reasonable fidelity valuable. We describe an extensible approach to creating these models, by mapping continuous signals into discrete events and avoiding explicit time integration.

**3:15 pm**

**BREAK**

**3:30 pm**
**22-4**

**Advancements in High-Speed Link Modeling and Simulation (Invited),** *Mike Peng Li, Masashi Shimanouchi, Hsinho Wu, Altera Corporation* **861**

This paper starts with reviewing the status of techniques and methods used in recent high-speed link simulation and modeling for signaling, integrated circuits, board circuits, and associated limitations and challenges, and then discusses new advancements that can overcome them.

**4:20 pm**
**22-5**

**Structure-Aware High-Dimensional Performance Modeling for Analog and Mixed-Signal Circuits,** *S. Sun, X. Li, C. Gu\*, Carnegie Mellon University, \*Intel Corp* **869**

Efficient high-dimensional performance modeling of nanoscale analog and mixed signal (AMS) circuits is challenging. In this paper, we propose a novel structure-aware modeling technique to accurately solve the model coefficients by exploiting the underlying structure of AMS circuits, and hence dramatically reduce the number of sampling points for performance modeling.

# Session 23 -- Analog Techniques II

Wednesday, 9/25, 3:25 pm
Fir Ballroom
**Session Chair:** Hasnain Lakdawala, Qualcomm

**Session Co-Chair**: Eric Naviasky, Cadence

3:25 pm **Introduction**

The session includes topics in time to digital converters, fast locking PLL's and double sampling sigma delta ADC's.

3:30 pm **A 148fs$_{rms}$Integrated Noise 4MHz Bandwidth All-Digital Second-Order ΔΣ Time-to-**
23-1 **Digital Converter Using Gated Switched-Ring Oscillator,** *W. Yu, K.S. Kim, S. Cho, KAIST* **873**

This paper presents an all-digital second-order ΔΣ time-to-digital converter(TDC) by using switched-ring oscillator (SRO) and gated switched-ring oscillator (GSRO). Unlike conventional multi-stage noise-shaping (MASH) TDC using the SRO, the proposed TDC does not require complex calibration to compensate for the error from frequency difference between the SROs. The prototype TDC achieves 148fsrms integrated noise and 80.4dB dynamic range in4MHz signal bandwidth at 400MS/s while consuming 6.55mW in a 65nm CMOS process.

3:55 pm **A 0.84ps-LSB 2.47mW Time-to-Digital Converter Using a Charge Pump and a**
23-2 **SAR-ADC,** *Zule Xu, Seungjong Lee, Masaya Miyahara, and Akira Matsuzawa, Tokyo Institute of Technology* **877**

We propose a 0.84ps-LSB, 2.47mW, 0.06mm² time-to-digital converter (TDC) using a charge pump and a SAR-ADC in 65nm CMOS. Sub-pico second time resolution is attainable by quantizing the time in charge domain. Low power consumption and small area are also feasible by using the SAR-ADC.

4:20 pm **A double-sampling cross noise-coupled Sigma Delta modulator with a reduced**
23-3 **amount of opamps,** *M. De Bock, P. Rombouts, UGhent* **881**

A second order double-sampling cross noise-coupled split-path Sigma Delta modulator is presented. The implementation of the noise-coupling is incorporated into the second integrator using a novel delaying feed-forward circuit. A prototype integrated in a 130nm CMOS technology achieves 77.8dBdynamic range and 71.4dB SNDR over a 5MHz bandwidth.

4:45 pm **A Novel OTA-Based Fast Lock PLL,** *Mezyad Amourah, Sandeep Krishnegowda, Morgan*
23-4 *Whately, Cypress Semiconductor* **885**

A novel fast lock scheme for phase-locked loops (PLLs). The proposed scheme uses a simple operational transconductance amplifier (OTA) to achieve significant reduction in PLL lock acquisition time without affecting PLL noise performance. The new fast lock schemes were implemented in multi-port SRAM chip to provide frequencies from 400MHz to 1.6GHz, The chip was fabricated using 65nm CMOS process. Silicon measurements across corner lots show significant reduction in PLL lock time, by a factor of 6.5X, over device operating conditions.

# Educational Session 7

Wednesday, 9/25, 1:30 pm
Cedar Ballroom
**Session Chair**: Mike Li, Altera
**Session Co-Chair:** Gerrit den Besten, NXP Semiconductors

E-7

1:30 pm –
3:00 pm

**Trends, Possibilities and Limitations of Photonic Integrated Circuits and Lasers,**

*John E. Bowers, University of California, Santa Barbara*     **889**

A number of important breakthroughs in the past decade have focused attention on Si as a photonic platform. We review here recent progress in this field, focusing on efforts to make lasers, amplifiers, modulators and photodetectors on or in silicon.  We also describe progress in silicon photonic integrated circuits. The impact active silicon photonic integrated circuits could have on interconnects, telecommunications and on silicon electronics is reviewed.

3:00 pm     **BREAK**

# Educational Session 8

Wednesday, 9/25, 3:30 pm
Cedar Ballroom
**Session Chair**: Jerry Jiang, Broadcom
**Session Co-Chair:** Eric Naviasky, Cadence

E-8

3:30 pm –
5:00 pm

**A/D Converter Circuit and Architecture Design for High-Speed Data Communication,**

*Boris Murmann, Stanford University*     **934**

A number of important breakthroughs in the past decade have focused attention on Si as a photonic platform. We review here recent progress in this field, focusing on efforts to make lasers, amplifiers, modulators and photodetectors on or in silicon.  We also describe progress in silicon photonic integrated circuits. The impact active silicon photonic integrated circuits could have on interconnects, telecommunications and on silicon electronics is reviewed.

# Welcome from the CICC Committee

Welcome to CICC 2013, our 35th annual Custom Integrated Circuits Conference – ▯▯▯ ▯▯▯▯▯▯▯▯▯▯▯▯▯▯▯▯▯▯▯▯▯▯▯▯▯▯▯▯▯▯▯▯▯▯▯▯▯▯▯▯▯▯. This year, our Technical Program Committee of over 80 experts has completely revamped CICC to make it more accessible to the audience. In addition to our continued emphasis on technical papers which address a broad spectrum of State-of-the-Art developments we have reorganized and revamped, what was already our strong suite, our educational sessions. Starting this year our educational and technical sessions run concurrently to allow attendees to choose their level of interest and understanding of the various technical subjects. Continuing with our tradition of educational emphasis, the Education Sessions are included as part of the conference registration fee.

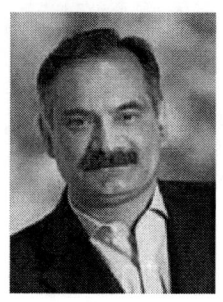

**Aurangzeb Khan**
General Chair

The eight Educational Sessions are distributed across the conference on Tuesday and Wednesday (Sept 24 -25). They feature a wide range of interesting topics by well recognized experts in the field. The sessions on Tuesday include Concurrent Design of ESD and Circuit performance by Prof. Albert Wang, UCR, Characterization and Modeling of Matching and Variability with Hans Tuinhout from NXP, DC to DC Converter Design with Prof. Philip Mok, HKUST, and Digital PLL Design with Prof. Salvatre Levantino, Politecnico di Milano. Wednesday's sessions include, Design of Nanoscale Paterning with Prof. Puneet Gupta, UCLA, Low Power Biomedical Chip Design with Brian Otis from Google, Photonics ICs and Lasers with Prof. John E. Bowers, UCSB, and High-Speed A/D Converter Design with Prof. Boris Murmann, Stanford.

**Philippe Jansen**
Conference Chair

The conference and Technical Sessions open on Monday morning with a Keynote by Prof. Mark Horowitz from Stanford University titled, "Digital Analog Design". In this not-to-be-missed event Prof. Horowitz focuses on the design of analog circuits by leveraging what we have learned about automating the design of digital circuits. The four Monday morning technical sessions focus on biomedical and sensing applications, high-speed wireline CDR, RF and technology issues beyond 14nm. The Forum after lunch, with a diverse set of expert panelists, focuses on 20nm design challenges and delves into analog/wireline design, tools and technology issues. This is in parallel with three technical sessions on power management, AMS verification, and wireless. The Panel on Monday afternoon has experts from across the world who will try to answer the question,

"Can Biomedical Electronic Startups Make Money"? At the end of the day on Monday, at the Welcome Reception, you can enjoy refreshments while exploring more than 20 Poster Presentations, visiting the Exhibitor Booths, and networking with your colleagues.

Be sure to join us for the Conference Luncheon on Tuesday where Prof. Joe Paradiso from the MIT Media Laboratory will present our Conference Luncheon Keynote titled, "Connecting with Emerging Nervous System of Ubiquitous Sensing." Three technical sessions and a parallel track of Educational Sessions highlight Tuesday. The technical sessions on Tuesday morning will focus on 3D and high power technologies, wireline transceivers and mm-Wave circuits. The luncheon talk will be followed by an exciting Panel Session titled, "Do You Need to Plug In to Get Your Fill of Bits?" trying to answer the question whether wireless or wired links will prevail for short distance ultra-high speed links. The technical sessions for Tuesday afternoon include testing for multi-Gbits/s electrical and optical I/Os and ADCs. Tuesday evening ends with another networking opportunity with refreshment and the opportunity to view over 20 Poster Presentations and Exhibitor Booths.

Wednesday provides another fully day of integrated Educational and Technical Sessions. The technical sessions on Monday morning include variation modeling for analog, SOC design and analog techniques. The afternoon technical sessions include memory, oversampled ADCs, mixed-signal simulation and delves further into analog techniques. This is in parallel with a set of four Educational Sessions.

# Steering Committee

**Jackie Snyder**
Marvell Semiconductor

**Rakesh Patel**
Consultant

**Tom Andre**
Everspin Technologies

**Aurangzeb Khan**
Altia Systems

**Ramesh Harjani**
University of Minnesota

**Philippe Jansen**
Texas Instruments

# Organizing Committee

**General Chair**
Aurangzeb Khan, Altia Systems

**Conference Chair**
Philippe Jansen, Texas Instruments

**Technical Program Chair**
Ramesh Harjani, University of Minnesota

**Educational Session Chair**
Howard Luong, Hong Kong University of
Science & Technology

**Exhibits Chair**
Trent McConaghy, Solido Design

**Panel Chair**
KimoTam, Analog Devices

**Sponsorship Chair**
Alessandro Picovaccari, Silicon Labs.

**Publicity Chair**
Ken Suyama, Epoch Microelectronics

**Best Paper Award Chair**
Ron Kapusta, Analog Design

**Treasurer**
Rakesh Patel, Consultant

# Technical Program Committee

*Analog Circuit Design*
Pavan Hanumolu, Oregon State University, Ken Suyama, Epoch, Eric Naviasky, Cadence, Ron Kapusta, Analog Devices, Yuji Nakajima, Renesas Electronics, Yusuf Haque, Crest Semiconductors, Xicheng Jiang, Broadcom, Don Thelen, ON Semiconductor, Alessandre Piovaccari, Silicon Laboratories, Mohammad Ranjbar, Cirrus Logic, John McNeill, Worchester Polytechnic Institute, Hasnain Lakdawala, Intel Corp.

*BioMedical Actuators, MEMs and Sensors*
Ed Lee, Alfred Mann Foundation, Emmanuel Quevy, Silicon Labs., Pedram Mohseni, Case Western Reserve University, Stephen O'Driscoll, UC Davis, Mourad El-Gamel, McGill University University, Christophe Antoine, Analog Devices, Patrick Chiang, Oregon State University and Hoi-Jun Yoo, KAIST

### IC Manufacturing

Rajiv Joshi, IBM T.J. Watson Research Center, Philippe Jansen, Texas Instruments, Takamaro Kikkawa, Hiroshima University; Ta Pen Guo, TSMC, Dinesh Somasekhar, Intel Corp. and Ramnath Venkatraman, LSI Corp.

### Simulation and Modelling

Larry Nagel, Omega Enterprises Consulting, Hidetoshi Onodera, Kyoto University, Colin McAndrew, Freescale Semiconductor, Trent McConaghy, Solido Design, Brian Chen, Agilent Technologies, Yu (Kevin) Cao, Arizona State University

### Memory

Chris H. Kim, University of Minnesota, Jean-Christophe Vial, Intel Mobile Communications, Toshiaki Kirihata, IBM, Koji Nii, Renesas Electronics, Tom Andre, Everspin, Muhammad Khellel, Intel Corp., Vikas Chandra, ARM

### System on Chip and 3D

Larry Clark, Arizona State University, Steve Witten, University of British Columbia, Visvesh Sathe, AMD, Rakesh Patel, Consultant, Aurangzeb Khan, Altia Systems, Rick Paul, Cisco Systems, Paul Billig, Consultant, Arif Rahman, Altera, Rick Williams, Oracle

### Power Management

Jerry Zheng, iWatt, Rajeevan Amirtharajah, UC Davis, Hoi Lee, University of Texas at Dallas, William McIntyre, Texas Instruments, Christophe Sandner, Infineon Technologies Austria

### Test, Debug & Reliability

Gordon Roberts, McGill University, Jackie Snyder, Marvell Semiconductor, Manoj Sachdev, University of Waterloo, Mike Li, Altera Corp., Takahiro Yamaguchi, Advantest

## *Wireless Devices*

⸤⸥⸤⸥⸤⸥⸤⸥⸤⸥John Rogers, Carleton University, Howard Luong, Hong Kong University of Science & Technology, Alireza Shirvani, RFMD!, Julian Tham, Broadcom, Andrea Mazzanti, Universita di Pavia, ⸤⸥⸤⸥⸤⸥⸤⸥⸤⸥ Ramesh Harjani, University of Minnesota, Foster Dai, Auburn University, Cicero Vaucher, NXP Semiconductors, Earl McCune, Consultant, Ehsan Afshari, Cornell University, ⸤⸥⸤⸥⸤⸥⸤⸥Rick Booth, Jonathan Borresmans, imec

## *Wireline Communications*

⸤⸥⸤⸥⸤⸥⸤⸥⸤⸥ Dennis Fischette, AMD, William Walker, Fujitsu Laboratories of America, Gerrit den Bestern, NXP Semiconductors, Kimo Tam, Analog Devices, Jaeha Kim, Seoul National University, ⸤⸥⸤⸥⸤⸥⸤⸥⸤⸥ Shahriar Mirabbasi, University of British Columbia, Afshin Momtaz, Broadcom, Elad Alon, University of California, Berkeley, Sam Palermo, Texas A&M University

**Microsystems for Biomedical and Sensing Applications**                                    **Session 2**

Chair: Christophe Antoine, Analog Devices
Co-Chair: Stephen O'Driscoll, University of California, Davis

Biomedical and sensor applications are growing apace and present a new set of challenges for integrated circuit and system designers. In this session, five papers representing advances in these research areas will be presented.

In the first paper, a biosensor interface IC as part of a miniaturized dielectric spectroscopy platform is presented that is capable of extracting dielectric readings of organic solvents in a wide measurement bandwidth from MHz to GHz using only the S21 information of a microfluidic dielectric sensor loaded with the solution-under-test.

Origami foldable implanted devices allow minimally invasive implantation surgery but their structure often necessitates communication between closely spaced integrated circuits. These ICs may move slowly over time due to tissue growth. The second paper presents a capacitive proximity communication link for implanted origami devices, which uses arrays of coupling plates to estimate alignment and adapt the proximity communication link to maximize efficiency for a given data rate.

Wireless inductive power delivery is the most effective method of power supply for many classes of biomedical implants. The received RF power must be converted to DC to supply the implanted electronics. A novel combined rectifier and regulator circuit that aims to improve RF-to-DC conversion efficiency is presented in the third paper.

For sensor electronic interfaces and embedded systems, measuring real-world quantities with minimal energy consumption has become an important axis of performance improvement. The fourth paper demonstrates a significant reduction in the power consumed by a temperature sensor without sacrificing the measurement accuracy.

Dead reckoning navigation based on inertial sensors has been hampered by drift accumulated over time and its associated long-term position inaccuracy. Rather than tackling inertial sensing drift directly, the fifth paper adopts a system-level approach and proposes to combine the inertial sensors with a ground reaction sensor array made of organic material. The sensor array is integrated into boots to detect bipedal locomotion patterns and uses the information as part of a gait-corrected inertial measurement unit (IMU).

**High-Speed Wireline Timing Recovery & PLLs**                                           **Session 3**

Chair: Samuel Palermo, Texas A&M University
Co-chair: Kimo Tam, Analog Devices, Inc.

Advances in high-performance clock generator circuits and timing recovery techniques are essential for the continued improvements in performance, power, and area demanded by current and future wireline communication systems. The papers presented in this section highlight developments in timing recovery techniques for ADC-based receivers and burst-mode systems, a new approach to realize PLLs with peaking-free transfer functions, and analysis and design comparisons of capacitor-multiplier and passive loop filters for low-area PLL implementations.

ADC-based receivers can be classified as phase-tracking and blind architectures. In the former, the VCO phase is controlled through a feedback loop so as to sample the received data in the middle of the data eye. In the latter, the received signal is sampled with a blind clock, i.e. not in a loop, and the data at the center is obtained by data processing techniques such as data interpolation and extrapolation. Our first paper compares these two architectures in terms of their design complexity and cost, and derives equations that relate the required ADC resolution to channel loss and to the characteristics of the FFE/DFE that follow the ADC.

Our second paper deals with burst-mode clock and data recovery (BMCDR), where it is required to recover data bursts with as few a number of bits as possible. The presented BMCDR is reconfigurable between data gating mode and phase tracking mode to achieve instantaneous (<1bit) phase-locking with jitter suppression for 10 GPON systems. Incorporating a 1/5-rate CDR with a 1:5 demultiplexer, it achieves 1.24pJ/bit energy efficiency with a recovered 2GHz clock jitter of $2.94ps_{rms}$ with 4MHz, $0.22UI_{pp}$ input data jitter.

A commonly-cited weakness of phase-locked loops against delay-locked loops is that PLLs accumulate jitter and therefore exhibit peaking in the frequency-domain transfer function and overshoots in the time-domain step response. The third paper describes a digital phase-locked loop that realizes a peaking-free jitter transfer directly in digital domain without requiring additional components. The prototype 9.2GHz digital PLL, fabricated in a 65nm CMOS, demonstrates a fast 1.58µs settling time with a 690kHz bandwidth, $3.48ps_{rms}$ divided clock jitter, and -120dBc/Hz phase noise at a 10MHz offset.

Our final paper evaluates the potential for leveraging capacitor-multiplier loop filters in PLLs in order to reduce area. Two PLLs operating at 8-12.2GHz are designed in 28nm CMOS, one with a capacitor multiplier-based active loop filter, and the other with a conventional passive loop filter. The authors show comparable $\sim 200fs_{rms}$ jitter performance for both PLLs, with the capacitor-multiplier PLL displaying similar area as that of state-of-the-art digital PLLs, while also offering power savings.

**RF Building Blocks**                                                                    **Session 4**

Chair: Andrea Mazzanti
CoChair:  Earl McCune

The never ending demands for costs reduction and improvements in performance and functionality of mobile communication systems drive continuous research and innovation on RF subsystems and building blocks. New technologies, architectures and circuit techniques are constantly emerging. This session presents recent advances on linearization techniques for digital PLL, integrated duplexer, T/R switches for smartphone and Ultra-Wide-Band programmable delay for beam-forming.

Spur reduction in digital PLL (DPLL) is a topic of intense research. The session starts with an invited paper presenting a new DPLL architecture based on a digital-to-time converter feedback divider. This technique allows cancellation of nonlinearities plus spur reduction with no penalty on noise level, along with reduced design complexity and low power consumption. The 65nm CMOS prototype covers the 2.9 - 4 GHz frequency range with -52 dBc in-band spur-free dynamic range and 4.5mW power dissipation.

The second and third papers of this session present respectively an integrated tunable duplexer and a multipath T/R switch for smartphone radios. The tunable duplexer is based on the electrical balance of a hybrid transformer and is integrated along with a direct conversion receiver. A novel calibration loop is proposed to achieve high TX/RX isolation by tracking antenna impedance variation. The single-pole multi-throw T/R switch is suited for quad-band GSM and multiple WCDMA. Fabricated in 180nm SOI CMOS it displays 1dB insertion loss, better than 40dB isolation with more than 8kV ESD robustness.

The last paper introduces a novel ladder filter with programmable gain and delay for Ultra-Wide-Band beam-forming. A 0.25µm BiCMOS prototype, designed for 3.1-10.6GHz UWB range, achieves a delay range of 80psec with 0.5psec resolution and gain variation from -30 to +10dB with 47mW power dissipation.

**Power Management**                                                                                    **Session 7**

**Chair:** Christoph Sandner, Infineon
**Co-Chair:** Rajeevan Amirtharajah, University of California, Davis

Power consumption is a major concern in nearly all systems, from data centers to portable devices and medical implants. Consequently, effective power management is increasingly important to designers working in a wide range of application domains. This session collects nine papers addressing the most recent advances in battery standards, circuits, and control strategies for power management.

Smart batteries increasingly power mobile devices. The session begins with an invited paper discussing an emerging battery interface standard that incorporates charging profiles and authentication features.

Solid-state lighting demands innovative power electronics to maintain energy efficiency. Two papers describe LED drivers that achieve greater than 90% efficiency. High voltage conversion also demands new approaches to achieve high speed and efficiency, as discussed in a paper focused on developing 100V gate drivers with sub-nanosecond switching speeds.

Low voltage applications continue to drive developments in different DC/DC conversion architectures. Four papers discuss inductor-based switching converters with digital control, switched-capacitor circuits that recycle charge, and efficient boost conversion for energy harvesting.

Managing power consumption in multicore processors remains challenging in many applications. The final paper in the session describes an integrated fast response power sensor circuit that targets this important class of integrated circuits.

**AMS Verification in Advanced Technologies**                    **Session 8**

Chair: Hidetoshi Onodera, Kyoto University
Co-Chair: Yu (Kevin) Cao, Arizona State University

Verification of Analog and Mixed-Signal(AMS) circuits and systems is increasingly challenging. This session explores novel AMS simulation and emulation techniques, and advanced reliability and performance issues with technology scaling.

Our first paper deals with efficient AMS design, verification, and testing methods that discretize the search spaces or discriminate possible cases. Applications to practical examples are demonstrated such as operational amplifier sizing, verification of correct start-up properties of coupled oscillators, and test suite derivation for fault screening in a phase interpolator.

The second paper discusses low-cost indirect performance sensors for on-chip analog analog self-healing. It adopts Bayesian model fusion to facilitate inexpensive-yet-accurate on-chip performance measurement. It has been applied to a 25 GHz VCO in a 32nm SOI process, and the self-healing has improved the parametric yield from 0 % to 69 %.

The third paper presents a fast FPGA emulation platform for adaptive analog circuit simulation and verification. A slowly adapted, background calibration technique for successive-approximation-register analog-to-digital converter has been demonstrated on the FPGA emulation platform such that a 3000x speedup over a software approach has achieved.

The fourth paper reviews circuit aging simulation approaches with focus on TSMC Model Interface (TMI). The limitations of aging models are also discussed. An example circuit under multi-waveform and multi-temperature stress is presented to illustrate reliability simulation flow through TMI.

Last three papers deal with modeling issues in advanced technologies, including ESD circuits, FinFETs, and gate stack resistance. Accurate and scalable behavioral modeling for a 3D field-programmable ESD circuits has enabled whole-chip ESD circuit design optimization and verification. Furthermore, a quasi-3D modeling of FinFETs has reduced a full 3D simulation to a set of 2D cut-plane simulations in ESD applications, yielding results 20x faster.

The final paper explains compact modeling of gate resistance with horizontal and vertical components, calibrated with silicon data from five technology generations. It demonstrates how the increase in the gate resistance constrains logic transistor performance during the scaling.

Wireline Transmitter and Receiver Design Techniques

Session 12

Chair: Dennis Fischette, Advanced Micro Devices, Inc.
Co-chair: Jaeha Kim, Seoul National University

Every year, designers of wireline transceivers are asked to do more with less. Ever higher data rates are demanded while at the same time power, area, and cost are expected to decrease. This session presents five notable wireline papers that have successfully tackled these challenges. They not only present novel circuit techniques, but also explore the thought process behind these designs.

Our first paper implements transmitter-side equalization in a power-efficient voltage-mode driver. The proposed transmitter decouples equalization and impedance control through pulse width modulation based de-emphasis, while providing selective frequency amplification beyond the Nyquist frequency. With 28-dB channel loss it achieves a horizontal eye opening of 0.3UI with BER < $10^{-12}$ while consuming only 16mW power at 5Gb/s.

Maintaining optimal impedance and output voltage level are challenges in current-steering transmitters. Our second paper introduces a new linear auto-tuned active line termination structure for continuous impedance tuning. The tunable active termination consists of an adaptive buffer bias current and load resistors and maintains accurate output voltage level ($\sigma$=1%) and impedance matching ($\sigma$=2%) in a 6 Gb/s current-steering transmitter with 3-tap pre-emphasis.

Our third paper (invited) explores design techniques for transceivers operating at high data rates in challenging backplane environments. The proposed transceiver employs a 5-tap feed-forward transmitter equalizer, two-stage peaking amplifier with active feedback, 15-tap receiver decision-feedback equalizer, and receive-side capacitive level-shifters to achieve error-free 28 Gb/s operation with 35-dB channel loss. Low-power operation at high data rates provides a particular challenge to receiver design. Our 4th paper (invited) focuses on low-power receiver topologies and analysis techniques, with emphasis on specifications and their impact on overall link performance. Also presented are techniques to properly characterize and measure these specifications, and measurements results for three links operating at 10Gb/s, 16Gb/s, and 20Gb/s.

Most clock and data recovery (CDR) schemes include both frequency and phase lock loops. During the lock process, these loops may interfere with one another, delaying phase lock. The proposed half-rate CDR reduces the number of feedback loops to one with a novel frequency detector employing clock phase selection. The design improves the CDR capture range by 3.6x while consuming only 8mW at 10Gb/s operation.

**Electrical and Photonic I/O Test and Debug**

**Session 15**

Chair: Mike Li, Altera
Co-Chair: Takahiro J. Yamaguchi, Advantest

This session addresses recent test challenges associated with high-speed I/Os implemented in electrical and photonic IC forms. In addition to conventional SERDES I/O test challenges, such as transmitter jitter generation and receiver jitter tolerance type tests, stress techniques in loop-back, known-good-die (KGD), and design-validation to debug to high-volume manufacturing, this session will include a discussion about similar photonic transmitter and receiver test and debug challenges. Moreover, this session will also cover the test and debug challenges of CMOS and silicon-photonic integration, either monolithically or heterogeneous, to provide overall cost-power-density-performance trade-offs and optimizations. Issues pertaining to reliability of these systems will also be discussed.

The first paper of this session will present new HVM test techniques to meet the ever increasing test complexity challenges in high-speed I/Os. Leading edge topics such as methodologies to test optical interfaces depending on their level of integration (Hybrid or Full). New innovative approaches in areas of: Complete No Touch Testing (NTT) methodology and systematic defect capture in Electrical High Speed IO circuits, will be presented.

The second paper presents TSV test circuitry which is implemented in the micro-IOs placed in between the fine pitch TSV arrays, capable of detecting and rejecting TSV defects prior to the stacking process. Another circuitry presented is for monitoring power noise caused by simultaneously switching of a 512 bit-wide I/Os.

The third paper looks at the verification and test techniques that can be deployed in silicon photonics solutions, and how it mirrors those that are used in conventional CMOS design.

The last paper will focus on the test and debug challenges associated with KGD, and with the heterous integration of electronics and CMOS photonics in a package format.

The never ending demands for costs reduction and improvements in performance and functionality of mobile communication systems drive continuous research and innovation on RF subsystems and building blocks. New technologies, architectures and circuit techniques are constantly emerging. This session presents recent advances on linearization techniques for digital PLL, integrated duplexer, T/R switches for smartphone and Ultra-Wide-Band programmable delay for beam-forming.

Spur reduction in digital PLL (DPLL) is a topic of intense research. The session starts with an invited paper presenting a new DPLL architecture based on a digital-to-time converter feedback divider. This technique allows cancellation of nonlinearities plus spur reduction with no penalty on noise level, along with reduced design complexity and low power consumption. The 65nm CMOS prototype covers the 2.9 - 4 GHz frequency range with -52 dBc in-band spur-free dynamic range and 4.5mW power dissipation.

The second and third papers of this session present respectively an integrated tunable duplexer and a multipath T/R switch for smartphone radios. The tunable duplexer is based on the electrical balance of a hybrid transformer and is integrated along with a direct conversion receiver. A novel calibration loop is proposed to achieve high TX/RX isolation by tracking antenna impedance variation. The single-pole multi-throw T/R switch is suited for quad-band GSM and multiple WCDMA. Fabricated in 180nm SOI CMOS it displays 1dB insertion loss, better than 40dB isolation with more than 8kV ESD robustness.

The last paper introduces a novel ladder filter with programmable gain and delay for Ultra-Wide-Band beam-forming. A 0.25μm BiCMOS prototype, designed for 3.1-10.6GHz UWB range, achieves a delay range of 80psec with 0.5psec resolution and gain variation from -30 to +10dB with 47mW power dissipation.

**Nyquist Rate A/D Converters**

**Session 16**

Chair: Mohammad Ranjbar
Co-chair: John McNeill

A/D converters enable a variety of applications ranging from low-power/low-speed sensing to high-speed wireless/wireline communications. Most often system specifications are so closely tied to ADC performance that incremental improvements are deemed too difficult without meaningful innovation on the ADC side. Due to this fact, a considerable worldwide effort has traditionally been dedicated to the advancement of ADC technology by numerous companies and R&D groups across the globe. This year CICC has received many high quality papers related to the topic. Papers presented in this session reflect some of the efforts on pushing the envelope of Nyquist rate ADCs.

The first three papers are about high-speed ADC implementations based on the time-interleaving technique. The first paper reports a 6-bit ADC using 32 way hierarchically time-interleaved SAR sub-ADCs to achieve 12.8 GS/s conversion rate. A cascode sample-and-hold circuit is further employed to extend the effective resolution bandwidths over 25 GHz. The second paper reports another 6-bit data converter based on 4 way time-interleaving of two-step sub-ADCs to achieve 10 Gs/s conversion speed. The use of a course-fine configuration along with partial activation of flash sub-ADCs allows keeping power consumption in check. The third paper presents a 4 GS/s 8-bit ADC based on time-interleaving of 4 higher resolution pipeline modules. A digital technique is utilized for correcting the errors caused by the static timing mismatch among the channels.

The design of a low-power 9-bit, 1 GS/s pipeline ADC is presented in the 4th paper. This work combines multiple design techniques such as double-sampled residue amplifier, 2 way interleaved pre-charged DACs, and a new residue amplifier nonlinearity calibration to achieve improved energy efficiency at a relatively high conversion speed.

The 5th paper presents a low-voltage 7-bit, 160 MS/s pipeline ADC using dynamic amplifiers to save power consumption. Furthermore the common-mode detection is employed in the amplifiers to increase the robustness against supply lowering.

The last three papers report SAR ADC designs that employ a variety of interesting techniques for improving the energy efficiency. Paper number 6 presents an 11-bit, 95 MS/s Asynchronous SAR in 65 nm, that employs a passive signal gain of 2x to reduce comparator noise impact which subsequently allows scaling the comparator power. A 12-bit, 1 MS/s SAR with 24 uW of power consumption is reported in the 7-th paper. In this work the use of a two-step switching scheme relaxes the capacitive DAC settling errors without requiring redundant elements. The last paper reports a 0.5 V, 10-bit SAR ADC based on a combination of non-binary redundant DAC and an optimized majority vote scheme to achieve a highly energy efficient conversion at 250 KS/s.

**Variation and Analog Modeling**                                    **Session 17**

Chair: Trent McConaghy
Co-Chair: Brian Chen

AMS, RF, and memory circuits can be highly susceptible to performance degradation caused by process variation, thermal noise, and other non-idealities. This session presents recent advances in modeling these effects, gaining insight into their operation, and techniques to mitigate them.

The session starts with an invited paper on thermal noise modeling in nano-scale MOSFETs. It includes measurement, characterization, modeling, and potential technology enhancement of thermal noise. Applications in low power, mixed-signal, and radio-frequency (RF) circuits are considered, and experimental results from 180 nm down to 40nm are presented.

The second paper describes a model-agnostic technique for simulating per-element distortion contributions. Nonlinearity of an element can be altered while maintaining the operating point and first-order terms by appropriately combining two instances of the nonlinear element with complementary scaling factors for incremental voltages above the operating points. The technique can be used in standard circuit simulators. It is demonstrated on a common source amp with nonlinear load, and a two-stage fully differential opamp.

Corner (best- and worst-case) models have been a mainstay of integrated circuit design for decades. Obviously they can be effective, especially for digital CMOS design. However, there are significant inaccuracies that arise when digital CMOS corner models are used for analog circuits, or any types of circuits or measures of circuit performance they were not targeted for. The third paper details what corner models can and cannot do, and shows their inadequacies for analog CMOS circuits.

The fourth and fifth paper examine SRAMs and variability. The fourth paper proposes an energy centric model of SRAM write operation. The model provides useful insight about energy and write error rate. The paper explores the model's use in evaluating write assist mechanisms, reducing memory error rates, and optimizing energy usage. The fifth paper examines read current (Iread) variability's dependence on transistor variability, by decomposing Iread variability into constituting pass-gate and pull down NMOS transistor variability; and reports a bottom-up model for Iread variability including feedback in stacked transistors.

The last paper characterizes mismatch in small metal fringe capacitors. It describes a test structure and measurements results in 32 nm SOI CMOS. Average standard deviations for these capacitances are 1.2% and 0.8%, respectively, confirming variance scaling according to Pelgrom's law.

**Energy Efficient SoC Design**                                    **Session 18**

Session Chair: Visvesh Sathe, Advanced Micro Devices
Session Co-Chair: Arif Rahman, Altera Corporation

The continued trend toward increased silicon integration in recent years has enabled large, complex SoC systems encompassing an ever-expanding range of capabilities. Maintaining energy-efficient operation continues to be at the heart of high-performance and low-power mobile applications alike, highlighting the crucial role of low-power design methodologies and techniques.

The papers presented in this session address these performance and efficiency challenges through a variety of different approaches. Accelerators, application specific processors and advances in clock generation and distribution are all discussed. Our first paper presents a Razor implementation of a loop-accelerator. Accelerators play a central role in modern SoC implementations, often providing a dramatically improved pareto performance-efficiency tradeoff compared to a general purpose implementation of commonly-performed operations.

The use of application-specific processors remains a continuing trend. Such processors feature architectures and data-paths that enable effective execution of a number of frequently executed tasks, such as our third, fourth and last paper in the areas of video-decoding, feature recognition and cryptography respectively. An emerging trend in application processors is to leverage computationally-relevant features of the application, as depicted in our second paper on a recognition and mining processor that addresses efficient computation at algorithmic, architectural, and circuit levels.

Our fifth paper presents an SoC used for an aerial microrobotic system application dubbed "RoboBees". Constrained by the target application, the authors present a clocking methodology that enables robust operation in the presence of supply variation both short-term (supply-noise) and long term (battery discharge).

As SoC implementations continue to increase in size, synchronization challenges become more severe. Distributing a single high frequency clock across the die in a low-skew fashion is increasingly difficult and dissipative. Our sixth paper addresses this problem through the design of a distributed clock generator using multiple ADPLL instances to achieve cross-design synchronization.

**Analog Techniques I**                                                                                      **Session 19**

Chair: Don Thelen, On Semiconductor
Co-chair: Xicheng Jiang, Broadcom Corporation

This year's CICC features two analog technique sessions. This session includes two invited papers and three regular papers that showcase a collection of analog techniques that enables high-performance analog designs.

Our session begins with Paper 19.1 from Oregon State University that presents a parallel gain enhancement technique for switched capacitor circuits. The proposed technique utilizes two dependent closed loop amplifiers to enhance the open loop DC gain of the main amplifier and to improve harmonic distortion performance. A proof of concept pipeline ADC in a 0.18 μm CMOS process using this technique achieves 75 dB SNDR, 91 dB SFDR, -87 dB THD at 20 MS/s. The ADC operates from a supply voltage of 1.3 V, consumes 5.9 mW, occupies 3.06 mm$^2$ and achieves a figure of merit of 65 fJ/CS.

The second paper (invited) from Analog Device introduces two circuit-level sampling techniques that allow the size of the input capacitor to be determined almost independently of the noise requirement. The first method broke the relationship between the sampling bandwidth and the dominant noise source. The second technique used active circuits and a second capacitor not driven by the input to cancel the noise sampled on the input capacitor. With the help of these circuit techniques, that the sampled thermal noise can be directly reduced by nearly 50% and also cancelled by more than 70% without change to the input capacitor.

The third paper from Oregon State University proposes a blind calibration algorithm which can be used to reduce harmonic distortion in ADCs. The proposed algorithm can be applied to any ADC architecture in general, and can be used to remove both even and odd harmonics. The prototyped VCO based first order sigma delta ADC with this algorithm improves the SNDR of the ADC by 39dB while improving SFDR by 45 dB.

The fourth paper from IBM features a tunable transmission line (t-line) structure with independent control of line inductance and capacitance. The t-line provides variable delay while maintaining relatively constant characteristic impedance using direct digital control through FET switches. Based on this structure, two CMOS 60 GHz phase shifters have been demonstrated in a 32 nm SOI process. Measured data from two phase shifter variants at 60 GHz showed phase changes of 175° and 185°, S21 losses of 3.5-7.1 dB and 6.1-7.6 dB, RMS phase errors of 2° and 3.2°, and areas of 0.073 mm$^2$ and 0.099 mm$^2$ respectively.

The final paper (invited) in this session comes from University of California, Los Angeles and presents charge steering technique for high-speed analog and mixed-signal circuits with low power consumption. The discrete-time nature of this design technique enables digital latching as well as multi-stage, nominally unstable op amps to perform in complex circuits. Issues associated with this design paradigm have been discussed and solutions have been proposed. Providing a fourfold power advantage over CML circuits, charge steering has been exploited in a 25-Gb/s CDR/deserializer dissipating 5 mW and a 10-bit 800-MHz ADC consuming 19 mW.

**Advanced Memory Topics**                                                    **Session 20**

Chair: Koji Niii
CoChair: Toshiaki Kirihata

This session covers scaling challenges, latest advances, and future trends on spin-torque MRAM, NAND, and logic-compatible flash, TCAM, and 6T/8T SRAM in advanced technology nodes.

The session starts with an invited paper presenting MRAM technology emerged from research and development into volume production within the last decade in the form of Toggle MRAM. Spin-Torque MRAM has reached the level of customer sampling, offering higher density and bandwidth. This paper describes the devices, fundamental circuit challenges, and applications of this evolving MTJ based memory.

The second invited paper discusses the scaling challenges of 2D and 3D NAND flash memory and then provides a state-of-the-art hybrid memory solution with storage class memory and NAND flash memory for the big data solid-state storage system. An error prediction low density parity check (LDPC) error correcting code (ECC) realizes an over 10 times extended lifetime. Proposed 3D TSV hybrid SSD achieves 11 times performance increase, 6.9 times endurance enhancement and 93% write energy reduction.

The third paper presents a high density 28nm 1R/1W 8T-SRAM with an effective scheme for design for testing (DFT). To achieve a smaller Macro area, a differential sense amplifier is introduced to read the data, where the reference voltage for reading 0/1 data is generated by unselected cell array. A screening test circuit for read disturb operation has also been proposed. Implemented 512kb macro achieves 3.16Mb/mm$^2$ density with 593ps read access time at 1.0V.

Next paper introduces a fully-pipelined tile-able 1MB SRAM IP with a 0.127μm$^2$ cell in a HKMG 28nm bulk planer CMOS technology. It has an area of 1.39mm$^2$/MB with 79.2% array efficiency. It operates with 2-cycle latency up to 1GHz. The no-repair hardware has a circuit limited yield of 99.92 and 53% at 100 and 850MHz, respectively with 0.75V VDD. A Data Retention Voltage of 0.42V has been measured.

The fifth paper demonstrates a bit-by-bit re-writable embedded flash memory in a generic 65nm logic process for moderate-density embedded non-volatile memory applications. The proposed 6T embedded flash memory cell improves the overall cell endurance by eliminating redundant program/erase cycles without disturbing cells in the unselected wordlines. A multi-story high voltage switch utilizes four boosted supply levels generated by a compact voltage doubler based on-chip negative charge pump.

The last paper indicates a memory sense-timing circuit with degraded VGS to emulate the behavior of weak memory tail-bits, improving Tail-Bit Tracking (TBT) across process, voltage and temperature. The TBT circuit generates timing for a TCAM search operation reducing Vmin by 200mV, and improving sense-time by 50%. Implemented in 32nm HKMG SOI process the 2Kx640b TCAM achieves 0.60V and 1G search/sec.

**Oversampled ADC's**                                                                                    **Session 21**

Chair: Eric Naviasky
Co-Chair Hasnain Lakdawala

Oversampled ADC's are steadily reaching higher effective bandwidths and resolutions at lower power dissipation levels through the use of clever design techniques. The use of continuous time architecture enhances the system level usability by reducing the need for anti-aliasing filters.

The first two papers make use of VCOs as combined integrators and quantizers. A VCO is well suited for this function in terms of low power and scalability to advanced process nodes but suffers from inherently poor linearity. The first of the two papers using VCO quantizers focuses on a low power, modest bandwidth convertor that uses a combination of 2 VCO's to provide integration, quantization, and an intrinsic data rotation to provide DEM. The second of the two papers reports on a wider bandwidth, higher power design that uses a dither based linearization of the VCO. This leads to a very impressive FOM of 90fJ/step.

The last two papers introduce new techniques for improving the linearity of more conventional continuous time ADC's. The first one corrects for the DAC errors by using reference shuffling rather than DAC element rotation in order to avoid the latency inherent in conventional DEM. This is all in a converter that is flexible enough to cover an order of magnitude in signal bandwidth and a factor of 3 in sampling rate.

The second paper of this group attacks another source of nonlinearity – the input transconductor – by introducing a nonlinear feedback D/A. This makes the design far more user friendly by removing the need for a high powered input buffer to drive the more traditional RC input section.

**AMS System Simulation Techniques**                                   **Session 22**

Session Chair: Larry Nagel, Omega Enterprises Consulting
Session Co-Chair: Colin McAndrew, Freescale Semiconductor

As integrated circuits become more complex and more difficult to design and verify prior to fabrication, evermore sophisticated simulation techniques are being developed to meet the design challenges, with particular emphasis on behavioral modeling and simulation. This session presents five papers describing new simulation techniques for analog mixed-signal systems.

The first paper is an invited paper on tools for automated behavioral modeling. This topic has been addressed by researchers for more than twenty years, but has proven to be both difficult and elusive. This paper presents algorithmic nonlinear macromodeling, which is studied from the perspective of projection in the state space defined by voltages and currents. The model accuracy and computational efficiency are compared to transistor-level models and linear models.

The second paper in the session presents a new and promising application of Volterra series models for event-driven simulation. Although Volterra series models have been used for many years to simulate the distortion of weakly-nonlinear circuits, this paper is the first to report a true event-driven simulator based on Volterra series models. The simulator capabilities are demonstrated by estimating the spectral regrowth of an RF power amplifier and characterizing the distortion-induced eye opening reduction of a continuous-time linear equalizer. Compared to a commercial SPICE simulator, this simulator achieves 300~1000X speed-up in performing time domain simulations with the same level of accuracy.

The third paper describes a Verilog piecewise-linear behavioral model for mixed-signal validation. The method for creating these models maps continuous signals into piecewise linear waveforms by creating analog events which contain a value and slope. By partitioning analog circuits into sub-blocks with mostly unidirectional ports, explicit time integration can be avoided, thus fitting well into an event-driven digital framework. This results in Verilog behavioral models that are pin-accurate, fast to simulate, and capture the key dynamics of analog circuits. A 2.5-1.8 V buck converter model and a 1 GHz PLL model are demonstrated.

The fourth paper is an invited paper which provides an overview of the simulation of high-speed data links. Important emerging requirements are determining process, voltage, and temperature variations, fully accounting for all the circuit blocks of the link, and closing the gap between modeling and measurements. This paper reviews the techniques used in recent data-link modeling and simulation such as behavioral, statistical, SPICE, and IBIS-AMI. The paper then presents new techniques that improve the accuracy and capability of high-speed data-link modeling and simulation.

The fifth and last paper in the session presents a structure-aware technique for modeling analog mixed-signal circuits which accurately determines the model coefficients by applying an efficient statistical algorithm which exploits the underlying structure of the circuit. This technique dramatically reduces the required number of sampling points and, hence, the computational cost. Several circuit examples demonstrate that structure-aware modeling achieves more than 2X runtime speedup over the traditional sparse regression technique without any reduction in accuracy.

**Analog Techniques II**                                                    **Session 23**

Chair Hasnain Lakdawala
Co-Chair: Eric Naviasky

The four papers in the session cover progress in the state of the art of analog circuits in data conversion and frequency generation.

The first two papers in the session detail advances in the TDC architectures. The first paper presents a all-digital second-order $\Delta\Sigma$ time-to-digital using switched-ring oscillator (SRO) and gated switched-ring oscillator (GSRO) that doesn't require extensive calibration to achieve a 148fsrms integrated noise and 80.4dB dynamic range in 4MHz signal bandwidth. The next paper describes a TDC that is designed using a charge pump and a low power SAR ADC yielding a 2.47mW design in 65nm CMOS with 0.84ps LSB.

The next paper presents a double sampled discrete time sigma delta ADC that uses noise coupling to using a novel delaying feedforward circuit. The design in 130nm achieves 77.8dB dynamic range in a 5MHz bandwidth.

The final paper in the session presents a fast locking 65nm CMOS PLL using a simple transconductor within the loop filter to improve the lock time by 6.5x,over a frequency range of 400MHz to 1.6GHz.

# Educational Sessions
*Tuesday, September 24 and Wednesday, September 25*

---

## Educational Session 1

Tuesday, 9/24, 9:00 am
Cedar Ballroom

**Session Chair**: Foster Dai, Auburn University
**Session Co-Chair:** Earl McCune, RF Communications Consulting

**E-1**
**9:00 am –**
**10:30 am**

**Concurrent Design of ESD Protection and Integrated Circuits for Optimization and Prediction,** *Albert Wang, University of California, Riverside*

As semiconductor technologies continuously advance into nano nodes, while integrated circuits (IC) become faster and more complex, on-chip ESD protection design quickly emerges as a huge IC design barrier nowadays. Major ESD design challenges include the followings: First, how to conduct simulation-based quantitative design to achieve ESD protection design optimization and prediction? Second, how to minimize ESD-induced parasitic effects that affect IC performance. Third, how to perform co-design of ESD protection and IC to achieve ESD protection and core circuit design optimization simultaneously. This lecture discusses critical aspects and techniques in practical ESD protection designs, including a mixed-mode ESD simulation-design method for design prediction, accurate RF ESD design characterization, complex ESD-IC interactions, ESD+IC co-design for whole-chip design optimization, etc. Real-world design examples will be presented.

**10:30 am**     **BREAK**

---

## Educational Session 2

Tuesday, 9/24, 11:00 am
Cedar Ballroom

**Session Chair**: Eric Naviasky, Cadence
**Session Co-Chair:** Gerrit den Besten, NXP Semiconductors

**E-2**
**11:00 am –**
**12:30 pm**

**Characterization of Matching, Variability, and Low-Frequency Noise for Mixed-Signal Technologies,** Hans Tuinhout, NXP

Parametric mismatches and low frequency noise are major performance limiters as well as notorious causes for redesigns of high performance mixed-signal (HPMS) circuits and systems. Consequently, it is extremely important to measure, analyze, interpret, model and document these effects for mixed-signal technologies.

Part one of this educational lecture discusses parametric mismatch benchmarks and variability characterization challenges for active and passive devices. Part two focuses on low frequency noise, in particular on the emerging challenge in this field, namely variability of 1/f noise and associated Random Telegraph Noise.

These topics will be exemplified with results from (Bi)CMOS technologies, ranging from the current HPMS cash cow technologies (140 to 250 nm minimum dimensions), up to more advanced 40 nm node devices which can be seen as the stepping stone to some of the ultimate challenges of sub 10 nm devices that will mark the end of the CMOS shrink roadmap.

---

## Educational Session 3

Tuesday, 9/24, 2:00 pm
Cedar Ballroom

**Session Chair**: Christoph Sandner, Infineon Technologies Austria AG
**Session Co-Chair:** Hoi Lee, University of Texas at Dallas

**E-3**
**2:00 pm –**
**3:30 pm**

**Single-Inductor-Multiple-Output DC-DC Converter Design,** Philip Mok, Hong Kong University of Science and Technology

Multiple well-regulated power supplies are essential for reducing power consumption and isolating the coupling noise between different functional blocks in VLSI design. With the increasing number of functional blocks in SoC applications, the need for a cost and efficiency effective solution of multiple power supplies is growing. Single-Inductor-Multiple-Output (SIMO) switching regulator, which provides several output voltages with only one inductor, is one of promising solutions and becomes a hot topic in DC-DC converter design due to the cost and volume reduction. However, with one inductor shared by all the outputs to accumulate and transfer energy from the input, cross regulation easily appears at outputs when a change in the inductive energy is induced by a load transient at one output. These unwanted voltage variations affect the performance or even the function of the loading devices. This talk will discuss the operation principle of SIMO switching converters and their design issues. To minimize cross regulation of a SIMO regulator, several control techniques will be presented and their pros and cons will be discussed.

**3:30 pm**      **BREAK**

---

## Educational Session 4

Tuesday, 9/24, 4:00 pm
Cedar Ballroom

**Session Chair**: Foster Dai, Auburn University
**Session Co-Chair:** Jonathan Borremans, IMEC

**E-4**
**4:00 pm –**
**5:30 pm**

**Advanced Digital Phase-Locked Loops,** Salvatore Levantino, Politecnico di Milano

This tutorial will introduce the fundamentals of digital phased-locked loops for wireless applications. After reviewing the basic architectures, the tutorial will analyze the mechanisms of generation of limit cycles, which manifest themselves as spurious tones in the output spectrum even when synthesizing integer-N channels. Then, loop-parameter settings and design strategies for spur elimination and phase-noise optimization will be derived. Next, we will move to the fractional-N case, in which quantization and nonlinearity add new sources of spur tones, and we will review the different design techniques which helps mitigate such impairments. Finally, examples of state-of-the-art implementations of frequency synthesizers and direct-FM modulators based on digital PLLs will be discussed.

---

## Educational Session 5

Wednesday, 9/25, 9:00 am
Cedar Ballroom

**Session Chair**: Gordon Roberts, McGill University
**Session Co-Chair:** Takamaro Kikkawa, Hiroshima University

**E-5**
**9:00 am –**
**10:30 am**

**Design for Nanoscale Patterning,** Puneet Gupta, UCLA

This tutorial explains how layout and circuit design interact with lithography choices. Lithography technology is rapidly evolving and has started to impose unusual restrictions on design layout. The tutorial will give a brief introduction to current and upcoming lithography technologies. We especially focus on multi-patterning technologies such as LELE double patterning and SADP. We will discuss design enablement of multi-patterning technologies, especially in context of cell-based digital designs. Models for electrical impact of lithography imperfections such as polysilicon/active rounding and overlay errors will be outlined. We will also briefly explore role of design in lithography technology development.

**10:30 am**     **BREAK**

---

## Educational Session 6

Wednesday, 9/25, 11:00 am
Cedar Ballroom

**Session Chair**: Howard Luong, Hong Kong University of Science & Technology
**Session Co-Chair:** Earl McCune, RF Communications Consulting

**E-6**
**11:00 am –**
**12:30 pm**

**Low Power Chip & System Design for Biomedical Applications,** Brian Otis, Google, Inc.

Advances in chip and system design will help define the next generation of wireless sensors for biomedical applications. This talk will investigate system and circuit design techniques for body-worn systems, implantable chips, and wireless sensors. These areas present unique challenges at the interface between the IC and the body that cannot be solved by technology scaling alone. Traditional circuit blocks, architectures, and even assembly techniques need to be questioned. Several future applications will demand thin-film realization and biocompatibility of complex systems. RFID-like techniques are highly useful for many of these emerging biomedical applications. High-Q RF resonators are useful for minimizing power and size of low power radios. We'll discuss a few examples of the above.

---

## Educational Session 7

Wednesday, 9/25, 1:30 pm
Cedar Ballroom

**Session Chair**: Mike Li, Altera
**Session Co-Chair:** Gerrit den Besten, NXP Semiconductors

**E-7**
**1:30 pm –**

**Trends, Possibilities and Limitations of Photonic Integrated Circuits and Lasers,** John E. Bowers, University of California, Santa Barbara

**3:00 pm**

A number of important breakthroughs in the past decade have focused attention on Si as a photonic platform. We review here recent progress in this field, focusing on efforts to make lasers, amplifiers, modulators and photodetectors on or in silicon. We also describe progress in silicon photonic integrated circuits. The impact active silicon photonic integrated circuits could have on interconnects, telecommunications and on silicon electronics is reviewed.

**3:00 pm       BREAK**

| Educational Session 8 |
|---|

Wednesday, 9/25, 3:30 pm
Cedar Ballroom

**Session Chair**: Jerry Jiang, Broadcom
**Session Co-Chair:** Eric Naviasky, Cadence

**E-8**          **A/D Converter Circuit and Architecture Design for High-Speed Data Communication.**
**3:30 pm –**    Boris Murmann, Stanford University
**5:00 pm**

As modern electrical and optical communication systems transition toward advanced modulation schemes, there exists a pressing need for power efficient A/D converters operating at tens of giga samples per second. Within this context, this tutorial will cover relevant circuit- and architecture-level design techniques for high-speed CMOS A/D converters. At the circuit level, we will discuss fundamental challenges in the design of track-and-hold circuits and voltage comparators, which will also include a review of clock jitter and metastability. At the architecture level, we consider tradeoffs in the design of time-interleaved SAR and flash converters as well as techniques for the estimation, system-level budgeting and calibration of circuit imperfections.

# A Broadband Biosensor Interface IC for Miniaturized Dielectric Spectroscopy from MHz to GHz

M. Bakhshiani, M. A. Suster, and P. Mohseni

Electrical Engineering and Computer Science Department, Case Western Reserve University

2123 Martin Luther King Jr. Drive, Glennan Building, Cleveland, OH 44106 USA

*Abstract*-**This paper describes a broadband biosensor interface IC as part of a miniaturized measurement platform currently under development for MHz-to-GHz dielectric spectroscopy (DS). Developed in 0.35μm 2P/4M RF CMOS, the IC measures the frequency-dependent $S_{21}$ magnitude and phase of a microfabricated microfluidic dielectric sensor, when the sensor is loaded with a solution-under-test (SUT). The IC architecture implements a modified frequency response analysis (FRA) method by first down-converting the biosensor response signal from the RF excitation frequency to an intermediate frequency (IF) of 1MHz using an LNA and active mixer, followed by down-converting the IF signal to dc using a coherent detector employing IF amplification stages with programmable gain, a passive mixer driven by in-phase ($I$) and quadrature-phase ($Q$) signals, and an active-RC lowpass filter (LPF). Dielectric readings of ethanol from the biosensor interfaced with the IC at five excitation frequencies in the range of 50MHz to 2GHz are in excellent agreement (error < 1%) with those from using a vector network analyzer (VNA) as the sensor readout. A reference measurement by an *Agilent* 85070E dielectric probe kit is also included to show proof-of-concept feasibility in miniaturization.**

## I. INTRODUCTION

Dielectric spectroscopy (DS), a method by which complex relative dielectric permittivity ($\varepsilon_r$) of a material is measured as a function of frequency, is a powerful monitoring technique with potential applications ranging from oil industry, to molecular and structural biology, to clinical diagnostics with blood and tissue analysis [1]. Existing instrumentation for DS incorporates a commercial dielectric probe kit (e.g., *Agilent* 85070E) that is interfaced with a benchtop vector network analyzer (VNA) to perform bulk-solution measurements with a large volume of the solution-under-test (SUT) over a very wide frequency range extended to several tens of GHz. Clearly, in order to translate DS studies from the lab bench to the field or the bedside, such an expensive and bulky measurement setup should be miniaturized into an autonomous, low-power, small-sized, and portable instrument to enable rapid, low-cost, and high-throughput measurements using only a small liquid volume (i.e., μL to nL).

Over the past several years, while many researchers have focused on developing miniaturized measurement platforms for biomolecular sensing applications (see [2] for an excellent review of the existent work), little effort has focused on miniaturizing the measurement setup for wideband DS. Daphtary and Sonkusale have previously reported a broadband current-mode interface IC that measured variations of a single sensing capacitor arising from changes in permittivity [3].

However, only simulation results were reported in a frequency range of 200MHz to 1GHz. Ghafar-Zadeh and Sawan have reported a hybrid microfluidic/CMOS sensor based on the charge-based capacitive measurement (CBCM) technique [4], but the measurements were all performed at dc, without demonstrating any spectroscopic measurements. Further, neither of these two works provided explicit dielectric readings from the capacitive sensor variations.

Helmy and colleagues have recently reported a miniaturized system that extracted the real part of $\varepsilon_r$ (i.e., $\varepsilon'_r$) of organic chemicals in 7 to 9GHz from the frequency shift of an LC-tank VCO within a frequency-synthesizer loop, when the capacitive sensing element was exposed to the SUT [2]. However, this approach is inherently unsuitable for applications in which the frequency range of interest might have to extend from MHz to GHz to capture the dielectric relaxation characteristics of an unknown material in β, δ and γ dispersion regions.

We have previously reported on the design, microfabrication, and testing of a microfluidic dielectric sensor for broadband DS studies in the range of MHz to GHz, requiring only an $S_{21}$ measurement for dielectric data extraction [5]. In this paper, we present an interface IC fabricated in 0.35μm RF CMOS that measures the $S_{21}$ magnitude and phase of the biosensor, when it is loaded with an SUT. Measured results from the IC in benchtop electrical tests, as well as dielectric readings of ethanol when the IC is interfaced with the biosensor are presented and compared with those from the biosensor interfaced with a VNA. A reference measurement by commercial DS equipment is also included to demonstrate proof-of-concept feasibility in miniaturization.

## II. SYSTEM ARCHITECTURE

Fig. 1 shows the system architecture of the biosensor interface IC and a simplified illustration of the experimental setup used in this work. The operation principle is that a single-tone sinusoidal excitation signal drives a microfluidic dielectric sensor based on a microstrip line with a centered gap as its parallel-plate capacitive sensing area [5]. As the SUT passes through this gap, the through-impedance of the sensor changes based on the dielectric permittivity of the liquid. The large through-impedance of the capacitive sensor structure dominates its series impedance, making it feasible to determine the impedance, and hence the permittivity of the liquid, using only an $S_{21}$ measurement.

978-1-4673-6145-3/13 $31.00 © 2013 IEEE

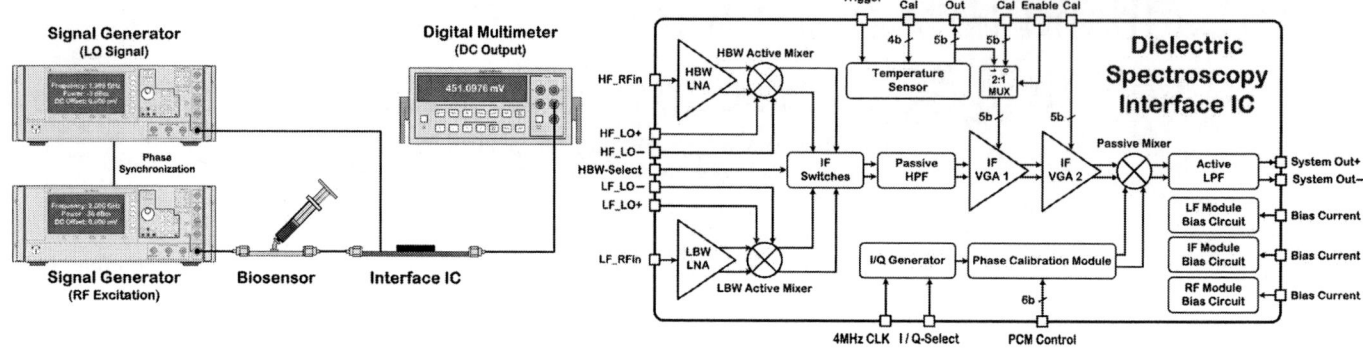

Fig. 1. System architecture of the dielectric spectroscopy interface IC and an illustration of the experimental setup.

The interface IC is therefore designed to accurately measure the $S_{21}$ magnitude and phase of the biosensor using a modified frequency response analysis (FRA) method, with the excitation frequency covering a wide range from MHz to GHz. In this broadband lock-in architecture, the biosensor response signal at the RF excitation frequency is first down-converted to an IF frequency of 1MHz using an LNA and an active mixer. The 1-MHz IF signal is next down-converted to dc using a coherent detector employing a passive highpass filter (HPF), two VGAs, a passive mixer driven by in-phase ($I$) and quadrature-phase ($Q$) signals, and a lowpass filter (LPF) to extract the dc component of the system output. It can be shown that this dc component is proportional to the imaginary (real) part of $S_{21}$ in $I$ ($Q$) mode of system operation. The dielectric permittivity of the SUT is then extracted from the $S_{21}$ measurement using a biosensor calibration algorithm that runs offline on a PC [5].

As seen in Fig. 1, the system architecture incorporates two parallel paths at the front-end for processing the biosensor response signal. The low-bandwidth (LBW) and high-bandwidth (HBW) paths are used for the excitation frequency range of ~14 to 200MHz and ~200MHz to 3GHz, respectively. Controlled by an external bit (*HBW-Select*), the IF Switches block routes the appropriate RF module output to the input of the coherent detector for the second down-conversion step. Since the two RF modules have independent power supplies, they can be turned on/off independently. This feature allows the user to save power by using the LBW RF module only, in case the experiment does not require high excitation frequencies. The IF Switches block has relaxed design constraints, given that it is placed *after* the active mixers and operates only statically.

After down-converting the biosensor response signal to 1MHz, a passive HPF removes the dc offset (and low-frequency noise) at the active mixer output, preventing the dc offset from saturating the subsequent VGAs. The gain of each VGA can be adjusted using an external 5b code (*Amp-Cal*) for signal-amplitude calibration. Further, the gain of the first VGA can additionally be adjusted by the 5b digital output of an on-chip temperature sensor to compensate for the effect of ambient temperature variation in the range of 0 to $60^{\circ}$C.

The amplified/filtered IF signal is next down-converted to dc via a passive mixer and a LPF. The 1-MHz, square-wave, $I/Q$ signals for the passive mixer are first derived by a digital $I/Q$ generator from an external 4-MHz clock and then phase-adjusted using a phase calibration module (PCM). Given the path delay experienced by the biosensor response signal at each excitation frequency, the PCM adds appropriate delay in the range of 0.8 to 50ns to the $I/Q$ signals (same delay for the two signals to preserve the $90^{\circ}$ phase shift) before they drive the passive mixer.

### III. INTEGRATED CIRCUIT ARCHITECTURE

Fig. 2 depicts the circuit architecture of the main building blocks in HBW RF, IF, and low-frequency (LF) modules of the biosensor interface IC. All circuitry are powered from 1.5V, but the control signals for the switches are at 3.3V.

The HBW LNA has an inductorless, wideband, common-gate–common-source (CG-CS) architecture with a single-ended input matched to ~50Ω [6], employing local feedback for $G_m$-boosting and a differential output that allows differential signaling throughout the rest of the system for canceling common-mode noise and even harmonics. Transistors $M_{2,3}$ implement a local feedback mechanism that boosts $gm_1$ by a factor of $(gm_2 + gm_3)/gm_4$, which in turn allows reducing the bias current of $M_1$ (600µA) for low-power operation and increasing its resistive load (450Ω) for enhanced gain. Similarly, current steering by $M_3$ (1.4mA) also allows reducing the bias current of $M_4$ (1.2mA) and hence increasing its resistive load. The bias currents of $M_{2,3}$ are set by tunable voltages $V_{b2,3}$, which allows controlling the $gm_1$-boosting factor and hence the LNA input impedance.

The LBW LNA employs a dual architecture based on a *pMOS* CG transistor to reduce the low-frequency noise, and with lower bias currents due to the reduced bandwidth (3-dB BW of 197MHz). Two capacitors (250fF) placed in parallel with the resistive loads further reduce the out-of-band high-frequency noise.

The HBW active mixer is a standard double-balanced Gilbert cell with resistive loads and source-degenerated RF input transistors (using two resistors) for enhanced linearity.

978-1-4673-6145-3/13 $31.00 © 2013 IEEE

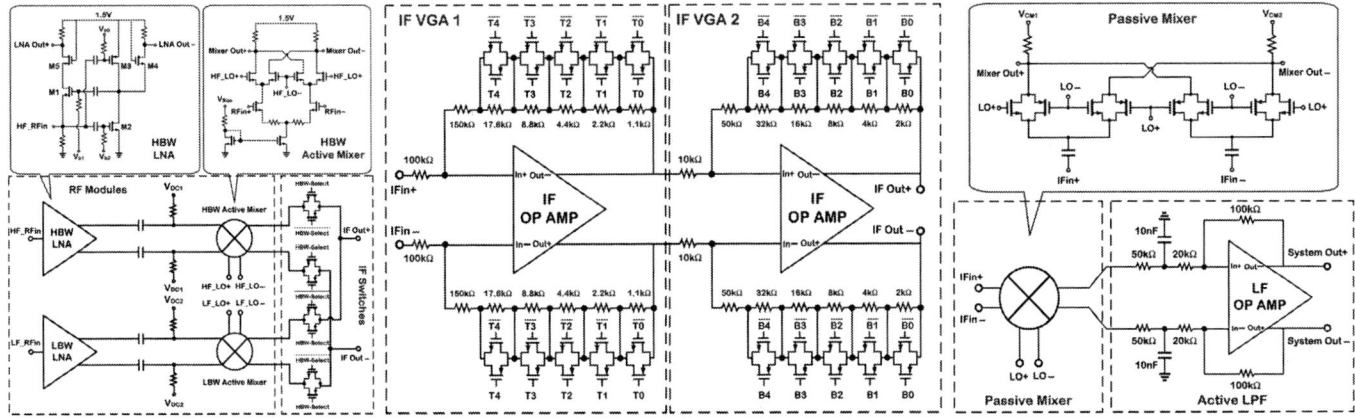

Fig. 2. Circuit architecture of the main building blocks in HBW RF (left), IF (middle), and LF (right) modules.

The bias current of the mixer (325µA), which is controlled by the external $V_{Bias}$, is set to be low enough to reduce the power consumption, while meeting the design constraints related to mixer bandwidth, linearity, and noise figure. The low bias current also allows using relatively large resistive loads (3kΩ) at the output for increased conversion gain without adding extra 1/f noise.

The LBW active mixer employs an identical architecture, but with lower bias current (165µA) due to the reduced bandwidth. This also allows further increasing the resistive loads (8kΩ) due to the higher available voltage headroom at the output. To maintain enhanced linearity, the sizing of the RF input transistors is reduced, obviating a need for resistive source degeneration.

The IF module comprises two 5b-programmable VGAs, with each gain stage based on a differential-input differential-output op-amp with resistive feedback. The op-amp employs a two-stage, telescopic, folded-cascode architecture with common-mode feedback, featuring a high open-loop gain (132.4dB) for enhanced linearity and accurate gain in closed-loop mode, rail-to-rail input/output dynamic range, and a unity-gain bandwidth of 34MHz for optimized power consumption and operation speed. The properly sized transmission gates in parallel with the five feedback resistors in each gain stage are driven by digital control signals at 3.3V for negligible ON resistance (compared to the LSB feedback resistor) for enhanced linearity in gain variation.

The LF module incorporates a passive mixer and a 1st-order active-RC LPF. The passive mixer, employing four identical transmission gates as its switches, is a double-balanced Gilbert cell with reduced dc offset and low-frequency noise at its output compared to an active counterpart. The double-balanced architecture also mitigates the 1-MHz LO feed-through at the output, relaxing the design constraints of the LPF. The LF op-amp of the LPF has an identical architecture to that of the IF op-amp, but with much lower unity-gain bandwidth (1.97MHz), and hence consumes less power. Using two external capacitors (10nF), the 3-dB bandwidth of the LPF is ~1.48kHz for a tradeoff between the total measurement time of the system for each excitation frequency (affected by the settling time of the LPF output) and noise bandwidth as well as the requisite filter attenuation at 1MHz (IF/LO feed-through) and 2MHz (up-converted IF) at the passive mixer output. The externally tunable bias voltages, $V_{CM1,2}$, are used to cancel out the dc offset of the LPF output.

## IV. MEASUREMENT RESULTS

A prototype chip was fabricated in 0.35µm 2P/4M RF CMOS measuring 3mm × 3mm, including the bonding pads (see Fig. 3.) The top plots in Fig. 4 depict the measured conversion gain and linearity plots of the HBW RF module (i.e., HBW LNA + active mixer). The conversion gain was 28.22dB at 200MHz and decreased to 9.92dB at 2.8GHz. This reduction in the RF conversion gain versus frequency was compensated by a gain increase in the IF module via amplitude calibration. With the RF signal at 1GHz, the 1-dB compression point was -21dBm, showing that the HBW RF module was highly linear with an RF input power < -31dBm (backing off by 10dB from the 1-dB compression point).

The bottom left plot depicts the measured linearity of the full system in $Q$ mode when the RF input amplitude at 2GHz was varied in the range of 0.1 to 6.5mV, showing that the full system was highly linear with an RF input amplitude < 5mV. The total system gain from RF to dc was 165.5V/V at 2GHz with maximum gain in the IF module. The bottom right plot depicts the dc system output in $I$ mode versus the RF input phase at 2GHz, demonstrating a sinusoidal response with no phase offset after phase calibration by the PCM.

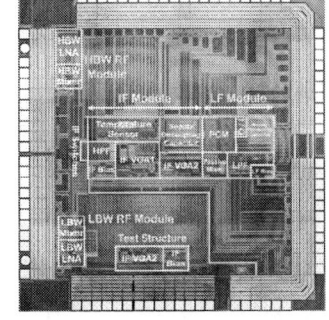

| Process | 0.35µm RF CMOS |
|---|---|
| Die Size | $3 \times 3 \text{mm}^2$ |
| Supply Voltage | 1.5V (Analog) |
| | 3.3V (Digital) |
| Operation Freq. | MHz – GHz |
| Scan Rate | $135 \text{ s}^{-1}$ |
| Dynamic Range | 88.5dB @ 50MHz |
| (BW = 3.2kHz) | 74.8dB @ 2GHz |

Fig. 3. Die micrograph of the fabricated chip.

Next, the IC was interfaced with the biosensor to extract the frequency-dependent dielectric permittivity of organic chemicals. Fig. 5 shows the measured $S_{21}$ magnitude and phase of the biosensor loaded with ethanol as the SUT and interfaced with a VNA (solid line, 14MHz to 3GHz) as well as the IC (blue square). The IC measurements were done at seven distinct excitation frequencies from 50MHz to 3GHz using the HBW RF module (except for the one at 50MHz that was done using the LBW RF module). As can be seen, the IC readings closely matched those of the VNA after a 1-point calibration to remove constant offset errors in magnitude and phase at each excitation frequency. Finally, Fig. 6 depicts the frequency-dependent $\varepsilon'_r$ of ethanol extracted from the $S_{21}$ measurements along with a bulk-solution reference measurement with *Agilent 85070E* dielectric probe kit (solid line). Only raw $S_{21}$ measurements by the IC (i.e., without 1-point calibration) were used to extract $\varepsilon'_r$ in this step. Dielectric readings of the "biosensor + IC" showed excellent agreement with those of the "biosensor + VNA", with an error < 1% up to 2GHz. The increase in error above 2GHz (e.g., 8.7% at 2.5GHz) was primarily because of the increased input noise level of the IC due to the reduction in RF conversion gain, as well as the sensitivity of the biosensor calibration algorithm to the measured $S_{21}$ data when extracting small values of dielectric permittivity. Table I provides a summary of the measured performance comparison with prior work.

## V. CONCLUSION

This paper reported on a broadband interface IC in 0.35μm RF CMOS that measures the $S_{21}$ magnitude and phase of a microfabricated microfluidic sensor for MHz-to-GHz miniaturized dielectric spectroscopy of organic chemicals. Dielectric readings of ethanol using the IC at five excitation frequencies in the range of 50MHz to 2GHz had an excellent match with those using a VNA (error < 1%). The next-generation IC will also feature on-chip generation of the excitation frequencies for a truly miniaturized standalone measurement platform for broadband dielectric spectroscopy.

## REFERENCES

[1] F. Kremer and A. Schonhals, *Broadband Dielectric Spectroscopy*, Springer-Verlag, Berlin, Germany, 2003.
[2] A. A. Helmy, et al., "A self-sustained CMOS microwave chemical sensor using a frequency synthesizer," *IEEE J. Solid-State Circuits*, vol. 47, no. 10, pp. 2467-2483, October 2012.
[3] M. Daphtary and S. Sonkusale, "Broadband capacitive sensor CMOS interface circuit for dielectric spectroscopy," in *Proc. IEEE Int. Symp. Circuits and Systems (ISCAS'06)*, pp. 4285-4288, 2006.
[4] E. Ghafar-Zadeh and M. Sawan, "A hybrid microfluidic/CMOS capacitive sensor dedicated to lab-on-chip applications," *IEEE Trans. Biomed. Circuits and Systems*, vol. 1, no. 4, pp. 270-277, December 2007.
[5] M. A. Suster and P. Mohseni, "An RF/microwave microfluidic sensor based on a center-gapped microstrip line for miniaturized dielectric spectroscopy," *IEEE MTT-S Int. Microwave Symp. (IMS) Dig.*, Seattle, WA, June 2-7, 2013, in press.
[6] H. Wang, L. Zhang, and Z. Yu, "A wideband inductorless LNA with local feedback and noise cancelling for low-power low-voltage applications," *IEEE Trans. Circuits and Systems – Part I*, vol. 57, no. 8, pp. 1993-2005, August 2010.

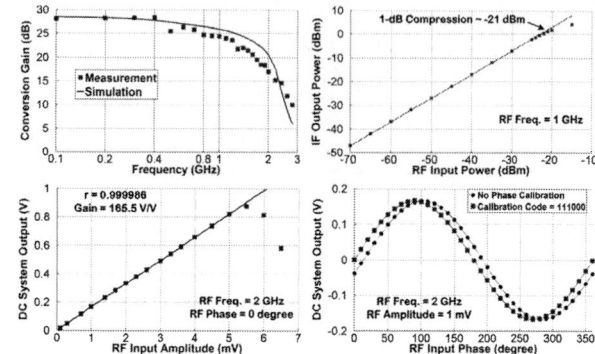

Fig. 4. Representative benchtop measured results. Conversion gain (top left) and linearity (top right) characterization of HBW RF module. DC system output *vs.* amplitude (bottom left) and phase (bottom right) of RF input signal.

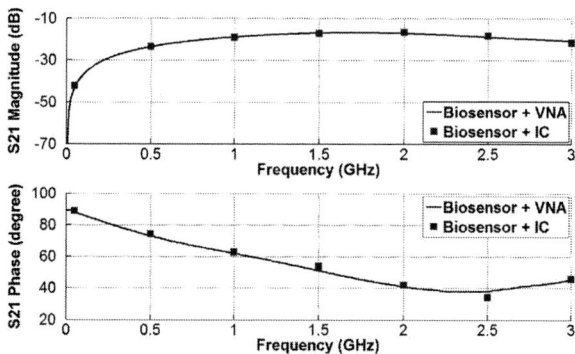

Fig. 5. Measured $S_{21}$ magnitude and phase *vs.* frequency of the biosensor loaded with ethanol as the SUT.

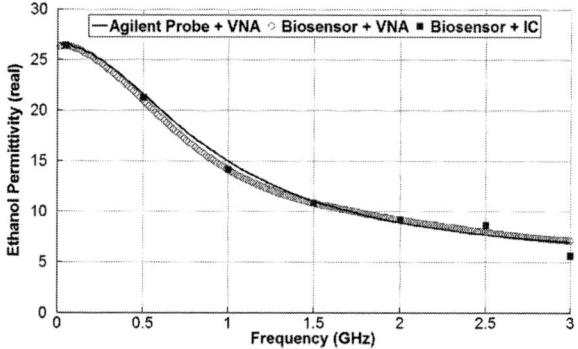

Fig. 6. Real part of complex relative dielectric permittivity ($\varepsilon'_r$) *vs.* frequency for ethanol extracted from $S_{21}$ measurements by the biosensor. Solid line depicts the measured curve with an *Agilent 85070E* dielectric probe kit.

TABLE I: PERFORMANCE COMPARISON

|  | **This Work** | [2] | [4] |
|---|---|---|---|
| **Functionality** | Measurement of dielectric permittivity of organic chemicals | | |
| **Methodology** | $S_{21}$ measurement (modified FRA) | Oscillation frequency shift | CBCM |
| **Capacitive Sensor Type** | μfabricated center-gapped μstrip line | On-chip interdigitated | On-chip interdigitated |
| **Operation Freq.** | MHz – GHz | GHz only | DC only |
| **Permittivity Range** | 1 – 30 | 1 – 30 | 1 – 30 |
| **Permittivity Error** | < 1% (50MHz – 2GHz) | < 3.5% (7 – 9GHz) | < 6% |
| **Total Power Consumption** | 4mW (w/ LBW RF) 9mW (w/ HBW RF) | 16.5mW @ 1.3V | N/A @ 1.8V |
| **Technology** | 0.35μm RF CMOS | 90nm CMOS | 0.18μm CMOS |

# Capacitive Proximity Communication with Distributed Alignment Sensing for Origami Biomedical Implants

Matthew Loh and Azita Emami-Neyestanak
California Institute of Technology
Pasadena, CA 91125

*Abstract*—Origami implant design is a 3D integration technique which addresses size and cost constraints in biomedical implants. A capacitive proximity interconnect that enables this technique is presented. The interconnect embeds an alignment sensor that measures link quality directly and simplifies its adaptation to alignment. The sensor and transceiver share functional blocks, saving power and area. Data rates from 10-60 Mbps are achieved over 4-12 $\mu$m of parylene-C, with efficiencies up to 0.180 pJ/bit.

## I. INTRODUCTION

Designers of medical implants face three primary challenges: size, cost and power consumption. At the same time, there is a desire to increase the capability of these implants. For example, recent developments in retinal prosthesis design have targeted as many as 1024 electrodes [1], [2] in order to achieve reasonable visual resolution. Such increases in capability require the development of specialized, highly-integrated system-on-chip designs, which can be cost-prohibitive for low-volume applications typical in the biomedical market. Even with highly-integrated designs, the size of the system required can pose an implantation challenge in small and delicate organs such as the eye; for example, if the design in [1] is scaled up to 1024 electrodes, approximately 8x8 mm$^2$ is required for the stimulator array alone, excluding power delivery, data telemetry and digital control. Compounding this problem, the high voltages ($\sim$5-10V) typically used to drive the desired stimulus currents prevent the use of finer-geometry process technology.

In order to work around size limits, large systems can be split into multiple chips and connected using 3D integration techniques. For example, they can be placed on a flexible, bio-compatible substrate such as Parylene-C [3]. This substrate can be folded compactly for implantation, then unfolded into its operating configuration inside the body, allowing minimally-invasive surgery [4]. Other applications of this Origami folding technique include conforming a retinal prosthesis to the back of the eye to improve electrode contact and make stimulation more effective. This concept can be further extended to address the high cost of developing custom SoC designs for each new implant; electronics can be partitioned into commonly-used functional blocks, mass-produced as ICs that are embedded into parylene library modules. Custom implants can be assembled from these (relatively) cheap modules.

To enable Origami implants, individual ICs will need to be able to communicate with each other wirelessly. Proximity communication [5] provides a compelling way to achieve this, thanks to its low cost of integration and the fact that many of the ICs in an Origami implant can be placed close to each other and face-to-face. However, existing approaches to proximity communication have targeted multi-Gbps links in high-performance computing [6] and memory-stacking [7] applications, and so have been designed under a different set of constraints than the relatively low data rates and ultra-low power consumption required by implants.

## II. PROXIMITY COMMUNICATION FOR ORIGAMI IMPLANTS

Since the Origami implant is deployed in-body, the alignment between the communicating chips is poorly controlled. Furthermore, the alignment will change over time due to patient movement and tissue growth. As a result, the proximity communication system needs to periodically sense its alignment and adapt itself to maximize power efficiency for a given data rate. Alignment sensing is also useful for providing feedback on the deployment status of the implant. Due to the tight power constraints on implants in sensitive organs such as the eye, this alignment-and-adaptation operation needs to be energy efficient and computationally simple.

Although capacitive coupling is more sensitive to separation between the chips than inductive coupling, capacitive plates exhibit less crosstalk and can be made smaller and denser. Additionally, inductive proximity communication tends to have higher power consumption, since it requires either a constant current drive [7] or imposes tight timing requirements at the receiver to capture current pulse inputs [8]. Given the requirements for extremely low power and the ability to sense alignment, a capacitive solution has been used.

The proximity interconnect is formed by capacitive coupling between plates in the pad-level metal of the two chips (Fig. 1). Since only one side of the link needs to be able to sense alignment, two different array types, sensor and target, are used. The target array contains only a transmitter and receiver, while the sensor array adds alignment sensing blocks. Embedding both alignment sensing and communication operations in the same (sensor) array eliminates the need to infer link quality from a separate alignment sensor,

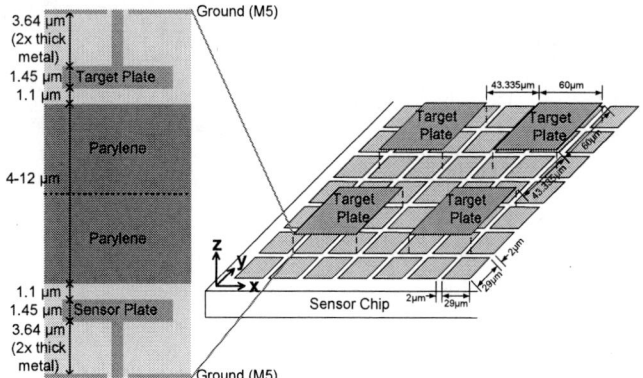

Fig. 1. Dielectric layers and sensor/target array arrangment (target chip outline not shown for clarity).

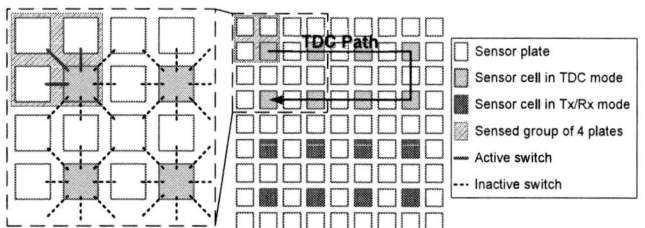

Fig. 2. Structure of the sensor array, showing TDC path for alignment sensing at indicated plates.

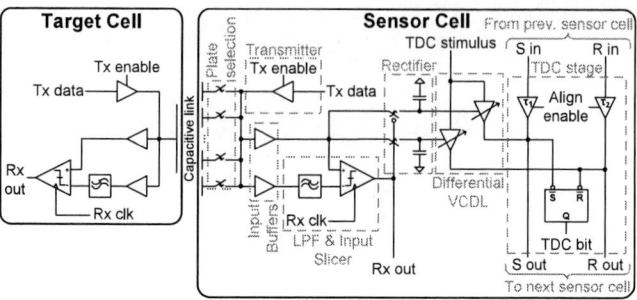

Fig. 3. Architecture of the sensor and target cells, with key functional blocks highlighted.

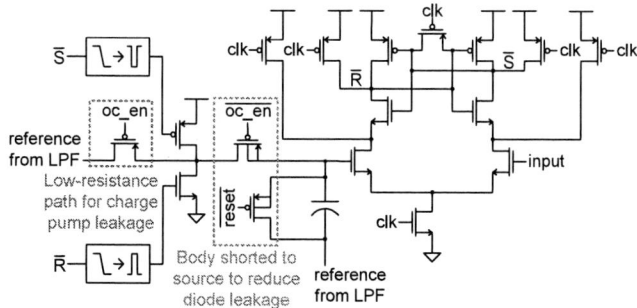

Fig. 4. Input slicer and offset compensation (SR latch not shown). 'Reset' zeroes the offset compensation capacitor, 'oc_en' is asserted during offset compensation calibration.

thus simplifying the alignment-and-adaptation operation. In order to increase granularity during alignment sensing and to maximize coupling capacitance for communication, the sensor array is composed of smaller plates that can be connected together in groups of four; each group of four sensor plates corresponds in size to a single target plate. The target plate pitch is $3\frac{1}{3}$ times the sensor plate pitch. Although this large spacing reduces signaling density, it minimizes crosstalk and avoids the need for differential crosstalk-mitigation schemes [6], allowing a lower-power single-ended design. Using a non-integer multiple of the sensor plate pitch prevents all the target plates from being simultaneously poorly aligned to the sensor, thus allowing the adaptation scheme some flexibility to optimize the array for a given data rate under different alignment conditions.

The dielectric between the two plates is composed of the passivation as well as a parylene sheet 4-12 $\mu$m thick, depending on the exact structure of the parylene module. Since the distance between the plates can be relatively large compared to the distance between each plate and its corresponding ground plane (Fig. 1), it is difficult to distinguish the capacitance between the sensor and the target plate from that between the sensor and the ground plane under the target plate. Instead, a stimulus (e.g. an alternating sequence) is applied to the target plate to electronically differentiate it from the ground plane. The amplitude received at the sensor is proportional to the amount of coupling between the plates, and this signal is rectified into a bias for a voltage-controlled delay line (VCDL). The output of the VCDL is in turn converted into a digital word via a time-to-digital converter (TDC).

## III. SYSTEM ARCHITECTURE

The structure of the sensor array is presented in Fig. 2. Each sensor cell can associate itself with one of 4 possible groups of 4 sensor plates, and shares these plates with its neighbours (the control logic ensures that no plate is connected to more than one cell at a time). In order to save power and area, the TDC stages are distributed through the sensor array, one stage per cell. During alignment sensing, 8 of these stages are connected to form a complete 3-bit TDC. By comparison, the structure of the target array is simple - each target plate is uniquely associated with a single target cell, which contains the transmitter and receiver for that plate

The architecture of the sensor and target array cells is presented in Fig. 3. Since the target cell is a subset of the sensor cell, subsequent discussion will focus on the design of the sensor cell.

Each sensor cell connects to one of 4 different groups of plates, so a set of switches is provided to select the correct group. Source-follower buffers are used to isolate the plates from the rest of the sensor cell blocks in order to minimize parasitic loading of the capacitive link. The transmitter is a simple tri-state buffer, modified with a leakage path to bias the source-follower buffers appropriately. These source follower buffers drive two distinct signal paths - the first contains no filtering, and is used by both the input slicer and the rectifier. The second path contains a low-pass filter, and is used to generate a reference voltage for the input slicer.

The input slicer itself is a strongARM latch used as a comparator (Fig. 4), followed by an SR latch. Chip-to-chip data rates in the 10's of Mbps are targeted, so latch gain is made fairly low in order to save power. Offset is a more

978-1-4673-6145-3/13 $31.00 © 2013 IEEE

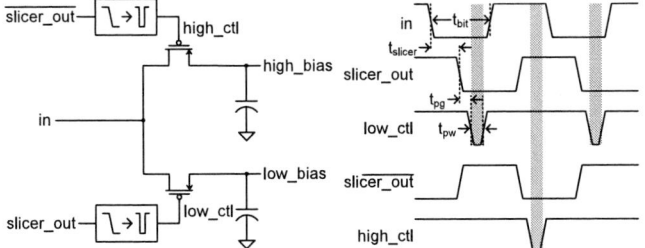

Fig. 5. Rectifier and associated timing diagram. To capture the correct value, $t_{slicer} + t_{pg} + t_{pw} < t_{bit}$.

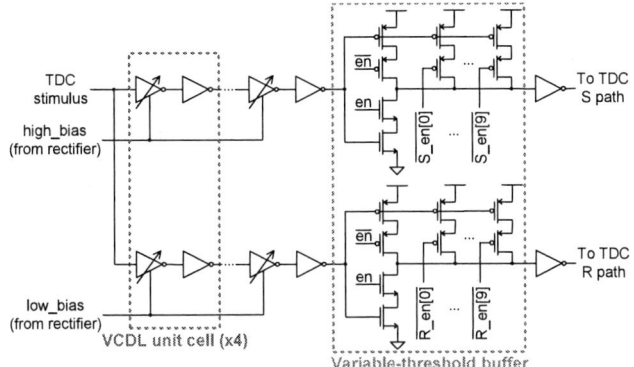

Fig. 6. Differential voltage-controlled delay line.

Fig. 7. Die micrograph, with sensor and target arrays marked.

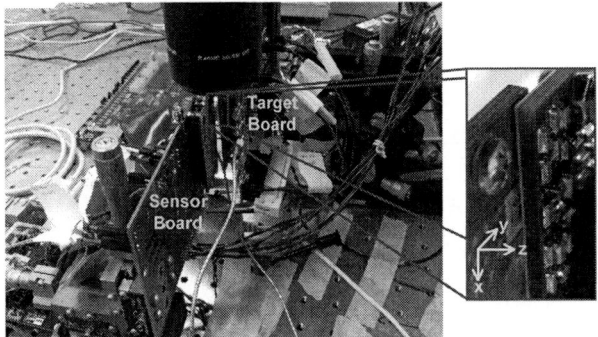

Fig. 8. Test setup. Inset: Detail of chips when brought into alignment.

## IV. MEASUREMENT RESULTS

Both a 6x4 cell (13x9 plate) sensor and a 4x3 cell target array were implemented in a single 65 nm bulk CMOS test chip (Fig. 7). The sensor plates measure 29x29 $\mu$m with 2 $\mu$m spacing, while the target plates are 60x60 $\mu$m. The sensor cell circuitry and digital logic for testing are designed to fit within a 2x2 set of sensor plates (62x62 $\mu$m). The target cell electronics are likewise designed to fit under a single target plate. Two test chips were mounted on a pair of micropositioners, and aligned visually under a microscope (Fig. 8). Slight over-torque was applied in order to cause the PCBs to flex against each other and ensure the test chips were reasonably close [5]. The alignment-sensing function of the sensor array was then used to correct any remaining alignment error. Tests were conducted with different thicknesses of parylene (4, 5, 6, 2x4, 2x5 and 2x6 $\mu$m, with $\pm 0.5$ $\mu$m tolerance) by adhering a single sheet of parylene to the sensor chip and adding a second sheet to the target, as necessary.

The alignment sensing was tested at various thicknesses of parylene, both for vertical (Z-axis) separation (Fig. 9) and in-plane (X/Y-axis) misalignment (Fig. 10). Vertical separation is measured directly from the appropriate sensor cell's TDC output. Since the amount of coupling capacitance is inversely proportional to the distance between plates, achieved resolution depends on the amount of separation and parylene between the chips, and varies from 1-4 $\mu$m/LSB. In-plane misalignment can be estimated by taking the ratio of adjacent sensor cell outputs; just as with vertical separation measurements, resolution degrades as coupling gets smaller. An in-plane alignment resolution between 3-20 $\mu$m is achieved.

When properly aligned, communication over all 12 channels available was demonstrated at data rates up to 60 Mbps/channel with BER $< 10^{-9}$. Input slicer offset compen-

pressing concern, since the received signal amplitudes can be small if the alignment is poor. Monte Carlo simulation of the input slicer suggests that it has a 3-sigma offset of $\sim \pm 50$ mV; in order to correct this to $<10$ mV within a limited power and area budget, charge pump-based offset compensation is added in series with the low-pass filter generating the reference voltage. Leakage is reduced through the use of a thick-oxide storage capacitor, triple-well devices to eliminate diode leakage through switches and the provision of a low-resistance path to shunt leakage from the charge-pump switches away from the storage capacitor. These measures limit charge pump leakage to about 1 mV/ms in the FF corner, and it can be refreshed periodically when the link is taken down to re-acquire chip-to-chip alignment.

The rectifier observes the input slicer output; when the output transitions, it generates pulses to control switches that shunt the high or low levels of the input signal onto the appropriate storage capacitor (Fig. 5). The delay from a transition in the input signal to the pulse must be short enough so that the pulse ends before a further transition in the input signal; to increase timing margin, the target cell transmitter is set to a low-frequency mode during alignment sensing.

The rectified voltages are used to bias a differential VCDL (Fig. 6). Offset error of the ADC formed by the VCDL and TDC is the most significant error affecting the assessment of link quality and alignment. This is corrected by a variable-threshold buffer at the output of VCDL, adjusted by observing the zeroth TDC output bit (prior to any TDC delay elements). The TDC is a straightforward vernier design. The differential delay is generated by adding a capacitive load to one signal path, so that $\tau_1 > \tau_2$. An SR-latch arbiter is used for symmetry.

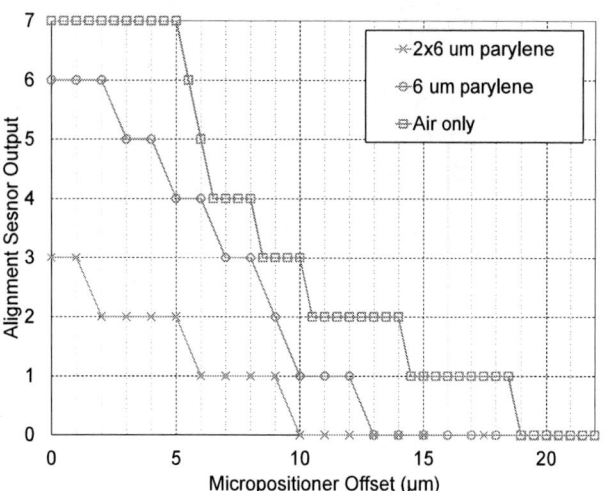

Fig. 9. Alignment sensor output under vertical (Z-axis) separation. Readings for micropositioner offsets $<2$ $\mu$m experience some non-linearity due to over-torquing of the two chips against each other. Coupling with air-only dielectric is good enough to cause the alignment sensor output to saturate.

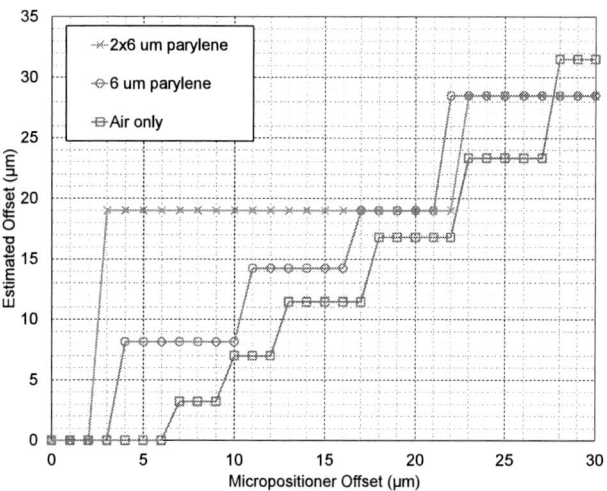

Fig. 10. Alignment sensor output under in-plane (X/Y-axis) misalignment. Resolution degrades substantially as parylene thickness is increased, due to the reduction in coupling between sensor and target plates.

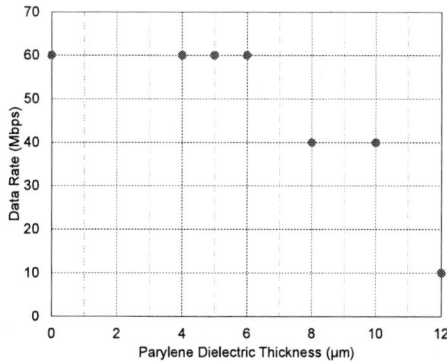

Fig. 11. Maximum data rates achievable (BER $< 10^{-9}$) under best-case alignment, for various thickness of parylene dielectric.

### TABLE I
#### PERFORMANCE SUMMARY

| | |
|---|---|
| Process | 65 nm bulk CMOS |
| Die Area | 1.6 mm x 2.4 mm |
| Sensor Array Area | 401 $\mu$m x 277 $\mu$m |
| Target Array Area | 370 $\mu$m x 267 $\mu$m |
| Data Rate | 12 x 60 Mbps |
| Transmitter & Input Buffer Supply | 1.0 V |
| Slicer, Rectifier, VCDL & TDC Supply | 0.7 V |
| **Power Dissipation (Sensor + Target)** | |
| Transceiver @ 12 x 60 Mbps | 100.9 $\mu$W + 27.7 $\mu$W |
| Alignment Sensor | 23.1 $\mu$W + 20.2 $\mu$W |
| **Figures-of-Merit** | |
| Power | 0.180 pJ/bit |
| Area (based on sensor array size) | 6446 Mbps/mm$^2$ |

sation was run for $\sim$10 $\mu$s (600 clock cycles @ 60 Mbps) every 2 ms, for a net data rate (over 12 channels) of $\sim$716 Mbps. Total power consumption at this data rate was 129 $\mu$W (Table I). Maximum achievable data rate is dependent on the amount of coupling between target and sensor plates; this is presented against parylene thickness in Fig. 11.

## V. CONCLUSION

A capacitive proximity interconnect for Origami biomedical implants has been developed and fabricated in 65 nm CMOS. The array embeds a distributed TDC-based chip-to-chip alignment sensor that provides direct information about link quality, enabling straightforward adaptation of the link to changing alignment conditions. Communication through up to 12 $\mu$m of parylene-C has been demonstrated with data rates from 10-60 Mbps and power efficiency of 0.180 pJ/bit.

## ACKNOWLEDGMENT

The authors thank M. Monge for helpful technical discussions, K. Potter for advice on the test setup, J. Park and Y. Liu for parylene fabrication, STMicroelectronics for chip fabrication and the NSF for funding support.

## REFERENCES

[1] E. Noorsal et al., "A neural stimulator frontend with high-voltage compliance and programmable pulse shape for epiretinal implants," *IEEE J. Solid-State Circuits*, vol. 47, no. 1, pp. 244–256, Jan. 2012.

[2] M. Monge et al., "A fully intraocular 0.0169mm$^2$/pixel 512-channel self-calibrating epiretinal prostehsis in 65nm CMOS," in *ISSCC Dig. Tech. Papers*, 2013, pp. 296–297.

[3] J.-C. Chang, R. Huang, and Y.-C. Tai, "High-density IC chip integration with parylene pocket," in *IEEE Int. Conf. on NEMS*, 2011, pp. 1067–1070.

[4] Y. Zhao, M. Nandra, and Y.-C. Tai, "A MEMS intraocular origami coil," in *Solid-State Sensors, Actuators and Microsystems Conf. (TRANSDUCERS)*, 2011, pp. 2172–2175.

[5] R. Drost et al., "Proximity communication," *IEEE J. Solid-State Circuits*, vol. 39, no. 9, pp. 1529–1535, Sep. 2004.

[6] D. Hopkins et al., "Circuit techniques to enable 430Gb/s/mm$^2$ proximity communication," in *ISSCC Dig. Tech. Papers*, 2007, pp. 368–609.

[7] N. Miura et al., "A 2.7Gb/s/mm$^2$ 0.9pJ/b/chip 1coil/channel ThruChip interface with coupled-resonator-based CDR for NAND flash memory stacking," in *ISSCC Dig. Tech. Papers*, 2011, pp. 490–492.

[8] ——, "A 1 Tb/s 3 W inductive-coupling transceiver for 3D-stacked interchip clock and data link," *IEEE J. Solid-State Circuits*, vol. 42, no. 1, pp. 111–122, Jan. 2007.

# An Active Rectifier/Regulator Combo Circuit for Powering Biomedical Implants

## Edward K.F. Lee

Alfred Mann Foundation, 25134 Rye Canyon Loop, Suite 200, Santa Clarita, CA, USA
edl@aemf.org

*Abstract-* **A circuit that combines an active rectifier and a linear regulator, which will be referred to as a rectulator, is proposed in this paper. The main transistor in the rectulator is used for both the rectification of the AC input ($V_{AC}$) and the regulation of the DC output ($V_O$) to reduce the overall voltage drop between $V_{AC}$ and $V_O$. The rectulator has a high power efficiency ($\eta_P$) and very good amplitude modulation (AM) rejection on $V_{AC}$ against AC field strength variations. A full-wave rectulator implemented in a 0.18μm CMOS process showed an AM rejection > 45dB for an AM frequency < 2kHz. A $\eta_P$ of 90.7% was measured for an input frequency of 5MHz, $V_O = 2.5V$ and an output power of 10.5mW.**

## 1. Introduction

In many biomedical implants, power is often delivered from an external controller (EC) outside the body, wirelessly across the skin, to the implants [1-3]. The EC usually consists of a coil driver driving a primary coil. The primary coil is inductively coupled to and powers a secondary coil that is located inside each implant. In some cases, the batteries inside the implants are used primarily for powering the implants. Nevertheless, these batteries are still required to be recharged by the EC [4]. The received signal on the secondary coil is an AC signal. It is necessary to convert the AC signal to a DC voltage using a rectifier [1-4]. Since the magnetic field strength and hence, the AC signal amplitude may vary due to changes in the distance between the EC and the implant, the DC voltage is typically regulated to the required supply voltages using linear regulators. Different integrated rectifier techniques have been proposed in the literature recently to improve the power efficiency of the conversion of the AC signal to DC voltages [5-8]. One popular technique is based on an active rectifier [5][7][8]. A conceptual design for powering an implant based on half-wave rectification followed by a linear regulator is illustrated in Fig. 1. The active rectifier consists of a switch $M_1$ and a comparator $CO_1$. When the received AC signal, $V_{AC}$, has a voltage higher than the unregulated DC voltage, $V_{UDC}$, $CO_1$ will turn on $M_1$. Ideally, $V_{UDC}$ will be charged to a value equal to the peak voltage of $V_{AC}$ given as $V_{ACP}$. However, due to the on-resistance of $M_1$ and the current flow through $M_1$ as well as the delay of $CO_1$, the maximum $V_{UDC}$ is $V_D$ lower from $V_{ACP}$. $V_D$ is typically in the range between 0.2V and 1V [6-8]. The linear regulator regulates $V_O$ by monitoring $V_O$ through the resistor voltage divider ($R_1$ and $R_2$) and comparing the voltage divider output to a reference input voltage, $V_{ref}$, using an error amplifier $A_E$. $V_O$ is then regulated by controlling the gate voltage of $M_2$. Typically, there is a minimum voltage drop (dropout voltage) given as $V_{DO}$ in the range of 0.2V to 0.4V between $V_{UDC}$ and $V_O$ required by the linear regulator. As a result, for a desired $V_O$, the minimum peak voltage requirement for $V_{AC}$, min[$V_{ACP}$], is equal to $V_O+V_{DO}+V_D$. One of the merits for characterizing rectifiers is the voltage conversion efficiency,

which is defined as $V_{UDC}/min[V_{ACP}]$. However, from a system perspective, the overall voltage conversion efficiency, $\eta_V$, should include $V_{DO}$ due to the linear regulator and should be redefined as $V_O/min[V_{ACP}]$. As a result, for $V_O < 2V$, $\eta_V$ is typically < 80%. Since the power efficiency of the rectifier is related to $V_D$ and the power efficiency of the regulator is almost directly related to $V_{DO}$ for an $A_E$ with negligible biasing current, the overall maximum power efficiency, $\eta_P$, will also be < 80% for $V_O < 2V$. $\eta_P$ is defined as the load power ($P_L$) dissipated on $R_L$ divided by the input power ($P_{in}$) from $V_{AC}$ at min[$V_{ACP}$]. In addition to $\eta_V$ and $\eta_P$, another concern for an implant is the limited space for incorporating off-chip components. In the conventional approach as shown in Fig. 1, it usually requires two off-chip bypass capacitors – $C_{B1}$ and $C_{B2}$. In this paper, a circuit called *rectulator* that combines an active <u>rect</u>ifier and a linear reg<u>ulator</u> for improving $\eta_V$ and $\eta_P$ is proposed. Only one bypass capacitor is needed at each output for the proposed circuit.

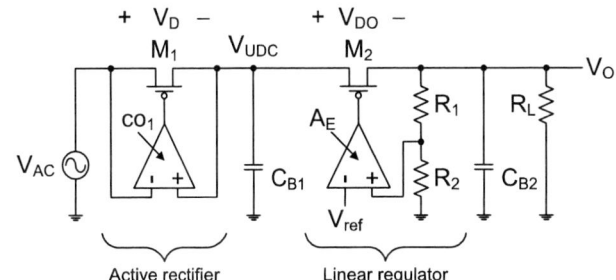

**Fig. 1: Conventional approach for powering an implant**

## 2. Proposed Active Rectifier/Regulator Combo Circuit

The basic concept of the proposed circuit based on a half-wave rectification configuration is illustrated in Fig. 2. Similar to the active rectifier in Fig. 1, $CO_1$ in Fig. 2 is used for detecting $V_{AC}$. If $V_{AC} > V_O$, $CO_1$ will turn on switch $S_1$ and turn off switch $S_2$, allowing the amplifier $A_E$ to control the gate voltage of $M_S$ such that $V_O$ is regulated to the desired output voltage given as $V_{ref}(1+R_1/R_2)$. If $V_{AC} < V_O$, $CO_1$ will turn off $S_1$ and turn on $S_2$. In this case, the gate voltage of $M_S$ is equal to $V_O$. $M_S$ will be off and there will be no current flow between $V_{AC}$ and $V_O$. Overall, the proposed circuit will behave like a linear regulator with a regulated $V_O$ except that the input is now an AC signal. Since there is only one transistor between $V_{AC}$ and $V_O$, the voltage drop between $V_{AC}$ and $V_O$ given as $V_S$ is a single source-to-drain voltage drop of $M_S$, which can have a minimum value typically in the range between 0.2V and 0.7V, depending on the drain current. Hence, $\eta_V$ and $\eta_P$ can be improved comparing to the design in Fig. 1. Furthermore, there is no intermediate DC voltage between $V_{AC}$ and $V_O$ and only a single off-chip bypass capacitor is needed for each output.

978-1-4673-6145-3/13 $31.00 © 2013 IEEE

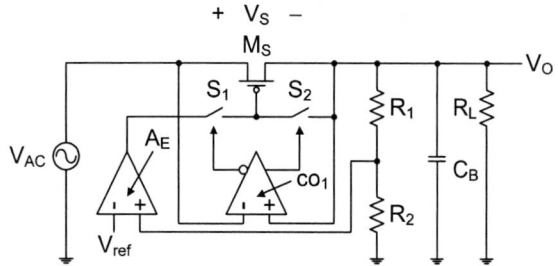

**Fig. 2: Conceptual half-wave rectulator design**

### 3. Full-Wave Rectulator Realization

It can be seen in Fig. 2 that $V_{AC}$ can range between $-V_{ACP}$ and $+V_{ACP}$. To minimize the voltage stress on $M_S$ for a practical implementation, a full-wave rectulator as shown in Fig. 3 is considered in this paper for demonstrating the proposed concept. Two cross-coupled NMOSs ($M_A$ and $M_B$) are used for rectifying $V_{AC}$ and generating two half-wave rectified signals – $V_{AC1}$ and $V_{AC2}$ on the two rectulator input nodes – A and B, respectively. The voltage swings on nodes A and B are now reduced to be between ~0V and ~$V_{ACP}$. The rectulators provide two output voltages – $V_{O1}$ and $V_{O2}$, which can be connected together as a single output if desired. Each rectulator consists of a switch driver, an error amplifier $A_E$, a comparator $CO_1$, a biasing circuit and a start-up comparator $CO_S$. The total resistance of $R_{1a}$ and $R_{1b}$ is equal to $R_1$ in Fig. 2. The design of the switch driver is shown in Fig. 4. It includes the main transistor $M_{S1}$, similar to $M_S$ in the half-wave rectulator design shown in Fig. 2 and the circuitry driving it. Notice that the source voltage of $M_{S1}$ can swing between ~0V and ~$V_{ACP}$. Therefore, it can have a voltage lower than $V_{O1}$. In order to prevent current flowing from $V_{O1}$ back to $V_{AC1}$ when $V_{AC1} < V_{O1}$, it is required to connect the gate of $M_{S1}$ to $V_{O1}$ through $M_3$ and $M_4$, and the bulk of $M_{S1}$ (node $N_W$) to $V_{O1}$ through $M_2$. The condition for $V_{AC1} < V_{O1}$ is detected using $CO_1$ in Fig. 3. The output signals of $CO_1$ are $C_{out}$ and $C_{outb}$ as shown in Fig. 4. When $V_{AC1} > V_{O1}$, $N_W$ is disconnected from $V_{O1}$ by switching off $M_2$ and $N_W$ is then connected to $V_{AC1}$ by switching on $M_1$. It can be seen that $N_W$ has a voltage swing between $V_{O1}$ and ~$V_{ACP}$. It is the highest potential in the rectulator and therefore, is used for biasing the bulks of the PMOSs to prevent any undesired current conduction due to the parasitic diodes of the PMOSs. Ideally, when $V_{AC1} > V_{O1}$, the gate of $M_{S1}$ is driven by $A_E$ as illustrated in Fig. 2. However, the supply voltage of $A_E$ and the biasing circuit are derived from $V_{O1}$ in the actual rectulator implementation and $V_{O1}$ can sometimes be regulated to a voltage much lower than $V_{ACP}$. As a result, $A_E$ may not have a sufficient output swing to control $M_{S1}$. To overcome this issue, transistors – $M_{5-8}$ are added between the output of $A_E$ ($A_{EO}$) and the gate of $M_{S1}$ as shown in Fig. 4. When $V_{AC1} > V_{O1}$, $M_{6-7}$ will be turned on by $CO_1$ and there will be a current flow between $M_5$ and $M_8$. $M_{5-8}$ will then act as an inverting gain stage with $M_5$ being an active load connected to $V_{AC1}$ such that this gain stage will have sufficient voltage swing to control $M_{S1}$. $M_{S1}$ has a much larger W/L ratio than the W/L ratio of $M_5$ to allow for a large output current. Fig. 5 shows the design of the comparator $CO_1$. The bulk voltage of the PMOSs,

$N_W$, is obtained from the switch driver. $N_W$ is also used for supplying the two inverters. $M_{10}$ and $M_{11}$ are used for comparing $V_{AC1}$ and $V_{O1}$. The biasing current of $M_{11}$, $I_{12}$, is set by $M_{12}$ through the biasing voltage $V_B$. The biasing current for $M_{10}$, $I_{13}$, is controlled through $M_{13-15}$. When $V_{AC1} < V_{O1}$, $I_{13}$ is approximately equal to $I_{12}$ since the gate of $M_{14}$ is connected to $M_{11}$ through $M_{16-17}$, which are turned on by the two inverters. When $V_{AC1} > V_{O1}$, the output values of the inverters will change. As a result, $M_{16-17}$ will be off and $M_{18-19}$ will be on. The gate voltage of $M_{12}$ now depends on $V_{AC1}$ via $M_{20}$. Hence, $I_{14}$ and $I_{13}$ will increase as $V_{AC1}$ decreases from its peak value, causing node X to switch to ground before $V_{AC1}$ equal to $V_{O1}$. As a result of the earlier switching, the switches in the switch driver (Fig. 4) will turn on or off slightly earlier to minimize the current flow back from $V_{O1}$ to $V_{AC1}$. This earlier switching is also used for compensating the delays due to the inverters as well as the comparison stage consisted of $M_{10-13}$. In this design, the comparator was optimized for an AC input with a frequency in a few MHz range.

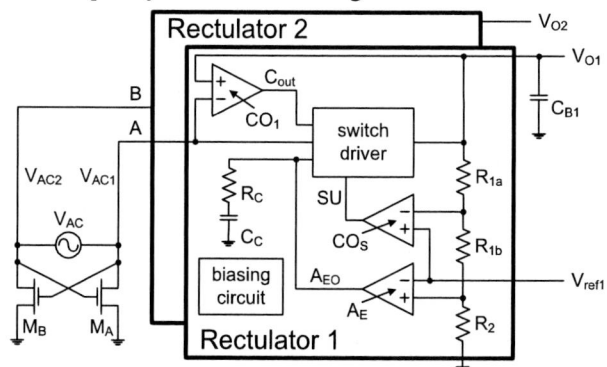

**Fig. 3: Full-wave rectulator architecture**

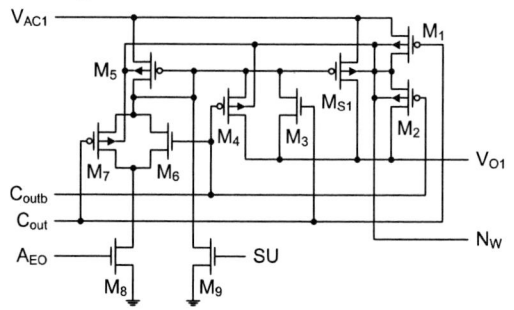

**Fig. 4: Switch driver design**

During initial start-up, $V_{O1}$ is charged up mainly by the parasitic diode associated with transistor $M_{S1}$ in Fig. 4. Since $M_{S1}$ may not have turned on properly, $V_{O1}$ may only charge up to < 1V in simulation. For a proper start-up, a simple and low speed comparator – $CO_S$ shown in Fig. 3 is added to compare $k V_{O1}$ to $V_{ref1}$ where k is set by $R_{1a}$, $R_{1b}$ and $R_2$. When $k V_{O1} < V_{ref1}$, the start-up signal – SU will have a logic high and the start-up transistor, $M_9$, shown in Fig. 4 will turn on. As a result, $M_{S1}$ will be turned on initially to charge up $V_{O1}$. When $k V_{O1} > V_{ref1}$, $M_9$ will turn off and $M_{S1}$ will be controlled by $A_E$ and $CO_1$ as discussed above. $A_E$ is realized using a simple folded-cascode amplifier with $R_C$ and $C_C$ (Fig. 3) added at the output for the overall frequency compensation of the rectulator.

978-1-4673-6145-3/13 $31.00 © 2013 IEEE

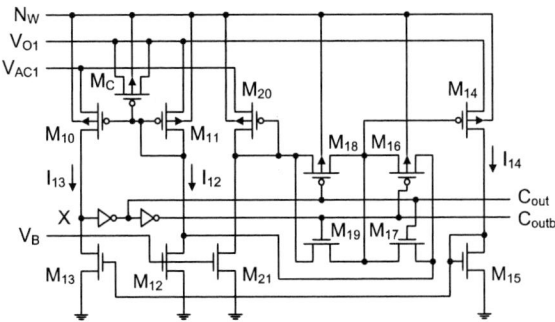

**Fig. 5: Comparator CO₁ design**

## 4. Implementation and Experimental Results

The proposed full-wave rectulator was implemented in a conventional 0.18μm CMOS process. Only 3.3V devices were used to provide wide input and output voltage ranges. The gain from $V_{ref1}$ to $V_{O1}$ (and from $V_{ref2}$ to $V_{O2}$) was set to 2. Fig. 6 shows the die photo and the layout of the rectulator. The active area is 350×200 μm². A 50Ω AC source connected to a 1:1 transformer followed by a 10Ω resistor was used as the input source for testing the rectulator. The input power of the rectulator, $P_{in}$, was measured using the 10Ω resistor. Each rectulator output was connected to a 470nF bypass capacitor.

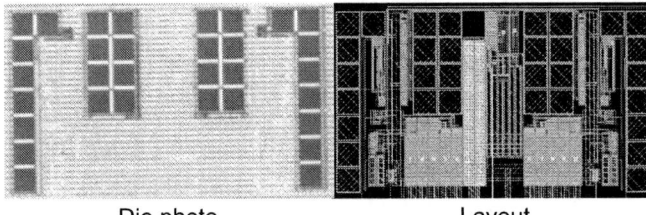

Die photo                    Layout
**Fig. 6: Die photo and layout of the full-wave rectulator**

Fig. 7 shows the output waveforms of the rectulator for $f_{in}$ = 5MHz, where $f_{in}$ is the input frequency of $V_{AC}$. $V_{ACP}$ had an amplitude just high enough for $V_{O1}$ and $V_{O2}$ to be regulated to 2V and 1.5V, respectively. A 1.2kΩ load resistor ($R_L$) was connected to each output for a total load power, $P_L$, of ~5.2mW. $V_{AC1}$ and $V_{AC2}$ were distorted differently due to the finite source impedance and different $P_L$'s at $V_{O1}$ and $V_{O2}$. The ripples on $V_{O1}$ and $V_{O2}$ were within 15mV. An AM modulation was applied to the AC input to demonstrate that the rectulator was capable of regulating $V_{O1}$ and $V_{O2}$ even with variations on the AC field strength. Fig. 8 shows the results for a 1MHz $V_{AC}$ with a 1.75V (2.65V < $V_{ACP}$ < 4.4V), 10kHz AM modulation. $V_{O1}$ was regulated to 2V. The peak-to-peak variations on $V_{O1}$ was ~9mV, corresponding to a ~45.8dB rejection on the AM modulation. Fig. 9 shows the measured AM rejection vs. frequency of the AM modulation ($f_{AM}$) for $f_{in}$ = 1MHz and $f_{in}$ = 5MHz. Over 45dB of AM rejection could be obtained for $f_{AM}$ < 2kHz. Degradations on AM rejection at high $f_{AM}$'s were due to the overall finite loop bandwidth of the rectulator. When measuring $\eta_V$ and $\eta_P$, both $V_{O1}$ and $V_{O2}$ were set to the same output value – $V_O$ for convenience and $V_{ACP}$ was then reduced to the minimum value (min[$V_{ACP}$]), which was determined by measuring $V_O$ until $V_O$ was reduced by

~0.5 – 1% from its nominal value set by $V_{ref1}$ and $V_{ref2}$. The measured $\eta_V$'s are shown in Fig. 10. The maximum $\eta_V$'s were measured to be 96.4% and 96.1% at $V_O$ = 3V for $f_{in}$ = 1MHz and $f_{in}$ = 5MHz, respectively. Fig. 11 and 12 show the measured $\eta_P$'s for different $P_L$'s and $V_O$'s for $f_{in}$ = 1MHz and $f_{in}$ = 5MHz, respectively. For $f_{in}$ = 1MHz, maximum $\eta_P$, max[$\eta_P$], was measured to be 93.3% at $P_L$ = 7mW and $V_O$ = 2V. For $f_{in}$ = 5MHz, max[$\eta_P$] was measured to be 90.7% at $P_L$ = 15.7mW and $V_O$ = 2.5V. For $V_O$ > 2V and $f_{in}$   5MHz, $\eta_P$ maintained over 80% for 5mW < $P_L$ < 40mW. For $V_O$ = 1.5V, the biasing circuit in the rectulator had limited supply voltage that limited the biasing current and hence, the speed of CO₁. As a result, $\eta_P$ began to degrade at $f_{in}$ ~ 5MHz. Fig. 13 shows $\eta_P$ vs. $f_{in}$ for different loading conditions. $\eta_P$ started to degrade for $f_{in}$ > 10MHz due to the limitations on the response time of CO₁. When the load current changes between 180μA and 12.5mA, a change on $V_O$ was measured to be ~0.75mV for $V_O$ = 2V with $f_{in}$ = 1MHz, corresponding to a load regulation of < 61μV/mA. In this case, $V_{ACP}$ was adjusted to a value ~100mV higher than the required min[$V_{ACP}$]. The settling time for 0.1% accuracy on $V_O$ was measured to be < 48μs for the same change in the load current. When $V_{AC}$ was initially applied to the rectulator (with a 2.2kΩ//470nF load at each output), a start-up time of < 350μs was measured.

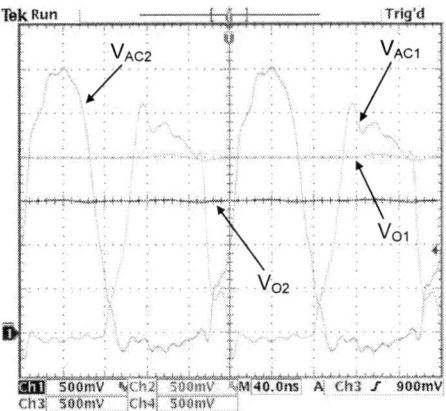

**Fig. 7: Output waveforms with $f_{in}$ = 5MHz and $P_L$ ~ 5.2mW**

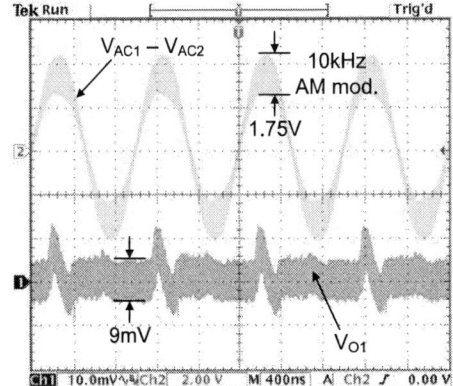

**Fig. 8: Output variations for $f_{in}$ = 1MHz with AM modulation**

Table I compares the performances of recent published rectifiers and the rectulator. $\eta_V$'s and $\eta_P$'s of other works are compatible with $\eta_V$'s and $\eta_P$'s of the rectulator. However,

978-1-4673-6145-3/13 $31.00 © 2013 IEEE

their output voltages were not regulated and there were no AM rejections. It can be argued that linear regulators may be added for achieving regulated DC outputs. However, the overall $\eta_V$'s and $\eta_P$'s will then degrade considerably. Hence, the rectulator has significant advantages when compared to the conventional approach that uses a rectifier followed by linear regulators.

## 5. Summary and Conclusion

In this paper, a circuit called rectulator that combines an active rectifier and a linear regulator is proposed for powering biomedical implants. The main transistors of the active rectifier and the linear regulator in a conventional design are combined into a single transistor in the rectulator to minimize the voltage drop between the AC input and the regulated DC output ($V_O$). As a result, the overall voltage conversion efficiency, $\eta_V$, and the maximum power efficiency, $\eta_P$, are improved. Based on a conventional 0.18μm CMOS process, a full-wave rectulator achieved a max[$\eta_V$] of 96.1% at $V_O = 3V$ and a max[$\eta_P$] of 90.7% at $V_O = 2.5V$ for a 5MHz AC input. It also provided > 45dB AM rejection on the AC input for an AM frequency of < 2kHz and a load regulation of < 61μV/mA.

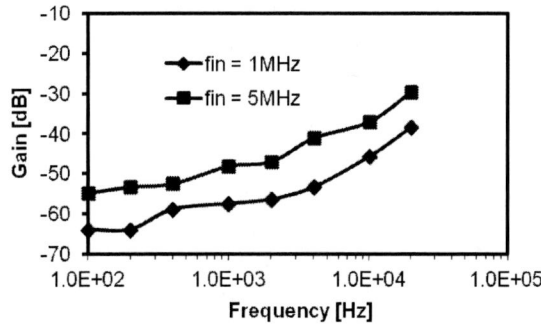

**Fig. 9: Measured AM rejection vs. $f_{AM}$ for $f_{in}$ = 1MHz & 5MHz**

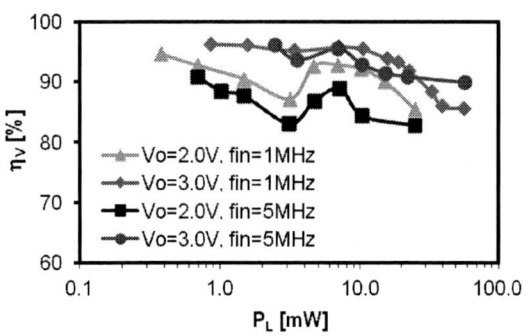

**Fig. 10: Measured $\eta_V$ vs. $P_L$ for $f_{in}$ = 1MHz & 5MHz**

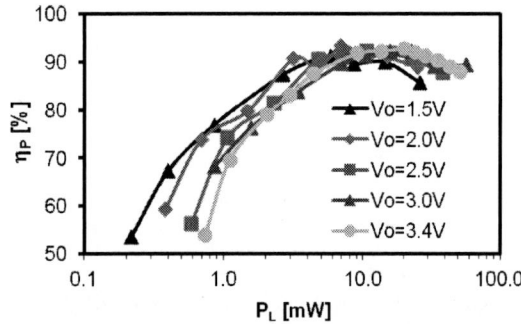

**Fig. 11: Measured $\eta_P$ vs. $P_L$ for $f_{in}$ = 1MHz**

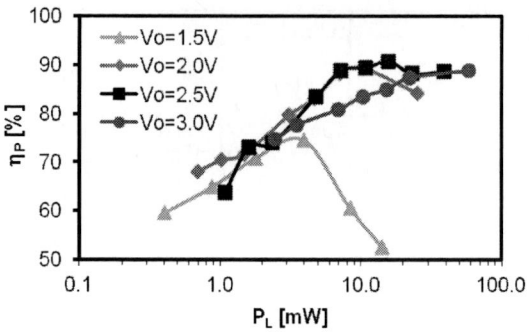

**Fig. 12: Measured $\eta_P$ vs. $P_L$ for $f_{in}$ = 5MHz**

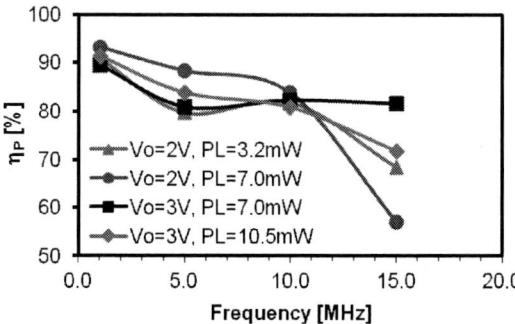

**Fig. 13: Measured $\eta_P$ vs. $f_{in}$ for different load conditions**

**Table I: Comparison with previous rectifier designs**

|  | [6] | [7] | [8] | This work |
|---|---|---|---|---|
| Technology | 0.18μm CMOS | 0.35μm CMOS | 0.5μm CMOS | 0.18μm CMOS |
| $V_O$ range (V) | 0.3 – 2 | 1.13 – 3.65 | 3.12 | 1.5 – 3.4 |
| Active area | ~0.45mm² | 0.121mm² | 0.18mm² | 0.07mm² |
| max[$P_L$] | 2mW | 24.5mW | 19.5mW | 58.1mW |
| $f_{in}$ (MHz) | 1 – 100 | 13.56 | 13.56 | 0.5 – 15 |
| $\eta_V$ (%) | 37.5 – 74 | 74 – 90 | ~68 – 96 | 56.9 – 96.4 |
| max[$\eta_P$] (%) @ $f_{in}$ (Hz) | 80 @ 10M | ~90* @ 13.56M | 80.2 @ 13.56M | 93.3 @ 1M 90.7 @ 5M 88.6 @ 10M 81.5 @ 15M |
| AM rejection | No | No | No | > 45dB for $f_{AM}$ < 2kHz |
| Load regulation | No | No | No | < 61μV/mA |

*simulated $\eta_P$

## References

[1] W. Biederman, et al., "A fully-integrated, miniaturized (0.125 mm²) 10.5 μW wireless neural sensor," IEEE JSSC, vol. 48, pp. 960-970, Apr. 13.

[2] D. Pivonka, et al., "A mm-sized wirelessly powered and remotely controlled locomotive implant," IEEE Tran. Bio. Cir. Syst., vol. 6, pp. 523-532, Dec. 12.

[3] Y. Liao, et al., "A 3μW CMOS glucose sensor for wireless contact-lens tear glucose monitoring," IEEE JSSC, vol. 47, pp. 335-344, Jan. 12.

[4] E. Lee, et al., "A biomedical implantable FES battery-powered micro-stimulator," IEEE TCAS I, vol. 56, pp. 2583-2596, Dec. 09.

[5] S. Guo and H. Lee, "An efficiency-enhanced integrated CMOS rectifier with comparator-controlled switches for transcutaneous powered implants," Proc. IEEE 2007 CICC, pp. 385-388, Sep. 07.

[6] S. Hashemi, M. Sawan and Y. Savaria, "A high-efficiency low-voltage CMOS rectifier for harvesting energy in implantable devices," IEEE Tran. Bio. Cir. Syst., vol. 6, pp. 326-335, Aug. 12.

[7] Y. Lu, W. Ki and J. Yi, "A 13.56MHz CMOS rectifier with switched-offset for reversion current control," Sym. on VLSI Cir., pp. 246-247, 11.

[8] H. Lee and M. Ghovanloo, "An integrated power-efficient active rectifier with offset-controlled high speed comparators for inductively powered applications," IEEE TCAS I, vol. 58, pp. 1749-1760, Aug. 11.

# 65nW CMOS Temperature Sensor for Ultra-Low Power Microsystems

Seokhyeon Jeong[1], Jae-yoon Sim[2], David Blaauw[1], Dennis Sylvester[1]

[1] University of Michigan, Ann Arbor, MI –[2]Pohang University of Science and Technology (POSTECH), Korea

seojeong@umich.edu, jysim@postech.ac.kr, blaauw@umich.edu, dmcs@umich.edu

**Abstract—We propose a temperature sensor using a novel process-invariant temperature sensing element and voltage to current converter for battery-operated ultra-low power micro systems. By introducing a new temperature-to-voltage sensing element that outputs only 75mV, the sensor achieves ultra-low power. The sensor was implemented in 180nm CMOS process and uses 0.09mm² of area. Measurements from test chips show 65nW power consumption, the lowest reported to date, with an inaccuracy of +1.3°C /-1.4°C across 0°C to 100°C after 2-point calibration.**

## I. INTRODUCTION

Ultra-low power wireless microsystems are emerging as a new class of computing that will find use in a wide range of applications. One common sensing modality in such systems is temperature, and various types of temperature sensors have been introduced. BJT-based sensors offer high accuracy at the expense of high power consumption (µW range) [1], making them unsuitable for miniaturized battery-powered applications [2]. MOSFET-based sensors consuming sub-µW have been reported but still consume over 100nW while still requiring an accurate external clock [3]-[5]. Such a time reference is not typically available in a wireless microsystem and can increase power consumption significantly. A temperature sensor that uses an on-chip time reference while consuming sub-µW has been reported [6]. However, it exhibits larger inaccuracy compared to others due to non-ideal characteristics of the reference clock. In this work, we propose new techniques to improve temperature inaccuracy while consuming extremely low power and energy.

The rest of the paper is structured as follows. Section II introduces the proposed topology for the low power sensor with detailed description and analysis. Section III presents measured results of the test chip. Finally, Section IV concludes the work.

## II. PROPOSED LOW POWER TOPOLOGY

### A. Overview of Approach

A conventional approach to generating a temperature-dependent current through a resistor ($I_R$) using a voltage source ($V_s$) is shown in Fig. 1(a). A current flowing across a resistor is mirrored through $M_1$ and $M_2$ to create a temperature-dependent voltages $V_H$ and $V_L$, which are used to generate pulse or frequency for further processing [3]-[5]. To reduce power consumption, either a very large resistor or very small voltage is required. Given a typical bandgap voltage reference of 1V, a resistance of >20MΩ is required to achieve sub-100nW power consumption, which is impractical in area-constrained microsystems [2,7].

In this work, a novel temperature sensing element that uses an ultra-low (75mV) voltage source is introduced. To further reduce

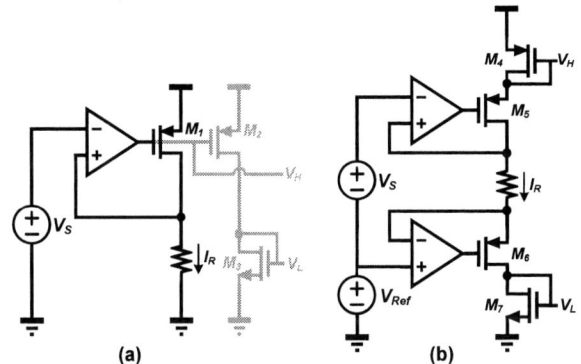

Figure 1. (a) Conventional scheme for voltage to current converter and (b) proposed scheme for low power operation. Process-invariant temperature sensing element is introduced to reduce output level of voltage source (Vs) to sub-100mV, enabling low-power operation.

power and obtain effective amplifier common mode voltage, a differential approach is used that eliminates the current mirror ($M_{1-2}$) as shown in Fig 1(b), reducing the required current by half. With these techniques, the sensor achieves ~3.4× lower power and ~11× lower energy per conversion as well as better inaccuracy compared to previous fully-integrated temperature sensors.

### B. Circuit description and analysis

Fig. 2 shows the block diagram of the proposed temperature sensor, containing three major components: 1) a temperature sensing core, 2) a current to frequency converter, and 3) a frequency to digital converter.

The temperature sensing core generates a temperature-sensitive proportional-to-absolute temperature (PTAT) current and a temperature-insensitive reference current by applying a voltage across a resistor ($V_R$). To achieve power consumption in the nW range, $V_R$ is reduced well below 100mV with the sensing element shown in Fig. 3. A key component of the sensing element is a 2T-based voltage reference, which generates linearly increasing output voltages and only consumes pA.

A prior implementation of the 2T voltage reference uses two different types of transistors (with highly disparate threshold voltages) to increase the reference voltage as much as possible [8]. However, such a high reference voltage increases power consumption in this application. Furthermore, the use of different threshold transistors causes its output voltage to vary widely across process, resulting in degraded linearity and sensing error.

In this work, we propose to use the same type of transistors for both top and bottom devices in the 2T reference to eliminate the threshold voltage term, achieving low variability. Also, the top transistor's gate is connected to the output to achieve higher

978-1-4673-6145-3/13 $31.00 © 2013 IEEE

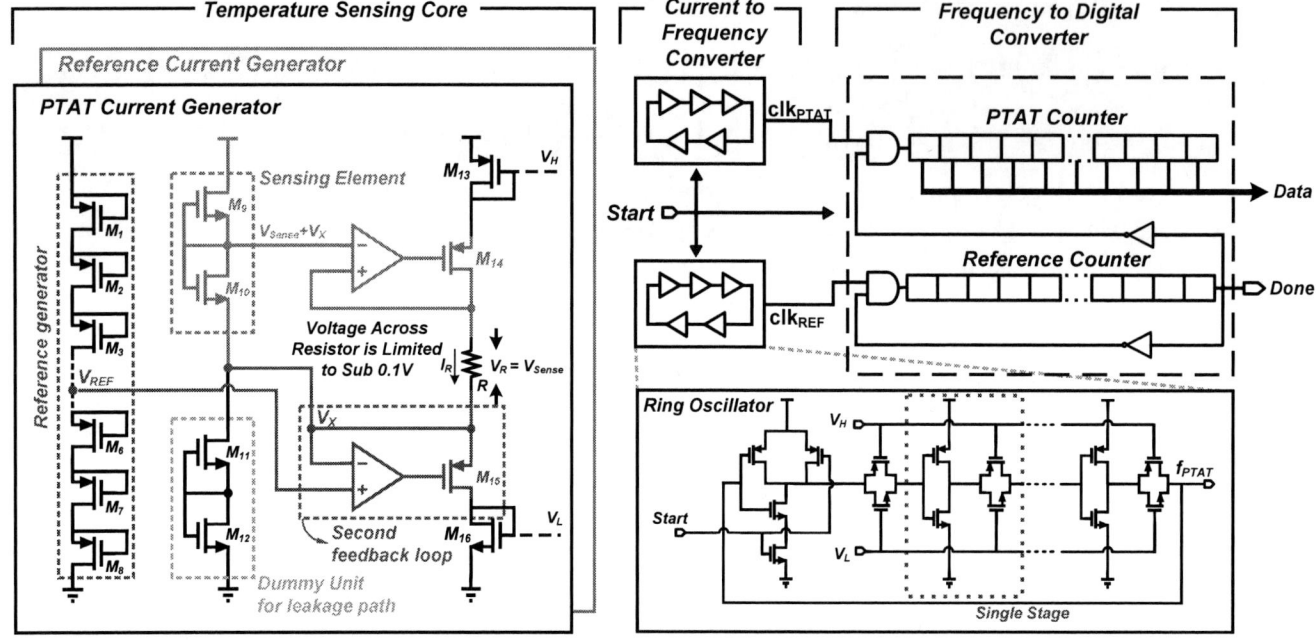

Figure 2. Structure of proposed temperature sensor architecture. To reduce power, voltage across the resistor is limited by using two feedback loops and bottom NMOS $M_{16}$ is added to reduce power further by eliminating additional mirror structure.

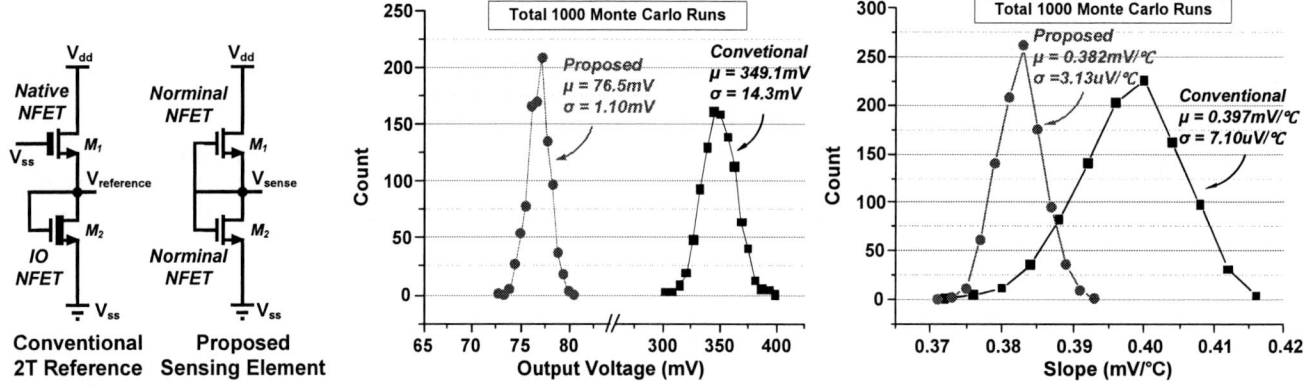

Figure 3. Circuit diagram of conventional 2T reference and proposed sensing element (left), simulated output voltage (center), and slope distribution (right) of conventional 2T and proposed structure. Compared to the conventional 2T scheme, new structure shows less variation in output voltage (1.7×) and slope (2.18×).

temperature sensitivity. The resulting output voltage can be expressed as

$$V_{sense} = m_2 V_T \ln\left(\frac{\mu_1 C_{ox1} W_1 L_2}{\mu_2 C_{ox2} W_2 L_1}\right) \quad (1)$$

which shows PTAT behavior due to the thermal voltage ($V_T$). In order to maintain highly linear PTAT behavior, the output voltage must exceed $3V_T$ (~75mV) to render subthreshold current independent of drain to source voltage ($V_{ds}$). Compared to the conventional 2T structure, the proposed structure's low output voltage enables a 5× reduction in resistor area for an equivalent current. Also, the proposed topology exhibits 1.7× lower output voltage variation ($\sigma/\mu$) and 2.2× less variation in slope (temperature coefficient) in Monte Carlo simulations (Fig.3).

The sensing element drives a conventional voltage to current converter to generate currents. A negative feedback loop consisting of an amplifier, transistor, and resistor duplicates the sensing element's output voltage across a resistor. Two different types of resistors along with two sensing elements are used to generate a temperature-insensitive reference current ($I_{REF}$) and temperature-sensitive PTAT current ($I_{PTAT}$) and generated currents can be modeled as follow.

$$\frac{V_o(1 + \alpha_{V1}T)}{R_o(1 + \alpha_{R1}T + \alpha_{R2}T^2)} = I_o(1 + \alpha_{I1}T + \alpha_{I2}T^2) \quad (2)$$

$$\alpha_{I1} = \alpha_{V1} - \alpha_{R1} \quad (3)$$

$$\alpha_{I2} = \sqrt{\alpha_{R1}^2 - \alpha_{R1}\alpha_V - \alpha_{R2}^2} \quad (4)$$

The first order temperature coefficient $\alpha_{V1}$ in (3) is a positive value set through sizing of the sensing element. To generate $I_{REF}$, a PTAT resistance is used to cancel $\alpha_{R1}$ with $\alpha_{V1}$. On the other hand, a complementary-to-absolute temperature (CTAT) resistor is desired for $I_{PTAT}$, in order to increase the temperature dependence. In the chosen process, an n$^+$ diffusion resistor and a high resistance poly resistor meet these requirements while minimizing the second order coefficient $\alpha_{I2}$.

The amplifier is designed for 105dB open-loop gain while

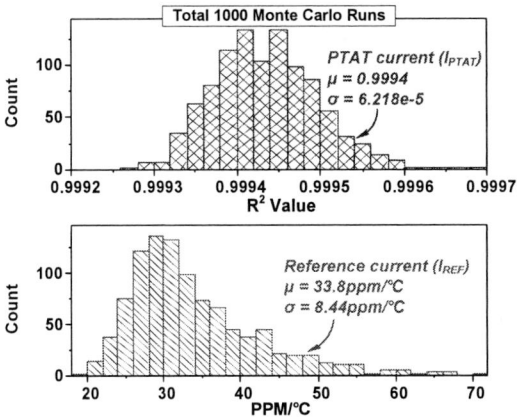

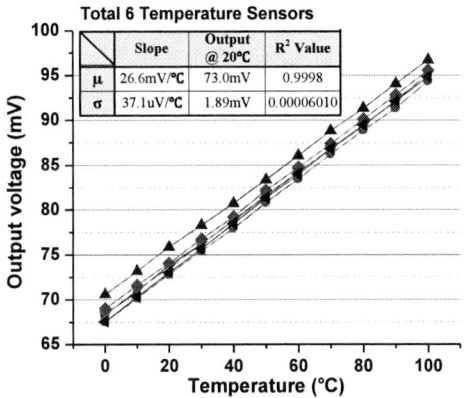

Figure 4. Effect of process variation and mismatch on PTAT linearity (top) and reference temperature coefficient (bottom).

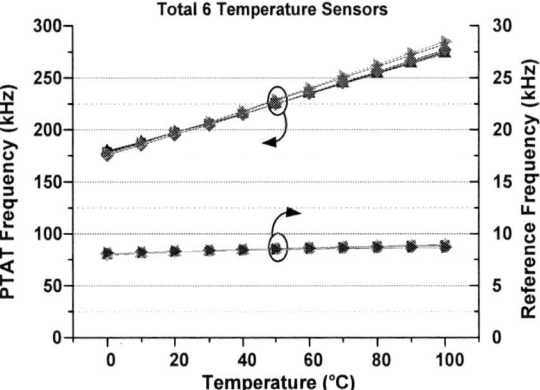

Figure 6. Measured PTAT and reference frequency across 6 dies.

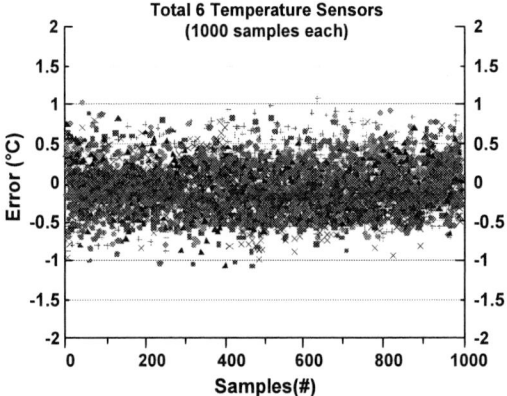

Figure 7. Measured temperature uncertainty at conversion rate of 32samples/sec. The sensor shows an error of 0.3°C (rms).

Figure 5. Measured output voltage of the sensing element.

consuming 136pW at room temperature (simulated). The high gain of the amplifier ensures that $V_R$ tracks $V_{sense}$. Due to the negligible amplifier power consumption, temperature sensing core power is dominated by $I_R$ (in the nA range). Conventionally, a current mirror is required to provide control voltages ($V_H$ and $V_L$) for the subsequent ring oscillators; however, 50% power saving can be achieved by avoiding such a current mirror. Thus, we introduce a second feedback loop to boost up the ground of the sensing element and the resistor ($V_x$) by $\sim V_{DD}/2$ and include a diode-connected transistor $M_{16}$ at the bottom of the stack. This ensures $M_{16}$ is saturated; otherwise, it becomes cutoff due to the sub-100mV output voltage from the sensing element. As a result, $V_H$ and $V_L$ are generated directly from $I_R$ without an additional mirror. This scheme also protects $I_R$ against supply variations by limiting $V_R$ to $V_{sense}$. Simulated line sensitivity shows that current changes by 0.974%/V in the 1.0−1.4V range. A duplicate sensing element serves as a dummy structure connected between $V_x$ and ground to provide a leakage path for the sensing element. This prevents $V_x$ from increasing at high temperature, which would introduce error in subsequent stages.

Amplifier offset and process variation of the resistors degrades sensor performance. Fig.4 shows the effects of process variation and mismatch on $I_{PTAT}$ and $I_{REF}$ in 1000 Monte Carlo simulations. The calculated resulting average error due to non-linearity is 0.3°C.

Voltage-controlled ring oscillators and asynchronous counters are used for digital temperature read-out. Oscillator frequencies are controlled by adjusting the resistance of transmission gates

($R_{TG}$) with voltages ($V_H$ and $V_L$) from the previous stage. For large $R_{TG}$ values, each stage delay will be determined by $I_{REF}$ and $I_{PTAT}$, regardless of the inverter delay temperature dependence. As a result, the two ring oscillators generate a PTAT frequency (clk$_{PTAT}$) and reference frequency (clk$_{REF}$). Each oscillator is connected to the first stage of an asynchronous counter. When *Start* signal is triggered, each counter counts upward until the reference counter becomes filled, setting *Done* signal. This stops both counters, and data is read from the PTAT counter. The counter is reset by the *Start* signal.

## III. MEASUREMENT RESULTS OF PROPOSED TEMPERATURE SENSOR TOPOLOGY

The proposed temperature sensor was implemented in 0.18μm CMOS in 0.09mm². Measurements were taken on 6 dies, taken from 3 different wafers in the same lot to observe the effect of process variation. The measured output voltage of the sensing element is shown in Fig. 5. The proposed circuit generates a process-independent slope and output value while maintaining high linearity. Average power consumption was only 10pW. Measured PTAT frequency ranges from 175kHz to 275kHz, leading to a resolution of 0.04°C/LSB with the 8-bit reference counter running at 8.4kHz (Fig 6.). However, as shown in Fig. 7 the effective resolution is limited by noise, resulting in 0.3°C (rms) at a conversion rate of 32 samples/sec. Resolution can be improved by having longer conversion time at the expense of energy. The sensor consumes 54nA at room temperature with

978-1-4673-6145-3/13 $31.00 © 2013 IEEE

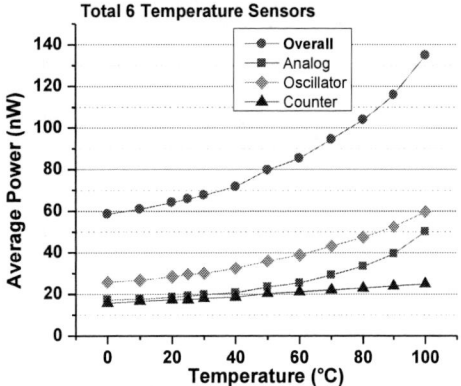

Figure 8. Measured average power of the sensor, including breakdown.

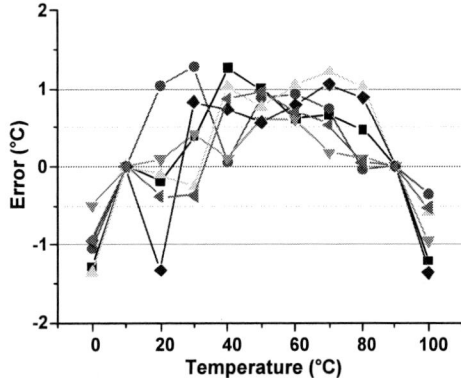

Figure 9. Measured temperature error over 6 dies.

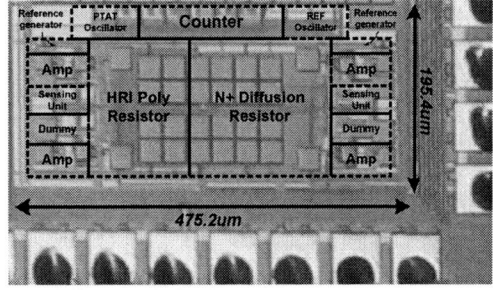

Figure 10. Die photo of the temperature sensor. The sensor occupies an area of 0.09mm² in 180nm CMOS.

TABLE I. COMPARISON WITH PREVIOUS WORK

| Parameters | This Work | [6] | [4] | [5] | [1] |
|---|---|---|---|---|---|
| Technology | 180nm | 180nm | 180nm | 180nm | 0.16μm |
| Type | MOSFET | MOSFET | MOSFET | MOSFET | BJT |
| Area | 0.09mm² | 0.05mm² | 0.042mm² | 0.032m² | 0.08mm² |
| Supply Voltage | 1.2V | 1V | 0.5V, 1V | 1.2V | 1.5V |
| Temperature Range | 0-100°C | 0-100°C | -10-30°C | 0-100°C | -55-125°C |
| Resolution | 0.3°C[1] | 0.3°C/LSB | 0.2°C/LSB | 0.3°C/LSB | 0.02°C[1] |
| Calibration | 2-point | 2-point | 2-point | 2-point | 1-point |
| Inaccuracy | +1.3°C/-1.4°C[2a] | +3°C/-1.6°C[2a] | +1°C/-0.8°C[2a] | +1°C/-0.8°C[2a] | 0.3°C[2b] |
| Relative Inaccuracy[3] | 2.7 | 4.6 | 4.5 | 1.8 | 0.2 |
| Fully Integrated | Yes | Yes | No | No | No |
| Power | 65nW | 220nW | 120nW[4] | 405nW[4] | 5.1uW[4] |
| Energy/Conversion | 2.0nJ | 22nJ | 3.6nJ[4] | 0.41nJ[4] | 27nJ[4] |

1. Degree rms
2a. Maximum error value, 2b. 3σ value
3. Relative Inaccuracy = Max error/Temperature range×100
4. Power or Energy for generating external clock not included

ultra-low power system. Without any external components, the sensor consumes only 2nJ for each conversion.

## REFERENCES

[1] K. Souri, Y. Chae, and K. Makinwa, "A CMOS Temperature Sensor with a Voltage-Calibrated Inaccuracy of ±0.15°C (3σ) from -55°C to 125°C," *IEEE ISSCC 2012*, pp. 208-209, Feb. 2012.

[2] Y. Lee, et al., "A Modular 1mm³ Die-Stacked Sensing Platform with Optical Communication and Multi-Modal Energy Harvesting," *IEEE ISSCC Dig. Tech. Papers*, pp. 402-403, Feb. 2012.

[3] A. Vaz, A. Ubarretxena, I. Zalbide, D. Pardo, H.solar, "A Full Passive UHF Tag With a Temperature Sensor Suitable for Human Body Temperature Monitoring," *IEEE Trans. on Circuits and Systems – part II*, vol. 57, no. 2, pp. 95-99, Jan. 2012.

[4] M.K. Law, A. Bermak, H.C. Luong, "A Sub-uW Embedded CMOS Temperature Sensor for RFID Food Monitoring Application," *IEEE J. Solid-State Circuits*, vol. 45, no. 6, pp. 1246-1255, Jun. 2010.

[5] M.K. Law, A. Bermak, "A 405-nW CMOS Temperature Sensor based on Linear MOS Operation," *IEEE Trans. on Circuits and Systems – partII*, vol. 56, no.12, pp. 894-895, Dec. 2009.

[6] Y.S. Lin, D. Blaauw, D. Sylvester, "An Ultra Low Power 1V, 220nW Temperature Sensor for Passive Wireless Applications," *IEEE Proc. Custom Integrated Circuits Conf.*, pp. 507-510, Sep. 2008.

[7] E. Y. Chow, S. Chakraborty, W. J. Chappell, P. P. Irazoquil, "Mixed-signal integrated circuits for self-contained sub-cubic millimeter biomedical implants," *IEEE ISSCC Dig. Tech. Papers*, pp. 236-237, Feb. 2010.

[8] M. Seok, G. Kim, D. Blaauw, D. Sylvester., "A Portable 2-Transistor Picowatt Temperature-Compensated Voltage Reference Operating at 0.5 V," *IEEE J. Solid-State Circuits*, vol. 47, no. 10, pp. 2534-2545, Oct. 2012.

supply voltage of 1.2V (57nA average). The overall power breakdown is given in Fig. 8 with the ring oscillator being the largest component. Fig. 9 shows the measured temperature sensor inaccuracy over 6 different test chips. After 2-point calibration at 10°C and 90°C, the measured error is +1.3°C/-1.4°C across 0 to 100°C. Table 1 provides a comparison with other low-power temperature sensors. The proposed design shows significantly better energy per conversion as well as relative inaccuracy compared to the previous fully-integrated temperature sensor. The sensor consumes lowest power even when compared to other designs that use high accuracy external clocks while achieving comparable resolution and inaccuracy. A die photo of the test chip is given in Fig.10.

## IV. CONCLUSION

In this work, we have designed and tested a novel temperature sensor that can be integrated into a battery-driven

# Design and Characterization of Electronic Sensing System for a 13 x 13 Biomechanical Ground Reaction Sensor Array

Q. Guo[1], M. A. Suster[2], R. Surapaneni[1], C. H. Mastrangelo[1] and D. J. Young[1]

[1]Electrical and Computer Engineering Department, the University of Utah, Salt Lake City, UT 84112 USA
[2]EECS Department, Case Western Reserve University, Cleveland, OH 44106 USA

*Abstract*— **This paper presents the design and characterization of an electronic sensing system interfaced with a high-density flexible biomechanical ground reaction sensor array (GRSA). The prototype system can be incorporated into a personal boot heel to measure real-time ground force, shear strain and sole deformation associated with a human bipedal locomotion, thus providing zero-velocity correction to an inertial measurement unit placed in a close proximity. This approach can greatly reduce inertial error accumulation over time and improve positioning accuracy. The electronic sensing system consists of a front-end multiplexer that can sequentially connect individual capacitive sensing nodes from a 13 x 13 GRSA to a capacitance-to-voltage converter followed by a 12-bit ADC sampled at 66.7 k-samples/sec, a digital timing & control unit, and a driving circuitry. The electronics were fabricated in XFAB 0.35 μm CMOS process and can achieve a gait ground velocity sensing resolution of 40 μm/sec while dissipating 3mW power.**

## I. INTRODUCTION

GPS-denied environments call for alternative position tracking solutions. MEMS technology has enabled system integration of miniature low-power inertial measurement units (IMUs) based on accelerometers and gyroscopes. However, these devices suffer from an excessive output drift over time, thus inadequate for a long-term accurate position tracking. It was recently demonstrated that a personal navigation system could be developed by employing high-resolution-gait-corrected IMUs [1]. The system combines a commercial IMU with a flexible error-correcting biomechanical ground reaction sensor array (GRSA). The IMU and GRSA are placed within the heel of a personal boot as depicted in Figure 1, where the IMU can measure inertial information while the biomechanical GRSA independently measures dynamic ground force, shear strain, and sole deformation associated with a ground locomotion gait. During a fraction of the mid-stance of a human bipedal locomotion, the velocity of the heel is zero [2] This critical information can be detected by measuring pressure contours or contours centroid movement captured by a GRSA placed between the heel and insole of a shoe depicted in Figure 1, and in turn provides a zero-velocity correction to the IMU, thus drastically reducing inertial error accumulation and improving positioning accuracy. To demonstrate the concept, a prototype navigation system incorporating a commercial heel-shaped pressure sensor array was implemented with an IMU mounted externally to a boot near the heel. A loop-closing walk test over 30 minutes was performed achieving a position error less than 4 meters [1]. Further improving position accuracy over an extended application time calls for a GRSA with an increased density. More data points available from a high-density array are expected to detect much smaller contact pressure profile change during the stationary contact of a heel, thus improving the accuracy of the zero-velocity estimation. This paper presents the design and characterization of an electronic sensing system interfaced with a prototype high-density (13 x 13) GRSA.

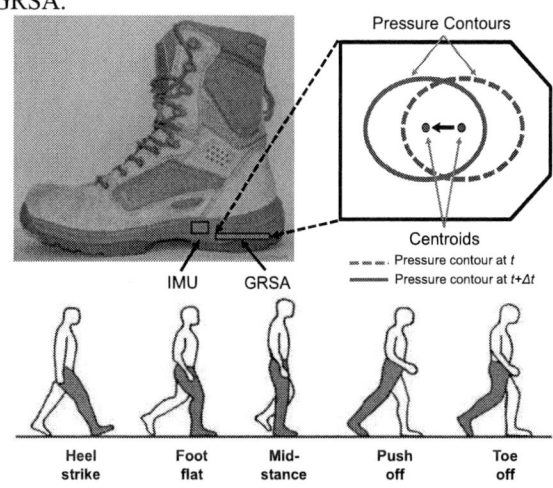

Fig. 1. Proposed personal navigation system employing high-resolution-gait-corrected IMU

## II. INTERFACE ELECTRONICS SYSTEM DESIGN

Figure 2 presents the interface electronic system design architecture for a 13 x 13 ground reaction sensor array implemented based on capacitive sensing schemes. The electronic circuitry consists of a front-end multiplexer that can sequentially connect individual sensing nodes in the GRSA to a capacitance-to-voltage (C/V) converter followed by a 12-bit ADC, a digital timing & control unit, and a driving circuitry. Each sensing node from the array can detect z-axis pressure as well as x/y axes shear force. An array area of 57 mm x 53 mm is selected to match with a typical boot heel dimension. A sensing node schematic with its corresponding electrical model is also illustrated in Figure 2. The sensor terminals can be dynamically configured by switches to achieve z-axis (single-ended) normal pressure and x/y-axes (differential) shear sensing. For navigation applications, in addition to the normal pressure sensing it is desirable to obtain shear force for slippage detection and shoe rotation estimation. A sensor pitch size of approximately 4 mm is chosen as a compromise

978-1-4673-6145-3/13 $31.00 © 2013 IEEE

between the array wiring complexity and electronic system dynamic range required for a ground velocity sensing resolution. The prototype sensor array is fabricated using kapton and PDMS materials embedding the sensors electrodes [3]. The fabricated sensing nodes exhibit a nominal capacitance value of approximately 0.8 pF and 1.6 pF for the differential mode and single-ended mode, respectively, with a capacitance change of 1% and 10% under a maximum load for the corresponding modes.

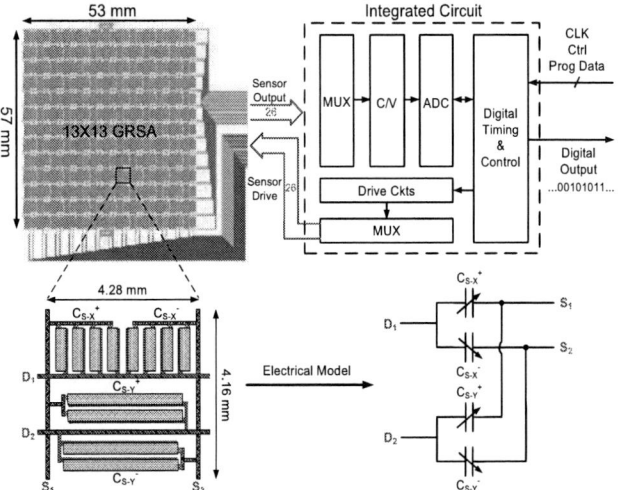

Fig. 2. Architecture of electronic sensing system design interfaced with 13 x 13 GRSA

Figure 3 presents the design of a front-end C/V converter proceeded by a multiplexer (MUX).

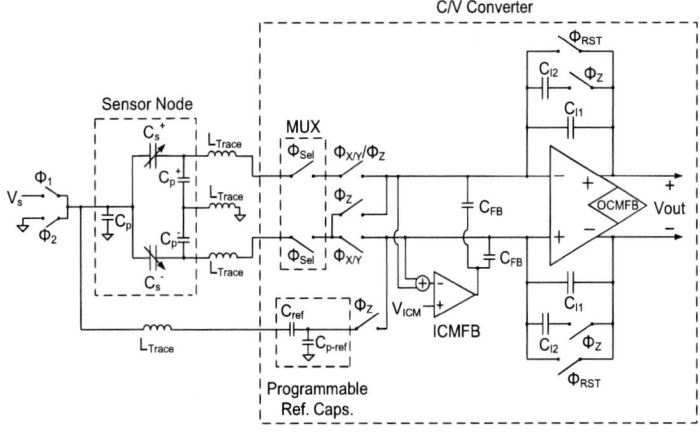

Fig. 3. Front-end interface electronic design architecture

A pair of differential sensing capacitors, $C_s^+$ and $C_s^-$, from the GRSA electrical model shown in Figure 2, is used as a sensing node model for simplicity. By properly configuring $\Phi_{X/Y}$ and $\Phi_Z$, the sensing electronics can interrogate each sensing node along z-axis or x/y-axes. When $\Phi_{X/Y}$ is closed, the sensor capacitance difference between $C_s^+$ and $C_s^-$ will be converted to an output voltage. When $\Phi_Z$ is closed, $C_s^+$ and $C_s^-$ will be combined together. The C/V converter will process the capacitance difference between the sum of the sensing capacitances and an on-chip programmable reference capacitor, $C_{ref}$. The capacitance difference thus represents the

normal pressure information. The programmable feature also enables the electronic system to be interfaced with a sensor array exhibiting a wide range of nominal sensor capacitance. The GRSA exhibits an inherent complex wiring and is connected to the interface electronic sensing system through long interconnect traces, introducing inductances modeled as $L_{trace}$ shown in Figure 3. These inductances can act as antennas to couple system clock edge transitions as interference signals to various locations, thus limiting system performance. An effective electrical shielding, therefore, becomes a critical part of the overall system integration to minimize interference coupling.

The capacitive sensor is interfaced with a switched-capacitor-based fully differential C/V converter. The converter is designed to alternatively operate between single-ended mode and differential mode to sequentially scan each sensor node in an array. A 10 msec array scanning time is chosen as a trade-off between power dissipation and number of frames needed to estimate a gait ground velocity, thus allocating 60 µsec to complete a sequential z, x, z and y sensing for each node. A stimulation voltage, $V_s$, with an amplitude of 2V is applied to the sensor, converting the capacitance difference to an output voltage, $V_{out}$, through the amplifier integrating capacitor. An input common-mode feedback (ICMFB) circuit is incorporated to ensure a propper conversion operation. The switchable integrating capacitor, $C_{I2}$, is connected in parallel with $C_{I1}$ in the single-ended mode. $C_{I1}$ and $C_{I2}$ are designed to be 0.7 pF, thus resulting in an integrating capacitor of 1.4 pF for the single-ended mode and 0.7 pF for the differential mode.

A fully differential amplifier consisting of a cascode transconductance stage followed by a Class A/B stage is chosen for the C/V converter design for its high gain, low noise and low power dissipation as shown in Figure 4.

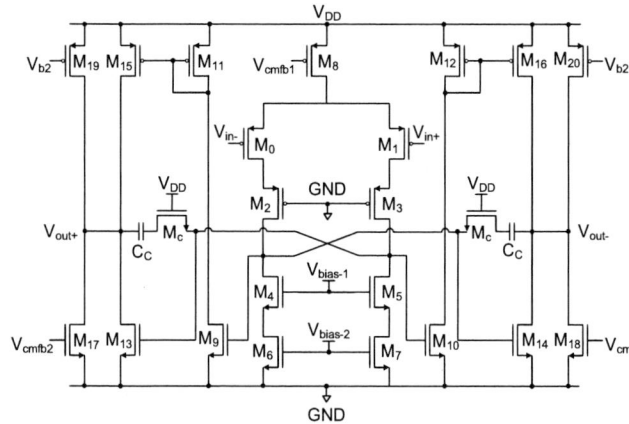

Fig. 4. Two-stage fully differential amplifier design architecture

The amplifier is designed to achieve an open-loop DC gain of 129 dB and a unity-gain frequency of 11.5 MHz resulting in a closed-loop bandwidth of approximately 300 kHz with a phase margin of 64 degrees, which satisfy the array scanning speed and settling requirements while dissipating 160 µA DC current from a 3V supply. The amplifier input-referred noise of 7

$nV/\sqrt{Hz}$ is designed to achieve a system output noise of 0.5 mV$_{rms}$, corresponding to an expected dynamic range of 70 dB with a capacitance sensing resolution of approximately 140 aF.

A 12-bit analog-to-digital converter (ADC) sampled at 66.7 k-sample per second is designed to digitize the output signal from the front-end C/V converter. The 12-bit dynamic range is chosen to ensure a certain system design safety margin. A ratio independent algorithmic ADC is selected because it can achieve the required resolution without calling for well-matched on-chip capacitors while consuming a small area and power dissipation.

## III. PROTOTYPE SYSTEM INTEGRATION AND MEASUREMNT RESULTS

The electronics were fabricated in XFAB 0.35 μm CMOS process. A fabricated ASIC occupies an area of 3.5 x 2.2 mm$^2$ and dissipates 3 mW power. The ASIC is then mounted into a socket on a custom-designed PCB and interefaced with a flexible high-density ground reaction sensor array to form a prototype system as shown in Figure 5.

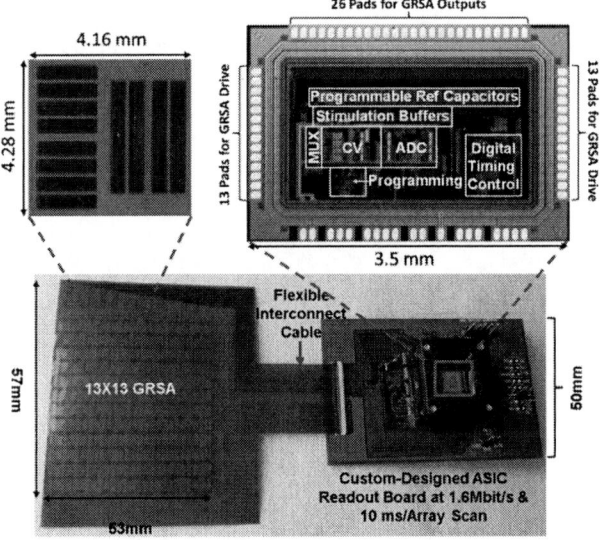

Fig. 5. Packaged prototype electronic detection system connected with fabricated GRSA

Two shielding layers, each consisting of a 2 mm-thick fabric as a dielectric layer and a grounded thin aluminum foil, are employed to sandwich the GRSA. For illustration purpose, the shielding layers are not shown in Figure 5.

Characterization of the prototype design reveals a system output noise of approximately 5 mV for the single-ended mode and 1 mV for the differential mode, caused by the electrical interferences coupling in the system due to long interconnect traces [4]. Individual sensing nodes from the sensor array were also tested under applied normal pressure. Assuming an individual carrying equipment weighs approximately 100 kg with a heel area of 30 cm$^2$, the maximum normal pressure applied on a GRSA inserted into a boot during a normal walking condition is thus around 320

KPa. Figure 6 presents a measured output voltage versus applied normal pressure from a typical sensing node in a GRSA. The output voltage increases with the pressure initially and then reaches saturation. Based on the output voltage change, the PDMS layer thickness shrinkage can be obtained, thus determining PDMS elastic modulus, which is also plotted in Figure 6.

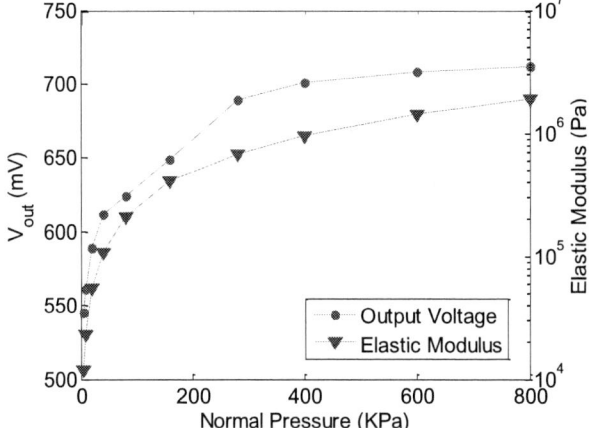

Fig. 6 Measured output voltage and elastic modulus vs. normal pressure

In order to detect a small pressure contour change or contour centroid movement to capture zero-velocity information associated with a human bipedal locomotion, it is necessary to sense small pressure variation centered around the nominal pressure bias point of 320 KPa, which calls for calculating the slope (or small-signal sensitivity) of the measured strain versus pressure profile. Figure 7 presents the resulting sensitivity plot, revealing that the PDMS small-signal elastic modulus is approximately 9 MPa around a normal pressure bias point of 320 KPa. This large elastic modulus will produce a small sensor capacitance change, which in turn demands an enhanced electronic system dynamic range for a given resolution requirement; hence an increased power dissipation. A real-time calibration technique to offset the nominal pressure bias effect can be considered to maintain the system performance requirements without increasing power dissipation.

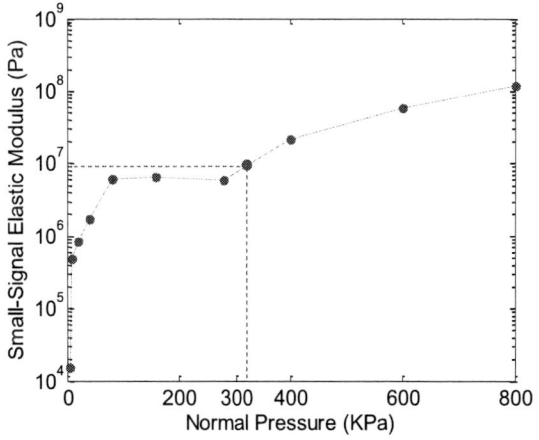

Fig. 7. PDMS small-signal elastic modulus vs. normal pressure

978-1-4673-6145-3/13 $31.00 © 2013 IEEE

To emulate a standing heel pressure profile under a walking condition, a normal force is applied onto a soft ball positioned over a ground reaction sensor array. Figure 8 presents the measured pressure profile.

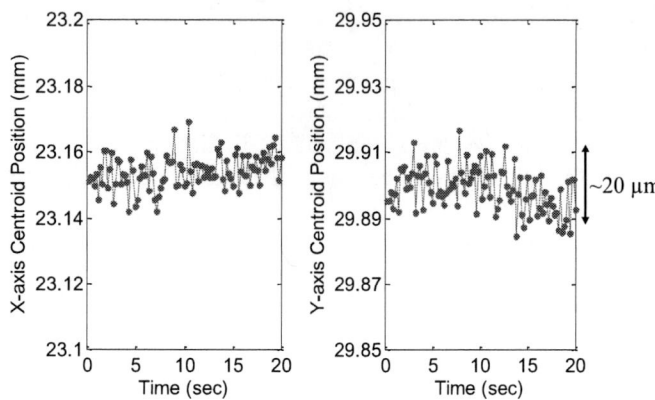

Fig. 9. GRSA pressure contour centroid position over time

## IV. CONCLUSION

A high-performance electronic sensing system has been designed and demonstrated with a flexible high-density ground reaction sensor array. Static pressure testing through a soft ball positioned over a prototype system indicates a pressure contour centroid position noise of 8 $\mu m_{RMS}$ averaged over 200 msec, thus corresponding to a gait ground velocity sensing resolution of 40 $\mu m$/sec. The demonstrated performance is limited by the system electrical interferences and saturated sensor PDMS material. A further enhanced performance can be expected by improving system output signal-to-noise/interference ratio as well as minimizing sensors pitch size.

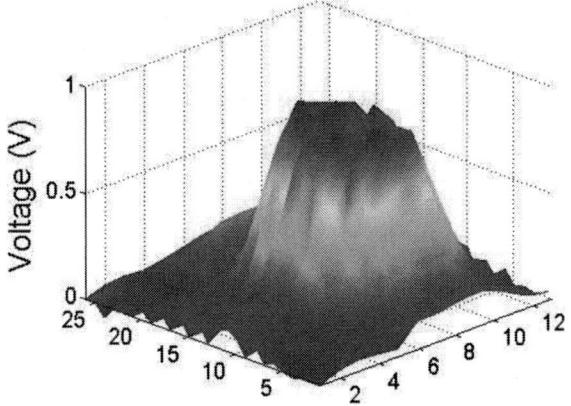

Fig. 8. GRSA output voltage (pressure) profile under normal force applied on a soft ball positioned over GRSA

The centroid position of the pressure contour can be determined by using the following expressions,

$$X_C = \frac{\sum_1^{n_y}(\sum_1^{n_x}(V_{xy}*n_x*\Delta l_x))}{\sum V_{xy}} \quad (1)$$

$$Y_C = \frac{\sum_1^{n_x}(\sum_1^{n_y}(V_{xy}*n_y*\Delta l_y))}{\sum V_{xy}} \quad (2)$$

where $x$ and $y$ represent the coordinates of a GRSA, $V_{xy}$ is the output voltage from a sensing node located at x and y location, $\Delta l_x$ and $\Delta l_y$ represent the sensor pitch size along the x-axis and y-axis, respectively, $n_x$ is the number of columns, and $n_y$ is the number of rows in a GRSA. The expressions indicate that noise and interference signals will limit the pressure contour centroid position accuracy. Further analysis reveals that the centroid position noise is proportional to $\frac{V_n}{V_{sig}} * \Delta l_{x/y}$, where $V_{sig}$ and $V_n$ represent the output voltage signal and noise, respectively, from a particular GRSA sensing node. Therefore, enhancing system output signal-to-noise ratio (SNR) and minimizing sensor pitch size are critical steps for achieving an improved centroid position accuracy.

Figure 9 presents the corresponding centroid position variation over a 20-second time frame for the pressure contour profile shown in Figure 8. It should be noted that (1) an array area slightly larger than the size of a softball used in the experiment is selected to compute the centroid position and (2) each data point in the plot presents an average value of 20 frames, equivalent to a time interval of 200 msec, which is adequate for estimating zero-velocity during a mid-stance of a human bipedal locomotion. The resultant centroid position variation is determined to be 8 $\mu m_{RMS}$, thus corresponding a gait ground velocity sensing resolution of 40 $\mu m$/sec. The demonstrated superior performance is expected to greatly improve personal navigation system performance.

## ACKNOWLEDGEMENT

This project has been sponsored by the U.S. Defense Advanced Research Project Agency under contract number: W31P4Q-08-C-0253.

## REFERENCE

[1] O. Bebek, M. A. Suster, S. Rajgopal, M. J. Fu, X. Huang, M. C. Cavusoglu, D. J. Young, M. Mehregany, A. J. van den Bogert, and C. H. Mastrangelo, "Personal Navigation via High-Resolution-Gait-Corrected Inertial Measurement Units," *IEEE Transactions on Instrumentation and Measurement*, vol. 59, No. 11, pp. 3018–3027, 2010.

[2] H. Lanshammar and L. Strandberg, "Horizontal floor reaction forces and heel movements during the initial stance phase," *Biomechanics VIII-B*, pp. 1123–1128, 1982.

[3] R. Surapaneni, Y. Xie, Q. Guo, D. J. Young and C. H. Mastrangelo, "A High-Resolution Flexible Tactile Imager System Based on Floating Comb Electrodes," *IEEE Sensors Conference*, 2012, pp. 208-211.

[4] Q. Guo, M. A. Suster, R. Surapaneni, C. Mastrangelo, D. J. Young, "Characterization of Electrical Interferences for Ground Reaction Sensor Cluster," *IEEE Sensors Conference*, 2012, pp. 596-599.

# Design Metrics for Blind ADC-Based Wireline Receivers

### (Invited Paper)

Ali Sheikholeslami[1] and Hirotaka Tamura[2]

[1]Department of Electrical and Computer Engineering, University of Toronto, Canada, [2]Fujitsu Laboratories Limited, Japan

*Abstract*—ADC-based receivers use an ADC in the front end to convert the incoming signal to digital where significant equalization can be done in digital domain. These receivers can be classified as phase-tracking and blind architectures. In the former, the VCO phase is controlled through a feedback loop so as to sample the received data in the middle of the data eye. In the latter, the received signal is sampled with a blind clock, i.e. not in a loop, and the data at the center is obtained by data processing techniques such as data interpolation and extrapolation. This paper compares the two architectures in terms of their design complexity and cost, and derives equations that relate the required ADC resolution to channel loss and to the characteristics of the FFE/DFE that follow the ADC.

## I. INTRODUCTION

As the data rates per single channel in chip-to-chip and backplane signaling march towards 100Gb/s, the main question still remains as whether the 100Gb/s receiver will be implemented as a binary receiver or as an ADC-based receiver. As shown in Fig. 1(a), a binary receiver consists of an analog equalizer at the front end followed by a Decision Feedback Equalizer (DFE). The equalized signal is then fed into a clock recovery (CR) unit that together with the slicer in the DFE form the clock and data recovery (CDR) block. This is in contrast with the ADC-based receivers, shown in Fig. 1(b)(c), where the received data is partially equalized in analog domain prior to being digitized by the ADC. The FFE, DFE, and the entire CDR are now implemented fully in digital. It is this capability of providing extensive equalization in digital domain that has given the ADC-based receivers an edge over their binary counterparts. However, this advantage comes at the price of power consumption, at least for now.

There are three main contributors to the power consumption in ADC-based receivers: the ADC, the digital processing, and the clock distribution network. Among the three, the digital processing power, and to some extent the clock distribution power, is expected to decrease over time as we move to smaller geometries. The ADC power, however, is expected to reduce at a lower rate, especially when a flash ADC is used to accommodate the high data rate. Since the ADC power grows exponentially with the ADC resolution, one remedy to the power issue is to devise architectures to minimize the required effective number of bits (ENOB) for the ADC. We will discuss this and other design considerations of the ADC-based CDRs in the second half of this paper. In the first half, we provide a primer on the basics of CDR for both binary and ADC-based receivers.

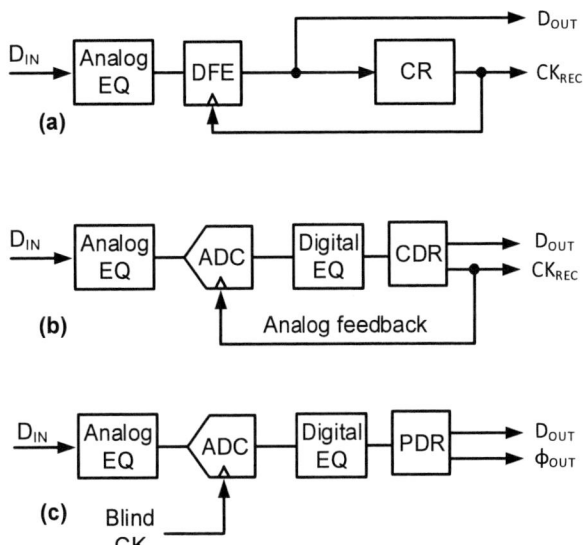

Fig. 1. Basic architectures for (a) binary receiver, (b) phase-tracking ADC-based receiver, and (c) blind ADC-based receiver

## II. CLOCK AND DATA RECOVERY PRIMER

### A. Binary CDR

A block diagram for a receiver architecture with the binary CDR is shown in Fig. 2(a). The receiver consists of a front-end analog equalizer, also known as continuous-time linear equalizer or CTLE, followed by a DFE and a CR unit. The analog equalizer partially compensates for the high-frequency attenuation introduced by the channel and hands over this partially-equalized signal to the DFE for further equalization. The equalized signal from the DFE is then sampled by a clean clock, which is provided by the CR unit.

The CR unit receives the equalized signal from the DFE and produces a clean clock. As shown in Fig. 2(b), The CR consists of a phase detector (PD), a charge pump (CP), a loop filter (LF) and a voltage-controlled oscillator (VCO). The PD compares the phase of its input with that of the recovered clock and accordingly produces early/late signals as shown in Fig. 2(c). The early/late signals control the flow of charge from the charge pump (CP) to the loop filter, and accordingly control the voltage of the VCO. Following a data transition (i.e. when $D_n \neq D_{n+1}$), an early signal is produced when the boundary data ($B_n$) matches $D_n$. Similarly, a late signal is produced when $B_n$ matches $D_{n+1}$. When there is no transition, both

978-1-4673-6145-3/13 $31.00 © 2013 IEEE

early and late signals remain inactive. Since the loop gain in this architecture is very high at low frequencies and very low at high-frequencies, the closed-loop system passes the low-frequency jitter of its input to the recovered clock (that is the VCO output) but attenuates the high-frequency jitter. The recovered clock is considered clean in the sense that it contains little high-frequency jitter. Once the clock recovery is complete, assuming a full-rate CDR, one of the falling (or rising) edge of the VCO is expected to align with the data transition and its rising (falling) edge to align with the center of the data. As such, the rising edge is used to sample the equalized signal to form the recovered data.

Alternatively, a CDR may use a phase interpolator (PI), instead of a VCO, to recover the clock. In this case, the PI receives a phase code from the loop in order to interpolate between two phases of a clock signal with a fixed frequency.

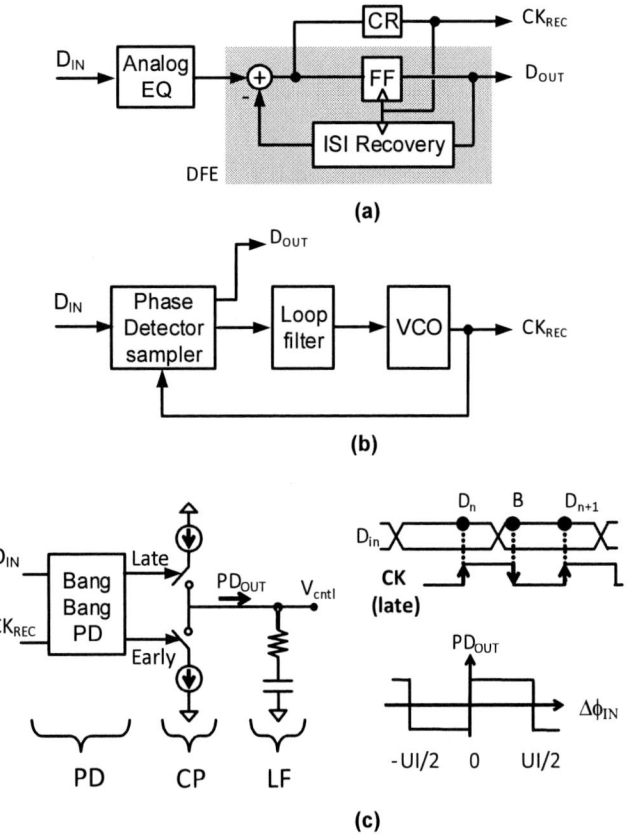

Fig. 2. (a) Binary receiver with CTLE and DFE, (b) bang-bang CDR, (c) bang-bang PD operation and its timing diagram

The binary CDR as described above is adequate for clock and data recovery if the linear equalizer and the DFE could open the eye to a sufficient level for the slicer. If this is not the case, i.e. when the channel is highly attenuative, more sophisticated equalization schemes become necessary, some of which are easy to implement in digital domain following an ADC.

## B. Phase-Tracking ADC-Based CDR

As shown in Fig. 1(b), a phase-tracking ADC-based CDR consists of an analog equalizer in the front end followed by an ADC that samples the equalized signal at the center of the eye. These samples are converted to digital where significant signal processing can be applied in digital domain. This architecture is commonly used in applications with high channel loss and dispersion, [1][2].

Including an ADC at the front end and utilizing mostly digital circuits in the backend has several advantages. First, as we mentioned earlier, this architecture allows for more sophisticated signal processing, and hence it could cover a wider range of channel attenuation. Second, the digital circuits can be designed using Verilog and can be easily ported to other technologies. This reduces the design time significantly compared to that of its alternative mostly-analog design. Third, as we move to more advanced technologies, the cost of digital circuits (in terms of area and power) continues to decrease while the cost of analog circuits is expected to remain the same, or at least not to decrease at the same rate. This implies that the cost of ADC-based CDR will continue to become more competitive over time.

Another step towards ease of design and better portability is to eliminate the feedback from the digital domain to the ADC, i.e. to eliminate the feedback that provides the recovered clock to the ADC. We discuss this architecture next.

## C. Blind ADC-based CDR

Fig. 3 shows the block diagrams of two architectures for a blind ADC-based CDR: the feed-forward architecture [3] and the DI-based architecture [4]. In both architectures, the sampling clock of the ADC is blind; i.e. it does not use any feedback from the digital backend. Nevertheless, the sampling clock is assumed to have a limited frequency offset, in the order of 100s or 1000s of ppm, with respect to the transmit clock. By using the blind clock, these architectures remove the feedback from the digital to the analog domain and hence greatly simplify the design as the ADC and the digital backend could be designed separately with minimum co-simulation.

The feed-forward architecture, Fig. 3(a), applies the blind clock (or its divided version in an actual implementation) to the digital backend. The digital backend consists of an FFE and DFE in addition to what we have called the Phase and Data Recovery (PDR) unit. We distinguish this block from a CDR as it does not recover *clock*; rather, it recovers the *phase* of the recovered clock in digits. The DI-based architecture (Fig. 3(b)) feeds back the recovered phase to a data interpolation (DI) block so as to reconstruct the received samples at the middle of the eye for ease of equalization by FFE and DFE. We will discuss this and the implications of not recovering the clock in the context of a 2x blind sampling in the next section.

## III. BLIND ADC-BASED CDR OPERATION

### A. 2x sampling

Fig. 4 shows the basic block diagram of a 2x blind ADC based receiver (the analog front end is not shown [5]). For

978-1-4673-6145-3/13 $31.00 © 2013 IEEE

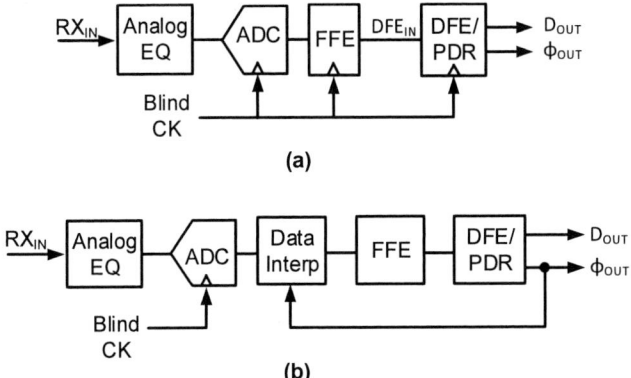

(a)

(b)

Fig. 3.  Two blind ADC-based architectures: (a) feed-forward, (b) DI-based

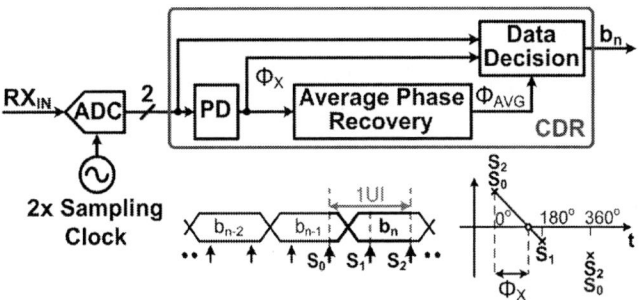

Fig. 4.  Basic architecture of a 2x blind ADC-based CDR [3]

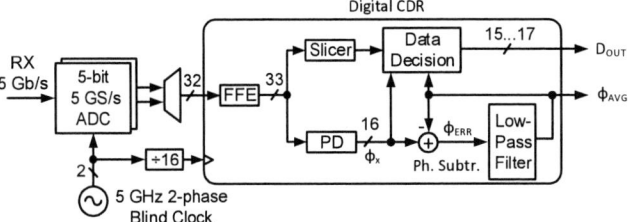

Fig. 5.  Detailed implementation of a 2x blind ADC-based CDR [3]

simplicity, we assume the received signal is sampled by a *single* ADC at the rate of 2x the bit rate. That is, the ADC takes two samples for every UI of data. These samples are denoted by $S_0, S_1$, and $S_2$ in the example timing diagram. In this example, there is a data transition (zero crossing) between $S_0$ and $S_1$. The role of the phase detector is to determine the timing of this zero crossing, denoted by $\phi_x$ on the diagram. Once we collect a sequence of $\phi_x$'s corresponding to several transitions, we will average them in digital domain to determine the average location of the transitions, denoted by $\phi_{AVG}$. This average phase represents, in digital, the falling edge location of the recovered clock, without a clock signal being present. Given the sequence of samples, $\phi_x$, and $\phi_{AVG}$, the Data Decision block then recovers the data, denoted by $b_n$.

In an actual implementation [3] in 65nm CMOS, shown in Fig. 5, two 5-bit 5GS/s ADCs are used to convert the received signal to two data streams of 5GS/s each. These two streams are then demuxed by a factor of 16 to provide 32 data streams of 312.5 MS/s where the digital logic can comfortably operate at 312.5MHz in 65nm CMOS. The details of FFE and its adaptation can be found in [6]. Here, we note that the PD takes 33 samples on each 312.5MHz clock period and produces up to 16 digital values for $\phi_x$ depending on the transition density of the received signal. The Data Decision block can output either 15, 16, or 17 bits depending on the relative position of $\phi_x$ and $\phi_{AVG}$.

As noted in Fig. 5, the blind ADC-based CDRs do not recover clock; instead, they recover the digital phase corresponding to a recovered clock. The recovered phase is used inside the Data Decision block for data recovery, and to help handle the frequency offset between the transmitter and receiver clocks using a flexible FIFO. We explain this in greater detail in the following.

In a typical application, such as in USB3.0, the recovered clock is used to write data (say 16 bits at a time) into a flexible FIFO shown in Fig. 6(a). A different clock, with possible frequency offset, coming from the core logic, reads the content of the FIFO. The FIFO is designed with enough depth so as to handle transient frequency offsets (both positive and negative) between the two clocks. In our implementation of the blind

CDR, we do not generate the recovered clock, so the question arises as how to handle the frequency offset using a FIFO. As shown in Fig. 6(b), we use the core clock for both writing and reading from the FIFO. However, we choose to write either 15, 16, or 17 bits at a time into the FIFO while we consistently read 16 bits from it at the output. This variable length in the input size allows for the effective bit rate at the input port to be different than that at the output, achieving the same effect as using of FIFO with two different clock frequencies but fixed data size.

One of the challenges of the blind ADC-based CDR as described above is the design of a corresponding DFE. Fig. 7 illustrates this challenge by comparing a one-tap DFE implantation in a phase-tracking versus blind architecture. In the former, the samples are taken in the middle of the eye where the ISI contribution from the previous bit is constant in amplitude (assuming only one post-cursor ISI for simplicity). In the latter, however, the ISI amplitude would depend on the phase of the sampling clock with respect to the center of the eye. And since this phase could change over time (because of any frequency offset), no fixed value could be used for the DFE coefficient. One design example [7] to address this challenge is shown in Fig. 8. Here, eight coefficient values are identified ($\alpha_1$ to $\alpha_8$) corresponding to eight relative locations of $\phi_{AVG}$ with respect to the sampling phase. The DFE Coefficient Selector then feeds two coefficients ($c_1$ and $c_2$) to the ISI Replica Generator. These coefficients correspond to two samples taken in one UI. The details of adaptive engine for this DFE are discussed in [8].

It is clear at this point that the feed-forward architecture of Fig. 3(a) requires a rather complex DFE architecture. This complexity can be avoided to a large extent by using the DI-based architecture of Fig. 3(b) where the use of data

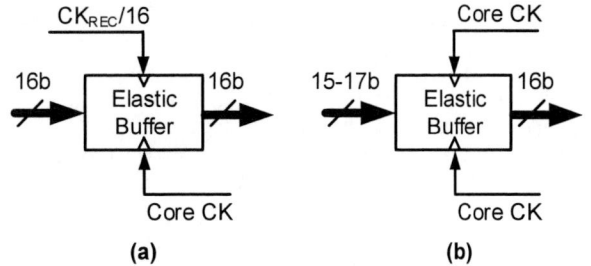

Fig. 6. Handling frequency offset in (a) USB3.0 and (b) 2x blind CDR

interpolation provides access to the data at the center and the edge. Therefore, a simpler, conventional DFE architecture can be used.

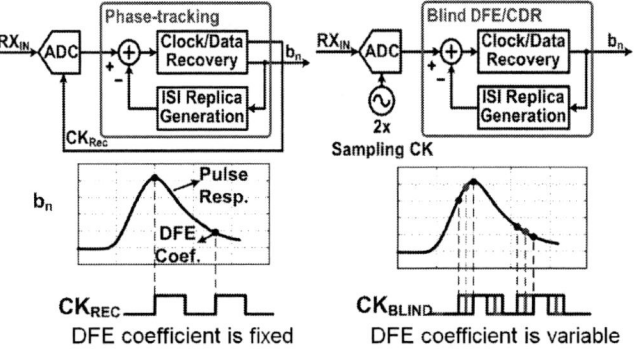

Fig. 7. DFE implementation in (a) phase tracking and (b) blind CDR [7]

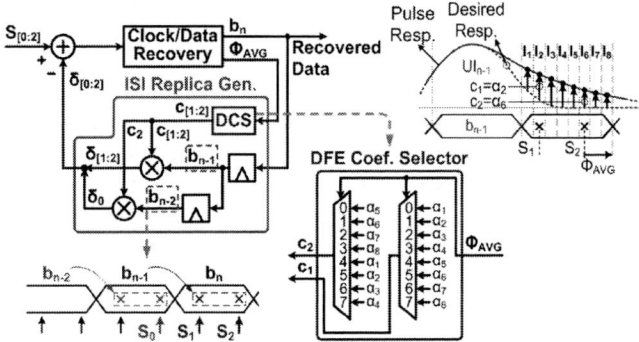

Fig. 8. Details of DFE implementation in 2x blind ADC-based CDR [7]

### B. 1.45x sampling

The main purpose of sampling the received signal at twice the baud rate is to estimate with good accuracy the location of its zero crossings. However, if we allow a small degradation in this accuracy, or equivalently in jitter tolerance, it is possible to reduce the sampling rate to below 2x, hence reducing power consumption or increasing the bit rate. In an example implementation in [9], the ADC-based receiver samples the incoming signal at 1.45x the baud rate yet offers a reasonable estimate of the zero crossings as evidenced by its simulation and measurement results.

Fig. 9 shows how conceptually an eye diagram (and hence a zero crossing location) appears as we fold samples taken at 1.45x the baud rate into a one UI interval. This sampling rate corresponds to approximately 0.7UI distance between adjacent samples, or equivalently to a total of 16 samples per 11UI. The block diagram shows a 6.875Gb/s received signal is sampled by four phases of a 2.5GHz blind clock to produce an aggregate 10GS/s for the digital CDR. The design uses a PD and a Data Decision block, which are similar to those used in [3], and a Data Compactor, where 16 samples are translated to nominally 11 bits (corresponding to 11UIs). Finally, a FIFO, with its diagram in Fig. 10, takes 10-12 bit words at its input and spits out 16-bit words using the retiming clock. Note that this FIFO absorbs the frequency offset between the blind clock and the clock embedded in the data similar to what we described in connection with Fig. 6.

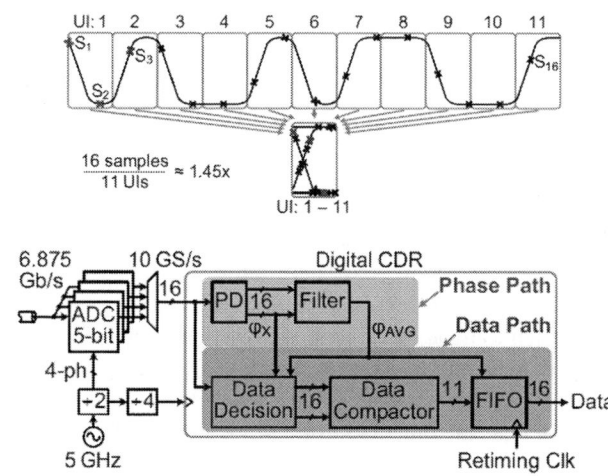

Fig. 9. 1.45x blind ADC-based CDR: (a) identifying the zero crossing from the samples, (b) implementation [9]

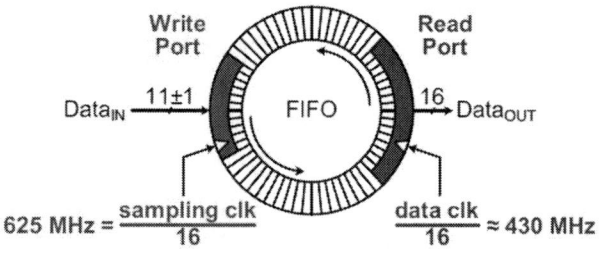

Fig. 10. FIFO implementation in 1.45x blind ADC-based CDR [9]

### C. 1x sampling

Baud-rate sampling allows for data and timing recovery in an architecture known as Mueller-Muller [10] if feedback is provided directly to the ADC clock [11]. However, when the ADC clock is blind, it is possible that the received signal is sampled directly on its zero crossings, and the data is lost completely (since there is only one sample per UI). Yet, by using an integrate and dump filter [5] in the analog domain prior to the ADC, Ting et al. [4] introduce intentional ISI in the data stream such that each bit is spilled over the adjacent bits. As a result, the data is contained not only in the center of the eye but also at zero crossings, and this makes 1x blind baud-rate CDR possible.

The block diagram of a 1x blind baud-rate CDR is shown in Fig. 11. Here the received signal is 10Gb/s and is sampled at 10GS/s using four time-interleaved 2.5GS/s ADCs. The ADCs are preceded by 1UI integrate and dump filters in analog and followed by a digital summer to form a 2UI I&D samples. These samples are then interpolated using the $\phi_{AVG}$ provided by the digital loop filter in the backend. A Mueller-Muller PD is used to estimate the phase, and a speculative 2-tap DFE is used to recover the bits. The measurement results provided in [4] confirm phase and data recovery operation with a high-frequency jitter tolerance of about 0.2UIpp even in the face of a 1000ppm of frequency offset, i.e. when the blind clock at the receiver samples the incoming signal at below the baud rate.

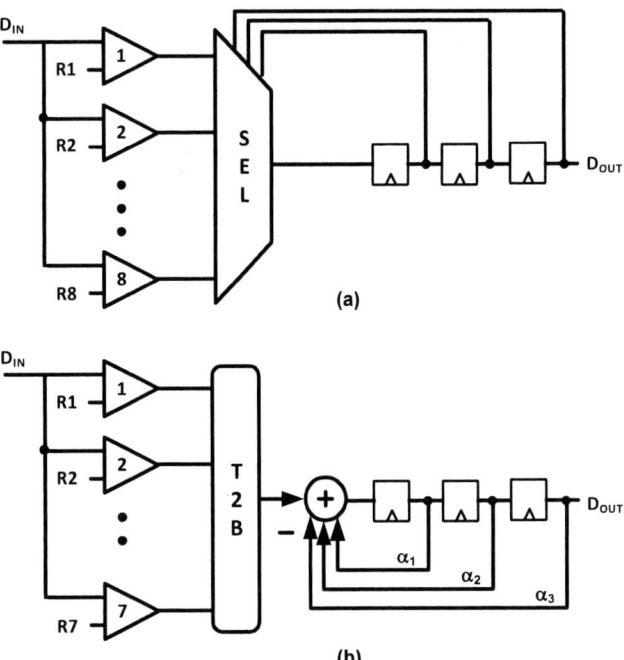

Fig. 12. (a) Speculative DFE in binary receivers, (b) ADC plus non-speculative DFE

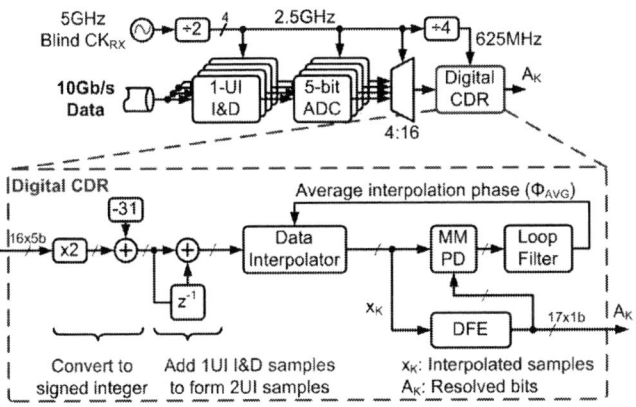

Fig. 11. Block diagram of a 1x blind ADC-based receiver [4]

## IV. BINARY VERSUS ADC-BASED: ARE THEY REALLY DIFFERENT?

Fig. 12(a) shows a simplified block diagram of a binary CDR with 3-bit speculative DFE. The received signal is compared against 8 reference levels, corresponding to the ISI associated with three consecutive bits. The output of one of the comparators is selected as the current bit based on the three previously recovered bits. Note that the reference levels in this architecture are not equidistant since the 1st, 2nd, and 3rd post-cursor ISI's are subject to the channel impulse response. However, once the previous bits are recovered, they are used directly to control the select operation, i.e. to select the current bit among 8 candidates.

This architecture is similar to an ADC-based receiver followed by a non-speculative DFE, as shown in Fig. 12(b). Here, the input is compared against 7 reference levels, but the reference levels are distributed uniformly in the input range as it is typical in a flash ADC. However, instead of only passing one bit to the next stage, all the thermometer

bits (corresponding to the full digital output of a 3-bit ADC) are fed to a non-speculative DFE. Note that even though the number of bits produced by the comparators is about the same in both cases, they carry different information because the bits correspond to different reference levels. In the former, the reference levels incorporate the ISI information; and hence the previous bits are used directly to pick one bit among 8. In the latter, the bits carry no information of the past ISI (as the reference levels have nothing to do with the ISI values). To include the ISI information, the ADC-based architecture multiplies the previous bits by the ISI values $(\alpha_1, \alpha_2, \alpha_3)$ and subtracts these digital values from the digital output of the ADC.

Given the comparison above, it is natural to suggest that perhaps the ADC-based architecture would be identical to that of a speculative DFE if the ADC reference levels are chosen non-uniformly, to be consistent with the speculative DFE case. This is indeed true but with one important caveat. The ADC based CDR (with equidistant reference levels) allows for digital FFE (i.e. FFE in digital domain) prior to DFE. This is not true for the speculative DFE because the input waveform shape is lost once we pass the set of comparators at the front end. Nevertheless, it is possible to think of novel architectures where the benefits of the two architectures can be combined [12].

The similarity between the two architectures will vanish to a large extent once we assume the sampling clock is blind. In this case, the binary CDR with speculative DFE, in its current architecture, will not be functional, as the ISI information needs updating depending on the sampling phase. The ADC-based CDR, on the other hand, functions well as it preserves

978-1-4673-6145-3/13 $31.00 © 2013 IEEE

the samples' magnitudes (except for the quantization error) of the incoming waveform for processing in the digital domain.

## V. DESIGN CONSIDERATIONS OF ADC-BASED CDRs

Fig. 13 shows the block diagram of a blind ADC-based CDR using data interpolation. An ADC digitizes the input data at a sufficient sampling rate. To compensate for the frequency-dependent signal loss in the channel, feed-forward equalizers (FFEs) are inserted in the signal path. The FFE can be analog, placed before the ADC, or digital, after the ADC, or both. Data at the decision timing (i.e., at the eye center) are then generated in the receiver by using interpolation in digital domain (Fig. 14). The interpolation ratio is obtained in the same manner as that in conventional PI-based receivers. A decision circuit with a decision feedback equalizer (DFE) performs a binary decision of the retimed data.

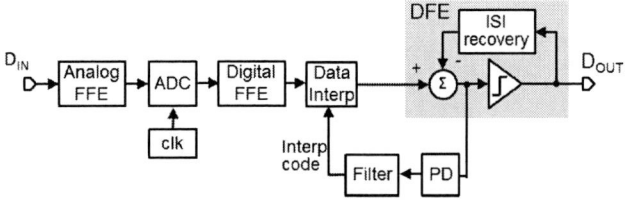

Fig. 13.   Blind ADC-based CDR using data interpolation

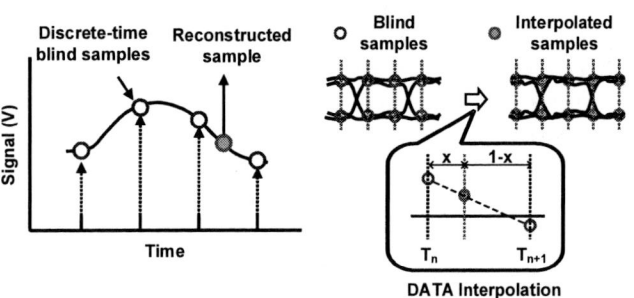

Fig. 14.   Reconstructing data by interpolation

The finite resolution of the ADC affects the CDR performance through two mechanisms. One is that the ADC quantization error propagates to an amplitude error in the eye-center data generated by the data interpolation. This amplitude error reduces the vertical eye opening. The other is that the ADC quantization error generates a phase error in the phase detection performed in the digital signal processing. The error in the phase detection results in CDRs phase tracking error, further decreasing the eye opening.

Since the digital FFE combined with the data interpolation is a linear operation on the input data, transfer from the ADC peak-to-peak quantization error $\delta V_{PP}$ to the retimed-data peak-to-peak amplitude error $\delta A_{PP}$ can be expressed as

$$\delta A_{PP} = G_e \delta V_{PP} \qquad (1)$$

where $G_e$ is the quantization-error amplification factor, and $\delta V_{PP} = LSB$. When a digital FFE is used, it amplifies the ADC quantization noise, resulting in $G_e$ greater than 1.

The value of $G_e$ is calculated in Appendix A. Assuming the channel attenuation in dB is proportional to frequency, we can write,

$$G_{ch}(f) = exp[-(ln10/10)L_0(f/f_b)] \qquad (2)$$

Here $L_0$ is the channel loss in dB at half the baud-rate frequency, $f_b$. A digital FFE equalizes this channel transfer function to that of Gaussian-filter characteristics, $G_{tot}(f)$,

$$G_{tot}(f) = exp[-(ln2/2)(f/f_{3dB})^2] \qquad (3)$$

Here $f_{3dB}$ represents the 3-dB cut-off frequency of the equalized channel. Fig. 15 shows plots of the frequency responses for the channel ($G_{ch}(f)$), equalizer ($G_{eq}(f)$), and the equalized channel ($G_{tot}(f)$) for four values of $L_0$. The value of $G_e$ increases by increasing the $f_{3dB}$ of the equalized channel.

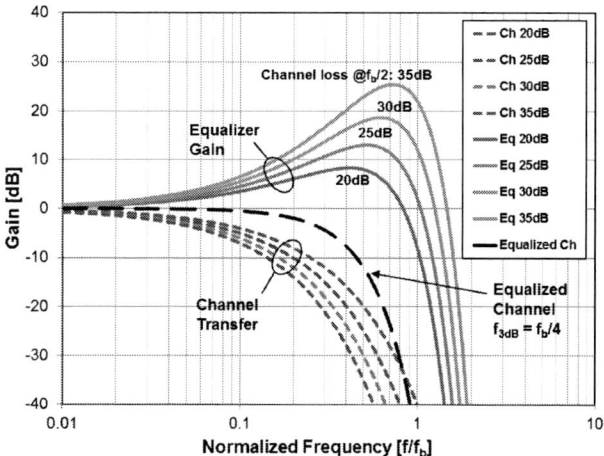

Fig. 15.   Channel, equalizer, and the equalized channel frequency response

The phase error comes from the dead zone in the phase detection. For example, 2x bang-bang phase detection produces 0.5UIpp of a dead zone if a blind clock signal is used. Since both the non-linearity of quantization and the bandwidth limitation of the data come into play, it is hard to express the error in a compact analytical form for more general multi-bit cases. If, however, the waveform is adequately bandwidth limited and can be approximated as a straight line between two adjacent samples, the dead zone for the first-order interpolation is easily calculated. The dead zone is given by the overlapping of the two dead zones from the first and second samples (Fig. 16), and the timing error corresponds to the maximum of the two dead zones. As a result, the timing error becomes:

$$\delta\theta/\Delta T = Min[1, Max[1 - \frac{\lfloor\lfloor x \rfloor\rfloor}{|x|}, \frac{\lfloor\lfloor x \rfloor\rfloor + 1}{|x|} - 1]] \qquad (4)$$

where $\Delta T$ is the sampling interval, $dV/dt$ is the slope of the input waveform, $x$ is the normalized slope ($= (dV/dt)/(LSB/\Delta T)$) and $\delta\theta$ is the dead zone error in terms of timing. The value of $\delta\theta/\Delta T$ oscillates between $1/x$ and

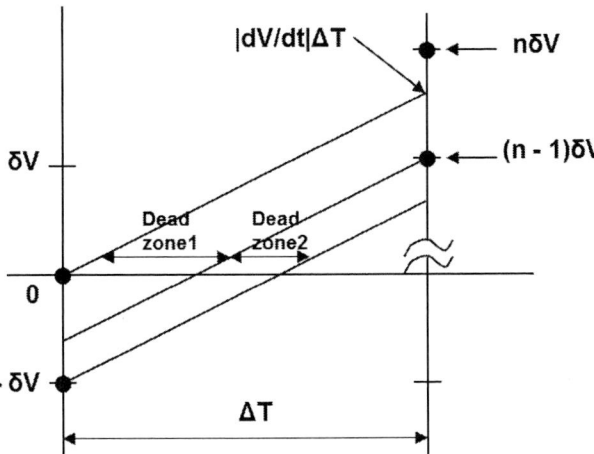

Fig. 16. Dead zones in phase detection

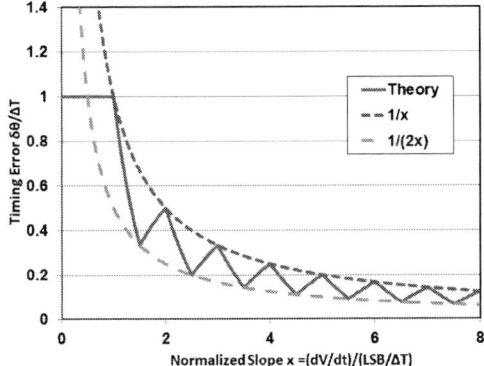

Fig. 17. Timing error as a function of normalized slope

$1/(2x)$ as shown in Fig. 17, and the worst-case timing error becomes

$$\delta\theta = Min[\Delta T, LSB/ \mid dV/dt \mid] \qquad (5)$$

Note that (5) is achieved by converting the ADC peak-to-peak quantization error, LSB, into the timing error using the slope $\mid dV/dt \mid$ as a timing-to-voltage conversion factor.

If the input waveform is well equalized by an analog FFE, the value of $\mid dV/dt \mid$ is on the order of $V_{pp}/UI$, where $V_{pp}$ is the peak-to-peak value of the input signal and UI is the unit interval. We assume that the amplitude $V_{pp}$ is adjusted by an automatic gain control such that $LSB \sim V_{PP}2^{-N}$, where N is the ADC number of bits. Then a rule of thumb expression is obtained from (5) as $\delta\theta \sim 2^{-N}$ [UI].

In theory it is possible to reduce the CDR tracking error to be less than the phase-detection dead zone given by (4) if means such as dithering is implemented. For example, adding a known amount of phase modulation would remove the dead zone error if the amount of the added phase exceeds the dead-zone width. The ADCs maximum quantization noise $\delta V_{max}$ that a blind data-interpolated ADC-based CDR can tolerate is calculated from an eye diagram drawn under a given bit-error rate (e.g. $10^{-12}$) criterion (Fig. 18). The condition for a correct signal reception in the CDR is given as

$$S_{eff} - \delta A_{ADC} - (\delta S/\delta t)\delta\theta > 0 \qquad (6)$$

where $S_{eff}$ is the effective signal strength that is given by the eye opening at the center of the eye, and $(\delta S/\delta t)$ is the slope of the eye opening at the eye center. These terms are calculated by the methods described in Appendix A and B, and the value of $\delta V_{max}$ is calculated from (6).

Increasing the 3-dB bandwidth of the equalized signal by using a digital FFE improves the effective signal amplitude but it enhances the quantization noise at the equalizer output. Due to the noise enhancement, the range of the 3-dB bandwidth of the equalized channel where the correct signal reception is possible becomes narrower when the ADC quantization noise $\delta A$ is increased. At a certain maximum value $\delta A_{max}$, the width of the correct operation range shrinks to zero (Fig. 19). We define the minimum required number of bits of the ADC as

$$\delta A_{max} = V_{pp}2^{-N_{min}} \qquad (7)$$

The minimum required number of bit $N_{min}$ increases with the channel loss $L_0$ (Fig. 20). The calculated curves matches the results obtained from behavioral simulations within $\pm 1$ bit (Fig. 21).

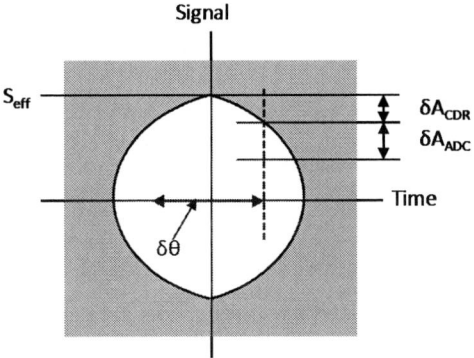

Fig. 18. Phase error and quantization error reduce the signal sense margin

ACKNOWLEDGMENT

The authors would like to acknowledge Masaya Kibune, Yanfei Chen, Mohammad Sadegh Jalali, and Clifford Ting for their contributions to this paper.

APPENDIX A

By using (2) and (3), the equalizer transfer function is given by

$$G_{eq} = \frac{G_{tot}}{G_{ch}} = exp[-\frac{ln2}{2}(\frac{f}{f_{3dB}})^2 + \frac{ln10}{10}L_0(\frac{f}{f_b})] \qquad (8)$$

The ADC quantization noise is treated as a white noise with the rms value given by $< \delta V^2 > = LSB^2/12$ and the bandwidth up to $f_b/2$. The rms value of the output of the equalizer $< A_{ADC}2 >$ is calculated as

$$< (\delta A_{ADC})^2 > = \int_0^{f_b/2} \mid G_{eq}(f) \mid^2 < \delta V^2 > df \qquad (9)$$

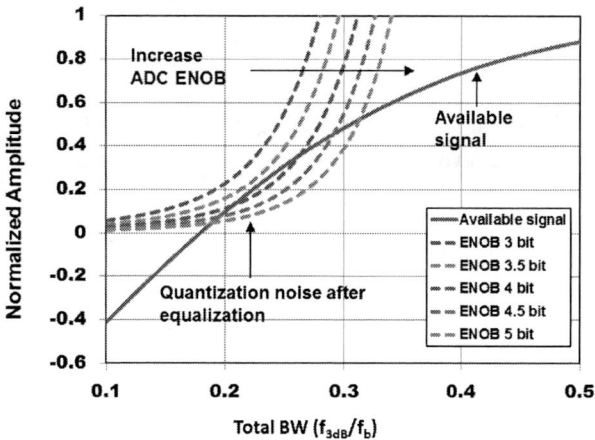

Fig. 19. Noise and signal versus equalized channel bandwidth

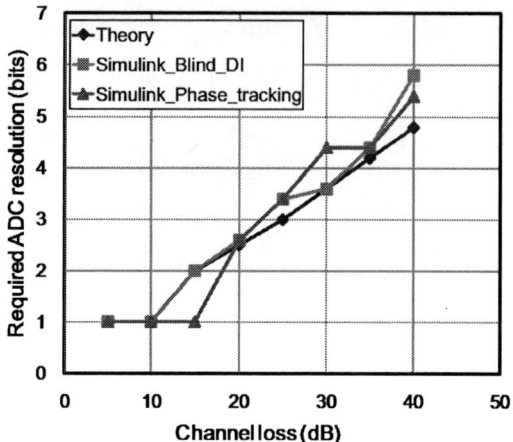

Fig. 21. Minimum ADC number of bits versus channel loss

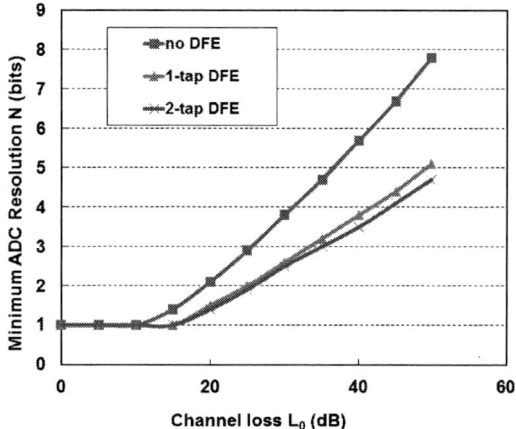

Fig. 20. Minimum ADC number of bits versus channel loss

The unit pulse response h(t) is calculated from the Gaussian filter step response s(t).

$$S(t) = 1 - 0.5 erfc(\pi f_{3dB}(2/ln2)^{1/2}t) \quad (13)$$

## REFERENCES

[1] D. Crivelli, et al., "A 40nm CMOS Single-Chip 50Gb/s DP-QPSK/BPSK Transceiver with Electronic Dispersion Compensation for Coherent Optical Channels," *ISSCC*, Dig. of Tech. Papers, pp. 338-329, Feb. 2012.

[2] J. Cao, et al., "A 40nm CMOS Single-Chip 50Gb/s DP-QPSK/BPSK Transceiver with Electronic Dispersion Compensation for Coherent Optical Channels," *IEEE J. of Solid-State Circuits*, Vol. 45, No. 6, pp. 1172-1185, June 2010.

[3] O. Tyshchenko, et al., "A 5Gb/s ADC-Based Feed-Forward CDR in 65nm CMOS," *IEEE J. of Solid-State Circuits*, Vol. 45, No. 6, pp. 1091-1098, June 2010.

[4] C. Ting, et al., "A Blind Baud-Rate ADC-Based CDR," *ISSCC*, Dig. of Tech. Papers, pp. 122-123, Feb. 2013.

[5] T. Tahmoureszadeh, et al., "A Combined Anti-Aliasing Filter and 2-tap FFE in 65-nm CMOS for 2x Blind 2-10 Gb/s ADC-Based Receivers," *IEEE Custom Integrated Circuits Conference*, pp. 1-4, Sep. 2010.

[6] H. Yamaguchi, et al., "A 5-Gb/s Transceiver with an ADC-Based Feed-Forward CDR and CMA Adaptive Equalizer in 65-nm CMOS," *ISSCC*, Dig. of Tech. Papers, pp. 168-169, Feb. 2010.

[7] S. Sarvari, et al., "A 5Gb/s Speculative DFE for 2x Blind ADC-based Receivers in 65-nm CMOS," *IEEE Symposium on VLSI Circuits*, Dig. of Tech. Papers, pp. 69-70, June 2010.

[8] B. Abiri, et al., "An Adaptation Engine for a 2x Blind ADC-Basead CDR in 65 nm CMOS ," *IEEE J. of Solid-State Circuits*, Vol. 46, No. 12, pp. 3140-3149, Dec. 2011.

[9] O. Tyshchenko, et al., "A Fractional-Sampling-Rate ADC-Based CDR with Feed-Forward Architecture in 65nm CMOS," *ISSCC*, Dig. of Tech. Papers, pp. 166-167, Feb. 2010.

[10] K. Mueller and M. Muller, "Timing Recovery in Digital Synchronous Data Receivers," *IEEE Trans. on Communications*, pp. 516-531, May 1976.

[11] M. Harwood, et al., et al., "A 12.5Gb/s SerDes in 65nm CMOS Using a Baud-Rate ADC with Digital Receiver Equalization and Clock Recovery," *ISSCC*, Dig. of Tech. Papers, pp. 436-437, Feb. 2007.

[12] J. Kim, et al., "Equalizer design and performance trade-offs in ADC-based serial links", *Custom Integrated Circuits Conference*, pp. 1-8, Sept. 2010.

Assuming the ADC quantization noise has a uniform distribution, this rms value can be multiplied by $2\sqrt{3}$ to provide the peak-to-peak value.

## APPENDIX B

For a given channel characteristics, let the unit pulse response be written as $h_k = h(kT)$, $(k \in Z)$, $T$ is the unit interval, and $h(0) = h_0$ is the main cursor where the unit pulse response has a peak. For Gaussian channel, the values of $h(kT)$ is non-negative and the maximum ISI is given as the sum of all post-cursor and pre-cursor taps. Thus we have

$$ISI_{max} = \sum_{i \neq 0} h_i = 1 - h_0 \quad (10)$$

$$S_{eff} = h_0 - ISI_{max} = 2h_0 - 1 \quad (11)$$

We assume that an ideal n-tap DFE makes the post cursor taps zero up to $h_n$. This decreases the ISImax by $\sum_{i=1}^{n} h_i$, producing

$$S_{eff} = 2h_0 + \sum_{i=1}^{n} h_i - 1 \quad (12)$$

# A 10Gbps, 1.24pJ/bit, Burst-Mode Clock and Data Recovery with Jitter Suppression

Ming-Chiuan Su[1], Wei-Zen Chen[1], Pei-Si Wu[2], Yu-Hsian Chen[2], Chao-Cheng Lee[2], and Shyh-Jye Jou[1]

Department of Electronics Engineering and Institute of Electronics, National Chiao Tung University[1]

Realtek Corp[2], Hsinchu, Taiwan, R. O. C.

*Abstract*—A 10Gbps, 1/5-rate burst mode clock and data recovery (BMCDR) circuit is proposed. The BMCDR is reconfigurable between data gating mode and phase tracking mode to achieve instantaneous phase-locking with jitter suppression for 10 GPON. Incorporating a 1/5-rate CDR with 1:5 demultiplexer, it achieves a high energy efficiency of 1.24pJ/bit. With a 4MHz, $0.22UI_{pp}$ input data jitter, the recovered clock jitter at 2GHz is $2.94ps_{rms}$. The prototype chip is fabricated in UMC 55nm CMOS technology. Chip size is $200 \times 150 \mu m^2$.

## I. INTRODUCTION

Burst mode clock and data recovery (BMCDR) plays a key role in gigabit passive optical network (GPON) receiver. It is required to recover data burst within tens to hundreds of bit time. Typically, instantaneous phase-locking in a BMCDR can be achieved by gating the voltage controlled oscillator (GVCO) in a PLL using input data [1]-[3]. Fig. 1 illustrates a burst-mode CDR architecture. It is based on the master/slave architecture for sampling clock generation, where the PLL with a replica GVCO (GVCO1) is used to initialize the sampling clock frequency of GVCO2, and then the input data gates GVCO2 for instantaneously phase-locking [7].

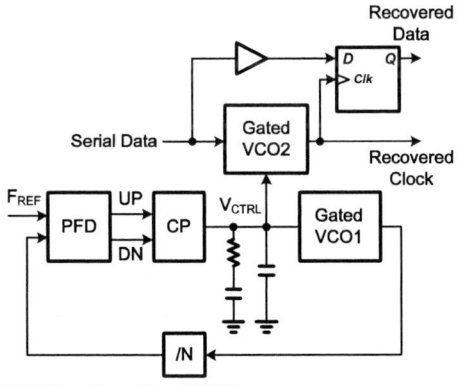

Fig. 1 BMCDR with replica GVCO

However, due to inevitable frequency mismatches between GVCO1 and GVCO2, and also frequency offset between transmitter and receiver sides, a longer stream of consecutive identical bits can lead to significant accumulated phase error in a phase gating CDR. As a consequence, it results in a higher jitter transfer associated with the input data jitter, and may cause higher bit error rate.

To overcome the aforementioned shortcomings, a dual mode BMCDR capable of switching between data gating mode and phase tracking mode is proposed in this paper. Frequency initialization is achieved by using GVCO-merged PLL with a local reference. As only one GVCO is used, it is free from frequency mismatch problem in a conventional master/slave architecture. Meanwhile, a lock detector is used to moderate the frequency locking condition so as to switch the frequency initialization mode into data gating and phase tracking mode through gating controller. The proposed GVCO can achieve phase alignment within tens of bit time. After phase alignment, the BMCDR will be configured as a phase tracking CDR loop to continuously track input data frequency, and also enhance jitter suppression capability. By dynamically reconfiguring the BMCDR during the locking process, it achieves instantaneous phase alignment and jitter suppression simultaneously while meeting the jitter tolerance mask of 10 GPON.

## II. PROPOSED BMCDR ARCHITECTURE

The proposed BMCDR is shown in Fig. 2. It consists of a 1/5-rate GVCO, a 1/5-rate phase detector (PD), a lock detector (LD), and a dual mode PLL. Compared to conventional full rate or half rate architecture [7][8], the GVCO frequency is greatly reduced to save power. On the other hand, as only one GVCO is adopted in this architecture, it is free from potential frequency mismatch issue in conventional dual GVCOs architectures [7]-[9]. Moderated by lock detector and gating controller, the PLL is switchable between data gating mode and phase tracking mode. Before enabling input data, the GVCO's frequency is brought to the vicinity of 1/5 data rate through a 2nd order PLL by disabling **lock (=0)** and **gate (=0)** signals. Monitored by the lock detector, the CDR would step into gating mode by enabling **lock (=1)** and **gate (=1)** signals after frequency initialization, and apply the input data to GVCO.

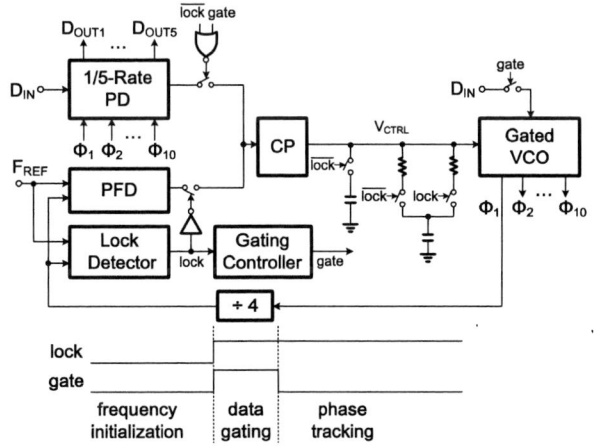

Fig. 2 Proposed dual mode BMCDR architecture

For instantaneous phase alignment, the gating controller enables data gating operation by setting **gate (=1)** signal at the onset of phase locking. Meanwhile, the CDR is opened loop by disabling the PFD and 1/5-rate PD to prevent the GVCO from being mutually pulled by the PLL during the gating operation. Moderated by a timer in the gating controller, the gating operation lasts for less than 10 ns to avoid frequency drift caused by potential charge leakage in the loop filter, while providing sufficient timing margin for injection locked at 10 Gb/s operation (~100 bits hits).

The CDR then resumes its closed loop operation by resetting **gate (=0)** signal **(lock =1),** and is reconfigured as a 1st order, 1/5 data rate PLL for continuously phase tracking. By activating the 1/5-rate PD, the CDR will keep tracking jitter and frequency of input data. Thus it can tolerate long run of consecutive identical bits by alleviating frequency mismatch between the transmitter and receiver side, and suppressing jitter through low pass filtering of PLL.

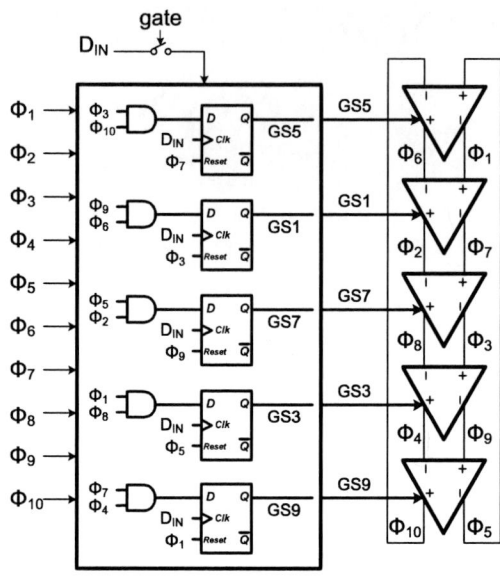

Fig. 3 GVCO architecture

## III. CIRCUIT IMPLEMENTATION

### A. Gated Voltage Controlled Oscillator

Fig. 3 shows the proposed GVCO architecture, which is composed of a 5 stage ring oscillator and a Selective Gating Generator. To comply with 1/5 rate operation, the selective gating generator is incorporated to pass the input data edge to the proper gating stage of GVCO. Fig. 4 illustrates the timing diagram of 1/5 rate selectively gating process. During the data gating mode, the input data continuously samples the multiphase output of GVCO ($\Phi_1$-$\Phi_{10}$). The Selective Gating Generator then provides the gating pulses (**GS1**, **GS3**, **GS5**, **GS7**, and **GS9**), which pick up the corresponding stages of GVCO that has the closest transition phases with $\mathbf{D_{IN}}$ [4]. Guiding by these gating pulses, $\mathbf{D_{IN}}$ is capable of adjusting GVCO output phase at 1/5 rate, drastically mitigating the speed and power requirement of BMCDR. When the GVCO is locked, $\mathbf{D_{IN}}$ will be aligned with $\mathbf{\Phi_1}$, $\mathbf{\Phi_3}$, $\mathbf{\Phi_5}$, $\mathbf{\Phi_7}$, $\mathbf{\Phi_9}$. Meanwhile, $\mathbf{\Phi_2}$, $\mathbf{\Phi_4}$, $\mathbf{\Phi_6}$, $\mathbf{\Phi_8}$, $\mathbf{\Phi_{10}}$, can be utilized for data recovery.

Fig. 5 depicts the circuit schematic of GVCO's delay cell. When **GSi =1 (i=1, 3, 5, 7, and 9),** the delay cell output will be reloaded to high for instantaneous phase synchronization. On the contrary, it will be configured as a closed loop oscillator when **GSi =0**. A cross coupled latch is placed in the middle of the delay cell to accelerate switching speed for reducing phase noise. On the other hand, it provides nearly rail to rail output swing so as to have higher disturbance immunity.

During data gating operation, both the PFD and the 1/5-rate PD are disabled to avoid pulling effect. To alleviate charge leakage issue, the data gating process only lasts during preamble, which is less than 100 bit times. Afterwards, the 1/5-rate PD will be activated to continuously track the input data. Compared to our previous work in [4], the GVCO in this design is reconfigurable as a typical VCO by disabling data gating path during phase tracking. By reconfiguring the loop filter and loop bandwidth during CDR mode, it provides the advantages of jitter suppression without sacrificing the locking speed and jitter tolerance. As the input data is retimed by 1/5-rate GVCO, it accomplishes BMCDR as well as 1:5 demultiplexer.

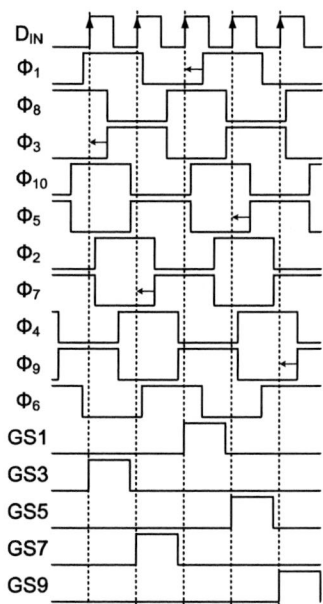

Fig. 4 GVCO timing diagram

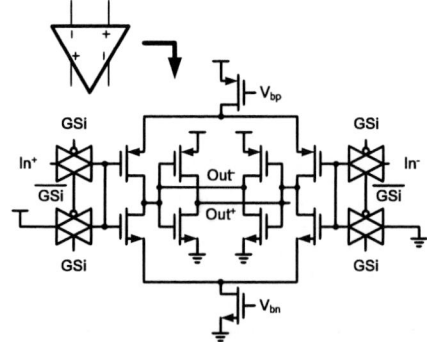

Fig. 5 GVCO's differential delay cell

978-1-4673-6145-3/13 $31.00 © 2013 IEEE

## B. 1/5-Rate Phase Detector

In a phase detector design, it must be able to detect data edges and compare them with clock phases to generate lead-lag information. Fig. 6 illustrates the 1/5-rate phase detector timing diagram, where the circles and solid dots on $D_{IN}$ respectively represent the sampling points by ($\Phi_1$, $\Phi_3$, $\Phi_5$, $\Phi_7$, $\Phi_9$) and ($\Phi_2$, $\Phi_4$, $\Phi_6$, $\Phi_8$, $\Phi_{10}$). By extending the Alexander phase detector to 5-way time interleaved architecture, the 1/5-rate phase detector can be implemented to generate UP/DN signals from three of the consecutive sampling points. Fig. 7 illustrates one out of the five sequentially toggled sub-PDs. The UP/DN information for charge pump is then derived by tallying the **UPi/DNi (i= 1, 2, 3, 4, and 5)** collected from the five sub-PDs. Thus in the locked state, ($\Phi_1$, $\Phi_3$, $\Phi_5$, $\Phi_7$, $\Phi_9$) are aligned to the data edge. Meanwhile, ($\Phi_2$, $\Phi_4$, $\Phi_6$, $\Phi_8$, $\Phi_{10}$) are served as sampling clock to recover and demultiplex input data.

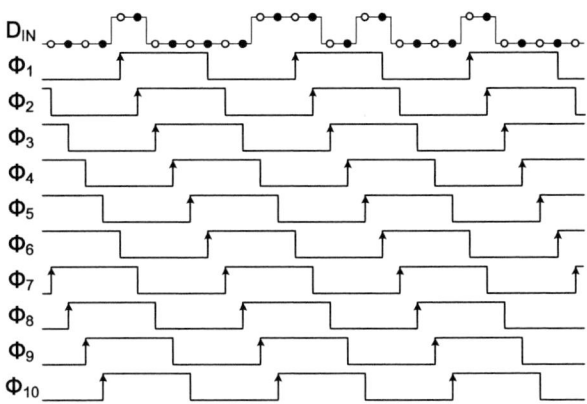

Fig. 6 The 1/5-rate phase detector timing diagram

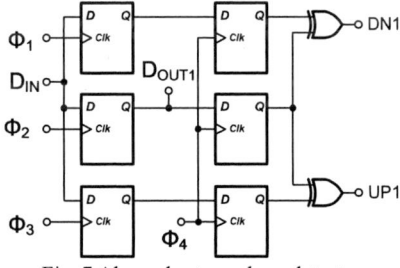

Fig. 7 Alexander-type phase detector

## IV. EXPERIMENTAL RESULTS

The 10 Gb/s, 1/5-rate BMCDR is fabricated in UMC 55nm SP CMOS technology. Fig. 8 shows the chip micrograph. The core area is only $200\times150\mu m^2$. The divider ratio of the PLL is 4, and the reference clock for frequency initialization is 500MHz. In order to better observe jitter suppression capability, an input data with jitter is applied to verify its performance. With a 4MHz, $220mUI_{pp}$ sinusoidal data jitter at 10 Gb/s, the measured recovered clock jitter is $2.94ps_{rms}$ at 2GHz, as is shown in Fig. 9. The measured recovered clock spectrum at 2GHz is also shown in Fig. 10.

Fig. 8 Chip micrograph

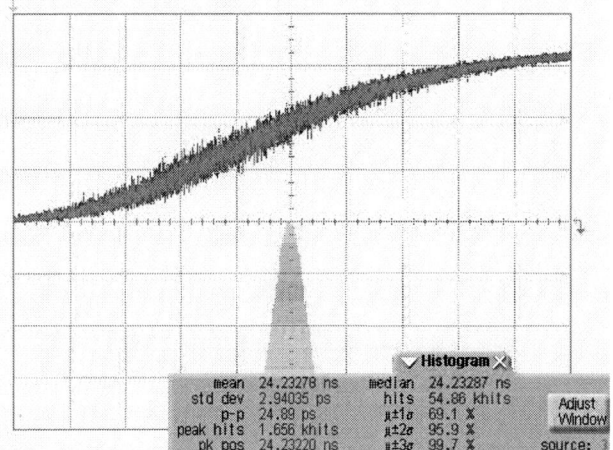

Fig. 9 Measured recovered clock jitter and (b) its histogram @ 2GHz (w/i 4MHz, $220mUI_{pp}$ sinusoidal jitter at 10 Gb/s input data)

Incorporating with 1/5-rate GVCO, it accomplishes BMCDR as well as 1:5 demultiplexer with $BER < 10^{-12}$. The jitter tolerance test reveals that it passes the 10 GPON jitter tolerance mask with $2^7$-1 PRBS input pattern, as is shown in Fig. 11. Fig. 12 shows the locking behavior for a 10 Gb/s input data. With a 10 Gb/s repeated data sequence of (00111 00100 11111 11101 11100 11011 11011 00101), one of the 1:5 demultiplexed data at 2 Gb/s is observed as (10100110···). The recovered data is captured at the first input data edge. The lock time is less than 1 UI. The latency including serial to parallel conversion and I/O buffer is about 5ns. Excluding the I/O buffers, the 10 Gb/s BMCDR and 1:5 demultiplexer consume only 12.4mW.

Table I summarizes the performance benchmark with the prior art. To the authors' best knowledge, it accomplishes BMCDR and 1:5 demultiplexer with the best energy efficiency of 1.24pJ/bit. On the other hand, the proposed architecture is also highly area efficient which only occupies $0.03mm^2$.

978-1-4673-6145-3/13 $31.00 © 2013 IEEE

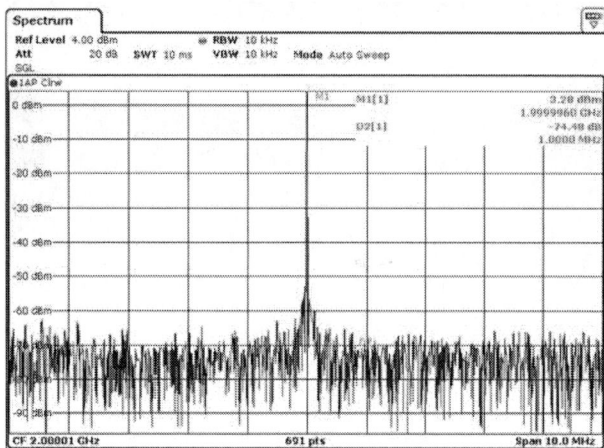

Fig. 10 Measured recovered clock spectrum @ 2GHz

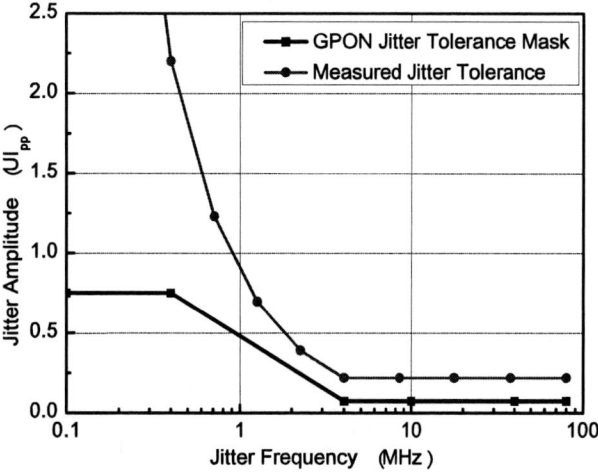

Fig. 11 Measured jitter tolerance with 10 GPON jitter tolerance mask (10 Gb/s, $2^7$-1 PRBS)

Table I Performance benchmark

|  | This Work | [5] | [6] |
|---|---|---|---|
| Process | 55nm | 65nm | 40nm |
| Supply | 1V | 1.2V | 1.1V |
| Method | Burst-mode | Burst-mode | Burst-mode |
| Data Rate | 10 Gb/s | 6 Gb/s | 5.184 Gb/s |
| Lock Time | <1 bit | <1 bit | <20 bits |
| Power | 1.24mW/Gb/s | 3.67mW/Gb/s | 2.4mW/Gb/s |

## V. CONCLUSIONS

A 1/5-rate BMCDR with jitter suppression is proposed. By reconfiguring the CDR between data gating mode and phase tracking mode, it achieves instantaneous phase locking and jitter suppression simultaneously. Thus severe trade-offs in conventional PLL-based or GVCO-based CDRs can be circumvented. By implementing selectively-gating technology, the operating speed of GVCO can be greatly reduced to save power consumption. To the authors' best knowledge, it accomplishes BMCDR and 1:5 demultiplexer with the best energy efficiency of 1.24pJ/bit. Fabricated in 55nm CMOS technology, the active area is 0.03mm$^2$ only.

## ACKNOWLEDGMENTS

The work is sponsored in part by NSC-101-2221-E-009-162, CIC and RealTek Corp, Taiwan.

## REFERENCES

[1] J. Terada, K. Nishimura, S. Kimura, H. Katsurai, N. Yoshimoto, and Y. Ohtomo, "A 10.3125Gb/s Burst-Mode CDR Circuit using a ΔΣ DAC," *IEEE ISSCC Digest of Technical Papers*, pp. 226-227, 2008.

[2] C. F. Liang and S. I. Liu, "A 20/10/5/2.5Gbps Power-Scaling Burst-Mode CDR Using GVCO/Div2/DFF Tri-mode Cells," *IEEE ISSCC Digest of Technical Papers*, pp. 224-225, 2008.

[3] S. L. J. Gierkink, "A 2.5 Gb/s Run-Length-Tolerant Burst-Mode CDR Based on a 1/8th-Rate Dual Pulse Ring Oscillator," *IEEE Journal of Solid-State Circuits*, vol. 43, no. 8, pp. 1763-1771, Aug. 2008.

[4] Y. H. Chen and W. Z. Chen, "A 0.6-7 Gbps, 1/7 Rate, Burst Mode Clock and Data Recovery Circuit and Demultiplexer," *IEEE Radio Frequency Integrated Circuits Symposium*, pp. 531-5334, 2012.

[5] B. Abiri, R. Shivnaraine, A. Sheikholeslami, H. Tamura, M. and Kibune, "A 1-to-6Gb/s Phase-Interpolator-Based Burst-Mode CDR in 65nm CMOS," *IEEE ISSCC Digest of Technical Papers*, pp. 154-155, 2011.

[6] K. Maruko, T. Sugioka, H. Hayashi, Z. Zhou, Y. Tsukuda, Y. Yagishita, H. Konishi, T. Ogata, H. Owa, T. Niki, K. Konda, M. Sato, H. Shiroshita, T. Ogura, T. Aoki, H, Kihara, and S. Tanaka, "A 1.296-to-5.184Gb/s Transceiver with 2.4mW(Gb/s) Burst-Mode CDR Using Dual-Edge Injection-Locked Oscillator," *IEEE ISSCC Digest of Technical Papers*, pp. 364-365, 2010.

[7] M. Banu and A. Dunlop, "A 660 Mb/s CMOS clock and data recovery circuit with instantaneous locking for NRZ data and burst-mode transmission," *IEEE ISSCC Digest of Technical Papers*, pp. 102-103, 1993.

[8] C. F. Liang, S. C. Hwu and S. I. Liu, "A 10Gbps Burst-Mode CDR Circuit in 0.18μm CMOS," *IEEE Custom Integrated Circuits Conference (CICC)*, pp. 599-602, 2006

[9] M. Nogawa, K. Nishimura, S. Kimura, T. Yoshida, T. Kawamura, M. Togashi, K. Kumozaki, and Y. Ohtomo, "A 10 Gb/s burst-mode CDR IC in 0.13μm CMOS," *IEEE ISSCC Digest of Technical Papers*, pp. 228–229, 2005

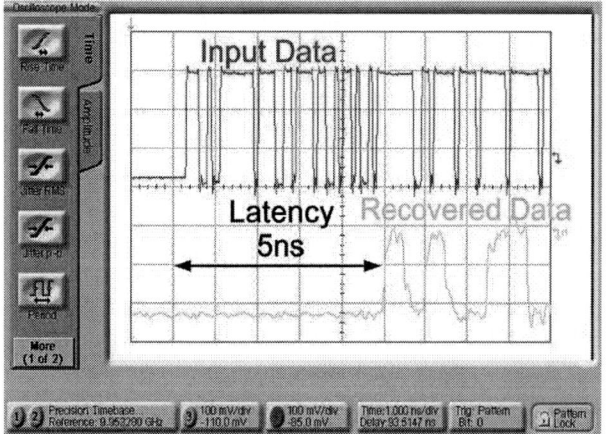

Fig. 12 Locking behavior (latency ~ 5ns, 10 Gb/s input data pattern = 00111_00100_11111_11101_11100_11011_11011_00101_···, 2 Gb/s recovered data pattern = 10100110···)

# A 9.2-GHz Digital Phase-Locked Loop with Peaking-Free Transfer Function

Sigang Ryu, Hwanseok Yeo, Yoontaek Lee, Seuk Son and Jaeha Kim

School of Electrical Engineering and Computer Science, Inter-university Semiconductor Research Center
Seoul National University, Seoul, Korea

*Abstract*- **This paper describes a digital phase-locked loop (PLL) that realizes a peaking-free jitter transfer. That is, the PLL's second-order transfer function does not have a closed-loop zero. Such a PLL does not exhibit overshoots in the phase step response and achieves fast settling. Unlike the previously-reported peaking-free PLLs, the proposed PLL implements the peaking-free loop filter directly in digital domain without requiring additional components. A time-to-digital converter (TDC) is implemented as a set of three binary phase-frequency detectors that oversample the timing error with time-varying offsets, achieving a linear TDC gain and PLL bandwidth insensitive to the jitter condition. And a 9.2-GHz digitally-controlled LC oscillator (DCO) with transformer-based tuning realizes a predictable DCO gain set by a ratio between two digitally-controlled currents. The prototype 9.2-GHz-output digital PLL fabricated in a 65nm CMOS demonstrates a fast settling time of 1.58-µs with 690-kHz bandwidth. The PLL has a 3.477-ps$_{rms}$ divided clock jitter and -120dBc/Hz phase noise at 10-MHz offset while dissipating 63.9-mW at a 1.2-V supply.**

## I. INTRODUCTION

A commonly-cited weakness of phase-locked loops (PLLs) against delay-locked loops (DLLs) is that PLLs accumulate jitter and therefore exhibit peaking in the frequency-domain transfer function and overshoots in the time-domain step response [1]. And it is widely regarded as the reason why PLLs typically show the poorer jitter and slower settling time compared to DLLs. The jitter accumulation behavior of a conventional second-order PLL is in fact due to the presence of a zero in its closed-loop transfer function. The popular use of the proportional-integral (PI) loop filter, or equivalently a series-RC filter with a charge pump, places a closed-loop zero in the vicinity of the PLL bandwidth frequency, of which position cannot be adjusted without affecting the bandwidth or stability of the PLL. This work presents a PLL without such an undesired closed-loop zero, by implementing a new loop filter in digital domain. The proposed PLL exhibits fast settling without overshoots and jitter transfer without peaking.

While peaking-free have been demonstrated in prior literature [2]-[4], the proposed PLL is a one that can be straightforwardly implemented in most digital PLLs simply by adopting a new digital loop filter without requiring an additional jitter-sensitive component. A key to eliminating the closed-loop zero is to make an alternative phase adjustment path bypassing the oscillator. For instance, the peaking-free analog PLL presented in [2] used an additional voltage-controlled delay line (VCDL). Also, the PLLs presented in [3] and [4] used a programmable divider in order to shift the phase of the feedback clock. However, these additional circuit

components placed in the noise-sensitive input or feedback clock paths can degrade the jitter performance. In contrast, the proposed PLL implements this alternative phase adjustment entirely within its digital loop filter. Therefore, the design can be easily adopted in many existing digital PLLs with a linear time-to-digital converter (TDC) and a digitally-controlled oscillator (DCO).

The presented digital PLL employs a low-cost linear TDC and a transformer-tuned LC-DCO, in order to realize predictable PLL characteristics against the variations in process, voltage, temperature (PVT), and jitter condition. The TDC consists of three bang-bang phase-frequency detectors (BB-PFDs) each triggered by an independently dithered clock phase and offset to achieve a linearized TDC gain that does not change with the external jitter condition. The DCO adopts a transformer-based tuning described in [5], in which the current flowing into the secondary coil can change the effective inductance of the primary coil and therefore the LC oscillation frequency. Compared to the varactor-based designs, the DCO realizes a more predictable tuning characteristic, set by a ratio between two digitally-controlled currents.

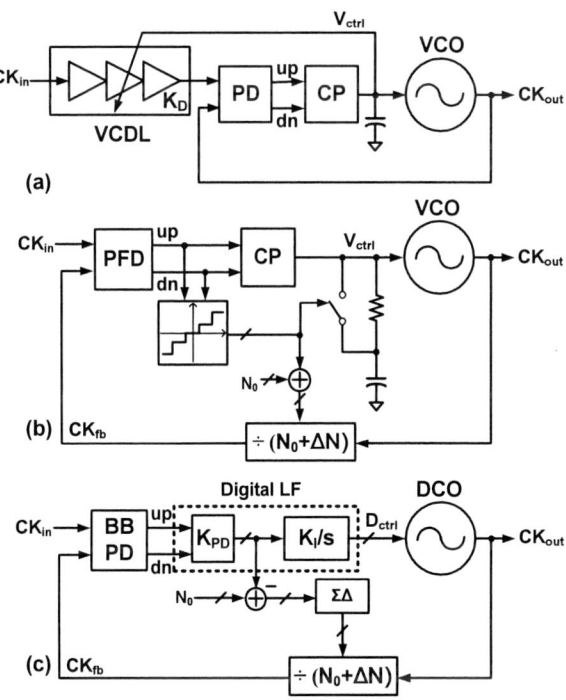

**Fig.1. The previously-published peaking-free PLLs: (a) the analog PLL with a cascaded VCDL [2], (b) the analog PLL with a programmable divider for phase error compensation [3], and (c) the digital equivalent to (b) in [4].**

978-1-4673-6145-3/13 $31.00 © 2013 IEEE          33

The paper is organized as follows. Section II presents the architecture of the proposed peaking-free digital PLL, while making comparisons with the previously-reported PLLs. Section III describes the implementation of the key circuit components, including the TDC and DCO. Finally, Section IV discusses the measurement results from the prototype chip fabricated in a 65nm CMOS and Section V concludes the paper.

## II. Peaking-Free Digital PLL Architecture

As stated in the introduction, the PI loop filter used in the conventional second-order PLLs yields a closed-loop zero, which causes peaking in the jitter transfer and overshoots in the transient response. The PI filter adjusts the control input to the oscillator according to the following equation:

$$V_{ctrl}(s) = (K_P + K_I/s) \cdot (\phi_{in} - \phi_{fb}) \tag{1}$$

where $\phi_{in}$ and $\phi_{fb}$ are the phases of the input and feedback clocks, and $K_P$ and $K_I$ are the proportional and integral gains of the loop filter, respectively. Here the control input is named $V_{ctrl}$ and expressed in s-domain assuming a voltage-controlled oscillator (VCO), but the equivalent z-domain expression can be derived for a DCO input, $D_{ctrl}$. With this PI filter, the resulting closed-loop transfer function of the PLL $H(s)$ is as follows:

$$H(s) = \frac{\phi_{out}(s)}{\phi_{in}(s)} = N \frac{K_P K_{VCO} s + K_I K_{VCO}}{s^2 + K_P K_{VCO} s + K_I K_{VCO}} \tag{2}$$

where $K_{VCO}$ is the VCO gain and $N$ is the dividing factor of the feedback clock divider. As seen in Eq. (2), $H(s)$ has a zero at $s=-K_I/K_P$. The presence of this zero makes $|H(s)|$ rise above $N$ near the cut-off frequency even when the PLL is well over-damped.

An analog PLL that realizes a closed-loop transfer function without such a zero was presented in [2]. Its block diagram of the PLL is shown in Fig. 1(a). The PLL has an additional VCDL in the input clock path, which shares the control input $V_{ctrl}$ with the VCO. Since the clock phase arriving at the phase detector (PD) input is offset by the VCDL delay, the equation governing $V_{ctrl}(s)$ becomes:

$$V_{ctrl}(s) = K_I/s \cdot (\phi_{in} - \phi_{fb} - K_D V_{ctrl}(s)) \tag{3}$$

where $K_D$ is the VCDL gain. Then, it can be shown that this PLL does not have a closed-loop zero:

$$H(s) = \frac{\phi_{out}(s)}{\phi_{in}(s)} = \frac{N \cdot K_I K_{VCO}}{s^2 + K_I K_D s + K_I K_{VCO}} \tag{4}$$

However, one difficulty associated with this design is that the

VCDL tuning range must be wide enough to provide the necessary phase compensation at all situations. A wide-range VCDL is typically sensitive to the supply/substrate noise and to the noise on $V_{ctrl}$.

On the other hand, the PLLs in [3] and [4] achieve the peaking-free jitter transfer by using a programmable divider instead of the VCDL to compensate the phase error. Fig. 1(b) and (c) depict the analog and digital versions of this PLL, respectively. For both designs, the feedback dividing factor $N$ is changed from its nominal value, $N_0$, depending on the phase detector outputs. Since the phase of the feedback clock ($CK_{fb}$) changes with the time-integral of the dividing factor's deviation from $N_0$, the PLL realizes an equivalent loop filter to Eq. (3), except that the feedback phase is adjusted instead of the input phase. However, one limitation is that the phase correction resolution can be too coarse especially when $N_0$ has a small value. It makes the programmable divider unsuitable for correcting a small phase error. To mitigate this, the design in Fig. 1(b) reverts back to a conventional second-order PLL when the phase error is small. The one in Fig. 1(c) uses a delta-sigma modulator to realize a fractional divider, but the PLL bandwidth becomes constrained to suppress the quantization noise.

The digital PLL proposed in this work realizes the filter response in Eq. (3) directly in the digital domain without requiring additional circuit blocks on the clock path. Fig. 2 illustrates the architecture of the proposed PLL, which consists largely of a TDC, a digital loop filter (LF), a DCO, and a fixed-ratio frequency divider. Basically, the phase compensation feedback is implemented inside the digital filter, by subtracting a $K_D$-scaled version of the oscillator control ($D_{ctrl}$) from the digitized phase error ($D_{err}$). In order to apply the feedback with the AC component of $D_{ctrl}$ only, a high-pass filter (HPF) with a sufficiently low cut-off frequency is inserted in the feedback path. Without this HPF, the PLL will exhibit a static phase offset that varies with the final $D_{ctrl}$ value. The added HPF does not alter the peaking-free closed-loop transfer function of the PLL as long as the cut-off frequency is sufficiently lower than $K_I K_D$ (e.g. 2-kHz in our design). Note that the finer phase correction than those in Fig. 1(b) and (c) is possible, since the resolution is now limited only by the resolution of the digital accumulator (the $K_I/s$ block in Fig. 2) and the subsequent DCO.

## III. Circuit Implementation

### A. Linearization of Bang-Bang PFD via Dithering

The proposed peaking-free PLL requires a linear TDC that can digitize the phase error information. The implemented TDC is a set of three bang-bang phase-frequency detectors (BB-PFDs), each of which measures the input clock phase compared to an independently dithered feedback clock phase. As depicted in Fig. 2, the TDC is composed of three BB-PFDs, an adder logic that aggregates their outputs, a delta-sigma modulator (DSM) that generates random dithering patterns, and phase-domain digital-to-analog converters (phase-DACs) that synthesize the BB-PFD triggering clocks with intentional clock dithers and offsets. Note that we have used the noise-

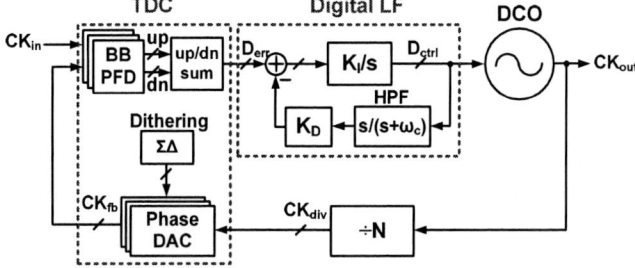

Fig.2. The architecture of the proposed digital PLL with peaking-free transfer function.

shaping property of a third-order multi-stage noise shaping (MASH) DSM to generate a pseudo-random dithering pattern [6].

The proposed TDC achieves a linear transfer characteristic that is insensitive to PVT and jitter conditions. The BB-PFDs oversampling the phase error at different offset positions recover the coarse information on the phase error, while the random dither added to each triggering clock phase recovers the fine information. The TDC linearization principle is illustrated in Fig. 3. With a sufficiently large dither, the overall TDC characteristic becomes a straight line whose slope is determined largely by the phase spacing between the BB-PFDs. Since this phase spacing is set as a fixed fraction of the clock period by the use of phase-interpolating DACs, the slope, corresponding to the effective TDC gain, will remain unchanged despite the change in the PVT or input/feedback clock jitter conditions.

Fig. 4 shows the schematics of two key circuit blocks of the TDC: BB-PFD and phase-DAC. The BB-PFD circuit is basically a linear PFD followed by two cascaded SR-latches, which provides binary information on the phase or frequency error [7]. Unlike the bang-bang phase-only detector, this BB-PFD can aid the frequency acquisition during the locking transient. On the other hand, the phase-DAC circuit in Fig. 4(b) consists of two multiplexers followed by an interpolating stage. The interpolation weights between the two selected phases (CK$_1$ and CK$_2$) are controlled by adjusting the bias currents, I$_{B1}$ and I$_{B2}$. The interpolating stage is similar to the one described in [8] except that its complementary structure can interpolate both the rising and falling edges of the input clocks.

### B. Digitally-Controlled LC Oscillator

To realize a DCO with a predictable control gain ($K_{DCO}$), a transformer-based tuning [5] is adopted instead of a varactor-based one. The circuit schematic of the LC-DCO is shown in Fig. 5. The DCO is composed of two differential oscillator cores coupled in a quadrature relationship. Each core has a transformer-capacitor tank whose effective resonant frequency ($\omega_{OSC}$) is tuned by the ratio of two currents, $I_1$ and $I_2$, each flowing through the primary and secondary coils of the transformer, respectively:

$$\omega_{osc} = \frac{\omega_0}{2Q} \cdot \frac{M}{L_1} \cdot \frac{I_2}{I_1} + \sqrt{\left(\frac{\omega_0}{2Q} \cdot \frac{M}{L_1} \cdot \frac{I_2}{I_1}\right)^2 + \omega_0^2} \quad (5)$$

where $\omega_0$ is the tank's resonant frequency without the secondary coil, $Q$ is the quality factor, and $L_1$, $L_2$ and $M$ are the self- and mutual- inductances of the transformer coils.

By controlling this current ratio $I_1/I_2$ with a current steering DAC, the DCO can realize a desired DCO gain $K_{DCO}$ insensitive to PVT variations without relying on the absolute characteristics of the circuit components. The designed LC-DCO has 10-bit resolution with 0.00637-%/step relative gain spanning a frequency range of 8.9~9.5-GHz.

## IV. MEASUREMENT RESULTS

The prototype of the described digital PLL was fabricated in a 65nm LP CMOS process. Its die photograph and the chip performance summary are shown Fig. 6. The PLL occupies the total active area of 0.55x0.68mm² and operates at a single nominal supply of 1.2V. From 143.75-MHz reference input, the PLL generates a 9.2-GHz output. The DCO tuning range is 8.9~9.5-GHz. The digital loop filter operates at a 143.75-MHz clock. The PLL consumes the total power of 63.9-mW. The DCO consumes the majority of the power of 51-mW.

The transients of the feedback clock phase in response to a step change in the input phase are shown in Fig. 7, which demonstrates the peaking-free and overshoot-less characteristics of the proposed digital PLL. The digital filter can be configured either as the proposed one or as a conventional PI-filter. Fig. 7(a) shows the response of a conventional second-order PLL with 670-kHz bandwidth and 82.5° phase margin while Fig. 7(b) shows that of the peaking-free PLL with 690-kHz bandwidth and 80° phase margin. As expected, the former exhibits an overshoot due to jitter accumulation while the latter does not. Note that the noise on the waveforms is due to the added dither to the feedback clock phase for TDC linearization. The measured settling times are 4.23-μs and 1.58-μs, respectively, showing 2.68x fast settling for the proposed peaking-free PLL.

Fig. 8 plots the measured TDC transfer curve, demonstrating its linear characteristic that is insensitive to the input clock jitter condition. This curve is measured by an on-chip built-in self-test (BIST) circuit, which measures the resulting static phase offset of the PLL when a deliberate offset is added to the TDC output code $D_{err}$. With an added dither of 912-ps$_{pp}$, the TDC has an effective gain of 266.8-ps/step, which does not change even when the input clock jitter varies from 5.4 to 14.3 and 24.6-ps$_{rms}$. With a reduced dither of 456-ps$_{pp}$, the TDC gain reduces to 57.5-ps/step, as expected.

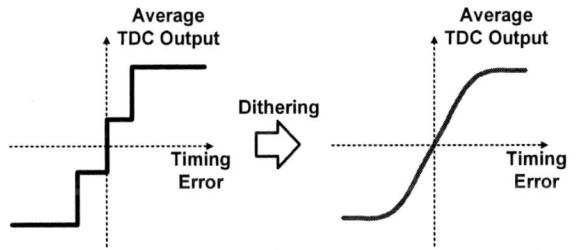

**Fig.3. Linearizeation of the TDC characteristics with the addition of random dither.**

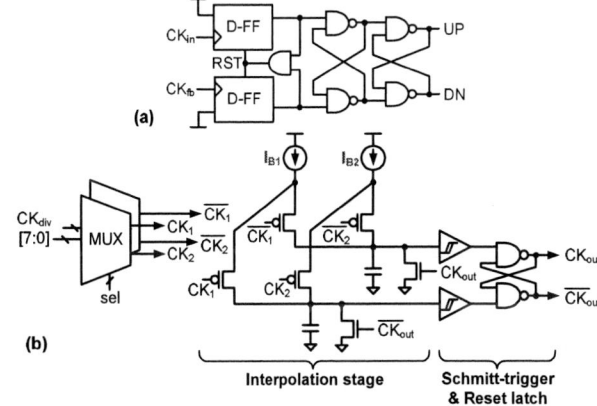

**Fig.4. (a) Bang-bang phase-frequency detector (b) phase interpolator.**

Fig. 9(a) and (b) show the measured jitter histograms of the 1.15-GHz divided-by-8 output clock and the feedback clock with the added dither, respectively. The divided output clock has the measured jitter of 3.477-ps$_{rms}$ and 24-ps$_{pp}$. Fig. 9(b) shows the 7 quantized phase positions visited by the dithering pattern. The spacing between them is 0.0208-UI or 147-ps, corresponding to the 1-LSB step of the phase-DAC. The plot is in fact measured with an added input jitter of 24.6-ps$_{rms}$. It shows that the dithering span is sufficiently large compared to the input jitter and the TDC gain can remain unchanged even with the input jitter variation.

## V. CONCLUSION

In this paper, a digital PLL with a peaking-free transfer function is presented. The necessary phase compensation feedback is realized entirely within the digital loop filter, without adding any noise-sensitive components on the input or feedback clock paths. The presented TDC achieves a linearized characteristic insensitive to the external jitter condition, and the LC-DCO design with a magnetic tuning achieves a well-controlled DCO gain set by a current ratio. The measurement results demonstrated the fast settling in the PLL's phase step without exhibiting an overshoot and the TDC gain invariant with the input jitter condition.

## ACKNOWLEDGMENTS

This research was supported by the KCC (Korea Communications Commission), Korea, under the R&D program supervised by the KCA (Korea Communications Agency) (KCA-2013- (12-911-01-102)).

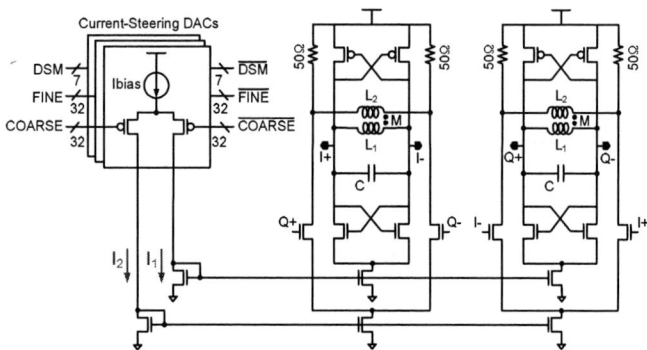

**Fig.5. Magnetically-tuned digitally-controlled LC oscillator.**

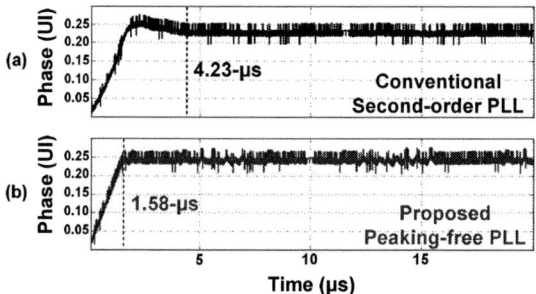

**Fig.7. The measured transient response of the PLL feedback phase to a 0.25-UI step input : (a) a conventional second-order PLL and (b) the proposed peaking-free PLL.**

## REFERENCES

[1] S. Sidiropoulos, et al., "A Semi-Digital DLL with Unlimited Phase Shift Capability and 0.08-400MHz Operating Range," in *Int'l Solid-State Circuits Conference(ISSCC) Dig. Tech. Papers*, pp. 332-333, Feb. 1997.

[2] T.–H. Lee, et al., "A 155-MHz Clock Recovery Delay- and Phase-Locked Loop," *IEEE J. Solid-State Circuits*, pp. 1736-1746, Dec. 1992.

[3] W. H. Chiu, et al., "A Dynamic Phase Error Compensation Technique for Fast-Locking Phase-Locked Loops," *IEEE J.Solid-State Circuits*, pp. 1137-1149, Sep. 2010.

[4] M. A. Ferris, et al., "A 14mW Fractional-N PLL Modulator With a Digital Phase Detector and Frequency Swithcing Scheme," *IEEE J.Solid-State Circuits*, pp. 2464-2471, Nov. 2008.

[5] Y. Tang, et al., "A 65nm CMOS Current-Controlled Oscillator with High Tuning Linearity for Wideband Polar Modulation," in *Custom Integrated Circuits Conference(CICC)*, pp. 1-4, Sep. 2012.

[6] J. Song, et al., "Spur-Free MASH Delta-Sigma Modulation," *IEEE Trans. Circuit and System-I*, pp. 2426-2437, Sep. 2010.

[7] T. Olsson, et al., "A Digitally Controlled PLL for SoC Applications," *IEEE J.Solid-State Circuits*, pp. 751-760, May. 2004.

[8] A. Agrawal, et al., "A 19Gb/s Serial Link Reciever with Both 4-Tap FFE and 5-Tap DFE Functions in 45nm SOI CMOS," in *Int'l Solid-State Circuits Conference(ISSCC) Dig. Tech. Papers*, pp. 134-135, Feb. 2012.

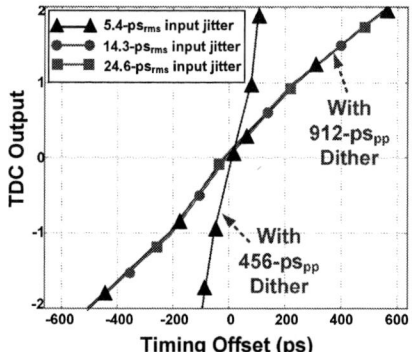

**Fig.8. The measured TDC transfer characteristic with various input jitter conditions and amount of dither.**

| Technology | 65nm 1P8M CMOS LP |
|---|---|
| Supply Voltage | 1.2 V |
| Input Frequency | 139 ~ 148.44 MHz |
| Output Frequency | 8.9 ~ 9.5 GHz |
| Bandwidth | 0.3 ~ 1.5 MHz |
| Clock Jitter(÷8) | 3.477 ps$_{rms}$ @ 1.15 GHz |
| | 24 ps$_{pp}$ |
| Phase Noise | -93 dBc/Hz @ 1MHz offset |
| | -120 dBc/Hz @ 10MHz offset |
| Settling Time | 1.58 µs @ 0.69MHz-BW |
| Power Dissipation | 63.9 mW |
| Active Area | 0.374 mm² |

**Fig.6. Die photo and performance summary.**

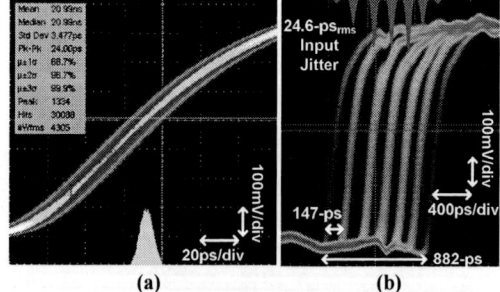

**Fig.9. Measured jitter histogram of (a) the 1.15-GHz divided by 8 output clock and (b) the 143.75-MHz dithered feedback output clock.**

978-1-4673-6145-3/13 $31.00 © 2013 IEEE

# A Sub-200 fs RMS Jitter Capacitor Multiplier Loop Filter-Based PLL in 28 nm CMOS for High-Speed Serial Communication Applications

Burak Çatlı, Ali Nazemi, Tamer Ali, Siavash Fallahi, Yang Liu, Jaehyup Kim, Mohammed Abdul−Latif,
Mahmoud Reza Ahmadi, Hassan Maarefi, Afshin Momtaz, Namik Kocaman

Broadcom Corporation, 5300 California Avenue, Irvine, CA 92617, USA, e-mail: burak@broadcom.com

*Abstract* **An 8.0 GHz to 12.2 GHz PLL with a capacitor multiplier-based active loop filter is designed in a 28 nm digital CMOS process. A passive loop filter-based version of the PLL is also implemented for comparison. While the PLL area is comparable to that of digital PLLs, the PLL performance is as good as that of an analog PLL that employs a passive loop filter. The capacitor multiplier-based active loop filter PLL has a jitter performance of 198 fs (rms), while its passive loop filter-based counterpart shows a jitter performance of 195 fs (rms). The PLL occupies 0.093 mm² and consumes 15.5 mA at 1.0V.**

## I. INTRODUCTION

The ever-increasing demand for higher data rates has expanded the scope of wireline applications, ranging from cellphones to cloud data centers. While the increasing data rates have made wireline applications more and more sophisticated to meet the tougher specifications, from customers perspective, the expectations have remained the same: compact form and low-power operation.

With its considerable area and power consumption, the PLL is one of the key building blocks that play an important role in the performance of wireline transceivers. Because of these factors, researchers have proposed using a digital PLL that has a smaller PLL loop filter area to make it more compact. Although some reported example designs have demonstrated a jitter performance of 190 fs (1 kHz to 10 MHz) [1], the operation frequencies remained around 5 GHz to 6 GHz or lower [2], and designs with higher frequencies around 10 GHz suffered from high jitter [3]. Nevertheless, other examples report the current limits of the digital PLLs, with a jitter performance as low as 295 fs in a SOI CMOS [4] or 345 fs with high power consumption [5].

In this paper, we investigate a technique that keeps analog PLLs still competitive in terms of area, while demonstrating excellent jitter performance. To reduce the area of the PLL, the loop filter was targeted, and the area covered by the integrator capacitance, $C_{int}$, (Fig.1a) was reduced significantly, based on a capacitor multiplier technique. To compare and evaluate the performance of the C-multiplier based PLL with its passive counterpart, two identical transceivers were implemented with two similar PLLs that can be distinguished from each other only by the filter type that they employ. In the following sections, we explain the concept of capacitor multiplier and show how it can be applied to a loop filter, addressing practical issues such as noise, swing, etc. Finally, we present the experimental results of the implemented PLL, taking its passive counterpart as a benchmark. To the best of

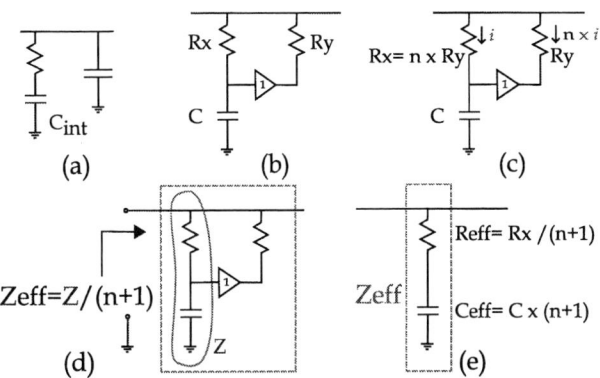

Fig. 1. Basic concept of the capacitor multiplier.

our knowledge, this study is the first comprehensive work on C-multiplier based loop filters at both the theory and implementation level with a passive counterpart comparison.

## II. THE LOOP FILTER

### A. Capacitor Multiplier

The capacitor multiplier concept used in this work was originally proposed by Larsson [6]. Fig. 1b shows how a simple series RC ($R_X$ and C) filter can be modified to obtain a capacitor multiplier. A unity gain buffer inserted between $R_X$ and an auxiliary resistor $R_Y$ can redefine the effective impedance of the filter Z. With the introduction of the buffer, the voltage drop on $R_X$ is copied on $R_Y$, and an additional current is drawn proportional to the resistance ratio $n=R_X/R_Y$ (Fig. 1c). This additional current effectively modifies the original impedance of the loop filter by $n+1$, $Z_{eff}=Z/(n+1)$ (Fig. 1d). The benefit of the technique is seen better if $Z_{eff}$ is decomposed to its equivalent R and C configuration that gives $n+1$ times less series resistance and $n+1$ times higher capacitance, which is the key feature that enables compact loop filter implementation (Fig. 1e).

### B. The Loop Filter Noise Analysis

The equivalent noise circuit for the loop filter is shown in Fig. 2. For the completeness of the analysis, shunt capacitor $C_P$ is also shown. In Fig. 2, $V_x$ and $V_y$ shows thermal noise of the resistors $R_X$ and $R_Y$, and $Vo$ shows the input-referred noise of the buffer. The noise contribution of each component to the noise at node P can be given as in the following:

978-1-4673-6145-3/13 $31.00 © 2013 IEEE          37

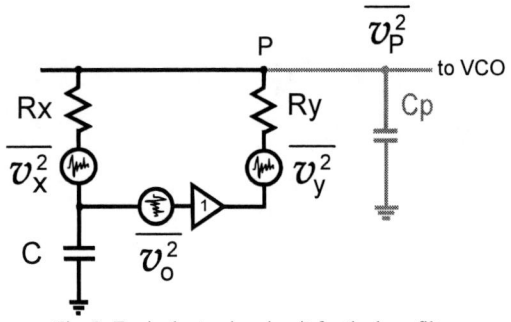

Fig. 2. Equivalent noise circuit for the loop filter.

$$v_{p,x} = \frac{v_x(-1 + sR_yC)}{s(R_yC_p + C(R_x + R_y + sR_xR_yC_p))} \quad (1)$$

$$v_{p,y} = -\frac{v_y(1 + sR_xC)}{s\left(R_yC_p + C(R_x + R_y + sR_xR_yC_p)\right)} \quad (2)$$

$$v_{p,o} = -\frac{v_o(1 + sR_xC)}{s\left(R_yC_p + C(R_x + R_y + sR_xR_yC_p)\right)} \quad (3)$$

where $v_{p,x}$, $v_{p,y}$, and $v_{p,o}$ are the noise voltages due to $R_x$, $R_y$ and the buffer. Thus, the total noise at node P can be expressed as in the following:

$$\overline{v_P^2} = \frac{v_x^2 + v_y^2 + v_o^2 + \omega^2C^2(v_x^2R_y^2 + R_x^2(v_y^2 + v_o^2))}{\omega^4R_x^2R_y^2C^2C_p^2 + \omega^2(R_yC_p + (R_x + R_y)C)^2} \quad (4)$$

The calculated loop filter noise is used in jitter estimation simulations. Table I shows a jitter estimation example for the PLLs with C-multiplier based and passive loop filters. In this example, PLL loop parameters (charge-pump current, VCO gain, etc.) are kept constant while the loop bandwidth is varied

Table I. Jitter estimation for the PLLs with C-multiplier based and passive loop filter.

| $C_{eff}$ [pF] | $R_{eff}$ [k$\Omega$] | BW [MHz] | PLL with Passive LF | | PLL with C-Multiplier LF | |
|---|---|---|---|---|---|---|
| | | | LF J-rms | PLL J-rms | LF J-rms | PLL J-rms |
| 300 | 1.8 | 2.8 | 67 fs | 188 fs | 101 fs | 203 fs |
| 300 | 2.4 | 4 | 67 fs | 185 fs | 90 fs | 194 fs |
| 300 | 3.2 | 5.6 | 68 fs | 187 fs | 82 fs | 192 fs |

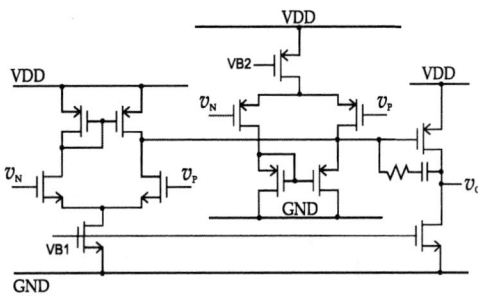

Fig. 3. Transistor level schematic of the rail-to-rail opamp.

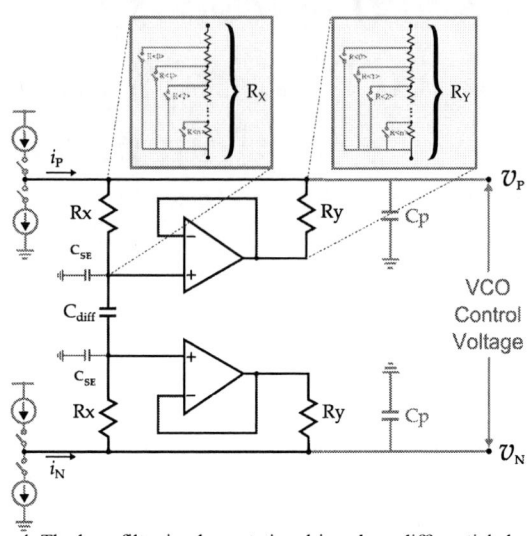

Fig. 4. The loop filter implementation driven by a differential charge pump.

by changing only the effective series resistance of the loop filters. For each loop filter configuration, the loop filter noise is calculated and added to estimation process. Table I shows the jitter generated by the filters and the total RMS jitter in each configuration. Although the total RMS jitter is a combination of many jitter components, we can still see from Table I that increasing the effective series resistance reduces the noise contribution of the loop filter to the point at which the total RMS jitters of both PLLs become equal, as it will also be shown experimentally in section IV.

C. Buffer Design

A Miller opamp based unity gain stage is designed as the buffer. Special attention is paid to minimize the opamp noise. One should note that the C-multiplier based filter should work properly in the control voltage range defined by the VCO and the charge pump circuit. Practically, this range requires a rail-to-rail buffer in this application. Thus, a complementary differential input stage is utilized as shown in Fig. 3. A constant-gm circuit is avoided to keep the area compact. The opamp is designed to achieve a GBW sufficiently larger (80 MHz, for the worst case) than the targeted PLL bandwidth (up to 10 MHz). This margin ensures that the opamp can still work as a unity gain amplifier in the PLL bandwidth range, and c-multiplier can continue to emulate the integrator capacitor.

D. Circuit Implementation

Fig. 4 shows the schematic of the loop filter driven by a differential charge pump. Because both the charge pump and the VCO are differential, two buffer stages are employed, and in addition to single-ended $C_{SE}$ capacitors, a considerable portion of the capacitor C is implemented differentially ($C_{diff}$). As shown in Fig. 4 both $R_x$ and $R_y$ are implemented programmable as different applications demand various PLL bandwidth, peaking, and noise requirements. To use the area

978-1-4673-6145-3/13 $31.00 © 2013 IEEE

Table II. Loop filter parameter set for reconfiguration, n=4 for each row.

| $R_X$ [kΩ] | $R_Y$ [kΩ] | $R_{eff}$ [kΩ] | $C_{eff}$ [pF] | BW [MHz] |
|---|---|---|---|---|
| 6 | 1.5 | 1.2 | 300 | 1.4 |
| 7 | 1.75 | 1.4 | 300 | 1.6 |
| 8 | 2 | 1.6 | 300 | 1.8 |
| 9 | 2.25 | 1.8 | 300 | 2 |
| 10 | 2.5 | 2 | 300 | 2.2 |
| 12 | 3 | 2.4 | 300 | 2.8 |
| 14 | 3.5 | 2.8 | 300 | 3.5 |
| 16 | 4 | 3.2 | 300 | 4 |
| 18 | 4.5 | 3.6 | 300 | 4.5 |

efficiently, the programmable resistor is varied by adding or removing unit resistors in series fashion. Table II shows the parameter set that can be used for the reconfiguration of the loop filter. For each row in Table II, $n=4$, and $R_{eff}$ scales with $R_X$, whereas C is always 300 pF ($[n+1]\times[2\times C_{diff}+C_{SE}]$). Various $n$ coefficients, however, can be realized by selecting $R_X$ and $R_Y$ independently. For example, if one picks the shaded resistor values in Table II, $n$ would be equal to 7, which would reconfigure the filter for the desired $R_{eff}$ (1.75 kΩ) and C (480 pF) values.

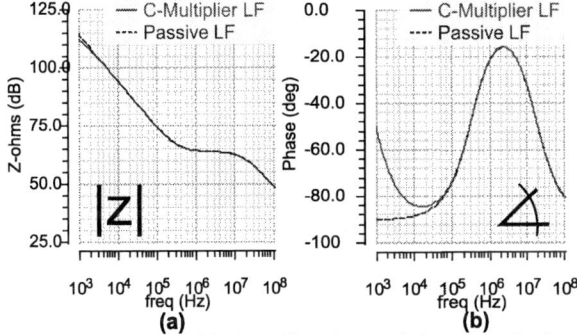

Fig. 5. Comparison of the loop filter characteristics: a) amplitude, b) phase.

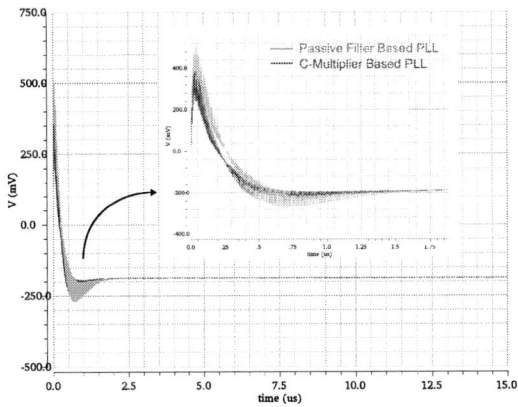

Fig. 6. VCO control voltage in locking phase for C-multiplier based and passive loop filter.

Fig. 5 shows the comparison of impedance characteristics of a C-multiplier based loop filter with a transistor level unity gain buffer and its passive counterpart for the parameter set

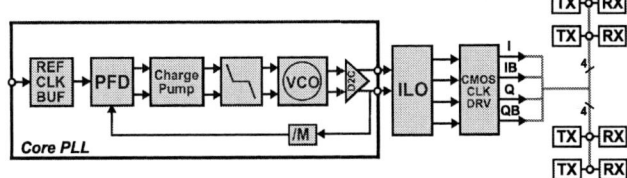

Fig. 7. Block diagram of the PLL.

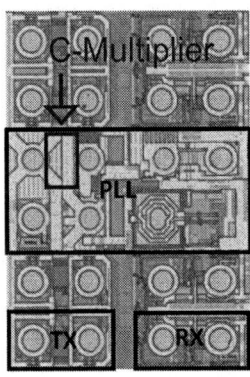

Fig. 8. Die of the PLL with TX and RX blocks of the transceiver.

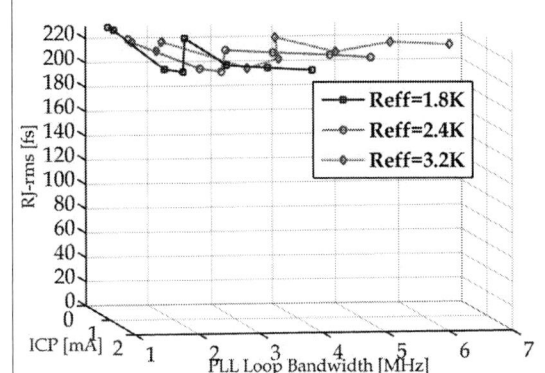

Fig. 9. Measured RMS jitter of the PLL for various $R_X$ and Charge-Pump currents.

given in the 3rd row of Table II. The C-multiplier based filter matches the expected characteristics over the PLL bandwidth and beyond.

The large signal characteristic of the loop filters should also be compared, especially for the locking phase in which the loop filter experiences large transients. This verification is especially important to guarantee proper locking behavior. Fig. 6 shows the VCO control voltage deviation in the locking phase for both the C-multiplier based and passive loop filter. Although the control voltage transients slightly deviate from each other, they still track together.

## III. PLL

The PLL consists of a phase and frequency detector, a charge pump, a loop filter, a 10 GHz LC VCO, a programmable integer divider, a 10 GHz injection locked oscillator, and 10 GHz CMOS clock buffers (Fig. 7). No CML logic is used in the entire PLL to reduce the power and the area. The VCO core is directly connected to the CMOS D2C (CML-to-CMOS converter) block for a compact

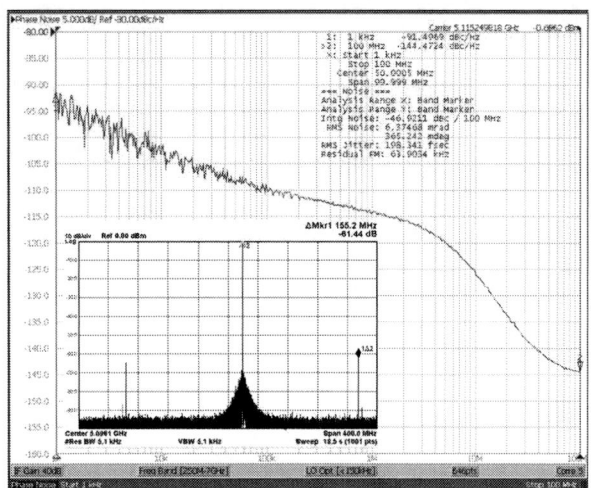

Fig. 10. Spectrum measurement at 10. 2 GHz at the output of the TX with a constant 1-0 pattern.

implementation. D2C drives both the integer-N divider block and the ILO that generates the full-rate quadrature clock. Tapered CMOS buffer stages that are employed at the output of the ILO draw 8 mA from a 1V supply at 10 GHz driving four RX-TX lanes.

## IV. MEASUREMENT RESULTS

The presented PLL was integrated in a SerDes transceiver chip in 28 nm CMOS. The clock performance was measured through the output of the designed full-rate transmitter. The presented PLL chip operates from 8 GHz to 12.2 GHz. Fig. 8 shows the die micrograph of the PLL section. The PLL consumes 15.5 mW from a 1V supply and occupies an area of 0.093 mm$^2$.

To see the jitter performance consistency under different loop bandwidth values, the PLL was measured around 10 GHz for several charge pump currents and all effective resistance values shown in Table II. Fig. 9 shows three sets of jitter measurements for different effective resistance values. In these measurements, charge pump current was varied from 0.1 mA to 1.5 mA for each measurement set, and a loop bandwidth range from 1.3 MHz to 6.5 MHz was measured. The peaking changed from 0.2 dB to 1 dB, while higher peaking was observed for lower loop bandwidths. Overall, RMS jitter changed between 198 fs and 230 fs, showing a stable jitter performance over a wide BW range. Fig. 10 shows corresponding spectrum measurements at 10.2 GHz with 198 fs jitter performance. The measured reference spur is

-61 dBc. The PLL performance was also measured at 12 GHz. The jitter increases slightly at 12 GHz to 207 fs, which can be attributed to the relatively higher phase noise of the VCO at this frequency. The measured reference spur is 58 dB below the carrier frequency. The PLL employs exactly the same PLL subblocks but the passive loop filter was also measured. A 195 fs RMS jitter performance was obtained for the passive version, which confirms the claim in Section II-B.

Table III compares the performance of the PLL with the state-of-the-art examples from the literature around 10 GHz. As can be seen, the designed PLL occupies an area as small as that of a digital PLL [5], consumes less power, and has the best jitter performance.

## V. CONCLUSIONS

A compact PLL with a capacitor multiplier-based loop filter is implemented in a wireline transceiver. The performance of the PLL exceeds or is comparable to that of recently reported digital PLLs or PLLs employed in wireline transceivers. Moreover, we present a comprehensive picture of capacitor multiplier based loop filters in PLLs, from theory to experimental results, and provide a description of its implementation and a comparison of its performance with its all-passive counterpart.

## REFERENCES

[1] C. W. Yao, et al., "A low spur fractional-N digital PLL for 802.11a/b/g/n/ac with 0.19 ps rms jitter," in Symp. VLSI Circuits (VLSIC) Dig., Jun. 2011, pp. 110–111.
[2] D. Park, et al., "A 14.2 mW 2.55-to-3 GHz Cascaded PLL With Reference Injection and 800 MHz Delta-Sigma Modulator in 0.13 um CMOS," in IEEE J. Solid-State Circuits, vol.47, no.12, pp.2989-2998, Dec. 2012
[3] S. Yang and W. Chen, "A 7.1mW 10GHz All-Digital Frequency Synthesizer with Dynamically Reconfigurable Digital Loop Filter in 90nm CMOS," ISSCC Dig. Tech.Papers, pp. 90-91, Feb. 2009.
[4] A. Goel, et al., "A compact 6 GHz to 12 GHz digital PLL with coupled dual-LC tank DCO," inVLSI Symp. Dig. Tech. Papers, Jun. 2010, pp.141–142
[5] A. Rylyakov, et al., "Bang-Bang Digital PLLs at 11 and 20GHz with sub-200fs Integrated Jitter for High-Speed Serial Communication Applications," ISSCC Dig.Tech. Papers, pp. 94-95, 2009.
[6] P. Larsson, "An offset-cancelled CMOS clock-recovery/demux with a half-rate linear phase detector for 2.5 Gbp/s optical communication," in IEEE Int. Solid-State Circuits Conf. Dig. Tech. Papers, Feb. 2001, pp.74–75.
[7] N. Kocaman, et al., "11.3Gb/s CMOS SONET-compliant transceiver for both RZ and NRZ applications," ISSCC Dig. Tech.Papers, pp.142-144, Feb. 2011
[8] J. Savoj et al.,"Design of high-speed wireline transceivers for backplane communications in 28nm CMOS," IEEE Custom Integrated Circuits Conference (CICC), 2012 IEEE , pp.1,4, 9-12 Sept. 2012
[9] G. Ono, et al., "A 10:4 MUX and 4:10 DEMUX gearbox LSI for 100-Gigabit Ethernet link," IEEE J. Solid-State Circuits, vol. 46, no. 12, pp. 3101–3112, Dec. 2011.

Table III. Performance summary of the PLL and state-of-the-art example PLLs operating around 10 GHz.

| | This Work | Yang [3] | Goel [4] | Rylyakov [5] | Kocaman [7] | Savoj [8] | Ono [9] |
|---|---|---|---|---|---|---|---|
| Jitter-RMS (fs) | **198/207** | 900[1] | 295 | 345 | 250[3] | 399[3] | 429[3] |
| Jitter Integration Range (Hz) | **1k–100M** | N/A | 7M–5.82G | 1k–10G | 50k–80M | N/A | 10k–100M |
| Output Carrier Frequency (GHz) | **10.2/12.0** | 9.92 | 11.65 | 11 | 11.3 | 13.1 | 12.89 |
| Tuning Range (GHz) | **8–12.2** | 9.75–10.17 | 7.89–11.64[2] | 8.1–11.8 | 8.5–11.3 | 8–13.1 | N/A |
| Reference (MHz) | **156.25** | 40 | N/A | 275 | 706.25 | N/A | 625 |
| Oscillator | **LC VCO** | LC DCO | LC DCO | LC DCO | LC VCO | LC VCO | LC VCO |
| Core Power Dissipation (mW) | **15.5/16.5** | 7.1 | 20.8 | 31 | N/A | N/A | N/A |
| Core Area (mm$^2$) | **0.093** | 0.352 | 0.111 | 0.088 | N/A | N/A | N/A |
| PLL Type | **Analog** | Digital | Digital | Digital | Analog | Analog | Analog |
| Technology (nm) | **28-CMOS** | 90-CMOS | 45-SOI | 65-CMOS | 65-CMOS | 28-CMOS | 65-CMOS |

[1]Includes 100 fs trigger jitter of instrument.  [2]Range is slightly higher for Push Mode.  [3] Jitter measured at TX output.

978-1-4673-6145-3/13 $31.00 © 2013 IEEE

# Nonlinearity Cancellation in Digital PLLs

## (Invited paper)

### Salvatore Levantino and Carlo Samori

### Politecnico di Milano, Italy

*Abstract*-**One decade after their introduction into wireless applications, digital fractional-N phase-locked loops are becoming a competitive solution for products. Their ultimate level of spurs is often bounded by the resolution and the linearity of the time-to-digital converter. Although methods for mitigating its nonlinearity have been proven effective in lowering spurs, they typically increase the level of random noise. By contrast, digital-PLL architectures based on digital-to-time converters enable nonlinearity cancellation and spur reduction with no penalty on noise level, while reducing design complexity and power consumption.**

## I. INTRODUCTION

The phase-locked loop (PLL) employed as local oscillator in wireless front-ends is a typical example of integrated circuit whose analog performance degrades with technology scaling. The power consumption of some of its building blocks, especially the charge pump, does not scale down and neither does the area devoted to the loop filter (if on chip), which may dominate the overall area occupation of the PLL. In analog charge-pump PLLs, the typical digiphase scheme, which is essential to suppress fractional spurs or $\Delta\Sigma$ quantization noise, requires analog correlators and digital-to-analog converters, which have limited accuracy of cancellation and high power consumption. By contrast, digital PLLs are more amenable to scaling: (i) They eliminate the charge pump and its power consumption, (ii) their filter being digital scales down with new nodes and it is reconfigurable, (iii) digiphase and other calibration algorithms are naturally implemented with less design complexity, better accuracy and limited area occupation and power consumption.

The two most used architectures of digital PLL are schematically shown in Fig. 1, in both cases the digital filter output drives a digitally-controlled oscillator (DCO): (a) The all-digital PLL (ADPLL) also referred to as phase-domain or divider-less PLL and (b) the digital $\Delta\Sigma$ fractional-N PLL. The ADPLL was first adopted in wireless applications in 2004 [1]. The second topology proposed in [2] is derived from the $\Delta\Sigma$ fractional-N charge-pump PLL [3], in which a time-to-digital converter (TDC) replaces the conventional phase-frequency detector (PFD) and the charge pump. As in analog fractional-N PLLs, the $\Delta\Sigma$ modulator quantizes the frequency control word (FCW) into an integer part, controlling the modulus of the integer-N divider. In the scheme in Fig. 1(b), a first-order digital $\Delta\Sigma$ modulator drives the divider modulus control. It contains a quantizer $Q$, which truncates the least significant bits (LSBs) of its input word. The quantization added by the $\Delta\Sigma$ produces noise/spurs and it is conventionally cancelled out at the TDC output by the digiphase scheme. The correlator $C$ provides an estimation of the gain of the divider/TDC cascade to perform accurate cancellation over PVT variations and realizes a least-mean square (LMS) algorithm.

Even though they may seem different at first glance, the two circuits in Fig. 1 are equivalent from the point of view of spectral purity. In both architectures, the levels of in-band noise and spurious tones are mainly set by the resolution and linearity of the TDC. Several techniques have been proposed both to refine TDC resolution and improve its linearity. Despite good results have been achieved, most of these solutions are very complex and power consuming, and often the reduction of the spur level is often paid with a growth of random phase noise.

By contrast, a recently proposed architecture of digital PLL follows a completely different path. It adopts a feedback divider based on a digital-to-time converter (DTC), which enables to reduce the number of TDC bits down to the limit case of a single-bit TDC. The advantage of the single-bit approach is twofold: (i) The resolution and linearity of the TDC are no more an issue, being the TDC itself a simple time-threshold. (ii) The fine resolution of the DTC can be achieved with less power consumption with respect to a TDC and, above all, the effects of DTC nonlinearity can be easily corrected in the digital domain by means of an automatic predistorsion algorithm and without adding dithering noise.

## II. PHASE-DOMAIN VS. $\Delta\Sigma$ FRACTIONAL-N PLL

The main metric of a PLL for wireless applications is spectral purity, i.e. its phase noise and spurious tones performance. While the dominant term of out-of-band noise comes from the digitally controlled oscillator (DCO) and is ultimately traded against its power dissipation, in-band noise and spurs (both in-band and out-of band tones) are mainly set by the TDC block. The latter is essentially an analog-to-digital converter (ADC) and the impact of its performance on PLL output spectrum can be synthesized as follows:

i)  The finite resolution of the TDC sets the ultimate limit for the level of in-band phase noise and it is responsible for the generation of fractional spurious tones.
ii)  The non-linearity of the TDC characteristic increases the level of fractional spurs and may cause spectrum folding of quantization noise.

From these standpoints, the two architectures are equivalent, as we will briefly recall by using intuitive arguments.

### A. Spurs due to TDC finite resolution

In analogy to any ADC, the finite time resolution, $\Delta t$, of the TDC is often taken into account by adding a uniformly distributed quantization noise with variance $\Delta t^2/12$ at its input. This noise is typically assumed to be white over the Nyquist

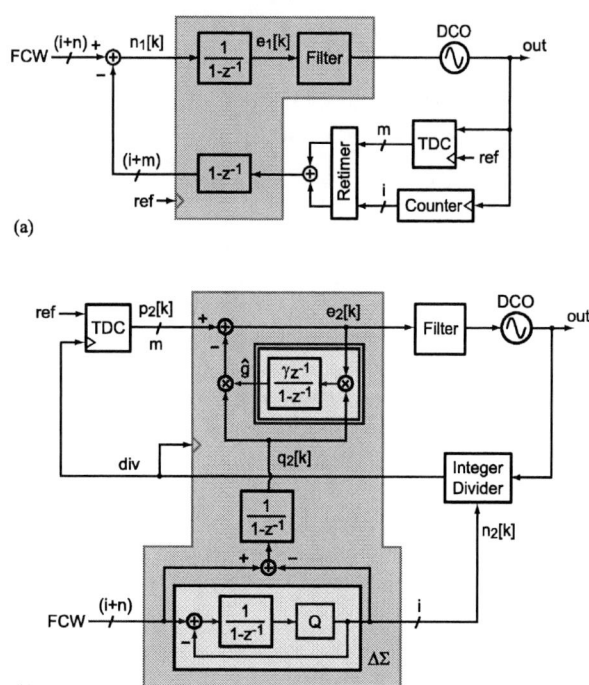

(a)

(b)

Fig. 1. Common architectures of digital PLL: (a) Phase-domain ADPLL, (b) $\Delta\Sigma$ Fractional-N PLL.

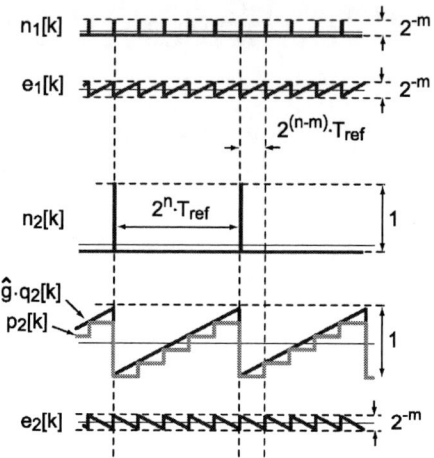

Fig. 2. Typical signal waveforms of the digital PLLs in Fig. 1 (Signals $n_1[k]$, $e_1[k]$, $p_2[k]$, $e_2[k]$ are to be intended as equivalent signals added to the respective closed-loop variables).

bandwidth, $f_{ref}$, being $f_{ref}$ the reference frequency. In both the architectures of digital PLL in Fig. 1, this noise is low-pass filtered by the loop and its spectrum is amplified by a factor of $FCW^2$, FCW being the frequency control word. The resulting in-band phase noise level is given by $(\pi^2/6f_{ref})(\Delta t/T_{dco})^2$, being $T_{dco} = 1/f_{dco}$ the DCO output period. Thus, a 1-GHz digital PLL employing 40-MHz reference and 10-bit TDC (i.e. about 1-ps resolution assuming a full-scale range of $T_{dco}$) should show a level of in-band phase noise below -134 dB/Hz. This figure is never achieved in practical circuits, not only because either thermal noise from the PLL building blocks or supply disturbances dominates over quantization noise, but also because the hypothesis of white quantization error is not necessarily true. In practice, quantization error may manifest itself also as concentrated tones, i.e. fractional spurs, when particular fractional channels near integer boundaries are synthesized. The level of these spurs depends on TDC time resolution.

Let us consider the phase-domain ADPLL and denote as $N$ the integer part of $FCW = f_{dco}/f_{ref}$. As the number of DCO periods within $T_{ref}$ is not integer, a time interval equal to a fraction of $T_{dco}$, $(FCW-N)\cdot T_{dco}$, must be measured by the TDC. Therefore, the TDC must accommodate an input range of $T_{dco}$, with some additional margin. This range, together with the time resolution determines the number, $m$, of the TDC bits. On the other hand, the number $n$ of fractional bits of FCW is set by the desired frequency resolution, and it is usually much larger than $m$. Hence, when a channel with resolution finer than that allowed by the TDC must be synthesized, the output frequency can converge to the wanted one (FCW times $f_{ref}$) only on average. The signals in Fig. 2 are obtained when FCW has only its LSB equal to 1 among its last $(n-m)$ bits, i.e. the

output frequency is close to an integer channel. Starting from synchronous rising edges of the reference and DCO signals, their time shift increases to $2^{-n}T_{dco}$ after $T_{ref}$, a value that the TDC cannot resolve. This time error piles up in the following cycles and it reaches one LSB of the TDC only after $2^{(n-m)}$ reference cycles. If we open the loop, the resulting error signal $n_1[k]$ is not zero, but it is a sequence of pulses of duration $T_{ref}$ (although its average must be zero, because of the cascaded accumulator). The sequence $n_1[k]$ at the input of the accumulator produces the sawtooth signal $e_1[k]$ at the filter input, whose output modulates the DCO tuning and it is responsible for fractional spurs at a frequency offset $f_{ref}/2^{(n-m)}$ from the carrier. If we close the loop, the same sequence $e_1[k]$ must be equivalently added to the variable $e_1$ and transferred to the output of the loop.

Moving to the $\Delta\Sigma$ fractional-N PLL in Fig. 2, let us assume that a first-order $\Delta\Sigma$ modulator is employed and that the resolutions of FCW and TDC are the same as in the previous example. The $\Delta\Sigma$ output $n_2[k]$ increments the division factor by one (corresponding to a time increment of $T_{dco}$) during one reference period every $2^n T_{ref}$ seconds, producing a phase error ramp much larger than in the ADPLL case. This ramp is however proportional to the accumulated $\Delta\Sigma$ quantization error $q_2[k]$, thus it can be cancelled out at TDC output. To obtain perfect cancellation, $q_2[k]$ has to be multiplied by the exact TDC gain which is estimated by an LMS algorithm for noise cancelling. However, also in this case the TDC time resolution ($2^{-m}T_{dco}$) limits the accuracy of cancellation. In fact, the number of TDC bits is much less than the number of bits of the signal to be subtracted from TDC output and this gives rise to the staircase $p_2[k]$ (when the loop is open). The time diagram of TDC output after correction $e_2[k]$ is identical to the signal $-e_1[k]$ obtained for the ADPLL topology and hence, it will produce an identical fractional spur with same amplitude and frequency offset from the carrier. It is possible to verify that the situation does not substantially change by employing higher-order $\Delta\Sigma$ modulators, since the limited TDC resolution still produces the same effect.

In both cases, the residual fractional spur cannot be filtered, as the loop-filter bandwidth cannot be reduced below the frequency resolution of the PLL. Under this hypothesis, the equivalent sequences $e_1[k]$ and $e_2[k]$ are transferred unaltered to the output of each loop, respectively. These error signals represent a sawtooth waveform of the time error, having peak-to-peak amplitude equal to TDC resolution, $\Delta t$. The amplitude of the first harmonic of this sawtooth signal is $\Delta t/\pi$. Hence, the doubled-sided phase-noise spectrum at the output of the PLL contains a spur of power $(\Delta t/T_{dco})^2$ at offset frequency $f_{ref}/2^{(n-m)}$ from the carrier.

According to this expression, a 4-bit TDC with $\Delta t = 2^{-4} \cdot T_{dco}$ would cause an in-band spur of about -24dBc, while a 10-bit TDC would produce a spur lower than about -60dBc. These estimated values match well simulation results simulations, as long as the level of random noise associated to the TDC input is much lower than the power of the spurs. Instead, when random noise power becomes comparable to spur power, the level of fractional spurs is lower than the expected figure. This phenomenon can be intuitively explained by thinking that random noise dithers the sawtooth signals and spreads their power over the whole spectrum.

In some cases, the dynamic range required to the TDC is even larger than $T_{dco}$ and additional bits are needed. If the $\Delta\Sigma$ modulator dithering the divider modulus in the scheme in Fig. 1(b) has an order larger than one, the time shift produced at the divider output would span a range larger than $T_{dco}$ and the TDC would be asked to have a wider dynamic range to convert it. For instance, a MASH 1-1-1 modulator would produce a peak-to-peak time shift of $7T_{dco}$ and, consequently, the TDC would be required to have 3 additional bits. Besides, one or two more bits may be needed to cover PVT spreads of the TDC. This wide relative range aggravates further the issue of TDC nonlinearity, which we will discuss in the following subsection.

### B. Spurs due to TDC nonlinearity

The adverse impact of nonlinearity on the generation of spurs in ADPLLs was originally discussed in [4]. The effects of nonlinearity are identical in both the digital-PLL topologies in Fig. 1. This is not surprising, given that the transfer function from the output of the TDC to the phase of the signal at PLL output is identical in both architectures. As in any ADC, the nonlinear conversion characteristic produces distortion tones that ultimately appear as spurs in the output spectrum. If we refer to the digital PLL in Fig. 1(b) and we assume a nonlinear characteristic of the TDC, the shape of the signal at TDC output departs from the ideal staircase $p_2[k]$. Conversely, the signal subtracted by the digiphase scheme is still a ramp. An example of those waveforms is shown in Fig. 3, assuming a TDC with sinusoidal nonlinearity and infinite resolution. As it is evident, the residual error $e[k]$ would not be zero even if the TDC had infinite number of bits. Thus, nonlinearity produces inaccurate cancellation of quantization noise and fractional spurs at the output.

In general, the evaluation of the level of the fractional spurs due to TDC nonlinearity is not straightforward. In the case of sinusoidal nonlinearity described in the example in

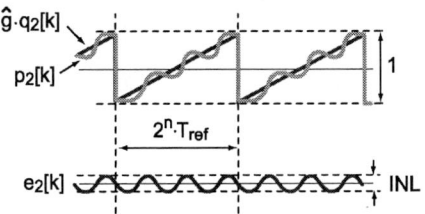

Fig. 3. Signal waveforms of digital PLL in Fig. 1(b), assuming a nonlinear TDC characteristic (Signals $p_2[k]$, $e_2[k]$ are to be intended as equivalent signals added to the respective closed-loop variables).

Fig. 3, the spur level equivalently referred to the TDC input has amplitude equal to $\Delta t_{INL}/2$, $\Delta t_{INL}$ being the integral nonlinearity (INL) of the TDC. Thus, the double-sided phase spectrum of the PLL is given by $(\pi^2/4)(\Delta t_{INL}/T_{dco})^2$. This expression, although obtained in a very simplified case, suggests that the level of INL must be kept below one LSB (i.e. $\Delta t_{INL}$ slightly lower than $\Delta t$) to guarantee a level of spurs due to nonlinearity comparable to the level of spurs due to finite resolution. Simulations and measurements confirm at a good extent this result even in practical cases.

For these reasons, it is not surprising that several solutions have been proposed in recent years to mitigate TDC nonlinearity. For instance, dynamic element matching (DEM) of TDC elements can trade linearity against noise. In practice, if the delay elements of the TDC are shuffled, the effect of their mismatches is no longer periodic and the energy of fractional spurs is spread across the spectrum [5]. Same result is achieved by means of the sliding scale technique, which consists in adding a dithering sequence to the input of the TDC and subtracting it at its output [4]. A more effective solution consist in shaping in frequency the effect of mismatches, so that their energy is concentrated at high frequency and it is filtered out by the loop filter. This behavior is obtained for instance in the TDC based on the gated-ring-oscillator (GRO), proposed in [6]. These solutions are effective to a certain extent to reduce the level of spurs due to nonlinearity. However, they are power consuming, substantially complicate design and, above all, reduce spur level at the expenses of increased noise, as it usually occurs with dithering techniques. A remarkable further step in this direction is a recent work [7], which discusses and mitigates the issues related to GRO design.

### III. DTC-BASED PLLs

As discussed so far, better performance in digital PLLs can be traded against increased design complexity, power consumption and area occupation of the TDC converter. An alternative path is to operate at architectural level. A step in this direction was done in [5]. In that work, a digital-to-time converter (DTC) in the feedback path enables a "true" fractional-N divider. The key advantages of this solution are the following ones: (i) the input range required to the TDC is narrowed, thus relaxing TDC design; (ii) the nonlinearity of the DTC can be easily corrected in digital domain without noise penalty, in contrast to TDC linearization techniques. Let

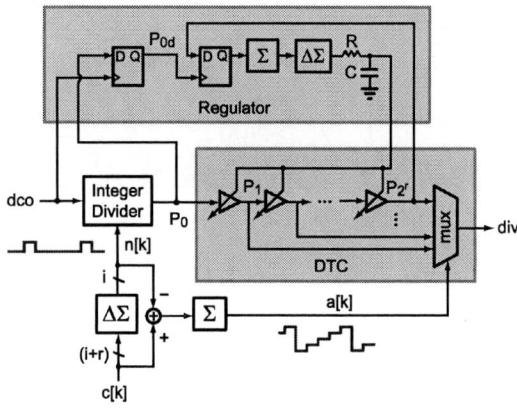

Fig. 4. Example of fractional-N divider.

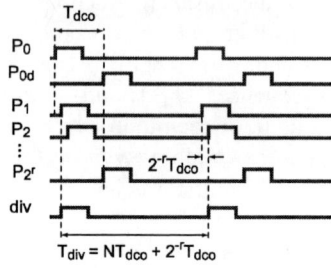

Fig. 5. Typical signal waveforms of the fractional-N divider in Fig. 4.

us analyze in details this architecture, starting from the realization of the fractional-N divider.

### A. DTC-based fractional-N divider

An example of realization of a fractional-N divider is shown in Fig. 4. It is the cascade of an integer-N divider and a regulated DTC (or phase interpolator). The DTC is a delay line with variable delay stages and a multiplexer selecting one of the $2^r$ phases generated. The dynamic range of the DTC is automatically adapted to be exactly equal to DCO period $T_{dco}$ by regulating the delay of the stages. Doing so, assuming identical delay stages, the phases will be offset in steps of $2^{-r}T_{dco}$. In practice, this scheme realizes a digitally-controlled delay-locked loop (DLL) [5]. Fig. 5 reports the relationship between the generated phases $P_i$. If the input $c[k]$ of the fractional-N divider in Fig. 4 is equal to $(N+2^{-r})$, being $N$ an integer number and $r$ the DTC number of bits, the output of the first-order $\Delta\Sigma$ is $N$ for $(2^r-1)$ clock cycles and it is $(N+1)$ for just one clock cycle (when $\Delta\Sigma$ overflows). The sequence $a[k]$ obtained by integration of $\Delta\Sigma$ error is used to control the DTC, i.e. to choose which one among the $2^r$ phases must be routed to the output. For the value of $c[k]$ chosen in this example, the DTC will shift the output of the integer-$N$ divider by $2^{-r}T_{dco}$ at each reference clock cycle. As it is evident from the signal waveforms in Fig. 5, the division factor is exactly equal to $(N+2^{-r})$. Hence, no quantization error is produced when the input control word $c[k]$ is within DTC resolution.

### B. Digital PLL with DTC-based fractional-N Divider

The digital PLL embedding this DTC-based fractional-N divider is shown in Fig. 6. The frequency resolution commonly needed by wireless standards (e.g. below 100 Hz for a 1-GHz signal) cannot be achieved by increasing the number of DTC bits, for practical implementation reasons. Thus, in analogy with a classical fractional-N synthesizer, a second $\Delta\Sigma$ modulator (with at least $n = 24$ fractional bits) is used to dither the input of the fractional-N divider. It quantizes FCW into $r$ fractional bits and the quantization error produced is cancelled out at TDC output by means of the usual digiphase scheme. Instead, the remaining $r$ fractional bits are cancelled out by means of the DTC, as described in the previous sub-section. Thus, in contrast to the classical

fractional-N synthesizer, the quantization error at TDC input is reduced by a factor of $2^r$ (i.e. peak amplitude is $2^{-r}T_{dco}$ instead of $T_{dco}$). Although the time resolution requirement of the TDC is unchanged, the narrower range required relaxes significantly its nonlinearity requirement. Moreover, a lower number of TDC bits enable a flash-type converter with a small number of elements and allows a practical and easy implementation of the above-discussed element-randomization technique to increase further its linearity.

### C. DTC nonlinearity correction

Apparently, this PLL architecture relaxes TDC linearity requirements only to shift this issue to the DTC block, which may represent an additional source of unwanted spurs. In fact, mismatches among the delay elements in Fig. 4 produce a non-uniformly-distributed delay among phases $P_i$ in Fig. 5, although the total delay of the line is forced to be $T_{dco}$. A similar source of inaccuracy is the delay mismatch of multiplexer gates of the DTC. If we assume for instance a delay error $t_i$ associated to the $i$-th phase $P_i$, the divider output would periodically contain this shift, every time $P_i$ is selected. This periodic modulation would appear at TDC output and produce spurs [with fundamental component at $f_{ref}/2^r$ from the carrier, if the division factor is set to $(N+2^{-r})$].

In contrast to TDC nonlinearity issue, the nonlinearity in the DTC can be completely cancelled out without requiring any dithering technique. The nonlinearity correction block in Fig. 6 operates in the background of normal operation and it estimates the errors associated to each phase $P_i$ and cancels it at TDC output. Instead of correcting the analog position of signal edges, the error associated to $P_i$ is cancelled in digital domain. As demonstrated in [8], the correction block is a multipath version of the adaptive filter employed to cancel quantization noise in the scheme in Fig. 1(b). If we neglect, for a moment, the feedback from the output $d[k]$, we can recognize $2^r$ correlators operating in parallel. The error associated to each phase is detected by the TDC and, by means of the $a[k]$ signal designating the DTC phase selected, it is routed through the $i$-th path of the adaptive filter. As in a standard LMS correlator, the $i$-th accumulator closed in feedback will converge to the $i$-th mismatch error of the delay line cancelling it. Note that only one correlator is active at a time. TDC nonlinearity is clearly still an issue, since it directly impacts on this cancellation, but the required INL must now be achieved along a much narrower range.

In the absence of the feedback branch from $d[k]$, the nonlinearity correction block would introduce a zero at dc in

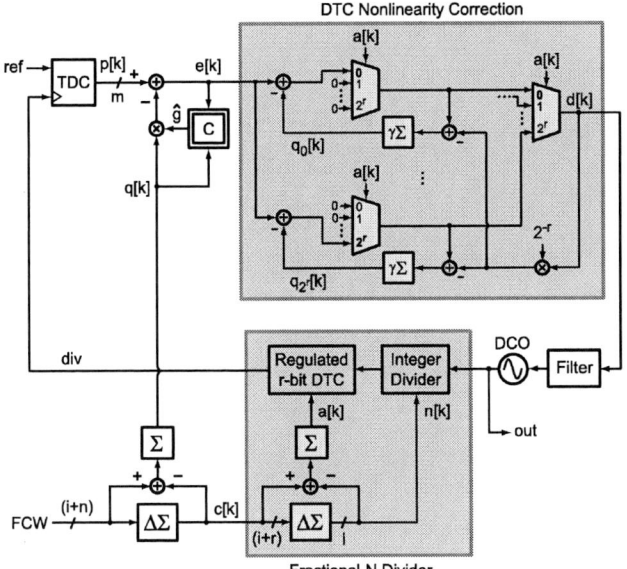

Fig. 6. Digital-time-converter-based digital PLL and background nonlinearity correction.

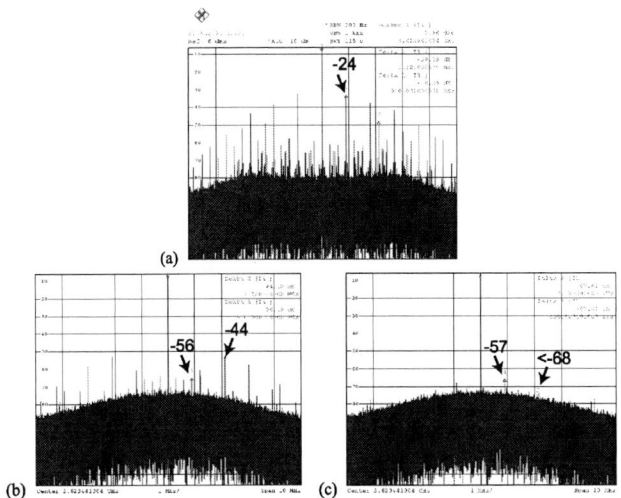

Fig. 7. Measured spectra of the DTC-based PLL in Fig. 6.

the PLL open-loop gain. In other words, a constant value for $e[k]$ would be not transferred to $d[k]$, because of the presence of the accumulators in feedback. This zero would make the PLL a type-I loop. This issue is solved by subtracting $d[k]$, multiplied by $2^{-r}$, at each accumulator input. In this way, in the case of a constant $e[k]$, the input of each accumulators has zero average value and no cancellation of $e[k]$ occurs. Adopting this solution, the final scheme of the linearizer block is multiplier-free, thus it allows low area and low power consumption despite using a data-path of 16 bits.

It is interesting to verify the quantitative impact of this alternative architecture of digital PLL on the output spectrum. A PLL adopting the scheme in Fig. 6 and operating around 3.6GHz was fabricated in a 65-nm CMOS process [5]. The PLL employs a 4-bit DTC, which reduces the peak-to-peak amplitude of the quantization error introduced in the fractional frequency divider to $T_{dco}/16$ (i.e., about 18ps), and a 4-bit TDC with range of about 45ps, which can handle with good margin and linearity this time error. Thanks to the reduced number of bits, the TDC embeds a simple shuffling technique of the delay elements. The measured output spectra for a critical near-integer fractional channel are shown in Fig. 7. When no algorithm is enabled, the signal waveforms are similar to those drawn in Fig. 2 before digiphase cancellation. In this case, thanks to the 4-bit DTC, the sawtooth time error has peak-to-peak amplitude equal to $T_{dco}/16$. According to the theoretical discussion in Sect. II-A, the expected power level of the fractional spur is $(1/16)^2$ (i.e., -24dBc), which matches the dominant measured spur in the spectrum in Fig. 7(a). When digiphase and TDC element scrambling are enabled, the primary spur drops to -56dBc and its harmonics to lower values, at the price of an increased level of in-band noise [See Fig. 7(b)]. However, a secondary spur still remains at -44dBc, which can be ascribed to DTC nonlinearity. This residual spur reduces below -68dBc without additional noise penalty, when the DTC nonlinearity correction is enabled [See Fig. 7(c)].

## IV. DTC-BASED PLL WITH SINGLE-BIT TDC

As we discussed so far the impact of TDC finite resolution and nonlinearity can be strongly attenuated at the price of increased noise level. Beside that, the TDC is a power hungry block and its dissipation for a given INL being about an exponential function of the number of bits. For these reasons, it seems very advantageous to further push the DTC-based architecture in Fig. 6 to its ultimate limit, i.e. to reduce the number of TDC bits to one and to increase DTC resolution. However, to make a fractional-N synthesizer based on a single-bit TDC feasible, three main issues must be clarified: (i) under what conditions the single-bit converter behaves as a linear block and produces no spurs, (ii) how to make a regulated DTC with large number of bits practically feasible at low power and (iii) how the DTC-nonlinearity-correction concept, discussed above, can be adapted to this case. Let us discuss those issues one by one.

### A. Linear behavior of single-bit TDC

Single-bit TDCs are often used in integer-N PLLs for clock and data recovery (CDR) circuits. In that context, they are often referred to as bang-bang (or lead-lag) phase detector. It is well known that the hard nonlinearity introduced in the loop by such an element gives rise to limit cycles that severely degrade output spectrum, both increasing spur content and/or producing high peaking. This is usually not an issue in PLLs for CDR applications; instead, it is unacceptable for wireless frequency synthesizers, where phase noise and spurs specifications are tighter and fractional-N synthesis is desired. A solution to the problem of limit cycles comes from an important property that has been widely discussed, also in recent literature [9], namely, under what condition a bang-bang phase detector can be modeled as a constant gain. In synthesis, when random jitter at TDC input, caused by flicker and thermal sources in the loop, is comparable or it even dominates over the deterministic component of time error, the single-bit TDC behaves as a linear block and the equivalent gain is inversely proportional to the random jitter at TDC input [9]. In other words, the linear approximation holds when

978-1-4673-6145-3/13 $31.00 © 2013 IEEE

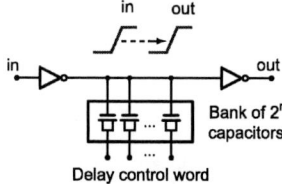

Fig. 8. Example of single-cell DTC implementation.

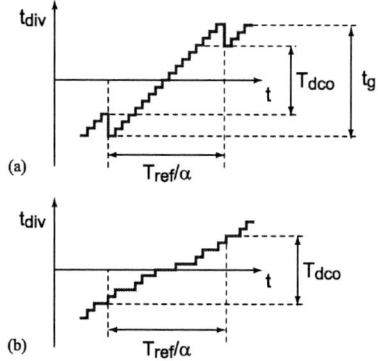

Fig. 10. Delay introduced by the single-cell DTC before and after gain adaptation.

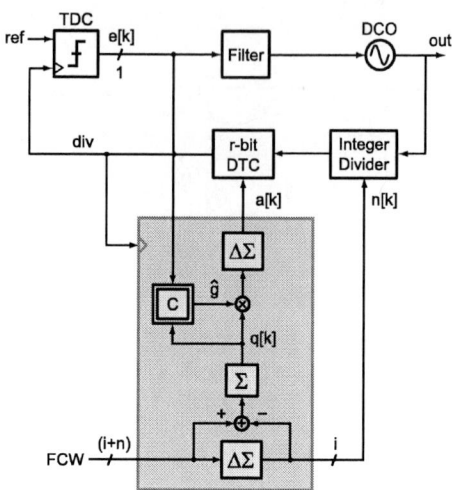

Fig. 9. Digital-time-converter-based PLL with single-bit TDC.

random noise, mainly DCO phase noise, dithers TDC input and scrambles limit cycles. We can call this condition *random-noise regime*, as opposed to the previous *limit-cycle regime*. In random noise regime, a spectrum with very good spur/noise performance can be achieved and it is well predicted by PLL linear model [10]. This property can be exploited in a relatively direct fashion in *integer-N* PLLs, but things drastically change in *fractional-N* PLLs. When in a fractional-N divider the modulus is incremented by one, the divider output time shift increases by $T_{dco}$, a figure much larger than the typical random jitter at TDC input (usually in the order of or less than 1ps rms in high-performance applications). We can say that in this case the TDC input would be overloaded and limit cycles would appear as huge spurs in the output spectrum. To recover random-noise regime, the amplitude of the deterministic quantization error introduced in the fractional-N frequency division must be reduced below the level of the integrated random noise. To do so, we must reduce the resolution of the DTC below random jitter. Therefore a PLL operating at 3.6GHz and targeting an output jitter of 400-500fs will require a DTC with resolution of about 300fs (about 10-bit DTC, assuming a range of $T_{dco}$).

*B. Single-cell DTC implementation and regulation*

Obviously, a flash-type approach like the one adopted in the DTC in Fig. 4 would be impractical at large number of bits. Looking more carefully to the problem, we note that just one phase is needed at the output of the DTC. The generation of $2^r$ phases at each clock cycles is useless and it consumes large power, which scales as the number of phases. Thus, the DTC can be reduced to the single digitally-controllable delay cell shown in Fig. 9 [11]. In a 65nm CMOS process, using non-minimum inversion-mode MOS capacitors, a resolution of about 340fs can be easily achieved and scaling of CMOS technology will enable finer time resolutions.

A main issue arises from the adoption of a single delay cell. The DTC range cannot be regulated as in the DLL scheme. The delay range of the DTC will considerably vary over PVT variations and will not match the required $T_{dco}$ value (or a multiple of it, when $\Delta\Sigma$ modulators with order higher than one are used to drive the fractional-N divider). This issue was solved in [11] with the digital regulation scheme in Fig. 9. A primary $\Delta\Sigma$ modulator (second-order) driven by FCW dithers the modulus of the integer-N divider. The $\Delta\Sigma$ quantization error is integrated to produce $q[k]$, multiplied by a gain $\hat{g}$, quantized at $r$ bits by means of a secondary $\Delta\Sigma$ and finally cancelled out via the DTC. The integration of the primary $\Delta\Sigma$ error is needed because the DTC controls the phase of the *div* signal, while the integer-N divider controls the frequency of the *div* signal. The multiplication by $\hat{g}$ adapts the amplitude of the quantization error to the dynamic range of the DTC, so that no error is produced.

In practice, the DTC output range $t_g$ is intentionally designed larger than $T_{dco}$ to cover PVT spreads. If the signal $q[k]$ were directly applied to the DTC, the waveform of the time shift, or equivalently the excess-phase ramp of the *div* signal, would be like the one in Fig. 10(a) (The unwrapped ramp is shown in this plot). The periodic discontinuities correspond to the overflow of the primary $\Delta\Sigma$, which increases the modulus of the integer-N divider by 1, and it is caused by the mismatch between $t_g$ and $T_{dco}$. To avoid the large fractional spurs produced by the DTC range mismatch, $q[k]$ is multiplied by $\hat{g}$. The value of $\hat{g}$ is regulated in the background in the same fashion we regulate the gain in the digiphase scheme: $q[k]$ is multiplied by the single-bit error signal $e[k]$ and integrated to get $\hat{g}$. This scheme realizes a sign-LMS estimation. The signal $\hat{g} \cdot q[k]$ is fed to the secondary $\Delta\Sigma$ modulator, whose output $a[k]$ drives the DTC. The resulting excess phase sketched in Fig. 10(b) is a continuous signal. The error with respect to an ideal linear ramp has energy mainly located at high frequency, which is filtered out by the loop.

Adopting the scheme in Fig. 9, the regulation of DTC range is performed in the digital domain and the multiplication inside the correlator block $C$ is realized by a simple digital

978-1-4673-6145-3/13 $31.00 © 2013 IEEE

multiplexer, as $e[k]$ is a single-bit signal. This makes this scheme extremely efficient if compared to the analog regulation performed in the DLL. This concept introduced in [11] has been successfully employed to reduce fractional spurs in analog charge-pump fractional-N synthesizers [12], [13].

### C. DTC linearization

As it has been so far discussed, the DTC is required to have large number of bits (in the order of 10 bit in wireless applications) and good linearity over this wide range (INL less than one LSB). The single-cell DTC sketched in Fig. 8 has three main sources of nonlinearity. The first one is the random mismatch among switched capacitors. A second mechanism is the nonlinearity of the *delay-vs-capacitive-load* relationship. As it is well known, the input-to-output delay of a simple stage, let say a CMOS inverter or a differential stage, is not a linear function of the output capacitance. A third source of mismatch arises from practical implementation reasons. The thermometric encoding of the DTC capacitor bank would guarantee better linearity than binary encoding, but it would require huge area for the unit capacitors and for the bit control lines. As in any DAC with high number of bits, a segmented approach must be used in practice. Instead of duplicating current legs, we can just make a segmentation of the capacitor bank of the DTC, realizing an MSB and an LSB bank (featuring $r$ and $s$ bits, respectively). The fine bank contains $2^s$ nominally identical capacitors with unit capacitance $C$, while the coarse bank contains $2^r$ nominally identical capacitors with unit capacitance $2^sC$. Obviously, in this segmented DTC, process spreads cause mismatch between the two banks and add another source of nonlinearity (*inter-bank*). While scrambling the capacitor elements could, in principle, attenuate the first source of nonlinearity, the second and the third sources occur even in the case of ideal matching among load capacitors of each bank. Therefore, a linearization technique completely different from either capacitance shuffling or any form of dithering must be devised.

Fig. 11 illustrates the digital-PLL block schematic, implementing the proposed linearization technique. To intuitively understand the working principle, let us first assume that the DTC MSBs has a delay range exactly equal to $T_{dco}$ and let us neglect the "nonlinearity correction" block in Fig. 11. The signal $q[k]$ is coarsely quantized to $r$ bits and fed to the DTC MSBs, while the quantization error produced is fed to the DTC LSBs. However, to solve the issue of mismatch between the two capacitor-banks, the DTC LSBs range must be regulated. Therefore, the regulation is performed as in the previous scheme in Fig. 9: the quantization error is multiplied by the adaptive gain $\hat{g}_1$, which is obtained by an LMS loop. Doing so, the range of the DTC LSBs is equivalently adapted to the LSB of the MSB section of the DTC, i.e. it is regulated to be equal to $2^{-r}T_{dco}$.

Let us now focus on the non-ideality of the coarse DTC and let us illustrate the "nonlinearity correction" block. This scheme implements the same concept already used in the multipath correction scheme in Fig. 6. The digital circuit correlate the use of the $i$-th level of the DTC MSBs with the error signal $e[k]$. At each clock cycle, only one of the correlators $C$ has non-zero inputs, and only its output, let us

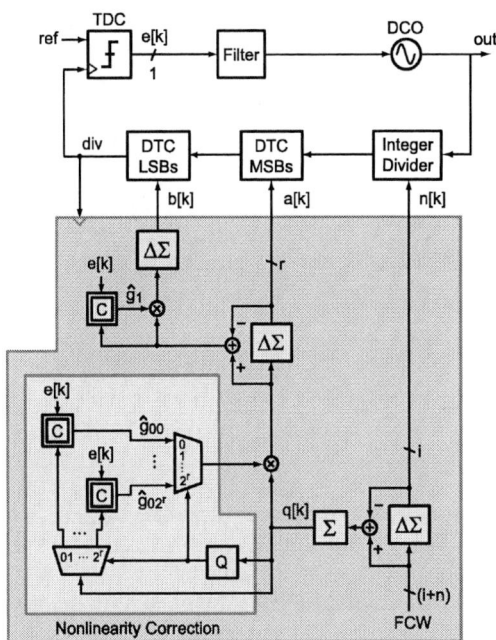

Fig. 11. Digital-to-time converter based PLL with single-bit TDC, segmented DTC and background nonlinearity correction.

say $\hat{g}_{0i}$, is used as correction gain of the quantization error $q[k]$. Any error produced by the $i$-th delay of the DTC MSBs is integrated and the correcting gain $\hat{g}_{0i}$ is estimated. When the value of $\hat{g}_{0i}$ is such to cancel exactly the error, the corresponding correlation goes to zero and $\hat{g}_{0i}$ will remain constant. As the error $e[k]$ is correlated to the MSBs levels of the DTC, this technique is able to correct both for mismatches among capacitors of the coarse bank and for nonlinearity of the delay-versus-capacitive-load characteristic. As highlighted in the previous sub-section, even in this case, the bank of $2^r$ correlators is multiplier free, thus results into a low-power and low-area implementation. The residual nonlinearity due to the mismatches of the DTC LSBs is usually negligible, but can be in principle corrected replicating the same scheme. The proposed technique is a multiple-adaptive-gain scheme, which pre-distorts the DTC control word and cancels its nonlinearity without adding dithering noise.

## V. EXPERIMENTAL RESULTS

The effectiveness of the proposed linearization technique can be verified by comparing the spur performance obtained enabling/disabling the nonlinearity-correction block. The digital PLL fabricated in a 65-nm CMOS technology and adopting the topology in Fig. 11 has been presented in [14]. A 10-bit DTC with 340-fs resolution is segmented in two 6-bit thermometric capacitors banks. The die photo is shown in Fig. 12 (The chip includes other digital algorithms for the direct FM modulation of the PLL, which are not here discussed). The circuit synthesizes frequencies in the 2.9-to-4.0-GHz range while reference frequency is 40MHz, loop bandwidth about 300kHz and power consumption 4.5mW.

The measured level of the in-band spur is -42dBc at 50-kHz, when a channel 50-kHz offset from the integer channel is synthesized (at about 3.6GHz) and when the DTC

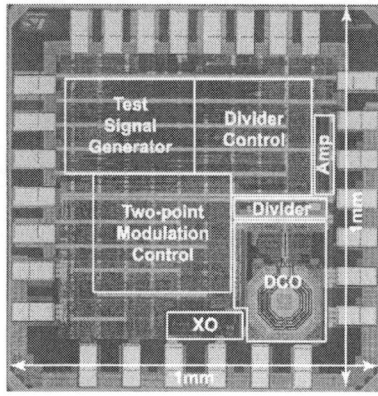

Fig. 12. Chip photograph.

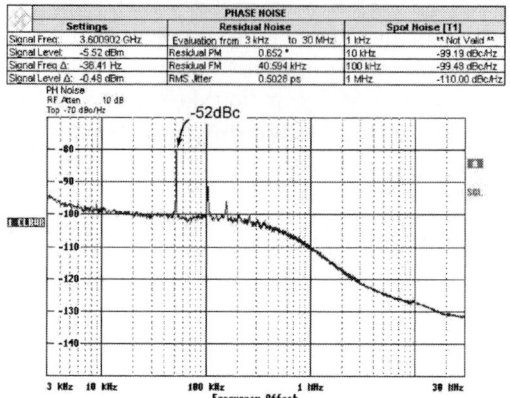

Fig. 13. Measured phase-noise spectrum.

linearization algorithm is disabled (only one gain is estimated for the MSBs'). Circuit simulations confirm this level of spurs, which is consistent with the simulated INL of the DTC of about 3ps. This value of INL is dominated by the nonlinearity of the delay-vs-capacitive-load characteristic.

When the nonlinearity-correction block illustrated in Fig. 11 is enabled, the level of the in-band spur decreases to -52dB at 50kHz (See measured spectrum in Fig. 13). The overhead of area occupation and power consumption of the extra 64 correlators needed for the correction algorithm is irrelevant (we can estimate less than 20% area overhead for the digital section and less than 0.2mW power), while in-band and out-of-band phase noise are unaffected. The residual level of spur is at the present time under investigation, as it would be expected to be lower than -65dBc. The fact that same level of spur (-52dBc) is measured even when the loop is opened suggests that it may result from spurious coupling of the reference signal. The lower level of in-band spurs (-57dBc), which was obtained in [5] adopting the architecture in Fig. 6, must not be ascribed to an intrinsic superior performance of that architecture, but most likely to the differential logic adopted in the whole integer-N divider, which strongly reduces the amount of supply disturbances and which is paid with much larger power consumption (80mW vs. 4.5mW).

## VI. CONCLUSIONS

Digital PLL architectures based on DTCs in feedback allow TDCs with reduced input range, lower number of bits, lower power consumption, and relaxed linearity requirement. A TDC with few bits is also amenable to implement scrambling techniques to improve linearity performance. By increasing further the resolution of the DTC, it is possible to realize a PLL based on a single-bit TDC and avoid the insurgence of limit cycles. The resolution and linearity requirements are in this way shifted from TDC to DTC. However, in contrast to TDC, the implementation of high-performance DTCs is greatly simplified by the adoption of digital calibration algorithms. A novel multipath version of adaptive filtering has been presented and demonstrated in silicon, which cancels the impact of DTC nonlinearities at very low power and without adding dithering noise.

## REFERENCES

[1] R. B. Staszewski et al, "All-digital TX Frequency Synthesizer and Discrete-Time Receiver for Bluetooth Radio in 130-nm CMOS," IEEE JSSC, Vol. SC-39, No. 12, pp. 2278–2291, 2004.

[2] C.-M. Hsu, M. Z. Straayer, and M. H. Perrott, "A Low-Noise Wide-BW 3.6-GHz Digital ΔΣ Fractional-N Frequency Synthesizer With a Noise-Shaping Time-to-Digital Converter and Quantization Noise Cancellation," IEEE JSSC, Vol. SC-43, No. 12, , pp. 2776–2786, 2008.

[3] A. L. Lacaita, S. Levantino, C. Samori, "Integrated Frequency Synthesizers for Wireless Systems," Cambridge University Press, 2007.

[4] E. Temporiti, C. Welti-Wu, D. Baldi, M. Cusmai, F. Svelto, "A 3.5 GHz Wideband ADPLL with Fractional Spur Suppression Through TDC dithering and Feedforward Compensation," IEEE JSSC, Vol. SC-45, No. 12, pp. 2723-2736, 2010.

[5] M. Zanuso, S. Levantino, C. Samori, and A. L. Lacaita, "A Wideband 3.6 GHz Digital ΔΣ Fractional-N PLL With Phase Interpolation Divider and Digital Spur Cancellation," IEEE JSSC, Vol. SC-46, pp. 627-638, No. 3, 2011.

[6] M. Z. Straayer, and M. H. Perrott, "A Multi-Path Gated Ring Oscillator TDC with First-Order Noise Shaping," IEEE JSSC, Vol. SC-44, No. 4, pp. 1089–1098, 2009.

[7] C. Yao, A. N. Willson, "A 2.8-3.2-GHz Fractional-N Digital PLL with ADC-assisted TDC and Inductively Coupled Fine-Tuning DCO," IEEE JSSC, Vol. SC-48, No. 3, pp. 698–710, 2013.

[8] C. Samori, M. Zanuso, S. Levantino, A. L. Lacaita, "Multipath Adaptive Cancellation of Divider Non-Linearity in Fractional-N PLLs," in Proc. of IEEE International Symposium on Circuits and Syst. ISCAS 2011, Rio de Janeiro, May 15-18 2011, pp.418-421.

[9] N. Da Dalt, "Markov Chains-Based Derivation of the Phase Detector Gain in Bang-Bang PLL," IEEE TCAS II, Vol. 53, No. 11, pp. 1195-1199, 2006.

[10] M. Zanuso, D. Tasca, S. Levantino, A. Donadel, C. Samori, and A. L. Lacaita, "Noise Analysis and Minimization in Bang-Bang Digital PLL," IEEE TCAS I, Vol. 56, pp. 835-839, No. 11, 2009.

[11] D. Tasca, M. Zanuso, G. Marzin, S. Levantino, C. Samori, A. L. Lacaita, "A 2.9-to-4.0GHz Fractional-N Digital PLL with Bang-Bang Phase Detector and 560 fs rms Integrated Jitter at 4.5mW Power," IEEE JSSC, Vol. SC-46, No. 12, pp. 2745–2758, 2011.

[12] S. Levantino, D. Tasca, G. Marzin, M. Zanuso, C. Samori, A. L. Lacaita, "A Wideband Fractional-N PLL with Suppressed Charge-Pump Noise and Automatic Loop Filter Calibration," in Proc. of IEEE Radio-Frequency Integrated Circuits Conf., RFIC 2012, Montreal, 17-19 Jun. 2012, pp. 177-180.

[13] Tsung-Kai Kao, Che-Fu Liang, Hsien-Hsiang Chiu, Michael Ashburn, "A Wideband Fractional-N Ring PLL with Fractional- Spur Suppression Using Spectrally Shaped Segmentation," in IEEE Int. Solid-State Circuits Conf. Dig. Tech. Papers, ISSCC 2013, San Francisco, 17-21 Feb. 2013, pp. 416–417.

[14] G. Marzin, S. Levantino, C. Samori, and A. L. Lacaita, "A 20 Mb/s Phase Modulator Based on a 3.6 GHz Digital PLL with -36 dB EVM at 5 mW Power," IEEE JSSC, Vol. SC-47, No. 12, pp. 2974-2988, 2012.

# Hybrid Transformer-Based Tunable Integrated Duplexer with Antenna Impedance Tracking Loop

Sherif H. Abdelhalem[1], Prasad S. Gudem[2], and Lawrence E. Larson[1,3]

[1]University of California at San Diego, La Jolla, CA 92093, [2]Qualcomm Inc., San Diego, CA 92121,
[3]Brown University, Providence, RI 02912

—Electrical balance between the antenna and the balance network impedances is crucial for achieving high isolation in a hybrid transformer duplexer. In this paper, an auto calibration loop for tuning a novel integrated balance network to track the antenna impedance variations is introduced. It achieves an isolation of more than 50 dB in the transmit and receive bands, with an antenna VSWR within 2:1, and between 1.7 and 2.2 GHz. The duplexer, along with a cascaded direct-conversion receiver, achieves a noise figure of 5.3 dB, a conversion gain of 45 dB and consumes 34 mA. The insertion loss in the transmit path was less than 3.8 dB. Implemented in a 65-nm CMOS process, the chip occupies an active area of 2.2 mm².

—Cellular phones, CMOS process, duplexers, LTE, tunable circuits and devices, WCDMA.

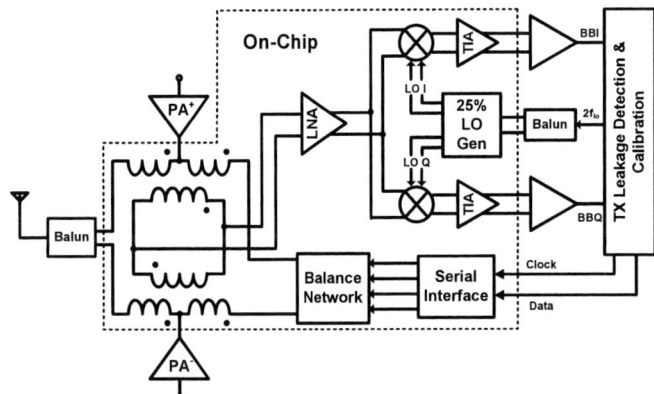

Fig. 1.   Integrated differential duplexer with antenna impedance tracking loop.

## I. INTRODUCTION

$\mathbf{F}$REQUENCY division duplex (FDD) transceivers require the simultaneous operation of the transmitter and receiver while sharing the same antenna. A duplexer, currently implemented as two highly selective off-chip SAW filters, one centered at the receive band and the other at the transmit band, separates the transmit and receive signals. Due to the finite isolation of the duplexer, the strong transmit signal leaks to the receiver input, desensitizing the receiver through several mechanisms: receive band noise, reciprocal mixing, cross-modulation distortion and second-order intermodulation. A high TX-RX isolation is required to limit these effects and keep the receiver noise and linearity requirements feasible.

The need for high-$Q$ resonators to implement these filters prohibits duplexer integration in a CMOS process. For each band of a modern multi-band transceiver, an off-chip duplexer is required. A multi-pole switch selects the appropriate duplexer based on the operating frequency. With over forty bands currently envisioned for mobile applications by the 3GPP, the system cost and complexity rises significantly. Replacing the bank of off-chip duplexers with a single integrated tunable duplexer would enable a fully integrated and reconfigurable multiband transceiver.

An integrated tunable duplexer based on the electrical balance of a hybrid transformer was introduced in [1]. Relying on electrical balance rather than frequency selectivity makes duplexer integration along with the rest of the CMOS RF IC possible, as matching and resolution are easier to achieve than high-$Q$ passives in a CMOS process. In [2] the concept was extended to a differential hybrid transformer architecture to

suppress the common-mode coupling and allow reliable high-power operation. In all these previous demonstrations of hybrid transformer duplexers, the antenna impedance was assumed to be 50 Ω and isolation would degrade significantly under realistic antenna impedance frequency dependence and variation in a mobile environment.

In this paper, we introduce a calibration loop that can detect the TX leakage generated due to imbalance in the duplexer and correct the balance network impedance to track antenna impedance variation and restore isolation. The paper is organized as follows: Section II introduces the duplexer auto-calibration loop design; the design of a novel high power balance network is proposed to achieve high isolation at the TX and RX bands simultaneously under antenna impedance mismatch. The measurement results of the system implemented in a 65-nm CMOS process is covered in Section III. Section IV concludes the paper.

## II. TRACKING LOOP DESIGN

The TX-RX isolation of a hybrid transformer duplexer similar to the one used in Fig. 1 can be derived using a simple S-matrix manipulation as done in [3]:

$$ISOL_{TX-RX}(dB) = 20\log\left|\Gamma_{ANT}(\omega) - \Gamma_{BAL}(\omega)\right| - 6 \qquad (1)$$

where:

$$\Gamma_{ANT}(\omega) = \frac{Z_{ANT}(\omega) - R_o}{Z_{ANT}(\omega) + R_o} \quad \text{and} \quad \Gamma_{BAL}(\omega) = \frac{Z_{BAL}(\omega) - R_o}{Z_{BAL}(\omega) + R_o} \qquad (2)$$

978-1-4673-6145-3/13 $31.00 © 2013 IEEE

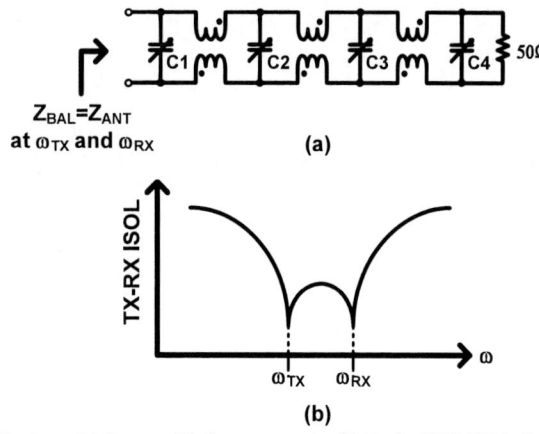

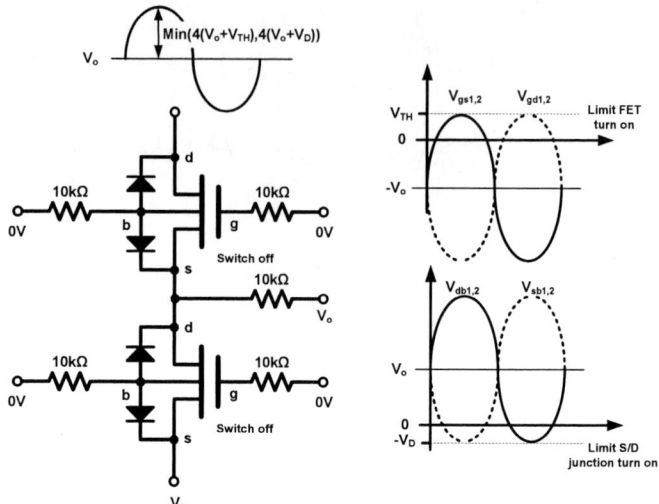

Fig. 2. (a) Proposed balance network. (b) Desired TX-RX isolation response.

Fig. 4. High voltage switch operation and waveforms.

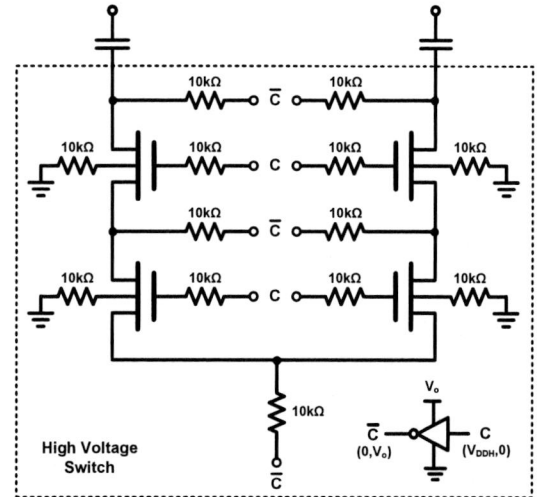

Fig. 3. Differential high voltage switched capacitor used in the balance network.

When $\Gamma_{ANT} \cong \Gamma_{BAL}$, high isolation can be achieved. In a mobile environment, besides the normal frequency dependence, $\Gamma_{ANT}$ would vary considerably due to variation in the surroundings and human interaction [4]. From (1), good TX-RX isolation can be maintained only if $\Gamma_{BAL}$ tracks the variations in $\Gamma_{ANT}$. Such variations are slow in nature and can be calibrated infrequently. Toward that goal, we propose the feedback loop shown in Fig. 1. A differential duplexer similar to the one used in [3] together with a direct-conversion receiver is utilized. The receiver can be the same one used for normal reception, or a special receiver with relaxed performance requirements can be used. The receiver downconverts the TX leakage using the TX LO, measures the amplitude of that leakage, and feeds back a correction signal to the balance network to restore the hybrid transformer balance and minimize the leakage.

*A. Balance Network Design*

The balance network should provide a wide range of impedances and also tolerate high power; meanwhile it should be highly linear as any intermodulation products generated will leak back to the LNA degrading isolation [3]. From (1), if $\Gamma_{ANT}$ and $\Gamma_{BAL}$ have the same frequency dependence, wide band isolation can be achieved. In practice, for a generic antenna the frequency dependence of $Z_{ANT}$ cannot be easily

matched by $Z_{BAL}$, limiting the achievable isolation bandwidth. A practical approach would be to match $Z_{ANT}$ and $Z_{BAL}$ at both the TX and RX frequencies, and only achieve high isolation at the transmit and receive bands, i.e.:

$$Z_{ANT}\left(\omega_{TX}\right) \cong Z_{BAL}\left(\omega_{TX}\right) \text{ and } Z_{ANT}\left(\omega_{RX}\right) \cong Z_{BAL}\left(\omega_{RX}\right) \quad (3)$$

Matching the real and imaginary parts of the impedance at two different frequencies requires four tuning parameters. A possible implementation of this tuner is a cascade of three $\pi$-sections with four high voltage highly linear switched capacitors, as shown in Fig. 2. The $\pi$-section can convert the 50 $\Omega$ termination up or down and, with a proper inductance and capacitance range, can cover a wide range of impedances between 1.7 and 2.2 GHz. The differential implementation makes use of coupled transformers to improve the $Q$-factor and reduce area, and a switched capacitor with stacked devices switches to tolerate high power and maintain high linearity. Limited by the achievable quality factor, a simulated VSWR range of 2:1 can be covered.

The high voltage switched capacitor unit is shown in Fig. 3. A potential problem arises when the switch is *off* and a large voltage swing appears across it. A stack of triple-well thick-oxide NMOS devices with RF floating gate and bulk allows the swing to almost split equally across them. A source/drain(S/D) bias ($V_o$) maintains the device and S/D to bulk junctions *off* to maintain high linearity. The operation is illustrated in Fig. 4, where stacking and S/D bias can allow for a maximum swing equal to:

$$V_{max} = 2n \times Min\left[\left(V_o + V_{th}\right),\left(V_o + V_D\right)\right] \quad (4)$$

where $n$ is number of stacked devices and $V_o$ is the S/D DC bias. One limit is dictated by the NMOS device turning *on* and the other is by the S/D-to-bulk junction turning *on*. The value of $V_o$ and $n$ is chosen so that $V_{gs}$ or $V_{gd}$ across any device will not exceed the oxide breakdown reliability limit under maximum transmit power of 27 dBm. Similar voltage enhancement techniques were used before in [5] and [6]. A trade-off between quality factor and tuning range exists; a wider device can reduce the on resistance and hence improve $Q$-factor, but also increases the parasitic capacitance and degrades tuning range.

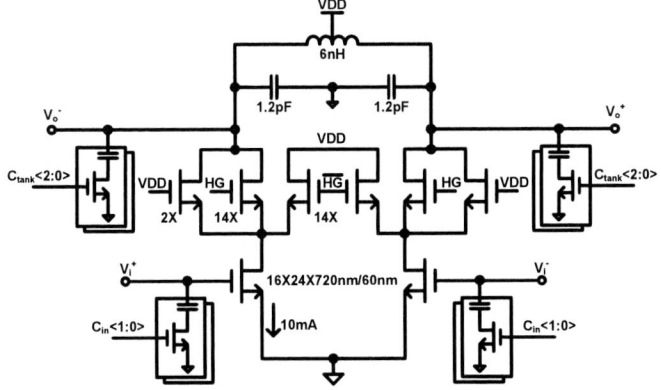

Fig. 5.  Noise matched LNA with gain programmability.

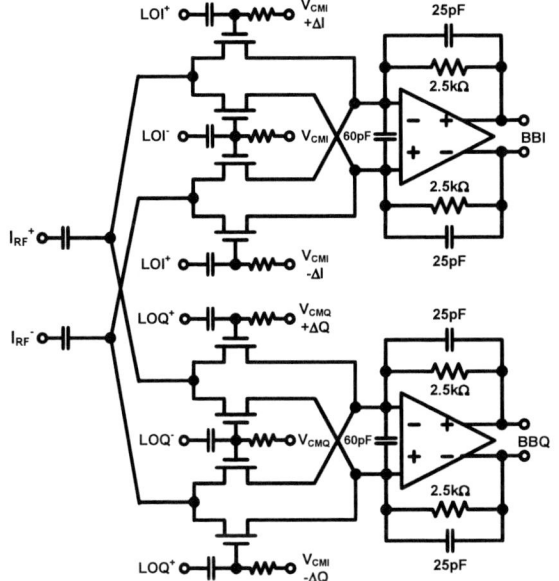

Fig. 6.  Current mode passive mixer with 25% LO duty cycle drive and output TIA.

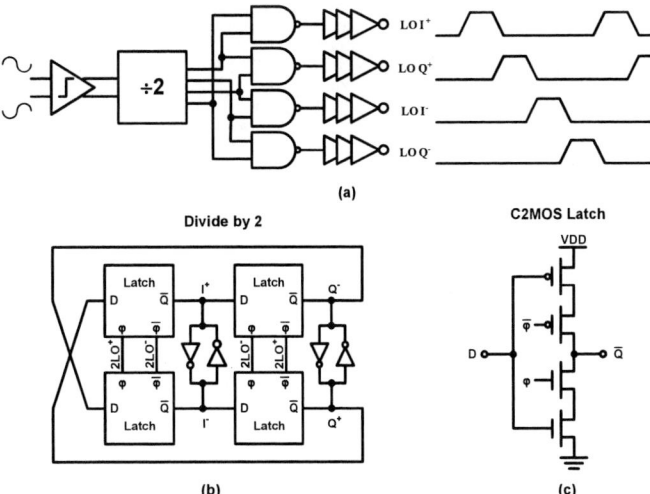

Fig. 7.  (a) 25% duty cycle LO generation. (b) Divide by 2 schematic. (c) Dynamic C²MOS latch used in divider.

## B. Direct-Conversion Receiver Design

A relaxed performance receiver could be used for the TX leakage detection. But in this implementation, a receiver with typical WCDMA/LTE performance requirements is designed to test the overall performance with the integrated duplexer. As shown in Fig. 1, the LNA acts as a transconductor that converts the input voltage to current for a current-mode passive mixer for downconversion. Finally a TIA provides some filtering and converts the signal back to voltage. The mixer is driven with a 25% duty cycle LO generated on-chip from twice the LO signal.

The LNA, shown in Fig. 5, is a noise-matched LNA [3]. The duplexer inductance resonates the LNA input capacitance to maximize the passive voltage gain in front of the LNA, thus minimizing its noise contribution. A switched capacitor at the input is used to tune that resonance across the operating frequency range. Two gain modes are provided using current steering in the cascode devices to reduce gain at high TX leakage so that the TIA output will not saturate. A tuned tank load is used to maximize the source impedance seen by the mixer at the RF frequency which helps minimize the TIA noise contribution and improves linearity [7].

Fig. 6 shows the schematic for the mixer and TIA. A current-mode passive mixer achieves high linearity without flicker noise, as no DC current flows in the switches. The LO is AC coupled to the mixer switches with the DC level offsets to calibrate the mismatches in order to minimize second-order intermodulation distortion. The TIA provides a low impedance across the channel bandwidth. At higher frequency (e.g. at the TX jammer offset), as the TIA gain decreases, the added capacitor at the TIA input reduces the input impedance to maintain good linearity. The TIA has a single pole at 2.5MHz to provide low-pass filtering and was designed using a two stage amplifier.

The 25% LO duty cycle reduces the interaction between the I and Q channels and provides 3dB higher gain [7]. The circuit for LO generation is shown in Fig. 7. The input sinusoidal signal at twice the required frequency is amplified and applied to a divide by 2 divider to generate the 50% duty cycle I and Q LO signals, which are anded to generate the 25% duty cycle LO. The divider is a simple master slave flip-flop implemented with dynamic C²MOS latches and connected in differential manner to generate the differential phases. Small cross-coupled inverters are used to guarantee differential phases and proper startup.

## III. MEASUREMENT RESULTS

The differential duplexer based on hybrid transformer with the newly proposed balanced network and cascaded with a direct conversion receiver was implemented in a 65nm CMOS process occupying an active area of 2.2 mm². Fig. 8 shows the die microphotograph. The die was packaged in a 40 pin QFN plastic package and mounted on an FR4 board for testing.

The calibration algorithm was implemented in Matlab using a genetic algorithm for global optimization of the balance network settings. The measured TX-RX isolation is shown in Fig. 9. More than 50 dB of isolation is achieved at the TX and RX frequencies concurrently. Fig. 10 shows the voltage conversion gain and noise figure in the RX path, and the TX insertion and return loss. Roughly 45 dB of gain was achieved with a cascaded noise figure around 5.3 dB. The TX insertion loss was less than 3.8 dB. Fig. 11 shows the IIP2 calibration

978-1-4673-6145-3/13 $31.00 © 2013 IEEE

Fig. 8.   Die microphotograph.

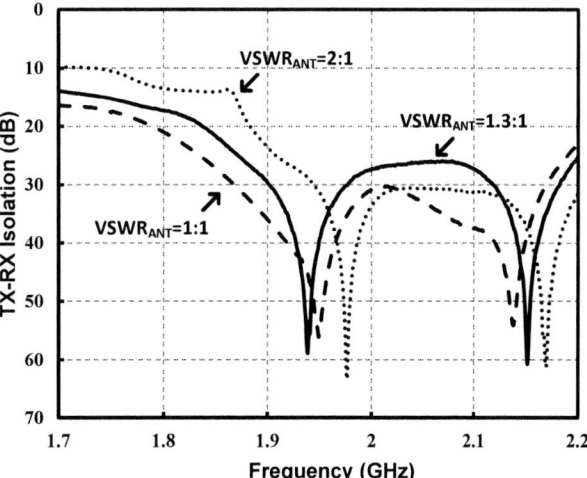

Fig. 9.   Measured TX-RX Isolation for different channels and VSWR$_{ANT}$.

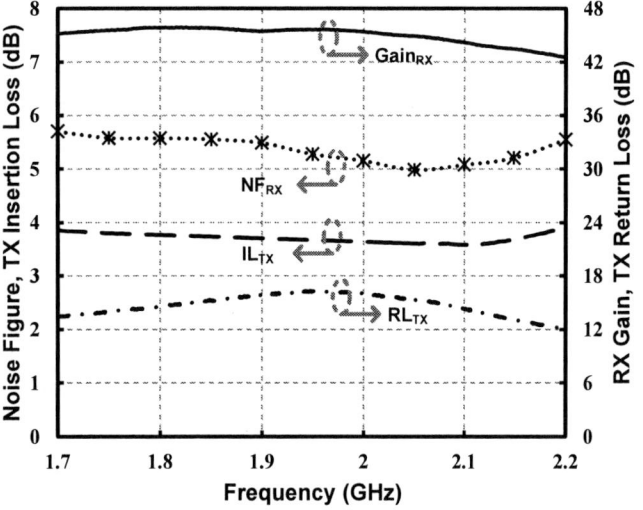

Fig. 10.   Measured RX gain, noise figure, TX insertion and return loss.

contours generated by sweeping the DC offset in the I and Q mixers shown in Fig. 6. An uncalibrated IIP2 of 40 dBm was measured, and more than 60 dBm can be achieved after calibration.

## IV.   CONCLUSION

An integrated duplexer that is tolerant to antenna impedance

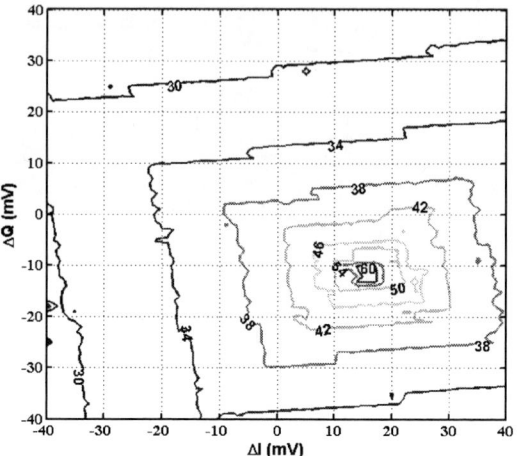

Fig. 11.   Receiver IIP2 calibration contours.

TABLE I
PERFORMANCE SUMMARY

| Frequency | 1.7–2.2 GHz |
|---|---|
| Isolation in TX Band | > 50 dB |
| Isolation in RX Band | > 50dB |
| Antenna VSWR Range | 2:1 |
| Max. TX Power | 27 dBm |
| TX Insertion Loss | 3.7 dB |
| Cascaded Noise Figure | 5.3 dB |
| Conversion Gain | 45 dB |
| IIP3 | -4.8 dBm |
| IIP2 | > 60 dBm |
| I and Q Quadrature Error | 0.5 dB, 1.2 degree |
| Power | 28 mA from 1.2V, 7 mA from 2.5V |
| Area | 2.2 mm$^2$ |

variations was introduced. By downconverting the TX leakage and feeding back a correction signal to a wide impedance range balance network, high isolation can be maintained. A novel balance network architecture was introduced to achieve high isolation at the TX and RX bands concurrently. Table I summarizes the achieved performance of the duplexer cascaded with a direct-conversion receiver.

## REFERENCES

[1]   M. Mikhemar, H. Darabi, and A. Abidi, "A tunable integrated duplexer with 50dB isolation in 40nm CMOS," *ISSCC Dig. Tech. Papers*, pp. 386-387, Feb. 2009.

[2]   S. Abdelhalem, P. Gudem, and L. Larson, "A tunable differential duplexer in 90nm CMOS," *RFIC Symp.*, pp. 101–104, Jun. 2012.

[3]   S. Abdelhalem, P. Gudem, and L. Larson, "Hybrid transformer-based tunable differential duplexer in a 90-nm CMOS process," *IEEE Trans. Microw. Theory Tech.*, vol. 61, no. 3, pp. 1316–1326, March 2013.

[4]   K.R. Boyle, Y. Yuan, and L.P. Ligthart, "Analysis of mobile phone antenna impedance variations with user proximity", *IEEE Trans. on Antennas and Propagation*, vol. 55, issue 2, pp. 364 –372, Feb. 2007.

[5]   Y. Yoon , H. Kim , Y. Park , M. Ahn , C.-H. Lee and J. Laskar, "A high-power and highly linear CMOS switched capacitor," *IEEE Microw. Wireless Compon. Lett.*, vol. 20,  no. 11,  pp. 619–621. Nov. 2010.

[6]   M. Ahn, H. Kim, C. Lee, and J. Laskar, "A 1.8-GHz 33-dBm P0.1-dB CMOS T/R switch using stacked FETs with feed-forward capacitors in a floated well structure," *IEEE Trans. Microw. Theory Tech.*, vol. 57, no. 11, pp. 2661–2670, 2009.

[7]   D. Kaczman et al., "A single–chip 10-band WCDMA/HSDPA 4-band GSM/EDGE SAW-less CMOS receiver with DigRF 3G interface and +90 dBm IIP2," *IEEE Journal of Solid-State Circuits*, vol. 44, no. 3, pp. 718–739, March 2009.

# A Smartphone SP10T T/R Switch in 180-nm SOI CMOS with 8kV+ ESD Protection by Co-Design

X. Shawn Wang[1], Xin Wang[2], Fei Lu[3], Li Wang[3], Rui Ma[3], Zongyu Dong[3], Li Sun[1], Albert Wang[3*], C. Patrick Yue[1, 4*], Dawn Wang[5] and Alvin Joseph[5]

[1] Dept. of ECE, University of California, Santa Barbara, CA 93106, USA, xiaowang@umail.ucsb.edu
[2] OmniVision Technologies, Santa Clara, CA 95054, USA,
[3] Dept. of EE, University of California, Riverside, CA 92521, USA, aw@ee.ucr.edu
[4] Hong Kong University of Science and Technology, Kowloon, Hong Kong SAR, eepatrick@ust.hk
[5] IBM Microelectronics, USA

*Abstract*—**This paper reports the first 8kV+ ESD-protected SP10T transmit/receive (T/R) antenna switch for quad-band (0.85/0.9/1.8/1.9-GHz) GSM and multiple W-CDMA smartphones fabricated in an 180-nm SOI CMOS. A novel physics-based switch-ESD co-design methodology is applied to ensure full-chip optimization for a SP10T test chip and its ESD protection circuit simultaneously.**

## I. INTRODUCTION

The demands for slim and light smartphones require ultra-compact RF front-end module (FEM) between antenna and multi-mode multi-band RF transceivers, where a single-pole multiple-throw (SPMT) T/R switches plays a key role. High-performance SPMT switches are normally designed in high-cost GaAs and SoS technologies [1, 2]. Recently, SPMT T/R switch design in SOI CMOS has gained research attentions due to its superior RF properties including low parasitic capacitance, high-Q passives, low substrate losses, high substrate isolation and CMOS integration [3, 4]. Compared to traditional GaAs pHEMT, SOI allows low battery voltage and high integration similar to bulk CMOS. Meanwhile, robust and whole-chip ESD protection is a challenge in SPMT switch designs due to adverse interactions between core switch circuit and ESD protection [5-7]. Particularly, the inherent ESD-induced parasitic capacitance ($C_{ESD}$) has such significant impacts on SPMT switch performance, including insertion loss (IL), isolation and linearity, that most recently reported SPMT switches had weak or no on-chip ESD protection [1-4], e.g., partial 0.7-kV ESD protection for a SP6T in SOI [3]. We report a fully integrated 8-kV ESD-protected SP10T switch in SOI CMOS and a new switch-ESD co-design methodology enabling whole-chip SPMT and ESD protection design optimization to address the challenge.

## II. SP10T SWITCH AND ESD CO-DESIGN

### A. SP10T T/R Switch Design

Fig. 1 shows the simplified topology for the new SP10T switch consisting of a transmitter (Tx) block covering the GSM 850/900, DCS and PCS bands, a GSM receiver (Rx) block and a transceiver (TRx) block dedicated for all W-CDMA bands. This SP10T switch features a compact series-shunt topology covering 10 GSM Tx/Rx and W-CDMA TRx channels. Each T/R path connects to antenna via a series branch, while the shunt branch to ground enhances inter-channel isolation. The linearity requirement is most stringent in GSM Tx mode as the

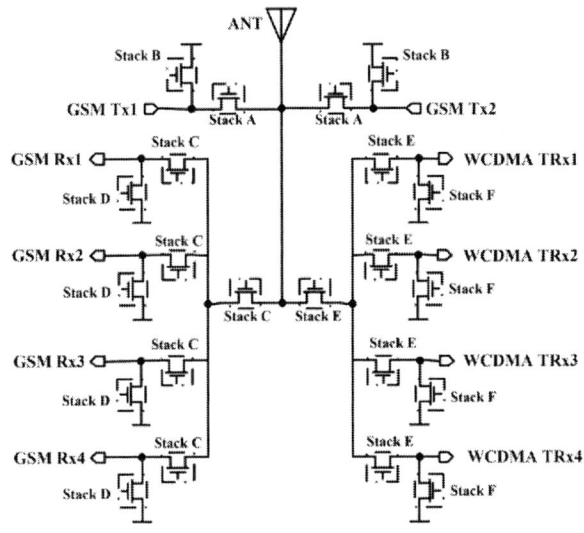

Fig. 1. A simplified SP10T T/R Switch topology.

off-state branches must handle output power up to 35 dBm, hence,

$$P_{Watt} = 10^{\frac{35dBm}{10}} \times 0.001 = 3.16\,W \tag{1}$$

$$V_{pk} = \sqrt{2 \times P_{Watt} \times Z_0} = 17.8\,V \tag{2}$$

$$V_{max} = V_{pk}\left(1 + \frac{VSWR-1}{VSWR+1}\right) = 28\,V \tag{3}$$

From Equations (1)-(3), the power in the GSM Tx paths may produce a peak AC voltage up to 28 V at VSWR 4:1, which requires stacking up 12 NMOS transistors in GSM Tx paths.

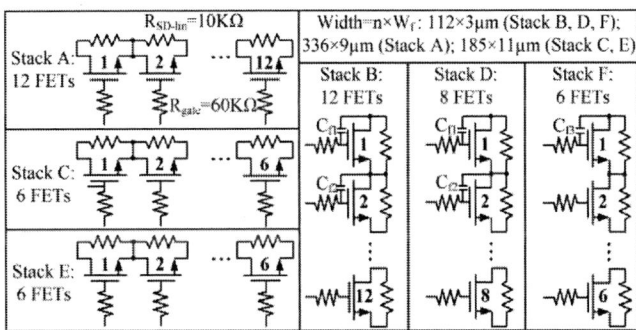

Fig. 2. Stacked FET topologies in series/shunt branches (All L=0.32 μm).

Similarly, a stack transistor topology is needed in all other GSM Rx and W-CDMA TRx channels with different stacking numbers as summarized in Fig. 2. The channel IL and isolation must also be considered in determining the transistor stack numbers and the transistor sizes. To improve the overall linearity, AC floating gate bias resistors and feed-forward capacitors (FFC, i.e., $C_f$) are used in all series and shunt channels. Particularly, the uneven distribution of voltage drop across transistors in the shunt paths may degrade the linearity and possibly cause breakdown to the top devices in the stacked transistors of the shunt branches at high power levels [8, 9]. Therefore, FFCs are used across the drain and gate of the top transistors in the shunt branches and their values must be carefully designed (i.e., $C_{f1}$=60 $fF$, $C_{f2}$=40 $fF$, $C_{f3}$=50 $fF$) to ensure close-to-even voltage drop across these transistors.

### B. Full ESD Protection Design

Robust full-chip ESD protection is designed for the SP10T switch aiming for 8-kV ESD protection level. Fig. 3 shows the ESD protection scheme for one GSM Rx channel. Per SOI CMOS features, an anti-parallel diode-string ESD protection topology, using gated STI diodes, is adopted in this design. To ensure ESD protection design optimization and prediction, as well as minimize ESD-induced parasitic effects, complete mixed-mode ESD design simulation was conducted [10]. The P+/NW ESD diode string is optimized for eight fingers of 60-μm wide, features ESD triggering voltage of $V_{t1}$ ~2V and achieves 8-kV ESD protection. Fig. 4 depicts simulated diode-string ESD protection structure, its transient

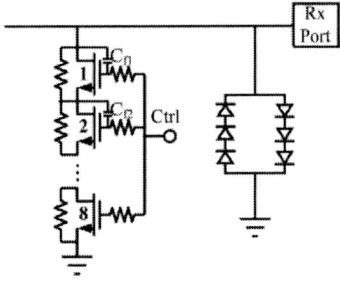

Fig. 3. ESD protection scheme uses anti-parallel diode-string structure.

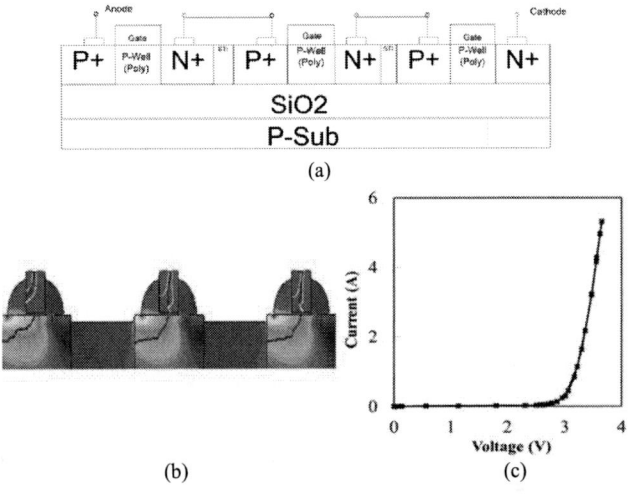

(a)

(b)　　　　　(c)

Fig. 4. Simulated diode-string ESD structures (a), transient ESD heating (b), and transient ESD discharging I-V curve (c) under 8-kV ESD stressing reveal ESD behavior details for ESD design optimization and prediction.

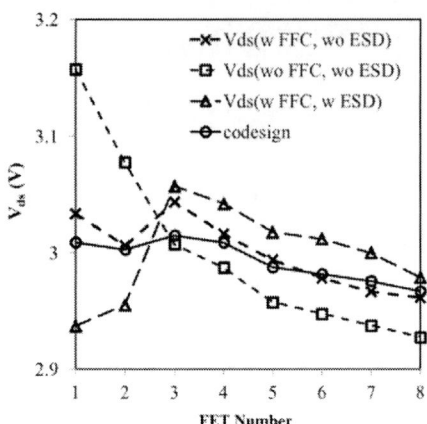

Fig. 5. ESD co-design ensures even voltage drop across the stacked FETs.

ESD heating effect (temperature contours and hot spot) and transient ESD discharging I-V curve under 8-kV ESD. Such function details are extremely critical to accurate ESD protection design with prediction, particularly for SOI ESD designs. Further, the ESD diode strings must be optimized for minimum ESD-induced $C_{ESD}$. In addition, the stacked transistors are also evaluated for their self ESD protection contribution in switch circuit design. Overall, 9-kV chip-level ESD protection is predicted with a margin of 1 kV due to self-protection by simulation.

### C. SP10T Switch and ESD Co-Design

ESD protection has inevitable negative impacts on RF IC performance [5-7], particularly for SPMT switches with robust full ESD protection because the large number of ESD units introduce significant parasitics. Specifically, referring to the ESD protection scheme shown in Fig. 3, the FFCs ($C_{f1}$ and $C_{f2}$) serve to balance the voltage drops across top FETS in the stack, which unfortunately is disrupted by the ESD-induced $C_{ESD}$ (~400$fF$) after the ESD diode-string is added. The parasitic $C_{ESD}$ can cause severe degradation in both linearity and IL. For example, the GSM Rx linearity is seriously degraded by ESD because the AC voltage is no longer evenly distributed across the top FETs due to the extra $C_{ESD}$ added to the accurately sized FFCs. Careful co-design is therefore conducted to resolve this design degradation problem, which resulted in revised FFC values from $C_{f1}$=60 $fF$ and $C_{f2}$=40 $fF$ without ESD protection to $C_{f1}$=55 $fF$ and $C_{f2}$=33 $fF$ with ESD protection. Fig. 5 compares the simulated voltage drop ($V_{ds}$) across the top FETs in a stacked shunt path with four cases: without FFCs and ESD, with FFCs but no ESD, with FFCs and ESD, and with ESD co-design. It clearly shows that FFCs can ensure the even $V_{ds}$ drop across top FETs, however, ESD may severely disrupt the evenly $V_{ds}$ distribution. Fortunately, a careful switch-ESD co-design almost recovered all the degradation caused by the ESD protection, which is a very important design philosophy in designing SPMT with robust full-chip ESD protection.

### III. MEASUREMENT AND DISCUSSIONS

This SP10T switch was fabricated in a commercial thick-film 180-nm SOI CMOS. Fig. 6 shows a die photo for the SP10T circuit with an area of 820 μm X 1000 μm including

978-1-4673-6145-3/13 $31.00 © 2013 IEEE

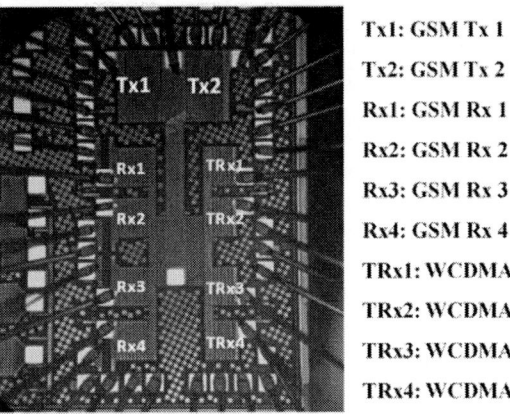

Tx1: GSM Tx 1

Tx2: GSM Tx 2

Rx1: GSM Rx 1

Rx2: GSM Rx 2

Rx3: GSM Rx 3

Rx4: GSM Rx 4

TRx1: WCDMA 1

TRx2: WCDMA 2

TRx3: WCDMA 3

TRx4: WCDMA 4

Fig. 6. A SP10T switch die photo.

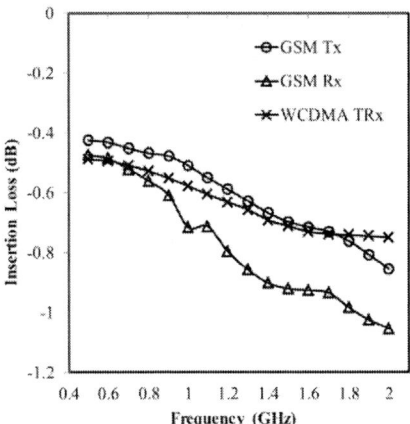

Fig. 7. Measured IL in all GSM and W-CDMA modes.

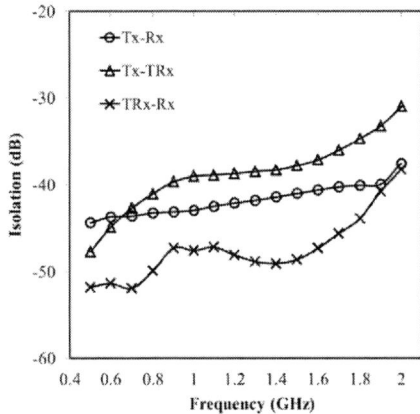

Fig. 8. Measured isolation across GSM and W-CDMA channels.

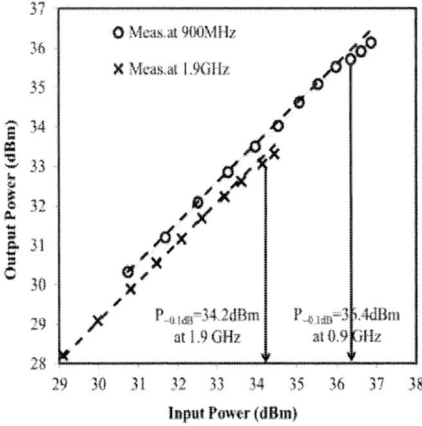

Fig. 9. Measured $P_{-0.1dB}$ at 900 MHz and 1.9 GHz.

ESD protection and pads. The floating-body (FB) SOI NMOS transistors in all channels have multiple fingers and the FFCs are metal-insulator-metal (MIM) capacitors using top metal with capacitance density of 2.05 $fF/\mu m^2$. Comprehensive measurements were conducted for both SP10T circuit and ESD protection. The SP10T die was mounted and bonded onto an FR-4 evaluation board with symmetric transmission lines from the pads to the SMA connectors and a straight thru-line of equal length to de-embed PCB trace losses.

*A. Small-Signal Characterization*

Small-signal SP10T circuit performance was measured using an Agilent E8364B Network Analyzer and the PCB thru-line losses were de-embedded. Fig. 7 shows the measured IL curves in GSM (Tx and Rx) and W-CDMA (TRx) modes, respectively, which includes bond-wire parasitics, however, removed PCB and SMA connector losses by de-embedding. The measured IL is around 0.48/0.8 dB for low/high bands in GSM Tx mode, 0.7/1.0 dB for low/high bands in GSM Rx modes, and 0.6/0.75 dB for low/high bands in W-CDMA modes, respectively. The measured IL at 2 GHz in the GSM Rx mode is 1.1 dB that is about 0.4 dB worse than that in the W-CDMA mode, which may be attributed to several possible factors including ESD-induced $C_{ESD}$ (~400$fF$) and bonding wire parasitic inductance. Fig. 8 shows measured isolation between Tx and Rx, as well as from TRx to Tx/Rx channels. It achieves better than 43/40-dB isolation for low-high bands between Tx and Rx channels in all GSM and W-CDMA modes.

*B. Large-Signal Characterization*

Large-signal SP10T circuit performance was characterized using an SMF 100A Signal Generator, R&S FSU Spectrum Analyzer and a 10-W amplifier to boost the signal. Fig. 9 presents the measured output power $P_{out}$ versus its input power $P_{in}$ in the GSM Tx mode, showing $P_{-0.1dB}$ of 36.4 dBm in low band and 34.2 dBm in high band as extracted. The main loss in the GSM Tx mode is due to large on-state resistance of the stacked channel ($R_{on}$ ~ 3.2$\Omega$ for the 12 stacked transistors) and the power leakage into other stacked paths in OFF state. Nonlinearity mainly comes from nonlinear off-state capacitance of the other stacks and ESD protection blocks. To investigate higher order nonlinearities in the SP10T T/R switch,

the 2$^{nd}$ and 3$^{rd}$ order harmonics were simulated. In GSM Tx mode, the 2$^{nd}$/3$^{rd}$ harmonics are 84/80 dBc below the fundamental tone with 35-dBm input. In W-CDMA TRx mode, the 2$^{nd}$/3$^{rd}$ harmonics are 86/79 dBc below the fundamental tone at 26-dBm input.

*C. ESD Protection Characterization*

Whole-chip ESD protection characterization was conducted for the SP10T circuit using transmission line

978-1-4673-6145-3/13 $31.00 © 2013 IEEE

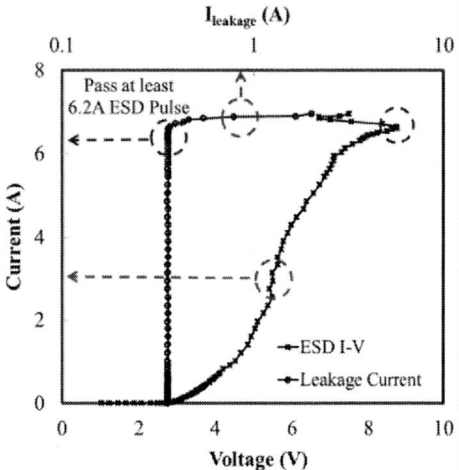

Fig. 10. Measured SP10T ESD performance at GSM Rx port by TLP testing shows it passed 6.2 A, i.e., about 9.3 kV.

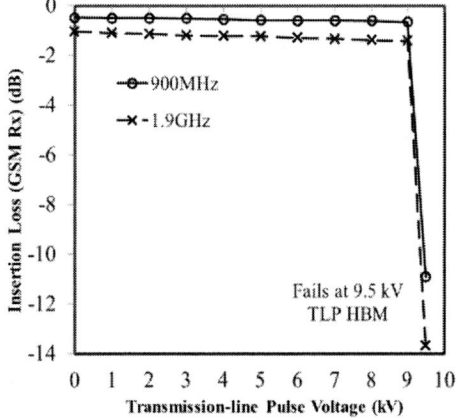

Fig. 11. Measured IL degradation for SP10T switch in GSM Rx mode after continuous ESD stressing by TLP test (passed 9 kV and failed at 9.5 kV).

pulsing (TLP) tester (Barth 4002) dedicated for HBM ESD evaluation. TLP testing is critical for chip level ESD evaluation because it provides instantenous I-V curve and exams leakage under ESD stressing. TLP testing was conducted by stepping up the ESD pulse levels with 100-ns pulse duration. After each TLP stress applied to the SP10T circuit pads, the SP10T circuit was inspected by performing regular small-signal and large-signal circuit testing to check if any ESD-induced circuit performance degradation occurred. Fig. 10 shows a typical I-V curve and leakage behavior for the SP10T under TLP ESD stressing, which reveals details of ESD diacharging behaviors at chip level. Fig. 10 shows that the SP10T chip can handle 6.2-A TLP current without leakage, i.e., about 9.3-kV ESD protection level. Fig. 11 presents measured SP10T circuit IL degradation at 0.9 and 1.9 GHz while stepping-up TLP ESD stressing with 100-ns pulse, which shows that the SP10T can survive at least 9-kV ESD stress in HBM mode, agreeing well Fig. 10. The full-chip ESD testing results, including ~8kV from diode-string ESD unit and a margin of ~1kV from self-protection, are well predicted by mixed-mode ESD co-design simulation at chip level.

Table I compares the measured key SP10T circuit specs with the relevant state-of-the-art SPMT switches reported.

This SOI CMOS SP10T switch performs comparable to SP6T switches designed in high-cost GaAs process [1]. This SP10T outperforms reported SP6T switch in 130-nm SOI CMOS and compares favorably to SP14T switch in 180-nm SOI CMOS [3, 4]. The favorable circuit performance comparison for this SP10T over other reported state of the art was achieved with robust 8-kV+ ESD protection, while others have very weak or no on-chip ESD protection.

## Conclusion

A high-linearity SP10T switch with full-chip 8kV+ ESD protection is reported for multi-band, multi-mode GSM/W-CDMA smartphones. Implemented in 180-nm SOI CMOS, this SP10T switch outperforms reported SPMT switches in SOI CMOS and performs comparable to SP6T switches designed in more expensive GaAs and SOS technologies. A new switch-ESD co-design technique plays a key role in achieving whole-chip design optimization for both switch circuit and on-chip ESD protection.

TABLE I
SP10T SWITCH PERFORMANCE COMPARISON

|  |  | [1] | | [3] | | [4] | | **This Work** | |
|---|---|---|---|---|---|---|---|---|---|
| SPXT | | SP6T | | SP6T | | SP14T | | **SP10T** | |
| $f$(GHz) | | 0.9 | 1.9 | 0.9 | 1.9 | 0.9 | 1.9 | **0.9** | **1.9** |
| $P_{-0.1dB}$ (dBm) | | 37 | 36 | 36 | 34 | 28 | | **36.4** | **34.2** |
| IL (dB) | Tx | 0.21 | 0.54 | 0.7 | 0.7 | 0.55 | 0.65 | **0.48** | **0.81** |
| | Rx | 0.33 | 0.65 | 0.8 | 0.8 | 1.0 | 1.0 | **0.71** | **1.03** |
| Tx-Rx Iso (dB) | | 37.2 | 31.7 | 40 | 30 | 37 | 31 | **43.1** | **40** |
| ESD HBM | | NA | | 700 V | | NA | | **9.3 kV** | |
| Area (mm²) | | 2.25 | | 1.51 | | 2.7 | | **1.23** | |
| Technology | | GaAs pHEMT | | 0.13-μm SOI | | 0.18-μm SOI | | **0.18-μm SOI** | |

## References

[1] H. Tosaka, T, Fujii, K. Miyakoshi, K. Ikenaka and M. Takahashi, "An antenna switch MMIC using E/D mode p-HEMT for GSM/DCS/PCS/WCDMA bands application," *Proc. IEEE RFIC Symp.*, pp. 519-522, 2003.

[2] D. Kelly, C. Brindle, C. Kemerling and M. Stuber, "The state-of-the-art of silicon on sapphire CMOS RF switches," *Proc. IEEE CSIC Symp.*, pp. 200-203, 2005.

[3] C. Tinella, et al, "0.13μm/spl mu/m CMOS SOI SP6T antenna switch for multi-standard handsets," *Proc. IEEE Silicon Monolithic Integrated Circuits in RF Systems*, pp. 58–61, 2006.

[4] Q. Chaudhry, et al, "A linear CMOS SOI SP14T antenna switch for cellular applications," *Proc. IEEE RFIC Symp.*, pp. 155–158, 2012.

[5] A. Wang, et al, "A review on RF ESD protection design", *IEEE Trans. Electron Devices, Vol. 52, No. 7*, pp. 1304-1311, July 2005.

[6] X. Wang, et al, "Post-Si programmable ESD protection circuit design: mechanisms and analysis", in press, *IEEE J. Solid-State Circuits*, 2013.

[7] L. Wang, et al, "A design technique overview on broadband RF ESD protection circuit designs", *Proc. IEEE MWSCAS*, pp.590-593, 2012.

[8] H. Xu and K. O. Kenneth, "A 31.3-dBm bulk CMOS T/R switch using stacked transistors with sub-design-rule channel length in floated p-wells," *IEEE J. Solid-State Circuits, Vol. 42*, pp. 2528–2534, Nov. 2007.

[9] K. Miyatsuji and D. Ueda, "A GaAs high power RF single pole dual throw switch IC for digital mobile communication system," *IEEE J. Solid-State Circuits, Vol. 30, No. 9*, pp. 979–983, Sep. 1995.

[10] H. Feng, et al, "A mixed-mode ESD protection circuit simulation-design methodology", *IEEE J. Solid-State Circuits, Vol. 38, No. 6*, pp. 995-1006, June 2003.

# A Lumped Component Programmable Delay Element for Ultra-Wideband Beamforming

Naga Rajesh and Shanthi Pavan, Indian Institute of Technology Madras, India

*Abstract*—We introduce a ladder filter based programmable time delay element for beamforming in Ultra-Wideband (UWB) systems. When compared to conventional methods based on the tapped delay line architecture, our technique achieves lower power dissipation, better area efficiency, and finer delay and gain resolution more efficiently. The proposed architecture is more scalable, has better parasitic absorption capability and highly programmable with delay and gain resolution dependent only on transconductor resolution. A prototype delay line designed for the 3.1-10.6 GHz UWB range achieves a delay range of 80 ps with 0.5 ps resolution and a gain range of -30 dB to +10 dB with 0.15 dB step. Fabricated in a 0.25 $\mu$m SiGe BiCMOS process, the delay element occupies an active area of 1 mm$^2$ and consumes 47 mW from a 2.5 V supply. A four antenna beamforming system using the delay element can achieve scanning range of $\pm 51^\circ$ with resolution of 0.86$^\circ$ for antenna spacing of 10 mm.

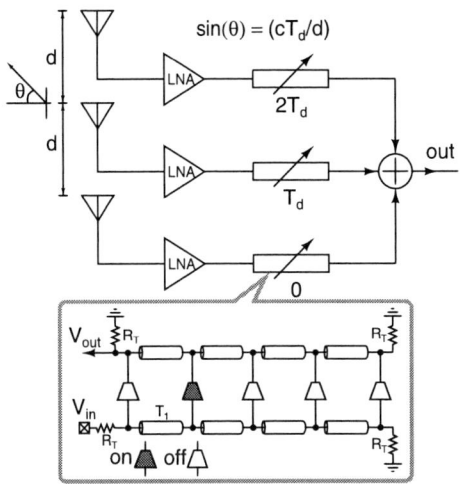

Fig. 1. An UWB multi-antenna beamforming receiver. The inset shows a commonly used digitally tunable delay element based on the traveling wave structure.

## I. INTRODUCTION

Pulse based Ultra-Wideband (UWB) radio technologies operating in unlicensed frequency spectrum of 3.1-10.6 GHz are used in radar and imaging. Using a multiantenna system with a variable delay in each path imparts spatial selectivity to the antenna array, there by enabling beamforming. The basic idea behind a wideband multi-antenna beamforming receiver is explained with the aid of Fig. 1, which shows a three antenna example. The individual channel outputs are combined after passing through front end low noise amplifiers and variable time delay circuits. It is easy to see that signals arriving at an angle $\theta = \sin^{-1}(cT_d/d)$ interfere constructively, thereby imparting spatial selectivity to the array. By varying the delays in the individual channels, the array can be electronically "steered". A common way of realizing the variable delay element is the tapped delay line based structure shown in the inset of Fig. 1 [1][2]. It is basically a traveling wave amplifier, with the output at the "anti-phase" end. The input and output delay lines are coupled using transconductors, only one of which is on at any time. Denoting the delay per section of the input (and output) lines by $T_1$, we see that the input-output delay can be varied digitally from 0 to $8T_1$ in steps of $2T_1$.

The traveling wave type variable delay element has several problems. In principle, the bandwidth of the delay line is infinite - in practice however, capacitive loading of the transconductors degrades bandwidth by reducing the cutoff frequency of the synthetic delay line. Double termination of both input and output lines, as well as line losses causes significant attenuation, needing increased gain (and power dissipation) in the transconductors. Further, it does not seem to be easy to achieve a small delay step without significantly increasing the area of the delay element *with more sections*.

In this work, we investigate a programmable delay element based on singly terminated LC ladder filters. The key features of the technique are as follows. Terminating the ladder on one end reduces attenuation, thereby reducing the power needed

to achieve a desired gain. Due to the lumped design, parasitic capacitances of the transconductors used to sense capacitor voltages can be absorbed into the ladder without effecting performance. Lumped element design also reduces the area needed to achieve a desired delay. Unlike methods based on transmission lines, fine delay tunability is obtained by varying bias currents in active elements - something that is easily doable. The technique and its implementation details form the subject of the rest of this paper, which is organized as follows. Section II, we discuss the basic idea and tradeoffs behind our technique. Circuit design details are given in Section III. Results from a prototype chip fabricated in a 0.25 $\mu$m SiGe BiCMOS process, are given in Section IV. Measurements show a tunable delay of 80 ps with a resolution of 0.5 ps and gain step of 0.15 dB. The IC, which has an active area of 1 mm$^2$, dissipates 47 mW from a 2.5 V supply. Section V concludes the paper.

## II. PROPOSED TECHNIQUE

Our technique is described using Fig. 2(a), which shows two singly terminated LC ladder filters. Fifth order filters are chosen to illustrate the principles, but our IC implementation uses 12$^{th}$ order ladders. The transfer function from $v_{in}$ to $v_1$ in Ladder-1 can be shown to be of the all-pole low pass kind and is denoted by $1/D(s)$. Further, analysis shows that the transfer functions to the capacitor voltages and inductor currents have numerator polynomials comprised of even and odd powers of $s$ respectively. Thus, by summing the state variables of an LC ladder network (with appropriately chosen weights), we see that one can realize a transfer function of the form $N(s)/D(s)$, where $N(s)$ can be *any* polynomial (with order $\leq$ the order of $D(s)$). By choosing $N(s)$ to approximate $e^{-sT_d}$,

978-1-4673-6145-3/13 $31.00 © 2013 IEEE

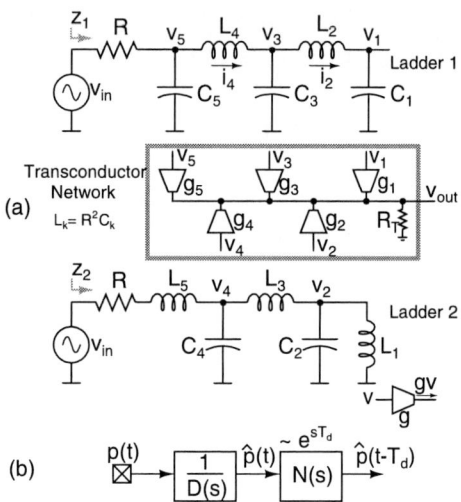

**(a)** $L_k = R^2 C_k$

**(b)**

Fig. 2. (a) Principle of operation of the delay element based on singly terminated LC ladders and (b) equivalent block diagram.

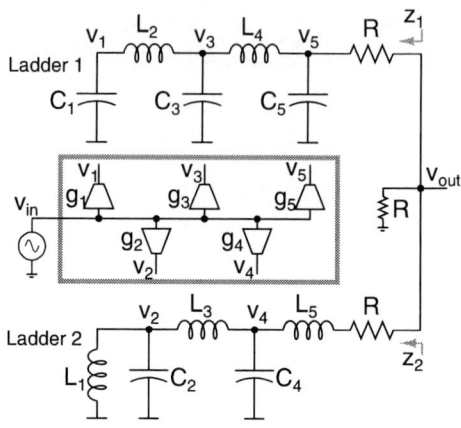

Fig. 3. The "companion form" of the system of Fig. 2 used in the actual implementation.

i.e $N(s) = 1 - sT_d + \frac{s^2}{2}T_d^2 \cdots$, a delay of $T_d$ can be realized. The delay can be made tunable by varying the coefficients by which the state variables are weighted.

From the discussion so far, it is seen that weighted summation of state variables allows the realization of programmable delay. However, sensing the ladder's inductor currents is problematic. Fortunately, this limitation can be overcome using a dual ladder (Ladder-2 in Fig. 2(a)). The inductor and capacitor values in the main and dual ladders are related by $L_k = R^2 C_k$, $k = 1, \cdots, 5$. Recall that a dual network is one in which the roles of voltage and current are interchanged - thus, capacitor voltages in Ladder-2 are proportional to inductor currents in Ladder-1. The transconductors (Fig. 2(a)) enclosed in the grey rectangle sense the capacitor voltages in the main and dual ladders and their currents are added into $R_T$ to produce $v_{out}$, which is the weighted sum of state variables of the LC ladder filter. Delay tunability is achieved by making the transconductors digitally programmable.

The operation of the system can be abstracted as shown in Fig. 2(b). The input pulse (say a Gaussian monocycle) is equivalently filtered by a low pass filter with transfer function $1/D(s)$ before being delayed by $T_d$. In order to not excessively distort the pulse shape, $D(s)$ must be chosen to have a flat group delay. Since a high order Bessel filter has a flat group delay with flatness extending well beyond the 3 dB bandwidth, the components of the LC ladder were chosen to realize a Bessel approximation.

While Fig. 2 illustrates the principle of our delay element, the actual implementation uses the so called "companion form" of the LC ladders. That is, rather than sense the capacitor voltages and convert them into currents which are then added, the input is converted into currents which are injected into $v_1, \cdots, v_5$ as shown in Fig. 3, and the output voltages of the two ladders are added. The addition can be performed passively, by using yet another interesting property of dual networks, namely the complementary impedance property. It can be shown [3] that in Fig. 3, $z_1(s) \parallel z_2(s) = R$. Thanks to this, the individual ladder outputs can be added using a resistor as shown in the figure. It is easy to see that the output voltages $v_{out}$ are identical in the circuits of Figs. 2 and 3. The practical

advantage of the companion form is the following. The output parasitic capacitance of the transconductors does not vary with the transconductor value. This, therefore, maintains the ladder frequency responses in spite of tap weight changes. The variation of input capacitance with tap weight is not an issue as the transconductors are driven by a buffer.

As mentioned earlier in this section, fifth order ladder filters are used to explain our technique - the actual realization uses twelfth order Bessel ladder filters. Apart from the use of singly terminated ladders (thereby avoiding signal attenuation in conventional delay cell designs), the use of LC ladders enable the absorption of transconductor parasitics into the ladder. The bandwidth of the Bessel filter was chosen to be 8 GHz - this was adequate to realize group delay flatness up to 12 GHz. Ideas similar to the one discussed above have been proposed for adaptive analog equalization at microwave frequencies [4].

## III. CIRCUIT DESIGN

The programmable UWB delay element was designed in a $0.25 \mu m$ SiGe BiCMOS process. The architecture of the test chip is shown in Fig. 4. The fully differential signal path uses bipolar devices for high speed operation. Amplifiers $A_1$ and $A_3$ are nominally identical. Only one of them is on at any time. $B_1$ is a buffer with a 50 Ω output impedance. The path $A_3 - B_1$ forms a bypass path that assists in de-embedding the response of the test setup. The output of $A_1$ drives emitter-followers that in turn drive the array of programmable transconductors that excite the main and dual ladders (denoted by Ladder 1 and Ladder 2 in the figure).

*Programmable Transconductors:* Fig. 5 shows a simplified schematic of the digitally programmable transconductor used in this work. It consists of two differential pairs, formed by $Q_{1,2}$ and $Q_{3,4}$. Only one of these is turned on at any given time, thereby enabling the realization of positive and negative transconductances. $I_1$ and $I_2$ are MOS current sources that can be digitally set by 5-bit control. The non-active differential pair also neutralizes the collector-base capacitance of the active pair. The differential pairs connect to $Q_{5,6}$, which form a cascode stage. This stage is placed close to the ladder to reduce routing capacitance at the transconductor output. The collectors of $Q_{1,2,3,4}$ connect to the emitters of $Q_{5,6}$ through coplanar strip lines (CPS). Their geometries are optimized to

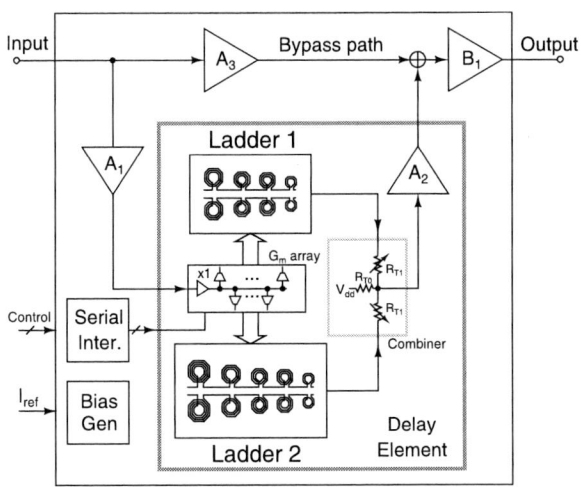

Fig. 4. The block diagram of prototype chip.

have a negligible effect on the group delay. Bleeder currents $I_b$ are used to maintain a sufficiently low impedance at the emitters of $Q_{5,6}$. These currents are only active for small tap weights (when $I_1/I_2$ are small).

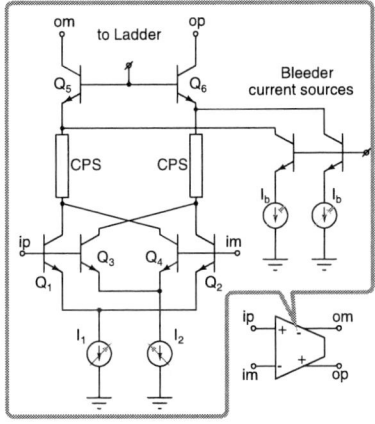

Fig. 5. Simplified schematic of the programmable transconductor cell.

*Ladder Filters:* The main and dual filters are realized using octagonal spiral inductors and MIM capacitors. The inductors are realized using $3\mu m$ thick top metal. The finite quality factors of the inductors are accounted for in the filter design by predistorting the filter transfer function to account for inductor losses. Accurate estimation of inductor parasitics and coupling between neighbouring inductors is important to achieve flat group delay. The Sonnet electromagnetic (EM) simulation environment is used to extract S-parameters and an equivalent 2-$\pi$ circuit model is determined as in [5]. The terminal parasitic capacitances of the inductors are absorbed into the ladder capacitors. As discussed in the previous section, the main and dual ladder networks form complementary impedances and their outputs are combined using a T-network of termination resistors. In our work, the terminations are implemented as fixed $65\,\Omega$ resistors in parallel with a tunable component, realized by varying the quiescent current through a diode. The reason to use a tunable termination is the following. In practice, process variations will cause a systematic shift of all capacitors in the ladder filters. If they increase by a

factor $\alpha$, it is straightforward to see that the shape of the frequency response of the LC ladder can be restored by reducing the termination resistance by $\sqrt{\alpha}$. The combined output of the ladders is processed by amplifier $A_2$, whose gain is programmable using a 3-bit word. The output of $A_2$ drives the output buffer $B_1$.

The test chip also consists of a bias generator and a serial interface that enables loading of tap gain settings from an external computer.

## IV. MEASUREMENT RESULTS

The test chip was fabricated in an IHP $0.25\,\mu m$ SiGe BiCMOS process through the Europractice program. The microphotograph of chip is shown in Fig.6.

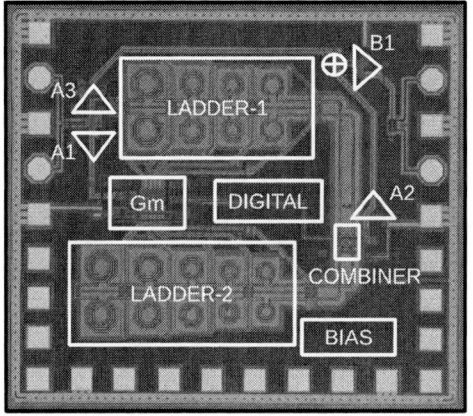

Fig. 6. Microphotograph of the chip.

The circuit was characterized by direct probing of the die using a CascadeMicrotech probe station. At the time of writing it was only possible to perform single-ended measurements, which were made using an Agilent E5071C two port network analyzer. The power and digital control signals were connected to the die using a multi-contact DC probe.

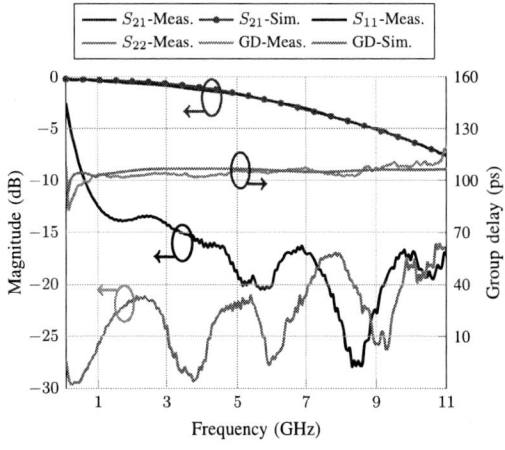

Fig. 7. Measured and simulated $S_{21}$ and group delay, and measured $S_{11}, S_{22}$.

The measured S-parameters of the delay element, when the nominal delay is set to 105 ps, are shown in Fig.7. The simulated group delay and $|S_{21}|$ are shown for comparison. The input reflection coefficient increases at low frequencies due to the following. The input common-mode voltage of the

978-1-4673-6145-3/13 $31.00 © 2013 IEEE

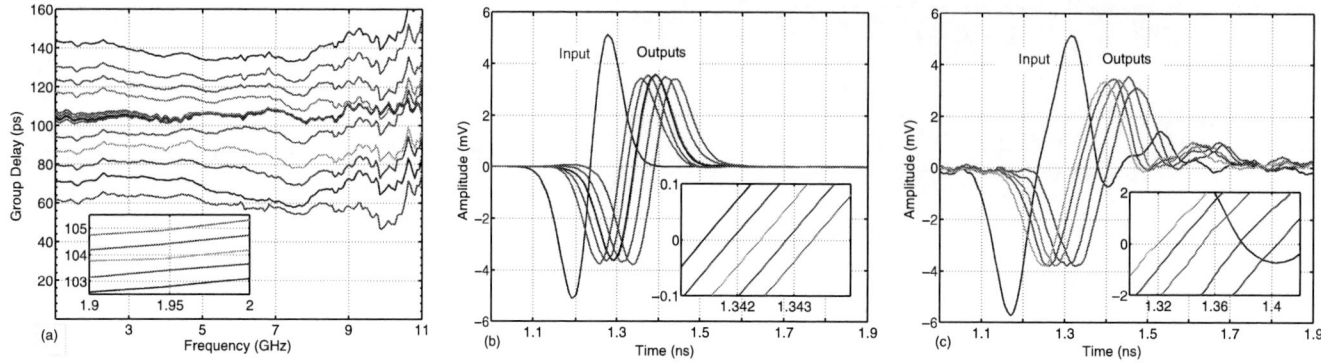

Fig. 8. (a) Measured group delay as the delay is tuned from minimum to maximum. (b) Pulse responses computed using frequency response data and (c) Measured time domain pulse responses.

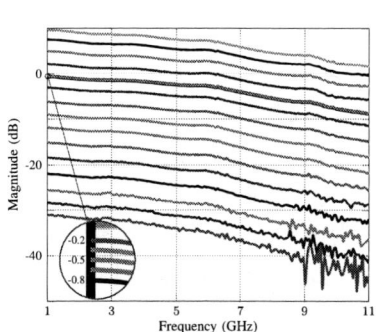

Fig. 9. Responses demonstrating tunable gain.

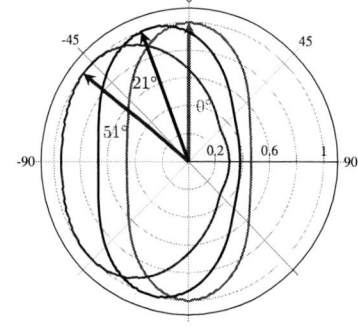

Fig. 10. Computed radiation pattern for 4-element array with spacing of 1 cm.

TABLE I    Performance summary and comparison

|  | This work | [1] |
|---|---|---|
| Delay range (ps) | 60 to 140 | 80 to 144 |
| Delay step | 0.5 ps | 4 ps |
| Gain range (dB) | -30 to 10 | 5 to 10 |
| Gain step (dB) | 0.15 | 1 |
| Active area (mm²) | 1 × 1 | 0.8 × 2 |
| Power | 47 mW @ G=10 dB 35 mW @ G=2 dB | 87.5 mW @ G=10 dB |

the chip is set by an on-chip bias generator, the output of which is capacitively bypassed to ground. The capacitor size is chosen to keep the impedance of the bias generator low in the desired band (3-11 GHz), causing $S_{11}$ to increase at low frequencies.

Fig. 8(a) shows the group delay variation from 60-140 ps. Only a few steps are shown in the interest of clarity. The inset demonstrates the 0.5 ps delay tunability. The time domain responses computed from the frequency responses are shown in Fig. 8(b). Measured pulse responses are shown in Fig.8(c) - due to reflections in the test setup, the input monopulse is distorted. However, the delay range and resolution are apparent. Again only few steps are shown for clarity and inset shows close-up view of zero crossing points demonstrating 80 ps delay range. The measured responses demonstrating gain tunability from -30 dB to 10 dB are shown in Fig. 9 and the inset shows the 0.15 dB step.

Fig.10 shows the computed radiation pattern using measured frequency response data for a four antenna beamformer with an element spacing of 1 cm. The beam-steering range is $\pm 51°$ and with a steering resolution of $0.86°$. The delay element, which occupies an active area of $1 \, mm^2$, dissipates 47 mW from a 2.5 V supply. The achieved performance is summarized in Table I and compared with results from [1].

## V. CONCLUSIONS

We proposed a lumped component programmable time delay element for UWB beamformers. The delay element,

based on singly terminated ladder filters exploited the concept of duality to enable voltage sensing, and the principle of complementary networks to achieve passive addition. These techniques allow transconductor parasitic capacitances to be part of the ladder filter, and enable low power dissipation. The efficacy of these techniques were demonstrated by measurements from a test chip.

## ACKNOWLEDGEMENT

This work was funded by the Department of Science and Technology, Government of India, through the Swarnajayanthi Project. The authors also thank Dr.H.Krishnaswamy of Columbia University for useful comments on the manuscript.

## REFERENCES

[1] J. Roderick, H. Krishnaswamy, K. Newton, and H. Hashemi, "Silicon-based ultra-wideband beam-forming," *IEEE Journal of Solid-State Circuits*, vol. 41, no. 8, pp. 1726 –1739, Aug. 2006.

[2] T.-S. Chu, J. Roderick, and H. Hashemi, "An integrated ultra-wideband timed array receiver in 0.13μm CMOS using a path-sharing true time delay architecture," *IEEE Journal of Solid-State Circuits*, vol. 42, no. 12, pp. 2834 –2850, Dec. 2007.

[3] E. Guillemin, *Introductory Circuit Theory.* Wiley, 1953.

[4] S. Pavan, "Power and area-efficient adaptive equalization at microwave frequencies," *IEEE Transactions on Circuits and Systems I: Regular Papers*, vol. 55, no. 6, pp. 1412 –1420, July. 2008.

[5] Y.-G. Ahn, S.-K. Kim, J.-H. Chun, and B.-S. Kim, "Efficient scalable modeling of double- π equivalent circuit for on-chip spiral inductors," *IEEE Transactions on Microwave Theory and Techniques*, vol. 57, no. 10, pp. 2289 –2300, Oct. 2009.

# The Past, Present, and Future of Design-Technology Co-Optimization

Greg Yeric[1], Brian Cline[1], Saurabh Sinha[1], David Pietromonaco[2], Vikas Chandra[2], and Rob Aitken[2]

Research and Development, ARM

[1]Austin, TX, USA, [2]San Jose, CA, USA

greg.yeric@arm.com

*Abstract*— Design-Technology Co-Optimization (DTCO) has evolved from early Design-for-Manufacture (DFM) needs into a multi-faceted, multi-lateral co-optimization below 20nm, where multiple patterning and FinFETs add significant complexities. Effective DTCO now involves end product metrics applied to a myriad of design-technology choices. This paper will highlight past and present examples of DTCO in practice for low-power SoC design, and examine a future of even more complexity that will drive a continued evolution in DTCO.

## I. INTRODUCTION

In past technology nodes, foundries' continued ability to deliver process node scaling meant everyone except the foundries could remain complacent regarding Moore's Law. As conventional transistor and lithography scaling began to hit significant difficulties, other avenues had to be exploited. Design-Technology Co-Optimization (DTCO) has become an increasingly key component enabling scaling entitlement for advanced process nodes. But like Design-for-Manufacturability (DFM) before it, the term means many things to many people, and in practice it is a moving target. Today's technology choices are increasingly complex, and reaching the optimal result necessarily involves multiple parties in early communication. By examining the evolution from past to present, we can better understand DTCO's positive effects and identify future opportunities.

## II. THE PAST: FROM OPTIMIZATION TO CO-OPTIMIZATION

The disaggregation of the semiconductor industry fostered the growth of independent fab, fabless, IP, EDA, and packaging companies. In simpler technology nodes, advance communication between these entities was not required to produce sufficient technology scaling. Fabs produced Process Design Kits (PDKs) including design rules and transistor models, and products scaled accordingly.

Beginning in earnest around the 90nm node, various yield concerns resulted in (DFM) initiatives [1], which attempted to prescribe restrictions to designers that would maximize yield. Communication was primarily unidirectional, from fab to designer, in the form of increasingly restrictive design rule checks (DRC). Designers (of physical IP, for instance), would then optimize their products within the bounds of the PDK. After IP was created, end product development would begin with synthesis, place and route flow.

This past dynamic is depicted in Figure 1. Two to three years before the year of production (YOP), possible technology choices are narrowed and the target process development begins. After the process stabilizes and is characterized, a full PDK is issued, allowing IP to be created and then used to implement early designs in the yield ramp stage. The primary scaling decisions have already occurred.

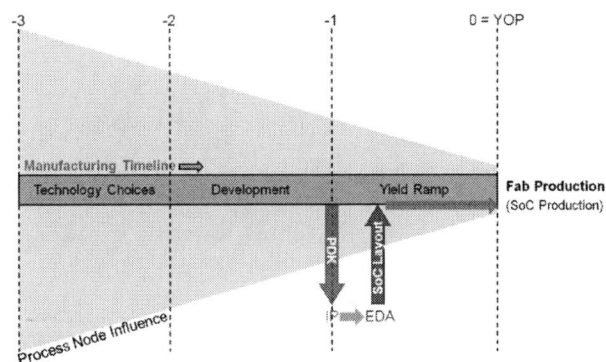

Figure 1: Manufacturing technology development timeline. Year of Production (YOP)=0. Conventional interactions with IP and EDA also noted.

### A. Distant past: DFM below 90nm

Restrictive design rules began to emerge in earnest in response to the 65nm node. Physical design engineers continued to rely on historical layout constructs which allowed them to minimize standard cell area, but many of these constructs became increasingly difficult to manufacture at 65nm and below in sub-wavelength lithography. Figure 2 shows layout for a flip-flop typical of the 65nm era, with key area-saving constructs (e.g. "outer channel" poly routes, which provide connectivity between disparate areas of the flop without using metal wires).

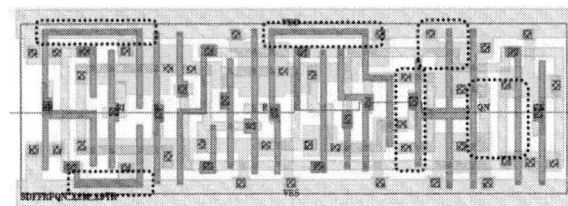

Figure 2: 65nm-era flip-flop. Key area-saving constructs are highlighted.

At 45/40nm and 32/28nm, with no wavelength scaling, increasingly stringent lithography restrictions made many of the area saving layout constructs of the past illegal. A 28nm-era flip-flop layout is shown in Figure 3. Comparing the similar circuit implementations in Figure 2 and Figure 3, the same layout required more area (horizontal gate pitches) to implement, as well as wiring on metal 2. This result can be

978-1-4673-6145-3/13 $31.00 © 2013 IEEE

considered a cost of regularity. A primary issue was dipole printing of 28nm gate pitches, requiring gate poly in only one preferred direction.

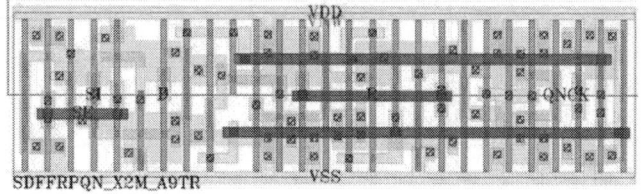

Figure 3: 28nm-era flip-flop. More gate pitches (area) required to implement the same function as compared to 65nm.

This eliminated all of the useful layout constructs shown in Figure 4, including out-bound poly routes, parallel gate connections, offset gate contacts, and non-uniform gate CD and non-uniform pitch poly. Thus, logic scaling from 65nm to 28nm could not scale to the entitlement predicted by the metal pitch and gate pitch scaling.

Figure 4: 65nm-era area-savings constructs that were lost to 28nm preferred gate orientation design rules.

### B. Past: Beyond Standard Cell Area

A problem with the state of DTCO in the 65nm-28nm era was that the final cost of the design restrictions was not known during their definition. IP designers could provide an estimate of flip-flop size increase, but the size of standard cells was not sufficient information, as it had been in the past. For example, each of the gate layer constructs shown in Figure 4 had to be replaced with metal wiring. The increase in metal within the cells could reduce yield according to critical area, but more importantly the pin access (ability of a router to connect to the pins of the standard cell) qualities of the standard cells had to be compromised. In many cases, such as the flip-flop shown in Figure 3, metal 2 wiring had to be employed (red lines) where none was required in previous technology nodes. Metal 2 use inside the cells further restricts routing and decreases implemented block utilization. However, getting to these answers requires complete library construction and well as optimized synthesis, place, and route, and the time line of Figure 1 did not allow for that.

Another example impetus in the evolving DTCO conversation came from the increased strain required to compensate for lack of physical gate length scaling below 90nm. A point was reached that was counter-intuitive at first glance: Larger standard cells produced smaller circuits. With strain, using a gate pitch larger than the minimum resulted in faster standard cells. This performance increase often more than compensated for the area penalty when implemented into logic blocks.

An example from 28nm test chip data is shown in Figure 5. Gate pitches P1-P5 represent minimum to continuously larger pitch. In this case, the intermediate pitches P3 and P4 resulted in performance gains high enough to offset the standard cell

size increase in many implementation cases. Higher drive transistors allow downsizing of gates and reduced repeater insertion, resulting in smaller implemented blocks.

However, given the constraints of implementing deep sub-wavelength lithography, decisions involving the gate pitch had been made much earlier, and in many cases could not be reversed. This issue helped underscore the value of using early learning from the implementation of logic blocks back to inform early technology development. In order to optimize the gate pitch, fabs must model and confirm in hardware the transistor-level characteristics (Miller capacitance, mobility) but without understanding the effect of technology choices through the product design flow may not result in the best end result.

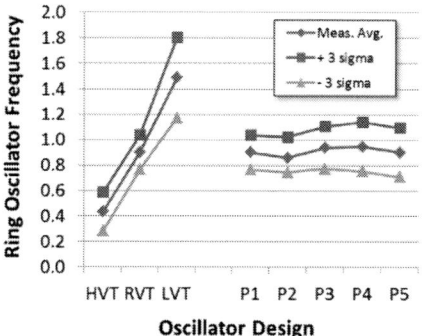

Figure 5: 28nm-era ring oscillator frequency as a function of gate pitch (right-side). Delay with varying VT type is shown in the left for comparison.

### C. Past: Deep sub-λ Lithography and IP

As the industry continued to scale pitch, the constant lithography wavelength (193nm) forced manufacturers to become much more aggressive with Optical-Proximity Correction (OPC) techniques, including Off-Axis Illumination (OAI) and sub-resolution assist features (SRAFs). This resulted in highly non-linear effects in layout, including significant influence of shapes beyond immediate proximity, which created significant challenges in the creation of standard design rules that would enable robust designs. This dynamic gave rise to the use of lithography simulators as additional design rule checks (Figure 6). By enabling such simulations of logic layout, a feedback loop with designers allowed inspection of local topologies and OPC interactions enabled tuning of OPC to provide higher density of common standard cell constructs.

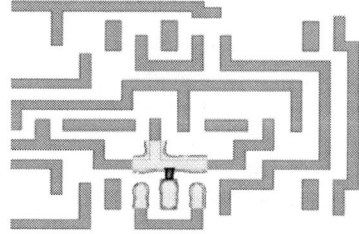

Figure 6: 28nm-era M1 layout and lithographic contour analysis example

Of course, no one can comprehend all possible end topologies and their printability. Thus, it became prudent to test printability schemes using end-user (IP design) layout styles much earlier in the technology development timeline.

Figure 7 shows a successful use case in test chip silicon, showing the identification of an OPC escape identified in implemented standard cell layout [2].

This placed further value on earlier two-way communication between fab and IP designer, before technology choices were set. This two-way collaboration, pushed into the technology development phase, is depicted in Figure 8 (as contrasted to Figure 1).

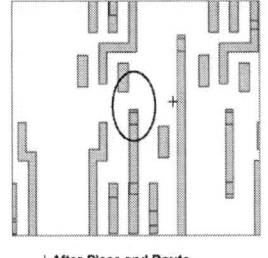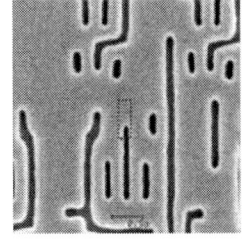

| After Place and Route          On Silicon

Figure 7:  Example of standard-cell like design-OPC interaction failure identified on a printability test wafer.

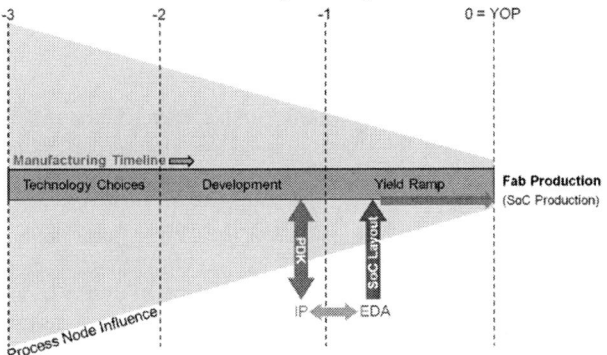

Figure 8:  Evolution of DTCO to include 2-way feedback during the technology development phase.

## III.   THE PRESENT: EARLY, MULTI-LATERAL DTCO

The technology below 28nm represents an inflection point in the application and benefit of DTCO. 20nm introduced double patterning of the metal layers, and the 16/14 nodes introduced FinFETs. Each of these issues added new levels of complexity into design/technology co-development which required early evaluation of higher level design metrics in order to achieve desired product scaling.

### A.  FinFETs

The transition to FinFETs below 20nm provided a key paradigm shift in DTCO. In previous technologies with relatively un-quantized device widths, a standard cell designer could receive M2 pitch and device characterization from a PDK and then independently determine which cell height(s) (in number of M2 tracks) represented the optimal end result for standard cell library(ies). The additional constraint of fitting a discrete number of fins within a cell changed this.

The existing paradigm can remain if the fin pitch equals the M2 pitch. However, in low-power standard cells there was significant pressure to reduce the fin pitch to below the M2 pitch. To understand this impetus, consider that not all fin tracks are available for active transistors. Power rail connections at the top and bottom of the cell force the removal of 1 fin each, and typically 2 additional fin tracks must be

removed in the center of the cell to accommodate gate input connections, which (as of today) are not allowed over the active diffusion regions. Because low power standard cell libraries have historically been 7-9 M2 tracks tall, if fin pitch equaled M2 pitch, there would be too few active fins available. This is illustrated in Figure 9. The drawing on the left shows the example of an 8-track standard cell, where only 4 active fins would remain. That results in only 2 fins per FET (PMOS and NMOS), which would not be acceptable for performance, but perhaps more importantly creates unacceptable quantization of device strength (only one device tuning option would remain, a -50% option). Figure 10 shows a circuit tuning example for a conventional, planar technology. Continuous (non-discrete) tuning of device widths allows the devices to achieve optimum power/performance. Discrete tuning between 1 and 2 fins only would render moot any real device tuning and result in higher power implementations.

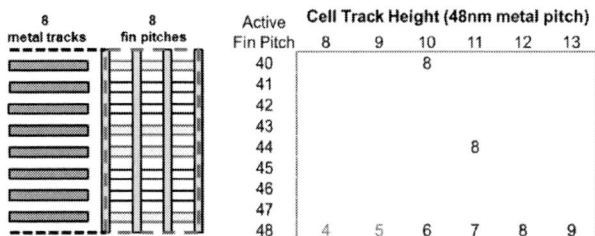

Figure 9:  With multi-gate devices, an integer number of fin pitches must fit within an integer number of metal tracks, defining the standard cell height (left). The table on the right lists the total number of active fins (total fins minus 4) in a standard cell for various fin pitches, if a solution exists.

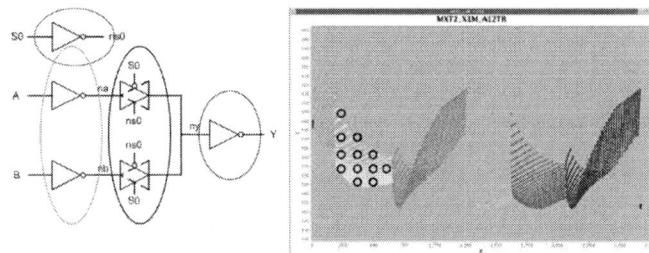

Figure 10:  Devices in schematic (left) are sized in minimum increments in order to find global circuit optimization. The right graph records cell delay as a function of tuning experiment, for the case of planar (non-multigate) technology.

As shown in the table on the right of Figure 9, there are limited solutions to the number of fins that can fit in a standard cell height, (only 2 integer solutions exist when fin pitch does not equal the metal pitch). One could populate the table with more solutions by allowing slight adjustments to specific fin locations. But this fine tuning is necessarily tightly linked to standard cell layout evaluation, including power rail construction, transistor contacting, etc., in order to determine exactly which fin configuration results in the best marriage to contacts and wire design rules for most cells.  Answers to sufficient accuracy must evaluate the aggregate effect on hundreds of key cells, many of which are quite complicated to construct (various And-Or-Invert, Flip-Flops, etc.).  The fin quantization era means the lead time required to create fairly complete cell libraries must be factored in while the fin patterning image is being determined, which is much earlier in the technology development cycle than before.

As an added complexity, multiple standard cell heights are typical for a technology. Generally speaking, there is a minimum size standard cell that is rational within a technology, and a larger cell size that represents the maximum performance available. SoC implementations can mix and match these minimum area and maximum performance cell libraries in various logic blocks to optimize power and performance. Thus, complex evaluations of IP become part of the discussion represented by Figure 8.

Fin patterning below 20nm does not escape problems with deep sub-wavelength lithography. For example, standard cell designers prefer to taper devices in order to optimize the power and performance. To support device taper below 20nm may involve additional mask layers (cuts). Thus, the situation in Figure 11 must be addressed: Increase standard cell area when taper is required (due to additional spaces between active regions of varying width) or increase wafer cost. Issues such as this bring the end customer more tightly into the discussion, because the right answer is a combination of two suppliers' metrics: Wafer cost and IP area.

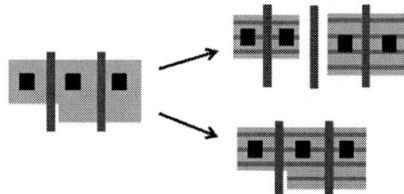

Figure 11: Device taper in older planar technology on the left, and choices for sub-20nm FinFET on the right.

Effects related to process, voltage and temperature (PVT) variability can also be significantly different with FinFETs. Fin width variation will affect circuits in new ways, because it should be more locally correlated than random dopant fluctuations and LER, but also because it directly affects drain-induced barrier lowering (DIBL) and sub-threshold slope [3] and should not be modeled by simple $V_T$ shift. Voltage scaling, above and below the nominal supply point, is an important design lever, and FinFETs can provide extra benefit in both directions. Additionally, MOSFET inverted temperature dependence (ITD) [4], which has been increasingly penal in recent nodes, may be ameliorated with proper FinFET construction.

### B. SRAM and Design assist

SRAM bitcells have historically been optimized with specific transistor size ratios that do not fit the integer world of FinFETs. The minimum area bitcell, with one fin each in the pull down, pass gate, and pull up devices in a bitcell (often referred to as a "111" bitcell), is inherently less stable than a traditionally-sized bitcell.

At the memory instance level, this means that design assist techniques which were optional in the past can be required with FinFETs below 20nm [5]. While FinFETs can provide better device matching than planar devices, owing to lower channel doping when properly constructed, Pelgrom's law makes life more difficult at each successive node [6].

Multiple bitcell types are usually offered (minimum area vs. higher performance options, for example). This is depicted in Figure 12, showing a "111" bitcell next to a "221" bitcell.

Additional larger cells can be more stable, but an additional complexity arises for the memory designer. Due to the discrete transistor sizing, some cell types will benefit from certain types of design assist but others could be degraded by the same assist method. Careful matching of bit cell with assist methodology is increasingly a key part of memory compiler design.

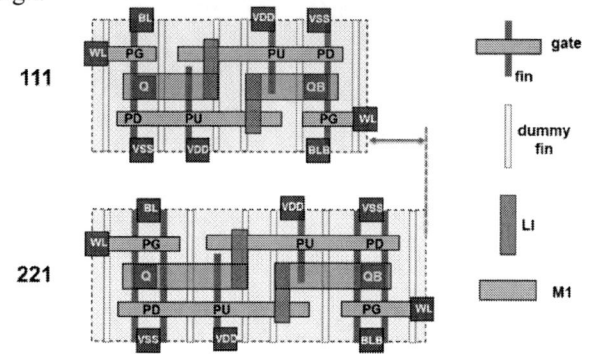

Figure 12: Minimum "111" fin bitcell compared to larger potentially more stable 221 bitcell.

### C. Multiple Patterning

Beyond 28nm, the required pitches forced the use of multiple patterning [7]. While a line/space grating can achieve the "peak" density offered by double patterning lithography (DPL), the average density in implemented standard cells ranges from this peak value toward the single-mask density. Figure 13 shows example layout of a cell with a high pin density and the set of shapes requiring resolution of a two-color conflict (where the two masks in double patterning are referred to as being different "colors"). In many cases such as this, there is no solution except to increase cell size, and the full entitlement of the pitch scaling is not achievable [8].

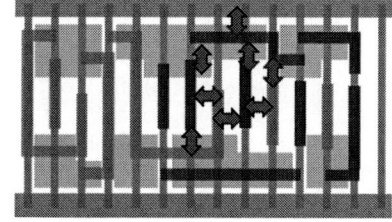

Figure 13: Example of complex DPL coloring conflict in standard cell.

Consider the simplified standard cell areas of Figure 14(a), in a two color layout. Shapes within a cell must not conflict with the power rail at the top and shapes from arbitrary neighboring cells on the left and right. That would result in a two-color loop conflict, often called an "odd cycle" [9]. However, maximally dense standard cell layout requires exactly this. One straightforward fix would be to add white space both vertically and horizontally as in Figure 14(b), but this is typically illegal due to contacting rules unless the cell width is extended by an additional poly pitch at every point where horizontal space needs to be added. Another option is to maintain the horizontal density but then require placement restrictions, as shown in Figure 14(c) [10]. This latter option may result in the best tradeoff, but that result requires comprehension of placer capabilities during technology definition, thus this scenario highlights the need for early EDA

978-1-4673-6145-3/13 $31.00 © 2013 IEEE

involvement in the technology definition, especially with regard to multiple patterning optimization.

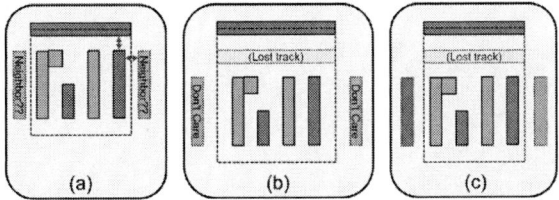

Figure 14: Standard cell DPL color conflict (a) and potential fixes. Fix (b) sacrifices area at left/right cell edges to make placement of abutting cells color-insensitive. Fix (c) does not increase cell size horizontally but requires restrictions on cell abutment combinations.

An additional issue at the 20nm node is the introduction of local interconnect (LI) layers between the transistors and M1. Contact layers became rectangular as dipole lithography was required to meet contacted gate pitch requirements. The rectangles provided a benefit in recapturing diffusion tabs (convex corners), one of the lost constructs shown in comparisons of Figure 2 and Figure 3, allowing transistor source/drain regions to be connected to the outside metal rails. In Figure 3 you see the diffusion tabs of Figure 2 replaced with metal tabs, which then blocked other routes from using those outer tracks. Additionally, by adding a second orthogonal local interconnect patterning layer, the process was able to support some of the key gate constructs without using M1, as depicted in Figure 15 [11].

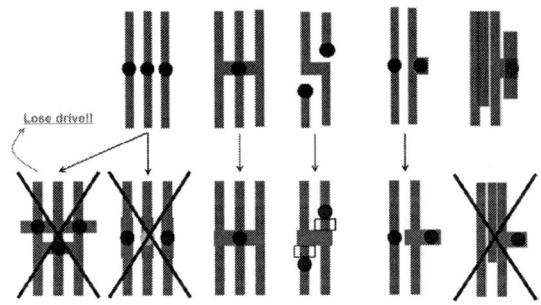

Figure 15: Key standard cell constructs, in 65nm-era 2D poly style (top) and in 20nm DLP/LI style (bottom).

The cost/benefit analysis of specific local interconnect and double patterning options requires understanding beyond simple pitch scaling, and even beyond the examination of a few standard cells. Consider the question of whether or not to change a layer from single to double patterning. If that were to add 2-3% to the wafer cost, then the resulting effect on final block area scaling must be determined to at least this accuracy, in order to provide meaningful feedback. Because these patterning details have varied effect on cell route-ability, one must build a fairly complete set of standard cell in order to be able to accurately calculate the aggregate result. Commensurately, the effects of these technology choices could not be evaluated on physical IP without a detailed understanding of EDA tool efficiency in these technologies. Parasitic variation due to misalignment of wires on different masks must also be taken into account [12]. These double-patterning issues further drove DTCO to the current paradigm depicted in Figure 16, where the development timeline includes multiple learning cycles involving fab, IP, and EDA.

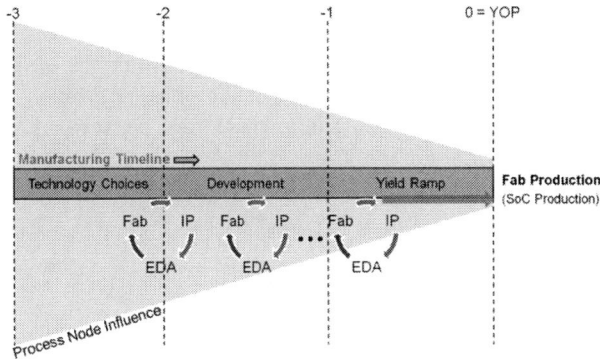

Figure 16: Contemporary DTCO, with multiple learning cycles involving close collaboration between fab, IP, and EDA.

DTCO of this nature, which begins more than 2 years prior to production, is currently targeted to the 10nm node. Given the lack of production-ready EUV at 10nm, at least in the development period, DPL cannot provide adequate pitch scaling, and therefore triple patterning lithography (TPL) may be required. Triple patterning coloring conflicts at 10nm cannot be contained/moderated by the solutions depicted in Figure 14, and furthermore TPL DRC is NP complete [13]. Thus, block level coloring (also known as decomposition) and TPL-aware placement and routing is a primary co-optimization concern. DRC, cell layouts, and decomposition algorithms must all be developed concurrently. To further complicate matters, multiple patterning must now be extended into the local routing layers, adding router algorithms to the above set of concerns. 10nm patterning, without EUV, promises to be significantly more complex than 20nm.

## IV. THE FUTURE: AN EXPANDED DTCO ECOSYSTEM

The increasingly complex technology choices discussed in the previous section represent merely the beginning of an inflection point in the semiconductor industry. As Figure 17 attempts to show qualitatively, FinFETs and DPL/TPL were just the beginning of intensified technological change. Continued attainment of scaling entitlement will require accelerated change and heterogeneity to the technology development landscape.

A prime example is the transistors themselves. The planar MOSFET was able to last 4 decades, but the silicon FinFET may last only 2 technology nodes. Active development for replacement devices includes mobility enhanced devices such as Quantum Well FETs (QWFETs), both horizontal and vertical nanowires (HNW, VNW), a host of 2D semiconductors, carbon nanotubes (CNT) and even non-field-effect devices. It is likely that multiple foundries will evaluate (different) multiple devices across multiple nodes in the near future.

The patterning roadmap portends a future of mixed-lithography choices. Triple patterning lithography (also known as LELELE) may coexist with self-aligned double patterning (SADP), and EUV, and then beyond the 10nm even multiple patterning versions of EUV. Augmentations at specific layers may come from direct write e-beam (DWEB) and/or directed self-assembly (DSA). All of these technologies must be evaluated in specific use cases in order to accurately understand the tradeoffs. Lithographic restrictions in the M1 layer will affect the efficiency of the place and route, which itself will be considering tradeoffs between gridding restrictions and runtime [14].

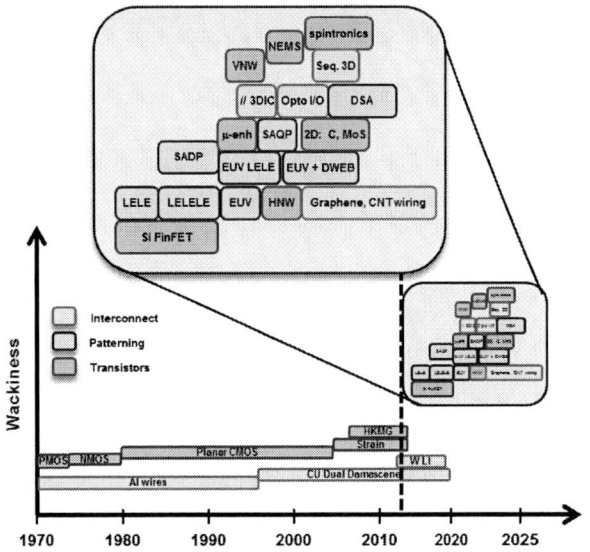

Figure 17: Qualitative assessment of past and future device complexity.

Cost of ownership is a primary concern regarding EUV, and that is driven by throughput, which is limited by source power. But simple wafer cost is not the complete view. The EUV source power issue is in a sense multiplicative, because Line Edge Roughness (LER) is inversely proportional to source power. If source power remains low, LER will remain higher, increasing variability that must be dealt with in design at the expense of circuit power, performance, and/or area.

EUV may present additional value at the design level by supporting fewer lithography-related design restrictions, possibly turning back the clock on some of the trends discussed above. A potentially significant example relates to M1 routing. As lithography progressed into deeper sub-wavelength nodes, at some point routers had to simply ignore M1. This is because the rules governing M1 became too complex, and routers are evaluated in large part by run time. In not allowing M1 route, block sizes can increase by up to 20%, or a significant fraction of a full technology node. If EUV were to allow M1 design rules to be simplified to the point where routers could once again utilize that layer, this would add to the EUV value statement.

If LER can be reduced, EUV could possibly be used to enable more complex active and poly shapes, which may allow for a reversal of the trend depicted in Figure 2 and Figure 3 and ease the pressure on the metal layers.

A possible design-technology limitation is EUV flare, which may need to be addressed via expanded, and possibly variable [15], transition regions between areas of differing pattern density. This will increasingly affect the density of

memory arrays, as parasitic limitations continue to push memory toward smaller instances, and smaller bitcell sub-array blocks within memory instances, as seen below:

Figure 18: Example large SRAM array with bitcells broken into sub-arrays and surrounded by up to 50% periphery (left). At right, array-periphery transition regions in M1.

In the following sections, we consider the inflection point of increasing technology complexity shown in Figure 17. We expect the following trends in DTCO to promote continued scaling.

*A. DTCO ecosystem expands along conventional axis*

The need to quantify end product metrics earlier and earlier in the process has added communication links that were previously non-existent. Research consortia and universities working on fundamental device and process R&D now regularly partner with fabless companies in order to quantify and validate fundamental technology choices. Technology evaluation through to the design level has also extended to equipment developers, as the processing challenges have become more intertwined with equipment characteristics. An example is given in Figure 19, showing ARM standard cell layout image tests on ASML EUV lithography tools.

Figure 19: Example of 10nm-node standard-cell like image on EUV printability test wafer. From [1]

End product cost is no longer neatly compartmentalized in fab and design buckets. The fab may consider many possible multiple patterning solutions, all with different costs, and different effects on IP scaling. Wafer cost differences might be rather straightforward to quantify, but their results on implemented designs are not. The choice between bulk and SOI wafers is one example. Improved device isolation and/or performance may compensate for an increase in wafer cost. Because specific patterning options have different effects on the scaling of specific cells, the design cost (in terms of total implemented chip area) is a statistical function of the micro-architecture, the standard cell library, and the performance target. This then necessitates end customer needs to be considered in order to accurately quantify options and choose the best technology path.

As an example, consider two design implementations of the same architecture and IP libraries, but with two different performance targets. These two implementations can arrive at very different conclusions regarding the optimum design-technology tradeoffs. Figure 20 illustrates this dynamic, using

978-1-4673-6145-3/13 $31.00 © 2013 IEEE

the example of varying transistor $V_T$ option. For lower frequency targets, cells with different performance (here via $V_T$ option), will converge to a minimum implementation area. As the performance target is increased, the synthesis, place and route flow will include larger drive strength cells (via transistor folding), and/or more buffer insertions. It will need to do this more aggressively with the lower performance cells (higher $V_T$). Thus, customer with a higher performance target can see a much larger benefit to any technology choice that increases the cell drive strength. This same differentiation would arise from reduced device size to accommodate staggered gate contacts as depicted in the lower left case of Figure 15. As shown in Figure 20, the difference in block area can be much greater than one or two additional mask steps, but is variable.

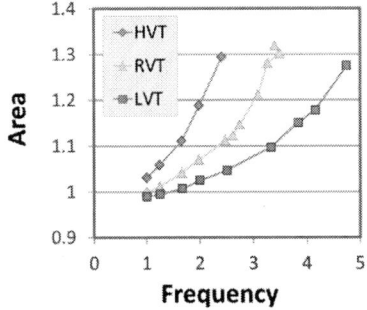

Figure 20: Implemented area as a function of target frequency and $V_T$ option.

In the future, results varying by product target may put pressure to bifurcate process choices in order to better meet criteria of different end customers. This is not a new paradigm, in the sense that older technologies provided distinctly different "HP" and "LP" transistor integrations. The difference for future technologies would be that this concept would extend to patterning choices and other fundamental process concepts.

### B. "More than Moore" brings more ecosystem expansion

Packaging, especially 3D IC technologies are expected to be a key enabler of future scaling progress. Opportunities lie in re-structuring memory and logic topology and in mixing different process technologies within the same final "IC". The value of the latter concept will be of increasing value, as supporting specific I/O requirements, including ESD, has become less and less attractive in advanced technology nodes. This might have occupied 10% of chip area in the past but can easily occupy 20% now, which will be compounded by the anticipated poor scaling of wafer cost into the future. Adding a lower-cost layer (that might allow better I/O performance anyway) is a tradeoff that will clearly be considered as cost-effective 3DIC comes online. Therefore, packaging entities will be an increasingly important part of future DTCO. Added device options, including opto-electronic I/O and embedded non-volatile memory could also be considered.

### C. Reliability as added DTCO dimension

As continued technology scaling faces increased difficulty in achieving desired progress, the DTCO ecosystem will need to continue to uncover additional areas of scaling opportunity. Reliability awareness should become an integral part of the complete technology offering to meet the power, performance

and area requirements. Design for reliability is a perfect example of DTCO, where close collaboration of innovation in devices, process, materials and circuit design can improve overall scaling as compared to past implementations (which generally simply guard-band around the unknown).

Reliability mechanisms which pose challenges to future scaling include Bias Temperature Instability (BTI), Time-dependent Dielectric Breakdown (TDDB), Hot Carrier Injection (HCI), Electromigration (EM), Soft errors and Random Telegraph Noise (RTN) [17]. Figure 21 depicts the increased pressure on reliability with continued technology scaling. A key enabler will be providing accurate reliability modeling in the early stages of technology assessment.

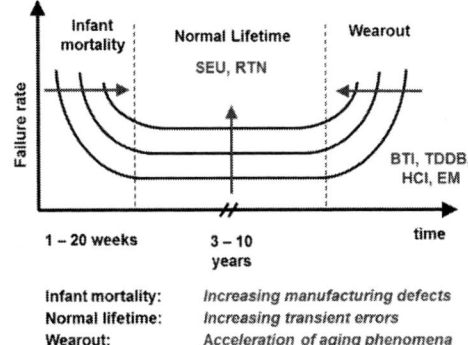

Figure 21: Reliability mechanisms becoming worse with technology scaling

Bias temperature instability, which relates to formation of traps in the gate oxide, is fairly well understood today, but may fundamentally change as entirely new channel materials are employed. Unlike other aging effects, BTI can be partially offset by "healing" – when the device is oppositely biased, some of the traps collapse. Designers can thus mitigate the effects of BTI by balancing bias states.

Time-dependent dielectric breakdown (TDDB), also known as soft oxide breakdown, is also trap related. FinFET devices can have improved TDDB due to vertical field reduction and increased barrier to tunneling [18], but 3D features must be carefully engineered and TDDB is still a critical reliability mechanism. With continued dimension scaling, in combination with pressure to employ lower dielectric constant films, increasing electric fields in the interconnect has extended TDDB concerns to the wiring [19].

Electromigration (EM) has moved from a problem of mild effect around high drive buffers to something that must be carefully checked throughout a design. Figure 22 shows that the maximum DC current allowed through local metal has not been keeping pace with device current scaling. This trend, which should intensify into the FinFET era, necessitates more robust power delivery network (PDN) design, wider metal in critical nets, and even in some cases strapping outputs in M2. All of these remedies necessarily increase block area.

While the soft error rate (SER) per bit cell has stabilized in recent technologies, and with FinFETs may actually improve as compared to planar, the rate per area has been increasing [20]. SER has conventionally been most important in SRAM arrays and dealt with via error correction. With continued scaling, the flip-flop SER has become comparable to that of

SRAMs [21]. However, protecting (hardening) flip-flops against soft error is challenging due to the fact that they are spatially distributed [22].

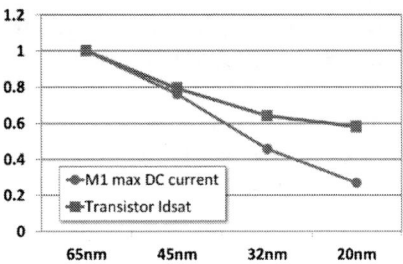

Figure 22: Relative scaling of metal maximum DC current versus $I_{DSAT}$ of inverter-sized transistor.

Random telegraph noise (RTN), which causes time-varying threshold voltage [23], is a future reliability concern. While this effect has historically not been significant in digital design, the trend for RTN variation, $\sim (L \times W)^{-1}$, is steeper than that for random variability, $\sim (L \times W)^{-0.5}$.

Mitigation of reliability effects extends past physical IP because many of the mechanisms are activity dependent and state dependent, which is complicated enough even before considering effects such as BTI healing. This activity dependence means consideration of workloads can result in substantial changes to product reliability prediction [24]. This will be yet another reason to add more understanding of end product into the DTCO evaluation.

### D. Continued importance of hardware optimization

As discussed above, design-technology choices can depend highly on power and performance targets. But beyond that, the choices for optimal logic design will vary as well. An example of this is heterogeneous multicore designs [25] where two or more entirely different, but software-compatible, classes of processor implementation are combined to provide hardware optimization for power and performance. Another example is the use of specialized graphics hardware (GPU) to provide higher performance for certain parallel compute tasks. Increasingly, system programmers and developers will have to make intelligent choices as to what specialized hardware to use for what tasks. Solving this problem, while managing software costs, will be key to improving hardware that has been optimized down to the process level for specific tasks.

## V. CONCLUSIONS

The semiconductor ecosystem has evolved from fab-centric process scaling to multi-lateral design-technology co-optimization as the scaling challenges have intensified. All evidence suggests that this trend will continue and intensify in the future.

### REFERENCES

[1] V. Pitchumani, "A hitchhiker's guide to the DFM universe", IEEE Asia Pacific Conf. Circuits and Systems (APCCAS) , 2006, pp. 1103-1106.

[2] S. Idgunji, V. Chandra, C. Pietrzyk, I. Iqbal, R. Aitken and G. Yeric, "An embedded process monitor test chip architecture", IEEE Intl. Conf. Microelectronic Test Structures, 2010, pp. 122-127.

[3] S. Sinha, B. Cline, G. Yeric, V. Chandra and Y. Cao, "Exploring sub-20nm FinFET design with predictive technology models", IEEE Design Automation Conference (DAC), 2012, pp. 283-288.

[4] A. Calimera, R. Bahar, E. Macii and M. Poncino, "Ensuring temperature-insensitivity of dual-Vt designs through ITD-aware synthesis", Intl. Workshop on Thermal Investigation of ICs and Systems (THERMINIC), 2008, pp. 31-36.

[5] V. Chandra, C. Pietrzyk and R. Aitken, "On the efficacy of write-assist techniques in low voltage nanoscale SRAMs", Design, Automation & Test in Europe (DATE), 2010, pp. 345-350.

[6] T. Matsukawa, Y. Liu, S.-I. O'uchi, K. Endo, et al., "Decomposition of on-current variability of nMOS FinFETs for prediction beyond 20 nm", IEEE. Trans. Electron Devices, v59 n8, Aug. 2012, pp. 2003-2010.

[7] J. Chen, W. Staud, and B. Arnold, "DFM challenges for 32nm node with double dipole lithography (DDL) and double patterning technology (DPT)", IEEE Symp. Semiconductor Manufacturing (ISSM), Sept. 2006, pp. 479-482.

[8] L. Liebmann, D. Pietromonaco, and M. Graf, "Decomposition-aware standard cell design flows to enable double-patterning technology", Proc. SPIE 7974, Apr. 2011, 79740K-1-12.

[9] J. Kye, Y. Ma, L. Yuan, Y. Deng and H. Levinson, "Lithography and Design Interaction – new paradigm for the technology architecture development", IEEE Custom Integrated Circuits Conference (CICC), Sept. 2012, pp. 1-4.

[10] R. Ghaida, K. Agarwal, S. Nassif, X. Yuan, L. Liebmann and P. Gupta, "Layout decomposition and legalization for double-patterning technology", IEEE. Trans. Computer-Aided Design of Integrated Circuits and Systems, v32 n2, Feb. 2013, pp. 202-215.

[11] G. Northrop, "Design technology co-optimization in technology definition for 22nm and beyond", IEEE Symp. on VLSI Technology (VLSI-T), June 2011, pp. 112-113.

[12] K. Jeong, A. Kahng, and R. Topaloglu, "Assessing Chip-Level Impact of Double Patterning Lithography", IEEE Intl. Symp. Quality Electronic Design (ISQED), March 2010, pp. 122-130.

[13] B. Yu, K. Yuan, B. Zhang, D. Ding, and D. Pan, "Layout decomposition for triple patterning lithography", IEEE Intl. Conf. Computer-Aided Design (ICCAD), Nov 2011, pp. 1-8.

[14] C.-T. Lin and Y.-L. Li, "Double patterning lithography aware gridless detailed routing with innovative conflict graph", IEEE Design Automation Conference (DAC), 2010, pp. 398-403.

[15] Fang, S.-Y., and Chang, Y-W., "Simultaneous flare level and flare variation minimization with dummification in EUVL", IEEE Design Automation Conference (DAC), 2012, pp. 1175-1180.

[16] M. van den Brink, "Continuing to shrink: Next-generation lithography - progress and prospects", IEEE Intl. Solid-State Circuits Confernece (ISSCC), Feb. 2013, paper 1.1

[17] R. Aitken, "Reliability Evaluation at the Device Level and its Impact on Design", IEEE Design Automation and Test in Europe (DATE), 2013.

[18] S. Ramey, A. Ashutosh, C. Auth, J. Clifford, et al., "Intrinsic Transistor Reliability Improvements from 22nm Tri-Gate Technology," IEEE Intl. Reliability Physics Symposium (IRPS), 2013, paper 4C.5.

[19] R. Achanta, P. McLaughlin and F. Chen, "Failure rates for interconnect dielectric breakdown: Trends determining technology reliability scaling limits", IEEE. Trans. Device and Materials Reliability, v11 n2, June 2011, pp. 273-277.

[20] N. Seifert, B. Gill, S. Jahinuzzaman, J. Basile, et al., "Soft error suceptibilities of 22 nm tri-gate devices", IEEE Trans. Nuclear Science, v59 n6, Dec. 2012, pp. 2666-2673.

[21] A. Oates, "Reliability challenges for the continued scaling of IC technologies", IEEE Custom Int. Circuits Conf. (CICC), 2012, pp. 1-4.

[22] S. Devarapalli, P. Zarkesh-Ha, and S. Suddarth, "SEU-hardened dual data rate flip-flop using C-elements", IEEE Intl. Symp. on Defect and Fault Tolerance in VLSI Systems (DFT), 2010, pp. 167-171.

[23] K. Takeuchi, T. Nagumo, K. Takeda, S. Asayama, et al., "Direct observation of RTN-induced SRAM failure by accelerated testing and its application to product reliability assessment," IEEE Symposium on VLSI Technology, 2010.

[24] E. Mintarno, V. Chandra, D. Pietromonaco, R. Aitken, and R. Dutton, "Workload Dependent NBTI and PBTI Analysis for a sub-45nm Commercial Microprocessor," IEEE Intl. Reliability Physics Symposium (IRPS), 2013, paper 3A.1.

[25] L. Lugini V. Petrucci and D. Mosse, "Online thread assignment for heterogenous multicore systems", Intl. Conf. Parallel Processing Workshops (ICPPW), 2012, pp. 538-544.

# From 2D-Planar to 3D- Non-Planar Device Architecture:
# A Scalable Path Forward?
## (Invited Paper)

Ghavam G. Shahidi

IBM T. J. Watson Research Center, Yorktown Heights, NY 10598

## Abstract

The microelectronics industry is in the process of transitioning from 2D-planar devices to 3D-non-planar (FinFET). In this paper, a metric is developed to assess the impact of scaling and device performance on chip (circuit) power as it is migrated node-to-node. The impact of node migration is assessed at product level as it is moved from 32 nm (2D-planar) to 22 nm (3D-non-planar device). Some of the limitations of the existing 22 nm 3D-device is reviewed that may explain some of the short comings in the product performance. Going forward it is critical that in two areas the FinFET needs to be improved: Multi-$V_T$ implementation and move away from the tapered fin shape.

## I - Introduction

Despite much talk of the "end-of scaling", it is fully expected that the CMOS technology scaling will continue through 14 nm, 10 nm, and beyond nodes. Lithography, the most important enabler of scaling, seems to be capable to print at the minimum dimensions of the upcoming nodes. One central question is the value of scaling, i.e. if there are any benefits in terms of cost and/or power-performance. Closely tied to the value question is the transistor performance in the upcoming nodes. There are many transistor requirements, challenges, and options not only to enable scaling, but to maximize the value of scaling. To sort through the scaling value proposition and tie it to the transistor, it is necessary to establish a metric for the various tradeoffs. In this paper, we use both scaling (i.e. chip per unit area) and drop in energy per operation vs. frequency (i.e. chip power at constant frequency) as the main benefits of scaling. As we will discuss, these are closely related, and are also tied to the transistor performance.

In the first part of this paper, we will discuss the value of scaling using historical data. We will discuss the range of possibilities for the upcoming nodes, and couple that to the expected transistor performance roadmap and the expected value of scaling in the future nodes. In the second part of the paper we

review the reasons behind the migration to 3D transistors (FinFET), and the challenges in its first implementation at 22 nm. We will argue that because of some fundamental issues (multi-threshold setting, 3D process challenges), there were compromises in the first implementation of FinFETs, and these challenges need to be overcome in order to enable their scalability to beyond 14 nm.

## II – Technology Scaling Value Proposition

CMOS node scaling has been in practice since the early 90's. It has two major components. One is Moore's Law, and the other part is Dennard's scaling theory. Moore's Law [1] refers to scaling the x- and the y-dimensions of the chip by 30% (shrink factor), and the area by 50% (Fig. 1a), and thus doubles the number of chips per wafer by each successive technology node. This benefit of Moore's Law is very well publicized.

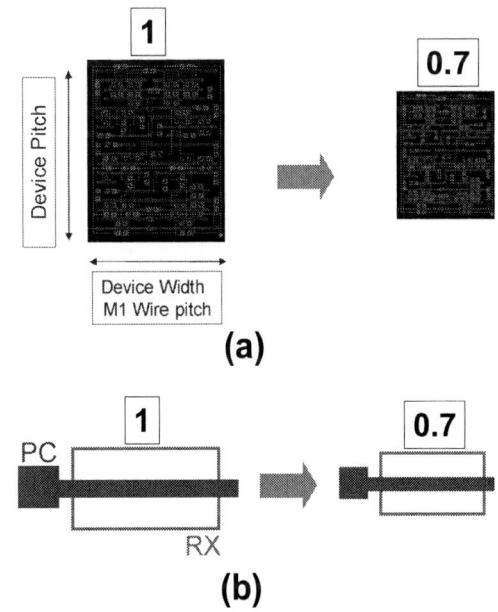

Figure 1 - Circuit (chip) and the device width shrink in Moore's Law

Another important side-benefit of the act of simple shrink (Moore's Law and the width scaling in the

978-1-4673-6145-3/13 $31.00 © 2013 IEEE

Dennard's scaling) is the drop in the chip power by the same scale factor (i.e. 30% drop in chip power). The reason behind this power drop is the drop in the total transistor width (and the capacitance) on the die by 30% (Fig. 1b).

Moore's Law does not make any reference to voltage and/or performance scaling. That is covered by Dennard's scaling theory, which in broad terms covers the improvement in device performance at every node. This is depicted in Figure 2, which is a plot of device performance (i.e. device-loaded ring oscillator frequency) vs. time. Since the early 90's through early 2010 the device performance has been improving by about 17% per year (about 35% per node) through combinations of Dennard scaling and introduction of other performance-improving effects such as strain. The improved performance can be traded off to run at a lower voltage (and power).

In this paper, the energy per operation (or chip power at constant frequency) is used as a metric for the scaling benefit. The combination of 30% device width scaling and 35% device improvement (traded off to run at lower power) result in slightly >70% drop in chip power at every node. A good example of the scaling benefit is Sony's Playstation 2 chips which were originally manufactured in 250 nm node and later scaled 3 times through 180 nm, 140 nm, and 90 nm [3]. Moving the chip through 3 nodes resulted in about 73% drop in power per node (Table 1).

| Technology Node | Year | Chip Area (mm$^2$) | Chip Power (W) |
|---|---|---|---|
| 250 nm | 98 | 518 | 23 |
| 90 nm | 04 | 87 | 0.5 |

**Table 1** - Sony Playstation 2 chip area and power (constant frequency) in 250 nm and after scaling for 3 nodes.

Device performance gain per node has been slowing down for the last 1-2 nodes (Fig. 2), and it is expected to be lower than the historical values (through 90 or 65 nm nodes). This reduction is driven by the slowdown in oxide and voltage scaling (as the operating voltage is getting close to the device threshold). The drop in the device performance and the slowdown in voltage scaling have resulted in less scaling benefit per node. For example in moving from node to node one can expect 40-50% drop in chip power (30% from the shrink and 10-20% from the device performance traded off for lower chip power). To illustrate this benefit, we use both Intel's Core-i7 power-frequency for 45 nm and 32 nm technologies (generation 1 and 2 of Core i7) and IBM's Power 7 server microprocessor.

In the case of Intel, moving products from node-to-node, they use a "Tick-Tock" approach [4], where the first product in the new node is a straight map of the previous generation ("Tick"), followed by new core and/or features in the new node ("Tock"). Figure 4 shows the data for all the 4 core and 8M L3 cache parts and the mobile parts (2 core-4 M cache) scaled to 4 core and 8M L3 (open circle and square). The lower envelope of the power-frequency data is a measure of the best that the technology can offer. We use representative parts on the envelope curve to estimate the power reduction per node at a given frequency. Using (direct mapped) parts i7-975 (in 45 nm) and i7-990x (in 32 nm), one can estimate an early technology benefit of ~33% in power reduction (both with no graphic cores). Later in 32 nm, Intel moved to Sandy Bridge core (Generation 2) with graphics ("Tock").

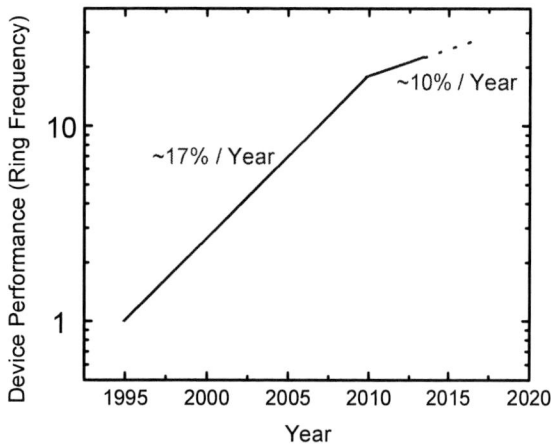

**Figure 2** - Device performance (i.e. ring frequency) through CMOS generations.

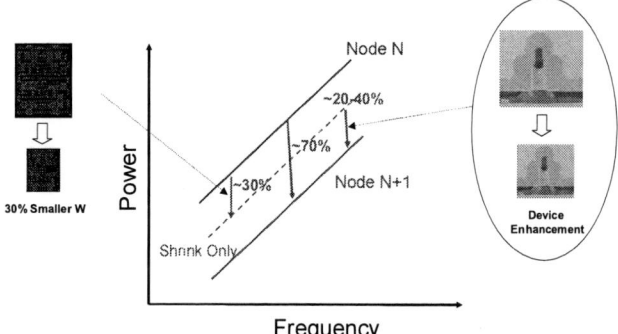

**Figure 3** - Power vs. frequency for a chip moving from node (N) to node (N+1)

Using later parts in 32 nm (with approximately the same core transistor count of 75-80 M), and taking into account the graphic core power in Sandy Bridge [14], one obtains about ~50% power reduction, node-to-node (using parts i7-870S in 45 nm and i7-2960XM in 32 nm, with ~12W allocated for the graphic core running @ 650 MHz).

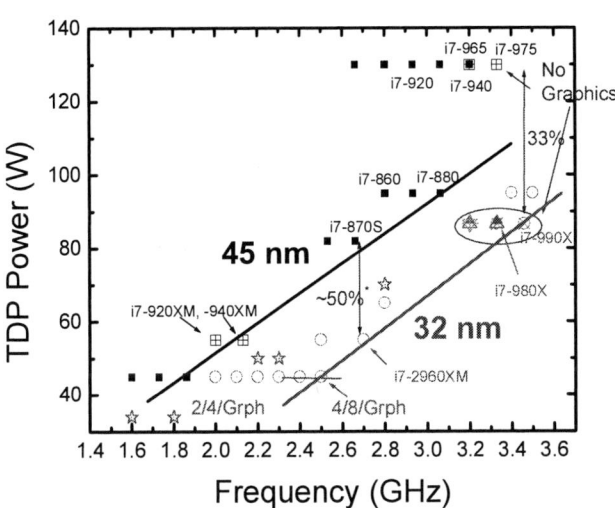

**Figure 4** - Intel Core-i7 (4 cores, 8MB L3 cache) power vs. frequency in 45 nm and 32 nm.

To illustrate the technology benefit, we also use the IBM's POWER7 (in 45 nm [5]) and 7+ (in 32 nm [6]), (Table 2). Chips in both technologies have the same area (576 mm$^2$) and the same total power. Using the published data, one obtains a normalized power (per transistor per GHz) of 1 POWER7 and 0.46 for POWER7+. This implies >50% reduction in the energy per operation when moving from 45 nm to 32 nm.

| Product | Technology Node | Transistors (B) | Normalized Power Per Transistor per GHz |
|---------|-----------------|-----------------|------------------------------------------|
| Power7 | 45 nm | 1.2 | 1 |
| Power7+ | 32 nm | 2.1 | 0.46 |

**Table 2** - IBM Server chips, POWER benefit from scaling, moving from 45 nm to 32nm.

Going forward, if there are breakthroughs in device performance (i.e. higher mobility substrates) then the benefit of scaling will improve. Similarly if there are changes in the 30% shrink factor, that will directly impact the power-performance value of scaling.

## III – Device Transition to 3D (FinFET)

One major requirement of scaling is that the FET must fit in the shrinking device pitch (contacted poly pitch or CPP). Table 3 shows the CPP for the last few nodes, and the expected CPP for the next few upcoming nodes.

| Node | CPP – Contacted Poly Pitch (nm) |
|------|----------------------------------|
| 32-28 | 110-130 |
| 22-20 | 90-100 |
| 14 | 65-80 |
| 10 | 55-65 |
| 7 | 48-56 |

**Table 3** - Contacted Poly Pitch (CPP) for the recent and upcoming nodes

One must be able to shrink L, contact size, and spacers so they all fit in the CPP. Figure 5 shows the value of L for the last few nodes. There was a large drop in L through the 90 nm node to about 40-45 nm, and then through the 32 nm, L has remained the same (at about 40 nm). This points to the challenge in scaling L by the approaches that have been in use through 32 nm. Clearly as CPP drops to below 100 nm, L must be reduced below 40 nm.

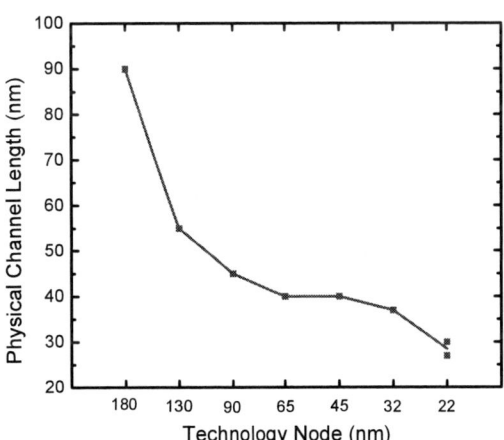

**Figure 5** - Physical channel length, 180 nm through 22 nm CMOS technologies

There are other device issues that need to be addressed as we scale forward. If one stays with the conventional 2D planar devices, it is expected that the channel doping would steadily increase at every node, to about low $10^{18}$'s for 22 nm (assuming conventional halo doping approaches). High doping will result in higher junction leakage and random dopant-caused $V_T$ fluctuations (RDF). RDF in turn would make the operation (or yield for large circuits) at low voltage (<0.7 V) very difficult (e.g. SRAM without assist features) [7]. To address L scaling, RDF, and other benefits, Intel introduced fully depleted FinFETs in 22 nm technology node in 2012. Benefits of FinFET are well known and documented (Table 4). Undoped channel results in the elimination of RDF and sharper sub-threshold. If the FinFET width is small enough, it will result in improved short channel effect. It is expected that the combination of sharper sub-threshold and lack of RDF would enable operation at low voltage (i.e. below 0.7-0.8 V).

| FinFET Property | Impact | Circuit Implication |
|---|---|---|
| Thin Body (8-12 nm) | Scale to Short L | Low Power, Fit in CPP |
| No Channel Doping | No RDF | $V_{MIN}$ (Low Voltage Operation) |

**Table 4** - Key features of an ideal FinFET and their resulting benefit.

FinFET is now on the roadmap of most major technology companies, and is the process of record for 14 nm technology node for many of them.

## IV – Multiple-$V_T$ Setting in 3D Fully-Depleted Devices

One of the requirements for a technology node is the ability to offer multiple threshold devices, in order to optimize power-frequency of the logic chip. The number of $V_T$'s can vary and be up to 4 in some designs. Usually the lowest $V_T$ is used in the performance sensitive critical paths. The highest $V_T$ is used in SRAM arrays [7], to minimize the stand-by leakage. Mid- $V_T$'s are used in the rest of the logic circuits. An ideal fully depleted device has an undoped channel while in conventional devices, $V_T$ is set by channel doping. So one logical question is how one would implement multiple thresholds in fully-depleted devices and its implications.

The approaches to set the device $V_T$ (before the advent of metal gate) has been to use channel doping and/or the poly work-function. If one uses channel doping and at the same time keeps the body fully-depleted, then the threshold becomes a function of the body thickness. This is illustrated in Fig. 6 [8]: When the device becomes fully depleted threshold voltage changes as a function of body (film) thickness. Scaling down the nodes increases the doping and therefore $V_T$ variation: As one scales the technology to the future

nodes, the fully-depleted device body thickness has to be reduced (to help with SCE). Then in order to obtain multiple $V_T$ (or the same shift in threshold) one has to apply a higher doping, which results higher $V_T$ variation (being on a steeper curve corresponding to higher doped fully-depleted device).

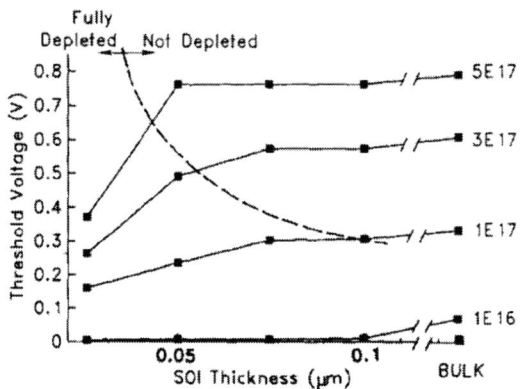

**Figure 6 -** $V_T$ as a function of body (i.e. SOI) thickness for $t_{ox}$ of 7 nm [Reproduced from reference 6]

With the advent of high-K and metal gate and the possibility of multiple work-function gate stack, the fully-depleted devices have received renewed attention. In this scenario, one would use the work-function to set the $V_T$ and obtain multiple $V_T$'s by simply adjusting the work-function of the metal gate. As of now, multiple work-function gate stacks for pFET and nFET have not been demonstrated in a late node technology. It is nevertheless a topic of intense research [9].

In the first implementation of the 22 nm FinFET technology, a combination of using the $V_T$ roll-off as well as channel doping was used [10,11]. High-K work-function was set to obtain the lowest $V_T$ of about 0.2 V (at shortest L of 27 nm). $V_T$-roll-off was used to get to the next one or two thresholds (about 0.25-0.3 V) by increasing channel length. Since the roll-off of the FinFET is minimal, to get to the highest $V_T$ (0.4 V) for the array devices, channel doping was used [10] (low $10^{18}$ dopant/cm$^3$). There are two implications of this approach: It forces one to use longer L for the mid- $V_T$ (i.e. higher power). In high $V_T$ devices (i.e. SRAM array), it introduces $V_T$ variation caused by the FinFET width variation, in addition to RDF. This new source of $V_T$ variation is unique to doped fully-depleted devices. It is expected that this variation will impact low voltage operation.

## V – Tapered FinFET and Spacer Formation

One feature of the 22 nm implementation of the FinFET is the tapered shape [11, 12]. In the middle of

978-1-4673-6145-3/13 $31.00 © 2013 IEEE

the fin the width is 8 nm (Fig. 7), while at the bottom of the fin the width is 16 nm [12]. The bottom of the fin is heavily doped to stop the source-drain punch-through. The short channel effect is dominated by the FinFET effective width at the bottom, estimated to be about

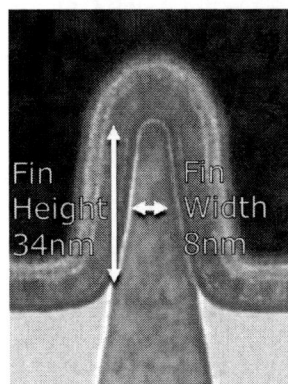

**Figure 7 -** Fin profile in 22 nm technology [Reproduced from reference 12]

14 nm, depending on the placement of the heavy doping punch-through-stop doping (at the bottom of the FinFET). As a measure of the SCE, the drain-induced barrier lowering (DIBL) at L=27 nm for a square FinFET of 14 nm width is about 86 mV/V while for a FinFET width of 8 nm, the DIBL is about <30 mV [13]. For a shorter L=18 nm, the impact on the DIBL of the tapered FINFET is larger (Fig. 8).

There are a number of motivations for the 22 nm implementation to be tapered. There are many challenges in FinFET processing and fabrication. Many of these challenges have been overcome and FinFET's are in high volume manufacturing. One area that has been particularly challenging and can impact the device performance is that of the spacer etch.

One of the most critical parameters in device design and CMOS node technology enablement is the placement of the extension (junction) relative to the gate edge. Too much overlap results in large gate to drain and source capacitance, which in turn (the $C_{gd}$) gets multiplied by the Miller effect, and impacts the performance. An under-lapped device will result in large parasitic source-drain resistance.

In traditional 2D devices, extension placement (i.e. implant) is done after the placement of a thin first spacer. The thickness, profile, and control of this spacer across pitch and various shapes is critical. Any variability in spacer thickness along the device width, or from device to device will result in large variability in device performance (i.e. switching times of the device). Spacer processing consists of depositing a thin layer of dielectric, followed by a very isotropic etch. Uniformity of the deposition and etch (along the device,

and across chip) are important in order to minimize the variability. Wherein lies the challenge to etch the spacer on a vertical-sidewall FinFET. In a vertical FinFET, the spacer on top of the Fin is cleared first, followed by the middle and then the lower part of the fin. In such a process, the FinFET receives non-uniform spacer etch along its width (Fig. 9) [14].

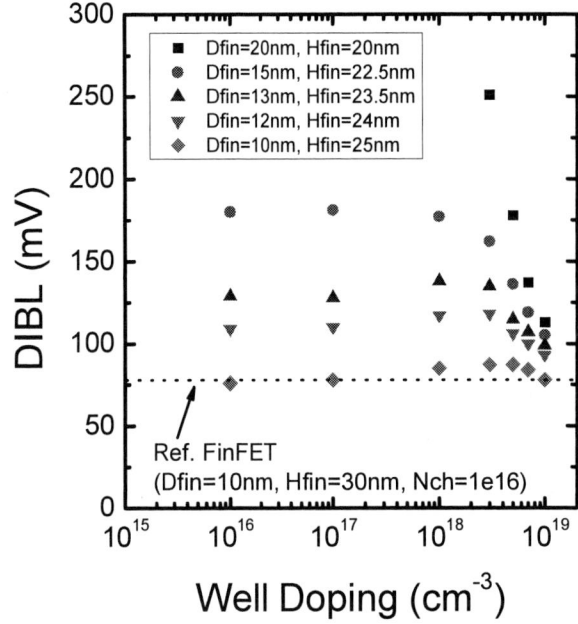

**Figure 8 -** DIBL vs. doping at various FinFET widths, for L=18 nm [Reproduced from reference 13]

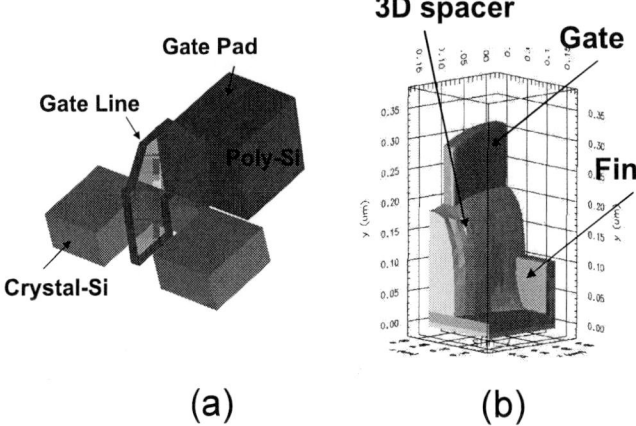

**Figure 9 –** Spacer etch on a vertical FinFET [Reproduced from reference 14]

One possible solution to address this issue is to make the fin tapered. With a slanted sidewall, other than the very top of the device, the rest of the FinFET receives a uniform spacer etch, although a long over-etch is required to clear the spacer. The very top of the Fin will have a thinner spacer, but since the Fin is thinnest on the top, it will also have better SCE in that device region, and it can handle a thinner spacer (and shorter L). The major implication of the slanted fin is that since the fin is wide at the bottom, the SCE is degraded. A degraded SCE would force a technology to be centered at longer L and result in higher capacitance. In the first implementation of the FinFET, the channel lengths were about 27-30 nm. This is a longer L as compared to other 20 nm technologies where the nominal L is 20 nm [15], or even in a 28 nm fully-depleted 2D-planar technology [16], where nominal L is 24 nm.

## VI – Performance Implication of Existing 22 nm FinFET Implementation

The FinFET holds great promise in terms of scaling to short L and operation at low voltage. Because of implementation challenges, a number of compromises have been made that impact the performance of the device as well as its ability to operate at low voltage .

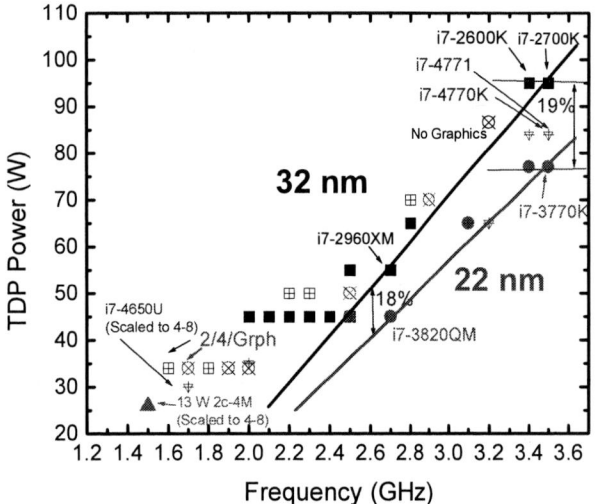

**Figure 10** – Intel Core-i7 in 32 nm (black symbols) and 22 nm (red symbols). Blue symbols are Haswell. Data are for 4C-8M or 2C-4M scaled to the 4C-8M.

To assess the impact on the chip performance, we use power-performance change for Intel core-i7 (4 cores and 8M L3 cache) as it moves from 32 nm to 22 nm ("Tick"), i.e. Generation 2 through Generation 4. In Fig.

10, black symbols are 32 nm i7-2x (Generation 2) parts (solid squares are 4 core and 8M and open symbols are the mobile parts, 2 core and 4M, scaled to 4C and 8M). Red symbols are 22 nm i7-3x (Generation 3) parts (solid red squares are 4C-8M and open symbols are the 2C-4M scaled to 4C-8M). The main difference in the "Tick" product migration between 32 and 22 nm is the graphic core [17]. The processor core was not much changed (about 80 M transistors in both technologies). The envelope of each data set is a measure of the best power-performance point in that technology. Using parts i7-2700 and i7-3770 (or i7-2960XM vs. i7-3820QM), node-to-node, the power performance improved by ~20%. For straight shrink with no device change, one expects ~30% change in power-performance. The fact that the gain is <30%, raises the question of whether there was an actual effective device degradation in going from 32 nm to 22 nm (caused possibly by larger effective capacitance and/or high variability ).

Intel Core-i7 had different graphic core in 32 nm and 22 nm. The graphic core power can be estimated using Xeon parts [18]. Taking that into account, actually degrades the node-to-node gain to <20%.

To the best of our knowledge, the nominal voltage for all the Intel processors in 22 nm is about 1.0 volt. In early 2013 Intel announced a low power (13 W TDP and 7 W SDP) version of Core-i7 [19]. The solid red triangle in Fig. 9 is the power-frequency point for Core-i7-3689Y, scaled from 2C-4M to 4C-8M (for comparison purpose). The low-power point falls on the existing data for Core-i7.

In mid-2013, Intel is announcing the Core-i7's with Haswell core, also called Generation 4 i7 (Core i7-4x). Not much is released about the Haswell core yet. It has been claimed that it may use a different (and presumably improved) device than the Ivy-bridge [20]. There is some speculation that it will have 10% more transistors, and configurable TDP and/or low power mode. Initial Haswell benchmark indicates 10-15% higher performance. Benchmarked chips were Ivy-Bridge i7-3770K @ 77W vs. i7-4770K @ 84W (i.e. i7-4770K running at 10% higher chip power [21]. Despite the differences in the "Tock" design, we place the power-frequency points for the Generation 4 (i7-4X) on the same graph (the blue triangles). The i7 4-core-8M Haswell core is within the 4-core-8M envelope for the earlier 22 nm i7's.

There is another implementation of a high performance 22 nm node technology, this based on 2D planar device [22]. This technology has faster devices (i.e. ring oscillators) than those reported in a technology based on 3D FinFET's [12]. It also has at least 13 level of metal, EDRAM, and the highest reliability grade in the industry. The earlier version of this technology (i.e. the 32 nm node) resulted in the largest microprocessor chips, running at the highest processor frequency, and very good gain node to node, as we described earlier (i.e. POWER 7+ vs. POWER 7). It is fully expected that any product using the IBM's 22 nm 2D technology will obtain 40-50% power reduction per transistor (at constant frequency) [23].

## VII – Requirements for Scalability

The product data comparison between 32 nm 2D planar technology and 22 nm 3D FinFET technology points to low node-to node gain. This is consistent with the modeled high capacitance in 3D structures [22], and with the long L (27-33 nm) and high variability that have been the result of doped fully-depleted devices. The deviations in the existing 22 nm FinFET implementation from the ideal FinFET (Table 4) is depicted in Table 5.

| FinFET 22 nm Implementation | Impact | Implication |
|---|---|---|
| Tapered Body – Wide at the bottom (16 nm) | Longer L | Higher C - Higher Power |
| Channel doping for the highest $V_T$ | RDF, $V_T$ variation caused by Fin width changes | High Voltage Operation |

Table 5 - Key features of actual FinFET technology in 22 nm and the resulting impact.

It is conceivable that the existing 22 nm FinFET implementation can be scaled to one more node: At 14nm, it is expected that the fin pitch will be at 40-45 nm (down from 60 nm in 22 nm FinFET). Keeping the FinFET height the same as in 22 nm with the same taper, will result in little L scaling. Furthermore spacer processing at the tighter fin and PC pitch will be more challenging. If one scales the FinFET height (and the width) by the scale factor, then the thinner FinFET will enable L scaling (but still a long L for the beyond 22 nm, similar to the same issue as in 22 nm). The challenge will be the parasitic gate capacitance at the bottom of the FinFET, where the heavily doped punch-through-stop bottom of the FinFET will create parasitic gate capacitance.

In addition to the excess capacitance caused by longer L and/or the heavily doped bottom of the FinFET, there is also large parasitic capacitance between gate-source and gate-drain [22]. In a 14 nm technology, at reduce fin pitch and reduced CPP, because of the increased parasitic capacitance and possible increase in variability, at the product level, it will be probably challenging to obtain performance gain

(i.e. power reduction) associated with the historical technology scaling.

In order to scale the 3D structure to future nodes and obtain performance, two fundamental issues need to be addressed. One is that of setting multiple $V_T$ with means other than doping and/or using the roll-off curve and going to longer L for higher $V_T$. If one uses doping to adjust the threshold, that would result in ever increasing doping levels and more variability as a result of dimensional changes. Methods to obtain different thresholds may include using multiple work-function metals in the gate stack, or using materials with different band gap than that of Si such as SiGe or Ge (in the pFET case) in some devices. In both cases they need to have low variability at the product level. The other issue is the use of tapered FinFET. It is difficult to scale this structure to many nodes beyond 22 nm. If one does not scale the FinFET height by the scale factor, then the wide bottom of the fin will pose challenge in scaling L. Without addressing these two issues, one is forced into a device which is similar to that in Table 4. A future technology implemented with these attributes will have low power-performance benefit and is far from the promised ideal FinFET of Table 3.

At present, the microelectronics industry for the most part is committed to implementing FinFET technology for 14nm (and beyond). Significant work is underway to address the above issues. In case some of these problems pose insurmountable challenges, alternative 2D-fully-depleted technology has been demonstrated [13]. Furthermore, there is some work indicating 2D-plannar devices may be scaleable [24] with good variability control [25].

## VIII – Conclusions

The introduction of the 22 nm 3D FinFET in 2012 has been a great engineering achievement. Nevertheless, due to a number of technology limitations, the implementation of the technology resulted in a tapered FinFET structure and doped FinFET's. The implication of these choices has been that the technology was centered along a longer L device and probably was impacted by variability (caused by doping especially in the SRAM devices). The power-drop, when migrating a product from node to node was less than expected even from a straight shrink.

If the existing structure is extended to 14 nm (with the same limitation ($V_T$ setting and use of tapered FinFET), then it is expected that the power-performance gain of the technology will suffer. Furthermore one can question the scalability of the existing implementation to 10 nm and beyond.

978-1-4673-6145-3/13 $31.00 © 2013 IEEE

To fully exploit the benefit of FinFET in future generations, it is critical that means are developed to set the $V_T$ by other than doping and move away from the tapered FinFET (i.e. introduce better etch technology for the spacer).

## References

[1] G. Moore, "Progress in digital integrated electronics", IEDM Tech. Dig., p. 11, 1975

[2] R. H. Dennard, et al, "Design of Ion-Implanted MOSFET's with Very Small Physical Dimensions," IEEE J. Solid-State Circuits SC-9, p. 256, 1974.

[3] K. Kutaragi, "Toward Future Computer Entertainment Systems", ISSCC Tech. Digest, p. 40, Feb. 2006.

[3] "Intel Tick Tock Model", http://www.intel.com/content/www/us/en/silicon-innovations/intel-tick-tock-model-general.html?wapkw=first+microprocessor.

[5] R Kalla, et al, "POWER7: IBM's Next Generation Server Procesor", Hot Chips 21, http://www.hotchips.org/wp-content/uploads/hc_archives/hc21/3_tues/HC21.25.8 00.ServerSystemsII-Epub/HC21.25.829.Kalla-IBM-POWER7NextGenerationServerProcessorv7display. pdf, Aug 2009,

[6] S. Taylor, "POWER7+™: IBM's Next Generation POWER Microprocessor " Hot Chips 24, http://www.hotchips.org/wp-content/uploads/hc_archives/hc24/HC24-8-DataCenter/HC24.29.815-Power7-Taylor-IBM-120828-Final.pdf, Aug. 2012.

[7] R Joshi et al, "Low voltage consideration for accurate yield predictions of VLSI Circuits", To be submitted to IEDM 2013.

[8] G. Shahidi et al, " A Room Temperature 0.1 µm CMOS on SOI", IEEE TED, Vol 41, Dec. 1994, p. 2405.

[9] R. Muralidhar, et al., "Meeting the Challenge of Multiple Threshold Voltages in Highly Scaled Undoped FinFETs", IEEE TED, Vol. 60, No. 3, p. 1276, 2013.

[10] M. Bohr, "Technology Insight: Silicon Technology Leadership for the Mobility Era ", http://www.intel.com/content/www/us/en/intel-developer-forum-idf/san-francisco/idf-2012-san-francisco.html. September 2012.

[11] C. Auth et al, "A 22nm High Performance and Low-Power CMOS Technology Featuring Fully-Depleted Tri-Gate Transistors, Self-Aligned Contacts and High Density MIM Capacitors", Proc. Of Symp. On VLSI Technology, 2012.

[12] K. Kuhn, " Technology Insight: CMOS and Beyond: Transistor Technology for the Mobile World", http://www.intel.com/content/www/us/en/intel-developer-forum-idf/beijing/IDF-2013-Beijing-special-attractions.html. April 2013.

[13] C-H Lin et al, "Non-Planar Device Architecture for 15nm Node: FinFET or Trigate?", IEEE SOI Conference, p.1, Oct. 2010.

[14] Y. Zhang, "Plasma Etching: Fundamentals, Applications, and Challenges", AVS Short Course, 2009.

[15] H. Shang et al, "High Performance Bulk Planar 20nm CMOS Technology for Low Power Mobile Applications", Proc. Of Symp. On VLSI Technology, p. 129, June 2012.

[16] N. Planes et al, "28nm FDSOI Technology Platform for High-Speed Low-Voltage Digital Applications", Proc. Of Symp. On VLSI Technology, p.133, June 2012.

[17] B. Heaney, "Designing a 22nm Intel® Architecture Multi-CPU and GPU", DAC 2012.

[18] Using Xeon parts E3-1275 and E3-1275V2 (in 32 nm and 22, with graphic cores) and E3-1270 and E3-12-70V2 (in 32 and 22, without the graphic cores), one can estimate the graphic processor power to be about 15W @ 850 MHz in 32 nm and 8W @ 650 MHz in 22 nm.

[19] A. L. Shimpi, " Intel Brings Core Down to 7W, Introduces a New Power Rating to Get There: Y-Series SKUs Demystified", http://www.anandtech.com/show/6655/intel-brings-core-down-to-7w-introduces-a-new-power-rating-to-get-there-yseries-skus-demystified, Jan. 2013.

[20] "IDF: Intel says Haswell won't use Ivy Bridge transistors", http://www.theinquirer.net/inquirer/news/2206077/idf-intel-says-haswell-wont-use-ivy-bridge-transistors.

[21] "Core i7-4770K: Haswell's Performance, Previewed", http://www.tomshardware.com/reviews/core-i7-4770k-haswell-performance,3461.html.

[22] S. Narasimha et al., "22nm High-Performance SOI Technology Featuring Dual-Embedded Stressors, Epi-Plate High-K Deep-Trench Embedded DRAM and Self-Aligned Via 15LM BEOL," IEDM Tech. Dig., pp. 415-418, Dec. 2012.

[23] J. Stuecheli, "Next Generation POWER Microprocessor", To be presented in Hot Chips 25, Aug. 2013.

[24] R. Muralidhar et al., "A Comparison of Short-Channel Control in Planar Bulk and Fully Depleted Devices", IEEE EDL, Vol. 33, No. 6, p. 776, 2012

[25] R. Rongenmoser et al, "Reducing Transistor Variability for Higher-Performance Lower-Power Chips", IEEE Micro, p.18, March 2013.

------------------------------------------------------------------

Draft submitted to CICC 2013 on 5/25/13. Revised version submitted on 7/15/13.

# Foundations for Scaling Beyond 14nm

Richard Schenker, Vivek Singh
Intel Corporation
Hillsboro, OR USA
richard.schenker@intel.com

## Abstract

**The path to extending Moore's Law beyond 14nm technology node will require a combination of advanced imaging, computation, patterning and design methods. Use of phase shift masks in combination with inverse lithography methods can enable imaging of complex, tight pitch patterns. Customizing designs to have asymmetric minimum metal pitch design rules can improve overall density. Pitch division methods like pitch quartering permit scaling beyond physical imaging limits from a single exposure. For example, metal test structures at 24nm pitch are generated using spacer based pitch quartering. Co-optimizing design and process allow application of pitch division to logic devices. Pitch division and Computation Lithography methods can be combined with EUV (Extreme Ultraviolet) lithography to further enhance scaling and design rule flexibility.**

## Introduction

Previously, scaling along Moore's Law relied on improvement in lithography equipment, occasionally by reducing the wavelength and frequently by improving the effective Numerical Aperture (NA). The wavelength used to define critical dimensions during much of the nineties was 248nm, with the industry switching to 193nm in the following decade. The introduction of immersion lithography allowed 193nm to continue to be the workhorse for semiconductor scaling. Moore's Law, however requires a significant innovation every two years. To make up for the divergence of lithography tool improvement rate versus the required rate of feature scaling, improved imaging technologies were applied such as Phase Shift Masks (PSMs), optical proximity correction (OPC), off-axis illumination (OAI) and Sub-Resolution Assist Features (SRAFs) [1-4].

Currently, the highest resolution lithography systems capable of high volume wafer manufacturing are 1.35NA, 193nm immersion systems. These tools have roughly equivalent resolution capabilities as tools used in the 32nm node. This paper describes some of the methods that can enable continued scaling of circuit density to the 14nm node and beyond.

## Computation Lithography

Computational Lithography comprises a broad set of techniques that use physics-based calculations to achieve greater performance from a given generation of lithography tooling. One advanced features of computation lithography is inverse lithography. An example of inverse lithography with pixilation phase shift masks is illustrated in Figure 1. In general, traditional OPC is unable to identify a converged mask solution for strong phase shift masks and complex design patterns. In this inverse lithography approach [5], the final solution is obtained iteratively using an optimization

scheme. The process starts with an initial phase assignment and the pattern is divided into pixels. The tool calculates a cost function after each iteration that characterizes both the image robustness and simulated edge pattern placement error. Mask pixels phase assignments are modified to improve the cost function during each iteration.

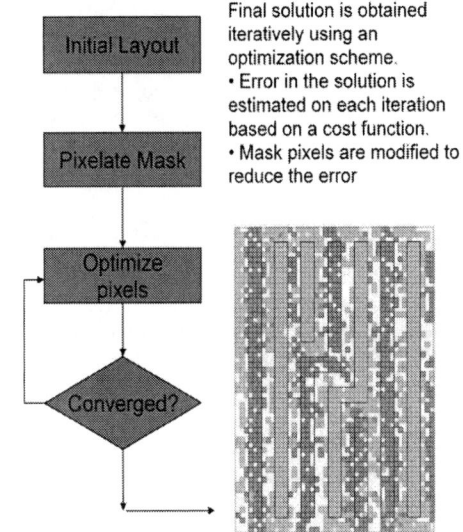

Fig. 1. Schematic of an inverse lithography approach applied to pixilation phase shift masks. Colors of mask represent either etcher or un-etched mask pixels (different phases).

Figures 2 and 3 show an examples of the inverse lithography method applied to pixelated phase shift masks.

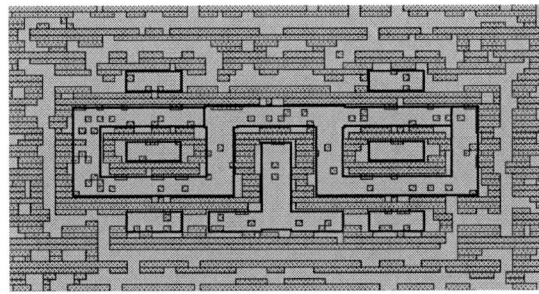

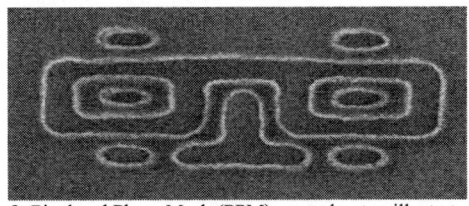

Fig. 2. Pixelated Phase Mask (PPM) example: top illustrates the mask pattern where blue/dark region represents etched glass (pi phase) and gray/light regions represent unetched glass (zero phase). Black lines are added for a reference to the desired pattern. Bottom shows wafer pattern from mask on left using a wavelength of 193nm, NA of 0.93 and cQuad illumination. Min pitch of pattern is 120nm (k1=0.29).

978-1-4673-6145-3/13 $31.00 © 2013 IEEE

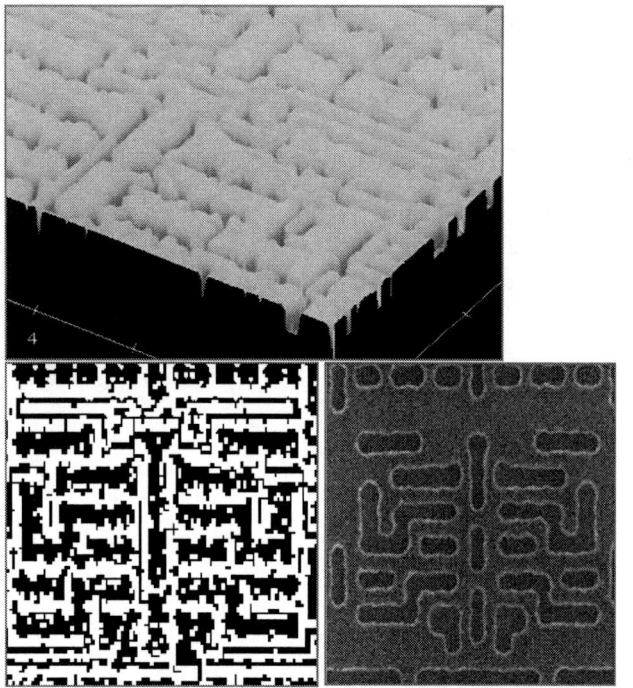

Fig. 3. Pixelated Phase Mask (PPM) example: Top is AFM scan of pixelated phase mask. Bottom left shows the mask design and bottom right shows the wafer lithography results (structure from first metal layer of a microporcessor).

The use of pixelated phase masks can enable high resolution imaging of complex two-dimensional patterns but requires complex process optimization of advanced mask making capabilities. [6-8] The techniques and tooling of inverse lithography and pixelated phase masks can be applied to other design styles, mask types and computational lithography methods.

## Pitch Division

Pitch division (pattern splitting) methods are used to enable scaling below 80nm pitch with 193nm lithography. One method, sometimes referred to as Litho-Etch Litho-Etch (LE-LE) splits a design pattern into two sets of patterns were each pattern can be resolved with a single lithography set and generates a final composite pattern with a minimum pitch as small as one-half the resolution limit of the lithography tool [9]. Another method, illustrated in Figure 4, shows a spacer based pitch quartering method [10]. This method generates a final pitch one-fourth the original lithography pattern without using additional lithography exposures. In general, one or more additional lithography steps are employed to remove unwanted features or to add additional breaks in the pitch quartered features.

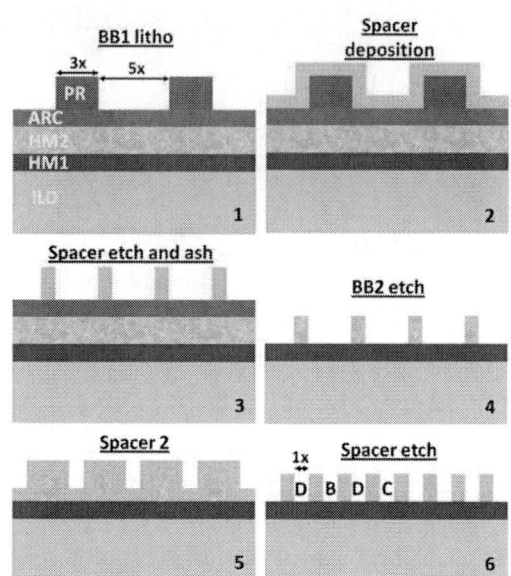

Fig. 4: Schematic of a pitch quartering flow using 2 conformal spacer depositions. The sequential processing steps are labeled 1-6.

A clear advantage of multiple pitch division is that the starting critical dimension and pitch can be well within the range of current 193nm wet lithography capability. For instance, the optically defined grating used in this work, called backbone 1 (BB1), has a CD 3x larger than the final metal wire CD. In addition, BB1 is printed at much looser (4x) pitch, as the quartering scheme can generate three additional trenches between any two backbone features.

Many parameters influence the ultimate uniformity of the resulting trenches. For instance, it is found that the four trenches resulting from a single BB1 fall into three independent trench categories: a B-trench located in the center of BB1, a C-trench which is centered between two BB1s, and two D-trenches that are positioned at BB2 locations, as labeled in Fig. 1. Different subsets of parameters affect the B, C, and D trenches. The D-trench for example, is insensitive to the spacer 2 thickness whereas this parameter directly affects the widths of the B- and C-trench. The ultimate trench profile can be corrected in a feed forward fashion by adjusting those parameters along the way to compensate for shifts observed at intermediate checks.

Figure 5 shows a 24nm pitch pattern in a hard mask with patterned breaks in some of the trenches. Figure 6 shows a 24nm pitch pattern in a low-k ILD with patterned breaks in the trenches and a Via. The pattern is part of a Via Chain test structure that showed electrical performance consistent with scaled resistance of earlier generation processes [11]. Figure 7 shows a TEM of metalized trenches and a Via in a low-k dielectric film connected to an underlying metal layer.

978-1-4673-6145-3/13 $31.00 © 2013 IEEE

Fig. 5. 24nm pitch features in a hard mask with patterned breaks in trench features.

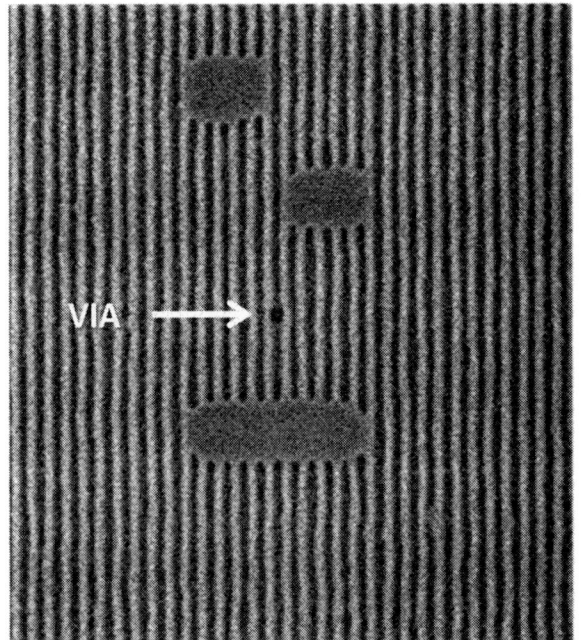

Fig. 6. Top-down image of a Via Chain test structure at 24nm pitch.

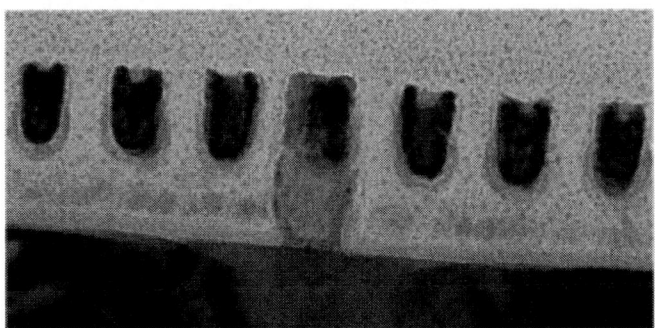

Fig. 7. Cross-section TEM image of dual-damascene electrically functional minimum size 12 nm-half-pitch structure post metallization.

## Process and Design Co-Optimization

Intel has reported several examples of process and design co-optimization for multiple technology nodes [12,13]. At the 65nm node, all active gates were orientated in a single direction to better utilize patterning methods like alternating phase shift masks for improved gate CD control. Different design rules were applied for minimum pitch, larger poly pitches and poly routing. For the 32nm and 45nm nodes, all poly was in one direction at one primary pitch. Trench contact local routing replaced orthogonal to gate poly routing. In order to extend the use of less expensive and more mature "dry" 193nm lithography for the 45nm node, trench contacts for source and drain connections replaced the tradition square contacts (see Figure 8).

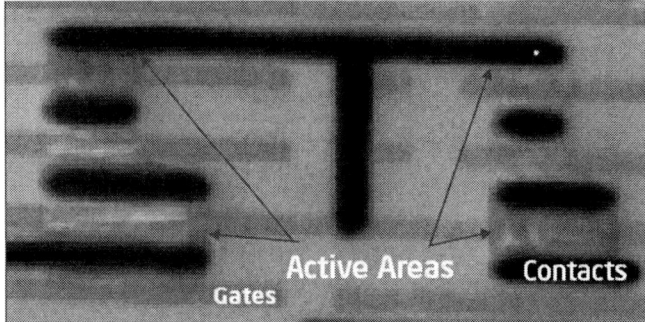

Fig. 8. 45nm node use of rectangular contacts. Extended use of "dry" 193nm lithography.

For the 22nm node, there is high value in being able to pattern complete metal layers down to 80nm pitch using a single 193nm lithography exposure. It was found that by use of multiple imaging enhancement method, 80nm single exposure pitch could be produced but only in one orientation simultaneously on the wafer. An analysis of the design found that equivalent density could be achieved if the non-preferred direction of the metal layer used looser than 80nm pitch. Figure 9 is an example of a 80nm pitch pattern in resist with orthogonal jogs.

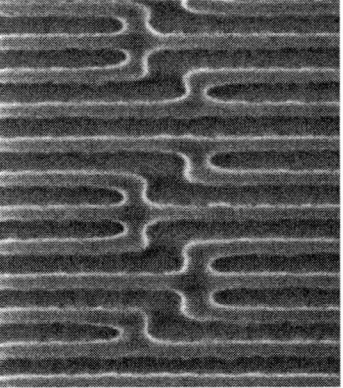

Fig. 9. 80nm pitch bi-directional patterning in photo-resist produced by a single 193nm exposure.

While the pitch quartering examples in this paper show regular arrays of features, the method can be modified to generate a variety of structures of use to logic designs. Figure 10 is an example of using spacer based pitch quartering method to generate patterns with variable widths or spacers. For layers that can utilize a single width features (such as gates or fins) ever other spacing between the fins can be modulated by either the original width or spacing of the starting lithography pattern. Likewise, for a damascene process applied to metal layers, every other metal width can be designed to be larger. This can be a useful feature for improving resistance of a metal or via feature.

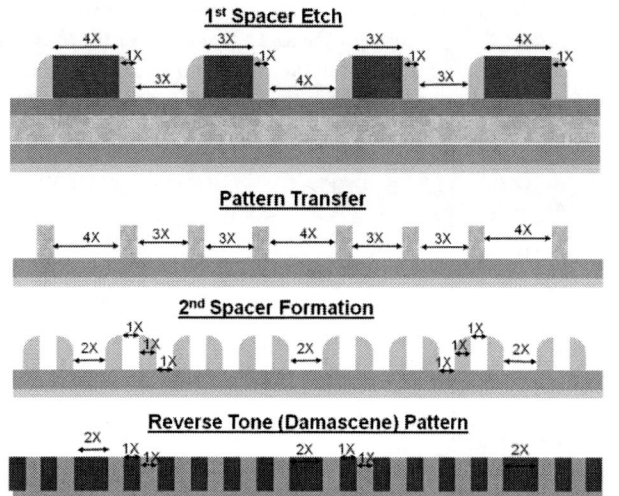

Fig. 10. Example application of spacer based pitch quartering patterning to generate structures with variable width or spacing.

## Pattern Splitting Lithography For Vias

## Complementary Lithography For Vias

Fig. 11. Example application of spacer based pitch quartering patterning to generate structures with variable width or spacing. Top image illustrates use of multiple 193nm exposures to define a single Via layer while the bottom case illustrates using EUV lithography to define Via layer. The under-lying grating layer could be defined with 193nm patterning with pitch division processing.

## Future Patterning Approaches

EUV with a 13nm wavelength has significantly better resolution than 193nm lithography. At this point, however, cost per EUV wafer exposure is significantly higher than a 193nm immersion exposure. To best utilize EUV resolution capabilities in a cost effective manner, complementary lithography [14] is a leading candidate for patterning beyond the 14nm node. Complementary lithography uses 193nm lithography with pitch division methods to generate primarily one-dimensional patterns and then uses EUV (or another high resolution lithography method) to pattern breaks in the one-dimensional features or Vias. Figure 11 illustrates the use of complementary lithography for Via layer patterning. In this case, 3 193nm

Via patterning steps are replaced by a single EUV patterning step.

## Conclusion

A combination of methods are needed to continue scaling beyond the 14nm node. Phase shift masks and computations enable extending the resolution of lithography tooling. Pitch division methods enable patterning scaling beyond the physical limits of a single 193nm lithography image. Process and design co-optimization enables designs to utilize the potential of advanced processing techniques. Selectively combining these techniques with high resolution lithography methods like EUV can provide a cost effective path for scaling.

## Acknowledgements

The results presented in this work couldn't be possible without the hard work and expertise of numerous members of Intel's Components Research, Portland Technology, Mask Shop and Computation Lithography groups. The contributions from Jasmeet Chawla, Curt Ward, Hui Jae Yoo, Kanwal Singh and Alan Myers were especially significant. The authors would also like to thank Yan Borodovsky for insightful discussion.

## References

[1] R. Socha et. al., "Simultaneous source mask optimization (SMO)," *Proc. SPIE* 5853, 2005.

[2] R. Socha et al., "Forbidden Pitches for 130nm lithography and below" *Proc. SPIE* 4000, 2000.

[3] R. Schenker et al., "Alt-PSM for 0.10-um and 0.13-um poly patterning" *Proc. SPIE* 4000, 2000.

[4] R. Schenker et al., "Alternating phase-shift masks for contact patterning" *Proc. SPIE* 5040, 2003.

[5] V. Singh et al., "Making a trillion pixels dance" *Proc. SPIE* 6924, 2008.

[6] Y. Borodovsky, Yan et al., "Pixelated phase mask as novel lithography RET" *Proc. SPIE* 6924, 2008.

[7] R. Schenker et al., "Integration of Pixelated Phase Masks for full chip random logic layers", *Proc. SPIE* 6924, 2008.

[8] J. Farnsworth et al., "Fabrication of defect free full-field pixelated phase mask", *Proc. SPIE* 6924, 2008.

[9] P. Wang et al., "Litho-Process-Litho for 2D 32nm hp Logic and DRAM Double Patterning", *Proc. SPIE* 7640, 2010.

[10] M. van Veenhuizen et al., "Demonstration of an electrically functional 34 nm metal pitch interconnect in ultralow-k ILD using spacer-based pitch quartering." *IEEE International Interconnect Technology Conference (IITC)*, 2012.

[11] J. Chawla et al., "Demonstration of a 12 nm-Half-Pitch Copper Ultralow-k Interconnect Process." *IEEE International Interconnect Technology Conference (IITC)*, 2013.

[12] C. Webb, "Intel design for manufacturing and evolution of design rules" *Proc. SPIE* 6925, 2008.

[13] R. Schenker et al., "The role of strong phase shift masks in Intel's DFM infrastructure development", *Proc. SPIE* 7641, 2010.

[14] Y. Borodovsky "ArF lithography extension for critical layer patterning", *LithoVision 2010*, San Jose, CA/USA 2010.

# BIF –Battery Interface Standard for Mobile Devices

Wolfgang Furtner, Stephan Schächer – Infineon Technologies AG, Germany
Markus Littow, Lionel Cimaz - ST-Ericsson, Finland/France
Pekka E. Leinonen – Nokia, Finland

## Abstract:

*The MIPI® Alliance Battery Interface (BIF) is the first comprehensive battery communication interface standard for mobile devices. MIPI BIF is a robust, scalable and cost-effective single-wire communication interface between the mobile terminal and smart or low cost batteries. It is suited for removable batteries as well as for embedded batteries. BIF improves mobile terminal safety and performance. It defines comprehensive battery monitoring and control functions such as temperature sensing and enables access to essential data for safe battery operation, e.g. charging parameters. BIF provides the communication layer for cryptographically secure battery authentication.*

## 1 INTRODUCTION

Mobile electronic devices with rechargeable batteries have become part of daily life. Billions of Lithium-Ion batteries are in use and have many different, mostly proprietary battery interfaces.

The lack of a commonly accepted battery interface standard has caused extra work and logistical effort throughout the industry. Mobile device manufacturers must coordinate, specify and maintain proprietary solutions from different parties in the ecosystem - themselves, mobile chipset suppliers, battery IC suppliers and battery pack manufacturers.

In 2010 several stakeholders in the mobile device industry started to develop a new industry standard for battery interfaces under the umbrella of the MIPI Alliance. The standard was named BIF which stands for Battery Interface (see [1] and [2]).

The MIPI Alliance is a global, collaborative organization comprised of companies that span the mobile ecosystem and are committed to defining and promoting interface specifications for mobile devices. The MIPI Alliance addresses the entire mobile device - from the antenna and peripherals to the modem and application processor.

### 1.1 Smart Batteries

Traditionally mobile device batteries have had analog interfaces. Such batteries often integrate an embedded resistive temperature sensor, next to the actual power supply connectors. Alternatively or additionally the type of the battery is indicated by identification resistors connected to battery pack terminals.

For modern highly sensitive battery technology it is very desirable to exchange more information between the mobile terminal host platform and the battery pack. Therefore "Smart Batteries" with digital communication are required.

Smart battery technology improves the safety of the end users significantly by providing access to reliable battery authentication, versatile sensing of operating conditions (e.g. multiple temperature sensors, stress sensors, etc.) and comprehensive battery related data sets (e.g. manufacturing parameters, charging recipes). In particular battery authentication with cryptographically strong algorithms improves end user safety by eliminating the use of potentially dangerous counterfeit batteries not complying with the required safety standards (e.g. [3]) or incompatible with the charging parameters of the mobile device.

### 1.2 Objectives of BIF

BIF is intended to support both traditional "analog" batteries and smart batteries. BIF defines the communication interface only. The actual power delivery interface and the mechanical parameters of the battery are out of its scope.

Support of low cost analog batteries is required for legacy designs, and for the simplest mobile devices. Most of the low cost batteries currently in the market have a built-in pull-down resistor in the battery pack. The measured value of the pull-down resistor usually represents certain capacity and chemistry information of the battery pack. For this function, the BIF standard supports measurement of a pull-down resistor value with a defined range and accuracy.

BIF minimizes the interface cost since it requires only one pin in addition to the power terminal of the battery. The mechanical pins of a typical battery pack have high reliability requirements due to a harsh operating environment, and take valuable space in the mobile device. These pins are a relatively high cost component in a mobile device.

BIF implements a simple multi-drop interface structure with one master device and one or more slave devices. It allows connecting multiple ICs on the same single wire bus. A smart battery may include multiple slave devices within the battery pack. The mobile device PCB may contain multiple slave devices, in addition to the master device.

The communication speed of the interface is dynamically scalable to match various available clock sources in the mobile device system under different operating conditions or data speed requirements.

A transceiver for a BIF master can be implemented with a serializer/de-serializer in hardware, or with software driving and sensing a general purpose I/O pin directly.

978-1-4673-6145-3/13 $31.00 © 2013 IEEE

A fast (approx. 1ms) battery pack presence detection is implemented without additional wires or contacts to inform the system immediately if the battery pack is disconnected. If the mechanical design of the battery pack connector assures that the communication pin is always the first one that disconnects, this can grant some time for system software to still perform critical shutdown actions.

While battery removals or longer contact breaks are detected and reported, short signal glitches due to contact instability, ESD or supply voltage bounces can be tolerated in communication.

BIF allows for a cost efficient implementation of data transceivers. A slave device can be built with an inexpensive and inaccurate clock source. The BIF protocol is constructed to cope with these inaccuracies.

With respect to the mobile device chipset BIF is designed for low voltage operation, supporting I/O voltages from 2.8V down to 1.1V. This enables interface implementation in the latest semiconductor processes.

Apart from physical layer and link layer protocol, BIF also defines higher level data structures. For certain standard functions (e.g. temperature sensor, non-volatile memory) standard register layouts are defined to enable the use of generic software driver in the systems.

BIF allows manufacturer specific functions in addition to the basic functions defined by the BIF specification. This enables slave device differentiation in the market and access to new innovative functions through the same unified interface.

BIF also takes care of storing non-volatile data at different phases of the battery pack production chain and during normal use of the battery pack.

## 2    ARCHITECTURE

BIF adds only one single wire (the battery communication line, BCL) to the two power connectors (VBAT, GND) of the battery pack. Communication signals are exchanged via BCL with reference to power ground (GND).

BCL carries all BIF related signaling including battery presence detection, analog battery identification, distinction between analog and smart batteries as well as data communication.

BIF data communication comprises of exchanges of data, address and command words, an in-band interrupt and wakeup from power-save modes.

There are two types of devices on a BIF bus, master and slave. There is only one master per BCL but there may be multiple slaves. Slaves may be located both inside the battery pack and on the mobile device side of the battery connector.

There are two types of slaves, primary slaves and secondary slaves. Primary slaves have a reserved address and may carry information about the other slaves found in a battery pack. Primary slaves need to have non-volatile memory in order to carry this information.

The BIF master device is typically placed in the power management IC (PMIC) as illustrated in Figure 1. Alternatively it can be placed on the digital baseband (BB) IC.

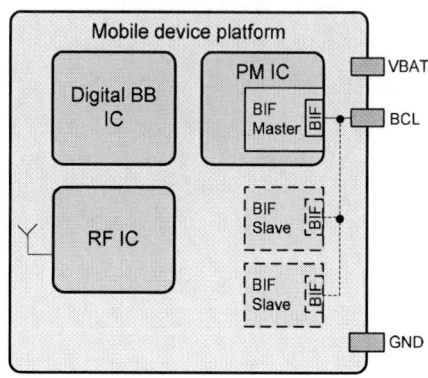

Fig. 1: BIF Master and Slave Devices in Mobile Terminal Host

BIF supports low cost battery packs (Figure 2) and smart battery packs (Figure 3), one connected at a time. Both of them have a pull-down resistor ($R_{ID}$) connected between the battery communication line (BCL) and GND. The value of $R_{ID}$ can be used to identify whether the battery is a smart or low cost type, and to identify the battery pack electrical characteristics for a low cost battery. The $R_{ID}$ can also be used for fast battery pack presence detection of both battery types. If a smart battery is disconnected, $R_{ID}$ also has the important role of pulling the BCL to GND and consequently putting the slave device(s) in power-down mode when the battery pack is removed from the mobile terminal.

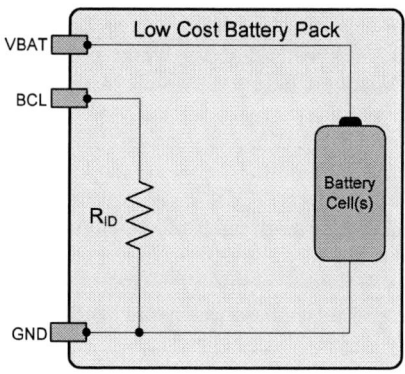

Fig. 2: Low Cost Battery Pack

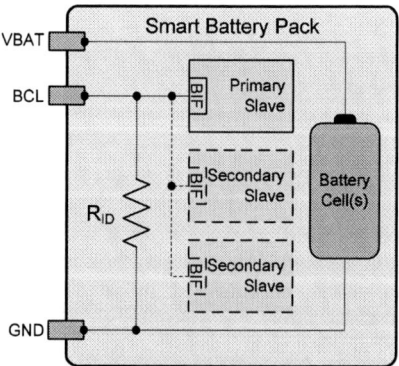

Fig. 3: Smart Battery Pack with BIF Slave Devices

A BIF master can address up to 256 slave devices

978-1-4673-6145-3/13 $31.00 © 2013 IEEE          82

connected to the BCL, using the so-called short 8-bit addressing. Each slave can have up to 64KByte of addressable memory space.

For basic standard compliance, master and slave devices have to implement the BIF physical and protocol layer. Higher levels of compliance require additional functions to be included.

## 3 PHYSICAL LAYER

The physical layer architecture is illustrated in Figure 4. The single communication line can be driven by each device with a switch to ground (TX). This is typically implemented with an NMOS transistor in open-drain configuration. A transmitter has to drive 1mA with a maximum residual voltage of 100mV. Moreover each device has a high impedance receiver circuit (RX) to sense the logical voltage level of the BCL. High levels are detected above 0.9V and low levels below 0.3V. A built in hysteresis of 50mV improves noise immunity.

The master provides a pull-up circuit ($R_{PU}$) to drive the line high, when no device is driving. The pull-up circuit may be implemented as a resistor or a current source. The signal high level is set by the host and is scalable from 1.1V to 2.8V. Minimum rise and fall times have been defined to limit potential electromagnetic interference (EMI) issues.

The master may implement fast presence detection and low cost battery identification ($R_{ID}$ value measurement). $R_{ID}$ value measurement typically requires an analog-to-digital converter (ADC) with a 10-bit resolution. The fast presence detection implementation can be done several ways. BIF specifies only timing parameters for this operation.

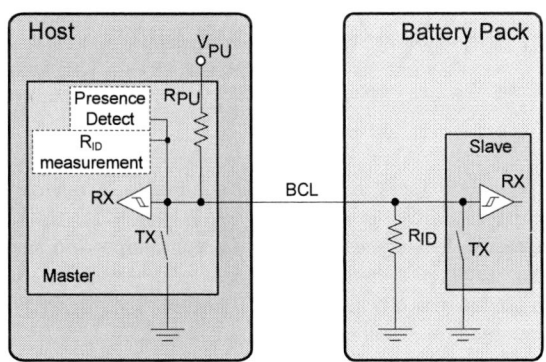

Fig. 4: BIF Physical Layer Interface

### 3.1 Low Supply Voltage PHY

When BIF is implemented in chipsets with low supply voltage (e.g. 1.8V or lower) it is recommended to use an active pull-up circuit such as depicted in Figure 5.

As long as BCL is pulled low the PMOS current mirror CMAB drives a defined current $I_{PU}=I_{MAX}$ into it. When no device is driving anymore, this current charges the line to a valid BIF high level voltage $V_{MAX}$. The operational amplifier throttles the current mirror such, that only the current necessary to maintain $V_{BIF}=V_{MAX}$ is driven into the line. This current is mainly determined by the value of $R_{ID}$ in the battery pack. For smart batteries

$R_{ID}$ is approximately 300KΩ and the current through it can be clearly distinguished from leakage. If the battery pack is removed the current reduces to very small values only caused by leakage on the host side.

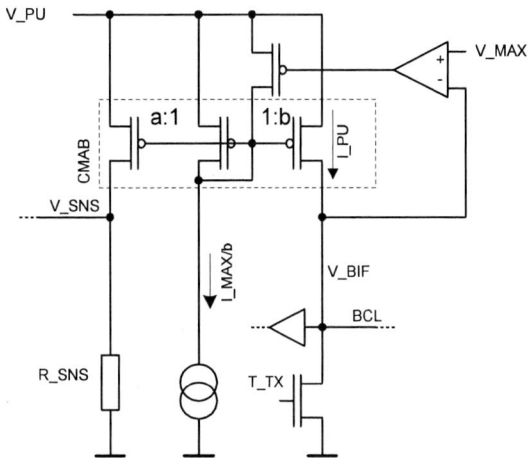

Fig. 5: Host Side PHY Implementation

The pull-up current $I_{PU}$ can thus be used for reliable detection of battery presence. In the circuit of Figure 5 it is mirrored into the leftmost branch of the current mirror with a ratio of a:b. With a resistor to ground ($R_{SNS}$) this current is converted to a voltage $V_{SNS}$.

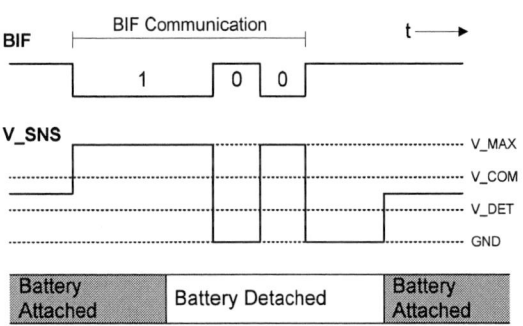

Fig. 6: Battery Presence Detection

This voltage is compared to fixed thresholds ($V_{COM}$, $V_{DET}$) to detect battery presence as depicted in Figure 6. It is not possible to detect battery presence when a transmitter is pulling the communication line low. However since the time for driving BCL low is limited by the protocol timing parameters it is assured that a detached battery can be detected within 1ms.

## 4 PROTOCOL

The BIF protocol is designed as a data transport interface. The actual battery applications, such as temperature measurement and authentication, make use of the protocol but do not interfere with it. Data transport and battery application usages are clearly separated.

One of the main objectives in BIF protocol design was to achieve flexibility while maintaining small silicon size. Hardware BIF transceivers can be implemented with approximately 1k gates in a typical CMOS process.

The BIF communication data rate is scalable between

3.27kbit/s – 250kbit/s (average). The minimum data rate was extended down to ~3kbit/s because in many systems there is a 32.768 kHz clock available due to a real time clock requirement. This enables BIF communication even in power save modes when only this clock is available in the host. The maximum data rate was limited to 250kbit/s to minimize the slave device receiver complexity. This was analyzed to be sufficient for the current use cases of BIF (e.g. allowing reasonably fast authentication). However higher speed definitions may be added to the BIF specification in the future.

BIF communication is always initiated by the master and is based on a data word. The master defines the communication speed in the beginning of every word and the addressed slave must use that speed in its response. So in theory, each transaction between the master and the slave could happen with a different speed.

### 4.1 Time Distance Coding

BIF signaling is based on the elapsed time between changes of signal level. This will be also referred to in the following as "Time Distance Coding". The signal level can be either High (= high voltage) or Low (= low voltage, close to GND). The time between polarity changes is classified into three duration classes. Short durations denote a logic $0_B$, long durations denote a logic $1_B$, and very long durations signal a STOP condition. Figure 7 shows the transmission of two subsequent BIF words. Words are separated by STOP codes and contain an odd number of bits. The STOP signal is only sent at a high voltage level, whereas $0_B$ and $1_B$ are not associated with a voltage level.

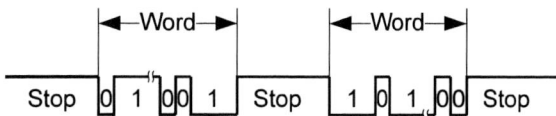

Fig. 7: BIF Communication

The duration of a logic $0_B$ transmission is called the BIF Time Base $\tau_{BIF}$. The symbol timing is specified relative to $\tau_{BIF}$. The three possible codes ($0_B$, $1_B$, STOP) are transmitted by modulating the duration between signal toggles as depicted in Figure 8. Their nominal time constants are integer multiples of the BIF Time Base ($1\tau_{BIF}$, $3\tau_{BIF}$, $5\tau_{BIF}$). A receiver normally uses two time thresholds - also integer multiples of the Time Base ($2\tau_{BIF}$, $4\tau_{BIF}$) – to distinguish between the three codes. The clear timing difference between the three symbols allows their robust distinction even in noisy environments and with asymmetric rise and fall times.

A transmitter modulates the durations with $\pm0.25\tau_{BIF}$ accuracy and a receiver has to guarantee reception of bits with timing deviations up to $\pm0.5\tau_{BIF}$ from the nominal value (Table 1). The BIF Time Base, $\tau_{BIF}$, is defined by the master for every data word with a training sequence. The master may choose any $\tau_{BIF}$ within the specified boundaries, and it may alter the $\tau_{BIF}$ used from data word to data word. A slave must support the full range of data rates specified by BIF. The transmitter of a word has to

assure proper stop timing for the stop signal preceding its transmission.

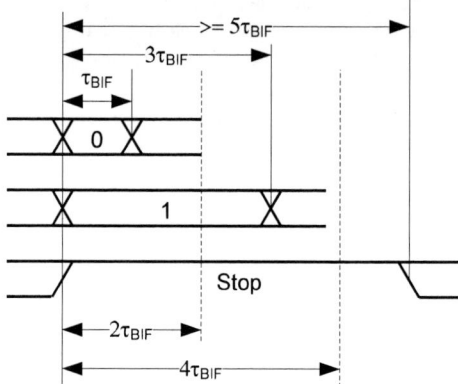

Fig. 8: BIF Signalling

| Code | Receiver | | Transmitter | |
|------|------|------|------|------|
| | Min. | Max. | Min. | Max. |
| 0 | $0.5\tau_{BIF}$ | $1.5\tau_{BIF}$ | $0.75\tau_{BIF}$ | $0.25\tau_{BIF}$ |
| 1 | $2.5\tau_{BIF}$ | $3.5\tau_{BIF}$ | $2.75\tau_{BIF}$ | $3.25\tau_{BIF}$ |
| STOP | $4.5\tau_{BIF}$ | $\infty$ | $4.75\tau_{BIF}$ | $\infty$ |

Tab. 1: BIF Timing

### 4.2 BIF Data Word

BIF defines data words of 17 bits in length as depicted in Figure 9. This requires the BCL to toggle 18 times per word. Each 17-bit data word consists of the following elements:

- Training Sequence, BCF → 2 bit
- Payload, D[9:0] → 10 bit
- Parity, P[3:0] → 4 bit
- Inversion → 1bit

A data word can carry a command, a device or register address, read data or write data. The training sequence bits are used to tell the communication speed and also tell whether the word is a Broadcast type (intended for all slaves) or a Unicast / Multicast type (intended for selected slaves).

The payload is the actual data to be transported. The use of 10 bits allows to distinguish between commands, devices addresses, register addresses and read or write data.

The parity bits (conforming to Hamming-15/CRC-4 coding) are used to detect possible communication errors. Strong detection of communication errors is important especially for a battery interface because of the physical connector on the line. Such connectors may be exposed to mechanical shocks and aging in a harsh environment.

The inversion bit is included due to the data dependent duration of BIF signaling. If more than half of the data word bits are $1_B$ ($3\tau_{BIF}$), all bits in the word are inverted and the inversion bit is set. This reduces the maximum and the average word length and makes the

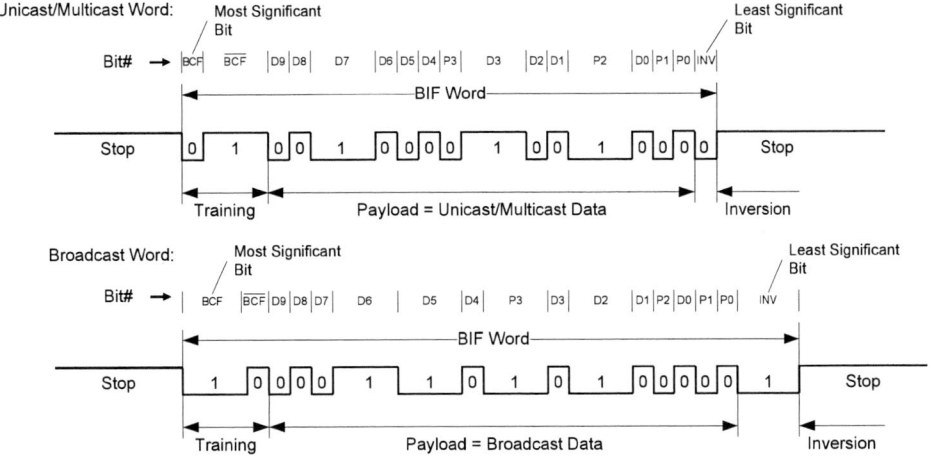

Fig. 9: BIF Data Words

word length less dependent on the actual data content. As a result it increases data bandwidth and decreases transmission power.

## 5   SLAVE DATA STRUCTURE

BIF provides up to 64Kbyte addressable register and memory space for each slave. Certain rules determine how this memory space shall be structured. Moreover, for certain standard functions, register layouts are mandated. This allows building generic software drivers. However BIF also provides support for vendor specific functions to enable slave device differentiation in the market

An example BIF slave data structure is illustrated in Figure 10. Memory locations can be used for different storage types such as RAM, ROM or reprogrammable NVM. The data map always starts with a fixed length 10-byte Device Descriptor Block (DDB-L1) defined by the MIPI Alliance. DDB-L1 contains generic device identification information like Manufacturer ID and Product ID. The last two bytes of the DDB-L1 tell the total size of the function directory following immediately after the DDB-L1.

### 5.1   Function Directory

A device contains one or multiple device functions (e.g. temperature sensor) each of them associated with certain registers and memory space. The function directory lists all programmable functions contained in the device. Each entry in the function directory consists of four bytes. The first two bytes contain the function type and version. This provides sufficient information for the host to select the correct software driver for that function. The last two bytes are a pointer to the first byte of the function capability section. Each function has a capability section whose content is function specific.

### 5.2   Functions

Functions consist of a capability section (static data), a register section (dynamic volatile data, readable and/or writable) and an optional section for non-volatile data, such as MIPI-defined data objects. The different sections of multiple functions may be grouped according to their access type.

The capability section provides essential information for the software driver about implemented features of the function. It also includes one or more pointers to the function registers. The function registers are used to control and monitor operations of a certain function.

A function can be either a MIPI BIF pre-defined type or manufacturer-defined function. The main idea is that the host software can find and identify both types of functions listed in the function directory. For MIPI-defined functions a generic software driver should be able to use capabilities and operate registers directly. For manufacturer-specific functions, an additional software driver provided by the manufacturer will be needed.

### 5.3   BIF Functions

MIPI BIF 1.0 pre-defines the following functions:

- Protocol
- Slave Control
- Non-Volatile Memory (NVM)
- Temperature Sensor
- Authentication

Only the protocol function is mandatory for a slave while the others are optional. The protocol function provides support for communication (e.g. word/error counting, reprogramming of device address).

The slave control function provides central control and status information for the tasks that are being executed on a slave device. Moreover it supports the in-band interrupt of BIF.

The NVM function allows reading and writing of a consecutive user NVM space of arbitrary size and independent of the physical organization of the memory. The user NVM can be used for multiple purposes such as storage of battery-related information or recording usage data in the field. Parts of the user NVM can be write-protected, to prevent intended or accidental altering by further manufacturing steps or in the field. For example, the slave IC manufacturer can write initial data, followed by the battery pack manufacturer writing additional data and finally the mobile device maker writing further

978-1-4673-6145-3/13 $31.00 © 2013 IEEE

content. After each of these steps the written data can be protected against overwriting by subsequent steps. The remaining NVM space may be used to store usage data in the field. In addition to the User NVM the NVM function also provides mechanisms to write and lock Manufacturer parameters (e.g. initial device address of a slave).

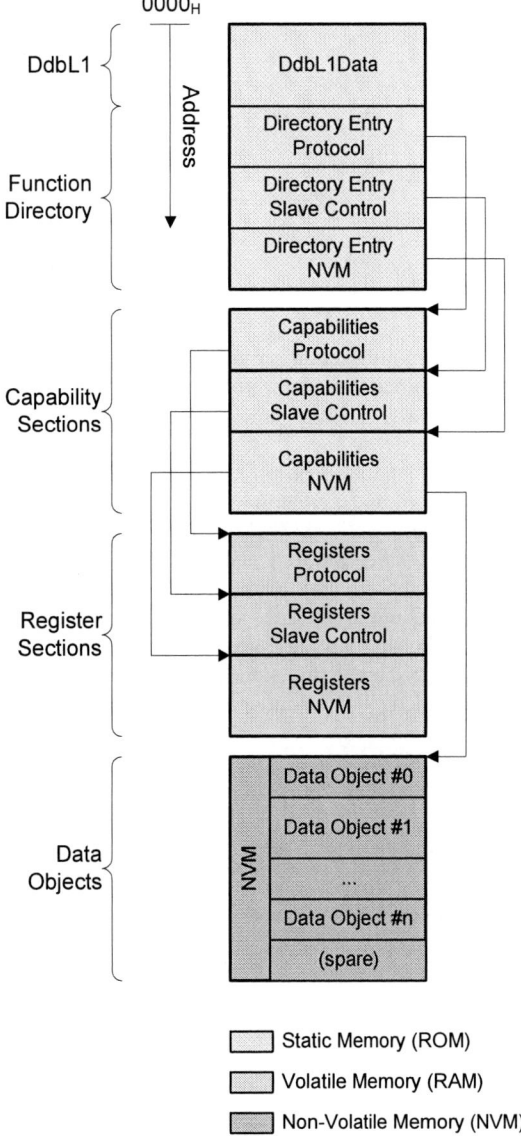

Fig. 10: Slave Data Structure

The temperature sensor function allows a single and periodic temperature measurement. It can generate an interrupt when the measurement is completed. Additionally it can monitor if temperature goes beyond an upper or a lower temperature threshold. It also provides information about the accuracy of the measurement.

### 5.4 Authentication

Authentication of the battery pack allows the mobile device to verify if the inserted battery is genuine, in order to validate if it is an authorized and quality controlled product of the respective manufacturer. This verification is done by proving that the battery pack possesses a certain secret (a "key") without actually communicating the key itself. A possible authentication scheme is shown in Figure 10. Essentially the host sends a random challenge, which is then encrypted with the secret key in the battery pack and sent back. To an outside observer the response is random too. The response is decrypted using a corresponding key on the host side and compared against the random challenge.

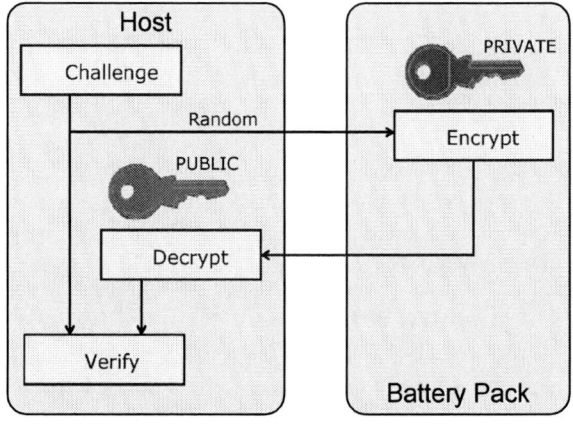

Fig. 11: Asymmetric Authentication

Conventional authentication algorithms have been based on symmetric cryptography. That means the same key is used both in the host and the battery pack. This imposes a security risk, since the key is typically easy to extract on the host side e.g. with a code debugger.

Modern authentication schemes employ asymmetric cryptography. A private key is stored on the battery side and protected against extraction by hardware means. A corresponding public key is used on the host side and can be freely accessible. Asymmetric cryptography is e.g. also employed in PGP encryption for emails.

The BIF authentication function defines a generic set of registers suited for any challenge-response-type authentication. BIF authentication requires at least an equivalent of 80-bit symmetric key strength according to [4]. The actual implementation of the authentication function is manufacturer-specific to allow multiple flavors of cryptographic algorithms.

### 5.5 BIF Objects

BIF also standardizes the structure of certain data objects stored in the User NVM. The idea is to provide a method for passing important parameters of the battery pack from battery pack makers to the mobile device. Reading and writing the objects is done through the NVM function. BIF defines the generic frame format for data objects including length information and data integrity protection using CRC-16. Moreover it defines a "secondary slave" object to store addressing information for secondary slaves within the User NVM of a primary slave.

Having a pre-defined battery data object structure and length defined in the beginning of the object enables a

978-1-4673-6145-3/13 $31.00 © 2013 IEEE

generic software driver to access and identify the actual data packet. MIPI-defined objects can be read and interpreted by the generic driver directly. For manufacturer defined objects the host will be able to interpret the data packet content after adding a manufacturer-specific software driver.

Typical battery data objects contain information about the battery model, capacity, chemistry, charging and discharging and aging parameters. More objects types will be standardized by BIF in future versions of the standard.

## 6   RULE BASED CHARGING

Battery charging requires particular attention to ensure safe operation for some battery types. This is specifically true for Lithium-based batteries widely used in portable devices today (see also [5]). While charging these batteries, appropriate constraints must be followed to control charging current and voltage precisely. The charging segment, defined as the charging voltage and current combination, depends generally on battery temperature and battery voltage. The charging segment is selected by a charging algorithm and applied on a precisely controllable CC/CV (constant current / constant voltage) charger.

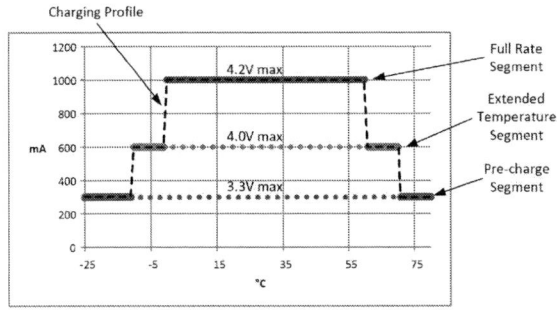

Fig. 12: Charging Profile

For a given battery, a charging profile, composed by a set of charging segments, can be established so that charging is safe over temperature and battery voltage (an example is shown in Figure 12). Disrespecting this charging profile may result in various defects such as accelerated aging of the battery, over-heating or even battery physical damage which can cause end-user injuries.

### 6.1   Limitations of Conventional Charging

A portable device charging subsystem is usually designed for a specific battery or family of batteries. It usually cannot guarantee safe operation or identical performance when used with batteries other than the design prototype. This inflexibility limits the choice of battery throughout the product life.

This strong link between a portable device and a battery type is a significant limitation, e.g. considering the effect of this product design on the multi-sourcing of batteries. Logically, each source of the battery would provide its essential charging profile to the charging subsystem and the product charging subsystem would

adjust appropriately. But with a fixed charging subsystem, the approach is quite the opposite: the charging subsystem sets the charging profile requirement for the battery sources. This often results in reduced charging performance and batteries from different sources are not utilized optimally. It could also result in higher battery unit cost from a given battery supplier because they may need to modify and customize their battery design to support the established charging profile for the charging subsystem of each portable device.

The strong attachment of a battery type to a portable device charging subsystem may also limit the use of newer or improved battery technology which requires a different charging profile. Once the portable device is widely available in the market to the end users, it usually cannot use the latest most advanced batteries or even adapt an updated charging profile (for example, improved safe charging rates). In the same way, the end user may not be able to effectively use higher capacity batteries in after-sales markets (probably it would take a longer time to charge than necessary for the new battery or perhaps operate with underutilized capacity).

In the desire to implement charging for various batteries and chemistries, classical software state machines have grown more and more complex and have become harder to maintain.

### 6.2   Charging with BIF

BIF provides the User NVM function and the standardized data object storage mechanism to clearly separate the charging control software running on the host from the actual battery related charging parameters stored in the battery pack (see [6]). Each battery pack may come with its own optimal charging recipe. The charging recipes may even be modified over the production life of a battery.

BIF proposes a "Rule Based Algorithm" for host side charging control and it specifies the battery related charging rules to be stored inside the battery pack. There may be additional system related charging rules stored on the host side. This algorithm is implemented on the host once when the mobile device is designed and can remain unchanged regardless of the battery qualified for the particular device later.

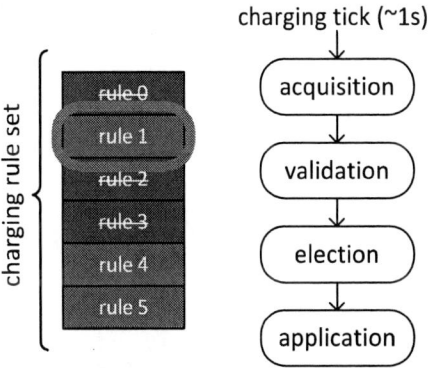

Fig. 13: Rule Based Charging

Figure 13 illustrates how Rule Based Charging is performed. In the acquisition phase the set of charging

978-1-4673-6145-3/13 $31.00 © 2013 IEEE                87

rules is read from the battery pack. In the validation phase all rules are checked if they are valid under the prevailing operating conditions (valid rules are marked in green and invalid ones in red in Figure 13). In the election phase the rule with the best charging performance is selected from the valid rules (encircled in blue in Figure 13). Since the rules are stored in prioritized order this is the topmost valid rule of the list. And finally in the application phase the charging circuit is programmed to the charging current or voltage stored in this rule. Each charging rule stored in User NVM contains the parameters listed in Table 2.

| Parameter | Description |
|---|---|
| $T_{min}$ | Minimum battery temperature. If temperature is lower than $T_{min}$, rule is invalid. |
| $T_{max}$ | Maximum battery temperature. If temperature is above $T_{max}$, rule is invalid. |
| $V_{min}$ | Minimum battery voltage. If voltage is lower than $V_{min}$, rule is invalid. |
| $V_{max}$ | Maximum battery voltage. If voltage is higher than $V_{max}$ (after hysteresis correction) rule is invalid. If rule applies, charger CV target is set to $V_{max}$. |
| $I_{min}$ | Termination current. If rule applies and charging current drops below $I_{min}$, rule is invalidated. |
| $I_{max}$. | CC limit. If rule applies, charger CC limit is set to $I_{max}$. |
| $t_{max}$ | Time out. If the rule is applied for a longer time than $t_{max}$, rule is invalidated and charger algorithm goes into error condition. |

Tab. 2: Charging Rule Parameters

## 7 BIF IMPLEMENTATIONS

Several manufacturers are sourcing BIF-compatible parts now and expect significant adoption in the first half of year 2014. Earliest adoption is expected in mobile phones and tablets, to be followed by cameras and ultrathin laptops. BIF interoperability testing is offered by UNH-IOL (University of New Hampshire Interoperability Laboratory). An example of a MIPI BIF compliant slave device implementation from Infineon Technologies is shown in Figure 14.

Fig. 14: BIF Device Origa2™

BIF slave devices are typically manufactured in CMOS processes nodes from 250 down to 65nm. Their digital circuit complexity ranges between a few kilo gates and several 100k gates. Often BIF slave ICs are dominated by analog circuits for sensors and power management. The quiescent current of BIF slaves should not exceed a few micro-amperes in order to present virtually no additional load to the battery. BIF slaves are preferably packaged in very small (max. 2mm width) unleaded packages of limited height (max. 0.7mm) to fit on the narrow PCBs inside batteries, often just a few millimeters wide. Wafer scale packages are in principle very desirable for BIF slave devices, but not always permitted by the battery-PCB manufacturing rules. Because of the PCB area limitations, BIF slaves aim for a high integration level and require only a few small external passive components. In particular they normally operate directly from the battery voltage to spare additional voltage regulation. Only the battery protection circuit which disconnects the battery cell in case of misuse remains separate for safety reasons.

## 8 CONCLUSION

MIPI BIF presents a comprehensive standardized interface solution for the management of rechargeable batteries used in mobile devices. It addresses all the functionality needed for safe and efficient battery operation from physical interface implementation to host control software design. It enables smart batteries with highly desirable features such as cryptographic authentication of batteries and rule based charging. MIPI BIF is supported and maintained by major stakeholders in the mobile device industry.

## 9 ACKNOWLEDGEMENTS

The authors would like to gratefully acknowledge the contribution of all MIPI Alliance members who participated in the creation of the BIF standard and related material used in this paper. A list of all MIPI Alliance members can be found in [7].

## 10 REFERENCES

[1] MIPI BIF data sheet, <http://www.mipi.org/sites/default/files/BIF data sheet final 022012_0.pdf>, MIPI Alliance, February 2012

[2] MIPI BIF Whitepaper, <http://www.mipi.org/sites/default/files /BIF_whitepaper 020112 final.pdf>, MIPI Alliance, February 2012

[3] IEEE1725-2006, IEEE Standard for Rechargeable Batteries for Cellular Telephones, Institute of Electrical and Electronic Engineers, 2006.

[4] SP 800-57, Recommendation for Key Management – Part 1: General (Revised), <http://csrc.nist.gov/publications/nistpubs/800-57/sp800-57-Part1-revised2_Mar08-2007.pdf>, National Institute of Standards and Technology, 8 March 2007

[5] "A Guide to the Safe Use of Secondary Lithium Ion Batteries in Notebook-type Personal Computers", Japan Electronics and Information Technology Industries Association and Battery Association of Japan, April 20, 2007

[6] MIPI BIF rule based charging Whitepaper, <http://mipi.org/sites /default/files/mipi_BIF_rule-based-charging_white-paper_1.pdf>, MIPI Alliance, 2013

[7] MIPI Alliance Member Directory, <http://www.mipi.org/member-directory>, MIPI Alliance, 2013

# A 40V 10W 93%-Efficiency Current-Accuracy-Enhanced Dimmable LED Driver with Adaptive Timing Difference Compensation for Solid-State Lighting Applications

Dongkyung Park and Hoi Lee

Department of Electrical Engineering

The University of Texas at Dallas, Richardson, TX 75080-3021, USA

*Abstract-* **This paper describes a floating buck dimmable LED driver for solid-state lighting applications. Adaptive timing difference compensation is proposed to enable the driver to achieve high accuracy of the average LED current, fast settling time, and high-frequency operation over a wide range of input voltages and number of LED loads. The power efficiency of the proposed LED driver is benefited from the capabilities of using synchronous rectification and having no sensing resistor in the power stage. The synchronous rectification under high input supply voltage is enabled by a proposed high-speed and low-power gate driver with pseudo-digital level shifters. Implemented in a 0.35μm 50V CMOS process, the proposed 40V LED driver can operate at 1MHz and achieve 93% peak power efficiency when driving up to 10 series-connected LEDs. It has only 2.8% current error from the average LED current of 345mA and settles within 8.5μs under different line and load variations. The performances of the proposed driver significantly outperform all state-of-the-art counterparts.**

## I. INTRODUCTION

Today's breakthroughs in high-brightness light emitting diodes (LEDs), that offer much better luminous efficacy and longer lifetime, have the potential to replace traditional incandescent technology in residential and commercial lighting applications for providing significant electricity saving [1]. In either AC- or battery-powered LED lighting system shown in Fig. 1, a DC-DC converter-based LED driver is needed to regulate current passing through LEDs under different input voltages and different numbers of series-connected output LEDs. The LED driver is required to (i) operate at high switching frequency for reducing the required values of passive components, and (ii) have an effective control scheme to precisely control LEDs' brightness with dimming control under different input and output conditions. However, there are usually tradeoffs between above performance requirements and the driver power efficiency, leading to sub-optimal LED driver design.

Peak-current control (PCC) and hysteretic-current control (HCC) are two common schemes to regulate the average LED current ($I_{avg}$) in DC-DC converter-based LED drivers [2] – [4]. Fig. 2 shows these control schemes in a floating buck topology. As the PCC (Fig. 2(a)) can only provide the peak inductor / LED current information to the control loop, the value of $I_{avg}$ can vary significantly under different input and output conditions. By regulating both peak and valley LED currents, HCC can offer much better LED current accuracy. However, a sensing resistor is required to place in series with the inductor for detecting both peak and valley LED currents (Fig. 2(b)).

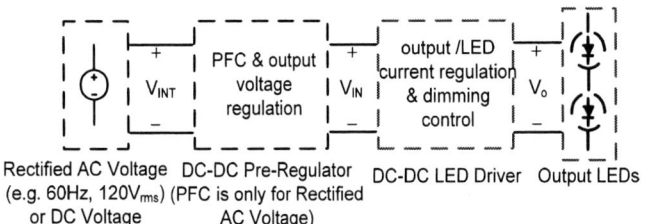

Fig. 1. Structure of a typical LED lighting system. (Noted that DC-DC pre-regulator can be bypassed if input voltages of the system are obtained from the battery).

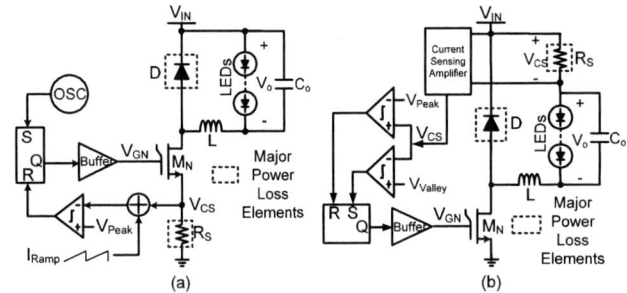

Fig. 2. Structures of traditional (a) peak-current-controlled and (b) hysteretic-current-controlled LED drivers.

As large LED current flows through the sensing resistor for the entire switching period, significant conduction power loss would be resulted in the driver. Recently, adaptive off-time control was reported to approximate the valley bound of LED current for improving the current accuracy [5]. However, the reported SAR algorithm to implement the adaptive off-time scheme leads to long calibration time and settling time of the LED current during dimming, limiting the maximum operation frequency of the driver. In addition, due to the lack of high-speed low-power high-side gate driver and level shifter, asynchronous power stage with a high-side power diode is commonly used in the high-input-supply LED drivers [2]. However, off-chip power Schottky diodes with a typical voltage drop of ~0.6V can greatly decrease the power efficiency of the LED driver especially under a small number of series-connected LEDs (small duty ratio of the converter due to low output voltage).

To mitigate above issues, a new control scheme: adaptive timing difference compensation (ATDC) is proposed to regulate $I_{avg}$ for achieving (i) high current accuracy, (ii) fast settling time in dimming, and (iii) high converter operation frequency. A high-speed low-power high-side gate driver is also developed to enable the proper operation of the

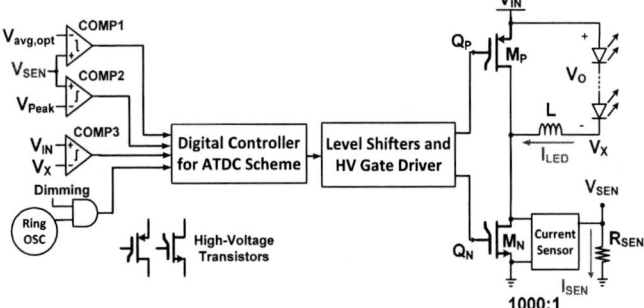

Fig. 3.  Structure of the proposed LED driver.

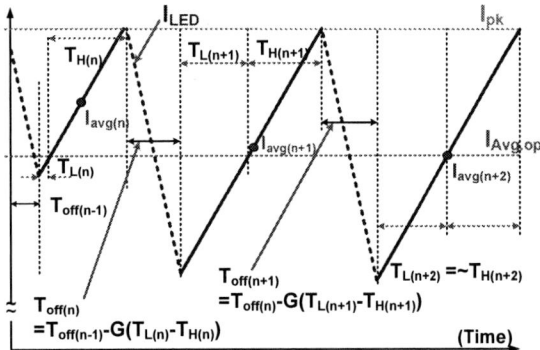

Fig. 4.  Concept of the proposed ATDC scheme.

synchronous rectifier in the driver power stage under high-frequency and high-input-voltage conditions. The proposed synchronous driver can minimize the conduction power loss for better power efficiency. Details of the proposed LED driver will be discussed in later sections.

## II. PROPOSED DIMMABLE LED DRIVER

### A. System Architecture and Operation Principle

Fig. 3 shows the structure of the proposed floating buck LED driver, which mainly consists of an on-chip low-side current sensor for detecting the peak LED current ($I_{pk}$); a synchronous rectifier (both high-voltage power transistors: low-side nMOS ($M_N$) and high-side pMOS ($M_P$)) enabled by a proposed high-speed low-power synchronous gate driver; a digital controller to realize the proposed adaptive timing-difference compensation (ATDC) for LED current regulation; and a dimming control block. As no power diode nor sensing resistor are used in the proposed power stage, significant conduction power loss can be saved compared with that in the conventional high-voltage LED drivers shown in Fig. 2. It should be noted that a sensed voltage $V_{SEN} = I_{SEN} \times R_{SEN} = (I_{LED} \cdot R_{SEN})/K$, where K = 1000 is used in the driver for providing the peak LED current information to the controller.

The average LED current ($I_{avg}$) in the floating buck converter is given as

$$I_{avg} = I_{pk} - \frac{\Delta I_L}{2} = I_{pk} - \frac{V_o}{2L}T_{off} = I_{pk} - \frac{V_{IN} - V_o}{2L}T_{on} \quad (1)$$

where $I_{pk}$ and $\Delta I_L$ are peak inductor current and inductor current ripple, respectively. L is the inductance used in the power stage; and $T_{off}$ and $T_{on}$ are off-time and on-time of power nMOS transistor $M_N$. From (1), if input voltage $V_{IN}$ of the driver changes or the number of output LED (thus output

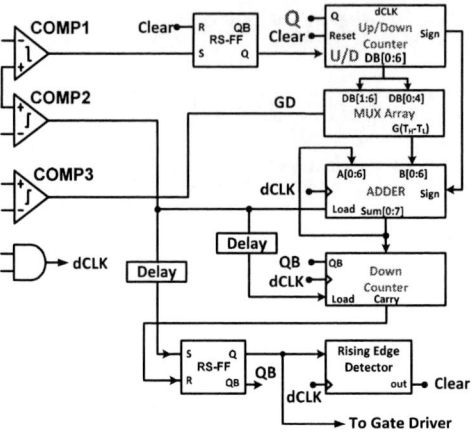

Fig. 5.  Block diagram of the proposed ATDC scheme.

voltage $V_o$) varies, then the inductor current ripple $\Delta I_L$ and thus $I_{avg}$ will be changed correspondingly, compromising the current accuracy. In order to maintain $I_{avg}$ to a constant value for high current accuracy under variations of $V_{IN}$ and $V_o$, $T_{off}$ should be adjusted adaptively. Fig. 4 illustrates the concept of the proposed ATDC scheme to adjust $T_{off}$. $T_H$ and $T_L$ are time durations that $I_{LED}$ is higher and lower than desired average current $I_{Avg,opt}$, respectively. ***In order to achieve $I_{avg} = I_{Avg,opt}$, we have to ensure $T_L = T_H$.*** The proposed ATDC will adjust $T_{off(n)}$ adaptively according to the following equation

$$T_{off(n)} = T_{off(n-1)} - G \cdot (T_{L(n)} - T_{H(n)}) \quad (2)$$

where 'G' is a gain constant dependent on the duty ratio of the driver and is realized by COMP3 in Fig. 3. As shown in Fig. 4, when $T_{L(n)} < T_{H(n)}$, it indicates that $I_{avg}$ is larger than $I_{Avg,opt}$ and the off-time $T_{off(n-1)}$ of power transistor $M_n$ in the previous switching cycle is smaller than the proper value. Therefore, the value $T_{off(n)}$ of this switching cycle will be increased according to (2). Keeping on this process for a few switching cycles, the difference between $T_L$ and $T_H$ will be converged to 0 such that $I_{avg}$ reaches $I_{Avg,opt}$. In case of $T_{L(n)} > T_{H(n)}$, $T_{off(n)}$ will be decreased based on (2) and eventually results in $T_L = T_H$. It should be noted that the proposed ATDC scheme is realized by a digital controller with a typical clock frequency (dCLK) of 160MHz. The digital controller is thus fast enough to enable off-time calculation in every switching cycle (1µs) of the converter under variations of input and output voltages. Additionally, the proposed ATDC scheme only needs a single low-side current sensor to detect the peak inductor / LED current $I_{pk}$ for determining $T_H$. As no high-voltage high-side current sensor is needed to provide the valley inductor current in the proposed LED driver, the current error from the high-side current sensing is eliminated.

### B. Circuit Implementations

Fig. 5 shows the digital controller that implements the proposed ATDC scheme. The controller mainly consists of an Up/Down (U/D) counter, an adder, a Down counter and RS latches. When the dimming signal is asserted, Q becomes logic "1", $I_{LED}$ ramps up and U/D counter starts to count up. When the sensed voltage $V_{SEN} > V_{Avg,opt}$ (= $I_{Avg,opt} \times R_{SEN}/1000$), the U/D counter starts to count down. As $V_{SEN}$ reaches $V_{Peak}$ (= $I_{pk} \times R_{SEN}/1000$), the nMOS power transistor is turned off

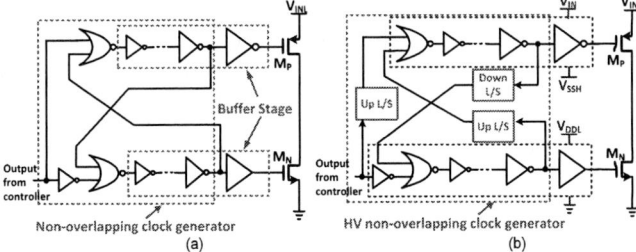

Fig. 8. Micrograph of the proposed dimmable LED driver.

Fig. 6. Structure of the (a) conventional low-voltage and (b) proposed high-voltage gate drivers.

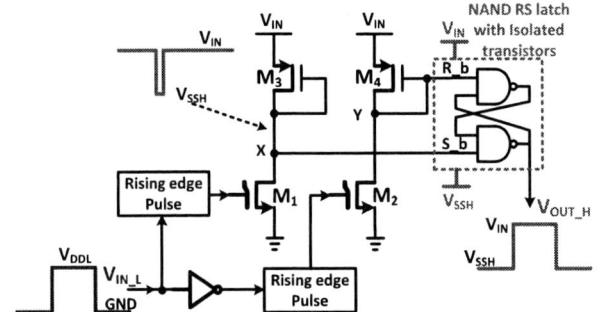

Fig. 7. Structure of the proposed pseudo-digital left shifter.

and the register DB[0:6] in the U/D counter stores the information of $T_H$ - $T_L$ in (2). The MUX array selects the value 'G' based on the converter duty ratio provided by $V_{IN}$ and $V_X$ information. The value $G \cdot (T_H - T_L)$ will then be added to the previous-cycle off-time to give the sum that is loaded to the Down counter. This sum determines the off-time of power transistor $M_n$ and in turns defines $T_{off}$ of the next switching cycle. The same process will continue until $T_L = T_H$ (i.e. $I_{avg} = I_{Avg,opt}$).

Fig. 6(a) shows the structure of a typical low-voltage gate driver that consists of a non-overlapping clock generator (NOCG) to generate proper dead-time during switching transitions of power transistors for minimizing the short-circuit power loss. However, as the input supply voltage of the proposed driver is as high as 40V, the structure in Fig. 6(a) is not appropriate to be used to generate a small differential high-side signal swing (1.5V in our design) referenced to the high common-mode supply voltage $V_{IN}$. A high-voltage gate driver shown in Fig. 6(b), that includes three additional level shifters compared with the low-voltage counterpart in Fig. 6(a), is thus developed. It is important to ensure these level shifters to have fast speed for minimizing the propagation delay of the gate driver.

Fig. 7 shows the structure of the proposed up level shifter. The core circuit $M_1$ – $M_4$ up-shifts the input signal from (0 – $V_{DDL}$) to ($V_{SSH}$ – $V_{DDH}$). Diode-connected transistors $M_3$ and $M_4$ lower the output impedance of the core circuit for delay minimization. Two short-pulse generators are placed at HV core-circuit inputs to reduce its turn-on time (<< on-time of the input signal). The core circuit is thus pseudo-digital in nature and the average current consumption of the level shifter can be significantly minimized. The SR latch in the isolated triple-well is used to recover the output signal of the core circuit to have the same on-time as the original input signal. Simulation results show that the proposed level shifter

achieves 0.62ns delay and dissipates 2.7μW/MHz for up shifting from (0 – 3V) to (38.5 – 40V). Compared with [6], both delay and power dissipation of the proposed level shifter are improved by 3.5 times and 30%, respectively.

## III. EXPERIMENTAL RESULTS

The proposed LED driver has been implemented in a 0.35μm 50V CMOS process. Fig. 8 shows the micrograph of the proposed driver. The chip area including all bonding pads is 3.47mm². Cree XB-D white LEDs are used for testing. The typical and maximum forward voltages of each Cree XB-D white LED are 2.9V and 3.5V, respectively.

The proposed LED driver was designed to operate with $V_{in}$ ranging from 10V to 40V with a maximum switching frequency of 1MHz. The driver delivers a typical average current of 345mA and drives a maximum 10 output series-connected LEDs. Fig. 9 shows different measured scope waveforms, indicating the proper operation of the proposed LED driver. Noted that differential gate swings for driving power transistors $M_P$ and $M_N$ are 1.5V and 3V, respectively. As the size $M_P$ and thus its input capacitance are much larger than those of $M_N$, a smaller differential signal swing for $M_P$ helps reduce its gate drive power loss that is proportional to the square of the differential gate swing. Fig. 10(a) shows that the proposed LED driver can achieve the maximum power efficiency of 92.5% under $V_{in} = 40V$ and the switching frequency of 1MHz. Fig. 10(b) shows the measured average LED current under different input and output conditions. The worst-case current error is 9.6mA (2.8% of the average LED current of 345mA) when the number of load LEDs changes from 5 to 10. Fig. 11 shows the settling behavior of the LED current after recovering from dimming. The worst-case settling time is only 8.5μs, which enables the fast dimming frequency of 10kHz and the wide dimming duty ratio from 20% to 100%. Fig. 12 shows the measured waveforms of the proposed LED driver under different duty ratios of the dimming signals.

Table I provides the performance comparisons of different LED drivers. Based on the reported work in [3], both the efficiency and current accuracy are degraded when the switching frequency of the driver is increased. The proposed LED driver provides a comparable peak power efficiency while operating at a much faster switching frequency and achieving much smaller current error under different input and output conditions compared with previous works from industry and literature. In addition, the proposed LED driver with ATDC scheme reduces the settling time of the LED current by 14 times compared with the previously-reported SAR-based adaptive off-time control [5].

978-1-4673-6145-3/13 $31.00 © 2013 IEEE       91

## IV. CONCLUSIONS

A new dimmable LED driver for the solid-state lighting applications is presented in this paper. With the proposed ATDC scheme and high-speed low-power synchronous gate driving under high input supply voltage, the proposed LED driver is proved to simultaneously achieve high power efficiency, high average LED current accuracy, and fast dimming (due to fast settling time and no calibration needed). The proposed LED driver is verified experimentally and the performances of the proposed DC-DC converter-based LED driver outperform those of state-of-the-art counterparts in lighting applications.

## ACKNOWLEDGMENTS

The authors thank for the financial supports from Semiconductor Research Corporation and U.S. National Science Foundation CAREER program.

## REFERENCES

[1] K. H. Loo, W. H. Lun, S. C. Tan, Y. M. Lai, and C. K. Tse, "On driving techniques for LEDs: toward a generalized methodology," *IEEE Trans. on Power Electronics*, vol. 24, no. 12, pp. 2967 – 2976, Dec. 2009.

[2] I. H. Oh, "An analysis of current accuracies in peak and hysteretic current controlled power LED drivers," in *Proc. IEEE Applied Power Electronics Conf.*, Feb. 2008, pp. 572 – 577.

[3] Diodes Inc. "ZXLD 1350: 30V 350mA LED driver with AEC-Q100," Datasheet, 2011.

[4] On Semiconductors Inc. "CAT 4201: 350mA high-efficiency step-down LED driver," Datasheet Rev. 8, May 2011.

[5] C. H. Liu, C. Y. Hsieh, Y. C. Hsieh, T. J. Tai, and K. H. Chen, "SAR-controlled adaptive off-time technique without sensing resistor for achieving high efficiency and accuracy LED lighting system," *IEEE Trans. Circuits Syst. I*, vol. 57, no. 6, pp. 1384 – 1394, Jun. 2010.

[6] Y. Moghe, T. Lehmann, and T. Piessens, "Nanosecond delay floating high voltage level shifter in a 0.35-μm HV-CMOS technology," *IEEE J. Solid-State Circuits*, vol. 46, no. 2, pp. 485 – 496, Feb. 2011.

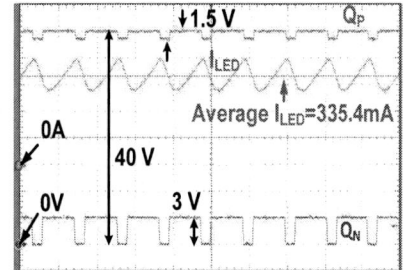

Fig. 9. Waveforms of the proposed LED driver under $V_{in}$ = 40V and 10 output LEDs.

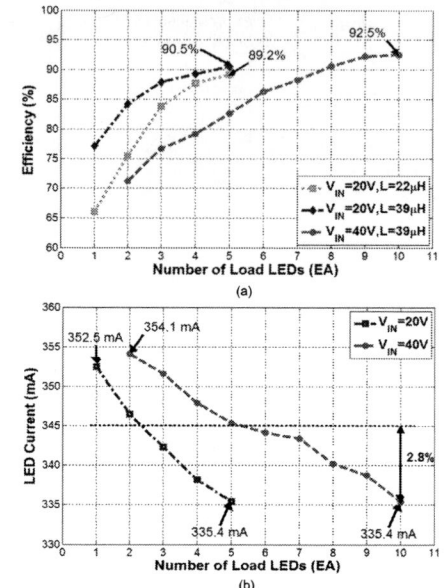

Fig. 10. (a) Efficiency and (b) current error of the proposed LED driver.

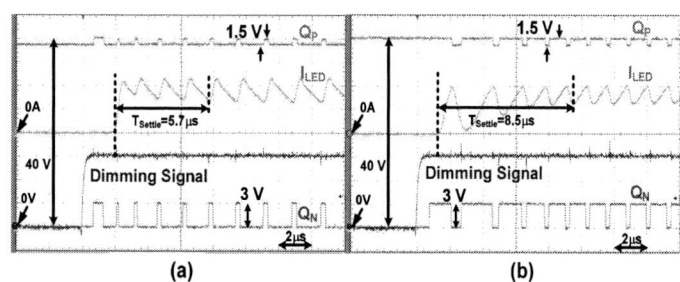

Fig. 11. Settling time after dimming under (a) $V_{in}$ = 10V and load = 2 LEDs, and (b) $V_{in}$ = 40V and load = 10 LEDs.

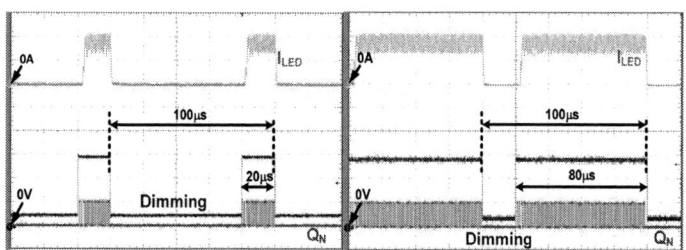

Fig. 12. Measured $Q_N$, LED current, and PWM dimming signal under different duty ratios.

TABLE I: PERFORMANCE COMPARISONS WITH PREVIOUS WORKS

| | ZXLD1350 2011 [3] | ZXLD1350 2011 [3] | CAT4201 2011 [4] | TCAS I 2010 [5] | This work |
|---|---|---|---|---|---|
| Switching Frequency (kHz) | 200 | 550 | 380 | ~188 | ~1000 |
| Input Voltage (V) | 6 - 30 | 6 - 30 | 6.5 - 36 | 8 – 40 | 10 – 40 |
| Max. LED Current (mA) | N. A. | N. A. | N. A. | 1500 | 500 |
| Typical Average Current (mA) | 350 | 350 | 300 | 720 | 345 |
| Worst-Case Settling Time (μs) | N. A.* | N. A.* | N. A.* | 120 | 8.5 |
| Peak Efficiency (%) | 95 | 92 | 90 (2 LEDs) | 94.3 | 93 |
| Current Error (mA) | 35 | 77 | N. A. | 14.5 (no. of LEDs: 5 - 8) | 9.6 (no. of LEDs: 5 - 10) 18.7 (no. of LEDs: 2 - 10) |
| Inductor (μH) | 100 | 47 | 22 – 56 | 33 | 10 – 39 |
| Max. No. of Drivable LEDs | ~ 8 | ~ 7 | ~ 9 | 8 | 10 |
| Fabrication Process | N. A.* | N. A.* | N. A.* | UMC 0.35-μm HV CMOS | AMS 0.35-μm HV CMOS |

* N. A. stands for not available

# A Stackable Switched-Capacitor DC/DC Converter IC for LED Drivers with 90% Efficiency

Chengrui Le[1], Mitchell Kline[2], Daniel L. Gerber[2], Seth R. Sanders[2], Peter R. Kinget[1]

[1]Columbia University, New York, NY 10027, USA
[2]University of California, Berkeley, Berkeley, CA 94720, USA

**Abstract**—A stackable switched-capacitor (SC) converter IC for a hybrid-SC-resonant LED driver using off-chip ceramic capacitors is presented in this paper. The IC can be configured to handle a range of input voltages through chip-stacking in the voltage domain. The tested driver delivers 17.6W at 470mA to the LED load with 90% peak efficiency from a rectified $120 V_{AC}$ line ($170 V_{DC}$), and maintains >85% efficiency over a rectified voltage range from $160 V_{DC}$ to $180 V_{DC}$.

## I. INTRODUCTION

As the cost of high brightness and high efficiency LEDs decreases, the need for efficient and miniaturized LED drivers is imminent. Traditionally, magnetics-based converters have dominated the moderate to high power (>100mW) applications, including those in lighting. However, since power magnetics are bulky, costly, and difficult to be integrated on die, switched capacitor conversion technology offers an attractive path to integration and passive component reduction. Furthermore, as outlined in [2,3,4], many switched capacitor topologies are able to realize medium or "high" voltage functionalities with basic low voltage elements, by organizing the elements in series. The work reported on in this paper takes advantage of this feature in utilizing a ladder topology, and further, by enabling stacking of chips, to effect a higher overall voltage solution than that available with either a single working device rating or with a single die.

Also, as outlined in [2,3], switched-capacitor conversion circuits are strategic in making effective use of the active device capabilities (i.e. overall V-A rating), and enjoy the enhanced energy and power densities of capacitors vis-à-vis magnetic elements. The presented DC/DC converter IC uses a hybrid resonant switched-capacitor strategy [1]. It offers all the advantages of the switched-capacitor ladder topology, while also permitting a clear path to soft-switching and lossless regulation, as is the typical case with the series resonant converter. In addition the multilevel switched capacitor LED driver prototype has a capacitive isolation barrier [1] offering galvanic isolation for the load. Taking advantage of the multilevel structure where the voltage rating of transistors and capacitors is substantially reduced, this work develops a highly integrated switched-capacitor standard module, where the power transistors, gate-drivers, level-shifters and internal DC-DC converters are integrated on chip using medium voltage rated devices. The module IC can then be "stacked" in the voltage domain to directly interface with the mains voltages.

The developed integrated circuit power conversion function has been prototyped in an Analog-Bipolar-CMOS-DMOS technology and achieves power levels in the 10-30W range. It

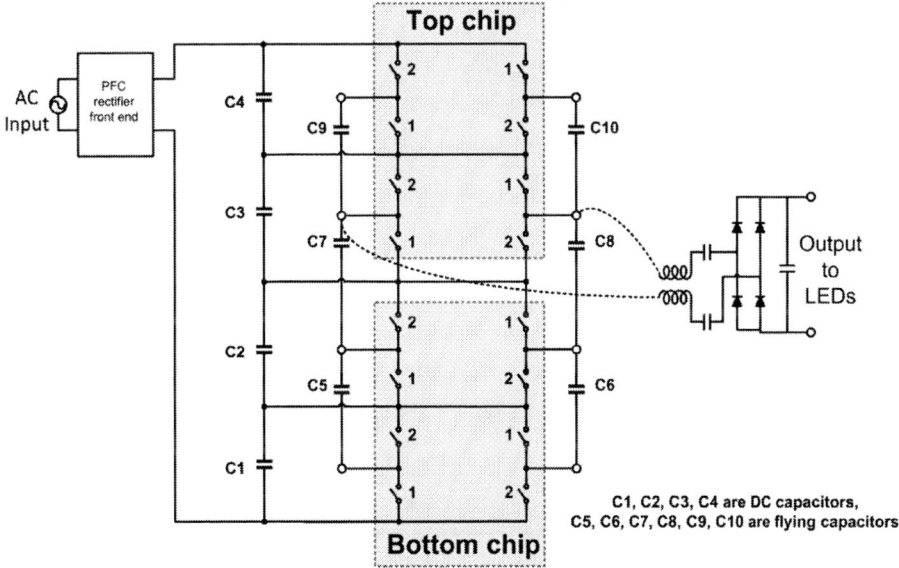

Fig. 1: Overall architecture of the proposed chip-stacking hybrid-switched-capacitor-resonant LED driver

significantly exceeds the ratings achieved in prior integrated switched-capacitor converter examples, e.g. those in [2,4,5,6].

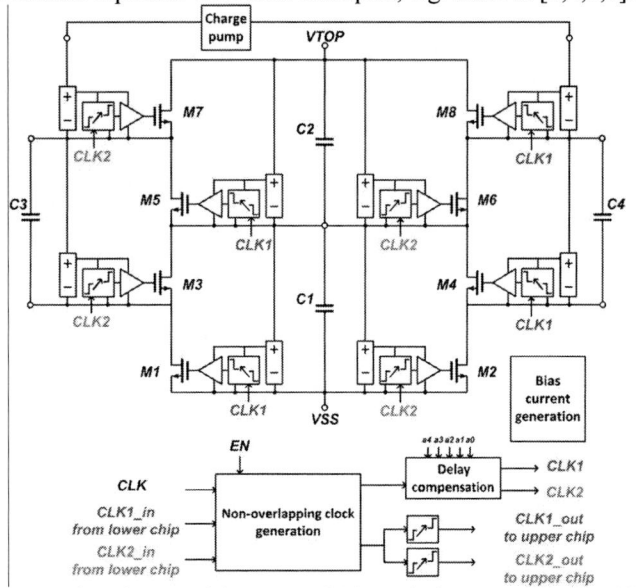

Fig. 2: Block diagram of the integrated 2:1 switched-capacitor converter IC

## II. ARCHITECTURE OF THE OFF-LINE PFC LED DRIVER

Fig. 1 shows the simplified architecture of the proposed hybrid-SC-resonant LED driver. The $120V_{RMS}$ AC line voltage is first rectified by a PFC rectifier front end and then passed through a switched-capacitor DC-DC converter to the LED load. The main body of the converter is a two-phase balanced 4:1 step-down power train. It down-converts the rectified DC voltage (nominally 170V) to a 900 kHz differential AC square-wave with a nominal amplitude of 42.5V, which gets fed through a resonant tank to the final output stage. Due to process breakdown voltage limitation, the power train is implemented with a stack of two integrated 2:1 switched-capacitor ICs connected in series in the voltage domain that will be discussed in detail in the following section. The balanced topology of the ladder can minimize high frequency common mode signal feed through to the output. With high voltage (e.g. 3kV) MLCC resonant capacitors, the resonant network can provide galvanic isolation. This approach is pursued here. Further, by controlling the switching frequency

of the switched-capacitor converter, we can change the impedance of the resonant tank and provide lossless regulation. Since the isolation capacitor in the resonant tank is rated for 3kV, the input of the resonant tank can be attached to any of the differential nodes in the switched-capacitor power train.

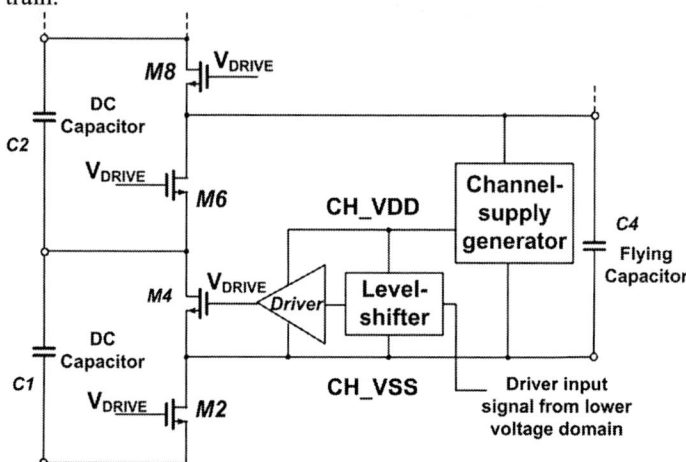

Fig. 3: Block diagram of the gate-driver channel

## III. STACKABLE SWITCHED-CAPACITOR CONVERTER IC

An integrated design of the switched-capacitor DC-DC converter's power train has been implemented with an Analog-Bipolar-CMOS-DMOS (ABCD) technology. Since the rectified DC voltage at the output of the PFC front end can peak to $180V_{DC}$, which exceeds the breakdown voltage limit of the ABCD process, a 'chip-stacking' topology is used to overcome the breakdown voltage limit. In the case shown in Fig. 1, for a nominal $170V_{DC}$ input, each capacitor is charged to 42.5V and the top chip's substrate sits on the bottom chip's highest voltage of 85V and the two chips communicate with each other by signal level-shifters.

Fig. 2 shows a simplified schematic of the stackable switched-capacitor converter IC. Each IC has 8 main switches with individual gate-driver channels. Selected by 'EN', the bottom chip uses the external clock input signal 'CLK' to generate two non-overlapping clock phases for the gate-drivers and level-shifts them to the top chip, while the top chip uses the 'CLK1_in' and 'CLK2_in' input clock signals shifted from the bottom chip for the gate-drivers.

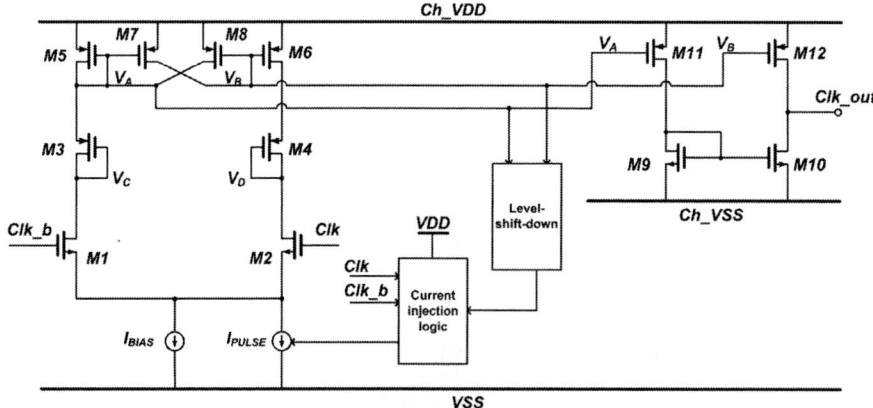

Fig. 4: Schematic of the level-shifter

978-1-4673-6145-3/13 $31.00 © 2013 IEEE

Each gate-driver channel is composed of a channel supply generator, a level shifter, and a gate-driver, as shown in Fig. 3. The channel-supply generator is a low-dropout (LDO) regulator that converts the 42.5V voltage drop across the DC or flying capacitors to a 5V voltage supply above the source of each main power switch which requires a $V_{GS}<5V$. The LDO can self-start, which simplifies the start-up control of the converter. Each level-shifter takes a gate-driving CLK signal from the 0-5V voltage domain and up-shifts it to each respective gate-driver channel. The schematic of the level-shifter is shown in Fig. 4, where nodes VDD and VSS are in the 5-0V voltage domain while Ch_VDD and Ch_VSS are in the flying gate-driver voltage domain. A differential pair M1 and M2 with a regenerative load M7 and M8 up-shift the signals to nodes $V_A$ and $V_B$, while the high voltage cascode devices M3 and M4 protect the low voltage devices. Two diode-connected PMOS transistors M5 and M6 are added to limit the swing of nodes $V_A$ and $V_B$, to speed up the transitioning of the up-shifted signals. Signals $V_A$ and $V_B$ are then fed through a down-shifting ladder back to the low voltage domain and control a current source $I_{PULSE}$. In steady state, the differential pair is biased by a small current $I_{BIAS}$ and has very low power dissipation. During the transition interval the current injection logic will pull a large pulse current $I_{PULSE}$ in parallel with $I_{BIAS}$ to speed up the flipping of $V_A$ and $V_B$, thus reducing the level-shifting delay.

Simulation shows that the best conversion efficiency can only be achieved when the gate-driving signals of each switch ($V_{DRIVE}$) are aligned within 6ns interval. So in order to better align the gate-driving signal, all level shifters inside the chip take their inputs from the lowest voltage domain and then up-shift them to the destination voltage domain in parallel. However, since the function of level-shifters is limited by the technology maximum voltage rating, the gate-driving signals of the top chip must go through two level-shifters in series and has a larger delay compared to the bottom chip. Thus a tunable delay element is added in each chip to compensate for the delay in the inter-chip signal paths, as shown in the bottom of Fig. 2. and in Fig. 5. The delay compensation is controlled by 5 digital input signals with maximum delay of 18ns.

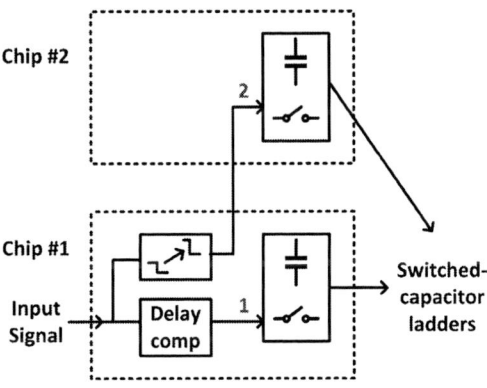

Fig. 5: Block diagram of the delay cancellation strategy

## IV. MEASUREMENT RESULTS

The stackable switched-capacitor converter chip has been fabricated in an Analog-Bipolar-CMOS-DMOS technology.

The converter die (Fig. 6) measures 14mm², including the ESD cells and the pads. The tested LED driver uses a split-winding 6.2 $\mu H$ inductor and two 10nF 3kV isolation capacitors as the differential resonant tank and the normal operating frequency is 0.9MHz. DC and flying capacitors are implemented with 1$\mu F$ capacitors. The load is a string of 12 1-Watt white LEDs. Fig. 7 (a) shows the output current regulated at 372mA with a rectified input DC voltage range from 160V to 180V, while efficiency goes from 85.6% to 90%. The output current range can vary from 63mA to 571mA using a 2.5kHz PWM dimming control with a duty cycle from <10% to 100%, while operating with an efficiency from 66% to 90%. (Fig. 7. (b)).

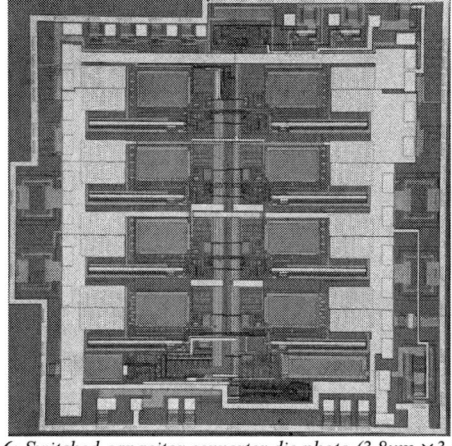

Fig. 6: Switched-capacitor converter die photo (3.8μm ×3.7μm)

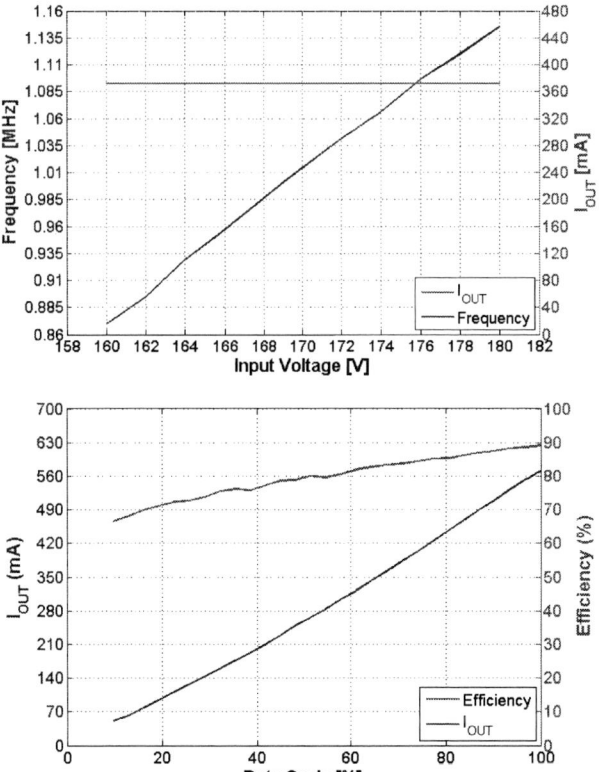

Fig. 7: Measurement performance of proposed LED driver with rectified 120V AC (170V DC) input. (a) Switching frequency and current regulation versus input voltage. (b) Current regulation and efficiency versus duty cycle.

The differential output voltage of a single stage of the switched-capacitor ladder is shown in Fig. 8. Zero voltage switching is evident from these waveforms. Fig. 9 shows the voltage waveforms of the flying nodes on one side of the balanced switched-capacitor ladder. Note that the well aligned gate drive signals on chip, and between chips, enables the nearly ideally aligned waveforms shown in Fig. 9. When no compensation is applied, the measured efficiency of the driver drops 5% due to the misalignment of the gate-driving signals. Misalignment of gate-driving signals can cause shoot-through current through flying capacitors and disturb zero-voltage switching, thus inducing more loss in the circuit.

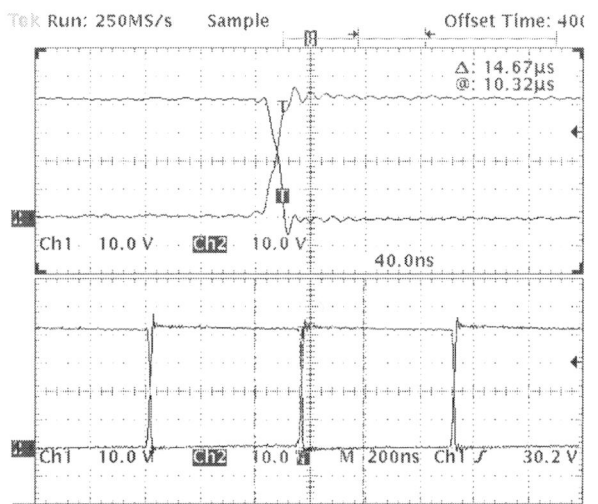

Fig. 8: Voltage waveforms of the differential node of the switched-capacitor ladder. The waveform on top is a zoomed view of the switching edge of the signals

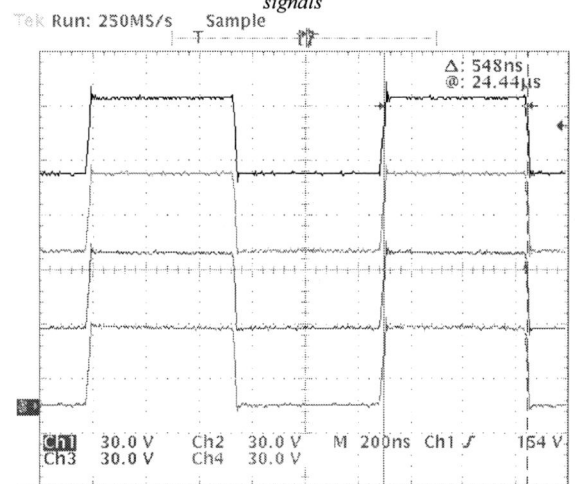

Fig. 9: Voltage waveforms of the flying nodes on one side of the switched-capacitor ladder

A photo of the tested DC/DC converter driving a string of 12 LEDs is shown in Fig. 11. The core area of the 2-layer test PDB measures 16cm$^2$. It can be further minimized by using 4-layer PCB and tighter layout.

## V. CONCLUSIONS

A stackable switched-capacitor (SC) DC/DC converter IC for a hybrid-SC-resonant LED was presented. Chip stacking allows the overall converter to exceed the process voltage ratings. The hybrid-SC-resonant topology offers near lossless regulation and galvanic isolation with small capacitors. The converter demonstrated a peak efficiency of 90% for a 17.6W output power which sets it apart as one of the highest power SC converters compared to the current state of the art.

Fig. 10: The two-chips-stacking DC/DC converter for LED driver

### ACKNOWLEDGMENTS

The authors thank Bijoy Chatterjee and Texas Instruments for silicon fabrication donation. Financial support was provided by the DOE under the ARPA-E ADEPT program, contract Metacapacitors DE-AR0000114. We also thank Ali Djabbari, Raj Subramoniam, Glen Wells, Jianyi Wu, Zhenyong Zhang, Gianpaolo Lisi, Jerry Socci, and Leigh Perona for their technical support.

### REFERENCES

[1] M. Kline I. Izyumin, B. Boser, and S. Sanders, "A transformerless galvanically isolated switched capacitor LED driver," *Proc. IEEE APEC*, Feb. 2012.
[2] M. Seeman and S. Sanders, "Analysis and optimization of switched-capacitor DC–DC converters," *IEEE Trans. Power Electron.* , vol. 23, no. 2, pp. 841–851, Mar. 2008.
[3] S. R. Sanders et al., "The Road to Fully Integrated DC–DC Conversion via the Switched-Capacitor Approach," *IEEE Trans. on Power Electron.*, vol. 28, no. 9, pp. 4146 – 4155, Sept. 2013.
[4] V. W. Ng and S. R. Sanders, "A high-efficiency wide-input-voltage range switched capacitor point-of-load DC–DC converter," *IEEE Trans. Power Electron.*, vol. 28, no. 9, Sept. 2013.
[5] H. Meyvaert, T. Van Breussegem, and M. Steyaert, "A 1.65 W fully integrated 90 m bulk CMOS intrinsic charge recycling capacitive DC–DC converter: Design & techniques for high power density," *Proc. IEEE Energy Conver. Congr. Expo.*, Phoenix, 2011, pp. 3234–3241.
[6] T. M. Anderson, F. Krismer, and J. W. Kolar, "A 4.6 W/mm$^2$ power density 86% efficiency on-chip switched capacitor DC-DC converter in 32 nm SOI CMOS," in Proc. IEEE Appl. Power Electron. Conf. (APEC), Mar. 2013.

# A 100V Gate Driver with Sub-Nanosecond-Delay Capacitive-Coupled Level Shifting and Dynamic Timing Control for ZVS-Based Synchronous Power Converters

Zhidong Liu and Hoi Lee

Department of Electrical Engineering

The University of Texas at Dallas, Richardson, TX 75080-3021, USA

*Abstract*-A high-voltage high-speed gate driver to enable synchronous rectifiers with zero-voltage-switching (ZVS) operation is presented in this paper. A capacitive-coupled level-shifter (CCLS) is developed to achieve negligible propagation delay and static current consumption. With only 1 off-chip capacitor, the proposed gate driver possesses strong driving capability and requires no external floating supply for the high-side driving. A dynamic timing control is also proposed not only to enable ZVS operation in the converter for minimizing the capacitive switching loss, but also to eliminate the converter short-circuit power loss. Implemented in a 0.5μm HV CMOS process, the proposed CCLS of the gate driver can shift up a 5V signal to the 100V DC rail with sub-nanosecond delay, improving the FoM by at least 29 times compared with that of state-of-the-art counterparts. The dynamic dead-time control properly enables ZVS operation in a synchronous buck converter under different input voltages (30V to 100V). The power losses of the high-voltage buck converter are thus greatly reduced under different load currents, achieving a maximum power efficiency improvement of 11.5%.

## I. INTRODUCTION

High-voltage (HV) power converters find wide applications in electronic vehicles, renewable energy systems, LED lighting systems, etc. To realize on-chip power transistors in HV power converters, thin-oxide HV devices are commonly used nowadays mainly due to advantages of significantly lower on resistance and threshold voltage than those of thick-oxide counterparts. The constraint of using thin-oxide power device however lies on driving the high-side power device because a $V_{GS}$ control signal usually in the range of $0 - 5V$ needs to be referenced to its source terminal. Therefore, a proper level-shifting technique should be used in the gate driver for the high-side driving to up-shift the gate control signal from low to high common-mode voltage, while still keeping the signal swing within the safety limit of the high-side device. Since the overall power dissipation and the propagation delay of the gate drivers are usually determined by the level shifters, developing a reliable, low-power and high-speed level shifter is crucial but challenging in the HV gate driver. In addition, high-performance gate drivers should also be able to (i) provide sufficient transient current to charge / discharge the gate terminals of power transistors for minimizing the propagation delay; (ii) offer a floating DC supply to enable high-side circuitry like logic circuitry and buffers using thin-oxide isolated-well low-voltage devices; and (iii) generate appropriate dead-time to minimize shoot-through currents during switching transitions of power

TABLE I: Performances of different level shifters

|  | Process Node L (μm) | $V_{DDH}$ (V) | Delay (ns) | FoM: D/(LV) (ns/μm·V) |
|---|---|---|---|---|
| [1] | CMOS 2 | 50 | 80 | 0.80 |
| [2] | HV CMOS 0.35 | 25 | 2.5 | 0.30 |
| [3] | HV CMOS 0.35 | 10 | 3.0 | 0.86 |
| [4] | HV CMOS 0.7 | 100 | 2000 | 28.6 |
| Proposed | **HV CMOS 0.5** | **100** | **~0.5** | **0.01** |

transistors and even enable zero-voltage-switching (ZVS) operation in the power converter for eliminating the capacitive switching power loss and switching noises under high-voltage condition.

Different level shifters have been previously reported for medium–voltage [1] – [3] and HV [4] applications. However, those reported level shifters suffer from either dissipating considerable static power [1], [2], requiring additional floating bias voltages that are usually obtained externally [3], or having long propagation delay from input to output [4]. In addition, an N-channel transistor is usually used as the high-side power device instead of the P-channel counterpart for saving the chip implementation area of the converter. However, high dV/dt and dI/dt noises at the switching node of the HV power converter might cause inherent reliability issue at the high-side gate drive for the N-channel power device [5]. Two additional off-chip components: a power diode and a capacitor are also needed to establish a floating supply in the high-side gate drive for the N-channel power device, increasing the board area and the system cost. Moreover, ZVS was previously reported to minimize the capacitive switching loss, switching noise and switching stress of HV power converters under the hard-switching operation [6]. An adaptive dead-time technique was previously reported to enable ZVS operation in a low-voltage buck converter [7]. However, its complicated circuit implementation involving a clocked sample-and-hold circuit and a V-to-I converter to generate dead-time for each power device would need large static current consumption for providing 10s of nano-second dead-time and is difficult to be extended to the high-side gate drives in HV power converters.

To address above difficulties and issues associated with gate driver design for HV power converters, this paper proposes a new gate driver for driving complementary-MOS power trains (high-side pMOS and low-side nMOS). It consists of a capacitive-coupled level shifter (CCLS) with only one off-chip capacitor as a built-in floating supply to

978-1-4673-6145-3/13 $31.00 © 2013 IEEE

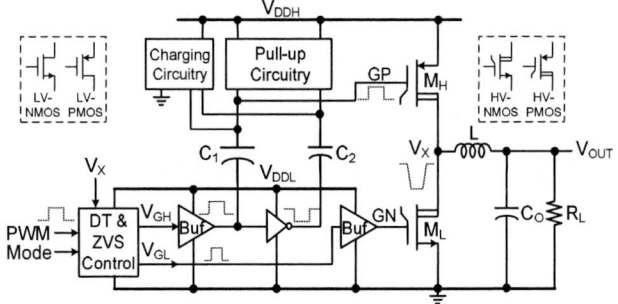

Fig. 1. Structure of proposed CCLS based gate driver with dynamic timing control for HV synchronous buck converters.

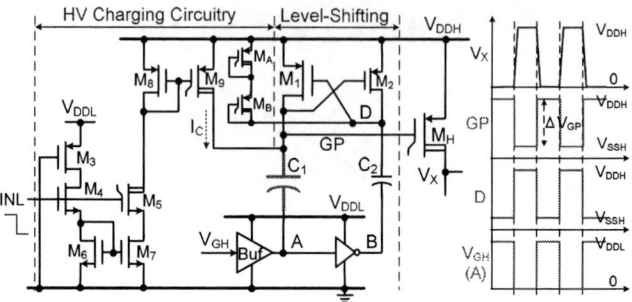

Fig. 2. Circuit and operation waveforms of the proposed CCLS with strong driving capability.

upshift the gate control signal for the high-side power pMOS with only sub-nanosecond delay. Table I provides the comparisons of different level shifters and a figure of merit (FoM) is adopted to gauge their performances [3]. The smaller the FoM is, the better the level shifter is. The proposed CCLS achieves ≥29 times improvement in FoM compared with previous counterparts. The proposed gate driver can also dynamically control the dead-time of power transistors to enable ZVS operation of HV power converter for saving different power losses in the power converter. The structure, operation principles and measurement results of the proposed gate driver will be discussed in detail in later sections.

## II. PROPOSED HIGH-VOLTAGE GATE DRIVER

The structure of the proposed HV gate driver for synchronous rectifiers is illustrated in a synchronous buck converter as shown in Fig. 1. The proposed driver consists of a capacitive-coupled level shifter (CCLS) formed by two capacitors $C_1$ and $C_2$, charging and pull-up circuitry, and an inverter providing the complementary signal at the level-shifter input. ZVS control block in the gate driver provides appropriate dead-time between $V_{GH}$ and $V_{GL}$ signals and correspondingly between GP and GN signals for enabling ZVS operation and minimizing the shoot-through current of power transistors $M_H$ and $M_N$. Two tapered buffers (Buf) in the gate driver are used to provide strong driving capability for ensuring fast gate-voltage switching of power transistors $M_H$ and $M_N$. It should be noted that $V_{DDL}$ is 5V and $V_{DDH}$ can be as high as 100V in our design. Power transistors $M_H$ and $M_L$ are realized by thin-oxide HV CMOS devices.

### A. Proposed CCLS with Built-In Floating Supply

Fig. 2 shows the schematic of the HV CCLS in which transistors $M_3 - M_9$, $M_A$ and $M_B$ form the HV charging circuitry to provide charging current for $C_1$ and $C_2$ in the start-up phase. These transistors are turned off in the steady state for reducing the power dissipation of the level shifter. The core of CCLS consists of cross-coupled transistors $M_1$ and $M_2$ with capacitors $C_1$ and $C_2$. The CCLS core is designed to be asymmetric: i.e. the sizes of $M_1$ and $C_1$ (the left branch) are much larger than those of $M_2$ and $C_2$ (the right branch). This is because large-size high-side power transistor $M_H$ is connected to the top plate of $C_1$ (at node GP) and a large-value $C_1$ (>> parasitic capacitance at node GP) is needed to maintain the gate voltage swing for $M_H$ to be close to $V_{DDL}$. $C_1$ acts as the built-in floating supply and is the only off-chip component in

the gate driver. The smaller sizes of $M_2$ and $C_2$ help reduce the chip area, while signals at D and B are still maintained complementary to those at GP and A, respectively.

In the CCLS, any signal variation at node A (or B) will be directly coupled to node GP (or D). The only delay element in the CCLS is the inverter between nodes A and B. Circuit simulations show that the delay from A to GP is negligible and that from A to B (or D) is about one gate delay, 0.5ns. Therefore, the proposed CCLS offers sub-nanosecond delay for translating voltage signal from low to high common-mode voltage. Moreover, due to the presence of $C_1$ and $C_2$, there is no static current consumption in the level shifter core. The dynamic power dissipation of the CCLS core is also minimal due to the relatively small voltage swing of only ~$V_{DDL}$ (5V in our design) at nodes A, B, GP and D.

The start-up mechanism of the CCLS is as follows. Signal INL in Fig. 2 is set to $V_{DDL}$ and provides current $I_C$ for charging $C_1$ when supply rail $V_{DDH}$ is being powered up initially. As $V_{GH}$ is equal to $V_{DDL}$ during start-up, $C_1$ will be charged to $V_{DDH}$ at node GP. Similarly, $C_2$ will be charged to $(V_{DDH} - V_{DDL})$ at node D via two diode-connected HV transistors $M_A$ and $M_B$. Transistor $M_2$ and $M_H$ is off during start-up. When $V_{DDH}$ finally reaches the desired value, signal INL is changed to 0. Transistors $M_3 - M_9$, $M_A$ and $M_B$ are turned off in the steady state to cut down static current consumption. Noted that, $M_5$ is a HV device in the cascode current mirror to sustain high voltage stress, while HV transistor $M_9$ has a higher threshold voltage than LV transistor $M_8$ for eliminating any undesirable leakage current of $M_9$ in the steady state.

### B. Operation and Design Considerations of CCLS

Since a large-size high-side power transistor $M_H$ is used to source large current to the power converter, a large parasitic capacitance $C_{GP}$ of $M_H$ is loaded at the output of the CCLS (node GP). So the charge in $C_1$ will be redistributed between $C_1$ and $C_{GP}$ during the switching instants and this determines the voltage at node GP. The operation of the charge redistribution is illustrated in Fig. 3. During the transition of $V_{GH}$ changing from $V_{DDL}$ to 0 in Fig. 3(a), transistor $M_{b2}$ in the last stage of the tapered buffer will be turned on to sink current from $C_1$. Both voltages at nodes A and GP will drop instantaneously with the intention to turn on power transistor $M_H$. The voltage swing at A ($\Delta V_A$) is $V_{DDL}$, while that at GP is $\Delta V_{GP} = V_{DDH} - V_{SSH}$, where $V_{SSH}$ is assumed to be the lowest voltage at GP. It is critical to design the value of $C_1$ such that

978-1-4673-6145-3/13 $31.00 © 2013 IEEE

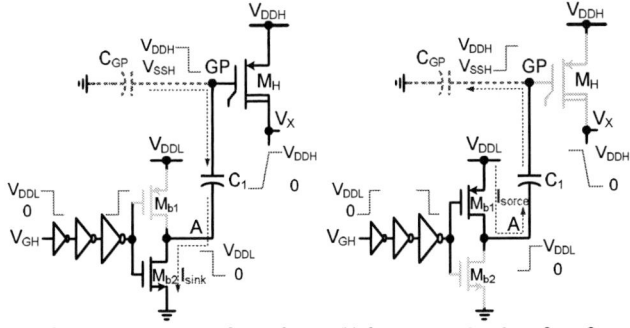

(a) Charge transfers from $C_{GP}$ to $C_1$     (b) Charge transfers from $C_1$ to $C_{GP}$

Fig. 3. The operation of coupling capacitor $C_1$ as the floating supply.

$\Delta V_{GP}$ is larger than the threshold voltage of $M_H$ for ensuring its proper turn on. The value of $\Delta V_{GP}$ can be obtained as

$$\Delta V_{GP} = \frac{C_1}{C_1 + C_{GP}} \Delta V_A \qquad . \qquad (1)$$

A similar charge redistribution (i.e. charge is transferred from $C_1$ to $C_{GP}$ as shown in Fig. 3(b)) also occurs during the transition of $V_{GH}$ changing from 0 to $V_{DDL}$ for turning off $M_H$, leading to the same equation of $\Delta V_{GP}$. Eq. (1) indicates that $\Delta V_{GP}$ is always smaller than $V_{DDL}$ and dependent on the relative sizes of $C_1$ and $C_{GP}$. The value of $C_1$ can be found as

$$C_1 = \frac{\Delta V_{GP}}{\Delta V_A - \Delta V_{GP}} C_{GP} \qquad . \qquad (2)$$

From (2), if $\Delta V_{GP} = 4.5V$ and $V_{DDL} = 5V$, $C_1$ should be at least 9 times larger than $C_{GP}$. In our design, $C_{GP}$ is about 500pF, and thus $C_1 = 5nF$ is chosen. Similarly, an on-chip capacitor $C_2$ of ~5pF is selected for obtaining a voltage swing of 4.5V at node D (Fig. 2), as the parasitic capacitance at node D is small.

In short, the operation of the CCLS in the steady state is as follows. When $V_{GH}$ input is $V_{DDL}$, voltages at nodes A, GP and D change to $V_{DDL}$, $V_{DDH}$, and ($V_{DDH} - \Delta V_{GP}$), respectively. Both $M_2$ and $M_H$ are thus turned off, while $M_1$ is on to connect $C_1$ to $V_{DDH}$ and allow voltage across $C_1$ to be ($V_{DDH} - V_{DDL}$). Similarly, when $V_{GH}$ input is 0, $M_1$ is off. Voltage at GP is equal to ($V_{DDH} - \Delta V_{GP}$) to turn on $M_2$ and $M_H$ and voltage across $C_2$ is ($V_{DDH} - V_{DDL}$). Since high voltages are developed across $C_1$ and $C_2$, pull-up transistors $M_1$ and $M_2$, tapered buffers and the inverter between nodes A and B in Fig. 2 can all be realized by LV transistors in the proposed CCLS.

### C. ZVS in HV Power Converters

In traditional hard-switching synchronous converters, the capacitive switching loss caused by charging / discharging the parasitic capacitance $C_X$ at node $V_X$ is significant under high-voltage condition. Substantial short-current power loss of the converter also occurs if no dead-time is provided by the gate driver.

ZVS technique is previously reported to minimize switching losses, and dV/dt and dI/dt noises by using inductor current to charge / discharge switching node $V_X$ during switching transitions [6], [7]. Fig. 4 shows the scheme and waveforms of ZVS operation in buck converters. The power losses associated with discharging $C_X$ will be eliminated if $M_L$ is turned on after $V_X$ going to zero, as $C_X$ is discharged by the peak inductor current during this transition period. Similarly, a

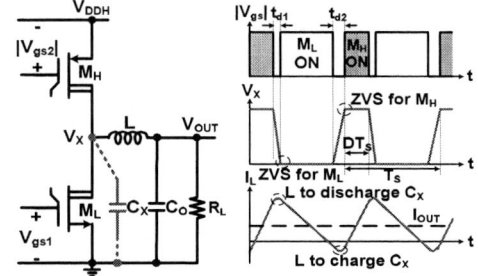

Fig. 4. Scheme and waveforms of ZVS operation in buck converters.

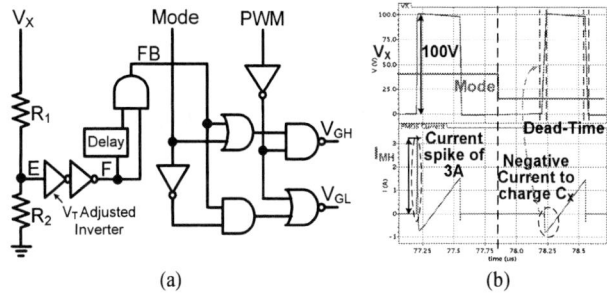

(a)         (b)

Fig. 5. Proposed dynamic ZVS control (a) circuit implementation and (b) simulation results of mode transition from hard switching to soft switching.

lossless low-to-high transition for $M_H$ at node $V_X$ can also be achieved by turning on $M_H$ after $V_X$ reaching $V_{DDH}$, as $C_X$ can charged by the negative inductor current during this dead-time period.

### D. Implementation of Dynamic ZVS Control

The circuit implementation of the proposed dynamic ZVS control is shown in Fig. 5(a). In this design, $V_X$ is monitored via a resistor divider realized by $R_1$ and $R_2$ to determine the zero-voltage state of power transistors. When $V_X$ is changing from 0 to $V_{DDH}$, the scaled-down of $V_X$ ($V_E$) indicates the zero-voltage state of $M_H$ if $V_E$ reaches a threshold voltage $V_T$ (signal F is logic 1). $V_T$ is set at 1V in our design to ensure optimum dead-time to be generated when $V_{DDH} = 30V$. When $V_{DDH}$ is larger than 30V, a controllable delay element in Fig. 5(a) will postpone the rising edge of signal FB automatically to increase the dead-time, such that appropriate dead-time between $M_H$ and $M_L$ as well as ZVS of $M_H$ can be simultaneously achieved when $V_{DDH}$ is larger than 30V. Similarly, when $V_X$ is changing from $V_{DDH}$ to 0, the dead-time is defined to ensure ZVS of $M_L$ if $V_E$ reaches $V_T$ to cause signals F and FB changing to logic 0. Gate drive timing is then adjusted by logic circuits based on signals FB and PWM signal. The proposed controller also consists of a mode selection, enabling the converter to perform either hard switching (Mode = "1") or soft switching (Mode = "0"). During hard-switching mode, the output drive signals $V_{GH}$ and $V_{GL}$ are identical with the PWM input with 2 gate delays. During soft-switching mode, gate drive timing is adjusted dynamically by signals PWM and FB to ensure ZVS operation under different input voltages through the detection of $V_X$ state and the tunable delay element. The simulated result of mode transition from hard switching to soft switching at $V_{DDH} = 100V$ is given in Fig. 5(b). A large current spike of 3A is generated at turn-on instant for charging $C_X$ during hard-switching mode. The soft switching enabled by dynamic ZVS

978-1-4673-6145-3/13 $31.00 © 2013 IEEE

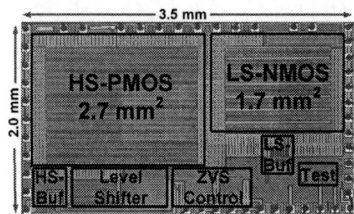

Fig. 6.  Micrograph of the proposed HV gate driver with power switches.

control minimizes the current spike as well as dV/dt and dI/dt noises. Due to the features of easy implementation and high-voltage compatibility, the proposed dynamic ZVS control can be extended to other converters as well.

## III.  MEASUREMENT RESULTS AND COMPARISONS

To verify the performances of the proposed high-voltage gate driver, it is fabricated together with two on-chip power transistors $M_H$ and $M_L$ (Fig. 1) in a 0.5μm HV CMOS process and the micrograph is shown in Fig. 6. The total chip area is 7.0mm² including all bonding pads, and the area for the gate drive circuit itself is less than 2.0mm².

Fig. 7 shows the measured input and output waveforms of the CCLS in the gate driver under $V_{DDH}$ = 80V. A voltage swing at GP is 4.5V with reference to $V_{DDH}$ = 80V. Although ~500pF capacitance is loaded at node GP due to gate capacitance of power transistor $M_H$, both rising-edge and fall-edge propagation delays from nodes A to GP are negligible based on the zoom-in waveforms. Hence, it proves that the delay of the CCLS is only the inverter delay between nodes A and B in Fig. 2 and is about 0.5ns. Table I provides performance comparisons of different level shifters. Fig. 8 shows the measured waveforms of the gate driver and the buck converter under different switching modes and supply voltages. Under hard switching with $V_{DDH}$ = 50V shown in Fig. 8(a), there are undesirable glitches on the rising/falling edges of $V_X$. As shown in Figs. 8(b) and 8(c), the glitches are eliminated by soft switching under different supply voltages. Additionally, both Figs. 8(b) and 8(c) show that GP changes to $V_{DDH}$ – 4.5V after $V_X$ increasing to $V_{DDH}$, and GN changes to 5V after $V_X$ decreasing to 0; therefore ZVS of $M_H$ and $M_L$ are achieved. The measured power efficiency of the synchronous buck converter with the proposed gate driver is given in Fig. 9. The converter power efficiency with the dynamic ZVS control is always larger than that with the hard-switching case for the entire load range from 50mA to 400mA. In particular, a maximum 11.5% power efficiency improvement is achieved in the ZVS buck converter.

## IV.  CONCLUSIONS

A 100V high-voltage gate driver enabled by CCLS and dynamic ZVS control is presented in this paper. The proposed CCLS achieves sub-nanosecond delay without static current consumption. CCLS also provides built-in high-side supply with only one off-chip capacitor and offers strong driving capability for power transistors. Proper gate drive timing is achieved by dynamic ZVS control in the driver to enable soft switching, thereby reducing switching losses and switching noises of power converters. The proposed gate driver with integrated power switches has been fabricated and verified

with silicon results. The proposed gate driver can significantly enhance the performances of high-voltage synchronous power converters.

## REFERENCES

[1]  M. J. Declerq, M. Schubert, and F. Clement, "5V-to-75 V CMOS output interface circuits," *IEEE Int. Solid-State Circuits Conference Dig. Tech. Papers*, Feb. 1993, pp. 162 - 163.

[2]  J. Buyle, V. De Gezelle, B. Bakeroot, and J. Doutreloigne "A new type of level-shifter for n-type high-side switches used in high-voltage switching ADSL line drivers," in *Proc. 2008 IEEE Int. Conf. Electronics, Circuits and Systems*, pp. 954 - 957.

[3]  Y. Moghe, T. Lehmann, and T. Piessens, "Nanosecond delay floating high voltage level shifters in a 0.35 μm HV-CMOS technology," *IEEE J. Solid-State Circuits*, vol. 46, no. 2, pp. 485 - 497. Feb. 2011.

[4]  J. Doutreloigne, "A fully integrated ultra-low-power high-voltage driver for bistable LCDs," in *Proc. IEEE Int. Symp. VLSI Design, Automation and Test*, Apr. 2006, pp. 1 - 4.

[5]  L. Balogh, "Design and application guide for high speed MOSFET gate drive circuits," *Texas Instruments Power Supply Design Seminar* (SEM - 1400), 2001.

[6]  C. Y. Chiang, and C. L. Chen, "Zero-voltage-switching control for a PWM buck converter under DCM/CCM boundary," *IEEE Trans. Power Electron.*, vol. 24, no. 9, pp. 2120 - 2126, Sep. 2009.

[7]  A. Stratakos, S. Sanders, and R. Brodersen, "A low-voltage CMOS dc-dc converter for a portable battery-operated system," in *Proc. IEEE Power Electronics Specialists Conf.*, 1994, pp. 619 - 626.

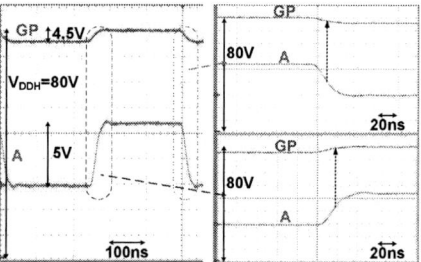

Fig. 7.  Measured waveforms of the CCLS in the gate driver at $V_{DDH}$ = 80V.

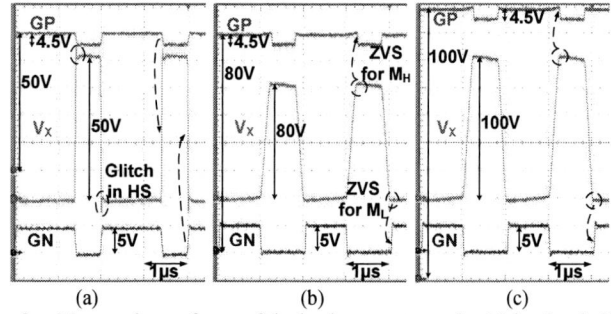

Fig. 8.  Measured waveforms of the buck converter under (a) hard switching and $V_{DDH}$ = 50V, (b) soft switching and $V_{DDH}$ = 80V, and (c) soft switching and $V_{DDH}$ = 100V.

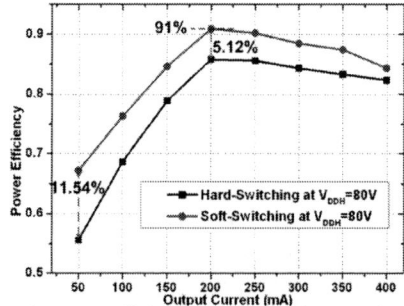

Fig. 9.  Measured power efficiency of the synchronous buck converter at $V_{DDH}$ = 80V under different switching modes.

# A Compact 120-MHz 1.8V/1.2V Dual-Output DC-DC Converter with Digital Control

Sakshi Arora, David K. Su, and Bruce A. Wooley
Stanford University, Stanford, California, U.S.A

*Abstract* — **A dual-output cascaded dc-dc buck converter is proposed to meet the challenge of miniaturization for embedded applications. A programmable switching frequency up to 120 MHz, output stage segmentation, and cascoding are used to achieve high power efficiency with small output inductors. A digital constant-off-time controller provides a fast load current step-response and the suppression of cross-regulation among multiple output voltages. Integrated in a 90-nm CMOS technology, an experimental prototype generates 1.8 V and 1.2 V from a 2.5-V input supply. Using only a PCB trace inductance of 4 nH, the 2.5-V to 1.8-V conversion is accomplished with the peak efficiency of 76% at a power density of 2.59 W/mm². The peak efficiency of the dual-output converter is 71.4% at 1.28 W/mm². These efficiency values increase to above 80% if 18-nH air-core inductors are used.**

## I. INTRODUCTION

To achieve high power density in battery-powered systems, as well as accommodate the reduced breakdown voltages in scaled CMOS technologies, systems-on-a-chip (SoCs) typically employ one or more supply voltages generated using dc-dc converters. To minimize the footprint of inductive dc-dc converters, which is typically dominated by the size of bulky off-chip inductors and capacitors, the operating switching frequency ($f_{sw}$) can be increased [1]-[3] by orders of magnitude above existing practice (< 10MHz). However, a higher $f_{sw}$ comes at the expense of increased switching losses ($P_{loss|sw}$) and the need for a fast, accurate controller.

Another challenge in SoC applications is that of achieving a high power density in a dc-dc converter module with multiple output voltages. In SoC applications, multiple voltages are typically derived from battery by connecting dc-dc converters in parallel. The cascading of dc-dc step-down (buck) converters has been explored [4], but is primarily seen as a two-step conversion process for implementing a low duty-ratio ($D = V_{out}/V_{in} < 20\%$), without a separate load current being drawn from the intermediary voltage. For example, in power management modules for laptops a front-end buck converter provides an intermediate voltage (a 5-V bus), and then several buck converters with $D > 20\%$ are connected in parallel without current otherwise being drawn from the front-end converter.

This work introduces a dc-dc converter employing a dual-output cascaded buck converter architecture with an $f_{sw}$ that is programmable up to 120 MHz to achieve high efficiency for output power from 150 mW to 1 W while using inductors with values in the nH range. Two fast digital constant-off-time control loops are used to suppress cross-regulation between the two outputs. Implemented in a 90-nm CMOS technology, an experimental prototype has an active area of 0.56 mm².

## II. DUAL-OUTPUT OVER-100MHZ POWER-PATH

This section describes a buck converter architecture based on a high-frequency power-path. It presents the challenges of achieving high efficiency while operating at high switching speeds and proposes an implementation for providing multiple output voltages, while specifically targeting the use of nH-level output inductors.

### A. Power losses in a high-frequency power-path

Fig. 1 shows the output stage of the proposed 2.5-V to 1.8-V buck converter, which has a total power loss, $P_{loss}$, given by

$$P_{loss} \approx \underbrace{i_{rms}^2 * r_{eff}}_{Conduction} + \underbrace{\alpha f_{sw}}_{Switching} \quad , \tag{1}$$

where,

$$i_{rms}^2 = \frac{\Delta i_L^2}{12} + I_{load}^2 \quad , \tag{2}$$

$$\Delta i_L = \frac{\left(\frac{V_{out}}{V_{in}}\right)(V_{in} - V_{out})}{L f_{sw}} \quad \text{and} \tag{3}$$

$$r_{eff} = ESR_L + D r_{high} + (1 - D) r_{low}; \, D{\sim}75\% \quad . \tag{4}$$

$\alpha$ is a proportionality constant, indicating that gate-drive switching, drain switching, diode conduction and shoot-through losses are, to first order, proportional to switching frequency. $\Delta i_L$ is the inductor ripple current. $V_{out}/V_{in}$ ( $= D$ ) is the conversion ratio. $r_{high}$ is the on-resistance of the high-side switch (comprising MP1, MP2 and MP3), and $r_{low}$ is that of the low-side switch (MN1).

Power losses from ripple conduction and transistor switching dominate the loss in high-frequency converters because of the small output inductor, $L_{out}$, and high $f_{sw}$.

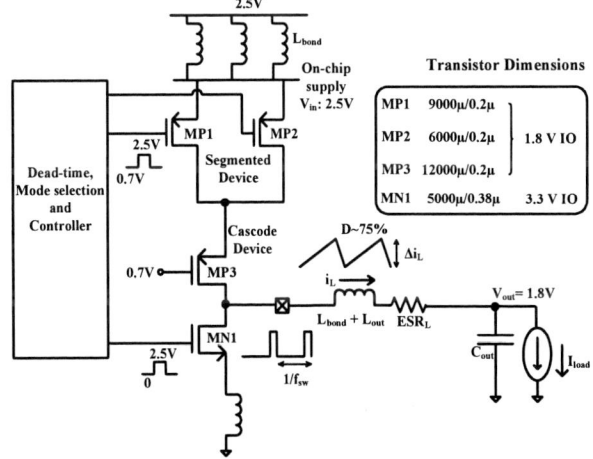

Fig. 1. 2.5-V to 1.8-V output stage with high-side switch segmentation.

978-1-4673-6145-3/13 $31.00 © 2013 IEEE

Constraining the value of the output inductor in Fig. 1 to a few nHs, so that small air-core inductors [1], bond-wires [2] or short PCB traces can be used, results in an increase in $\Delta i_L$. An increase in $f_{sw}$ (to over 100MHz) can be used to reduce $\Delta i_L$, but this increases $P_{loss|sw}$. Both $P_{loss|sw}$ and ripple conduction losses are significant for light loads (small $I_{load}$), where efficiency is critical for prolonged battery life in portable electronics.

To reduce the gate-drive switching losses, a cascode of 1.8-V thick-oxide transistors [5] is used as the high-side switch in Fig. 1 in order to withstand the 2.5-V input supply. Driving the gates of the high-side switch transistors MP1 and MP2 with levels of 0.7 V and 2.5 V maintains device reliability while reducing the gate-drive switching losses. It is apparent from (4) that MN1 does not add significant conduction loss because the duty ratio D is approximately 75%. As a result, its width can be small to reduce $C_{gg}$ of MN1.

### B. Programmable modes for high efficiency

Maximizing the efficiency across a wide load current range requires adjusting the power transistor widths as well as the switching frequency. Programmable modes are therefore implemented to enable adjustment of (1) the gate-capacitance ($C_{gg}$) and $r_{on}$ of the segmented high-side switch [6] and (2) $f_{sw}$ as functions of $I_{load}$. For example, to increase the light-load efficiency, a segment of the high-side switch (MP2 in Fig. 1) is disabled to reduce $C_{gg}$ and thus reduce $P_{loss|sw}$. Furthermore, somewhat counter intuitively, $f_{sw}$ is increased at light loads to reduce conduction losses due to $\Delta i_L$, which is the dominant loss mechanism under a light load.

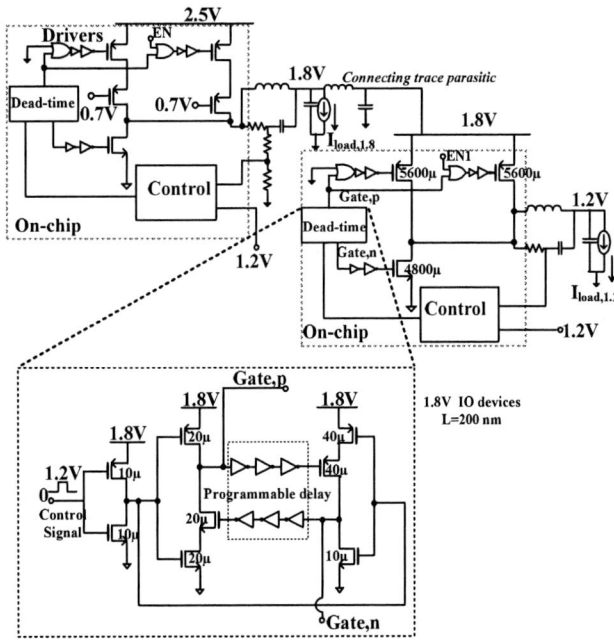

Fig. 2. Schematic of the proposed cascaded buck converter.

### C. Cascaded high-frequency dual-output buck converter

The techniques described above help to achieve high efficiency with a small inductor for the 1.8-V output, but not for generating a 1.2-V output directly from 2.5 V, primarily because $\Delta i_L$ increases significantly with reduced $V_{out}/V_{in}$ ($\Delta i_L$ is maximum at $V_{out}/V_{in} = 0.5$). To achieve high efficiency for low $V_{out}/V_{in}$ and a small output inductor, this work proposes using a

cascade of buck converters employing the dual-output, high-frequency architecture shown in Fig. 2.

Rather than using two converters operating in parallel to generate outputs of 1.8 V and 1.2 V from 2.5 V, the converters in Fig. 2 are connected in a 2.5-V–1.8-V–1.2-V cascade. In this work, unlike [4], the first stage not only acts as the supply for the next stage but is also used to supply power to a separate load at 1.8 V. Small nH-level inductors are used in both converter stages, thus minimizing the overall solution size. For the 1.2-V output, operating with a 1.8-V input increases $V_{out}/V_{in}$, thereby reducing conduction losses due to $\Delta i_L$. Furthermore, the 1.8-V input removes the need for a cascoded high-side switch in second converter and also lowers $P_{loss|sw}$.

If a very large load current is drawn from the 1.2-V output, or a large $L_{out}$ (of the order of µHs) is used, load conduction losses dominate and series loss in the first (2.5-V to 1.8-V) stage prohibits the use of cascaded stages. However, in terms of reducing ripple and switching losses, the proposed cascaded converter design provides an advantage that outweighs the increased series loss due to intermediate stage. For nH-level output inductors and output power up to 1 W, it is observed from simulations that

$$\eta_{2.5-1.8}\eta_{1.8-1.2} > \eta_{2.5-1.2} \ , \qquad (5)$$

where $\eta_{(Vin-Vout)}$ is the efficiency of the converter with input voltage $V_{in}$ and output voltage, $V_{out}$. The dual-output cascaded high-frequency architecture thus addresses the challenges specifically presented by small L (reduces $\Delta i_L$) and high $f_{sw}$ (reduces $P_{loss|sw}$).

The use of a programmable dead-time circuit in the architecture of Fig. 2 avoids shoot-through switching loss. However, since output power is drawn from both stages in this architecture, cross-regulation becomes an additional concern that requires a fast controller. The design of a suitable controller is considered in section III.

### D. Layout

To support a large $I_{load}$, wide power MOSFETs are used (Fig. 1 and Fig. 2) and the IO pads are placed in the middle of the MOSFET fingers to minimize the effective metal resistance and conduction losses. Multiple bond wires are placed evenly on the output pads that overlay the power MOSFETS to reduce the metal resistance and the effective resistance of bond wires due to reduced skin depth at high frequencies. Supply and ground bond-wire inductance induced ripple is reduced with an on-chip bypass MOS capacitor that occupies a die area of 0.06 mm$^2$.

### III. High-Frequency Digital Constant-Off-Time Control

A high $f_{sw}$ presents challenges in the controller design due to the small switching period (1/ $f_{sw}$), which is comparable to logic, chip interconnect and PCB parasitic delays. Moreover, a high degree of controller flexibility is needed for tuning $f_{sw}$ with $I_{load}$, as described in section II. In this design, a fully digital controller is used. Two state-machine-based, fast digital constant-off-time (DCOT) controllers (Fig. 3) regulate the output levels. A COT controller [7] is a type of hysteretic controller where the off time is fixed. Unlike a PWM controller, a COT controller does not require a compensation network, making it simpler, with low area and a fast transient response. The digital COT adds flexibility to the COT controller, and compared to a digital

978-1-4673-6145-3/13 $31.00 © 2013 IEEE

controller employing a high-speed ADC and a DPWM block, it requires less computational speed, and thus lower dynamic power.

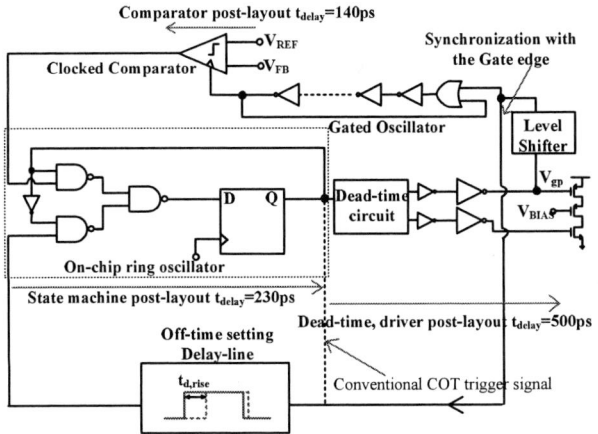

Fig. 3. Block diagram of the digital constant-off-time controller.

The state machine in Fig. 3 operates at ~1 GHz, while the comparator is clocked at ~500 MHz; the clocks are provided by free running ring-oscillators. $t_{off}$ is set by the rising edge delay of a programmable delay line so that $f_{sw}$ can be adjusted depending on $I_{load}$. The delay line is designed so that the falling edge has minimum delay and does not impact the state-machine timing.

Shown in Fig. 4 are the external load and feedback components that need to be considered in the design of a high-frequency buck converter. When a small output inductor (L in Fig. 4) is used, high-order harmonics appear at the output of such a converter because the parasitic inductance $L_{par,C}$, is comparable to the output inductor. Therefore, an RC network is used to emulate the output inductor current and generate a synthetic saw-tooth ripple voltage at node FB [8]. The feedback PCB trace is laid out to be short enough to ensure that the parasitic trace pole does not make the loop unstable.

Continuous-time comparators used in conventional hysteretic and constant-off-time controllers impact the switching frequency at high switching frequencies. Therefore, this DCOT controller uses the fast dynamic comparator shown in Fig. 5, which compares $V_{fb}$ to the reference voltage $V_{ref}$ (= 1.2 V). The comparator comprises a resistively loaded high-speed pre-amplifier (gain ~ 3) followed by a regenerative latch. In both of the cascaded converter stages used in this dual-output converter design, with a high $f_{sw}$ and $V_{out}/V_{in}$ ~ 0.7, the on-time ($t_{on}$) of the high-side switch in the controllers is longer than the off-time ($t_{off}$), which can be as low as 1 ns. Therefore, to ensure that the comparator propagation delay does not alter $f_{sw}$, $t_{on}$ is set by the comparator, while $t_{off}$ is constant.

To ensure loop stability, the comparator is synchronized to the falling edge of $V_{gp}$ (in Fig. 3), and the timing constraints are

$$T_{comp} < \frac{t_{on}}{2} \quad , \tag{7}$$

$$T_{sm} << min(T_{comp}, t_{off}) \quad . \tag{8}$$

$T_{comp}$ and $T_{sm}$ are the comparator and state-machine clock periods, respectively. The feed-forward path in the DCOT controller ensures that the converter has good line regulation, but there is a finite droop in the output voltage due to load current, given by

$$\overline{v_{OUT}} = \overline{v_{sw}} - I_{out} * R_L \approx kV_{ref} - I_{out} * R_L \quad , \tag{9}$$

where $R_L$ is the series resistance of the output inductor. $k = 1$ for the 1.2-V output and $k = 1.5$ for the 1.8-V output.

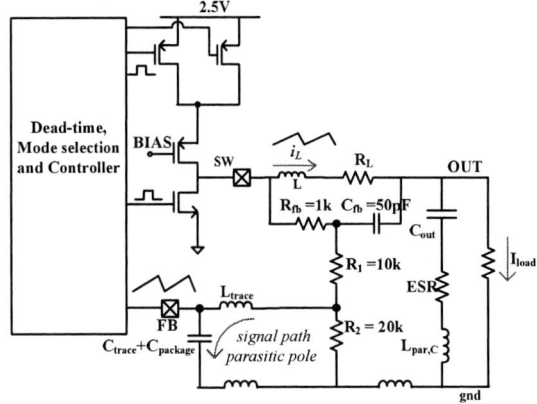

Fig. 4. External load and feedback components (2.5 V-1.8 V).

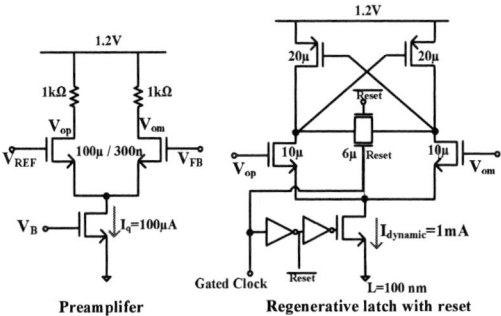

Fig. 5. Dynamic comparator with core devices.

IV. EXPERIMENTAL RESULTS

Fig. 6 shows a die photo of the experimental prototype, which has an active area of 0.56 mm². The two digital controllers draw a total of 7 mA of current from the 1.2-V supply. The measured 50% $I_{load}$ step-response and the cross-regulation of both outputs with external 18-nH inductors and 150-nF capacitors are shown in Fig. 7. Plots of the measured efficiency for the 2.5-V to 1.8-V converter output using just a parasitic PCB trace inductance and various surface-mount air-core inductors are shown in Fig. 8. The converter achieves a peak efficiency of 82.9% with an 18-nH air-core inductor and 76% with a parasitic PCB trace (~4 nH) at a power density of 2.59 W/mm². This is the highest reported power density for an integrated buck converter without a discrete external inductor.

Fig. 9 shows the efficiency measured when connecting the two converters in both cascade and parallel. The peak efficiency of the cascaded converter, which occurs at a power density of 1.28 W/mm², is 80% with an 18-nH inductor and 71.4% using only the PCB trace inductance; these values are approximately 5% higher than achieved with parallel-connected converters.

The key parameters for comparing high-frequency inductive dc-dc converters are power efficiency, power density, $V_{in}$ and $V_{out}/V_{in}$. As described in Section II, a low $V_{out}/V_{in}$ lowers the efficiency of a converter when employing a small output inductor. A higher input voltage increases the switching losses, thereby lowering the efficiency. Table I compares the prototype dual-output cascaded converter presented in this work with recently published high-frequency inductive buck converters. For a 1.2-V output, the prototype achieves a higher efficiency and power density than a previously reported converter with a similar conversion ratio close to 50% [2]. For conversion ratios close to 70%, as reported in [1] and [3], a comparable efficiency is achieved at a higher power density and for a higher $V_{in}$.

Table I

COMPARISON OF RECENTLY PUBLISHED INDUCTIVE MINIATURIZED HIGH-FREQUENCY BUCK CONVERTERS

|  | [1] | [2] | [3] | Present Work | |
|---|---|---|---|---|---|
| Process | 90-nm CMOS | 130-nm CMOS | 130-nm CMOS | 90-nm CMOS | |
| $V_{in}$ | 1.2V | 2.6V | 1.2V | 2.5V | |
| $V_{out}$ ($V_{out}/V_{in}$) | 0.9V (75%) | 1.2V (46%) | 0.9V (75%) | 1.8V (72%) and 1.2V (48%) | |
| Inductor (nH) | 27.2 air-core | 15.6 On-chip | ~8 Bondwire | ~4 parasitic PCB trace (18nH air-core) | |
| Capacitor (nF) | 2.5 | 12.7 | 3.7 | 150 | |
| Peak efficiency | 83.2% | 58% | 82.4% | 2.5V-1.8V 76% (83%) | 2.5V-1.8V-1.2V 71.4% (80%) |
| Power density at peak efficiency (W/mm²) | 1.6 | 0.07 | 0.25 | 2.59 (1.3) | 1.28 |
| Efficiency at peak power density | 82.5% @1.98 W/mm² | 48% @0.213 W/mm² | 80% @ 0.96 W/mm² | 76% (80%) @2.59 W/mm² | 67% (76%) @1.67 W/mm² |
| Die area | 0.14 | 1.67 | 1.25 | 0.375 | 0.56 |
| Switching frequency | 233 MHz | 30Hz-300MHz | 100 MHz | 40MHz - 120MHz | |

## V. CONCLUSION

A dual-output, cascaded buck converter using only nH-level inductors in both stages has been designed and integrated in a 90-nm CMOS technology. The experimental prototype operates at a power density as high as 2 W/mm² with power efficiency on the order of 80%. A fully digital controller with a constant off-time is used to achieve a fast control, thereby suppressing cross-regulation between the two output voltages. The proposed converter is amenable to CMOS scaling, making it an attractive option for SoC applications requiring multiple supply levels.

## VI. ACKNOWLEDGEMENTS

This work was supported by the Stanford RAD initiative. The fabrication of the experimental circuit was made possible by the TSMC University Shuttle Program.

## REFERENCES

[1] P. Hazucha, et al., "A 233-MHz 80%-87% efficient four-phase dc-dc converter utilizing air-core inductors on package," *IEEE J. of Solid-State Circuits*, vol.40, pp. 838- 845, April 2005

[2] M. Wens, et al., "An 800mW fully-integrated 130nm CMOS DC-DC step-down multi-phase converter, with on-chip spiral inductors and capacitors," *Energy Conversion Congress and Exposition*, pp. 20-24 Sept. 2009.

[3] C. Huang, et al., "An 82.4% efficiency package-bondwire-based four-phase fully integrated buck converter with flying capacitor for area reduction," *IEEE ISSCC Dig. Tech. Papers*, pp. 17-21, Feb. 2013.

[4] Brian R. Pelly, "Cascaded buck converter circuit with reduced power loss," U.S. Patent 5006782, Apr 9, 1991.

[5] S. Bandyopadhyay, et al., "20µA to 100mA DC-DC converter with 2.8 to 4.2V battery supply for portable applications in 45nm CMOS," *IEEE ISSCC Dig. Tech. Papers*, pp. 20-24, Feb. 2011.

[6] W. T. Ng, et al., "Output Stages for integrated dc-dc converters and power ICs," *IEEE Conf. on Electron Devices and Solid-State Circuits*,

pp.91-94, 20-22 Dec. 2007.

[7] L. Cheng, et al., "A constant off-time controlled boost converter with adaptive current sensing technique," *Proceedings of the ESSCIRC*, pp.443,446, Sept. 2011.

[8] K.D.T Ngo, et al., "Synthetic-ripple modulator for synchronous buck converter," *Power Electronics Letters, IEEE*, vol.3, no.4, pp.148- 151, Dec. 2005.

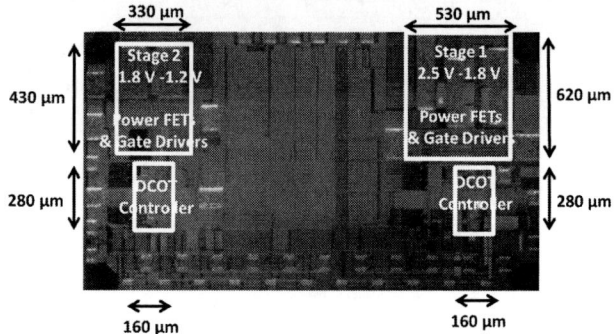

Fig. 6. Die photo

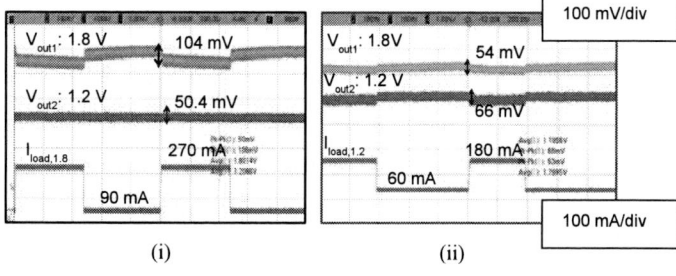

Fig. 7. Measured load current step-response of both stages (i) Impact of step for $I_{load,1.8}$, and (ii) Impact of step for $I_{load,1.2}$.

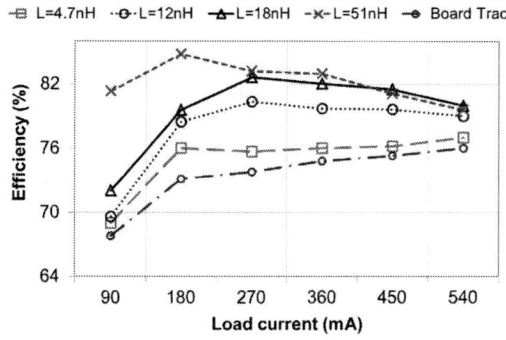

Fig. 8. Measured efficiency of the 2.5-V to 1.8-V converter.

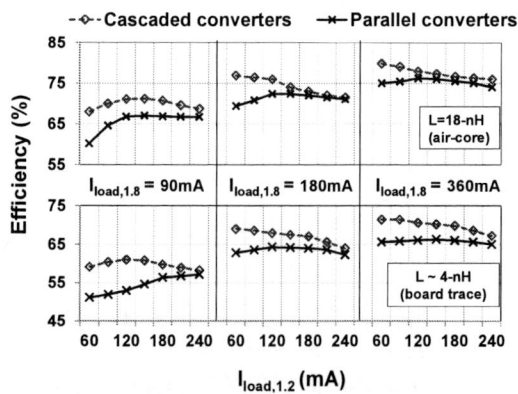

Fig. 9. Measured efficiency of the dual-output 1.8V/1.2V converter.

# A Monolithic Digitally Controlled Ripple-Based DC-DC Converter with Digital Inductor Current Sensor

Man Pun Chan and Philip K. T. Mok

Department of Electronic and Computer Engineering
The Hong Kong University of Science and Technology
Clear Water Bay, Kowloon, Hong Kong, China

*Abstract*—A ripple-based control scheme for DC-DC converters is presented which can fast load-transient responses by sensing inductor current and using it as feedback ripples. Conventionally, analog RC inductor current sensors can be used for that purpose. However, the RC passive components are too bulky to integrate on-chip, and digital controllers cannot use the analog ripples for the control purpose unless extra ADCs are available to quantize them. Therefore, this paper proposes a digital inductor current sensor that does not require the extra ADCs and occupies a small chip area. A ripple-based digital controller has been implemented to demonstrate how the digital sensor can be utilized. Both the digital sensor and controller are fully synthesizable with a UMC 0.13-μm digital CMOS process. Their total chip areas are 220μm×220μm. Measurements results show that a 2MHz buck converter achieves load-transient responses of 10μs by using the digital controller. The peak efficiency is 91% at 100mA of the loading current.

## I. INTRODUCTION

Nowadays, the ever lighter and more powerful portable devices, like smartphones, tablets and ultra-light notebooks, are often used as multimedia, entertainment and communication devices. With all these capabilities, there is a high demand for more computing power from these devices. To cope with the demand, their processors are evolving from single-core to multi-core [1–2] which requires low voltage (0.8V to 1.3V) operation and high loading current. These impose stringent requirements on monolithic DC-DC converters, especially their controllers' physical sizes and load-transient response. Among the different control schemes developed for the controllers, a ripple-based control scheme has the advantages of simple circuit implementations and fast load-transient responses. Some well-known control schemes could be considered as different variations of the ripple-based control scheme, for example, hysteretic control, constant on/off-time control and $V^2$ control [3].

Several ripple-based analog controllers [4–5] have demonstrated that fast load-transient responses can be achieved by sensing the inductor current and using it as "feedback ripples". To sense the inductor current, the controllers use analog RC inductor current sensors that are similar to the one shown in Fig. 1(a). It is basically an RC filter that is used to filter the inductor voltage $V_L$. The analog inductor current ripples $V_s$ can be obtained if the time constant $R_sC_s$ of the filter meets the condition as follows:

$$f_{sw} \gg \frac{1}{R_sC_s} \qquad (1)$$

where $f_{sw}$ is the switching frequency of the DC-DC converter.

From (1), the typical value of $C_s$ ranges from 10pF to 20pF, given that $R_s = 200k\Omega$ and $f_{sw} = 1MHz$ to 4MHz. Although the analog sensor can still be integrated on chip, it consumes large chip area. Alternatively, digital inductor current sensors, which are similar to the one shown in Fig. 1(b) [6], can be used. An ADC is used to quantize the inductor voltage as $V_L[n]$, and then a digital filter is applied to generate digital inductor current ripples as $V_s[n]$. Consequently, the use of passive components can be avoided and the digital sensor can be easily incorporated into digital controllers that offer many advantages over analog controllers. For instance, the digital controllers have the advantages of re-programmability, noise immunity and easy on-chip integration with other digital SoCs [7–10].

However, there are drawbacks of using the digital sensor. Firstly, the extra ADC that is needed to quantize the inductor voltage occupies more chip areas, consumes extra power and introduces greater quantization noise. Secondly, the inductor value L is needed for generating the digital ripples, as reported in [6]. This limits the flexibility of using the digital sensor with different specifications of DC-DC converters. To address these issues, this paper proposes a digital inductor current sensor in which the extra ADC and the inductor value are not required. A ripple-based digital controller has also been implemented to demonstrate how the digital sensor can be used for control purposes.

This paper is organized as follows. Section II first analyzes how the analog sensor operates. Then, the proposed digital sensor can be derived from the analysis. Section III gives the design and implementation of the proposed digital sensor and the ripple-based digital controller. Measurement results and conclusions are given in Section IV and Section V, respectively.

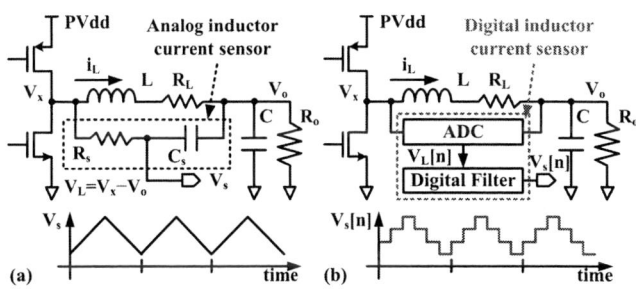

Fig. 1. A DC-DC converter with (a) an analog RC inductor current sensor and (b) a digital inductor current sensor

## II. ANALYSIS OF THE ANALOG AND DIGITAL INDUCTOR CURRENT SENSOR

The analog inductor current sensor is isolated from the DC-DC converter for analysis, as shown in Fig. 2(a). The switching node voltage $V_x(t)$ and the output voltage $V_o(t)$ are assumed to be two uncorrelated signal sources, exciting the RC network in two different terminals.

We have,

$$V_x(t) - V_s(t) = i_s(t)R_s \tag{2}$$

$$i_s(t) = C_s \frac{d(V_s(t) - V_o(t))}{dt} \tag{3}$$

Combining (2) and (3),

$$V_x(t) - V_s(t) = C_s \frac{d(V_s(t) - V_o(t))}{dt} R_s \tag{4}$$

Applying Laplace transform to (4),

$$V_s(s) = \frac{1}{(sR_sC_s + 1)}V_x(s) + \frac{sC_sR_s}{(sR_sC_s + 1)}V_o(s) \tag{5}$$

It can be seen from (5) that the analog inductor current ripple $V_s(s)$ consists of two parts. One is a low-pass filter (LPF) applying to $V_x(s)$. The other is a high-pass filter (HPF) applying to $V_o(s)$. In order to avoid using the passive components $C_s$ and $R_s$, the HPF and the LPF in (5) can be implemented by two corresponding digital filters. With this implementation, a digital inductor current sensor can be constructed and it is depicted in Fig. 2(b). Firstly, two ADCs are used to quantize the signals $V_x(t)$ and $V_o(t)$. Then, the digital LPF $L_1(z)$ and HPF $L_2(z)$ are used to filter the quantized signals $V_x[n]$ and $V_o[n]$. Finally, the two digital signals $i_{L1}[n]$ and $i_{L2}[n]$ are added together to generate the digital inductor current ripple as $i_{Lsen}[n]$. The digital algorithm is summarized in the following equation:

$$i_{Lsen}(z) = L_1(z)V_x(z) + L_2(z)V_o(z) \tag{6}$$

where $L_1(z) = A_{vx}\frac{1}{(1-\alpha_1 z^{-1})}$, $L_2(z) = A_{vo}\frac{1-z^{-1}}{(1-\alpha_2 z^{-1})}$

The parameters $A_{vx}/A_{vo}$ and $\alpha_1/\alpha_2$ are dc gains and pole locations of the filters, respectively. Their designs will be covered in the subsequent section. It seems that the digital sensor requires two ADCs to operate. However, $V_o[n]$ is always available in most digital controllers. Therefore, if $V_x[n]$ can also be obtained without using the ADC, then the digital sensor occupies a small chip area and can easily be incorporated into digital controllers.

## III. DESIGN AND IMPLEMENTATION OF THE PROPOSED DIGITAL SENSOR AND CONTROLLER

### A. Implementation of the Proposed Digital Sensor

Fig. 3 shows a ripple-based digital controller for a buck converter with 2MHz switching frequency. The digital controller has incorporated a digital inductor current sensor, which is similar to the one shown in Fig. 2(b). However, the difference is that only one ADC is used to quantize the output voltage $V_o$ and no ADC is required for quantizing the switching node voltage $V_x$. Instead, the internal signal duty[n] is used by the digital sensor. This is possible because $V_x$ and duty[n] are equivalent signals, i.e., $V_x$ = PVdd (Ground) corresponds to duty[n] = logic 1(logic 0).Therefore, applying the filter $L_1(z)$ to duty[n] is equivalent to apply to the quantized switching node voltage $V_x[n]$. As a result, the ADC for quantizing $V_x(t)$ can be eliminated.

The following shows how $L_1(z)$ and $L_2(z)$ are implemented by digital circuits. Rewriting $L_1(z)$ in (6) in time domain,

$$L_1(z) = \frac{i_{L1}[z]}{duty[z]} = A_{vx}\frac{1}{1-\alpha_1 z^{-1}} \tag{7}$$

$$i_{L1}[n] = A_{vx} \times duty[n] + \alpha_1 i_{L1}[n-1] \tag{8}$$

Similarly, rewriting $L_2(z)$ in (6) in time domain,

$$L_2(z) = \frac{i_{L2}[z]}{v_o[z]} = A_{vo}\frac{1-z^{-1}}{1-\alpha_2 z^{-1}} \tag{9}$$

$$i_{L2}[n] = A_{vo}(v_o[n] - v_o[n-1]) - \alpha_2 i_{L2}[n-1] \tag{10}$$

According to (8) and (10), the digital implementations of the filters $L_1(z)$ and $L_2(z)$ are illustrated in Fig. 4. It can be seen

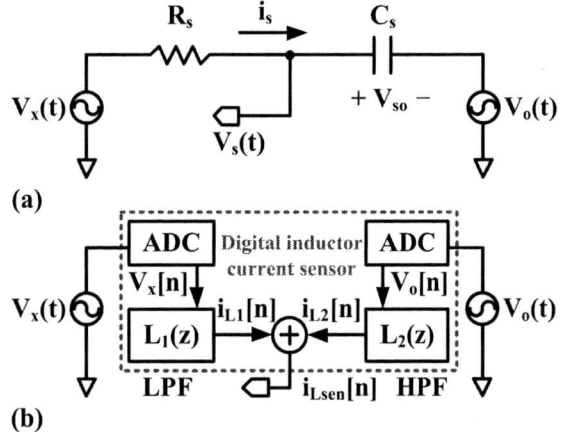

**(a)**

**(b)**

Fig. 2. Signal models of (a) an analog inductor current sensor and (b) digital inductor current sensor

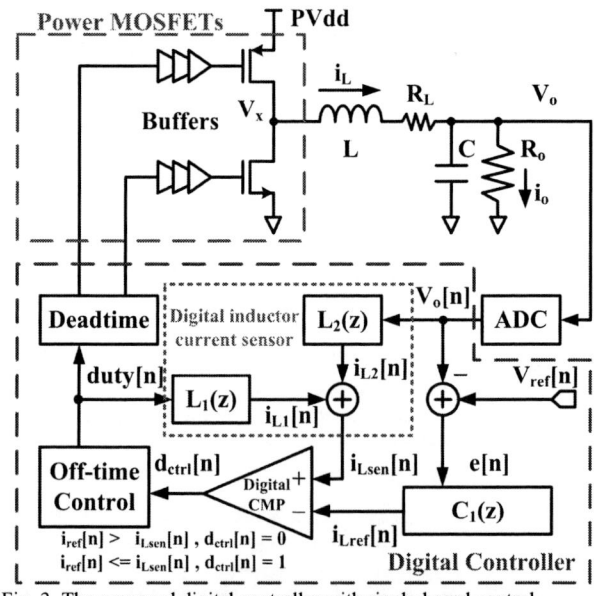

Fig. 3. The proposed digital controller with ripple-based control

that the filters only consist of adders, shifters and delay units, which occupy a small chip area. The shifters are used for multiplication. Since $L_1(z)$, as shown in Fig. 4(a), is a LPF, it integrates duty[n] to generate the digital inductor current $i_{L1}[n]$. In this case, the integration of duty[n] is linear if the pole location of $L_1(z)$ is much lower than the switching frequency $f_{sw}$ of the buck converter. This condition is similar to the one stated in (1). For example, if $f_{sw}$ = 2MHz, $\alpha_1$ = 0.9375 will be chosen. The corresponding pole location of $L_1(z)$ is 89.7kHz.

For the HPF $L_2(z)$ shown in Fig. 4(b), its main function is to speed up the load-transient response of the digital controller. During the load-transient, the sudden change in the output voltage $V_o[n]$ will "couple" through $L_2(z)$ and the digital controller can respond to the load-transient rapidly without being slowed down by the voltage-loop compensation $C_1(z)$, as shown in Fig. 3. The pole location of $L_2(z)$ is 6.24kHz, which corresponds to $\alpha_2$=0.9375.

### B. Operation Principle of the Proposed Digital Controller

The steady-state operation of the ripple-based digital controller is depicted in Fig. 5. The digital ripple $i_{Lsen}[n]$, generated from the digital sensor, is compared with the inductor reference current $i_{Lref}[n]$. The off-time $T_{off}$ of the power MOSFETs is fixed as $T_{off}$ = $N_1 T_c$, where $N_1$ is the number of clock cycles and $T_c$ is the digital clock period. As a result, the output voltage $V_o[n]$ is regulated by varying the on-time $T_{on}$, which is defined to be from the end of the off-time to the moment where $i_{Lsen}[n]$ "hits" $i_{Lref}[n]$.

Since $i_{Lref}[n]$ is generated from the error voltage $e[n]$ between $V_{ref}[n]$ and $V_o[n]$, as shown in Fig. 3, the on-time will change accordingly to enforce $e[n]$ being equal to zero. For example, if $V_o[n]$ is less than $V_{ref}[n]$, $e[n]$ will increase and so will $i_{Lref}[n]$. Therefore, $i_{Lsen}[n]$ will take a longer period of time

to "hit" $i_{Lref}[n]$ and $T_{on}$ will become larger. In this case, the inductor current has a longer time to charge up the output capacitor and hence increase the output voltage. This completes the negative feedback loop and regulates the output voltage back to $V_{ref}[n]$.

## IV. MEASUREMENT RESULTS

Table I shows the parameters of the buck converter and the proposed digital controller, which are designed with a 0.13μm digital CMOS process. Fig. 6 shows a chip photo of the buck converter, consisting of the digital controller, the power MOSFETs and a Successive Approximation (SA) ADC. Since both the digital controller and digital sensor are implemented by Verilog code and synthesized together, they only occupy a small chip area of 220μm×220μm. This area can be further scaled down with a more advanced digital CMOS process.

Fig. 7 shows mixed-signal waveforms during steady-state operation of the digital sensor. All the voltage/current labels in the figure correspond to the labels shown in Fig. 3. The buck converter operates at 2MHz with $V_o$ = 1.13V and loading current $i_o$ = 100mA. The switching node voltage $V_x$ indicates that the input voltage PVdd of the buck converter is equal to 2V. The digital ripple $i_{Lsen}[n]$ is represented by 12-bit hexadecimal numbers. It can be seen that $i_{Lsen}[n]$ increases/decreases as the actual inductor current $i_L$ ramps up/down. If the value of $i_{Lsen}[n]$ is plotted against time, the

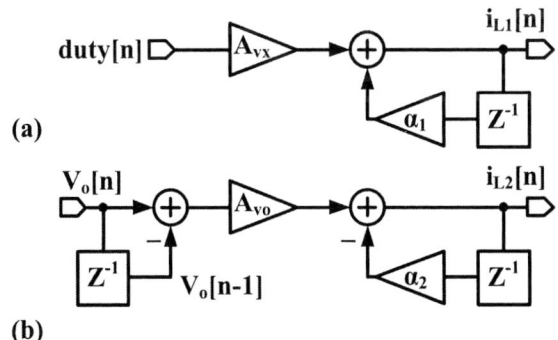

**(a)**

**(b)**

Fig. 4. Digital implementation of (a) $L_1(z)$ and (b) $L_2(z)$

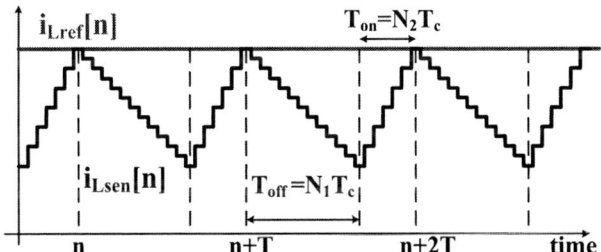

Fig. 5. Steady-state operation of the proposed digital controller with ripple-based control

Table I : Parameters of the buck converter and the digital controllers

| Technology | UMC 0.13μm CMOS |
|---|---|
| **Power Stage Parameters** | |
| Input Voltage (PVdd) | 2.0V |
| Output Voltage ($V_o$) | 0.8–1.3V |
| Output Capacitor (C) | 4.7μF |
| Output Inductor (L) | 4.7μH |
| Switching Frequency ($f_{sw}$) | ~2MHz |
| **Digital Controller Parameters** | |
| Digital Voltage (DVdd) | 1.2V |
| SA ADC resolutions | 8-bit |
| $i_{Lsen}[n]$ resolutions | 12-bit |
| Digital Clock Frequency ($F_c$=1/$T_c$) | ~50MHz |

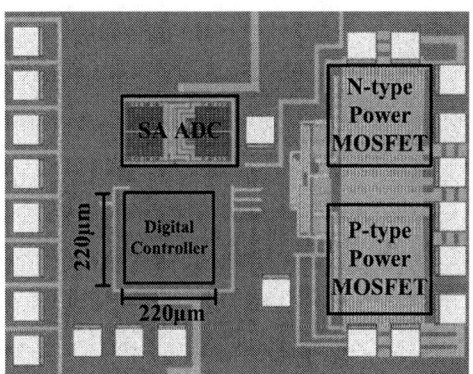

Fig. 6. Chip photo of the buck converter and the proposed digital controller

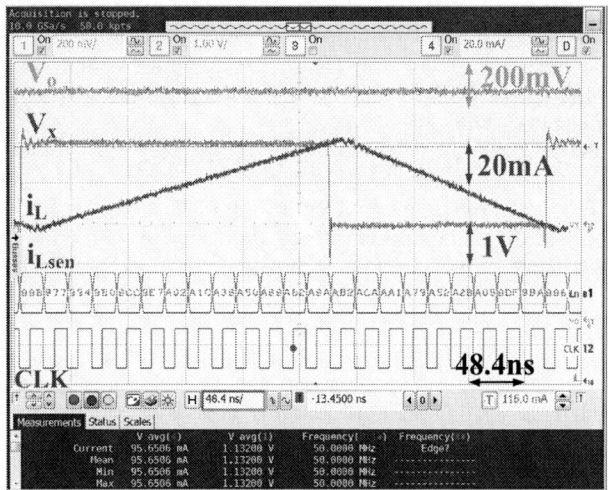

Fig. 7. Steady-state measurement of the proposed digital inductor current senor

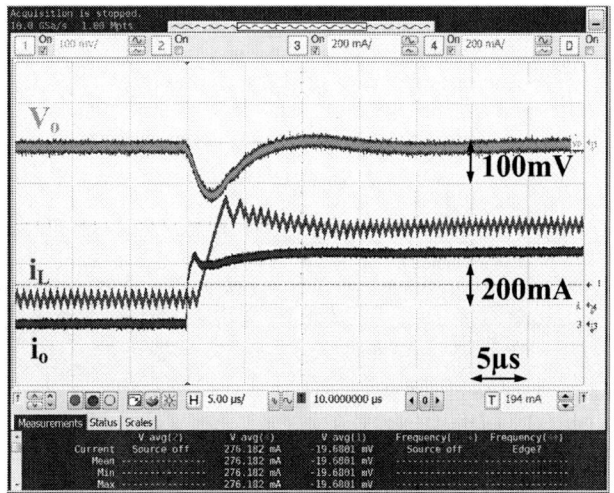

Fig. 8. Load-transient response of the proposed digital controller

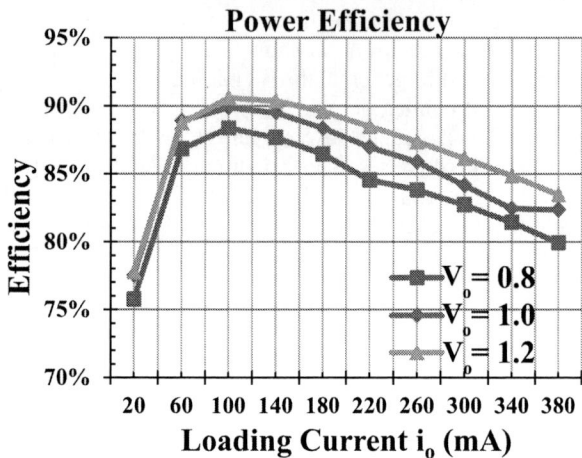

Fig. 9. Power efficiency of the buck converter with different output voltages

waveform will be the same as that in Fig. 5. The value of $i_{Lsen}[n]$ updates at each falling edge of the digital clock (CLK), which is equal to 50MHz.

Fig. 8 shows the load-transient response of the buck converter. The loading current $i_o$ steps from 30mA to 400mA in less than 500ns. The response time of the output voltage $V_o$ is about 10μs with 2MHz switching frequency of the converter. The undershoot voltage is about 130mV during the transient. Fig. 9 shows the power efficiency of the buck converter with different output voltages, given that PVdd = 2.0V. The peak efficiency is about 91% at 100mA of the loading current.

## V. CONCLUSIONS

This paper presents a digital inductor current sensor that is able to generate digital ripples without knowledge of the inductor value or an extra ADC. This allows the digital sensor to be fully incorporated with the digital controller, which utilizes a ripple-based control scheme for achieving fast load-transient responses. It should be pointed out that the digital sensor and controller are fully synthesizable in Verilog code.

Therefore, they occupy a small chip area and can be easily integrated with other digital SoCs. Also, the synthesizable codes are portable to any digital CMOS processes/foundries.

## ACKNOWLEDGEMENT

This work was partially supported by the Research Grant Council of Hong Kong SAR Government, China, under project No. 617308.

## REFERENCES

[1] Samsung, "Samsung Exynos 4 Quad User's Manual." [Online]. Available: http://www.samsung.com/global/business/semiconductor/file/product/Exynos_4_Quad_User_Manaul_Public_REV100-0.pdf.

[2] Texas Instruments, "Multicore DSP+ARM KeyStone II System-on-Chip Data Manual." [Online]. Available: http://www.ti.com/product/66ak2e05#doctype6.

[3] R. Redl and J. Sun, "Ripple-Based Control of Switching Regulators : An Overview," *IEEE Trans. Power Electron.*, vol. 24, no. 12, pp. 2669–2680, Dec. 2009.

[4] J. Fan, X. Li, J. Park, and A. Huang, "A Monolithic Buck Converter Using Differentially Enhanced Duty Ripple Control," in *Proc. IEEE Custom Integrated Circuits Conf.*, Sep. 2009, pp. 527–530.

[5] F. Su and W.-H. Ki, "Digitally Assisted Quasi-$V^2$ Hysteretic Buck Converter with Fixed Frequency and without Using Large-ESR Capacitor," in *ISSCC Dig. Tech. Papers*, Feb. 2009, pp. 446–447,447a.

[6] K.-Y. Cheng, F. Yu, F. C. Lee, and P. Mattavelli, "Digital Enhanced $V^2$-Type Constant On-Time Control Using Inductor Current Ramp Estimation for a Buck Converter With Low-ESR Capacitors," *IEEE Trans. Power Electron.*, vol. 28, no. 3, pp. 1241–1252, Mar. 2013.

[7] J. Song, G. Yoon, and C. Kim, "An Efficient Adaptive Digital DC-DC Converter with Dual Loop Controls for Fast Dynamic Voltage Scaling," in *Proc. IEEE Custom Integrated Circuits Conf.*, Sep. 2006, pp. 253–256.

[8] J. Shi, Y.-C. Hsu, E. Soenen, A. Roth, and J. Gaither, "A Wide-Range DC/DC Converter with 2nd Order Digital Compensation and Direct Battery Connection in 40nm CMOS," in *Proc. IEEE Custom Integrated Circuits Conf.*, Sep. 2011, pp. 1–4.

[9] M. P. Chan and P. K. T. Mok, "Design and Implementation of Fully Integrated Digitally Controlled Current-Mode Buck Converter," *IEEE Trans. Circuits Syst. I, Reg. Papers*, vol. 58, no. 8, pp. 1980–1991, Aug. 2011.

[10] M. P. Chan and P. K. T. Mok, "On-Chip Digital Inductor Current Sensor for Monolithic Digitally Controlled DC-DC Converter," *IEEE Trans. Circuits Syst. I, Reg. Papers*, to be published.

# A Fully Integrated Battery-Connected Switched-Capacitor 4:1 Voltage Regulator with 70% Peak Efficiency Using Bottom-Plate Charge Recycling

Tao Tong, Xuan Zhang, Wonyoung Kim, David Brooks, Gu-Yeon Wei

School of Engineering and Applied Sciences, Harvard University, Cambridge, MA, U.S.A
E-mail: taotong@seas.harvard.edu

*Abstract*—**This work presents a switched-capacitor (SC) DC-DC voltage regulator that converts a 3.7V battery voltage down to ~0.8V in order to power the 'brain' SoC of a flapping-wing microrobotic bee. A cascade of two 2:1 SC converters offers high efficiency for a 4:1 conversion ratio. A charge recycling technique reduces the flying capacitor's bottom-plate parasitic loss by 50% and overall conversion efficiency reaches 70%. The output droop is less than 10% of the nominal output voltage for a worst-case 47mA load step.**

## I. INTRODUCTION

In the aerial microrobotic bee application [1], the on-board battery (~3.7V) is the only source of energy. A digital SoC, which works as the 'brain' of the robotic bee, operates at low voltages (~0.8V or less). While a voltage regulator is required to bridge the voltage difference, the stringent weight and area requirements of the robotic bee make the regulator design challenging. First, the regulator needs to be fully integrated along with the SoC without using any external components in order to minimize weight and area. Second, the regulator must directly connect to the battery and support a high (4:1) step down ratio. Third, high conversion efficiency is important to achieve long flight times for the robotic bee.

SC converters are well suited for this application from weight and area perspectives since they only require capacitors and MOS transistors. On-chip MOS capacitors with density as high as $10nF/mm^2$ are available in digital CMOS processes [2]. However, choosing the right topology is important. Single-stage SC converters suffer from power switch voltage breakdown and high bottom-plate parasitic loss when the conversion ratio and input voltage are high [2][4][5]. One solution has been to cascade thick-oxide transistors to avoid transistor break down in 3:1 SC converters [2][5], but this degrades conversion efficiency. Novel switching techniques have also been shown to mitigate flying capacitor bottom-plate parasitic loss [4][8]. Unfortunately, these issues get worse in single-stage 4:1 SC converters.

This paper presents a fully integrated two-stage SC regulator to address these challenges. The proposed two-stage topology simplifies the overall design and implements several techniques to improve conversion efficiency: (1) it uses the appropriate flavor of transistors (thin oxide and think-oxide transistors) in each stage; (2) it applies a charge recycling technique to mitigate bottom-plate parasitic loss; and (3) it employs separate low-boundary feedback controls to regulate the each stage's output to desired levels. Lastly, the two-stage topology provides an intermediate voltage for use by other parts of the microrobotic bee.

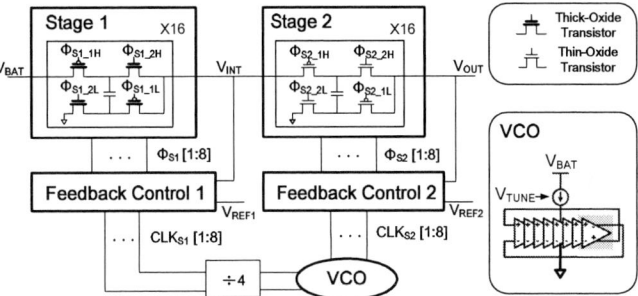

Fig. 1: Block diagram of proposed two-stage SC converter.

## II. PROPOSED TWO-STAGE CONVERTER

### A. Two-Stage Structure

Fig. 1 illustrates the system block diagram of the proposed SC converter. The design cascades two 2:1 SC stages, which are implemented and optimized for different purposes. The first stage converts the 3.7V battery voltage down to a 1.8V intermediate voltage ($V_{INT}$). To handle the 1.8V swing, this stage uses thick-oxide transistors available in the process. The second stage converts the intermediate 1.8V down to ~0.8V for the final output ($V_{OUT}$) using thin-oxide transistors. Each stage also includes identical, but separate feedback control loops, discussed later.

The two SC stages are nearly identical except for the type of transistors and sizing. Each SC stage implements a multi-phase topology to reduce voltage ripple. Sixteen modules operate off both edges of eight interleaved clock phases. A multi-phase current-starved pseudo-differential VCO generates the clock edges and operates directly off of the battery to guarantee proper start-up. To ensure there is always a balanced number of modules in operation, pairs of modules operate 180° out-of-phase off of one shared clock phase. SC converters have two basic phases of operation, thoroughly discussed in [2]. In one phase, energy drawn from the input charges the flying capacitor up and flows to the load. In the other phase, energy stored on the capacitor during the previous phase flows to the load. The power switches operate with stacked voltage domains similar to [3] and [6]. Taking the first-stage as an example, switches driven by $\Phi_{S1\_1H}$ and $\Phi_{S1\_2H}$ operate in the high voltage domain (between $V_{INT}$ and $V_{BAT}$) while switches driven by $\Phi_{S1\_1L}$ and $\Phi_{S1\_2L}$ operate in the low voltage domain (between ground and $V_{INT}$).

The maximum switching frequencies of the two stages are also different. The first-stage maximum switching frequency is one quarter of that in the second stage. By doing this, the two stages occupy similar chip area and have similar conversion efficiencies, resulting in optimal overall efficiency and power density for the regulator. By optimizing the two stages separately, the first stage connects to the high battery voltage,

978-1-4673-6145-3/13 $31.00 © 2013 IEEE

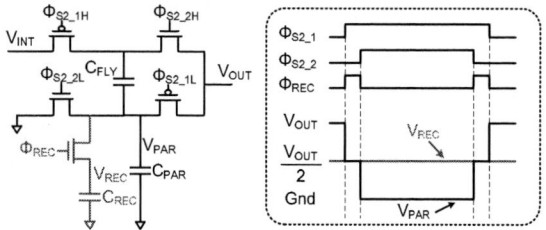

Fig. 2: Charge recycling technique and timing diagrams

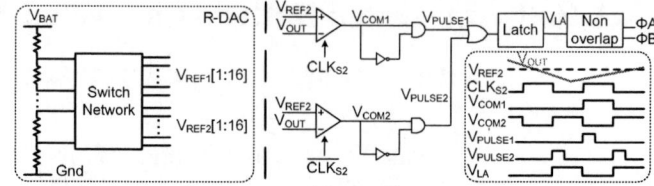

Fig. 3: Feedback control loop diagram (second stage)

but is decoupled from output load transients. The higher switching frequency of the second stage enables higher closed-loop bandwidth for fast output load transient response.

Cascading two 2:1 SC stages offers other advantages. $V_{INT}$ and $V_{OUT}$ can serve as stacked supply voltages for the switch drivers in each stage such that no additional voltage rail is required. Also, the bottom plate parasitic loss is lower, compared to single-stage 4:1 SC converters, which we further reduce via a charge recycling technique described below.

### B. Bottom-Plate Charge Recycling

A dominant source of efficiency loss in SC converters comes from switching the bottom-plate parasitic capacitance associated with the flying capacitor ($C_{FLY}$). All of the flying capacitors in this design rely on bulk MOS transistors, which usually have non-negligible bottom-plate parasitic capacitance (~2% in this technology, ~5% in [4]). Each stage implements circuitry that combines two-step charging/discharging with charge recycling, as illustrated in Fig. 2 for the second stage. $C_{PAR}$ is the parasitic bottom-plate capacitor of $C_{FLY}$. By adding an additional recycling capacitor, $C_{REC}$, the proposed technique avoids using an external voltage source. The two-step charging/discharging occurs during the converter's dead time to recycle charge, reduce losses, and improve conversion efficiency.

The charge recycling operation is as follows. Assume $C_{REC} \gg C_{PAR}$ and $V_{REC}$ starts out at $V_{OUT}/2$. When discharging $C_{PAR}$, $C_{PAR}$ first transfers charge to $C_{REC}$ through the additional switch controlled by $\Phi_{REC}$. In this process, $C_{PAR}$ discharges from $V_{OUT}$ to $V_{OUT}/2$. Then, the switch $\Phi_{REC}$ turns off and $C_{PAR}$ fully discharges to gnd. The amount of charge transferred from $C_{PAR}$ to $C_{REC}$ is $C_{PAR}V_{OUT}/2$, which is stored on $C_{REC}$ and is recycled in the charging phase. When charging $C_{PAR}$, $C_{PAR}$ first charges up from gnd to $V_{OUT}/2$ via $C_{REC}$. In this period, $C_{REC}$ transfers $Q = C_{PAR}V_{OUT}/2$ to $C_{PAR}$, which is the same amount of charge that $C_{REC}$ gets from $C_{PAR}$ in the discharging process. $C_{PAR}$ then disconnects from $C_{REC}$ and fully charges up to $V_{OUT}$. From an energy perspective, $V_{OUT}$ only needs to provide $E = C_{PAR}V_{OUT}^2/2$ in this charging process, which is half of the energy otherwise required. It is important to note that $V_{REC}$ eventually settles to $V_{OUT}/2$ regardless of its initial voltage, because this is the only balanced state where the energy stored on $C_{REC}$ when discharging $C_{PAR}$ matches the energy that $C_{REC}$ loses when charging $C_{PAR}$.

The above recycling process assumes $C_{REC} \gg C_{PAR}$. Thanks to the converter's multi-phase operation, $C_{REC}$ can be shared by all of the phases and $C_{REC}$ only needs to be larger than the parasitic capacitance in one phase, achieved with negligible penalty. In this implementation, $C_{REC}$ is 2% of the total flying capacitance.

### C. Low-Boundary Feedback Control

Closed-loop operation regulates $V_{OUT}$ and $V_{INT}$ to desired voltage levels. Each stage implements the same low-boundary feedback control loop illustrated in Fig. 3 [3]. Since the feedback toplogy is the same in both stages, the following illustration uses the second stage as an example. Pairs of the interleaved modules share separate feedback paths, i.e., there are a total of eight feedback paths in the 2nd stage. In each feedback path, two comparators operate off of complimentary clocks generated by the VCO. The comparators compare $V_{OUT}$ with a reference voltage, $V_{REF2}$, on the rising and falling edges of the clock. If $V_{OUT}$ is smaller than $V_{REF2}$, $V_{LA}$ switches either from low to high or high to low, depending on its previous state. $V_{LA}$ then propagates through to control the power switches and switch the state of the SC converter. This action increases the output voltage $V_{OUT}$. If $V_{OUT}$ is larger than $V_{REF2}$, $V_{LA}$ remains in its previous state. The power swiches do not switch and $V_{OUT}$ decreases until the SC converter reacts.

A resistor DAC (R-DAC), shown in Fig. 3, provides separate reference voltages to the 16 comparators via a switch network that connects each individual comparator to the resistor ladder separately. By doing do, we can use the R-DAC to calibrate comparator offsets. The switch network also generates 16 separate reference voltages for the first SC stage. Calibrating comparator offsets improves steady-stage voltage ripple and conversion efficiency.

## III. MEASUREMENT RESULTS

The two-stage SC converter was fabricated in TSMC's 40nm CMOS technology. The chip was tested in two modes: open- and closed-loop operation. In open-loop operation, the output voltage and output power can be tuned by changing the switching frequency, $F_{sw}$, of the converter via the VCO. In closed-loop operation, the VCO frequency is set to its maximum and the feedback control loop adjusts the effective switching frequency of the converter to regulate the output.

In open-loop operation, there is a relationship between the switching frequency and the output voltage and power. Shown in Fig. 4(a), higher output power requires high switching frequency to deliver energy more frequently. However, when switching frequency increases, there is less time for the switched capacitor circuit to settle in each cycle. Because of this incomplete charge transfer, the energy that is delivered from input to output in each cycle decreases as switching frequency increases. Hence, switching frequency increases super linearly with output power. Switching frequency, and thus switching loss, increases faster than the delivered power. Fig. 4(b) shows that higher output voltages also require higher switching frequencies. As the output voltage increases, there is less energy that can be delivered from input to output in each cycle [2]. So, switching frequency and switching loss increase faster than $V_{OUT}$ increases.

978-1-4673-6145-3/13 $31.00 © 2013 IEEE

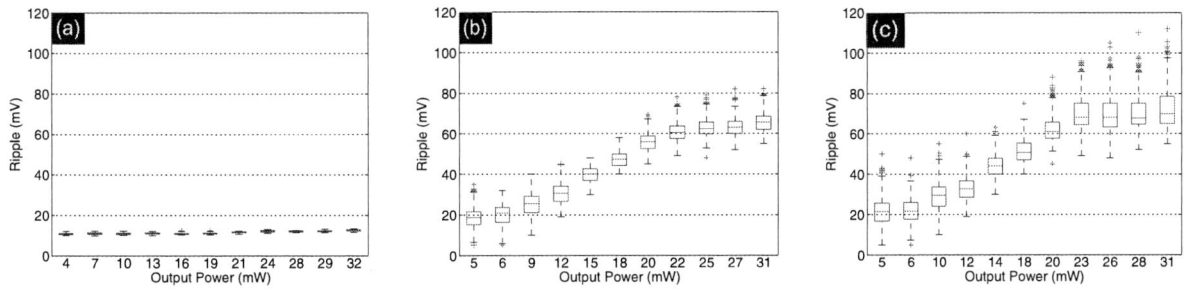

Fig. 5: Measured output voltage ripple @ $V_{BAT}$=3.8V, $V_{OUT\_AVE}$= ~800mV in (a) open-loop operation, (b) closed-loop operation with calibrated comparators, and (c) closed-loop operation with un-calibrated comparators

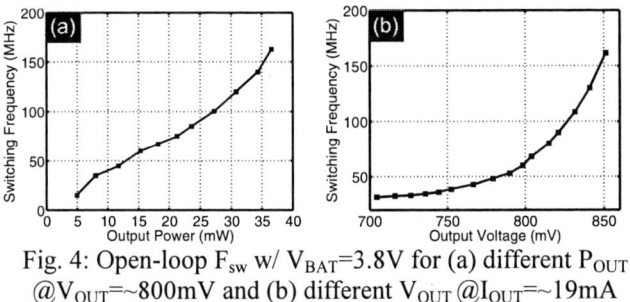

Fig. 4: Open-loop $F_{sw}$ w/ $V_{BAT}$=3.8V for (a) different $P_{OUT}$ @$V_{OUT}$=~800mV and (b) different $V_{OUT}$ @$I_{OUT}$=~19mA

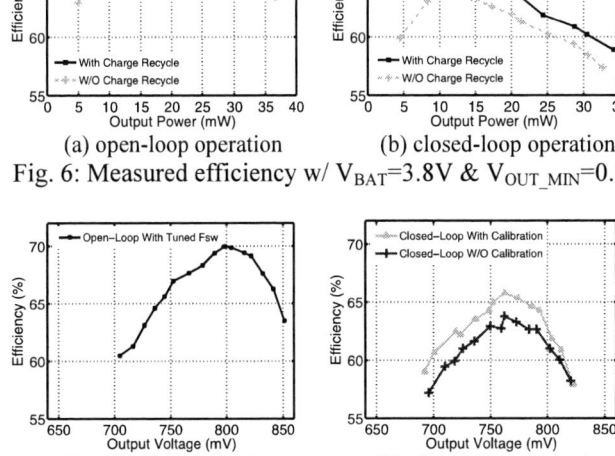

(a) open-loop operation (b) closed-loop operation

Fig. 6: Measured efficiency w/ $V_{BAT}$=3.8V & $V_{OUT\_MIN}$=0.8V

(a) open-loop operation (b) closed-loop operation

Fig. 7: Measured efficiency w/ $V_{BAT}$=3.8V & $I_{OUT}$=~19mA

This section presents the experimentally measured results as follows: Section III.A first compares steady-state voltage ripple for open- and closed-loop modes of operation. Then, Section III.B presents conversion efficiency results versus $V_{OUT}$ and $P_{OUT}$. The transient responses in open- and closed-loop modes of operation are discussed in Section III.C. Finally, Section D provides a summary of test chip characteristics and compares it to prior work.

*A. Voltage ripple*

The box plots in Fig. 5 compare the measured steady-state output voltage ripple across a range of output power conditions for the SC converter in open- and closed-loop operation. In open-lop operation, we manually tuned the VCO frequency to keep $V_{OUT}$ at ~800mV for each power level. In closed-loop operation, the feedback loop keeps the output voltage at ~800mV. Steady-state ripple in open-loop operation is small (~10mV) due to the interleaved design with constant switching frequency. In contrast, closed-loop ripple is generally higher due to the cycle-skipping nature of the feedback topology. In each cycle, the feedback controller must determine whether the converter should switch or not. As a result, the instantaneous switching frequency can vary widely from cycle to cycle. Delay through the feedback loop further exacerbates the ripple, because the control loop must react to the output decreasing below the reference voltage. The longer the feedback delay is, the larger the ripple is. Measurement results show that closed-loop ripple increases with output power since larger load currents discharge the output voltage more quickly. Comparing Figs. 5(b) and (c), calibration helps to reduce voltage ripple by minimizing inconsistent switching thresholds across all of the comparators in the multiple feedback paths. In all subsequent plots, the comparators are always calibrated unless noted otherwise.

*B. Conversion efficiency*

In SC converters, the major sources of efficiency loss are due to linear charge redistribution loss, bottom-plate parasitic loss, other switching loss, and voltage ripple overhead. The minimum output voltage is used to calculate conversion efficiency, because the worst-case speed of the digital load circuits depends on the lowest transient voltage condition.

Fig. 6 plots efficiency measurements for both open- and closed-loop operation. In Fig. 6(a), open-loop efficiency reaches a peak of 70% at $P_{OUT}$=15mW. The efficiency rolls off for higher output power, because switching frequency and switching losses increase faster than the delivered power. Efficiency also rolls off for lower output power, because of static overheads. Comparing Figs. 6(a) and (b), closed-loop efficiency is generally lower than open-loop efficiency, because of larger voltage ripple. Fig. 6 also shows charge recycling consistently improves conversion efficiency by ~2%. Charge recycling is always on for all subsequent plots.

Fig. 7 plots conversion efficiency across different output voltage levels and exhibits the characteristic efficiency versus voltage curve of SC converters. In open-loop operation, the output voltage is set by tuning $F_{SW}$. In closed-loop operation, changing the reference voltage regulates the output voltage to different levels. Conversion efficiency rolls off as output voltage decreases due to linear charge redistribution loss and rolls off as output voltage increases due to higher switching loss.

978-1-4673-6145-3/13 $31.00 © 2013 IEEE    111

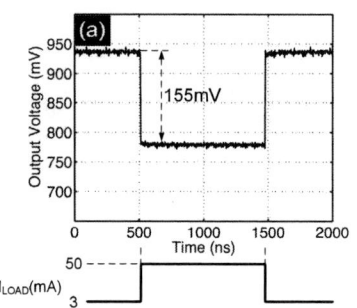 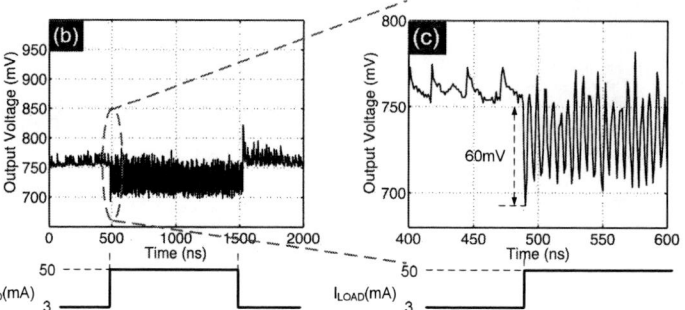 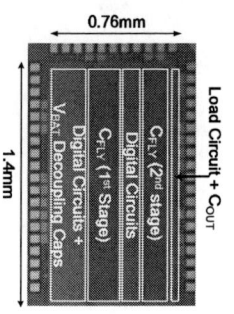

Fig. 9: Transient response (a) open-loop with maximum F$_{SW}$, (b) closed-loop, and (c) zoom-in of (b)     Fig. 10: Die Photo

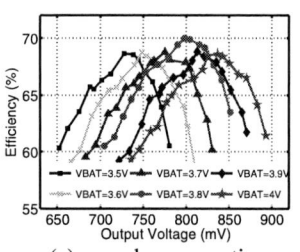 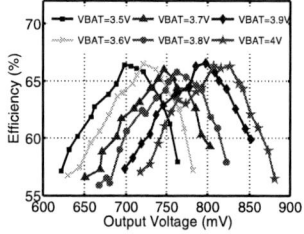

(a) open-loop operation     (b) closed-loop operation

Fig. 8: Measured efficiency with different V$_{BAT}$ and V$_{OUT\_MIN}$

|  | [5] | [3] | This Work |
|---|---|---|---|
| Technology | 65nm | 90nm | 40nm |
| Input voltage | 3V-4V | 3V-3.9V | 3.5V-4V |
| C$_{FLY}$/C$_{FILTER}$ | 3.88nF/0 | 2nF/3.2nF | 2.24nF/0.4nF |
| Efficiency (η) Peak eff. Conversion ratio | 73% 74.3% 3:1 | 74% 77% 2:1 | 66% 70% 4:1 |
| Power density (mW/nF)@η | 31.3 (121mW/3.88nF) | 28.8 (150mW/5.2nF) | 13.3 (35mW/2.64nF) |
| Load step/C$_{TOTAL}$ (mA/nF)@t$_{rise}$ | 41.7@50ps | 5.8@25ns | 17.8@100ps |
| Output droop | 76mV | 30mV | 60mV |

Table. 1 Comparison to prior art.

Comparing the three curves in Fig. 7, open-loop operation consistently achieves higher conversion efficiency since it has the smallest voltage ripple. Calibration improves efficiency, as expected, since it reduces voltage ripple in closed-loop operation. The efficiency in closed-loop operation peaks at a lower output voltage compared with that in open-loop operation again because of voltage ripple and because the minimum output voltage is used to calculate efficiency.

Fig. 8 summarizes conversion efficiency versus output voltage for different battery voltages (V$_{BAT}$). First, conversion efficiency is higher for open-loop operation, consistent with previous results above. Second, conversion efficiency peaks at higher output voltages when V$_{BAT}$ is higher since charge redistribution loss and switching loss are both related to V$_{OUT}$/V$_{BAT}$ [2].

*C. Transient response*

Fig. 9 presents the SC converter's measured response to 47mA output load transients using an on-die load circuit with rise and fall times of ~100ps. As seen in Fig. 9(a), when the SC converter runs in open-loop with maximum switching frequency, a 3mA to 50mA load step causes V$_{OUT}$ to drop by 155mV. When running in closed-loop with the nominal output voltage set to 750mV, however, the control loop quickly reacts and the voltage droop caused by the load current step is much smaller. In fact, the ~60mV droop in Fig. 9(c) is mostly due to the larger steady-stage voltage ripple previously seen with respect to higher output power.

*D. Test chip summary*

The silicon area, shown by the micrograph in Fig. 10, was not optimized for power density but was governed by the pads and circuitry added for testing. Flying capacitors and output filter capacitors, which occupy half of the overall area, total 2.64nF. Table. 1 compares this work to prior art SC converters. The 70% peak efficiency of this design is comparable to the efficiencies in [3] and [5], but for a higher 4:1 conversion ratio.

## IV. CONCLUSIONS

This paper demonstrates a fully integrated battery-connected switched capacitor converter for the brain SoC of a microbotic bee. The two-stage topology, with bottom-plate charge recycling, offers high conversion efficiency for the high 4:1 conversion ratio. While closed-loop regulation provides fast transient response, it also exhibits larger stead-state voltage ripple, which results in efficiency drop compared to open-loop operation. This tradeoff motivates exploring an adaptive clocking strategy to improve overall system efficiency as described in [7].

## ACKNOWLEDGMENTS

This work was supported in part by the NSF Expeditions in Computing Award #: CCF-0926148. The authors thank the TSMC university shuttle program for chip fabrication.

## REFERENCES

[1] M. Karpelson et al., Energetics of flapping-wing robotic insects: Towards autonomous hovering flight, *IROS*, 2010.

[2] H.-P. Le et al., Design techniques for fully Integrated switched-capacitor DC-DC converters, *JSSC*, Sep. 2011.

[3] T. Van Breussegem et al., Monolithic capacitive DC-DC converter with single boundary–multiphase control and…, *JSSC*, July 2011.

[4] Y.K. Ramadass et al., Voltage scalable switched capacitor DC-DC converter…, *PESC*, 2007.

[5] H.-P. Le et al., "A sub-ns response fully integrated battery-connected switched-capacitor voltage regulator…, *ISSCC*, 2013.

[6] W. Kim et al., A fully-integrated 3-level DC/DC converter for nanosecond-scale DVFS, *JSSC*, Jan. 2012.

[7] X. Zhang et al., Supply-noise resilient adaptive clocking for battery-powered aerial microrobotic system-on-chip in 40nm CMOS, submitted to *CICC*, 2013.

[8] T. Andersen et al., A 4.6 W/mm$^2$ power density 86% efficiency on-chip switched capacitor DC-DC converter in 32nm SOI CMOS, *APEC* 2013.

# A 110nA Synchronous Boost Regulator with Autonomous Bias Gating for Energy Harvesting

Khondker Z. Ahmed and Saibal Mukhopadhyay

School of Electrical and Computer Engineering, Georgia Institute of Technology, Atlanta, GA 30332 USA

Phone:+1.404.894.2688, Fax:+1.404.894.4641, E-mail: khondker@gatech.edu, saibal@ece.gatech.edu

*Abstract*–An autonomously bias gated synchronous boost regulator consuming 110nA at 1V is demonstrated in 130nm CMOS. The IC generates regulated 1V output from 30mV input, starts up autonomously (battery-less) at 265mV, and regulates output ranging from 0.78V-3.3V. The peak efficiency is 83% with 10μA and 85% with 10mA load.

## I. INTRODUCTION

Advanced transducers can produce appreciable (10s of μW) power from small energy fields [1]. Effective energy harvesting from these power sources requires boost regulators with low operating input voltage, low standby current and high efficiency [2-5]. Boost regulators with digital control blocks with sub-μA bias currents have been presented [2-3]. However, designing regulators with all-analog control blocks and sub-μA bias remains challenging as low bias current of the analog blocks degrade the regulator's performance.

This paper presents *autonomous bias gating* to achieve ultra-low standby current in pulse frequency modulated (PFM) boost converter with all-analog control (Fig.1). The Feedback Comparator (FC) and the Reference Generator (REF) are required during all the modes of operation of a PFM regulator; but the Oscillator (OSC), the Zero Current Comparator (ZCC), and the Current Limit Comparator (CLC) are only required during the charging (active) mode of the output and are not essential in the discharging (idle) mode. The proposed regulator gates the bias currents of OSC, ZCC, and CLC by turning off digital switches (bias gating) during the discharging mode. The bias gating is performed *autonomously* using the already available feedback signal in the regulator. The autonomous gating of the bias currents during the idle mode reduces standby and average operating current, while still maintaining performance during active mode.

The autonomously bias gated PFM boost regulator is demonstrated in 130nm CMOS technology. The bias currents of the non-bias gated blocks are reduced to 10s of nA region. A high duty cycle analog oscillator enables boosting from very low input voltage. The measurements demonstrate a total bias current of 110nA at 1V, a sub-300mV all-electronic autonomous battery-less startup voltage, and high (> 80%) efficiency at both heavy (10mA) and light (10μA) load conditions.

## II. DESIGN OF THE BOOST REGULATOR

Fig. 1 shows the functional block diagram of the proposed synchronous PFM boost regulator. The regulator can start-up and operate either with an external battery or a capacitor with stored energy connected to $V_{BAT}$ (assisted mode) or with $V_{OUT}$ connected to $V_{BAT}$ (self-powered or battery-less mode). The main power stage and drivers are designed with 3.3V devices

to enable wide output range (0.78V to 3.3V). A $V_{DD}$-Select Comparator circuit is designed that helps regulate output above the battery voltage. The control blocks are designed with low voltage (1.2V) devices to reduce area/power.

The FC senses the output through the external resistor divider, compares it with internally generated reference, and generates the EN signal. When $V_{FB}$ falls below $V_{REF}$ (i.e., $V_{OUT}$ falls below the regulation level), EN goes high (active mode), turning on the oscillator. The OSC signal then propagates through the power FET drivers and enables the inductor to build up current and charge the output capacitor. Once output rises above the regulation point, EN goes low (idle mode). During this mode, no switching occurs and output discharges under the given load. During the idle mode, the bias currents of OSC, ZCC, and CLC (shaded in Fig. 1) are cut off (bias gated) using the EN signal. Additionally, within the active mode, the OSC circuit is always partially bias gated, and the ZCC is bias gated when the pFET is off, thus reducing the average bias current during active mode. The REF, FC, and

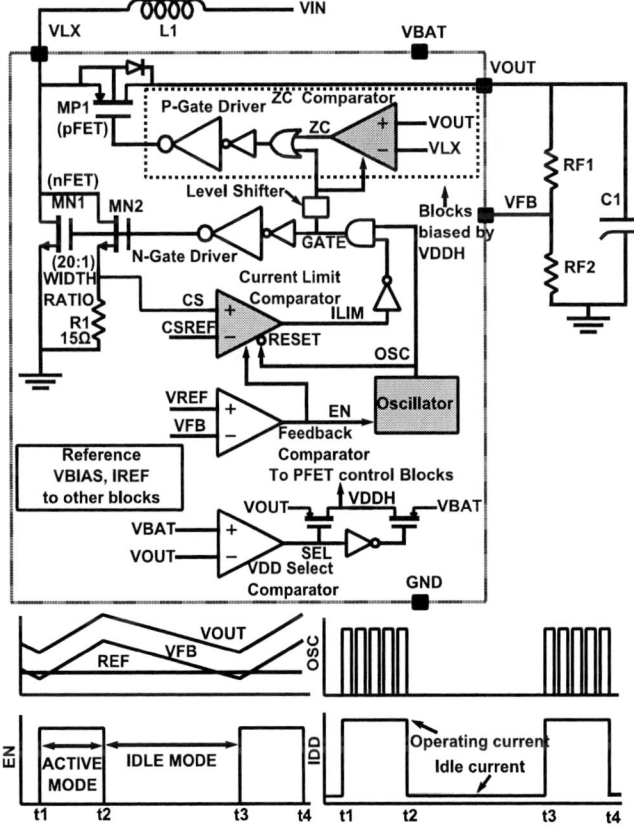

Fig.1: Functional block diagram of the proposed synchronous boost regulator. Beside the overall low bias current design, the bias currents of the shaded blocks are cut off during idle mode (EN low), to further reduce converter loss.

978-1-4673-6145-3/13 $31.00 © 2013 IEEE    113

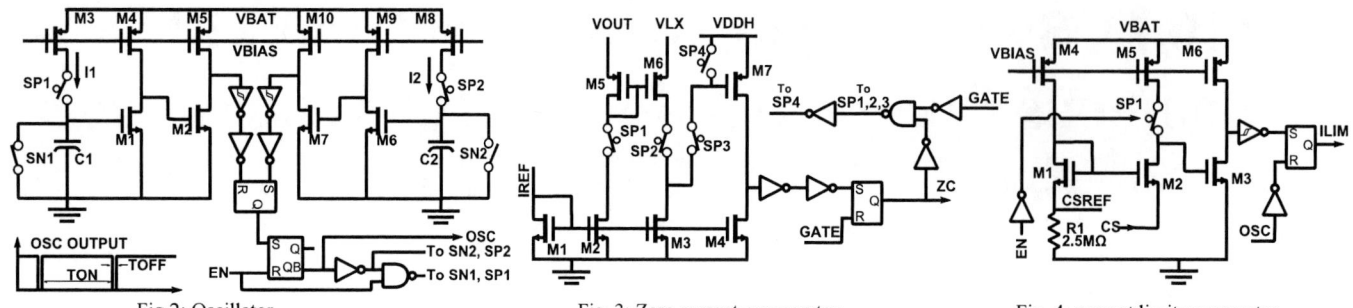

Fig 2: Oscillator      Fig. 3: Zero current comparator      Fig. 4: current limit comparator

Fig 5: Reference circuit      Fig. 6: Feedback comparator      Fig. 7: $V_{DD}$-Select Comparator

Table I: Bias currents for key blocks

|  | Idle mode (bias gated) (nA) | Active mode[1] (non-bias gated) (nA) |
|---|---|---|
| OSC | 3 | 215 |
| ZCC | 9 | 27 |
| CLC | 9 | 18 |
| FBC | 18 | 18 |
| REF | 25 | 25 |
| V_sel | 30 | 30 |
| Driver[2] | 16 | 16 |

[1] Excluding switching current.
[2] Leakage current.

$V_{DD}$-Select Comparator are not bias gated since they remain essential in all modes. The operating current of the chip depends on the time in the active versus the idle mode and is the weighted average of the two current levels (Fig.1).

### A. Design of the Bias Gated Blocks

Fig. 2 shows the high-duty cycle analog *oscillator* (OSC) block with two symmetric half-circuits having $I_1/I_2$=2.5 and $C_2/C_1$=28, (4.2pF and 0.15pF *on-chip* capacitors) resulting an overall on-time to off-time ratio of 70. In the active mode, at the beginning of $T_{OFF}$, switch $SP_1$ closes and $SN_1$ opens and $I_1$ charges $C_1$. When the voltage across $C_1$ crosses $M_1$ threshold, a reset (R) edge is created and $T_{OFF}$ ends. This also enables $I_2$ to charge $C_2$ and $T_{ON}$ begins. Similarly, a positive edge at set (S) ends the $T_{ON}$, after which OSC goes to low state and the next $T_{OFF}$ repeats. The latches ensure that the positive pulse is not terminated by the asynchronous EN signal, i.e., if EN goes low while OSC is high, the circuit goes into the idle mode only after the continuing $T_{ON}$ period elapses itself. Without this masking, in certain conditions, the OSC may produce unwanted frequency and duty cycle. *During active mode, only one half of the circuit is operative while the other half is bias gated, and during idle mode both halves are bias gated by means of $SP_{1,2}$ and $MN_{1,2}$ switches.*

The *Zero Current Comparator (ZCC)*, shown in Fig. 3, generates the output (ZC) to turn off the pFET ($MP_1$ in Fig. 1) preventing reverse current through the inductor. The GATE signal is generated by OSC and $I_{LIM}$, and serves as the bias gating signal for the ZCC. When GATE goes low, the pFET turns on, and ZCC starts comparing $V_{LX}$ with $V_{OUT}$ and generates output (ZC goes high) when the current falls below 10mA (a designed positive offset to compensate for delay and variation). The ZC signal turns the pFET off and the remaining current flows through the body diode of the pFET. Except the diode branch ($M_1$), all currents are cut off (bias gated) when either ZC or GATE is high. *This occurs during the idle*

*mode and within the active mode when pFET is off.* Switches $SP_{1-4}$ provide isolation from the dynamic $V_{LX}$ node and provide correct initial state for the ZC signal. The diode-connected device ($M_1$) is not bias gated to reduce circuit settling time at the beginning of each ZCC active phase.

The *Current Limit Comparator (CLC)* circuit is essential to clamp inrush and over current through the inductor. As shown in Fig. 4, it compares inductor current level [via CS, sensed at nFET $MN_2$ (Fig.1)] with the local reference, $CS_{REF}$. When CS goes above $CS_{REF}$, $I_{LIM}$ changes state to high and turns off the nFET ($MN_{1-2}$ in Fig. 1). This circuit is not required during idle mode and hence the bias current is cut off using EN signal and switch $SP_1$. *The current through the diode connected device ($M_1$) is not cut off during idle mode to keep the $CS_{REF}$ available and the amplifier ready at the beginning of next EN pulse to minimize the settling time.*

### B. Design of the Non-Bias Gated Control Blocks

An open loop, diode based, supply independent voltage *Reference Generator* (REF) is used (Fig.5 [6]). The gate voltage of the diode-connected device ($M_3$) is shared to replicate bias currents all blocks ($V_{BIAS}$ in Fig. 2-4). The elimination of the closed loop reference and temperature correction is justified by the fact that in many operations the variation of ambient temperature is small. As shown in Fig. 6, the *Feedback Comparator* (FC) compares a fraction of output voltage, $V_{FB}$ with internal reference $V_{REF}$. The output (EN) goes high when $V_{FB}$ goes below $V_{REF}$. The circuit is designed with hysteresis to regulate the valley of the output voltage.

The battery voltage and the output regulation range are decoupled to ensure wide output range in the battery assisted mode. To achieve this goal, the *$V_{DD}$-Select Comparator* is designed to select the higher voltage between the external $V_{BAT}$ and the generated $V_{OUT}$ as the biasing voltage for all high voltage circuits ($V_{DDH}$) (Fig.7 and Fig. 1). This circuit ensures that the pFET driver receives the highest potential of the

978-1-4673-6145-3/13 $31.00 © 2013 IEEE      114

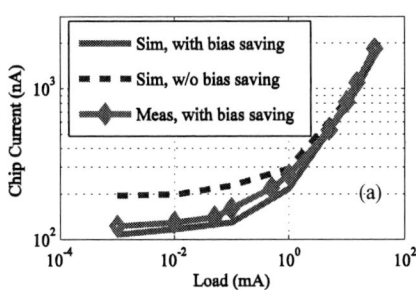 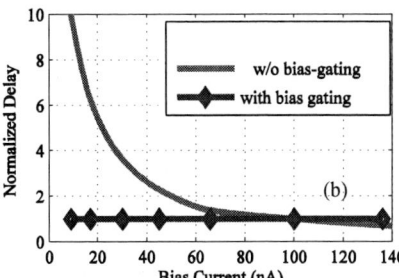 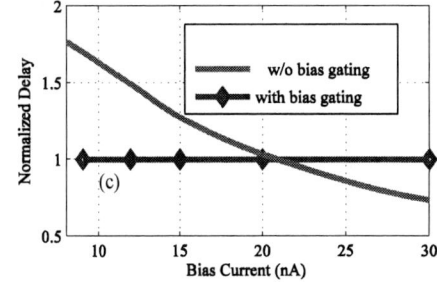

Fig.8: The effect bias gating over (a) total chip current measured data matches with simulation, (b) delay of ZCC (simulation) and (c) delay of CLC (simulation). For the results w/o bias gating, the results were obtained by re-designing the circuits at various bias currents. The results with bias gating consider that the current during active and idle modes are constant but the average bias current varies with the activity factor, and different points denote different activity factors.

whole system to turn the transistor completely off. The circuit is not required in the self-powered mode (battery-less mode, $V_{OUT}$ shorted to $V_{BAT}$), however, then the operating range of output is limited to 1.2V due to the use of low-voltage devices.

### C. Impact of Bias Gating on Power

The simulated bias currents of the individual blocks are shown in Table I. Simulated chip currents with and without the bias gating are shown in Fig. 8(a). At higher load active to idle time ratio increases, thereby increasing the chip current. The peak consumption is limited by the continuous switching condition (i.e. no idle time). With bias gating, the initial current consumption is much lower (~1.5X); however, at higher load the extra switching energy associated with the gate and parasitic capacitances of the bias gating transistors becomes important. The circuits are designed to ensure the bias gating remains energy-efficient until 40mA of load, which is beyond the maximum current capability of the converter.

### D. Impact of Bias Gating on Efficiency and Regulation

The bias gating helps reduce overall current while meeting specific delay requirements to manage efficiency and functionality of the booster. A higher delay in the ZCC results in negative inductor current (i.e., discharges the output) which degrades the efficiency. The delay of CLC circuit causes overshoot in the inductor current resulting in higher output ripple and potential malfunction. The bias gating doesn't affect the delay since the decisions are generated only in active mode (higher bias current). If the standby power reduction is to be achieved solely by reducing the bias current (in all conditions), the delays of these blocks would have been degraded significantly as illustrated in Fig. 8(b) and (c). At maximum bias current the marginal performance overhead of the bias-gated circuit is due to the added gating circuitry. A high wake-up time for ZCC and CLC at the beginning of each active cycle can negatively impact the performance. The proposed design minimizes the wake-up time by not bias gating the diode connected devices of these blocks.

The design needs to ensure that the additional delay in the feedback path due to the overall current reduction does not degrade stability. The load pole, introduced by variable load (0 to 30mA) and output capacitor (100μF), varies from the origin to 50Hz. With the designed bias current, the feedback path, including the bond pad capacitance and delays of the FB, OSC and drivers, introduce another pole at ~30kHz. The bias gating in the OSC doesn't affect the pole location any further because

with or without the bias gating, the circuit has the same delay (capacitors' initial conditions are same in both cases). Since the second pole is far from the load pole, the transient stability is not affected and the converter behaves like a single pole system with the dominant load pole.

### III. TEST-CHIP AND MEASUREMENT RESULTS

Fig. 9(a) shows the die-photo of the converter. The fundamental oscillation frequency of the converter is 100kHz. The measured nFET and pFET resistances of the power stage are 400mΩ and 900mΩ, respectively. Test board shown in Fig. 9(b) is used to measure performance parameters of the packaged (LCC44) die mounted on a socket. Parasitic resistances include the board trace (in power loop) (20mΩ) and the socket resistance (25mΩ). External inductor (100μH, $R_{ESR}$=240mΩ) and input/output capacitors (both, 100μF, $R_{ESR}$=450mΩ) are used. All measurements are performed using the socket and the board; hence results include all associated (parasitic and ESR) losses. Table II compares the primary features of the regulator with prior works and shows the bias current advantage of this design. Fig 10(a) shows the all-electronic autonomous (battery-less) startup of the regulator, showing the converter bootstraps itself from 265mV at the output. The zoom in view of startup phase, in Fig. 10(b), shows the operation of the current limit protection scheme. Since the regulator is biased by its own generated output, the delay of the current limit circuit decreases with higher output voltage hence the threshold is seen to move slightly downward. Fig. 10(c) shows a regulated output at 1V from 30mV input, delivering 10μA while toggling between the idle and active modes. Fig. 10(d)

Table II: Comparison chart with other works

| Item | [2] | [3] | [4] | [5] | This work |
|---|---|---|---|---|---|
| Min. $V_{IN}$ (w/ $V_{OUT}$=1V) | 20mV | 100mV | 25mV | 30mV | **30mV** |
| Min. self start | 600mV | 330mV | 35mV[1] | 40mV[2] | **265mV** |
| Bias current | 1.1μA | 150nA[3] | N/A | N/A | **110nA[4]** |
| $V_{OUT}$ | 1.4V | 3V | 1.8V | 2V | **0.78V- 3.3V** |
| Input range | 20mV-250mV | 100mV-VOUT | 25mV-VOUT | 50mV-VOUT | **30mV-VOUT** |
| Peak efficiency | 75% | 92% | 60% | 61% | **85%** |
| Topology | Sync. | Sync. | Sync. | Async. | **Sync.** |
| Max Pout | 175μW | ~12mW | 300μW | N/A | **33mW** |
| Process | 0.13μm | N/A | 0.35μm | 0.13μm | **0.13μm** |
| Area | 0.12mm² | N/A | 4.2mm² | 0.09mm² | **0.2mm²** |

[1]Requires Motion Activated Sensor. [2]Requires Negative-VTH MOSFET.
[3]Total bias current is 330nA, the charger (boost) consumes 150nA.
[4]Total bias current in 110nA, the booster without $V_{DD}$_select is 80nA.

978-1-4673-6145-3/13 $31.00 © 2013 IEEE

shows the highest conversion ratio (62:1) operation while boosting from 50mV input to 3.1V output. Measured minimum $V_{IN}$ for boosting is 8mV, generating a 0.45V unregulated output (not shown here). Fig. 11 shows the load and the line regulation characteristics. The measured maximum ripple (at highest input and lowest load) is 125mV. Fig. 12(a) shows the measured current consumption of the chip with different load. The measurement shows that for a given output voltage, lower input voltage reaches maximum switching condition (peak chip current) at lower load to compensate for the higher conversion ratio. With higher $V_{IN}$, the measured minimum chip current at the lightest load is ~120nA and increases to ~2.2µA at maximum load. Fig. 12(b) shows the efficiency of the converter ($V_{OUT}$=1.2V). The efficiency increases at high load current and lower conversion ratio. The measured peak efficiency is 85% with 10mA and 83% with 10µA of load current.

## IV. CONCLUSION

This paper demonstrates the application of bias gating in reducing stand-by and operating power of a boost regulator. A PFM boost regulator with autonomous bias gating is presented where bias currents of selected analog circuit blocks are cut off during the output discharging phase to reduce standby current consumption. The demonstrated test-chip in 130nm CMOS achieves very small bias current (110nA at 1V) and

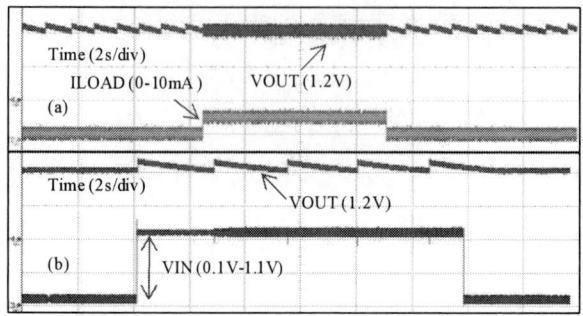

Fig. 11: (a) load regulation (0.25%/mA) and (b) Line regulation (5%/V)

maintains high efficiency across wide load, input, and output region. The maximum delivered peak output power is 33mW. The presented regulator is effective for energy harvesting systems such as wireless sensor nodes which exhibit long intervals of low-power mode but also moderately high peak power.

***Acknowledgement:*** This work is supported in part by National Science Foundation and ONR Young Investigator Program.

### REFERENCES

[1]. eTEG HV56; Nextreme Thermal Solutions, Accessed Dec. 2012.
[2]. E. Carlson, et al., IEEE JSSC, vol. 45, no 4, pp. 741-750, April 2010.
[3]. K. Kadirvel, et al., ISSCC, pp.106-108, 19-23, Feb. 2012.
[4]. Y. Ramadass, et al., IEEE JSSC, vol. 40, no. 1, pp. 296-297, Jan. 2011.
[5]. J. Im, et al., ISSCC, pp.104-106, 19-23, Feb. 2012.
[6]. B. Razavi, Design of Analog CMOS IC, McGraw-Hill, 2001.

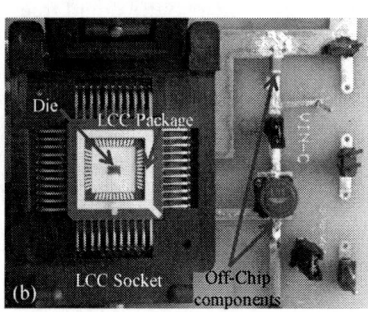

Fig. 9: (a) Die photo, and (b) the board (package, socket and external components are also shown) used for measurements.

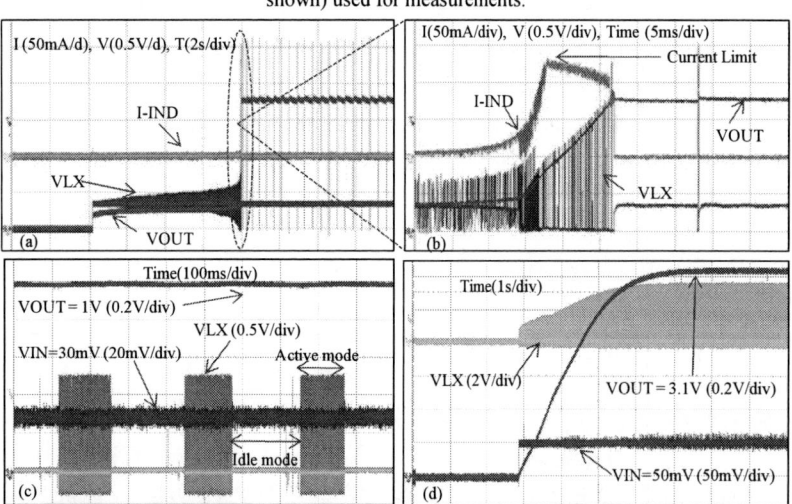

Fig. 10: Operation waveforms (a) Autonomous (battery-less) startup with 265mV at output, (b) current limiting during start up event, (c) regulated output with 10µA load current from 30mV input after battery assisted start-up showing idle mode and active mode of operations and (d) battery assisted startup from 50mV input to 3.1V output.

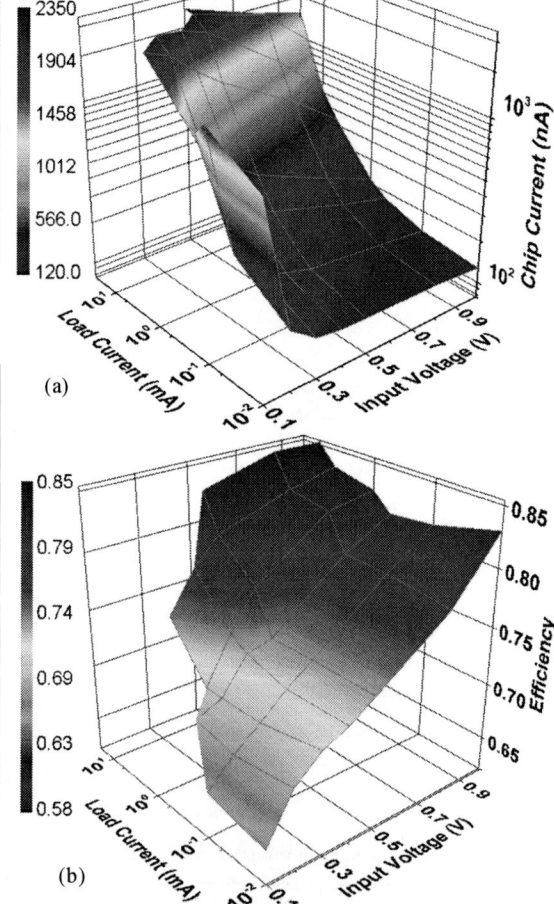

Figure 12: (a) Chip current consumption and (b) Efficiency measurement with different load and input conditions. ($V_{OUT}$=1.2V.)

# A Power Sensor with 80ns Response Time for Power Management in Microprocessors

Srikar Bhagavatula, and Byunghoo Jung

Electrical and Computer Engineering, Purdue University. E-mail: {sbhagava, jungb}@purdue.edu

*Abstract*—A real-time, on-chip power sensor that estimates load currents and on-chip temperatures concurrently is presented. It occupies an area of 0.01mm² in 0.13μm CMOS technology. With a simplified 1-point calibration and a response time of 80ns, it shows improvements in input dynamic range by 10×, response time by 6× and sensitivity by 3× over previous such sensors. A current reference with a measured temperature coefficient 91ppm/°C (-20°C to 120°C) is presented. This reference is used for online calibration of the power sensor to enable greater tolerance to PVT variations and aging effects.

## I. INTRODUCTION

Energy consumption has emerged as a critical design criterion due to increasing relevance of battery-powered mobile devices like tablets, smartphones, laptops etc. Continued rise in the density of thermal power dissipation due to increasing device densities also leads to higher cooling costs and reliability issues. Power management techniques such as clock gating, power gating, and Dynamic Voltage and Frequency Scaling (DVFS) have been proposed to reduce system power consumption while mitigating the effect on performance. Such techniques rely on real-time feedback from workload-based simulation models or from thermal sensors ([1]-[5]) for power estimation ([6]) and power management.

However, simulation models lack the detail to track variations in operating conditions and computation of power from temperature values is a resource intensive process. Combined with the inherent delay associated with the conversion of electric power in to heat dissipation, such techniques have very poor overall response times. In addition, due to lateral heat spread across the chip surface, thermal sensors also lacks the spatial resolution required for fine-grained power management. A current sensor described in [7] introduces a small resistance in series and samples the voltage drop across the resistor to estimate power accurately. However, it has poor temporal resolution (Hz-KHz) and a large area overhead (1mm²). A sensor with a fast response time and a low area overhead was presented in [8]. However,

This material is based upon work supported by NSF under grant no: CCF -0964634.

estimation required measurement of on-chip temperature and even with a two-point calibration, estimates may be vulnerable to aging defects. It also has limited dynamic range which, although sufficient for estimation of average power, may not be sufficient to estimate spikes in power consumption. In addition, methods described in [7] and [8] require an external current source for calibration, which increases the test complexity.

This paper presents an on-chip power sensor for estimating power dissipation by utilizing the voltage drop across sleep-transistors. A novel online calibration scheme along with an on-chip reference current has been proposed which results in significant improvements to variation tolerance and input dynamic range.

## II. ARCHITECTURE

### A. Sensor

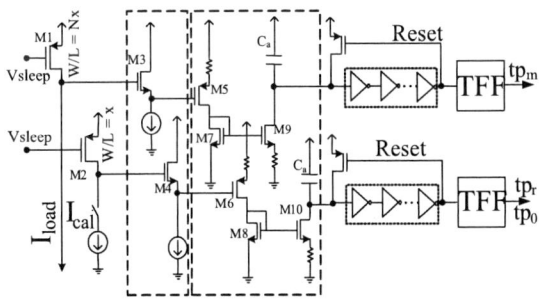

Fig. 1: Architecture of the power sensor

Fig. 1 shows the architecture of the proposed on-chip power sensor. The sleep transistor acts as a small series resistor in the on-state with a resistance of $R_{ON}$ and produces an IR-drop proportional to the load current ($I_{load}$) as it operates in the ohmic region. This voltage is amplified (with a transconductance of $G_m$) and converted to a current ($I_{chg}$). $I_{chg}$ is used to discharge a capacitor ($C_a$) which is reset to Vdd with some delay after it reaches an inverter threshold ($V_{t,inv}$). The output of inverter chain is used to trigger a flip-flop producing a waveform with a pulse-rate proportional to $I_{load}$ as follows:

$$I_{load} = \frac{2 \cdot C_a \cdot V_{t,inv}}{R_{ON} \cdot G_m} \cdot \left(\frac{1}{tp_m} - \frac{1}{tp_0}\right) \quad (1)$$

where $tp_m$ is the output time-period at a given load current and $tp_{0,m}$ is the zero error, measured as the output time-period at zero load current.

This design is replicated in a second branch with one change: The sleep transistor is scaled down by a factor of "N" to reduce the area and power overhead of the sensor. The ON-resistance of the series resistor in this path now equals $N \times R_{ON}$, therefore, the current required to produce the same IR-drop as the main-branch is reduced by N-times. To ensure good matching with the primary branch, all components in the two branches are designed in an interdigital, common-centroid layout.

When the replica branch is loaded with a calibration current $I_{cal}$, the ratio of the two currents $I_{load}$ and $I_{cal}$ is given by

$$\frac{I_{load}}{I_{cal}} = N \cdot \frac{\left(\frac{1}{tp_m} - \frac{1}{tp_0}\right)}{\left(\frac{1}{tp_r} - \frac{1}{tp_0}\right)} \quad (2)$$

where $tp_m$ is the output time-period of the primary branch, $tp_r$ is the output time-period of the replica branch and $tp_0$ is the zero-load output time-period.

As PVT variations, supply noise and aging in the two branches are highly correlated, the ratio of their outputs is tolerant to such effects. Consequently, this sensor can provide variation tolerant estimates of the load current without needing to know the on-chip temperatures. This strategy of measuring current as a ratio of sensor outputs also eases the constraints on linearity of the sensor gain, enabling design with an improved input dynamic range and a higher sensitivity. In addition, increased gain of the sensor coupled with greater noise immunity results in a faster response time.

In addition to serving as a means for calibration, the replica branch can also provide an added functionality as a concurrent temperature sensor. When $I_{cal}$ is switched off from the replica branch, $I_{chg}$ depends on the threshold voltages of n and p channel transistors in the gain stages of the sensor [8]. This relationship is approximately linear. As a result, the output pulse rate at zero load, $1/tp_0$, increases linearly with temperature.

### B. Current reference

A compact, low-power, aging-tolerant current reference with low temperature co-efficient (TC) is therefore needed to make the sensor tolerant to any variations. Some current reference circuits have been reported earlier with good line-regulations and low power overheads [9]-[10]. However, with footprints in excess of $0.1mm^2$,

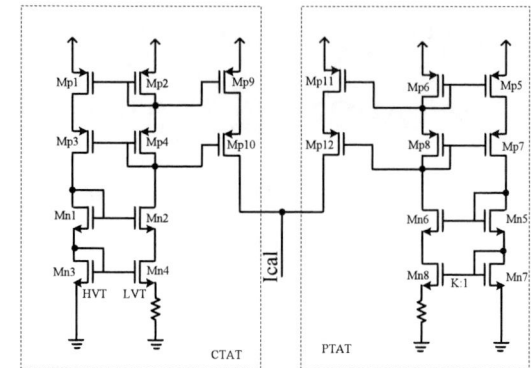

Fig. 2: Schematic of the calibration current source.

such designs increase system costs significantly, especially when multiple instances of such current sources are needed. On the other hand, reference currents local to a given sensor are preferred to reduce routing overheads. Fig. 2 shows the schematic of an on-chip CMOS calibration current source which overcomes these challenges. This current source consists of four branches biased in weak-inversion, two of which generate a Complementary to Absolute Temperature (CTAT) current, while the other two generate a Proportional to Absolute Temperature (PTAT) current. In both the pairs, the top three transistor pairs (Mp1-Mn2, Mp5- Mn6) are matched. In the section generating CTAT current, Mn3 is chosen to have higher than nominal threshold voltage and a low-threshold voltage is chosen for Mn4. In the section generating PTAT current, Mn7 and Mn8 are sized k:1 but have the same threshold voltage. Summation of the currents in these two sections is given as follows:

$$I_{cal} = \frac{V_{t3} - V_{t4}}{R_1} + \frac{n\phi_t}{R_2} \cdot ln(k) \quad (3)$$

where n is the subthreshold slope factor and $\phi_t$ is the thermal voltage providing the PTAT component in the reference current. Therefore, a current source with a low Temperature Coefficient (TC) can be obtained by suitable scaling of k, $R_1$ and $R_2$. This circuit is also inherently resilient to aging defects as the aging defects have been shown to be proportional to applied electric fields [11]. Although tolerant to variations in ambient conditions, its value may be susceptible to process variations, hence, one measurement at any temperature is necessary.

### III. Calibration

From (2), a variation tolerant estimate of $I_{load}$ can be obtained if we can calibrate the scaling ratio, N and know the value of the calibration current, $I_{cal}$. To calibrate

N, a copy of the reference current $\alpha I_{cal}$, which can be measured externally is created for a one-time, one-point calibration. As this copy is used as a load to the main branch, it is scaled up by a factor $\alpha$ to ensure that the IR-drop due to this copy is large enough to reduce measurement and systematic (linearity) errors. As this current source will be used only once, it does not contribute to the power overhead of the sensor. With the $\alpha I_{cal}$ as a load to the primary branch and $I_{cal}$ as the load to the replica branch of the sensor, output pulse rates ($1/tp_{m,1}$, $1/tp_{r,1}$) are measured. In order to correct for zero error, output time-periods in the two branches at zero load current ($tp_{0,m}$, $tp_{0,r}$) are also measured, with their ratio expressed as $\eta$. Scaling ratio N is given by

$$N = \left(\frac{\alpha I_{cal}}{I_{cal}}\right) \cdot \left(\frac{1}{tp_{r,1}} - \frac{1}{tp_0}\right) / \left(\frac{1}{tp_{m,1}} - \frac{1}{\eta \cdot tp_0}\right) \quad (4)$$

Thereafter, the value of $I_{load}$ at any given condition is estimated as follows:

$$I_{load} = (\alpha I_{cal}) \cdot \frac{\frac{1}{tp_{r,1}} - \frac{1}{tp_0}}{\frac{1}{tp_{m,1}} - \frac{1}{\eta \cdot tp_0}} \cdot \frac{\frac{1}{tp_m} - \frac{1}{\eta \cdot tp_0}}{\frac{1}{tp_r} - \frac{1}{tp_0}} \quad (5)$$

Thus, a one-time, one-point calibration is sufficient to achieve variation resilient current estimates. If the added functionality of the sensor to measure temperature is to be utilized, a simple, two-point calibration at two different temperatures is needed to evaluate the equation of the line relating $1/tp_0$ to chip temperature 'T'.

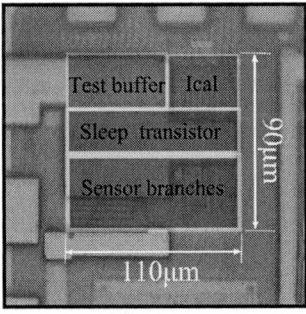

Fig. 3: Die Microphotograph.

## IV. MEASUREMENT RESULTS

This sensor occupies an active on-chip area of $110\mu$m $\times$ $90\mu$m as shown in Fig. 3 and consumes $180\mu$A at 1V supply. Output pulse-rate was measured for loads from 0 to 25mA at various temperatures ranging from -23$^o$C to 100$^o$C. To ensure fast response times, the output pulse-rate was designed to be higher than 50MHz. Averaging the outputs over a time window (80ns) and estimating current as a ratio of the two branches further improves

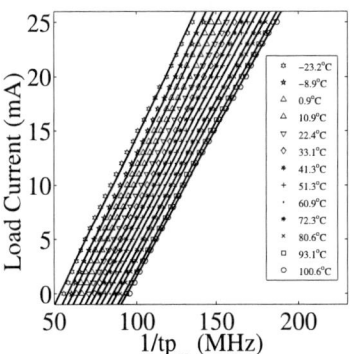

Fig. 4: Measured output of power sensor for loads up to 25mA at temperatures from -23$^o$C to 100$^o$C.

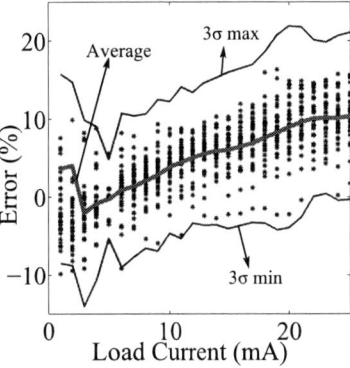

Fig. 5: Distribution of estimation errors after on-line calibration.

estimation accuracy. Fig. 4 shows that the output pulse-rate varies linearly with current for loads up to 25mA, which is equivalent to an IR-drop in excess of 150mV. At all given temperatures, the sensitivity was higher than 2.8GHz/A. Fig. 5 shows that the average error in current estimation across 3 samples and 13 different temperatures was less than $\pm 8.25\%$ with a 3-$\sigma$ error $\leq$ $\pm 15\%$. Its input dynamic range was in excess of 20mA across 39 points of measurement, which is equivalent to $0.1V_{dd}$, normalizing for sleep-transistor size and supply voltage. As sleep transistors are scaled with $I_{load}$ to meet IR-drop specifications, this dynamic range ensures that the sensor design can be replicated for any other block on a given chip without significant redesign.

The circuit for generating $I_{cal}$ occupies an on-chip area of $30\mu$m $\times$ $50\mu$m and when measured through a test pad, $I_{cal}$ showed a PSRR better than -25.8dB. Fig. 6 shows that the temperature coefficient of the current reference across three samples is less than 91ppm/$^o$C. Fig. 7 shows that the sensor can provide an additional func-

978-1-4673-6145-3/13 $31.00 © 2013 IEEE

TABLE I: Performance summary and comparison with state-of-the-art

| Metric | [1] | [2] | [3] | [4] | [5] | [8] | [7] | This work |
|---|---|---|---|---|---|---|---|---|
| Technology | 32nm | 0.16$\mu$m | 0.18$\mu$m | 0.13$\mu$m | 32nm | 45nm | 0.13$\mu$m | 0.13$\mu$m |
| Power overhead | 3.78mW | 5.1$\mu$W | 30$\mu$W | 1.2mW | 1.6mW | 120$\mu$W | 82$\mu$W | 180 $\mu$W |
| Area overhead | 0.02mm$^2$ | 0.08mm$^2$ | 0.18mm$^2$ | 0.16mm$^2$ | 0.02mm$^2$ | 0.02mm$^2$ | 1.1mm$^2$ | 0.01mm$^2$ |
| Conversion time | 10-100$\mu$s | 5.3ms | 12.5$\mu$s | 0.2ms | 1ms | 0.5$\mu$s | 5ms | 0.08$\mu$s |
| Temperature inaccuracy (3-$\sigma$) | 4.5$^o$C | $\pm$0.15$^o$C | $\pm$0.5$^o$C | $\pm$2.3$^o$C | 5$^o$C | $\pm$4.05$^o$C | - | $\pm$3$^o$C |
| Current inaccuracy | - | - | - | - | - | $\pm$10% | $\pm$0.03% | $\pm$8.25% |
| Energy per Conversion | 37.8nJ | 26.5nJ | 0.375nJ | 240nJ | 1.6$\mu$J | 0.06nJ | 410nJ | 0.014nJ |
| Input Dynamic Range | -10-110$^o$C | -55-125$^o$C | 0-100$^o$C | 0-100$^o$C | -10-110$^o$C | 0-0.01V$_{dd}$ (22-85$^o$C) | 0-1A (-40-85$^o$C) | 0-0.1V$_{dd}$ (-20-100$^o$C) |

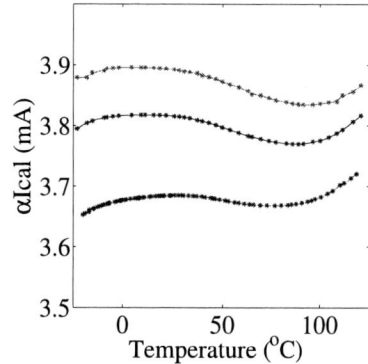

Fig. 6: Measured reference current (-20$^o$C to 120$^o$C, 3 samples).

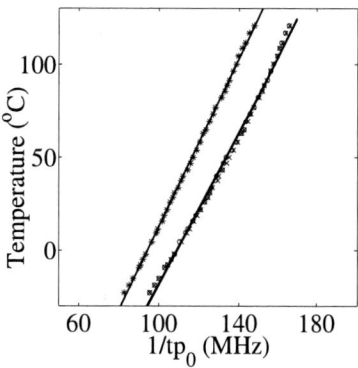

Fig. 7: Linear variation of 1/tp$_0$ with tempearture (-20$^o$C to 120$^o$C, 3 samples).

tionality as a temperature sensor, as 1/tp$_0$ can be used to estimate temperature with an error within $\pm$0.7$^o$C and a 3-$\sigma$ error $\leq$ $\pm$3$^o$C from $-20^o$C to 120$^o$C which is sufficient for thermal management in microprocessors [1]. Table. I summarizes and compares the test results with the state of the art sensors for power management.

## V. CONCLUSION

An on-chip power sensor with an 80ns response time has been presented. A compact, low TC reference current source for online calibration of the sensor has also been presented which improves the resilience to PVT variations and aging defects significantly, and leads to 10$\times$ improvement in input dynamic range. With low overheads, wide dynamic range, and a fast response time, this sensor paves the way for a smarter, fine-grained power management in microprocessors.

## REFERENCES

[1] J. Shor, K. Luria and D. Zilberman, "Ratioametric BJT-based thermal sensor in 32nm and 22nm technologies," *ISSCC Dig. Tech. Papers*, pp. 210-212, Feb. 2012.

[2] K. Souri, Y. Chae and K. Makinwa, "A CMOS temperature sensor with a voltage-calibrated inaccuracy of $\pm$0.15$^o$C (3$\sigma$) from -55 to 125$^o$C," *ISSCC Dig. Tech. Papers*, pp. 208-210, Feb. 2012.

[3] C. Wu, W. Chan, T. Lin, "A 80kS/s 36$\mu$W resistor-based temperature sensor using BGR-free SAR ADC with a unevenly-weighted resistor string in 0.18$\mu$m CMOS," *Symp. On VLSI Circuits*, pp.222-223, June 2011.

[4] K. Woo et al., "Dual DLL-based cmos all-digital temperature sensor for microprocessor thermal monitoring," *ISSCC Dig. Tech. Papers*, pp. 68-69, 69a, Feb. 2009.

[5] Y. W. Li et al., "A 1.05V 1.6mW 0.45$^o$C 3$\sigma$-resolution $\Sigma\Delta$-based temperature sensor with parasitic-resistance compensation in 32nm cmos," *ISSCC Dig. Tech. Papers*, pp. 340-341, 341a, Feb. 2009.

[6] H. Wang et al., "Runtime power estimator calibration for high-performance microprocessors," *Proc. Design Automation and Test in Europe*, pp. 352-357, Mar. 2012.

[7] S. H. Shalmany, D. Draxelmayr, K. A. Makinwa, "A micropower battery current sensor with $\pm$0.03% (3$\sigma$) inaccuracy from -40 to +85$^o$C," *ISSCC Dig. Tech.. Papers*, pp. 386-387, 387a, Feb. 2013.

[8] S. Bhagavatula and B. Jung, "A low power real-time on-chip power sensor in 45nm SOI," *IEEE Trans. on Circuits and Systems I*, vol. 59, pp. 1577-1588, July 2012.

[9] G. Serrano and P. Hasler, "A precision low-tc wide-range cmos current reference," *IEEE J. Solid-State Circuits*, vol. 43(2), pp. 558-565, Feb. 2008.

[10] B. D. Yang et al., "An accurate current reference using temperature and process compensation current mirror," *Proc. IEEE A-SSCC*, pp.241-244, Nov. 2009.

[11] R. Thewes et al., "Device reliability in analog CMOS applications," *IEDM Dig. Tech.*, pp.81-84, Dec. 1999.

# Discretization and Discrimination Methods for Design, Verification, and Testing of Analog/Mixed-Signal Circuits

Jaeha Kim, Jiho Lee, Do-Gyoon Song, Taehwan Kim, Kyung-Hoon Kim, Seobin Jung, Sangho Youn

School of Electrical and Computer Engineering, Inter-university Semiconductor Research Center
Seoul National University, Seoul, Korea

*Abstract-* **This paper describes how the difficult problems of designing, verifying, and testing analog circuits in presence of variability can be converted to easier ones by discretizing the search spaces or discriminating one case from another. For instance, discretizing the continuous design space of analog circuits enables the use of an efficient, predictive global circuit optimizer. Also, discretizing the initial condition space of a circuit enables one to establish its global convergence property over the entire space by exploring only a small number of samples. Lastly, discriminating the test responses of the circuit with and without a fault in consideration of the underlying statistical distribution provides a formal guide on how to quantify the fault coverage of analog/mixed-signal circuit tests. It is noteworthy that it is variability that introduces the cross-correlations in the performance metrics, convergence behaviors, and test responses between two nearby candidates in consideration and therefore enables the use of discretization and discrimination methods listed in this paper. The proposed methods are demonstrated on the practical examples of sizing an operational amplifier, verifying the correct start-up of a coupled ring oscillator, and composing a test suite for screening faults in a digitally-controlled phase interpolator circuit.**

## I. INTRODUCTION

Many problems related to analog circuits look difficult or even mysterious when they address continuous quantities or spaces. For instance, designing an analog circuit involves determining the optimal transistor sizes, which can take continuous values. Verifying whether an oscillator always starts a proper oscillation involves trying out all of its possible initial conditions. Determining the acceptable range of a circuit's test response for screening bad parts during production seems like a decision relying on heuristics only. This paper aims to show that these problems can be converted to those dealing with discrete spaces with a finite number of samples, especially in the presence of variability in the circuits. As a result, these seemingly difficult problems in designing, verifying, and testing analog circuits can be solved more efficiently with the simpler algorithms.

It is noteworthy that variability is the key to converting the continuous search space into a discrete one. The variability underlying in modern integrated circuits may include global variations in process, voltage, and temperature (PVT) conditions as well as local device mismatches in their threshold voltage, oxide thickness, doping densities, etc. While taking variability into account typically incurs substantial computational costs in many analog tools, it is this variability that helps us decide how far the discretized points

can be placed in a continuous search space. Variability introduces cross-correlation among the neighboring candidate points, implying that by exploring one discrete point in the search space, it is possible to infer some knowledge on the other points in the neighborhood and therefore to perform a search while considering only a finite number of samples on a discretized grid. It is interesting that the more variability the circuit has, the coarser the discretized grid can be, making the search process easier with the less candidate points to explore.

This paper will demonstrate how these discretization and discrimination methods leveraging variability can help ease many problems in analog circuits. For instance, discretizing the continuous design space of analog circuits enables the use of an efficient, predictive global search algorithm. Also, discretizing the initial condition space enables one to establish the global convergence property over the entire space by exploring only a small number of discrete points, again with the aid of the predictive global search algorithm. Finally, discriminating the test responses of the parts with and without a fault leads to a way of quantifying the fault coverage of analog/mixed-signal circuit tests.

The rest of the paper is organized as follows. Section II lays the foundation for discretizing the continuous search spaces when variability introduces cross-correlations among the

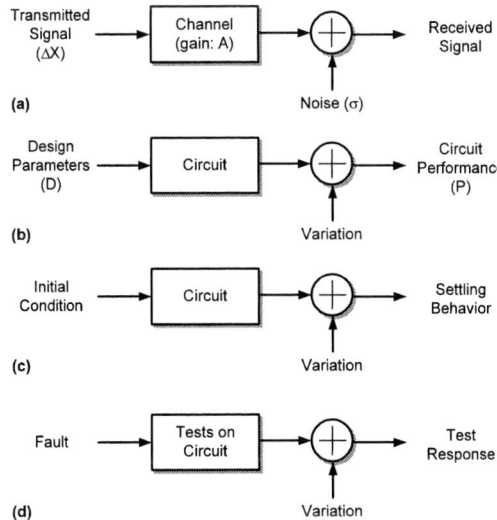

Fig. 1. The additive noise channel model analogies for the design, verification, and testing problems in analog circuits.

978-1-4673-6145-3/13 $31.00 © 2013 IEEE

neighboring candidate points. Section III, IV, and V then show how these discretization and discrimination methods can be applied to the design, verification, and testing of analog circuits, respectively. Finally, Section VI concludes the paper.

## II. DISCRETIZATION OF CONTINUOUS SPACES LEVERAGING VARIABILITY-INDUCED CORRELATIONS

The aforementioned problems in analog can be cast into the problems of telling a difference between two designs, initial conditions, or test responses. For instance, in case of finding the optimal design ($D$) that maximizes the circuit's performance ($P$), it is important to note that a designer does not have a full control of $P$ with $D$ in presence of variability, as illustrated in Fig. 1(b). That is, the same design may possess different performances from sample to sample. It means that unless one design is sufficient different from another, one cannot distinguish the two based their performance metrics. An analogy can be found with a noisy communication channel [2],[4] shown in Fig. 1(a). Due to the presence of noise added to the output of the channel, there is a limit on the amount of information that can be conveyed with the input signal with a swing $\Delta X$. Assuming that the added noise is white Gaussian with the variance of $\sigma^2$ and the channel has a gain of $A$, the peak-to-peak input signal swing $\Delta X$ must be larger than $\sigma\sqrt{12}/A$ in order to transmit one-bit of information across this channel, according to the channel capacity theorem. One bit of information corresponds to telling a difference between two possible input symbols.

The similar analogies can be made for the problems of verifying a circuit's convergence property and quantifying the fault coverage of a given circuit test, as depicted in Fig. 1(c) and (d). Both the circuit's settling behavior due to different initial conditions and circuit's test response to a fault are subject to global and local variations. It implies that unless two initial conditions are sufficiently different from each other, the trial on the second condition does not provide much information regarding the convergence behavior of the circuit

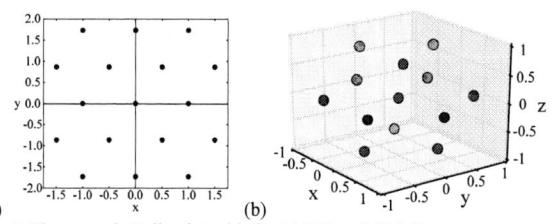

(a)                                           (b)
**Fig. 2.** The example Polka-dot grids for (a) 2-D and (b) 3-D spaces.

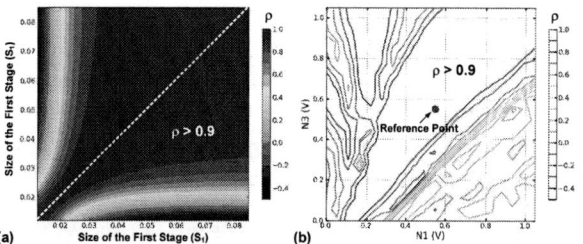

**Fig. 3.** Cross-correlations introduced by variability: (a) among the performance metrics of different circuit designs and (b) among the settling times for different initial conditions.

beyond the first one. It also implies that in order for a test to cover a fault and the amount of change in the test response must be sufficiently large compared to the underlying uncertainty due to variability.

This observation justifies the use of discretization or discrimination methods for addressing the design, verification, and test problems in analog circuits. For the design and verification problems, it means that a set of discrete design or state points can capture the true response of the circuit without loss of information. For the testing problem, it means that the methods to discriminate discrete symbols in communication can be used to quantify the fault coverage of a given test suite. In all cases, the necessary difference between the designs, initial conditions, and test responses is governed by variability.

To efficiently fill a continuous space with a set of discrete points, we have previously introduced the isotropic, Polka-dot grid [4]. Examples of the Polka-dot grids for 2-D and 3-D cases are illustrated in Fig. 2. The points on a Polka-dot grid can be derived from the center-points of constant-radius hyper-spheres that are maximally packed in the space. The grid can also be regarded as an extension of the hexagonal close packed lattice found in chemistry.

The Polka-dot grid has a number of favorable properties compared to a classical Cartesian grid. First, each point on the grid has uniform distances to its nearest neighbors. Second, the angle formed with two adjacent vectors originating from a given point is always equal to 60° regardless of the dimension. Third, the number of nearest neighbors scales mildly with the dimension $N$, as $N(N+1)$.

It can be shown that the basis vectors $\{\vec{e_i}\}$'s constituting the described Polka-dot grid must satisfy the following property:

$$\vec{e_i} \cdot \vec{e_j} = \begin{cases} \|\vec{e_i}\|^2 = 1 & \text{if } i = j \\ 1/2 & \text{if } i \neq j \end{cases} \quad (1)$$

Then, it follows that for an $N$-dimensional space with its Cartesian basis vectors denoted as $\vec{c_1}, \vec{c_2}, ..., \vec{c_N}$, the $n$-th Polka-dot basis vector $\vec{e_n}$ can be found recursively from $\vec{e_1}, \vec{e_2}, ..., \vec{e_{n-1}}$ using the following rule:

$$\vec{e_n} = \frac{1}{n}\sum_{i=1}^{n-1}\vec{e_i} + \sqrt{\frac{n+1}{2n}}\vec{c_n}. \quad (2)$$

On the other hand, the proper spacing of the Polka-dot grid can be determined based on the cross-correlation among the neighboring points. Note that this cross-correlation is introduced by variability; that is, the performance metrics, settling behaviors, and test responses of two nearby candidates on the grid share the similar tendencies with the common fluctuations in the variability condition. In simulation, this cross-correlation can be characterized by using common random Monte-Carlo samples for the candidate points. Fig. 3 illustrates the cross-correlations found in the design and verification problems, which are discussed in further detail in Section III and IV, respectively.

With proper spacing, the neighboring points on the Polka-dot grid become sufficiently correlated to provide accurate predictions on the points in the neighborhood. A conservative way to quantify the necessary cross-correlation is to impose an

978-1-4673-6145-3/13 $31.00 © 2013 IEEE

upper-bound on the model prediction error. Using the correlation-coefficient model of $\rho(d) = (1/\rho_1)^{-d^2/d_1^2}$ being a function of the distance $d$, where $\rho_1$ and $d_1$ are the correlation coefficient and distance between two nearest neighbors, respectively, the mean squared error (MSE) of the best linear unbiased predictor can be found [5],[7]:

$$MSE = \sigma^2 \left[1 - r^T R^{-1} r\right] \qquad (3)$$

where $r$ is a row vector of the correlation coefficients between the test point and its neighbors and $R$ is a matrix of the correlation coefficients among the neighboring points. $\sigma^2$ is the variance of the point's response distribution.

Fig. 4 plots the correlation coefficients ($\rho$'s) required between the nearest neighbors on the Polka-dot grid in order to achieve a model prediction error less than 10, 20, and 30% of $\sigma$, respectively. Note that even with 30% of error, the overall standard deviation of the estimated distribution increases only by 4.4% due to the root-mean-square way of summing the independent contributions. The plot illustrates that $\rho$ of 0.9 is sufficient in most cases and the requirement can be further relaxed for high-dimensional spaces since the number of nearest neighbors increases with the dimension.

### III. DISCRETIZATION FOR OPTIMAL CIRCUIT SIZING

The first case to demonstrate the proposed discretization method with is the problem of optimally sizing analog circuits. Analog circuit optimizers, that can automatically size the transistors in a circuit according to a prescribed cost function, can be an effective productivity tool for analog, yet the reality is that their adoption by circuit designers has been rather slow. We find the main cause in the fact that most traditional analog circuit optimizers look for the optimal solution in a continuous, high-dimensional design space [3]. For instance, the algorithms such as simulated annealing, convex optimization, and genetic/evolutionary algorithms tend to repeatedly evaluate the similar design points without knowing whether a similar design in the vicinity has been explored in the prior iterations. This is particularly the case when refining a solution to a certain precision or when finding a global optimum in presence of surface roughness and/or local optima. Furthermore, the ever-increasing complexities and variabilities in the deeply-scaled devices aggravate the situation with these

optimizers.

One can realize a more efficient circuit optimizer by adopting a model-predictive global search algorithm in [5] and using to find the optimum from a finite set of design candidates on a discrete grid. The outline of the proposed algorithm is depicted in Fig. 5. First, the continuous design space of the circuit is discretized into a set of design candidates, where the spacing is properly chosen to capture the response surface without loss of information. Second, the algorithm starts by evaluating a small number of randomly chosen pilot samples and characterizing the statistical distributions of their performance metrics. Third, based on the statistics gathered from the pilot samples, a predictive response surface model is constructed. Such a model can be built using radial-basis function interpolation [6] or stochastic kriging [7]. These models can provide the predictions on the statistical distribution of the unexplored design candidates. Fourth, using these predictions, the algorithm evaluates the probability of discovering a better solution than the currently known best. Then it selects the most promising design candidate for evaluation and updates the response surface model based on the outcome. This process repeats until the probability drops below a certain threshold (e.g. 0.1%).

Combining the predictive global search algorithm with the use of discrete grids solves many problems of using each approach alone. For instance, the use of discrete grids guarantees that the optimizer will not waste efforts in overly

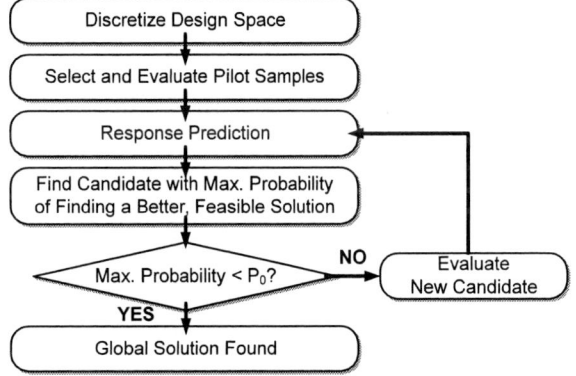

**Fig. 5.** A yield-aware, predictive global search algorithm that finds the optimal design from a discretized design space.

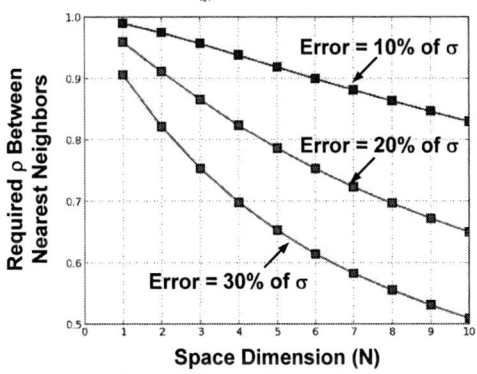

**Fig. 4.** The required correlation coefficients ($\rho$'s) among the nearest neighbors on the Polka-dot grid in order to achieve the model prediction error less than 10, 20, and 30% of the distribution's standard deviation ($\sigma$).

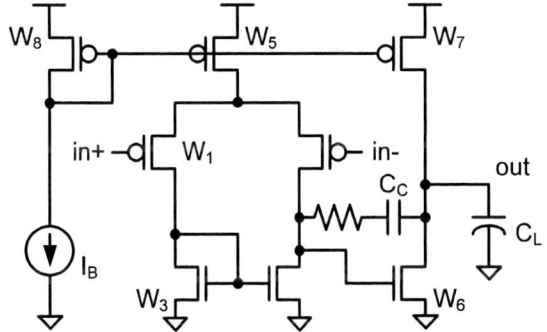

**Fig. 6.** A two-stage operational amplifier with zero compensation.

978-1-4673-6145-3/13 $31.00 © 2013 IEEE          123

exploring one particular area of the design space. Also, the predictive global optimizer can mitigate the scalability issue commonly encountered in optimizers searching a discretized space [4]. Yet another advantage is that the algorithm finds the optimal solution fairly early in the process, rather than reaching it at the very last step, enabling quick interactions between the optimizer and designer.

The described circuit optimization algorithm is demonstrated on a two-stage operational amplifier with zero compensation shown in Fig. 6. The objective is to maximize the open-loop bandwidth of the amplifier while satisfying the constraints on the open-loop gain, phase margin, power-supply rejection ratio, common-mode rejection ratio, input-referred noise, and power dissipation. Considering of the symmetry and matching requirements, the final independent variables chosen are the widths of the transistors $M_1$, $M_3$, $M_5$, $M_6$, and $M_8$, as well as the bias current $I_B$ and the compensation capacitor $C_c$. With these 7 design variables, the number of candidate points in the discretized design space is more than 800,000.

The process of this optimizer finding the solution for this operational amplifier is illustrated in Fig. 7. Note that the optimizer terminates the search after 250 iterations but the optimum is found at the 140-th iteration. The subsequent evaluations are to confirm that this solution was indeed the global optimum. The probability of improving the current solution decreases with iterations in general but may show some fluctuations, especially when the predicted responses do not agree with the simulated outcomes. As the optimizer learns more about the circuit's response surface, the probability of improvement converges towards zero. The search is initiated with 20 pilot samples and each design point is evaluated with 50 Monte-Carlo samples. It has been confirmed that the choice of pilot samples does not alter the

final solution as long as they span all the dimensions of the design space.

## IV. Discretization for Verifying Global Convergence

Global convergence failures refer to functional failures in which a system, although operating well in most circumstances, converges to undesired behavior when starting from certain initial conditions [8]. The start-up failures encountered in ring oscillators is a typical example of global convergence failures. For instance, a three-stage coupled ring oscillator shown in Fig. 8 may exhibit differential-mode oscillation, common-mode oscillation, or DC equilibrium depending on its initial conditions. Other examples of global convergence failures can be found in phase-locked loops (PLLs) and DC-DC converters. These failures are largely due to the complex, nonlinear interactions among the circuit components, resulting in multiple equilibrium states to which the system can converge.

Verifying a circuit's correct convergence to the desired operating mode requires examining the final states that are reachable by all possible initial conditions of the circuit for all possible variability conditions. It is a very challenging task. The prior work in literature [8]-[10] have focused mainly on establishing the circuit's convergence over the initial condition space, each taking an approach in simulation, stability analysis, and formal verification, respectively. However, the main difficulty lies in the fact that all parts of the continuous space of initial conditions must be explored and examined. Considering variability in addition to the initial conditions only seems to aggravate the problem.

However, it is noteworthy that global convergence can be

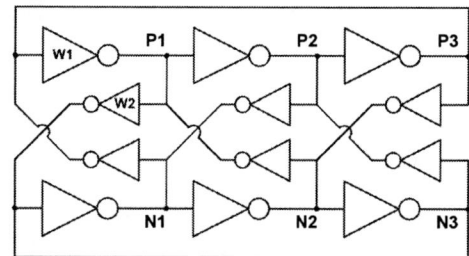

**Fig. 8.** A 3-stage coupled ring oscillator that can exhibit false oscillation behaviors depending on its initial state and variability condition.

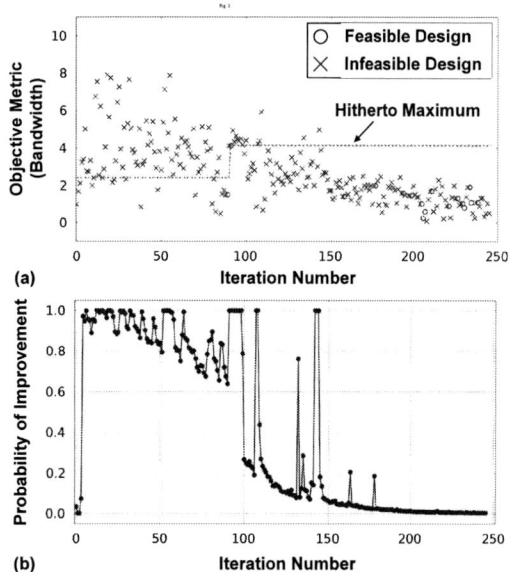

**Fig. 7.** The search process of the predictive global circuit optimizer illustrated for the two-stage operational amplifier example: (a) progression in the objective metric (i.e. bandwidth) and (b) progression in the probability of improving the current solution.

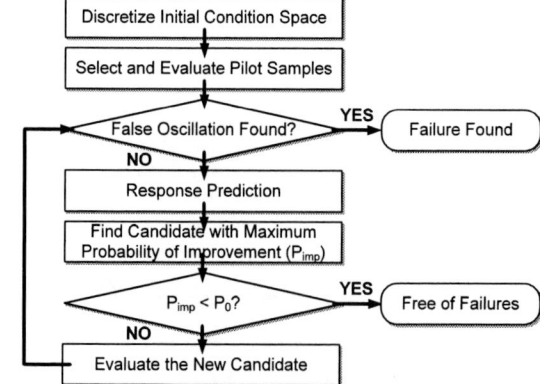

**Fig. 9.** A two-stage operational amplifier with zero compensation.

978-1-4673-6145-3/13 $31.00 © 2013 IEEE          124

more easily verified when variability is taken into consideration. It is again due to the correlations introduced by variability. For two closely-spaced initial conditions, the circuit is expected to have the similar tendency for change in the circuit characteristic $P$ with respect to variability. It implies that the continuous space of the circuit's initial condition can be effectively covered by a finite number of discrete samples. Based on this observation, this paper briefly outlines a simulation-based approach presented in [12].

Recall that Fig. 3(b) illustrated the cross-correlation found among the settling times of the 3-stage coupled ring oscillator in Fig. 8 starting from different initial conditions. To aid visualization, only the initial values on the nodes P1 and N1 are varied while those of the other nodes P2, P3, N2, and N3 were fixed. The plot shows that the correlation coefficient higher than 0.9 is observed when the initial condition is within the distance of 0.2 from a reference point of P1=0.55 and N1=0.55. Therefore, according to our discussion in Section II, we can discretize the initial condition space in steps of 0.2-V.

The proposed variability-aware, global convergence analysis algorithm uses a predictive global optimizer similar to the one described in the previous section. Based on an observation that the circuit's settling time increases with its initial condition's distance to the equilibrium, the algorithm searches for an initial condition that has the longest settling time. Such an effort can lead to the discovery of a problematic initial state that converges to a wrong equilibrium. The search is conducted again over a discretized state space leveraging the cross-correlations introduced by variability.

Fig. 9 outlines the proposed algorithm for variability-aware, global convergence analysis of oscillators. The algorithm looks for a problematic initial state among a finite set of candidates, using the settling time as a guidance metric. As explained earlier, a predictive global optimizer searches for an initial state candidate that globally maximizes the settling time, by estimating the probability of the unevaluated candidates yielding the longer settling times than the currently known maximum and evaluating new candidates until this probability drops below a certain threshold. Concurrently with this search process, our global-convergence verifier checks whether the initial state under evaluation exhibits a false start-up behavior, by means of Monte-Carlo simulation. If the initial state exhibits a false start-up behavior, the algorithm reports a failure. If the optimizer reaches to a global maximum without discovering a false start-up behavior, the algorithm concludes that the circuit is free of start-up failures.

The algorithm is demonstrated on a two-stage coupled ring

oscillator in Fig. 10 [12]. It is known that this oscillator can exhibit false oscillation behaviors if the main-to-auxiliary buffer ratio lies between 0.4 and 2.0. However, due to variability, the oscillator is not guaranteed free of failures even when it is sized with the ratio outside this range [11]. Assuming a 45-nm CMOS technology, the oscillator's 4-dimensional initial condition space is discretized with 0.2-V spacing and each selected initial condition is evaluated with a 1,000-sample Monte-Carlo simulation. For different size configurations, the algorithm is able to report a possible failure (e.g. a DC equilibrium) or conclude that the circuit is free of failures by evaluating only at most 68 candidates out of 356 total.

Fig. 11 and Fig. 12 illustrate the search process of the proposed variability-aware global convergence analysis algorithm. To aid visualization, the search space was reduced to a two-dimensional space spanning the initial conditions of A2 and B1 nodes only, while the other nodes A1 and B2 are assumed fixed at 0.9 and 0.2V, respectively. Fig. 11 is the true distribution of the settling time across this reduced state space. The oscillator starts normally when the initial condition is (A2, B1) = (0, 1.0V), which corresponds to the point where the settling time is the minimum. On other hand, the oscillator fails to oscillate when it starts from (A2, B1) = (1.0V, 0), which corresponds to the area where the settling time keeps increasing without a bound.

Fig. 12 shows the progression of the estimates on the settling times as the number of explored initial condition samples increases. Fig. 12(a) shows the initially estimated response surface made with only three pilot samples. Among the three samples, the one marked as a green dot has the longest settling time and is the hitherto solution. Based on model-based predictions, the algorithm next suggests that the point marked

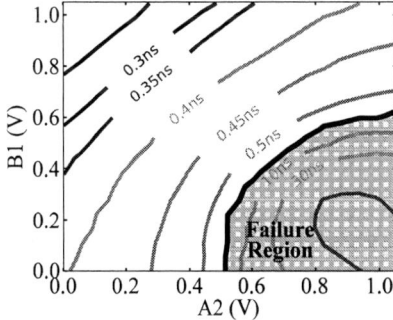

**Fig. 11.** The true response surface of the settling time as a function of the initial values of the nodes A2 and B1.

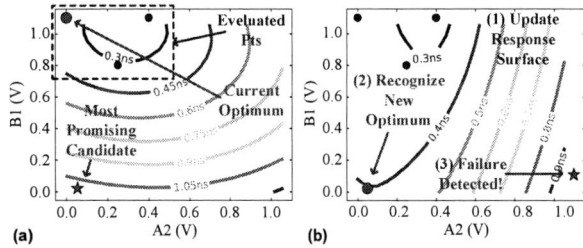

**Fig. 12.** Progression of the settling time response surface as the more candidate points are evaluated.

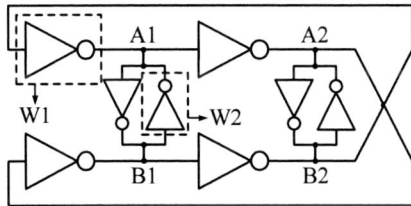

**Fig. 10.** A two-stage coupled ring oscillator whose global convergence properties were analyzed in [9],[11].

978-1-4673-6145-3/13 $31.00 © 2013 IEEE          125

by a red star is the most likely initial condition to yield the longer settling time, or possibly a start-up failure. Therefore, the algorithm evaluates this initial condition with 1,000 Monte-Carlo samples. However, this condition is not problematic and the search process continues with the updated response surface, as shown in Fig. 12(b). This time, the algorithm suggests the red star in the lower-right corner as a new initial condition to explore and it reveals a start-up failure.

## V. DISCRIMINATION FOR QUANTIFYING ANALOG FAULT COVERAGE

Variability can ease another difficult problem in analog: quantifying a test's fault coverage. As IC parts that fail due to defects in analog/mixed-signal circuits increase and the resulted penalties and costs rise sharply, there is a strong demand for ways to compose an efficient analog/mixed-signal test suite that achieves high fault coverage. A basic pre-requisite to any test generation method is a way to quantify the fault coverage of a given test. The heart of this problem is to tell whether the test makes enough difference in its response when the chip has a particular defect or fault. However, this very notion of 'enough difference' has not been well defined for analog tests due to their continuous nature.

The urgency is with the sharply rising number of mixed-signal ICs escaping the production screening, even with catastrophic defects. It is largely due to the fast-spreading use of digital calibration loops in analog systems [1]. For instance, a state-of-the-art high-speed I/O interface may use various digital calibration loops for timing generation/recovery, impedance control, equalizer adaptation, transmitter swing adjustment, receiver bandwidth control, etc. These digital calibration loops offer ease of programmability and process migration, but can significantly reduce the visibility of the internal defects. To address this, some have resorted to exhaustive tests, e.g. trying all possible input codes for testing a digital-to-analog converter (DAC), but found that they incur excessive time and cost.

This section will highlight the results from [14], which describes a way of quantitatively measuring the fault coverage of an analog test suite by leveraging variability. Simply put, we say that a test can tell the existence of a fault when it causes large enough shift in the test's response that can be distinguished from the statistical variations of the circuit due to variability. This formal definition of fault coverage can alleviate designers from having to define the tolerance range for each of the test responses, either based on design specifications or using heuristics.

Fig. 13 illustrates the proposed way of discriminating faulty parts from fault-free parts by their deviation in the statistical test responses. The example circuit is a 5-bit current digital-to-analog converter (DAC) shown in Fig. 13(a), which produces a digitally-programmable current using a set of five binary-weighted switched current sources. When a stuck-short fault is introduced to the transistor M2 (i.e. M2 stays on regardless of its gate voltage), the circuit may produce a different level of current than without the fault. For instance, as shown in Fig. 13(b), the test with the input code CD[4:0]=00001 has a large shift due to the fault while the one with 00011 has a small shift. So in this case, we say the former covers the fault while the latter doesn't. The required shift in the circuit's statistical distribution depends on the circuit's variability and target screening error probability. In case when the statistical distributions are estimated with finite-sample Monte-Carlo simulations, additional margin may be allocated to accommodate the uncertainty in the sample statistics [14].

Like other methods described in the paper, the correlations among the test responses can be utilized in multiple ways to improve the effectiveness of this approach. First, the cross-correlations among the multiple test responses in a given suite can be used to enhance the effective coverage of the test suite

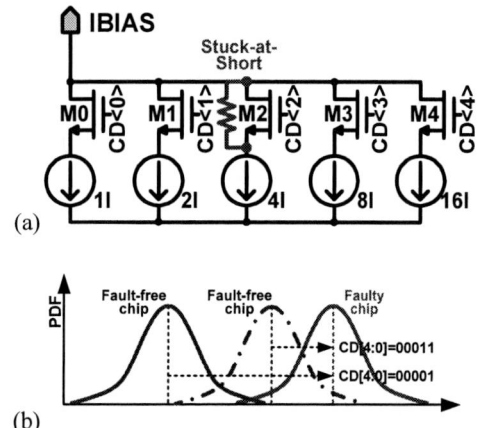

**Fig. 13.** Discriminating the fault-free and faulty parts of a 5-bit current digital-to-analog converter (DAC) circuit: (a) the schematic of the circuit with a stuck-short fault introduced at the transistor M2, and (b) the resulting change in the statistical distribution when different test codes are applied.

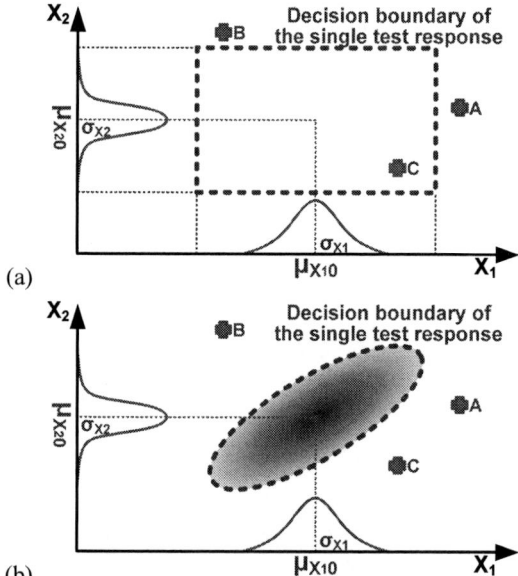

**Fig. 14.** Leveraging cross-correlation among the test responses when quantifying the fault coverage: (a) without considering the cross-correlation, the decision boundary is a rectangular region which cannot discriminate the fault C; (b) with the cross-correlation, the decision boundary is a tilted ellipse which can discriminate the fault C.

[13]. Second, the cross-correlations between the circuit responses with and without a fault can be exploited to estimate their statistical distributions with a small number of Monte-Carlo samples. A good example is the use of a control-variate method [15].

Fig. 14 illustrates how the cross-correlation among multiple test responses can improve the fault coverage. Suppose that two tests are conducted, whose responses are labeled $x_1$ and $x_2$, respectively. The statistical distributions shown along the axes visualize those of the fault-free parts with the means $\mu_{x1}$ and $\mu_{x2}$ and variances $\sigma_{x1}$ and $\sigma_{x2}$, respectively. When a fault is introduced, the circuit's test response may be shifted. For instance, the three points A, B, and C marked on the plot denote the shifts in the mean responses due to three different faults introduced, respectively. When the fault coverage is measured by checking whether any of the individual tests reveals the existence of the faults, the corresponding decision boundary is a rectangular box as depicted in Fig. 14(a). The two tests can cover the faults A and B, but not C since its response still lies inside the box. On the other hand, when the fault coverage is defined with the joint statistics of the two test responses, the decision boundary encloses a smaller region as shown in Fig. 14(b). Hence, the fault coverage of the test suite is improved, in this case covering the fault C as well. The general expression for the decision boundary including the proper margin is derived in [14].

On the other hand, a control variate method can be utilized to estimate the statistics of the fault-free and fault-induced parts with reduced numbers of Monte-Carlo samples. Despite the injected faults, the fault-free and fault-induced parts are likely to have strongly correlated responses with respect to common fluctuations in the variability conditions. Since a control-variate method [15] is a way to estimate the statistical parameter of one random variable $X$ with high confidence when it is strongly correlated with another random variable $Y$, one can use the method for finding the statistical distribution of the fault-induced parts once that of the fault-free, nominal part is obtained. For instance, when estimating the expectation $E[X]$, the method instead estimates the expectation of $Z$, where $Z$ is defined as $Z = X - \alpha(Y - E[Y])$. It can be shown that this linear estimator is unbiased, i.e., $E[Z]=E[X]$, and its variance $Var[Z]$ can be reduced by a factor $1-\rho^2$ compared to $Var[X]$ when $\alpha$ is equal to $Cov[X,Y]/Var[Y]$. Here, $\rho$ is the correlation coefficient between $X$ and $Y$. For instance, if $\rho$ =0.9, the achievable variable reduction is ~5.3×, which translates to a 2.3× reduction in the number of Monte-Carlo samples for the same level of accuracy and confidence. In addition to the expectation, the control-variate method can be also applied to estimating other statistics such as variance and covariance.

The proposed fault-coverage analysis algorithm is demonstrated using the catastrophic stuck-open and stuck-short fault models in [16], although the algorithm is not limited to them. These fault models consider a MOS transistor as an ideal switch and model its defect as the switch being permanently open or shorted, respectively. Being transistor-level fault models, they can be readily applied to analog circuits. For instance, for a digitally-controlled phase interpolator shown in Fig. 15, the circuit has 76 transistors in total and the number of single-fault cases is 152, enumerating a stuck-open and stuck-short faults for each transistor. In general, the number of fault cases will grow only linearly with the number of transistors in the circuit.

First, the described fault-coverage analysis is demonstrated on a purely analog circuit, like a two-stage operational amplifier in Fig. 6. It is to show that achieving high fault coverage for such linear circuits is easy, explaining why quantifying fault coverage for analog has been overlooked for a long time. The test suite consists of six common specification tests for an operational amplifier, including the ones that measure the power consumption, input-referred offset, input common-mode voltage range, DC gain, unity-gain frequency, and phase margin. With total of 8 individual transistors in the circuit, there are 16 stuck-short or stuck-open faults to be considered. Table I lists the analyzed fault coverages of these tests, showing that each test indeed has nearly 100% coverage over the 16 possible stuck-short or stuck-open faults. In fact, 100% coverage can be achieved only with the input common-mode voltage range, DC gain, unit gain frequency, and phase margin measurements.

The second example that demonstrates the need for the proposed fault-coverage analysis is a digitally-controlled phase interpolator (PI) circuit shown in Fig. 15. The circuit is widely used in many clock-and-data recovery (CDR) circuits and synthesizes an arbitrary clock phase according to a digital

| | Power | Input Refered Offset | Imput Common Mode Range | DC gain | Unit Gain Frequency | Phase Margin |
|---|---|---|---|---|---|---|
| M0O | 1 | 1 | 1 | 1 | 1 | 1 |
| M1O | 1 | 1 | 1 | 1 | 1 | 1 |
| M2O | 1 | 1 | 1 | 1 | 1 | 1 |
| M3O | 0 | 1 | 1 | 1 | 1 | 1 |
| M4O | 1 | 1 | 1 | 1 | 1 | 1 |
| M5O | 1 | 1 | 1 | 1 | 1 | 1 |
| M6O | 1 | 1 | 1 | 1 | 1 | 1 |
| M7O | 1 | 1 | 1 | 1 | 1 | 1 |
| M0S | 1 | 0 | 1 | 1 | 1 | 1 |
| M1S | 1 | 1 | 1 | 1 | 1 | 1 |
| M2S | 1 | 1 | 1 | 1 | 1 | 1 |
| M3S | 1 | 1 | 1 | 1 | 1 | 1 |
| M4S | 1 | 1 | 1 | 1 | 1 | 1 |
| M5S | 1 | 1 | 1 | 1 | 1 | 1 |
| M6S | 1 | 1 | 1 | 1 | 1 | 1 |
| M7S | 1 | 1 | 1 | 1 | 1 | 1 |

**Table I.** The fault coverage of the six specification tests on a two-stage operational amplifier.

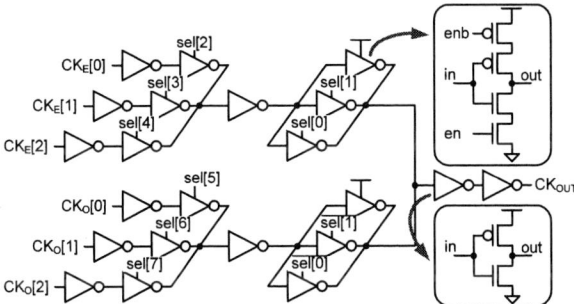

**Fig. 15.** A digitally-controlled phase interpolator circuit.

978-1-4673-6145-3/13 $31.00 © 2013 IEEE

input $D_{in}$. In this particular implementation, the first-stage multiplexers select two adjacent clock phases from 6 evenly-spaced ones and the second-stage interpolator generates a middle phase in-between controlled in 18 steps. This PI circuit was in fact reported as one of the possible causes of defect escapes in the modern DRAM's production screening since it is practically difficult to try all the input codes due to time and cost constraints.

The fault coverage report on the tests that measure the rising- and falling-transition delays of the PI circuit for various input code settings is listed in Table II. Unlike the case with the operational amplifier, each test covers only a small number of faults, due to the digital and nonlinear nature of the PI circuit. Without taking the cross-correlation among the test responses into account, it takes at least 8 test cases to achieve a fault coverage of 92.5%. Note that the faults in the transistors associated with low interpolation weights are particularly difficult to cover, since their differential responses are hard to distinguish from the underlying statistical variations of the PI delays. However, when the cross-correlations are included in the analysis, it is possible to cover 100% of these faults with only 5 test cases. It is noteworthy that the catastrophic faults in this 8-bit digitally-controlled circuit could be covered with only 5 tests, when the analysis is guided by the statistical information due to variability.

## VI. CONCLUSION

This paper described ways to leverage variability and the resulting correlations when addressing the problems of designing, verifying, and testing analog circuits. On the

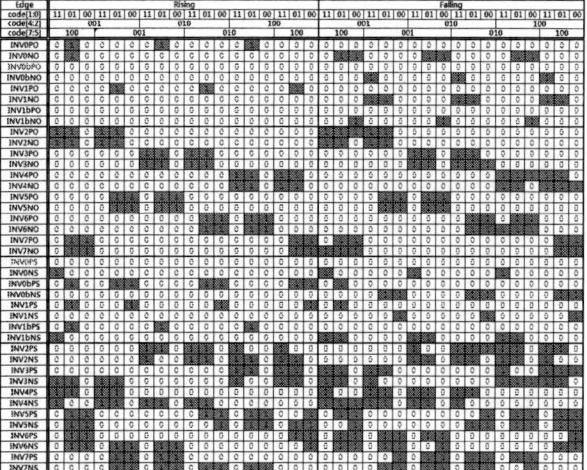

**Table II.** The fault coverage of the delay measurement tests on the digitally-controlled phase interpolator in Fig. 15.

practical examples of sizing an operational amplifier, verifying the global convergence of a coupled ring oscillator, and screening faults in a digitally-controlled phase interpolator circuit, it was demonstrated that discretization and discrimination methods can significantly simplify the problems, which are otherwise very difficult with the existing continuous methods. It is believed that this observation opens the opportunity for various discrete methods to help solve analog circuit problems.

### ACKNOWLEDGMENT

This work was supported by Semiconductor Research Corporation (SRC) under GRC Task 1836.068 and the authors would like to thank the industry liaisons at Texas Instruments, Inc. and Intel Corp.

### REFERENCES

[1] B. Murmann, "Digitally Assisted Analog Circuits," *IEEE Micro*, Mar. 2006, pp. 38–47.

[2] T. M. Cover and J. A. Thomas, *Elements of Information Theory*, 2nd Ed., New York: Wiley-Interscience, 2006.

[3] G. Gielen and R. Rutenbar, "Computer-Aided Design of Analog and Mixed-Signal Integrated Circuits," *Proceedings of IEEE*, pp. 1825-1852, Dec. 2000.

[4] S. Jung, et al., "Variability-Aware, Discrete Optimization for Analog Circuits," *in Proc. IEEE/ACM Design Automation Conference (DAC)*, pp. 536-541, Jun. 2012.

[5] D. R. Jones, et al., "Efficient Global Optimization of Expensive Black-Box Functions," *J. of Global Optimization*, pp. 455-492, Dec. 1998.

[6] H.-M. Gutmann, "A Radial Basis Function Method for Global Optimization," *J. of Global Optimization*, pp. 201-227, Mar. 2001.

[7] B. Ankenman, et al., "Stochastic Kriging for Simulation Metamodeling," *Operations Research*, pp. 371-382, Mar.-Apr. 2010.

[8] S. Youn, et al., "Global Convergence Analysis of Mixed-Signal Systems," *in Proc. IEEE/ACM Design Automation Conference (DAC)*, pp. 498-503, Jun. 2011.

[9] M. R. Greenstreet and S. Yang, "Verifying Start-up Conditions for a Ring Oscillator", *in Proc. Great Lakes Symp. on VLSI*, May 2008.

[10] S. K Tiwary, et al., "iSpice: A Boolean Satisfiability Based Approach to Formally Verifying Analog Circuits", *in Proc. of Formal Verification of Analog Circuits*, July 2008.

[11] S. Steinhorst and L. Hedrich, "Trajectory-Directed Discrete State Space Modeling for Formal Verification of Nonlinear Analog Circuits," *in Proc. IEEE/ACM Int'l. Conf. Computer-Aided Design (ICCAD)*, pp. 202-209, Nov. 2012.

[12] T. Kim, et al., "Verifying Start-Up Failures in Coupled Ring Oscillators in Presence of Variability Using Predictive Global Optimization," *submitted to IEEE/ACM Int'l Conf. on Computer-Aided Design (ICCAD)*, 2013.

[13] F. Liu and S. Ozev, "Statistical Test Development for Analog Circuits Under High Process Variations," *IEEE Trans. on Computer-Aided Design of Integrated Circuits and Systems*, Aug. 2007, pp. 1465–1477.

[14] K.-H. Kim and J. Kim, "Fault Coverage Analysis on Analog/Mixed-Signal Circuits Based on Statistical Dissimilarity," *submitted to IEEE Trans. on Very Large Scale Integration Systems*, 2013.

[15] D. Hocevar, et al., "A Study of Variance Reduction Techniques for Estimating Circuit Yields," *IEEE Trans. Computer-Aided Design of Integrated Circuits and Systems*, pp. 180-192, Jul. 1983.

[16] M. L. Bushnell and V. D. Agrawal, *Frontiers in Electronic Testing*, vol. 17, Boston: Kluwer Academic Publishers, 2002.

# Indirect Performance Sensing for On-Chip Analog Self-Healing via Bayesian Model Fusion

S. Sun[1], F. Wang[1], S. Yaldiz[1], X. Li[1], L. Pileggi[1], A. Natarajan[2,3], M. Ferriss[2], J. Plouchart[2], B. Sadhu[2], B. Parker[2], A. Valdes-Garcia[2], M. Sanduleanu[2], J. Tierno[2], D. Friedman[2]

[1]Electrical & Computer Engineering Department, Carnegie Mellon University, Pittsburgh, PA, USA, 15213
[2]IBM T. J. Watson Research Center, Yorktown Heights, NY, USA, 10598
[3]Electrical & Computer Engineering Department, Oregon State University, Corvallis, OR, USA, 97331

*Abstract*—On-chip analog self-healing requires low-cost sensors to accurately measure various performance metrics. In this paper, we propose a novel approach of indirect performance sensing based upon Bayesian model fusion (BMF) to facilitate inexpensive-yet-accurate on-chip performance measurement. A 25GHz differential Colpitts voltage-controlled oscillator (VCO) designed in a 32nm CMOS SOI process is used to validate the proposed indirect performance sensing and self-healing methodology. Our silicon measurement results demonstrate that the parametric yield of the VCO is improved from 0% to 69.17% for a wafer after the proposed self-healing is applied.

## I. INTRODUCTION

With the aggressive scaling of nanoscale integrated circuit (IC) technology, large-scale process variation is a critical issue for today's analog ICs [1]-[3]. As the traditional over-design technique becomes impractical, on-chip self-healing has emerged as a promising methodology to address the variability issue [4]-[5]. The key idea of self-healing is to actively monitor the post-manufacturing circuit performance metrics and then adaptively adjust a number of tuning knobs (e.g., bias current) in order to meet the given performance specifications.

To practically implement on-chip self-healing, a large number of analog performance metrics must be measured accurately and inexpensively by on-chip sensors. Such a measurement task, however, is not trivial, because many analog performance metrics (e.g., phase noise) cannot be easily measured by on-chip sensors. For this reason, the idea of indirect performance sensing has been recently proposed [6]-[7] where the *performance of interest* (PoI) is not directly measured by an on-chip sensor. Instead, it is accurately predicted from a set of other performance metrics, referred to as the *performances of measurement* (PoM), that are highly correlated with PoI and are easy to measure.

Towards this goal, indirect sensor modeling is a critical task where the objective is to build a mathematical model to capture the correlation between PoI and PoM so that PoI can be predicted from PoM. These models must be repeatedly re-calibrated to accommodate the process shift associated with manufacturing lines. While efficient modeling methods have been explored in the literature [6], the aforementioned model re-calibration issue has not been extensively studied yet. Hence, there is a need to develop a new methodology to facilitate efficient model re-calibration with low cost (i.e., requiring few additional measurement data). As such, the overhead of indirect performance sensing and, eventually, the overhead of analog self-healing can be minimized.

In this paper, a novel Bayesian model fusion (BMF) technique is proposed to address the model re-calibration problem and, hence, make analog self-healing of practical utility. The key idea of BMF is to combine the old (i.e., before process shift) indirect sensor model with very few new (i.e., after process shift) measurement data to generate a new model that is aligned with the new process conditions. Mathematically, the old model is encoded as prior knowledge, and a Bayesian inference is derived to optimally fit the new model by maximum-a-posteriori (MAP) estimation.

Furthermore, an analog self-healing flow is developed where the indirect sensor models are extracted by the proposed BMF technique. The proposed self-healing flow is validated for a 25GHz differential Colpitts VCO designed in a 32nm CMOS SOI process. Our silicon measurement results demonstrate that the parametric yield of the VCO is improved from 0% to 69.17% for a wafer after self-healing is applied.

The remainder of this paper is organized as follows. In Section II, we review indirect performance sensing and then derive the BMF method in Section III. The proposed BMF-based self-healing flow is described and validated for a VCO example in Section IV. Finally, we conclude in Section V.

## II. INDIRECT PERFORMANCE SENSING

We denote a PoI as $f$ and the PoM as:

$$\mathbf{x} = \begin{bmatrix} x_1 & x_2 & \cdots & x_M \end{bmatrix}^T, \tag{1}$$

where $M$ stands for the number of performance metrics belonging to PoM. The objective of indirect performance sensing is to accurately predict the PoI $f$ from the PoM $\mathbf{x}$ that is highly correlated with $f$ and can be easily measured by on-chip sensors. To this end, we need to approximate $f$ as an analytical function of $\mathbf{x}$:

$$f(\mathbf{x}) = \sum_{k=1}^{K} \alpha_k \cdot b_k(\mathbf{x}), \tag{2}$$

where $\{b_k(\mathbf{x}); k = 1, 2, \ldots, K\}$ contains the basis functions (e.g., linear and quadratic polynomials), $\{\alpha_k; k = 1, 2, \ldots, K\}$ contains the model coefficients, and $K$ is the total number of basis functions. Once the model $f(\mathbf{x})$ in (2) is determined, it can be used to predict the PoI $f$ from the on-chip measurements of the PoM $\mathbf{x}$.

In general, a number of silicon chips are required to fit the model $f(\mathbf{x})$ in (2). The PoM (i.e., $\mathbf{x}$) values of these chips are measured by on-chip sensors and the corresponding PoI (i.e., $f$) values are measured by an off-chip tester. These measurement data are then used to determine the model

978-1-4673-6145-3/13 $31.00 © 2013 IEEE

coefficients $\{\alpha_k; \; k = 1, 2, ..., K\}$ in (2) by ordinary least-squares (OLS) fitting [8]. More details about the traditional OLS method can be found in [8].

## III. BAYESIAN MODEL FUSION

Our proposed BMF method aims to efficiently re-calibrate the indirect sensor model $f(\mathbf{x})$ in (2) with consideration of process shift. Towards this goal, we consider two different models: the old model $f_{OLD}(\mathbf{x})$ and the new model $f_{NEW}(\mathbf{x})$:

$$f_{OLD}(\mathbf{x}) = \sum_{k=1}^{K} \alpha_{OLD,k} \cdot b_k(\mathbf{x}) + \varepsilon_{OLD} \qquad (3)$$

$$f_{NEW}(\mathbf{x}) = \sum_{k=1}^{K} \alpha_{NEW,k} \cdot b_k(\mathbf{x}) + \varepsilon_{NEW}, \qquad (4)$$

where $\{\alpha_{OLD,k}; \; k = 1, 2, ..., K\}$ and $\{\alpha_{NEW,k}; \; k = 1, 2, ..., K\}$ contain the old and new model coefficients respectively, and $\varepsilon_{OLD}$ and $\varepsilon_{NEW}$ denote the modeling error associated with the old and new models respectively.

The old model $f_{OLD}(\mathbf{x})$ in (3) is fitted before process shift occurs. Hence, we assume that the old model coefficients $\{\alpha_{OLD,k}; \; k = 1, 2, ..., K\}$ are already known, before fitting the new model $f_{NEW}(\mathbf{x})$ in (4). The objective of BMF is to accurately determine the new model coefficients $\{\alpha_{NEW,k}; \; k = 1, 2, ..., K\}$ by combining the old model coefficients $\{\alpha_{OLD,k}; \; k = 1, 2, ..., K\}$ with very few new measurement data.

Our proposed BMF method consists of two major steps: (i) statistically extracting the prior knowledge from the old model coefficients $\{\alpha_{OLD,k}; \; k = 1, 2, ..., K\}$ and encoding it as a prior distribution, and (ii) optimally determining the new model coefficients $\{\alpha_{NEW,k}; \; k = 1, 2, ..., K\}$ by MAP estimation. In what follows, we will describe these two steps in detail.

### A. Prior Knowledge Definition

Since the two models $f_{OLD}(\mathbf{x})$ and $f_{NEW}(\mathbf{x})$ in (3)-(4) both approximate the mathematical mapping from PoM to PoI, we expect that the model coefficients $\{\alpha_{OLD,k}; \; k = 1, 2, ..., K\}$ and $\{\alpha_{NEW,k}; \; k = 1, 2, ..., K\}$ are similar. On the other hand, $f_{OLD}(\mathbf{x})$ and $f_{NEW}(\mathbf{x})$ cannot be exactly identical due to process shift. To statistically encode the "common" information between $f_{OLD}(\mathbf{x})$ and $f_{NEW}(\mathbf{x})$, we define a Gaussian distribution as our prior distribution for each new model coefficient $\alpha_{NEW,k}$:

$$pdf(\alpha_{NEW,k}) \sim Gauss(\alpha_{OLD,k}, \lambda^2 \cdot \alpha_{OLD,k}^2) \quad (k = 1, 2, \cdots, K), \quad (5)$$

where $\alpha_{OLD,k}$ and $\lambda^2 \cdot \alpha_{OLD,k}^2$ are the mean and variance of the Gaussian distribution respectively, and $\lambda$ is a parameter that can be determined by cross-validation [8]-[9].

The prior distribution in (5) has a two-fold meaning. First, the Gaussian distribution $pdf(\alpha_{NEW,k})$ is peaked at its mean value $\alpha_{NEW,k} = \alpha_{OLD,k}$, implying that the old coefficient $\alpha_{OLD,k}$ and the new coefficient $\alpha_{NEW,k}$ are likely to be similar. In other words, since the Gaussian distribution $pdf(\alpha_{NEW,k})$ exponentially decays with $(\alpha_{NEW,k} - \alpha_{OLD,k})^2$, it is unlikely to observe a new coefficient $\alpha_{NEW,k}$ that is extremely different from the old coefficient $\alpha_{OLD,k}$. Second, the standard deviation of the prior distribution $pdf(\alpha_{NEW,k})$ is proportional to $|\alpha_{OLD,k}|$. It means that the absolute difference between the new coefficient $\alpha_{NEW,k}$ and the old coefficient $\alpha_{OLD,k}$ can be large (or small), if the magnitude of the old coefficient $|\alpha_{OLD,k}|$ is large (or small). Restating in words, each new coefficient $\alpha_{NEW,k}$ has been provided with a relatively equal opportunity to deviate from the corresponding old coefficient $\alpha_{OLD,k}$.

To complete the definition of the prior distribution for all new model coefficients $\{\alpha_{NEW,k}; \; k = 1, 2, ..., K\}$, we further assume that these coefficients are statistically independent and their joint distribution is represented as:

$$pdf(\mathbf{\alpha}_{NEW}) = \prod_{k=1}^{K} pdf(\alpha_{NEW,k}), \qquad (6)$$

where $\mathbf{\alpha}_{NEW} \in \mathfrak{R}^K$ is a vector containing all new coefficients $\{\alpha_{NEW,k}; \; k = 1, 2, ..., K\}$. The independence assumption in (6) simply implies that we do not know the correlation information among these coefficients as our prior knowledge. The correlation information will be learned from the new measurement data, when the posterior distribution is calculated by MAP estimation in the next sub-section.

### B. Maximum-A-Posteriori Estimation

Once the prior distribution is defined, we collect a few new measurement data $\{(\mathbf{x}^{(n)}, f_{NEW}^{(n)}); \; n = 1, 2, ..., N\}$, where $\mathbf{x}^{(n)}$ and $f_{NEW}^{(n)}$ are the values of $\mathbf{x}$ and $f_{NEW}(\mathbf{x})$ at the $n$-th data point respectively after process shift occurs. These new measurement data can tell us additional information about the process shift and, hence, help us to determine the new coefficients $\{\alpha_{NEW,k}; \; k = 1, 2, ..., K\}$.

Based on Bayes' theorem [8]-[9], the uncertainties of the new coefficients $\{\alpha_{NEW,k}; \; k = 1, 2, ..., K\}$ after knowing the data $\{(\mathbf{x}^{(n)}, f_{NEW}^{(n)}); \; n = 1, 2, ..., N\}$ can be mathematically described by the following posterior distribution:

$$pdf(\mathbf{\alpha}_{NEW} | \mathbf{f}_{NEW}) \propto pdf(\mathbf{\alpha}_{NEW}) \cdot pdf(\mathbf{f}_{NEW} | \mathbf{\alpha}_{NEW}), \qquad (7)$$

where $\mathbf{f}_{NEW} \in \mathfrak{R}^N$ is a vector with the new data $\{f_{NEW}^{(n)}; \; n = 1, 2, ..., N\}$. In (7), the prior distribution $pdf(\mathbf{\alpha}_{NEW})$ is defined by (6). The conditional distribution $pdf(\mathbf{f}_{NEW} | \mathbf{\alpha}_{NEW})$ is referred to as the likelihood function. It measures the probability of observing the new data $\{(\mathbf{x}^{(n)}, f_{NEW}^{(n)}); \; n = 1, 2, ..., N\}$.

To derive the likelihood function $pdf(\mathbf{f}_{NEW} | \mathbf{\alpha}_{NEW})$, we assume that the modeling error $\varepsilon_{NEW}$ in (4) can be represented as a random variable with zero-mean Gaussian distribution:

$$pdf(\varepsilon_{NEW}) \sim Gauss(0, \sigma_0^2), \qquad (8)$$

where the standard deviation $\sigma_0$ indicates the magnitude of the modeling error. Similar to the parameter $\lambda$ in (5), the value of $\sigma_0$ can be determined by cross-validation [8]-[9]. Since the modeling error associated with the $n$-th data point $(\mathbf{x}^{(n)}, f_{NEW}^{(n)})$ is simply one sampling point of the random variable $\varepsilon_{NEW}$ that follows the Gaussian distribution in (8), the probability of observing the $n$-th data point is:

$$pdf(f_{NEW}^{(n)} | \mathbf{\alpha}_{NEW}) = \frac{1}{\sqrt{2\pi} \cdot \sigma_0} \cdot$$
$$\exp\left\{-\frac{1}{2 \cdot \sigma_0^2} \cdot \left[f_{NEW}^{(n)} - \sum_{k=1}^{K} \alpha_{NEW,k} \cdot b_k(\mathbf{x}^{(n)})\right]^2\right\} \cdot \qquad (9)$$

Note that the likelihood function $pdf(f_{NEW}^{(n)} | \mathbf{\alpha}_{NEW})$ in (9) depends on the model coefficients $\{\alpha_{NEW,k}; \; k = 1, 2, ..., K\}$.

Assuming that all data points $\{(\mathbf{x}^{(n)}, f_{NEW}^{(n)}); n = 1, 2, ..., N\}$ are independently generated, we can write the likelihood function $pdf(\mathbf{f}_{NEW} | \boldsymbol{\alpha}_{NEW})$ as:

$$pdf\left(\mathbf{f}_{NEW} | \boldsymbol{\alpha}_{NEW}\right) = \prod_{n=1}^{N} pdf\left(f_{NEW}^{(n)} | \boldsymbol{\alpha}_{NEW}\right). \quad (10)$$

After the new data $\{(\mathbf{x}^{(n)}, f_{NEW}^{(n)}); n = 1, 2, ..., N\}$ are available, the new coefficients $\{\alpha_{NEW,k}; k = 1, 2, ..., K\}$ can be modeled by the probability density function $pdf(\boldsymbol{\alpha}_{NEW} | \mathbf{f}_{NEW})$ (i.e., the posterior distribution) in (7). Depending on the shape of the posterior distribution $pdf(\boldsymbol{\alpha}_{NEW} | \mathbf{f}_{NEW})$, the new coefficients $\{\alpha_{NEW,k}; k = 1, 2, ..., K\}$ do not take all possible values with equal probability. If the posterior distribution $pdf(\boldsymbol{\alpha}_{NEW} | \mathbf{f}_{NEW})$ reaches its maximum value at $\{\alpha^{*}_{NEW,k}; k = 1, 2, ..., K\}$, these values $\{\alpha^{*}_{NEW,k}; k = 1, 2, ..., K\}$ are the optimal estimation of the new coefficients, since these coefficient values are most likely to occur. Such a method is referred to as the MAP estimation in the literature [8]-[9].

The aforementioned MAP estimation can be formulated as an optimization problem:

$$\underset{\boldsymbol{\alpha}_{NEW}}{\text{maximize}} \quad pdf\left(\boldsymbol{\alpha}_{NEW} | \mathbf{f}_{NEW}\right). \quad (11)$$

Substituting (5)-(7) and (9)-(10) into (11) and taking the logarithm for the merit function result in the following equivalent optimization formulation for MAP:

$$\underset{\boldsymbol{\alpha}_{NEW}}{\text{minimize}} \quad \frac{\sigma_0^2}{\lambda^2} \sum_{k=1}^{K} \left(\frac{\alpha_{NEW,k} - \alpha_{OLD,k}}{\alpha_{OLD,k}}\right)^2 +$$
$$+ \sum_{n=1}^{N} \left[f_{NEW}^{(n)} - \sum_{k=1}^{K} \alpha_{NEW,k} \cdot b_k\left(\mathbf{x}^{(n)}\right)\right]^2 . \quad (12)$$

Studying (12) would reveal several important properties. First, the optimization in (12) is convex and, hence, can be solved both efficiently and robustly. Second, only the ratio $\sigma_0/\lambda$ is required to solve the new model coefficients $\{\alpha_{NEW,k}; k = 1, 2, ..., K\}$ from (12). Hence, we only need to determine the ratio $\sigma_0/\lambda$, instead of the individual parameters $\sigma_0$ and $\lambda$, by cross-validation. Once the new model coefficients $\{\alpha_{NEW,k}; k = 1, 2, ..., K\}$ are found from (12), an updated indirect sensor model $f_{NEW}(\mathbf{x})$ in (4) is generated to match the new process condition after process shift occurs.

## IV. SELF-HEALING FLOW AND EXPERIMENTAL RESULTS

In this section, we describe our proposed self-healing flow for a 25GHz differential Colpitts VCO example designed in a 32nm CMOS SOI process. Fig. 1 shows the simplified schematic of the VCO. It consists of a cross-coupled differential pair connected to two common-gate Colpitts oscillators. The capacitor at the output is tunable so that the VCO frequency can be centered at different frequency bands. The bias voltage $V_b$ is controlled by a DAC for self-healing. More details about the VCO design can be found in [10].

For the VCO shown in Fig. 1, phase noise is an important performance of interest. Accurately measuring the phase noise at 25GHz is not trivial. Hence, indirect performance sensing is applied for on-chip phase noise measurement. Towards this goal, we identify four PoM metrics that are highly correlated with phase noise, as summarized in Table I. A quadratic model is fitted to predict the PoI (i.e., phase noise) from the PoM. In other words, the indirect sensor model $f(\mathbf{x})$ in (2) is a quadratic polynomial function. Once the model coefficients of $f(\mathbf{x})$ are determined, the quadratic model $f(\mathbf{x})$ is used for on-chip phase noise estimation that is required for self-healing.

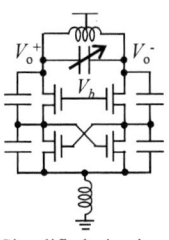

Fig. 1. Simplified circuit schematic is shown for a Colpitts VCO.

TABLE I
PoI AND PoM OF INDIRECT PHASE NOISE SENSOR FOR SELF-HEALING

| | Performance Metric |
|---|---|
| PoI | Phase Noise |
| PoM | Oscillation Frequency |
| | Oscillation Amplitude |
| | Bias Current |
| | Bias Voltage |

In this VCO example, the objective of self-healing is to find the optimal bias voltage to minimize phase noise. To this end, we implement a brute-force search algorithm where all possible bias voltages are set by the DAC and the phase noise values are estimated by the quadratic model $f(\mathbf{x})$ based on PoM. Next, the bias voltage corresponding to the minimum phase noise is chosen as the optimal solution.

Off-chip measurement data are collected from two silicon wafers and these data are used to perform an off-chip data analysis to validate the aforementioned self-healing flow. The first and second wafers contain 55 and 61 functional VCOs, respectively. Without self-healing, the parametric yield for these two wafers is 76.4% and 0%, respectively. Note that there is a significant process shift between these two wafers, since the yield of the second wafer is substantially different from that of the first wafer.

In this example, since the parametric yield of the second wafer is extremely low, we want to apply self-healing to improve its yield. For testing and comparison purposes, four different self-healing methods are implemented:

- **Ideal**: The optimal bias voltage is determined by directly measuring the phase noise with an off-chip tester for all bias voltages. This approach is not considered as on-chip self-healing; however, it provides the upper bound of the yield improvement that can be achieved by self-healing.

- **OLS-1**: Traditional OLS is applied to fit the indirect sensor model based on the phase noise data measured by an off-chip tester for a few VCOs on the second wafer. In this case, no data from the first wafer is used. Next, the indirect sensor model is applied to the aforementioned self-healing flow to heal all VCOs on the second wafer.

- **OLS-2**: Traditional OLS is applied to fit the indirect sensor model based on the phase noise data measured by an off-chip tester for all VCOs on the first wafer and a few VCOs on the second wafer. The indirect sensor model is then applied to self-heal all VCOs on the second wafer.

- **BMF**: An indirect sensor model is first fitted based on the measurement data from the first wafer. Next, the proposed BMF algorithm is applied to re-calibrate the old model and generate a new model based on the phase noise data

978-1-4673-6145-3/13 $31.00 © 2013 IEEE

measured by an off-chip tester for a few VCOs on the second wafer. The new model is applied to self-heal all VCOs on the second wafer.

Fig. 2 shows the parametric yield of the second wafer that is achieved by four different self-healing methods. Table II further summarizes the off-chip measurement cost for self-healing. Studying Fig. 2 and Table II reveals several important observations. First, BMF requires substantially less VCOs measured by an off-chip tester to build the indirect phase noise sensor than the traditional OLS method. In this example, BMF needs to measure one VCO only, while both OLS-1 and OLS-2 require to measure four VCOs (4×) to achieve similar yield.

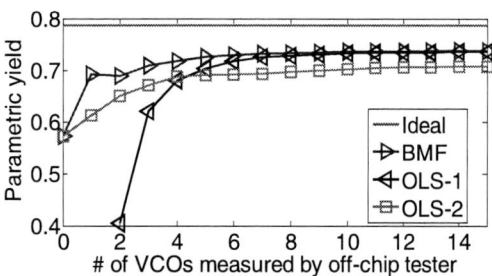

Fig. 2. Post-self-healing parametric yield of the second wafer is shown as a function of the number of VCOs from the second wafer measured by an off-chip tester.

Table II
MEASUREMENT COST AND PARAMETRIC YIELD FOR SELF-HEALING

| Self-Healing Method | # of Measured VCOs from Second Wafer | Parametric Yield for Second Wafer |
|---|---|---|
| Ideal | 61 | 78.69% |
| OLS-1 | 4 | 68.06% |
| OLS-2 | 4 | 68.78% |
| BMF | 1 | 69.17% |

Second, studying the BMF results in Fig. 2, we notice that if no measurement data is collected from the second wafer (i.e., the number of VCOs from the second wafer measured by an off-chip tester is zero) and the self-healing of the second wafer is performed with the indirect phase noise model fitted from the first wafer, the post-self-healing parametric yield of the second wafer is 59% only. Once a single VCO is measured for the second wafer, the indirect phase noise model is re-calibrated by BMF and the post-self-healing yield is increased to 69.17%. It, in turn, demonstrates that the aforementioned model re-calibration is a critical step for yield enhancement.

Third, the OLS-2 results in Fig. 2 demonstrate that if all measurement data from the first wafer are directly used by OLS, the fitted indirect sensor model is biased and does not accurately capture the shifted process condition associated with the second wafer. For this reason, the post-self-healing yield of OLS-2 remains lower than that of OLS-1, even if a large number of VCOs are measured for the second wafer. On the other hand, BMF does not suffer from such a bias issue, as it appropriately exploits the useful information from the first wafer through Bayesian inference.

Finally, the post-self-healing parametric yield of BMF is close to that of the "ideal" case. Compared to BMF, "ideal" self-healing increases the yield from 69.17% to 78.69%. It implies that the modeling error of our proposed BMF method

is reasonably small, even if a single VCO is measured from the second wafer. Fig. 3 further shows the scatter plot between the actual phase noise measured by an off-chip tester and the predicted phase noise estimated by BMF where the average modeling error is 0.56dBc/Hz.

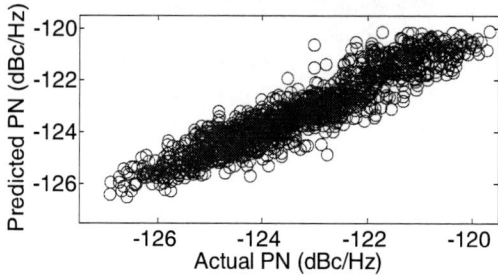

Fig. 3. Scatter plot is shown for the actual phase noise (measured by an off-chip tester) and the predicted phase noise (estimated by the indirect phase noise sensor) for all VCOs on the second wafer.

## V. CONCLUSION

In this paper, we propose a novel BMF technique to efficiently re-calibrate indirect performance sensors to accommodate the process shift associated with manufacturing lines. Towards this goal, a Bayesian inference is developed to accurately update the indirect sensor model by combining the old model coefficients with very few new measurement data. The proposed indirect performance sensing based on BMF is incorporated into an on-chip self-healing flow and is validated by off-chip data analysis for a 25GHz VCO designed in a 32nm CMOS SOI process.

## VI. ACKNOWLEDGMENTS

This work is sponsored by the DARPA HEALICS (Self-Healing Mixed-Signal Integrated Circuits) program under Air Force Research Laboratory (AFRL) contract FA8650-09-C-7924. The views expressed are those of the author and do not reflect the official policy or position of the Department of Defense or the U.S. Government.

## VII. REFERENCES

[1] S. Nassif, "Modeling and analysis of manufacturing variations," *IEEE CICC*, pp. 223-228, 2001.

[2] X. Li, et al., *Statistical Performance Modeling and Optimization*, Now Publishers, 2007.

[3] Semiconductor Industry Associate, *International Technology Roadmap for Semiconductors*, 2011.

[4] G. Keskin, et al., "Statistical modeling and post manufacturing configuration for scaled analog CMOS," *IEEE CICC*, 2010.

[5] C. Chien, et al., "Dual-control self-healing architecture for high performance radio SoC's," *IEEE Design & Test*, 2013, to be published.

[6] S. Yaldiz, et al., "Indirect phase noise sensing for self-healing voltage controlled oscillators," *IEEE CICC*, 2011.

[7] B. Sadhu, et al., "A linearized, low-phase-noise VCO-based 25-GHz PLL with autonomic biasing," *IEEE JSSC*, vol. 48, no. 5, pp. 1138-1150, May 2013.

[8] C. Bishop, *Pattern Recognition and Machine Learning*, Prentice Hall, 2007.

[9] X. Li, et al., "Efficient parametric yield estimation of analog/mixed-signal circuits via Bayesian model fusion," *IEEE ICCAD*, pp. 627-634, 2012.

[10] J. Plouchart, et al., "A 23.5GHz PLL with an adaptively biased VCO in 32nm SOI-CMOS," *IEEE TCAS-I*, 2013, to be published.

# Fast FPGA Emulation of Background-Calibrated SAR ADC with Internal Redundancy Dithering

Guanhua Wang and Yun Chiu

Analog and Mixed-Signal Lab, Texas Analog Center of Excellence
University of Texas at Dallas, Richardson, TX, USA

*Abstract*—A custom FPGA emulation platform for the verification of a slowly adapted, background calibration technique for successive-approximation-register (SAR) analog-to-digital converter (ADC) is demonstrated in an Altera DE4 board. The internal redundancy of a sub-binary SAR is exploited for the identification of ten leading bit weights in a 14.5-bit SAR ADC using pseudorandom bit sequence (PRBS) injection with background correlation. Experimental results reveal that the FPGA emulation achieves a 3000× speedup for the same simulation executed on a general-purpose microprocessor.

## I. INTRODUCTION

On-chip deployment of cost-efficient, adaptive digital algorithms to compensate for deterministic and random analog impairments presents an important trend for A/MS/RF circuit design in scaled process nodes. The flip side is that it also presents a difficult challenge for the simulation/verification of such systems incorporating adaptation loops with convergence times some $10^6$ times slower than the on-chip circuit time conconstants [1]. Transistor-level verification of the resulting mixed-signal system is simply not feasible; and even behavioral simulation proves to be difficult and painful due to the extremely long simulation time when executed on a general-purpose microprocessor. A critical need arises in this scenario to explore alternative fast simulation techniques that can provide sufficient accuracy while capturing the dynamics of the closed-loop interaction between the analog and digital building blocks in pre-silicon design phase.

**Table 1  Convergence speed of background-calibrated pipeline ADC**

| Work | Resolution [bits] | Linearity [dB] | Convergence time [samples] |
|------|------|------|------|
| [6] | 15 | 90 | 134 M |
| [7] | 15 | 93 | 268 M |
| [8] | 15 | 96 | 225 M |
| [9] | 15 | 98 | 400 M |

## II. MATLAB SIMULATION VS. FPGA EMULATION

Behavioral simulations coded in MATLAB or C language are traditionally used to model A/MS/RF circuits for algorithmic level investigations. However, the simulation performance is limited to a few hundred cycles per second when executed on a general-purpose microprocessor [2]. In adaptive analog circuits, a large number of iterations are often required before the circuit settles to a steady state. For example, the conver-

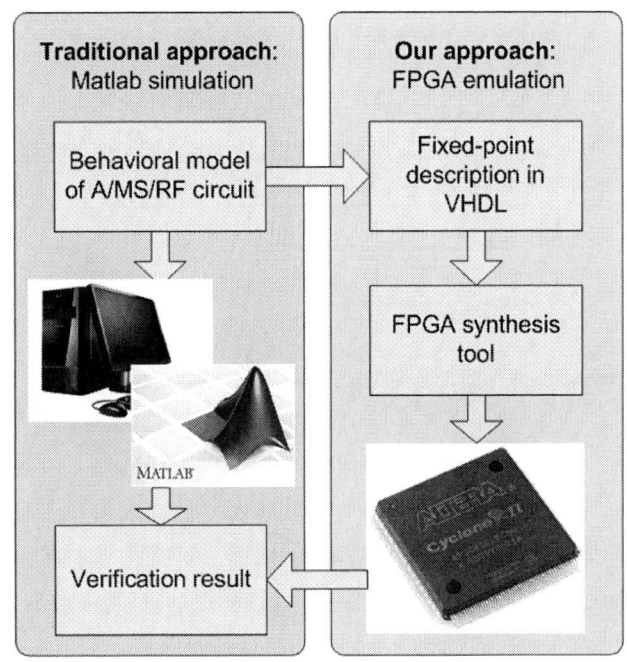

**Figure 1.** MATLAB simulation vs. FPGA emulation of adaptive analog circuits

gence time of a few recent pipeline ADC works employing correlation-based calibration is summarized in Table 1. With the traditional approach, it can take days or even weeks to complete a simulation, which greatly limits the efficiency for circuits and algorithm characterization. FPGA-based emulation provides a solution in such scenarios, and has been widely adopted for ASIC design, communication system modeling, and digital signal processing [3]. Various works have shown that FPGA can provide up to three orders of magnitude acceleration in simulation times [4], [5].

This paper describes the approach and experimental results of an FPGA emulation platform for fast behavioral simulation of adaptive analog circuits (Fig. 1). In this work, the behavioral model of a 14.5-bit sub-binary SAR ADC with background bit-weight calibration is described in VHDL and then synthesized and downloaded into an Altera DE4 board. After the emulation is done, the adaptation results are sent out from the FPGA to a PC, where optional post-emulation data analysis can be performed. Experimental results show that multi-billion adaptation cycles can be executed in hours with a 50-MHz system clock; in contrast, the same simulation will consume years in MATLAB executed on a general-purpose 3-GHz microprocessor. The acceleration factor is 3000×.

978-1-4673-6145-3/13 $31.00 © 2013 IEEE

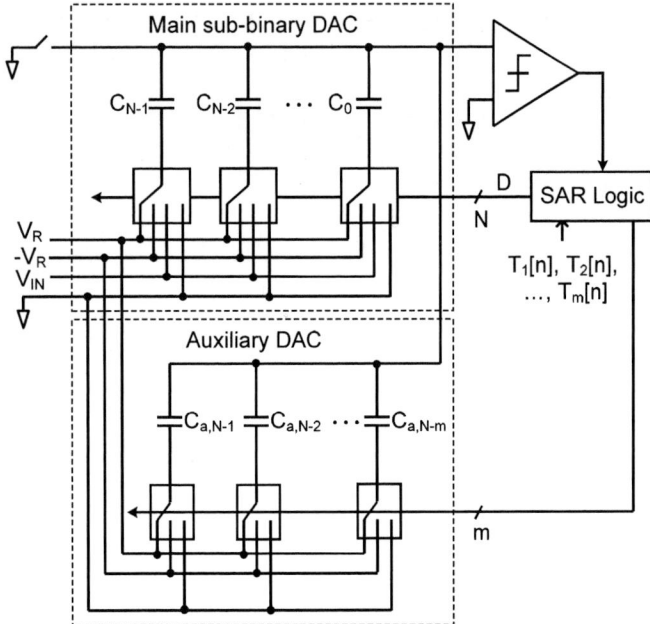

**Figure 2.** Proposed SAR architecture with an auxiliary DAC to generate the dither for the decision thresholds during bit cycling

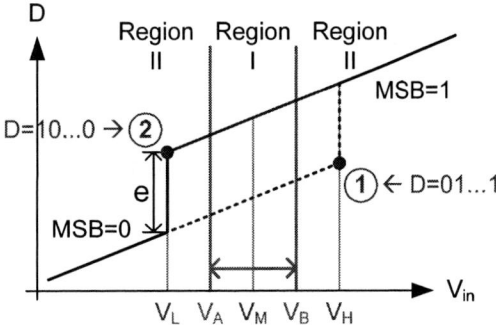

**Figure 3.** A transfer curve illustration of the bit-weight calibration. Two new bit-decision thresholds $V_A$ and $V_B$ are introduced within the redundancy region (bounded by $V_L$ and $V_H$). The ADC outcomes are dithered by the PRBS for inputs falling into Region I.

## III. CASE STUDY – SAR ADC CALIBRATION

### A. Sub-Binary SAR Redundancy

*In situ* identification of key circuit parameters, e.g., inter-stage gain or DAC mismatch coefficients, using PRBS injection and digital-domain correlation is a popular background calibration technique for high-performance pipeline ADC [6]–[9]. The core calibration algorithm usually relies on exploiting the built-in redundancy of the underlying analog circuits, which was originally invented to combat comparator offset and other analog non-idealities.

A form of inherent redundancy also exists in the sub-binary SAR ADC as the one shown in Fig. 2, where a DAC capacitor is sized slightly larger than half of the value of its more-significant neighbor, i.e., $C_{N-2}$ is slightly larger than $0.5C_{N-1}$ and so on. In the sub-binary case, the raw conversion curve is discontinuous as illustrated in Fig. 3, where the over-

lapped curves from $V_L$ to $V_H$ form a redundancy region – any input falling into this region can be potentially digitized to two distinct raw outcomes depending on which bit-decision threshold is used, $V_H$, $V_L$, or $V_M$, which corresponds to the DAC code $01...1$ (code ①), $10...0$ (code ②), or tying all capacitors to the common mode, respectively, as shown in Fig. 3.

It has been shown that the redundancy makes the conversion process insensitive to small dynamic errors due to comparator noise, reference bouncing, and DAC incomplete settling [10]. In this work, we will utilize the same redundancy to perform background bit-weight calibration for the SAR ADC. For this purpose, we first note that neglecting quantization noise the input can be accurately represented as a weighted sum of the raw bits in a sub-binary structure even in the presence of capacitor mismatch error [10]

$$d_{out} = \left\lfloor \frac{V_{in}}{V_R} \right\rfloor = \sum_{i=0}^{N-1} \frac{C_i}{C_{tot}} (2D_i - 1) = \sum_{i=0}^{N-1} W_i (2D_i - 1), \quad (1)$$

where $d_{out}$ is the final digital output, $C_{tot} = \sum_{i=0}^{N-1} C_i + C_0$ is the total DAC capacitance, $W_i = C_i/C_{tot}$ is the weight corresponding to the $i^{th}$ DAC capacitor, and $D_i$ is the $i^{th}$ raw bit. Thus, calibrating the DAC mismatch error is equivalent to identifying all the bit weights $\{W_i\}$.

### B. Internal Redundancy Dithering (IRD)

Fig. 3 shows conceptually how a bit-weight can be identified. Using the MSB for example, in the presence of capacitor mismatch, the vertical distance $e$ between the two segments denoted as MSB = 0 and MSB = 1 of the transfer curve deviates from its ideal value. According to (1), a proper MSB weight would restore the linearity of the ADC. To identify the optimal weight, the bit-decision threshold can be dithered inside the redundancy region, e.g., $V_M$ (the assumed threshold) is left- and right-shifted by a small amount to yield two new thresholds $V_A$ and $V_B$ to randomize the ADC outcomes for inputs falling inside the window of dithering. The effect of this is a random hopping between the two redundant conversion curves inside Region I. If the MSB weight is ideal, this hopping exhibits no effect on $d_{out}$; while a non-ideal weight will result in a noticeable conversion error in $d_{out}$ correlated with the PRBS. For inputs residing outside of the window (i.e., in Region II), the dither will produce zero correlation as the transfer curve is unique in Region II. Mathematically,

$$\overline{d_{out} \cdot T_1} = \frac{\sum_{n=1}^{M} d_{out}[n] \cdot T_1[n]}{M} = \begin{cases} \dfrac{e}{2} & V_{in}[n] \in \text{Region I} \\ 0 & V_{in}[n] \in \text{Region II} \end{cases}, \quad (2)$$

where $T_1[n] = \{-1, 1\}$ is a zero-mean PRBS and $M$ is the block size used to estimate the correlation.

### C. IRD Multi-Bit-Weight Calibration

The calibration algorithm can be easily extended to multiple bits as shown in Fig. 4. To identify $m$ weights, $m$ zero-mean PRBS $T_1[n], ..., T_m[n]$ are injected during the first $m$ bit cycles for every sample, respectively. The ADC output is cor-

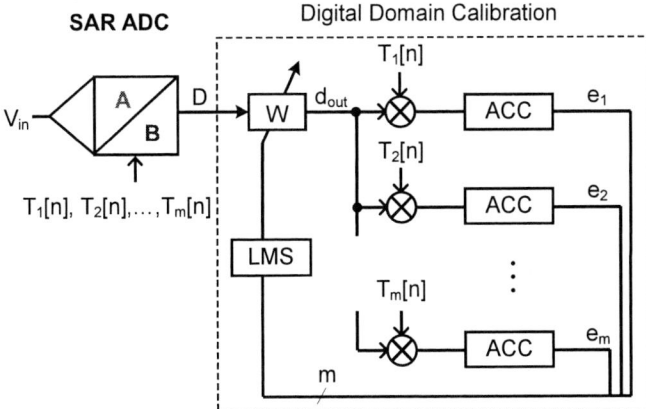

**Figure 4.** IRD calibration of *m* leading bit weights in a SAR ADC

related with each of $\{T_l\}$, fed to a digital accumulator (ACC in Fig. 4), and followed by a least-mean-square (LMS) weight updater. The iterative update equations are

$$
\begin{aligned}
e_l(j) &= \sum_{n=jM+1}^{(j+1)M} d_{out}[n] \cdot T_l[n], \quad l = 1...m, \\
W_l(j) &= W_l(j-1) - \mu \cdot e_l(j), \quad l = 1...m.
\end{aligned}
\tag{3}
$$

where $W_l(j)$ denotes the $l^{th}$ weight on the $j^{th}$ update instance and $\mu$ is the step size.

Fig. 2 shows one implementation of the SAR ADC to accommodate the PRBS injection. In addition to the main sub-binary DAC, an auxiliary DAC is used to set the dither thresholds in the redundancy region. For example, during the MSB decision the capacitor $C_{a,N-1}$ is connected to $V_R$ if $T_1[n] = 1$, otherwise to $-V_R$. In the next bit-decision cycle, $C_{a,N-1}$ is switched to the AC ground while $C_{a,N-2}$ is connected to $V_R$ if $T_1[n] = 1$, otherwise to $-V_R$. The conversion proceeds in a similar fashion for the rest of the bits.

Since the auxiliary DAC is to instrument small offsets to the bit-decision thresholds, its total capacitance and area are negligible relative to the main DAC. In addition, as the exact amount of dither is not required to be precise, the matching accuracy is relaxed for the auxiliary DAC (as long as the resulting offset does not exceed the redundancy boundary). Alternatively, low-rank capacitors of the main DAC can be exploited to implement the dither. However, a dedicated auxiliary DAC tends to minimize the digital logic effort involved.

Relative to the calibration technique reported in [10], the proposed calibration retains the full ADC conversion speed while all bit weights are still learned simultaneously. In addition, IRD does not consume any signal dynamic range nor is it necessary to remove the PRBS from the ADC output. However, the convergence speed is much slower as expected due to the correlation nature of the calibration algorithm [11].

## IV. EXPERIMENTAL RESULTS

The IRD calibration algorithm was emulated in an Altera DE4 board for a 14.5-bit sub-binary SAR ADC with 16 raw bits. The setup of the emulation platform is illustrated in Fig. 5, which includes a bit-accurate description of the digital pro-

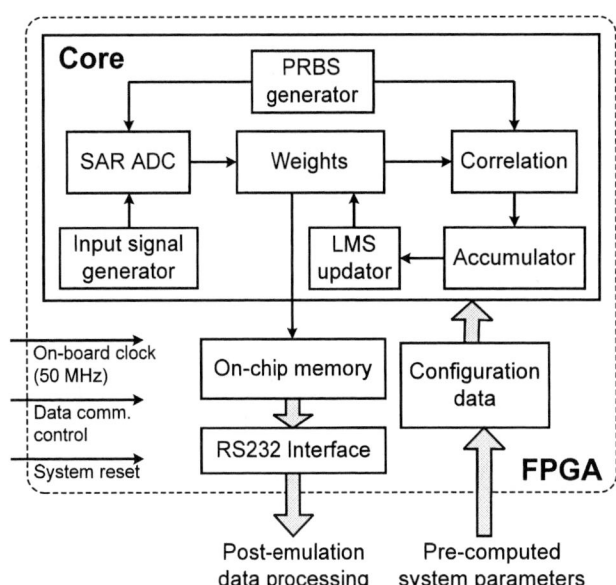

**Figure 5.** Setup of the FPGA emulation platform for SAR ADC

cessing engine and a fixed-point behavioral model of the analog circuits. The DAC capacitor mismatch was set to 1% for all 16 capacitors, and the ten leading bit weights were calibrated. The FPGA was clocked at 50 MHz and the block size for LMS update was set to one million in the emulation. We also chose to utilize 80% of the redundancy region shown in Fig. 3 for dithering, i.e., $V_B - V_A = 0.8*(V_H - V_L)$. An auxiliary 10-bit DAC as the one shown in Fig. 2 was used to implement the dither. As the unit capacitor in this design was set to 1 fF for kT/C noise consideration, the smallest capacitor in the auxiliary DAC was determined to be 2.4 fF, which can be realized in modern CMOS processes with ease [10], [12].

Fig. 6 shows the emulated learning curves of the spurious-free dynamic range (SFDR) and signal-to-noise plus distortion ratio (SNDR) of the SAR ADC. The convergence time for the calibration is approximately 500 billion samples – it will take 5000 seconds for a 100-MS/s ADC in reality to produce these many samples – which no doubt precludes the possibility of MATLAB simulation. The recorded runtimes for the FPGA

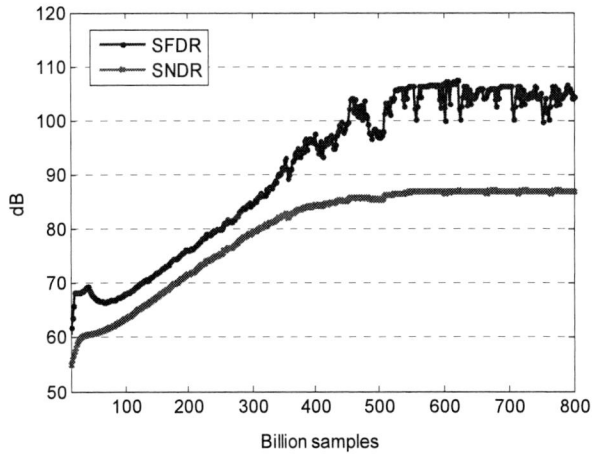

**Figure 6.** The emulated SFDR and SNDR learning curves

978-1-4673-6145-3/13 $31.00 © 2013 IEEE

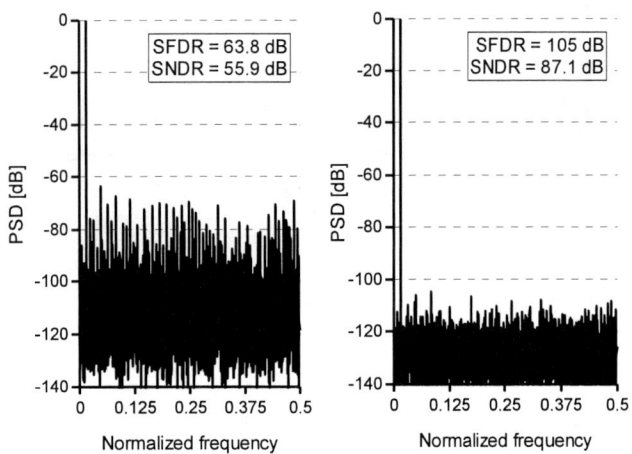

**Figure 7**. The emulated ADC output spectrum before (left) and after (right) calibration

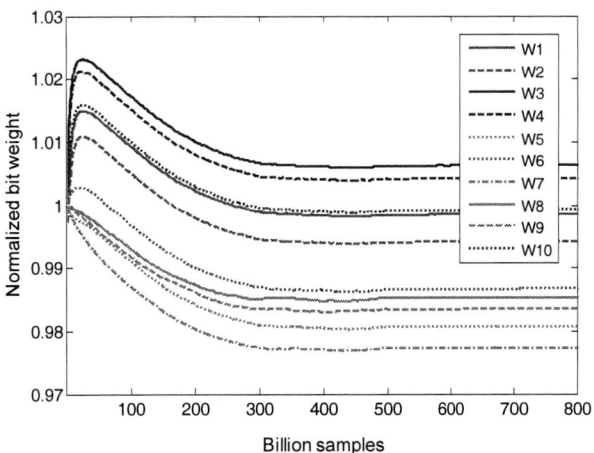

**Figure 8**. The learning curves for the ten leading bit weights

emulation and MATLAB simulation are compared in Table 2. The speedup factor is 3000× for this particular experiment.

Fig. 7 shows the emulated ADC output spectrum before and after calibration. The SFDR and SNDR are improved by more than 40 dB and 30 dB, respectively. The effective number of bits (ENOB) of the SAR ADC after calibration is 14.2, a 5.2-bit improvement from that before calibration.

Fig. 8 shows the learning curves of the ten leading bit weights. The weights are normalized to their ideal values in order to facilitate comparison. We note a slight negative bias of the ten learned weights, i.e., the mean value of the weights is slightly less than unity, which is caused by the mismatch among the six least-significant DAC capacitors that is modeled but not corrected in the emulation.

## V. CONCLUSION

In summary, we have presented a fast FPGA emulation platform for adaptive analog circuit simulation and verification. The runtime of the emulation is 3000× faster than its counterpart MATLAB simulation executed on a desktop computer. A technique for the digital background calibration of SAR ADC is successfully verified using the FPGA emulation platform. The technique exploits the internal redundancy of a sub-binary DAC to glean mismatch information and the calibration overhead is low due to the simplicity of the PRBS injection and digital processing. The ADC throughput is fully retained in this approach relative to an earlier technique at the cost of a much longer convergence time for the calibration.

### Table 2 Performance comparison

|  | MATLAB simulation | FPGA emulation |
|---|---|---|
| Number of samples | 100 million | |
| Clock/CPU frequency | 3.07 GHz | 50 MHz |
| Runtime | 30 hours | 36 seconds |

## REFERENCES

[1] M. Courtoy, "Rapid system prototyping for real-time design validation," *Proc. Ninth International Workshop on Rapid System Prototyping*, pp. 108-112, 1998.

[2] S. Banerjee and T. Gupta, "Design Aware Scheduling of Dynamic Testbench Controlled Design Element Accesses in FPGA-based HW/SW Co-simulation Systems for Fast Functional Verification," *IEEE Asia Symposium on Quality Electronic Design*, pp. 175-181, 2010.

[3] C. Chang *et al.*, "Implementation of BEE: A Real-Time Large-Scale Hardware Emulation Engine," *Proc. 2003 ACM/SIGDA 11th Int. Symp. on Field-Programmable Gate Arrays (FPGA 03), ACM Press*, pp. 91-99, 2003.

[4] J.P. Durbano *et al.*, "FPGA-Based Acceleration of the 3D Finite-Difference Time-Domain Method," *Proc. 12th Ann. IEEE Symp. Field-Programmable Custom Computing Machines (FCCM 04), IEEE CS Press*, pp. 156-163, 2004.

[5] W. Chen *et al.*, "An FPGA Implementation of the Two Dimensional Finite-Difference Time-Domain (FDTD) Algorithm," *Proc. 2004 ACM/SIGDA 12th Int'l Symp. Field-Programmable Gate Arrays (FPGA 04), ACM Press*, pp. 213-222, 2004.

[6] E. J. Siragusa and I. Galton, "A digitally enhanced 1.8V 15b 40MS/s CMOS pipelined ADC," *IEEE J. Solid-State Circuits*, vol. 39, 2126-2138, 2004.

[7] H.-C. Liu, Z.-M. Lee, and J.-T. Wu, "A 15b 20MS/s CMOS pipelined ADC with digital background calibration," *IEEE International Solid-State Circuits Conference (ISSCC), Digest of Technical Papers*, pp. 454-455, 2004.

[8] K. Nair and R. Harjani, "A 96dB SFDR 50MS/s digitally enhanced CMOS pipeline A/D converter," *IEEE International Solid-State Circuits Conference (ISSCC), Digest of Technical Papers*, pp. 456-457, 2004.

[9] Y.-S. Shu and B.-S. Song, "A 15b linear, 20MS/s, 1.5b/stage pipelined ADC digitally calibrated with signal-dependent dithering," *IEEE Sym. VLSI Circuits, Dig. Tech. Papers*, pp. 218-219, 2006.

[10] W. Liu, P. Huang, and Y. Chiu, "A 12b 22.5/45MS/s 3.0mW 0.059mm² CMOS SAR ADC achieving over 90dB SFDR," *IEEE International Solid-State Circuits Conference (ISSCC), Digest of Technical Papers*, pp. 380-381, 2010.

[11] J. McNeill, M. Coln, and B. Larivee, "A split-ADC architecture for deterministic digital background calibration of a 16b 1MS/s ADC," *IEEE International Solid-State Circuits Conference (ISSCC), Digest of Technical Papers*, pp. 276-278, 2005.

[12] W. Liu, P. Huang, and Y. Chiu, "A 12-bit 50-MS/s 3.3-mW SAR ADC with background digital calibration," *Proc. IEEE Custom Integrated Circuits Conference (CICC)*, pp. 1-4, 2012.

# Circuit Reliability Simulation Using TMI2

Min-Chie Jeng, Cheng Hsiao, Ke-Wei Su, Chung-Kai Lin

Taiwan Semiconductor Manufacturing Company,
168, Park Ave. 2, Hsinchu Science Park, Hsin-chu County, Taiwan 308-44, ROC.

*Abstract*—**Using simulation to assess the impacts of various reliability mechanisms to circuit performance has become prevail for advanced technologies due to smaller headroom (=Vdd-Vth) and less design margins. This paper reviews existing circuit aging simulation approaches with focus on TMI. The limitations of aging models are also discussed so that reliability simulations can be executed more correctly with the right expectation. An example circuit under multi-waveform and multi-temperature stress is presented to illustrate reliability simulation flow through TMI.**

## I. INTRODUCTION

Device degradation due to hot-carrier injection (HCI) and bias-temperature instability (BTI) will cause circuit performance to drift over time [1][2]. It's very important to ensure a circuit can still function over the planned lifetime of continuous operation even when device characteristics may have been shifted from their fresh states by HCI, BTI, or other reliability mechanisms. Every device in a circuit degrades differently because the amount of degradation is highly dependent on the voltage waveforms applied to that device. It's difficult to estimate the circuit performance drift due to device degradations even for a small-sized circuit. This motivated the advent of circuit reliability simulator and the aging models. The first public-domain circuit reliability simulator, BERT, was created in late 1990's by U.C. Berkeley [3]. Even though BERT was developed about 20 years ago, almost all existing reliability simulators still follow BERT's simulation flow.

A reliability simulator must be used together with a circuit simulator such as SPICE because we need the circuit simulator to provide the voltage waveforms of each device in order to calculate the degradations accumulated over time. The model equations and parameters associated with reliability simulators are referred to as aging models. Aging model parameters usually are divided into two parts: one part specifies the degradation-bias relationship and the rest describes how the SPICE model parameters should be adjusted corresponding to the degradations. The degradations can be any electrical parameters such as drain currents (e.g. Idsat and Idlin) or threshold voltages (Vth). Analog parameters such as transconductance (Gm) and channel conductance (Gds) are seldom used for degradation targets because they are difficult to characterize accurately after stress. It's crucial for aging

models to capture the tracking among all the degradation targets under various stress conditions to ensure correct simulated circuit degradation trends.

To create an aging model, devices are stressed under different bias conditions and/or different temperatures. The degradations versus time are recorded. At each monitoring time points during the stress, I-V or C-V characteristics may also be measured in order to find the relationship between SPICE model parameters and the degradations. This process is shown in Fig. 1. Usually the degradation versus time follows a power-law (approximate to $t^n$) relationship under a fixed stress condition. Each degradation mechanism has its own characteristic slope when plotted in the log-log scale. Even for the same mechanism, the slope can be different under different stress conditions. For each degradation parameter (e.g. Idsat, Idlin, Vth), the slope may also be different. This makes the aging model development and parameter extraction extremely complicated and cumbersome especially when the number of electric parameters of interest increases. To simplify the work, most existing aging models use a fixed slope for each reliability mechanism and assume that all the mechanisms are mutually independent so that the degradations due to each mechanism is additive. Some of these approximations are too primitive and impose limitations on the accuracy of existing aging models (compared to SPICE models).

The first step to create an aging model is to find the relationship between the degradations and the biases so that the degradations can be expressed as

$$D = A(V, I, T) \times t^n \qquad (1)$$

where D is the degradation targets such as Idsat and Vth, A(V, I, T) is the relationship to be determined. For example, Idsat degradation due to HCI is often written as

$$Didsat = \left[ \frac{Ids}{HW} \left( \frac{Isub}{Ids} \right)^m \times t \right]^n \qquad (2)$$

where W is to the width of the transistor, m and H are technology-dependent parameters to be extracted from the

978-1-4673-6145-3/13 $31.00 © 2013 IEEE      137

stress data, Isub is the substrate current, and Ids is the drain current. In BERT, the above equation is re-written as

$$Didsat = (Age)^n \qquad (3)$$

where

$$Age = \frac{Ids}{HW}\left(\frac{Isub}{Ids}\right)^m \times t \qquad (4)$$

The "age" concept greatly simplifies the implementation of degradation accumulation in simulators because it transforms the degradation rates under different biases into a unified age so that the age can be summed up and transforms back to obtain the total degradation. The restriction here is that "n" must be constant for all biases. If the slope is not constant or if the degradation doesn't follow the power law, the "age" approach will fail.

The second step is to find the SPICE model parameter changes corresponding to a given age. This is typically done by re-extracting some of the key SPICE model parameters based on the degraded I-V and C-V characteristics collected during the stress as shown in Fig. 1.

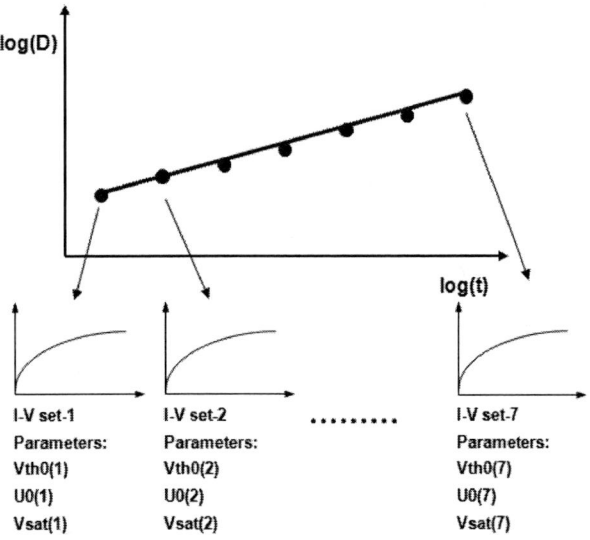

Fig. 1.    Degradation increases with stress time. Different sets of SPICE model parameters are extracted from every I-V/C-V curve taken during the stress.

Unlike standard models in SPICE, different circuit reliability simulators implement different proprietary aging models. Creating simulator-dependent aging models are very costly and time consuming let alone the confusion caused by the different simulation results among these simulators even though all these proprietary aging models are created from the same data. This raised the need for a unified aging model for all simulators. One solution is to go through the CMC (Compact Model Council) standardization procedures. However, reliability is highly process dependent and the detailed information usually cannot be shared to other companies. It's difficult to come up with a standard. This paper describes how we implemented a unified aging model through TSMC Model Interface (TMI) - the industry standard model interface for circuit simulators. With TMI, the aging model equations can be kept proprietary to protect company's sensitive information and yet all circuit simulators can use the same set of aging model parameters. Circuit reliability simulation with TMI aging models doesn't require additional license or cost as long as the circuit simulators support TMI.

In Section-2, we describe the general circuit reliability simulation flow in existing simulators. We also highlight what aging model and reliability simulations can and cannot do so that designers know how to interpret the reliability simulation results correctly. Section-3 describes how we implement the aging models in TMI. Section-4 gives a TMI aging simulation example for an inverter chain under multiple-waveform stress at different temperatures. Besides using TMI to simulate reliability mechanisms such as HCI and BTI, which gradually degrade device performance, we can also use TMI to calculate Time Dependent Dielectric Breakdown (TDDB) information after SPICE simulation. Topics related to RF and SRAM degradation simulations are also touched. Section-5 concludes this paper.

## II.    GENERAL CIRCUIT AGING SIMULATION FLOW

A typical circuit aging simulation flow is shown in Fig. 2. It starts with a regular SPICE transient simulation. The degradations for all or selected devices accumulated during this transient period (e.g. 1m seconds) are calculated and projected to a future time (e.g. 10 years) assuming that the same waveforms are repeated over the period of time to be projected. This degradation integration process can be done after SPICE simulation is completed through a post-processor as in Cadence's RelXpert or can be done on the fly during SPICE simulation as in Synopsys' MOSRA and TMI. The model parameters for each device are updated depending on the amount of degradations occurred. The circuit is re-simulated with updated model parameters to project the aged circuit behavior. Because devices degrade gradually over time, the waveforms applied to the devices will also drift. Linearly projecting the degradations calculated from the fresh state to a long future time may cause some error especially if the devices degrade quickly. In this case, we can divide the degradation projection into a sequence of shorter steps at the cost of more SPICE simulations. For example, we can project the degradation from fresh state to 1 month first; re-calculate the degradation rates at 1 month and project to 1 year; then project to 5 years, and finally to the desired circuit lifetime such as 10 years. The output waveforms of a fresh and 10-year old ring oscillator are shown in Fig. 3 as an example.

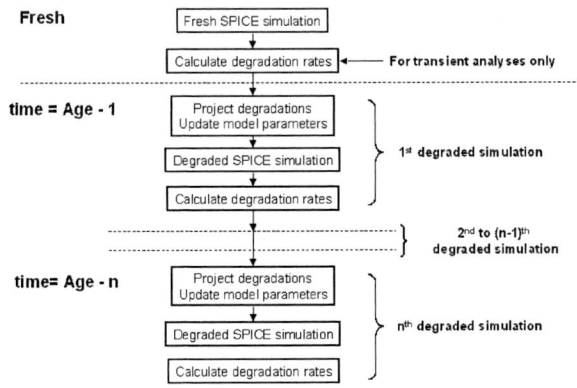

Fig. 2.    Typical aging simulation flow.

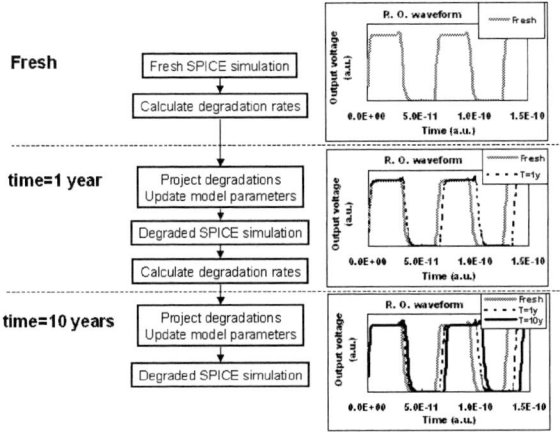

Fig. 3.    The simulation results of an 11-stage ring oscillator. The output waveform is shift for 1 year and 10 year stress.

Although this circuit aging simulation flow is logical and straightforward, we need to be aware of some of its restrictions. First of all, the degradation can only be integrated in transient analyses. If we would like to simulate a DC stress, we still need to run a transient analysis with DC biases. Furthermore, device degradation is a strong function of voltages. For some circuits, the spikes of SPICE simulated waveforms may contribute to a significant portion of the total degradations. Therefore, the accuracy of SPICE models and the transient integration algorithms (such as trapezoidal or gear) used will affect the aging results. We should always use the consistent simulator options in circuit aging simulations to avoid contaminating the simulated degradation results.

Most importantly, we should realize that aging models and SPICE models are fundamentally different in nature. SPICE models represent the performance of a technology. No matter how the process is tuned or improved to enhance yields, improve reliability, or for other reasons, the device performance will still be retained to be close where SPICE models stand. Otherwise, SPICE models must be revised to reflect the technology changes. Aging models on the other hand, represent the worst possible reliability performance of

the technology. It is perfectly legitimate and welcome to improve the device lifetime much better than what the aging models predict. Therefore, it should not be a surprise if you find that your circuits degrade less than what the aging simulation predicts.

Like other device characteristics, degradation is also a statistical phenomenon. If we stress two identical devices with exactly the same electrical characteristics and under the same stress conditions, we will most likely get different degradations. Furthermore, devices at the fast corner do not necessarily will have smaller lifetimes than those at the slow corner, although some correlation between SPICE model corners and reliability corners may exist. Most existing aging models do not include degradation variation effect, which means aging model will give the same degradation for two identical devices under the same stress. Without the variation effect, aging models will not be able to simulate the mismatch drift due to stress. Therefore it's inappropriate to use aging models to simulate Vccmin drift for SRAM cell aging. One of the reasons for not including the variation effect in aging models is the huge amount of effort needed to collect data for statistical aging model creation. In addition to the mismatch, it's also difficult to accurately simulate aging effect for analog parameters also mainly because of the difficulty in collecting stable and reliable data in both fresh and stressed devices to create a statistically meaningful model for analog applications. Unless we innovate more effective ways to characterize device stress, these aging model limitations will continue to exist.

Because aging simulation involves SPICE simulation, its capability is also dictated by SPICE model restrictions. For example, SPICE models are usually only validated up to 20% over-drive (1.2*Vdd) with a temperature range of –40C to 125C. Even though the aging models may be created from stress data at higher biases and temperatures, we should not simulate device and circuit degradations outside of SPICE model's range.

A full-fledged aging model takes long time to create. A simpler alternative to simulate circuit-aging effect is the End-Of-Life (EOL) model approach. EOL model can be considered as a special corner model, in which we assume all devices reach the end of life condition (for example, Idsat degrades by 10%). This assumption may not be realistic, but EOL model can provide some circuit performance trend after devices degrade. Fig. 4 shows an example of EOL model with Idsat degraded 10% relative to other SPICE model corners.

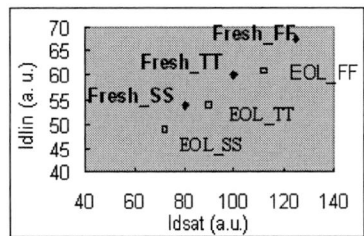

Fig. 4.    EOL models at 10% idsat degradation for TT, SS, and FF corners.

978-1-4673-6145-3/13 $31.00 © 2013 IEEE          139

EOL model is conceptually very different from aging model because it lets all devices degrade simultaneously to the EOL condition irrelevant to the waveforms and temperatures applied to the devices. This makes EOL models more suitable for memories (such as SRAM pull-up device) than for logic circuits because devices in memories are identical and are under similar stresses. If EOL model is to be used for logic circuits, we can apply the EOL models only to critical devices instead of letting all devices in the circuit reaching EOL, which will make EOL models more practical. We can extend EOL models to include variability to simulate random reliability distribution as shown in Fig. 5. Using a Gaussian distribution to represent reliability variability is somewhat artificial here because we don't have sufficient data to verify it's correctness, but EOL plus variability can provide us a quick way to analyze the impact of reliability variation to circuit performance [6], such as SRAM Vccmin, at least to the first order.

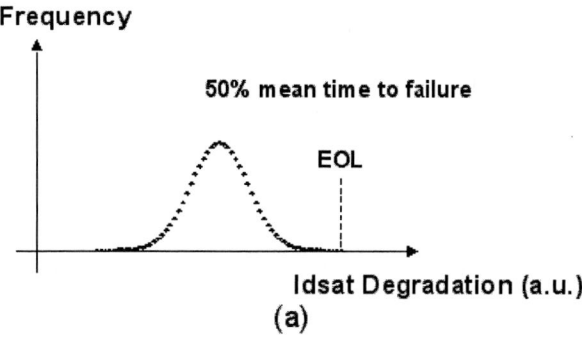

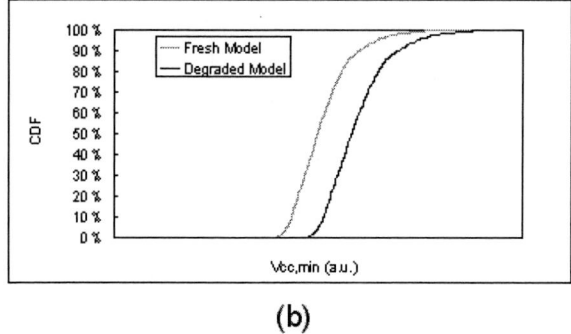

Fig. 5.     (a) An example of EOL model with variability, the tail of idsat degradation is 10%. (b) Vcc,min shift caused by EOL model with variation capability.

Although aging model has more than twenty years of development history, it becomes prevail only recently when the headroom of MOSFET's is reduced below 0.5V. A small Vth drift such as 50mV, was considered tolerable in the past, will reduce the headroom by more than 10% nowadays and may cause circuits to cease function. Circuit designers need aging models to simulate circuit performance drift with time extending the circuit sign-off conditions from PVT (Process, Voltage, Temperature) to PVTT (Time). However, as mentioned earlier, aging model cannot achieve the same level

of accuracy as that of SPICE models. Simulated relative shift and degradation trends are more meaningful than the absolute degradation numbers. That is, aging models are most appropriate to analyze the impact of over-drive to circuit lifetime relative to the normal operating condition. For example, if the simulated circuit lifetime under normal and over-drive conditions are 20 years and 4 years, respectively. We should interpret that over-drive will reduce the circuit lifetime by approximately a factor of 5 rather than taking the simulated lifetime numbers literally. Aging simulation can also help us to identify the most vulnerable devices in the circuits or the weak spots in the critical paths. Usually by enforcing a few of the most degraded devices in a circuit, the lifetime of the circuit can be greatly enhanced at a minimum cost and re-design effort.

Aging simulation should be performed on complete circuits instead of on individual functional blocks (IP's) because degradation is strongly related to the voltages across a device. For example, a NAND gate will be stressed differently in different circuits. We cannot apply aging simulations to a NAND gate to define its lifetime unless we are certain about the waveforms to be used. Nevertheless, we can use aging simulations to compare the relative reliability strength of an IP under different configurations. One example is shown in Fig. 6. In this example; a two-input NAND is used to as an inverter with three possible configurations. Using aging models, we found that Configuration-C has the best lifetime because the voltage is divided between the two NMOS during switching. In many I/O circuits, low-voltage core devices are often cascaded to interface with external high-voltages as shown in Fig. 7 [5]. This is another good application of aging models to help check that the core devices can sustain the possible over-stress.

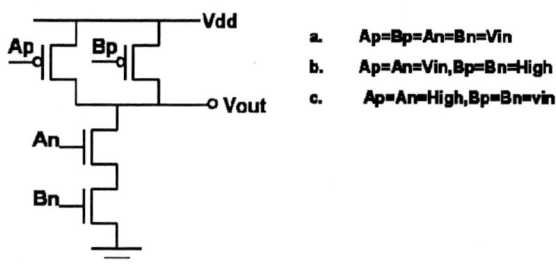

Fig. 6.     An inverter can be implemented using a 2-input NAND gate with the configurations a, b and c. Aging model could help to know configuration c have best lifetime.

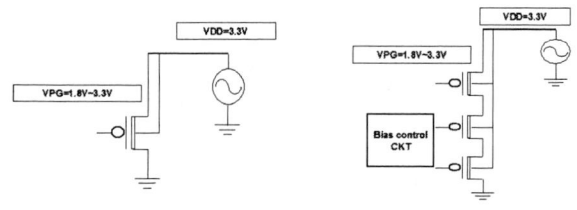

Fig. 7.     (a) Single device structure has poor reliability. (b) Multi-cascaded devices structure has better reliability due to voltage division.

## III. TMI-BASED AGING MODEL

TMI has been elected by CMC as the industry standard interface specifications between models and circuit simulators. The original goal of TMI is to augment SPICE built-in models to include layout-effects. We further extended it to enable circuit-aging simulations. TMI attaches itself to simulators through a shared library and communicate with simulators at the machine code level as shown in Fig. 8. TMI specifications allow model developers to access most of the information pertinent to MOSFETs in simulators. Because TMI is an external extension to simulators, model enhancements can be added/modified without the need to update simulators. Many commercial circuit simulators have already complied with TMI specifications; hence automatically have the TMI aging simulation capability. TMI aging model has the advantage of using a unified set of model parameters for all simulators without extra license cost. The aging simulation results are consistent among all simulators. TMI is flexible enough to define aging model equations, degradation integration algorithms, the mapping between degradation and SPICE model parameters, and information to be output, etc. Existing aging models and simulation flow in BERT can also be re-implemented in TMI, but we decided not to do so because many of the restrictions will also be carried over. In this section, we describe the existing aging model implementation and our implementation in TMI.

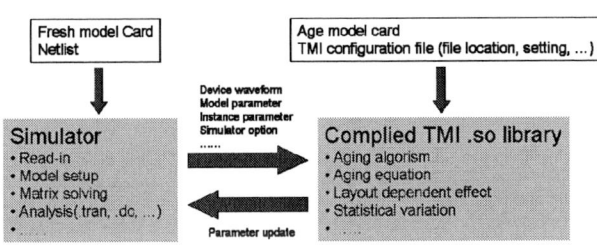

Fig. 8.     TMI interacting with circuit simulator flow.

Aging simulation involves two procedures. The first procedure calculates the degradation rates based on current SPICE simulation and projects the degradation to a future time. In BERT, this procedure is done through the "age" of the devices. We didn't adopt the "age" approach used in BERT, which requires degradation to follow a fix-sloped power-law behavior. We observed that most degradation versus time data has a one-to-one correspondence between the degradation D and the degradation rate dD/dt. Usually dD/dt decreases with D. By characterizing the relationship between D and dD/dt, we can calculate, integrate, and project the degradations directly without resorting to age. We will use an example to illustrate TMI aging model approach. For simplicity, let's assume that a device is repetitively stressed by a 3-step pulse shown in Fig. 9(a) and the degradation versus time characteristics is shown in Fig. 9(b). The amount of degradation accumulated for each pulse period (equivalent to dD/dt) depends on the initial

degradation where the stress starts. During SPICE simulation, TMI aging model calculates dD/dt for several initial degradation values and constructs dD/dt versus D relationship as shown in Fig. 9(c). Re-plotting 1/(dD/dt) versus D shown in Fig. 9(d), the area under the curve is equal to the total stress time, hence the degradation versus time under a periodic waveform stress can be obtained as shown in Fig. 9(e). This TMI aging model approach doesn't have any restrictions on the degradation versus time characteristics as long as the one-to-one relationship between D and dD/dt holds.

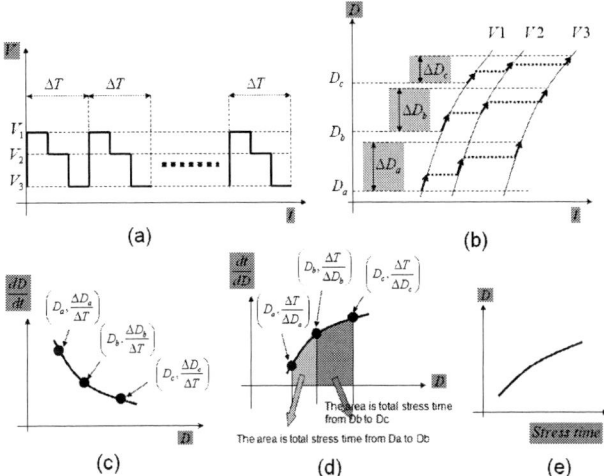

Fig. 9.     An example illustrates how TMI calculates and projects the degradation.

The second procedure in aging simulation involves SPICE model parameter adjustment. Fig. 10 shows how this is done in most commercial aging simulators. SPICE model parameters versus age are extracted and fitted to create aging models as shown in Fig. 1. During simulation, these aged SPICE model parameters are combined to calculate the degraded device characteristics. When the parameters versus age are smooth and trends are consistent, this approach works fine. However, if the SPICE model parameters are scattered over ages as illustrated in Fig. 10, the calculated degradation may not be monotonic versus age. In our experience, this situation occurred quite often because SPICE model parameters are usually not fully independent. To avoid this issue, we innovated a different approach as shown in Fig. 11. Instead of extracting many sets of SPICE model parameters for different degradations, we create a mapping matrix between degradations and model parameters. This mapping matrix is part of the aging model. The fitting error between the simulated degradation and the measured degradation is assured to be within a pre-determined spec before we release the aging model. For a given set of projected degradation targets (Idlin, Idsat, and Vth in our case), TMI calculates the corresponding SPICE model parameters from this mapping matrix on the fly. This process is very efficient and transparent to users. The degradation integration and projection algorithm is described in Fig. 11. The detailed description of degradation to parameter mapping is beyond the scope of this paper and will not be discussed here.

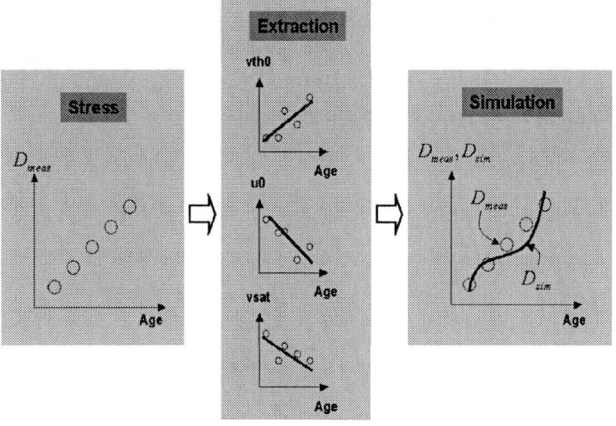

Fig. 10. Degradation simulation flow in most commercial simulators. The simulated degradation may not match measurement well.

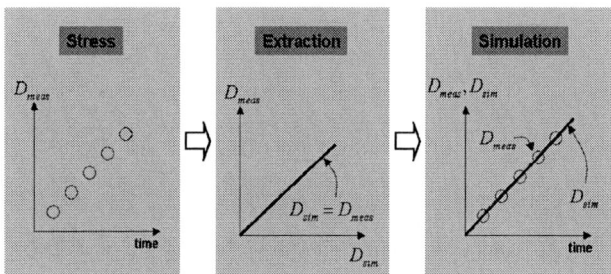

Fig. 11. Degradation simulation flow in TMI. Simulated degradation can match measurement well.

We also studied the impact of reliability on RF device characteristics. In this experiment, we first measured the I-V/C-V curves and s-parameters of a fresh MOSFET. A base-band model was created from the I-V and C-V curves. An RF model was built on top of the base-band model plus a parasitic substrate network to fit the s-parameters. Then the device was under HCI and BTI stress until Idsat degraded to around 10%. We measured I-V/C-V and s-parameters of this degraded device again, re-extracted the base-band model, and plugged in this new base-band model into the RF model without changing any parasitic component in the substrate network. We found that this revised RF model can fit the s-parameters of the degraded device well as shown in Fig. 12. This indicates that HCI and BTI only affect the intrinsic MOSFET characteristics, not the parasitic components, which agrees with our expectation. This is the first time we verified our speculation on RF device degradation experimentally. It's a relief for aging model providers to know that aging models created for base-band can be directly wrapped into RF models for simulating RF circuit aging behavior.

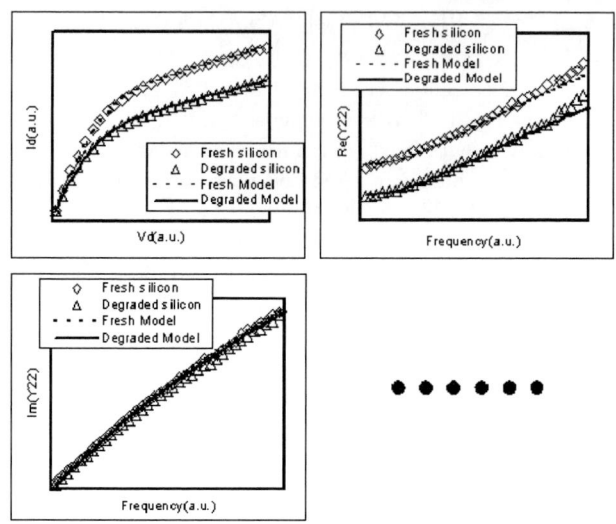

Fig. 12. DC and RF characteristics before and after stress. Only base-band model parameters were adjusted after stress.

## IV. TMI AGING SIMULATION EXAMPLES

Circuit aging simulation involves degradation projection from current time to future times. In order to correctly project future circuit performance, the simulation conditions must mimic the actual long-term operating conditions of the circuit as much as possible. For example, if an outdoor equipment is operated under a different waveform for each season. Then we will need to design multi-waveform and multi-temperature SPICE simulations to represent the different climate of each season and integrate all the degradations accumulated under all the waveforms as a whole and do the projection. In this section, we will demonstrate the TMI aging simulation of an 11-stage inverter chain under multiple-waveform and multi-temperature stress.

The waveforms we use for this example are given in Fig. 13(a). Each waveform is associated with a different temperature. It's difficult to achieve this in one transient simulation in SPICE. We separated each waveform into one SPICE simulation and instructed TMI to inform SPICE not to reset the device information of previous simulations when we start another waveform simulation so that the degradations can be accumulated for all waveforms. When the fresh-state simulations of all waveforms are completed, TMI will treat them as one unit, project the degradation for a future time, calculate the SPICE model parameter changes for each degraded device based on the mapping matrix, and simulate the aged circuit performance. The delay of the inverter chain versus operation time is shown in Fig. 13(b), which can be fitted well by a straight line in the log-log space. In fact, when the device degradation is small (say less than 10%), most digital circuit performance drifts follows the device-level degradation trend. For this example, we could have projected the degradation directly from the fresh state to 20 years in one step without the intermediate projection steps to save simulation time. In a separated simulation exercise, we considered each waveform as an individual case and summed up the degradations due to each waveform at 5 years. The result is very close to that in Fig. 13(b) initially until the device degradations become large and the two results start to deviate. This indicates that a careful selection of representative stress waveforms for your circuit is important in circuit aging simulations.

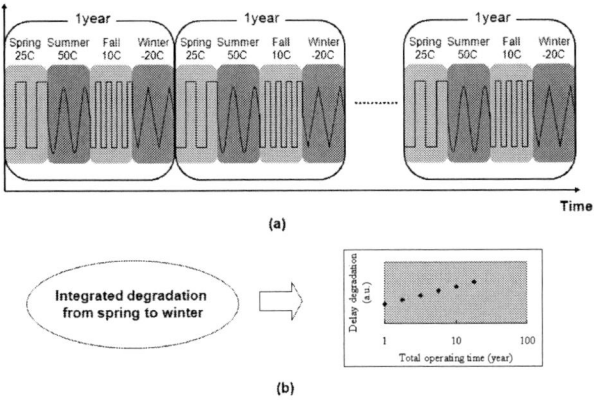

Fig. 13.     (a) Operating condition vs. time for an 11-stage inverter chain. (b) Delay degradation vs. total degradation time.

In this example, we didn't consider the BTI recovery effect, thus the simulated delay shift can be too pessimistic. It's not difficult to implement the recovery effect in TMI or in other circuit aging simulators. The main obstacle lies in how to characterize the recovery effect accurately because every time when we pause the stress to monitor the device characteristics shift, BTI recovery has occurred and the degradations have been partially recovered. It's extremely difficult to accurately quantify the amount of recovery. Instead of spending effort to characterize the recovery effect and develop the recovery model and algorithms in TMI, we implemented a scaling factor multiplied with the degradations serving to compensate the averaged device recovery. The default value of this scaling factor is 1, which means no recovery effect. A value of 0 is equivalent to full recovery. Any value between 0 and 1 represents partial recovery. The amount of recovery is highly related to the frequency, duty cycle, and the magnitude of bias change, etc. There is no simple rule to determine a suitable value for each device in a circuit. We are still looking for ways to automatically set the values based on waveforms applied to

devices. However, if the devices switch on and off frequently in a circuit as those in a ring oscillator, we can set the scaling factor to a small number like 0.2 or smaller. We also implemented a similar scaling factor for HCI. By setting the scaling factor to 0, we can turn off each reliability mechanism individually to study the impact of different reliability mechanism on your circuits. This important information can be easily obtained by simulation, but not by silicon.

Besides simulating the gradual circuit performance drift due to HCI and BTI, TMI can also serve as a convenient reliability calculator because TMI can communicate with SPICE to access most device information including the geometry, temperature, and the voltage waveforms applied to devices. After the circuit simulator completes a transient analysis, TMI can calculate the TDDB lifetime of each device type, calculate the projected lifetime of each device without running another SPICE simulation, sort the device degradations based on the criteria specified in a configuration file, etc.

## V.    CONCLUSION

We described how the aging models are implemented in TMI. A multiple-waveform and multiple-temperature stress example on an 11-stage ring oscillator is given as an example to illustrate the simulation flow. The different between existing aging models and TMI approach is highlighted. The restrictions of aging models and simulations are also discussed so that circuit aging simulation results can be interpreted more correctly.

## REFERENCES

[1]  P. Fang, J. Tao, J.F. Chen, and C. Hu. Design in hot-carrier reliability for high performance logic applications. In IEEE Custom Integrated Circuits Conference, pages 525–531, 1998.

[2]  M. A. Alam and S. Mahapatra. A comprehensive model of PMOS NBTI degradation. Microelectronics Reliability, Volume 45 Issue 1, pages 71 – 81, 2005.

[3]  R.H. Tu, E. Rosenbaum, W.Y. Chan, C.C. Li, E. Minami, K. Quader, P.K. Ko, Chenming Hu. Berkeley reliability tools - BERT. IEEE Transactions on Computer-Aided Design of Integrated Circuits and Systems,  Volume 45 Issue 10, pages 1524–1533, 1993.

[4]  E. Grossar, M. Stucchi, K. Maex, W. Dehaene. Read stability and write-ability analysis of SRAM cells for nanometer technologies. IEEE J. Solid-State Circuits, Volume 41 Issue 11, pages 2577-2588, Nov. 2006.

[5]  ]K.B. Klaassen. Digitally Controlled Absolute Voltage Division. IEEE Transactions on Iizstrumenratioiz and Measurement, Volume 24 Issue 2, pages 106-1 12, June 1975.

[6]  J.C. Lin, A.S. Oates, H.C. Tseng, Y.P. Liao, T.H. Chung, K.C. Huang, P.Y. Tong, S.H. Yau, Y.F Wang. Prediction and control of nbti - induced sram vccmin drift. IEEE International Electron Devices Meeting, pages 1–4, Dec. 2006.

# Scalable Behavior Modeling for 3D Field-Programmable ESD Protection Structures

L. Wang[1], X. Wang[2], Z. T. Shi[3], R. Ma[1], C. Zhang[1], Z. Dong[1], F. Lu[1], H. Zhao[3] and A. Wang[1]

[1]Dept. of Electrical Engineering, University of California, Riverside, USA, aw@ee.ucr.edu

[2]Omnivision Technologies, USA; [3]Marvell Semiconductor, USA

*Abstract*-**This paper reports new accurate and scalable behavioral modeling for novel 3D field-programmable ESD protection circuits using Verilog-A, which enables post-Si on-chip and in-system ESD protection design simulation and verification. New field-programmable ESD protection devices were fabricated in CMOS-compatible processes utilizing SONOS and nano crystal dots structures. The ESD behavior models were developed from ESD testing results and verified in SPICE circuit simulation.**

## I. INTRODUCTION

Electrostatic discharge (ESD) failure is a major reliability problem to integrated circuits (IC) [1, 2]. Whole-chip ESD protection circuit simulation is essential to chip-level ESD protection design synthesis, optimization, verification and prediction. Today, trial-and-error approaches still dominate in practical ESD circuit designs due to lack of accurate ESD device modeling technique and full-chip ESD simulation tools [3-5]. Several ESD device models were reported for traditional diode and MOSFET type ESD structures [6-8]. However, due to complex ESD behaviors, particularly the electro-thermal-process-device-circuit-layout coupling effects, existing ESD models have limited accuracy in describing complex ESD physics, such as thermal boundary condition and snapback I-V behavior [1-5]. On the other hand, all traditional ESD structures have fixed ESD-critical parameters [2], including ESD triggering voltage ($V_{t1}$), after Si design and fabrication are done. However, the measured ESD-critical parameters may be different from simulation design due to inevitable process, voltage and temperature (PVT) variations in fabrication. In addition, IC scaling down leads to ESD Design Window shrinking, new programmable ESD protection mechanisms hence become essential to accurate ESD protection designs [5]. Overall, field-programmable ESD protection mechanisms and structures are needed to fine-tune key ESD-critical parameters in post-Si field designs [9-11]. We devised two novel 3D field-programmable ESD protection solutions using silicon-oxide-nitride-oxide-silicon (SONOS) and nano crystal dots (NCD) structures in CMOS-compatible processes, and report accurate Verilog-A ESD behavior models for the new non-traditional ESD protection structures.

## II. FIELD-PROGRAMMABLE ESD PROTECTION

### A. SONOS ESD Protection

Fig. 1 shows new 3D SONOS ESD structure (a) and a typical ESD protection circuit scheme (b). The novel ESD protection mechanism follows: with an embedded floating SONOS gate, extra charges are stored in SONOS where the charge density can be controlled by gate programming. This modifies threshold voltage ($V_{th}$) of a MOEFET, hence

resulting in change of ESD triggering voltage ($\Delta V_{t1}$). As showed in Fig.1b, SONOS ESD structure is connected as a ggNMOS ESD device with a field-programming control to its gate. Fig. 2a shows measured ESD I-V curves for a SONOS ESD device with W10μm/L0.15μm by transmission line pulsing (TLP) ESD testing [9]. The desired snapback ESD I-V curves are readily observed. It also clearly shows that erasing operation reduces the ESD $V_{t1}$, while programming increases the ESD $V_{t1}$.

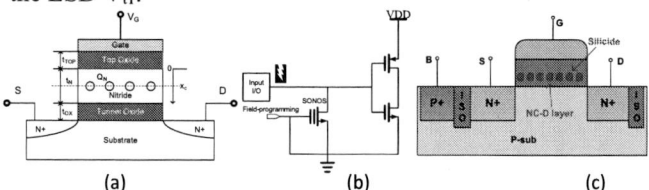

Fig. 1 SONOS (a) and NCD (c) field-programmable ESD protection structure concepts, and ESD protection circuit scheme (b).

### B. NCD ESD Protection

Fig. 1c depicts the new 3D NCD ESD structure and its ESD protection circuit scheme [10]. In principle, an NCD ESD structure is an MOSFET containing a layer of nano crystal dots, which is connected as a ggNMOS ESD structure. The ESD $V_{t1}$ programmability for NCD ESD devices is realized by charging/de-charging the NCD layer with free carriers through gate control. Fig. 2b shows measured I-V curves for a sample NCD ESD device (W100/L2μm) before and after programming by TLP testing. It clearly shows that the desired snapback ESD I-V behavior that delivers a wide $\Delta V_{t1} \sim 2V$ by programming the embedded nano crystal dots.

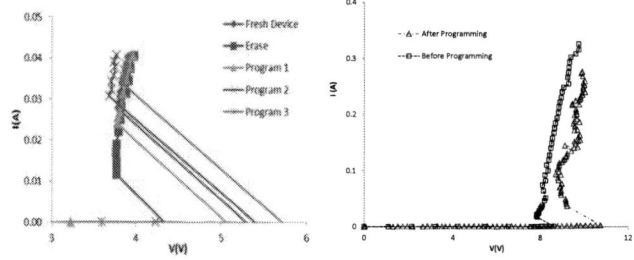

Fig. 2 Measured ESD I-V curves by TLP for SONOS (a) and NCD (b) field-programmable ESD protection structures.

## III. NEW SCALABLE ESD BEHAVIOR MODELING

To fully understand ESD protection mechanisms and further to enable whole-chip ESD circuit design simulation and verification, accurate ESD device modeling is required, which is very challenging due to the extremely complex ESD discharging behaviors. This is particularly true to model any non-traditional ESD structures, such as new SONOS and NCD

978-1-4673-6145-3/13 $31.00 © 2013 IEEE

ESD structures. The first step for ESD modeling is to understand basic ESD discharging functions, such as snapback ESD I-V curve shown in Fig. 3, which depicts the ESD-critical parameters include triggering ($V_{t1}$, $I_{t1}$), holding ($V_h$, $I_h$), ESD discharging resistance ($R_{ON}$) and ESD thermal failure threshold ($V_{t2}$, $I_{t2}$) [1]. Behavioral ESD device modeling enables accurate description of ESD device behaviors without being limited by the complex, and often unknown, ESD device physics. It overcomes the difficulties to extract complicated parameters based on high current and thermal physics. It is relatively easy to achieve a semi-physical model to describe high current ESD device behaviors using behavior language Verilog-A. It is empirical since scalability with device dimensions for key ESD-critical parameters ($V_{t1}$, $V_h$) can be established by matching TLP testing with design splits. In addition, behavioral ESD modeling also contribute to ease the convergence problem often caused by avalanche breakdown.

A new ESD behavior modeling technique is developed for the novel SONOS and NCD programmable ESD structures, which can accurately describe ESD discharging I-V behaviors piece-wisely and delivers reliable behavior ESD device models. The new ESD behavior modeling technique utilizes Verilog-A to describe complicated ESD discharging behavior by ESD-critical parameters extracted from ESD I-V curves by TLP testing, which is then verified in SPICE circuit simulation. Fig.4 shows a flow chart to develop scalable behavior ESD device models for the new SONOS and NCD ESD devices. Firstly, the relationship between ESD-critical parameters and design splits must be established. Comprehensive TLP testing, featuring ESD pulse rise time $t_1 \sim 10ns$ and duration $t_d \sim 100ns$, plays a key role in this step. After extracting ESD-critical parameters from TLP testing results and analysis, a scalability model per device dimensions can be set up. Next, the TLP

curves are divided into several section-wise segments according to ESD functions including: device off, device triggering, device snapback and ESD discharging, which may be repeating depending upon the complexity of measured ESD discharging curve for a given ESD device as shown in Fig. 5. Each segment of the TLP testing curve can be modeled by a formula to describe the corresponding ESD function in Verilog-A and the fitting parameters can be extracted from the TLP curve directly correlated with device dimensions. For example, as shown in Fig. 5, Function_off describes the device-off region of a TLP curve. An accurate scalable ESD behavior model is then obtained, which must be verified by comparing SPICE circuit simulation with TLP testing.

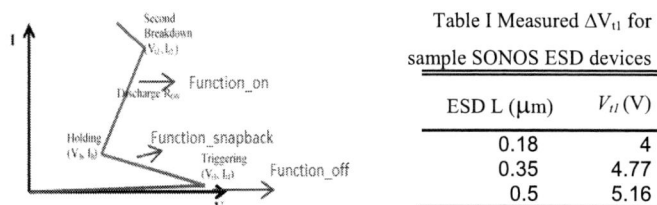

Table I Measured $\Delta V_{t1}$ for sample SONOS ESD devices

| ESD L ($\mu$m) | $V_{t1}$ (V) |
|---|---|
| 0.18 | 4 |
| 0.35 | 4.77 |
| 0.5 | 5.16 |

Fig. 5 Partition TLP curve into sections per ESD discharging functions.

### A. SONOS ESD Device Model Method

Developing accurate and scalable ESD device models requires carefully correlating the measured results with ESD design features, including field-programmable ESD device sizes, layout patterns, SONOS material properties. Table I compares measured $V_{t1}$ for sample SONOS ESD devices. Fig. 6 depicts the measured $V_{t1}$ versus SONOS ESD channel length (L) relationship, showing desired programmable ESD $\Delta V_{t1}$ by designs. Equations 1 is extracted to fit into measured ESD behaviors for Verilog-A modeling. Similarly, all other ESD-critical parameters ($I_{t1}$, $V_h$, $I_h$, $R_{ON}$, $V_{t2}$, $I_{t2}$) can be extracted from TLP testing and described in Verilog-A. ESD models were extracted for the field-programmable ESD structures with varying design dimensions in this work, which leads to scalable behavior ESD models.

$$V_{t1} = 4.53(L - 0.18) + 4 \qquad (1)$$

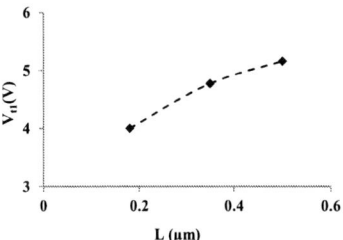

Fig. 6 Measured $V_{t1}$ versus device size for sample SONOS ESD structures.

A W10/L0.5$\mu$m SONOS ESD device is selected as an example to describe how the behavior model is generated. Fig. 7 shows that the measured TLP curve is divided into multiple sections according to its complex ESD discharging behaviors, with its fitting formulas for different I-V sections listed below.

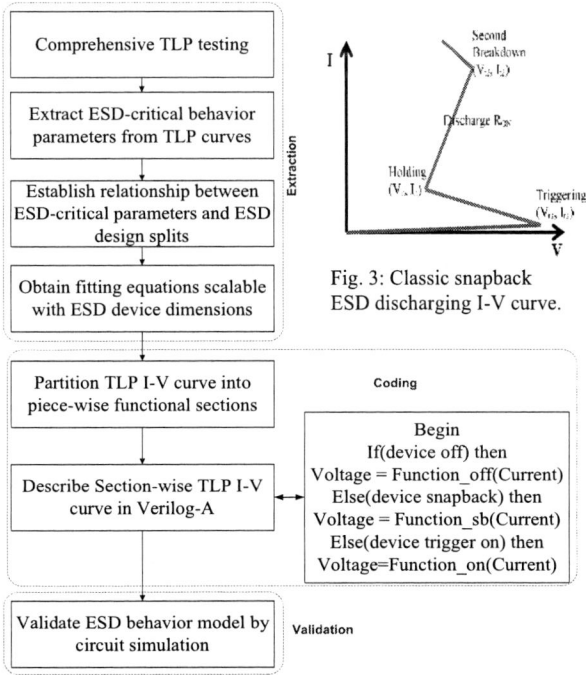

Fig. 3: Classic snapback ESD discharging I-V curve.

**Coding**

Begin
If(device off) then
Voltage = Function_off(Current)
Else(device snapback) then
Voltage = Function_sb(Current)
Else(device trigger on) then
Voltage=Function_on(Current)

Fig. 4 A flow chart of ESD behavior modeling for new field-programmable SONOS and NCD ESD protection structures by Verilog-A.

978-1-4673-6145-3/13 $31.00 © 2013 IEEE          145

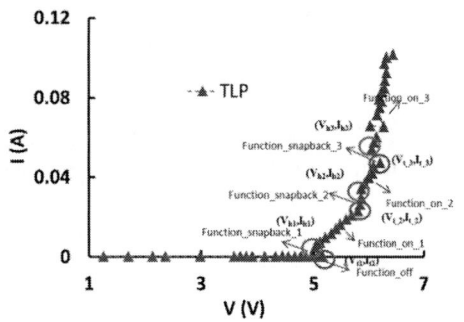

Fig. 7: TLP curve partition for a W10/L0.5μm SONOS ESD device.

If ($V<V_{t1}$)

$V <+ A*I$;    (Function_off)

Else if ($V<V_{h1}$)

$V <+ B+C*I$;    (Function_snapback_1)

Else if ($V<V_{t\_2}$)

$V <+ D+E*I$;    (Function_on_1)

Else if ($V<V_{h2}$)

$V <+ F+G*I$;    (Function_snapback_2)

Else if ($V<V_{t\_3}$)

$V <+ H+I*I$;    (Function_on_2)

Else if ($V<V_{h3}$)

$V <+ J+K*I$;    (Function_snapback_3)

Else if ($V<V_{t2}$)

$V <+ L+M*I$;    (Function_on_3)

Parameter A is achieved from $I_{t1}$, which can be extracted from measured TLP curve and $V_{t1}$ that is obtained from equation 1, correlated with device channel length. Similarly, fitting parameters B~M can be obtained either from correlated with device dimension or extracted from TLP curve.

### B. NCD ESD Device Model Method

NCD ESD device behavior models are generated similarly. Tables II compares measured $V_{t1}$ for sample NCD ESD devices. Fig. 8 depicts the measured relationship between ESD Vt1 and L of NCD ESD devices, which shows desired programmable ESD $\Delta V_{t1}$ by design variation. Equations 2 is obtained to fit into measured ESD behavior model parameters using Verilog-A.

Table II Measured $\Delta V_{t1}$ for sample NCD ESD devices

| ESD L (μm) | $V_{t1}$ (V) |
|---|---|
| 1 | 7.9 |
| 2 | 8.6 |
| 3 | 8.95 |

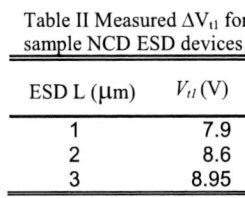 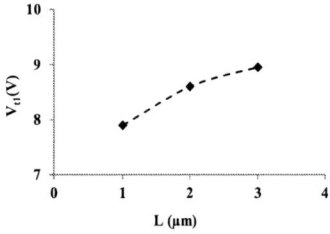

Fig. 8: Measured $V_{t1}$ versus device size for sample NCD ESD structures.

$$V_{t1} = 0.7(L-1)+7.9 \qquad (2)$$

A W/L=100/1μm NCD ESD device is used to show ESD behavior modeling procedures. Fig. 9 shows its TLP curve that can be divided into three ESD functional sections with the fitting formulas given below for model parameter extraction.

If ($V<V_{t1}$)

$V <+ A*I$;    (Function_off)

Else if ($V<V_h$)

$V <+ B+C*I$;    (Function_snapback)

Else if ($V<V_{t2}$)

$V <+ D+E*I$;    (Function_on)

Parameter A can be obtained from $I_{t1}$ that is extracted from measured TLP curve and $V_{t1}$ that is obtained from equation 2, correlated with device channel length. Similarly, fitting parameters B~E were generated either by correlating with device dimension or extraction from TLP curve.

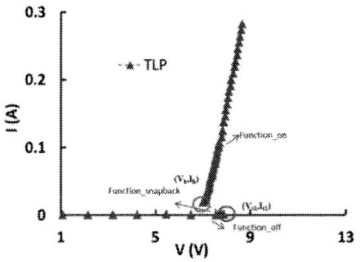

Fig. 9: TLP curve partition for a W100/L1μm NCD ESD device.

## IV. ESD BEHAVIOR MODEL VERIFICATION

The newly developed scalable ESD behavior models for SONOS and NCD ESD structures were fully verified by SPICE circuit simulation and TLP testing. Fig. 10 shows a schematic for ESD circuit simulation using the ESD behavior models developed. ESD circuit simulation was conducted using HBM ESD standard and compared with TLP testing for a group of new SONOS and NCD ESD structures.

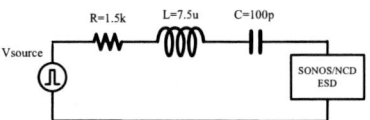

Fig. 10 ESD circuit simulation schematic.

### A. SONOS ESD Device Model Verification

Many new SONOS ESD devices with varying layout dimensions were used for behavior model extraction and verification. Figs. 11-12 show ESD I-V curves from TLP testing and SPICE circuit simulation using the new ESD behavior models developed for two sample ESD devices (W/L=10/0.18μm, 10/0.5μm). Clearly, simulated ESD circuit function using the new behavior models matches TLP testing curve very well. Further, good agreement between TLP testing and SPICE simulation holds for SONOS ESD devices with varying dimensions, confirming that the new ESD behavior modeling technique developed is not only accurate, but also scalable to layout variation, which is very important for practical on-chip ESD protection circuit designs because the actual ESD design dimensions may vary to meet specific circuit requirements including programmable ESD $\Delta V_{t1}$ and ESD protection level, etc.

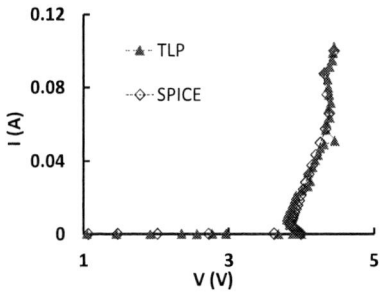

Fig. 11 SPICE circuit simulation for a W10/L0.18μm SONOS ESD device using ESD behavior model matches TLP testing curve well.

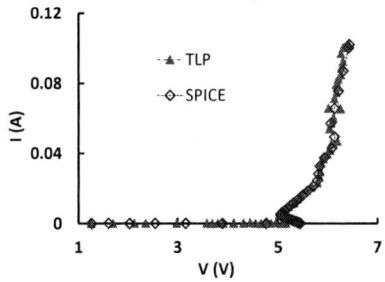

Fig. 12 SPICE circuit simulation for a W10/L0.5μm SONOS ESD device using ESD behavior model matches TLP testing curve well.

### B. NCD ESD Device Model Verification

Similarly, new NCD ESD device of different design splits was studied for modeling verification. Figs. 13-14 compare ESD I-V curves from TLP testing with SPICE circuit simulation using new ESD behavior model for two sample NCD devices (W/L=100/1μm, 100/2μm), which again show excellent agreement and accuracy, as well as scalability of the new ESD behavior models. Further, to enable field design variations, behavior models for ESD devices under different programming conditions are developed. Fig. 15 compares TLP and SPICE simulation curves for a sample NCD device after programming, which shows good agreement (the same device in Fig. 14 tested before-programming). Similarly, SONOS ESD behavior models were verified before and after ESD programming. It clearly shows that the new scalable ESD behavior modeling technique allows to develop accurate ESD device models for the novel field-programmable ESD structures, which enables whole-chip ESD circuit simulation, synthesis, optimization, verification and prediction in designs.

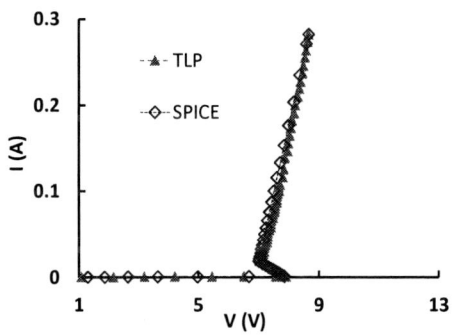

Fig. 13 SPICE circuit simulation for a W100/L1μm NCD ESD device using ESD behavior model matches TLP testing curve well (Fresh).

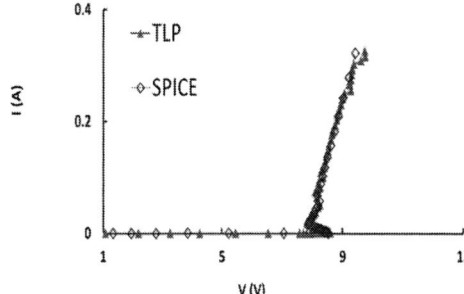

Fig. 14 SPICE circuit simulation for a W100/L2μm NCD ESD device using ESD behavior model matches TLP testing curve well (before programming).

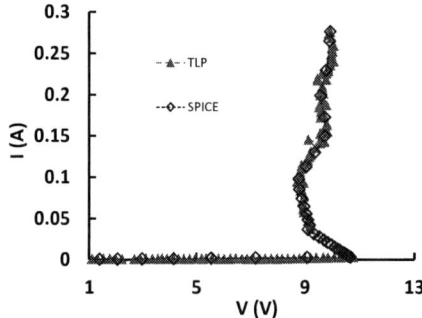

Fig. 15 SPICE simulation for a W100/L2μm NCD ESD device (Fig. 14) using behavior model matches TLP testing curve well (after programming).

## V. CONCLUSION

We report a new scalable ESD behavioral modeling technique using Verilog-A to develop accurate ESD behavior models for novel 3D field-programmable ESD protection structures, including SONOS and NCD ESD devices. The new ESD behavior modeling technique was fully verified by SPICE circuit simulation and TLP testing, which will enable whole-chip ESD circuit design optimization and verification.

### REFERENCES

[1] A. Wang, *On-Chip ESD Protection for ICs*, Kluwer, 2002.

[2] A. Wang, et al, "A review on RF ESD protection design", *IEEE TED*, p1304, July 2005.

[3] R. Zhan, et al, "ESDInspector: A new layout-level ESD protection circuitry design verification tool using a smart-parametric checking mechanism", *IEEE Trans. CAD of ICs & Sys.*, p1421, Oct. 2004.

[4] R. Zhan, et al, "ESDExtractor: A new technology-independent CAD tool for arbitrary ESD protection device extraction", *IEEE Trans. CAD of ICs & Sys.*, p1362, Oct. 2003.

[5] L. Lin, et al, "Whole-chip ESD protection design verification by CAD", *Proc. EOS/ESD Symp*, p28, 2009.

[6] A. Amerasekera, et al, "Modeling MOS snapback and parasitic bipolar action for circuit-level ESD and high current simulations," *IEEE Circuits and Devices Magazine*, p7, 1997.

[7] J. Li, et al, "Compact modeling of on-chip ESD protection devices using Verilog-A," *IEEE Trans. CAD of ICs & Sys.*, p1047, 2006.

[8] W. Li, et al, "A scalable Verilog-A modeling method for ESD protection devices," *Proc. IEEE EOS/ESD Symp.*, p1, 2010.

[9] J. Liu, et al, "Field Programmable SONOS ESD Protection Design", *Proc. IEEE CICC*, p1, 2012.

[10] Z. Shi, et al, "Programmable on-chip ESD protection using nano crystal dots mechanism and structures", *IEEE Trans. Nanotech.*, p884, Sept. 2012.

[11] X. Wang, et al, "Post-Si programmable ESD protection circuit design: mechanisms and analysis", *IEEE JSSC, V48, N5*, pp.1237, May 2013.

# Quasi-3D method: time-efficient TCAD and Mixed-Mode Simulations on finFET Technologies

Geert Hellings, Shih-Hung Chen, Dimitri Linten
imec, Leuven, Belgium
e-mail:geert.hellings@imec.be

Mirko Sholz
imec, Leuven, Belgium
and University of Brussels, Dept.
Electrical Engineering
Brussels, Belgium

Guido Groeseneken
imec, Leuven, Belgium
and University of Leuven, Dept.
Electrical Engineering
Leuven, Belgium

*Abstract*—— The Quasi-3D allows to drastically speed up TCAD and mixed-mode simulations of finFET technologies, by solving on well-chosen 2D finFET cross sections. The method accurately reproduces important transistor metrics requiring only 1/20$^{th}$ of the simulation time.

*Keywords—TCAD, Modeling, Mixed-Mode simulations, bulk finFET, 3D simulations.*

## I. INTRODUCTION

With the introduction of the 22nm node, finFET technologies are quickly becoming industry standard. The main reason for this change in transistor architecture is the improved electrostatic gate control, compared to planar devices with the same footprint [1]. While SOI-based finFETs [2] have long co-existed with their bulk finFET counterparts, it seems the latter will be dominating at least the next few technology nodes [3, 4]. In a bulk finFET technology, the fin structures are fabricated starting from a bulk Si substrate. As such, the fins are directly connected to Si substrate (e.g. fig. 1).

Finite-element simulations have since long been used for technology exploration and optimization. However, the three-dimensional nature of the finFET introduces a dilemma.

On the one hand, simulating the entire 3D structure allows accurate modeling of short channel effects and fin-substrate interactions (e.g. drain-to-substrate leakage), at the expense of simulation time in comparison to a 2D simulation. This difference can be quite large (slowdowns of 10-100× are common). Consequently, circuit oriented mixed-mode and/or transient simulations, which are often already time-consuming using 2D models, become very cumbersome.

On the other hand, simulating 2D cut planes or a 2D projection of the finFET will have a positive effect on simulation times, but risks omitting critical effects. Typically, such simulations yield reliable results only for events which are confined to the top part of the Si fin (e.g. ON-state current). An example of such an approach is modeling the finFET as a dual-gate structure [5] or as a cylindrical nanowire: both allow using a 2D simulation domain, but are unable to capture short channel effects and fin-substrate interactions in a realistic way, especially in bulk finFET technologies.

When one is requiring mixed mode simulations, this dilemma often becomes particularly persistent. In this work, the simulation of an ESD-pulse (Electrostatic Discharge) is

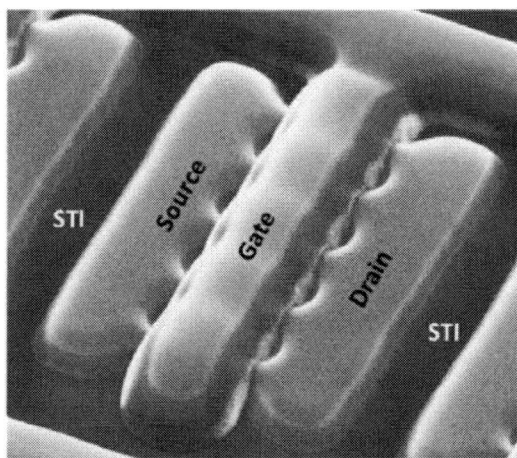

Figure 1: Electron Microscopy image of a typical finFET in the 65-nm technology used in this work.

used as an example of such a mixed mode simulation. Since many ESD-related effects occur far below the gate oxide, one has to consider such a large 3D volume that a transient or mixed-mode simulation becomes unpractical. A fast simulation setup would be particularly valuable to investigate trends, e.g. the effect of layout parameters on ESD performance.

In this work, a method is developed which allows simulating those essential 3D effects in bulk finFETs, using well-chosen 2D cross sections. In section III, this method is compared to slower, full-3D simulations. Its results are also benchmarked against measured transistor data. In section IV, this proposed method is used to simulate the response of a gated diode in a 65nm finFET technology, to an ESD event

## II. QUASI-3D METHOD

This section introduces the Quasi-3D method and discusses the simulation setup.

### A. Quasi-3D method summary

As mentioned in the introduction, a classical 2D approximation of the finFET would be a dual-gate structure. This approach is shown schematically in fig. 2(a): a plane A horizontally cuts the fin, yielding a dual gate structure. This

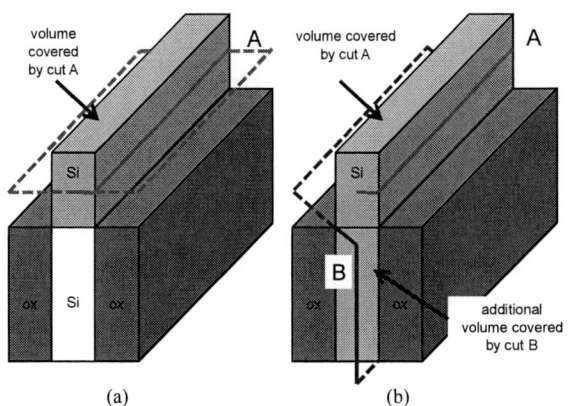

(a)                                    (b)

Figure 2 Schematic image showing 2 methods to convert a 3D finFET structure to its 2D equivalent: (a) classical approach, cutting the fin horizontally, yielding a dual gate structure. (b) Quasi-3D approach, which includes a second cut B. Note that in (b) the volume below the actual fin is also modeled.

dual gate structure covers a volume located at the top of the fin, delimited by a gate electrode left and right (not shown). The volume in between the STI oxide is not taken into account in such a simulation.

The new method introduces a second, kinked cut plane B shown in fig. 2(b). The largest part of this cut plane runs vertically through the middle of the fin. As such, it allows also covering the additional Si volume below the actual fin. the top part of this volume can still be influenced to some extent by the gate electrode – an effect which quickly diminishes at greater depths. To capture this effect, the plane is kinked 45° (see fig.2(b) ), making sure this second cross section (B) ends up in the gate electrode on the top.

The result is a set of two cut planes, which can be placed in parallel to represent the full finFET. To this end, the current in both cuts should be normalized considering that cut A covers the top part of the fin, yielding $W_{norm}=2 \times H_{fin}$ (with $H_{fin}$ = fin height). For cut B, $W_{norm}=W_{fin}$ (with $W_{fin}$ = fin width). This normalization is used throughout this paper.

III.  BENCHMARKING TO A 65-NM FINFET TECHNOLOGY

This section benchmarks the simulation results obtained using the Quasi3D method with those of a full 3D simulation and measurements, using a $L_G$ = 65nm finFET technology.

A. Simulation setup

This method was applied to a $L_G$ = 65nm finFET technology, starting from a calibrated 3D TCAD setup containing both the specifics of the fabrication process and relevant electrical models. A 3D model of p-well gated diodes in this technology is shown in fig. 3(a). Applying the Quasi-3D method to this structure yields two 2D cuts. The first arises from cutting the fin horizontally, with the plane A. This results in a dual gate structure, since the fin is delimited by a gate electrode on both sides. Because of symmetry, only one half of

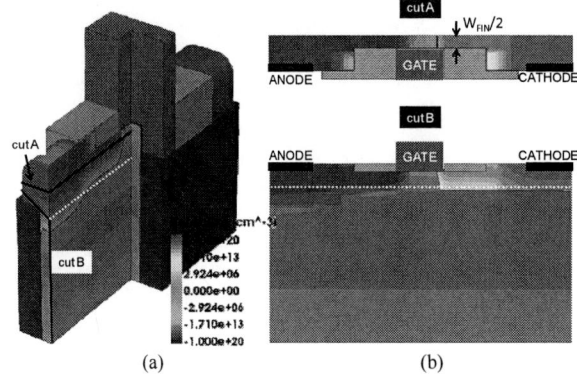

(a)                                    (b)

Figure 3: (a) 3D simulation of a finFET gated p-well diode, showing active doping concentration. (b) Quasi-3D equivalent of this structure, resulting from two cuts along cut A (parallel to the wafer plane) and cut B (vertical along the fin, with a 45° kink).

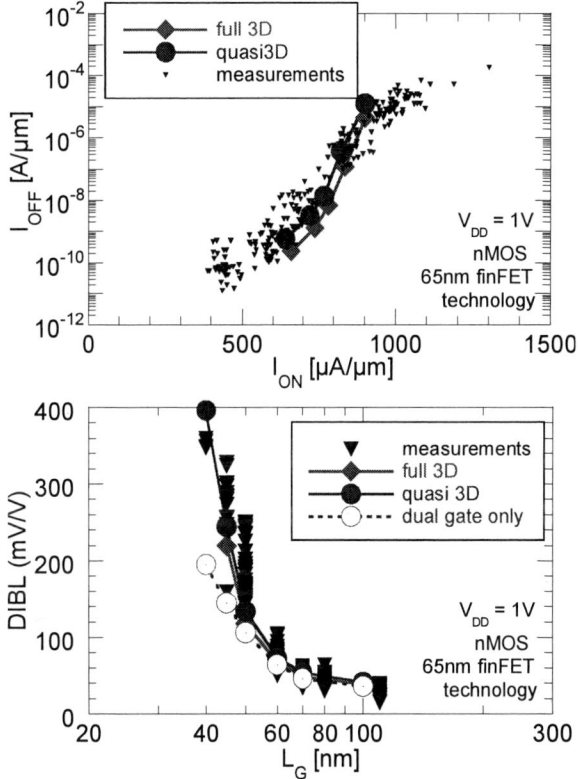

Figure 4: Comparison between full 3D TCAD simulations and the Quasi-3D method presented in this work and measured values for (a) $I_{ON}$-$I_{OFF}$ and (b) DIBL vs. $L_G$ on nFETs.

this dual gate structure is simulated, yielding the upper structure – cut A in fig. 3(a). The second arises from cutting the fin along the kinked cut plane B, yielding the bottom structure in fig. 3(b). Placing cut A and cut B in parallel, one obtains the quasi-3D equivalent of the gated diode in this technology.

978-1-4673-6145-3/13 $31.00 © 2013 IEEE

## B. Benchmarking

The method was applied to nFET core transistors and compared to experimental results. The gate length $L_G$ was varied from $L_G = 40$ to $110$ nm on the test chip and between $45$ and $90$ nm in simulations. The result of this comparison is plotted in fig.4(a-b). Results obtained with the Quasi-3D simulations are very close to those from full 3D simulations, using the classical calibrated TCAD deck arising from process simulations: The $I_{ON}$-$I_{OFF}$ comparison is shown in fig. 4(a). Fig. 4(b) shows DIBL as a function of gate length, a common metric for the presence of short channel effects (Drain-Induced Barrier Lowering [mV/V]).

Fig. 4(b) also clearly shows the need to include cut B in the simulations, as simulations that only consider cut A (=dual gate only), yield much lower DIBL values than are observed in measurements.

Finally, the simulation time is compared for the full 3D TCAD and the Quasi-3D simulations using a commercial TCAD simulator [6]. Note that the meshing in the 3D TCAD was heavily optimized to increase speed. For this case (full $I_D$-$V_G$ of a finFET nFET for low and high $V_D$), the simulation time is reduced from 30 minutes to 1.5 minutes using the Quasi-3D method, yielding results 20× faster.

## IV.  MIXED-MODE ESD SIMULATIONS

This section gives some application examples of the Quasi-3D method using ESD-oriented mixed mode simulations. The response of an ESD protection device to an HBM pulse (Human Body Model) is simulated electrically (full 3D and Quasi-3D) and compared to measurements.

### A. Pwell gated diode for ESD protection

A gated pwell diode shown in fig. 3 was imported in to a mixed-mode simulation tool [7], allowing to evaluate the Quasi-3D method under ESD conditions. To this end, both cuts were connected in parallel and subjected to an HBM pulse to trace the devices' response. Measurements were conducted on a gated diode, consisting of 400 parallel fins. Consequently, the width of the simulated cuts was adjusted to match this configuration. This yields an effective width of $400 \times 60$ nm for cut A (2× the fin height per fin), and $400 \times 24$ nm for cut B (for a fin width of 24nm). A schematic of the mixed-mode setup is shown in fig. 4(a).

The results obtained from these simulations were compared to TLP measurements on a $L_G$=45nm gated diode in the finFET technology. Measured devices show a failure current $IT_2 = 2$A, and average $R_{ON}$ around 0.75 Ω. The measured, simulated Quasi-3D and simulated full 3D gated diodes show very similar characteristics, as shown in fig. 4(b). Remaining differences (currents in excess of 1.5A ) between the full 3D and the measured data are due to the fact that no heating was simulated. Secondly, for these high currents, a small offset (about 0.2Ω) in $R_{series}$ exists between the full3D and quasi-3D simulations. In such cases where this level of accuracy is required, these effects can be calibrated for.

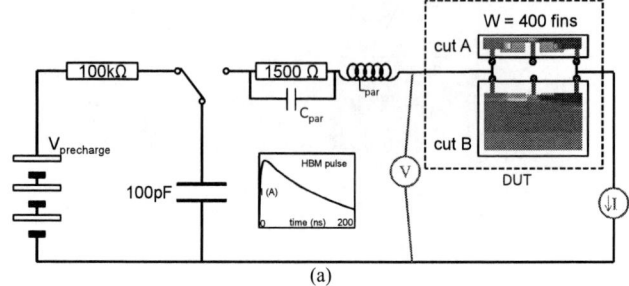

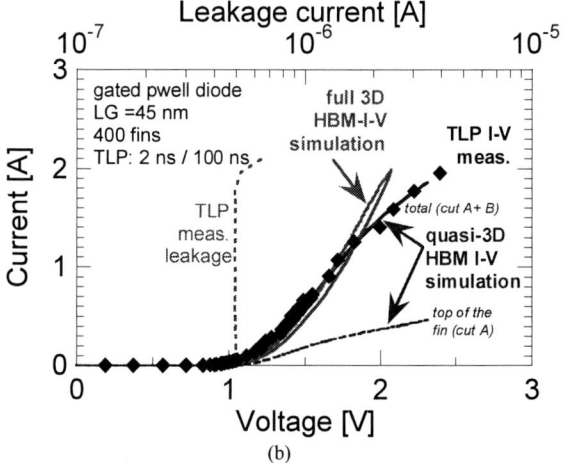

Figure 4 (a) schematic of the simulation setup: a fin-based gated diode, consisting of 400 fins is connected to ESD pulse source in the mixed mode simulator. (b) Comparison between measured TLP characteristics of this gated diode and the simulations (TLP and HMB I-V). Clearly, the dual-gate only approach is insufficient for ESD-related simulations (mixed-mode simulation time: 1'42")

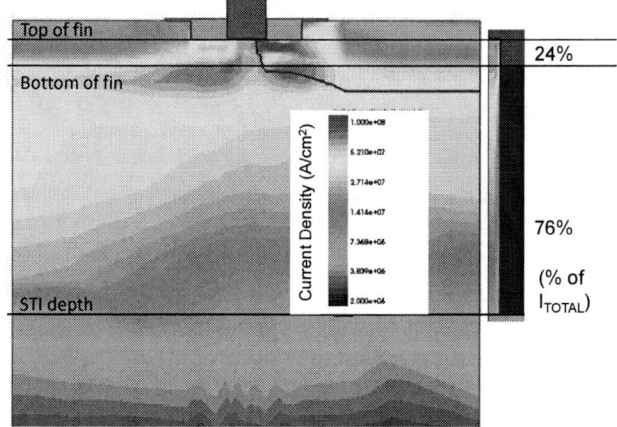

Figure 5: Current distribution of the fin-based gated diode at the peak of the ESD pulse, showing the current distribution through the fin (full 3D Mixed-Mode simulations). A duals gate-based approach would only consider the top part of the fin, carrying less than a quarter of the total current.

An interesting observation from these simulations is that the bottom part of the fin (captured by but B) carries most of the ESD current. The full 3D and the Quasi 3D simulations predict 76% and 83% (respectively) of the current to be in this area respectively. This signifies that an approach where only the top part of the fin is modeled (cfr. dual gate) would be completely insufficient to simulate ESD-related characteristics (see also fig. 5). Ignoring the bottom volume would mean that one assumes all ESD current flows in the top of the fin.

## V. CONCLUSIONS

In this work, a method is presented to reduce a full 3D simulation of a finFET to a set of 2D cut planes. This Quasi-3D equivalent was shown to accurately reproduce important core transistor characteristics ($I_{ON}$, $I_{OFF}$, DIBL), while delivering 20× faster simulations. Because it considers the full fin (also the part below the gate), this method is shown to produce more accurate results than the classical 2D dual-gate approximation of the finFET.

Secondly, the method is applied to mixed-mode ESD simulations. TLP measurements on a fin-based gated diode were accurately reproduced and show agreement with the full 3D TCAD. Hence, the Quasi-3D method enables a more time-efficient development using mixed-mode simulations of e.g. ESD protection strategies.

## ACKNOWLEDGMENT

The authors gratefully acknowledge the support from imec's LOGIC-INSITE partners and Angstrom Design Automation.

## REFERENCES

[1] T. Chiarella et al., Benchmarking SOI and bulk FinFET alternatives for PLANAR CMOS scaling succession, Solid-State Electronics, Vol. 54, nb. 9, pp. 855-860, 2010.

[2] M. Poljak et al, *"Technological constrains of bulk FinFET structure in comparison with SOI FinFET,"* Semiconductor Device Research Symposium, vol., no., pp.1-2, 2007

[3] C.-H. Jan et al., A 22nm SoC Platform Technology Featuring 3-D Tri-Gate and High-k/Metal Gate, Optimized for Ultra Low Power, High Performance and High Density SoC Applications, *International Electron Devices Meeting – IEDM*, pp.44-47, 2012.

[4] M.J.H. van Dal, Demonstration of scaled Ge p-channel FinFETs integrated on Si, *International Electron Devices Meeting – IEDM*, pp.521-524, 2012.

[5] D. Tremouilles et al., "Understanding the optimization of sub-45nm FinFET devices for ESD applications," *29th Electrical Overstress/Electrostatic Discharge Symposium*, pp.7A.5-1-7A.5-8, 2007.

[6] Sentaurus suite v2012-06, Synopsys inc., Mountain View, CA, USA.

[7] Decimm, v5.5.0, Angstrom Design Automation Ltd., San Jose, CA, USA.

# Gate Stack Resistance and Limits to CMOS Logic Performance

R. A. Wachnik*, S. Lee, L. H. Pan*, N. Lu, H. Li*, R. Bingert **, M. Randall*, S. Springer, C. Putnam

IBM Semiconductor Research and Development Center, Essex Junction, VT 05452
*IBM Semiconductor Research and Development Center, Hopewell Junction, NY 12533
**ST Microelectronics, Hopewell Junction, NY 12533 USA

*Abstract*-The input resistance of CMOS circuits is a measurable limit on the performance of typical static CMOS logic gates. A survey of measured data from five generations of CMOS technology including polysilicon oxynitride gate first stacks (P-SiON), high-K metal gate first stacks (GF), and high-K replacement metal gate stacks (RMG) shows a trend of increasing gate resistance. We show DC and RF measurements may be analyzed to determine horizontal and vertical components of gate resistance in terms of scalable parameters and the sum of these components may be represented by a compact scalable equation representing total gate resistance. Gate resistance increases at advanced nodes and affects typical logic performance of a 20nm replacement gate technology.

## I. INTRODUCTION

For silicon integrated circuit technologies with minimum Field Effect Transistor (FET) gate length less than ~40nm the semiconductor industry has migrated from polysilicon electrodes on oxynitride gate stacks to metal electrodes on high-K dielectrics to control short channel effects [1, 2, 3, 4]. Silicided polysilicon disposed on oxynitride gate stacks have been replaced by gate first stacks comprising silicide, polysilicon, and work function metals disposed on hafnia based dielectrics or by replacement gate stacks comprised of fill, barrier, and work function metals disposed on hafnia based dielectrics. Gate first patterning schemes offer the advantage of compact layout. Gate last or replacement gate schemes offer advantageous placement of thermal treatment during the process sequence. Scaling of device dimensions has increased the relative contribution of parasitic elements in determining drive current and capacitance. Gate electrode resistance contributes to the input resistance of radio frequency (rf) circuits and this contribution may be extracted from high frequency power measurements using s-parameters [5, 6, 7]. Industry standard rf compact FET model formulations for rf circuit design include support for gate resistance [8] but these standard formulations do not incorporate scaling of vertical resistance components. This work will show these vertical components are detectable in many gate stacks below the 65nm node and introduce significant deviations from simple model formulations. A simple physical model may be used to understand observations within and across technologies.

## II. EXPERIMENTAL DATA

The wafers were fabricated using silicon integrated circuit technologies as described in the references [1, 2, 3, 4, 5, 9]

and the specific wafers used for these studies were carefully screened for uniform and well controlled physical dimensions and electrical parameters. With the exception of the 90nm gate first SOI technology [9], all wafers used high-k metal gate integration schemes. Multi-finger field effect transistors of near minimum channel length with scaled device width were embedded in high frequency test structures. S-parameter measurements using open-short de-embedding at the frequencies of interest may be converted to a hybrid parameter representation within which the real part of the $H_{11}$ parameter

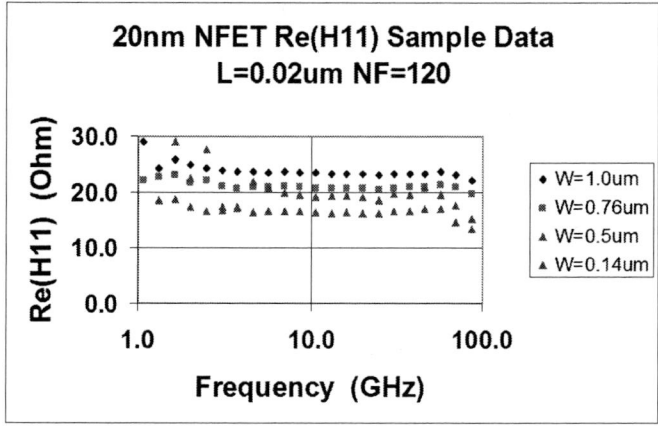

(a)

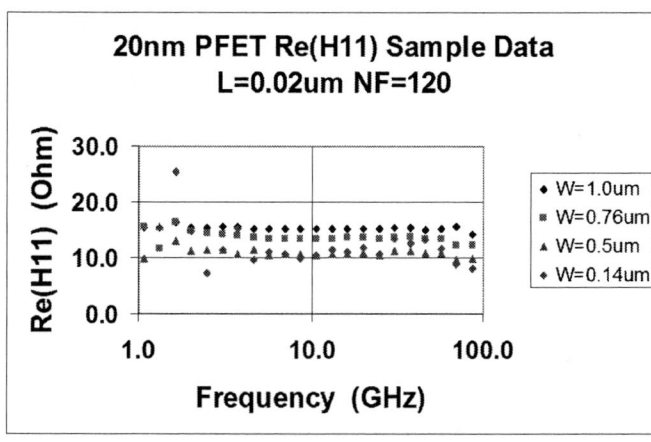

(b)

Figure 1. Re($H_{11}$) from s-parameter measurements from four NFETs (a) and four PFETs (b). The active area width or Wfinger is shown in the legend. The nominal total width is NF*Wfinger.

represents the gate resistance and a portion of the channel and the source-drain resistance. Gate metal sheet resistance and source drain resistance elements were determined from separate measurements on similar structures. After de-embedding, transforming to hybrid parameters, and taking the real part of $H_{11}$, typical measured resistance data for 20nm n-type transistors and p-type transistors are shown in Fig. 1 as a function of frequency. Measured values from four different widths of device active area (i.e. four finger lengths) from a typical chip are shown in the figures and it can be seen that the value of $Re(H_{11})$ is nearly constant between a few GHz and a several tens of GHz. The gate length and number of fingers is given in the figure and the nominal total device width is computed by the product of finger length and number of fingers.

Similar analysis of s-parameter data from a 32nm SOI technology [2] shown in Fig. 3 may also be interpreted with the gap between $Re(H_{11})$ and the silicide resistance as representing a vertical resistance. Note this resistance may be modulated by reducing doping in the silicon region.

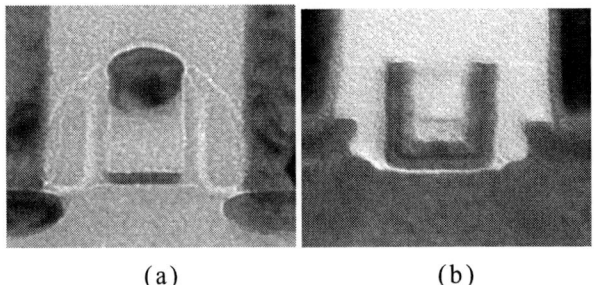

(a)                    (b)

Figure 4. The 32nm gate first hiK metal gate stack (a) has a dark metal silicide region on top and a dark work function metal region on the bottom [1]. The 20nm gate last or replacement gate stack (b) has a U shaped region of work function metals with a lighter core of metal fill [2]

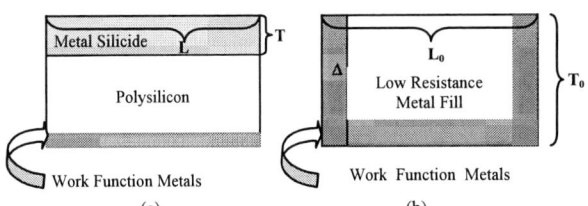

(a)                    (b)

Figure 5. Compare the schematic drawings above with the transmission electron micrographs in Figure 4. For a gate first stack (a) current is conducted by the metal silicide and characterized by a sheet resistance $r_{sh} = \rho_{MSi}/T$. For a replacement gate stack (b), the work function metals have uniform thickness $\Delta$.

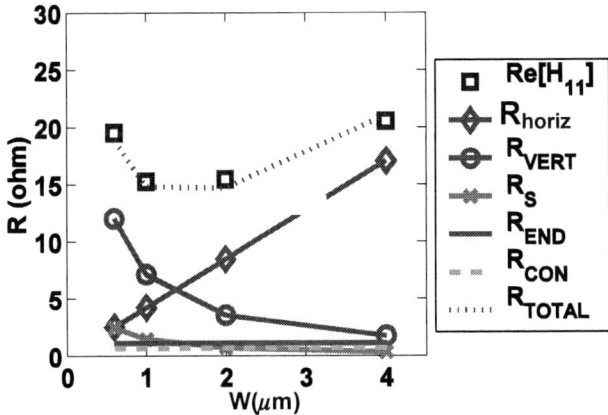

Figure 2. $Re(H_{11})$ for 90nm SOI NFETs with P-SiON gate stack, having 20 fingers with four different finger widths (0.6, 1.0, 2.0, 4.0 μm), measured at $V_{gs}$=0.6V, $V_{ds}$=0.7V. See text.

Fig. 2 shows typical data with $V_{gs}$=0.6V $V_{ds}$=0.7V for a 90nm SOI technology [1] which uses a silicided polysilicon oxynitride gate stack. The gate resistance exceeds the value of $R_{horiz}$ predicted from measured silicide sheet resistance. This excess resistance, inversely scaling with transistor gate finger width, may be assigned to a vertical gate ($R_{vert}$), channel and source/drain resistance ($R_S$). The components of gate contact ($R_{con}$) and end ($R_{end}$) resistance with optimized layout are shown to be smaller.

The relationship between material parameters, integration scheme, and resistance of gate stacks is evident from inspection of the transmission electron micrographs in Fig. 4 and redrawn as a cartoon in Fig. 5. A high K metal gate scheme as shown in the left side of Fig. 4 has a silicon layer between the metal work function layers and a top silicide layer. The sheet resistance is dominated by the low resistance silicide layer. A well devised process will minimize the resistance of the silicon to metal Schottky barriers by maximizing doping atom concentration at the interfaces. A replacement gate scheme as shown in the right side of Fig. 4 has a liner comprised of work function metals and diffusion barriers. The sheet resistance is dominated by the fill which is chosen to have a much lower resistance when compared to the barrier and work function metals. The area fraction in cross section of low resistance metal has been the standard simple model for representing the resistance of multi-level metallization and this model should apply to the horizontal component of a replacement gate scheme. It is difficult to predict with confidence the resistance experienced by a vertical current traversing the thin work function metals and associated boundaries and this work addresses decomposing resistance measurements to determine these components.

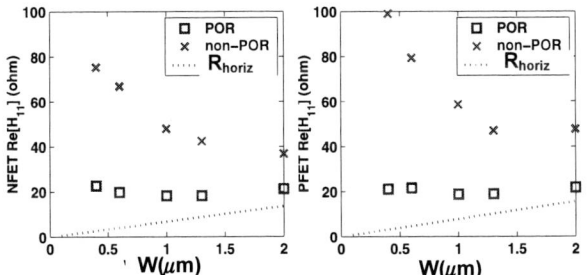

Fig. 3. $Re(H_{11})$ vs finger width for GF (having 20 fingers, measured at $V_{gs}$=0.9V, $V_{ds}$=0V) for two different process flows is compared to the $R_{horiz}$ predicted from sheet resistance data. Non-POR means reduced doping in gate silicon.

978-1-4673-6145-3/13 $31.00 © 2013 IEEE          153

## III. SIMPLE SCHEMATIC MODEL

Gate electrode resistance contributes to the input resistance of radio frequency (rf) circuits and this contribution may be extracted from high frequency power measurements using s-parameters [5, 6, 7]. We have developed a schematic transistor model for a multi-finger transistor, and this analysis may be generalized to FINFETs [10]. The gate resistance tree is shown in Fig. 5 and the gate resistance can be represented by

$$R_{g,tot} = R_{horiz} + R_{end} + R_{vert} + r_{wire} \quad . \qquad (1)$$

$$R_{g,tot} = \frac{r_{sh}W}{3N_{gcon}^2 n_f L} + \frac{r_{end}}{N_{gcon} n_f} + \frac{r_v}{n_f W} + r_{wire} \quad .(2)$$

Here, $W$ is a transistor's effective width, $L$ is transistor channel length, $r_{sh}$ is gate sheet resistance when measured by a horizontal current flow, $N_{gcon}$ is gate contact number (either 1 or 2), $n_f$ is the number of fingers, $r_{end}$ represents gate resistance outside FET width region as well as, if present, the resistance of a discrete contact at one end of a poly finger, $r_v$ is gate vertical resistance (in units of $\Omega$-m), and $r_{wire}$ represents an effective resistance of a metal wire or a poly that connects the $n_f$ poly fingers together. Fig. 5 illustrates gate resistance vs. transistor width $W$ for two ratios of $r_v/r_{sh}$. Note that the total gate resistance $R_{g,tot}$ displays a minimum at $W=\mathrm{sqrt}(3N_{gcon}L\,r_v/r_{sh})$ and for $W$ above this minimum, $R_{g,tot}$ approaches the limiting case predicted from sheet resistance $r_{sh}$ alone. Below this minimum in $R_{g,tot}$ the vertical resistance is dominant. Note in Fig. 3 that for the non POR process Re($H_{11}$) decrease with increasing Wfinger.

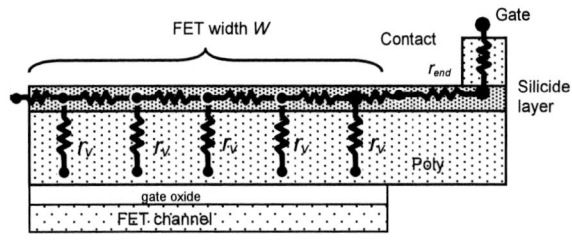

Figure 6. A network representation of distributed gate resistance of a FET width $W$, showing both horizontal and vertical resistive elements.

## IV. CROSS TECHNOLOGY COMPARISON

The analysis in figures 2 and 3 can be repeated to compare chips chosen from several technologies fabricated on bulk wafers but with differing gate stack integration schemes. A comparison of cold s-parameter data between 20nm RMG in Fig. 8 (top) [3], 32nm GF in Fig. 8 (middle) [4], and P-SiON in Fig. 8 (bottom) [5] is summarized in table I. The bias used, $V_{gs}$=0V and $V_{ds}$=0V, minimizes the influence of channel resistance. It can be seen from the Fig. 2, 3, and 8 that $R_{vert}(W)$ is very roughly comparable to $R_{horiz}(W)$. The horizontal

component (yellow triangles) can be measured using sheet resistance structures which mimic the gate electrode layout and topography. Low resistance silicide is a key technology objective for gate first schemes and low resistance metal fill is a

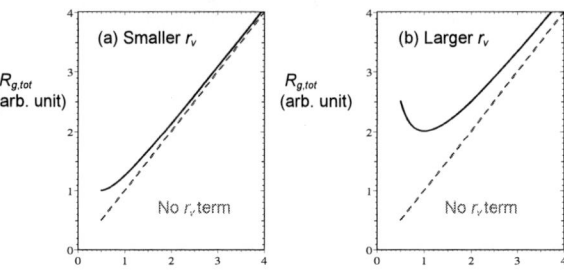

Fig. 7. FET gate resistance $R_{g,tot}$ vs. FET width $W$ for two cases of contact resistance $r_v$: (a) $r_v = r_{sh}/12L$; (b) $r_v = r_{sh}/3L$. In the plot, $n_f = 1$, $N_{gcon}=1$, and $r_{end}=r_{wire}=0$.

key objective for replacement gate schemes. The vertical gate resistance component (light blue crosses) is more complex and is again inferred from the difference between measured resistance Re($H_{11}$) (indicated by dark blue diamonds) and the horizontal resistance after correcting for additional.

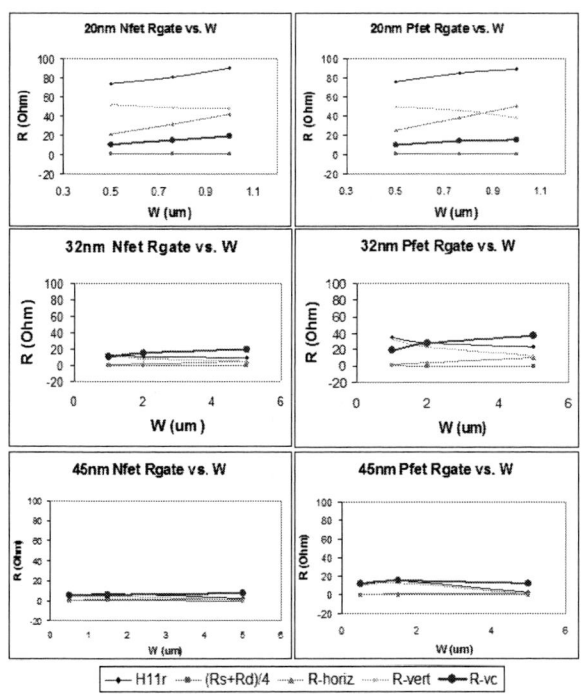

Fig. 8. Gate resistance components for devices from ten 20nm RMG (top), six 32nm GF (middle), and three 45nm P-SiON (bottom) chip. Dark blue diamonds are Re($H_{11}$), light blue crosses are $R_{vert}$, and dark purple dots are $r_{vc}$ in $\Omega$-$\mu m^2$.

small parasitic elements. For replacement gate, the vertical component of the gate stack resistance may be represented by a simple contact resistance model. This simple assumption is supported by the near constant value of the normalized vertical component $r_{vc}$ plotted in $\Omega$-$\mu m^2$

978-1-4673-6145-3/13 $31.00 © 2013 IEEE

## V. DISCUSSION

Early modeling of input resistance of CMOS device focused on those device elements which dominated rf behavior, especially frequency dependant non-quasi-static behavior related to the response time of carriers in the channel. For high performance logic devices at 65nm through 20nm the physical electrode resistance dominates $Re(H_{11})$ at high frequencies and the scaling of the electrode resistance provides insight into the components of the gate resistance tree in Fig. 6. The material parameters from Fig. 8 are summarized in Table I. The resistances $Re(H_{11})$, $R_{vert}$, and $R_{horiz}$ increase as node feature is reduced and $R_{vert} \sim R_{horiz}$ for the technologies in Fig. 8. The contact resistivity $r_{vc}$ is roughly constant across technology although this can be altered by doping or inter-layers with large bandgaps. Sheet resistance $r_{sh}$ shows a tendency to increase at smaller technology nodes. Table II shows the delay of a FO1 inverter ring as a function of gate resistance. Gate resistance constricts performance of a 20nm technology and gate-fill material limits achievable $r_{sh}$. The effect of $r_{sh}$ is more significant for wide devices.

## VI. SUMMARY

Gate resistance for several integration schemes may be accurately represented by a network comprised of vertical and horizontal elements. The values scale with cross section or interface area and gate resistance may be efficiently evaluated with a closed form expression. With continued scaling, gate resistance increases and constrains logic transistor performance.

## TABLE I
TYPICAL VERTICAL and HORIZONTAL COMPONENTS of gate resistance for 20nm RMG, 32nm GF, and 45nm P-SiON gate stacks.
Note $r_{vc}$ is roughly constant and $r_{sh}$ increases with node.

| Node - gate | NFET | NFET | PFET | PFET |
|---|---|---|---|---|
| | $r_{vc} \Omega$-$\mu m^2$ | $r_{sh} \Omega$/sq | $r_{vc} \Omega$-$\mu m^2$ | $r_{sh} \Omega$/sq |
| 20bulk-RMG | ~15 | 50 | ~14 | 66 |
| 32bulk - GF | ~15 | 17 | ~30 | 12 |
| 45bulk P-SiON | ~15 | 11 | ~15 | 11 |

## TABLE II
Simulated ring oscillator delay increase using a PSP model and 20nm technology net list. Delay increase in % for $20\Omega$/sq increase in gate resistance for each of 3 widths of an FO1 ring oscillator normalized so the increase at 200nm is negligible. For 20nm gate stacks $R_{g,tot}$ limits performance.

| Width $W_n = W_p$ | % delay increase |
|---|---|
| 200nm | 0 |
| 400nm | 1.5% |
| 600nm | 2.2% |

## ACKNOWLEDGMENT

The authors thank P. M. Solomon, K. Jenkins, and M.H. Na for review and comments. This work has been supported by the independent Bulk CMOS and SOI technology development projects at the IBM Microelectronics, Div. Semiconductor Research & Development Center, Hopewell Junction, NY 12533.

## REFERENCES

[1] S. J. Lee *et. al.* "Advanced modeling and optimization of high performance 32nm HKMG SOI CMOS for RF/analog SoC applications", Honolulu, *2012 Symposium on VLSI Technology (VLSIT)*, 2012 IEEE Conference Publications.

[2] H. Shang *et. al.* "High Performance Bulk Planar 20nm CMOS Technology for Low Power Mobile Applications", Honolulu, 2012 *Symposium on VLSI Technology (VLSIT)*, 2012 IEEE Conference Publications.

[3] F. Arnaud *et. al.* "32nm General Purpose Bulk CMOS Technology for High Performance Applications at Low Voltage", San Francisco *International Electron Device Meeting Technical Digest*, 2005 IEEE Conference Publications.

[4] J. Yuan *et. al.* "A 45nm Low Power Bulk Technology Featuring Carbon Co-implantation and Laser Anneal on 45 degree rotated Substrate Honolulu *2008 Symposium on VLSI Technology (VLSIT)*, 2012 IEEE Conference Publications.

[5] X. Jin, J. Ou, C-H Chen, W. Liu, M.J. Deen, P.R. Gray, C. Hu, "An effective Gate Resistance Model for CMOS RF and Noise Modeling," in *International Electron Device Meeting Technical Digest*, 1998 IEEE Conference Publications.

[6] H.W. Lin, S.S. Chung, S.C. Wong, G.W. Huang, "An Accurate RF CMOS Gate Resistance Model Compatible with HSPICE," *Proc. IEEE 2004 Int. Conf. on Microelectronic Test Structures*, 2004, vol. 17, pp. 227-230.

[7] M. Kang, I.M. Kang, Y.H. Jung, H Shin, "Separate Extraction of Gate Resistance Components in RF MOSFETs" *IEEE TED* vol. 54 no. 6 pp 1459-1463 2007.

[8] BSIM group, http://www-device.eecs.berkeley.edu/bsim/

[9] S. J. Lee*, L. Wagner, B. Jagannathan, S. Csutak, J. Pekarik*, N. Zamdmer, M. Breitwisch, R. Ramachandran, and G. Freeman, "Record RF Performance of Sub-46 nm Lgate NFETs in Microprocessor SOI CMOS Technologies" Wash, DC., *International Electron Device Meeting Technical Digest*, 2005 IEEE Conference Publications.

[10] N. Lu, T. B. Hook, J. B. Johnson, C. Wermer, C. Putnam, R. A. Wachnik "An Efficient and Accurate Schematic Transistor Model of FinFET Parasitic Elements", National Harbor, MD accepted *2013 Nanotech Workshop on Compact Modeling*, available online.

# IIP2 and HR Calibration for an 8-Phase Harmonic Recombination Receiver in 28nm

Barend van Liempd[1], Jonathan Borremans[1], Sungwoo Cha[2], Ewout Martens[1], Hans Suys[1] and Jan Craninckx[1]

[1] imec, Leuven, Belgium
[2] Renesas Electronics Corporation, Takasaki, Japan

***Abstract* – Fully integrated CMOS receivers achieve high linearity and low noise due to harmonic recombination, but suffer from limited IIP2 and harmonic rejection due to mismatch and inaccuracies. This paper presents an 8-phase harmonic recombination receiver with independent IIP2, HR3 and HR5 calibration techniques. Calibrated >80dBm IIP2, >70dB HR3 and >75dB HR5 are measured.**

## I. INTRODUCTION

Recent instantiations of direct-conversion wideband software defined radio (SDR) receivers [1]-[4] have demonstrated stringent performance for noise, linearity, filtering and re-configurability with similar area and current consumption as narrow-band circuits. Leveraging the high $f_T$ of <32nm scaled CMOS [5], low-power 8-phase passive mixers reject signals at the 3rd- and 5th-order harmonics, critical to SAW-less wideband receivers below 3GHz [1].

In FDD systems, transmitter (Tx) leakage degrades the receiver (Rx) sensitivity via 2nd-order intermodulation distortion due to mismatches in the mixers. Hence, IIP2 calibration is desired.

In a zero-IF Rx architecture with 4-phase passive mixers, mismatch in each I- and Q-channel mixer may be cancelled independently. However, 4-phase passive mixers do not reject signals at the odd harmonic frequencies of the LO [6]. When 8-phase passive mixers are used, I- and Q-paths are recombined from 4 mixer outputs, complicating independent quadrature mismatch cancellation for IIP2 calibration [7].

This paper details calibration techniques to optimize IIP2 and 3rd/5th-order harmonic rejection (HR3/5) independently in an 8-phase harmonic recombination receiver. A prototype is implemented in a 28nm process. Independently calibrated IIP2 (>80dBm) and HR3/5 (>70dB/>75dB) performance and robustness vs. supply variation is demonstrated.

## II. RECEIVER ARCHITECTURE & CALIBRATION TECHNIQUES

The proposed harmonic recombination receiver architecture [1] is shown in Fig. 1: low- and high-band (LB/HB) LNAs are used for 0.4-3GHz and 3-6GHz inputs, respectively. In the LB path, 4 passive mixers are driven by a 12.5% duty-cycled 8-phase LO, while the HB path has 2 mixers driven by a 25% duty-cycled 4-phase LO. The LB path omits the coupling capacitors between LNA and mixer to improve out-of-band filtering. Both LO signals are generated by a divider chain, assuming a frequency synthesizer with e.g. a dual-VCO 6-12GHz PLL is available.

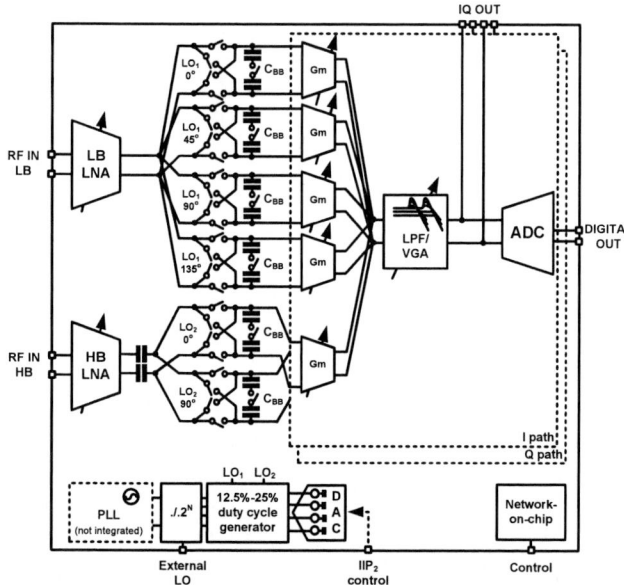

Fig. 1 Harmonic recombination receiver architecture [1].

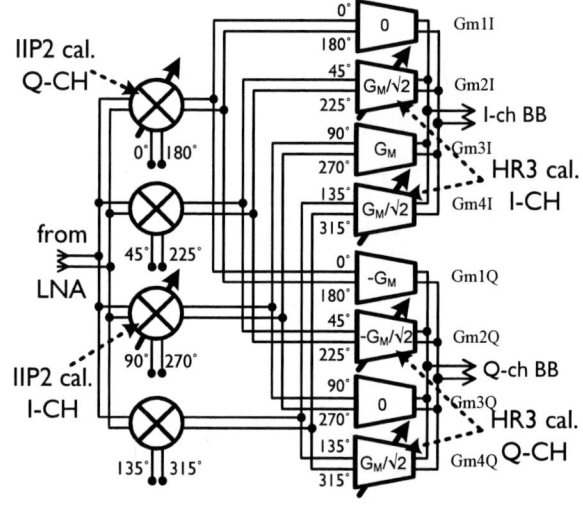

Fig. 2 HR3 and IIP2 calibration approach in the 8-phase Rx (LB).

In the baseband section, all mixers are followed by inverter-based variable gain transconductors, whose output currents are combined to provide harmonic recombination (Fig. 2). That drives trans-impedance amplifiers (TIAs) which provide 0.5-50MHz tunable bandwidth and are shared between LB and HB paths. A 11b >400MHz ADC finishes the Rx chain [8].

978-1-4673-6145-3/13 $31.00 © 2013 IEEE

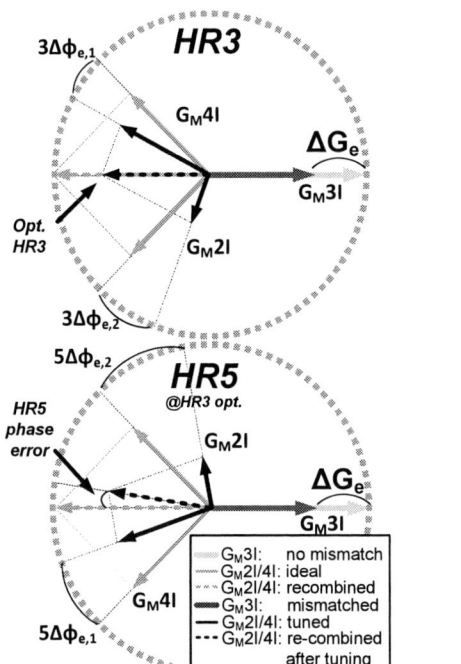

Fig. 3 Phasor diagram of $3^{rd}$- & $5^{th}$-order harmonics (LB path, I-channel).

## A. Calibration of $3^{rd}/5^{th}$-order Harmonic Rejection

In an 8-phase Rx system, I and Q signals are reconstructed out of 0°/45°/90°/135° down-converted paths by proper combination of these signals with the right weighting factor. In the voltage-mode Rx in Fig. 2, this recombination is achieved by current summation through Gm cells.

Fig. 3 shows the phasor representation of the recombination of the $3^{rd}$- and $5^{th}$-order harmonics at the mixer outputs (I-channel only). The diagrams in Fig. 3 assume the Gm3I phasor is the phase-reference.

If done well, it can be shown that $1^{st}$-order signal vectors constructively add up, while $3^{rd}$ and $5^{th}$ harmonics of the input signal are cancelled. However, if phase and amplitude of the different paths are not perfect due to mismatch, the harmonic recombination is not perfect. Mismatch can be caused by e.g. differences in the LO duty-cycle or mismatch between the Gm-cells used for harmonic recombination.

Fig. 3 shows that when *only* the magnitude of Gm2I and Gm4I phasors are tuned, the combined gain and phase error in all three phasors may be corrected to achieve very high $3^{rd}$- or $5^{th}$-order harmonic rejection (HR3, HR5).

Phase mismatch increases at higher harmonics, e.g. $\Delta\varphi_e$ in Fig. 3 is multiplied by 3 for the $3^{rd}$ harmonic and 5 for the $5^{th}$ harmonic. As a result, optimizing HR3 does not necessarily improve HR5, because the tuned Gm2I/4I magnitudes cannot compensate two phase differences. Vice-versa, when tuning for HR5, the HR3 is not optimized necessarily.

In the prototype presented here, this scheme is implemented by tuning the Gm-cells in a coarse-fine fashion, achieving 10b resolution.

## B. Calibration of IIP2 for 8-Phase Passive Mixers

Mismatches in mixer devices govern IIP2. Without any calibration, simulations for the 8-phase mixer Rx chain over

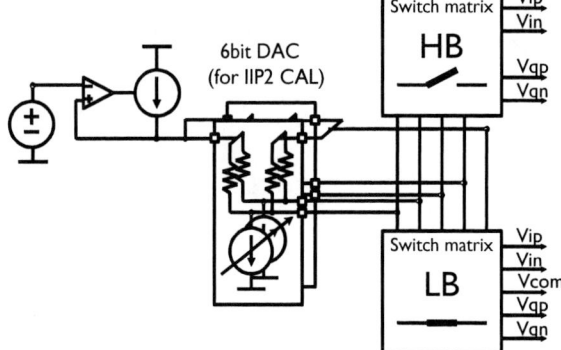

Fig. 4 Calibration circuitry for IIP2, re-used by both LB & HB path.

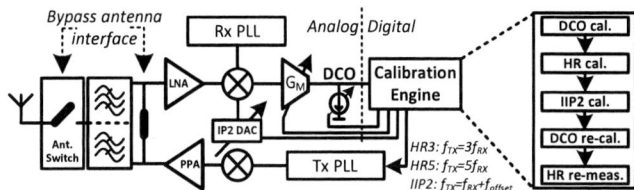

Fig. 5 Tx loopback for DCO, HR and IIP2 calibration in a practical system.

$3\sigma$ corners predict +55dBm IIP2. To improve $6\sigma$ IIP2, calibration may be provided through gate and/or common-mode voltage tuning [9] or via current injection [10].

A known IIP2 calibration approach used with 4-phase quadrature mixers allows to independently calibrate the I- and Q-channel mixers [9],[11]. The HB Rx chain in this work utilizes this approach.

However, for the four mixers in an 8-phase system, tuning the gate voltage for each mixer would add significant circuitry overhead and calibration may be cumbersome. Instead, Fig. 2 shows the approach for the LB Rx chain, allowing independent I- and Q-channel IIP2 calibration by tuning only 2 out of 4 mixers.

In the proposed scheme, the output of the 0°-path is only used for the I-path harmonic recombination, while the 90°-path mixer is only used for the Q-path harmonic recombination. When these *independent* mixers are calibrated only, I- and Q-channel calibration is independent as well, cancelling the *collective* mismatches in all mixers.

Fig. 4 shows the DACs used in the IIP2 calibration system. Two differential 6-bit DACs tune the gate voltages differentially, compensating $V_{th}$-mismatch in either the I- or Q-channel independent mixer for either LB or HB. Two switch matrices allow the same DACs to be used by both the LB and HB paths. $V_{com}$ is the common-mode bias voltage, used to bias the non-calibrated mixers.

## C. Practical implementation

Fig. 5 shows how HR and IIP2 calibration can be implemented in a practical system through loopback, using the Tx PLL. To calibrate HR3 or HR5, this work uses a brute-force calibration, tuning Gm2I, Gm4I, Gm2Q and Gm4Q to minimize Rx output power when a harmonic is applied to the Rx. For IIP2, similarly, a two-tone is generated by the Tx and applied to the LNA input, while a binary search algorithm minimizes the Rx output power.

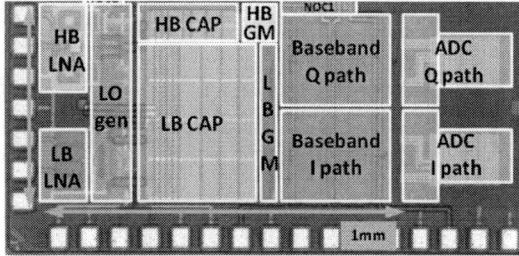

Fig. 6 Photograph of the 28nm CMOS prototype chip.

## III. MEASUREMENT RESULTS

Fig. 6 shows the prototype HR receiver, implemented in 28nm CMOS and occupying 0.6mm$^2$. The receiver operates from 0.4-6GHz and consumes 20-40mW from a single 0.9V supply, including 12.5%/25% duty-cycle generation and mixer drivers. For the measurements presented here, the ADC is bypassed.

Fig. 7 reports NF and OB-IIP3 across frequency, showing a minimum NF of 1.8dB. An OB-IIP3 of +3 to +5dBm can also be achieved.

In the following measurement results, the calibration steps are consecutively executed as shown in Fig. 5:

1. DC-offset (DCO) calibration (also available [1]) to provide a good starting DC condition for the Rx,

2. either HR3 or HR5 calibration,

3. IIP2 calibration (using a two-tone at $f_{offset}$ = 30MHz),

4. a second DCO calibration, removing any DCO caused by previous calibration steps.

After these steps, HR3 and HR5 are re-measured to verify whether the calibrated performance degrades when performing subsequent calibration steps.

Fig. 8 shows IIP2 at 870MHz vs. full I and Q calibration DAC codes. The orthogonal optima illustrate the independency of I and Q calibration for the 8-phase LB-path when compensating the offset at only 2 out of 4 mixers.

Similarly, Fig. 9 shows HR3 at 870MHz, sweeping the Gm2I/4I cells around a coarse setting to illustrate I- and Q-path independent tuning. Since all phasors are compensated independently, I and Q do not cross-influence each other and very high HR3 or HR5 can be achieved in a 2-dimensional search space.

Fig. 10 reports >80dBm IIP2 after binary search calibration (finding the optimum in Fig. 8), achieved over the LB frequency range. Non-calibrated IIP2 is >55dB, confirming 3σ corner simulations. A good IIP2 is reported, even when the coupling capacitors in the front-end are omitted and IM2 products from the LNA are not filtered.

Fig. 11 and Fig. 12 report >70dB HR3 and >75dB HR5 after IIP2 calibration (finding the optimum in Fig. 9), illustrating relative independency of IIP2 and HR calibration techniques. This is to be expected, since the mixer gain does not change by differential compensation.

As predicted, calibrating either HR3 or HR5 results in worse rejection for the other harmonic at higher frequencies, indicating that phase mismatch between the LO pulses limits recombination accuracy. Finally, the results were verified on a second (randomly selected) sample.

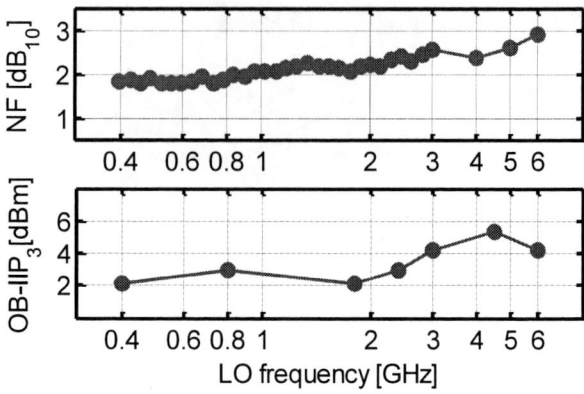

Fig. 7 Measured NF and OB-IIP3 (20MHz offset).

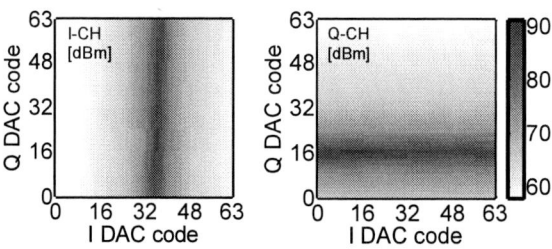

Fig. 8 Measured I- & Q-channel IIP2 versus DAC code at 870MHz.

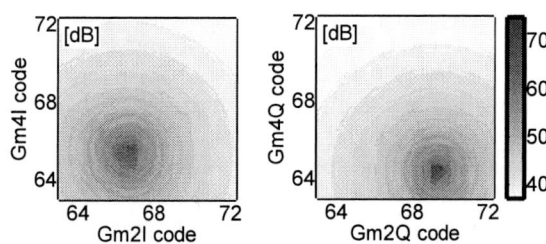

Fig. 9 Measured HR3 versus tuning code at 870MHz (Gm3I=Gm1Q=64).

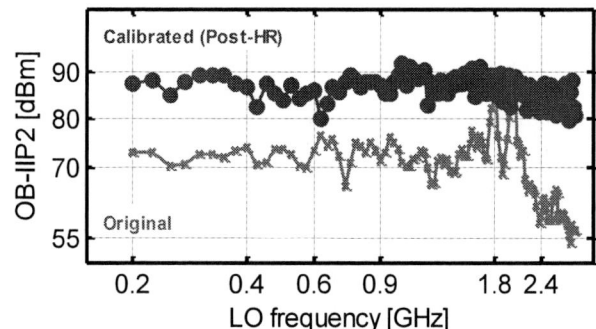

Fig. 10 Calibrated IIP2 for the 8-phase Rx path.

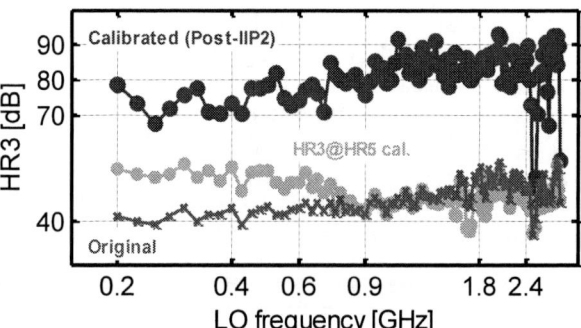

Fig. 11 Calibrated HR3 for the 8-phase (LB) Rx path.

978-1-4673-6145-3/13 $31.00 © 2013 IEEE

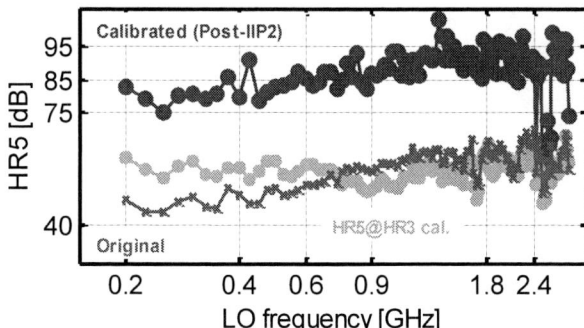

Fig. 12 Calibrated HR5 for the 8-phase (LB) Rx path.

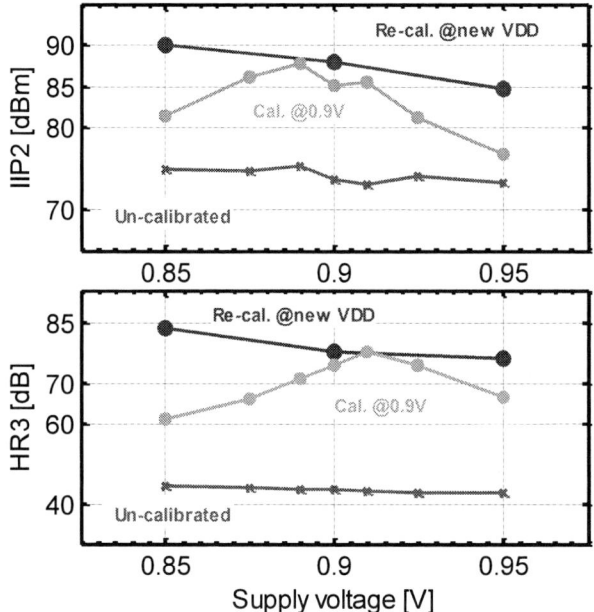

Fig. 13 IIP2 and HR3 for 0.85-0.95V supply at 870MHz.

TABLE I.    CALIBRATED IIP2, HR3 AND HR5 COMPARED

| Specification | This work | | IIP2 [11] | HR3 [4] |
|---|---|---|---|---|
| | LB | HB | | |
| Frequency [GHz] | 0.4-3 | 3-6 | 0.4-6 | 0.4-0.9 |
| Power [mW] | <40 | <35 | 60 | 60 |
| Max. gain [dB] | 70 | 58 | 70 | 34.4 |
| NF [dB] | 1.8-2.4 | 2.5-3 | 3.2-4.5 | 4 |
| OB-IIP3 [dBm] | 3 | 5 | 10 | 16 |
| OB-IIP2 [dBm] | >80 (>55¹) | >80 [1] | >70 | 56¹ |
| HR3 [dB] | >70² (>40¹) | N/A | N/A | >60² (>80³) |
| HR5 [dB] | >75² (>50¹) | N/A | N/A | >65² (>80³) |
| # of phases | 8 | 4 | 4 | 8 |
| Area [mm²] | 0.6 | | 1.4 | 1.0 |
| Supply [V] | 0.9 | | 2.5/1.2 | 1.2 |
| Technology | 28nm CMOS | | 40nm CMOS | 65nm CMOS |

¹Non-calibrated, ²Analog only, ³Using Digital Interference Cancellation

## IV. CONCLUSIONS

This work demonstrates that IIP2, HR3 and HR5 can be robustly and independently calibrated up to >80dBm, >70dB and >75dB, respectively, in an 8-phase Rx with minimal hardware and complexity. The calibration techniques have been shown to be relatively insensitive to supply variations.

## REFERENCES

[1] J. Borremans, et al., "A 0.9V Low-Power 0.4-6GHz Linear SDR Receiver in 28nm CMOS," to appear in *VLSI Dig. of Tech. Papers*, June 2013.

[2] D. Murphy, et al., "A Blocker-Tolerant, Noise-Cancelling Receiver Suitable for Wideband Wireless Applications," *J. Solid-State Circuits*, Vol. 47, No. 12, pp. 2943-2963, December 2012.

[3] Chi-Yao Yu, et al., "A SAW-Less GSM/GPRS/EDGE Receiver Embedded in 65-nm SoC," *J. Solid-State Circuits*, Vol. 46, No. 12, pp. 3047-3060, December 2011.

[4] Z. Ru et al, "Digitally Enhanced Software-Defined Radio Receiver Robust to Out-of-Band Interference," *IEEE J. Solid-State Circuits*, vol.44, pp. 3359-3375, December 2009.

[5] M. Yang, et al., "RF and Mixed-Signal Performances of a Low Cost 28nm Low-Power CMOS Technology for Wireless System-on-Chip Applications," *Symp. on VLSI Technology*, pp. 40-41, June 2011.

[6] J. Weldon et al, "A 1.75-GHz Highly Integrated Narrow-Band CMOS Transmitter With Harmonic-Rejection Mixers," *J. Solid-State Circuits*, Vol.36, pp. 2003-2015, December 2001.

[7] A. Molnar, et al., "Impedance, Filtering and Noise in N-Phase Passive CMOS Mixers," *Custom Int. Circuits Conference*, pp.1-8, September 2012.

[8] B. Verbruggen, et al., "A 2.1mW 11b 410 MS/s Dynamic Pipelined SAR ADC with Background Calibration in 28nm Digital CMOS," to appear in *VLSI Dig. of Tech. Papers*, June 2013.

[9] D. Kaczman, et al., "Mixer Circuits for Second-Order Intercept Point Calibration," US Patent US20110201296A1, Filed April 2011.

[10] I. Elahi, et al., "IIP2 Calibration by Injecting DC offset at the Mixer in a Wireless Receiver," *Trans. On Circuits and Systems II*, Vol. 54, pp. 1135-1139, December 2007.

[11] J. Borremans, et al., "A 40 nm CMOS 0.4–6 GHz Receiver Resilient to Out-of-Band Blockers," *J. Solid-State Circuits*, Vol.46, No.7, pp. 1659-1671, July 2011.

Un-calibrated, >40dB (HR3) and >50dB (HR5) is measured, while calibration works best up to 2.5GHz. Beyond these frequencies, the mismatch appears to be too large to correct. Nevertheless, blockers at harmonic frequencies above 6GHz are generally not considered as an issue.

Fig. 12 shows the sensitivity of calibrated IIP2 and HR3 to supply voltage variations. First, IIP2 and HR3 are calibrated at 0.9V, then the supply is varied and both numbers are measured at the same calibration settings. Second, IIP2 and HR3 are re-calibrated at the extreme ends of the supply range.

Without re-calibration, IIP2 remains above 75dBm and HR3 does not drop below 60dB when the supply is swept from 0.85V to 0.95V. After re-calibration across the supply range, IIP2 and HR3 reach 85dBm and 75dB, respectively (at this specific frequency).

Table 1 compares calibrated IIP2 and HR3 results with state-of-the-art. Both the presented analog-only calibration scheme for HR3 and IIP2 for 8-phase harmonic recombination receiver compare favorably to state-of-the-art.

# Advances in the Design of Wideband Receivers

David Murphy, *Member, IEEE*, Mohyee Mikhemar, *Member, IEEE*, Ahmad Mirzaei, *Member, IEEE*, and Hooman Darabi, *Member, IEEE*

*Abstract*—To be practical, wideband receivers must tolerate large out-of-band blockers, which can desensitize the receiver through gain compression or reciprocal mixing with LO phase noise. This paper reviews how a new noise-cancelling receiver architecture – that utilises 3 important circuit innovations – mitigates gain compression without compromising noise figure. While the architecture is still susceptible to reciprocal mixing, it is shown how a recently proposed reciprocal mixing cancelling technique (if incorporated into the receiver) can eliminate the need for a dramatic rise in LOGEN current.

## I. INTRODUCTION

The past few years have seen significant progress towards a practical and fully integrated realization of a Software Defined Radio (SDR) [1]. This is particulary true of the receiver portion of the system, which has benefitted from the application of 3 important circuit innovations. The first is the development of noise-cancelling topologies [2] [3], which facilitate the design of wideband, low-noise and matched receivers. The second is use of passive-mixers driven by non-overlapping clocks, which enable low-noise and highly linear downconversion [4] [5] [6] [7] [8]. The final important innovation is the replacement of conventional LNAs that provide RF voltage gain with RF-transconductances that directly downconvert RF current [9] [10] [11]. Employing some or all of these techniques has resulted in dramatic performance improvements in wideband receivers [12] [13] [6] [7] [8] [14] [15]. This paper reviews two recent works that continues this good progress and overcomes two major challenges in wideband blocker-tolerant receiver design, specifically issues related to gain compression and reciprocal mixing [16] [17] [18].

We will first discuss the *frequency-translational noise-cancelling receiver (FTNC-RX)* architecture [16] [17], which makes use of the 3 circuit innovations mentioned above, to limit the effects of gain compression without degrading noise figure. In particular, we will explore the key role of the passive-mixers employed in the design, and we will introduce a simple calibration scheme that exploits the reciprocity of their input and output terminals. This scheme can optimise noise-cancelling regardless of any antenna variations or path mismatch. New insights are also provided: we will show how the FTNC-RX is advantageous even in applications where a 50Ω match is not required, and how elimination of the matching requirement can further boost the linearity of the FTNC-RX.

A drawback of the FTNC-RX, indeed all mixer-based receivers that eschew passive filtering, is that reciprocal mixing of LO phase noise with large unwanted blockers can place additive noise in the RX band. This suggests that an ultra low

phase noise LO chain is required. Unfortunately, a dramatic reduction in the contribution of VCO phase noise, which dominates for close-in blockers, can only be accomplished by increasing power. However, as we will discuss, this limitation of VCO design can be overcome by utilizing a promising reciprocal mixing cancellation technique that is perfectly suited for use in the FTNC-RX [18].

## II. WIDEBAND RECEIVER CHALLENGES

The out-of-band continuous-wave (CW) blocker scenarios for two cellular bands are shown in Fig. 1. These scenarios, which are among the most challenging of all the wireless standards, show that an RX designed for use at these frequencies must be able to tolerant 0dBm blockers as close 20MHz from the wanted signal in the case of E-GSM900 or 80MHz in the case of PCS1900. A conventional narrowband RX will employ

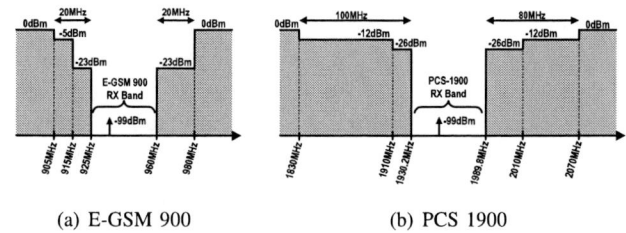

(a) E-GSM 900          (b) PCS 1900

Fig. 1. Out-of-blocker profiles.

off-chip filtering to suppress these blockers followed by a tuned LNA, which is designed to provide enough RF voltage gain such that the noise contribution of the downconverison mixers and baseband amplifiers is suppressed. A wideband receiver cannot make use of fixed RF filtering and, so, these large 0dBm blockers can easily compress the gain of a such design (Fig. 2).

Moreover, without upfront filtering, any blocker present (whether or not it causes gain compression) will be down-converted along with the wanted signal. When the blocker mixes with LO phase noise, it deposits additive noise in the receive channel proportional to the blocker amplitude (Fig. 3). Thus, for a perfectly linear wideband receiver to maintain the same noise figure (NF) as an equivalent narrowband receiver, its LO phase noise must reduced by one dB for every dB of filter attenuation that is removed at the blocker frequency.

The two new techniques covered in this paper seek to alleviate these unwanted sources of performance degradation.

## III. THE NOISE-CANCELLATION RECEIVER

The recently proposed FTNC-RX architecture can tolerate challenging 0dBm out-of-band blockers without significant

All authors are with Broadcom Corporation, Irvine, CA 92617, USA.

978-1-4673-6145-3/13 $31.00 © 2013 IEEE

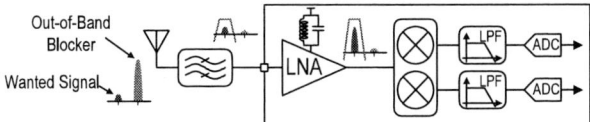

(a) A narrowband direct-conversion receiver typically employs both off-chip and on-chip passive RF filtering to attenuate any unwanted signals.

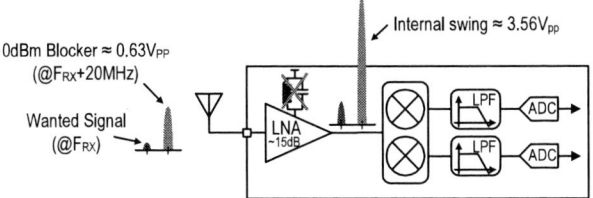

(b) A wideband receiver cannot employ passive RF filtering and, therefore, a large blocker will saturate a conventional front-end design.

Fig. 2.    Effect of an out-of-band blocker on narrowband and wideband receivers.

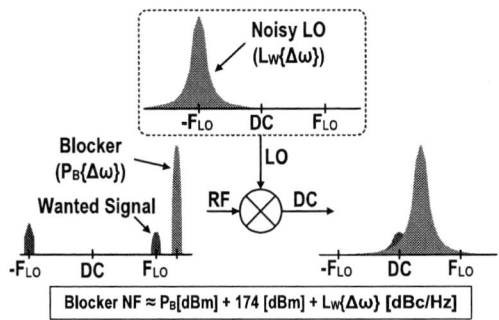

Fig. 3.    Noise degradation due to reciprocal mixing in a linear mixer.

gain compression, and without a substantial noise-figure degradation. The architecture leverages 3 important (and relatively recent) circuit techniques, which are reviewed in this section followed by discussion of the FTNC-RX architecture itself.

### A. Important Innovations

*1) Noise-Cancelling:* Noise-cancelling, as applied to an LNA [2] [3], is able to deliver both wideband impedance matching and a low noise factor. A conceptual model is shown in Fig. 4, which demonstrates how the noise of the matching resistor can be eliminated by taking a voltage measurement at the RF input *and* a measurement of the current flowing through the matching resistor. Most noise-cancelling LNA (NC-LNA) topologies apply this gain at RF and, so, any out-of-band blockers can easily desensitize the LNA. However, as demonstrated in [19] [20], there is no requirement to apply voltage gain at RF provided two separate down-conversion paths are used.

*2) Passive-Mixer-Based Receivers:* The primary requirement of an LNA is to provide enough voltage gain to overcome the noise contribution of the downcoversion and baseband stages. Too much gain, however, and linearity will suffer. This trade-off has been relaxed in recent years because of the adoption of passive-mixers. Compared to their active counterparts, passive mixers are extremely linear, and have the potential to introduced little additional noise. Indeed, in

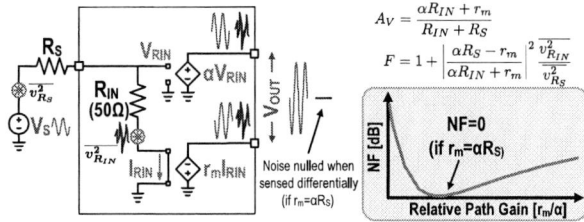

Fig. 4.    A conceptual model of a noise-cancelling LNA.

certain applications, passive mixers can be connected directly to the antenna (see Fig. 5) [5] [6] [7] [8], RF voltage gain is avoided completely, and a reasonable noise factor is still retained. In such a setup (ignoring the matching network and assuming the RF node is purely resistive), the noise figure of an $M$ phase mixer is given by:

$$F = \left(1 + \frac{R_{SW}}{R_S} + \frac{\overline{v_{BB}^2}}{M(4kTR_S)}\right)\frac{1}{CG^2} \qquad (1)$$

where is $R_{SW}$ is the resistance of mixer switches, $\overline{v_{BB}^2}$ is the input referred noise of each baseband TIA, and $CG$ is the RF-to-IF current conversion gain of the mixer given by $CG = \text{sinc}(\pi/M)$. This implies that the noise factor can be made arbitrarily low if the switches are made very large, and the baseband noise is minimised. In practice, GHz mixer-first receivers are generally limited to 3dB and above. The

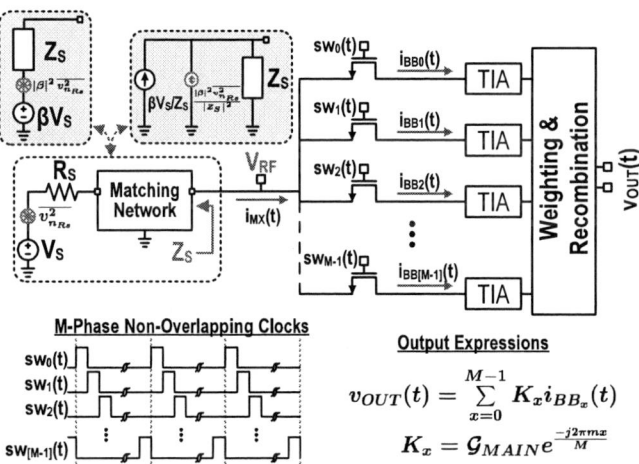

Fig. 5.    An arbitrary $M$-phase passive mixer.

reciprocity of passive-mixers can also be utilized to boost the large-signal linearity of a given design. This happens, when a passive mixer is configured to operate as a N-path filter (see Fig. 6) [21] [22]. If the baseband outputs of the mixer are connected to lowpass filters (either an explicit RC circuit or the input impedance of the TIA), these impedances will effectively upconvert to RF and present an RF input impedance that has a high-Q bandpass characteristic. This feature has been used to filter unwanted out-of-band in a variety of both narrowband and wideband designs [23] [24] [25] [26] [6] [7] [14] [15].

*3) RF Transconductance:* In [9] and later in [10], [11], [27], the LNA is replaced with a linear transconductance, which drives a current mode passive mixer (Fig 7). In such

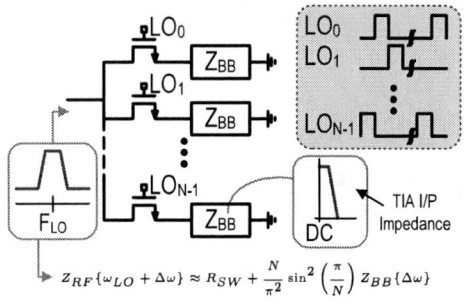

$$Z_{RF}\{\omega_{LO} + \Delta\omega\} \approx R_{SW} + \frac{N}{\pi^2}\sin^2\left(\frac{\pi}{N}\right)Z_{BB}\{\Delta\omega\}$$

Fig. 6. The mixer-first receiver; RF input impedance is combination of mixer switch resistance ($R_{SW}$) and upconverted TIA input impedance ($Z_{BB}$).

an approach, RF voltage gain can be avoided by employing large passive mixer switches and ensuring the input impedance of the transimpedance amplifiers (TIAs) are close to zero, thereby preventing the amplification of blockers and limiting gain compression effects, but still allowing a low noise figure in narrowband receivers. For a wideband design, however, the design of the RF transconductance will require some performance compromise. No single-ended $G_M$ cell can simultaneously provide low-noise and wideband matching. If differential inputs are used, a differential common-gate topology with partial noise-cancelling (such as the one employed in [14], [15]) can be employed, but even in this case only moderate noise performance can be expected. Differential inputs also require a wideband balun, which, depending on the bandwidth requirements, may not be practical.

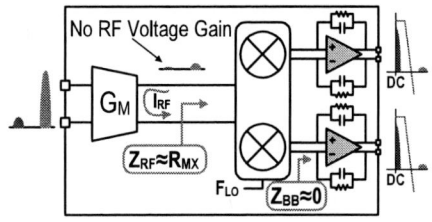

Fig. 7. An RX employing an RF transconductance.

### B. The Noise-Cancelling Receiver

*1) Architecture Overview:* By drawing on the 3 techniques listed above, we can better understand the new FTNC-RX architecture [16] [17]. Figure 8 shows the evolution of the FTNC-RX from the simple noise-cancelling model. Instead of converting the current measurement to a voltage at RF, a passive mixer immediately downconverts the RF current to baseband. A TIA then converts any current in the receive-band to voltage. This path is essentially a mixer first receiver (like that shown in Fig. 5). The voltage measurement is provided by an auxiliary path, where an RF-transconductance converts the RF node voltage to a current, which is then downconverted by another passive mixer. Another TIA then converts any in-band current to voltage. This path can be viewed as an RF-transconductance-based receiver (like that shown in Fig. 7), but with an $G_M$-cell that does not have to provide matching.

As shown in Fig. 9, if high-gain baseband operational amplifiers are employed, the input terminal of both TIAs

appear as virtual grounds. Additionally, if large switches are used in the passive mixers, the impedance looking into the RF terminal of each mixer is small and, therefore, no RF voltage gain is experienced. While large out-of-band blockers will result in a large current flowing into the mixers, particulary in the auxiliary path, this current once downconverted will be shunted to ground through large capacitors at the inputs of the TIAs.

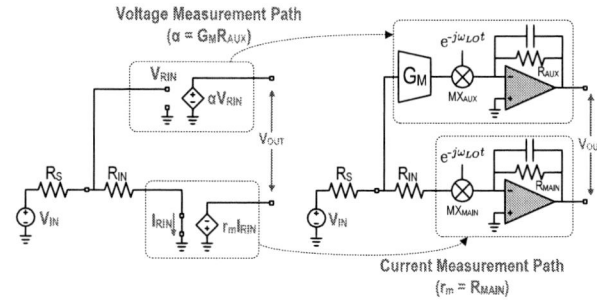

Fig. 8. Evolution of the proposed frequency translational noise-cancelling receiver (FTNC-RX).

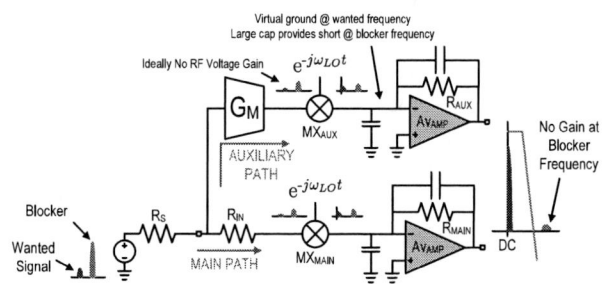

Fig. 9. The proposed noise-cancelling receiver.

The FTNC-RX preserves the beneficial features of other noise-cancelling topologies. Shown in Fig. 10 is an $M$-phase model of the FTNC-RX with the noise sources in the system explicitly shown. In the main path, noise associated with mixer switches $(\overline{v_{R_{SW}}^2})$ appears in series with the matching resistor noise $(\overline{v_{R_{IN}}^2})$ and is cancelled. In addition, the low frequency noise associated with the main path TIAs $(\overline{v_{BB}^2})$ will be upconverted by the passive mixers switches appear at the RF terminal and, so, will also be cancelled. In the auxiliary path, the noise associated with the passive mixer switches and TIAs will contribute negligibly if the $G_M$ cell has a large output impedance. Therefore assuming a purely resistive RF node, and the gain of both paths is appropriately set the noise figure of the receiver is given by:

$$F = \left(1 + \frac{\overline{v_{GM}^2}}{4kTR_S}\right)\frac{1}{CG^2}, \qquad (2)$$

where $\overline{v_{GM}^2}$ is the input referred noise of the $G_M$-cell. As this $G_M$-cell does not provide matching, we use a simple class-AB cell, which is extremely linear. It is sized large enough to ensure a noise figure less than 2dB.

*2) Fabricated FTNC-RX Results:* The fabricated FTNC-RX is shown in Fig. 11. The two passive mixers are configured

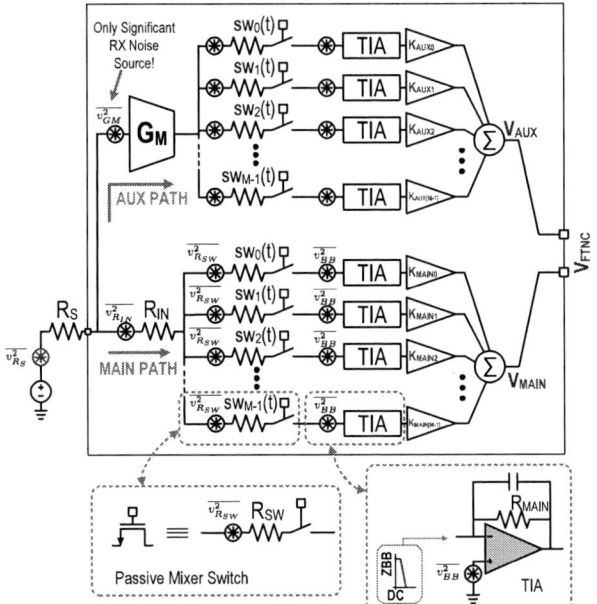

Fig. 10. Simplified FTNC-RX model highlighting main noise sources. All noise sources, with the exception of the $G_M$ noise, can be nulled or contribute negligibly.

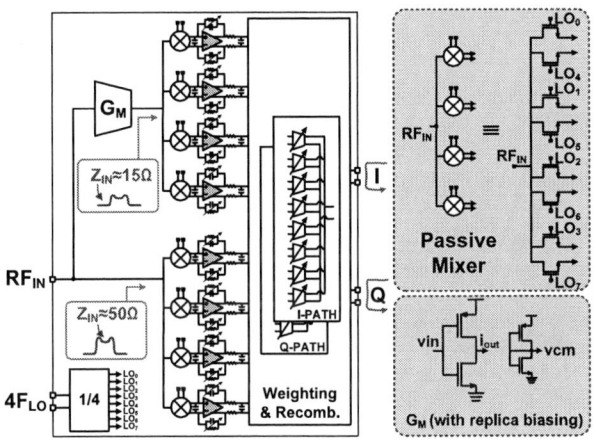

Fig. 11. The complete FTNC-RX fabricated in 40nm.

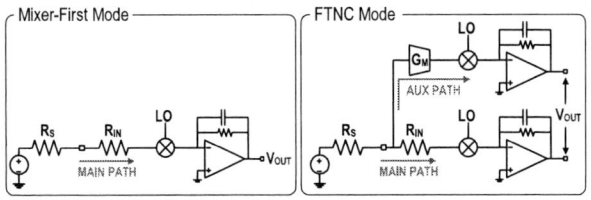

(a) Modes of operation of the FTNC-RX.

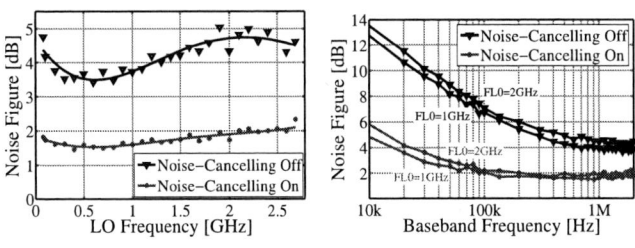

(b) Measured noise figure versus receive frequency. (c) Measured noise figure versus baseband frequency.

Fig. 12. Measured nose figure of the FTNC-RX in two modes of operation.

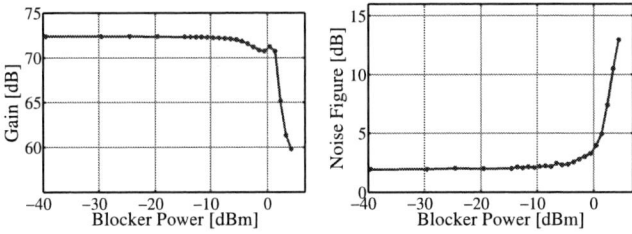

Fig. 13. Measured receiver gain and noise figure assuming a wanted signal at 1.5GHz accompanied by a 1.58GHz continuous-wave blocker.

to operate as implicit N-path filters, since the mixer switches and input impedances of the TIAs are sized to give some light filtering at the RF input terminals. The measured noise figure in the two modes of operation is shown in Fig. 12. In the low power mode, the auxiliary path is powered down and the receiver behaves like a mixer-first receiver. In the low-noise mode, the auxiliary path is powered up and the noise drops below 2dB across most of the band.

Fig. 13 shows the blocker performance of the receiver and clearly demonstrates that in the absence of passive RF filtering, 0dBm blocker tolerance need not come at the expense of a low noise figure.

*3) Noise-Cancellation Optimisation:* Since noise cancellation takes place after downconversion, the phase and magnitude of both paths can be programmed to optimise noise-cancellation in the presence of variations in the antenna impedance or the variations in the two downconversion paths. In an FTNC-RX with a large value for $M$ (see Fig. 14(a)), the optimum noise-cancelling condition is given by:

$$\mathcal{G}_{MAIN} = G_M \mathcal{G}_{MAIN} Z_S \tag{3}$$

where $\mathcal{G}_{MAIN}$ is the V-to-I baseband gain in the main path, $\mathcal{G}_{AUX}$ is the V-to-I baseband gain in the auxiliary path and the $Z_S$ is the antenna's effective impedance. Fig. 14 shows how varying the phase and magnitude of the two paths can minimise the noise figure. The question then arises as to how this optimum point can be found in the presence path mismatch and antenna variations? The most straightforward approach is to inject small currents into the output terminals of the main path TIAs, as shown in Fig. 15. This injected current will cause a voltage at the output of the main path, but will also be upconverted (like TIA noise) to the RF input node and will a generate voltage at the output of the auxiliary path. Referring to Fig. 15 and assuming $M$ is large, the conversion gain of the injected current sources to the output of the main path is:

$$TF_{MAIN}\{\Delta\omega\} = \frac{-\mathcal{G}_{MAIN}\frac{I_{inj}\{\Delta\omega\}}{G_{M_{TIA}}}}{Z_S\{\omega_c + \Delta\omega\} + R_{SW} + Z_{BB}\{\Delta\omega\}/M}, \tag{4}$$

where $G_{M_{TIA}}$ is the transconductance of the baseband amplifiers. While the conversion gain of the same current sources to the auxiliary path output is:

$$TF_{AUX}\{\Delta\omega\} = \frac{-\mathcal{G}_{AUX}G_M Z_S\{\omega_c + \Delta\omega\}\frac{I_{inj}\{\Delta\omega\}}{G_{M_{TIA}}}}{Z_S\{\omega_c + \Delta\omega\} + R_{SW} + Z_{BB}\{\Delta\omega\}/M}. \tag{5}$$

Note that the ratio between the two outputs is $\mathcal{G}_{MAIN} = G_M \mathcal{G}_{AUX} Z_S$, which is the optimised noise cancelling condition. Accordingly, actuating the circuit with small baseband currents and sensing the outputs can be used to calibrate the receiver.

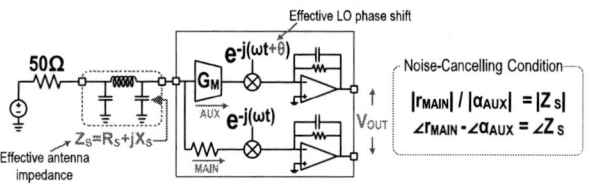

(a) Optimum noise cancelling conditions depends on the effective antenna impedance ($r_{MAIN} = -\mathcal{G}_{MAIN}$, $\alpha = -G_M \mathcal{G}_{AUX}$).

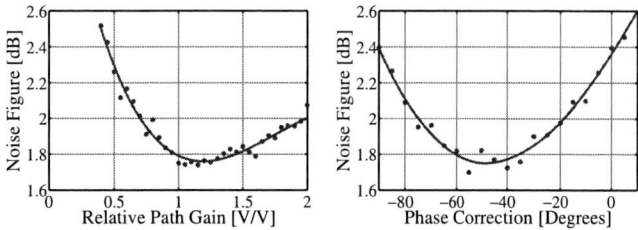

(b) Measured effect of gain correction on noise figure.     (c) Measured effect of phase correction on noise figure.

Fig. 14. Noise-cancelling optimization via baseband phase and magnitude correction.

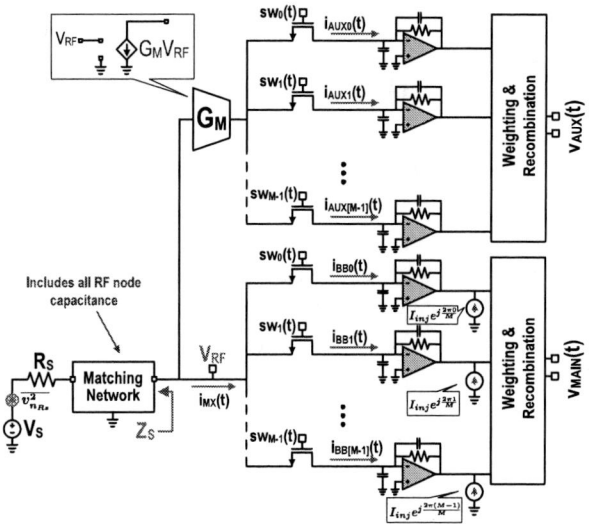

Fig. 15. Injected small baseband currents (i.e. $I_{inj}$) into the TIA output nodes can be use to calibrate the FTNC-RX

*4) Distortion Cancellation:* The reader may have noticed that the noise figure of the FTNC-RX is the same as an identical receiver, but with the main path removed (i.e. a RF-transconductance-based receiver with an unmatched $G_M$-cell). This would imply that the FTNC-RX only advantage over such a topology is noiseless wideband matching. This is, indeed, the primary advantage of the topology. However, since a noise-cancelling receiver is also a distortion-cancelling receiver [3] there are some linearity advantages as well. When the FTNC-RX is optimally configured, distortion originating from the main-path is mostly cancelled, while the presence of a matched main-path reduces the inband voltage swing at

the RF port by 6dB. The reduction for out-of-band signals is even larger, since the main-path can be configured to operate as an N-path filter. This protects the $G_M$-cell from large input swings and increases the blocker tolerance of the receiver.

The small-signal linearity of the receiver also benefits from the presence of the main path. Consider the simplified model shown in Fig. 16(a), which for simplicity ignores frequency conversion effects. In the same way as noise in the main path appears at the input terminal, so does distortion. We can model this distortion as:

$$v_{RF}(t) = \beta_1 v_S(t) + \beta_2 v_S^2(t) + \beta_3 v_S^3(t) + \dots, \quad (6)$$

where $\beta_1 = R_{IN}/(R_{IN} + R_S)$. Now the output of the main path is given by

$$v_{MAIN}(t) = -\frac{r_m}{R_S}\left(v_S(t) - v_{RF}(t)\right), \quad (7)$$

while the auxiliary path output is given by

$$v_{AUX}(t) = \alpha_1 v_{RF}(t) + \alpha_2 v_{RF}^2(t) + \alpha_3 v_{RF}^3(t) + \dots, \quad (8)$$

Solving for $v_{FTNC}(t) = v_{AUX}(t) - v_{MAIN}(t)$ under the noise-cancelling condition ($r_m = \alpha_1 R_S$) gives

$$v_{FTNC}(t) = \alpha_1 v_S(t) + \sum_{n=2}^{\infty} \alpha_n \left(\sum_{m=1}^{\infty} \beta_m v_S^m(t)\right)^n, \quad (9)$$

and so the input referred IIP3 is calculated as

$$A_{IIP3(FTNC)} = \sqrt{\frac{4}{3}\left|\frac{\alpha_1}{2\beta_1\beta_2\alpha_2 + \beta_1^3\alpha_3}\right|}. \quad (10)$$

Assuming the second order components are small (i.e. $\beta_2 \approx 0$ and/or $\alpha_2 \approx 0$), the IIP3 of the FTNC-RX simplifies to

$$A_{IIP3(FTNC)} \approx \left(\frac{R_{IN} + R_S}{R_{IN}}\right)^{3/2} A_{IIP3(AUX)}, \quad (11)$$

where $A_{IIP3(AUX)}$ is the IIP3 of the auxiliary path assuming the main-path has been removed. This implies that two things: First, distortion introduced by the main-path is largely cancelled and, second, the distortion of auxiliary path is reduced by factor that depends on the input impedance and is always less than an equivalent unmatched receiver composed of only the auxiliary path (in the case of a perfectly matched FTNC-RX, the IIP3 will be up to 9dB better). Interestingly, as $R_{IN}$ is reduced the noise factor remains unchanged but the linearity of the receiver improves. This effect is seen in Fig. 16(b), which plots the IIP3 of a verilog-A model of the FTNC-RX assuming the strongly nonlinear elements noted in Fig. 16(a). Therefore, even if the matching requirement is removed, the main path *should* be retained and, if practical, its input impedance should be lowered.

## IV. RECIPROCAL MIXING CANCELLATION

Even without upfront passive filtering, the FTNC-RX can tolerate 0dBm blockers without significant gain compression. It does not, however, inhibit the reciprocal mixing of a blocker with LO phase noise. Although both passive-mixers in the FTNC-RX can be configured to operate as implicit N-path filters, such filtering is based on mixing and *cannot* reduce

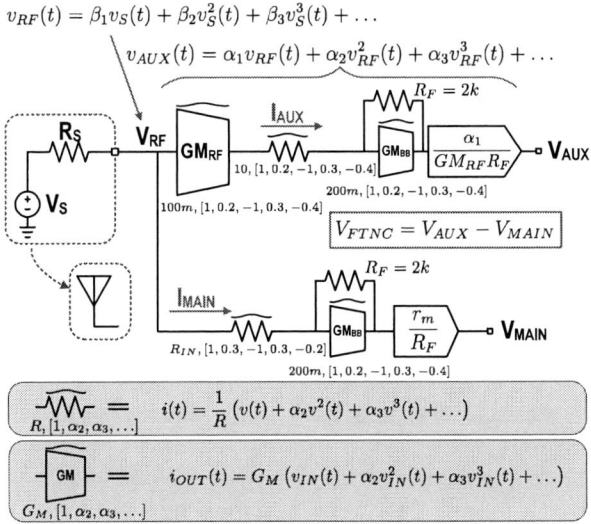

$$v_{RF}(t) = \beta_1 v_S(t) + \beta_2 v_S^2(t) + \beta_3 v_S^3(t) + \dots$$

$$v_{AUX}(t) = \alpha_1 v_{RF}(t) + \alpha_2 v_{RF}^2(t) + \alpha_3 v_{RF}^3(t) + \dots$$

$$\widetilde{\text{---WW---}} = i(t) = \frac{1}{R}\left(v(t) + \alpha_2 v^2(t) + \alpha_3 v^3(t) + \dots\right)$$
$R, [1, \alpha_2, \alpha_3, \dots]$

$$\boxed{\text{GM}} = i_{OUT}(t) = G_M\left(v_{IN}(t) + \alpha_2 v_{IN}^2(t) + \alpha_3 v_{IN}^3(t) + \dots\right)$$
$G_M, [1, \alpha_2, \alpha_3, \dots]$

(a) Simplified FTNC-RX testbench (VerilogA).

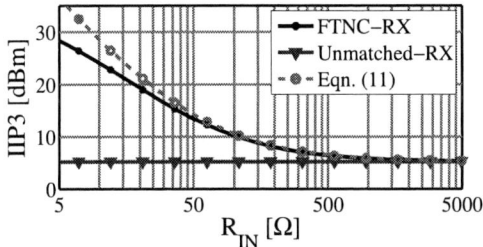

(b) IIP3 simulated results.

Fig. 16. Linearity advantages of FTNC-RX over an unmatched RF-transconductance based receiver.

reciprocal mixing effects. Therefore, the architecture requires a very low noise LOGEN chain. It has been shown that very low noise LO division and buffering can be accomplished through retiming [16] [17], however, VCO phase noise will typically dominate the process for blockers located at moderate offsets. This section first shows why the trade-off between phase noise and power consumption in a VCO is fundamentally limited, before a reciprocal mixing technique is discussed that circumvents this trade-off.

### A. Limitations of VCO Performance

Despite intensive research, oscillator performance metrics have not improved a great deal over the past fifteen years. Indeed the best Figure of Merit ($FOM$) currently reported was published back in 2001 [28]. To understand why, let's rewrite the expression for $FOM$ in terms of oscillator efficiency $\eta$ and noise factor $F$:

$$FOM = \frac{\left(\frac{\omega_c}{\Delta\omega}\right)^2}{\mathcal{L}\{\Delta\omega\}P_{DC[mW]}} = \frac{2\eta Q^2}{kTF}10^{-3}, \quad (12)$$

where $\eta = P_{TANK[mW]}/P_{DC[mW]}$, the power dissipated in the tank is $P_{TANK[mW]} = A_c^2/(2R_p)$, the oscillation amplitude is $A_c$, and the equivalent tank loss is given by $R_p$. Leeson's well-known expression for phase noise [29] has been

used in the simplification:

$$\mathcal{L}\{\Delta\omega\} = \frac{4kTFR_p}{A_c^2}\left(\frac{1}{2Q}\right)^2\left(\frac{\omega_c}{\Delta\omega}\right)^2. \quad (13)$$

This equation for $FOM$ depends on only three variables: quality factor ($Q$), efficiency ($\eta$) and noise factor ($F$). Since $Q$ is set for a given process, a circuit designer can only optimise $\eta$ and $F$. Now consider the generalised $LC$ oscillator shown in Fig. 17(a), which consists of a lossy $LC$ resonator and a memoryless nonlinear conductance. Assume the inputs and outputs of the conductance can be connected to any internal nodes within the passive structure. Since the conductance is memoryless, the PSD of its thermal noise is at best proportional to its instantaneous conductance (i.e. $\overline{i_{gm}^2(t)} = 4kT\gamma gm(t)$, where $\gamma = 2/3$ in long-channel CMOS).

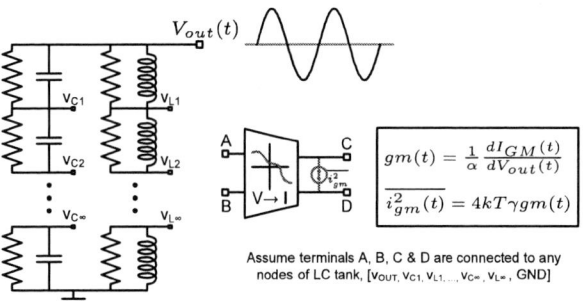

(a) A generic LC oscillator with an arbitrary conductance.

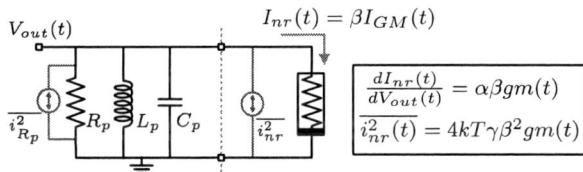

(b) The negative-gm $LC$ oscillator model [30].

Fig. 17. The $LC$ Oscillator.

We can redraw this circuit in the form of Fig. 17(b) [30], which is a lossy $LC$ tank in parallel with a memoryless nonlinear negative resistance, where $\alpha$ and $\beta$ are used in the transformation, each having a maximum possible value of 1, which corresponds to the case where both the inputs and outputs are across the entire tank. Given this setup, Bank's general result [31] [32] [30] tells us that the noise contributed by the tank loss will be equal to the noise contributed by the negative resistance, but in this case scaled by the proportional constant $\beta\gamma/\alpha$. This is regardless of the specific IV characteristic of the negative resistance. Minimising $F$ implies that $\alpha$ should be set to its maximum of 1. Furthermore maximising $\eta/F$ implies setting $\beta = 1$. With this optimum connection arrangement, the noise factor is fundamentally limited to $F = 1 + \gamma$.

A standard $LC$ oscillator (unlike a Colpitts) employs this optimum connection arrangement and achieves an efficiency of 63.67% when driven rail-to-rail, which is just short of the maximum possible efficiency, i.e. 100%[1]. Fig. 18(b) plots the

[1]An ideal class-C oscillator is theoretically capable of 100% efficiency, limited only by practical design issues [32] [33]

maximum achievable $FOM$ of the standard topology (assuming the output amplitude is maximised) and the maximum achievable $FOM$ of any $LC$ oscillator. Remarkably, for a given $Q$, the theoretically best possible $FOM$ is only 2dB better than the standard topology.

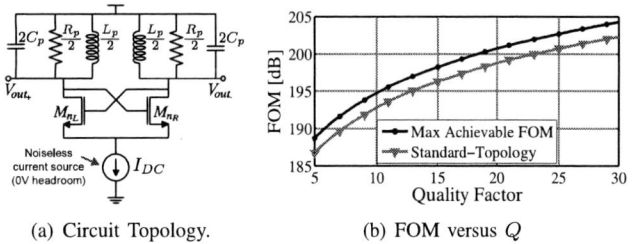

(a) Circuit Topology.  (b) FOM versus $Q$

Fig. 18.  The ideal figure of merit for the standard $LC$ topology.

While practical design issues will prevent this ideal $FOM$ for being obtained, it does demonstrate that the opportunities for circuit innovation are fundamentally limited. In order to significantly improve VCO phase noise, one must move to a technology with lower resistivity metal or burn more power. In the context of wideband receivers, the penalty for removing passive filtering is thus significant: Assuming reciprocal mixing of VCO phase noise dominates, if 10dB of RF filtering is removed, VCO phase noise must improve by 10dB in order to attain the same blocker NF. This corresponds to a 10X increase in power.

### B. The Proposed Reciprocal Mixing Cancellation Technique

To avoid this dramatic rise in VCO power, we can employ a recently proposed reciprocal mixing cancellation technique [18]. The proposed technique makes use of the inherent symmetry of phase noise of an LO signal. As shown in Fig. 19(a), when a noisy LO is mixed with a narrowband blocker, it yields a symmetrical reciprocal-mixing profile around the blocker beat frequency $\Delta f_b$. The tail of reciprocal-mixing profile around DC is in-band and is indistinguishable from the desired signal, while the image of the in-band reciprocal-mixing component is located at $2\Delta f_b$. If this image is frequency-shifted by $2\Delta f_b$ and properly scaled, the noise due to reciprocal mixing can be cancelled by simple subtraction. After cancellation, the inband noise is ideally independent of both VCO phase noise and blocker power.

A limiter-based realisation of this technique, which is suitable for CW blockers, is shown in 19(b). This topology makes use of the fact that when a band-limited sinusoid that is modulated with a small single-sideband is passed through a limiter, the resulting output is a phase-modulated waveform with symmetric sidebands with a carrier-to-sideband ratio that is 6dB less than the input. Therefore, if the downconverted image of reciprocal mixing noise is directed to an auxiliary path, high-passed filter and past through a limiter, the reciprocal mixing image can be shifted to DC. This signal is then use to cancel the noise due to reciprocal mixing in conventional signal path.

A fabricated 40nm version of this topology is shown in 19(c). To evaluate the performance, the receiver was tuned

to 2.3GHz and a CW blocker was applied at 20MHz away. The resulting receiver noise figure versus blocker power is plotted in Fig. 19(d), which demonstrates that the post-cancellation NF improves by up to 19dB. The receiver small signal NF was measured at 2.4dB, which increased to 9.5dB at the presence of a -15dBm blocker after cancellation, limited by the noise of the auxiliary path. Ultimately, the receiver is limited by gain compression effects for higher blocker powers, but it is important to note that the fundamental trade-off between VCO power and noise due to reciprocal mixing has been broken. Note also that the auxiliary path input is the current shunted through the large capacitors at the inputs of signal path TIAs and, therefore, this technique can be directly applied to the FTNC-RX.

More sophisticated realisations of this technique, which can cancel noise due to the reciprocal mixing of modulated blockers, are possible. Such techniques utilise low-frequency carrier-recovery circuitry in the auxiliary path instead of simple limiters, but are outside the scope of this work.

## V. CONCLUSION

The deleterious effects of blockers on wideband receivers can be mitigated. Gain compression can be overcome with the use of passive-mixer-based architectures, while noise performance need not be degraded provided two downconversion paths are employed. Furthermore, the noise due to reciprocal-mixing of LO phase noise with unfiltered blockers can be cancelled, which suggests that LOGEN current does not have to be dramatically increased when passive filtering is removed.

These circuit techniques have the potential to make a single wideband receiver a practical alternative to multiple narrowband receivers.

## REFERENCES

[1] J. Mitola, "The software radio architecture," *IEEE Commun. Mag.*, vol. 33, no. 5, pp. 26–38, May 1995.

[2] F. Bruccoleri, E. Klumperink, and B. Nauta, "Noise cancelling in wideband CMOS LNAs," in *IEEE ISSCC Dig. 2002*, Feb. 2002, pp. 406–407.

[3] ——, "Wide-band CMOS low-noise amplifier exploiting thermal noise canceling," *IEEE J. Solid-State Circuits*, vol. 39, no. 2, pp. 275–282, Feb. 2004.

[4] S. Zhou and M.-C. Chang, "A cmos passive mixer with low flicker noise for low-power direct-conversion receiver," *IEEE J. Solid-State Circuits*, vol. 40, no. 5, pp. 1084–1093, May 2005.

[5] M. Soer, E. Klumperink, Z. Ru, F. van Vliet, and B. Nauta, "A 0.2-to-2.0GHz 65nm CMOS receiver without LNA achieving >11dBm IIP3 and <6.5 dB NF," in *IEEE ISSCC Dig. 2009*, Feb. 2009, pp. 222–223,223a.

[6] C. Andrews and A. Molnar, "A passive-mixer-first receiver with baseband-controlled RF impedance matching, <6dB NF, and >27dBm wideband IIP3," in *IEEE ISSCC Dig. 2010*, Feb. 2010, pp. 46–47.

[7] ——, "A passive mixer-first receiver with digitally controlled and widely tunable RF interface," *IEEE J. Solid-State Circuits*, vol. 45, no. 12, pp. 2696–2708, Dec. 2010.

[8] ——, "Implications of passive mixer transparency for impedance matching and noise figure in passive mixer-first receivers," *IEEE Trans. Circuits Syst. I*, vol. 57, no. 12, pp. 3092–3103, Dec. 2010.

[9] E. Sacchi, I. Bietti, S. Erba, L. Tee, P. Vilmercati, and R. Castello, "A 15 mW, 70 kHz 1/f corner direct conversion CMOS receiver," in *Proc. CICC 2003*, Sept. 2003, pp. 459–462.

[10] Z. Ru, E. Klumperink, G. Wienk, and B. Nauta, "A software-defined radio receiver architecture robust to out-of-band interference," in *IEEE ISSCC Dig. 2009*, Feb. 2009, pp. 230–231,231a.

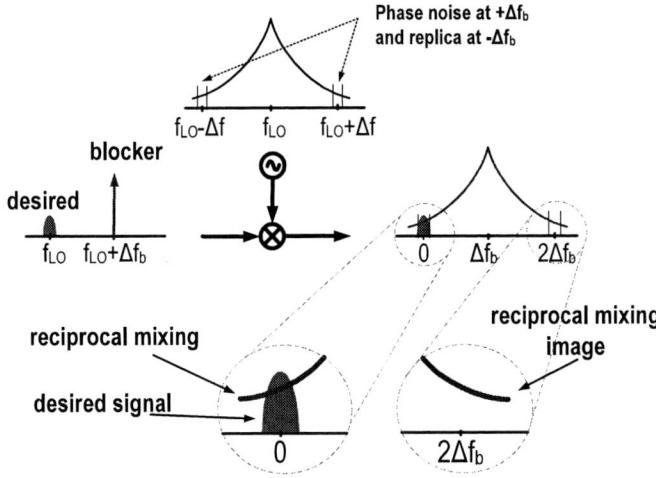

(a) Image of Reciprocal Mixing Product.

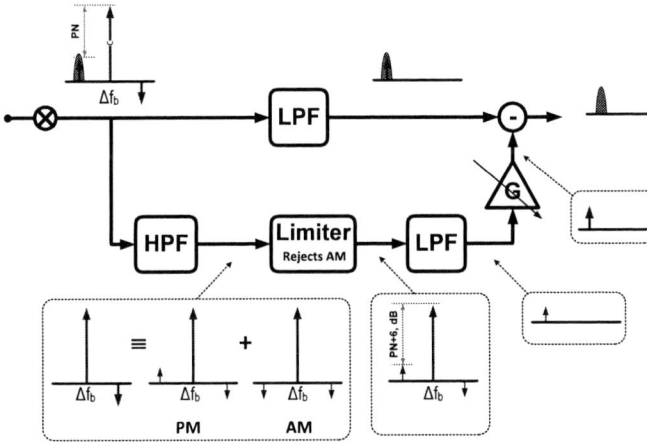

(b) Proposed Architecture

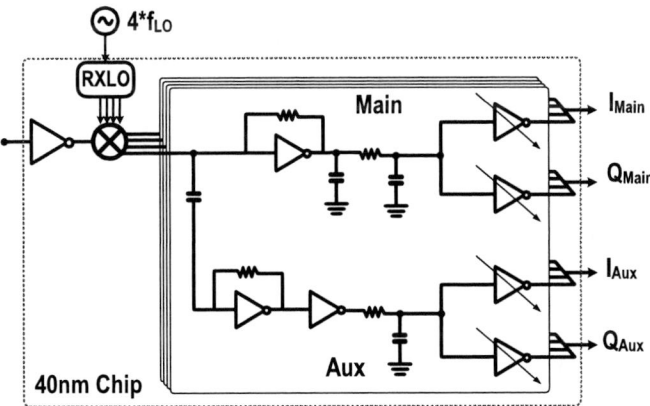

(c) Circuit Schematic

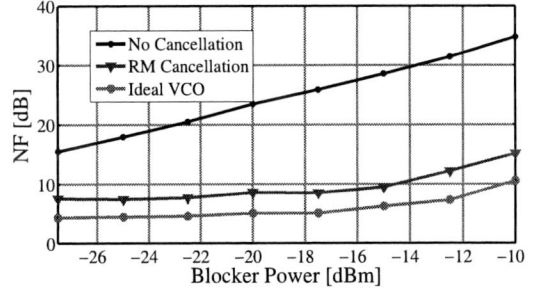

(d) Measured Cancellation

Fig. 19. A Reciprocal Mixing Cancellation Technique.

[11] Z. Ru, N. Moseley, E. Klumperink, and B. Nauta, "Digitally enhanced software-defined radio receiver robust to out-of-band interference," *IEEE J. Solid-State Circuits*, vol. 44, no. 12, pp. 3359–3375, Dec. 2009.

[12] R. Bagheri, A. Mirzaei, S. Chehrazi, M. Heidari, M. Lee, M. Mikhemar, M. Tang, and A. Abidi, "An 800MHz to 5GHz software-defined radio receiver in 90nm CMOS," in *IEEE ISSCC Dig. 2006*, Feb. 2006, pp. 1932–1941.

[13] R. Bagheri, A. Mirzaei, S. Chehrazi, M. Heidari, M. Lee, M. Mikhemar, W. Tang, and A. Abidi, "An 800-mhz ndash;6-ghz software-defined wireless receiver in 90-nm cmos," *IEEE J. Solid-State Circuits*, vol. 41, no. 12, pp. 2860–2876, Dec. 2006.

[14] J. Borremans, G. Mandal, V. Giannini, T. Sano, M. Ingels, B. Verbruggen, and J. Craninckx, "A 40nm CMOS highly linear 0.4-to-6GHz receiver resilient to 0dBm out-of-band blockers," in *IEEE ISSCC Dig. 2011*, Feb. 2011, pp. 62–64.

[15] J. Borremans, G. Mandal, V. Giannini, B. Debaillie, M. Ingels, T. Sano, B. Verbruggen, and J. Craninckx, "A 40 nm CMOS 0.4-6GHz receiver resilient to out-of-band blockers," *IEEE J. Solid-State Circuits*, vol. 46, no. 7, pp. 1659–1671, July 2011.

[16] D. Murphy, A. Hafez, A. Mirzaei, M. Mikhemar, H. Darabi, M.-C. Chang, and A. Abidi, "A blocker-tolerant wideband noise-cancelling receiver with a 2dB noise figure," in *IEEE ISSCC Dig. 2012*, Feb. 2012, pp. 74–75.

[17] D. Murphy, H. Darabi, A. Abidi, A. A. Hafez, A. Mirzaei, M. Mikhemar, and M.-C. F. Chang, "A blocker-tolerant, noise-cancelling receiver suitable for wideband wireless applications," *IEEE J. Solid-State Circuits*, vol. 47, no. 12, pp. 2943–2963, Dec. 2012.

[18] M. Mikhemar, D. Murphy, A. Mirzaei, and H. Darabi, "A phase noise and spur filtering technique using reciprocal-mixing cancellation," in *IEEE ISSCC Dig. 2013*, Feb. 2013, pp. 86–87.

[19] S. Blaakmeer, E. Klumperink, D. Leenaerts, and B. Nauta, "A wideband balun LNA I/Q-mixer combination in 65nm CMOS," in *IEEE ISSCC Dig. 2008*, Feb. 2008, pp. 326–617.

[20] ——, "The blixer, a wideband balun-LNA-I/Q-mixer topology," *IEEE J. Solid-State Circuits*, vol. 43, no. 12, pp. 2706–2715, Dec. 2008.

[21] L. Franks and I. Sandberg, "An alternative approach to the realizations of network functions: N-path filter," *Bell Syst. Tech. J.*, pp. 1321–1350, 1960.

[22] D. von Grunigen, R. Sigg, J. Schmid, G. Moschytz, and H. Melchior, "An integrated CMOS switched-capacitor bandpass filter based on N-path and frequency-sampling principles," *IEEE J. Solid-State Circuits*, vol. 18, no. 6, pp. 753–761, Dec. 1983.

[23] A. Mirzaei, H. Darabi, and D. Murphy, "A low-power process-scalable superheterodyne receiver with integrated high-Q filters," in *IEEE ISSCC Dig. 2010*, 2011, pp. 60–62.

[24] ——, "A low-power process-scalable super-heterodyne receiver with integrated high- filters," *IEEE J. Solid-State Circuits*, vol. 46, no. 12, pp. 2920–2932, Dec 2011.

[25] A. Mirzaei, A. Yazdi, Z. Zhou, E. Chang, P. Suri, and H. Darabi, "A 65nm CMOS quad-band SAW-less receiver for GSM/GPRS/EDGE," in *Proc. Symp. VLSI Circuits 2010*, June 2010, pp. 179–180.

[26] A. Mirzaei, H. Darabi, A. Yazdi, Z. Zhou, E. Chang, and P. Suri, "A 65 nm CMOS quad-band SAW-less receiver SoC for GSM/GPRS/EDGE," *IEEE J. Solid-State Circuits*, vol. 46, no. 4, pp. 950–964, Apr. 2011.

[27] Z. Ru, "Frequency translation techniques for interference-robust software-defined radio receivers," Ph.D. dissertation, University of Twente, Enschede, November 2009.

[28] E. Hegazi, H. Sjoland, and A. A. Abidi, "A filtering technique to lower LC oscillator phase noise," *IEEE J. Solid-State Circuits*, vol. 36, no. 12, pp. 1921–1930, Dec. 2001.

[29] D. B. Leeson, "A simple model of feedback oscillator noise spectrum," *Proc. IEEE*, vol. 54, no. 2, pp. 329–330, Feb. 1966.

[30] D. Murphy, J. Rael, and A. Abidi, "Phase noise in LC oscillators: A phasor-based analysis of a general result and of loaded Q," *IEEE Trans. Circuits Syst. I, Fundam. Theory Applicat.*, vol. 57, no. 6, pp. 1187–1203, 2010.

[31] J. Bank, "A harmonic-oscillator design methodology based on describing functions," Ph.D. dissertation, Chalmers University of Technology, Sweden, 2006.

[32] A. Mazzanti and P. Andreani, "Class-C harmonic CMOS VCOs, with a general result on phase noise," *IEEE J. Solid-State Circuits*, vol. 43, no. 12, pp. 2716–2729, Dec. 2008.

[33] ——, "A 1.4mW 4.90-to-5.65GHz class-C CMOS VCO with an average FoM of 194.5dBc/Hz," *Proc. Int. Solid-State Circuits Conf. (ISSCC)*, pp. 474–629, Feb 2008.

# An Asymmetric Dual-Channel Reconfigurable Receiver for GNSS in 180nm CMOS

Nan Qi, Baoyong Chi, Yang Xu, Zhou Chen, Jun Xie, Yang Xu, Zheng Song, Zhihua Wang

Institute of Microelectronics, Tsinghua University, Beijing, 100084, China

*Abstract*-A fully integrated dual-channel reconfigurable receiver supporting all the GNSS (GPS, Compass, GLONASS, Galileo) signals in 180nm CMOS is presented. The two channels share the frequency synthesizer and RF front-end circuits, but employ separate IF strips to support simultaneous dual-constellation reception. In order to save the power of digital baseband, one of the two IF strips can be configured to a dual-conversion mode, which lowers the highest signal frequency and thus the sampling rate. The two asymmetric channels are designed with different bandwidths, covering from 2.2MHz to 20MHz for both civil and high precision applications. Besides, I/Q mismatch calibration is introduced into each down-conversion to improve the image rejection ratio (IRR). The highly integrated receiver also integrates on-chip LDOs, crystal oscillator, AFC, AGC and DCOC modules. Thanks to the flexible frequency plan and scalable IF circuits, the typical dual-channel power consumption can be reduced to 45mW. The receiver finally achieves 2.5dB noise figure, 40dB minimum IRR, 55dB dynamic range with 1dB gain steps and -57dBm input referred in-band 1dB compression point. The result of collaboration with digital baseband shows that the chip achieves positioning with >40dB carrier to noise ratio (CNR).

## I. INTRODUCTION

As China and the European Union accelerate their construction of Compass and Galileo system in these years, the Global Navigation Satellite System (GNSS) comes to a new era characterized by multi-constellation interoperability and high precision positioning. On one hand, there are plenty of satellites in the sky, serving for different constellations including GPS, Compass, GLONASS and Galileo. However, the number of observable satellites from one system is often insufficient to accomplish the positioning. Therefore, GNSS interoperation is demanded, in which signals from different systems (two systems in this work) are received and processed simultaneously, collaborating to achieve faster positioning and improved reliability.

On the other hand, high precision GNSS signals (GPS L5, Compass B2 and Galileo E5) are open to the public recently, which are modulated with high data rate PRN codes and occupy wide bandwidths. Higher process gain can be achieved when de-spreading, which helps to increase the final signal-to-noise ratio. Compared to traditional GPS radios, the GNSS analog front-end (AFE) should provide wider bandwidth to cover the main lobe of the satellite signals.

Recently there are several works focused in the related area. Reference [1] simply adopts two independent receivers in a single chip, but the chip area and power consumption are inevitably huge. Besides, LO cross-talk and frequency pulling issues might be troublesome. Reference [2] shares the LO between the two channels, and adopts a flexible frequency

plan to optimize the power consumption. However, there are still modes with high IF frequency and sampling rates, which not only increases the complexity of analog IF circuits but also burdens the digital baseband (DBB). Reference [3] employs only one analog channel to receive dual-constellation signals, and separates them by complex filtering in the DBB. It does save power in the AFE but complicates the processing in the digital domain, especially under a high sampling rate. Moreover, it doesn't support wideband reception for high precision signals.

This paper presents a fully integrated dual-channel reconfigurable receiver, which is capable of receiving single system or simultaneously dual-system GNSS signals with narrow or wide bandwidth. The two channels share the frequency synthesizer and RF front-end circuits, but employ separate asymmetric IF-strips. To save power both in AFE and DBB, the two IF-strips are designed with different passband modes, and one of them can be configured to the dual-conversion architecture. The analog baseband circuit employs power-scalable op-amp arrays to regulate the power consumption across various operation modes. I/Q mismatch calibrations are introduced into each down-conversion to improve the image rejection ratio. The receiver has been implemented in 180nm CMOS, and tested both individually and cooperated with digital baseband. Measurement results show that the chip achieves positioning with low power, low noise figure and high image rejection ratio.

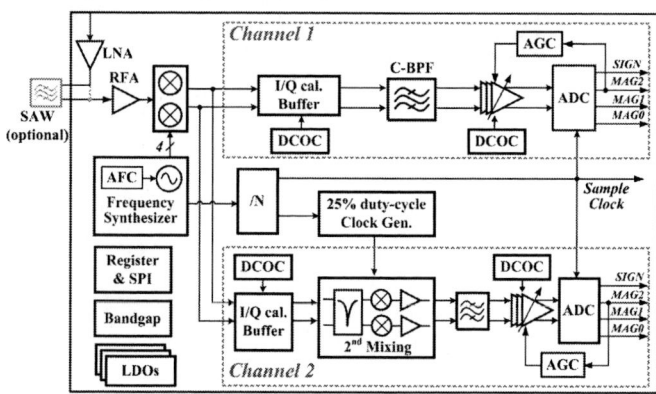

Fig. 1.    Block diagram of the GNSS receiver.

## II. RECEIVER ARCHITECTURE

### A. Receiver Architecture

As shown in Fig. 1, the dual-channel receiver shares RF circuits and employs two independent IF-paths that established after the down-conversion. Only one frequency synthesizer is used for the two channels in order to avoid LO cross talk and save power. By default, the whole receiver is configured to the single-conversion low-IF type to receive dual-constellation signals simultaneously. Besides, the second channel can be reconfigured to a dual-conversion mode, in which a passive poly-phase filter (PPF), a passive mixer and a calibration buffer would be inserted. In the dual-conversion mode, the first image band is rejected by the PPF, while the second one suppressed by the complex bandpass filter (C-BPF).

The RF section of the receiver consists of a single-ended LNA, a single-to-differential radio frequency amplifier (RFA), and a quadrature down-conversion mixer. After frequency mixing, the signal path is divided into two IF-branches, each of which includes an I/Q calibration buffer, a C-BPF, a tri-stage PGA and a multi-bit switchable ADC. The PGA has 45dB dynamic range with 1dB gain step and a digital assisted AGC loop. An external SAW can be inserted between LNA and RFA in case of strong out-of-band interferences, but the cascaded noise figure will get 0.5dB worsen (assuming the insertion loss of SAW is 2dB).

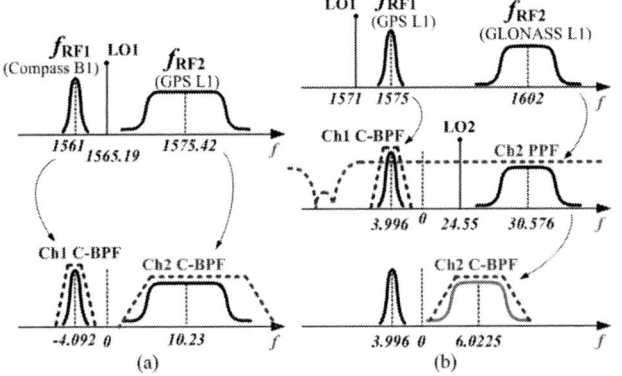

Fig. 2 Dual-channel frequency plan:(a) single-conversion, (b) dual-conversion

### B. Frequency Plan

As shown in Fig. 2, the frequency plan for the receiver is flexible for different operation modes. In single channel modes, the receiver is simply a low-IF single-conversion type. In dual-constellation modes when the two RF signals locates near (Fig. 2a), the LO is set to the middle of them, and the C-BPF directly picks up each positive (or negative) corresponding IF signal. Besides, when the frequency gap between them is large (Fig. 2b), LO1 is firstly selected to convert $f_{RF1}$ to sufficiently low, while leaving the other

channel's frequency high. Then LO2 further shifts down the IF signal in channel2 in the second conversion to a reasonable frequency. In the whole process, the C-BPF in each channel takes charge of band selection, while the PPF is responsible for the first image rejection.

The proposed receiver supports five dual-system and all the single-system modes of the current GNSS. Since the two channels are asymmetric, wideband (>10MHz) signals are always processed in the second channel. The frequency plan for dual-system modes is listed in table I. Since the single-channel modes can be easily calculated referring to low-IF conversion method, they are not listed here.

### C. Dynamic Range and AGC

The analog front-end can provide 55dB dynamic range to cover the temperature induced input strength variation, passive or active antenna application shifting and in-band jammers that 20dB above the noise floor. An AGC loop is employed to regulate the PGA's output, which monitors the highest magnitude output (MAG2) of the ADC, accumulates its value within a long term, and compares it to the predetermined threshold. For 2-bit ADC the optimum duty-cycle is 30.8%, while for 3 and 4-bit modes it would be 23.5% and 18.8% respectively.

### III. BUILDING BLOCKS

### A. RF Front-end

The RF Front-end consists of a LNA, a RFA and a quadrature Mixer. The LNA adopts the common source topology with its degeneration inductor realized by bond wires. Besides, the capacitor array is utilized at the output node, compensating the resonation frequency shift induced by temperature or process variation. The RFA is realized with a CG-CS [4] topology for the single-to-differential conversion, which achieves the wideband input matching that covers both L1 and L2 bands. Moreover, the noise cancellation scheme helps to improve the radio's noise performance, especially when the antenna directly feeds signals into the RFA for higher linearity. A Gilbert type mixer down-converts RF signals to analog baseband with quadrature outputs. Current bleeding technique has been utilized to reduce the switching transistor's current and thus its flicker noise contribution.

### B. I/Q Mismatch Calibration Buffer

## TABLE I
### DUAL-CONVERSION MODES SUPPORTED BY THE GNSS RECEIVER

| | Mode | Channel | RF (MHz) | LO1 (MHz) | IF1 (MHz) | LO2 (MHz) | IF2 (MHz) | BW (MHz) | Sample Clock (MHz) |
|---|---|---|---|---|---|---|---|---|---|
| 1 | GPS L1 (Galileo E1) | Ch1 | 1575.42 | $96f_{ref}$ | 3.996 | $f_{LO1}/110$ | - | 2.2 (4.2) | $f_{ref}$ |
| | Compass B1 | Ch2 | 1561.098 | | -10.326 | | 3.96 | 4.2 | |
| 2 | GPS L1 (Galileo E1) | Ch1 | 1575.42 | $96f_{ref}$ | 3.996 | $f_{LO1}/64$ | - | 2.2 (4.2) | $2f_{ref}$ |
| | GLONASS L1 | Ch2 | 1602 | | 30.576 | | 6 | 10 | |
| 3 | Compass B1 | Ch1 | 1561.098 | 1565.19 | -4.092 | - | - | 4.2 | $f_{LO1}/32$ |
| | GPS L1w (Galileo E1) | Ch2 | 1575.42 | | 10.23 | - | - | 18 | |
| 4 | GLONASS L2 | Ch1 | 1246 | 1239.876 | 6.124 | - | - | 10 | $f_{LO1}/28$ |
| | GPS L2 | Ch2 | 1227.6 | | -12.276 | - | - | 18 | |
| 5 | Compass B2 | Ch1 | 1207.14 | 1217.37 | -10.23 | - | - | 4.2 | $f_{LO1}/28$ |
| | GPS L2 | Ch2 | 1227.6 | | 10.23 | - | - | 18 | |

Note: $f_{ref}$=16.369MHz for typical modes.

An I/Q calibration buffer is placed prior to the IF C-BPF to improve the image rejection ratio (IRR). The buffer is realized in a current mode to improve linearity, in which the V-I and I-V conversion is distributed into two stages (Fig. 3). The trans-conductor (or $g_m$) stage is built up by the combination of forward-path array and cross-coupled array. The forward-path array takes charge of I/Q amplitude error calibration, while the cross-coupled one is responsible for phase error calibration. The $g_m$-cell in the array is realized in independent-biased inverters, whose input gates are controlled by separate switches. The buffer can provide 6 or 12 dB voltage gain by adjusting $R_f$ in the TIA stage.

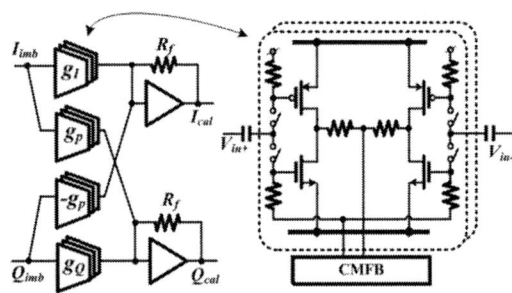

Fig. 3.    Circuit implementations of I/Q mismatch calibration module.

### C.    The Second Down-Conversion

The second channel of the proposed GNSS receiver can be configured to the dual-conversion architecture (Fig. 4), in which a PPF, a $g_m$-TIA buffer and a passive mixer are inserted into the signal chain. A 2$^{nd}$-order PPF with 8MHz BW is employed to reject the image band for the first conversion. The phase sequence of I/Q inputs can be switched, in order to select positive or negative band rejection. The PPF shares a RC-tuner with the C-BPF, which automatically tune the capacitance to overcome the effects of the temperature and process variation. A current mode mixer is adopted for higher linearity, which is regulated by a 25%-duty-cycle clock. Compared to the 50% duty-cycle, the non-overlapping clock can provide higher conversion gain and linearity. The feedback resistor in the TIA includes a fixed part and a variable part, which is used to calibrate amplitude imbalances between I/Q signals. The second down-conversion can be disabled when bypassing signals directly to the C-BPF.

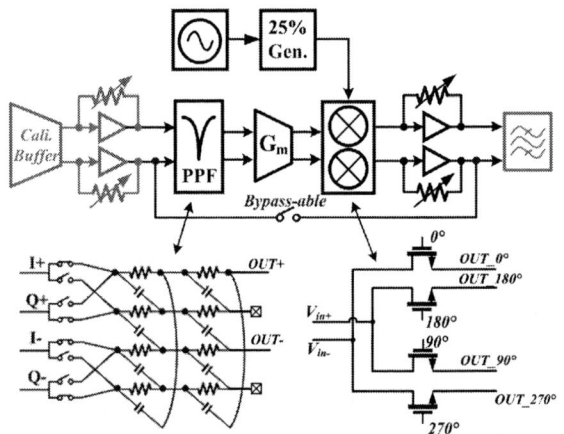

Fig. 4.    Circuit implementations of the 2$^{nd}$ conversion in channel-2.

## IV. MEASUREMENT RESULTS

The GNSS receiver has been implemented in 180nm CMOS, which occupies 3x3.5mm$^2$ area (Fig. 5). The dies have been mounted on PCB directly and measurements have been carried out on each building block and the whole receiver.

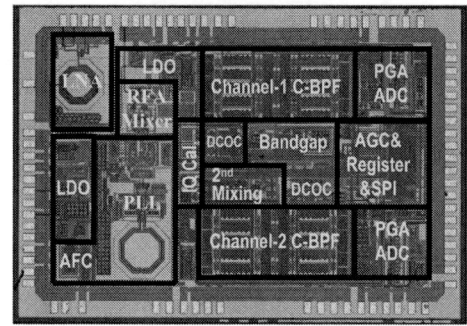

Fig. 5.    Microphotograph of the presented dual-channel GNSS receiver.

### A.    Image Rejection Ratio (IRR)

The IRR is tested twice with different test tones for the two image bands in each down-conversion. In the 1$^{st}$ conversion, the image and desired signal locate at both sides of the 1$^{st}$ LO. They are deliberately set asymmetric to the LO for identification. The tested IRR is compared before and after I/Q calibration (Fig. 6). In the 2$^{nd}$ conversion, the RF image signal is selected to generate an image at the 2$^{nd}$ IF, which would not suppressed by the PPF. The tested desired and calibrated image signal are also shown below.

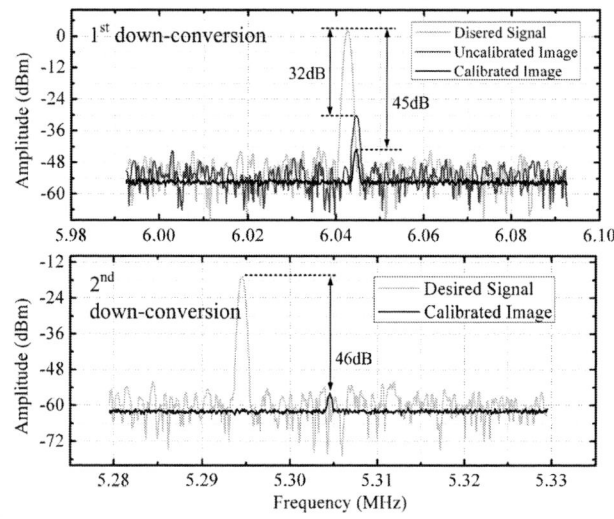

Fig. 6.    Tested IRR in each of the conversion.

### B.    Cascaded Frequency Response

The frequency response of the whole receiver is tested by sweeping the RF input frequency, and monitoring the PGA outputs on a spectrum analyzer. The C-BPF has the narrowest bandwidth in the signal chain, thus it dominates the cascaded frequency response. Measurement results shown in Fig. 7 include four operation modes in each channel, with the bandwidth covering from 4.2MHz to 20MHz.

978-1-4673-6145-3/13 $31.00 © 2013 IEEE

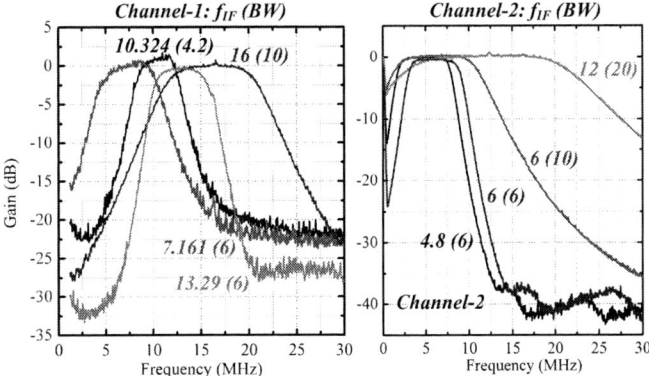

Fig. 7. Tested frequency response of the cascaded AFE.

### C. Fractional-N Frequency Synthesizer

The 20bit fractional-N frequency synthesizer is designed to cover both L1 and L2 bands. Fig. 10 shows the phase noise in different modes, which settles around -90dBc/Hz at 100kHz and -110dBc/Hz at 1MHz offset respectively. Since frequency of the testing point is $2f_{LO}$, the actual LO phase noise should be 4-5dB better.

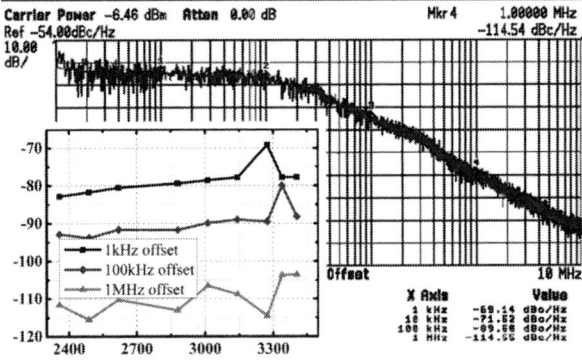

Fig. 8. Tested LO phase noise at $2f_{LO}$ in different modes.

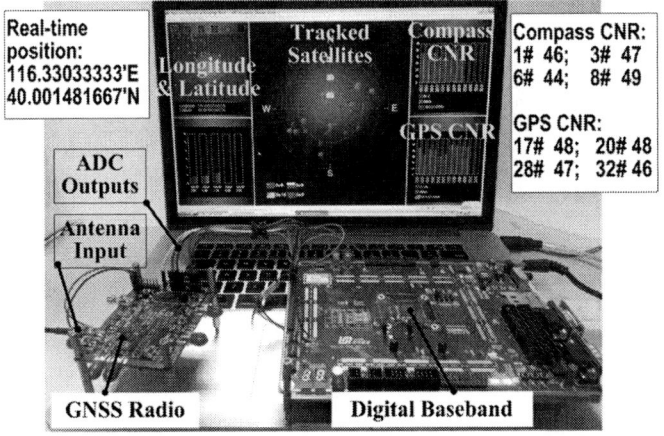

Fig. 9. The measurement platform and testing results of the co-operation.

### D. Co-Operation with Digital Baseband (DBB)

Experiments have been carried out, in which the GNSS radio receives GPS/Compass dual-system signals from an active antenna. The ADC in each channel outputs 3-bit digitized data to the DBB with the AGC loop enabled. Fig. 9 shows the real-time tracked GPS (green) and Compass (red) satellites, as well as the CNR of each satellite and the longitude/latitude information. This verifies the functionality of the GNSS radio when co-operating with DBB in the dual-system mode.

### E. Performance Comparison

Table II summarizes the performance of the presented receiver, and makes a comparison with the-state-of-the-arts. It could be seen that our receiver optimizes the dual-channel DC power while achieving the same or better performances in the other specifications.

TABLE II
COMPARISION WITH PREVIOUS PUBLISHED GNSS RECEIVERS

| Index | This work | [1] | [2] | [3] |
|---|---|---|---|---|
| Technology | 180nm | 180nm | 65nm | 40nm |
| Constellations | GNSS | GNSS | GNSS | GNSS |
| NF(dB) | 2.5dB | 2.5dB | 2.2dB | 2.1dB |
| P-1 dB (dBm) | -57 | -58 | -61 | - |
| IRR (dB) | 45 | 30dB | ~50 | 33dB |
| Phase Noise@ 1MHz(dBc/Hz) | $-114@2f_{LO}$ | -112 | -118 | -94 |
| Dual-channel Power (mW) | 45 | 90* | 60 | single chan. 8.9mA** |
| Sample rate (MHz) | 32.738 GPS/Glonass | - | 65.476 GPS/Glonass | 66.192 GPS/Glonass |

*Estimated from reference paper, **Supply voltage is 1.8V.

### V. CONCLUSION

A fully integrated dual-channel multi-constellation reconfigurable GNSS receiver is implemented in 180nm CMOS. The two channels share the RF front-end circuits and the frequency synthesizer, and employ separate asymmetric IF-strips that switchable between single and dual-conversion. Flexible frequency plan helps to satisfy all the dual-system modes, while keeping the sample-rate low. Moreover, I/Q mismatch calibration is carried out to achieve more than 40dB IRR in each down-conversion. The receiver achieves 2.5dB NF, 45dB IRR, while consuming 45mW power for two channels.

### ACKNOWLEDGEMENTS

This work was supported in part by the National Science and Technology Major Projects of China under Grant 2012ZX03004007 and in part by the National Natural Science Foundation of China under Grant 61020106006, 61076029, 61222405, JCYJ20120616142625998.

### REFERENCES

[1] D. Chen et al., "Reconfigurable dual-channel multiband RF receiver for GPS/Galileo/BD-2 systems," in IEEE Trans. Microw. Theory Tech., vol. 60, no. 11, pp. 3491–3501, Nov. 2012.

[2] Nan Qi, et al., "A Dual-Channel GPS/Compass/Galileo/GLONASS Reconfigurable GNSS Receiver in 65nm CMOS," in CICC 2011.

[3] C. G. Tan et al., "A universal GNSS (GPS/Galileo/Glonass/Beidou) SoC with a 0.25mm² radio in 40nm CMOS," in IEEE ISSCC 2013, pp. 334-335.

[4] Stephan C. Blaakmeer, et al., "Wideband Balun-LNA With Simultaneous Output Balancing, Noise-Cancelling and Distortion-Cancelling," in IEEE J. Solid-State Circuits, vol. 43, no. 6, pp. 1341–1350, June 2008.

# A 5-GHz 11.6-mW CMOS Receiver for IEEE 802.11a Applications

Aliakbar Homayoun and Behzad Razavi
Electrical Engineering Department
University of California, Los Angeles

## Abstract

A direct-conversion receiver employs a 1-to-6 transformer as a low-noise amplifier along with passive mixers and non-invasive baseband filters. Realized in 65-nm CMOS technology, the receiver provides an average noise figure of 5.3 dB and a sensitivity of −70 dBm at a data rate of 54 Mb/s. The prototype draws 11.6 mW from a 1-V supply and occupies an active area of 0.18 mm².

## I. INTRODUCTION

While advances in the art have considerably reduced the power consumption of RF oscillators, frequency dividers, and analog-to-digital converters, the main receiver (RX) chain in 5-GHz systems draws a disproportionately high power, e.g., about 46 mW in [1]. It is therefore desirable to develop low-power RX front ends and baseband filters for WiFi applications.

This paper introduces a complete 5-GHz CMOS receiver that meets the 11a sensitivity, blocking, and filtering requirements while consuming 11.6 mW. This fourfold reduction in power is achieved through the use of a transformer as a low-noise amplifier (LNA), passive mixers, and "non-invasive" baseband filtering [2].

Section II introduces the receiver architecture and Section III elaborates on the design of the transformer. Section IV deals with the interface between the transformer and the mixers and its effect on the RX input matching. Section V describes the baseband channel-select filters and Section VI presents the experimental results.

## II. RECEIVER ARCHITECTURE

With the choice of passive mixers in a receiver, the power consumption arises from three other building blocks: The LNA, the local oscillator (LO) buffers, and the baseband filters, with the last typically dissipating the most [1]. As shown in Fig. 1, we implement the LNA by means of a transformer, thus obtaining voltage gain and ensuring input matching. The small passive mixer devices require an LO buffer power of 0.4 mW (Section IV). We also exploit non-invasive filtering to realize a fourth-order elliptic response with a more relaxed power-linearity-noise trade-off than that of conventional filters.

By virtue of its high turns ratio, the transformer in Fig. 1 exhibits a relatively high output impedance, approximating a current source. Operating with 25%-duty-cycle LOs, the

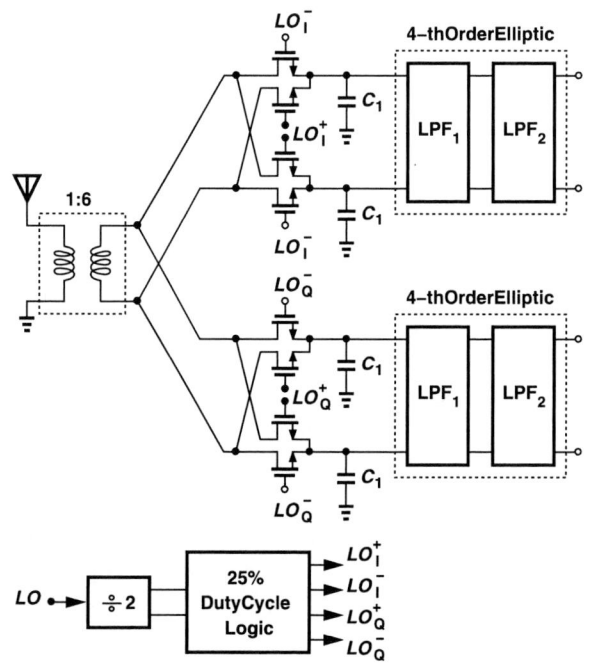

Fig. 1. Receiver architecture.

switches can therefore be viewed as current-driven mixers, thus contributing less noise than voltage-driven topologies [3].

We should highlight two advantages of our approach over the LNA-less receiver in [4]. First, the input matching inherent in our receiver provides a robust interface with the antenna in the presence of long external traces. Second, in addition to saving power, our front end benefits from a higher linearity.

## III. TRANSFORMERS AS LNAS

A low-noise amplifier provides voltage gain and proper input matching but it need not draw supply current. This work explores the possibility of using a high-turns-ratio transformer for this purpose and co-designing it with passive mixers so as to achieve an acceptable noise figure.

The 1-to-6 transformer is realized as shown in Fig. 2, with a one-turn primary in metal 8 and a six-turn secondary in metal 9. Different from planar [5] or other stacked [6] structures, this geometry exhibits a more favorable trade-off between the insertion loss and the loaded voltage gain. As the number of turns in the secondary increases, the voltage gain rises but

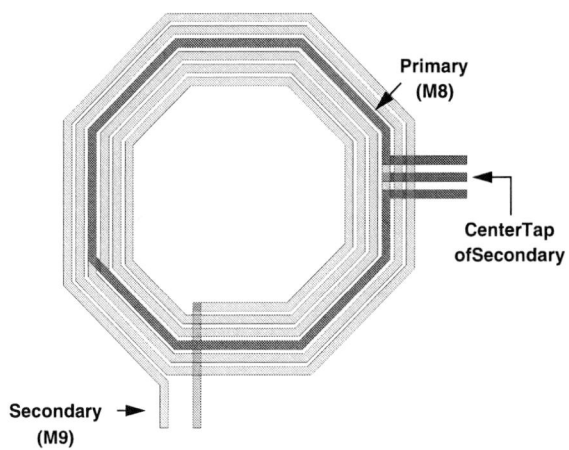

Fig. 2. Transformer geometry.

flattens out because the outer turns begin to have negligible coupling with the primary. The choice of the geometry also depends on the input impedance of the passive mixers and is thus finalized in conjunction with their design.

According to HFSS simulations, the above transformer has an insertion loss of 2.4 dB and a loaded voltage gain of 12 dB at 5.5 GHz. The outer diameters of the primary and the secondary are 146 $\mu$m and 170 $\mu$m, respectively.

## IV. MIXER DESIGN

Driven by a 50- antenna, the transformer presents an output impedance of 800 . Thus, the quadrature passive mixers in Fig. 3 must be designed for an overall input resistance equal to this value. Since the input impedance of current-driven mixers depends on the source impedance [7], we model the interface as shown in Fig. 3, where $I_T$ and $Z_T$ represent the

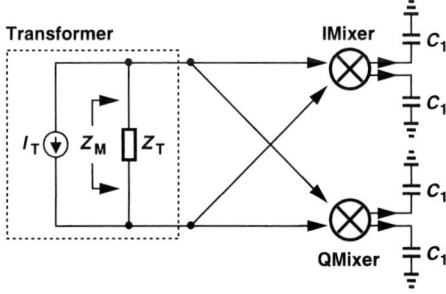

Fig. 3. Transformer-mixer interface.

transformer over a wide bandwidth and $Z_M$ denotes the composite impedance resulting from $Z_T$ and the input impedance of the I and Q mixers. With a baseband capacitive load of $C_1$, $Z_M$ can be simplified to [7]:

$$
Z_M(\omega) = R_{sw}\|Z_T(\omega) + \left[\frac{Z_T(\omega)}{Z_T(\omega) + R_{sw}}\right]^2 \div
$$
$$
\sum_{k=-\infty}^{\infty} \frac{1}{(4k+1)^2[Z_T(\omega + 4k\omega_{LO}) + R_{sw}]}, \quad (1)
$$

where $R_{sw}$ is the switch on-resistance. Due to the bandpass nature of $Z_T$, the summation on the right-hand side must be

carried out for about 14 terms. Ideally, in the range of 5 to 6 GHz, we must have $Re\{Z_M(\omega)\} \approx Z_T(\omega)/2 \approx 400$ and $Im\{Z_M(\omega)\} \approx 0$. With a choice of $W/L = 10\ \mu$m/ 60 nm for the switches, we obtain an $S_{11}$ of $-12$ dB in this band. The LO buffers driving eight such switches draw a total power of $fCV_{DD}^2 \approx 0.4$ mW at 6 GHz.

Simulations indicate that the "zero-power" RF front end consisting of the transformer and the mixers exhibits a noise figure of 4.5 dB and an input $P_{1dB}$ of $-5.2$ dBm at 5.5 GHz. For a target RX NF of less than 6 dB, all of the subsequent stages must contribute no more than 1.5 dB, demanding additional circuit techniques.

## V. FILTER DESIGN

In the 11a standard, the adjacent and alternate adjacent channels can be higher than the desired channel by 16 dB and 32 dB, respectively. The baseband filters must therefore provide a sharp roll-off to reduce these channels to well below the desired signal level − unless the baseband ADCs offer a dynamic range wide enough and a sampling rate high enough to handle partially-attenuated blockers.

Figure 4 shows the realization of the fourth-order elliptic filter. The circuit consists of two second-order sections, each

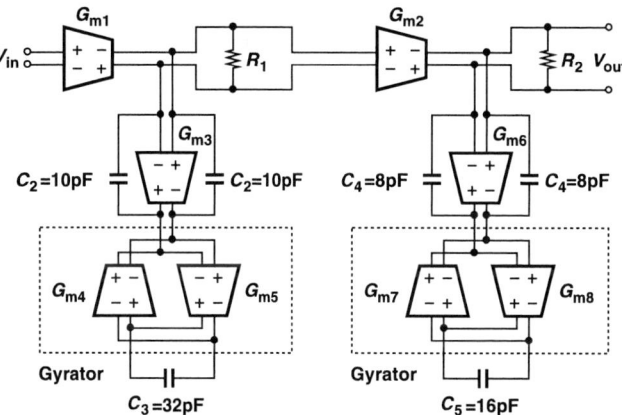

Fig. 4. Fourth-order elliptic low-pass filter.

formed as a $G_m$ cell and a frequency-selective load [2]. Created by $G_{m3}$-$G_{m5}$, $G_{m6}$-$G_{m8}$, and the capacitors, the loads remain *open* in the passband, contributing small noise and nonlinearity to the desired signal, and act as a short circuit in the stopband. This stands in contrast to conventional filters that process the desired signal and the blockers in the same stage and hence add considerable noise and nonlinearity.

The gyrators in Fig. 4 transform their load capacitors to an inductor, which then creates a resonance in each integrator. Proper choice of these resonance frequencies shapes the frequency response of the overall filter, including its passband ripple and stopband rejection. The fourth-order filter exhibits an input-referred noise voltage of 2 nV/$\sqrt{\text{Hz}}$ at 5 MHz, an in-channel $IIP_3$ of 193 mV$_{rms}$ and a voltage gain of 39 dB while consuming 4.3 mW. The filter voltage gain is programmable in steps of 2 to 3 dB for a total range of 43 dB.

978-1-4673-6145-3/13 $31.00 © 2013 IEEE

## VI. EXPERIMENTAL RESULTS

The receiver of Fig. 1 has been fabricated in 65-nm digital CMOS technology. Figure 5 shows the die photograph. The

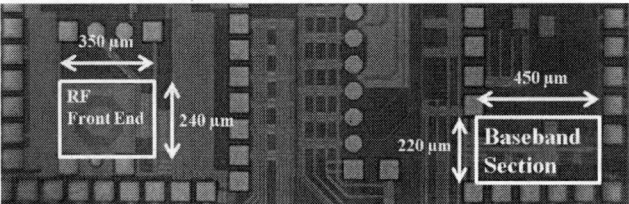

Fig. 5. Die photograph.

RF section occupies 350 $\mu$m $\times$ 240 $\mu$m and the baseband section 450 $\mu$m $\times$ 220 $\mu$m.[1] The circuit has been characterized in a chip-on-board assembly with a 1-V supply.

Figure 6 plots the measured noise figure of the complete receiver as a function of the baseband frequency. The average noise figure is about 5.3 dB.

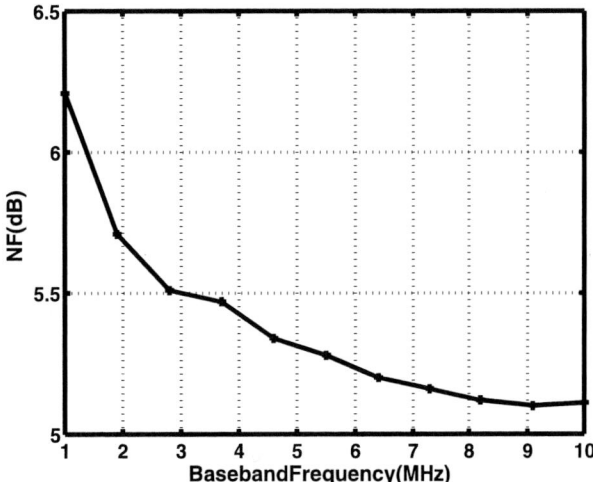

Fig. 6. Measured noise figure.

The sensitivity of the receiver is measured with the aid of Agilent's N5182 MXG vector signal generator and N9020A MXA signal analyzer, which respectively apply a 64-QAM signal and sense the baseband outputs to construct the signal constellation. Figure 7 shows the results for a −65-dBm 5.7-GHz input. The error vector magnitude (EVM) is equal to −28 dB, exceeding the 11a specification. (For an input level of −70 dBm, an EVM of −23.4 dB is measured.)

Figure 8 plots the $S_{11}$ from 5 to 6 GHz, measured at each input frequency, while the mixers switch at the corresponding LO frequency. It is expected that a slightly larger transformer can yield $S_{11} = -10$ dB at the lowest 11a frequency, 5.15 GHz.

Figure 9 plots the measured receiver transfer function, revealing a passband peaking of 1 dB and a rejection of 22 dB at 20 MHz and 43 dB at 40 MHz.[2] Owing to the finite output

---

[1] Due to limited silicon area, the receiver layout is decomposed and placed within other unrelated circuits, but all of the connections are present on the chip.

[2] In this measurement a first-order RC section follows each output on the PCB.

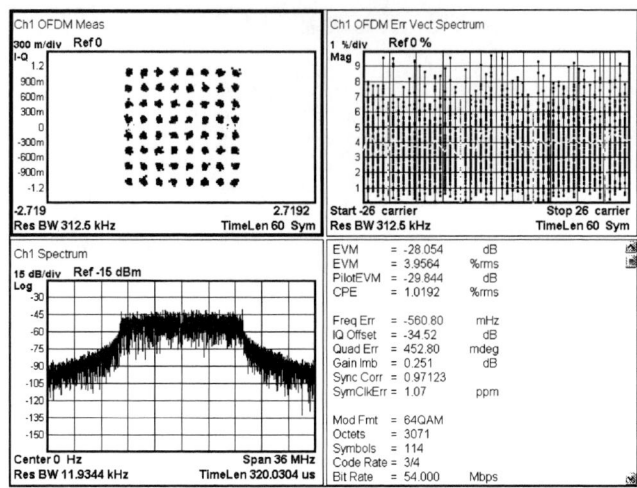

Fig. 7. Measured EVM at $P_{in} = -65$ dBm.

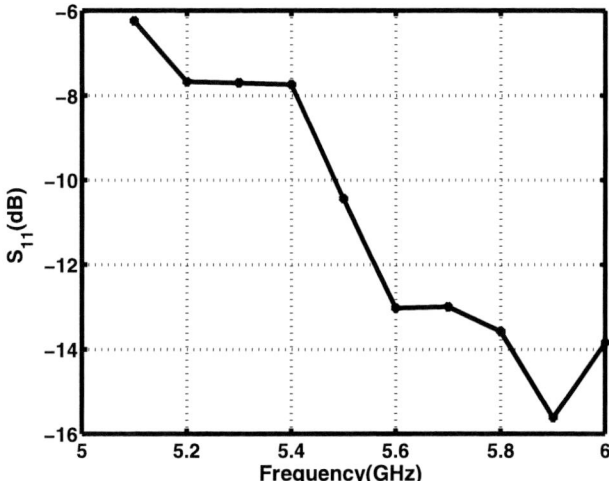

Fig. 8. Measured input return loss.

resistance of the $G_m$ cells, the filter does not exhibit the deep notches that are characteristic of elliptic transfer functions. The

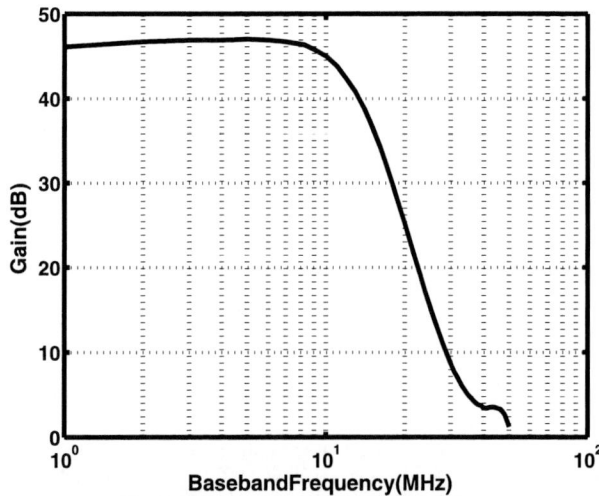

Fig. 9. Measured receiver transfer function.

performance of the baseband filter is ultimately tested when a large blocker accompanies a small desired signal. In such a case, the filter must remain sufficiently selective and linear so that the desired signal does not experience compression. Figure 10 plots the measured passband gain as a function of the power of an RF blocker in the adjacent or alternate adjacent channel.

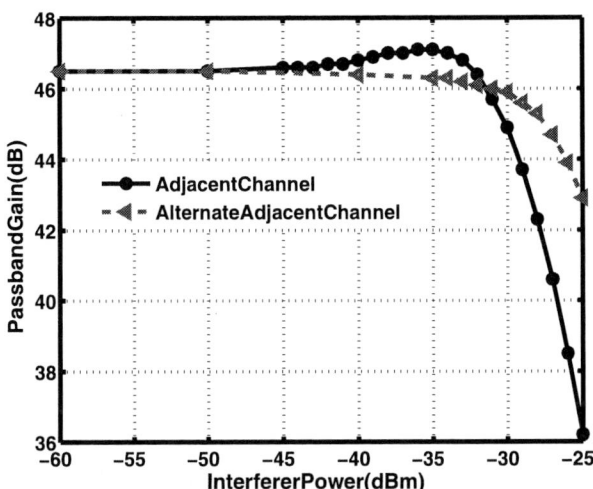

Fig. 10. Measured passband gain in the presence of a blocker.

The filter nonlinearity resulting from a blocker may also corrupt the 11a 64-QAM OFDM signal by creating cross modulation among the sub-channels. This effect is characterized by setting the RF input signal level 3 dB above the sensitivity, applying a blocker, and raising its level until the EVM falls to −23 dB. Figure 11 plots the relative blocker level as a function of the frequency offset with respect to the desired signal center frequency.

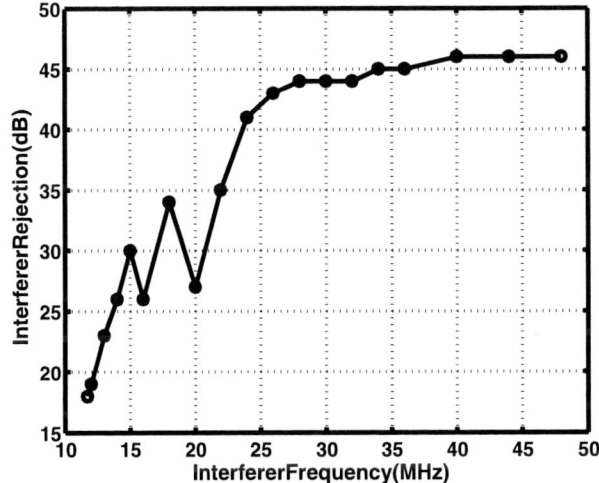

Fig. 11. Measured interferer rejection.

Table 1 summarizes the receiver performance and compares it to that of prior art.

Table 1. Comparison with state-of-the-art.

|  | ThisWork | [1] | [8] | [9] |
|---|---|---|---|---|
| Frequency(GHz) | 5.1–5.9 | 5.15–5.35 | 4.9–5.95 | 5.1–5.9 |
| NF(dB) | 5.3 | 8.0 | 4.4 | 5.5 |
| $IIP_3$(dBm) | +2.6 | −11.2 | +5 | +16 |
| Gain(dB) | 5–48 | 14–94.5 | 8–74 | 19–89 |
| Sensitivity(dBm) at54Mb/s | −70 | NA | NA | −75.5 |
| Power(mW) | 11.6 | 46 | 108* | 72.7** |
| LNA | 0 | 11.7 |  |  |
| Mixers | 0 | 9.8 |  |  |
| LOBuffers | 0.4 | 10.8 |  |  |
| Filters,VGAs | 10 | 13.7 |  |  |
| Divider/ 25%Logic | 1.2 |  |  |  |
| CMOSProcess | 65nm | 0.18m | 0.18m | 0.13m |
| Area() mm$^2$ | 0.183 | NA | NA | NA |

*IncludingADC.

**WithoutLOBuffer.

## Acknowledgment

The authors wish to thank TSMC's University Shuttle Program for chip fabrication. This work was supported by Realtek Semiconductor.

## REFERENCES

[1] L. L. L. Kan et al., "A 1-V 86-mW RX 53-mW TX single-chip CMOS transceiver for WLAN IEEE 802.11a," *IEEE J. Solid-State Circuits,* vol. 42, no. 9, pp. 1986–1998, Sep. 2007.

[2] A. Zolfaghari and B. Razavi, "A low-power 2.4-GHz transmitter/receiver CMOS IC," *IEEE J. Solid-State Circuits,* vol. 38, no. 2, pp. 176–183, Feb. 2003.

[3] D. Kaczman et al., "A single-chip 10-band WCDMA/HSDPA 4-band GSM/EDGE SAW-less CMOS receiver with DigRF 3G interface and +90 dBm IIP2," *IEEE J. Solid-State Circuits,* vol. 44, no. 3, pp. 718–739, Mar. 2009.

[4] M. Soer, et al., "A 0.2-to-2.0GHz 65nm CMOS receiver without LNA achieving >11dBm IIP3 and <6.5 dB NF," *ISSCC Dig. of Tech. Papers,* pp. 222–223, Feb. 2009.

[5] J. R. Long, "Monolithic transformers for silicon RF IC design," *IEEE J. Solid-State Circuits,* vol. 35, no. 9, pp. 1368–1383, Sept. 2000.

[6] A. Zolfaghari, A. Chan, and B. Razavi, "Stacked inductors and transformers in CMOS technology," *IEEE J. Solid-State Circuits,* vol. 36, no. 4, pp. 620–628, Apr. 2001.

[7] A. Mirzaei and H. Darabi, "Analysis of imperfections on performance of 4-phase passive-mixer-based high-Q bandpass filters in SAW-less receivers," *IEEE Trans. Circuits Syst. I, Reg. Papers,* vol 58, no.5, pp. 879–892, May 2011.

[8] T. Maeda et al., "Low-power-consumption direct-conversion CMOS transceiver for multi-standard 5-GHz wireless LAN systems with channel bandwidths of 5-20 MHz," *IEEE J. Solid-State Circuits,* vol. 41, no. 2, pp. 375–383, Feb. 2006.

[9] K. Lim et al., "A 2x2 MIMO tri-band dual-mode direct-conversion CMOS transceiver for worldwide WiMAX/WLAN applications," *IEEE J. Solid-State Circuits,* vol. 46, no. 7, pp. 1648–1658, Jul. 2011.

# An Adaptive Predistorter for Wireless LAN RFSoC with embedded PA and T/R switch in 55nm CMOS

Khurram Muhammad, Ming-Cho Chen, Kai-Hung Wang, Kuang-Ping Ma, Yu-Lin Hiseh, Wei-Show Hsu, Yuan-Yu Fu, Meng-Chang Lee, Shuo-Yuan Hsiao, Chih-Ming Hung

MStar Semiconductor Inc., Chupei, Hsinchu Hsien, Taiwan, R.O.C
E-mail: khurram.muhammad@mstarsemi.com

## Abstract

We present an adaptive predistortion system for a WLAN transceiver in 55nm CMOS. The forward DSP path utilizes complex gain predistortion while the APD module in the feedback path computes AMAM and AMPM coefficients by comparing ideal transmit signal with the distorted signal from the receiver. This module operates with various calibration signals generated on-chip in addition to TX data. Measurement results show improvement of EVM by 1dB with the proposed approach. Improvement of P1dB of more than 3dB was obtained using fully automatic processing. The total solution utilizes 120k gates.

Keywords: VLSI, CMOS, DSP, Predistortion

## Introduction

Recent WLAN radio SoCs have integrated WLAN PA that can easily output 18 dBm of linear output power. Efficient and linear PA is highly desirable for low-cost smart phones. Higher margin to specs allows trading current consumption with linearity. Hence, there is a high interest in deployment of PA predistortion in future generations of WLAN radios [1-4]. The AMAM and AMPM distortion of the PA can be digitally compensated, however, it can vary with process and temperature. This variation can be reduced by factory calibration using open-loop methods based on process and temperature sensing; however, in order to obtain superior performance over all operating conditions and reduce the cost of calibration time, adaptive predistortion is highly desired. This paper presents an adaptive predistorter for WLAN transceiver based on complex gain estimation using the receive path while coping with IQ imbalance and DC offsets. The goal is to have fully automated calibration at power on as well as fine coefficient tuning in the background while transmitting payload data. The MODEM, DFE embedding the adaptive predistorter, transceiver and a class-AB type PA were all combined in a single chip WLAN radio in 55nm digital CMOS 1P7M process. The PA integrates a balun and operates at 3.3V using IO transistors for cascode device to protect the gm device from high voltage stress.

## Digital Frontend

Fig. 1 shows a block diagram of the overall predistortion solution. Initially, an FPGA board was used and included the WLAN MODEM together with the digital front end (DFE) embedding the predistorter for the transmitter. The validated solution was fabricated in the final RFSoC. The DFE contains PA predistortion circuit based on complex gain predistortion; hence, PA distortion is modeled as AMAM and AMPM distortion as a function of signal envelope [3,4]. The Predistortion block (PD) compensates a backed-off input signal to emphasize it in the region where the PA is compressed. This block is followed by transmitter distortion correction circuits that add image signal and LO feed-through (LOFT) pre-compensation. After up-sampling by 4, the 80MHz output is passed through to the DACs and upconverted in the analog sections of the transceiver. The RF output of the transceiver is amplified by the WLAN PA. The output of the PA goes to T/R switch and parasitically couples into the RX where it is down-converted and digitized using the RX path.

The TESTSIGS block generates various test signals such as ramp, sawtooth and complex exponential waveforms in order to synthesize a calibration signal. The output from this block or the modulation waveform generated from the MODEM can be used as an input to the DFE. The receiver ADC output signal contains distortions from the transmitter as well as the receiver. These distortions are removed in the IQ mismatch estimator (IQME)

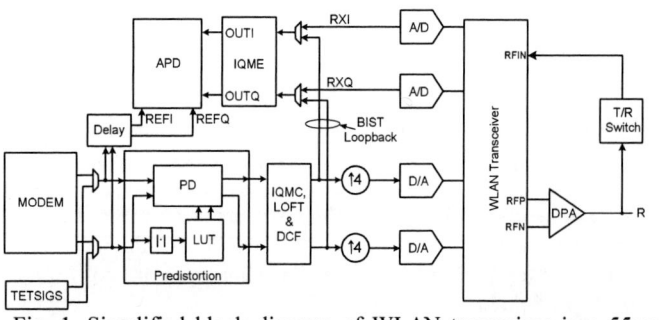

Fig. 1. Simplified block diagram of WLAN transceiver in a 55nm connectivity RFSoC.

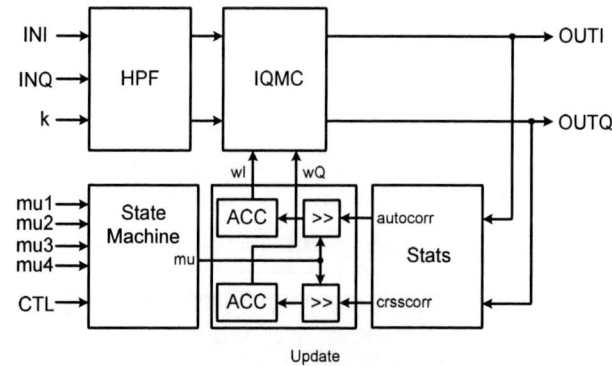

Fig. 2. IQ imbalance estimation and correction.

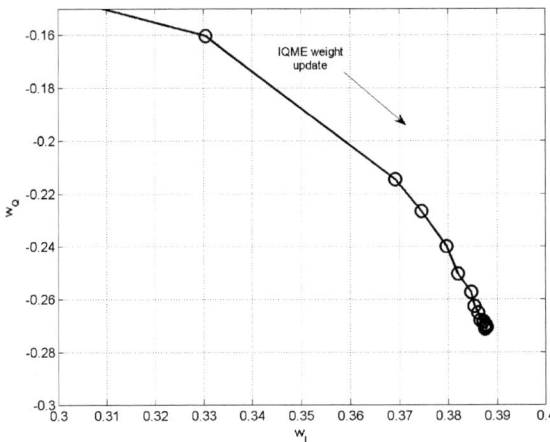

Fig. 3. Simulated response of IQME settling with 32 iterations

and corrector shown in Fig. 2. This block implements a programmable digital high pass filter (HPF) to remove the DC offset with a programmable corner frequency ranging from 100kHz to 20MHz to allow fast settling through "gear-shifting". It is followed by IQ mismatch compensator (IQMC) circuit which subtracts the estimated image signal. Next, the difference between auto-correlation of I and Q outputs of IQMC as well as the cross-correlation between these signals are computed and independently accumulated similar which reduces the convergence noise over ref [5]. A state machine selects mu value for LMS update. The outputs of the autocorrelation and cross-correlation accumulators are used to update the weight by "gear-shifting" mu to achieve faster convergence. When the algorithm converges, the difference in auto-correlation as well as the cross-correlation between I and Q signals at the IQMC output are forced to zeros, thereby eliminating the image signal.

In addition to calibration using test signals, payload data based background fine-tuning of predistortion coefficients is also supported. This is a key novelty and differentiates the work presented in this paper with prior art [1-4]. The digital transmit signal is also the ideal TX signal; hence, it is used as a reference signal by the learning algorithm in the adaptive predistortion (APD) block that follows IQME in Fig. 1. The reference transmit signal is delayed to match the analog loopback delay before comparing with the RX signal. WLAN is a TDD system, hence, in RX mode the transmitter is powered off when IQME computes RX imbalance. The converged weight values are used to eliminate the RX imbalance and also stored in memory.

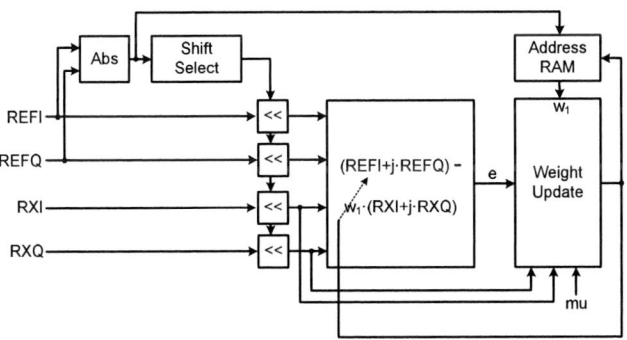

Fig. 4. APD block implementing complex gain estimation.

In the TX mode, the transmitter is powered on. The LO feed-through (LOFT) and TX image due to TX mixer imbalance appear in addition to the RX impairments at the RX ADC output. The difference between I and Q auto-correlations and cross-correlation now indicates the combined IQ imbalance due to the RX and TX. When the IQME settles down, OUTI and OUTQ (see Fig. 1) have the same power and are orthogonal to each other. Hence, the image signal is eliminated. The TX image signal is generally around 40dB below the TX signal with 25% duty cycle LO. Similarly, the RX image is also around 40dB below the RX signal. Hence, the RX introduces an uncorrected distortion (image of image) that is 80dB below the TX signal and does not influence the APD coefficients significantly. In case, TX IQ imbalance estimation is desired, it can be extracted from the two estimates obtained, or may be computed directly by using the corrected RX signal instead of the ADC output in the TX mode. The only application for highly accurate TX IQ balance is low-IF operation of the transmitter otherwise 40dB of IQ balance is sufficient for high SNR signal conditions.

Fig. 3 shows the simulated response of the IQME settling within 32 iterations. Each iteration takes as many clock cycles as the number of samples selected in the "Stats" block for averaging. 16 iterations also produce acceptable quality results. The clock runs at 40MHz and good estimate can be computed with in 25usec. The IQME block utilizes 15k gates.

**Adaptive Predistorter**

Once the RX and TX impairments are removed, compression in the PA needs to be estimated in the APD block. The time-aligned reference TX signal is divided by the feedback signal using a complex divider. This operation directly estimates the AMAM and AMPM compression. FIG. 4 shows the circuit used to compute the complex gain. The entire dynamic range of the reference signal magnitude is divided into 32 equally spaced intervals unlike [4]. Each interval has an associated complex gain which is an estimate of AMAM and AMPM compression for the signal falling in that interval.

To compute this, first, the reference signal envelope (magnitude) is calculated in Abs block. Next, the interval in which the reference signal sample falls in is computed and Address RAM block selects the address in RAM corresponding to the APD weight for this interval. The LUT contains 32 weights, only one of which is relevant to the sample being processed. The time-aligned sample from the feedback path is the distorted version of this reference sample. The selected weight becomes the weight used by an LMS update algorithm. The error signal is defined as the difference between the reference signal sample (REFI + $j$ REFQ) and the weight times the compensated RX signal sample (RXI + $j$ RXQ). This weight is updated using a LMS update equation to drive the error signal towards zero in a mean square sense whenever the reference sample falls in its interval. As the reference signal traverses the signal envelope values in different regions in the total dynamic range, corresponding weights get selected from the RAM and

978-1-4673-6145-3/13 $31.00 © 2013 IEEE         177

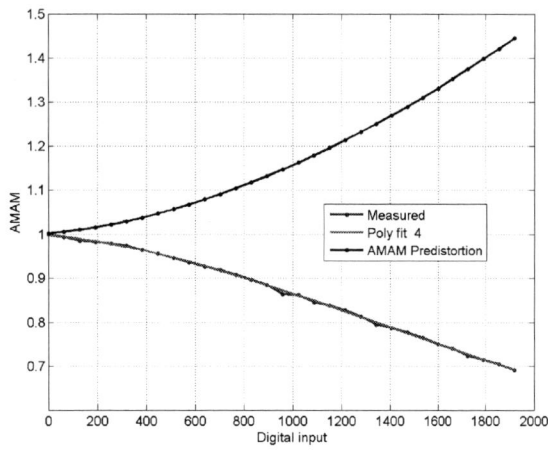

Fig. 6. Pre-empahsis for PA predistortion

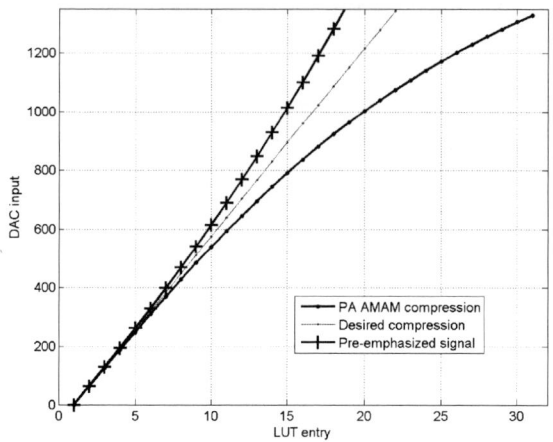

Fig. 5. Predistortion approach used in this work.

are updated. Hence, all 32 weights are updated and converge together.

For initial calibration (at power up) the TESTSIGs block is programmed to generate a sawtooth waveform. Hence, each of the 32 intervals dividing the dynamic range of the PA is visited equally and almost the same number of points is presented to the APD which estimates the complex gain for the interval. This achieves faster overall convergence. If, for example, we use the modulation waveform as the training signal, complex gain can be computed while the payload data is being transmitted; however, different regions of the total dynamic range see different number of points and it takes longer to train all 32 weights as some are more accurate than others. This is because the peak values are visited less than average or small values while predistortion compensation is required more in the compressed region where peaks occur.

When the signal traverse the lowest intervals where PA is linear, convergence takes longer due to very small increments in weight updates. This is corrected by normalizing the input based on the envelope of the reference signal and shown by the left shift operation in Fig. 3 before the error calculation. The weight update operation produces the update applied to the

selected weight from the 32 entries of the LUT. When the reference signal drops in magnitude, all four inputs are shifted left by an equal amount to allow the use of the entire dynamic range available with the use of 12-bit data path. This not only reduces the impact of quantization noise in the update operation, but also maintains the error signal to be large enough and is similar to the use of normalized LMS algorithm. The advantage is maintaining uniform rate of convergence for low magnitude sample points. The predistortion approach shown in Fig. 5 offers an advantage that the pre-emphasis in the linear region of the PA is known for weights where the SNR of the feedback signal is worst. Therefore, average of weights in the linear region of PA can be used to calculate the loop gain of this system and applied to the lowest weight values.

At lowest values the error due to quantization noise becomes high. This situation is exacerbated by the low SNR of the feedback signal and produces poor estimates. In contrast to [3,4] we chose to predistort the PA using the curve shown in Fig. 6 which provides no compensation in the linear region while it pre-emphasis in the signal in the compressed region. Higher the magnitude of the envelope, the higher is the pre-emphasis. Therefore, the strategy is not to interfere with the transmit operation where PA is linear and only modify the signal when PA starts to compress. Hence, quantization noise due to PD block is avoided in the region where PA is linear. The operation of complex divider of Fig. 4 is shown in Fig. 7 for an example test signal. As the signal enters the interval corresponding to a particular APD coefficient (weight), the error quickly goes towards zero within a few iterations. The adaptive predistorter uses 48k gates in this implementation.

### Measurement Results

The PA occupies only 0.5mm² of die area in 55nm CMOS and can output up to 18dBm of linear modulated output power using a 3.3V power supply while consuming 238 mA current. The raw EVM at 18dBm output power for 64QAM modulation can be improved from -25dB w/o predistortion to -26dB with predistortion. 3dB of P1dB improvement can be obtained using precise predistortion. Fig. 9 shows spectral plot of the predistorted signal that meets specification by a wide margin.

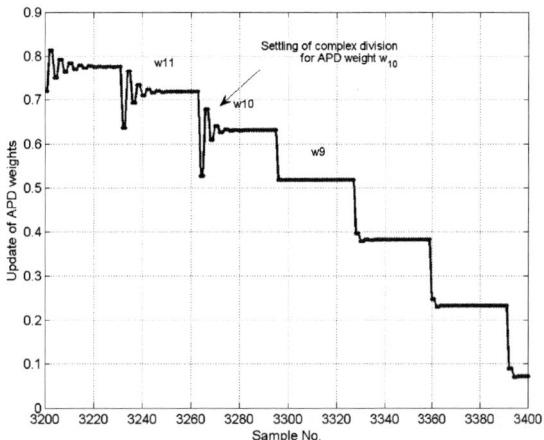

Fig. 7. Settling of complex division as calibration signal sweeps the dynamic range of the reference signal.

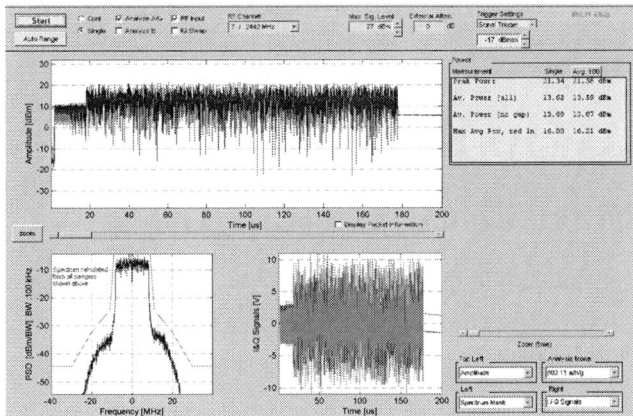

Fig. 8. Measured results with predistorted of 64QAM modulation

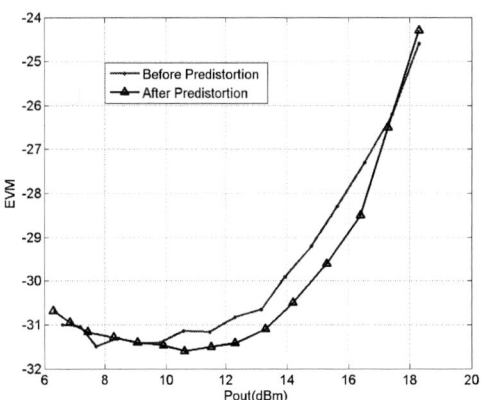

Fig. 9. Measured EVM versus Pout.

If higher Pout is desired, due to high peak-to-average of OFDM signal of about 9.6dB, modulation peaks get pre-emphasized rapidly and can easily clip and degrade EVM. Hence, more output power is not easy to obtain and requires very precise pre-emphasis. This is further explained by observing the pre-emphasis shown in Fig. 5. If we wish to increase the output power by increasing the digital input to the predistorter PD in Fig. 1, the peak-average signal will also increase by the same proportion. The change in pre-emphasis for the RMS signal is very small; however, the pre-emphasis increases rapidly for higher signal magnitude as shown in Fig. 6 and can easily clip the output. For PAR of 9 dB, the voltage signal at peak excursion is 2.82 times the voltage signal at RMS. Therefore, very little room exists for increasing Pout for a PA designed to produce linear output power without use of predistortion. Therefore, predistortion can now be used to improve the linearity of the PA trading off PA bias to reduce its current consumption at peak output powers when predistortion can be enabled. For lower output powers, predistortion circuits are bypassed and disabled and do not consume any current.

Fig. 8 shows the measured EVM of the PA with and without predistortion. The PA is originally very linear and meets the WLAN specification; however, EVM can be improved by about 1dB easily above 14dBm-17dBm output power. EVM improvement is better than the results reported in [1] and consistent with those reported in [2]. Limitations are seen at lower and highest end of power due to different reasons. At lowest powers, the quantization effects are visible; however, this is the linear region of the PA where EVM cannot be further improved. RX signal in linear region has low SNR. Conveniently, the proposed predistortion scheme performs no pre-emphasis in the linear region and can be fully bypassed. At highest Pout, combined residual errors limited the improvement. This shows that there still are challenges lingering in the use of adaptive predistortion on the handset side as only simple approaches can be used to justify the current consumption overhead of predistortion.

Fig. 10 shows the die photo of the RFSoC which integrates the WLAN MODEM, transceiver, PA and T/R switch in addition to other connectivity radios.

## Conclusion

Adaptive predistortion allows improvement of PA linearity consistently over temperature and frequency and reduces factory calibration time. This work presents PA distortion estimation that uses the RX path of WLAN radio to compute updates to AMAM and AMPM compensation applied in the forward path. Measurement results show 1dB improvement of EVM and up to 3dB improvement in 1dB compression point. The solution utilizes 120k gates in 55nm CMOS process. The proposed approach works in the presence of other RX path impairments and shows the challenges that still linger against fully automatic self-calibration in modern transmitters.

## References

[1] D.H.Kwon, H. Li, Y. Chang, R. Tseng, Y. Chiu, "Digitally equalized CMOS transmitter front-end with integrated power amplifier", IEEE JSSC, Vol.45, No.8, pp.1602-1614, Aug. 2010.

[2] C.-J. Chang et. al., "A CMOS transceiver with internal PA and digital pre-distortion for WLAN 802.11a/b/g/n applications", 2010 IEEE RFIC, pp.435-438.

[3] C.D. Presti, F. Carrara, A. Scuderi et. al., "A 25 dBm digitally modulated CMOS power amplifier for WCDMA/EDGE/OFDM with adaptive digital predistortion and efficient power control," IEEE JSSC, Vol. 44, No. 87, pp. 1883-1896, July 2009.

[4] J. Mehta, V. Zoicas, O. Eliezer et. al. , "An efficient linearization scheme for a digital polar EDGE transmitter," IEEE Trans. Circuits & Systems-II: Express Briefs, Vol. 57, No. 3, pp.193-197, Mar 2010.

[5] I. Elahi, K. Muhammad, P.T. Balsara, "I/Q mismatch compensation using adaptive decorrelation in a low-IF receiver in 90nm CMOS process," IEEE JSSC, Vol. 41, pp.395-404, Feb. 2006.

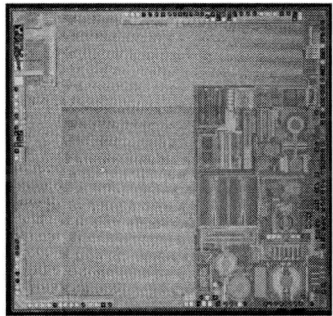

Fig. 10. Die photograph of RFSoC with embedded MODEM, DFE, WLAN PA, balun and T/R switch in 55nm digital CMOS process.

978-1-4673-6145-3/13 $31.00 © 2013 IEEE

# A 116nW Multi-Band Wake-Up Receiver with 31-bit Correlator and Interference Rejection

Seunghyun Oh, Nathan E. Roberts, and David D. Wentzloff

University of Michigan, Ann Arbor, MI, 48109, USA

*Abstract* — This paper presents a 116nW wake-up radio complete with crystal reference, interference compensation, and baseband processing, such that a selectable 31-bit code is required to toggle a wake-up signal. The front-end operates over a broad frequency range, tuned by an off-chip band-select filter and matching network, and is demonstrated in the 402-405MHz MICS band and the 915MHz and 2.4GHz ISM bands with sensitivities of -45.5dBm, -43.4dBm, and -43.2dBm, respectively. Additionally, the baseband processor implements automatic threshold feedback to detect the presence of interferers and dynamically adjust the receiver's sensitivity, mitigating the jamming problem inherent to previous energy-detection wake-up radios. The wake-up radio has a raw OOK chip-rate of 12.5kbps, an active area of 0.35mm$^2$ and operates using a 1.2V supply for the crystal reference and RF demodulation, and a 0.5V supply for subthreshold baseband processing.

## I. INTRODUCTION

Wireless sensor nodes spend most of the time in an ultra-low-power sleep state with their radios off to conserve energy. This presents a problem when remotely waking up and synchronizing to these nodes. Wake-up radios (WRX) are a viable solution [1-4], but only if their active power is below the sleep power of the node, otherwise the WRX power dominates and dictates the lifetime of the node. With digital sleep power being reported in the nW range, this presents a significant challenge to WRX design. A simple method for reducing the power of a WRX is to reduce sensitivity, which is tolerable for short-range communication and when the primary goal is a lifetime of multiple years [3]. For example, with a receiver sensitivity of -40dBm, 6m communication at 400MHz is theoretically possible with only 0dBm transmit power. This range and power level is suitable for a broad set of personal and internet of things (IoT) applications.

WRXs use energy detection architectures to keep power low; however, any signal at the proper frequency can trigger a false wake-up of these radios, and false wake-ups result in significant amounts of wasted energy on the node. In order to prevent this, a WRX must have enough local processing to differentiate a wake-up event from interference without use of the node's main processor. The proposed WRX addresses these issues.

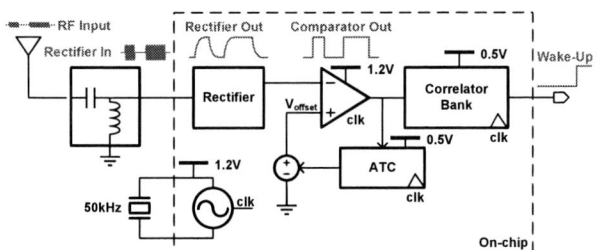

Fig. 1. Architecture of the WRX with rectifier, baseband processor, and crystal oscillator on-chip

A 116nW WRX complete with crystal reference, interference compensation, and all the necessary baseband processing is presented. To operate in the nW power range, the power hungry LNA is replaced with a high-sensitivity, passive RF rectifier, tuned by an off-chip band select filter and matching network. To prevent false wake-ups the WRX does two things: first, it must receive a selectable 31-bit OOK-modulated CDMA code, and second, the baseband processor implements automatic threshold feedback to detect the presence of interferers and dynamically adjusts the receiver's sensitivity, which mitigates the jamming problem found in previous energy-detection radios [1-4]. Operation of the WRX has been demonstrated in the 402-405MHz MICS band and the 915MHz and 2.4GHz ISM bands. Section II will introduce the architecture of the WRX. Section III will explain the major circuit blocks of the WRX in detail, and Section IV will present measurement results. Finally, Section V will conclude the paper.

## II. SYSTEM ARCHITECTURE

Fig. 1 shows a block diagram of the WRX. From the antenna and band-select filter, the RF signal passes through an input matching network that filters and boosts the signal before going on-chip. A 30-stage rectifier down-converts the RF signal to baseband, which is then sensed by a comparator clocked at 4X the chip-rate. The input offset voltage of the comparator is controlled by the ATC (Automatic Threshold Controller) which is used to overcome interferers. A bank of 124 correlators (31 shifts of the code, 4x oversampled) continuously compare the

978-1-4673-6145-3/13 $31.00 © 2013 IEEE

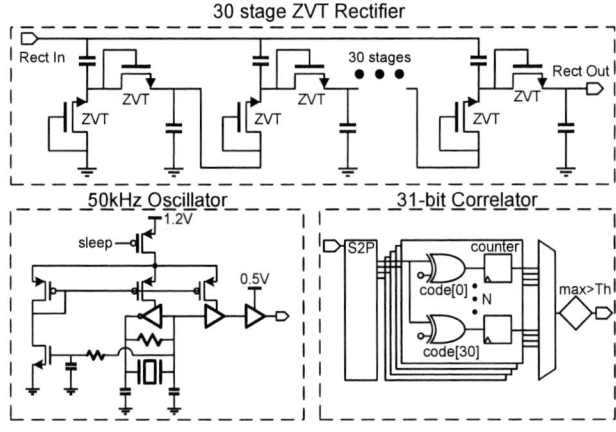

Fig. 2. Schematic of the rectifier, crystal oscillator, and parallel correlators

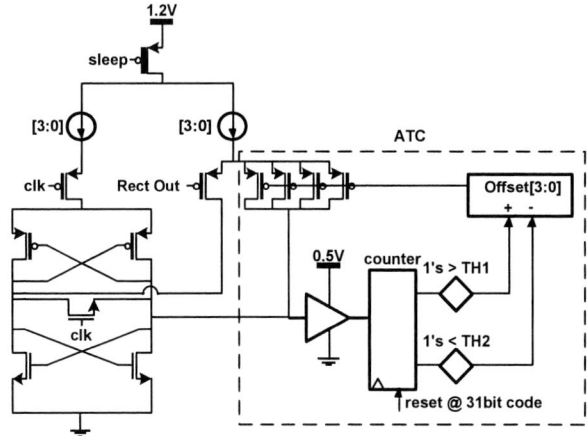

Fig. 3. Schematic of the clocked comparator and automatic threshold control (ATC)

received chip sequence with a programmable wake-up code, and toggles the wake-up signal only when a correlation result exceeds a user-programmable threshold. The reference clock for the receiver is generated using an off-chip 50kHz crystal with an integrated oscillator. The oscillator and comparator operate from a 1.2V supply while all the digital logic operates in subthreshold at 0.5V.

## III. CIRCUIT DESCRIPTION

In this section, the major circuit blocks of the WRX are explained in detail. All circuits include a thick-oxide PMOS header to reduce sleep-mode power at the expense of slightly higher active power.

### A. Off-chip matching network

For the WRX, a 2-element off-chip matching network was used and provided a passive 5dB voltage boost. For example, at 400MHz the input impedance of the chip was measured on a network analyzer to be 23-j35Ω, and a 12pF series capacitor and a 15.7nH shunt inductor were used in this case. The Q factor of the input impedance is low, due to a voltage limiter that prevents the rectified voltage from exceeding the breakdown voltage of the FETs, so a broadband matching network could be implemented. Devices like BAW or FBAR resonators can also be used to tune to the desired frequency of operation

### B. RF rectifier

An RF rectifier, shown in Fig. 2, replaces a traditional LNA to save significant power. The rectifier's structure is the same as the Dickson Multiplier, with the exception that all transistors operate in the sub-threshold regime (at low Rx power levels). The output voltage calculation is therefore different [5]. This subthreshold rectifier uses

zero-threshold transistors and 30 stages to achieve sufficient RF gain with fast charging time.

### C. Crystal oscillator

A 50kHz crystal oscillator in Fig. 2 [6] serves as the reference clock of the WRX. An off-chip crystal is used, and the oscillator's primary amplifier is an inverter with resistive feedback. When the oscillator circuit first turns on, the transconductance of the primary amplifier is much greater than the critical transconductance of the crystal in order to start oscillation and increase amplitude. As the amplitude increases the DC level of the oscillation drops, which is used in feedback to starve the primary amplifier until it settles to 38nW while sustaining oscillations.

### D. Comparator with ATC

The clocked comparator, shown in Fig. 3, applies regenerative feedback clocked by the 50kHz oscillator [7]. Two separate current biases are each controlled by 4-bit binary-weighted current DACs. The programmable 4-bit binary-weighted input threshold of the comparator is controlled by the ATC which dynamically adjusts the offset voltage to overcome interference signals. The ATC monitors the samples coming from the comparator output for one 31-bit code period. If the number of 1's is greater than a programmable value (indicating an interferer is present), then the ATC will increase the comparator's threshold to bring the sensitivity of the receiver above that of the interfering signal. When the number of 0's at the output of the comparator reaches a separate programmable value (indicating the interferer is gone), the ATC then reduces the threshold to increase the sensitivity of the receiver. Hysteresis is added between these values to eliminate limit-cycles.

978-1-4673-6145-3/13 $31.00 © 2013 IEEE

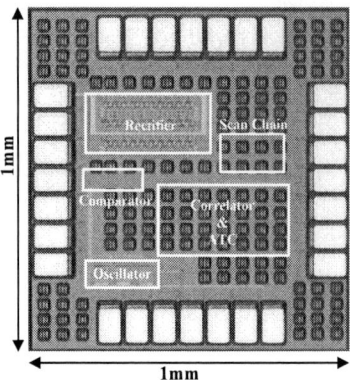

Fig. 4. Die photo

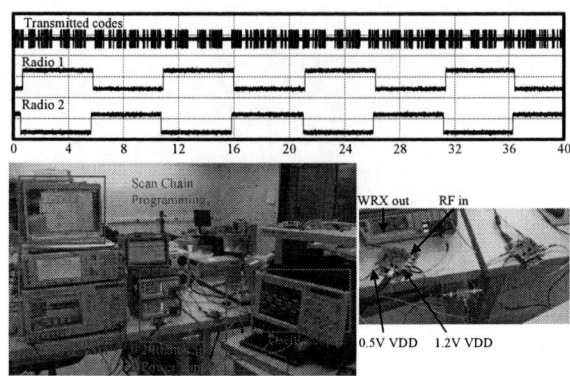

Fig. 5. Measurement with 2 codes and 2 receivers.

## E. Correlators

Four banks of 31 correlators, each shifted by one sample, work simultaneously to account for any phase shift between the transmitter and WRX. Each correlator takes two samples per chip from the 4X sampled comparator output when correlating with the wake-up code. Therefore, each 31-bit code has 62 total comparisons. This is used to synchronize the WRX to the chip boundary. In addition, each correlator simultaneously compares with every possible shift of the code, in order to align to the code boundary. A programmable correlator threshold determines the number of correct sample points needed before the wake-up code is toggled. A lower correlator threshold means fewer bits have to match the code, improving sensitivity, but leads to more false wake-ups. A higher correlator threshold prevents false wake-ups, but also reduces the sensitivity of the receiver.

## IV. MEASUREMENTS

The IC is fabricated in an IBM 130nm CMOS process. Fig. 4 shows the die photo of the fabricated chip. The total size is 1mm x 1mm, and the wake-up receiver occupies 0.35mm$^2$ without pads.

Fig. 5 shows the measurement setup. It also demonstrates the WRX's ability to accept only the programmed code and reject others. In this setup an arbitrary waveform generator (AWG) and vector signal generator (VSG) were transmitting two different codes back to back. The signal was then split and sent to two different WRXs that were each programmed with two different codes. The top trace shows the transmitted OOK signal and the bottom traces show that each WRX toggles its wake-up signal when receiving its own code, but not when receiving the other code.

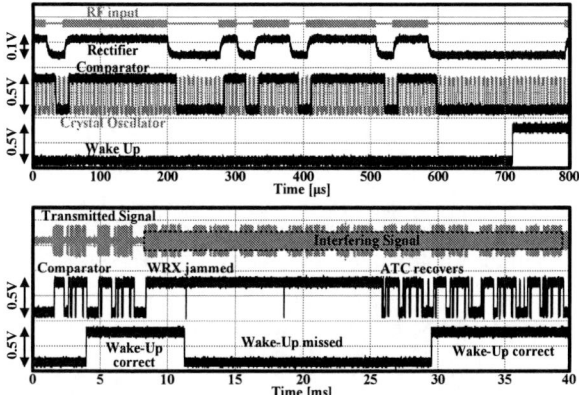

Fig. 6. Transient response of the WRX in normal operation, and in the presence of an interferer

Detailed transient operation of the WRX receiving a 31-bit code is shown in Fig. 6. The WRX automatically synchronizes to the incoming bit stream. The top two traces show the RF input signal and the RF rectifier converting the signal to baseband. The third trace shows the output of the comparator being clocked at 4X the data-rate by the local oscillator and the final trace is the wake-up signal being toggled by the correlator. The WRX is capable of CDMA by selecting different codes used by the correlator block.

If an interfering signal is strong enough to exceed the comparator threshold (saturating the bit-slicer), then the ATC increases the comparator's threshold until it is above the interfering signal. A transient of this operation can be seen in Fig. 6. The top signal is the received RF signal, which is jammed by a 2.4GHz tone at 8ms. With the interferer present, the comparator initially outputs 1's so that the receiver cannot receive the code. After 15ms, the ATC has raised the threshold of the comparator above that of the interferer, and the WRX regains synchronization.

978-1-4673-6145-3/13 $31.00 © 2013 IEEE            182

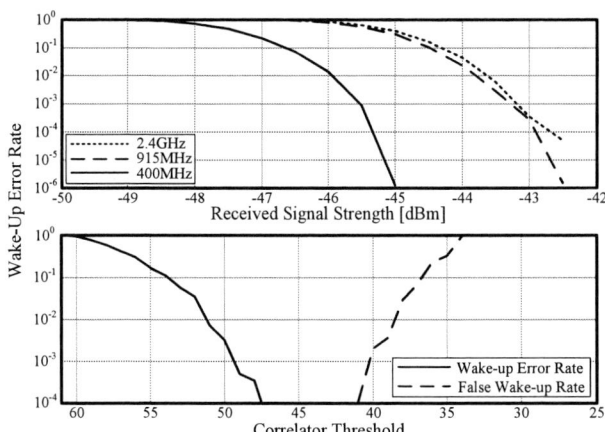

Fig. 7. Wake-up error rate vs. signal strength and correlator threshold

Table I
Power breakdown and receiver specs

| Power Breakdown [nW] | | Receiver Specs | |
|---|---|---|---|
| RF Rectifier | 0 | Energy/bit | 9.28pJ |
| Comparator | 8.4 | Energy/wakeup | 287.7pJ |
| Digital Logic | 69.5 | Max signal level | -15dBm |
| Crystal Oscillator | 38.4 | Max interferer level | -20dBm |
| TOTAL | 116.3 | Code length | 31 |
| Sleep [pW] | 20 | # of pre-defined codes | 8 |

Table II
Comparison with other state of the art work

| | This Work | | | [1] | [2] | [3] |
|---|---|---|---|---|---|---|
| Power [µW] | 0.116 | | | 52 | 45 | 0.098 |
| Sleep [pW] | 20 | | | N/A | N/A | 11 |
| Frequency [MHz] | 403 | 915 | 2400 | 2000 | 5800 | 915 |
| Data-rate [kbps] | 12.5/31 | | | 100 | 14 | 100 |
| Sensitivity [dBm] | -45 | -43 | -41 | -72 | -45 | -41 |
| SIR [dB] | 3.3 | 1.7 | 1.7 | N/A | N/A | N/A |
| Die Area [mm²] | 0.35 | | | 0.1 | N/A | 0.03 |
| VDD [V] | 1.2\|0.5 | | | 0.5 | 3.0 ~ 3.6 | 1.2 |
| Process [nm] | 130 | | | 90 | 130 | 130 |

The top of Fig. 7 shows the chip error rate (BER) curves for the 403MHz, 915MHz, and 2.4GHz bands. Sensitivity is best in the 403MHz range. The bottom of Fig. 7 shows BER as the correlator threshold is varied. The measurements were taken using a -40dBm signal in the 2.4GHz band. The figure also shows the impact this threshold has on false wake-ups. From these two data sets, the correlator threshold can be set to maximize sensitivity while minimizing the possibility of a false wake-up.

The WRX has an active power of 116nW with a sleep power of 18pW, afforded by thick-oxide headers. A full power breakdown can be found in Table I. The digital baseband processing consumes the majority of the power in the WRX. The measured performance is summarized and compared with other WRXs in Table II.

## V. CONCLUSION

This paper introduced a 116nW WRX that uses CDMA codes to provide interference rejection from both in-band and out-of-band interferers. In addition, with programmable wake-up codes, the WRX can be used in a network with similar WRXs and be able to uniquely wake up. With reduced sensitivity specifications, the use of a zero-power RF energy harvester was used as the RF front end of the receiver and subthreshold design was implemented to keep entire radio in the nanowatt power region. With power that is less than a typical sensor node's sleep power, the WRX is not the energy dominant circuit when the node is asleep and can provide false wake up rejection, making it a very suitable synchronization technique for sensor nodes.

## ACKNOWLEDGEMENT

This material is based upon work supported by the National Science Foundation under Grant No. CNS-1035303.

## REFERENCES

[1] N.M. Pletcher, et. al, "A 2GHz 52µW Wake-Up Receiver with -72dBm Sensitivity Using Uncertain-IF Architecture, " ISSCC Dig. Tech. Papers, pp. 524-525, Feb. 2008

[2] Jeongki Choi, et. al, "An interference-aware 5.8GHz wake-up radio for ETCS, " ISSCC Dig. Tech. Papers, pp. 446-448, Feb. 2012

[3] N.E. Roberts, and D.D. Wentzloff, "A 98nW Wake-Up Radio for Wireless Body Area Networks," Radio Frequency Integrated Circuits Symposium, pp.373-376, June 2012

[4] X. Huang, et. al, "A 2.4GHz/915MHz 51µW Wake-up Receiver with Offset and Noise Suppression, " ISSCC Dig. Tech. Papers, pp. 222-223, Feb. 2010

[5] Seunghyun Oh, and D.D.Wentzloff, "A −32dBm Sensitivity RF Power Harvester in 130nm CMOS," Radio Frequency Integrated Circuits Symposium, pp.483-486, June 2012

[6] Hector Ivan Oporta, "An Ultra-low Power Frequency Reference for Timekeeping Applications," Master's Thesis, Oregon State University, December, 2008

[7] B. Razavi, and B.A. Wooley, "Design techniques for high-speed, high-resolution comparators," Solid-State Circuits, IEEE Journal of, vol.27, no.12, pp.1916-1926, Dec. 1992

# A 11μW Sub-pJ/bit Reconfigurable Transceiver for mm-Sized Wireless Implants

Anatoly Yakovlev, Jihoon Jang, Daniel Pivonka, and Ada Poon

Stanford University, Stanford, CA

*Abstract* — **A wirelessly powered 11 μW transceiver for implantable sensors has been designed and demonstrated through 35 mm of porcine heart tissue. The prototype occupies 1 mm × 1 mm in 65nm CMOS with an external receive antenna. The IC consists of a rectifier, regulator, demodulator, modulator, controller, and sensor interface. The forward link transfers power and data on a 1.32 GHz carrier using low-depth ASK modulation that minimizes impact on power delivery and achieves from 4 to 20 Mbps with 0.3 pJ/bit at 4 Mbps. The backscattering link modulates the antenna impedance with a configurable load for operation in diverse biological environments and achieves 2 Mbps at 0.7 pJ/bit. The device supports TDMA, allowing for simultaneous operation of multiple sensors.**

## I. INTRODUCTION

Wireless implantable sensors can greatly enhance a variety of existing diagnostic procedures and open new possibilities for emerging noninvasive techniques. Intracardiac mapping, which is currently accomplished with a point-by-point manual catheter measurement, can instead use distributed implantable probes that capture real-time localized action potentials (APs) as they propagate through the heart [1]. This information can dramatically improve the treatment of cardiac arrhythmias.

Existing transceivers are not sufficiently small or robust to serve the needs of sensing applications such as intracardiac mapping. Batteries are too large for these devices, and wireless powering provides a very limited power budget. As a further complication, tissue depth and composition varies for each treatment, and changes over time. This especially impacts data recovery from the reverse backscatter link because the optimal antenna impedance to maximize reflected energy changes with the environment [2].

To overcome the challenges associated with power management and deep-tissue implantation, this paper presents a new miniaturized transceiver design that is both power-efficient and reconfigurable for operation in variable biological conditions. With the exception of an external antenna, the entire transceiver is fully integrated on a 1 mm × 1 mm IC fabricated in 65 nm GP CMOS process. Because the device relies on continuous RF power transfer for operation, the forward link must be efficient and have minimal impact on power delivery. To achieve this, an asynchronous, low-depth amplitude modulation scheme with data encoded in the pulse-width (ASK-PW) was designed and implemented. The reverse link operates via backscattered pulses, and the IC can adjust the antenna loading and the pulse width to optimize for robust operation with high power efficiency. Both the forward and reverse links have a reconfigurable data rate and support multiple devices using time domain multiple access (TDMA).

The organization of this paper is as follows. Section II describes the high-level system design of the external transceiver and the IC architecture. Section III presents the implant transceiver design and its implementation. The experimental results are discussed in section IV, and the paper is concluded in section V.

## II. SYSTEM DESIGN

The purpose of the chip was to demonstrate a fully functioning miniaturized transceiver. The external transceiver is similar to [3], and consists of an amplitude modulated carrier, a power amplifier, and a direct conversion receiver. The implantable transceiver is fully-integrated except for an external antenna, and the chip architecture is diagrammed in Fig. 1. The IC consists of power management circuitry, transceiver circuitry, a digital controller, and an analog sensor interface. The implantable transceiver has been designed to operate with a variety of antenna types with varying sizes, and the device is tested with three antennas: a 1 mm × 1 mm loop, a 2 mm × 2 mm loop, and a 2.5-turn 800 μm-long coil antenna with an 800 μm diameter.

Wireless powering is essential for miniaturization, and is integral in the design of the system. The RF carrier is converted to DC voltage with a 4-stage charge-pump connected rectifier. The output of the last stage is approximately 1.2 V under normal conditions, which is then regulated to a stable 0.7 V supply based on a bandgap voltage reference. The rectifier was sized to drive up to 10 μA for the sensors in addition to powering the transceiver and the power management circuitry. To simplify circuit design, a low dropout regulator is used, although a switching regulator would result in higher efficiency. The power management circuitry consumes approximately 4.3 μW.

It is important that both forward and reverse data transfer have minimal impact on power delivery because of the very limited power budget. In the forward link, we implemented amplitude shift keying (ASK) with low modulation depth (minimum of 9%), and encoded data in the pulse width (PW). This method is inherently asynchronous, allowing for simple on-chip clock and data recovering circuitry while minimally

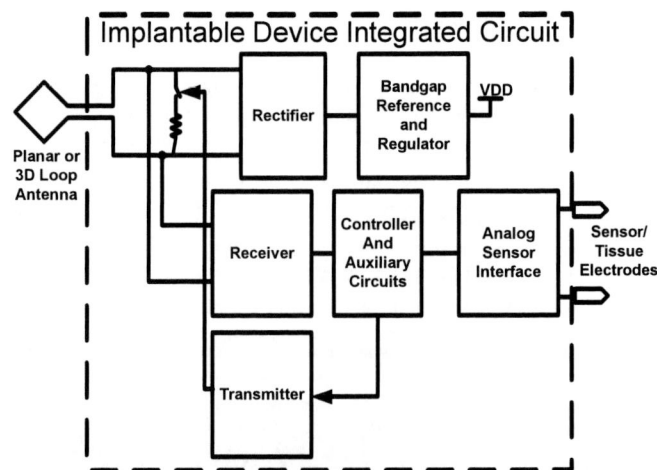

Fig. 1: Transceiver chip architecture and communication protocol timing diagram

perturbing the power carrier. Reverse data transfer is accomplished with a backscattering link, in which energy pulses are reflected by modulating the load on the antenna. The pulses encode data, and a short preamble allows for decoding by the external transceiver without a precise clock on the implant saving space and power.

The communication protocol supports multiple devices using TDMA, with each device having a unique programmable ID. Each forward packet contains the device ID and configuration data, and the corresponding device responds with a reverse packet containing a preamble, followed by 5 data bits. As will be described in the following section, the implanted transceiver can configure the modulating load, the backscattered pulse width, and both forward and reverse data rates to accommodate diverse operating conditions.

III. TRANSCEIVER ARCHITECTURE

Batteryless implantable devices are forward link limited and require robust, power-efficient transceiver design. A custom ASK-PW modulation scheme was implemented in the forward link, which minimizes impact on power delivery and allows for asynchronous clock and data recovery and therefore no power-hungry synchronization circuitry. The reverse link relies on backscattered energy, and uses reconfigurable circuitry to robustly operate in diverse conditions. This section describes the circuit architecture for these techniques.

A. Asynchronous Clock and Data Receiver

A high-level description of the data receiver is shown in Fig. 2. The demodulator provides both the clock signal for the digital controller and decodes incoming data. The low modulation depth and fluctuating input power make it impossible to use a fixed reference voltage for the ASK threshold detector. Instead, a dynamic reference voltage is generated concurrently with envelope detection. The demodulator interface with the antenna consists of a rectifier with a small time constant to recover the RF signal envelope and an RC low pass filter to average the envelope as shown in Fig. 3. These two signals are input to a comparator to generate the digital signal $V_{out1}$. This signal is buffered to produce a digital clock. $V_{out1}$ is also integrated and compared to a threshold to decode the data. With this implementation, long pulses produce high output and short pulses produce low output. The demodulated data is captured on the falling edge of the clock by a low-power digital controller, which configures the transmitter and the low power sensing circuit.

Clock and data signals are recovered from the envelope and the dynamic reference, which are first input to a comparator to generate the full-swing digital signal $V_{out1}$. This comparator consists of two differential amplifier stages followed by a Schmitt-trigger inverter as shown in Fig. 4. Two low-power differential amplifiers ensure that the gain remains high for a wide range of common-mode input voltages, which vary depending on input power. The Schmitt-trigger inverter reduces the crowbar current due to slow transitions of the amplifier output, and it also decreases noise sensitivity. The resulting digital signal is both buffered to generate the clock and integrated to decode the data as shown in Fig. 5.

The integrator consists of a skewed inverter with a capacitive load to provide slow rising and fast falling edges. This capacitance defines the pulse width that causes the data to transition from low to high, and therefore sets the minimum

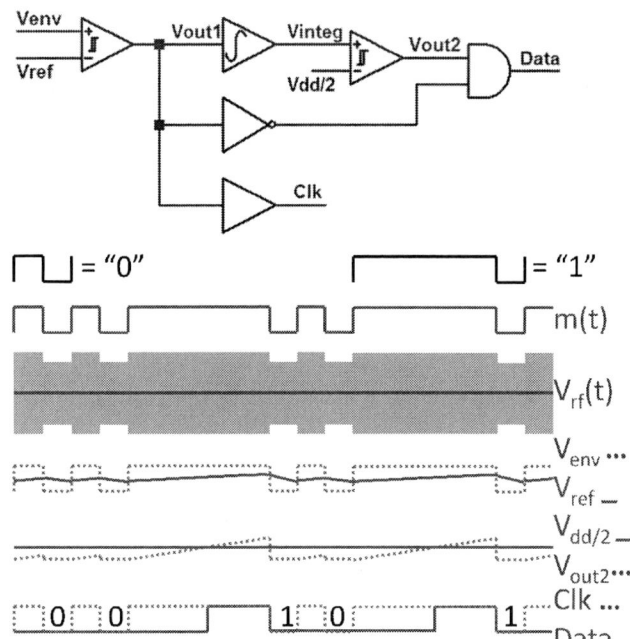

Fig. 2: Demodulator and description of operation

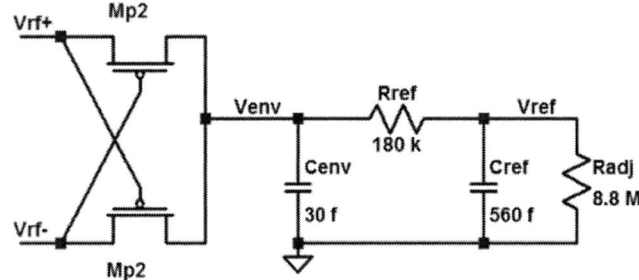

Fig. 3: Envelop detection and avering circuit

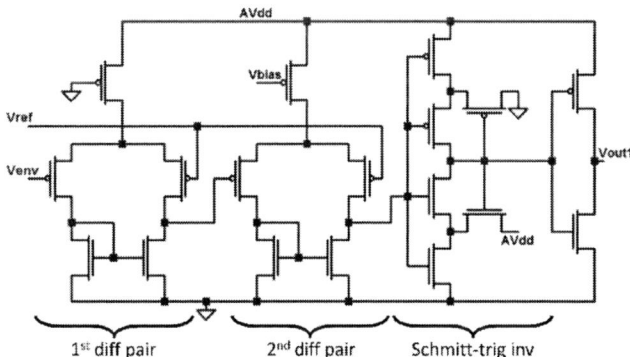

Fig. 4: First comparator that converts envelope into digital signal

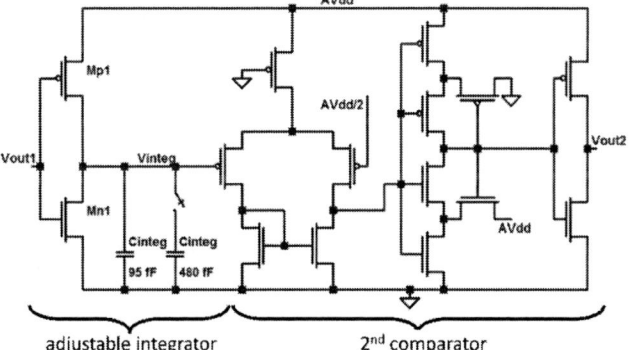

Fig. 5: Adjustable integrator for forward data rate control and second comparator for data decoding

and maximum data rates. As shown in the figure, the integrator capacitive load can be adjusted to configure the forward data rate. In the current implementation the load capacitance can be set to 95 fF or 475 fF, which corresponds to a forward data rate of either 20 Mbps or 4 Mbps, respectively. On the falling edge of each incoming pulse, data is captured from a comparator that compares the integrated result with a fixed reference at $V_{dd}/2$. This comparator consists of a single differential pair followed by a Schmitt-trigger inverter. The entire clock and data recovery circuit draws a current of 1.7 µA at 4 Mbps, achieving 0.3 pJ/bit energy efficiency which outperforms comparable systems by a factor more than 20 [4].

## B. Reconfigurable Backscattering Link

Backscattering links are highly power-efficient though they can suffer from a high bit error rate (BER) when implanted because of varying tissue composition, which also changes over time. To mitigate the effects of the unpredictability in the operating environment, the reverse link has been designed with a reconfigurable modulating load, data rate, and backscattered pulse width.

Analysis in [5] provides a method for determining the optimal load to minimize BER for binary backscattering modulation. Two key objectives must be satisfied: maximizing the average backscattered power per bit – max $\{\sigma_1 + \sigma_2\}$, and maximizing the Euclidean distance between the reflection coefficients on the Smith chart corresponding to the matched and mismatched loads – max $\{|\Gamma_1 - \Gamma_{2,k}|\}$. In the above $\sigma_1$ and $\sigma_2$ correspond to the termination-dependent implant antenna radar cross-section and $\Gamma_1$ and $\Gamma_{2,k}$ are the reflection coefficients corresponding to the implant antenna being match terminated and terminated with the $k^{th}$ modulating load, respectively. The first objective requires that the selected modulating load maximizes Euclidian distance from structural antenna mode $A_s$ which is a parameter defined by the antenna geometry and its physical composition. The second objective demands maximizing Euclidean distance between the matched reflection coefficient $\Gamma_1$ and the mismatched reflection coefficient $\Gamma_{2,k}$ when the antenna is terminated with the $k^{th}$ modulating load. In both cases, the Euclidean distance is maximized when the reflection coefficients are separated by a straight line collinear with the Smith chart center.

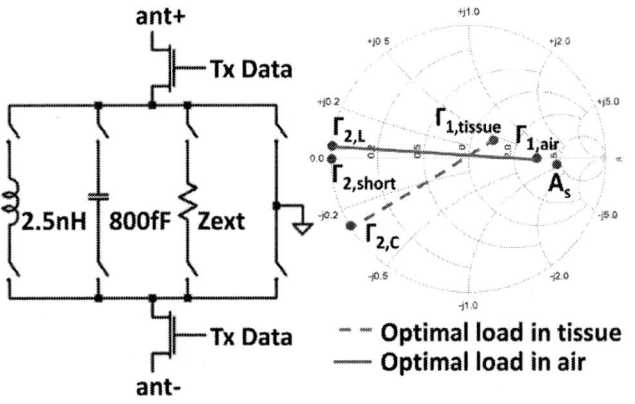

Fig. 6: Configurable load for backscatter link and Smith chart representation

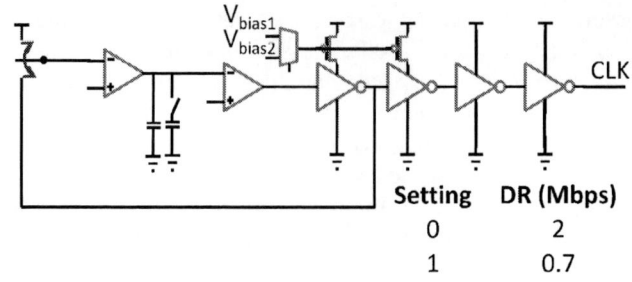

| Setting | DR (Mbps) |
|---------|-----------|
| 0       | 2         |
| 1       | 0.7       |

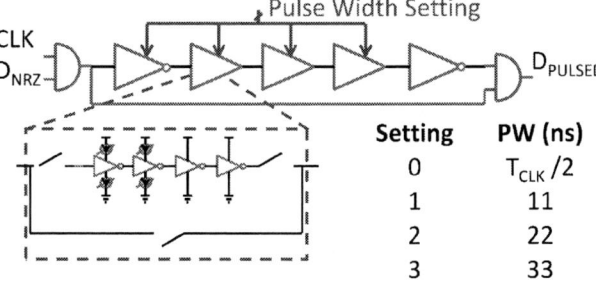

| Setting | PW (ns)        |
|---------|----------------|
| 0       | $T_{CLK}/2$    |
| 1       | 11             |
| 2       | 22             |
| 3       | 33             |

Fig. 7: Reconfigurable clock for data rate adjustment and pulse width control

To satisfy these objectives, the IC was designed to have four loads to accommodate a variety of conditions as shown in Fig. 6. The pass gates and the modulation transistor are sized to balance the loading on the RF path while minimizing the parasitic effective series resistance (ESR). The fourth load can be provided externally if necessary. Fig. 6 shows the Smith chart at 1.32 GHz for the device with the 3D coil antenna in air and implanted in a porcine heart with the structural antenna mode $A_s$, reflection coefficient $\Gamma_1$ corresponding to antenna impedances in both air and tissue (matching was designed for tissue), and reflection coefficients $\Gamma_{2,k}$ corresponding to the three loads for backscatter modulation. It is clear from the chart that the capacitive load and the inductive load should be used for the device in tissue and in air, respectively.

To accommodate multiple devices for intracardiac mapping with TDMA, the data rate must be sufficiently high. Some devices may have less available power and others may need more frequent polling, justifying the need for reverse data rate configurability. This configurability is accomplished by adjusting the clock frequency to set data rates to either 0.7 Mbps or 2 Mbps as shown in Fig. 7. The decoder is unaffected by this frequency change because each data packet contains a preamble which can be used to synchronize bit recovery.

Furthermore, the on-chip clock is only enabled for the reverse data transmission to save power. Additionally, the backscattered pulses can be set to 11, 22, 33 ns, or half of the clock period with a circuit shown in Fig. 7. As explained in [6], shorter pulses minimize the time that the RF path is mismatched and thus not harvesting power. However, very short pulses are spectrally inefficient and result in reduced signal quality as the higher frequencies are filtered by the antennas, matching network, or other elements in the reverse link. Therefore, adjustability of the pulse width allows for optimization of the data rate while maintaining an adequate BER. The transmitter consumes 1.4 µW at 2 Mbps including the lost energy that is reflected and thus not harvested, achieving 0.7 pJ/bit energy efficiency which outperforms comparable systems by a factor more than 25 [6].

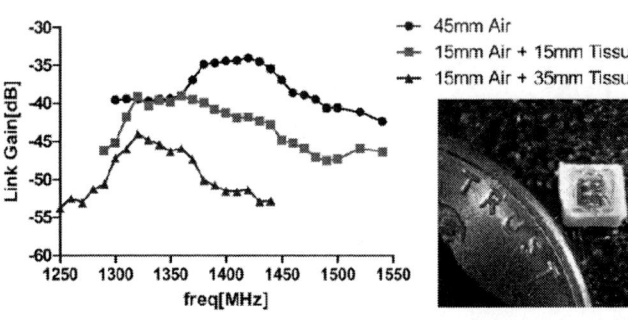

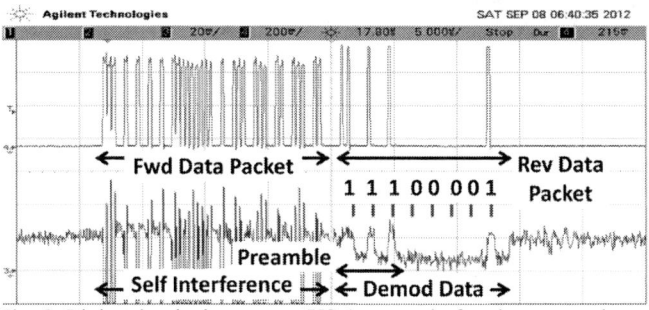

Fig. 8: Link gain, device next to US 1 cent coin for size comparison, and successful demodulation of forward and reverse data link

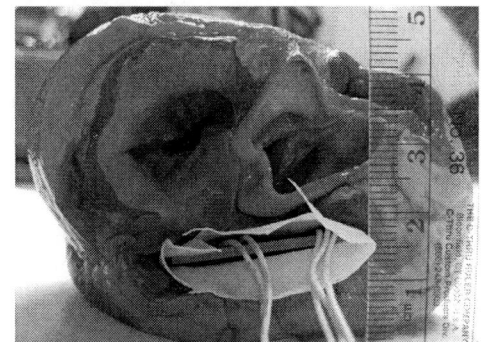

Fig. 9: Experimental setup of the device inside porcine heart

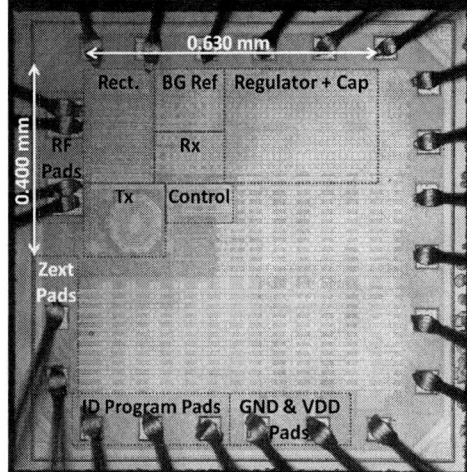

Fig. 10: IC annotated micrograph

Table 1: IC Performance Summary

|  | Receiver | Transmitter | Power Breakdown | |
|---|---|---|---|---|
| Modulation | Config. ASK + PWM | Config. Pulsed LSK | Bandgap Ref | 2.3µW |
| Carrier freq | 1.32 – 2.14 GHz | 1.32 – 2.14 GHz | Regulator | 2µW |
| Data Rate | 4 or 20 Mbps | 0.7 or 2 Mbps | Demodulator | 1.2µW |
| Power | 1.2 µW @ 4 Mbps | 1.4 µW @ 2 Mbps | Modulator | 1.4µW |
| Energy/bit | 0.3 pJ/bit | 0.7 pJ/bit | Controller | 3.8µW |
| Other | Mod. Depth ≥ 9% | Conf. Loads & PW | Total | 10.7µW |

## IV. EXPERIMENTAL RESULTS

The power and data links were evaluated over air and at two different depths of tissue for three different antennas: a 1 mm × 1 mm loop, a 2 mm × 2 mm loop, and a 2.5-turn 800 µm-long coil antenna with an 800 µm diameter. The resonance of the antennas in air varied from 1.4 GHz to 2.1 GHz. The antenna and the IC were encapsulated in low-density RF transparent epoxy. The link gain for three different environments is plotted in Fig. 8 for the 3D coil antenna, illustrating the effect of propagation medium surrounding the tag antenna. Deeper implantation results in reduced link gain and shifts the optimal operating point to lower frequencies. The link gain is approximately 2 dB and 6 dB better for the 1 mm × 1 mm and 2 mm × 2 mm antennas, respectively. Similarly, both forward and reverse data links were evaluated through air and tissue. The experimental setup for tissue tests is shown in Fig. 9. The modulating load had affected the link performance, and the received signal was strongest with the capacitive modulating load in tissue, as expected from the Smith chart in Fig. 6. The direct conversion receiver was constructed with off-the-shelf equipment, and it required a circulator to suppress self-interference. Fig. 8 shows the successful demodulation of reverse data without the use of any post-processing. Fig. 10 shows the annotated chip micrograph, and the IC performance is summarized in Table 1.

## IV. CONCLUSION

This work demonstrates a miniaturized, power-efficient transceiver architecture with a reconfigurable link for operation in diverse biological environments. The total power consumption of the device was 10.7 µW, and it operated successfully through 35 mm of porcine heart tissue. Forward transfer achieves up to 20 Mpbs with 0.3 pJ/bit efficiency at 4 Mpbs. The backscattering link operates at up to 2 Mbps with an efficiency of 0.7 pJ/bit. Additionally, multiple transceivers can operate simultaneously using TDMA and a unique programmable ID for each device.

## ACKNOWLEDGEMENTS

The authors would like to acknowledge Yuji Tanabe and the RAD program. The fabrication of this chip was made possible by the TSMC University Shuttle Program.

## REFERENCES

[1] Kneeland, P. P., et al., Trends in catheter ablation for atrial fibrillation in the United States. J. Hosp. Med.

[2] Griffin, J.D., et al., "Complete Link Budgets for Backscatter-Radio and RFID Systems," APM, Apr. 2009

[3] D. Pivonka, et al., "A mm-sized wireless powered and remotely controlled locomotive implant," TBCAS, 2012

[4] A. Ghenim, et al., "A full digital low power DPSK demodulator and clock recovery circuit for high data rate neural implants," ICECS, Dec. 2010

[5] A. Bletsas, et al., "Improving backscatter radio tag efficiency," MTT, June 2010

[6] Mark, M., et al., "A 1mm3 2Mbps 330fJ/b transponder for implanted neural sensors," VLSI, June 2011

# An All-Digital Time Difference Hold-and-Replication Circuit utilizing a Dual Pulse Ring Oscillator

Tetsuya Iizuka[†], Teruki Someya[†], Toru Nakura[‡] and Kunihiro Asada[‡]

[†]Department of Electrical Engineering and Information Systems, University of Tokyo

7-3-1 Hongo, Bunkyo-ku, Tokyo 113-8656, Japan

[‡]VLSI Design and Education Center, University of Tokyo

2-11-16 Yayoi, Bunkyo-ku, Tokyo 113-0032, Japan

Email: iizuka@vdec.u-tokyo.ac.jp

*Abstract*— This paper presents a time-domain analog signal hold-and-replication circuit which holds an input time interval of two signal transitions and replicates it any number of times. The proposed Time Difference Hold-and-Replication (TDHR) circuit utilizes a dual pulse ring oscillator to hold the input time interval without any time deterioration due to mismatches. The proposed TDHR circuit is implemented in a 65nm standard CMOS technology only with digital cells in the standard-cell library. It realizes 100ps to 1.2ns time interval hold and replication with the maximum error of ±50ps and consumes 0.51mW while occupying $40 \times 60 \mu m^2$ area. This TDHR circuit is also used for an input time amplification and a TDC resolution enhancement application of the proposed TDHR is also demonstrated in this paper.

## I. INTRODUCTION

The improvement of VLSI process technologies enables us to integrate a large number of transistors on a single chip, and significantly improves the circuit performance. Since a time-domain resolution is becoming superior to a voltage-domain resolution due to the high-speed transistors and the reduced supply voltage especially in nano-scale CMOS processes[1], several novel time-domain circuit architectures like Time-to-Digital Converter (TDC), All-Digital PLL (ADPLL), etc. have been proposed and widely used. Moreover, several papers which utilize time-domain analog signal have been proposed recently. A pulse-width-controlled PLL proposed in [2] tunes its oscillation frequency directly with a PFD output pulse width, which is a time-domain analog signal, and achieves a highly area-efficient PLL implementation. [3] proposes time-domain analog and digital mixed signal processing scheme for area and energy efficient LDPC (low-density parity-check) decoder.

As shown by these examples, several time-domain circuit architectures have been developed continuously with good correspondence with the voltage-domain circuits, like TDC in time-domain corresponds to Analog-to-Digital Converter (ADC) in voltage-domain. However, there has been no proper circuit to hold the time-domain signals, whereas the S/H circuit is one of the most primitive components in voltage-domain. Especially in the time-domain analog signal processing architectures, a circuit to hold the time-domain signals will be one

of essential components to keep or memorize the processed information. In this paper, we propose a hold-and-replication circuit for time-domain signals which is an input time interval of two signal edge transitions. The proposed circuit holds an input time interval of two signal transitions and replicates it any number of times. Using this circuit in front of TDCs enables a stable conversion of time-difference signals, and it can also be used to realize several time-domain signal operations like comparison, summation, subtraction and so on, to open up new time-domain signal processing. The proposed Time Difference Hold-and-Replication (TDHR) circuit utilizes a Dual Pulse Ring Oscillator (DPRO)[4] to keep the input time interval as a time difference between two pulses propagating on the ring, and is implemented only using digital cells in the standard-cell library.

## II. PROPOSED TDHR ARCHITECTURE AND IMPLEMENTATION

Figure 1 illustrates a block diagram of the proposed TDHR circuit and its timing diagram. During a stand-by state, this circuit stays in the stand-by (S/B) mode without any active blocks and waits for a time interval input. Once two rise transitions with $T_{in}$ time difference are injected to IN1 and IN2 ports as shown in Fig. 1, these transitions are first converted to pulses through a Pulse Generator (PG) and the pulse for IN2 propagates to DPRO with $T_{in} + T_{offset}$ delay after the pulse for IN1 to keep the enough time difference between two pulses.

To hold the time difference between two pulses, we use buffer-ring-based circuit as illustrated in Fig. 2. Ideally, a simple buffer ring circuit shown in Fig. 2 (a) can be used to keep the pulse propagation on the ring. However, a delay mismatch between rising and falling transitions of the buffer due to the process variability, and/or an irregularity of the MUX circuit often cause the pulse width shrinking or expanding along with the pulse propagation. This results in the pulse disappearing and fails to hold the time difference during sufficient period. To prevent this pulse disappearing, DPRO circuit illustrated in Fig. 2 (b) is employed in the proposed TDHR circuit. The DPRO consists of NAND gates and inverters so that the rising edge of the pulse is propagated through the gates while the

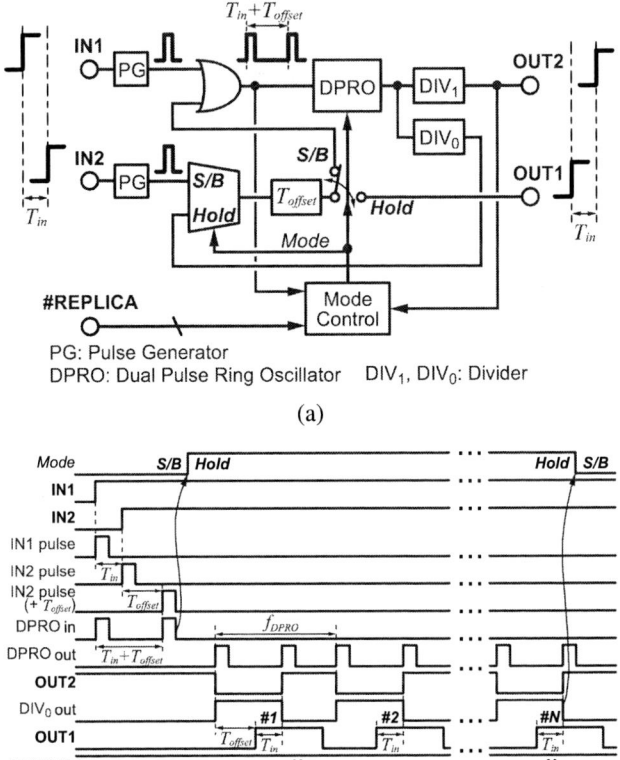

(a)

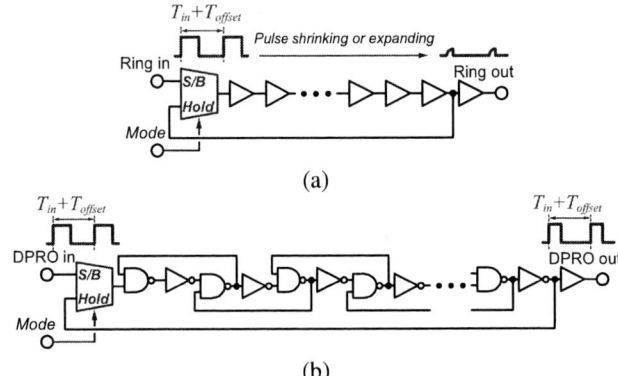

(a)

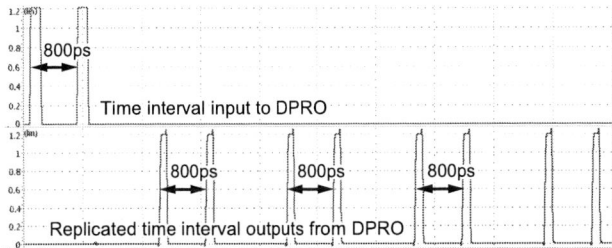

(b)

Fig. 2. A schematic diagrams of (a) a simple buffer ring and (b) a dual pulse ring oscillator (DPRO).

(b)

Fig. 1. (a) A block diagram of the proposed time-difference hold-and-replication circuit architecture and (b) its timing diagram.

Fig. 3. Simulated waveforms of a time interval input to and a replicated time interval output from DPRO.

falling edge of the pulse is generated in each delay stage[4]. Thanks to this property, DPRO keeps a pulse propagating on the ring infinitely, whereas a pulse width gradually shrinks or expands when it propagates on a simple buffer-based ring. The proposed TDHR circuit uses this DPRO to hold the input time difference by injecting the two successive pulses with $T_{in} + T_{offset}$ delay difference. Figure 3 shows simulated waveforms of a time interval input to and a replicated time interval output from DPRO. In this example, we inject 800ps time interval between two rising transition of the pulses and the output successive pulses replicate the 800ps time interval repeatedly. The DPRO keeps this time interval without any process-induced mismatch effects since these two pulses propagate through exactly the same paths during the hold mode.

After these two pulses are injected to DPRO, the TDHR automatically enters into the hold mode by switching the Mode signal to high. Then the repeated output of two successive pulses from the DPRO is fed into two dividers, $DIV_0$ and $DIV_1$. During the S/B mode, outputs of $DIV_0$ and $DIV_1$ are fixed to 0 and 1, respectively. Thus, during the hold mode, the rise transitions of $DIV_0$ and $DIV_1$ outputs are synchronized with $(2n-1)$-th and $2n$-th $(n = 1, 2, \ldots)$ rise transitions of DPRO output, respectively. Using this architecture, we can

replicate the time difference of two successive pulses on DPRO as two rise edge transitions with the same time difference. Since this time difference still includes the intentional offset delay $T_{offset}$, $DIV_0$ output is fed into the delay offset block to cancel the offset. We can again prevent the mismatch effects by sharing the same circuit block for both adding and canceling the offset delay. This TDHR can replicate the time interval any number of times and its output frequency is equal to the DPRO frequency $f_{DPRO}$. In the prototype implementation of this paper, it terminates the time interval output and return to the S/B mode after the required number of replica outputs is counted.

### III. APPLICATION TO TIME-TO-DIGITAL CONVERSION

One of the promising applications of this TDHR circuit is placing it in front of a TDC to enhance a time resolution. After holding the input time interval using the proposed TDHR circuit, the repetitive replicated time interval outputs can be fed into the TDC as a pulse train. We can use a gated ring oscillator (GRO)[5] or gated delay line (GDL) based TDC to enhance its time resolution with this pulse train as proposed in [6]. A simplified block diagram of the GRO-based TDC is illustrated in Fig. 4 (a). As shown in Fig. 4 (b), quantization error in a single pulse is carried on to the next pulse and the effective time resolution is enhanced to $4 \times T_{res}$ by injecting four enable pulses in this example, where $T_{res}$ is the original time resolution which is equal to a delay of a single stage

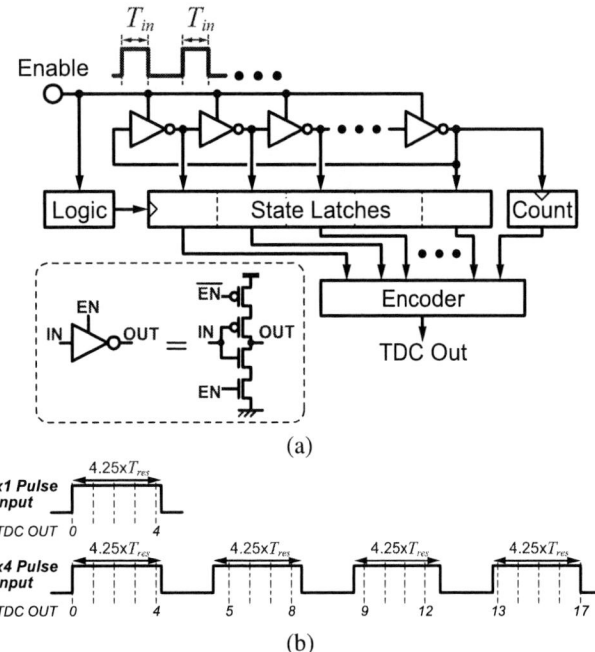

(a)

(b)

Fig. 4. (a) A simplified block diagram of the GRO-based TDC and (b) timing diagrams of the resolution enhancement with multiple enable pulse input.

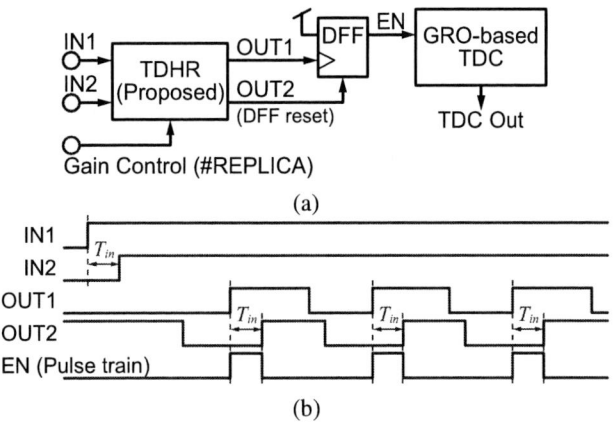

(a)

(b)

Fig. 5. (a) A block diagram of the resolution-enhanced TDC with TDHR circuit and (b) its simplified timing diagram.

inverter in GRO. Therefore, the GRO-based TDC treats the pulse-train input as a single pulse having a width multiplied by the number of pulses and we can use the proposed TDHR circuit as a pulse-train time amplifier (PTTA)[6]. Since the TDHR circuit can replicate the input time interval any number of times, this can be used as PTTA with arbitrary gain. A block and timing diagrams of the resolution-enhanced TDC with TDHR circuit is illustrated in Fig. 5. Replicated time difference outputs are converted to a pulse train using a DFF, then fed into the GRO-based TDC. As mentioned above, the gain of PTTA can be controlled arbitrarily by tuning the number of the replica outputs.

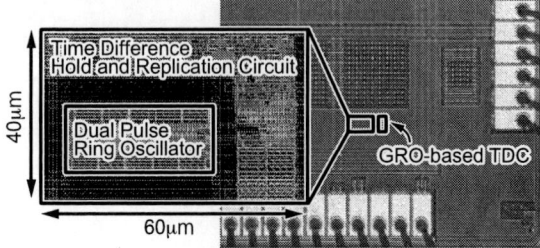

Fig. 6. A chip micrograph and a layout of the TDHR circuit.

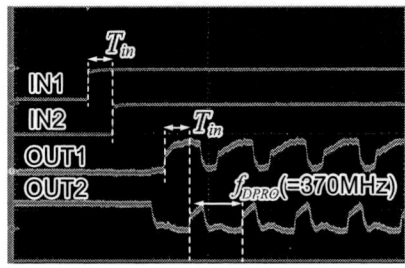

Fig. 7. Input and output waveforms of the TDHR circuit.

## IV. MEASUREMENT RESULTS

A prototype of the TDHR circuit is fabricated in a 65nm standard CMOS technology only with digital cells in the standard-cell library and occupies $40 \times 60 \mu m^2$ area as shown in Fig. 6. The TDHR core consumes a total of 0.51mW power during the hold mode from 0.9V and 1.2V supplies for the DPRO and other peripheral circuits, respectively. Figure 7 shows measured waveforms of the time interval input and output of the TDHR circuit, and verifies that the proposed TDHR circuit replicates the time interval input any number of times as explained in the previous section. These waveforms also show that the output frequency which is equal to the DPRO frequency is about 370MHz in this implementation.

Figure 8 plots the measured output time interval dependence on the number of the replica outputs up to 32 times, and verifies stable time interval outputs from the TDHR. The measured output time interval errors from the input are shown in Fig. 9. As shown in these graphs, this prototype TDHR accepts 100ps to 1.2ns time interval input with the maximum error of ±50ps.

To demonstrate a TDC resolution enhancement application of this TDHR, a 9-stage GRO-based TDC is also implemented on the test chip and its occupation area is $40 \times 25 \mu m^2$. First, we evaluated the characteristics of TDHR circuit as a PTTA. When we inject a time interval input $T_{in}$ into the TDHR circuit and replicate it $N$ times, the sum of the replicated output time interval becomes $N \times T_{in}$ ideally. Figure 10 plots the error between ideal $N \times T_{in}$ and the measured sum of the time interval outputs. In the cases of the time interval larger than 200ps, the TDHR circuit achieves below ±10% time amplification errors and demonstrates up to 32× time amplification. Figure 11 shows the normalized GRO-based

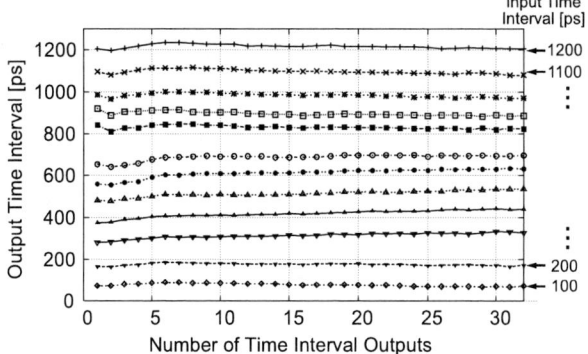

Fig. 8. Measured output time interval dependence on the number of the replica outputs.

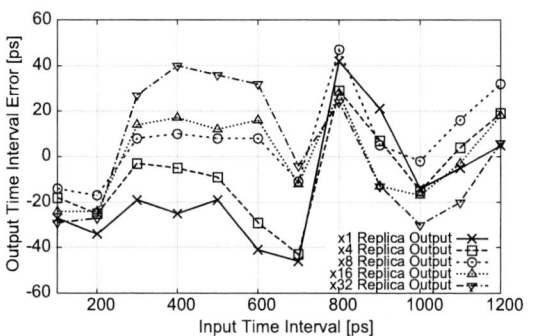

Fig. 9. Measured time replication error depending on the input time interval by changing the number of the replica outputs.

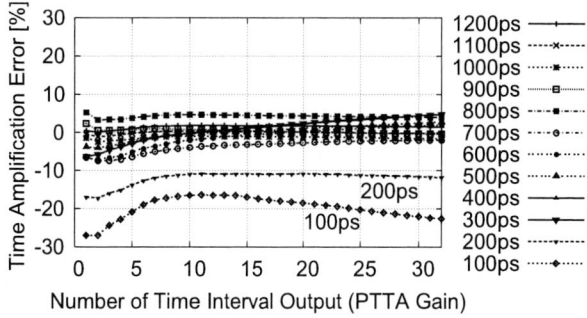

Fig. 10. Measured time amplification error depending on the PTTA gain factor.

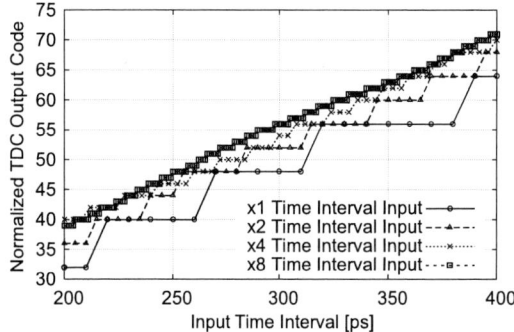

Fig. 11. Measurement results of a GRO-based TDC resolution enhancement using the proposed TDHR circuit.

TDC outputs of each amplification setting and verifies that the proposed TDHR effectively enhances the time resolution of the TDC by replicating the input time interval 2 to 8 times.

## V. Conclusions

This paper proposed a time-domain analog signal hold-and-replication circuit which holds an input time interval of two signal transitions and replicates it any number of times. The proposed TDHR circuit holds the input time interval without any time deterioration due to mismatches by utilizing a dual pulse ring oscillator, and was implemented only using digital cells in the standard-cell library. The measurement results with 65nm CMOS implementation showed that it realizes 100ps to 1.2ns time interval hold and replication with the maximum error of ±50ps and consumes 0.51mW while occupying $40 \times 60 \mu m^2$ area. This TDHR circuit is also used for an input time amplification as a PTTA and a TDC resolution enhancement application of the proposed TDHR circuit was also demonstrated in this paper.

## Acknowledgments

The VLSI chip in this study has been fabricated in the chip fabrication program of VLSI Design and Education Center(VDEC), the University of Tokyo in collaboration with STARC, e-Shuttle, Inc., and Fujitsu Ltd.

## References

[1] R. B. Staszewski *et al.*, "All-Digital TX Frequency Synthesizer and Discrete-Time Receiver for Bluetooth Radio in 130-nm CMOS," *IEEE JSSC*, vol. 39, no.12, pp. 2278–2291, 2004.

[2] T. Nakura and K. Asada, "Low Pass Filter-Less Pulse Width Controlled PLL Using Time to Soft Thermometer Code Converter," *IEICE Trans. on Electronics*, pp. 297–302, 2012.

[3] D. Miyashita *et al.*, "A 10.4pJ/b (32, 8) LDPC Decoder with Time-Domain Analog and Digital Mixed-Signal Processing," *IEEE ISSCC Dig. Tech. Papers*, pp. 420–421, 2013.

[4] D. Park and S.-H. Cho, "A 14.2mW 2.55-to-3GHz cascaded PLL with reference injection, 800MHz delta-sigma modulator and 255fs$_{rms}$ integrated jitter in 0.13$\mu$m CMOS," *IEEE ISSCC Dig. Tech. Papers*, pp. 344–346, 2012.

[5] M. Z. Straayer and M. H. Perrott, "A Multi-Path Gated Ring Oscillator TDC With First-Order Noise Shaping," *IEEE JSSC*, vol. 44, no. 4, pp. 1089–1098, 2009.

[6] K.-S. Kim *et al.*, "A 7b, 3.75ps Resolution Two-Step Time-to-Digital Converter in 65nm CMOS Using Pulse-Train Time Amplifier," *IEEE Symp. VLSI Circuits Dig. Tech. Papers*, pp. 192-193, 2012.

# A 15-Bit Binary-Weighted Current-Steering DAC with Ordered Element Matching

Tao Zeng, Kevin Townsend, Jingbo Duan[*], and Degang Chen
Department of Electrical and Computer Engineering, Iowa State University, Ames, IA.
[*]Broadcom Corporation, San Jose, CA.
Email: zt123@iastate.edu

*Abstract-* **Device variability has become one of the fundamental challenges to high-resolution and high-accuracy DACs in nanometer and emerging processes. This paper introduces a 15-bit binary-weighted current-steering DAC in a standard 130nm CMOS technology, which utilizes a new random mismatch compensation theory called ordered element matching to improve the static linearity performance with the presence of large variability. The chip's core area is less than 0.42mm$^2$, among which the 7-bit MSB current source area is well within 0.021mm$^2$. Measurement results have shown that the DAC's DNL and INL can be reduced from 9.85LSB and 17.41LSB to 0.34LSB and 0.77LSB, respectively.**

## I. Introduction

High-resolution and high-accuracy DACs can be widely found in various medical, instrumentation, and test and measurement applications. For these types of circuits, device matching is one of the most critical design parameters. As IC technology continues to evolve, the minimum feature size is quickly approaching nanometer scale. In these technology nodes and beyond, significant variability, due to process, supply voltage, temperature, and stress, imposes grand challenges to achieving accurate device matching. With the emerging materials and devices that may provide an alternative to CMOS, variability is no less.

Traditional matching techniques can compensate random mismatch errors to certain degrees, but they possess some disadvantages. For example, trimming [1] suffers high cost in terms of test equipments and time; calibration [2] requires complicated compensation circuitry; switching sequence adjustment [3] offers no improvements to DNL; and dynamic element matching (DEM) [4] limits its most applications to sigma-delta data converters.

A totally different approach called ordered element matching (OEM) was developed and rigorously proven using order statistics [5]. It reduces random variations significantly by grouping the complementary ordered circuit components. After repeating multiple OEM operations to a unary-weighted element array with the presence of large variability, a well-matched binary-weighted array can be generated. Additionally, incorporating outlier elimination strategy can be highly effective in matching performance enhancement.

In this paper, a 15-bit current-steering DAC is designed and fabricated in a standard 130nm CMOS technology to demonstrate the significant linearity improvements illustrated by OEM in [5]. The DAC has 7-8 segmentation, where OEM is continuously applied to the 7-bit unary-weighted MSB

array. By doing so, a 7-bit binary-weighted MSB array is formed at the end. The 8-bit LSB array has a conventional binary-weighted structure, therefore yielding an overall 15-bit binary-weighted DAC. The chip's active area is less than 0.42mm$^2$, among which the 7-bit MSB array only consumes 0.021mm$^2$. Measurement results show that the DNL can be reduced from 9.85LSB to 0.34LSB, whereas the INL can be reduced from 17.41LSB to 0.77LSB.

This paper is organized as follows. In Section II, the OEM operation is illustrated, while the 15-bit DAC architecture and its related circuit implementation are shown in Section III. Followed by those, measurement results are provided in Section IV. Finally, conclusion is drawn in Section V.

## II. Ordered Element Matching

Random mismatch in CMOS devices is due to inherent variations in the semiconductor process. It is by far the largest source of error degrading the performance of high-resolution and high-accuracy DACs. Based on the standard mismatch model [6], for a MOSFET in saturation with an overdrive voltage $V_{gs}$-$V_t$, the relative variance of the drain current is given by:

$$\left(\frac{\sigma_I}{I}\right)^2 = \frac{A_\beta^2 + 4A_{Vt}^2 / \left(V_{gs} - V_t\right)^2}{2 \cdot W \cdot L} \tag{1}$$

where $A_\beta$ and $A_{Vt}$ are process mismatch parameters, and $W \cdot L$ is gate area. Similar formulas for capacitor and resistor mismatch errors also show variance inversely proportional to area [7] [8]. This leads to the basis of the widely used rule of thumb: quadrupling area for every factor-of-two reduction in mismatch errors.

Nevertheless, instead of varying the design variables listed in (1), we can combine a pair of complementary ordered components in a population to reduce standard deviation of the mismatch errors. This process is called OEM. From statistical analysis, the random variations can be reduced by a factor of at least 6.5 for a sample population size greater than 64 with one-time OEM iteration. The detailed study is available in [5].

To take advantage of the significant variance reduction offered by OEM, we can create a binary-weighted array, achieving system level matching out of a mismatched population. A 3-bit unary-weighted array is taken as an example for the whole process illustration. Since outlier elimination is proven to be effective in boosting the matching

978-1-4673-6145-3/13 $31.00 © 2013 IEEE

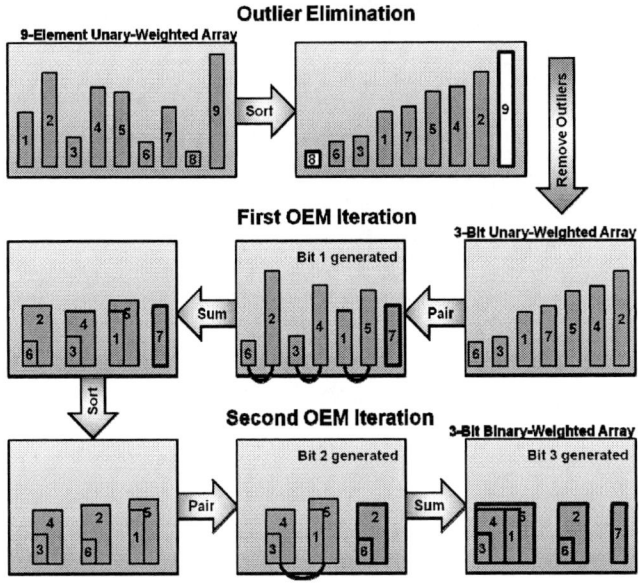

Fig. 1. 3-bit binary-weighted array generation based on outlier elimination and two OEM iterations from a 9-element unary-weighted array

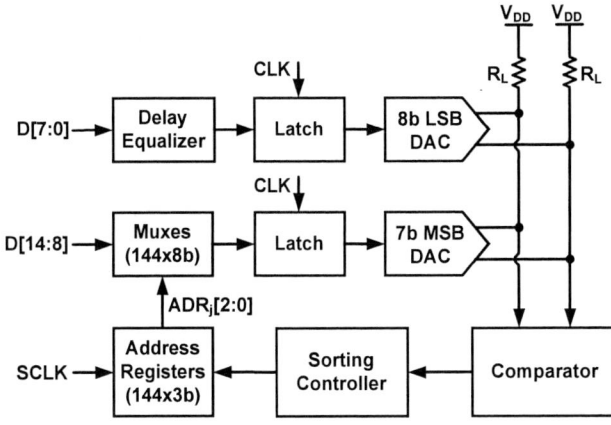

Fig. 2. 15-bit binary-weighted DAC architecture

performance [5], we will start with 9 unary mismatched components. Fig. 1 shows the outlier elimination step, which simply truncates the sorted element array to remove two outliers. The area of each rectangle in the figure denotes the value of each element with random mismatch error. Followed by that, we can apply two OEM iterations to the remaining elements.

The specific steps proceed as follows. First, all elements are sorted in ascending order. Then, the complementary ordered elements are paired, and a single element in the middle is left alone. Finally, two elements in each pair are summed together, and the singleton is moved to the end of the array. Thus, we have generated a new 2-bit unary-weighted array with each element nearly twice the value of the original elements, and a 1-bit binary weighted array. The random variations in the new elements are reduced significantly by the OEM theory. If we continue this process to the new unary-weighted array, only

three elements are left at the end, and they differ by a factor of 2. Furthermore, the parameter variations are rapidly diminishing compared to the previous step. In general, N-1 OEM iterations are required to convert an N-bit unary-weighted array into an N-bit binary-weighted array.

## III. DAC ARCHITECTURE

OEM implementation necessitates two key functionalities, which are (1) to obtain the relative orders of circuit components and (2) to connect all elements in a binary-weighted fashion according to the results of OEM operations. Since [3] has already proposed an on-chip solution for ranking components, we will not repeat the circuits in this work but rather replace them with off-chip electronics and a FPGA board. It should be pointed out that the on-chip digital sorting controller can be much simpler in our case because of the binary-weighted operation. On the other hand, each component needs to be connected to one of the (N+1)-bit lines, i.e., the actual N-bit input plus one extra bit to indicate outlier throwaway. This requires, for each element, an (N+1)-to-1 digital mux and a $\log_2(N+1)$-bit register to store the mux address code.

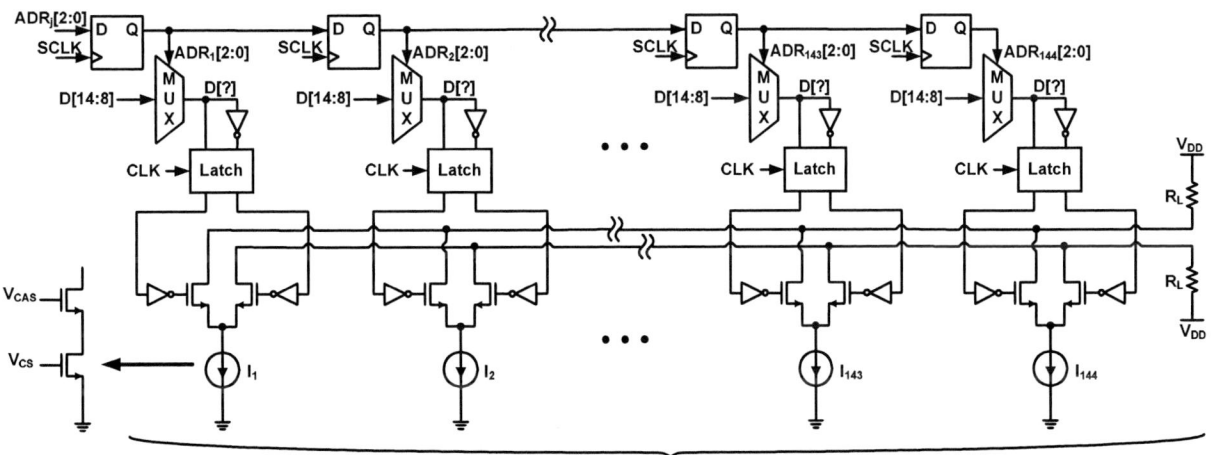

**127 Out of 144 Equal Current Sources Encoded into 7 Binary-Weighted MSB Bits**

Fig. 3. Circuit implementation of 7-bit MSB array

978-1-4673-6145-3/13 $31.00 © 2013 IEEE

TABLE I
ADDRESS CODE MAPPING DURING NORMAL CONVERSION PHASE

| Address code | Number of elements | Digital input lines |
|---|---|---|
| 000 | 17 | Don't Care |
| 001 | 1 | D[8] |
| 010 | 2 | D[9] |
| 011 | 4 | D[10] |
| 100 | 8 | D[11] |
| 101 | 16 | D[12] |
| 110 | 32 | D[13] |
| 111 | 64 | D[14] |

TABLE II
ADDRESS CODE MAPPING DURING COMPARISON PHASE

| Address code | Number of elements | Digital input lines |
|---|---|---|
| 001 | $k^a$ | D[8] |
| 010 | $k^a$ | D[9] |
| 110 | $72-k^a$ | D[13] |
| 111 | $72-k^a$ | D[14] |

[a]The variable k depends on the stages of OEM iterations, where k = 1, 2, 4, 8, 16, and 32.

In this work, we choose N=7 and create 127 plus 17 elements to produce a 12×12 array of cells, which demand 144 8-to-1 muxes and 3-bit registers. Fig. 2 shows a block diagram of this implementation for the 15-bit current-steering DAC with 7-8 segmentation. Meanwhile, Fig. 3 gives the detailed circuit blocks of the 7-bit MSB array.

As shown in Fig. 3, the outputs of the back registers are connected to the inputs of the front registers. Then, we can sequentially load 144 address codes into the DAC. Table I illustrates the corresponding address code and number of elements for different bit lines during the normal conversion phase. For example, if 010 is assigned to two current sources, D[9] will be selected to control them. With these arrangements, the 7-bit MSB DAC will operate in a binary-weighted manner.

When we are in the current source comparison phase, 144 specific address codes will be loaded. To be more instructive, a variable k is introduced here. The address codes to the 144 current sources and the corresponding digital bit lines are shown in Table II. Our goal is to compare the current sources with address code 001 to those with 010. The variable k is determined by the stages of OEM operations. Based on the theory, we need 6 OEM iterations in order to generate 7 binary-weighted current sources out of 144 unit current sources. Thus, k can be set to 1, 2, 4, 8, 16, and 32.

Suppose that we are at the first time OEM iteration, then k = 1. We set D[8] and D[9] to be 1 and 0, respectively. D[13] and D[14] are assigned to 1 and 0, respectively. Furthermore, all the other digital input bits are 0. From this setup, each output side will have 72 MSB current sources flowing. We can then store the resulting differential current output. Followed by that, we will adjust D[8] and D[9] to be 0 and 1, respectively, and keep all the other inputs as the same as before. Consequently, we have swapped the output sides for the current sources with address codes 001 and 010. Subtracting the previously stored value from the present value, we can have current difference doubled between the comparing current sources, thus obtaining the relative ranks. This process can be repeated for any other current sources at any stages of

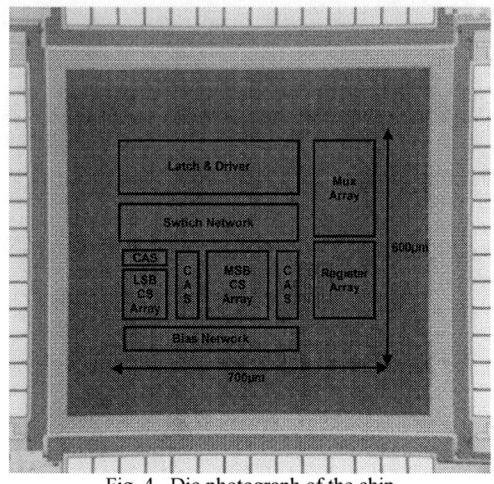

Fig. 4. Die photograph of the chip

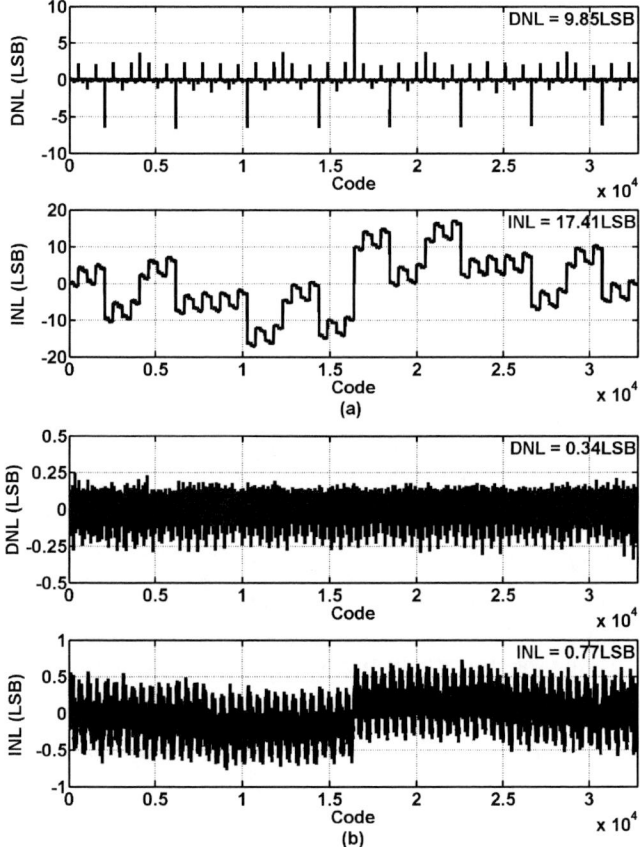

Fig. 5. Static linearity performance of the 15-bit binary-weighted DAC (a) before and (b) after OEM iterations

OEM iterations. Therefore, the complete current source orders can be easily attained by an efficient merge-sort algorithm implemented in the FPGA.

It should be pointed out that the described operation above is similar to the on-chip implementation shown in [3] but with different digital control mechanisms. Therefore, the current comparator can be directly used in our case. Furthermore, thanks to the binary-weighted operation, the sorting controller can be much simpler here since there are only 7 possible MSB routing address codes and it requires no thermometer decoder.

978-1-4673-6145-3/13 $31.00 © 2013 IEEE

TABLE III
MATCHING PERFORMANCE COMPARISON WITH STATE OF THE ART

| Specifications | Self-Cal. [2] | SSPA [3] | OEM |
|---|---|---|---|
| Resolution | 14-bit | 14-bit | 15-bit |
| Structure | 6b Unary MSB<br>8b Binary LSB | 7b Unary MSB<br>7b Binary LSB | 7b Binary MSB<br>8b Binary LSB |
| Technique | Calibrate<br>MSB current<br>values | Adjust<br>MSB switching<br>sequence | Regroup<br>MSB current<br>sources |
| Process | CMOS 130nm | CMOS 180nm | CMOS 130nm |
| Power supply | 1.5V | 1.8V | 1.2V |
| Full current | 10mA | 16mA | 5mA |
| DNL | 5LSB(before)<br>0.34LSB(after) | 0.63LSB(before)<br>0.56LSB(after) | 9.85LSB(before)<br>0.34LSB(after) |
| INL | 9LSB(before)<br>0.43LSB(after) | 1.37LSB(before)<br>0.76LSB(after) | 17.41LSB(before)<br>0.77LSB(after) |
| Area | 0.1mm$^{2\,a}$ | 2mm$^{2\,b}$ | 0.42mm$^2$ |

[a]The area of 16-bit sigma-delta ADC was not taken into account.
[b]The area of sorting controller is taken out of the total chip area.

## IV. MEASUREMENT RESULTS

The 15-bit binary-weighted current-steering DAC is implemented in a 130nm digital CMOS process with 1.2V supply voltage. The full-scale current is 5mA driving 50$\Omega$ load resistors. Fig. 4 shows the die photograph of the chip. The active area is less than 0.42mm$^2$, among which the 7-bit MSB current source area is well within 0.021mm$^2$.

With random binary group assignments to the 144 MSB unit current sources, the DAC's DNL and INL are 9.85LSB and 17.41LSB, respectively. However, after loading address codes obtained by OEM iterations, the DNL and INL can be reduced to 0.34LSB and 0.77LSB, respectively. Fig. 5 plots the static linearity performance before and after OEM iterations.

Table III shows the performance advancements compared to state-of-the-art matching techniques, e.g., [2] and [3]. Since we did not implement the sorting controller on chip, the area is taken out of the total chip area for [3] to maintain a fair comparison. On the other hand, the current comparator is so small that it does not affect the overall area consumption. Meanwhile, it is worth mentioning that the authors did not account for the area of 16-bit sigma-delta ADC in [2].

Fig. 6 shows the measured DAC's output spectrum performance before and after OEM iterations with 0.4MHz signal frequency and 10MHz sampling frequency. The SFDR can be increased from 67dB to 84dB.

## V. CONCLUSION

In order to confirm the OEM theory on silicon, a 15-bit binary-weighted current-steering DAC is fabricated in a 130nm digital CMOS process. The active area of the chip is less than 0.42mm$^2$. More importantly, the MSB current source area is well within 0.021mm$^2$. Experimental results have shown that the DAC's DNL and INL can be both reduced significantly. The new matching technique only demands the component orders, thus requiring a comparator and some digital circuitry. Such implementation scales well with IC technologies, which may offer one alternative solution to random mismatch errors in the variability-excessive processes.

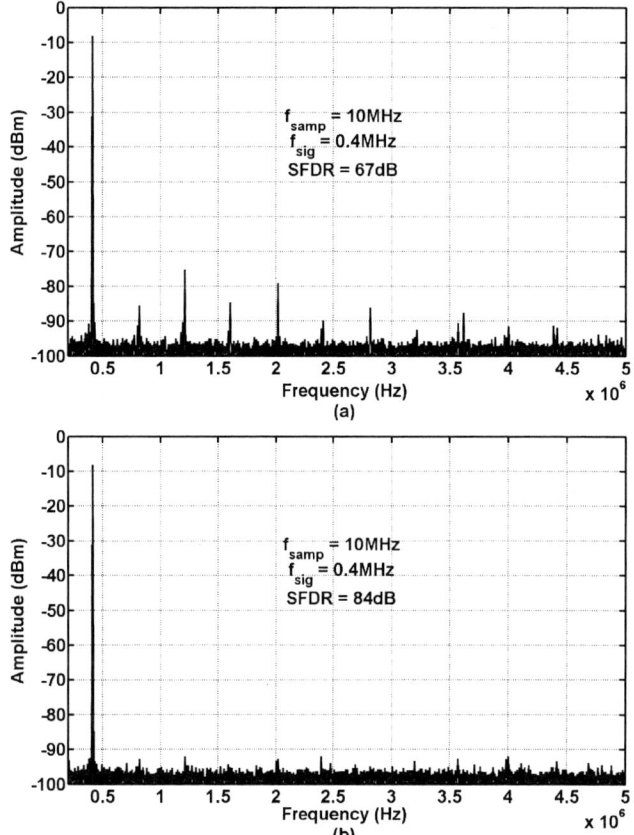

Fig. 6. Output spectrum of the 15-bit binary-weighted DAC (a) before and (b) after OEM iterations with $f_{samp}$ = 10MHz and $f_{sig}$ = 0.4MHz

## ACKNOWLEDGMENTS

The authors would like to thank Yuan Ji and Yan Duan for their valuable supports on this project.

## REFERENCES

[1] D. Marche, Y. Savaria, and Y. Gagnon, "Laser fine-tuneable deep-submicrometer CMOS 14-bit DAC," *IEEE Trans. Circuits Syst. I, Reg. Papers*, vol. 55, no. 8, pp. 2157–2165, Sep. 2008.

[2] Y. Cong and R. L. Geiger, "A 1.5-V 14-bit 100-MS/s self-calibrated DAC," *IEEE J. Solid-State Circuits*, vol. 38, no.12, pp. 2051 -2060, Dec. 2003.

[3] T. Chen and G. Gielen, "A 14-bit 200-MHz current-steering DAC with switching-sequence post-adjustment calibration," *IEEE J. Solid-state Circuits*, vol. 42, no. 11, pp. 2386-2394, Nov. 2007.

[4] R. E. Radke, A. Eshraghi and T. S. Fiez, "A 14-bit current-mode ΣΔ DAC based upon rotated data weighted averaging," *IEEE J. Solid-state Circuits*, vol. 35, no. 8, pp. 1074-1084, Aug. 2000.

[5] T. Zeng and D. Chen, "An order-statistics based matching strategy for circuit components in data converters," *IEEE Trans. Circuits Syst. I, Reg. Papers*, vol. 60, no. 1, pp. 11-24, Jan. 2013.

[6] M. J. M. Pelgrom, A. C. J. Duimaijer, and A. P. G. Welbers, "Matching properties of MOS transistors," *IEEE J. Solid-State Circuits*, vol. 24, no. 5, pp. 1433–1439, Oct. 1989.

[7] J. B. Shyu, G. C. Temes, and F. Krummenacher, "Random error effects in matched MOS capacitors and current sources," *IEEE J. Solid-state Circuits*, vol. 19, no. 6, pp. 948-956, Dec. 1984.

[8] W. A. Lane and G. T. Wrixon, "The design of thin-film polysilicon resistors for analog IC applications," *IEEE Trans. Electron Devices*, vol. 36, no. 4, pp. 738-744, Apr. 1989.

# A 500 MS/s 76dB SNDR Continuous Time Delta Sigma Modulator with 10MHz Signal Bandwidth in $0.18\mu$m CMOS

Rune Kaald*[†], Bjørnar Hernes[†], Christian Holdø[†], Frode Telstø[†], and Ivar Løkken[†]

*Department of Electronics and Telecommunications,
Norwegian University of Science and Technology
[†]Hittite Microwave Norway

*Abstract*—A 5th order continuous time delta sigma modulator is designed in $0.18\mu$m CMOS. At a sampling rate of 500MHz it achieves 76dB SNDR over a 10MHz bandwidth consuming 58mW. 5th order noiseshaping is realized with 4 opamp based RC integrators and a VCO realizing an integrator and a 4 bit quantizer. A THD of -82.3dBc is achieved without calibration of feedback DACs. A high speed capacitive implementation of excess loop delay compensation, together with a method for reducing the switching activity of the output codes from the VCO are proposed.

## I. Introduction

With the never-ending downscaling of process technology VCO based quantizers have gained a lot of attention. Trading resolution in time versus voltage is beneficial when the supply voltage decreases and speed increases. One of the major power consuming blocks in a multibit high speed conventional single loop continuous time delta sigma (CTDS) modulator is the quantizer. A voltage controlled ring oscillator with phase quantization and comparison offers very power efficient quantization and integration [1]. However, due to its digital nature and being an integrator with voltage input, compensating for excess loop delay becomes more problematic. In this work we address this issue by proposing a capacitive implementation of the excess loop delay compensation and summation of feedforward signals. Further, we introduce a method for reducing code-dependent switching activity causing inter-symbol-interference and we suggest ways to achieve high performance.

This paper is organized as follows: section II describes the modulator architecture. Section III describes circuit implementation specific details. Section IV shows the experimental results and finally section V draws the conclusion.

## II. Modulator Architecture

To compensate for excess loop delay (i.e. the time it takes from the quantizer toggles to the feedack DACs are updated) a direct feedback path to the quantizer is necessary to ensure correct impulse response of the loop filter. When a VCO is the last integrator, a differentiation of the quantized feedback signal with summation of signals from the loop filter prior to its input is necessary. In prior work, this has been achieved using passive summation based on current steering DACs and resistors [1], [2]. While having low complexity and potentially lower power, the parasitic pole arising from routing and device

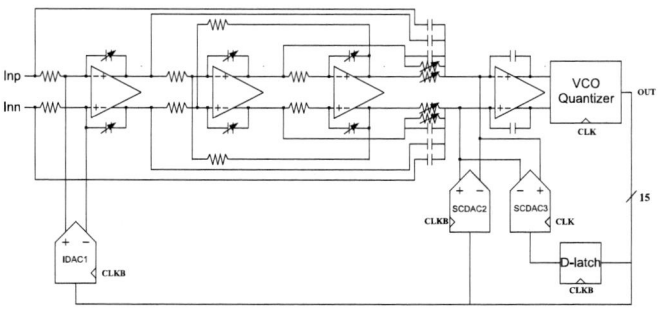

Fig. 1. Modulator architecture

capacitances, reduces the loop filters' phase margin, degrading the stability of the modulator. An active solution based on an opamp with resistive feedback network may still not achieve the necessary wide bandwidth, due to a pole being formed at virtual ground.

The architecture of this work is shown in Fig. 1. The 5th order loop filter has one in-band optimized resonator for optimal loop gain considering VCO non-idealities. In this loop filter topology the summing point is shifted back to the integrator prior to the VCO. This enables capacitive summation of feedforward, input and feedback paths from the VCO quantizer. The feedback signals however must be differentiated to cancel integration. In Fig. 3b the switched capacitor DAC structure is shown. This structure has several advantages; the digital inputs are directly connected to the switches ensuring very low setup time, a single switch for the clock at the output enables low clock network complexity and small parasitic load. Since only the difference in the digital input codes leads to charge being transmitted, the required slew-rate for the amplifier is low. Because there is no reset of the unit capacitors, the input is differentiated. To generate the zero order path, a second DAC is used for discrete time differentiation. They operate on different clock phases such that only one is connected to the amplifier at a time. The benefit of this is that the DAC will not load the amplifier when it is not in use, improving the feedback factor. Whereas a current steering DAC has delay from both the latch and current switches, the proposed structure has a delay mainly determined by the bandwidth of the amplifier. This is very important since the delay from the DAC can cause a CTDS modulator to become unstable. Adjusting the position of a clock edge is power intensive and adds to the complexity.

When operating the VCO quantizer at one fourth of the modulator sampling frequency, the output of the phase quantizers follow an alternating pattern as shown to the right in Fig. 2. Current steering NRZ pulse type DACs are prone to inter-symbol-interference. To reduce signal-dependent switching distortion in IDAC1, alternating switching multiplexers are added to restore an almost minimum-switching thermometer code (a simplified illustration in the block diagram to the left in Fig. 2). This principle is illustrated with state diagrams to the right in Fig. 2. A constant input is input to the VCO quantizer to generate a ramp output function. A 9 level version is shown here for illustrative purposes, while a 15 level was implemented.

The loop filter is synthesized for half a clock period delay. The additional switching delay of IDAC1 is compensated for, but not for SCDAC2 and 3. While the $2^{nd}$ to $5^{th}$ order paths could be set to have a full clock cycle delay the bandwidth requirements of this path would be greatly increased. The out-of-band gain of the noise transfer function (NTF) is set to 2. While a higher out of band gain allows for higher suppression of noise, it also degrades the stability at high input levels and induces more down folding of out-of-band quantization noise due to clock jitter, amplifier and DAC non-linearity. Another concern regarding the out-of-band gain is the loop filter's unity gain frequency. A lower out-of-band gain implies a lower unity gain frequency of the loop filter. Excess loop delay will then have less impact on the phase margin. A higher order loop filter can realize the same loop gain as a lower order one, but with a much lower out-of-band gain. Together with more degrees of freedom with regards to state scaling these are important advantages for a higher order loop filter.

Non-linearity originating in the DAC coupled to the first integrator is the main cause of distortion for a multibit continuous time delta sigma ADC. For a DAC having non-return-to-zero (NRZ) output waveform, not only is the static accuracy important but also the dynamic accuracy. At high sampling frequencies (relative to the process node), the non-linearity contribution from the element switching can be dominant. The switching non-linearity scales with input signal. When mismatch-shaping techniques like Data-Weighted Averaging (DWA) is employed, the contribution from the static non-linearity is shaped, but the dynamic is not. With the exception of [3], achieving high dynamic linearity has proven to be difficult, as was pointed out in [1] with DACs having NRZ output waveform. In this work, high linearity is achieved through the use of large sized current sources. The integrator bandwidth is set to 140MHz for the main DAC. This allows for higher unit element current, which improves the matching and provides a higher gain in the first integrator, justifying the use of a higher power consumption for the modulator front-end. The full scale input voltage is $3V_{ppd}$ while the modulator internal full scale is $2V_{ppd}$, the feedforward capacitor to the $4^{th}$ integrator can then be reduced further improving the

corresponding feedback factor.

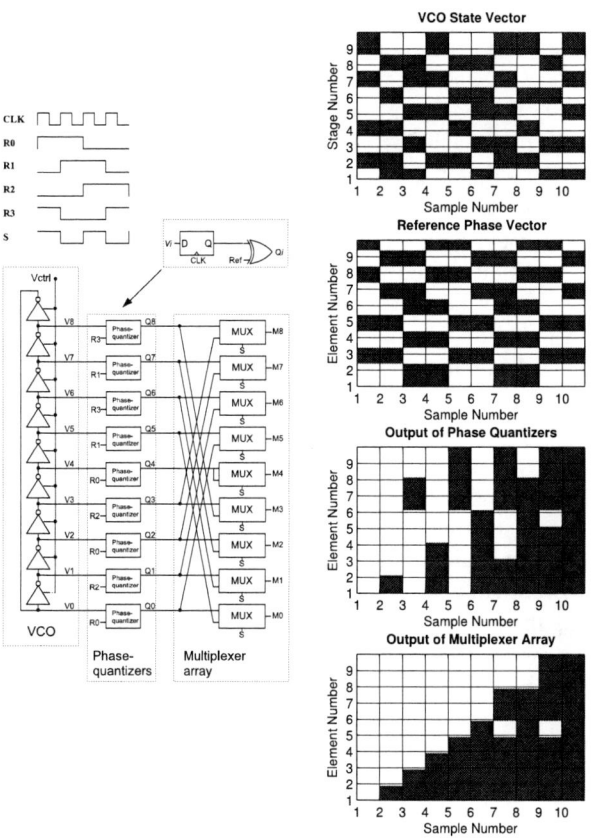

Fig. 2.   VCO quantizer. Left; Block diagram. Right; State diagram

## III. CIRCUIT IMPLEMENTATION DETAILS

The front-end consists of a current steering DAC feeding into a wide-band folded-cascode active-RC integrator. The input common mode voltage is set nominally to 1.3V to facilitate a high overdrive voltage of the IDAC1 NMOS unit current sources and full-swing logic SR-latch switch drivers [3] for high static and dynamic linearity, respectively. To avoid systematic mismatch in the differential output current waveform from the DAC, two inverters are used for buffering the inputs of each SR-latch. The high input common mode voltage puts a burden on the cascoded PMOS common mode current sources used for the DC biasing of the DAC. These must be biased with a low overdrive voltage to provide sufficient output impedance over process, voltage, temperature (PVT) and input common mode variations. This adds significantly to the overall noise budget. However, this design's SNDR is limited by the mismatch in the IDAC1's unit cells. Since there is no calibration of the DACs mismatch, the current sources of IDAC1 are sized large to achieve a mismatch below 0.04% over PVT variations. The IDAC1's current sources are further cascoded to shield their large parasitic capacitance and improve output impedance. In the layout, each current cell is split into four segments in a common centroid fashion. The segmented unit cells are further interleaved to reduce abrupt changes in the DNL characteristic. Current noise from the band gap reference are filtered with a rather large NMOS decoupling capacitor connected to the biasing transistor for the unit elements. The power routing for the unit cells are sized such that each unit cell sees an equal $I \times R$ drop.

Excess phase shift of an active-RC integrator is not only determined by the finite bandwidth of the operational amplifier but also by the phase shift in the input resistors. Due to the short distance between the poly layer and substrate, poly resistors exhibit a parasitic capacitance to substrate. When a high resistive value is used to reduce the required capacitive load of the amplifier, the pole formed by the resistance and parasitic capacitance can easily dominate the excess parasitic phase shift of an integrator. To mitigate this, floating N-wells are placed below the resistors in the second to the fourth integrator to create a series capacitance to the substrate, effectively halving the resistors parasitic capacitance. Since the input referred noise from these integrators are heavily attenuated by the large gain of the first integrator, higher resistances are allowed. High resistive values are not only beneficial for reducing the power consumption of the amplifiers realizing the integrators but also for increasing the gain of the amplifiers. The resistor tuning network used for the input resistors in the $4^{\text{th}}$ integrator is shown in Fig. 3a. To ensure binary coding, as well as a high linearity and an improved frequency response, a current mode binary switchable R2R ladder in shunt with a fixed resistor is employed. Most important for the tunable resistors at this location in the modulator is the excess phase shift. Compared to the conventional tuning resistor method where a series of resistors are switched to the output depending on the tuning code, this approach has a much lower phase shift.

The $2^{\text{nd}}$ and $3^{\text{rd}}$ integrator employ power efficient and low noise telescopic cascode opamps with a relatively low gain of 30dB and unity-gain bandwidth around 500MHz. Thanks to the feedforward coupling from the modulator input to the $4^{\text{th}}$ integrator, the signal swing at these states are very low allowing for single stage amplifiers. Integrator 4 has the highest requirements for speed and needs higher opamp gain due to a feedback factor of 0.25. A Miller compensated cascaded NMOS differential pair achieves the desired unity gain-bandwidth of 2GHz. While this unity gain-bandwidth is higher than needed, it was deliberately over-designed since parasitic phase shift in the first order path affects the overall stability and that this was our first design with this type of ADC architecture. The bias currents in this amplifier are high to push the parasitic poles high enough up in frequency. At a slightly lower unity gain-bandwidth target, a substantial amount of the power consumption could be saved. All passives are made tunable to cover a $\pm40\%$ RC product variation and a $\pm20\%$ variation around the nominal sampling frequency of 500MHz.

The VCO employed here is the same as in [2], and has a separate power supply due to poor PSRR. A high VCO frequency gain results in a better "virtual ground" (lower

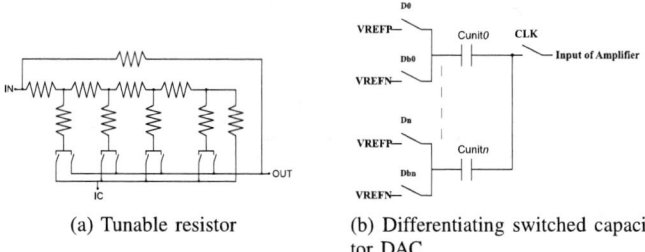

(a) Tunable resistor  (b) Differentiating switched capacitor DAC

Fig. 3.

VCO input swing), and thus lowers quantization noise down-folding and nonlinearity contribution due to the tuning devices intrinsic nonlinearity. In addition, the input capacitors to the fourth integrator can be lowered, resulting in power saving. However, in this design the frequency gain was chosen to be around 240MHz/V to reduce the VCO's power supply sensitivity. Sense amplifier based flip-flops are used for reading out the delay cells' state. The XOR ports and multiplexers are realized with pass gate logic, which for these two types of ports can be realized with compact area, high speed and low power. The modulator core is laid out with focus on achieving minimum routing distance between the VCO quantizer and the feedback DACs to ensure a low digital delay.

## IV. Experimental Results

With a 2MHz input signal the prototype modulator achieves a maximum SNDR of 76dB (Fig. 4). A sweep of input level measuring SNR/SNDR is shown in Fig. 5. The dynamic range, being defined as the input range where SNR>0, is 80dB. The spurs located around 125MHz (Fs/4) are caused by kickback from the VCO, and parasitic coupling from the phase references into the VCO. However, due to being out of band, this doesn't affect the behavior or performance. The out-of-band gain was measured to be 3.3. From post-layout simulations this was revealed to be due to insufficient decoupling of the VCO power supply and the reference voltages. This limited the performance and stability of the modulator at input levels higher than -6dBFS. A micrograph of the modulator core is shown in Fig. 6, the modulator core occupies $1.2\text{mm}^2$. At a nominal supply voltage of 1.8V, the power consumption for the modulator is 55mW analog and 3mW digital.

For power efficiency evaluation, a figure-of-merit defined as $FOM_1 = SNDR + 10 \cdot log_{10} \frac{BW}{Power}$ is used, achieving 158.4dB. This FOM is better suited for thermal noise limited ADC designs[4]. With the commonly employed figure-of-merit, defined as $FOM_2 = \frac{Power}{2^{ENOB} \cdot 2 \cdot BW}$, the efficiency is 567fJ/conversion. In table I this work is compared to state of the art designs using NRZ type current steering main DAC and having a bandwidth higher than 5MHz.

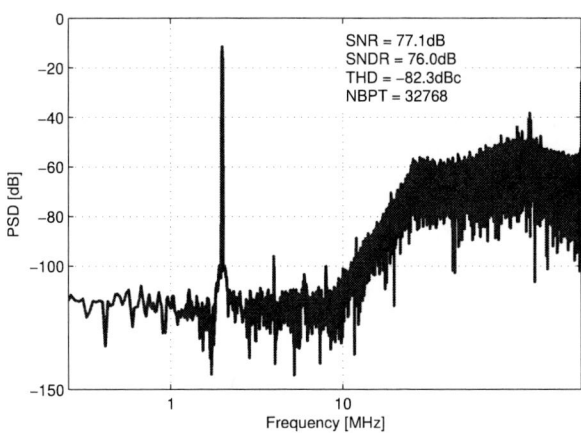

Fig. 4.  Measured spectrum for a 2MHz input signal at -6.7dBFS

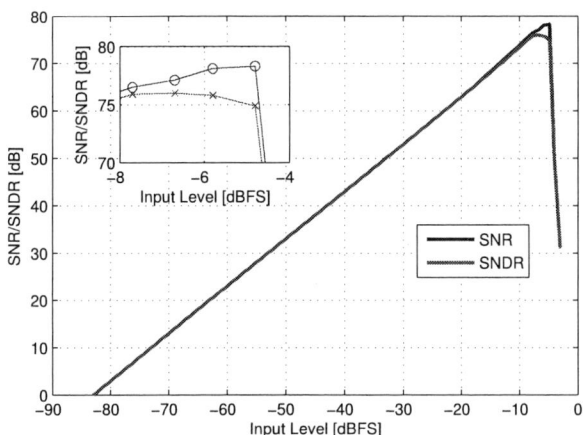

Fig. 5.  Input level versus SNR/SNDR

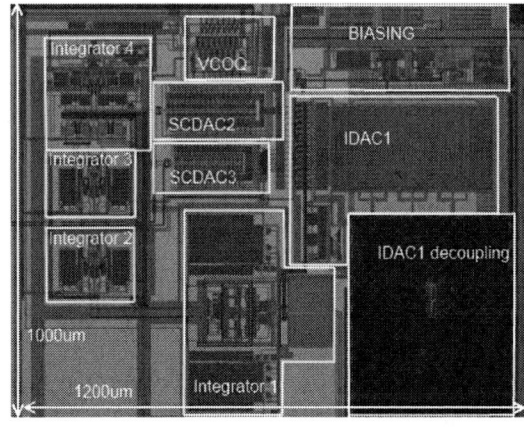

Fig. 6.  Micrograph of the modulator core

## TABLE I
### PERFORMANCE COMPARISON

| Design | Feature Size (nm) | SNR/SNDR/THD (dB/dB/dBc) | BW (MHz) | Power (mW) | FOM$_1$/FOM$_2$ (dB/fJ) |
|--------|-------------------|--------------------------|----------|------------|-------------------------|
| This   | 180 | 78.4/76/-82.3 | 10 | 58 | 158.4/567 |
| [5]    | 180 | 67/65/-69.3* | 16 | 47.6 | 150.3/1002 |
| [6]    | 130 | 76/74/-78 | 20 | 20 | 164/122 |
| [1]    | 130 | 81.2/78.1/-81* | 20 | 87 | 161.7/331 |
| [7]    | 180 | 67.2/64.2/-67.2* | 15 | 20.7 | 152.8/520 |
| [3]    | 180 | 84/82/-87 | 10 | 100 | 162/486 |
| [8]    | 110 | 68.5/62.5/-69.4* | 10 | 5.32 | 155.2/244 |

\*: Estimated from SNDR/SNR difference at measured peak SNDR

achieved very efficiently with the use of a VCO. We propose a capacitive implementation for excess loop delay compensation using high speed differentiating switched capacitor DACs. A switching method to reduce the alternating output from the VCO quantizer is used, which together with careful DAC layout and a low distortion loop filter topology achieves a THD of -82.3dBc at the peak SNDR of 76dB. To our best knowledge, for CTSD modulators having NRZ type current steering main feedback DAC, the THD achieved is better than any other designs relying on good matching.

## REFERENCES

[1] M. Park and M. Perrott, "A 78 dB SNDR 87 mW 20 MHz Bandwidth Continuous-Time ADC With VCO-Based Integrator and Quantizer Implemented in 0.13μm CMOS," *Solid-State Circuits, IEEE Journal of*, vol. 44, no. 12, pp. 3344 –3358, dec. 2009.

[2] M. Straayer and M. Perrott, "A 12-Bit, 10-MHz Bandwidth, Continuous-Time ADC With a 5-Bit, 950-MS/s VCO-Based Quantizer," *Solid-State Circuits, IEEE Journal of*, vol. 43, no. 4, pp. 805 –814, april 2008.

[3] W. Yang, W. Schofield, H. Shibata, S. Korrapati, A. Shaikh, N. Abaskharoun, and D. Ribner, "A 100mW 10MHz-BW CT Sigma Delta Modulator with 87dB DR and 91dBc IMD," in *Solid-State Circuits Conference, 2008. ISSCC 2008. Digest of Technical Papers. IEEE International*, Feb. 2008, pp. 498–631.

[4] R. Schreier and G. C. Temes, *Understanding Delta-Sigma Data Converters*. Piscataway, NJ: IEEE Press, 2005.

[5] V. Singh, N. Krishnapura, S. Pavan, B. Vigraham, N. Nigania, and D. Behera, "A 16MHz BW 75dB DR CT Delta Sigma ADC compensated for more than one cycle excess loop delay," in *Custom Integrated Circuits Conference (CICC), 2011 IEEE*, sept. 2011, pp. 1 –4.

[6] G. Mitteregger, C. Ebner, S. Mechnig, T. Blon, C. Holuigue, and E. Romani, "A 20-mW 640-MHz CMOS Continuous-Time Delta Sigma ADC With 20-MHz Signal Bandwidth, 80-db Dynamic Range and 12-bit ENOB," *Solid-State Circuits, IEEE Journal of*, vol. 41, no. 12, pp. 2641–2649, Dec. 2006.

[7] K. Reddy and S. Pavan, "A 20.7mW continuous-time delta sigma modulator with 15MHz bandwidth and 70 dB dynamic range," in *Solid-State Circuits Conference, 2008. ESSCIRC 2008. 34th European*, sept. 2008, pp. 210 –213.

[8] K. Matsukawa, Y. Mitani, M. Takayama, K. Obata, S. Dosho, and A. Matsuzawa, "A Fifth-Order Continuous-Time Delta-Sigma Modulator With Single-Opamp Resonator," *Solid-State Circuits, IEEE Journal of*, vol. 45, no. 4, pp. 697 –706, april 2010.

## V. CONCLUSION

In this paper we show the design and measurements of a 5$^{th}$ order feed-forward based CTDS modulator with 10MHz signal-bandwidth in 0.18μm CMOS. At 500MS/s the power consumption is 58mW from a 1.8V supply. Quantization is

978-1-4673-6145-3/13 $31.00 © 2013 IEEE

# A 0.1-3GHz Cell-Based Fractional-N All Digital Phase-Locked Loop Using ΔΣ Noise-Shaped Phase Detector

Yao-Chia Liu[1], Wei-Zen Chen[1], Mao-Hsuan Chou[2], Tsung-Hsien Tsai[2], Yen-Wei Lee[1], Min-Shueh Yuan[2]

Department of Electronics Engineering, National Chiao Tung University[1], TSMC[2], Hsinchu, Taiwan

*Abstract* - A 0.1-3 GHz, cell-based, fractional-N ADPLL with ΔΣ noise-shaped phase detector is presented. By dithering the reference phase and quantization phase error through an additional feedback path, linear phase detection and zero stabilization are accomplished without resort to sophisticated time to digital converter (TDC). The measured rms jitter from a 3GHz carrier is 1.9 ps with a multiplication factor of 60. Implemented in TSMC 40nm general purpose superb CMOS technology, the chip size is 280um x 240um.

Keywords: TDC, ΔΣ phase detector, fractional-N ADPLL

## I. INTRODUCTION

Fractional-N frequency synthesizers are widely utilized in wireless and wireline communication systems for its higher frequency resolution and feasibility for direct digital modulation. As the VLSI technologies advance into the nano-meter CMOS arena, all digital PLL (ADPLL) based frequency synthesizers are expected to provide several advantages over conventional semi-digital counterparts, such as easier portability for technology migration and facilitating digital control for performance enhancement. Additionally, by replacing analog charge pump and passive loop filter with digital circuitries, they circumvent voltage headroom issues and are more cost effective for system chip integrations.

In a fractional-N ADPLL, a highly linear phase frequency detector is required to resolve the phase difference between reference input and feedback signal. Conventionally, a time-to-digital converter (TDC) based PFD is adopted [1-2], and it requires calibration [1] or phase-locking schemes [2] to maintain the accuracy of timing reference ladder. Instead of directly increasing the resolution of TDC, a fractional-N PLL using Bang-Bang PD is proposed by introducing an additional phase minimizing loop [3]. As it requires a multi-modulous divider for phase interpolation, the implementation is critical for multi-GHz frequency range operation by cell-based design. Some other noise shaping techniques are proposed recently to improve the PD resolution [4], however, analog intensive designs are still required.

This paper presents a fractional-N ADPLL with ΔΣ noise-shaped phase detector. By circumventing the design complexity of analog intensive linear PD, it is synthesizable using the standard cell library, and is also attractive for technology migration in nanometer CMOS technology.

## II. ARCHITECTURE AND IMPLEMENTATION

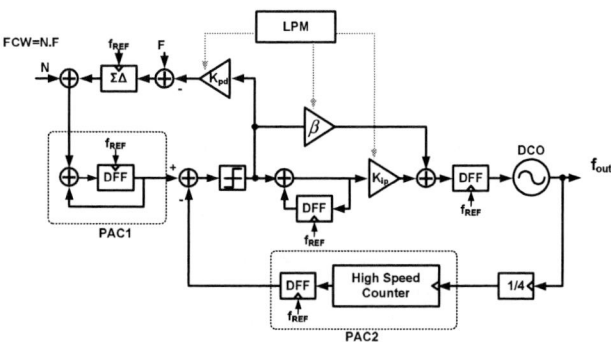

Fig. 1 TDC-less fractional-N all digital PLL architecture

Fig. 1 illustrates the fractional-N ADPLL architecture, which comprises of a dual mode phase frequency detector (DPD), a locking process monitor (LPM), a ring oscillator based digital controlled oscillator (DCO), and phase accumulators PAC1 and PAC2 for the accumulation of reference phase and DCO phase respectively. The frequency multiplication factor of the ADPLL is composed of an integer part ($N$) and a fractional part ($0.F$), which is implemented by interpolating the accumulated reference phase through a 2nd order ΔΣ modulator. To realize linear phase detection through a bang-bang type phase detector, the PD output feeds back to the input of the reference phase accumulator through $K_{pd}$, forming a ΔΣ noise shaping loop of the quantization error during the locking process. As both the PAC1 and PAC2 are ripple counter based architectures, they facilitate cell-based automatic synthesis for the digital design. The detail operations are explained as follows.

978-1-4673-6145-3/13 $31.00 © 2013 IEEE

### A. Fractional-N Frequency Synthesis and ΔΣ Noise-Shaped Phase Detector

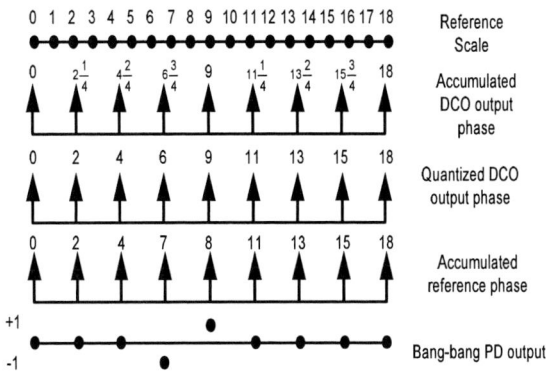

Fig.2 Timing diagrams for fractional-N frequency synthesis ( N = 2.25)

The fractional-N frequency synthesis is achieved by dithering the reference phase accumulator through a 2nd order ΔΣ modulator. Fig. 2 illustrates an example for fractional-N frequency synthesis with a multiplication factor of N.F=2.25. Here the 1st row shows the reference timing scale. The second row is the ideal DCO output phase in the locked state. As the DCO output phase is accumulated by a ripple counter, it will be truncated to integer when sampled by the bang-bang PD for phase comparison. The timing misalignment between reference and DCO output will inevitably lead to quantization error, as is shown in the third row. By feeding back the quantization phase error through the reference feedback path and combining it with ΔΣ phase interpolator, the reference phase accumulator is dithered as shown in PAC1. Thus the bang-bang PD output is randomized and the quantization noise is high pass shaped with a zero mean, as is shown in the 4th row.

The quantization noise of the reference phase and PD is then suppressed by PAC1, forward path integrator, and the low pass nature of DCO. By simulation, the proposed ΔΣ noise-shaped PD provides an effective phase resolution of 6 bits. Compared to the prior art [5], no additional digital to phase converter is required to reduce fractional phase error. Another advantage comes with the new architecture is to dither quantized phase error and reference phase through ΔΣ modulator simultaneously. Thus fixed pattern noise relevant to deterministic jitter can be diminished.

### B. Loop Filter

Typically, ADPLL employs PI loop filter consisting of integral and proportional paths [1-5]. In the locking state, a quantized phase error will inevitably disturb DCO through the proportional path, and thus lead to spurious tones. In contrast to the PI loop filter, only integral path remains in this design in the lock state, and the proportional path is replaced by the additional negative feedback path in the reference phase accumulator. From the perspective of loop stabilization, it adjusts the instantaneous frequency of reference rather than DCO according to the PD output to diminish phase error. Thus the DCO frequency is much less disturbed to reduce frequency spurs. In combination with ΔΣ noise-shaped PD, the reference feedback loop provides a stabilization zero as well as noise shaping advantage of the PD quantization error.

### C. Locking Process Monitor (LPM)

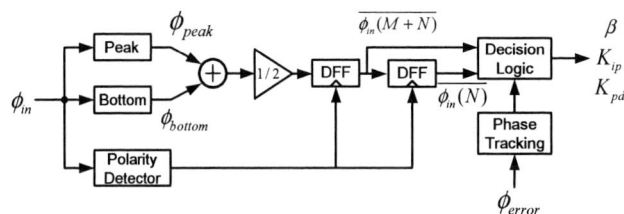

Fig.3 Locking Process Monitor structure

The loop dynamics of the ADPLL is moderated by a locking process monitor (LPM) to accelerate phase locking. Fig. 3 illustrates the circuit architecture of LPM. During frequency initialization, the DPD is operated in linear mode. Meanwhile, the loop filter is configured as a simple forward path $\beta$ dynamically adjusts for high speed locking( $\beta > 0$ ) and the feedback path is disabled (Kpd = 0). The whole system becomes a first order frequency locked loop so as to speed up the updating process of DCO.

As the loop approaches frequency locked, the output of Bang-Bang phase detector will be brought to the vicinity of the targeting frequency. Afterwards, the DPD is operated in the binary mode. The feedback path from PD output to the reference phase accumulator is activated, and the loop filter is reconfigured as an integrator to suppress the high pass noise-shaped quantization noise ( $\beta = 0$ ). The LPM continuously monitors the locking process by sensing the mean value of filter output through a peak and bottom detector, where the peak ( $\phi_{peak}$ ) and bottom ( $\phi_{bottom}$ ) values of $\phi_{in}$ is captured when the polarity of phase error is inverted. As the difference between consecutive samples $\overline{\phi_{in}(N+M)}$ and $\overline{\phi_{in}(N)}$ diminishes gradually, the $K_{ip}$ and $K_{pd}$ will be adaptively adjusted as well to accelerate phase locking without sacrificing its noise performance [6].

### D. Digital Control Oscillator

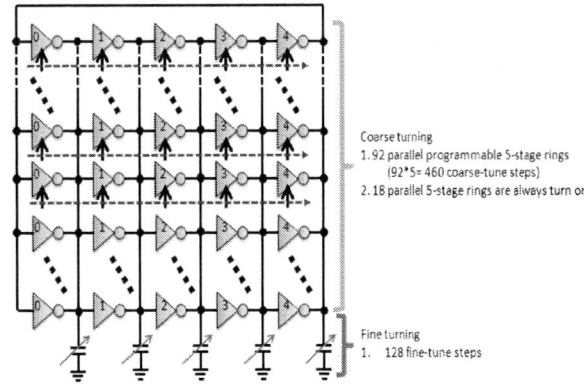

Fig. 4  DCO Architecture

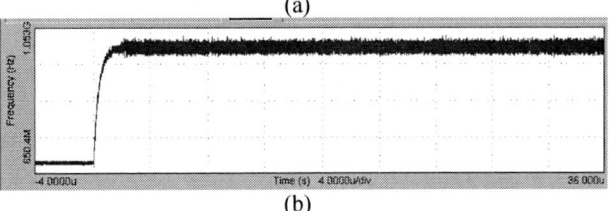

(a)

(b)

Fig.5 Measured locking behavior (a)locking detector signal
(b)frequency hopping dynamic

The circuit schematic of the DCO is shown in Fig. 4, which is basically a 5 stage ring oscillator. It composes of 9 bits inverter banks for coarse tuning followed by 7 bits capacitor bank for fine tuning. Part of the inverter banks are always turned on to maintain a stable oscillation. With the aid of 10 bits $\Delta\Sigma$ interpolation, the frequency resolution is around 800Hz per LSB. Incorporating with 3bits post divider after the DCO, it is capable of delivering 30 X tuning range covering from 0.1-3 GHz.

## III. EXPERIMENTAL RESULTS

Fig. 5 illustrates the ADPLL locking behavior. In Fig. 5(a), the blue line represents the reset signal and the red line stands for built in locking detector signal, the delay between these two curves only 4.16μs. Fig.5 (b) shows the frequency tracking process. Incorporating dynamic loop filter, the locking time is less than 5 μsec. The measured rms jitter at 0.1 GHz and 3 GHz are 2.69 ps and 1.9 ps respectively, as are shown in Fig. 6 (a) and (b). Fig. 7 shows the measured output phase noise performance. The phase noise is -94dBc/Hz at 1 MHz offset from a 1 GHz carrier. The photograph of the implemented test chip

and ADPLL layout are shown in Fig. 8 (a) and (b). The measurement results are summarized in Table I. Table II provides a comparison of this work with the prior art. By incorporating the proposed $\Delta\Sigma$ noise-shaped phase detector, the proposed ADPLL is digitally synthesizable and highly area-efficient, while capable of covering wide frequency range.

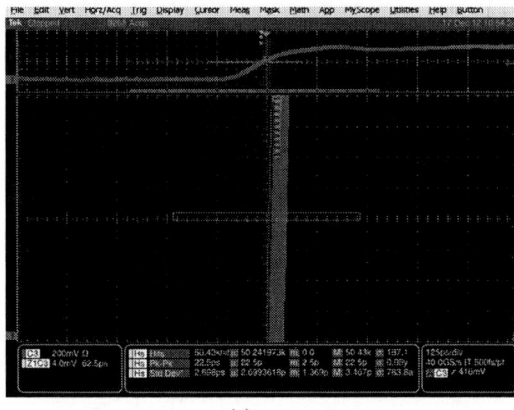

(a)

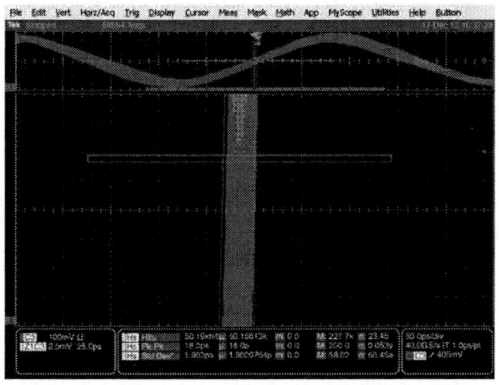

(b)

Fig.6 Measured clock jitter at (a) 0.1  GHz  (b) 3 GHz

TABLE I. PERFORMANCE SUMAARY OF THE ADPLL

| Technology | 45nm GP CMOS |
|---|---|
| Output Frequency | 0.1~3GHz |
| rms jitter @ 3 GHz | 1.9 ps |
| In-band phase noise | -94dBc/Hz |
| Out of band phase noise | -114dBc@10MHz |
| Power | 9.1mW |
| Area | 280um x 240um |
| Locking Time | <5 μsec |

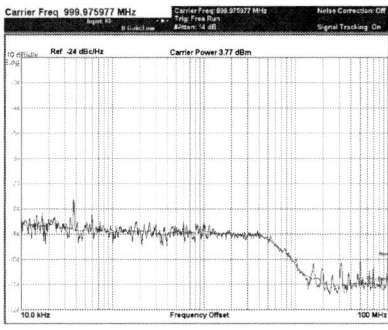

Fig. 7 Measured phase noise @ 1GHz

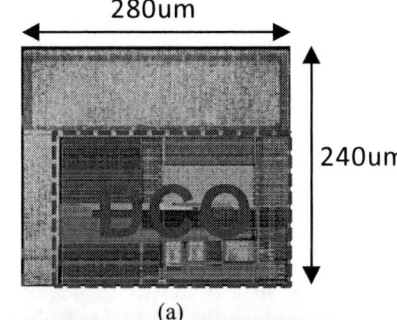

(a)

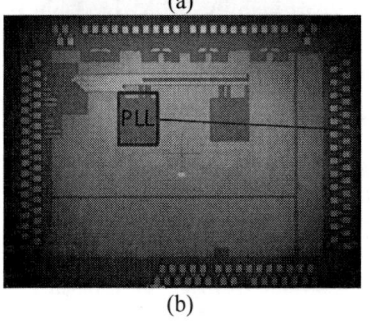

(b)

Fig. 8 (a) ADPLL layout  (b) Test chip photograph

## TABLE II PERFORMANCE COMPARISON WITH OTHER REPORTED PLLS

|  | [3] | [8] | [9] | [10] | This work |
|---|---|---|---|---|---|
| Technology (nm) | 130 | 130 | 90 | 65 | 45 |
| Power [mW] | 14 | 12.8 | 9.6 | 12 | 9.1 |
| Frequency range [GHz] | 1.5~2.25 | 1.9~3.8 | 3.57~4.35 | 2.29~2.92 | 0.1~3 |
| PN10MHz [dBc/Hz] | -108@1MHz | -117@1MHz | -130 | -135 | -114 |
| Active area(mm^2) | 0.7 | 0.087 | 0.34 | 0.24 | 0.067 |
| Locking Time | NA | <22 μ | NA | <30 μ | <5 μ |

## IV. Conclusion

A fractional-N all-digital PLL is implemented through standard cell design flow. Without resort to analog intensive phase detector, the phase error can be $\Delta\Sigma$ noise-shaped by the additional feedback loop on the reference integral path. A locking process monitor is incorporated to accelerate phase locking while maintaining its noise performance. The experimental prototype is digitally synthesizable and highly area efficient with wide output frequency range.

## Acknowledgment

This project is supported by NSC under contract 101-2220-E-009-015. The authors would like to thank TSMC university shuttle program for chip fabrication.

## REFERENCE

[1] Robert Bodan Stazweski, "Spur free all-digital PLL in 65nm for mobile phones," *IEEE ISSCC*, pp. 52-53, 2011.

[2] Frank Opteynde, " A 40 nm CMOS all digital fractional-N synthesizer without requiring calibration," *IEEE ISSCC*, pp. 346-347, 2012.

[3] M. Ferriss and M. P. Flynn, "A 14mW Fractional-N PLL modulator with an enhanced digital phase detector and frequency switching scheme, " ISSCC Dig. Tech. Papers, pp. 353–353, 2007.

[4] Jon-Phil Hong, et al, "A 0.004mm², 250 μW, $\Delta\Sigma$ TDC with time difference accumulator and a 0.12 mm² 2.5 mW bang-bang digital PLL using PRNG for low-power SoC applications,"*IEEE ISSCC*, pp. 240-241, 2012.

[5] Nenad Pavlovic, and Jos Bergervoet, "A 5.3 GHz digital-to-time converter based fractional-N all-digital PLL," *IEEE ISSCC*, pp. 54-55, 2011.

[6] Song-Yu Yang, Wei-Zen Chen, and Tai-You Lu "A 7.1mW 10GHz all-digital frequency synthesizer with dynamically reconfigurable digital loop filter in 90nm CMOS, " *IEEE JSSC*, pp. 578-586, 2010.

[7] J.A. Tierno, et al., "A Wide Power Supply Range, All Static CMOS All Digital PLL in 65nm SOI," IEEE J. Solid-state Circuits, vol. 43, no., 1 pp. 42-51, 2008

[8] Jaewook Shin and Hyunchol Shin et al., "A 1.9–3.8 GHz Fractional-N PLL Frequency Synthesizer With Fast Auto-Calibration of Loop Bandwidth and VCO Frequency," *IEEE JSSC*, pp. 665-675, 2012.

[9] Mi-Jeong Park; Byonghoon Mhin; Seong-Do Kim;Moon-Yang Park; Hyunku Yu "A 4-GHz all digital fractional-N PLL with low-power TDC and big phase-error compensation," IEEE Custom Integrated Circuits Conference, 2011, pp. 1-4.

[10] Liangge Xu, et al.,"A 2.4-GHz low-power all-digital phase-locked loop," IEEE Custom Integrated Circuits Conference, 2009, pp. 331-334.

# A Direct-Battery Hookup, Fully Integrated Stereo Headphone Module with 82 mW Output Power and 110 dB PSRR

Khaled Abdelfattah[1], Sherif Galal[1], Iuri Mehr[1], Alex Jianzhong Chen[1], Ahmet Tekin[2], Xicheng Jiang[1], Todd L. Brooks[1]

[1]Broadcom Irvine, CA; [2]Semtech, San Jose, CA

khaled@broadcom.com

*Abstract* — **A complete stereo ground-referenced headphone module that supports direct battery hookup is integrated on a 40 nm mobile baseband SoC. Several techniques were employed to guarantee the reliability of the module circuitry under high output swing and limited safe operating regions in this low-voltage technology. Additional techniques to reduce area and enable low-cost integration were also employed. The module delivers 3.24 Vpp to a 16Ω load (82 mW) and achieves 100 dB dynamic range (DR), 110 dB PSRR, and 84 dB THD+N with an area of 0.675 mm² on the SoC.**

## I. INTRODUCTION

The integration of audio and power management functions with the baseband processing and radio functions is becoming more critical for reducing the footprint and minimizing the cost of a cellular SoC [1-2]. This level of integration presents a unique set of challenges for device reliability, cost, and performance. For example, in 40 nm CMOS, the I/O devices can withstand only 3.3V across the gate-to-source or gate-to-drain junctions. This constraint can potentially limit how much power an audio system can deliver without sacrificing device reliability.

In order to attain the lowest implementation cost in a cellular system, it is desirable to reduce the PCB footprint and the number of I/O pins. Hence, integrating analog functions and high-speed complex digital functions in a fine geometry CMOS process can achieve this goal. However, fine geometry has higher cost per unit area; therefore, a key challenge for integration is to minimize the analog die area. Another important aspect for a cellular audio system is to allow direct battery (3.1V–4.5V) hookup. This requirement can pose significant challenges regarding device reliability, which will be addressed in this paper. The battery connection can also affect the overall performance of the audio module if it does not possess sufficient power supply rejection. Since the module shares the same battery connection with the cellular RF PA, which draws high current when it is on, a high power supply rejection (PSRR) is required to suppress the time-division duplex (TDD) bursts due to the periodic PA turn-on and turn-off times [5].

This paper describes a Class-AB headphone module in 40 nm that delivers a peak power of 82 mW with 1.25 mA per channel quiescent current. This high output power is achieved without sacrificing device reliability.

The driver achieves very high PSRR of 110 dB, a peak THD+N of 84 dB and a dynamic range (DR) of 100 dB. It also maintains 80 dB THD+N at output power levels up to 75 mW. Techniques and design choices to maintain the devices in the safe operating regions as well as to save analog die area will be discussed in the following sections.

## II. ARCHITECTURE

Fig. 1 depicts the complete stereo headphone module including two ground-referenced headphone channels powered from a shared power management unit (PMU). Each channel is driven by 6.5 MHz, 3-bit oversampled data generated by a digital sigma-delta modulator. The PMU is powered off the battery directly and contains two low dropout regulators and an inverting charge pump. The output of LDO2 (VDD2 = 2.8V) serves as the supply for the DAC and the front end of the driver, whereas the output of LDO1 (VDD1 = 2.8V) powers the charge pump and the output stage of the driver. The negative output of the charge pump (VSS = -2V) enables the driver to provide ground-referenced outputs while eliminating large external capacitors. Overcurrent detection is integrated into each channel to flag excessive current due to short-circuit events and improper insertion of the headphone jack. The module is powered down when an overcurrent condition is detected to avoid depleting the battery or damaging the chip metallization. Pop/click suppression circuitry is integrated into each channel to minimize the audible pops generated

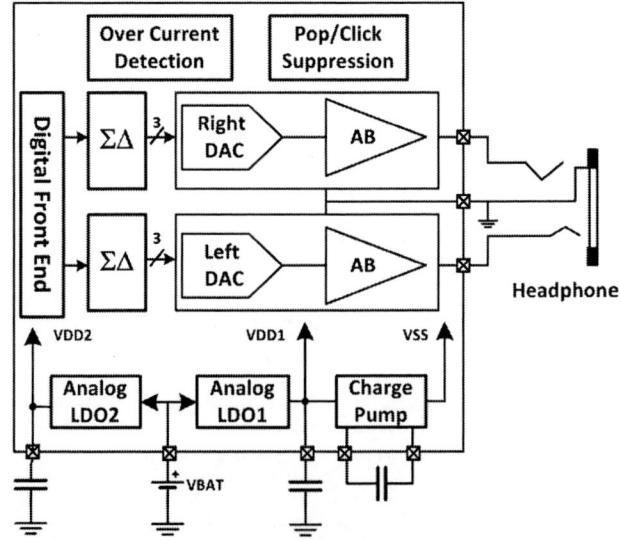

Fig. 1. Stereo Headphone Module

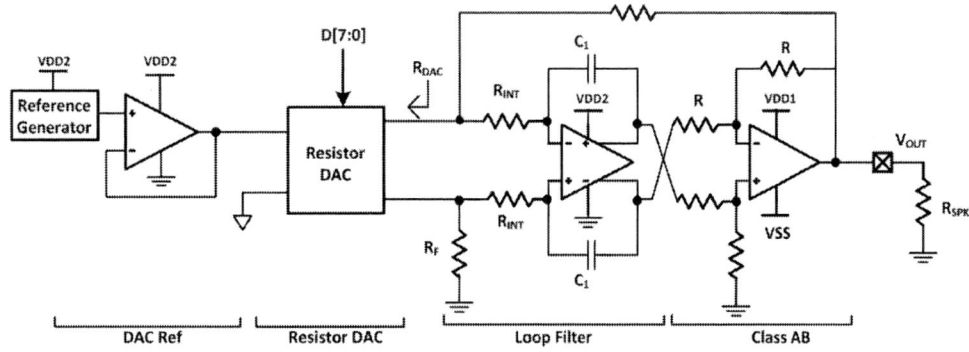

Fig. 2. Detailed Channel Architecture

during power-up and power-down. This is accomplished by sequentially turning individual blocks on and off [4].

Fig. 2 shows the architecture of a single headphone channel. A resistor DAC is driven by 9-level, shuffled, oversampled data running at 6.5 MHz. A reference buffer powered from VDD2 is used to generate the DAC reference. The DAC output feeds into a first-order loop to drive the headphone load. In addition to providing the driving capability, the loop acts as a low-pass filter for the out-of-band DAC quantization noise.

The closed-loop feedback circuit contains an integrator, followed by a Class-AB driver, and a feedback path that is resistive overall. To enhance the PSRR of the entire module, LDO2 (on Fig. 1) is used to power the loop filter and the DAC reference buffer, while the Class-AB driver is powered from LDO1 and the charge pump (CP). The LDO1 output voltage ripple caused by charge-pump switching is attenuated by the loop gain. LDO1 helps to maintain a high module PSRR by rejecting the effects of TDD transmission bursts on the battery.

## III. DESIGN CHALLENGES

Integrating this module with other baseband functions in 40 nm CMOS presents significant challenges that can affect audio performance. A major challenge is due to the limited safe operating range of the devices in this 40 nm technology. This means that for any I/O device to operate safely, the gate-to-drain and gate-to-source junction voltage should not exceed 3.3V. Several techniques were employed to help achieve the target performance without sacrificing design reliability.

Additionally, several design choices and techniques were adopted to help save die area. In the following subsections, more information on maintaining device reliability and reducing area is presented.

### A. Reliability Challenges

The Class-AB driver used in this design is shown in Fig. 3. It consists of a folded cascode input stage with a floating battery ($M_1$ and $M_2$) and a push-pull second stage ($M_p$ and $M_n$) with Miller compensation. A quiescent current ($I_Q$) bias control circuit [5] is used to generate the bias voltages for $M_1$ and $M_2$ to accurately set the quiescent current in the output transistors.

If the Class-AB driver is powered directly from the battery, the output devices will have to operate with a much higher rail-to-rail supply. This poses even more challenges to maintain device reliability. To relax this requirement, LDO1 is used to generate the Class-AB positive supply from the battery, thus reducing the maximum voltage across the Class-AB output transistors. With this arrangement, the driver's rail-to-rail supply is 4.8V which is necessary to generate the required output power. Since the rail-to-rail supply is still higher than the technology limit for the gate-to-source and gate-to-drain junction breakdown voltages, protection mechanisms for these junctions are employed.

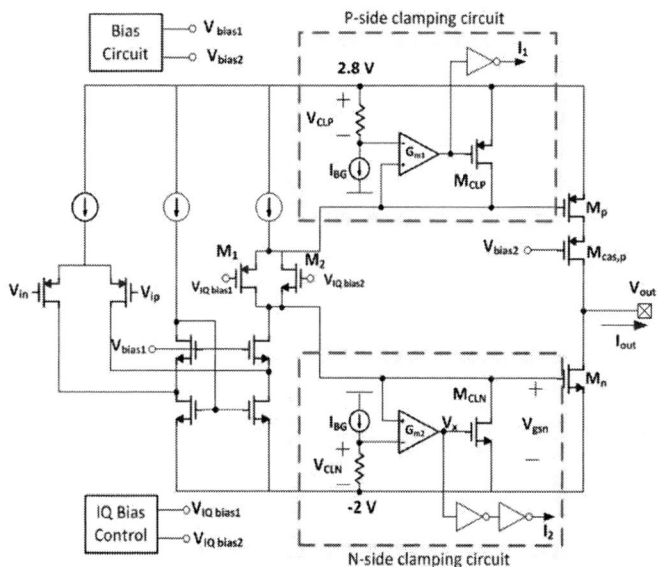

Fig. 3. Class-AB Amplifier

During normal operation, $V_{gd}$ of $M_p$ can exceed 3.3V. This reliability challenge is solved using $M_{cas,p}$. This will clamp the drain of $M_p$ relative to the cascode device-gate bias during negative output signal peaks. During positive signal peaks, $M_{casp,p}$ is in the linear region and will not cause reliability concerns. Even in the absence of the cascode device on the negative side, the peak gate-to-drain voltage of $M_n$ is less than 3.3V.

Sizes for $M_p$ and $M_n$ are chosen to support the maximum load current during normal headphone operation. However, if a faulty headphone is used with a significantly lower resistance, then $M_p$ and $M_n$ will flow higher load

978-1-4673-6145-3/13 $31.00 © 2013 IEEE            205

current and eventually break down because the gate-to-source junction voltage will exceed 3.3V.

A solution to this issue is to use the P-side and N-side clamping circuits as shown in Fig. 3. For the sake of illustration, only the N-side is discussed. During driver operation, the gate-to-source voltage of $M_n$ ($V_{gsn}$) is compared against a reference ($V_{CLN}$). If the load current is within the normal range, then $V_{gsn}$ is smaller than $V_{CLN}$ and the clamping circuit is deactivated by pulling the gate of $M_{CLN}$ ($V_x$) to the negative rail. As the load current through $M_n$ increases due to a reduced load resistance, $V_{gsn}$ approaches $V_{CLN}$ and the clamping circuit is activated forcing $V_{gsn} = V_{CLN}$. Fig. 4 shows the simulation result of the clamping circuit. The output current is swept from –200 mA to 200 mA, while monitoring $V_x$, $V_{gsn}$, and $V_{out}$.

The gate voltages of both $M_{CLP}$ and $M_{CLN}$ are digitized to binary levels using skewed inverters to generate signals

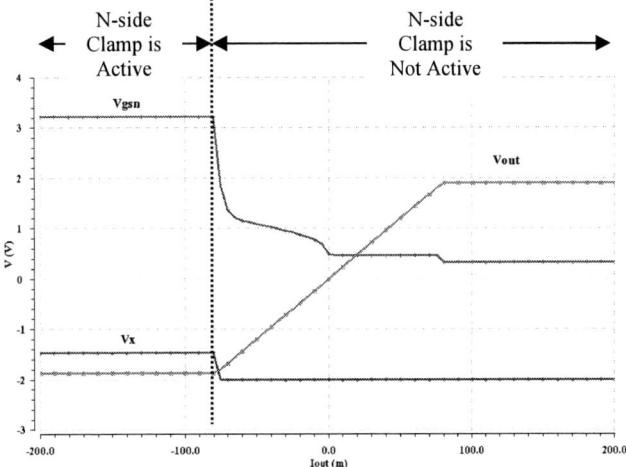

Fig. 4.   Simulation Results for the Clamping Circuit

($I_1$ and $I_2$) to indicate that either clamp is activated as shown in Fig. 3.

Relying solely on the clamping circuit to limit the output current is not sufficient because it only protects the output devices when their $V_{gs}$ approaches a certain limit. Due to process and temperature variation, the load current caused by a faulty speaker can increase above the reliable limit without the clamps being activated. Although this situation poses no breakdown threat for the output devices, the excessive output current can eventually damage the chip metallization due to electromigration. Therefore, an overcurrent detection function is included to monitor the output current and compare it to a reference current on the chip.

Fig. 5(a) shows the overcurrent detection scheme on the P-side. A similar scheme is also used on the N-side. The circuit has an accurate current mirror (K:1) that uses an amplifier to equalize $V_{ds}$ of the main device and the mirrored device. The mirrored current gets compared against $I_{REF}$. If the mirrored current exceeds $I_{REF}$, then $I_3$ is pulled high and an overcurrent condition is detected. The N-side overcurrent detector works the same way to generate the $I_4$ signal. Finally, the $I_1$ and $I_2$ signals in Fig. 3 are

combined with $I_3$ and $I_4$, as shown in Fig. 5(b), to generate an interrupt signal (IRQ) that is sent to a digital state machine to shut off the headphone module.

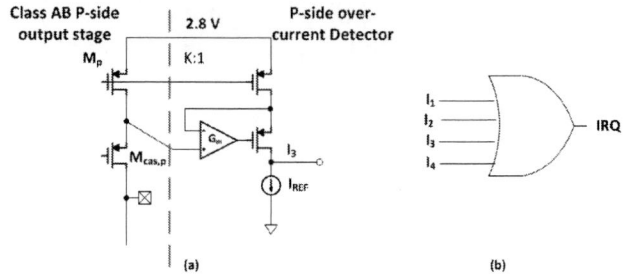

Fig. 5.   (a) Overcurrent Detector Circuit , (b) Logic Function to Generate Final Interrupt

### B.   Area-Saving Techniques

As shown in Fig.2, a resistor DAC is used because of its inherent linearity. However, the resistor DAC, which is illustrated in Fig. 6(a), can be very expensive to use. In the figure, eight units are connected in parallel to realize a 9-level resistor DAC; each unit has a pair of resistors and four switches. Each of the resistors in each unit is 8x larger than the overall required $R_{DAC}$ resistance, which is determined by the noise requirement. Hence, a large area corresponding to a total resistance value of 64 x $R_{DAC}$ is required to realize an equivalent resistance with a value of 1 x $R_{DAC}$. To reduce this area in our implementation while keeping the total equivalent resistance the same, as shown in Fig. 6(b), the total DAC resistor ($R_{DAC}$) is partitioned into a series

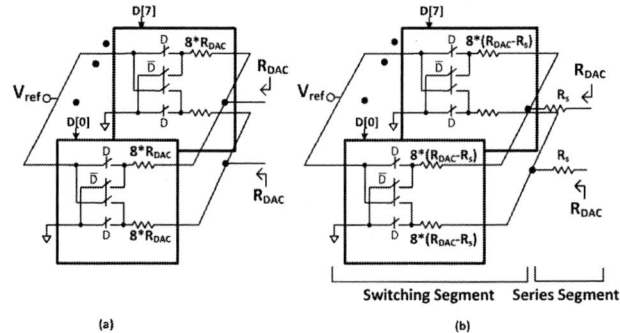

Fig. 6.   Resistor DAC Area Savings: (a) Without Partitioning, (b) With Partitioning

segment that has a single resistor of value $R_S$ and a switching segment with an equivalent resistance equal to ($R_{DAC} - R_S$). Therefore, the total DAC resistance is kept the same and equals $R_{DAC}$. The equivalent resistance of the switching segment is used to implement the 9-level DAC with a unit resistance equal to 8 x ($R_{DAC} - R_S$). This allowed a much smaller area, corresponding to a reduced total resistance of {64 x ($R_{DAC} - R_S$) + $R_S$} to realize the same equivalent DAC resistance of 1 x $R_{DAC}$.

Additional area savings can be achieved by lowering the integrating capacitor in Fig. 2. The loop bandwidth is given by $1/R_{eq}C_1$, where $R_{eq}$ is given by:

$$R_{eq} = R_{INT} + R_F + \frac{R_{INT}R_F}{R_{DAC}} \qquad (1)$$

The resistor T-network formed by $R_{DAC}$, $R_{INT}$, and $R_F$ is used to produce a high equivalent resistance ($R_{eq}$) which allows the capacitor $C_1$ to scale down while maintaining the same loop bandwidth. This bandwidth needs to be low enough to attenuate out-of-band DAC noise and sufficiently high to preserve THD and PSRR performance.

In order to save additional area, the charge pump provides only the negative supply for the Class-AB stage, while an LDO provides the positive supply. This also reduces the pin count and the number of external capacitors on the board.

## IV. EXPERIMENTAL RESULTS

The module was fabricated in 40 nm CMOS. Fig. 8 shows a plot of the measured A-weighted THD+N versus the output power for a 1 kHz signal with a 16Ω load. The

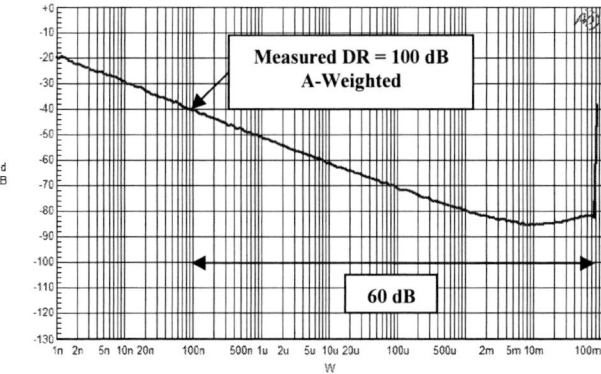

Fig. 8. A-weighted THD+N Versus Output Power (16Ω Load)

1% THD power level is 82 mW, and the peak THD+N (at $P_{out} = 10$ mW) is 84 dB.

The figure also shows that the system achieves a dynamic range equal to 100 dB. The THD+N exceeds 80 dB for output power levels up to 75 mW. The measured quiescent current per channel is 1.25 mA. Fig. 9 shows 110 dB measured PSRR at 217 Hz, which is the TDD burst frequency. Fig. 10 shows the die micrograph for the stereo headphone module, which measures 0.675 mm².

All measurements and comparisons with state-of-the-art Class-AB and Class-G drivers are summarized in Table 1.

## REFERENCES

[1] S. Wen et al. "A 5.2 mW, 0.0016% THD up to 20 kHz, Ground-Referenced Audio Decoder with PSRR-Enhanced Class-AB 16Ω Headphone Amplifiers," *2012 IEEE Symposium on VLSI Circuits*, pp. 20-21, 2012.

[2] M. Hammes et al. "A GSM Baseband Radio in 0.13 um CMOS with Fully-Integrated Power Management," *ISSCC Dig. Tech Papers*, pp. 264-265, 2007.

[3] V. Dhanasekaran et al. "Design of Three-Stage Class-AB 16Ω Headphone Driver Capable of Handling a Wide Range of Load Capacitance," *IEEE Journal of solid-State Circuits*, Volume 44, Issue 6, pp. 1734-1744, June 2009.

[4] J. Chen et al. "A 62 mW Stereo Class-G Headphone Driver with 108dB Dynamic Range and 600uA/channel Quiescent Current," *ISSCC Dig. Tech Papers*, pp. 182-183, 2013.

[5] S. Galal et al. "A 60 mW Class-G headphone Driver for Portable Battery-Powered Devices," *IEEE journal of Solid state Circuits*, Volume 47, Issue 8, pp. 1921-1934, August 2012.

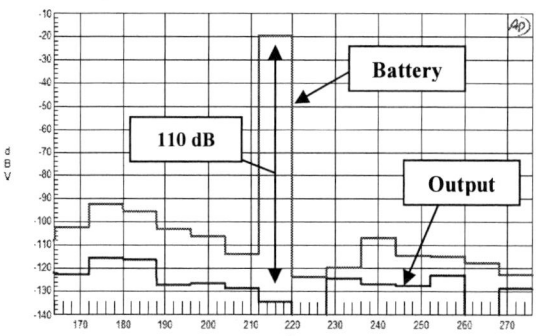

Fig. 9. PSRR of the Complete Headphone Module at 217 Hz Power (16Ω Load)

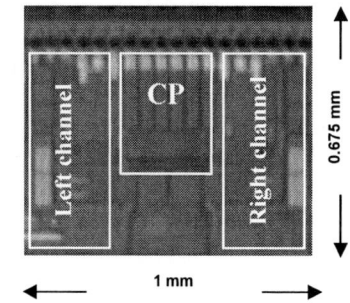

Fig. 10. Die Micrograph

Table 1: Performance Summary and Comparison

|  | [1] | [3] | [4] | [5] | This Work |
|---|---|---|---|---|---|
| **Process** | 40 nm | 130 nm | 180 nm | 180 nm | 40 nm |
| **Peak output voltage ($V_{pp}$)** | 1.4 | 1.6 | 2.81 | 2.77 | 3.24 |
| **Output power (mW)** | 15.6 | 20 | 62 | 60 | 82 |
| **Integrated CP/DAC** | Yes/Yes | No/No | No/No | Yes/No | Yes/Yes |
| **Quiescent current/channel ($I_{Q1}$) (mA) (with DAC reference buffer current and CP included)** | 1.17 | – | – | – | 1.25 |
| **Quiescent current/channel ($I_{Q2}$) (mA) (without DAC reference buffer nor CP current)** | – | 1.46 | 0.6 | 1.15 | 0.84 |
| **Dynamic range (A-weight) (dB)** | 100 | 92 | 108 | 111 | 100 |
| **Peak THD+N (dB)** | –91 | –84 | –88 | –95 | –84 |
| **PSRR ($f = 217$ Hz, 200 m$V_{pp}$) (dB)** | 90 | – | - | 120 | 110 |
| **Class-AB or Class-G** | AB | AB | G | G | AB |

# A Fast-Locking Digital DLL with a High Resolution Time-to-Digital Converter

Dandan Zhang [1,2], Hai-gang Yang [1], Zhujia Chen [1], Wei Li [1], Zhihong Huang [1,2], Lijiang Gao [1], Wenrui Zhu [1,2]

[1] Institute of Electronics, Chinese Academy of Sciences, Beijing 100190, China

[2] University of the Chinese Academy of Sciences, Beijing 100049, China

*Abstract*-A fast-locking digital delay-locked loop (DLL) is presented in this paper. By adopting a novel high resolution Time-to-Digital Converter (TDC), the time for generating fine-tuned codes is reduced to two clock cycles. Thus the total locking time is greatly reduced to 8 input reference clock cycles and remarkably shortened by 80% to 94.6% compared to previous closed-loop architectures. The proposed DLL has been fabricated in a 0.13μm CMOS technology and operates from 80MHz to 450MHz. The measured RMS and peak-to-peak jitters are 2.3ps and 10ps respectively.

Keywords: fast-lock, closed-loop DLL, TDC, multiphase

## I. INTRODUCTION

Delay-Locked Loops (DLLs) have been widely used in high performance digital system synchronization [1], high-speed clock/data recovery (CDR) [2] and DRAM interfaces application [3] for their fast locking time, unconditional stability, and excellent jitter performance. DLLs can be categorized as three types: analog, mixed-signal and digital. Analog and mixed-signal DLLs could achieve better jitter and skew performance, yet digital DLL is more suitable in SOC systems due to its insensitivity to the process variation and power management. Besides, the digital DLLs have a significant advantage of short locking time which means short start-up time or response time for digital systems. Although digital DLLs have the merits of shorter locking time, they are still not fast enough to satisfy some high performance digital system requirements.

Some approaches are proposed to improve the locking time of digital DLLs. Ref. [4] adopts a sequential search algorithm such as the counter-controller scheme which is composed of several simple D-type flip-flops. The realization of DLL is easy, but the locking time increases exponentially as the number of control bits increases. To accelerate the locking process, Ref. [5] provides a binary search algorithm and the locking time increases linearly with the number of control bits, but the locking time is still too long. Ref. [6] introduces a two-stage architecture with a combination of TDC and shift registers to achieve a shorter locking time than [5]. The TDC circuit estimates the input clock period and completes coarse-stage locking in only one clock period. Owing to the low time resolution of TDC, the DLL adopts conventional binary search algorithm approach to generate fine codes. However, binary search algorithm can only tune the fine codes once in one reference clock period. Furthermore, the fine tuning time will increase as the number of fine codes increases. Ref. [7] adopts a clock-synchronized-delay (CSD) to achieve a shorter locking time of fine stage. However, it cannot track temperature or supply variations once it is in lock because of its open-loop architecture. Consequently, a high

resolution TDC is greatly desired for fine codes generation to shorten the total locking time while ensuring a high time resolution performance of DLL. Additionally, a closed-loop architecture is also desired for higher process voltage temperature (PVT) robustness.

This paper proposes a fast-locking closed-loop DLL with a high resolution TDC. The high resolution TDC can generate fine codes in two input reference clock cycles, which greatly reduces the fine-locking time. The total locking time is remarkably reduced compared to [6]. Furthermore, it can work stably against PVT variations due to its closed-loop architecture, thus achieves higher PVT robustness than [7]. The measured locking time is 8 input reference clock cycles and remarkably reduced by 80% to 94.6% compared to previous closed-loop DLLs. The proposed DLL has been realized in a 0.13μm CMOS technology and operates from 80MHz to 450MHz. The measured RMS and peak-to-peak jitters are 2.3ps and 10ps respectively.

## II. PROPOSED DLL ARCHITECTURE

The architecture of the proposed DLL is shown in Fig 1. It is composed of a digital-controlled delay line (DCDL), a coarse-code generator (CCG), a fine-code generator (FCG), and a phase detector (PD). The DCDL comprises four same hierarchy delay units (HDUs). The four HDUs are controlled by same control codes, so the proposed DLL can provide four accurate orthogonal output clocks. Each of the HDUs is composed of a coarse delay unit (CDU) and a fine delay unit (FDU). The CCG is composed of a coarse-tuned TDC [8] and a coarse-code register (CCR), which is used to complete coarse delay line tuning. Unlike conventional DLL structure [8], which adopts a BSR to generate fine codes, a high resolution fine-tuned TDC and a fine-code BSR are adopted in the FCG to complete fine delay line tuning. Furthermore, the fine-code BSR can also retune the fine codes in the presence of PVT variations.

In this design, the locking procedure of the proposed DLL is divided into 2 stages: coarse locking and fine locking. The coarse locking stage is divided into two steps. In the first step, coarse-tuned TDC generates coarse-tuned control codes Q [15:0] by measuring input reference clock period between two sequential rising edges of the input reference clock. Then in the second step, the CCR loads the coarse-tuned control codes to DCDL. After this stage, the delay difference between input reference clock CLK_REF and output feedback clock CLK360 is less than a delay step of a coarse delay line, thus the coarse locking of DLL is completed.

In the second stage, the fine locking is accomplished by a high resolution fine-tuned TDC which utilizes the same delay

978-1-4673-6145-3/13 $31.00 © 2013 IEEE     208

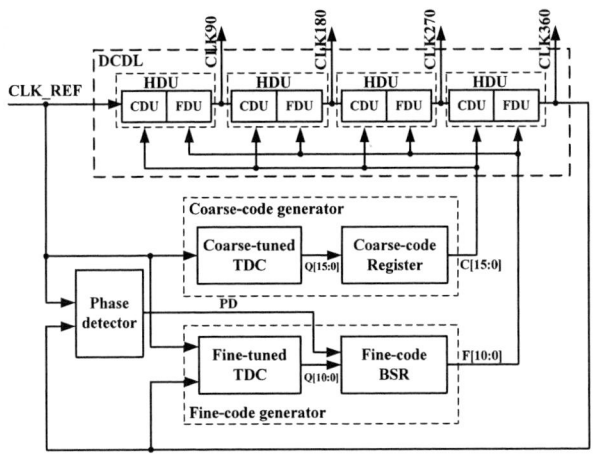

Fig. 1 Architecture of proposed DLL

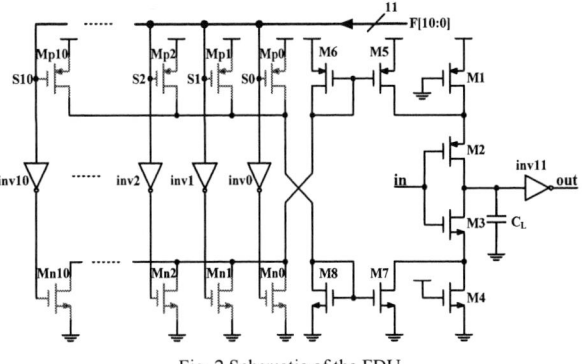

Fig. 2 Schematic of the FDU

cells as the fine delay line. This stage is also divided into two steps. In the first step, the proposed high-resolution fine-tuned TDC firstly produces fine-tuned control codes Q [10:0] according to the remained phase difference between reference clock CLK_REF and feedback clock CLK360. Then in the second step, the fine-code BSR loads the fine-tuned control codes to DCDL. After this step, the difference between CLK_REF and CLK360 is less than the resolution of DLL. In other words, the DLL is locked after the second step.

Theoretically, each step operation could simply need one clock cycle, thus the total locking time is 4 clock periods similarly as [7]. Practically, control codes taking completely effect to delay the DCDL is not immediate but at the next clock cycle. For achieving a high delay precision, we choose that each step costs a time of two clock cycles in the proposed DLL. Thus the total locking procedure costs 8 clock cycles. The proposed DLL could get locked at the 9th clock cycle. In some non-ideal conditions, there may be some deviation of fine-codes generated by the high-resolution TDC due to PVT variations. If it happens, the fine-code BSR can retune the fine control codes until the DLL gets locked. In this case, the total locking time will be a little longer.

## III. CIRCUIT IMPLEMENTATION

### A. Digital Control Delay Line (DCDL)

The DCDL is composed of a coarse delay line which consists of four same CDUs, and a fine delay line which consists of four same FDUs. For achieving a wide operating frequency range, the delay step of the coarse delay line is long. As a result, the time resolution is low. But the FDU can compensate for resolution problem. Fig. 2 shows the schematic of FDU, which comprises a current-starved inverter and a delay-fixed inverter (inv11). The current-starved inverter is controlled by a digitally adjustable current mirror pair which is tuned by controlling transistor arrays (Mp0-Mp10, Mn0-Mn10). The transistors M1 and M2 are always turned on to provide an intrinsic current for the current-starved inverter. The parallel-connection controlling transistors Mp0-Mp10 and Mn0-Mn10 are controlled by control bits S0-S10 respectively. When the transistors Mp0-Mp10 and Mn0-Mn10 are turned on, the drain-source currents of M7 and M8 increase. Then M5 and M6 mirror the currents of M7 and M8, and increase the charge/discharge current of the current-starved inverter. The delay time of the current-starved inverter will decreases following the

charge/discharge current with a relation of:

$$\Delta t \propto \frac{C_L}{\Delta I} \qquad (1)$$

Where $\Delta t$ and $\Delta I$ represent the variation of the delay time and charge/discharge current of the current-starved inverter respectively, $C_L$ is the parasitic capacitances at the output node of current-starved inverter. By properly choosing the sizes of transistors in FDU, the DCDL can obtain a high resolution of less than 15ps by simulating.

### B. Proposed High Resolution Fine-tuned TDC

As mentioned above, the locking procedure of the proposed DLL includes coarse locking stage and fine locking stage. The coarse locking stage is carried out by the coarse-code generator, which consists of a coarse-tuned TDC and a coarse-code register. The coarse-tuned TDC [8] generates coarse codes by measuring input reference clock period between two sequential rising edges of the input reference clock. The coarse-tuned TDC could generate coarse codes in two input reference clock cycles, which greatly shortens the coarse locking time. However, the delay resolution of the coarse-tuned TDC equals to the delay step of a coarse delay line, which is too low to be used for generating the fine-code, so a high resolution TDC is newly introduced to generate fine codes in this paper.

Fig. 3(a) and (b) show the block diagram and timing diagram of proposed high-resolution fine-tuned TDC circuit respectively. It consists of 11 fine-tuned TDC cells. Each of fine-tuned TDC cells is composed of two FDUs and a D-type flip-flop (DFF), in which the FDU is adopted to achieve a high resolution. Timing resolution and power consumption are in a trade-off relationship. One of the two FDUs in a fine-tuned TDC cell is regarded as datum cell with a minimum delay by setting all control codes F [10:0] to zero. The other is regarded as variable cell with m-stage delay by setting m control codes to high, where the designed value of m is from 0 to 10.

The fine codes generation is carried out following the coarse codes generation. After the coarse locking, the feedback clock CLK360 leads a little reference clock CLK_REF. The operation procedure of the fine-tuned TDC is as follows: Firstly, the input reference clock CLK_REF passes through the datum cell in 11 fine-tuned TDC cells in parallel and has a delay of the minimum time $T_{min}$. Simultaneously, feedback delay clock CLK360 passes through the variable cell with different delay in 11 fine-tuned TDC cells in parallel and has a delay of 11 different times $T_{min} + \Delta t$ , $\cdots$ ,

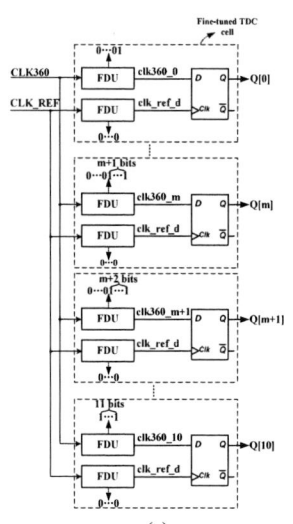 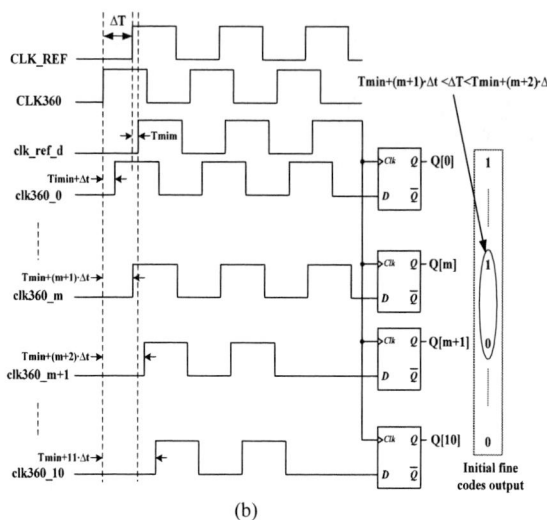

(a)                      (b)

Fig. 3 (a) Block diagram of high resolution TDC circuit, (b) Timing diagram of the proposed high resolution TDC

$T_{\min} + (m+1)\cdot \Delta t$, $T_{\min} + (m+2)\cdot \Delta t$, $T_{\min} + 11\cdot \Delta t$. Secondly, the DFF detects the phase difference between delayed reference clock signal clk_ref_d and delayed feedback clock signal clk360_m in parallel to generate "1" or "0". If signal clk_ref_d leads clk360_m, the DFF outputs "0", if signal clk_ref_d lags clk360_m, the DFF outputs "1". If Q [m] is 1 and Q [m+1] is 0, it means the time difference $\Delta T$ between clk_ref_d and clk360_m satisfies:

$$T_{\min} + (m+1)\cdot \Delta t < \Delta T < T_{\min} + (m+2)\cdot \Delta t \qquad (2)$$

Then the fine-tuned control codes of the fine delay line is settled just to Q [0···01···1], where the number of "1" is m while "0" occupies the leaving position of the 11 bits, the fine locking will be completed by loading the fine-tuned control codes to the fine delay line. And the DLL will get locked then.

The high resolution TDC can not only generate the fine codes in two input reference clock cycles which quickens the locking process, but also make it possible to achieve high time resolution. In addition, the fine-code BSR can retune the fine-tuned control codes against PVT variation until the DLL gets locked.

## IV. MEASUREMENT RESULTS

The proposed fast-locking DLL is fabricated in a 0.13μm CMOS technology. Fig. 4 shows the chip micrograph of the proposed DLL. The core area is 0.2 mm×0.4 mm.

Measurement results show that the DLL achieves a locking frequency range of 80MHz to 450MHz. All measurements described further are at the square root of this frequency range (≈180MHz). Fig. 5 shows the locking behavior of the proposed DLL at 180MHz. From Fig.5 we can see that the locking time of the proposed DLL is 8 input reference clock cycles. Fig. 6 shows the measured four-phase outputs when the proposed DLL works at 180MHz.

Fig. 7 shows the measured output jitter at 180MHz. The RMS and peak-to-peak jitters are 2.3ps and 10ps respectively. Fig. 8 shows the measured p-p and RMS jitters at different frequencies, which indicates that the variation range of p-p jitters and RMS jitters are 8.75ps ~ 14ps and 2ps ~ 3.4ps at the operation frequency range from 80MHz to 450MHz respectively.

Table I compares the proposed circuit with previously reported designs. From Table I we can see that [7] achieves

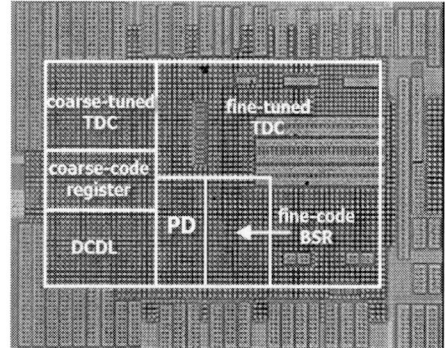

Fig. 4 Chip micrograph of the proposed DLL

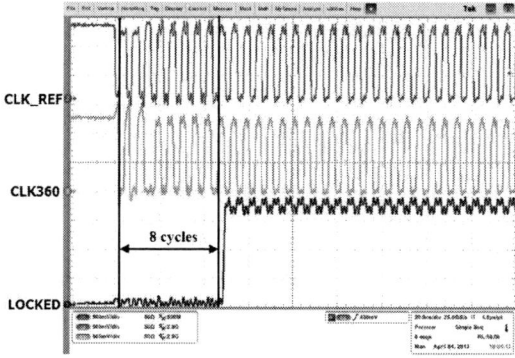

Fig. 5 Locking behavior of the proposed DLL at 180MHz

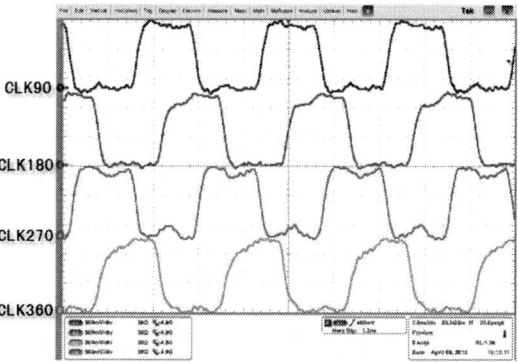

Fig. 6 Measured four-phase outputs at 180 MHz

978-1-4673-6145-3/13 $31.00 © 2013 IEEE      210

the fastest locking time, however, its open-loop architecture may not track PVT variations. Compared with other closed-loop DLLs, the proposed DLL shortens the locking time by 80% to 94.6%. As well, the locking time advantage is owing to the proposed high resolution fine-tuned TDC.

## V. CONCLUSION

A fast-locking closed-loop Delay-Locked Loop is newly introduced in this paper. By adopting a novel high resolution Time-to-Digital Converter, the time for generating fine-tuned codes is reduced to two input reference clock cycles, which greatly reduces the fine-locking time. The total locking time is greatly reduced to 8 input clock cycles and remarkably reduced by 80% to 94.6% compared to previous closed-loop architecture. And the closed-loop architecture can work stably against PVT variations. The proposed DLL has been realized in a 0.13μm CMOS technology and operates from 80MHz to 450MHz. The measured RMS and peak-to-peak jitters are 2.3ps and 10ps respectively.

## ACKNOWLEDGEMENTS

The authors gratefully acknowledge National Science and Technology Major Project of China (2013ZX03006004) and National Natural Science Foundation of China (61106025).

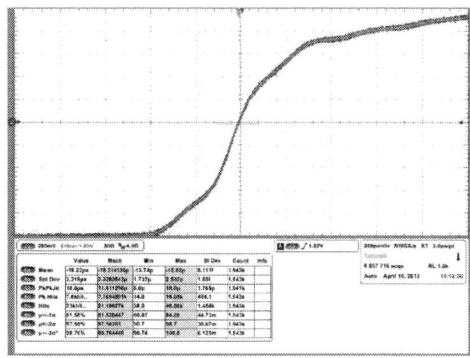

Fig. 7 Measured output jitter at 180 MHz

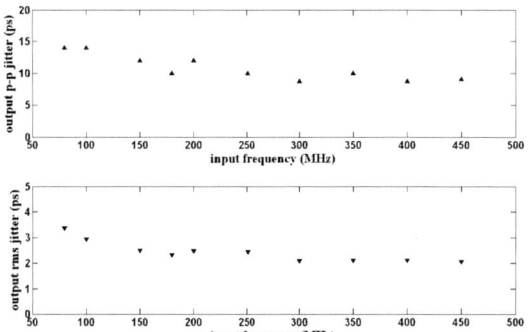

Fig. 8 Measured p-p and rms jitters versus input frequencies

TABLE I PERFORMANCE COMPARISON

|  | [7] CICC'10 | [9] JSSC'12 | [10] TSCAS'12 | [11] TSCAS'10 | [12] CICC'07 | This work |
|---|---|---|---|---|---|---|
| Process (μm) | 0.18 | 0.044 | 0.18 | 0.13 | 0.13 | 0.13 |
| Supply (V) | 1.8 | 1.5 | 1.8 | 1.5 | 1.2 | 1.5 |
| Type | open-loop | open-loop | closed-loop | closed-loop | closed-loop | closed-loop |
| Locking time (clock cycles) | 3 ~ 10 | 150 | 51 | 42 | 40 | 8 |
| Pk-Pk jitter (ps) | 14 @1GHz | 45.6 @1.08GHz | 20.9 @800MHz | 60 @30MHz | 40 | 10 @180MHz |
| RMS jitter (ps) | N/A | 6.13 @1.08GHz | 2.81 @800MHz | N/A | 6.88 | 2.3 @180MHz |
| Time resolution (ps) | 12 | N/A | N/A | <10 | 10 | 15 |
| Frequency range (MHz) | 100 ~ 1000 | 1100 | 400 ~ 800 | 30 ~ 1000 | 333 ~ 800 | 80 ~ 450 |
| Power (mW) | 64 @1GHz | 4.1 @667MHz | 19 @800MHz | 1.5 @30MHz | 16 @800MHz | 26 @180MHz |

## REFERENCES

[1] M.-H. Hsieh, L.-H. Chen, and S.-I. Liu, "A 6.7MHz-to-1.24GHz 0.0318 mm2 Fast-Locking All-Digital DLL in 90nm COMS," IEEE Int. Solid-State Circuit Conf, pp. 244-246, Feb. 2012.

[2] H. H. Chang, R. J. Yang, and S.-I. Liu, "Low Jitter and Multirate Clock and Data Recovery Circuit Using a MSADLL for Chip-to-Chip Interconnection," IEEE Trans. on Circuit Syst. I,Reg. Papers, vol. 51, pp. 2356-2364, Dec. 2004.

[3] L. H., D. K. S., H. T. C., Z. J., J. M. G., and I. T., "A 2.5 V CMOS Delay-Locked Loop for an 18 Mbit, 500 Megatyte/s DRAM," IEEE J. Solid-State Circuits, vol. 29, Dec. 1994.

[4] B.-S. Kim and L.-S. Kim, "100MHz All-Digital Delay Locked Loop for Low Power Application," IEEE Electronics Letters, vol. 34, Sep. 1998.

[5] R.-J. Yang and S.-I. Liu, "A 40-550MHz Harmonic-Free All-Digital Delay-Locked Loop Using a Variable SAR Algorithm," IEEE J. Solid-State Circuits, vol. 42, pp. 361-373, Feb. 2007.

[6] K.-H. Cheng, Y.-L. Lo, and W. F. Yu, "A Mixed-Mode Delay-Locked Loop for Wide-Range Operation and Multiphase Outputs," IEEE Int. Symp. Circuits and Systems (ISCAS), vol. 2, pp. 196-199, May. 2003.

[7] M.-J. Kim and L.-S. Kim, "A 100MHz-to-1GHz Open-Loop ADDLL with Fast Lock-Time for Mobile Application," IEEE Custom Integrated Circuits Conference (CICC), pp. 1-4, Sept. 2010.

[8] C. Zhujia, Y. Haigang, L. Fei, and W. Yu, "A Fast-Locking All-Digital Delay-Locked Loop for Phase/Delay Generation in an FPGA," Journal of Semiconductors, vol. 32, pp. 139-146, Oct. 2011.

[9] L. Hyun-Woo, C. Hoon, S. Beom-Ju, K. Kyung-Hoon, K. Kyung-Whan, K. Jaeil, et al., "A 1.0-ns/1.0-V Delay-Locked Loop With Racing Mode and Countered CAS Latency Controller for DRAM Interfaces," Solid-State Circuits, IEEE Journal of, vol. 47, pp. 1436-1447, 2012.

[10] R. Kyungho, J. Dong-Hoon, and J. Seong-Ook, "A DLL With Dual Edge Triggered Phase Detector for Fast Lock and Low Jitter Clock Generator," Circuits and Systems I: Regular Papers, IEEE Transactions on, vol. 59, pp. 1860-1870, 2012.

[11] L. Wang L and C. H. L, "An Implementation of Fast-Locking and Wide-Range 11-bit Reversible SAR DLL," IEEE Transactions on Circuits System II: Express Briefs (TSCAS-II), vol. 57, p. 421, 2010.

[12] J.-H. Bae, J.-H. Seo, H.-S. Yeo, J.-W. Kim, J.-Y. Sim, and a. H.-J. Park, "An All-Digital 90-Degree Phase-Shift DLL with Loop-Embedded DCC for 1.6Gbps DDR Interface," IEEE Custom Integrated Circuits Conference (CICC), pp. 373-376, 2007.

# A Stochastic Sampling Time-to-Digital Converter with Tunable 180-770fs Resolution, INL less than 0.6LSB, and Selectable Dynamic Range Offset

James S. Tandon[1], Takahiro J. Yamaguchi[1,2], Satoshi Komatsu[1], Kunihiro Asada[1]

VDEC-D2T, The University of Tokyo, Tokyo, Japan 113-8656 [1]

Advantest Laboratories, Ltd. 48-2, Matsubara, Kamiayashi, Aoba-ku, Sendai, Miyagi, Japan [2]

*Abstract*—We introduce a stochastic time-to-digital converter (TDC) that has 180-770fs tunable resolution, less than 0.6LSB INL, and selectable dynamic range offset. Previous arbiter-based TDCs have fine resolution but small dynamic range which is difficult to calibrate. Our approach uses comparators as decision elements to precisely control dynamic range offset.

*Index Terms*—stochastic, comparator, comparator group, offset voltage, time-to-digital converter, TDC, variation

## I. INTRODUCTION

As digital and mixed signal circuit designs move to higher and higher frequencies, timing accuracy becomes essential for correct operation. To measure and compensate for small timing errors on the order of femtoseconds to picoseconds, high resolution time-to-digital converters (TDCs) are used. Increasing timing jitter can cause bit error rate to increase in high speed serial links [1]; detecting phase error between two clock edges accurately can affect phase lock loop performance [2], [3]; timing error can introduce distortion in analog-to-digital converters [4]; these are just a few examples where high resolution time-to-digital conversion can be applied. Achieving subpicosecond resolution in CMOS requires advanced design techiques because it is less than the delay of a single inverter. Process variation has a strong effect on the accuracy of high resolution time-to-digital converters as small variation in a single digital gate delay can cause significant nonlinear features at subpicosecond resolution.

Several different architectures have been explored in recent literature for sub-inverter delay, time-to-digital conversion. The time residue amplifier was used in a two-step conversion process to realize high resolution, but it requires a complex design and calibration to ensure proper operation [5]. Vernier delay line methods can achieve sub-inverter delay but transistor mismatch needs mitigation [6]. Stochastic arbiter based converters harness transistor mismatch to achieve good linearity and fine resolution but have a very small dynamic range[2], [7]. Another deterministic converter uses mismatched SR-latches to induce timing bias in individual arbiters; it achieves a better dynamic range than stochastic arbiters, however, process variation can introduce nonlinear features in the transfer function [8].

This paper describes a time-to-digital converter that uses stochastic comparators instead of traditional stochastic arbiters as the decision circuits. While stochastic comparators have been applied to very finite range analog-to-digital conversion [9], this is the first time they are applied in a stochastic time-to-digital converter. After describing the background theory

and the advantages of using comparators, we describe our test circuit and test setup. Next, we detail our results with a discussion of their significance. Finally we conclude the paper.

## II. STOCHASTIC COMPARATOR TDC

A stochastic, comparator-based time-to-digital converter harnesses process variation to develop the time-to-digital transfer function. This section describes the comparator as a timing decision circuit. Next, it develops the theory for how timing and comparator voltage offset are related. Then it describes the method for tuning the center time offset and the method for tuning the resolution.

### A. Comparator Decision Circuit

The decision circuit of a stochastic time-to-digital converter takes two input step functions, then outputs a 'one' or 'zero' depending on which input arrived first. Our circuit uses a clocked comparator with a differential input clock, a single ended input $V_{IN}$, and a reference voltage $V_{LVL}$. The comparator schematic is shown in figure 1. When using a comparator for subpicosecond measurement, the slope of the input signal and the slope of the reference clock become significant. This is shown in figure 2. Given a clock signal with period $T_{period}$, when the clock signal transitions, the following function is calculated and latched at time $n = \lfloor t/T_{period} \rfloor$:

$$OUT[n] = (V_{IN}(t) > V_{LVL} + V_{OS}) \qquad (1)$$

where $OUT[n]$ is the digital output of the comparator latched at clock cycle $n$ and $V_{OS}$ is the voltage offset of the comparator. This decision element determines whether the signal $V_{IN}(t)$ is greater than $V_{LVL} + V_{OS}$ before the clock latches, or after.

### B. Stochastic Comparator Group

The stochastic comparator group forms the core of the time-to-digital converter. Process variation introduces a different voltage offset bias for each comparator. This offset voltage is a random process which approximates a normal distribution when enough comparators are implemented. If a comparator $i$, has a random offset voltage, $V_{OS,i}$, then the timing of the comparator decision circuit will be affected. Figure 2 shows how a voltage offset, $V_{OS}$, relates to a time offset, $T_{OS}$. These can be placed in a symmetric layout to form a stochastic comparator group.

The outputs of the stochastic comparator group are encoded with a summing encoder as shown in figure 3. The offset voltage transfer function of a comparator group follows a normal distribution function:

$$F(v) = \frac{N_{comp}}{\sqrt{2\pi}} \int_{-\infty}^{x} \exp\left(-z^2/2\right) dz \qquad (2)$$

$$x = \frac{v - \mu_{V_{OS}}}{\sigma_{V_{OS}}}$$

where $N_{comp}$ is the number of comparators, $\mu_{V_{OS}}$ is the mean of the offset voltage from $V_{LVL}$, and $\sigma_{V_{OS}}$ is the standard deviation of the offset voltage. A step function, which will have a finite slope in practice, is applied to $V_{IN}(t)$ of the TDC, so:

$$v = \frac{dV_{IN}(t)}{dt}(t - t_0) + v_0 \qquad (3)$$

where $v_0 = V_{LVL}$. If equation (3) is substituted into (2) and simplified, then:

$$x = \frac{t - \mu_{T_{OS}}}{\sigma_{T_{OS}}} \qquad (4)$$

$$\mu_{T_{OS}} = \frac{V_{LVL} - \mu_{V_{OS}}}{dV_{IN}(t)/dt}$$

$$\sigma_{T_{OS}} = \frac{\sigma_{V_{OS}}}{dV_{IN}(t)/dt}$$

and we have a tranfer equation for the distribution function $F(t)$ to characterize the output of the time-to-digital converter.

The dynamic range of the stochastic TDC is defined as the largest region of the normal CDF transfer function that approximates a linear function within a defined error tolerance. For our circuit, we define our tolerance to be INL < 0.6LSB.

### C. $V_{LVL}$ versus Dynamic Range Offset

The average offset voltage, $\mu_{V_{OS}}$, is defined relative to the reference voltage, $V_{LVL}$ of the comparator. Therefore the offset time $\mu_{T_{OS}}$ shifts relative to $V_{LVL}$; This can be seen in equation (4). There is a benefit because it is possible to control the dynamic range offset of the TDC by controling $V_{LVL}$. The dynamic range time offset of the TDC is proportional to:

$$T_{dyn,OS} = \frac{V_{LVL}}{dV_{IN}(t)/dt}. \qquad (5)$$

Thus, by controlling $V_{LVL}$, it is possible to extend the measurable dynamic range by shifting the mean of the normal distribution, $\mu_{T_{OS}}$.

### D. $V_{IN}$ Slope versus Time

The dynamic range of the stochastic TDC is directly proportional to the standard deviation of the stochastic comparator group offset voltage. We also can see from equation (4) that the time offset distribution is inversely proportional to $\frac{dV_{IN}(t)}{dt}$. Therefore, it is possible to control the resolution as well. A troublesome side effect of changing the resolution is that the dynamic range time offset $T_{dyn,OS}$ will shift as well. However, this can be mitigated by shifting it with $V_{LVL}$.

Using stochastic comparators affords several advantages to the designer. First, it is easier to analyse and characterise

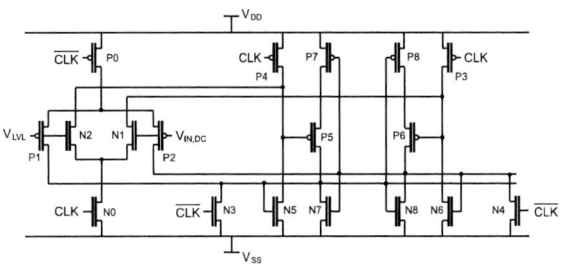

Fig. 1. Schematic of the comparator used in the stochastic time-to-digital converter.

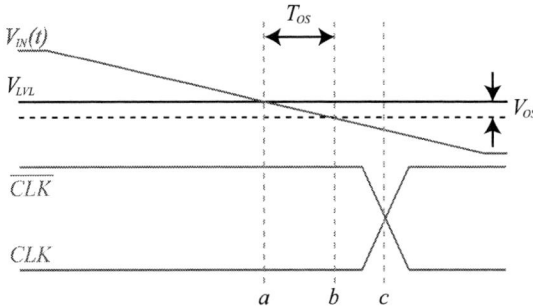

Fig. 2. Timing diagram that shows how the both the time offset and slope of $V_{IN}$ and the clock signal affect the detected time difference. Input $V_{IN}(t)$ crosses $V_{LVL}$ at time (a), but the comparator offset voltage, $V_{OS}$ will shift the time offset $T_{OS}$ to time (b). The comparator output will be determined by whether time (b) occurs before or after time (c).

using the slope of the input step function. Second, the input time offset can be precisely calibrated by changing an input reference voltage level. Third, the adjustable dynamic range time offset affords the ability to measure larger, total dynamic range by tuning $V_{LVL}$. Finally, converter resolution can be changed by controlling the input slope.

## III. EXPERIMENTAL SETUP

This section describes the experimental set up and the test algorithms used to test the TDC. Our test chip integrated 63 comparators and it was fabricated in the eShuttle 65 nanometer process; the die photo is shown in figure 4. A 25 MHz differential clock and the signal $V_{IN}$ were provided by a bit error rate tester (Agilent ParBERT 81250). The $V_{IN}$ input signal time offset was controlled by a mechanical phase shifter (Waka 01X0557-00) that swept a range of 141 picoseconds

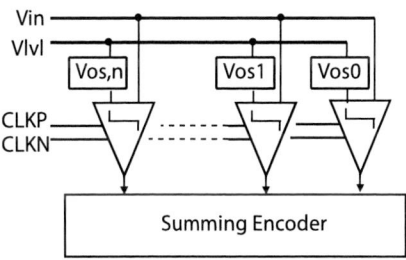

Fig. 3. Schematic of the proposed time-to-digital converter that uses stochastic comparators. The result is encoded by summing the comparator outputs.

978-1-4673-6145-3/13 $31.00 © 2013 IEEE

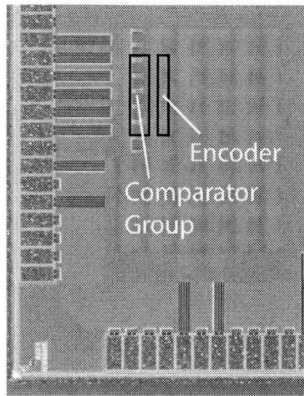

Fig. 4. Photograph of our test chip with the stochastic comparator group under test with 63 comparators.

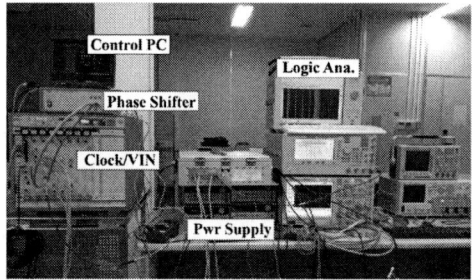

Fig. 5. Photograph of our test bench. An RF shield box was used to limit spurious interference.

in 200 femtosecond intervals. The reference voltage $V_{LVL}$ was provided by a DC power supply (Agilent 6612C) and was stepped from 590mV to 610mV in 5mV increments. The stochastic, comparator-based time-to-digital converter output was acquired by a logic analyzer (Agilent 16822A). The control PC used LabVIEW (National Instruments) to manage the test apparatus. The slope from the ParBERT was measured using an oscilloscope (Tektronix DSA71254B) to verify the input function, $V_{IN}(t)$, delivered to the TDC. The test setup is shown in figure 5.

Using this testing strategy, the controllability of the resolution, and the controllability of the TDC's dynamic range time offset is demonstrated. Additionally, it is shown that the TDC can be used to determine the input slope of $V_{IN}$.

## IV. RESULTS & DISCUSSION

The results of the implemented stochastic time-to-digital converter are shown here. The input slope was measured by an oscilloscope for each $V_{IN}$ input function that was provided to the TDC. Two example slopes are shown in figure 6. Then the TDC transfer function was measured in 200 femtosecond steps for five reference voltage levels input to $V_{LVL}$ (only three of five transfer functions are shown). The measured transfer functions are shown in figure 7 and figure 8 for two sample slopes of $dV_{IN}/dt = 3.03$mV/ps and $dV_{IN}/dt = 13.2$mV/ps. The measured time resolution was from 180fs to 770fs in our experiment, but it is possible to decrease the resolution further by providing a more gradual input slope. There is a

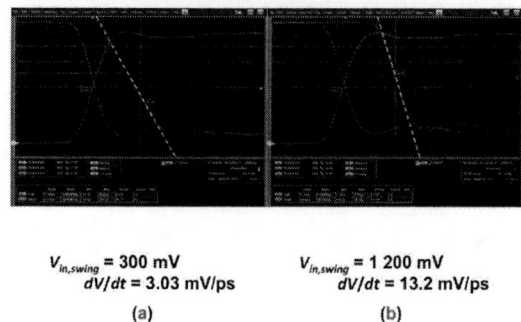

$V_{in,swing} = 300$ mV
$dV/dt = 3.03$ mV/ps
(a)

$V_{in,swing} = 1\,200$ mV
$dV/dt = 13.2$ mV/ps
(b)

Fig. 6. The reference waveforms captured by oscilloscope to determine the slope delivered to $V_{IN}(t)$. By varying the input peak-to-peak waveform from the ParBERT, the slope was controlled. The scale of the input function in green was changed to attain a better slope measurement. Detected slopes were (a) 3.03mV/ps and (b) 13.2mV/ps.

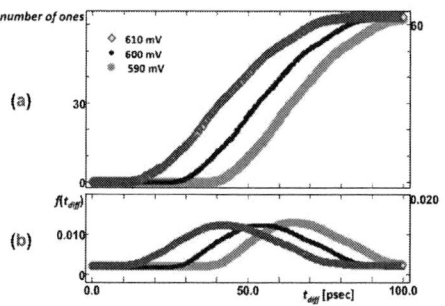

Fig. 7. Time-to-digital converter transfer functions with a scope measured $dV_{IN}(t)/dt = 3.03$mV/ps for $V_{LVL} = 590, 600, 650$mV (three of five transfer functions are shown). The detected resolution for this slope is 499fs.

fundamental tradeoff: a higher slope implies higher resolution, but reduces the dynamic range, and the tunability of the dynamic range offset. Reducing the slope will decrease the resolution, but increase the dynamic range.

We see from figures 7 and 8 that changing the input $V_{LVL}$ changes the center time offset of the transfer function. The probability density functions were estimated using the procedure specified in [10]. By measuring the change in $\mu_{T_{OS}}$ relative to $V_{LVL}$, it is possible to capture the slope of the input edge by using our TDC. Two sample slopes of 3.03mV/ps and 13.2mV/ps that are measured by the circuit are shown in figure 9. The correlation coefficient between the oscilliscope measurements and our TDC measurements for five slopes is shown to be 0.9991 in figure 10. The integral nonlinearity is shown to be less than 0.6LSB for an input slope of 13.2mV/ps when $V_{LVL} = 700mV$ in figure 11.

## V. CONCLUSION

We introduced a stochastic sampling time-to-digital converter (TDC) that has 180-770fs measured tunable resolution, integral nonlinearity less than 0.6LSB INL, and a selectable dynamic range offset. Previous stochastic, arbiter-based TDCs have fine resolution, but very small dynamic range that are difficult to calibrate. Our approach uses stochastic, clocked comparators as decision elements which allow for precise

978-1-4673-6145-3/13 $31.00 © 2013 IEEE

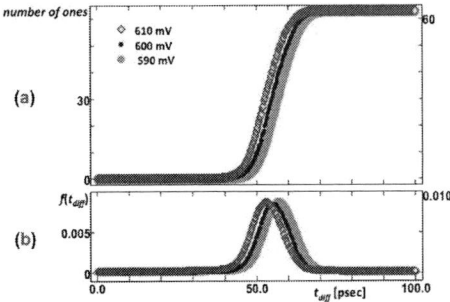

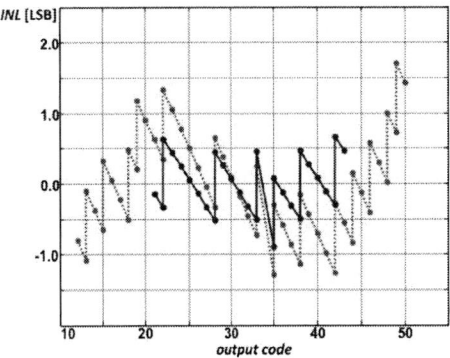

Fig. 8. Time-to-digital converter transfer functions with a scope measured $dV_{IN}(t)/dt = 13.2$mV/ps for $V_{LVL} = 590, 600, 650$mV (three of five transfer functions are shown). The detected resolution for this slope is 183fs.

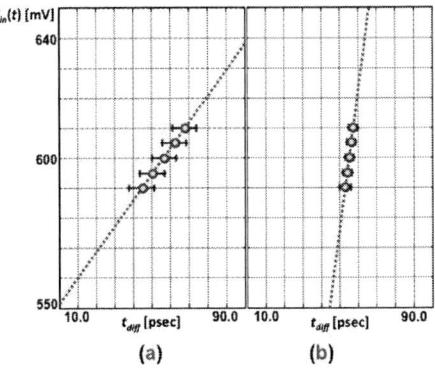

Fig. 9. The slope measured by mean time offset shift $\mu_{T_{OS}}$ relative to changing $V_{LVL}$.

Fig. 11. The measured integral nonlinearity for each output code when the linear region is selected as $2\sigma$ (red), and when the linear region is selected as $\sigma$. The smaller region shows an INL of no more than 0.6LSB.

in the chip fabrication program of VLSI Design and Education Center (VDEC) at the University of Tokyo in collaboration with e-Shuttle, Fujitsu, Synopsys, Cadence Design Systems, Mentor Graphics, and Agilent Technologies Japan.

### REFERENCES

[1] J. Yu and F. Dai, "On-chip jitter measurement using vernier ring time-to-digital converter," in *Proc. IEEE Asian Test Symp.*, Dec. 2010, pp. 167–170.

[2] V. Kratyuk, P. Hanumolu, K. Ok, U. Moon, and K. Mayaram, "A digital PLL with a stochastic time-to-digital converter," *IEEE Trans. Circuits Syst. I*, vol. 56, no. 8, pp. 1612–1621, Aug. 2009.

[3] A. Samarah, , A. Carusone, and E. Rogers, "A digital phase-locked loop with calibrated coarse and stochastic fine tdc," in *Proc. IEEE Custom Integrated Circuits Conf.*, 2012, pp. 1–4.

[4] C. Fan and J. Wu, "Jitter measurement and compensation for analog-to-digital converters," *IEEE Trans. Instrum. Meas.*, vol. 58, no. 11, pp. 3874–3884, Nov. 2009.

[5] M. Lee and A. Abidi, "A 9b, 1.25ps resolution coarsefine time-to-digital converter in 90 nm cmos that amplies a time residue," *IEEE J. Solid-State Circuits*, vol. 43, no. 4, pp. 769–777, Apr. 2008.

[6] P. Dudek, S. Szczepanski, and J. Hatfield, "A high-resolution cmos time-to-digital converter utilizing a vernier delay line," *IEEE J. Solid-State Circuits*, vol. 35, no. 2, pp. 240–247, Feb. 2000.

[7] S. Pellerano, P. Madoglio, and Y. Palaskas, "A 4.75-ghz fractional frequency divider-by-1.25 with tdc-based all-digital spur calibration in 45-nm cmos," *IEEE J. Solid-State Circuits*, vol. 44, no. 12, pp. 3422–3433, Dec. 2009.

[8] Y. J. Yamaguchi, S. Komatsu, M. Abbas, K. Asada, N. Khanh, and J. Tandon, "CMOS flash TDC with 0.84-1.3ps resolution using standard cells," in *Proc. IEEE Radio Frequency Integrated Circuits Symp.*, Jun. 2012, pp. 527–530.

[9] S. Weaver, B. Hershberg, D. Knierim, and U.-K. Moon, "A 6b stochastic flash analog-to-digital converter without calibration or reference ladder," in *Proc. IEEE Asian Solid-State Circuits Conf.*, Nov. 2008, pp. 373–376.

[10] T. J. Yamaguchi, K. Asada, K. Niitsu, M. Abbas, S. Komatsu, H. Kobayashi, and J. Moreira, "A new procedure for measuring high-accuracy probability density functions," in *Proc. IEEE Asian Test Symposium*, Niigata, Japan, Nov. 2012, pp. 1–2.

control of dynamic range offset by shifting comparator reference voltage $V_{LVL}$. We verified this precise control by characterizing the TDC for multiple reference voltages to find their dynamic range offsets. Then we derived the slope of the comparator input $V_{IN}$. Our TDC measured slope results show a cross correlation of 0.9991 with oscilloscope measured results.

## VI. ACKNOWLEDGEMENT

This research project has been supported by Advantest Corporation. The VLSI chip in this study has been fabricated

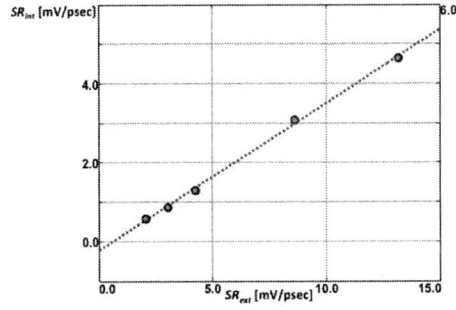

Fig. 10. Slope detected by the TDC transfer function mean offset shifts ($SR_{int}$) versus slope detected by oscilloscope ($SR_{ext}$). The detected cross-correlation coefficient is 0.9991.

# A 50 $\mu$W/Ch Artifacts-Insensitive Neural Recorder Using Frequency-Shaping Technique

### Jian Xu and Zhi Yang

Department of Electrical and Computer Engineering, National University of Singapore, Singapore
Email:{elexjian, eleyangz}@nus.edu.sg; Homepage: http://www.ece.nus.edu.sg/stfpage/eleyangz/

*Abstract*—**This paper presents a frequency-shaping (FS) neural recording interface that can inherently reject electrode offset, 5-10 times increase input impedance, 4.5-bit extend system dynamic range (*DR*), and provide much more tolerance to motion artifacts and 50/60 Hz power noise interferences. It is supposed to be more suitable for long-term brain-machine-interface (BMI) experiments. To achieve the mentioned performance above, the proposed architecture adopts an auto-zero offset calibration to avoid system saturation, a delayed-signaling noise cancellation to attenuate *kT/C* noise, and an automatical data-splitting technique to reduce input-referred noise at low frequencies. Measured at a 40 kHz sampling clock and $\pm$ 0.6 V supply, the recorder consumes 50 $\mu$W/ch, including 22 $\mu$W for FS amplifier, 12 $\mu$W for gain-stage amplifier, 12 $\mu$W for buffer, and 4 $\mu$W for successive approximation register (SAR) analog-to-digital converter (ADC). The designed SAR ADC achieves an effective-number-of-bit (*ENOB*) of 11-bit in a 160 kHz bandwidth. In addition, the recorder has a 3 pF input capacitance and 15.5-bit (11-bit+4.5-bit) system *DR* due to the utilization of FS technique. The designed chip occupies 0.76 mm$^2$/ch in a 0.13 $\mu$m CMOS process.**

## I. INTRODUCTION

Neural recording systems are useful tools in both neuroscience research and clinical diagnosis. After decades of engineering efforts on circuit design and trade-offs towards performance optimization [1]–[3], several faced challenges are needed to overcome in order to meet the demand of chronic high-density recording experiments. Firstly, given the electrode material, an increased recording density needs a proportionally increased amplifier input impedance to avoid degeneration in signal quality. The main reason is that the electrode impedance tends to increase due to the elicited adverse tissue response, resulting in electrical isolation from target neurons [4]. Secondly, a full-spectrum recording from very low frequency (<1 Hz) to several kHz is desired to simultaneously achieve local field potentials (LFPs) and extra-cellular spikes. It is further complicated by requiring removing electrode offset, which is achieved at a cost of compromised system specifications. Thirdly, motion artifacts and 50/60 Hz power noise interferences that appear in BMI experiments can cause electronics saturation, which increases the requirement on system *DR*. Thus, a piece of electronics robust to artifacts and interferences is also in great demand.

In response to the faced challenges, there are in general two circuit design trends [1]–[3], [5]–[10]. One is exploring elegant analog techniques to further optimize system performance [2], [3], [5], [6], [8]–[10]. This trend is a mainstream approach that has reached physical limits. For example, these architectures

need a high-pass filter consisting of feedback resistor $R_f$ and capacitor $C_f$ to remove intrinsic electrode offset. Because low frequency neural activities extend to sub-1 Hz, a filter corner frequency ($1/(2\pi R_f C_f)$) of less than 1 Hz is needed to avoid distortion, which translates to large $R_f$ and $C_f$. By operating transistors in cut-off region, $R_f$ achieves near $10^{12}$ $\Omega$. However, a further increase in $R_f$ is difficult and undesired due to leakage currents on bias circuits, induced nonlinearities to amplification, and tuning efforts required to stabilize the amplifier. Hence, $C_f$ is accordingly designed in sub-pF range, which causes that the input capacitance is around tens of pF. As a result, the input impedance reaches 10 G$\Omega$ at 1 Hz or 10 M$\Omega$ at 1 kHz. The other one is exploring digital compensation techniques for certain system parameter improvements [1], [7]. [1] has reported a open-loop amplifier with sophisticated digital compensations (off-chip) to gain 1 G$\Omega$ resistive input impedance. Paid costs are on signal distortion and system instability due to the open-loop configuration and electrode interface mismatch. Also, it is difficult to instrument on-chip digital modules that can reliably cancel electrode offset and perform unsupervised compensations. In a sense, current mainstream efforts rely on circuit design techniques and technology advancements that are difficult to support a further growth in recording density and provide more tolerance to motion artifacts and power noise interferences.

As a principled contribution of this work, this paper presents a new artifacts-insensitive system architecture for long-term high-density BMI experiments. From carefully designed experiments and analyses, it is demonstrated that the proposed architecture is capable of the following advantages. a) Higher input impedance: the input capacitance can be potentially reduced to a few pF or even less. b) Full-spectrum recording: the proposed architecture can simultaneously record signals from sub-1Hz to several kHz without assuming complex circuits. c) Extended system *DR*: the system *DR* is 4.5-bit extended compared with the conventional architecture without consuming much power and area cost. d) Improved tolerance to electrode surface degeneration and insensitive to artifacts. f) Suitability to deep submicron CMOS processes: under the proposed architecture, a sub-1 Hz high-pass filter to remove electrode offset is not required. Consequently, it avoids using MOS-bipolar pseudo-resistors with extremely high resistance, which is challenging to implement in deep submicron CMOS processes due to leakage current on bias circuits. g) Reliable, real-time signal processing: an exponential component and polynomial component (EC-PC)

978-1-4673-6145-3/13 $31.00 © 2013 IEEE

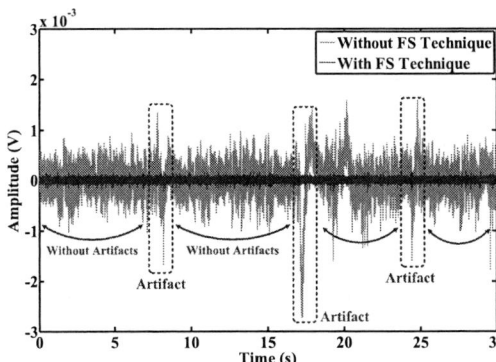

Fig. 1. Neural amplifier output without (gray) and with (blue) FS technique. Waveforms are offline recorded from an in-*vivo* preparation and feed to circuits input. Both output waveforms are normalized to the input and plotted in the same scale.

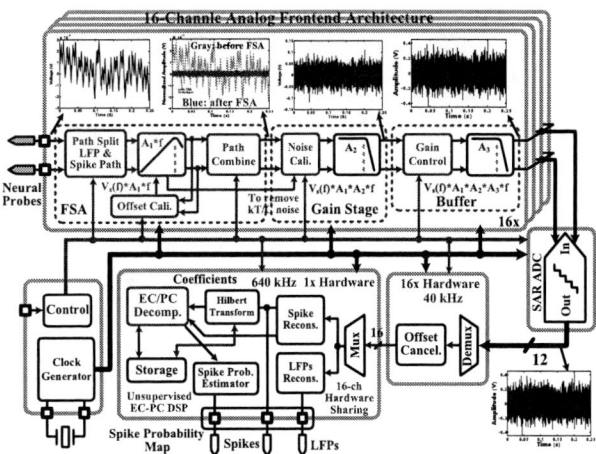

Fig. 2. The proposed FS neural recording and EC-PC signal processing system architecture. The prototype can simultaneously output field potentials, extra-cellular spike data, and spiking activity map.

spike processing framework [11] is proposed to analyze the existence probability of extra-cellular spikes.

## II. FREQUENCY-SHAPING SYSTEM ARCHITECTURE

### A. System Dynamic Range Reduction

neural data follow a $1/f^{1-3}$ spectrum. At low frequencies, neural signals, artifacts, and interferences are in millivolt range; while at high frequencies, signals require microvolt resolution for processing, such as alignment, feature extraction, classification, *etc*. To accommodate both low frequency and high frequency activities, a wide data acquisition *DR* (1 $\mu$V to >10 mV) is needed that is challenging and cost ineffective for circuit implementation. As a compromise, a majority of recording systems accommodate about 10-bit data *DR*, possibly switching between two modes to either record LFPs or extra-cellular spikes separately. Moreover, these systems are not suitable for recording large artifacts which appear more frequently in less constrained recording environment.

To study system *DR*, neural amplifier output data with and without FS technique are both plotted in Fig. 5. Tentatively assuming that the FSA is noise free, the relaxation on system *DR* can be visualized by comparing the waveform peaks, suggesting one to two orders extended system *DR*. With the extended *DR*, it is possible to allow less constrained recording environment and remove both artifacts and interferences on-chip. As shown in Fig. 5 (a), the FS technique provides a 4.5-bit extended system *DR* which is averaged from 167 in-*vivo* sequences.

### B. Proposed Recording Architecture Based on FS Technique

As shown in Fig.2, the proposed system consists of FS amplifier (FSA), ADC, integrator, filter, EC-PC decomposition engine, and spiking activity map generator. The circuits take neural data from the electrodes as input and can output three data streams simultaneously: 1-300 Hz LFPs sampled at 1.25 kHz, 0.5-10 kHz spike data sampled at 40 kHz, and a spiking activity map for information decoding.

The closed-loop gain of FSA is $2\pi(C_{in}/C_f)*(f/f_S)$, where $f$ and $f_S$ are signal frequency and sampling frequency, respectively; $C_{in}/C_f$ is the capacitor ratio as shown in Fig. 3. Due to

the FS technique, the architecture inherently rejects electrode offset without requiring a sub-1Hz high-pass filter and also allows a substantially improved input impedance. To alleviate the contradiction between sampling capacitor size and sampled noise, three noise removal strategies are used. First, amplifier virtual ground is employed for sampling signals, which largely removes the supposed $kT/C$ noise on $C_{in}$. Second, a delayed-signaling scheme is developed to cancel $kT/C$ noise on $C_f$. Third, a new path-splitting technique that automatically splits input data into two streams (LFP path and spike path) is used to reduce the input-referred noise at low frequencies.

The ADC output is fed to the digital part for data analysis. To undo FS modulation, a matched digital integrator is designed with offset-cancellation to avoid saturation. After integration, the word-length of each data sample extends from 12-bit to 16-bit. Due to the path-splitting technique, 31 out of every 32 samples are used to reconstruct extra-cellular spikes and the remaining 1 sample is used to reconstruct LFPs. As 40 kHz clock oversamples neural spikes several times, a periodically missing sample can be recovered (<-60 dB distortion). 16-channel LFPs each sampled at 1.25 kHz are band-pass filtered at 1-300 Hz and then serialized to an output port at a data rate of 320 Kbps. A programmable band-pass filter is used to obtain spike data, where the out-of-band rejection is over 64 dB and in-band ripples are less than 0.08 dB. Besides, an EC-PC data decomposition algorithm is implemented to predict the presence of neural spikes at any time and a spiking activity map for decoding, where the algorithm has been detailed in [11]. 16-channel spike data and their corresponding spiking activity map are serialized and output through two ports at 7.68 Mbps and 160 Kbps, respectively.

## III. CIRCUIT IMPLEMENTATION

### A. Frequency-Shaping Amplifier

The FS frontend circuit schematic is shown in Fig. 3. Several circuit techniques are added for performance enhancement: firstly, input data are split into two streams with gains set by

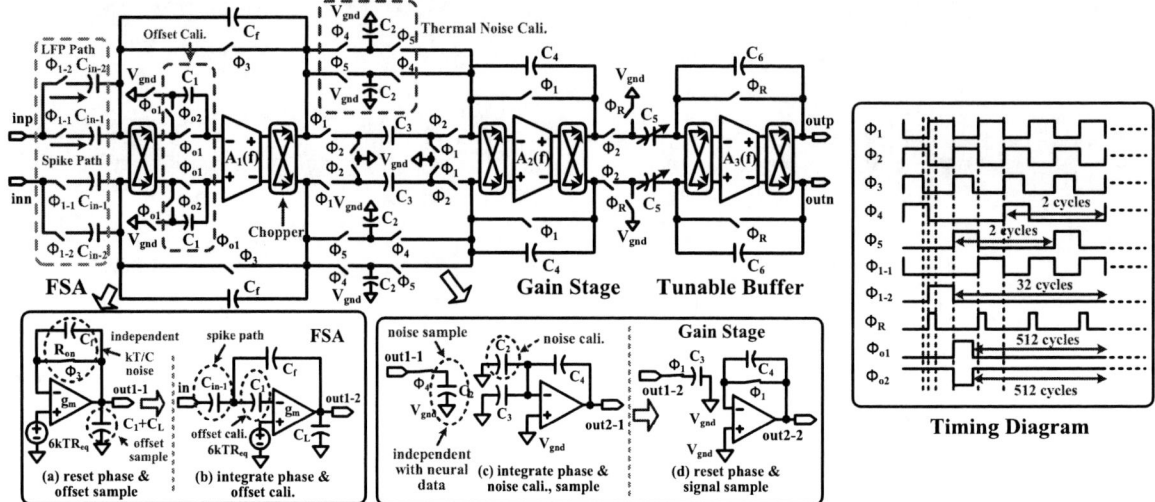

Fig. 3. The proposed FS neural amplifier implementation with offset calibration and noise cancellation circuits. It inherently rejects electrode offset and has improved amplifier input impedance, dynamic range, and provides more tolerance to electrode encapsulation, artifacts, and interferences.

$C_{in-1}/C_f$ (extra-cellular spikes), $C_{in-2}/C_f$ (LFPs), and with sampling clocks $\Phi_{1-1}$, $\Phi_{1-2}$ (both generated from a 40 kHz clock $\Phi_S$: $\Phi_{1-1}\bigcup\Phi_{1-2}=\Phi_S$, $\Phi_{1-1}\bigcap\Phi_{1-2}=0$). The scheme allows the flexibility to selectively record extra-cellular spikes, LFPs, or both for noise optimization and power saving. Secondly, an auto-zero circuit controlled by two non-overlapped clocks $\Phi_{o1}$ and $\Phi_{o2}$ at $1/512$ $\Phi_S$ is designed. This circuit is to attenuate the power at the chopper frequency thus avoiding circuit saturation. Thirdly, a novel delayed-signaling scheme is used to sample the appeared $kT/C$ noise and charge injection on $C_f$ ($\Phi_4$ and $\Phi_5$ as shown in Fig. 3), which are subtracted at the second stage for cancellation. Thus, an aggressive reduction of $C_{in}(C_f)$ doesn't proportionally elevate the noise floor.

### B. SAR ADC with On-Chip Calibrations

Two SAR ADCs clocked at 320 kHz are shared by 16 recording channels through multiplexing. As neural data exhibit a $1/f^{1-3}$ frequency dependency and are frequently contaminated by large artifacts and interferences, to simultaneously record neural data, motion artifacts, and power noise interferences, an ADC is required to have a wide *DR*. In this work, an 11-bit *ENOB* ADC is designed in Fig. 4. The FS architecture compresses neural data *DR* by 4.5-bit, consequently the system *DR* is extended to 15.5-bit. To achieve the targeted precision at a low power budget, a split-capacitor SAR ADC structure is chosen with circuit techniques summarized as follows. Firstly, a time-domain comparator consisting of two differential 14-stage voltage-controlled delay lines (VCDLs) and a binary phase detector is implemented to achieve a sensitivity of 4 ns/mV with a 10 fF load, which is automatically calibrated at each stage. Secondly, a three-stage auto-zero preamplifier and a digital background calibration are used to suppress noise, offset, and mismatch. Thirdly, a monotonic switching procedure is employed to give an extra 1-bit comparison. Hence, the total number of unit capacitors is reduced and the unit capacitance can be increased accordingly.

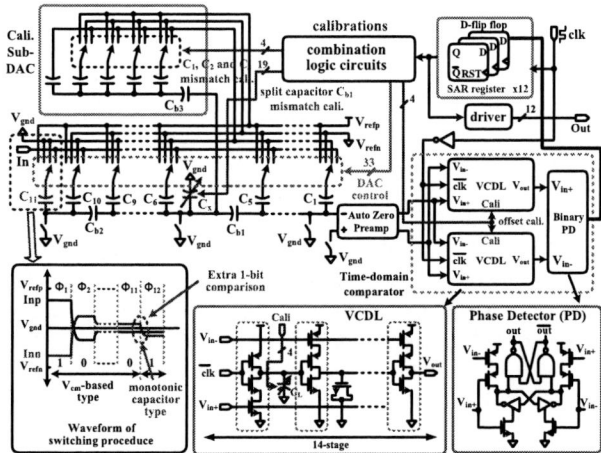

Fig. 4. 12-bit SAR ADC circuit schematic with power-efficient capacitor switching procedure and on-chip calibrations.

## IV. MEASUREMENT RESULTS

Fig. 5 summarizes the measured results of the designed FSA and SAR ADC. Measured results show that the designed amplifier has the FS function and the gain error is below 0.5%. The measured total input-referred noise for LFPs and spike recording is 13 $\mu$Vrms and 7 $\mu$Vrms, respectively. As metioned in Section II-A, in-*vivo* neural data follow a $1/f^{1-3}$ spectrum, and neural signals at low frequencies are in millivolt range while neural signals at high frequencies are in tens of microvolt range. Thus, under the measured input-referred noise, the proposed FSA architecture can still achieve high enough *SNDR* at all the frequencies to reconstruct good signal-quality neural data.The total power consumption of the proposed frontend amplifier is measured to be 46 $\mu$W, including 22 $\mu$W for the FSA, 12 $\mu$W for the gain stage and 12 $\mu$W for the buffer. The ADC is measured to have a 68 dB *SNDR* on average with ą0.3-bit channel-to-channel variation. The total power consumption of the 8-channel ADC is 30 $\mu$W with 160 kHz signal bandwidth

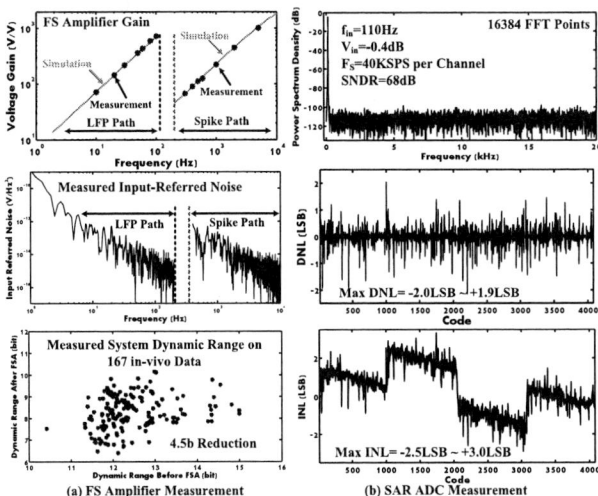

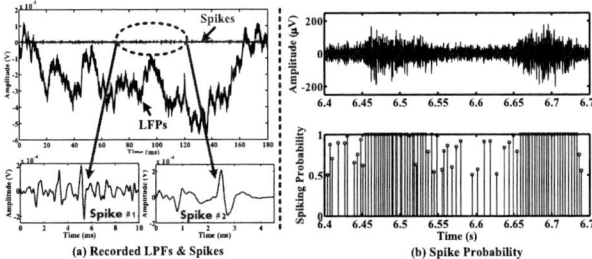

Fig. 5. Measurement results of the designed FS amplifier and SAR ADC.

(a) FS Amplifier Measurement     (b) SAR ADC Measurement

Fig. 6. In-*vivo* recording and neural signal processing experiments. (a) Recorded LFPs and extra-cellular spikes from micro wire array in rat hippocampus. (b) Spike probability map to indicate the presence of neural spikes [11].

| | Parameter | This Work | [3] | [5] | [7] |
|---|---|---|---|---|---|
| Chip | Process (μm) | 0.13 | 0.18 | 0.5 | 0.35 |
| | Number of Ch | 16 | 32 | 100 | 96 |
| | Supply (V) | ±0.6 | A:1.8 D:1.0 | 3.3 | 1.2 |
| | System DR (dB) | 97.8 | 63.7 | 69 | <60 |
| | FOM$_{dB}$ (fJ/conv.-step) | 26.3 | 63.1 | 5861.4 | >8321.8 |
| | Total Power (μW/Ch) | 50 | 10.1 | 135 | 68 |
| Amp | Amplifier Gain | Up to 6000*2πf/ f$_s$ (V/V) | (49-66) dB | 39.8 dB | N/A |
| | Input Impedance | 3pF | 14pF | 20pF | 20 pF |
| | Noise (μVrms) | 13 @ (0.5-300Hz) 7 @ (0.5-10kHz) | 5.4-11.2 @ (10Hz-65kHz) | 5.1 @ (1Hz-5kHz) | 4.3 @ (10Hz-5kHz) |
| | Total Power (μW/Ch) | 46 | 6.5 | 42.2 | 40.7 |
| SAR ADC | f$_s$ (kHz/Ch) | 40 | 31.25 | 15 | 31.25 |
| | SNDR (dB) | 70 | 47.8 | N/A | 60.26 |
| | ENOB (bit) | 11.3 | 7.65 | N/A | 9.72 |
| | DNL (LSB) | -2.0-+1.9 | ±0.4 | N/A | ±0.25 |
| | INL (LSB) | -2.5-+3.0 | ±0.4 | N/A | ±0.25 |
| | Total Power (μW/Ch) | 3.75 | 0.47 | N/A | 1.1 |
| | FOM (fJ/conv.-step) | 37 | 75 | N/A | 41.8 |

Fig. 7. Chip micrograph and performance summary.

(15.5-bit). In addition, its *FOM* based on system *DR* reaches 26.3 fJ/conversion-step, which is lowest in comparison with other neural recording designs. Overall, the proposed architecture shows multifold advantages over the conventional one, as a replacement to better support future long-term, high-density recording experiments.

## ACKNOWLEDGMENT

This work is supported by Singapore grants R-263-000- 699-305, R-263-000-A32-305 and R-263-000-A29-133.

## REFERENCES

[1] R. Muller, S. Gambini, and J. Rabaey, "A 0.013 mm$^2$, 5μW, DC-coupled neural signal acquisition IC with 0.5 V supply," *IEEE J. Solid-State Circuits*, vol. 47, no. 1, pp. 232–243, Jan. 2012.

[2] M. Azin, D. Guggenmos, S. Barbay, R. Nudo, and P. Mohseni, "A battery-powered activity-dependent intracortical microstimulation IC for brain-machine-brain interface," *IEEE J. Solid-State Circuits*, vol. 46, no. 4, pp. 731–745, Apr. 2011.

[3] R. Wattanapanitch and R. Sarpeshkar, "A low-power 32-channel digitally programmable neural recording integrated circuit," *IEEE Trans. Biomed. Circuits Syst.*, vol. 5, no. 6, pp. 592–601, Dec. 2011.

[4] I. Stevenson and K. Kording, "How advances in neural recording affect data analysis," *Nature Neuroscience*, vol. 14, no. 2, pp. 139–142, Feb. 2011.

[5] R. Harrison, P. Watkins, R. Kier, R. Lovejoy, D. Black, B. Greger, and F. Solzbacher, "A low-power integrated circuit for a wireless 100-electrode neural recording system," *IEEE J. Solid-State Circuits*, vol. 42, no. 1, pp. 123–133, Jan. 2007.

[6] M. Chae, W. Liu, Z. Yang, T. Chen, J. Kim, M. Sivaprakasam, and M. Yuce, "A 128-channel 6 mW wireless neural recording IC with on-the-fly spike sorting and UWB transmitter," in *IEEE Int. Solid-State Circuits Conf. Dig. Tech. Papers*, 2008, pp. 146–147.

[7] H. Gao, R. Walker, P. Nuyujukian, K. Makinwa, K. Shenoy, B. Murmannn, and T. Meng, "HermesE: A 96-channel full data rate direct neural interface in 0.13 μm CMOS," *IEEE J. Solid-State Circuits*, vol. 47, no. 4, pp. 1043–1055, Apr. 2012.

[8] J. Lee, H. Rhew, D. Kipke, and M. Flynn, "A 64-channel programmable closed-loop neurostimulator with 8 channel neural amplifier and logarithmic ADC," *IEEE J. Solid-State Circuits*, vol. 45, no. 9, pp. 1935–1945, Sep. 2010.

[9] Z. Xiao, C. Tang, C. Dougherty, and R. Bashirullah, "A 20 μW neural recording tag with supply-current-modulated AFE in 0.13 μm CMOS," in *IEEE Int. Solid-State Circuits Conf. Dig. Tech. Papers*, 2010, pp. 122–123.

[10] F. Zhang, J. Holleman, and B. Otis, "Design of ultra-low power biopotential amplifiers for biosignal acquisition applications," *IEEE Trans. Biomed. Circuits Syst.*, vol. 6, no. 4, pp. 344–355, Aug. 2012.

[11] Z. Yang, W. Liu, M. R. Keshtkaran, Y. Zhou, J. Xu, V. Pikov, C. Guan, and Y. Lian, "A new EC-PC threshold estimation method for in-*vivo* neural spike detection," *J. Neural Eng.*, vol. 9, no. 4, pp. 1–16, Aug. 2012.

and a figure-of-merit (*FOM*) of 37-56 fJ/conversion-step. Fig. 6 (a) plots both recorded LFPs and extra-cellular spikes from a live rat. Fig. 6 (b) shows on-chip signal processing results on spike probability map, which is used to indicate the presence of neural spikes. During the testing, the chip is connected to a credit card size board, which only requires one single USB cable for powering and data storage. The chip is fabricated in a 0.13 μm CMOS with microphotograph, performance summary and comparison as shown in Fig. 7. The core area is 0.76 mm$^2$/ch and the total power consumption of the recorder is 50 μW/ch. To evaluate the recorder performance, an $FOM_{DR}$ based on system *DR* is used

$$FOM_{DR} = \frac{Power_{total}}{2BW * 2^{(DR-1.76)/6.02}}, \quad (1)$$

where $Power_{total}$ is the total power consumption of the recorder including neural amplifier, buffer and ADC; *BW* is bandwidth; *DR* is system dynamic range.

## V. CONCLUSION

A new circuit architecture for neural recording has been proposed as a principled contribution to allow major performance improvement. It inherently rejects electrode offset and has more tolerance to electrode encapsulation, motion artifacts, and interferences. By using several novel circuit techniques, $kT/C$ noise has been greatly attenuated, and our design provides the smallest input capacitance (3 pF), and the widest system *DR*

# A Bipolar >40-V Driver in 45-nm SOI CMOS Technology

Yousr Ismail, Chang-Jin "CJ" Kim, and Chih-Kong Ken Yang
University of California, Los Angeles
Los Angeles, CA 90095, USA
{yousr, yang}@ee.ucla.edu

*Abstract—* This paper presents a high-voltage driver in nanometer-scale, low-voltage SOI CMOS technology well beyond the voltage limits of standard devices. The drive level is near the voltage-tolerance limit of the body insulator. A novel, bidirectional, switched-capacitor output stage that combines both voltage-conversion and pulse-drive is introduced. The two-level driver is implemented in 45-nm SOI CMOS technology and uses only process-compliant devices. It achieves a maximum output drive of 44 V and occupies an area of 600 μm x 350 μm. The output drive resistance depends on the pumping frequency and is equal to 36 KΩ at a current consumption of 28 mA drawn from a 1.5-V supply.

## I. INTRODUCTION

Many applications, such as, MEMS, automotives, and displays, require driving capacitive or large resistive loads with voltages in the range of several tens of volts. These applications often require extensive signal processing and feedback control. Because of the limited voltage-handling capability of state-of-the-art CMOS technology, designs often opt for either: an older technology node, thus sacrificing performance for higher voltage-handling capability [1], a bulky multi-chip solution, or an expensive high-voltage module [2]. If such target voltages could be implemented in fine-linewidth CMOS, this can facilitate a means towards more compact and lower cost, fully-integrated solutions.

In this paper, we describe a high-voltage driver design that provides up to ±44 V drive, without exceeding components tolerances of a 45-nm SOI CMOS technology. The proposed solution is enabled by two key-factors: 1) the high voltage-handling capability of the insulator substrate, and 2) a circuit architecture that leverages the voltage-tolerance inherent to a charge pumping action.

The paper is organized as follows, section II explains the driver architecture, section III discusses the circuit implementation, section IV provides measurement results, and section V concludes the paper.

## II. HIGH-VOLTAGE DRIVE ARCHITECTURE

One common realization of an on-chip, high-voltage drive uses charge pump(s) to generate the desired DC drive level(s) as depicted in Fig. 1(a). A high-voltage-tolerant output stage

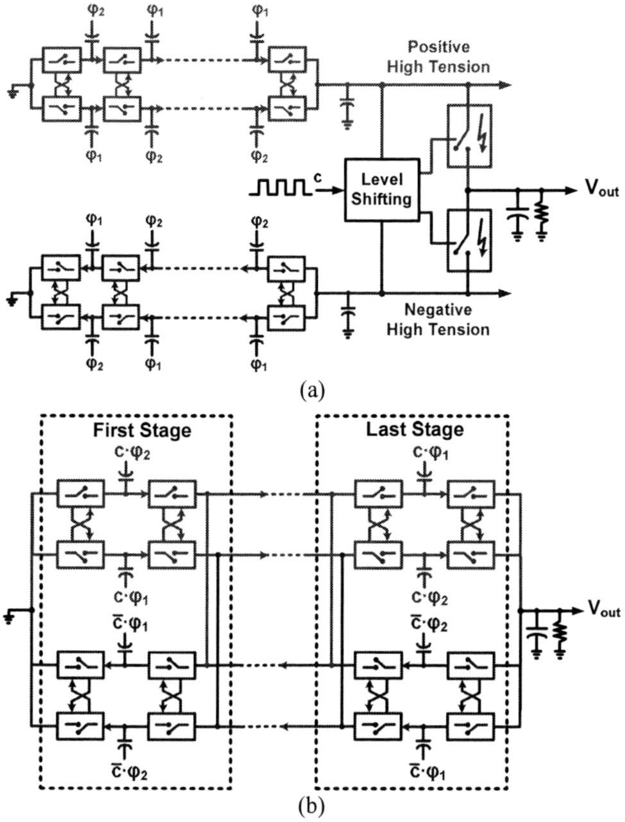

Fig. 1 (a) Conventional drive scheme (b) Proposed drive scheme.

is then necessary to perform the switching at the output. The design of the output stage typically requires switches with a high blocking-voltage and a pre-driver circuit. The pre-driver provides the low- and high-side gate drive signals necessary to maintain the switches gate-oxide reliability. The output switches can be implemented using high-voltage LDMOS devices [3], but they incur an extra cost and are unlikely to handle tens of volts in a fine-linewidth technology. Alternatively, stacking low-voltage devices extends the switch blocking-voltage but requires multiple gate controls. In which case, the pre-driver needs to supply gate control signals for all the individual stacked devices and to guarantee their reliability during switching transitions. Device stacking has been recently implemented up to only four or five devices because of the associated pre-driver complexity [4-5].

---

This work has been supported in part by DARPA award number W31P4Q-10-1-0008P00005.

978-1-4673-6145-3/13 $31.00 © 2013 IEEE

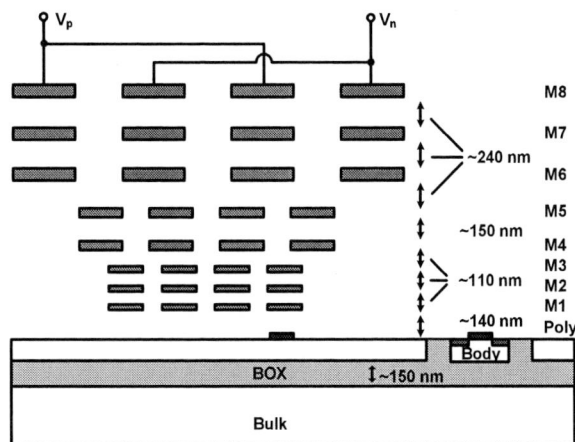

Fig. 2 Approximate process cross-section.

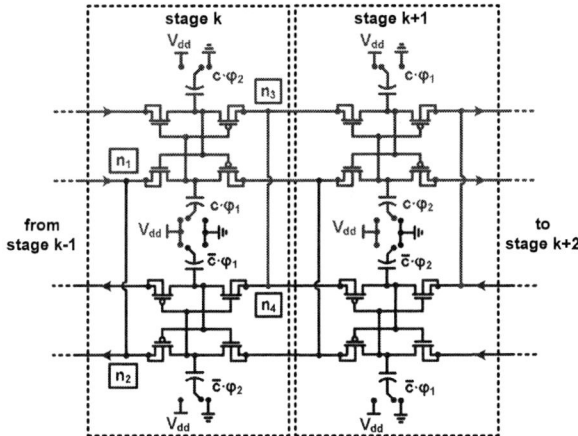

Fig. 3 Circuit implementation of two successive driver stages.

In this work we propose a drive architecture that directly uses the charge pump output as the drive output. The design relies on the voltage-tolerant nature of charge pump circuits, and presents a two-level, bidirectional pumping stage, as shown in Fig. 1(b). The driver is made up of a cascade of stages, with each stage comprised of two charge-transfer paths, namely, to and from the load. A control signal gates the clocks in each charge-transfer path to ensure a mutually exclusive up/down operation. Because of having two separate charge-transfer paths and the complementary nature of the drive levels, only one path is active in all stages at a time. Thus, while the capacitors of the active path are continually refreshed, those of the idle path are slowly discharging. This will result in voltage stresses across the transistors of the idle path. To alleviate this issue, both charge-transfer paths in each stage are bootstrapped to one another, as further explained in section III.

The proposed implementation can use nominal technology devices if each charge pump stage is ensured to meet the device voltage tolerances. A direct consequence of this design is the ability to stack tens of devices reliably in a modular and scalable fashion, and with well-controlled voltage-transition amplitudes. Because increasing the number of stacked devices does not incur an extra design complexity, the stack height becomes limited by the technology voltage-handling capability, which in our case is the breakdown voltage of the buried oxide (BOX) layer. The driver output resistance scales proportionally with the number of stages but can be controlled through the pumping frequency. The implemented driver for >40V comprises a cascade of 48 stages.

## III. IMPLEMENTATION DETAILS

### A. Technology and Passive Elements

Different voltage tolerances exist for different device types within a technology node. We chose a 45-nm SOI technology that provides 2.5-nm, thick gate-oxide devices with a 1.5 V ± 10% voltage tolerance. An SOI technology is chosen instead of bulk CMOS because its BOX breakdown limit is considerably higher than that of a well-substrate junction [6]. The higher voltage-handling capability of SOI enables us to demonstrate the extensive device stacking necessary to achieve the target voltages. Another advantage of SOI technology is that it does not rely on reverse biased junctions for device isolation as is the case with bulk technology. This convenient feature makes generating negative voltages similar to generating positive ones. An approximate cross-section of the process is shown in Fig. 2.

Metal spacing is a critical parameter in determining the reliability of low-k dielectrics. The time-dependent dielectric breakdown (TDDB) behavior is a function of the metal spacing and area [7], and both need to be considered in calculations. The minimum horizontal spacing between wires is 70 nm. Metal capacitors with minimum finger spacing are estimated to sustain ~15 V based on a TDDB cumulative failure rate of less than 0.1% over 10 years at 85°C. The TDDB calculations are based on a Weibull distribution, a field accelerated √E-model, an Arrhenius temperature dependence, and a Poisson area scaling model. The BOX thickness is ~150 nm and is estimated to sustain ~51 V based on the same criteria. Note that the only areas of the BOX that are exposed to voltage stress are the transistors' active areas. No data for the BOX reliability was available, so its TDDB estimates are based on the low-k dielectric parameters.

To minimize area, different stages use capacitors with variable finger spacing depending on the voltage requirement of the stage. For the first 16 stages, minimally-spaced finger capacitors from the technology library with a capacitance density of 2.3 fF/um² are used. The finger spacing is double-spaced (1.3 fF/um²) and triple-spaced (0.75 fF/um²) for the second and third 16 stages, respectively. Because different metal layers have different spacing rules, both fringing and parallel plate components contribute to the total capacitance of the minimally-spaced capacitor. However, for the doubly- and triply-spaced capacitors, the vertical metal spacing becomes smaller than or comparable to the finger spacing, and capacitors with fringing-only fields are devised. Thus, all the fingers of the doubly- and triply-spaced capacitors are vertically aligned and share the same finger width and spacing. It was also shown in [7] that increasing the finger spacing by only 50%, under the same field, results in a TDDB improvement by an order of magnitude. Consequently, we expect that the overall driver reliability is dominated by the TDDB failure rates of the minimally space capacitors.

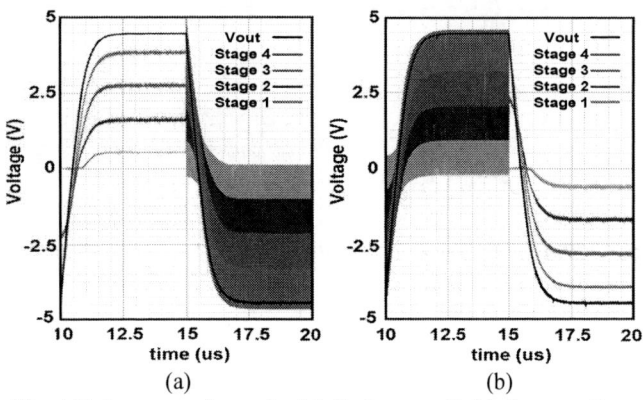

(a) (b)

Fig. 4 Voltage waveforms for (a) discharge path (b) charge path.

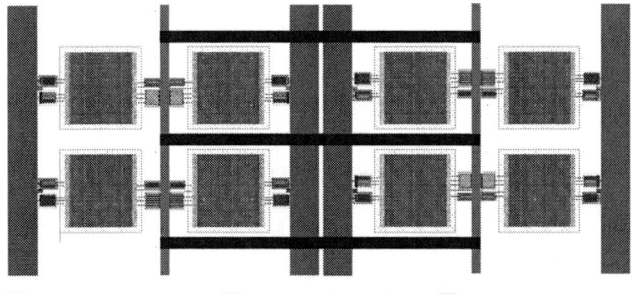

■ Low-Voltage Rails ■ High-Voltage Rails ■ Bootstrapping nets

Fig.5 Layout of two successive driver stages.

## B. Circuit Implementation

The basic building unit of the driver is a cross-coupled inverter pair configured as a two-phase voltage doubler (TPVD) [8]. The TPVD provides an elegant implementation for the clock-boosting required for efficient charge-transfer switches. Only thick-oxide devices are used, and device reliability is guaranteed as long as the pumping clock amplitudes are within the transistor rated voltage tolerance. Each driver stage uses two identical, yet, oppositely-oriented doublers for bilateral charge transfer. Figure 3 depicts the circuit schematic of two successive driver stages, indicating the transistors' bulk connectivity. For the stages closest to the output, and as the capacitors in the idle path start discharging, the devices connected to the output node experience a large voltage stress. To remedy this issue, nodes n1 and n3 within a stage are connected to nodes n2 and n4, respectively. This connectivity ensures that the node voltages in the idle path continuously follow those of the active path. As a result, the idle-path capacitors are automatically refreshed from the active-path capacitors. This behavior is better depicted by observing the simulation waveforms of a representative four-stage driver shown in Fig. 4. During the output positive half-cycle, only the clocks of the charge-path are active, nonetheless, the voltages of the discharge-path capacitors (Fig. 4a) are shown to track the voltages of their charge-path counterparts (Fig. 4b). Similarly, during the output negative half-cycle, only the clocks of the discharge-path are active, nonetheless, the voltages of the charge-path capacitors (Fig. 4b) are shown to track the voltages of their discharge-path counterparts (Fig. 4a).

The blocks are laid out while carefully considering the necessary lateral and vertical clearances between low- and high-voltage signals. Supply rails and control wires are separated from the high-voltage interconnects by a sufficient number of metal layers whenever crossing. The layout of two successive driver stages is shown in Fig. 5.

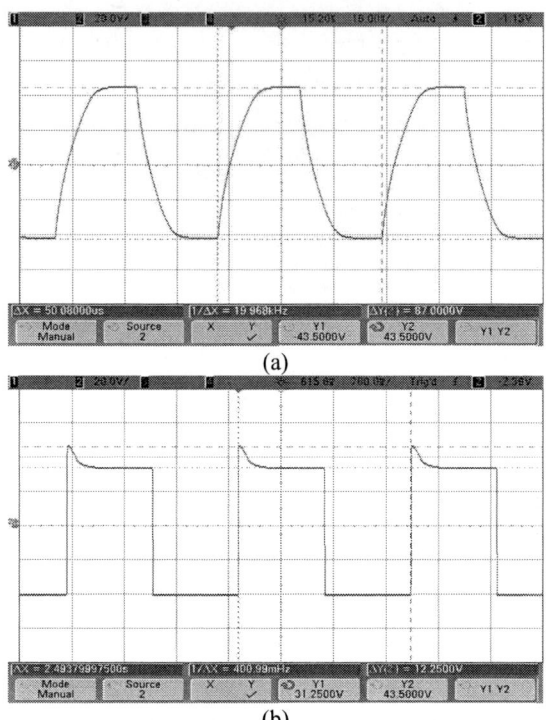

(a)

(b)

Fig. 6 Measured square waveforms at (a) 20 KHz (b) 400 mHz.

## IV. MEASUREMENT RESULTS

The driver is characterized over supply voltages ranging from 0.9 V to 1.65 V, and pumping frequencies ranging from 100 MHz to 1 GHz. The circuit is tested for both a square-wave output, and a DC output at different load currents.

Fig. 6(a) shows a measured 20-KHz, 44-V, bipolar, square waveform at the driver output (appears slightly lower due to the oscilloscope loading). The driver operates from a 1.65-V supply and a 450-MHz pumping frequency with a 35-pF capacitive load. Interestingly, when the driver is run at very low drive frequencies, the positive output level settles to the lower steady state value of 32 V as shown in Fig. 6(b). This is an artifact of the slow settling time constant of back-gate effect in SOI CMOS [9], and limits drive frequencies to remain above 10 Hz. Because transistor bodies that are closer to the driver output are floating at a much higher potential than that of the chip substrate, the substrate and the BOX form a back-gate and an inversion layer is formed above the BOX. The formed back-channel causes reverse conduction losses and leads to a lower absolute output voltage. Note that the effect is only observed for positive output voltages.

Because of the back-gate effect, the maximum achievable DC response is characterized for the negative drive level. Fig. 7 shows the driver DC characteristics at a 450-MHz pumping frequency and a range of supply voltages. The pump is capable of providing a 40-V output for load currents as high as 100 μA. Fig. 8 shows similar DC characteristics at a 1.5-V

978-1-4673-6145-3/13 $31.00 © 2013 IEEE  222

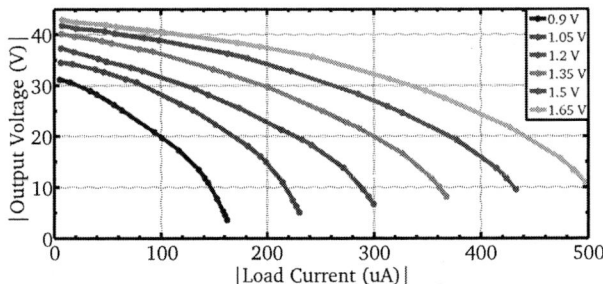

Fig.7 Driver DC characteristics at a 450-MHz pumping frequency.

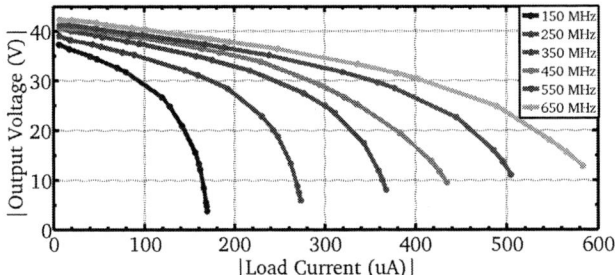

Fig.8 Driver DC characteristics at a 1.5-V supply voltage.

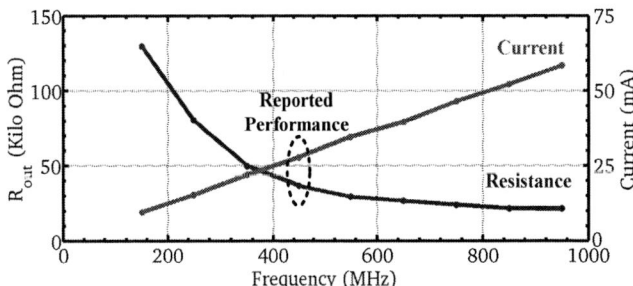

Fig. 9 Output resistance and current consumption at a 1.5-V supply.

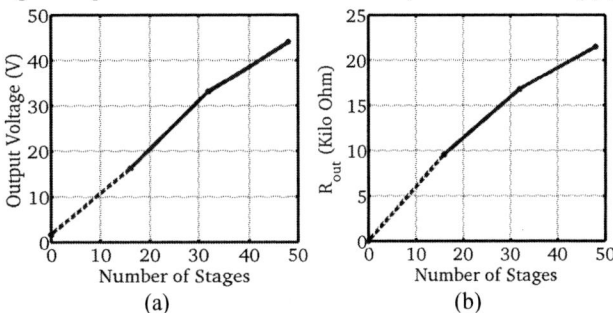

Fig. 10 (a) Max. voltage (b) Min. resistance vs. number of stages.

supply voltage and a range of pumping frequencies. The output drive resistance and current consumption are plotted versus the pumping frequency as shown in Fig. 9. The measured supply current represents the driver switching power consumption at no load current. The output resistance drops with higher pumping frequency and is eventually limited by the switches series on-resistance, whereas, the current consumption increases linearly with frequency as expected of CMOS logic. The driver performance is reported for the 36-K$\Omega$ output resistance data-point, at which, the driver consumes 28 mA from a 1.5-V supply voltage. The driver achieves a minimum output resistance of 21.5 K$\Omega$ for pumping frequencies higher than 800 MHz. The maximum achievable drive voltage and the minimum achievable output resistance are plotted versus the number of driver stages, and are shown in Fig. 10(a) and Fig. 10(b), respectively. The ripple voltage is found to be 1.2 mV at a load capacitance of 35 pF and a pumping frequency of 450 MHz.

The die photo and driver layout are shown in Fig. 11. The circuit occupies 600 µm x 350 µm, and achieves the highest voltage drive reported to date in a CMOS technology while adopting an all-low-voltage device implementation.

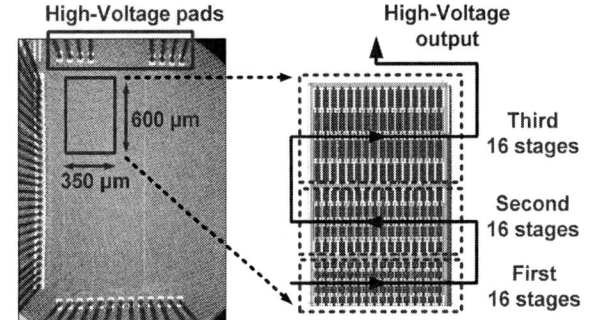

Fig. 11 Die photo and driver layout.

## V. CONCLUSION

A bidirectional-charge-pump-based output stage has been implemented in a 45-nm SOI CMOS process. The driver demonstrates the feasibility for >40V drive in a standard CMOS technology. The proposed driver leverages two properties: (1) the inherent voltage-tolerance of charge pump circuits, and (2) the extended voltage-handling of SOI CMOS technology. Attractive features of the driver include a modular gate drive that enables extended device stacking, and a frequency-controlled drive resistance. The proposed driver can prove beneficial in applications that need high-voltage switching, yet require an advanced technology-node implementation to fully harvest CMOS scaling benefits.

## REFERENCES

[1] A. Emira et al., "50V All-PMOS charge pumps using low-voltage capacitors," *IEEE Transactions on Industrial Electronics*, "in press".

[2] R-A. Bianchi, C. Raynaud, F. Blanchet, F. Monsieur, O. Noblanc, "High voltage devices in advanced CMOS technologies," *IEEE Custom Integrated Circuits Conference*, 2009, pp. 363-370.

[3] P. Favrat et al., "A 1.5 V supplied, CMOS ASIC for the actuation of an electrostatic micromotor," *IEEE, The Ninth Annual International Workshop on Micro Electro Mechanical Systems*, 1996, pp. 25-31.

[4] B. Serneels, T. Piessens, M. Stepert, W. Dehaene, "A high-voltage output driver in a standard 2.5 V 0.25 µm CMOS technology," *IEEE International Solid-State Circuits Conference*, 2004, pp. 146-148.

[5] B. Serneels, E. Geukens, B. De Muer, T. Piessens, "A 1.5W 10V-output Class-D amplifier using a boosted supply from a single 3.3V input in standard 1.8V/3.3V 0.18µm CMOS," *IEEE International Solid-State Circuits Conference*, 2012, pp. 94-96.

[6] M. Hoque, T. McNutt, J. Zhang, A. Mantooth, M. Mojarradi, "A high voltage Dickson charge pump in SOI CMOS," *IEEE Custom Integrated Circuits Conference*, 2003, pp.493-496.

[7] F. Chen, et al. "The effect of metal area and line spacing on TDDB characteristics of 45nm Low-k SiCOH dielectrics," *IEEE International Reliability Physics Symposium*, 2007, pp.382-389.

[8] R. Pelliconi, D. Iezzi, A. Baroni, M. Pasotti, P. Rolandi, "Power efficient charge pump in deep submicron standard CMOS technology," *IEEE European Solid-State Circuits Conference*, 2001, pp. 73-76.

[9] K. Yallup, B. Lane, S. Edwards, "Back gate effects in thick film SOI CMOS devices," *IEEE International SOI Conference*, 1991. pp. 48-49.

978-1-4673-6145-3/13 $31.00 © 2013 IEEE

# High-Sensitivity Photodetection Sensor Front-End, Detecting Organophosphourous Compounds for Food Safety

Lei Wan[1], Yajie Qin[1], Patrick Chiang[1,2], Guoping Chen[1], Ran Liu[1], Zhiliang Hong[1]

[1]State Key Laboratory of ASIC & System, Fudan University, Shanghai, 201203, China

[2]Oregon State University, Corvallis, OR, 97331, USA

{yajieqin, gpchenapple}@fudan.edu.cn

*Abstract* —A high-sensitivity, high dynamic range photodetection sensor front-end is presented, suitable for low-cost hand-held food safety systems. This sensor front-end for detecting organophosphorus (OP) compounds incorporates an on-chip deep N-well photodetector, pulse width modulation (PWM), and a folded reference. Designed in a 0.18um process, measurement results show an input optical power dynamic range of 71dB, a sensitivity of 3.6nW/cm² (0.77pA), and a power consumption of 14.5uW. OP compound detection experiments demonstrate a limit of detection (LOD) of 0.16u mol/L, comparable to that of a commercial spectrophotometer.

*Index Terms* — optical sensor, organophosphorus compounds, low power, high sensitivity, CMOS sensor.

## I. INTRODUCTION

With the continued advancement of globalization and population growth, environmental and food security issues have recently received significant attention. In the past decade, the use of organophosphorus (OP) compounds as pesticides has proliferated in agricultural farming [1-2], aggravating the pollution of food, water and vegetables. Consequently, the development of a convenient, real-time OP compound detection method would be of great interest. Previously, OP compounds have been detected using electrochemical sensing, piezoelectric detection or photoluminescence [3]. Although these previous methods have shown sufficient accuracy, they require expensive instruments/reagents and time-consuming pre-treatments, and therefore are not suitable for hand-held, real-time field detection. In this work, a compact photodetection system is proposed that can detect hydrolyzed OP compounds within fruit, grains and water (Fig. 1).

Commercial photodetection systems [4] use a charge-coupled device (CCD) as their detector in order to achieve high sensitivity. While a CCD exhibits good image quality, it is not compatible with the CMOS process and is therefore difficult to integrate with other circuits on a single chip. For example, readout circuits and photodiodes are compatible with CMOS processes and therefore easy to be integrated on-chip. Previous works in CMOS typically utilize an ADC to quantize the output voltage of the transimpedance amplifier (TIA) [5], resulting in several disadvantages. For example, a capacitive SAR ADC can be low power, but has large area overhead. On the other hand, a sigma-delta ADC can achieve better resolution, but consumes significant power.

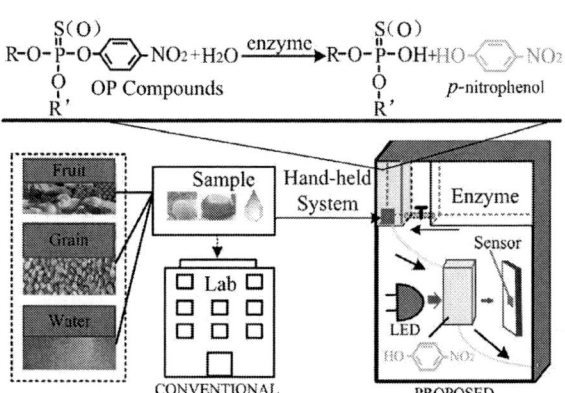

Fig. 1: Proposed food safety field detection system using hand-held sensor, compared with traditional lab detection

Pulse-width modulation (PWM) is another method for analog quantization [6], which converts the input photocurrent to a pulse width. Comparators instead of an ADC are used to convert the output voltage of the TIA into a PWM signal. In this way, the complexity and power consumption of the system are reduced. However, the reference voltage is typically fixed, such that the conversion requires excessively long integration time when the input optical intensity is small.

The optical detection sensor presented in this paper demonstrates several improvements compared with previously reported systems. First, the optical detection circuit here employs a folded reference to achieve both large dynamic range and high sensitivity at the same time. Second, the photocurrent is converted into a PWM signal, achieving high sensitivity and low power compared with a conventional time-to-digital converter (TDC). Furthermore, this pulse width modulation can easily be sent by a low power wireless transceiver, such as an ultra wideband radio [7]. Hence, the high-resolution time quantization can be performed by the remote wireless base station, further reducing the power of the sensor node. Finally, the photodiode can be manufactured in a standard CMOS process, and integrated with other peripheral circuits for a complete single sensor-on-chip solution.

The paper is organized as follows: In Section II, the operating principle of the chip and the folded reference, PWM-based optical detection circuit architecture is introduced. In Section III, the measurement setup and the chip test results are described. Section IV provides the experimental results and comparisons with other previous works. Finally, the conclusion is given in Section V.

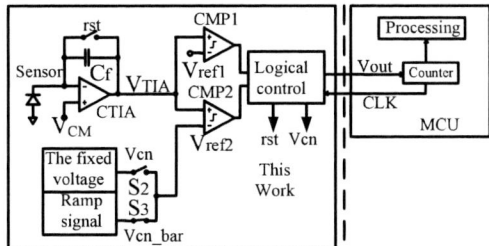

Fig. 2: Architecture of the folded-reference detector

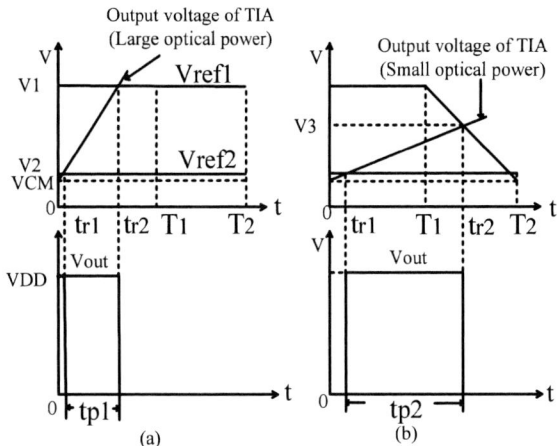

Fig. 3: Operating principle of the folded-reference detector when input optical power is: (a) large or (b) small

## II. FOLDED-REFERENCE, PWM-BASED DETECTION

### A. Architecture

The architecture of the folded-reference detector, shown in Fig. 2, consists of a TIA with a T-type low leakage current switch, two comparators, a ramp generator, and several logic circuits. The reference voltage of *CMP1* is a fixed voltage and *CMP2* is a folded reference. If the input optical power is large, the photocurrent is large as well, and therefore the output voltage of the TIA reaches $V_1$ in a short time duration. In contrast, when the input optical power is small, the photocurrent is also small, and the voltage $V_{ref1}$ becomes a folded reference voltage similar to Fig. 3 (b), thereby decreasing the integration time. Finally, an external *MCU* is used to measure the pulse width and calculate the photocurrent that is correlated to the input optical power.

### B. Operating principle

One major problem for conventional PWM-based optical detection systems is when the received input optical power is too small, the detector takes excessive time to perform quantization. In order to address this problem, this work utilizes a folded-reference voltage rather than a fixed reference in order to decrease integration time. Specifically, during $0 \sim T_1$, $V_{ref1}$ is a fixed voltage, similar to a conventional PWM based system. When the output voltage of the TIA is larger (less) than $V_1$ at $T_1$, $V_{ref1}$ is a fixed voltage (ramp signal) during $T_1 \sim T_2$. Fig. 3 describes these two conditions mentioned above.

When $V_{ref1}$ is a folded-reference, the expression is as follows below, with equation (1). The output voltage of the TIA across the two reference voltages $V_{ref2}$ and $V_{ref1}$ at times $t_{r1}$ and $t_{r2}$, respectively, is turned into a PWM signal by two comparators and proceeding logical gates. The photocurrent can be expressed as (2), where $C_f$ is the integration capacitor. The integration time $T_i$ required to cross the two different voltages is expressed in (3).

$$V_{ref1} = \begin{cases} V_1 \cdots\cdots\cdots\cdots\cdots\cdots (0 < t_{r2} \leq T_1) \\ (t_{r2} - T_1) \cdot \dfrac{V_{CM} - V_1}{T_2 - T_1} + V_1 \cdots (T_1 < t_{r2} \leq T_2) \end{cases} \quad (1)$$

$$I_{photo} = C_f \cdot \left(V_{ref1} - V_{ref2}\right) / \left(t_{r2} - t_{r1}\right) \quad (2)$$

$$T_i = \left(V_{ref1} - V_{ref2}\right) \cdot C_f / I_{photo} \quad (3)$$

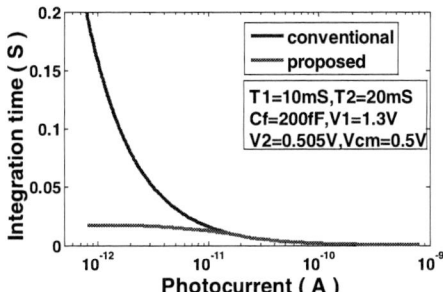

Fig. 4: Photocurrent versus integration time

If $V_{ref1}$ is a fixed voltage when the photocurrent $I_{photo}$ is small, the difference between $V_{ref1}$ and $V_{ref2}$ is quite large, and $T_i$ can be very large. However, if a folded reference is adopted, such as in expression (1), the integration time $T_i$ can be significantly reduced. Simulation results of two different integration times for both a conventional fixed reference and the proposed folded reference. As can be seen in Fig. 4, when the photocurrent is lower than 10pA, the integration time of the proposed reference ascends very slowly compared with the fixed reference.

There are some additional benefits that can be obtained by using this improved photocurrent sensor front-end. First, the use of two comparators instead of an entire ADC simplifies the architecture and lowers the power consumption. Second, the offsets from the op-amp, clock feed through and switch charge injection are eliminated using correlated double sampling. Third, the proposed folded-reference voltage $V_{ref1}$ increases the dynamic range (DR) of the input optical power, as it permits a smaller input optical power compared with conventional detection.

978-1-4673-6145-3/13 $31.00 © 2013 IEEE

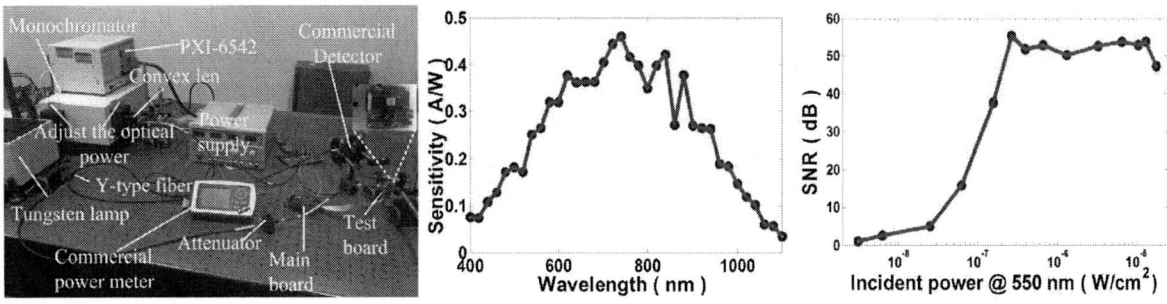

Fig. 5: (a) Measurement environment setup (b) Measured sensitivity of the n-well/p-sub photodiode sensor
(c) SNR of the chip with increasing incident power at 550nm and a 20ms exposure time

## III. RESULTS AND DISCUSSION

The measurement setup of the chip is shown in Fig. 5 (a). A tungsten lamp is used as the light source, exhibiting a wide spectral range and stability of the output optical power. The tungsten lamp is connected to a monochromator that can adjust the input and output optical power on to the chip. After the light is focused using a convex lens, the output light of the monochromator is sent to a Y-type optical fiber. One end of the Y-type optical fiber is connected to the detector input of a commercial optical power meter (Thorlabs PM100D) for quantizing the input incident power. The other end of the Y-type fiber is connected through an attenuator to the input of the proposed sensor chip. After calibrating the ratio of the outputs of the two optical fibers, the amount of optical power attenuated by the attenuator can be obtained. The clock is generated by a microcontroller (MSP430F149, TI) and the 1.8V power supply is generated by a low dropout regulator (TPS74701, TI). For test purposes, the output PWM signal is sent to a signal acquisition board (PXI-6542, NI) and the data is later post-processed in Matlab.

The optical detection chip was implemented in a 0.18um CMOS process, shown in the bottom-right corner of Fig. 6. The power consumption of the entire sensor is 14.5uW. The input optical power, centered at 550nm, is generated by the tungsten lamp and the monochromator. The photocurrent was calculated by measuring the pulse width versus the input incident power, as shown in Fig. 6.

The lowest power that can be detected by the chip at a signal-to-noise ratio (SNR) of 0.8dB is 3.6nW/cm$^2$. A total of 25 optical powers were measured from 3.6nW/cm$^2$ to 13.9uW/cm$^2$, for a total dynamic range of 71.7dB.

The chip was run at 50Hz with an exposure time of 20ms. A 10MHz clock signal in the signal acquisition board was used to digitize the PWM signal of the optical sensor. For an integration capacitor of 200fF, the measured photodiode responsivity was 0.27A/W at 550nm, approximately 62% of the maximum sensitivity at 760nm.

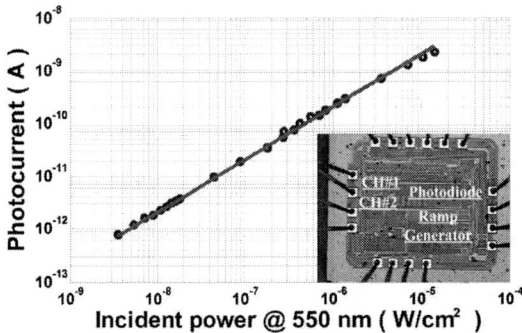

Fig. 6: Photocurrent versus incident power at 550nm with a 20ms exposure time; chip microphotograph

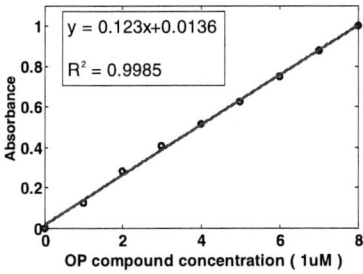

Fig. 7: Absorbance with increasing OP concentration

The dark signal was measured for an integration time of 9.7s, from 0.5V to 1.3V, with an integration capacitor of 200fF. This implies a dark signal of 82.4mV/s, at the output of the TIA, with an input-referred dark current of 16.5fA. Considering the pixel area of $3.84 \times 10^{-4}$cm$^2$, this equates to a dark current intensity of 0.043nA/cm$^2$.

The measured sensitivity of the N-well/p-sub photodiode is shown in Fig. 5(b). The peak responsivity wavelength was between 600nm to 700nm. More than 50 pulse width samples were collected at a particular optical power and measured by the power meter. Their mean and standard deviation were computed and plotted in Fig. 5(c), where SNR is defined as their mean divided by their standard deviation.

TABLE-1 shows a comparison between our work and prior works. It can be seen that the proposed work exhibits better sensitivity with less power consumption and larger dynamic range. Hence, these characteristics make the detector suitable for low-cost, hand-held optical sensor applications.

## IV. OP COMPOUND DETECTION

### A. Principle of Detection

The basic principle of OP compound detection occurs by measuring the absorbance of yellow *p*-nitrophenol, which is the hydrolysis product of OP compounds that are normally colorless. The absorbance of the liquid ($A_s$) can be expressed as follows:

$$A_s = \lg\left(I_s / I_t\right) = c\varepsilon_s l \qquad (4)$$

Where c is the concentration of the yellow *p*-nitrophenol, $\varepsilon_s$ is the absorptive coefficient, $l$ is the optical path, and $I_s$ and $I_t$ are the power of the incident light and the transmitted light, respectively. According to equation (4), the absorbance of the liquid is linearly correlated to the *p*-nitrophenol concentration, which is linear to the OP compound concentration (one mole of OP compound can be hydrolyzed into a mole of *p*-nitrophenol).

### B. Detection Results

A blue LED emitting at 405nm was used as the light source. The transmitted light passing through OP solutions was collected by the sensor chip. Nine different concentration solutions of methyl parathion (one type of OP compound) from 0 to $8\times10^{-6}$ mol/L are measured by the chip. The relationship between absorbance and OP concentrations is shown in Fig. 7, and the results show a significant linear relationship up to 8uM ($R^2$=0.9985). The limit of detection (LOD) is estimated at $0.16\times10^{-6}$M, which is three times that of the standard deviation of the absorbance measurement for the same buffer without MP.

TABLE-2 compares the characteristics of this work versus previously published work. The measured LOD in this paper is comparable to previous works that use bulk, discrete expensive instruments.

## V. CONCLUSION

The proposed photodetection sensor introduces pulse width modulation with a folded reference, achieving a large dynamic range, high sensitivity, and low power consumption. Hence, the proposed sensor chip is suitable for future low-cost, hand-held food safety and agricultural pesticide monitoring.

## ACKNOWLEDGEMENTS

The authors are grateful to the 863 programs of China under grant 2011AA100701, 2011AA100706-2, NSFC program under grant 61076027, NSFC program under grant 61177021, and the State Key Laboratory of ASIC & System, Fudan University, for funding this work.

**TABLE I:**
**COMPARISON OF LOW-INTENSITY CMOS SENSORS**

| Parameter | This work | [5] TBCAS' 2011 | [8]ISSCC' 2011 |
|---|---|---|---|
| Detector | n-well/p-sub | n-well/p-sub | GaAs detector |
| Die Size (mm$^2$) | 0.8×1 | 3×3 | 0.9×1.4 |
| Pixel Size(um$^2$) | 240×160 | 20.1×20.1 | NA |
| Technology | 0.18um CMOS | 0.5um CMOS | 0.18um CMOS |
| DR | 71dB | 48dB | 77dB |
| Sensitivity | 3.6nW/cm$^2$ (0.77pA) | 4nW/cm$^2$ | 4pA |
| Integration Time | 20mS | 14.3mS | 0.5mS |
| Power consumption | 14.5uW | 2.37mW (Array) | 2.4mW |

**TABLE II:**
**COMPARISON OF THE LIMIT OF DETECTION OF OP COMPOUNDS**

| Parameter | This work | [1]Sensors'2012 | [2] Biosensors andBioelec'2006 |
|---|---|---|---|
| Detection method | Optical absorption(@405nm) | Optical absorption(@400nm) | Optical absorption(@410nm) |
| Instrument | Chip (CMOS) | PIN-FET | Spectrometer (CCD) |
| LOD | 0.16uM | 4uM | 0.3uM |

## REFERENCES

[1] W.Lan, G.Chen, F.Cuiet. al."Development of a novel optical biosensor for detection of organophoshorus pesticides based on methyl parathion hydrolase immobilized by metal-chelate affinity," *Sensors*, pp. 8477-8490, 2012.

[2] J.Kumar, S.Jha, J.Seo et. al. "Optical microbial biosensor for detection of methyl parathionpesticide using Flavobacterium sp. whole cells adsorbedon glass fiber filters as disposable biocomponent," *Biosensors and Bioelectronics*, vol. 21, issue 11, May 2006,pp. 2100-2105.

[3] Li, L., Zhou S.S., Jin L.Xet. al. "Enantiomeric separation of organophosphoruspesticides by high-performance liquid chromatography, gas chromatography and capillaryelectrophoresis and their applications to environmental fate and toxicity assays," J. Chromatogr. B2010, 878, 1264–1276.

[4] www.ideaoptics.com/Products/PContent.aspx?pd=PG4000

[5] Murari K., Etienne-Cummings R., Thakor N.V., Cauwenberghs, G., "A CMOS In-Pixel CTIA High-Sensitivity Fluorescence Imager," *Biomedical Circuits and Systems, IEEE Transactions on*, pp. 449-458, Volume:5 , Issue: 5, 2011.

[6] C.Xu, S.Chaoet al, "A new correlated double sampling (CDS) technique for low voltage design environment in advanced CMOS technology," Solid-state circuit conference, ESSCIRC 2002, pp.117-120.

[7] C.Hu, Chiang P.Y., K.Hu et al; "A 90nm-CMOS, 500Mbps, fully-integrated IR-UWB transceiver using pulse injection-locking for receiver phase synchronization," Radio Frequency Integrated Circuits Symposium (RFIC), 2010 IEEE, pp.201-204, 2010.

[8] Heitz R.T., Barkin D.B. et. al. "A low noise current readout architecture for fluorescence detection in living subjects," *Solid-State Circuits Conference Digest of Technical Papers (ISSCC)*, pp. 308- 310, 2011.

# A 16-Channel, 359 $\mu$W, Parallel Neural Recording System Using Walsh-Hadamard Coding

Vahid Majidzadeh, Alexandre Schmid, and Yusuf Leblebici

Microelectronic Systems Laboratory (LSM), Swiss Federal Institute of Technology (EPFL), CH-1015 Lausanne

Email: vahid.majidzadeh@a3.epfl.ch

*Abstract*—**Application of an algebraic coding to a multi-channel parallel neural recording system is presented. The Walsh-Hadamard coding enables back-end hardware sharing between recording channels, using a linear and orthogonal superposition of the analog inputs. Moreover, this technique preserves the temporal information of the channels in contrast to the conventional architectures which use time-multiplexed ADC. In the proposed architecture a single ADC operates on a superposed signal, thereby the dynamic range of the ADC is effectively shared between channels benefiting from the sparsity characteristics of the channels. A 16-channel parallel recording system is implemented as a proof of concept. The system is implemented in a 0.18 $\mu$m CMOS technology and occupies 1.99 mm$^2$ of silicon area. The input-referred noise of a single channel integrated from 10 Hz to 100 kHz equals to 4.1 $\mu$V$_{rms}$, and the effective power consumption of each channel is measured at 22.4 $\mu$W from a 1.2 V power supply, which results in a system level NEF of 5.6.**

## I. INTRODUCTION

Action potentials and spikes are generated as electro-chemical operation of individual neuron's membrane. While individual recording is indispensable for understanding the behavior of single neuron, neuroscientists devote significant attention to studying the behavior of population of neurons. These studies target a better understanding of the complex brain network and its cognitive functions.

The number of recording channels is predicted to exceed thousand in a near future. Although extensive research and development have been conducted to improve the performance of the recording from individual channels [1], considerable improvements are still needed at the system level to support recording from large number of channels.

The multichannel recording system in [2] utilizes a single SAR ADC and a time-multiplexing technique to digitize the neural signals at the cost of losing some temporal information of the channels. A compressive modulation scheme based on a double sampling analog memory and a switched-capacitor delta circuit is presented in [3], which enables reducing the dynamic range requirement of the ADC and the data rate.

The recent architectures which are proposed in [4] [5] maintain the temporal information of the channels using a dedicated successive approximation register (SAR) ADC in each channel.

The frequency devision multiplexing (FDM) technique is proposed in [6]. This technique enables parallel recording from multiple channels. However, the overhead of the generation of the frequency references is considerable, since two precise crystal oscillators along with a multi-output frequency synthesizer are required to generate accurate frequency spacing between channels.

In this paper, the Walsh-Hadamard coding is presented as an alternative technique. A single ADC is used to digitize the orthogonal superposition of the channels. The modulation codes are generated using low power digital logics, which mitigates the need for power hungry frequency synthesizer.

This paper is organized as follows. Section II presents a brief introduction to Walsh-Hadamard coding technique. Section III introduces the system architecture embedding the Walsh-Hadamrd coding. In Section IV brief circuit implementation are discussed. Section V presents measurement results of the proposed architecture. Finally, Section VI concludes the paper.

## II. INTRODUCTION TO WALSH-HADAMARD CODING

The Walsh-Hadamard transform is a general class of Fourier transform that performs an orthogonal, symmetric, and linear operation on a vector consisting of $2^n$ data [7]. Indeed, this transform decomposes an arbitrary input vector signal to Walsh-Hadamard bases. The base matrix has a size of two and defined as:

$$H_1 \triangleq \frac{1}{\sqrt{2}} \begin{bmatrix} 1 & 1 \\ 1 & -1 \end{bmatrix} \tag{1}$$

The Hadamard matrix of size $n$ is defined as a recursive Kronecker product of two matrices:

$$H_n = H_1 \otimes H_{n-1} = \begin{bmatrix} H_{n-1} & H_{n-1} \\ H_{n-1} & -H_{n-1} \end{bmatrix} \tag{2}$$

The Hadamard matrix can also be obtained using binary representation of its element located in row $m$ and column $k$ as:

$$h(m,k) = (-1)^{\sum_{i=0}^{n-1} k_i m_i} = \prod_{i=0}^{n-1} (-1)^{k_i m_i} \tag{3}$$

where

$$k = \sum_{i=0}^{n-1} k_i 2^i = (k_{n-1} k_{n-2} \dots k_1 k_0)_2 \qquad (k_i = 0, 1)$$

$$m = \sum_{i=0}^{n-1} m_i 2^i = (m_{n-1} m_{n-2} \dots m_1 m_0)_2 \quad (m_i = 0, 1) \tag{4}$$

This binary representation is used to implement the digital logic generating the Walsh-Hadamard matrix coefficients. The

Walsh-Hadamard matrix presents interesting properties for vector-based signal processing. $H_n$ is symmetric and orthogonal, and can be used for linear and independent superposition of multiple signals.

An example of using the Walsh-Hadamard transform in multi-channel recording is shown in Fig. 1. The N-channel signal is shown as vector X. Linear and independent superposition of channels is achieved by multiplying X by $H_n$, where $n = log_2^N$. Since all matrix elements are equal to $\pm 1$, the multiplication operation is performed by exchanging the polarity in a differential circuit implementation. At the receiver side, the inverse of $H_n$ is used to reconstruct the original channel signal. Since $H_n$ is orthogonal, no systematic crosstalk between channels is introduced. Moreover, in contrast to techniques such as compressed sensing [8] Walsh-Hadamard coding provides a loss-less reconstruction.

## III. SYSTEM ARCHITECTURE

The architecture of the 16-channel neural recording system implementing the Walsh-Hadamard coding technique is shown in Fig. 2. Each channel is composed of a low-noise amplifier (LNA), a biquad low-pass filter (LPF) to avoid aliasing, a track and hold (T/H) circuit which operates at Nyquist rate and holds the input signal for a period of 16 successful modulation and sub-conversion cycles of the ADC, and a modulator which is controlled by the Walsh-Hadamard code generator block. Each channel performs modulation in current domain using a dedicated code generator logic which uses the channel index number for successive code generation, e.g., channel number one use the code generator h(m,1) and channel number k uses the code generator h(m,k). The outputs of channels are all summed up in current domain using wired addition, and further signal processing is performed through shared far-end hardware in voltage domain.

The variable gain amplifier (VGA) provides a programmable voltage gain ranging from 27.6 dB to 37.1 dB in three steps as well as serves as an on-chip ADC driver. A single 10-bit SAR ADC simultaneously converts the superposition of all channels. The required ADC resolution ($B_{ADC}$) is determined based on the required resolution of each individual channel ($B_{Channel}$) [9]:

$$B_{ADC} = B_{Channel} + log_2 \sqrt{N} \qquad (5)$$

A serial interface is used to reduce the pin count overhead of

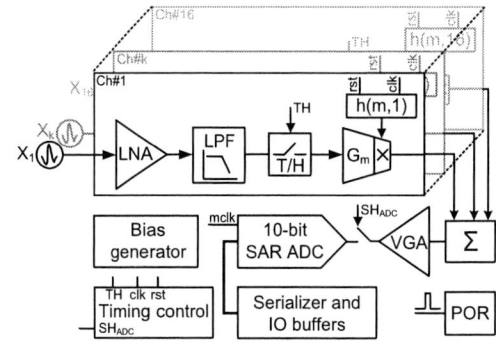

Fig. 2. Architecture of the 16-channel neural recording system implementing the Walsh-Hadamard coding technique.

the system. A power-on-reset (POR) circuit resets the ADC and timing control block at the rising of the supply voltage.

The timing diagram of the system is presented in Fig. 3. An external master clock (mclk) of 10 MHz is used to generate all required timings as well as to drive the SAR. The ADC's sample and hold signal SH is generated at a rate of 1/20th of the master clock cycle. This maximizes the sampling time of the ADC and mitigates the need for linear booststrap switches for sampling. The track and hold signal (TH) which samples the channel's output is generated at 1/17th of the SH in order to support the 16 consecutive modulation and sub-conversion operations. Modulation is performed at each rising edge of the SH and is followed by an analog to digital sub-conversion that occurs when the SH is in low state. Each sub-conversion lasts for 10 master clock cycles. After completion of 16 modulation and sub-conversion operation, the code generator is reset to its initial value and is ready to start a new conversion.

## IV. CIRCUIT IMPLEMENTATION

The front-end low-noise amplifier (LNA) is implemented using the capacitive coupling architecture, which enables rail-to-rail offset rejection. Bandwidth limitation is performed at the filter stage where the power is scaled down due to the relaxed noise requirements. The front-end LNA provides accurate gain of 22.6 dB and exhibits bandwidths of 11.5 Hz$\leq f \leq$ 712 kHz. The input referred noise is measured at 3.7 $\mu V_{rms}$ with current consumption equal to 11.2 $\mu A$. PSRR and CMRR are measured at 72.2 dB and 78 dB, respectively.

A lowpass filter (LPF) is required to avoid aliasing and limit the noise bandwidth. Fig. 4(a) shows the biquad filter schematic. Biquad operation and comlex poles are synthesized using positice feedback [10]. Non-existense of parasitic poles

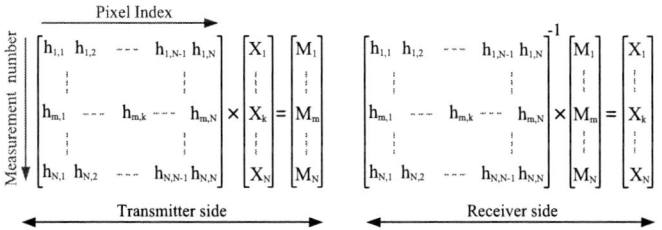

Fig. 1. Modulation and demodulation in a multi-channel recording system using Walsh-Hadamard coding.

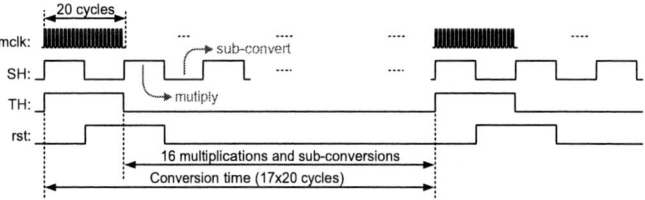

Fig. 3. Timing operation of the 16-channel neural recording system.

978-1-4673-6145-3/13 $31.00 © 2013 IEEE          229

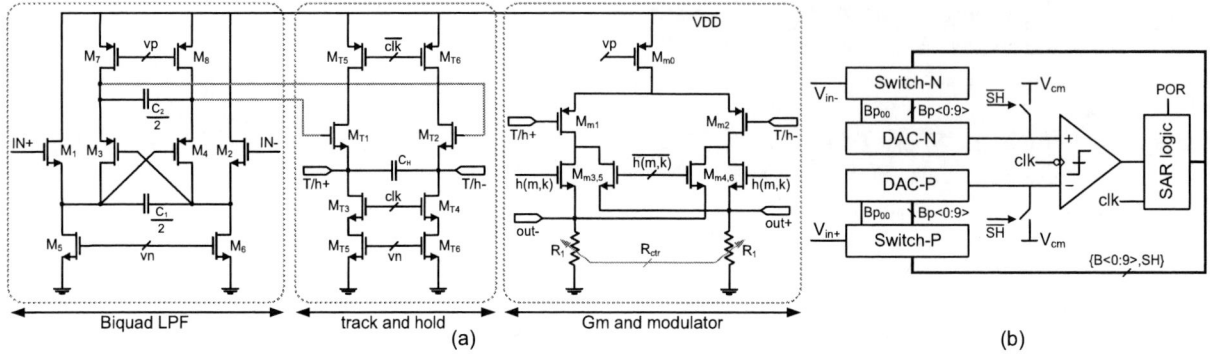

Fig. 4. Circuit schematic of the source-follower Biquad filter, track and hold, and Walsh-Hadamard modulator.

reduce the power consumption in this architecture compared with conventional architectures. The filter bandwidth is tunable over the frequency span of 7 kHz to 56 kHz, using a tuning bias current ranging from 10 nA to 100 nA.

The circuit schematic of the track and hold circuit is shown in Fig. 4(a). A source-follower buffer is used to track the input signal when the clock signal is in high state. During the negative alternate of the clock signal, the tail current source $M_{T5-6}$ and devices $M_{T7-8}$ operate in cut-off region, and the input signal is saved on the hold capacitor $C_H$. The proposed track and hold circuit only uses NMOS and PMOS switches and does not need either floating switch or clock boosting techniques.

The Walsh-Hadamard modulation is performed in current domain using a Gilbert-cell multiplier circuit which is shown in Fig. 4(a). The input voltage is converted to current, and further modulated in accordance to the coefficient by swapping the differential output currents. The output currents of all channels are superposed by a wired connection, and a passive polysilicon resistor converts the superposed current to voltage mode. The resistor is tunable in three steps using a two-bit digital control word. Fig. 5(a) presents the resistor control circuitry.

The binary representation of (4) is utilized to synthesize the Walsh code generator which is shown in Fig. 5(b). A 4-bit binary counter generates the time-varying modulation coefficient applied in each sub-conversion. The logic is identical in all channels, with the exception of the binary digital word $(k_3 \cdots k_0)$ which reflects the channel number and is a specific parameter to each channel.

A variable gain amplifier preceding the SAR-ADC maximizes the dynamic range. The VGA gain is distributed along two stages in order to optimize the area and power consumption. A 10-bit SAR-ADC with sampling rate of 500 kS/s is specified in order to achieve a resolution of 8-bit and sampling rate of 29.4 kS/s for each individual channel. The architecture of the 10-bit SAR ADC is presented in Fig. 4(b). The input signal is sampled on the bottom plate of the switched-capacitor DAC in order to mitigate the offset requirements of the comparator. Two sub-DACs, each with 5-bit resolution and an attenuator capacitor are used to implement a 10-bit DAC.

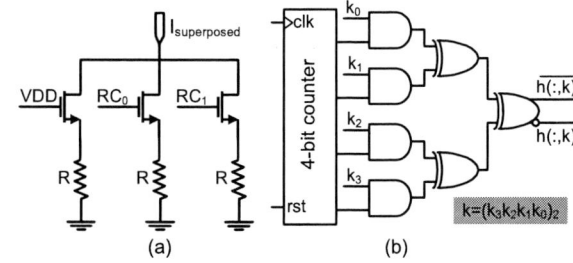

Fig. 5. (a) Resistor control circuitry, (b) digital logic implementing the Walsh-Hadamard sequence for each channel.

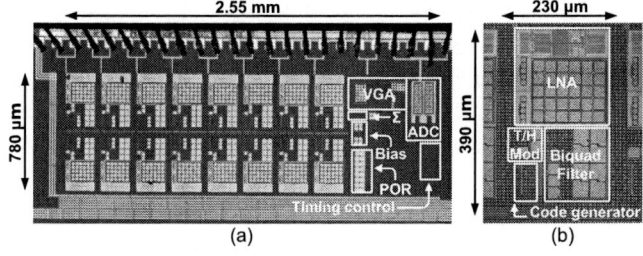

Fig. 6. (a) Chip photograph of the 16-channel neural recording system using the Walsh-Hadamard coding technique, (b) photograph with zoom into an individual channel.

This architecture reduces the input capacitance of the ADC and facilitates the design of the on-chip ADC driver . Supply rails serve as reference voltages of the ADC and the on-chip reference buffers are removed, which reduces the power consumption of the ADC.

## V. MEASUREMENT RESULTS

The 16-channel neural recording IC is implemented in a 0.18 $\mu$m CMOS technology. Fig. 6(a) shows the chip photograph which occupies a silicon area of 1.99 mm$^2$. A photograph with zoom into an individual channel is shown in Fig. 6(b). Each channel occupies a silicon area of 0.09 mm$^2$, which improves the spacial resolution of the recording system.

The modulated data is serially captured, and subsequently demodulated in the Matlab software. In order to verify the system functionality and performance, three channels are se-

978-1-4673-6145-3/13 $31.00 © 2013 IEEE

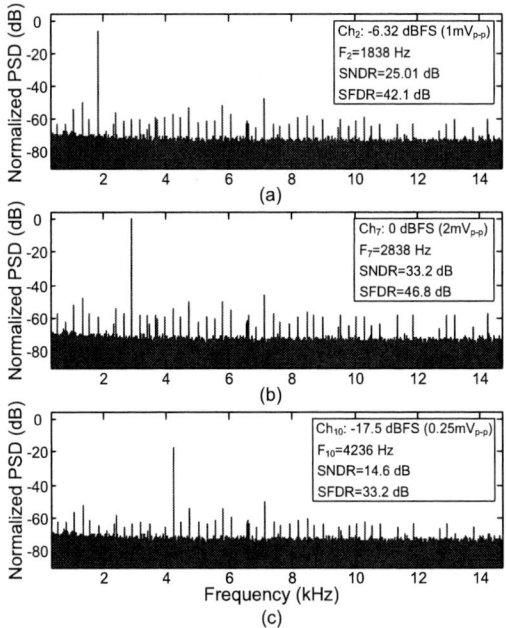

Fig. 7. (a) Demodulated output spectrum of channel number two, (b) demodulated output spectrum of channel number seven, (c) demodulated output spectrum of channel number ten.

lected and connected to independent signal sources. On-board passive voltage divider circuits at the input of the channels provide an attenuation equal to 26 dB, including a 50 $\Omega$ termination. Channel number two (CH$_2$) is fed using a single-tone sinusoid with a frequency and amplitude equal to 1838 Hz and 20 mV$_{p-p}$, respectively. Channel number 7 (CH$_7$) is stimulated using a single-tone sinusoid with a frequency and amplitude equal to 2838 Hz and 40 mV$_{p-p}$, respectively. Finally, channel number ten (CH$_{10}$) is stimulated with a single-tone sinusoid with a amplitude measured at 5 mV$_{p-p}$ and frequency measured at 4 kHz. The reference input and rest of the channels are connected to ground. Fig. 7 presents the demodulated output spectrum of the aforementioned channels. All spectrums are normalized with respect to CH$_7$. CH$_2$ achieves SNDR and SFDR equal to 25.01 dB and 42.1 dB, respectively, and exhibits a gain error equal to -0.32 dB. CH$_7$ which is named the strongest channel shows a SNDR equal to 33.2 dB and SFDR equal to 46.8 dB. The gain error of CH$_{10}$ is measured at +0.56 dB with SNDR and SFDR measured at 14.6 dB and 33.2 dB, respectively. The spurs observed in the spectrum originate from intermodulation of the channels due to the nonlinearity of the VGA driving the 10-bit SAR ADC. The nonlinearity of the VGA which is shared between all channels introduces error that appears as spurs in the output spectrum of all demodulated channels. The weak channels performance are affected more significantly from nonlinearity, as nonlinearity is mainly introduced by strong channels.

Table I shows the summary of the performance and comparison with the state of the art works. Channel coding technique enables parallel recording using a single ADC. The proposed architecture reports better trade-off between noise and power

TABLE I
SUMMARY OF THE PERFORMANCE AND COMPARISON WITH THE STATE OF THE ART WORKS.

| Parameter | This work | [4] | [5] |
|---|---|---|---|
| Architecture | Single ADC | SAR-ADC/ch | SAR-ADC/ch |
| Feature | Coding | Multiplying ADC | Windowed Integrator |
| Gain | 41-60 dB | 54-60 dB | 56 dB |
| Bandwidth (Hz) | $11.5 \leq f \leq 7\text{-}56$ k | $10 \leq f \leq 5$ k | $280 \leq f \leq 10$ k |
| Input noise | 4.1 $\mu V_{rms}$ | 6.5 $\mu V_{rms}$ | 2.2 $\mu V_{rms}$ |
| Power/ch | 22.4 $\mu W$ | 12.5 $\mu W$ | 68 $\mu W$ |
| NEF | 5.6 | 7.2 | 6.4 |
| Area/ch | 0.09 mm$^2$ | 0.09 mm$^2$ | 0.26 mm$^2$ |
| Technology | 0.18 $\mu m$ | 0.13 $\mu m$ | 0.13 $\mu m$ |
| Supply voltage | 1.2 V | 1.2 V | 1.2 V |

consumption (NEF), and occupies a competitive silicon area of 0.09 mm$^2$ in each recording channel.

## VI. CONCLUSION

Walsh-Hadamard coding enables parallel recording from multiple channels using a single ADC. The linear and orthogonal combination of channels maps the spacial information of the channels to the temporal information of a superposed signal. Further processing of the superposed signal is carried out using a shared far-end hardware architecture. The shared hardware architecture enables the effective usage of the dynamic range of the ADC between recording channels. Moreover, the spacial resolution of the recording sites is improved by moving the shared signal processing hardware to the outside of the sensor plane.

## REFERENCES

[1] R. Muller, S. Gambini, and J. Rabaey, "A 0.013mm$^2$ 5$\mu$W DC-DC-coupled neural signal acquisition IC with 0.5V supply," ISSCC Dig. Tech. Papers, pp. 302-303, Feb. 2011.

[2] R. R. Harrison, et al., "A Low-Power Integrated Circuit for a Wireless 100-Electrode Neural Recording System," IEEE J. Solid-State Circuits, vol. 42, no. 1, pp. 123-133, June 2007.

[3] J. N. Y. Aziz, et al., "256-Channel Neural Recording and Delta Compression Microsystem With 3D Electrodes," IEEE J. Solid-State Circuits, vol. 44, no. 3, pp. 995-1005, Mar. 2009.

[4] K. Abdelhalim, and R. Genov, "915-MHz wireless 64-channel neural recording SoC with programmable mixed-Signal FIR filters," Proc. European Solid State Circuit Conf., pp. 223-226, Sept. 2011.

[5] H. Gao, R. M. Walker, P. Nuyujukian, K. A. Mikanawa, K. V. Shenoy, B. Murmann, and T. H. Meng, "HermesE: A 96-channel full data rate direct neural interface in 0.13 $\mu$m CMOS," IEEE J. Solid-State Circuits, vol. 47, no. 4, pp. 1043-1055, April. 2012.

[6] A. R. Tavakkol, A. M. Sodagar, and M. H. Refan, "New architecture for wireless implantable neural recording microsystems based on frequency-division multiplexing," in Proc. IEEE Conf. Engineering in Medicine and Biology Society, (EMBC'10), pp. 6449-6452, 2010.

[7] R. E. Blahut, "Algebraic Codes for Data Transmission," Cambridge University press, 2003.

[8] F. Chen, A. Chnadrakasan, V. Stojanović, "Design and Analysis of a Hardware-Efficient Compressed Sensing Architecture for Data Compression in Wireless Sensors," IEEE J. Solid-State Circuits, vol. 47, no. 3, pp. 744-756, March 2012.

[9] V. Majidzadeh, A. Schmid, and Y. Leblebici, "Energy Efficient Low-Noise Neural Recording Amplifier with Enhanced Noise Efficiency Factor," IEEE Trans. on Biomedical Circuits and Systems, vol. 5, no. 3, pp 262-271, June 2011.

[10] S. Amico, M. Conta, and A. Baschirotto, "A 4.1-mW 10-MHz fourth-order source-follower-based continuous-time filter with 79-dB DR," IEEE J. Solid-State Circuits, vol. 41, no. 12, pp. 2713-2719, Dec. 2006.

# Analysis of Deviation from Pelgrom Scaling Law in $V_{th}$ Variability of Pocket-implanted MOSFET

Kiyohiko Sakakibara, Yaichiro Miura, Toshio Kumamoto, and Susumu Tanimoto

Renesas Electronics Corporation

*Abstract*- **This paper analyzes cause of deviation from Pelgrom scaling law in threshold voltage ($V_{th}$) variability of pocket-implanted long channel MOSFET. It has been reported that this deviation from Pelgrom scaling law becomes remarkable in 65nm and beyond technologies. It is clarified that deviation from Pelgrom scaling law is attributed to increasing behavior of offset-voltage variability $\sigma(\Delta I/g_m)$ in weak and moderate inversion regions. It is found that this increasing behavior of $\sigma(\Delta I/g_m)$ can be completely eliminated by using both-side (BS) ring gate structure. This means that deviation from Pelgrom scaling law is caused by subthreshold hump.**

## I. INTRODUCTION

Transistor variability plays an important role in technology scalability, low-voltage operation, and stable yield in mass production. In order to realize low-cost devices, fine gate pattern MOSFET is indispensable. Pocket implant (halo) technology has been generally used to achieve nanometer scaled gate length MOSFET. It was reported that pocket-implant technology affects analog performances [1]-[5]. Recently, deviation from Pelgrom scaling law in $V_{th}$ variability of pocket-implanted long channel MOSFET was reported especially in 65nm and beyond technologies [3]-[5]. However, cause of the deviation has not been clarified. On the other hand, STI structure is generally used in nanometer scaled channel width MOSFET. In STI structure, avoiding the subthreshold hump becomes more difficult [6]. It was verified that current mismatch behavior in weak and moderate inversion regions is affected by the subthreshold hump [7][8].

In this work, cause of deviation from Pelgrom scaling law of pocket-implanted long channel MOSFET will be clarified from analysis of offset-voltage variability $\sigma(\Delta I/g_m)$.

## II. DEVIATION FROM PELGROM SCALING LAW IN Vth VARIABILITY

Deviation from Pelgrom scaling law in $V_{th}$ variability $\sigma(\Delta V_{th})$ of pocket-implanted long channel MOSFET can be observed also in CMOS technology before 65nm. $\sigma(\Delta V_{th})(LW)^{1/2}$ versus L plot in our 90nm technology is shown in Fig. 1. STI fabrication process is experimentally changed. $V_{th}$ values of HVT and LVT devices are 550mV and 260mV at L=0.15μm, respectively. $V_{th}$ value is defined as the gate voltage required to drive a current that is equal to IL/W=10nA ($V_{th}$@10nA). HVT and LVT devices share the same pocket implant, but have different well implant. 5μm channel width devices are measured in saturation condition.

Deviation behavior depends on channel profile. The deviation becomes remarkable in LVT devices as shown in Fig. 1. Namely, as a difference of doping concentration between pocket and the adjacent well becomes larger, the deviation in long devices becomes remarkable. This trend is observed in many companies according to the paper [4].

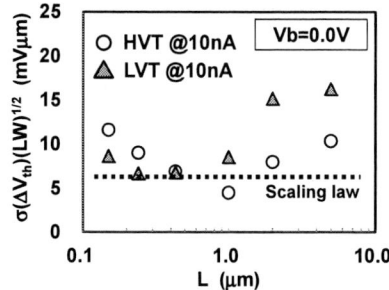

Fig. 1. Measurement results of $\sigma(\Delta V_{th})(LW)^{1/2}$ versus L (gate length) in 90nm N-channel MOSFET at Vb=0.0V. Deviation from scaling law becomes larger especially in LVT long devices.

It is verified that deviation from scaling law depends on bulk-bias level. In Fig. 2, various bulk-bias dependence can be seen according to gate length and channel profile. In HVT short devices, bulk-bias dependence of deviation from scaling law becomes remarkable. While, in HVT long devices and LVT devices, levels of $\sigma(\Delta V_{th})(LW)^{1/2}$ hardly depend on bulk-bias level.

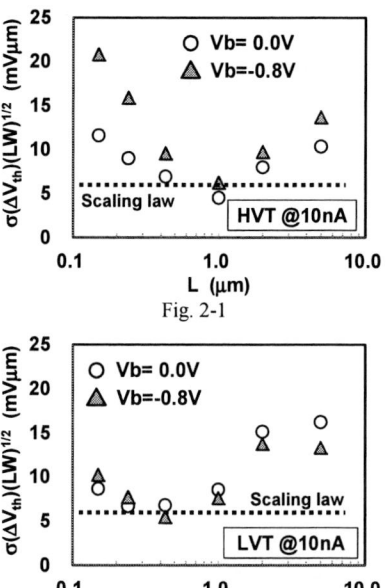

Fig. 2-1

Fig. 2-2

Fig. 2. Bulk-bias dependence of $\sigma(\Delta V_{th})(LW)^{1/2}$ versus L in HVT (Fig. 2-1) and LVT (Fig. 2-2) devices at Vth@10nA. In HVT short devices, deviation from scaling law becomes remarkable by applying bulk-bias.

It is also verified that deviation from scaling law depends on $V_{th}$ criteria-current level, as in Fig. 3. When $V_{th}$ value is defined as the gate voltage of a current level of IL/W=10μA, deviation from scaling law is eliminated in both HVT and LVT devices, respectively.

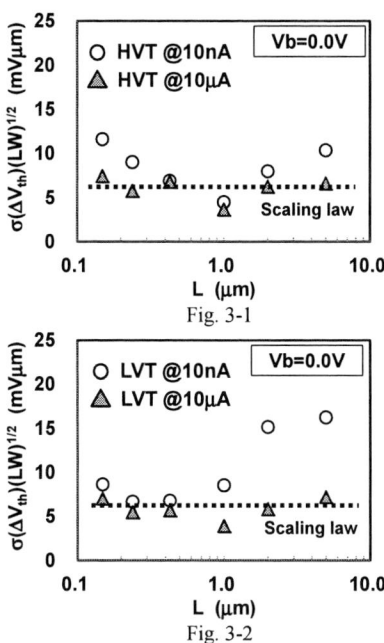

Fig. 3-1

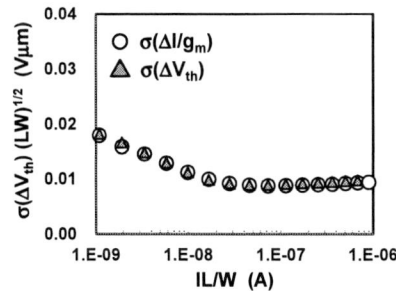

Fig. 3-2

Fig. 3. $V_{th}$ criteria-current level dependence of $\sigma(\Delta V_{th})(LW)^{1/2}$ versus L in HVT (Fig. 3-1) and LVT (Fig. 3-2) devices at Vb=0.0V. By selecting $V_{th}$ criteria-current level in strong inversion region, deviation from scaling law is removed in both HVT and LVT devices.

### III. ANALYSIS OF $\sigma(\Delta I/g_m)(LW)^{1/2}$ VERSUS IL/W PLOT

In order to clarify mechanism of the deviation from scaling law, we focus on $\sigma(\Delta I/g_m)(LW)^{1/2}$ versus IL/W plot. When $V_{th}$ value is locally defined as a gate voltage required to drive each current of IL/W, value of $\sigma(\Delta V_{th})$ becomes equal with value of $\sigma(\Delta I/g_m)$ in all inversion regions shown in Fig. 4. Thus, the deviation behavior from scaling law can be analyzed by using $\sigma(\Delta I/g_m)(LW)^{1/2}$ versus IL/W plot at each gate length.

Fig. 4. Comparison between $\sigma(\Delta V_{th})$ and $\sigma(\Delta I/g_m)$ versus IL/W.

The deviation behavior from scaling law which depends on channel profile, bulk-bias level, and Vth criteria-current level is caused by an increasing behavior of $\sigma(\Delta I/g_m)(LW)^{1/2}$ versus IL/W plot in weak and moderate inversion regions. $\sigma(\Delta I/g_m)(LW)^{1/2}$ versus IL/W plots of HVT and LVT devices at L=0.15μm and L=2μm are depicted in Fig. 5. As shown in all plots of Fig. 5, $\sigma(\Delta I/g_m)(LW)^{1/2}$ in weak and moderate inversion regions increases with decrease in current level. The remarkable deviation from scaling law of LVT long devices in Fig. 1 is caused by a higher level of $\sigma(\Delta I/g_m)(LW)^{1/2}$ at

IL/W=10nA in Fig. 5-4. The bulk-bias dependence of the deviation from scaling law in HVT short devices in Fig. 2-1 is attributed to that of $\sigma(\Delta I/g_m)(LW)^{1/2}$ at IL/W=10nA in Fig. 5-1. The removal of the deviation in $V_{th}$@10μA, which is shown in Fig. 3, is attributed to nearly same levels of $\sigma(\Delta I/g_m)(LW)^{1/2}$ in strong inversion region at IL/W=10μA in Fig. 5. Namely, the deviation mechanism can be interpreted by analyzing the increasing behavior in weak and moderate inversion regions of $\sigma(\Delta I/g_m)(LW)^{1/2}$ versus IL/W plot.

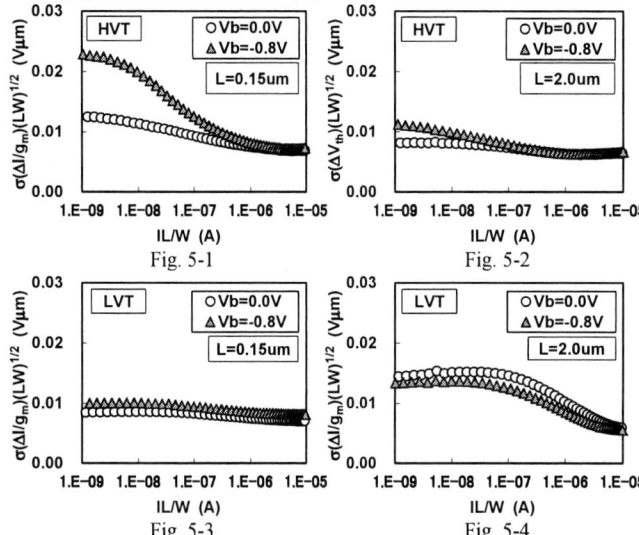

Fig. 5-1  Fig. 5-2

Fig. 5-3  Fig. 5-4

Fig. 5. Measurement results of $\sigma(\Delta I/g_m)(LW)^{1/2}$ versus IL/W in both HVT and LVT devices. Both dependence of gate length and bulk-bias are indicated.

We have reported that an increasing behavior in weak and moderate inversion regions of $\sigma(\Delta I/g_m)(LW)^{1/2}$ versus IL/W plot is closely related to the subthreshold hump [7][8]. $\sigma(\Delta I/g_m)(LW)^{1/2}$ behavior depends on presence of hump as shown in Fig. 6. MOSFET without pocket implant is measured. $\sigma(\Delta I/g_m)(LW)^{1/2}$ monotonically decreases with decrease in current level when MOSFET does not have hump. On the other hand, $\sigma(\Delta I/g_m)(LW)^{1/2}$ increases with decrease in current level when MOSFET has hump. This increasing behavior depends on wafer, temperature, and bulk-bias level [8]. This increasing behavior in Fig. 6 is quite similar with that in Fig. 5. By using ring-gate method [8], influence of hump on behavior of $\sigma(\Delta I/g_m)(LW)^{1/2}$ in weak and moderate inversion regions can be eliminated. In strong inversion region, $\sigma(\Delta I/g_m)(LW)^{1/2}$ level hardly depends on presence of hump.

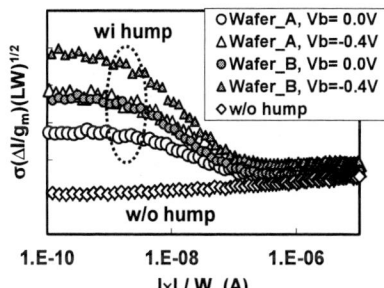

Fig. 6. Measurement results of $\sigma(\Delta I/g_m)(LW)^{1/2}$ in N-channel MOSFET without pocket implant. When MOSFET has hump, $\sigma(\Delta I/g_m)(LW)^{1/2}$ increases with decrease in current level, and depends on wafer and bulk-bias level.

978-1-4673-6145-3/13 $31.00 © 2013 IEEE

## IV. $\sigma(\Delta I/g_m)$ BEHAVIOR ACCORDING TO RING-GATE STRUCTURE

In MOSFET without pocket implant, hump is caused by existence of plural potential barriers in carrier excitation from source to channel. Potential diagram is schematically indicated in Fig. 7. In weak inversion region, current level is determined by potential barrier height $\phi_b$ between source and channel.

When STI is used, it is difficult to eliminate variation of parasitic current near STI channel edge caused by STI surface bump [8]. This situation can be schematically shown as in Fig. 8-1. There are potential barriers in source-side areas enclosed in dotted squares. Potential barrier height of $\phi_b^{edi}$ (i=1 or 2) near STI channel edge is lower than that of $\phi_b^{cnt}$ in the channel center. This is due to a gate electric field enhancement caused by STI surface bump. An equivalent circuit can be regarded as in Fig. 8-2.

An increasing behavior of $\sigma(\Delta I/g_m)(LW)^{1/2}$ versus IL/W plot is caused by variation of the parasitic current level. As shape of STI surface bump differs at every channel edge, current level near STI channel edge differs at every MOSFET.

When ring gate is used, potential barrier height at source side can be controlled into only one level of $\phi_b^S$. This situation is depicted in Fig. 9-1. Channel length is different according to current path. Current level in weak inversion region is determined by the same level of $\phi_b^S$ regardless of current path. Variation of $\phi_b^S$ is caused by random dopant fluctuation.

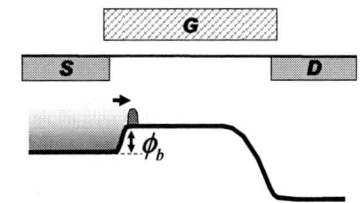

Fig. 7. Schematic potential diagram in MOSFET without pocket implant

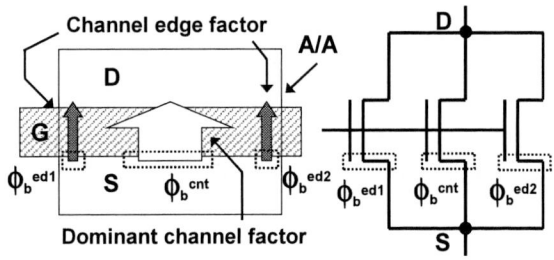

Fig. 8-1       Fig. 8-2
Fig. 8. Schematic MOSFET layout and current factors (Fig. 8-1), and equivalent circuit (Fig. 8-2).

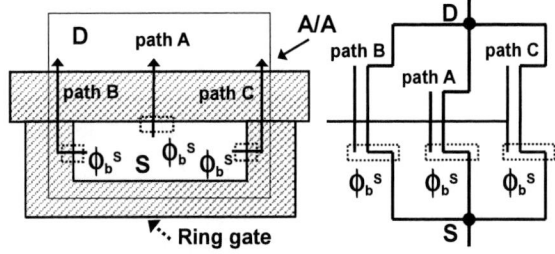

Fig. 9-1       Fig. 9-2
Fig. 9. Schematic MOSFET layout and current factors (Fig. 9-1), and equivalent circuit (Fig. 9-2). Ring gate is used at source side.

Thus, monotonic decreasing behavior in weak and moderate inversion regions can be obtained in $\sigma(\Delta I/g_m)(LW)^{1/2}$ versus IL/W plot [8].

In pocket-implanted MOSFET, two kinds of potential barrier should be considered along current path at both sides of source and drain. Schematic potential diagram is depicted in Fig. 10-1. An equivalent circuit can be written in series connection of MOSFET with different $V_{th}$ levels in Fig. 10-2.

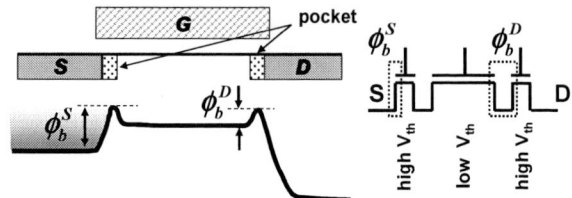

Fig. 10-1       Fig. 10-2
Fig. 10. Schematic potential diagram in pocket-implanted MOSFET (Fig. 10-1), and equivalent circuit along current path (Fig. 10-2).

Also in pocket-implanted long channel MOSFET, an increasing behavior of $\sigma(\Delta I/g_m)(LW)^{1/2}$ can be drastically improved by using ring gate. MOSFET fabricated in another pocket-implant condition is measured. Dose of pocket implant is lower than that in Fig. 1. $\sigma(\Delta I/g_m)(LW)^{1/2}$ behavior is measured in three types of gate structure; normal gate structure, source-side (SS) ring-gate structure, and both-side (BS) ring-gate structure. SS ring-gate structure is the same with that shown in Fig. 9. Gate length is selected in 2μm.

In normal gate structure, an increasing behavior of $\sigma(\Delta I/g_m)(LW)^{1/2}$ can be seen, as shown in Fig. 11-1. Deviation from scaling law is caused by this increasing behavior. In this channel profile condition, bulk-bias dependence can be seen remarkably. It is supposed that this behavior is attributed to variation of parasitic current level near STI channel edge. This situation is schematically shown in Fig. 11-2. It is supposed that the potential barrier height of $\phi_b^{Sei}$ or $\phi_b^{Dei}$ (i=1 or 2) near STI channel edge differs at every channel edge.

By using SS ring-gate structure, bulk-bias dependence can be removed, as in Fig. 12-1. Potential barrier height of carrier excitation at source side can be controlled into one level, as in Fig. 12-2. Thus it can be judged that the bulk-bias dependence, which is shown in Fig. 11-1, is caused by variation of potential barrier height of $\phi_b^{Se1}$ or $\phi_b^{Se2}$. In Fig. 12-1, an increasing behavior is not removed completely. This increasing behavior might be caused by variation of potential barrier height of $\phi_b^{De1}$ or $\phi_b^{De2}$ at drain side.

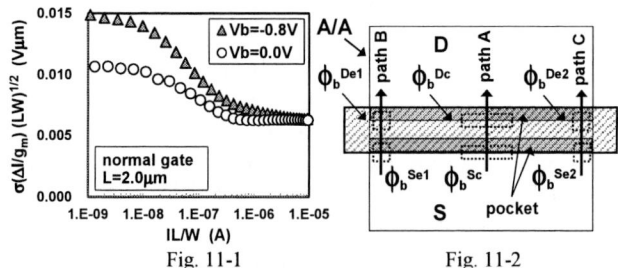

Fig. 11-1       Fig. 11-2
Fig. 11. $\sigma(\Delta I/g_m)$ versus IL/W plot (Fig. 11-1), and schematic MOSFET layout and current factors (Fig. 11-2) in normal gate structure. Hatched regions in channel indicate pocket regions. Along each current path, there are two kinds of potential barrier at source side and drain side, respectively.

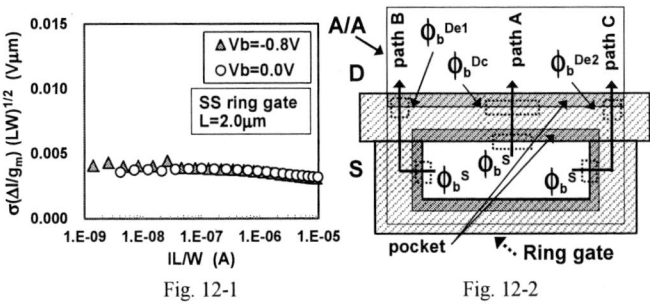

Fig. 12-1          Fig. 12-2

Fig. 12. $\sigma(\Delta I/g_m)$ versus IL/W plot (Fig. 12-1), and schematic MOSFET layout and current factors (Fig. 12-2) in Source-Side (SS) ring-gate structure. Unlike case of normal gate structure, bulk-bias dependence is removed.

By using BS ring gate structure, an increasing behavior of $\sigma(\Delta I/g_m)(LW)^{1/2}$ can be completely eliminated shown in Fig. 13-1. In order to remove influence of STI channel edge, overlapped length $\Delta x$ of ring gate on active area (A/A) should be enough long. Thus, potential barrier height $\phi_b^{D}$ of carrier excitation from well to pocket at drain side can be controlled into one level, regardless of current path, as in Fig. 13-2. As a result, monotonic decreasing behavior of $\sigma(\Delta I/g_m)(LW)^{1/2}$ in weak and moderate inversion regions can be obtained.

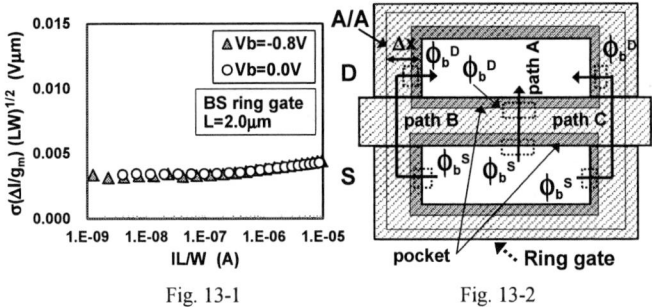

Fig. 13-1          Fig. 13-2

Fig. 13. $\sigma(\Delta I/g_m)$ versus IL/W plot (Fig. 13-1), and schematic MOSFET layout and current factors (Fig. 13-2) in Both-Side (BS) ring-gate structure. $\sigma(\Delta I/g_m)$ monotonically decreases with decrease in current level.

From the measurement results in Fig. 11, 12, and 13, it is clarified that the $\sigma(\Delta I/g_m)$ increasing behavior equivalent to the deviation from Pelgrom scaling law is caused by variation of current level near STI channel edge. The variation of current level is caused by the subthreshold hump.

## V. DISCUSSION

It should be noted that bulk-bias dependence of the deviation from scaling law in Fig. 2 is caused by variation of potential barrier height at source side. Potential barrier height at drain side does not depend on bulk-bias level. This is because potential barrier height at drain side is determined by difference of doping concentration between pocket and well in the same channel region. Thus, it can be regarded in Fig. 2 that $\sigma(\Delta V_{th})(LW)^{1/2}$ levels of HVT short devices are mainly determined by variation of potential barrier height at source side. While $\sigma(\Delta V_{th})(LW)^{1/2}$ levels of HVT long devices and LVT devices are mainly determined by variation of potential barrier height at drain side.

The remarkable deviation of LVT long devices in Fig. 1 can be interpreted by a difference of potential barrier height at drain side. As mentioned before, a difference of doping concentration between pocket and the adjacent well in LVT long devices is larger than that in HVT long devices. This means that potential barrier at drain side of LVT long devices is higher than that of HVT long devices. From variation of this higher potential barrier near STI channel edge, the remarkable deviation in LVT long devices is caused. On the other hand, in LVT short devices, distance between pockets of both sides becomes short. Doping concentration of the adjacent well becomes higher by adding that of pocket at the opposite side. Difference of doping concentration between pocket and the adjacent well in LVT short devices becomes smaller. This means that potential barrier height at drain side in LVT short devices becomes lower than that of LVT long devices. As a result, deviation from scaling law in LVT short devices becomes smaller than that in LVT long devices.

## VI. CONCLUSION

We have clarified for the first time that deviation from Pelgrom scaling law in $\sigma(\Delta V_{th})$ of pocket-implanted long channel MOSFET is caused by the subthreshold hump. The similar deviation is observed in many companies according to the paper [4]. The deviation from scaling law depends on both $V_{th}$ criteria-current level and bulk-bias level. This is due to an increasing behavior of $\sigma(\Delta I/g_m)(LW)^{1/2}$ in weak and moderate inversion regions. We have found that the increasing behavior can be completely eliminated by using both-side ring gate structure. This means that the increasing behavior is caused by variation of current level near STI channel edge. This current level variation is caused by the subthreshold hump. When current mismatch characteristics are discussed by $V_{th}$ whose criteria current level is in weak or moderate inversion regions, behavior of $\sigma(\Delta I/g_m)(LW)^{1/2}$ versus IL/W plot should be noted.

## REFERENCES

[1] K. Cao, W. Liu, X. Jin, K. Vasanth, K. Green, J. Krick, T. Vrotsos, C. Hu, "Modeling of Pocket Implanted MOSFETs for Anomalous Analog Behavior," in IEDM Tech. Dig., pp. 171-174, 1999.

[2] T. Tanaka, T. Usuki, T. Futatsugi, Y. Momiyama, T. Sugii, "Vth Fluctuation Induced by Statistical Variation of Pocket Dopant Profile," in IEDM Tech. Dig., pp. 271-272, 2000.

[3] J. B. Johnson, T. B. Hook, Y. Lee, "Analysis and Modeling of Threshold Voltage Mismatch for CMOS at 65 nm and Beyond," IEEE Electron Device Letters, Vol. 29, No. 7, pp. 802-804, July 2008.

[4] T. B. Hook, J. B. Johnson, A. Cathignol, A. Cros, G. Ghibaudo, "Comment on Channel Length and Threshold Voltage Dependence of a Transistor Mismatch in a 32-nm HKMG Technology," IEEE Trans. on Electron Devices, Vol. 58, No. 4, pp. 1255-1256, April 2011.

[5] C. Mezzomo, A. Bajolet, A. Cathignol, R. Frenza, G. Ghibaudo, "Characterization and Modeling of Transistor Variability in Advanced CMOS Technologies," IEEE Trans. on Electron Devices, Vol. 58, No. 8, pp. 2235-2248, August 2011.

[6] A. Mizumura, T. Ohishi, N. Yokoyama, M. Nonaka, S. Tanaka, H. Ammo, "A Study of 90nm MOSFET Subthreshold Hump Characteristics Using Newly Developed MOSFET Array Test Structure," in ICMTS Tech. Dig., pp. 39-42, 2005.

[7] K. Sakakibara, K. Arimoto, "An Accurate Prediction Model of Temperature Dependent Current Mismatch in All Inversion and Influence of Subthreshold Hump on Mismatch Characteristics," in Ext. Abstract of SSDM, pp. 829-830, 2011.

[8] K. Sakakibara, T. Kumamoto, K. Arimoto, "Impact of Subthreshold Hump on Bulk-bias Dependence of Offset Voltage Variability in Weak and Moderate Inversion Regions," in CICC Tech. Dig., 6-5, 2012.

# Low Power ARM® Cortex™-M0 CPU and SRAM Using Deeply Depleted Channel (DDC) Transistors with V_DD Scaling and Body Bias

V. Agrawal, N. Kepler, D. Kidd, G. Krishnan, S. Leshner, T. Bakishev, D. Zhao, P. Ranade, R. Roy, M. Wojko, L. Clark, R. Rogenmoser, M. Hori*, T. Ema*, S. Moriwaki**, T. Tsuruta**, T. Yamada**, J. Mitani**, S. Wakayama**

SuVolta Inc., 130-D Knowles Drive, Los Gatos, California 95032
*Fujitsu Semiconductor Limited, 1500 Mizono, Tado, Kuwana Mié, Japan
**Fujitsu Semiconductor Limited, 50 Fuchigami, Akiruno, Tokyo, Japan

*Abstract* — **An SoC with ARM® Cortex™-M0 CPU cores and SRAMs is implemented in both 65nm baseline and Deeply Depleted Channel™ (DDC) technologies. DDC technology demonstrates more than 50% active and static power reduction for the CPU cores at matched 350 MHz speed via $V_{DD}$ scaling and body biasing. Alternatively, DDC technology demonstrates 35% speed increase at matched power. The results hold across process corners and temperature, with appropriate body bias selection. DDC technology also increases SRAM static noise margin (SNM), reduces 8Mb $V_{DDmin}$ by 150 mV, reduces SRAM active leakage by 50% while maintaining Iread, and reduces SRAM retention leakage by 5x.**

## I. INTRODUCTION

Power dissipation has become the foremost architectural limiter for mobile SOCs, constraining both thermal envelope and battery life [1]. Supply voltage ($V_{DD}$) scaling used to be the most effective design lever to reduce power. In deep submicron technology nodes; however, increasing threshold voltage ($V_T$) variation caused by random dopant fluctuation limits the ability to reduce $V_T$, in turn limiting $V_{DD}$ scaling. $V_{DD}$ scaling has also been limited by poor SRAM $V_{DDmin}$ yield caused primarily by $V_T$ mismatch.

The Deeply Depleted Channel (DDC) transistor reduces threshold voltage ($V_T$) variation by half and provides a significantly stronger body effect than conventional transistors. Consequently, the nominal DDC $V_{DD}$ is lowered from 1.2 V to 0.9 V in the 65nm baseline process without loss in performance [2]. In addition, DDC SRAMs have greatly improved static noise margin (SNM) and $V_{DDmin}$ [3].

In this work we leverage the reduced $V_T$ variation, increased low voltage performance, and enhanced body effect of the DDC transistor to achieve increased performance to power ratio on a synthesizable ARM Cortex-M0 CPU. A corner-based design methodology is used. Separate well tap networks control the n-channel and p-channel body voltages independently. DDC design corners assume the use of body bias to pull-in the as-fabricated product toward target. The efficacy of body bias in mitigating both process (e.g., corner) and environment (e.g., temperature) effects is demonstrated. This work also demonstrates that the DDC transistor increases SRAM static noise margin (SNM), reduces $V_{DDmin}$, reduces SRAM leakage while maintaining $I_{read}$, and reduces SRAM retention leakage.

## II. TRANSISTOR

The DDC transistor structure used in this work is described in [2] and [3]. Transistor variation is dramatically improved by the un-doped channel and the highly doped screening layer, which reduce RDF and depletion depth ($W_{dep}$) variation, respectively. Fig. 1a shows the DDC reduction in global $V_T$ variation for an SRAM layout with typical LOD and well proximity effects [4]. Local $V_T$ variation is illustrated in Fig. 1b by SRAM mismatch improvement of 40% and 60% for DDC p-channel and n-channel transistors, respectively. For SRAM n-channel transistors extremely low mismatch of less than 15 mV is repeatedly demonstrated.

Fig. 2 quantifies the increased n-channel $I_{EFF}$ of the DDC transistors relative to the baseline process. Improved low voltage performance is achieved by increased $I_{EFF}$ [5] with lower $V_{DD}$, with particular improvement on stacked transistor structures, e.g., NAND and NOR gates.

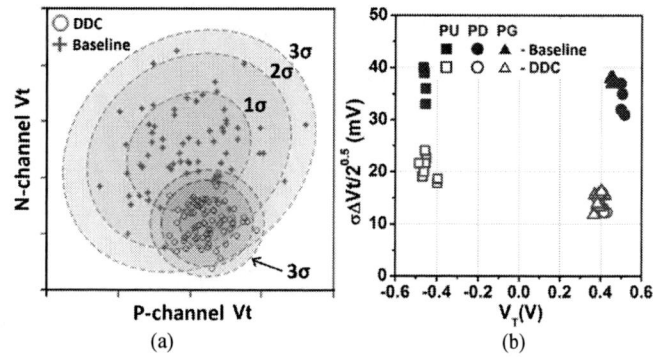

Fig.1. DDC and baseline $V_T$ distributions for (a) global within-wafer $V_T$ variation, and (b) SRAM transistor matching of the pull-up (PU), pull-down (PD), and pass-gate (PG) transistors.

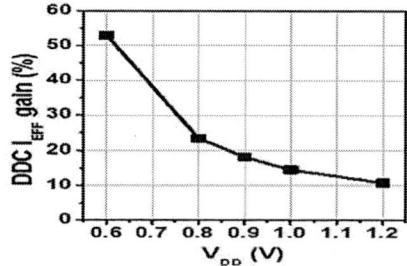

Fig. 2. DDC n-channel $I_{EFF}$ improvement compared to baseline.

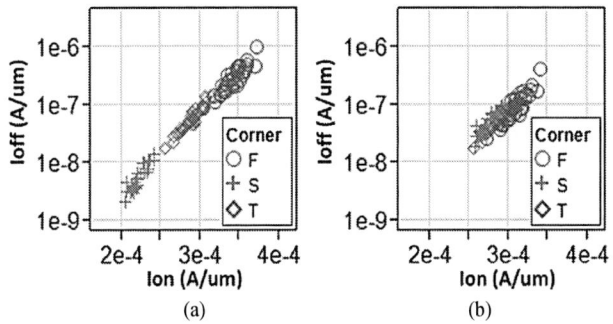

(a)                              (b)

Fig. 3. The DDC transistor's strong body coefficient allows effective use of body bias to mitigate systematic process variation as illustrated by p-channel $I_{off}$ vs. $I_{on}$ at fast, target and slow (F, T, S) corners with (a) nominal -0.3 $V_{SB}$ and (b) -0.6 $V_{SB}$ on fast and 0V $V_{SB}$ on slow silicon to collapse the corners.

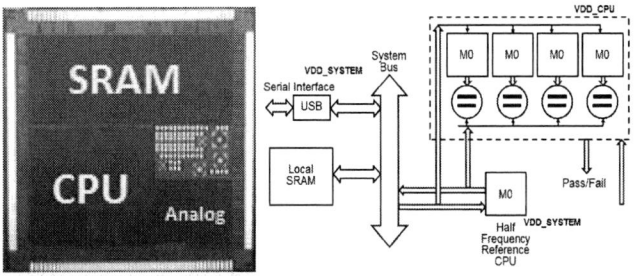

Fig.4 (a) Die Photo (b) Micro-architecture of ARM Cortex-M0 subsystem

The DDC transistor provides twice the body-bias response of the conventional transistor [2] because the highly doped screening layer that determines the channel depletion depth also produces a strong body coefficient that enables a greater control of performance and power, as well as circuit-level variability mitigation [6]. Fig. 3 shows the body bias response of p-channel transistors fabricated at fast (F), target (T), and slow (S) process conditions. The fast and slow distributions at the nominal back bias of 0.3 V (Fig. 3a) are pulled substantially towards the target behavior using 0 V back bias for slow and 0.6 V back bias for fast (Fig. 3b). The n-channel response is similar.

### III. DIGITAL CIRCUITS AND LOGIC

In this work an SoC was developed with ARM Cortex-M0 CPUs implemented using 12-track cell libraries and a conventional corner-based design approach. For the DDC designs, corner library timings were characterized with appropriate back bias selected for each PVT corner. Back bias of 100 mV is used for the slow (SS) corner and 600 mV for the fast (FF) corner in order to achieve the technology goal of half power at matched performance. Well taps are inserted in the place and route floor plan in columns to minimize routing congestion, and routed using minimum width bias wires. Thereby, only 2 out of 300 vertical metal-2 tracks available between maximally spaced substrate taps are consumed by the low-level bias network.

For test, body bias and individual CPU supply voltages are applied externally to 208-pin LFQFP packaged test die. Test die were fabricated on both the baseline and DDC processes and at the SS and FF process corners by biasing transistor channel length and doping skews.

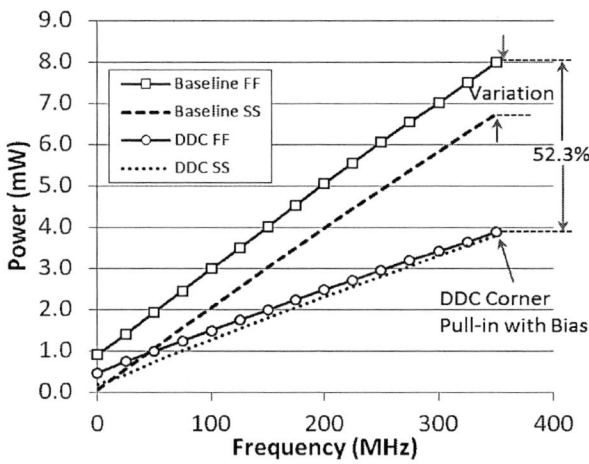

Fig.5 Measured DDC and baseline power at 85°C plotted vs. CPU frequency. Body bias pulls the corners together for DDC parts. DDC CPU matches baseline performance at SS.

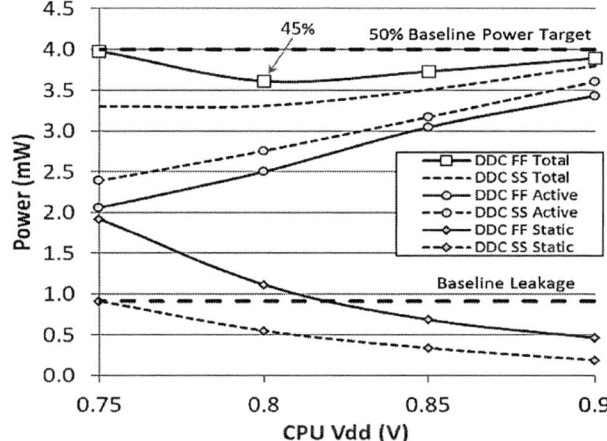

Fig.6 Measured DDC $V_{DD}$ scaled CPU (350MHz @ 85°C) showing the best power result at 45% achieved at 0.8 V compared to baseline.

The ARM Cortex-M0 CPUs are tested by comparing full frequency test CPUs against a half frequency reference CPU as described in [7]. Test die were measured with increasing clock rates until a CPU failed to match reference operation.

The SS corner baseline CPU achieves an operating frequency of 350 MHz at 85°C. For the DDC parts, body bias voltages were chosen per die to minimize leakage while meeting the baseline SS performance. At nominal voltages, silicon measurements at matched frequency show FF DDC CPU power at 52.3% of the FF baseline in Fig. 5. Corner pull-in provides nearly coincident DDC SS and FF responses.

Fig. 6 shows the $V_{DD}$ scaling response at 85°C for DDC CPUs running at the baseline performance level of 350MHz. $V_{DD}$ is reduced in 50 mV increments while an optimized body bias voltage is applied to minimize leakage while maintaining 350MHz operation. The FF DDC CPU achieves 45% of the baseline power of the FF baseline CPU at 85°C, with $V_{DD}$ of 0.8 V. It is notable that as VDD is reduced, the amount of reverse body bias must be reduced as well to maintain gate overdrive. This results in an increase in leakage power at lower $V_{DD}$ while the active power is reduced,

978-1-4673-6145-3/13 $31.00 © 2013 IEEE          237

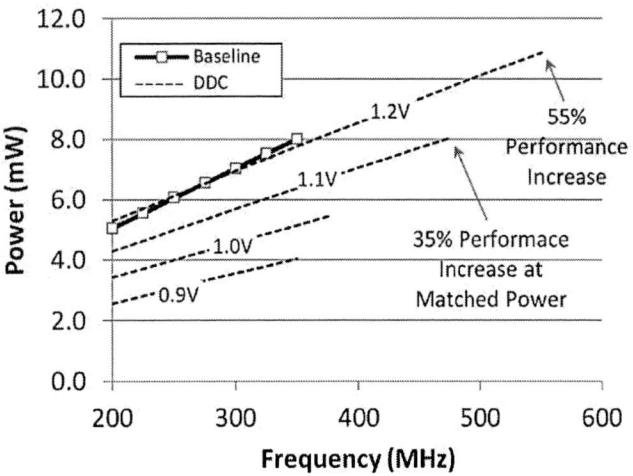

Fig.7 Measured DDC SS CPU has 35% higher performance than baseline SS CPU at equal total FF power and 55% higher performance at matched voltage, at $V_{DD}$ = 1.1 V and 1.2 V, respectively.

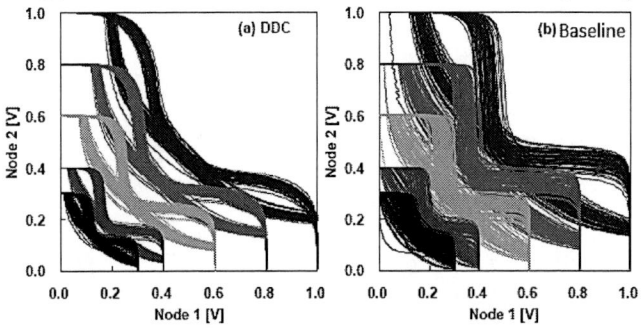

Fig. 8. Superposed butterfly curve of SRAM on (a) DDC and (b) baseline.

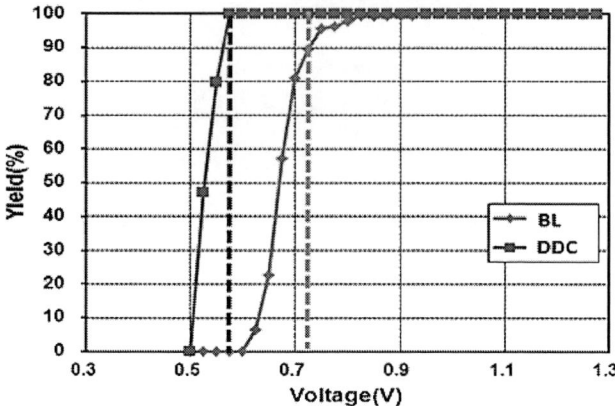

Fig.9 Measured $V_{DDmin}$ for 8Mb SRAM across process corners at 125°C

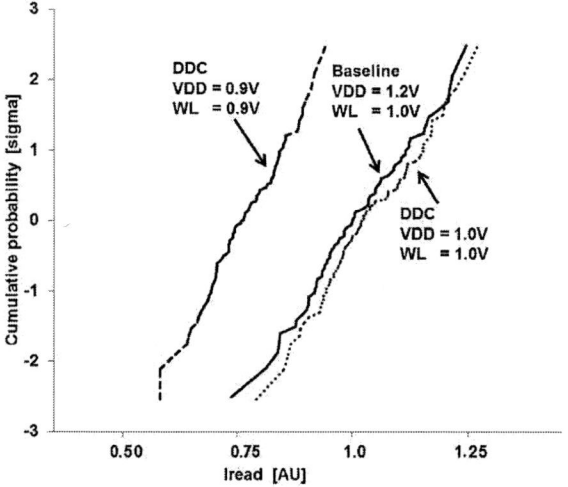

Fig.10 Read current distributions for baseline and DDC SRAM macros. DDC's enhanced stability allows full rail wordline, giving a drive current boost. DDC matches read current at 1.0V, vs. baseline 1.2V.

forming a minimum total power at 0.8 V for this implementation. CPUs designed for other performance or leakage levels will have a different optimum point.

The DDC process is reliable to the 1.2 V baseline $V_{DD}$ [3], allowing high frequency 'turbo' operation modes. DDC SS CPU operates at 550 MHz with nominal (100mV) back bias and $V_{DD}$ of 1.2 V, a 55% increase over baseline performance at the same voltage. At 1.1 V, the SS DDC CPU operates 35%, faster than the baseline with matched power at the FF corner. Fig. 7 quantifies DDC FF CPU power when operated at the maximum frequencies for SS parts at the same voltage.

## IV. SRAM

The stability of the SRAM bitcell is an important functional limiter in the advanced technology nodes. The static noise margin (SNM) is a measure of the cell's ability to retain its state. Silicon data in Fig. 8 shows the spread in the DDC $V_T$ variation is tight, resulting in a better "eye" opening even at $V_{DD}$ = 0.6 V.

This improved process control results in a substantially lower $V_{DDmin}$. Using standard 6T SRAM cells and conventional read or write-assist techniques on an 8Mb SRAM array, we achieved a $V_{DDmin}$ of 600 mV at 125°C across process corners (an improvement of 150 mV over baseline) as shown in Fig. 9.

For an SRAM bitcell in a given technology, the read current is reduced as the square of the $V_{DD}$ reduction. This 25% (0.9 V/1.2 V) reduction in VDD implies a 45% reduction in read current. Fig. 10 shows measured bitcell read current for DDC and baseline. In the 8Mb design, the baseline bitcell requires a wordline voltage that is lower than $V_{DD}$ to improve read stability; while the DDC bitcell's improved matching enhances stability and does not require the reduced wordline voltage. This helps to recover the read current which is reduced by 24%, rather than by 45%.

The speed of an SRAM read operation is determined primarily by the read current of the SRAM bitcell and the sense amplifier offset, neglecting contributions from peripheral logic delay, which are shown to be equal in section III. The discharge time required to reliably sense a small differential on the bitline, $t_{discharge}$ = $C\Delta V/I$ (1). This required voltage differential is generally the sense amplifier offset with additional safety margin. Simulation results for sense amplifier offset using a stratified sampling, factorial pseudo-

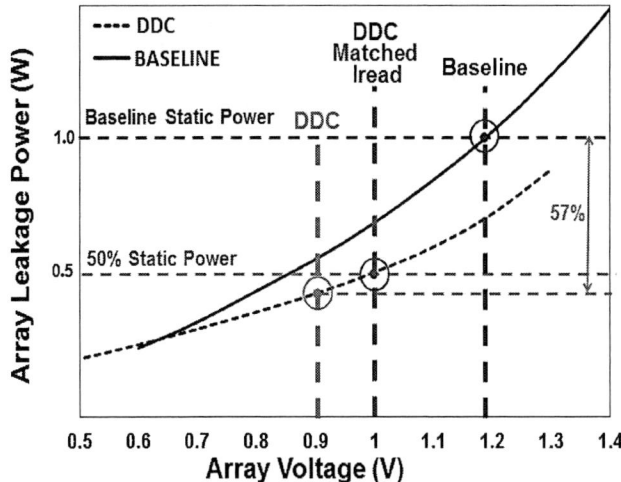

Fig.11 Measured 6-T SRAM Leakage for 1Mb SRAM at FF 85 °C.

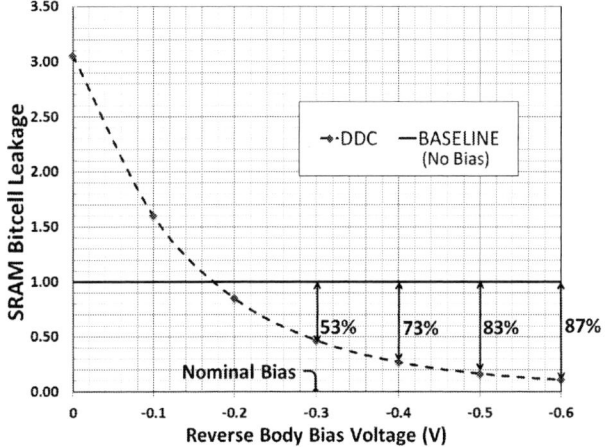

Fig.12 Measured 6-T SRAM Leakage for 1Mb SRAM at FF 0.6 V 85°C.

Monte-Carlo simulation shows approximately 1.6x sense amplifier offset voltage advantage due to better transistor matching (A$_{VT}$) with DDC transistors. Assuming matched capacitance, the discharge time for DDC is given from (1) with 63% $\Delta$V and 76% I$_{read}$ versus the baseline. This results in a 17% improvement of the bitline signal development time for a differentially sensed read. Additionally, the measured data in Fig. 10 also shows the DDC read current is matched to the baseline read current when V$_{DD}$ is 1.0 V.

Similarly, for a single-ended sense amplifier design the time required to discharge a capacitance is inversely proportional to the read current and directly proportional to the voltage on the bit line t$_{discharge}$ = CV/I (2). This gives matched delay for DDC and baseline, as V$_{DD}$ is reduced to 75% (0.9 V/1.2 V), and read current is reduced to 76%, essentially keeping the overall delay constant.

At nominal voltage the DDC SRAM bitcell dissipates 57% less leakage power at 0.9 V and 50% less leakage power at 1.0 V (with I$_{read}$ matched to the baseline) as shown in Fig. 11. Silicon measurements have also shown DDC SRAM instances at V$_{DD}$ = 0.9 V consume 43% less dynamic power than those

fabricated on the baseline process at 1.2 V for the same performance at 0.9 V.

In retention mode, DDC SRAM shows improvement with 10% lower data retention voltage compared to baseline due to better device mismatch. Fig.12 shows DDC SRAM bitcell leakage power is less than 50% of the FF baseline at a conservative data retention voltage of 0.6 V using nominal back bias. Further increasing reverse bias lowers retention leakage by more than 5x, demonstrating that the junction leakage component is not limiting, even at high reverse body bias.

V. CONCLUSION

An SoC with ARM® Cortex™-M0 CPU cores and SRAMs is implemented in both 65nm baseline and Deeply Depleted Channel (DDC) technologies. DDC technology demonstrates more than 50% active and static power reduction for the CPU cores at matched 350 MHz speed via V$_{DD}$ scaling and body biasing. Alternatively, DDC technology demonstrates 35% speed increase at matched power. The results hold across process corners and temperature, with appropriate body bias selection. DDC technology also increases SRAM static noise margin (SNM), reduces 8Mb V$_{DDmin}$ by 150 mV, reduces SRAM active leakage by 50% while maintaining Iread, and reduces SRAM retention leakage by 5x.

**Acknowledgments**

The authors thank and acknowledge Dave Flynn from ARM for his design and measurement insights.

**References**

[1] T. Mudge, "Power: A First-class Architectural Design Constraint," *IEEE Computer*, Vol. 34, pp. 52-58, April, 2001.

[2] L. Clark, *et al.*, "A Highly Integrated 65nm SoC Process with Enhanced Power/Performance of Digital and Analog Circuits," *Proc. IEDM*, pp. 14.4.1 – 14.4.4, Dec., 2012.

[3] K. Fujita, *et al.*, "Advanced Channel Engineering Achieving Aggressive Reduction of V$_T$ Variation for Ultra-Low-Power Applications," *Proc. IEDM*, pp. 749-750, Dec 2011.

[4] J. Faricelli, "Layout-dependent Proximity Effects in deep Nanoscale CMOS," *Proc. CICC*, Sept. 2010, pp. 1 – 8.

[5] K. von Arnim, et al., "An Effective Switching Current Methodology to Predict the Performance of Complex Digital Circuits," Proc. IEDM, 2007, pp. 483 – 486.

[6] J. Tschanz, J. Kao, S. Narendra, et. al., "Adaptive body bias for reducing impacts of die-to-die and within-die parameter variations on microprocessor frequency and leakage," *IEEE J. of Solid-state Circuits*, vol. 37, no. 11, pp. 1396–1402, Nov. 2002.

[7] D. Flynn, "High Performance State Retention with Power Gating applied to CPU Subsystems—Design Approaches and Silicon Evaluation, Proc. Hot Chips 24, Aug., 2012.

# Highly Efficient CMOS Rectifier Assisted by Symmetric and Voltage-Boost PV-Cell Structures for Synergistic Ambient Energy Harvesting

Koji Kotani

Department of Electronics, Graduate School of Engineering, Tohoku University
6-6-05 Aza-Aoba, Aramaki, Aoba-ku, Sendai 980-8579, Japan
Phone: +81-22-795-7122, Fax: +81-22-263-9396
E-mail: kotani@ecei.tohoku.ac.jp

*Abstract*– A highly efficient CMOS rectifier assisted by symmetric PV cells was developed as an example of possible realization of the synergistic ambient energy harvesting concept. In addition, output-voltage-boosted PV cell structures were found to effectively improve the efficiency of this rectifier. As a result, under typical indoor lighting conditions, 30% power conversion efficiency was achieved at -20 dBm of 920 MHz RF input and 100 kΩ loading condition, efficiency 4 times larger than a conventional rectifier without PV assistance.

*Keywords*: energy harvesting, rectifier, and photovoltaic cell

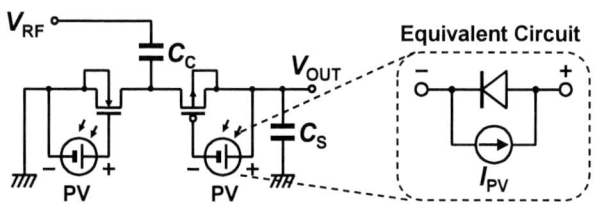

Fig. 1 PV-assisted CMOS rectifier.

## I. Introduction

There are various energy sources in the environment, such as light, heat, electromagnetic waves, vibration (kinetic energy), and others. Energy harvesting from such ambient energy sources is one of the key technologies for powering portable wireless devices, such as remote sensors, wireless tags, and biomedical implants [1]. I have already proposed a synergistic ambient energy harvesting concept, in which multiple environmental energy sources available at the same time and at the same place are effectively utilized together in a cooperative manner [2]. One example of this concept is the photovoltaic (PV) assisted rectifier, in which a CMOS rectifier for ambient radio wave energy harvesting is equipped with PV cells for compensating the threshold voltages (Vths) of MOSFETs and enhancing the power conversion efficiency (PCE) [2]. It was demonstrated that light energy can effectively support energy harvesting from ambient radio waves. However, the achieved PCE was not sufficient because the photo-generated voltages for nMOS biasing and pMOS biasing in the CMOS rectifier was unbalanced and varied with the change in the operating conditions. In order to increase the PCE, CMOS rectifiers assisted by symmetric PV-cell structures and by voltage-boost PV-cell structures are newly proposed in this study.

## II. Previous Implementation and Problems

Fig. 1 shows the PV-assisted CMOS rectifier circuit. The rectifier has a voltage doubler configuration composed of diode-connected nMOS and pMOS transistors. In order to increase the PCE, the compensation of Vths is essential. Instead of using the output voltage of the rectifier itself [3-5], PV cells composed of a pn junction are inserted into the gate bias paths and compensate Vths of MOSFETs by photo-generated voltage. An equivalent circuit of PV cell is also shown in Fig. 1. Different from the conventional solar energy generation by PV cells, output loads of PV cells in this case are high impedance MOS gates. Photo current generated by the pn junction flows in the pn junction itself and forward bias voltage required for the current to flow is developed as an output voltage. In a previous implementation [2], the same p+-diffusion / n-well / p-substrate structure was used for both nMOS biasing and pMOS biasing as shown in Fig. 2. Since both PV cells have to be integrated in the same substrate, making perfect electrical isolation between them impossible, the operation of these PV cells becomes unbalanced. More specifically, in the p+-diffusion / n-well / p-substrate structure, the p+-diffusion / n-well diode ① generates photo current for nMOS biasing, while the n-well / p-substrate diode ② mainly does so for pMOS biasing. In addition, in the pMOS biasing, photo-generated current flows in the p+-diffusion / n-well diode ① rather than in the n-well / p-substrate diode ② and a parasitic bipolar transistor action occurs when the voltage of the p+-diffusion electrode connected to the output of the rectifier increases according to the rectifier operation. As a result, the output voltage of the PV cell connected to the pMOS is much higher than that of nMOS and the difference further increases when the rectifier output voltage increases (Measured output voltage are shown in Fig. 5). Since the increase in bias voltage only in the pMOS does not increase the efficiency of the rectifier, or rather the excess bias voltage causes rectification loss due to the reverse leakage current, this imbalance spoils the PV assistance effect, and resultantly the improvement in PCE was limited in the previous implementation. It is important to note that we cannot control

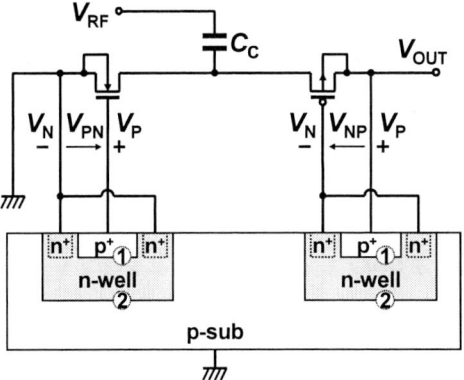

Fig.2 Previous PV-cell implementation for PV-assisted CMOS rectifier.

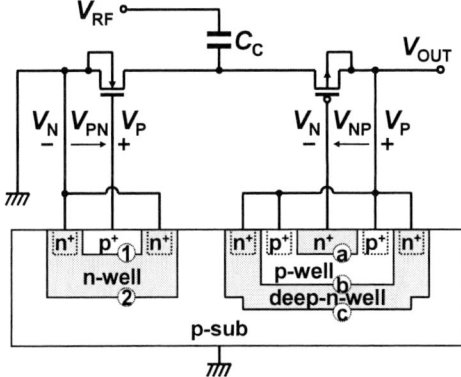

Fig.3 Symmetric PV cell structures implemented in this study.

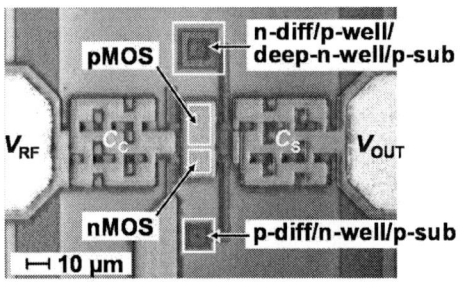

Fig.4 Photomicrograph of the CMOS rectifier test circuit.

output voltage of PV cells by increasing or decreasing PV cell area since the open-circuited output voltage of a PV cell is almost determined by the PV cell structure (impurity doping profile) and does not depend on the PV cell area.

## III. CMOS Rectifier with Symmetric PV Cells

In order to solve the problem, a symmetric (not the "same") PV cell design was applied to the newly developed rectifier, as shown in Fig. 3. For nMOS biasing, the p$^+$-diffusion / n-well / p-substrate structure, which was the same as that in the previous implementation, was employed. Although both the p$^+$-diffusion

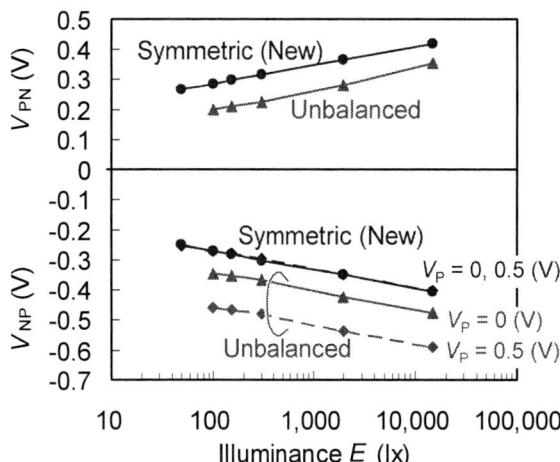

Fig. 5 Output voltage of a PV cell.

/ n-well diode ① and the n-well / p-substrate diode ② generate photo current, only the former works as a PV cell since the latter is shorted and therefore disabled. For pMOS biasing, an n$^+$-diffusion / p-well / deep-n-well / p-substrate structure was newly implemented by using triple well technology. Among the three pn junctions, only the n$^+$-diffusion / p-well diode ⓐ acts as a PV cell for biasing pMOS. The p-well / deep-n-well diode ⓑ is shorted and therefore can be ignored. The deep-n-well / p-substrate diode ⓒ generates photo current, but it is compensated for by a fraction of the output current of the rectifier. Photo-generated current of PV cell is much smaller than the output load current of the rectifier and therefore the compensation current can be ignored from the viewpoint of rectifier efficiency. As a result of this symmetrical configuration, both for nMOS biasing and pMOS biasing, junctions between surface diffusion and a well located underneath (①and ⓐ, respectively) act as PV cells. Physical dimensions and characteristics of these PV cells, such as impurity doping profiles and saturation currents, are almost the same and therefore balanced biasing can be realized and improved efficiency is expected.

A CMOS rectifier having symmetric PV cells was designed. Fig. 4 shows a photomicrograph of the test circuit fabricated by a 0.18-μm 5-ML CMOS process. The channel width / length of the nMOS transistor were 3.6 μm and 0.18 μm, respectively, and those of the pMOS transistor were 10.8 μm and 0.18 μm, respectively. The PV cell area (diodes ① and ⓐ) was 7 × 7 μm². The transistor area was covered by the top two metal layers (ML4 and ML5) for shading. Both coupling capacitor $Cc$ and output smoothing capacitor $Cs$ were designed to be 1.13 pF. A fluorescent light was used as the light source to emulate normal indoor environment. Illuminance was measured by photo radiometer (Delta OHM HD2302) with irradiance probe LP471RAD.

Fig. 5 shows measured photo-generated output voltages of the PV cells. Results for previous unbalanced PV cells as well as

978-1-4673-6145-3/13 $31.00 © 2013 IEEE

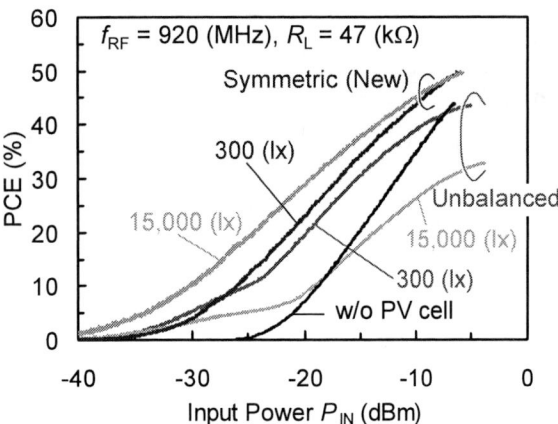

Fig.6 Power conversion efficiency (PCE) as a function of $P_{IN}$.

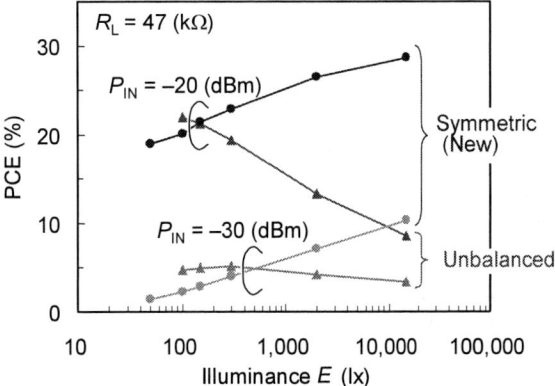

Fig.7 Power conversion efficiency (PCE) as a function of $E$.

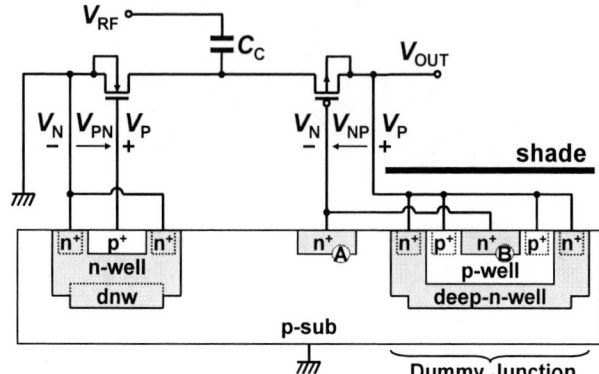

Fig.8 Voltage-boost PV cell structures.

newly-developed symmetric PV cells are shown. nMOS biasing voltage $V_{PN}$ and pMOS biasing voltage $V_{NP}$ were plotted in the upper and lower parts as a function of illuminance $E$. Different from the unbalanced PV cells, almost the same level of dc bias voltage was generated for $V_{PN}$ and $V_{NP}$, and $V_{NP}$ was independent of positive terminal voltage $V_P$ in the newly developed balanced biasing with symmetric PV cells. Increased $V_{PN}$ of the balanced design in spite of using the same PV cell structure was due to the revised layout design that reduced the shading effect.

PCE was then evaluated by on-wafer measurements. Measurements were performed at an RF input frequency of 920 MHz and an output load resistance of 47 kΩ. A vector network analyzer (Agilent N5242A) was used to apply RF power to the rectifier and input power $P_{IN}$ was calculated by,

$$P_{IN} = P_{SRC}(1 - |S11|^2),$$

where $P_{SRC}$ ans $S11$ are the output source power of the network analyzer and the measured reflection coefficient, respectively.

PCE measurement results are shown in Fig. 6. When the illuminance was 300 lx, 22.8% of PCE was achieved at a -20 dBm input by the newly developed rectifier with symmetric PV cells, which is 2.3 times larger than the conventional rectifier without PV cell assistance. When illuminance increased up to 15,000 lx, the PCE increased in the newly developed rectifier,

although it was severely degraded in the previous rectifier with unbalanced PV biasing.

PCE dependence on the illuminance is more clearly shown in Fig. 7. In the newly developed rectifier, PCE simply increased with an increase in illuminance, whereas it decreased in the previous unbalanced implementation. Under the illuminance levels from several hundred lx to a few thousand lx, which are typical indoor lighting conditions, much higher PCEs were achieved for the newly developed rectifier.

## IV. CMOS Rectifier with Voltage-Boost PV Cells

Since the PV cell characteristics are primarily determined by physical parameters, such as dopant profiles of the junctions, the photo-generated voltage of PV cells cannot be increased by increasing the PV cell area. This is because by increasing the PV cell area, photo generated current increases but forward pn-junction current also increases and resultantly, the same forward bias voltage develops (see the equivalent circuit shown in Fig. 1). In addition, although series connection of PV cells to increase the output voltage is possible [6], two electrically-isolated series-connected PV cells cannot be integrated in the same substrate. Based on the physical investigation of the PV cell operation mechanism, I have newly developed an advanced PV cell structures focusing on rectifier assistance applications.

Fig. 8 shows the advanced PV cell structures. For nMOS biasing, a deep n-well was added at the bottom of the n-well. Increased dopant concentration at the bottom of the n-well acts as a diffusion barrier for photo-generated holes. Holes are reflected back to the p⁺-diffusion and photo-generated current increases and resultantly, the output voltage increases. For pMOS biasing, a hybrid PV cell composed of n⁺-diffusion / p-substrate junction Ⓐ generating photo-current and shaded n⁺-diffusion / p-well / deep-n-well / p-substrate structure acting as a dummy pn-junction was introduced. Since the photo-carrier generation region of the n⁺-diffusion / p-substrate junction Ⓐ is much deeper than the n⁺-diffusion / p-well junction Ⓑ, a much larger photo current can be generated. The photo current flows in both n⁺-diffusion / p-substrate junction Ⓐ and the n⁺-diffusion / p-well junction Ⓑ when $V_P = 0$ V and flows only in the n⁺-diffusion / p-well junction Ⓑ when $V_P$ becomes positive

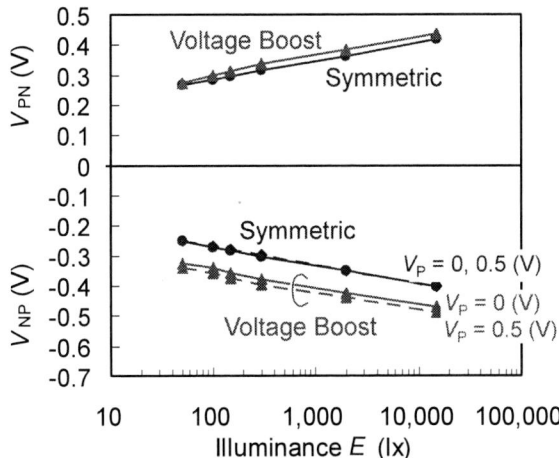

Fig.9 Measured output voltage of the PV cell.

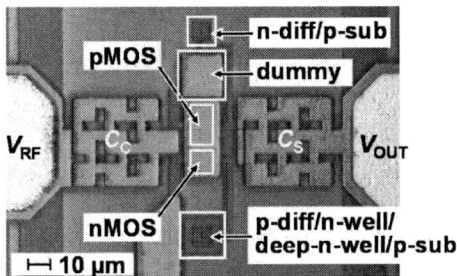

Fig.10 Photomicrograph of the CMOS rectifier test circuit with voltage-boost PV cells.

according to the $V_{OUT}$ generation since the n$^+$-diffusion / p-substrate junction Ⓐ is reverse biased at this condition. Since the saturation currents for these two junctions are almost the same level, the output voltage is not greatly affected by the $V_P$ conditions. Of course, symmetric hybrid PV-cell structure also for nMOS biasing is preferable, it is impossible in the bulk CMOS technology due to the same reason for the impossibility of the isolated series-connected PV cells.

Fig. 9 shows measured output voltages of the advanced PV cells in comparison with the symmetric PV cells. Both $V_{PN}$ and $V_{NP}$ increased over the entire illuminance range. $V_{NP}$ variation caused by $V_P$ change was kept small.

Figs. 10 and 11 show photomicrograph and measured PCEs of the rectifier with the advanced voltage-boost PV cells, respectively. The output load was set at 100 kΩ in this experiment. As compared with the rectifier with the symmetric PV cells, relative PCE improvement of at least 10% or more was achieved under typical indoor lighting conditions. When the illuminance level increased too much, however, the PCE decreased due to the small imbalance in the advanced PV cells

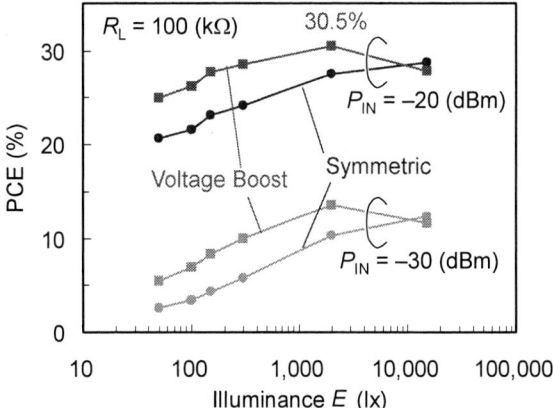

Fig.11 Power conversion efficiency (PCE) as a function of $E$.

for nMOS biasing and pMOS biasing.

## V. Conclusion

A highly efficient PV-assisted CMOS rectifier circuit was developed. Based on a physics-based investigation, PV cell structures were optimized. As a result, improved PCEs of the rectifier were achieved by the symmetric PV-cell structures and the voltage-boost PV-cell structures. It was demonstrated that the optimum collaboration of different energy sources can effectively enhance the efficiency of ambient energy harvesting.

## Acknowledgment

This research was supported by JSPS KAKENHI Grant Number 23360144 and 23656227. The VLSI chip in this study was fabricated in the chip fabrication program of VDEC, the University of Tokyo, in collaboration with Rohm Corporation and Toppan Printing Corporation.

## References

[1] "Special Section on Energy-Harvesting/Scavenging Circuits and Systems" *IEEE Trans. Circuits Syst. II*, Vol.58, No.12, 2011.

[2] K. Kotani, Takumi Bando, and Yuki Sasaki., "Photovoltaic-Assisted CMOS Rectifier Circuit for Synergistic Energy Harvesting from Ambient Radio Wave," *IEEE Asian Solid-State Circuits Conference*, pp. 329-332, 2012.

[3] K. Kotani and T. Ito, "Self-Vth-Cancellation High-Efficiency CMOS Rectifier Circuit for UHF RFIDs," *IEICE Trans. Electron.*, Vol. E92-C, No. 1, pp. 153-160, 2009.

[4] K. Kotani, A. Sasaki, and T. Ito, "High-Efficiency Differential-Drive CMOS Rectifier for UHF RFIDs," *IEEE J. Solid-State Circuits*, Vol. 44, No. 11, pp. 3011-3018, 2009.

[5] G. Papotto, F. Carrara, and G. Palmisano, "A 90-nm CMOS Threshold-Compensated RF Energy Harvester," *IEEE J. Solid-State Circuits*, Vol. 46 , No. 9, pp. 1985-1997, 2011.

[6] F. Horiguchi, "Integration of Series-Connected On-Chip Solar Battery in a Triple-Well CMOS LSI," *IEEE Trans. Electron Devices*, Vol. 59, No. 6, pp. 1580-1584, 2012.

# A Slew-Rate Based Process Monitor and Bi-directional Body Bias Circuit for Adaptive Body Biasing in SoC Applications

Sang-Soo Lee, Edward Boling, Augustine Kuo, and Robert Rogenmoser

SuVolta Inc., 130-D Knowles Drive, Los Gatos, California 95032

*Abstract* — **A process monitor based on slew-rate measurement has been applied to a body bias control system to detect the process corners and adjust the body bias voltage necessary to meet the power and performance requirements for CMOS circuits. The process monitor consists of N- and P- type slew generators, pulse generator, pulse extender, counter and control circuits. A new analog pulse extender has been developed to increase the pulse width for easy characterization of process corners in practical systems. Circuit description, design considerations, and measured results of the process monitor and body bias generator circuits in 55nm CMOS process are presented. Active circuit area for the process monitor and body bias generator are 0.023 mm$^2$ and 0.013 mm$^2$ and power dissipation for process monitor and bias generator are 400 μW and 55 μW using a 0.9 V supply. Measured silicon results match with simulation results and show correct operation of the circuits.**

## I. INTRODUCTION

With continued scaling in sub-65 nm process technologies, the effect of process variations on the parametric yield of VLSI circuits has increased significantly in recent years. The process variations directly impact the transistor characteristics as the threshold voltage of the transistor varies a lot over process corners. The net result is wide variations of the performance and power dissipation of the VLSI circuits. In order to compensate for the process variation, body biasing techniques can be used as part of performance & power management techniques in SoC implementations. Body biasing provides an effective knob to alter the delay and leakage of the circuit by modulating the threshold voltage of the transistors. Forward body bias (FBB) is applied to slow transistors and reverse body bias (RBB) is applied to fast but leaky transistors.

Body biasing techniques to optimize circuit and chip performance has been reported in the literatures [1-5]. The traditional approach is to apply a fixed forward or reverse body bias to alter the threshold voltage of the transistors depending upon the mode (standby or active) of operation. A more recent approach to body biasing involves dynamic or adaptive body biasing in a control loop so that the body bias voltage is adjusted based on environmental changes such as temperature and aging. This adaptive body bias (ABB) technique is becoming increasingly important in mobile processors [5]. The ABB is known to be an efficient way to reduce manufacturing variation and thereby improve yield. Implementing the ABB in a SoC, however, requires additional circuitry including

process monitor, body bias generator, and a controller as shown in Fig. 1. The controller could be built based on either a critical path replica based system [1] or a look-up table based system [3]. Regardless of the controller design approaches, the process monitor and body bias generator are essential blocks to implement the ABB.

A typical process monitor to detect the manufacturing process corner is based on delay measurement of a circuit such as a ring oscillator. However, purely delay-based process monitoring systems prove to be insufficient for optimizing both performance (delay) and power dissipation at the same time. Using the slew-rate in conjunction with the delay has been proposed in [6] to determine the optimal body bias voltage for both PMOS and NMOS transistors of the circuit. Fig. 2 shows the process monitor used in [6]. The circuit requires a series of signal conversion circuits consisting of slew-to-pulse, pulse to voltage, and voltage-to-frequency converters. The final frequency output is then compared with a reference frequency in a closed loop control system to adjust the body bias voltage. A key limitation of this approach is the frequency variation of VCO from process corners. In the conventional process monitor, as the VCO is sensitive to process corner variation, the frequency output includes not only the variation due to slew

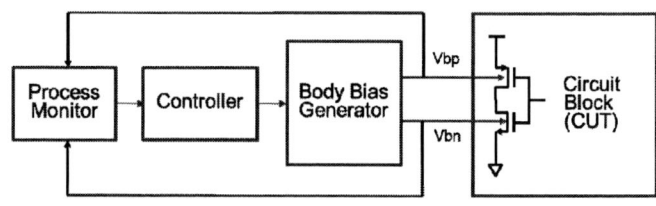

Fig. 1. A generic adaptive body bias (ABB) architecture.

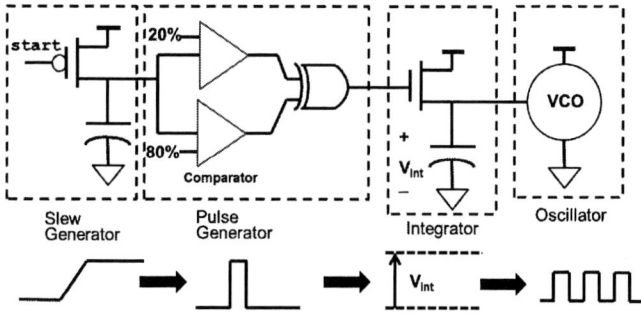

Fig. 2. A conventional process monitor based on slew signal input to frequency output conversion.

978-1-4673-6145-3/13 $31.00 © 2013 IEEE          244

signal but also the variation from the VCO itself. Therefore, this approach has intrinsic limitation on accurate measurement of the actual slew-rate of the signal.

This paper presents a novel approach to produce a process monitor based on a pulse extender circuit. Fig. 3 shows the overall architecture of the new process monitor and Fig. 4 shows a simplified schematic of the pulse extender circuit. The pulse extender replaces the integrator and the oscillator of the conventional process monitor. A key advantage of this approach is obtaining a linear slew-to-pulse output response depending only on the input slew signal since the effect of process corners on the pulse extender circuit can be easily calibrated out in the controller design. The proposed pulse extender circuit is described in Section II. Section III discusses the body bias generator used in this paper. Measured results are presented in Section IV. Section V concludes the paper with discussion.

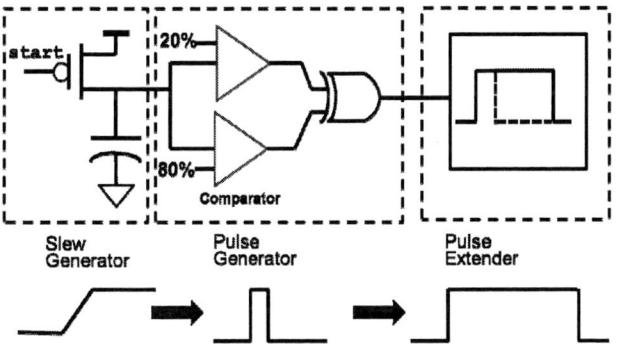

Fig. 3. A new proposed process monitor based on pulse extender.

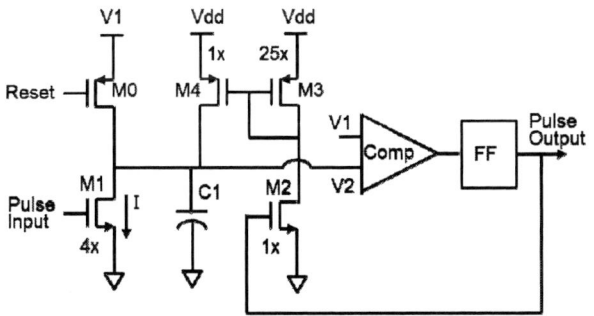

Fig. 4. A new pulse extender circuit.

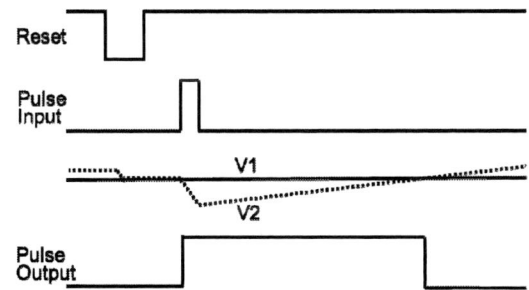

Fig. 5. Timing diagram of the pulse extender circuit.

## II. PULSE EXTENDER CIRCUIT

To replace the VCO in the conventional process monitor, an ADC can be used to convert the integrator output voltage into a digital code for the controller to compare the digital code against a reference code. However, an ADC would be larger in area and add more complexity compared to the traditional approach of using the VCO. In this paper, a novel approach of using the pulse output directly is presented. A key issue of using the pulse signal directly to measure the process corner variation is due to the pulse width being too short to implement a controller reliably in the time window allocated to do the required calculations. In this paper, the short pulse issue is solved by extending the pulse width by 100 times. Detailed operation of the pulse extender is described below.

Referring to Fig. 4 and Fig. 5, the voltage on the capacitor C1 is initially charged to V1 during the reset period. The output voltage from the flip-flop is designed to be low so that M2 is initially off. Therefore, there will be no current into the capacitor from M4 during the reset period. When the reset state is released and the input pulse comes in, the voltage V2 on the capacitor decreases linearly with a rate of I/C1. The transistors M1-M4 are sized so that the current from M4 is a lot lower than that of M1. As the voltage on V2 is discharged, the comparator forces the FF output to change its state to high turning M2 on. This causes M4 to carry I/100 based on the transistor size ratio for M1 & M2, and M3 & M4. Therefore, the capacitor voltage V2 is rising linearly with a rate of I/100C. As the V2 recovers its original voltage level and crosses over V1, the comparator output changes its state to make the FF output back to zero. This results in a 100 times wider pulse from the pulse extender output. Although not shown in Fig. 4, three input pulse ranges (0.5X, 1X, and 2X) have been designed to cover a wide range of input pulse width.

The 100x proportionality depends on the matching of transistors M1 & M2 and M3 & M4, not the actual values of the currents or capacitance. Thus, the transistor mismatches are affecting the overall gain (or scaling) of the pulse width. However, the gain and offset of the circuit are calibrated by measuring and comparing the actual signal output with the reference output from known pulse inputs in a controller design. The calibration procedure is done during the ATE test time for each die. This essentially removes process dependent corner variations of the circuit and results in the process monitor that produces an output relying only on the input signal.

## III. BODY BIAS GENERATOR

A conventional body bias generator in VLSI circuits is typically built with a programmable gain stage followed by a drive amplifier as reported in [7]. This circuit has limited ABB range and requires a separate supply voltage to produce RBB voltage outside the range between VDD and GND. To generate a body bias voltage higher than VDD or lower than GND from a single supply voltage, a charge pump circuit is commonly used as stand-alone DC-DC converters. A key issue is how to build an efficient body bias generator circuit and minimize power and area overhead. Fig. 6 shows the proposed bi-directional PMOS body bias generator circuit adopted in this paper. An NMOS body bias generator circuit has been

978-1-4673-6145-3/13 $31.00 © 2013 IEEE          245

implemented similarly but not shown. The new circuit utilizes a programmable 6-bit DAC to generate a 0.9 V bi-directional ABB voltage range (600 mV RBB to 300 mV FBB), an amplifier to produce the control voltage/current for a controlled oscillator, and a symmetric charge pump to rapidly charge up the load capacitor in both phases (CLK & CLKB) of the clock.

The circuit implemented in this paper works as follows. First, DAC1 is used to set the amount of maximum FBB voltage, 300 mV. With VDD = 0.9 V, VREF = VDD – I1·R1 = 0.9 V – 0.3 V = 0.6V. Second, DAC2 is used to select the amount of current to develop the desired voltage across R2 so that the final output voltage VBBP = VREF + I2·R2 = VDD – I1·R1 + I2·R2. The voltage drop across R1 & R2 can be adjusted to achieve the required RBB. For example, if the voltage drop on R1 is 150 mV and the drop on R2 is 750 mV, then the RBB voltage is VBBP = VDD – I1·R1 + I2·R2 = 0.9V – 0.15V + 0.75V = 1.5 V for 600mV RBB on the PMOS transistor. An important feature of this circuit is that the FBB and RBB levels can be controlled independently by programming DAC1 and DAC2. Consequently, bi-directional body bias can be efficiently generated in different parts of the SoC circuit block. The gm block compares the fraction of VBBP against VREF to produce the control voltage for the current controlled oscillator in a negative feedback loop. A charge pump [8] was used to create a charging signal at both phases of CLK and CLKB.

Fig. 7 shows the die photo and layout of the proposed PMOS & NMOS bias generators as well as the process monitor. In a typical SoC implementation, the area and power overhead of the body bias generators depends on the amount of load current the body bias circuit has to support. The load current is a strong function of the junction leakage of P+ S/D for PMOS and N+ S/D for NMOS. In the 16 mm$^2$ test chip, the area and power overhead of the bias and process monitor circuits was less than 0.5% and 4 mW, respectively.

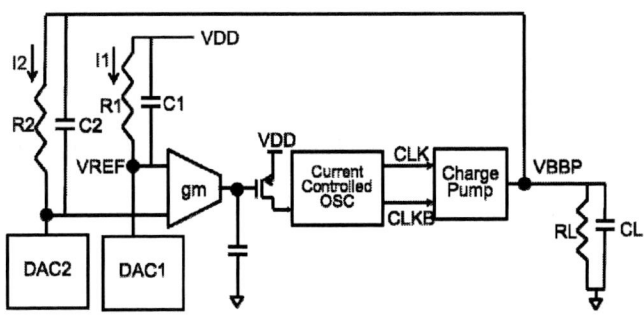

Fig. 6. Proposed bi-directional PMOS body bias (VBBP) generator circuit.

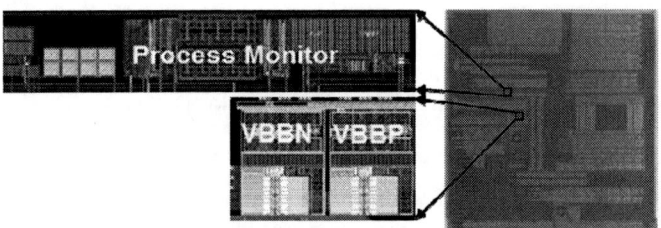

Fig. 7. Die photo and layout of the process monitor and body bias generator.

## IV. MEASURED RESULTS

The process monitor and body bias generator circuits have been fabricated in a 55-nm CMOS test chip based on DDC technology [9] that features a 2 times stronger body effect than the 55-nm CMOS baseline technology. In the test chip, the on-chip slew-signal is converted to a short pulse that is eventually converted to a long pulse output by the pulse extender circuit. Fig. 8 shows a comparison plot of simulated and silicon measured pulse widths for different slew-rate input over three ranges. Range bits are used to choose the appropriate range depending on the input pulse width. For small pulse widths, 0.5X range is used, and for large pulse widths, 2X range is used. There is an excellent agreement between the measured and simulated results with the error less than 3% for the entire input signal range up to 200 ns. Fig. 9 shows a representative transient measurement result of the pulse extender circuit for the rising input slew signal and the output pulse. Additional pulse output measurements on corner dies were performed to get a pulse width of 4.3µs for TT, 3.6µs for FF, and 5.4µs for SS die. These pulse output data were compared with the I$_{on}$ measurement data of corners transistors. The transistor size is W/L = 0.5µm / 60nm and I$_{on}$ is defined as the drain current when Vds=Vgs=Vdd. Fig. 10 shows a comparison plot of pulse output and scaled version of the 1/ I$_{on}$ for 3 corners. The close correlation between the pulse width and the corresponding transistor characteristics demonstrates the efficacy of the proposed process monitor.

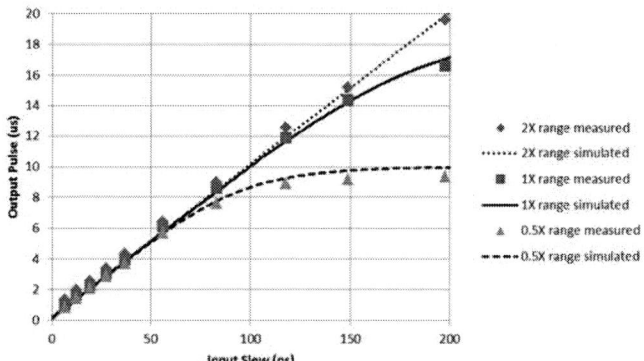

Fig. 8. Measured & simulated results of the pulse output vs. slew input.

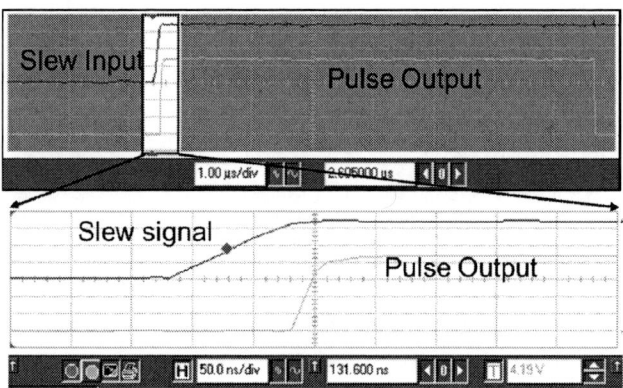

Fig. 9. Measured results of the pulse extender circuit from 70ns slew input to7us pulse output.

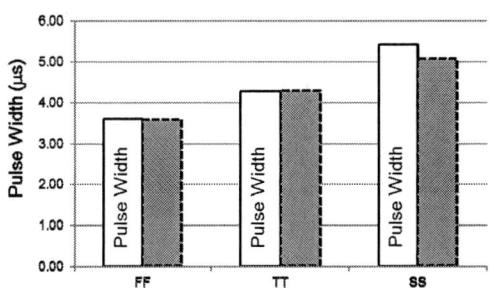

Fig. 10. Correlation between the measured pulse width and the normalized value of transistor's $1/I_{on}$ (dotted box) for 3 corners.

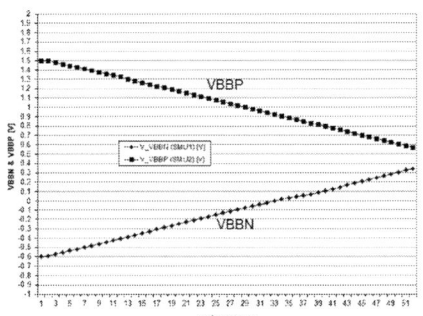

Fig. 11. Measured bi-directional body bias generator output voltages for VBBP & VBBN versus DAC code.

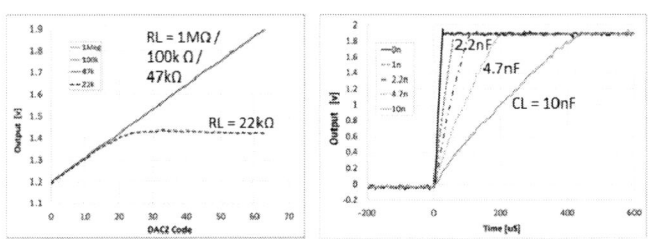

Fig. 12. Measured result of VBBP circuit for (a) different load resistance (b) different load capacitance.

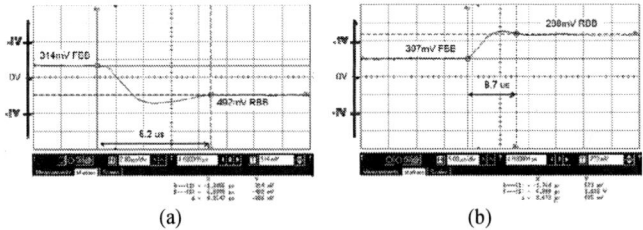

Fig. 13. Measured transient result of body bias generator. (a)VBBN output from 314 mV FBB to 492 mV RBB (b) VBBP output from 397 mV FBB to 288 mV RBB.

Fig. 11 shows the measured output voltages for VBBP & VBBN versus DAC code. The linear monotonic response confirms the correct operation of the bi-directional body bias circuit from 300mV FBB to 600mV RBB. Fig. 12(a) shows a measured VBBP output as a function of load resistance. The design target for this circuit is a maximum of 20 µA DC loading current and 4 nF capacitive load. The 20 µA current is two times the simulated total leakage current through the body and the capacitive load was calculated from the simulation for 1 million NAND2 gates enclosed by deep n-well structure. Fig. 12(b) shows a transient response of the VBBP generator

confirming the loop stability with 100 kΩ load in parallel with different capacitive loads. The transient measurements indicate a maximum charging current of 55µA. The average slew rate is 10 mV/µsec during the 200 µsec pump-up period for 4.7 nF load. Simulations indicate a ripple voltage of less than 1 mV during regulation with a 4 nF load capacitance. Measurements show 5 mV of ripple owing to test system limitation. Fig. 13(a) shows an additional measured transient waveform of body bias generator for the VBBN output going from 314 mV FBB to 493 mV RBB and Fig. 13(b) shows a measured transient VBBP output waveform changing from 397 mV FBB to 288 mV RBB. These transient measurement results agree well with the simulated results and confirm the correct operation of the bi-directional body bias generator.

## V. CONCLUSION

Low power requirements for mobile SoC demand ABB as a standard power management and yield improvement vehicle. A new approach of a process monitor based on a pulse extender circuit has been presented. Additionally a new bi-directional body bias generator circuit has been presented to produce independently programmable FBB and RBB voltages. Measured results agree with the simulated results within 3% confirming correct operation of the presented circuits. The process monitor and body bias circuits introduced in this paper are being integrated with a controller to form an onchip ABB system. The onchip ABB system requires minimal overhead in power and area (less than 0.5% area overhead and less than 4mW power overhead during active mode in the test chip), facilitating its use in many VLSI circuits.

## REFERENCES

[1] J. Tschanz, J. Kao, S. Narendra, et al., "Adaptive body bias for reducing impacts of die-to-die and within-die parameter variations on microprocessor frequency and leakage," *IEEE J. of Solid-state Circuits*, vol. 37, no. 11, pp. 1396–1402, Nov. 2002.

[2] J. Kao, M. Miyazaki, A. Chandrakasan, "A 175-mV multiply-accumulate unit using an adaptive supply voltage and body bias architecture," *IEEE J. of Solid-state Circuits*, vol. 37, no. 11, pp. 1545–1554, Nov. 2002.

[3] B. Choi, Y.Shin, "Lookup table-based adaptive body biasing of multiple macros," *IEEE Int. Symposium on Quality Electronic Design*, 2007, pp. 533-538.

[4] S. Kumar, C. Kim, S. Sapatnekar, "Body bias voltage computations for process and temperature compensation," *IEEE Trans. on VLSI Systems*, vol. 16, no. 3, pp. 249–262, Mar. 2008.

[5] G. Gammie, A. Wang, M. Chau, et al., "A 45nm 3.5G baseband-and-multimedia application processor using adaptive body-bias and ultra-low-power techniques," *ISSCC Dig. Tech. Papers*, 2008, pp. 258-259.

[6] A. Ghosh, R. Rao, J. Kim, et al., "On-chip process variation detection using slew-rate monitoring circuit," *IEEE Int. Conference on VLSI Design*, 2008, pp. 143-147.

[7] J. Tschanz, N. Kim, S. Dighe, et al., "Adaptive frequency and biasing techniques for tolerance to dynamic temperature-voltage variations and aging," *ISSCC Dig. Tech. Papers*, 2007, pp. 292-293.

[8] R. Pelliconi, D. Iezzi, A. Baroni, et al., "Power efficient charge pump in deep submicron standard CMOS technology," *IEEE J. of Solid-state Circuits*, vol. 38, no. 6, pp. 1068–1071, Jun. 2003.

[9] L. Clark, D. Zhao, T. Bakhishev, et al., "A highly integrated 65-nm SoC process with enhanced power/performance of digital and analog circuits," *IEDM Tech.Dig.*, 2012, pp.335-338.

# Comparison of Modeling Approaches through Hierarchical Behavioral Modeling of a GNSS Receiver Front-end

Zhimiao Chen* Yifan Wang* Joern Driesen[†] Fabio Garzia[0]
Stefan Koehler[0] Frank Henkel[†] Ralf Wunderlich* and Stefan Heinen*

*Chair of Integrated Analog Circuits and RF Systems, RWTH Aachen, 52074, Aachen, Germany, ias@rwth-aachen.de
[†]IMST GmbH, 47475, Kamp-Lintfort, Germany, driesen@imst.de
[0] Fraunhofer-Institut fuer Integrierte Schaltungen IIS, Nuernberg, Germany, fabio.garzia@iis.fraunhofer.de

*Abstract*—This paper analyzes and compares the mixed-signal modeling approaches (conservative, timed data flow, and event-driven, as well as the baseband modeling approach) through the hierarchical behavioral modeling of a GNSS (Global Navigation Satellite System) receiver front-end. Based on the result of the comparison, one hierarchical modeling flow is finally derived comprising multi-modeling approaches with reduced manual modeling effort.

## I. INTRODUCTION

In the past decades, the analysis and simulation of complex analog and mixed-signal circuits based on SPICE requires huge size of calculations exceeding the power of current computers.

One solution is to reduce the order of SPICE models, trading the model accuracy against the simulation speed. It still requires further discussion in the application of this method for non-LTI (Linear Time Invariant) systems[1].

Another idea is to provide high efficient simulation and verification flow to limit the size of circuits for the simulation at each hierarchy level[2] and therefore increase the simulation performance. Highly abstracted behavioral models are often used to describe the circuit functions at block level, and then as executable specifications at system level. Choosing proper modeling approaches and proposing the modeling flow are mainly discussed in this work.

### A. Existing Modeling Approaches

The modeling approaches are mainly classified as: conservative, timed data flow, and event-driven.

For conservative approach, the circuit behaviors are described based on ordinary differential equations (ODE) according to Kirchhoff's voltage and current law. This approach is implemented in VerilogA in this work.

Timed data flow approach, like wreal based RVM (Real Value Modeling) in VerilogAMS[3], or TDF (Timed Data Flow) in SystemC-AMS[4], digitizes analog circuit behaviors, sampling the continuous signal to be discrete data flow with a specific time step complying with Nyquist theorem.

This work is part of the NAPA project (BMBF16M3190F), which is sponsored by Germany Federal Ministry of Education and Research.

Abandoning iterative calculations for the convergent numerical solutions of ODE, this method brings a significant simulation speed increase.

Event-driven models realized in SystemC or SystemVerilog require coming events to be triggered. Inner clocks can be used to provide timed events as timed data flow approach. Or the analog waveform is described as a set of parameterized basis s-domain functions without time-step integration[5] when the analog behaviors of the modules can be regarded as LTI.

### B. Baseband Modeling

The baseband equivalent model for RF parts is a good choice to trade simulation accuracy for speed. The basic baseband equivalence[6] is shown as (1).

$$s(t) = I(t)cos(2\pi f_c t) - Q(t)sin(2\pi f_c t) \qquad (1)$$

The processing of high-frequency $s(t)$ is reduced to the processing of lower rate in-phase part $I(t)$, quadrature part $Q(t)$, and constant carrier frequency $f_c$.

This basic method requires the clear baseband part without distortions of carrier in $I(t)$ and $Q(t)$, so an extended solution is needed to describe the influence of higher order harmonics of the carrier. In this work, the baseband equivalence is from approximated Taylor expansion[7]:

$$\begin{aligned} s_{RF}(t) = DC(t) &+ I_1(t)cos(2\pi f_c t) - Q_1(t)sin(2\pi f_c t) \\ &+ I_2(t)cos(4\pi f_c t) - Q_2(t)sin(4\pi f_c t) \\ &+ I_3(t)cos(6\pi f_c t) - Q_3(t)sin(6\pi f_c t) + \cdots \quad (2) \end{aligned}$$

## II. COMPARISON OF MODELING APPROACHES

To the best of authors' knowledge, it still lacks an analytical comparison of the mentioned approaches in modeling accuracy and simulation speed. Many demonstrated simulation results are more affected by the other factors like the chosen mathematical model, the order of model matrix, the size of memory, et al. The comparison in this work is based on examples from behavioral modeling of a GNSS receiver front-end.

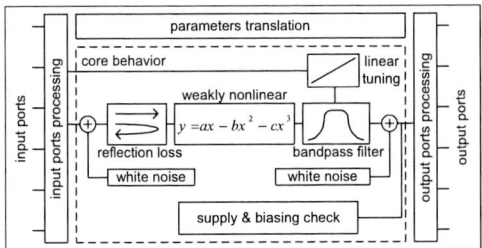

Fig. 1.   Modeling basic behaviors of LNA

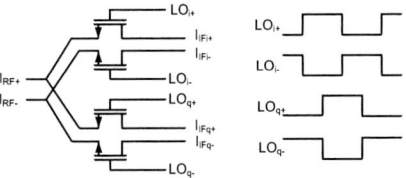

Fig. 2.   Structure of passive mixer

## A. Modeling Accuracy

In this work, the discussion of modeling accuracy is focused on the realization using modeling approaches under the same mathematical model with the same coefficients. Some specific examples in RF front-end are used to strengthen the discussion.

*1) LNA:* The main behaviors to be modeled in LNA (Low Noise Amplifier) are shown in Fig.1 as input port matching, input reference noise, first order nonlinearity, and bandpass characteristics.

For the matching behavior, the calculations are bidirectional to consider both output and input port specifications. It can be implemented in conservative model, but not in digitized timed data flow and event-driven approaches, whose calculations follow the direction of signal flow. In these two approaches, ports between modules are assumed as already matched.

For noise behavior, colored noise is considered as the Gaussian white noise with specific filter. For the mentioned three modeling approaches, Gaussian white noise is implemented using Gaussian random number generation. To get precise noise power, the event-driven model requires preknowledge of the time step.

Weakly nonlinear behavior in this work is modeled using the mathematical representation as (3).

$$y = ax - bx^2 - cx^3 - \cdots \tag{3}$$

where a, b, c are the constants and derived from the measurements of circuit nonlinear parameters. This behavior described in the mentioned three modeling approaches share the same performance for model accuracy. It involves only numerical calculations. For baseband model, the clamping behavior (when ignoring higher order harmonics in (3)) will bring in extra distortions[7].

In behavioral model, the filtering characteristic is described by corresponding Laplace transfer function. It is supported in VerilogA as conservative model. In timed data flow or event-driven modeling approaches, the z-domain equivalence is used instead. And this continuous to discrete conversion requires fixed time step. Baseband modeling requires pre-calculation of the baseband equivalence of the passband filter characteristics.

*2) Passive Mixer:* Current-driven passive mixer (shown in Fig.2) is usually used for low power design. It is critical to model the periodic current switching controlled by LO signals[8]. Conservative model can accurately describe this behavior. Timed data flow and event-driven approaches can only describe the numerical relation between output and input,

which is a function of LO signals. The latter can catch every transition points. Baseband model cannot simulate this behavior without pre-calculation based on the knowledge of certain switch pattern of LO signals.

*3) PLL:* For the modeling of PLL (Phase Lock Loop), the most critical aspect is the performance of phase noise. The modeling of phase noise is from the modeling of jitter in VCO (Voltage Control Oscillator) and the closed loop transfer function. VerilogA conservative model is able to describe PLL with single-tone VCO output or square wave output (composed by enough harmonics). The time step of VCO is forced small enough to quantize jitter accurately in time domain. Timed data flow model has the similar problem. For the PLL with square wave output, event-driven model can accurately catch the transition points and support better jitter modeling.

## B. Simulation Speed

Under the same computation environment, the simulation speed of modeling approaches is determined by the total amount of calculations ($C_{total}$). The number of modules ($N$), the number of callbacks for each module in a certain system time ($N_{CB,i}$), and the amount of calculations per callback($C_{CB,i}$) are used to estimate $C_{total}$ for each modeling approach.

$$C_{total} = \sum_i^N N_{CB,i} \cdot C_{CB,i} \tag{4}$$

For conservative model, $N$ equals to the number of circuit modules. $N_{CB,i}$ is determined by the maximum frequency of signals passing through modules. Due to iterative calculations of ODE, $C_{CB,i}$ is times the amount of calculations directly from the mathematical model.

For timed data flow mode, $N$ equals to the number of circuit modules. $N_{CB,i}$ is estimated through the assignment of time step, which is small enough to avoid spectrum alias. Without the iterative calculation, $C_{CB,i}$ for timed data flow model is in the same level as the calculations of the mathematical model.

For event-driven model, when the module is processed with fixed time step, a virtual inner clock is required and it therefore increases the number of modules. $N_{CB,i}$ is determined by the coming event (the time step of virtual clock, or the data rate of input signals). In the example of PLL, the output of VCO is sampled only at transition points instead of Nyquist sampling in the other two methods. Smaller $N_{CB,i}$ can be reached. $C_{CB,i}$ is in the same level as the mathematical model.

For baseband model, $N_{CB,i}$ is determined by the maximum frequency of baseband signals, which is much smaller than the passband counterpart. $C_{CB,i}$ is increased by at least 7 times because one RF signal is expanded up to 7 baseband signals in this work.

It is summarized as:

$$N^{(3)} \approx N^{(BB)} \geqq N^{(2)} = N^{(1)}$$
$$N_{\text{CB},i}^{(1)} \approx N_{\text{CB},i}^{(2)} \geqq N_{\text{CB},i}^{(3)} \gg N_{\text{CB},i}^{(BB)}$$
$$C_{\text{CB},i}^{(1)} \gg C_{\text{CB},i}^{(BB)} > C_{\text{CB},i}^{(2)} = C_{\text{CB},i}^{(3)}$$
$$C_{total}^{(1)} \gg C_{total}^{(2)} \approx C_{total}^{(3)} \gg C_{total}^{(BB)} \qquad (5)$$

where superscripts denotes conservative model $(1)$, timed data flow model $(2)$, event-driven model $(3)$, and baseband model developed in event-driven method $(BB)$ respectively. In the case of PLL, $C_{total}^{(2)} > C_{total}^{(3)}$

## III. MODELING FLOW

An ideal design and verification flow requires fast and accurate behavioral models. Based on the discussion above, neither of the mentioned modeling approaches is optimum in both modeling accuracy and simulation speed. On the other hand, these requirements are relaxed in a hierarchical flow, since the concerns at each hierarchy level are different. At block level, model accuracy is the most important factor to check the performance of limited size circuits. At subsystem or system level, the simulation speed becomes critical for the functional verification of huge size system.

For the behavioral modeling of the GNSS receiver front-end in this work, there are some more challenges.

Firstly, it requires long system time simulation to collect enough samples for functional verification due to the low message rate of GNSS signals. Secondly, the GNSS signals are always overwhelmed by noise at the RF front-end, therefore it would need an overall system simulation including the de-spreading parts at digital baseband.

Moreover, it requires the flow providing the encapsulated, user-friendly, and flexible toplevel behavioral models for industrial cooperation in various verification tasks.

The proposed modeling flow is given in Fig.3. VerilogA model is realized in conservative approach, VerilogAMS is

for timed data flow method, and SystemC model uses event-driven approach.

The models are developed from parametric basic behaviors and specifications from circuits. The subsystem level and top level models are generated recursively from these basic behaviors and circuit schematics. An automatic tool assists this process by extracting hierarchical information from circuit schematic to generate SystemC pin-accurate models[9].

In this way, the use of more than one modeling approach does not lead to the big increase of modeling effort. The duplicate work by hand only occurs when developing parametric basic behaviors, which can be reused in different modules and other projects. The match of easy test basic behaviors indicates the match of complex modules in different modeling approaches.

To further speed up simulations, the baseband model in SystemC is also used to test digital design concept as VerilogAMS and SystemC event-driven toplevel models. Virtual GNSS signals transmitter and one user-friendly interactive control panel are also implemented.

## IV. SIMULATION RESULTS

The comparison of model accuracy for modeling approaches are demonstrated at first in Fig.4 for the example of weakly nonlinear behavior in LNA. It demonstrates the match among different modeling approaches as previous analyzed.

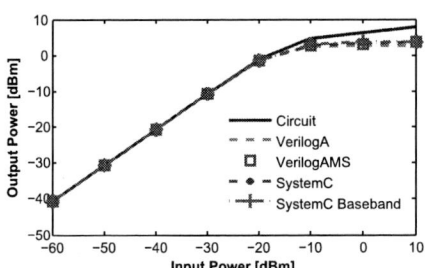

Fig. 4.    Compression curves for nonlinear behavior of LNA

Fig.5 compares the realization of bandpass behavior in VerilogA conservative model and VerilogAMS wreal model with the circuit SPICE model. The conservative model is more close to the circuit model than converted z-domain realization.

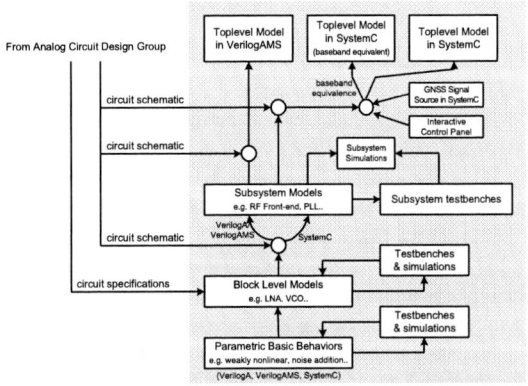

Fig. 3.    Proposed hierarchical behavioral modeling flow

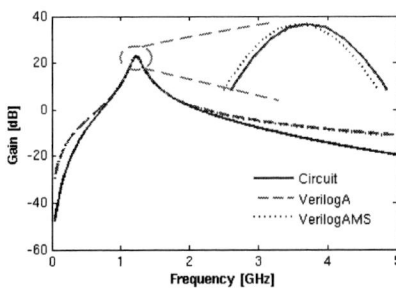

Fig. 5.    Modeling bandpass behavior in different approaches

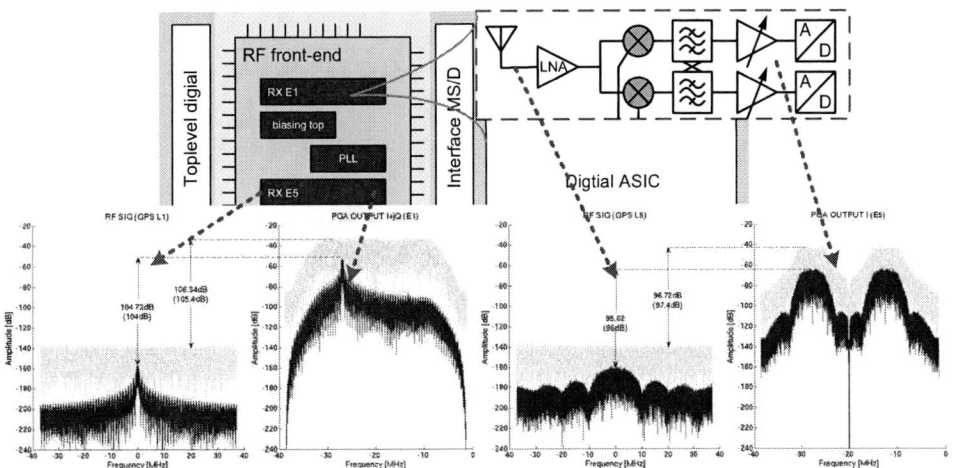

Fig. 6. Spectrum of GNSS Signal Flow at RF Receiver Front-end from SystemC Baseband Model

As shown in Fig.6 (SystemC baseband case), by comparing the input and output spectrum of RF front-end for E1 path (1575.42MHz band) and for E5 path (1176.42MHz band), two gains are measured for GNSS signals from non-noise mode simulation (the dark one) and for noised signals (gray background). The former is the power gain $G_p$, and the difference of the two values is NF (noise figure) of the front-end.

Comparing these measurements with the assigned value as Table I, it reveals that the three toplevel models are in the same level of accuracy.

| | Spec Value | VAMS Model | SC Model | SC Model (baseband) |
|---|---|---|---|---|
| $G_p$ (E1 path) | 104dB | 104.39dB | 104.46dB | 104.72dB |
| NF (E1 path) | 1.4dB | 1.82dB | 1.32dB | 1.62dB |
| $G_p$ (E5 path) | 96dB | 96.10dB | 96.55dB | 96.62dB |
| NF (E5 path) | 1.4dB | 1.612dB | 1.3dB | 1.1dB |

TABLE I
COMPARISON OF TOPLEVEL MODEL ACCURACY

The simulation speeds of modeling approaches are compared in Table II. VerilogA model is estimated to have 100x to 1000x simulation speed in comparison with SPICE circuit model[4].

| | VA Model | VAMS Model | SC Model | SC Model (baseband) |
|---|---|---|---|---|
| PLL | 5733s/100us | 31s/100us | 3s/100us | - |
| RX E1 | 137s/10us | 33s/10us | 3.8s/10us | 0.007s/10us |
| Toplevel | - | 9800s/1ms | 2700s/1ms | 6s/1ms |

TABLE II
COMPARISON OF SIMULATION SPEED OF TOPLEVEL MODELS

## V. CONCLUSION

Based on the comparison of conservative, timed data flow and event-driven modeling approaches at model accuracy and simulation speed, a hierarchical modeling flow is proposed to link the relatively accurate VerilogA model and fast SystemC baseband model without much modeling effort increase. The simulation results prove the analytical comparison in return.

## VI. ACKNOWLEDGMENT

The authors would like to thank Cadence Academic Network (CAN) for the software support used in this work.

## REFERENCES

[1] R. A. Rutenbar, G. G.E.Gielen, and J. Roychowdhury, "Hierarchical Modeling, Optimization, and Synthesis for System-Level Analog and RF Designs," *Proceedings of the IEEE*, vol. 95, pp. 640–669, March 2007.

[2] K. Kundert and H. Chang, "Verification of Complex Analog Integrated Circuits," in *IEEE Custom Integrated Circuits Conference (CICC)*, 2006, pp. 177–184.

[3] W. Hartong and S. Cranston. (2009) Real Value Modeling for Mixed Signal Simulation. [Online]. Available: http://www.cadence.com/rl/Resources/application_notes/real_number_appNote.pdf

[4] M. Barnasconi. (2010) Whitepaper: SystemC AMS Extensions: Solving the Need for Speed. [Online]. Available: http://www.accellera.org/community/articles/amsspeed/

[5] J.-E. Jang, M.-J. Park, D. Lee, and J. Kim, "True event-driven simulation of analog/mixed-signal behaviors in SystemVerilog: A decision-feedback equalizing (DFE) receiver example," in *IEEE Custom Integrated Circuits Conference (CICC)*, 2012, pp. 1–4.

[6] J. Chen. Modeling RF Systems. [Online]. Available: http://www.cktsim.org/modeling/modeling-rf-systems.pdf

[7] Z. Chen, Y. Wang, and S. Heinen, "Baseband Modeling of a GNSS Receiver Front-end using SystemC/-AMS," in *8th Conference on Ph.D. Research in Microelectronics and Electronics (PRIME)*, 2012, pp. 1–4.

[8] A. Mirzaei, H. Darabi, J. C. Leete, and Y. Chang, "Analysis and Optimization of Direct-Conversion Receivers With 25Mixers," *IEEE Transactions on Cicuits and Systems*, vol. 57, pp. 2353–2366, September 2010.

[9] Y. Wang, Z. Chen, and S. Heinen, "Hierarchical generation of pin accurate SystemC models based on RF circuit schematics," in *IEEE International Behavioral Modeling and Simulation Conference (BMAS)*, 2010, pp. 25–30.

978-1-4673-6145-3/13 $31.00 © 2013 IEEE

# Pulse Amplification Based Dynamic Synchronizers with Metastability Measurement using Capacitance De-rating

Bharan Giridhar, Matthew Fojtik, David Fick, Dennis Sylvester, David Blaauw

University of Michigan, Ann Arbor, MI 48109

**Abstract — We present dynamic buffer based synchronizers where only *pulses* rather than stable intermediate voltages cause metastability. We exploit this unique feature by amplifying such pulses to improve MTBF by >$10^6 \times$ over jamb latches and double flip-flops at 2GHz in 65nm CMOS. The synchronizers incur single-cycle latency with a MTBF of ~$2 \times 10^{11}$ years. A new method to experimentally measure metastability on chip is also proposed and used to evaluate synchronizer performance.**

## I. INTRODUCTION

Message-passing and shared-memory based multicore processors have risen in popularity and often employ independent DVFS of the cores to improve energy efficiency [1–2]. As a result, fast and reliable on-chip communication is a key challenge in these systems and has spurred extensive research to reduce the occurrence of metastability during synchronization. Previous hardware approaches [3–5] have addressed this by either increasing synchronization latency or constraining the relationship between the clock frequencies. This paper presents dynamic buffer based synchronizers where only *pulses*, rather than stable intermediate voltages, generate metastability events due to the one-sided operation of dynamic gates. This unique feature enables the key advantage that mean time between failures (MTBF) can be significantly improved by amplifying such pulses using skewed inverters in the synchronization path. In a 65nm test chip (FO4 delay = 11ps), this approach improves MTBF by ~$10^6 \times$ over jamb latches and ~$5 \times 10^7 \times$ over double flip-flops (2-FFs) at 2GHz. The synchronizers incur single cycle synchronization latency with an MTBF of up to ~$2 \times 10^{11}$ years. In addition, a new on-chip metastability measurement method using capacitance de-rating is also demonstrated and used to evaluate synchronizer performance.

## II. SYNCHRONIZER DESIGN

Typically, a synchronizer uses two series-connected flip-flops (Fig. 1, inset – double-flop synchronizer). The 1st FF (DFF1) has a finite probability of sampling the input (DCK1 from clock domain CK1) during a transition and becoming metastable due to the asynchronous relationship of the two clock domains (CK1 and CK2). This event can cause an arbitrarily slow transition at Q1, which in turn can cause the 2nd FF (DFF2) to become metastable during the subsequent cycle. This output (Q2) can be interpreted inconsistently by different downstream gates, potentially causing a functional

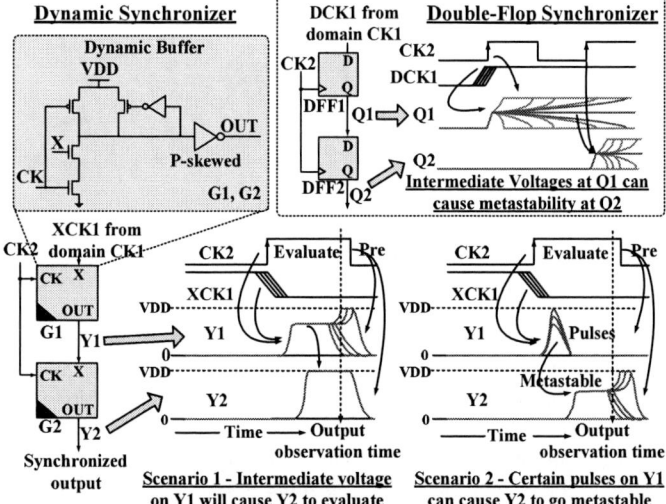

Figure 1. Metastability in dynamic synchronizers is only caused by pulses (scenario 2), in contrast with double-flop based synchronizers (inset).

failure. By increasing resolution time, the metastability probability reduces exponentially, giving rise to a fundamental latency/robustness tradeoff [7].

Dynamic buffers, however, exhibit one-sided evaluation. Fig. 1 shows that a similar intermediate signal voltage at Y1 causes the final dynamic buffer G2 to fully evaluate (scenario 1), avoiding metastability. Instead, metastability occurs when dynamic buffer G1 generates a pulse at Y1 (resulting from its keeper) that causes partial evaluation in G2 and metastability at output Y2 (scenario 2). We make the key observation that such pulses, unique to dynamic logic, can be amplified using skewed inverters, thereby reducing metastability without adding synchronization cycle latency.

Fig. 2 shows how pulses generated at Y1 by dynamic buffer G1 are amplified using properly skewed inverters. If the pulse at the input of dynamic buffer G2 is within a specific height/width range, its output will fail to evaluate to a rail voltage by the required time, causing a metastability event at Y2. By providing amplification, the range of pulses at the output of G1 causing metastability in G2 is compressed. This in turn compresses the window (range) of data-to-clock alignments at the input of G1 that yield metastability (metastability window). The amplifying inverters are skewed

978-1-4673-6145-3/13 $31.00 © 2013 IEEE

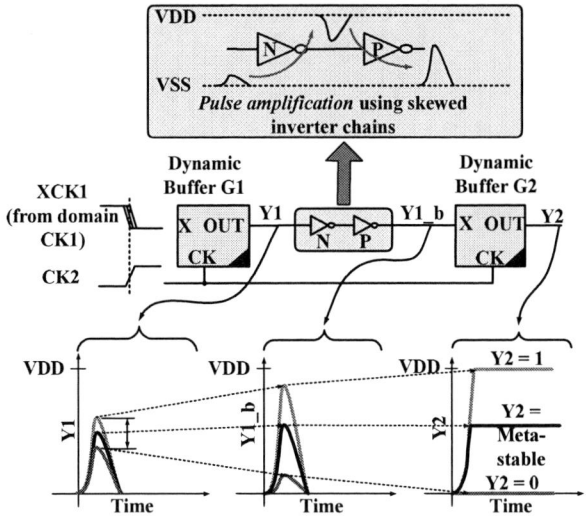

Figure 2. Pulses in dynamic synchronizers amplified using skewed inverters.

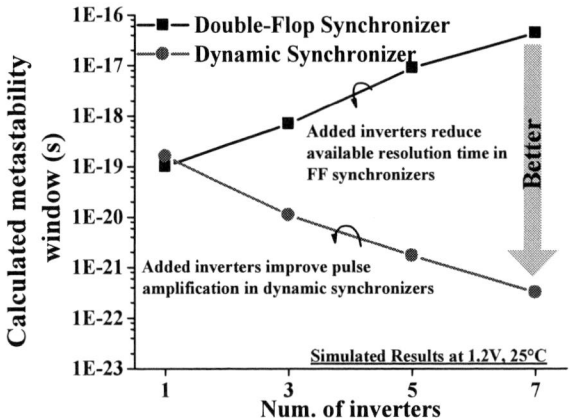

Figure 3. Pulse amplification in dynamic synchronizers contrasted with FF-based synchronizers.

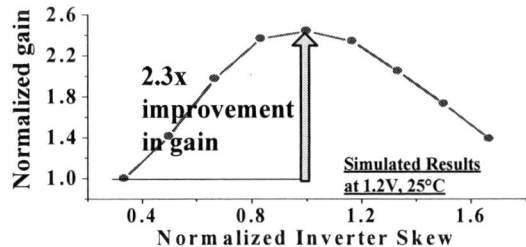

Figure 4. Properly skewing inverters in a 3-inverter chain by aligning their DC transfer functions with input pulse height improves stage gain by 2.3×.

by aligning their DC transfer function with the input pulse height to maximize gain (Fig. 4).

Based on simulation, adding skewed inverters improves MTBF by ~$2 \times 10^3 \times$ (Fig. 3) in a 2-stage synchronizer. In addition to inserting inverters, additional dynamic buffers, also clocked by CK2, can be inserted since they function much like inverters, providing gain for propagated pulses without adding cycle latency.

Fig. 3 also shows that inserting additional inverters in FF-based synchronizers does not improve metastability. A

properly skewed inverter chain with metastable input will drive its output to rail. However, since the metastable input can still resolve in either direction, the inverter output can still switch back at a later time, creating metastability in the capturing FF (in contrast to dynamic synchronizers with one-sided evaluation). Thus, inverter insertion only delays the metastability event and hereby worsens MTBF by reducing available resolution time (due to the additional delay of the inserted inverters).

Fig. 5 shows the circuit details of a dynamic synchronizer stage. The full keeper improves gain at the dynamic node by 13.3× (simulated). Cutoff device M1 prevents short-circuit current during precharge. Gate length in the inverters is increased to 70nm ($L_{min}$ = 60nm), improving gain by 30%.

## III. ON-CHIP METASTABILITY MEASUREMENT

To experimentally characterize MTBF, we propose a new measurement method where DUTS are de-rated (slowed) by connecting their internal nodes to selectable MIM capacitors (Fig. 5). By increasing node capacitance, gain-bandwidth product is reduced and the metastability window increases. Such windows are then measured and results are finally extrapolated to the actual metastability window under native (self-loaded) conditions. The proposed extrapolation based measurement has the disadvantage that it amplifies measurement error. However, to date, no known method to *directly* measure metastability in silicon exists and instead current methods [8, 9] rely on fitting of models. Our analysis shows that the results of the proposed method are within a reasonable margin of error (Section IV). Slowing the gates also requires resolution time (RT) to be de-rated (increased) due to the slower transitions of nodes to their steady-state values. The de-rating of RT and capacitance must be coordinated to obtain a linear dependence, which is critical to facilitate accurate extrapolation.

Fig. 6 shows that linearly scaling RT with capacitance does

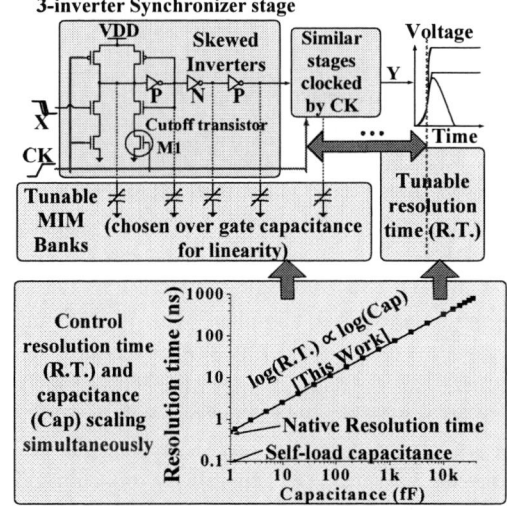

Figure 5. Measurement method for determining intrinsic metastability window using capacitance de-rating.

978-1-4673-6145-3/13 $31.00 © 2013 IEEE

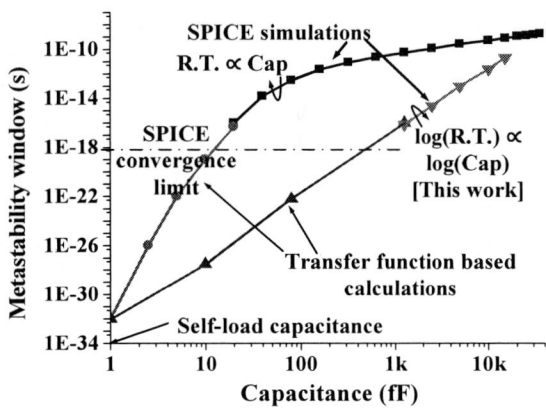

Figure 6. Scaling capacitance and RT with log-log proportionality results in linear dependence of metastability window on capacitance, enabling accurate extrapolation.

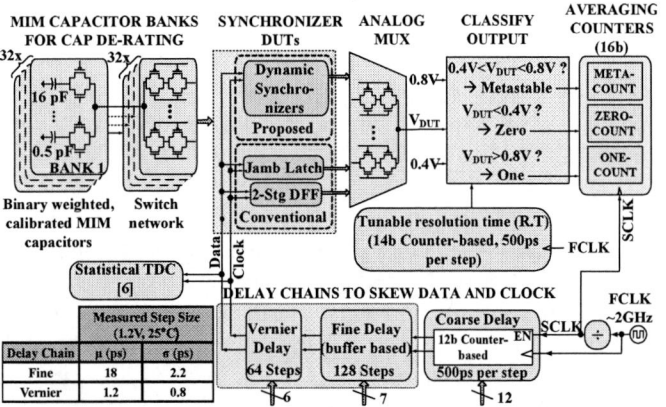

Figure 7. Block diagram of the test harness for on-chip metastability measurement using capacitance de-rating, implemented in 65nm CMOS.

not yield a linear change in the metastability window. Instead, we find that log-log proportionality between RT and capacitance provides linearity and enables accurate extrapolation to native RT (500ps) under self-loading conditions (1fF for the simulation in Fig. 6). This result is confirmed by SPICE simulations, transfer function-based calculations, and silicon measurements.

The test harness to measure metastability on chip using this de-rating scheme is shown in Fig. 7. The data-to-clock alignment was controlled using a 3-stage delay chain: 1) a counter-based coarse delay chain with measured steps of 0.5ns; 2) a fine delay chain with measured steps of ~18ps; and 3) a Vernier delay chain with measured mean resolution of 1.2ps. A statistical TDC [6] averaging $10^6$ results was used to measure the data-to-clock alignment with 1ps accuracy. The DUTs (at 1.2V VDD) were de-rated by connecting their nodes to calibrated, binary-weighted, selectable MIM capacitors. All switches were double-stacked to remove leakage effects. Two comparators were used to flag a metastable event by comparing the selected DUT output to off-chip references (0.8V and 0.4V) that define the metastable voltage range. Averaging counters recorded the number of 1, 0, and metastable events over several trials. Data was increasingly

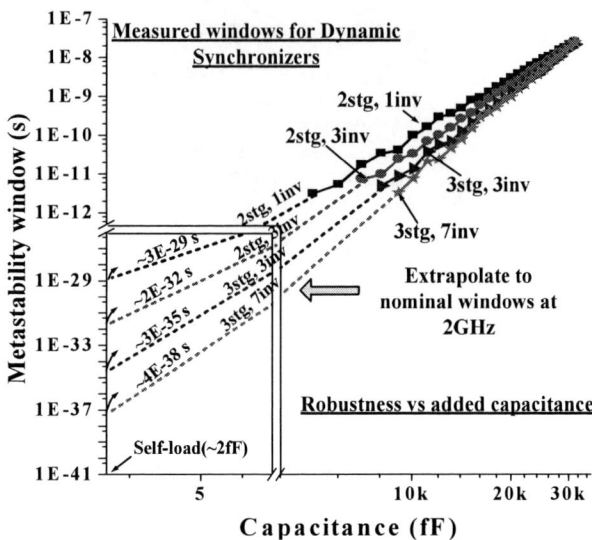

Figure 8. Measured metastability windows for several dynamic synchronizer configurations under de-rated conditions at 1.2V, 25°C.

delayed with respect to clock and the metastability window was defined as the time range when metastability count dominated over 0- and 1-counts.

## IV. MEASURED RESULTS

Fourteen dynamic synchronizer configurations differing in the number of stages and inverters per stage were tested at 2GHz and compared to jamb latch and 2-FF synchronizers.

Fig. 8 shows measured metastability windows for several dynamic synchronizer configurations (de-rated conditions) along with their extrapolated windows at native conditions (~2fF, 500ps RT). Fig. 9 corroborates the extrapolation approach by measuring windows using distinct RT / capacitance scaling ratios; results converge to a relatively small range at native conditions, as desired.

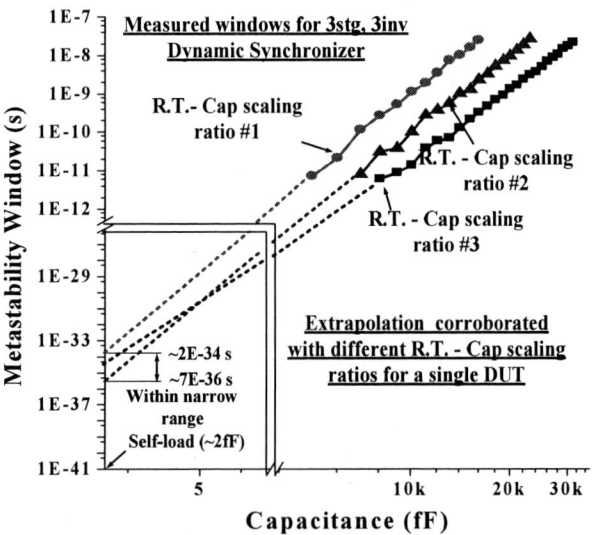

Figure 9. Measured windows for the 3-stage, 3-inverter dynamic synchronizer (at 1.2V, 25°C) with distinct RT / capacitance scaling ratios converge to a small range, corroborating the extrapolation.

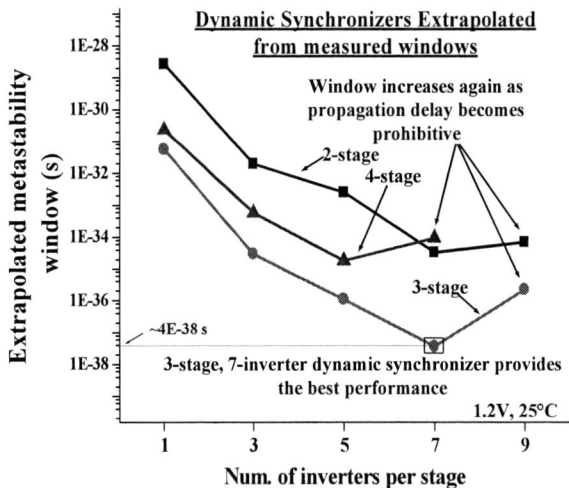

Figure 10. Measured results for all configurations are extrapolated to obtain nominal windows at 2GHz under self-load conditions. The 3-stage, 7-inverter configuration gives the best metastability window of ~4e-38s at native conditions.

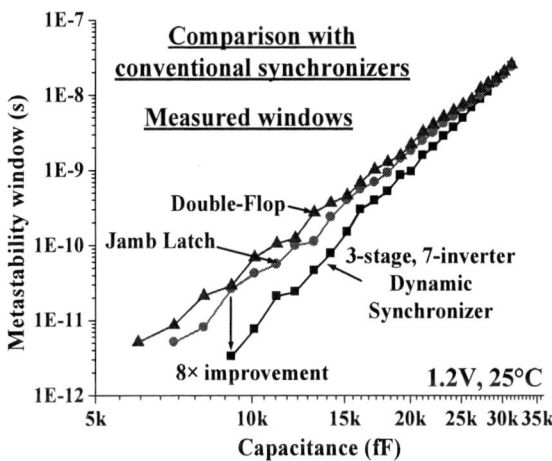

Figure 11. At 25°C, the 3-stage, 7-inverter configuration improves MTBF by 8× over jamb latch at 9.1pF load, 307ns RT; this translates to an improvement of ~1e6× at native conditions (2fF self-load, 500ps RT).

Extrapolated windows for all measured configurations are shown in Fig. 10, confirming that metastability reduces as inverters/dynamic buffers are inserted until their propagation delay becomes prohibitive. The 3-stage, 7-inverter synchronizer provides the best performance and MTBF improvement of 8× over the jamb latch (Fig. 11) at the smallest measureable de-rating condition (9.1pF loading, identical RT of 307ns). This translates to an improvement of ~$1\times10^{6}$× at native conditions.

Fig. 12 shows that the extrapolation error due to measurement and fit limitations is relatively small compared to improvement over jamb latch. This synchronizer has a MTBF of ~$2\times10^{11}$ years at 2GHz, mapping to a system failure rate of ~$8.7\times10^{-4}$/year for a CMP with $10^{3}$ synchronized signals at 2.5GHz (jamb latch rate = ~55.8/year). Fig. 13 shows the die micrograph and performance summary.

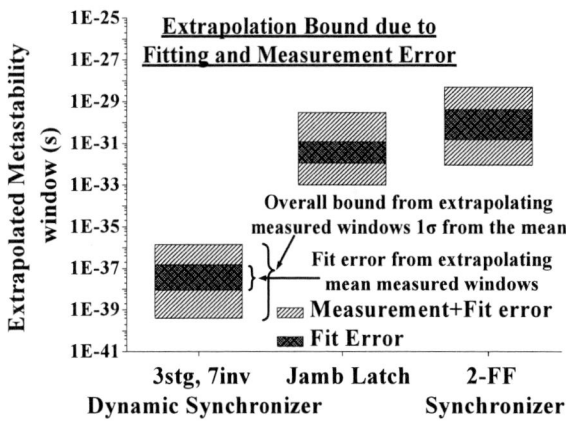

Figure 12. Extrapolation error due to fitting and measurement limitations is relatively small, compared to improvement over jamb latch.

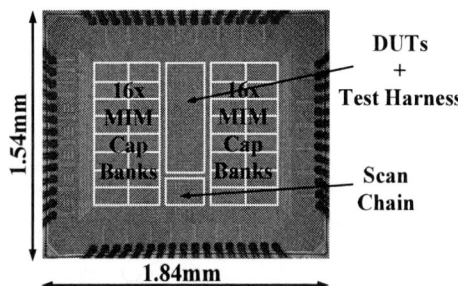

| Synchronizer | Normalized MTBF at 2GHz | Power (1.2V, 25°C, simulated) | Area |
|---|---|---|---|
| 3-stage, 7-inverter Dynamic Sync. | ~5E7× (This work) | 106 μW | 52 μm² |
| 2-FF | 1× | 60 μW | 24 μm² |
| Jamb Latch | ~50× | 24 μW | 8 μm² |

Figure 13. Die micrograph and performance summary in 65nm CMOS (FO4 delay = 11ps).

## V. CONCLUSION

This paper presented dynamic buffer based synchronizers where only pulses, rather than stable intermediate voltages, cause metastability. This unique characteristic was exploited by amplifying such pulses using skewed inverters and additional dynamic buffers to improve MTBF by ~$10^{6}$× over jamb latches and ~$5\times10^{7}$× over double flip-flops at 2GHz in 65nm CMOS. The synchronizers incur single cycle latency and were evaluated using a new on-chip metastability measurement method based on capacitance de-rating.

### REFERENCES

[1] J. Howard et al., ISSCC, 2010, pp.108-109.
[2] S. Dighe et al., JSSC, Jan. 2011, pp.184-193.
[3] Z. Yu et al., ISSCC, 2012, pp.64-66.
[4] A. Shibayama et al., Symp. VLSI Circuits, 2007, pp.158-159.
[5] P. Caputa et al., ISSCC, 2006, pp.1765-1774.
[6] D. Fick et al., ISSCC, 2010, pp.188-189.
[7] R. Ginosar, ASYNC, 2003, pp.89-96.
[8] S. Beer et al., ISCAS 2011, pp.2593-2596.
[9] J. Zhou et al., JSSC, Feb. 2008, pp.550-557.

# FireBird: PowerPC e200 Based SoC for High Temperature Operation

Radisav Cojbasic*, Ömer Cogal*, Pascal Meinerzhagen†, Christian Senning†, Conor Slater‡,
Thomas Maeder‡, Andreas Burg† and Yusuf Leblebici*
*Microelectronic Systems Laboratory (LSM),
†Telecommunications Circuits Laboratory (TCL),
‡Laboratory of Microengineering for Manufacturing 2 (LPM2),
Swiss Federal Institute of Technology Lausanne (EPFL), Station 11, 1015 Lausanne, Switzerland
email: radisav.cojbasic@epfl.ch

*Abstract*—**PowerPC Architecture microcontrollers are commonly used in embedded applications. In this work we present FireBird, the first PowerPC based SoC for reliable operation beyond 200°C. Designing SoCs for reliable operation at high temperatures is a significant challenge, due to increased static leakage current, reduced carrier mobility, and increased electromigration. To alleviate the consequences of high temperatures, this paper proposes to customize a PowerPC e200 based SoC by using a dynamically reconfigurable clock frequency, exhaustive clock gating, and electromigration-resistant power supply rings. A 20x9 mm$^2$ chip implementing this design has been fabricated in 0.35 μm CMOS technology. The custom testing procedure showed the expected maximum operating frequency reduction from 38 MHz at room-temperature to 30 MHz at 200°C, which illustrates the importance of an adaptable clock frequency under temperature variation. At 200°C, the maximum power dissipation at 3.3 V supply voltage was 1.2 W and the idle state static leakage current was 3.4 mA. Silicon measurements proved that this design outperforms PowerPC based SoCs available in the high-temperature microcontrollers market which are not operational at temperatures above 125°C.**

## I. INTRODUCTION

There is a growing demand for SoCs integrating one or more microprocessors, on-chip memories, and a variety of sensors and actuators on a single chip that is capable to operate at high temperatures [1]. For example, automotive applications such as ECU (*Engine Control Unit*) require SoCs that reliably operate in high-temperature environments [2]. Modern aircraft industry also demands microcontrollers and electric motors to operate at high temperatures in order to replace hydraulics. Fully integrated solutions allow better reliability than presently used hydraulic structures which exploit many interconnects, plugs and complex mounting procedures [3]. Downhole oil and gas drilling is the oldest and largest application domain of high-temperature electronics (T > 150°C). In this application, the operating temperature is a function of the underground depth of the well (the typical geothermal gradient is 25°C/km). Advances in technology have motivated the industry to drill deeper, thus there is a constant urgent demand for high-temperature microelectronic devices, particularly microcontrollers that are used for control applications.

In 1991, the PowerPC alliance (IBM Corporation, Motorola, Inc., and Apple Computer Corporation) defined the Power

(*Performance Optimization With Enhanced RISC*) Architecture [4] [5]. A PowerPC is a general-purpose processor based on a superscalar design [5]. The Power Architecture supports a family of processors that spans a wide range of system and application needs. It also provides a stable base for software, allowing applications that run on one PowerPC processor to run consistently on any other PowerPC processor. Operating systems can be moved from one processor implementation to another by making only a few minor changes [5] [6].

Already in 1994, PowerPC was used in aerospace communication systems as presented in [6]. In 2001, Honeywell presented a radiation hardened PowerPC e603 based single board computer for avionics applications that operates up to 80°C [7]. The PowerPC e200 based MCP5554 microcontroller by Freescale is often used for real-time control applications and it operates up to 125°C.

As commercially available PowerPC based SoCs are not operational at temperatures above 125°C, the main motivation for this work was to design the first PowerPC based SoC that operates beyond 200°C. To the best of our knowledge, there are not even any previous reports on experimental PowerPC processors operating beyond 125°C. Nevertheless, Table I compares the features and performance of the best commercial PowerPC based SoCs (reaching the highest temperature so far) with this work. Maximum operating temperature, frequency, and power consumption of the aforementioned products, are taken from product datasheets.

TABLE I
POWERPC BASED SOCS FOR HIGH-TEMPERATURE OPERATION.

| Model | MPC5554 | PC7447A | FireBird |
|---|---|---|---|
| Manufacturer | Freescale | e2v Semiconductors | EPFL |
| $T_{jMAX}$ [°C] | 125 | 125 | 225 |
| $f_{MAX}$ [MHz] | 132 | 1330[CPU], 166[BUS] | 38 |
| PowerPC core | e200 | Dhrystone | e200 |
| Cache Size [kB] | 32 | 64 | 32 |
| SRAM Size [kB] | 64 | - | 64 |
| Power Consumption [W] | - | 18 | 1.45 |

978-1-4673-6145-3/13 $31.00 © 2013 IEEE

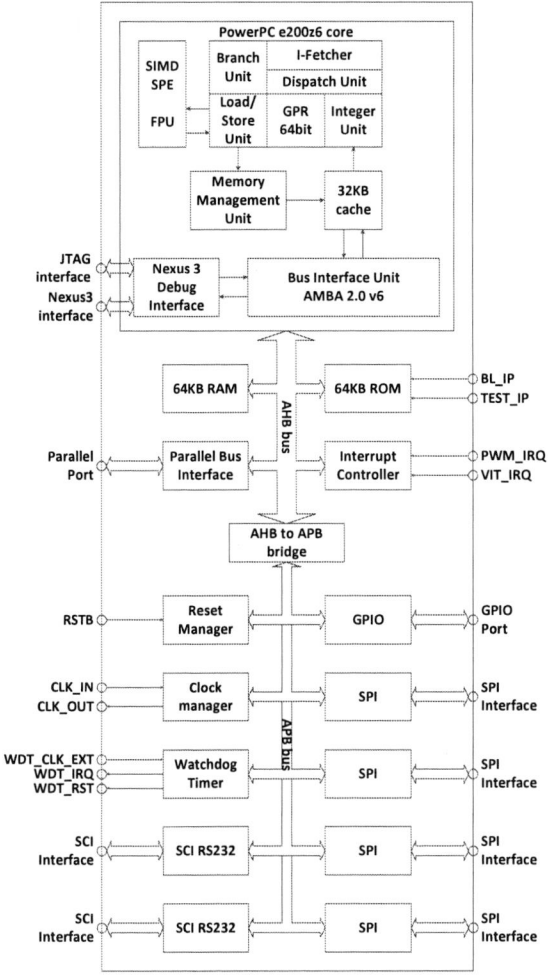

Fig. 1. The Top Level Architecture.

## II. TOP LEVEL ARCHITECTURE

Core e200 is a single-issue 32bit Book E Power Architecture [8] compliant core. Its seven pipeline stages feature a 64bit GPR (*General-Purpose Register*) file, a BPU (*Branch Processing Unit*), an LSU (*Load/Store Unit*) and an MMU (*Memory Management Unit*) with 32-entry fully associative TLB (*Translation Look-aside Buffer*) and multiple page-sizes support. The core e200 uses 32kB of 8-way set-associative unified cache. Accelerated processing is supported for integer and single-precision floating-point operations using 64bit operands.

The core e200 addresses memories and peripherals using the ARM Ltd. AMBA (*Advanced Microcontroller Bus Architecture*) and it is the only AMBA master in the system. The AMBA consists of the AHB (*Advanced High-performance Bus*) and APB (*Advanced Peripheral Bus*) buses that are connected with a bridge. The AHB addresses directly the RAM and ROM controllers, interrupt controller, parallel port and the AHB/APB bridge. The AHB/APB bridge is master to the APB and it addresses all the remaining peripherals in the system.

Apart from the PowerPC e200 core and AMBA system, the FireBird SoC has 64kB of both RAM and ROM. The ROM contains compiled boot-up instructions which download the RAM content. The RAM is divided into 32kB instruction and data sectors. The interrupt controller handles eight internal and two external interrupt sources. A reset manager ensures proper system power-up and external reset signal detection. The clock manager allows dynamic frequency reduction while operating at high temperatures. It also provides four external buffered clock sources. Communication interfaces include four SPI masters and two RS232 UARTs. Both interfaces are configurable through internal registers. Finally, a parallel port (programmable interface to external memory) and general-purpose input/output port are available for debugging and application-specific purposes.

The top level architecture of the FireBird SoC is presented in Fig. 1. The main features could be defined as following:

- 32bit PowerPC Core (e200z6 Architecture).
- Bus Interface Unit (AMBA 2.0 v6).
- Memory Management Unit.
- External Memory Interface (Parallel Port).
- Embedded on-chip Memories (64kB RAM and ROM).
- 32kB Cache Memory (Parity detection security).
- JTAG/Nexus3 Debug Interface.
- Two External Interrupt Sources.
- Clock Manager (4 configurable clock sources).
- Reset Manager.
- Two RS232 UART Interfaces.
- Four SPI Master Interfaces.
- 16bit GPIO Port (8bit Input and Output Ports).
- Voltage levels LVTTL compliant (3.3V).

## III. DESIGN IMPLEMENTATION

FireBird SoC design consists of 8 million transistors that are found exclusively in standard-cells and SRAM modules, since it does not infer any analog blocks. It was fabricated in a 0.35 µm CMOS process optimized for low-leakage high-temperature operation. The process provides only four metal layers including the top metal. Limited metal stack led to a reduced standard-cell density in order to be able to close the routing of the entire design and resulted in a large die size of $20 \times 9 \, mm^2$. This design uses as much as 256 pins and it is still core limited. The 142 I/O pins are presented in Fig. 1. The remaining pins were dedicated to a power supply, assuming a supply voltage of 3.3 V both for the I/O and the core power rings.

The core e200 accounts for the major part of the standard-cells and it is equivalent to 320 kGates (where a driving-one 2-input NAND is taken as reference gate) or 1.2 million transistors. The 64kB SRAM are implemented in 4x 16kB SRAM macros and 32kB of Cache memory are implemented in 24x SRAM macros. Overall, those 96kB SRAM count 4.8 million transistors. The 64kB on-chip ROM is implemented in a single block containing 2 million transistors. The FireBird SoC layout and the chip microphotograph are presented in Fig. 2 and Fig. 3, respectively.

978-1-4673-6145-3/13 $31.00 © 2013 IEEE

Fig. 2. The FireBird SoC Layout.

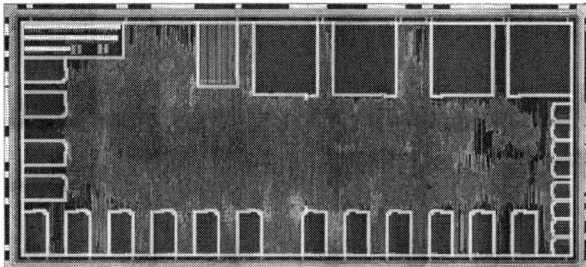

Fig. 3. The FireBird SoC Microphotograph.

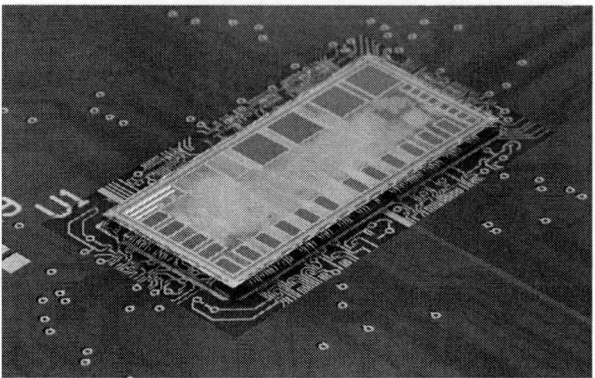

Fig. 4. The FireBird SoC Chip-On-Board Setup.

## A. Design for high-temperature Operation

When applied externally, a thermally-stable clock signal with a constant frequency can jeopardize the proper execution of the critical paths, since the maximum operating frequency is degraded due to the reduced carrier mobility at high temperatures [9]. The clock manager supports dynamically re-configurable frequencies of the clock signal and the capability to internally scale-down the clock frequency. The control of frequency at high temperatures is implemented in software.

FireBird SoC is designed to keep the active power consumption as low as possible over the entire operation range. The goal is to eliminate switching activity in registers when related outputs are not used by fetched instruction or selected operation mode. The switching activity is eliminated when the corresponding clock signal is idle. The clock is exhaustively gated. Depending on the operation modes and fetched instructions, all the major PowerPC core parts could be restricted from the clock, including the cache controller, MMU, integer and floating-point units. Hence, independently of the operating temperature, the active power consumption is held as low as possible.

The custom triple-rail power distribution network was designed to equally distribute power supply current. The power supply distribution network wrapped each memory module to guarantee robust power supply and prevent the IR drop.

## B. Design Summary

The post-place-and-route simulations showed maximum frequency of 33 MHz at 175°C (at 2.75 V supply voltage in the worst case corner). The transistor spice models are categorized up to this temperature. The critical-path of this design was the on-chip SRAM access. It starts from the MMU registers and it ends in the SRAM controller.

## IV. MEASUREMENTS RESULTS

Once fabricated, the FireBird SoC had to pass room-temperature tests. Tests included BIST (*Built-In Self-Test*) procedure execution. The BIST instructions were stored in the ROM, so the core e200 was capable to enter test-mode directly after power-up. Tests assumed proper March C algorithm [10] execution on the entire addresses range of the 64kB SRAM. It has been observed that March C algorithm provides the highest fault coverage with a reasonably low software complexity (making exclusively read and write accesses to the SRAM). Two RS232 UARTs were tested by sending hard-coded string from the ROM. Four SPI masters were tested by communicating in loop-back configuration.

The reset manger was not tested during the BIST operation, since it had a single functionality (reset generation) that was anyway executed before entering the test-mode. The interrupt controller was disabled during the test-mode, but it was used immediately after, during the boot-up RS232 code download that initialized SRAM. Also, there were no particular tests for the GPIO, but this port was used by all the other tests to send out test results.

While running in test-mode at 3.3 V supply, the Fire-Bird SoC was capable to operate at $f_{MAX} = 38$ MHz with $I_{DD} = 430$ mA. Room temperature tests were done on three setups and all had similar results.

## A. High-temperature Measurements

The high-temperature measurement setup consisted of an alumina plate ($18 \times 8$ mm$^2$, 250 μm thick) with a positive temperature coefficient (PTC) thick film resistor (PTC 2611, Elector Science Laboratories Inc., USA) mounted on top of the FireBird SoC naked die using Sylgard 527 silicone gel [11]. The die was bonded directly to the PCB using COB (*Chip On Board*) techniques. The COB setup is presented in Fig. 4. The temperature was set by a PID controller which controlled the power supply (Agilent E3631A) of our heater using a PC. The program set the voltage across the PTC resistor and measured the current through the resistor. The heater was calibrated using the IR camera (SC655, FLIR Systems Inc., USA). The *p-n* junction temperature, $T_j$, was measured using an on-chip diode.

978-1-4673-6145-3/13 $31.00 © 2013 IEEE

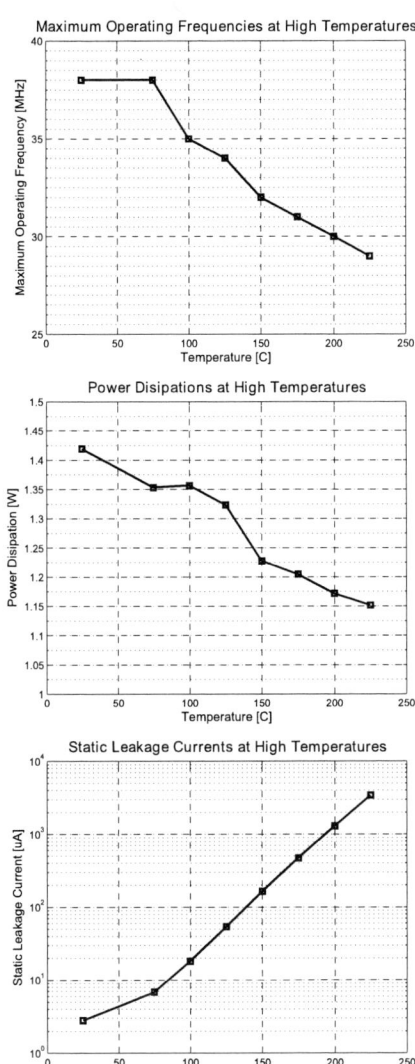

Fig. 5.  High Temperature Measurement Results.

The goal of high-temperature measurements was to determine the maximum operating frequency, related power consumption and static leakage current at the given temperature. The high-temperature measurement results are presented in Fig. 5. Measurement results verify our prediction that the maximum operating frequency degrades with elevated temperatures. As the post-place-and-route simulation reported 33 MHz at 175°C in the worst case corner, higher maximum operating frequency at room-temperature was expected and the result of 38 MHz is in line with our prediction. Maximum operating frequency degradation trend follows the reduction of carrier mobility with respect to temperature [9].

On one hand, the static leakage current increases exponentially with increased temperature. Thus static power dissipation becomes a significant part of the total power consumption at high temperatures and it is negligible at room-temperature, so the total power dissipation of 1.42 W can be referred to as the highest active power consumption of the FireBird SoC. On the other hand, as the maximum operating frequency is degraded with increased temperature, and the

active power consumption is reduced. The peak of the power consumption is when $T_j$ temperatures are in between room-temperature and 125°C. The static power increases slower than the active power consumption decreases, thus the measurement results indicate lower total power consumption with elevated temperature. Such behavior increases reliability and proves electromigration-resistant design, considering that the power supply current has the maximum value at the room-temperature.

## V. Conclusion

We designed, manufactured and measured FireBird, a PowerPC e200 based SoC for operation beyond 200°C. Besides a powerful core and bus, the system features on-chip memories and several communication interfaces. The FireBird SoC was designed to resist high-temperature effects in silicon semiconductors, such as electromigration, increased leakage current and reduced carrier mobility. Silicon measurement results confirmed excellent performance at high temperatures that outperform existing PowerPC based SoCs. FireBird is able to reliably operate at 200°C, a temperature which has never been achieved by previous PowerPC based SoCs.

## Acknowledgment

The authors would like to thank Sylvain Hauser for designing very complex COB, Nikola Katic for designing diode temperature-sensing setup, and Michael Zervas and Davide Sacchetto for wafer dicing.

## References

[1] P. Neudeck, R. Okojie, and L.-Y. Chen, "High-temperature electronics - a role for wide bandgap semiconductors?" *Proceedings of the IEEE*, vol. 90, no. 6, pp. 1065 – 1076, jun 2002.

[2] A. Laudenbach and M. Glesner, "Vlsi system design for automotive control," *Solid-State Circuits, IEEE Journal of*, vol. 27, no. 7, pp. 1050–1056, 1992.

[3] S. Marwedel, N. Fischer, and H. Salzwedel, "Improving performance and reliability assessments of avionics systems," in *Digital Avionics Systems Conference (DASC), 2011 IEEE/AIAA 30th*, 2011, pp. 7D1–1–7D1–7.

[4] P. D. Hester and B. Filip, "Preface," *IBM Journal of Research and Development*, vol. 38, no. 5, pp. 490–491, 1994.

[5] M. Vaden, L. J. Merkel, C. Moore, T. Potter, and R. J. Reese, "Design considerations for the powerpc 601 microprocessor," *IBM Journal of Research and Development*, vol. 38, no. 5, pp. 605–620, 1994.

[6] A. Jackson, "A new family of microcontrollers simplify aerospace communications systems," in *Aerospace Applications Conference, 1996. Proceedings., 1996 IEEE*, vol. 1, 1996, pp. 403–414 vol.1.

[7] G. Brown, "Radiation hardened powerpc 603etm based single board computer," in *Digital Avionics Systems, 2001. DASC. 20th Conference*, vol. 2, 2001, pp. 8C1/1–8C1/12 vol.2.

[8] K. Diefendorff and E. Silha, "The powerpc user instruction set architecture," *Micro, IEEE*, vol. 14, no. 5, pp. 30–, 1994.

[9] M. Darwish, J. Lentz, M. Pinto, P. Zeitzoff, T. Krutsick, and H.-H. Vuong, "An improved electron and hole mobility model for general purpose device simulation," *Electron Devices, IEEE Transactions on*, vol. 44, no. 9, pp. 1529–1538, 1997.

[10] J. Zhao, S. Irrinki, M. Puri, and F. Lombardi, "Testing sram-based content addressable memories," *Computers, IEEE Transactions on*, vol. 49, no. 10, pp. 1054–1063, 2000.

[11] T. Maeder, B. Jiang, F. Vecchio, C. Slater, G. Farine, and P. Ryser, "Ceramic hotplates based on thick-film and ltcc technology," in *Proceedings of 35th International Conference of IMAPS - CPMT IEEE Poland*, 2011, pp. 143–147.

978-1-4673-6145-3/13 $31.00 © 2013 IEEE

# A 1/10000 Lower Error Rate Achievable SSD Controller with Message-Passing Error Correcting Code Architecture and Parity Area Combined Scheme

Kai Li, Mitsuyoshi Ito and Atsushi Esumi

Siglead Inc.
1-38-10-203 Nakagawachuo Tsuzuki-ku
Yokohama-city Kanagawa 224-0003, Japan

*Abstract*-a new Error Correcting Code (ECC) solution aiming to improve the reliability of NAND flash memory (NAND) is proposed. Implemented in Solid-State Drive (SSD) controller IC (Fabricated in TSMC65LP), it is confirmed that more than 1/10000 lower error rate without increasing redundant bits (parity) of conventional BCH code, and 1.7x longer endurance of SSD can be achieved. This solution consists of two independent techniques: MP-ECC, a Message-Passing ECC architecture; and PAC, a Parity Area Combined ECC scheme.

## Introduction

The emerging Solid-State Drive (SSD) market strongly relies on the success of NAND flash memory (NAND), which is achieved by continued cost reduction. However, the reliability of NAND dramatically decreases due to the shrinkage of process nodes and the storage of more bits per cell [1]. So far, the reliability of SSD is maintained by Error Correcting Codes (ECC); and the conventional ECC for SSD controller is BCH code [2]. To address the continuous decline of NAND reliability, normal solution goes to simply increase the ECC strength to correct more errors. But larger ECC strength implies larger circuit area and more redundant bits (parity). So, such solution given by simply increasing ECC strength becomes not feasible and practicable. People also think of using more powerful ECC such as LDPC codes [3], but one still has to face challenges such as Log Likelihood Ratio (LLR) generation, error floor, ECC performance estimation, latency and circuit size [4]. Other related works introduce channel codes to reduce error rate or improve the longevity [5] [6]. Nevertheless, feasible ECC with stronger error correction performance without increasing circuit and parity size is extremely necessary for maintaining NAND/SSD reliability. In this work, a sophisticated ECC solution is proposed to achieve this purpose. It consists of two independent techniques: MP-ECC, an improved ECC architecture enabling effective message-passing algorithm among different ECC blocks; and PAC, the Parity Area Combined scheme. Fabricated in TSMC65LP, the proposed techniques have been equipped in SSD controller prototype IC. The measurement results demonstrate that this ECC solution achieves 1/10000 lower

error rate and 1.7x longer SSD endurance comparing with conventional BCH ECC.

## Proposed ECC Solution for SSD

### A. Error Rate Definition

We define 2 different error rates.

*rBER* (Raw Bit Error Rate) denotes the error rate before applying ECC:

$$rBER = \frac{Count\ of\ error\ bits}{Total\ read\ bits} \quad (1)$$

*uPER* (Uncorrectable Page Error Rate) denotes the error rate after applying ECC.

$$uPER = \frac{Count\ of\ uncorrectable\ pages}{Total\ read\ bits} \quad (2)$$

It is well known that as data storage products, SSDs shall guarantee certain uPER, such as $10^{-16}$ to satisfy the required reliability.

### B. MP-ECC: a Message-Passing ECC Architecture

Fig.1 shows MP-ECC's basic idea: unlike conventional BCH architecture, where BCH blocks are independent with each other; MP-ECC architecture enables BCH block within the same page to be overlapped. For example, for 4 different BCH blocks (BCH1~BCH4) within a page, part of BCH1 is also part of BCH2. This architecture enables an effective Message-Passing algorithm: if the decoding of BCH2 is failed, but the decoding of BCH1 is successful, then the overlapped parts can be replaced by decoded BCH1. Thus, for replaced BCH2, the overlapped area becomes error-free, hence the total error within updated BCH2 is reduced. We can then decode the updated BCH2 again. This replacement is essentially an effective message passing scheme. As a result, the probability of successful decoding of BCH2 becomes higher than before. Moreover, it can be observed that in the last BCH (such as BCH4), the overlapped area is smaller than other BCHs – only left side of BCH4 is overlapped. As a result, the error correction capability of BCH4 is weaker than others. The total uPER of MP-ECC will then be affected. Here, we propose a type of MP-ECC to address this problem: the idea is to make the ECC strength of last BCH to be slightly stronger (e.g. add extra 3-bit correctable). It is easy to proof that MP-ECC has the following features:

- The same guarantee: if conventional BCH can correct all errors within a page, MP-ECC can also actually correct the same errors;
- Better performance: for the same rBER, MP-ECC can outperform conventional BCH with the same parity size.
- Same circuit size: MP-ECC can be implemented in the same level circuit size.

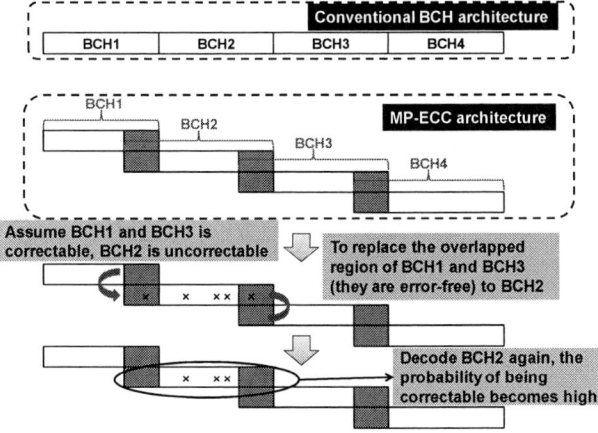

Fig.1 MP-ECC architecture and its decoding scheme

### C. PAC: Parity Area Combined Scheme

Fig.2 shows the background of PAC idea: in NAND, the error event differs from different pages (Error Asymmetry).

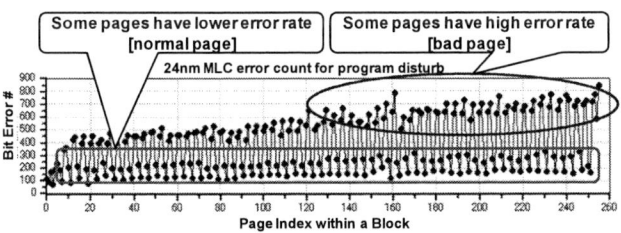

Fig.2 Normal page, Bad page in NAND

Within one block, some pages have higher rBER, we call these pages "bad pages"; while most pages have lower rBER, we call these pages "normal pages". Basically for example, in MLC NAND, upper pages have higher rBER than lower pages. On the other hand, only one kind of BCH is normally used in SSD controller, i.e. all BCH blocks have the same ECC strength. Thus the uPER is essentially determined by "bad pages". For those "normal pages", the ECC strength designed for bad pages is much stronger than being necessary. We only need a weaker BCH on normal pages for specified uPER; and part of parity area of normal page can be used by other bad pages. As a result, bad pages can be equipped with a much stronger ECC by sharing the area of normal pages. This is the basic idea of PAC, a Parity Area Combined scheme.

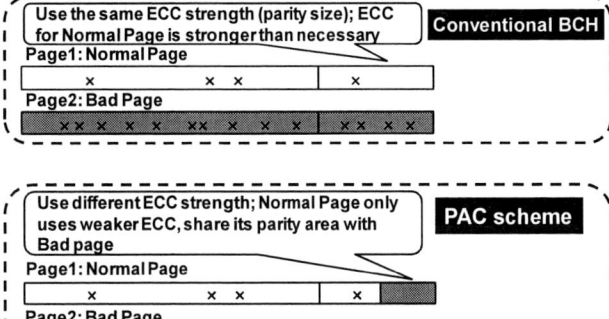

Fig.3 PAC scheme and the difference with conventional BCH

There are many ways to implement PAC scheme. Fig.4 shows the one adopted in our SSD controller test sample. Firstly, a programmable BCH ECC engine is implemented. This enables different type of BCH to be set with different ECC strength and codeword length. In the implementation, the programmable ECC strength ranges from 8-bit to 80-bit, and the codeword length could be 512 bytes or 1024 bytes. We can then flexibly set the normal pages to be covered by a "weak" BCH, such as 24-bit/1KB, meaning that for each 1KB BCH block, up-to 24 bits error can be correctable; for bad pages, we set a "strong" BCH, such as 40-bit/1KB, we call it "main-BCH". Note that main-BCH could be different for good pages and bad pages. Secondly, for those bad pages, we further choose part of data area to be as a new BCH codeword, we call it "Retry-BCH". For example, a 32-bit/512B BCH can be used, and its parity can be restored in other normal pages. When decoding of main-BCH in bad page failed, the system enters into a Retry mode and Retry-BCH decoding is then performed; if it is successfully decoded, the part of data (which is error free) is then used to replace the area of corresponding main-BCH, and a decoding will be done again (Retry). Due to that part of data is replaced by error-free bits, the total error bits in this main-BCH will be reduced, and the probability of successful decoding in Retry mode becomes higher.

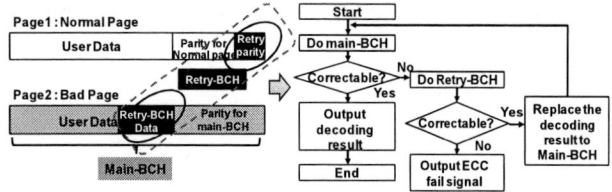

Fig.4 PAC scheme and its decoding chart

## Measurement Results

MP-ECC and PAC can either be used independently or together to improve the ECC performance. Fig.5 shows the measurement results in different case. Here ECC strength is set to be relatively weak to enable practical measurement. These results demonstrate remarkable improvement on SSD's reliability and endurance using proposed ECC solution.

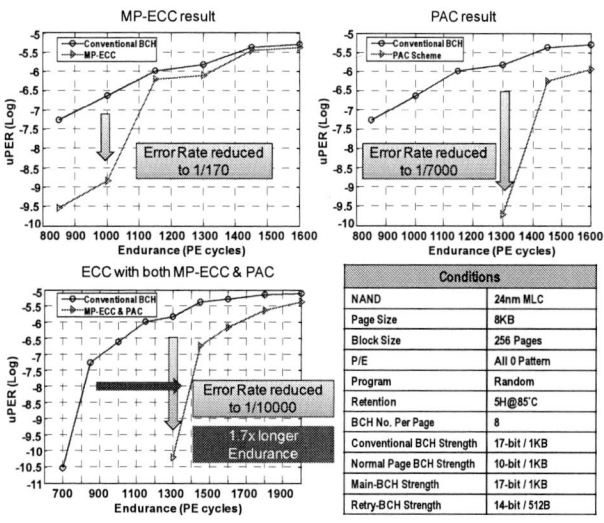

Fig.5 Measurement results

Although both of MP-ECC and PAC discussed here are based on BCH code, they can of course work with other ECC such as LDPC. Fig.6 shows the simulation results regarding LDPC performance

The proposed ECC solution has been implemented into a SSD controller prototype IC fabricated in TSMC65LP. This SATA III controller has die size of 5mm x 5mm, supports up-to 1TB SSD capacity. Fig.7 shows the die photograph and brief feature diagram.

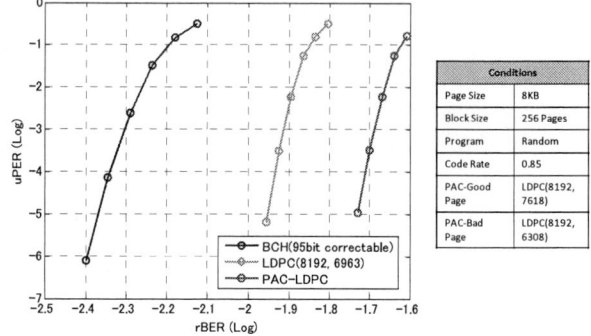

Fig.6 Simulation results of PAC-LDPC performance

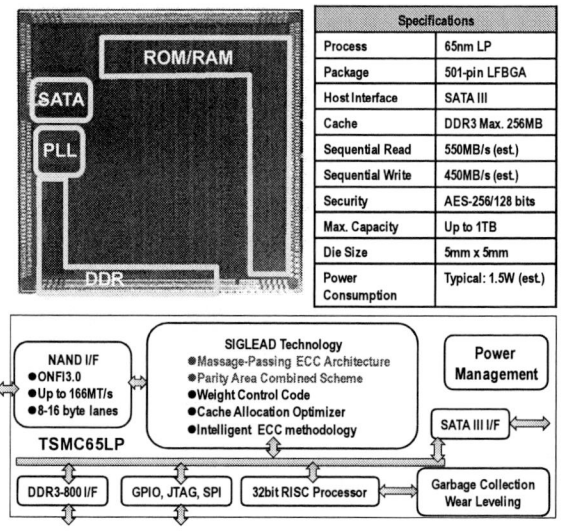

Fig.7 SSD controller die photograph, specifications and brief diagram

## References

[1] K. Takeuchi, "NAND successful as a media for SSD", ISSCC, Tutorial T-7, 2008

[2] S. Lin and D. Costello, "Error control coding", 2nd Edition, Prentice hall, 2004.

[3] D. MacKay, "Information theory, inference, and learning algorithms", Cambridge University Press, 2003.

[4] A. Yu, "Improving NAND reliability with Low-Density Parity Check codes (LDPC)", Flash Memory Summit, 2011.

[5] S. Tanakamaru et. al, "95% Lower bit error rate, 43% lower power intelligent solid-state drives (SSDs) with asymmetric coding and stripe pattern elimination algorithm", ISSCC Dig. Tech. Papers, Feb, 2011.

[6] A. Jiang, J. Bruck, "Joint coding for flash memory storage", Proc. IEEE ISIT, 2008.

# 45pW ESD Clamp Circuit for Ultra-Low Power Applications

Yen-Po Chen[1], Yoonmyung Lee[1], Jae-Yoon Sim[2], Massimo Alioto[3], David Blaauw[1], and Dennis Sylvester[1]

[1]University of Michigan, Ann Arbor, [2]Pohang University of Science and Technology, Pohang, Korea, [3]University of Siena, Siena, Italy

*Abstract-* **Novel ultra low-leakage ESD power clamp designs for wireless sensor applications are proposed and implemented in 0.18μm CMOS. Using new biasing structures to limit both subthreshold leakage and GIDL, the proposed designs consume as little as 43pW at 25°C and 119nW at 125°C with 4500V HBM level and 400V MM level protection, marking an 18−139× leakage reduction over conventional ESD clamps.**

## I. INTRODUCTION

Robustness against electrostatic discharge (ESD) is a critical reliability issue in advanced CMOS technologies. To prevent circuit damage due to ESD events (which can expose the circuit to kV range voltages), ESD clamp circuits are typically incorporated in supply pad library cells. These circuits use extremely wide devices (100s of μm) and thus exhibit leakage currents of 10nA to 10μA (at 25°C and 125°C, respectively) despite the use of various low power approaches [1-4]. Recently, there has been increased interest in ultra-low power wireless sensor node systems [5, 6] with constrained battery sizes and system standby power budgets as low as 10-100nW. Considering the need for multiple power pads, these systems cannot use existing ESD structures due to their high leakage, thereby compromising their reliability. To address this challenge, we propose three ultra-low leakage ESD circuits that use special biasing structures to reduce subthreshold leakage and gate-induced drain leakage (GIDL) while maintaining ESD protection. In 180nm silicon test chip results, we demonstrate 10s of pA (nA) operation at room temperature (125°C), which is a >100× improvement over prior state of the art.

## II. PROPOSED ESD TECHNIQUES

A standard commercial ESD clamp circuit is shown in

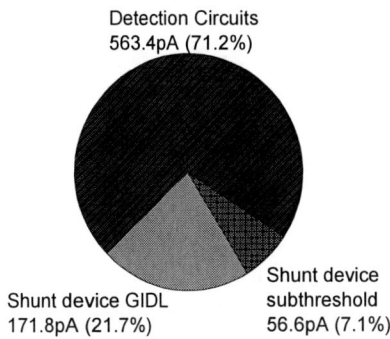

Fig. 2. Power breakdown of a standard ESD clamp circuit.

Fig. 1 and consists of an RC filter and inverter to detect the ESD event, as well as a large MOSFET to remove electrostatic charge. All transistors are thick-oxide high Vt devices. When a high voltage is applied to the supply rail due to an ESD event, transistor M2 turns on, pulling up the detection node and allowing the electrostatic charge to be dissipated through the large M4 shunt device. Waveforms for a 7kV Human-Body Model discharge are shown in Fig. 4. The key parameters associated with achieving high voltage protection are M4 size and the speed at which the detection node is pulled up. After the charge is dissipated, the resistor pulls up the inverter input to turn off the clamp.

Fig. 2 gives the simulated power breakdown of this conventional design, with two major components: 1) Detection circuits, and particularly, pull-up device M2, which dominates leakage as it is sized up to speed detection and also exhibits poorer subthreshold slope compared to NMOS; 2) the large shunting device M4. Due to the high supply voltage

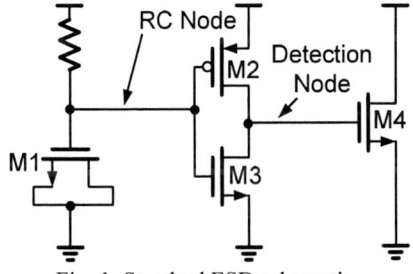

Fig. 1. Standard ESD schematic.

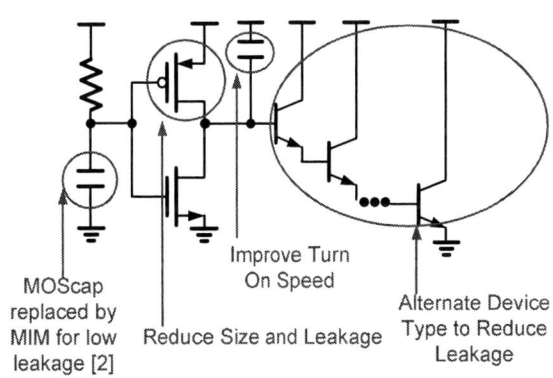

Fig. 3. The modified BJT based structure.

978-1-4673-6145-3/13 $31.00 © 2013 IEEE

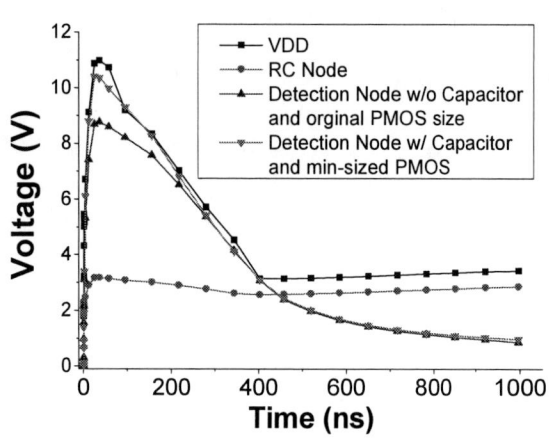

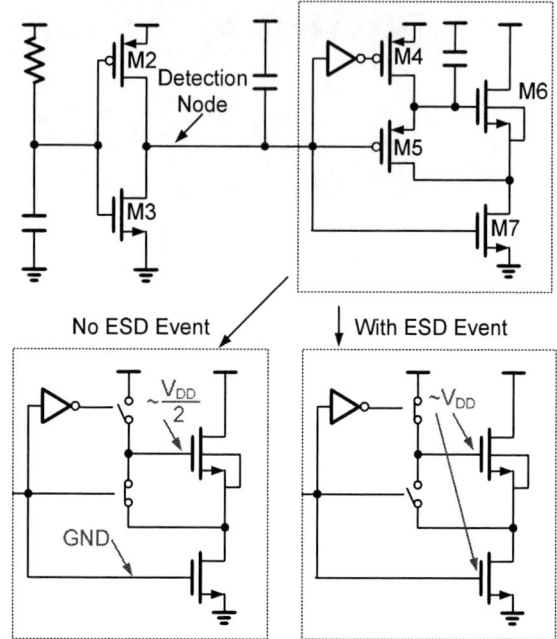

Fig. 4. Simulation waveform of the modified BJT based structure.

Fig. 5. Proposed GIDL reduction scheme.

($\geq$1.8V), GIDL of M5 is larger than its subthreshold leakage.

To reduce these leakage sources, we propose and test three circuit structures. The first and most straightforward approach is shown in Fig. 3. To address M2 leakage, an assisting capacitor is added. At the onset of an ESD event, the supply voltage rises rapidly and this assisting capacitor couples the detection node up, allowing the PMOS to be down-sized (near min-size), while maintaining the same effective turn-on speed and ESD robustness. Simulated waveforms of the detection node in Fig. 4 show that the assisting capacitor with downsized M2 slightly improves response time. Note that although leakage through the MOS capacitor in this technology is small (<2pA), for a scalable low-leakage approach, a MIMCAP is used in the RC filter (as in [2]). To limit M4 leakage we employ a BJT, which provides lower off-current than MOSFETs. However, in standard CMOS technologies only parasitic BJTs with small current gains are available, making it necessary to use a Darlington-like structure.

Overall, these modifications offer a 10× (104×) leakage reduction at 25°C (125°C) (silicon measurements below). However, the parasitic BJTs introduce several technology scaling concerns that make MOS-based solutions preferable. In particular, from simulations the base-emitter current gain drops from 25 in 180nm to 5 in 65nm. Also, bipolar clamp

snapback voltage decreases with technology scaling more rapidly than MOSFETs [8], reducing effectiveness for ESD protection. We therefore also propose two MOS-based structures that offer similar leakage reduction gains with better scalability and improved density. A well-known approach to reduce MOSFET leakage is stacking, which yields a 2.9× subthreshold leakage reduction in 180nm CMOS. However, as noted earlier, GIDL dominates leakage in the shunt device and hence stacking alone only reduces total leakage by 17%.

The first method to address GIDL in an MOS shunt device is shown in Fig. 5 and has similarity with [7]. When there is no ESD event the gate and source of M6 are shorted and the stacked shunt transistors M6 and M7 act as a voltage divider. As a result, the key GIDL parameter $V_{dg}$ is reduced by half for both transistors, lowering GIDL by 5.4×. When an ESD event occurs, the two MOS shunts fully turn on to remove the electrostatic charge. The same concept can be extended to a stack of 3 devices; simulations across temperature in Fig. 6 show temperature stability across a wide range (-20°C to 125°C). The 3-stack structure provides minimum leakage for this approach (denoted GIDL-1). Further extending the

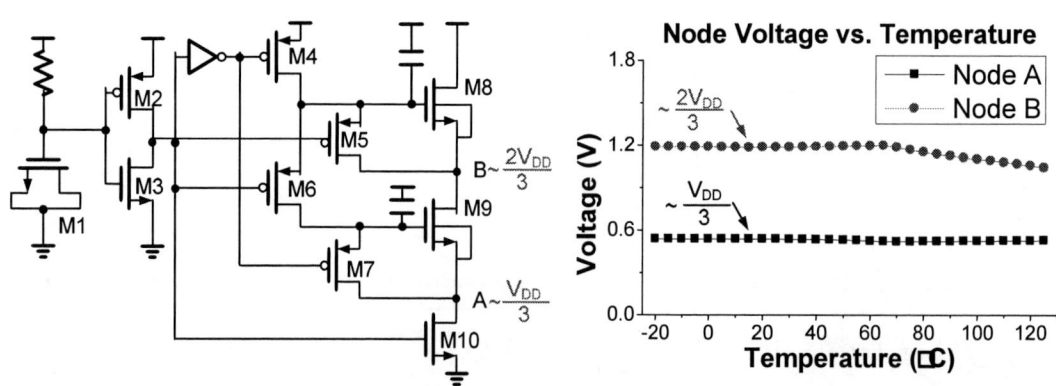

Fig. 6. GIDL reduction scheme for 3-stack (GIDL-1) with simulated internal node voltages across temperature at 1.8V.

978-1-4673-6145-3/13 $31.00 © 2013 IEEE

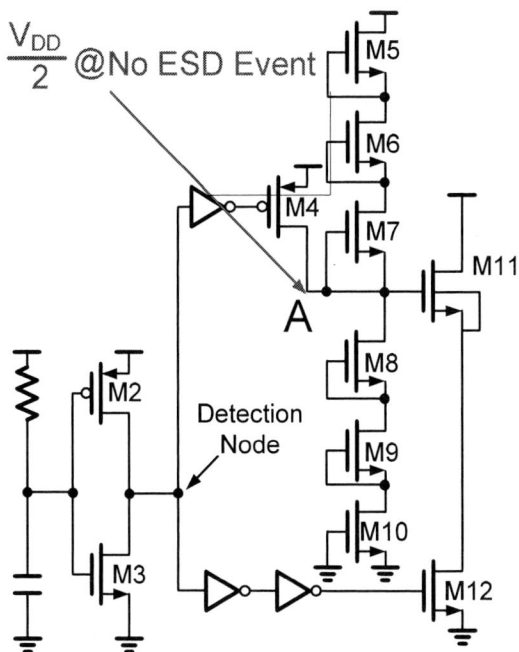

Fig. 7. Leakage-based GIDL reduction method (GIDL-2)

method to a 4-stack degrades shunt on-current, requiring device up-sizing for sufficient ESD protection and leading to higher leakage.

The second GIDL reduction approach (denoted GIDL-2) is given in Fig. 7. In this structure, a bias voltage of approximately VDD/2 is generated by a diode stack (M5-

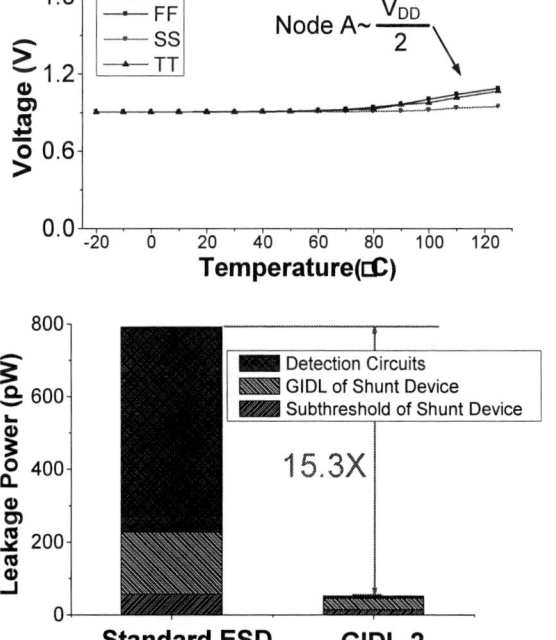

Fig. 8. Simulated internal node voltage across temperature and corners as well as leakage power breakdown of GIDL-2.

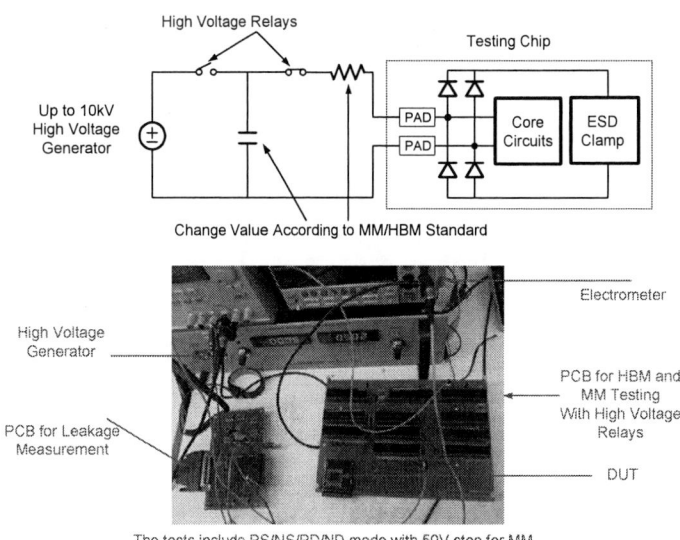

Fig. 9. Testing setup with high voltage generator for human body model (HBM) and machine model (MM).

M10), which is then applied to the topmost stacked output device (M11) to reduce GIDL in M11 and M12. Since there is no need for leaky PMOS switches in GIDL-2, total transistor area and overall leakage is reduced. Note that diode-connected NMOS M5-M10 have minimum W (with increased L) since they only need to overcome the subthreshold leakage of M4 and gate leakage of M11 to maintain VDD/2 at node A. As a result, the diode stack leakage is negligible. Simulations across temperature/process show the stability of node A voltage (Fig. 8). During an ESD event node A is charged to VDD through M4 and then slowly discharges to VDD/2 through the diode stack. During this relaxation time (350☐s in simulation) the ESD clamp experiences substantial GIDL. However, since ESD events are rare, the impact on total energy is minimal and the low quiescent current of the structure far outweighs it. Simulated leakage power breakdown of GIDL-2 is shown in Fig. 8, showing a 15.3 − 115× reduction (25 − 125°C) compared to a conventional commercial clamp.

III. MEASUREMENT REASULTS

The three proposed ESD structures (BJT, GIDL-1, GIDL-2) and a commercial ESD clamp circuit (baseline) were fabricated in a standard 180nm CMOS process. In addition, an ESD structure using smaller devices and offering a lower protection level was integrated with a mm-scale microsystem [5] to meet its nW system power budget. The human body model (HBM) and machine model (MM) are evaluated on the ESD structures (Fig. 9). Device leakage current is measured after each discharge of the HBM or MM test. We use a conventional definition of failure, namely the smallest voltage at which either 1) the structure exhibits a 30% increase in leakage, or 2) an analog block connected to the ESD pads functionally fails.

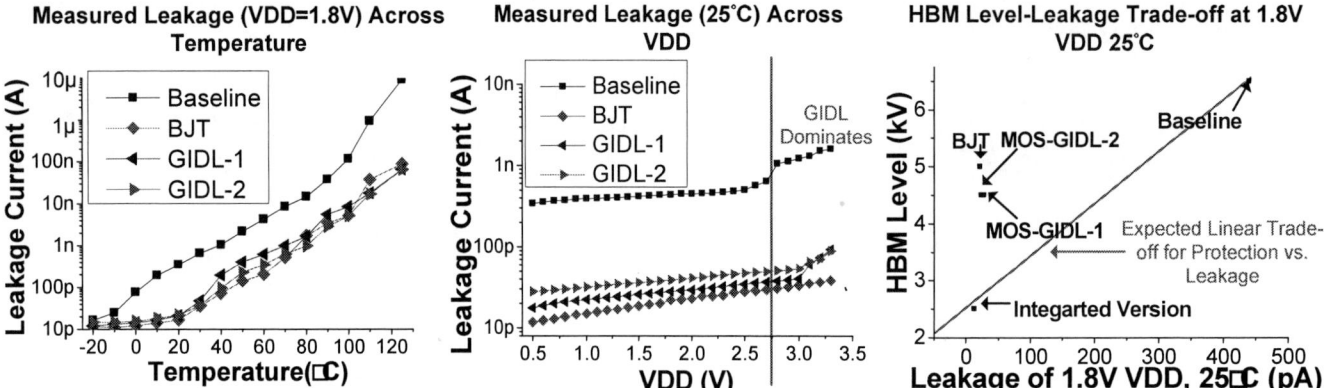

Fig.10. Measured leakage results across temperature and power supply

Fig.11.Measured scatter plot of baseline and 3 proposed structures

The measured leakage of each structure across temperature and VDD is shown in Fig. 10. The proposed clamps have lower leakage than the baseline design throughout the temperature range of 0°C to 125°C and VDD from 0.5V to 3.3V. The BJT structure has the lowest leakage (22pA) at room temperature, a 20× reduction over the baseline. At 125°C, GIDL-1 and GIDL-2 structures consume 67.8nA and 66nA, respectively, compared to 16.52µA for the baseline. A scatter plot showing ESD protection and leakage (25°C) of the 4 measured structures is also given in Fig. 11. The expected linear trend between protection level and leakage highlights the gains achieved by the proposed structures beyond straightforward device down-sizing. A histogram of leakage current for GIDL-2 at 85°C and 1.8V across 20 measured dies from one wafer is shown in Fig. 12. Nearly all dies consume

1.6−2.1nA with average leakage of 1.91nA and standard deviation of 317pA. The integrated version shows 13pA leakage at 25°C with 2.5kV HBM level and 300V MM level. Table 1 provides a summary table including a comparison of HBM and MM levels of the proposed structures to both the literature and measured baseline pads. Overall the proposed GIDL-2 structure provides 18−139× leakage reduction over commercial ESD clamps with 70-100% of ESD protection levels while avoiding special devices such as SCR. Die photos are given in Fig. 13.

### REFERENCES

[1] P.-Y. Chiu et al., ICICDT, 2011.
[2] M.-D. Ker et al., ISSCC, 2006.
[3] M.-D. Ker et al., ISCAS, 2009.
[4] C.-T. Wang et al., JSSC, 2009.
[5] Y. Lee et al., ISSCC, 2012.
[6] TI, MSP430F20x1, 2011.
[7] S. Bang et al., CICC, 2012.
[8] J. Li et al., EOS/ESD, 2009

Table 1. Summary table of proposed structures and related work.

| ESD Structure | Technology | Area (µm²) | HBM Level (kV) | MM Level (V) | Leakage 1.8V, 25°C | Leakage 1.8V, 125°C |
|---|---|---|---|---|---|---|
| Baseline Commercial Clamp | 0.18µm | 17500 | 6.5 | 400 | 440pA | 9.18µA |
| BJT | 0.18µm | 67200 | 5.0 | 350 | **22pA** | 88.1nA |
| GIDL-1 | 0.18µm | 67200 | 4.5 | 400 | **28pA** | **67.8nA** |
| GIDL-2 | 0.18µm | 44800 | 4.5 | 400 | **24pA** | **66nA** |
| Integrated Version For mm3 system [5] | 0.18µm | 35000 | 2.5 | 300 | 13pA | 41nA |
| [1] | 65nm | N/A | 4.0 | 350 | 358nA (1V) | 1.91µA (1V) |
| [2]* | 0.13µm | N/A | 6.5 | 400 | N/A | N/A |
| [3]* | 65nm | N/A | >8.0 | 750 | 228nA (1V) | 3.14µA (1V) |
| [4]* | 65nm | 1029 (7891)** | 7.0 | 325 | 96nA (1V) | 1.02µA (1V) |

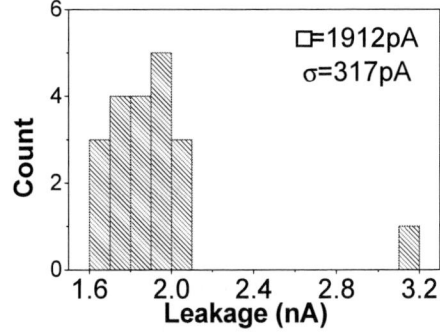

Fig.12. Measured histogram of leakage for GIDL-2 across 20 measured dies, 85°C and 1.8V.

Fig. 13. Die photo. The BJT, GIDL-1, and GIDL-2 version are shown in the left, and the integrated version is shown in middle. The complete mm3 system is shown at right and the commercial clamp is measured in the same run on a different die.

# A 1.14mW 750kb/s FM-UWB Transmitter with 8-FSK Subcarrier Modulation

Fei Chen[1], Yu Li[1], Deyuan Lin[1], Huiying Zhuo[1], Woogeun Rhee[1], Jongjin Kim[2], Dongwook Kim[2], and Zhihua Wang[1]

[1]Institute of Microelectronics, Tsinghua University, Beijing, China
[2]Samsung Advanced Institute of Technology, Suwon, Korea

*Abstract*—A noninvasive energy-efficient FM-UWB transmitter is implemented in 65nm CMOS for stereo hearing aid applications. Different form the conventional FM-UWB transmitter, the proposed transmitter employs an 8-FSK subcarrier modulation method to enhance data rate by three times with slight increment of hardware complexity. The 8-FSK subcarrier modulation is achieved by a fast-settling PLL along with a pre-tuned relaxation VCO. The FM-UWB signal generated by an FLL-assisted ring VCO and a single-stage class AB power amplifier meets the FCC spectrum mask. The 3.5-4GHz 750kb/s transmitter consumes 1.14mW from a 1V supply, achieving an energy efficiency of 1.5nJ/bit.

## I. INTRODUCTION

As more than two-thirds of the seniors in the world are facing hearing impairment problems, the wireless hearing aid device market has grown rapidly. Linking the left ear and the right ear, the emerging stereo hearing aid device helps the user promptly locate the source of sound. The required data rate of the stereo hearing aid device is higher than that of the conventional one. In addition, low peak transmission power could be an important factor for noninvasive wearable devices.

The frequency-modulated ultra-wideband (FM-UWB) transceiver, ratified as an optional PHY of the IEEE 802.15.6 standard for low data rate body area networks (BANs) [1], features a constant envelope and a steep spectral roll-off of the transmitted output. As shown in Fig. 1, the transmitter first modulates the baseband data with frequency-shift-keying (FSK) to generate an intermediate frequency (IF) subcarrier. The subcarrier modulates an RF oscillator with wideband FM modulation, resulting in UWB spectrum. Compared to the impulse radio ultra-wideband (IR-UWB) system which transmits an impulse with a high peak-to-average power ratio (commonly >100 for low data rate applications), the FM-UWB system exhibits a peak power which is typically a few percent of that of the IR-UWB system. Hence, the FM-UWB system is more suitable for the around-the-head audio application than the pulse based UWB system [2].

Conventional FM-UWB transmitters use 2-FSK subcarrier modulation [3]-[5] with a typical data rate of around 100 kb/s. To further increase the data rate without degrading the energy efficiency significantly, an 8-FSK FM-UWB technique is proposed to triple the bit rate per symbol period. The conventional 2-FSK modulation shown in Fig. 2(a) represents 1 bit by selecting one of the two subcarrier frequencies per symbol period, while the proposed 8-FSK modulation as shown in Fig. 2(b) represents 3 bits per symbol period by

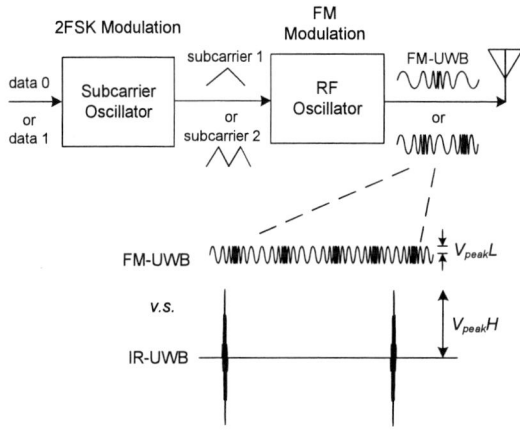

Fig. 1. Conventional FM-UWB transmitter system.

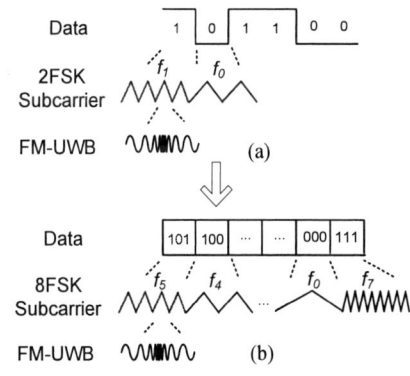

Fig. 2. Modulation principle for FM-UWB transmitter: (a) conventional 2-FSK FM UWB, and (b) proposed 8-FSK FM UWB.

selecting one of the eight subcarrier frequencies. In the proposed architecture, power hungry blocks such as the RF voltage-controlled oscillator (VCO) and the power amplifier (PA) are not changed for the 8-FSK subcarrier modulation, high energy efficiency is achieved. The system can be reconfigurable for 2-FSK/4-FSK/8-FSK modes. As we target <5mW power consumption for the 750kb/s transceiver and assume 3.8mW for the receiver [6], this work consider <1.2mW power consumption for the transmitter system.

In this paper, we present a 750kb/s FM-UWB transmitter by employing an 8-FSK modulation method. A fast-settling

978-1-4673-6145-3/13 $31.00 © 2013 IEEE

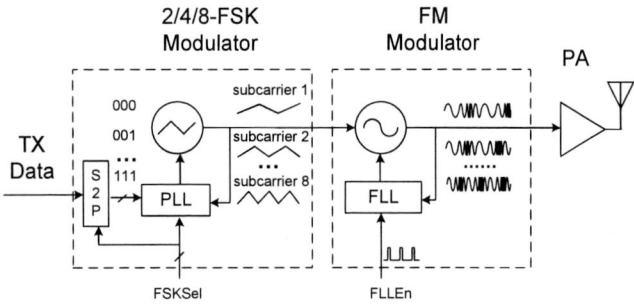

Fig. 3. Proposed 8-FSK FM-UWB transmitter.

TABLE I
SUBCARRIER FREQUENCY ALLOCATION

| Subcarr freq | 13 MHz | 14 MHz | 15 MHz | 16 MHz | 17 MHz | 18 MHz | 19 MHz | 20 MHz |
|---|---|---|---|---|---|---|---|---|
| 8-FSK | 000 | 001 | 010 | 011 | 100 | 101 | 110 | 111 |
| 4-FSK | | 00 | | 01 | | 10 | | 11 |
| 2-FSK | | | | 0 | | | | 1 |
| fracN x 4MHz | 3.25 | 3.5 | 3.75 | 4 | 4.25 | 4.5 | 4.75 | 5 |

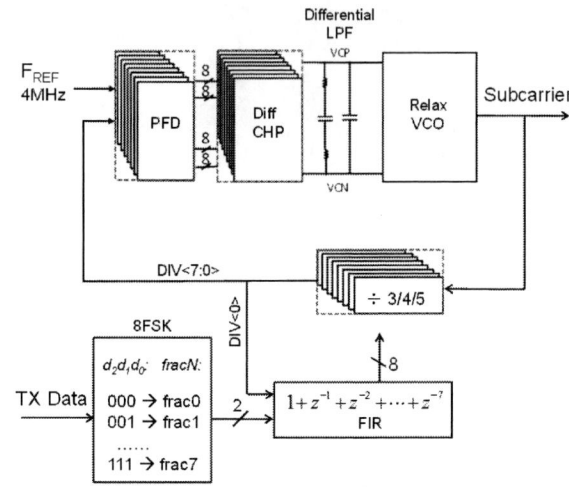

Fig. 4. The PLL based 8-FSK modulator for subcarrier generation.

phase-locked loop (PLL) for 8-FSK subcarrier modulation and a current-starved ring VCO for wideband FM modulation are designed. A gated frequency-locked loop (FLL) is used to periodically calibrate the center frequency of the ring VCO, and a class AB power amplifier outputs the UWB signal to a 50Ω antenna with a spectrum under the FCC mask.

## II. TRANSMITTER ARCHITECTURE

Fig. 3 shows a simplified block diagram of the proposed transmitter. It consists of an FSK modulator for subcarrier generation, an FM modulator for UWB signal generation, and a power amplifier to transfer power to an antenna. The baseband data is first processed by a PLL-based FSK modulator to generate the subcarrier. The modulator can be configured to 2-FSK for 250kb/s data rate, 4-FSK for 500kb/s, or 8-FSK for 750kb/s. The subcarrier frequencies in each case are allocated as in Table I. For example, the 8-FSK modulation maps a 3-bit symbol to one of the subcarrier frequencies which range from 13MHz to 20MHz with a frequency step of 1MHz. The wider frequency separation than that used in [4]-[5] makes the receiver achieve a better bit error rate (BER) performance [7]. The eight subcarrier frequencies can be also utilized to enable multi-user access in the 2-FSK or 4-FSK mode.

The triangular subcarrier directly modulates the ring VCO to get a spread RF spectrum in the range from 3.5GHz to 4GHz. The change of the instantaneous frequency of the ring VCO output reflects the change of the waveform of the subcarrier. In such a way, the subcarrier information is coded into the RF signal. Since the modulation index is 250MHz/20MHz > 10, this wideband FM-modulation results in a flat spectrum shape with high spectrum efficiency [8]. The center frequency of the ring VCO is calibrated by a gated FLL to prevent the undesired frequency drift due to PVT

variations. The slow center frequency shift does not require the FLL to work all the time. In this work, the FLL is enabled by a clock of 10% duty cycle to save 90% of the power if an always-on FLL is used.

The output stage is a class AB power amplifier, which is impedance matched to a 50Ω antenna. The amplified signal has a power spectral density limit of −41.3dBm/MHz. With a bandwidth of 500MHz in this design, the integrated output power is about −14dBm.

## III. CIRCUIT DESIGN

The FSK modulator is implemented by an 8-modulo fractional-N PLL with a 3/4/5 multi-modulus divider as shown in Fig. 4. The data controls the division ratio to make the PLL lock to a desired subcarrier frequency $f_{sub} = f_{ref} * div$. The divison ratio for each subcarrier is shown in the bottom row of Table I. An 8-tap hybrid finite-impulse response (FIR) filter is designed to reduce fractional spurs generated by periodic operation of the multi-modulus divider in the 8-modulo fractional-N PLL [9].

Although the PLL settling time to shift one subcarrier frequency to another is not a concern with low data rate such as 100kb/s [4], it can cause a noticeable delay before the desired subcarrier frequency is reached. The delay degrades the BER performance in the receiver, especially in the case where the frequency is shifted from the lowest frequency of 13MHz to the highest frequency of 20MHz. To alleviate the problem, a switched capacitor array is designed in the relaxation VCO. The output frequency is controlled directly by the input data as shown in Fig. 5. When new data $d_2 d_1 d_0$ come, the first two bits $d_2 d_1$ switch the capacitor array to instantly shift the frequency close to the desired one. The PLL only requires a short settling time to lock the frequency to the precise one determined by the division ratio. The comparison of the settling time is also shown in Fig. 5, where $t_1$ and $t_2$ are the settling times without and with direct capacitor control respectively, and $\Delta t$ is the shortened time during the frequency shift.

978-1-4673-6145-3/13 $31.00 © 2013 IEEE

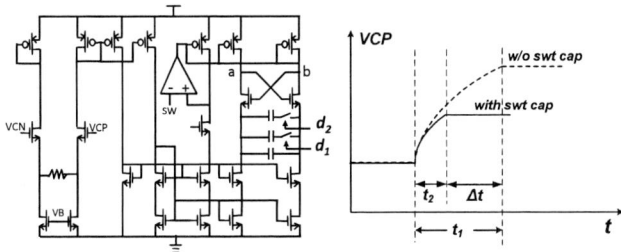

Fig. 5. (a) The relaxation VCO , and (b) fast PLL settling.

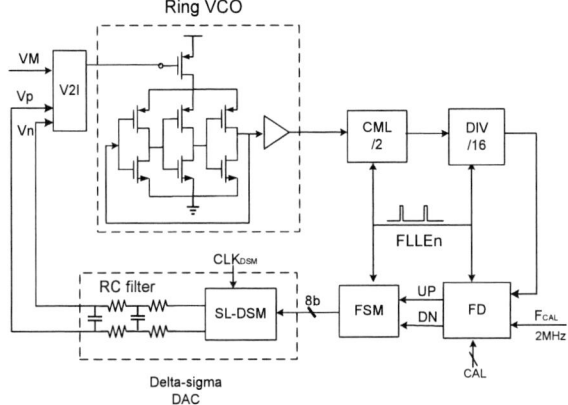

Fig. 6. The Ring VCO and the gated FLL for calibration.

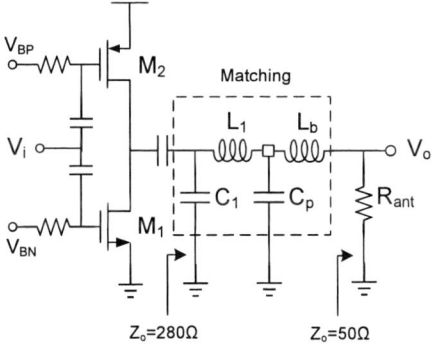

Fig. 7. The class AB power amplifier.

The FM modulator, as shown in Fig. 6, consists of a ring VCO and a gated FLL. The ring VCO has dual control paths. One path is the fast modulation path with a high VCO gain to generate the FM-UWB signal, and the other path is the slow calibration path with a low gain to overcome the undesired center frequency shift due to PVT variations. In the modulation path, the subcarrier voltage is first converted to a current to modulate the current-starved ring VCO. In the calibration path, the FLL compares the divided VCO center frequency with the reference in the frequency detector (FD). A finite state machine (FSM) sets the control word, which is converted to an analog differential voltage for calibration. Since the common-mode-logic (CML) divider in the FLL is power consuming, a 10% gating mechanism is applied to save power. The ring VCO shows a phase noise of –74dBc/Hz at

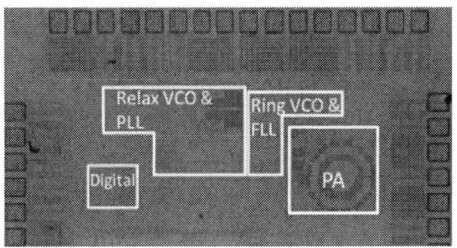

Fig. 8 Chip micrograph

1MHz offset, which satisfies the relaxed phase noise requirement of the FM-UWB system due to the wideband nature [8].

The output stage is a class AB amplifier as shown in Fig. 7. To save die area, the push-pull structure is selected since it does not require an RF choke or a harmonic resonator. The on-chip inductor $L_1$ and capacitor $C_1$ together with the bonding wire inductance $L_b$ and pad capacitance $C_p$ form a two-section matching network to convert the antenna load of 50Ω to a load of 280Ω for the push-pull transistors to improve the power efficiency. The PA has an output 1dB point at –6.5dBm, which is sufficient for the output power of –14dBm.

## IV. MEASUREMENT RESULTS

The prototype transmitter is fabricated in 65nm CMOS. A chip micrograph is shown in Fig. 8. The active area is 0.65μm x 0.35μm.

The generated subcarrier spectra with 2-FSK, 4-FSK, and 8-FSK modulation under the same symbol rate of 250kbaud are shown in Fig. 9. The marker 1 in each subplot is at 20MHz, and the marker 2 shows the subcarrier frequency separation of 4MHz, 2MHz, and 1MHz respectively.

The measured tuning range of the ring VCO in the modulation path, as shown in Fig. 10, is from 2.5GHz to 4.5GHz with the VCO gain of 2.2GHz/V. The calibration range of the center frequency is 3.4GHz to 4GHz with the gain of 225MHz/V. The calibration resolution is $600\text{MHz}/2^8$ = 2.3MHz, where 8 is the word length of the FSM in FLL.

Fig. 11 shows the measured FM-UWB spectrum. The spectrum occupies 3.5GHz to 4GHz with a fast roll off. The shape of the spectrum is mainly flat except for a small decline at the high end, which is caused by the non-ideal frequency response of the cable and connectors.

The transmitter system consumes the power of 1.14mW from a 1V supply, in which 0.31mW is consumed by the FSK modulator including the relaxation VCO and the PLL, 0.37mW by the FM modulator including the ring VCO and the gated FLL, and 0.46mW by the PA. The energy efficiency of 1.5nJ/bit is achieved for the 750kbps data rate with 8-FSK subcarrier modulation. The performance comparison with other FM-UWB transmitters is given in Table II, showing that the proposed transmitter offers superior energy efficiency.

## V. CONCLUSION

A 1.14mW FM-UWB transmitter with 8-FSK subcarrier modulation is implemented in 65nm CMOS. A fast-settling PLL is used for subcarrier modulation and a ring VCO with a

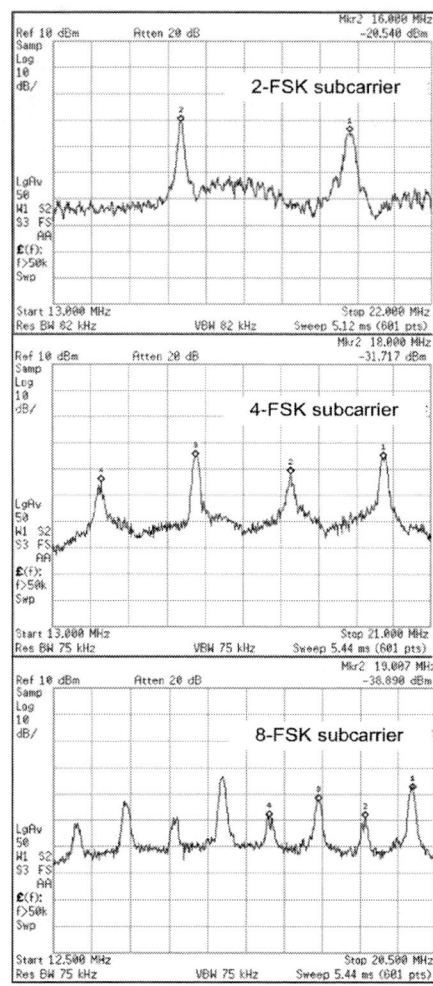

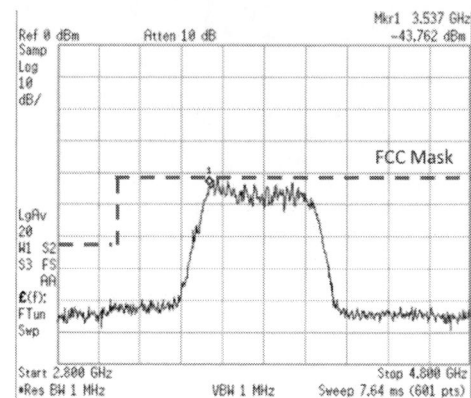

Fig. 11. Measured transmitted FM-UWB spectrum.

TABLE II
FM- UWB TRANSMITTER PERFORMANCE COMPARISON

|  | ICUWB 2009 [3] | TCAS-I 2012 [4] | JSSC 2011 [5] | This work |
|---|---|---|---|---|
| Subcarrier Modulation | 2-FSK | 2-FSK | 2-FSK | 8-FSK |
| RF freq | 6.25-8.25 GHz | 3.5-4 GHz | 3.75-4.25 GHz | 3.5-4 GHz |
| RF VCO | LC VCO | LC VCO | Ring VCO | Ring VCO |
| Freq Calibration | N.A + PLL | PLL+FLL | FLL+FLL | PLL+FLL |
| Power Consumption | 4.6 mW | 8.7 mW | 0.9 mW / 1.1mW | 1.14 mW |
| Data Rate | 100 kb/s | 100 kb/s | 100 kb/s | 750 kb/s |
| Energy Efficiency | 46nJ/bit | 87nJ/bit | 9nJ/bit | 1.5nJ/bit |
| Technology | 0.13um CMOS | 0.18um CMOS | 90nm CMOS | 65nm CMOS |
| Active Area | 0.06 mm² | 1.2 mm² | 0.1 mm² | 0.2 mm² |

Fig. 9. Measured subcarrier spectrum under 250 kbaud symbol rate.

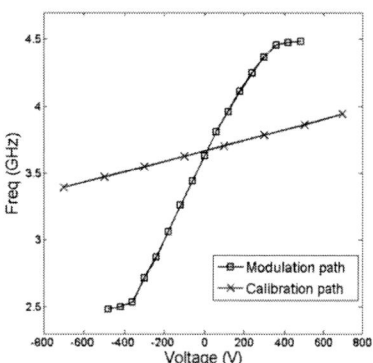

Fig. 10. Measured ring VCO tuning range.

gated FLL is designed for wideband FM modulation. A class AB power amplifier outputs the power under the FCC mask. Compared to the conventional transmitter with 2-FSK subcarrier modulation, the proposed transmitter reached a much higher data rate of 750kbps and obtains a much lower energy efficiency of 1.5nJ/bit.

REFERENCES

[1] Wireless Body Area Networks (WBAN), IEEE 802.15 Task Group 6 [Online]. Available: http://www.ieee802.org/15/pub/TG6.html.
[2] X. Wang et al., "A meter-range UWB transceiver chipset for around-the-head audio streaming," in IEEE ISSCC Dig. Tech. Papers, pp. 450-451, Feb. 2012.
[3] M. Detratti, E. Perez, J. Gerrits, and M. Lobeira, "A 4.6mW 6.25–8.25 GHz RF transmitter IC for FM-UWB applications," in Proc. IEEE ICUWB, pp. 180–184, Sep. 2009.
[4] B. Zhou et al., "A gated FM-UWB system with data-driven front-end power control," IEEE Trans. Circuits Syst. I, vol. 59, no. 6, pp. 1348-1358, June 2012.
[5] N. Saputra and J. R. Long, "A fully-integrated, short-range, low data rate FM-UWB transmitter in 90 nm CMOS," IEEE J. Solid-State Circuits, vol. 46, no. 7, pp. 1627-1635, July 2011.
[6] F Chen, W Zhang, W Rhee, J Kim, D Kim, and Z Wang, "A 3.8mW, 3.5–4GHz regenerative FM-UWB receiver with enhanced linearity by utilizing a wideband LNA and dual bandpass filters," IEEE RFIT Symp, pp. 150 – 152, Nov. 21-23, 2012.
[7] S. Haykin, Communication Systems, Wiley, 2009.
[8] J. F. M. Gerrits, M.H.L. Kouwenhoven, P. R. Van Der Meer, J. R. Farseroru, and J. R. Long, "Principle of FMUWB Principles and limitations of ultra-wideband FM communications systems," EURASIP J. on App. Signal Processing, vol. 2005, pp. 382-396, Mar. 2005.
[9] X. Yu, Y. Sun, W. Rhee, and Z. Wang, "An FIR-embedded noise filtering method for ΔΣ fractional-N PLL clock generators," in IEEE J. Solid-State Circuits, vol. 44, pp. 2426-2436, Sept. 2009.

# A 2.4 GHz Energy-Efficient 18-Mbps FSK Transmitter in 0.18 μm CMOS

Jingjing Chen, Weiyang Liu, Peng Feng, Haiyong Wang, and Nanjian Wu*

Institute of Semiconductors, Chinese Academy of Sciences, Beijing 100083, China

*nanjian@red.semi.ac.cn

*Abstract*—This paper presents a 2.4 GHz energy-efficient phase-locked loop (PLL)-based transmitter (TX) integrated in 0.18-μm CMOS technology. By using Twin-VCO transmission scheme, the data rate of the transmitter is free of loop bandwidth of PLL with stable carrier frequency. The frequency presetting technique is adopted to not only reduce the lock-in time of the PLL but also minimize the VCO gain ($K_{VCO}$) variation. Measured results show that the frequency precision of the Twin-VCO relative to center carrier is less than 21 ppm. The lock-in time of the PLL is less than 7 μs. The TX dissipates 11.6 mW when the output power is 0 dBm at a data rate of 18 Mbps and achieves an energy efficiency less than 0.64 nJ/bit.

## I. INTRODUCTION

With the rapid development of wireless networks applications, such as wireless personal network (WPAN), wireless body area network (WBAN) and wireless sensor network (WSN), it is desirable to implement an energy-efficient transmitter with high data rate and fast frequency settling time. In order to realize the energy-efficient transmitter, frequency shift keying is commonly adopted due to its low complexity and constant-envelope modulation characteristic, which allows the use of a power-efficient nonlinear power amplifier (PA).

Many architectures have been proposed for FSK transmitters. The mixer-based architecture is highly flexible and can be used for various modulations, but the power consumption and the cost is high due to the use of digital-to-analog converter (DAC), filter, mixer, PA and frequency synthesizer. Therefore, this architecture is unsuitable for low-power applications.

For the PLL-based transmitter architecture, there are four major approaches. One way is to apply the modulation data to the divider [1]. This approach is simple and accurate, but the PLL acts as a low-pass filter on the data signal, thus the data rate is limited by the loop bandwidth. The second way is to directly apply the modulation data to the VCO which is locked by PLL. In this case the low frequency components of modulation data will be corrupted since the PLL acts as a high-pass filter. The third type of the PLL-based TX is two-point modulation which combines the benefits of the two previous techniques by applying the baseband data to both the divider and the VCO [2], [3]. However, the requirement on perfect match in two-point modulation raises the complexity and the power consumption of the TX. The fourth type of the PLL-based TX is open-loop VCO modulation [4], [5], the main advantage of this topology is that the data rate is not constrained by the loop bandwidth. However, since the VCO is released from PLL loop during data transmission, the VCO suffers from frequency drift due to leakage current, undesired noise and environment temperature.

This paper proposes a novel PLL-based TX architecture. The proposed transmitter combines the advantages of both the open-loop VCO modulation and the closed-loop modulation by overcoming the problems of frequency drift and the limited transmission data rate. Meanwhile, the PLL-based TX utilizes fast lock-in technique in order to save energy consumption because of short settling time. The rest of this paper is organized as follows. In section II, the architecture of the transmitter is presented. Section III describes the details of the building blocks. Finally, the measured results and the conclusions are given in section IV and section V, respectively.

## II. TRANSMITTER ARCHITECTURE

### A. Architecture

Fig. 1 shows the architecture of the proposed PLL-based TX. The TX consists of a PLL and a variable gain power amplifier (VGPA). The PLL includes phase frequency detector (PFD), charge pump (CP), 3rd low-pass filter (LPF), presetting module, Twin-VCO, dual mode prescaler (DMP) and digital processor. The non-volatile memory (NVM) is used to store the presetting data for the presetting module. The PLL-based TX with Twin-VCO can operate in two modes: 1) generate quadrature carrier for receiver 2) transmit mode with high transmission data rate which freeing of loop bandwidth of PLL. In this paper, we focus on the second mode.

The Twin-VCO consists of two identical VCOs: iVCO and qVCO. During the receive mode, the iVCO and qVCO couple and generate quadrature frequency for the receiver, as shown in Fig. 2 (a). During the transmit mode, the iVCO and qVCO are decoupled each other. The iVCO is locked to a desired carrier frequency by the PLL while the qVCO is unlashed from PLL loop, as shown in Fig. 2 (b). Since the two VCOs share the same

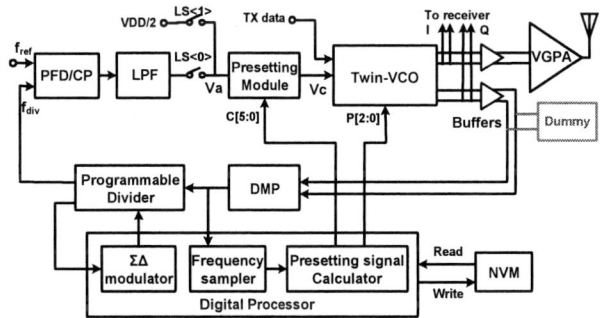

Fig. 1 Proposed PLL-based FSK TX architecture

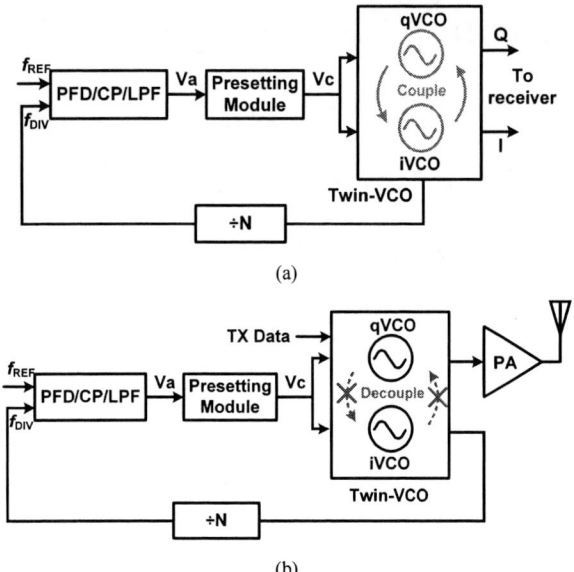

(a)

(b)

Fig. 2 PLL-based TX operate in (a) receive mode (b) transmit mode

control voltage, the frequency of the qVCO will closely track the frequency of the iVCO. So, we name "Twin-VCO" standing for the iVCO and the qVCO. The qVCO is used to transmit high rate baseband data by direct modulation because the data rate is free of loop bandwidth.

### B. Frequency Precision

The frequency precision of the Twin-VCO in transmit mode is defined as $\Delta f/f_c$, where $\Delta f$ is the the frequency deviation of the Twin-VCO and $f_c$ is the locked carrier frequency of iVCO. The frequency precision of Twin-VCO between the iVCO and qVCO is in proportion to deviation of the VCO gain, which can be expressed as:

$$\Delta K = K_I - K_Q = K_p \left( K_{iVCO} - K_{qVCO} \right). \quad (1)$$

Where $K_I$ and $K_Q$ are the total VCO gain of iVCO and qVCO, respectively. $K_P$ is the voltage gain of the presetting module in Fig. 1. $K_{iVCO}$ and $K_{qVCO}$ are VCO gain of iVCO and qVCO, respectively. Eq. (1) shows that if a small $K_P$ is chosen, then the deviation between the $K_{iVCO}$ and the $K_{qVCO}$ due to all kinds of mismatches can be further reduced. In this paper, we design $K_P$ with a small value, so $\Delta K$ is also small. Another technique to reduce the intrinsic frequency difference of the Twin-VCO is by careful layout with high symmetry.

### C. Frequency Presetting Method

It is important to reduce the lock-in time of the PLL because the energy consumption of the TX can be reduced with short lock-in time. The lock-in time of the PLL can be expressed as [6]:

$$T_{lock-in} = \frac{1}{\zeta \omega_n} \ln \left( \frac{\Delta F}{\varepsilon \sqrt{1 - \zeta^2}} \right). \quad (2)$$

Where $\varepsilon$ is the acceptable frequency error, $\Delta F$ is the initial frequency error, $\zeta$ is the damping factor and $\omega_n$ is the natural

frequency of the PLL. If the initial frequency error $\Delta f$ is small, then the lock-in time can be greatly reduced. In this design, the frequency presetting technique is adopted to preset the frequency of the VCO with a small initial frequency error $\Delta f$ relative to the target frequency by changing the VCO cap-array control word P[2:0] and the presetting word C[5:0], as shown in Fig. 1.

When the PLL starts up, the loop switch is open and the control voltage of the VCO is biased to VDD/2, then the frequency of the VCO versus P[2:0] and C[5:0] is automatically measured by the frequency sampler at five sampling points for every P signal. The sampling frequency will be stored in the NVM. Next, the values of the signals P and C for the target frequency can be calculated by linear interpolation. Thus the target frequency can be preset accurately by the digital signals P and C and finely tuned by the output voltage of the loop filter $V_a$. In this way the PLL can be settled down in a very short time.

### III.  CIRCUIT IMPLEMENTATIONS

#### A.  Twin-VCO

Fig. 3 shows the schematic of the proposed Twin-VCO. The Twin-VCO adopts complementary nMOS and pMOS cross-coupled negative-gm topology. A 3-bit binary switch-capacitor array is used to achieve a tunning range from 2.3 GHz to 2.7 GHz.

During the receive mode, the Twin-VCO can be configured to couple in quadrature by turning on the coupling current source $I_c$. Interposing a phase-shift into the coupling path can be used to eliminate bimodal oscillation behavior inherent in conventional QVCO [7]. The phase noise of the QVCO can also be reduced. In this paper, a parallel passive R-C network is added between the source terminals of the coupling transistors (MP1-MP2, MP3-MP4) to interpose additional phase shift in the coupling network.

During the transmit mode, the iVCO and the qVCO are decoupled by turning off the coupling current source $I_c$, thus the two VCOs run separately sharing the same control voltage and keep the same carrier frequency as identical as possible. The cascode transistors (MP5-MP8) are used to enhance isolation when the two VCO are decoupled each other.

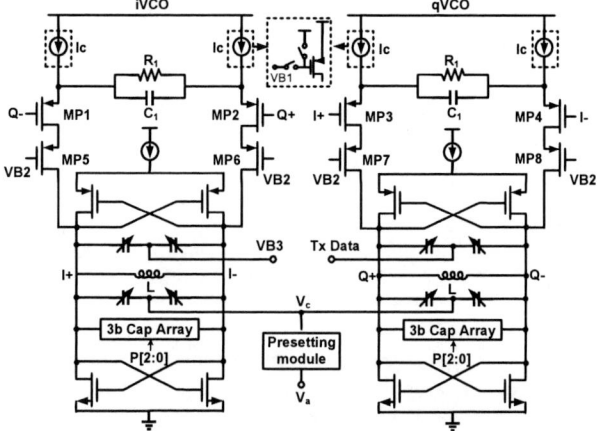

Fig. 3 Proposed Twin-VCO

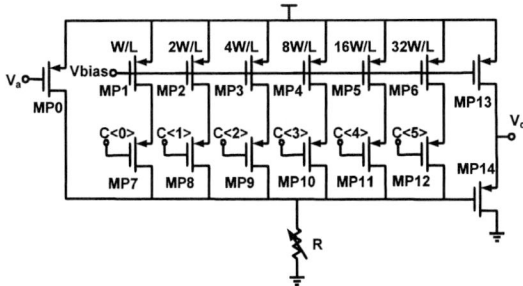

Fig. 4 Schematic of the presetting module

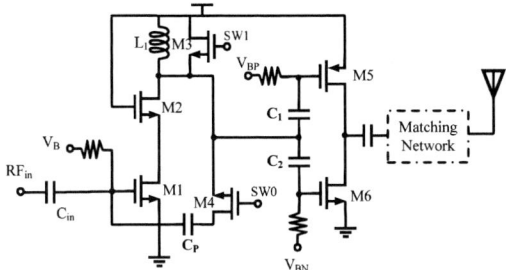

Fig. 5 Schematic of the VGPA

### B. Presetting module with Small Kp

Fig. 4 shows the schematic of the presetting module in Fig. 1 [8]. MP1-MP6 form a series of digital controlled current sources with a two-fold step. When a digital control word C[5:0] is applied to the presetting module, the module generate a control voltage $V_c$ by the source follower (MP13-MP14). The presetting word C[5:0] together with the switch-capacitor array control word P[2:0] can accurately preset the frequency of the VCO with a small initial frequency error relative to the desired frequency. The Va from the output of LPF finely tunes the frequency of the VCO by controlling the current of MP0. Small Kp value is implemented by the presetting module. The voltage gain $K_P$ from Va to Vc is $g_{m0}$, where $g_{m0}$ is the transconductance of MP0. In this design, a small value of $g_{m0}R$ (<0.1) is chosen to reduce the deviation between $K_{iVCO}$ and $K_{qVCO}$ as described in part B of section II.

### C. Variable gain Power Amplifier

The constant envelope of FSK modulation allows optimization of PA for high efficiency. Fig. 5 shows the schematic of the proposed VGPA. It consists of a driver stage and an output stage. The driver stage is a cascode amplifier and the output stage is a Class B inverter-type amplifier. The digital controlled switches SW0 and SW1 are used for gain control with three gain modes (high, medium, and low). It can provide a variable gain range of 20 dB and a maximum output power of 0 dBm.

## IV. MEASURED RESULTS

The proposed PLL-based TX is implemented in a 0.18 µm CMOS process. The chip microphotograph is shown in Fig. 6. The chip core area is 1 mm×1.5 mm.

Fig. 7 shows measured output frequency of the Twin-VCO for different digital presetting word C[5:0] and P[2:0] when $V_a$

Fig. 6 Microphotograph of the chip

is biased at VDD/2. The curves shows good linearity between the output frequency and the presetting word C[5:0].

The measured PLL lock-in time is less than 7µs with a frequency step from 2.4 GHz to 2.424 GHz, as shown in Fig. 8. The measured qVCO phase noise is -111 dBc/Hz@1-MHz while the iVCO is locked. Fig. 9 shows the measured frequency precision between the qVCO and iVCO while iVCO is locked by the PLL. It can be learned that the frequency precision is below 21 ppm from 2.4 GHz to 2.5 GHz.

The transmitter output spectrum with FSK modulation index of 0.5 (MSK) at 18 Mbps data rate is shown in Fig. 10. The PLL and the PA consumes 6.2 mW and 5.4 mW respectively while delivering 0 dBm output power. The energy efficiency of the proposed TX is less than 0.64 nJ/bit.

Table I summarizes major performance of the proposed TX and compares it with other designs. Fig. 11 shows the performance of the energy-per-bit versus output power which compares with recently published FSK transmitters.

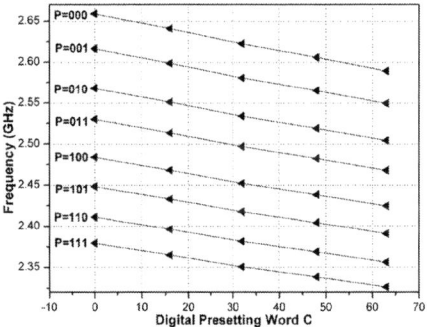

Fig. 7 Measured output frequency versus presetting word C[5:0]

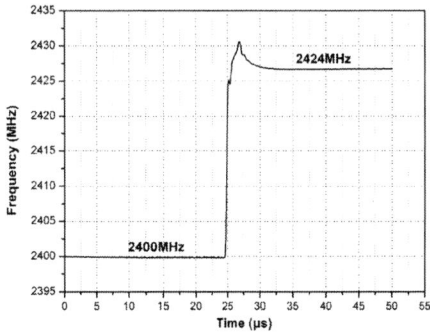

Fig. 8 Measured lock-in time while frequency-hopping

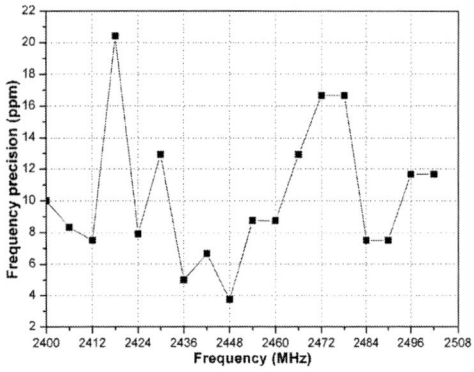

Fig. 9 Measured frequency precision of the Twin-VCO

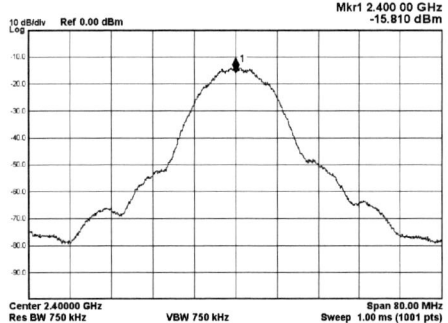

Fig. 10 Measured output spectrum at 18 Mbps MSK data

TABLE I.    TRANSMITTER PERFORMANCE COMPARISON

|  | This work | [3] | [4] |
|---|---|---|---|
| Modulation scheme | FSK | GFSK/OQPSK | GFSK |
| Technology | 0.18 μm | 90 nm | 0.18 μm |
| Data rate | 18 Mbps | 2 Mbps | 1 Mbps |
| PLL settling time | 7 μs | 40 μs | 94μs |
| Output power | 0 dBm | -1 dBm | 0 dBm |
| Power dissipation | 11.6 mW | 5.4 mW | 19.3 mW |
| Energy efficiency | 0.64 nJ/b | 2.7 nJ/b | 19.3 nJ/b |

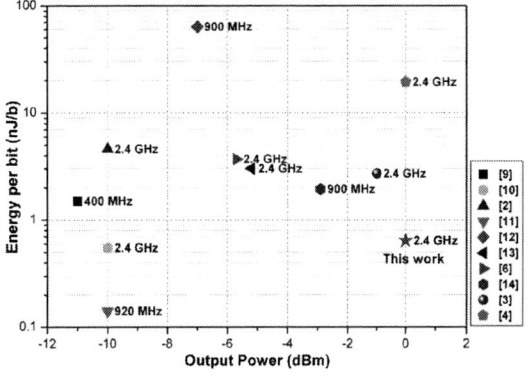

Fig. 11 Energy-per-bit versus output power with frequency annotated

## V.    CONCLUSION

A novel PLL-based TX with fast lock-in time has been presented in this paper. The proposed TX breaks the transmission data rate limitation by loop bandwidth in conventional closed-loop modulation architecture while still achieving frequency precision of less than 21 ppm at 2.4GHz carrier frequency. The PLL settling time is less than 7 μs. The TX consumes 11.6 mW for 18 Mbps MSK modulation data at 0 dBm output power.

## ACKNOWLEDGMENT

This work is financially supported by the National Key Technology Research and Development Program of the Ministry of Science and Technology of China under Grant No.2012BAH20B02 and the National Science and Technology Major Projects of the Ministry of Science and Technology of China under Grant No. 2012ZX03004007-002.

## REFERENCES

[1]  S. Pamarti, L. Jansson, and I. Galton, "A wideband 2.4-GHz delta-sigma fractional-N PLL with 1-Mb/s in-loop modulation," *IEEE J.Solid-State Circuits*, vol. 39, no. 1, pp. 49–62, Jan. 2004.

[2]  A. C. Wong *et al.*, "A 1 V 5 mA Multimode IEEE 802.15.6/Bluetooth Low-Energy WBAN Transceiver for Biotelemetry Applications," *IEEE J. Solid-State Circuits*, vol. 48, no.1, pp. 186–198, Jan. 2013.

[3]  Yao-Hong Liu *et al*, "A 2.7nJ/b multi-standard 2.3/2.4GHz polar transmitter for wireless sensor networks," *ISSCC Dig. Tech Papers*, pp. 448-450, Feb. 2012.

[4]  C. P. Chen *et al.*, "A low-power 2.4-GHz CMOS GFSK transceiver with a digital demodulator using time-to digital conversion," IEEE Trans. Circuits Syst. I, Reg. Papers, vol. 56, no. 12, pp. 2738–2748, Dec. 2009.

[5]  Xiaozhou Yan, Xiaofei Kuang,Nanjian Wu, "A smart frequency presetting technique for fast lock-in LC-PLL frequency synthesizer," *ISCAS*, pp. 1525-1528. May 2009.

[6]  J. Masuch, M. Delgado-Restituto, "A Sub-10 nJ/b +1.9-dBm Output Power FSK Transmitter for Body Area Network Applications," *IEEE Trans. Microw. Theory Tech.*, vol. 60, no. 5, pp. 1413–1423, 2012.

[7]  B. Razavi, RF Microelectronics, Second Edition, Prentice Hall 2011.

[8]  X. F. Kuang and N.J. Wu, "A Fast-Settling PLL Frequency Synthesizer with Direct Frequency Presetting," *ISSCC Dig. Tech papers*, pp. 204-205, Feb. 2006.

[9]  Yao-Hong Liu, and Tsung-Hsien Lin,  "A Wideband PLL-Based G/FSK Transmitter in 0.18 μm CMOS,"*IEEE J. Solid-State Circuits*, vol. 44, no. 9, pp. 2452–2462, 2009.

[10]  A. Paidimarri, P. M. Nadeau, P. P. Mercier, and A. P. Chandrakasan, "A 2.4 GHz Multi-Channel FBAR-based Transmitter With an Integrated Pulse-Shaping Power Amplifier," *IEEE J. Solid-State Circuits*, vol. 48, no. 4, pp. 1042–1054, 2013.

[11]  J. Bae, L. Yan, and H. Yoo, "A low energy injection-locked FSK transceiver with frequency-to-amplitude conversion for body sensor applications," *IEEE J. Solid-State Circuits*, vol. 46, no. 4, pp. 928–937, 2011.

[12]  A. C. Wong et al, "A 1 v, micropower system-on-chip for vital-sign monitoring in wireless body sensor networks," in *ISSCC Dig. Tech. Papers*, 2008, pp. 138–602.

[13]  B. Cook, A. Berny, A. Molnar, S. Lanzisera, and K. Pister, "Low-power 2.4-GHz transceiver with passive RX front-end and 400-mV supply," *IEEE J. Solid-State Circuits*, vol. 41, no. 12, pp. 2757–2766, Dec. 2006.

[14]  N. Panitantum, K. Mayaram, and T. S. Fiez, "A 900-MHz low-power transmitter with fast frequency calibration for wireless sensor networks," in Proc. *IEEE Custom Integr.Circuits Conf.*, Sep. 2008, pp. 595–598.

# A 60GHz, Linear, Direct Down-Conversion Mixer with mm-Wave Tunability in 32nm CMOS SOI

M. A. T. Sanduleanu[1], A. Valdes-Garcia[1], Y. Liu[1], B. Parker[1], Shlomo Shlafman[2],
Benny Sheinman[2], Danny Elad[2], Scott Reynolds[1] and Daniel Friedman[1]

[1]IBM T. J. Watson Research Center, Yorktown Heights, NY, USA email: masandul@us.ibm.com
[2]IBM Haifa R&D, Mount Carmel, Haifa, Israel

*Abstract*- **The gain/linearity trade-off is exploited to achieve the best linearity performance of a mm-Wave down-conversion system. The achieved linearity (IIP3) for the whole down-conversion chain is better than 11.06dBm for 5.8dB gain at 60GHz. Gain can be adjusted from -15dB to 11dB in 1dB steps depending on the signal level. By adjusting phase matching at RF input and the common-mode impedance of the mixer with a variable transmission line, the input phase imbalance can be corrected and the second-order distortion (HD2) can be reduced. The LO/RF isolation is 43dB and the LO/IF isolation is better than 82dB. The down-converter occupies 1.38mm$^2$ in 32nm CMOS SOI and consumes 19.2mW from 1V supply. The power consumption of the mixer itself is 4mW @ 1V supply.**

## I. INTRODUCTION

For phased-array receivers, in addition to the beam-steering capability, one well known advantage over individual single-channel receivers is the improvement of the SNR [1]. Whether it adopts a power combiner or an amplitude combiner, a phased-array receiver does not improve the noise factor (or noise figure) from an individual array channel, but shows the same noise factor (or noise figure). Although the noise figure remains unchanged, the output SNR ($SNR_N$) increases by a factor of N (array size) in the array receiver $SNR_N = N*SNR_1$. In a homodyne or a super-heterodyne phased-array receiver with many elements (N), after RF power combining, the mixer has to handle very large signals at its input. This then, requires extremely high values for the IIP3 and 1-dB compression of the total down-conversion chain. As the combined signal for a phased-array system with many elements is already large, one can trade off the noise figure of the mixer function for better linearity.

The paper presents a linear direct down-conversion system with integrated sensors and digital infrastructure. The high linearity of the mixer cell is facilitated by a novel linearization method without penalty on extra voltage headroom. As the linearity and the input impedance are virtually decoupled from each other, the differential impedance can be optimized to match a large range of different signal sources. The LO is applied single-ended and switches on and off the transconductor pair. It allows low voltage operation and large swing at the output nodes. Key for operation is the gain/linearity trade-off. The load of the mixer, as well the following VGA stage, have variable gain. For a given signal level, gain is adjusted for best linearity. The price paid for linearity is the worsening of the NF compared to a double balanced Gilbert cell.

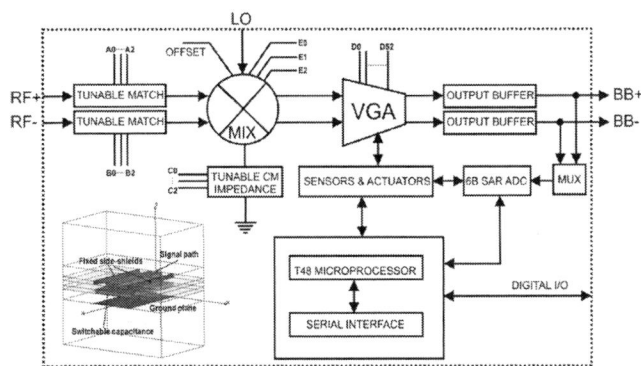

Fig. 1. Mixer architecture with digital infrastructure and tunable T-Line.

## II. Down-Conversion Mixer Architecture

Fig. 1 shows a block diagram for the implemented 60 GHz down-conversion system. At the input, the matching circuits use a variable transmission line (T-line) as an adjustable stub. In order to reduce the input common-mode, the phase difference between the two differential inputs ( ideally 180°) can be adjusted, thus maximizing the fundamental component. The T-Line has fixed side shields, a ground plane and a switchable capacitance layer in form of a grating structure connected either to ground or left floating [2]. The grating layer is divided in sections, individually controlled using MOS switches. The common-mode impedance of the mixer is tunable as well, by using the same variable T-Line structure.

The mixer is followed by a VGA and a buffer with 50Ω output impedance. The power gain of the mixer/VGA tandem is adjustable from -15dB to +11dB in small steps of 1dB. The gain of the VGA is distributed over two stages and for the same power gain, the gain distribution setting can be found for the best linearity performance. A peak detector incorporated in the mixer itself, correlates the signal level with linearity performance and is used as a linearity sensor. The offset adjustment, nulls the dc difference between the two differential outputs of the mixer mitigating the effect of even order distortion components. It prevents, as well, saturation of the VGA stages in high gain mode.

An 8-bit microcontroller and a 6-bit SAR ADC are integrated in the system. These components were not employed for the measurements presented in this work but could be employed for closed-loop, on-chip calibration. For measurement purposes the internal registers of the actuators (gain, bias controls, and tunable t-lines) are programmed from an on-chip serial interface.

978-1-4673-6145-3/13 $31.00 © 2013 IEEE

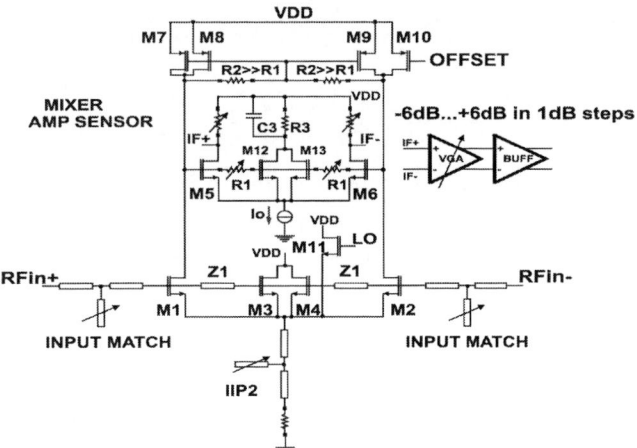

Fig. 2. Circuit diagram: mixer, amp-sensor and VGA.

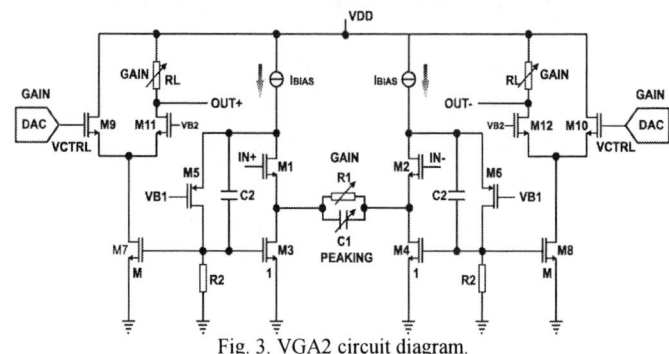

Fig. 3. VGA2 circuit diagram.

## III. Circuit Implementation

Fig.2 shows the circuit diagram of the mixer and the first stage of the VGA (VGA1). The input stage including M1...M4 together with the impedance Z1 provides simultaneous input matching and linearization. For small input differential signals, the gates of M3/M4 and their source are virtual ground and no signal flows at their drains. At large signal levels, the source of M1...M4 is not anymore virtual ground and the distortion components will flow in M3/M4 transistors. The impedance Z1 consists of a transmission line in series with a $10\Omega$ resistor. The crux of the linearization technique is the decoupling between input matching and linearity (the linear range does not depend on the choice of Z1). Moreover, this technique does not require extra voltage headroom when compared to a simple differential pair. Hence, the drain of the input pair could have larger voltage swing. The common-mode impedance is realized with a resistor, two fixed transmission lines and a variable T-Line acting as an adjustable capacitance stub. The common-mode current of the linearized pair, flows in M7...M10. The LO transistor M11 switches ON/OFF the linearized pair without the need of extra voltage headroom like in a Gilbert mixer. The differential current flows through two adjustable resistors R1 connected in series. This then, generates an input voltage for the first VGA stage (VGA1) with M5, M12, M13 and M6. The output load of VGA1 is adjustable as well.

The IF output is applied to a second gain stage (VGA2) with a gain control range from -6dB to +6dB in 1dB fine steps. VGA1 provides an extra 10dB of gain for the IF chain. The gate of M10 is connected to an offset control signal (OFFSET) generated after measuring the offset at the output of VGA2. The drain current of the middle transistors in VGA1 is low-pass filtered with R3, C3 and the resulting voltage is a measure of the input amplitude applied to VGA1.

The second VGA is shown in Fig.3. It consists of a super source follower stage M1...M6, a peaking load R1, C1 and an output stage M9...M12 with extra gain adjustment. The current $I_{BIAS}$ is forced in the transistors M1 and M2, facilitated by a local feedback loop with M5, R2, M3 and M6, R2, M4. As the gate-source voltages of M1 and M2 are constant the input

differential voltage applied between IN+ and IN- will be amplified with a voltage gain of 1 at the gain peaking load R1∥C1. A tunable capacitor C1, realized as a switching network of capacitors, provides high frequency peaking for the V/I converter bandwidth control. The PMOS transistors M5 and M6 are dc level shifters and are bypassed by capacitors C2 at higher frequencies. The current in M7, M8 is an amplified replica (factor M) of the current flowing in M3, M4 and R1∥C1. The DAC controls the amount of current flowing to the OUT+ and OUT- nodes in the differential pair M9, M11 and M12, M10. For low gain, most of the current is dumped to VDD (M9 and M10 are completely open).

## IV. Measurements Results

The down-converter implemented in a 32nm CMOS SOI process from IBM consumes 19.2mW (RF+analog+digital) from a 1V power supply. The mixer part consumes only 4mW. For one-tone measurements (Fig.4) a Magic Tee and an external amplifier are used to compensate the losses of cables and phase shifters. The same construction is used in the LO path. Although, the LO is applied single-ended to the mixer, an on-chip differential transmission line is used for LO distribution with one end terminated on a dummy load. This minimizes the LO coupling to RF input. For two-tone measurements (Fig.5), a power combiner generates the two required tones from two external generators. Three signal generators locked to an external 10MHz reference provide the LO signal and the two RF tones.

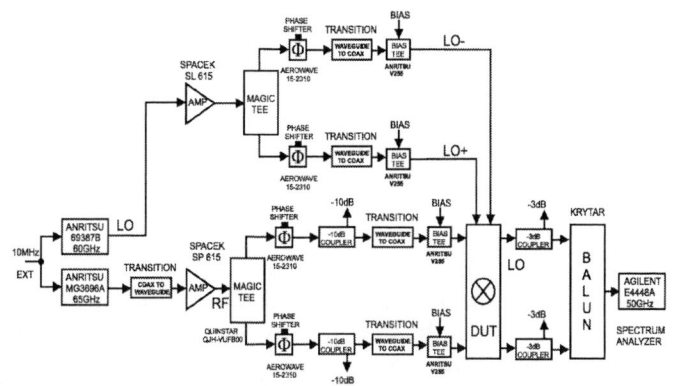

Fig. 4. One-tone measurement setup.

978-1-4673-6145-3/13 $31.00 © 2013 IEEE 276

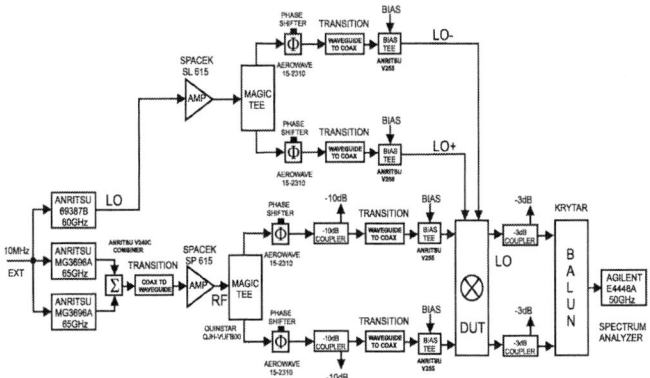

Fig. 5. Two-tone measurement setup.

The power gain of the mixer/VGA tandem is adjustable from -15dB to +11dB (Fig.5) in small steps of 1dB. The IF bandwidth varies from 3.5GHz to 7GHz (Fig.5). The LO is at 59GHz.

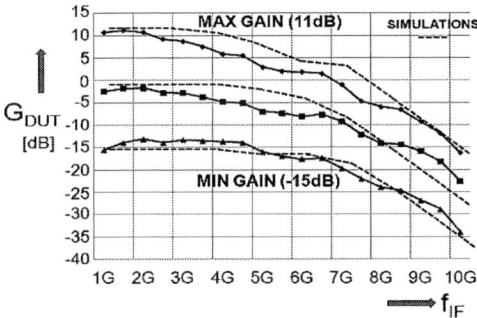

Fig. 6. Mixer+VGA tandem power gain

The gain-linearity trade-off is shown in Fig.7 and Fig.8. P1 represents the power of the fundamental and P3 is the power of the third harmonic. At maximum gain, the P3OI is -3.4dBm and at minimum gain P3OI is 8.4dBm. For intermediate gain levels, (the same total power gain), gain can be distributed differently between the mixer output, VGA1 and VGA2. As shown in Fig.8, for more gain in the last IF stages, the P3OI is 13.4dBm. When the mixer output has more gain, the P3OI is 0.32dBm. The two-tone measurement from Fig.9 shows consistent results with the one-tone behavior. Close to the maximum gain (10.6dB) the IIP3 is 1.49dBm whilst, at minimum gain (-14.5dB), the IIP3 is 13.3dBm ($f_{IF}$=1GHz).

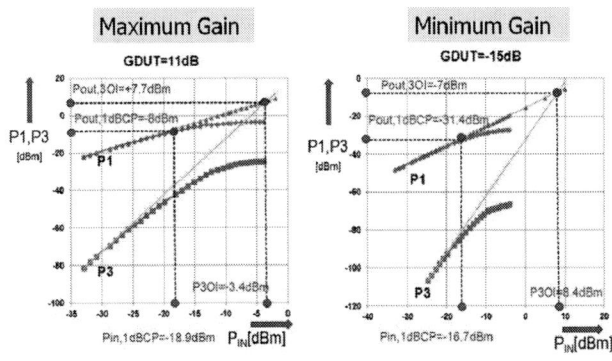

Fig. 7. One-tone measurement for maximum and minimum gain

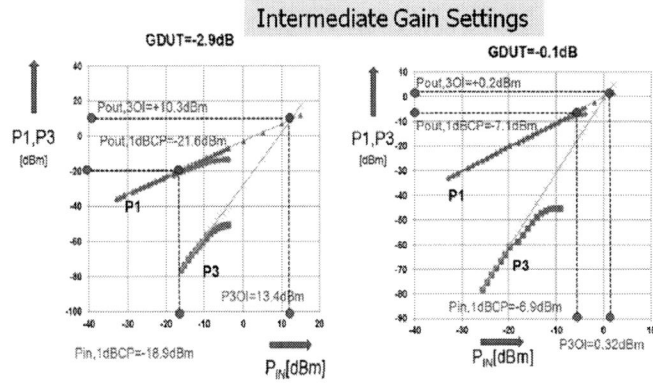

Fig. 8. One-tone measurement for two intermediate gain settings

As gain can be distributed in different ways, for the same gain (5.8dB) the best measured linearity is IIP3=11dBm (Fig.10) but can be as low as 5dBm for different gain distribution.

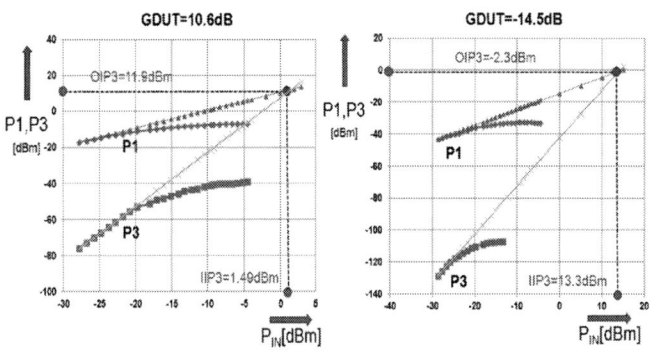

Fig. 9. Two-tone measurement for maximum and minimum gain.

In order to test the on-chip, tunable match circuit at the input of the mixer, the off-chip phase shifters are tuned to maximum imbalance. This introduces a common-mode component at the differential input and the result is the reduction of the fundamental at the input and output. Adjusting the phase imbalance in the matching circuit, the fundamental (HD1) is improved by almost 4dB (Fig.11). By adjusting the common-mode impedance of the mixer, HD2 is reduced by 3dB.

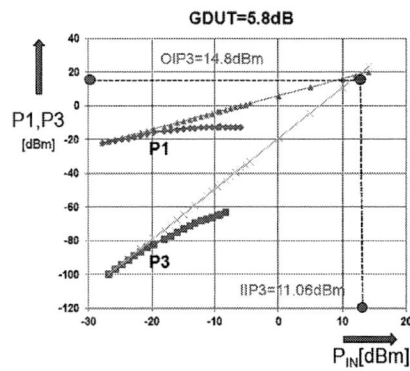

Fig. 10. Two-tone measurement ($G_{DUT}$=5.8dB @ best IIP3).

978-1-4673-6145-3/13 $31.00 © 2013 IEEE        277

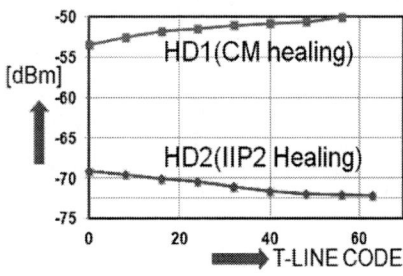

Fig. 11. HD1(Common-mode) and HD2(IIP2) healing.

The down-converter implemented in a 32nm CMOS SOI process from IBM occupies 2.3x0.6mm² real estate. The chip photomicrograph is presented in Fig.12. Table I shows the performance summary of the circuit.

| POWER GAIN (MIX+VGA) | -15dB to 11dB |
|---|---|
| LO POWER (PLO) | -2dBm |
| Pin,1dBCP (G=11dB) | -18.9dBm |
| Pin,1dBCP (G=-15dB) | -16.7dBm |
| IIP3 (G=11dB) | 1.49dBm |
| IIP3 (G=-14.5dB) | 13.3dBm |
| IIP3 (G=5.8dB) | 11.06dBm |
| NF (@ GDUT=11dB) | 17dB |
| LO→RF Isolation | 43dB |
| LO→IF Isolation | >82dB |
| TOTAL POWER (VDD=1V) | 19.2mW |
| TOTAL AREA | 2.3x0.6 mm² |

Table I. Measurements Summary

The LO applied at the input of the mixer is -2dBm. The LO/RF isolation is 43dB and the LO/IF isolation is >82dB. The simulated noise figure of the mixer/VGA tandem is 17dB for $G_{DUT}$=5.8dB. Table II shows a benchmark with other CMOS mm-Wave down-converters found in literature.

| PUB | TECH | GDUT[dB] | IIP3[dBm] | P[mW] | PLO[dBm] | Freq. [GHz] |
|---|---|---|---|---|---|---|
| This Work | 32nm SOI | -15...+11 | +11.06 @ G=5.8dB | 19@1V | -2 | 60 |
| μ-wave Let. 05/06 | 130nm | +6 | +4.5 | 97@3.3V | 5 | 76 |
| μ-wave Let. 03/09 | 65nm | -8 | 2.5 | 6 | 4 | 76 |
| RFIC 2010 | 130nm | +18 | -12 | 50.2 | 2 | 50.8-60.6 |
| [4] | 90nm | 3-5 | +11 | 93@3V | +6 | 40 |
| [3] | 65nm | +33.7 | -19 | 16.9 | -2 | 77 |
| RFIC 2012 | 65nm | +13dB | ~-2.4 | 14 | N/A | 49-67 |
| RFIC 2012 | 90nm | -13.5 | -10.5 | 21 | 4 | 30-65 |
| RFIC 2012 | 40nm | 27.6 | -22 | 28@0.9V | N/A | 60 |

Table II. Performance comparison

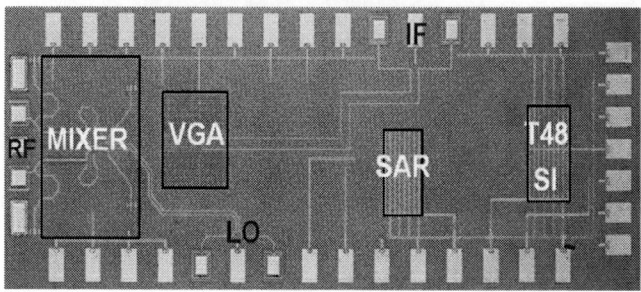

Fig. 12. Chip photo.

Based on our search, the presented down-converter has the highest ever reported IIP3 @ mm-Waves (f>30GHz) in a CMOS technology. The IIP3 value of 11dB reported in [4] is realized at a lower frequency (40GHz) and at 3V power supply.

## V. CONCLUSION

A 60GHz direct down-conversion mixer with integrated sensors and digital infrastructure was presented. It was realized in a 32nm CMOS SOI process (IBM). Applied to large phased-array systems with RF power combining, the gain/linearity trade-off is exploited to achieve the best linearity performance. The high linearity of the mixer cell is facilitated by a novel linearization method without the need of extra voltage headroom. The LO is applied single-ended and switches on and off a linearized transconductor pair. The down-converter occupies 1.38mm² in 32nm CMOS SOI and consumes 19.2mW from 1V supply. The mixer alone consumes only 4mA. The measured linearity (IIP3) for the whole down-conversion chain is better than 11.06dBm for 5.8dB gain at 60GHz. Gain can be adjusted from -15dB to 11dB in 1dB steps depending on the signal level. The LO/RF isolation is 43dB and the LO/IF isolation is better than 82dB. By adjusting phase matching at RF input and the common-mode impedance of the mixer with a variable transmission line, the input phase imbalance can be corrected and the second-order distortion (HD2) can be reduced.

## ACKNOWLEDGMENTS

This work was conducted under the DARPA HEALICs program and was partially funded by DARPA under AFRL contract # FA8650-09-C-7924. The views expressed are those of the author and do not reflect the official policy or position of the Department of Defense or the U.S. Government.

## REFERENCES

[1] J. Kim, et al., "Improvement of Noise Performance in Phased-Array Receivers," in *ETRI Journal*, Vol.33, Number 2, April 2011, pp.176-182.

[2] Shlomo Shlafman, et al., "Variable Transmission Lines: Structure and Compact Modeling," in *IEEE COMCAS*, Tel-Aviv, Israel, Nov. 2011.

[3] Chun-Hsing Li, et al., "16.9-mW 33.7-dB Gain mm-Wave Receiver Front-End in 65 nm CMOS," in *IEEE SiRF*, Santa-Clara, US, Jan. 2012, pp.179-182.

[4] Jeng-Han Tsai, et al.,"A 25-75GHz Broadband Gilbert-Cell Mixer Using 90-nm CMOS Technology," in IEEE Microwave and Wireless Components Letter Vol.17, April 2007, pp.247-249.

# A Fully Integrated Highly Linear Receiver with Automatic IP$_2$ Calibration Schemes for Multi-Standard Applications

Ashkan Borna[1], Yanjie Wang[2], Chris Hull[2], Hua Wang[3], and Ali Niknejad[1]

[1] Berkeley Wireless Research Center, UC Berkeley, Berkeley, CA, USA, 94704,

[2] Intel Corporation, Hillsboro, OR, USA, [3] Georgia Institute of Technology, Atlanta, GA, USA, 30332

*Abstract* - **This paper presents an entire receiver chain from RF to baseband with fully integrated self-calibration circuitries for suppressing the 2$^{nd}$-order intermodulation (IM$_2$) distortions in Homodyne receivers for multi-standard applications. All the potential sources for IM$_2$ generation are identified and tackled independently in the proposed receiver by architectural and calibration techniques, which results in a very robust IP$_2$ enhancement with independency of the amplitude and the frequency of the blockers. The prototype receiver implemented in a 90 nm CMOS process achieves at least 10 dB improvement on the receiver IP$_2$ performance at high-power blockers and less than 100 μs calibration cycle for whole receiver chain. It potentially provides truly SAW-Less, fully integrated, and frequency-agile receivers.**

*Index Terms* - **Direct down-conversion receiver, multi-standard radio, IP$_2$ enhancement, automatic calibration.**

## I. INTRODUCTION

Even order distortion is a well-known concern in RF wireless receivers [1]. Homodyne receivers and low-IF receivers, in particular, are susceptible to even-order nonlinearities, since they produce low-frequency intermodulation distortions which will corrupt the desired signal after down-conversion mixing and thereby degrade the effective signal-to-noise (SNR) ratio. Conventionally, this issue is alleviated by using high quality factor off-chip filters, e.g. SAW filters. However, these external filters significantly increase the system cost and complexity, and they are mostly tuned to limited blocker frequencies. Removing the external filters, if possible, will enable hardware sharing between different standards to achieve frequency-agile and multi-standard radios. Therefore, it is highly critical to develop on-chip calibration schemes to enhance the receiver even-order linearity (primarily the 2$^{nd}$ order intermodulation intercept point IP$_2$) and minimize the corruptions from the blockers. To date, most reported IP$_2$ calibration schemes achieve linearization by using a low frequency compensation current after the down-conversion mixers [2]-[3]. These designs typically assume a certain blocker frequency and amplitude for calibration, inject large complex tones, and use least-squares algorithm to minimize the IM$_2$. Such techniques, while effective on the calibrated blocker tone, generally cannot suppress the distortions by blockers at other frequencies and powers.

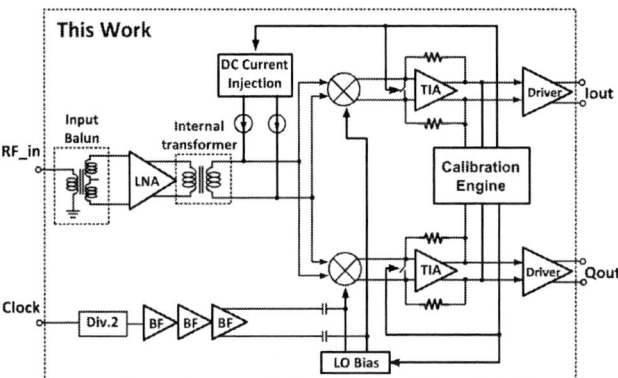

Fig. 1. Proposed fully integrated highly linear IP$_2$ receiver system.

This presents a serious limitation, since it is difficult to anticipate the blocker profile a priori and the blocker may often subject to change. In order to address this challenge, we propose a new self-calibrating scheme to suppress the 2$^{nd}$ order intermodulation distortions.

The proposed scheme uses only simple dc injection, detection, and logic for calibration. Most importantly, the method employs independent calibration for each IM$_2$ generation mechanism and achieves the overall IM$_2$ suppression insensitive to the specific blocker profile.

The paper is organized as follows. In section II the general IM$_2$ sources are investigated based on which our IP$_2$ calibration scheme is proposed. Section III examines these IM$_2$ sources in a canonical direct down-conversion receiver together with an implementation example in a 90nm CMOS. The measurement results are presented in Section IV to demonstrate the functionality of the proposed scheme.

## II. IM$_2$ GENERATION MECHANISMS

The IM$_2$ generation mechanisms in a canonical receiver can be summarized as follows.

1) LO and RF self-mixing. This is mainly due to the coupling and leakage between the mixer's LO and RF ports. This effect generally can be made negligible by using orthogonal routing of RF and LO ports and double-balanced mixer topologies.

2) LNA nonlinearity. This mechanism comprises of two parts. One is the IM$_2$ components directly generated by the LNA and the other one is due to the unbalanced output signals of the LNA when a differential mixer is used.

978-1-4673-6145-3/13 $31.00 © 2013 IEEE

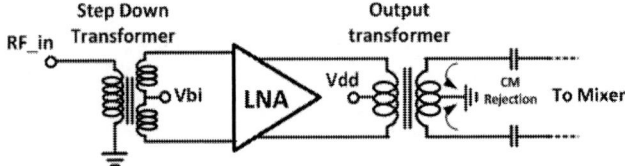

Fig. 2. Simplified schematic of input and output stage of LNA.

3) Mixer nonlinearities. This mechanism can be summarized as

    a) Switching pair mismatch,
    b) LO waveform dependency,
    c) Interaction with baseband mismatches.

Note that the interaction with baseband occurs in the passive mixers due to the bilateral property of this architecture. The active mixers buffer between the IF and RF domains and hence the baseband mismatches have little effect on the $IP_2$. However, there is a large signal at twice the frequency of the LO at the RF node in the active mixers [1]. This signal can be down converted to the baseband by the mismatches in the switches and exacerbates the sensitivity to the mismatch in the mixers' switches. Therefore, passive mixers outperform the active counterparts in addition to noise, power and linearity ($P_{1dB}$ and $IP_3$) [4, 5].

## III. PROPOSED RECEIVER CIRCUIT DESCRIPTION

The simplified schematic of the proposed receiver system is shown in Fig. 1 consisting of integrated input balun, broad-band LNA, mixers, self-calibration circuits, and transimpedance amplifiers (TIA) with output drivers for output matching. Based on our analysis, the dominant $IM_2$ generation mechanisms are the LNA imbalanced output currents, the switching pair mismatches, the LO waveforms and the baseband mismatches for the case of passive mixers, all of which are taken care of in the design through architectural and calibration techniques.

### A. LNA Design

The LNA generates $IM_2$ contents in baseband through two mechanisms.

First are the $IM_2$ products at its output. The magnitude of these signals due to the LNA can be relatively high. However, because of the ac-coupling capacitor after the LNA as shown in Fig. 2, this effect can be made quite negligible. There is a trade-off in selecting the magnitude of this ac-coupling capacitor in current mode receivers that are sensitive to the LNA load. It can easily increase the load to the LNA by as much as 50% if one deploys a small capacitor and it could introduce a considerable parasitic capacitance to the ground for large capacitors. The RF impedance at the output node of the LNA plays an important role in the $IP_2$ metrics of the receiver. It has been analyzed quite extensively in [1] and the

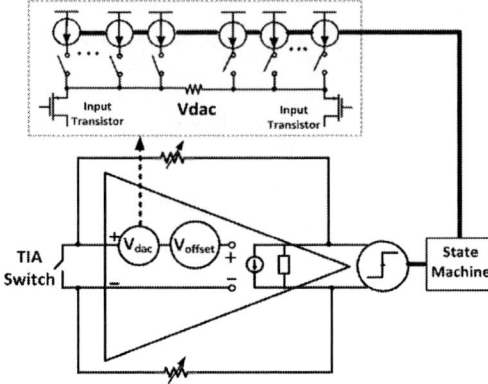

Fig. 3. Simplified schematic of baseband calibration circuitry incorporated within the TIA.

conclusion is the lower the parasitic capacitance at that node, the better the $IP_2$ metrics. One option is to tune out the capacitance at that node via an inductor. Another option is to deploy a transformer at that node which has another important benefit.

The second mechanism arises from the unbalanced RF signals at the input/output of the LNA. The ac-coupling capacitor can block the $IM_2$ content of the LNA. However, if the output current of the LNA becomes unbalanced, that could cause generation of $IM_2$ content. The unbalanced signals can be modeled as balanced signals plus a common mode signal and the common mode to differential mechanisms can generate the $IM_2$ at the output of the mixer. At the input of the LNA, the on-chip balun equalizes RF signals. Moreover, the secondary to primary ratio of this balun can be set to reduce the magnitude of the input current to the LNA, which leads to increase in the saturation power of the LNA. The common-gate architecture has a high $P_{1dB}$ and hence supplies balanced outputs while processing high power blockers.

A transformer with center-taps on both primary and secondary sides has been placed at the output of the LNA as shown in Fig. 2. This would suppress the common-mode content at the output of the LNA even further and hence has a two-fold advantage of balancing the RF signals and tuning-out the parasitic capacitors at the output nodes. The primary to secondary ratio has been set to increase the current gain of the transformer. Since the passive mixers have a high $P_{1dB}$, higher current gain would increase the sensitivity of the current-mode receiver and will not desensitize the mixers.

### B. Mixer Design

Mixer is the dominant block responsible for $IM_2$ generation. The mismatch in the mixers can be due to the transistor geometry, the threshold voltage variations, the LO waveform and the baseband related mismatches.

For the width mismatches, the simplest approach is to increase the size of the mixer switches. This has a two-fold advantage, a reduction in the mismatch of the mixer switches

978-1-4673-6145-3/13 $31.00 © 2013 IEEE

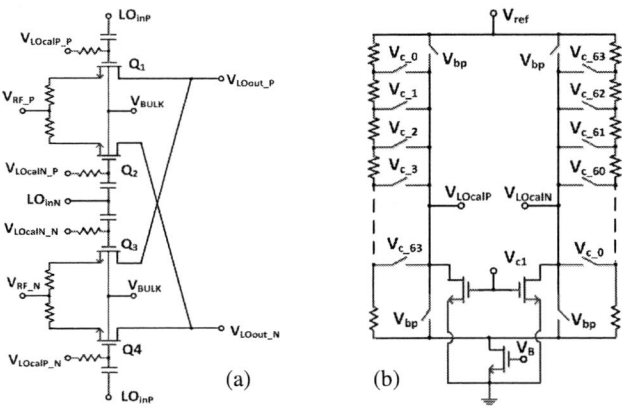

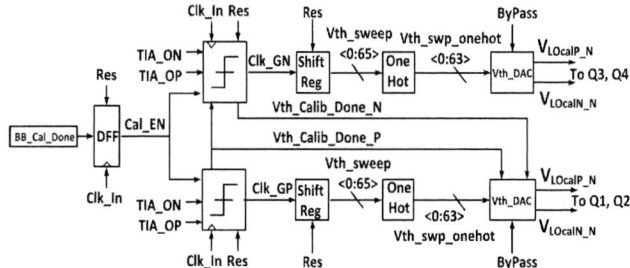

Fig. 4. (a) schematic of mixer with bias nodes, (b) simplified schematic of the threshold DAC.

and lowering the mixer load on the LNA, resulting in an increase in linearity and power handling ability. The disadvantage comes through the increase in the power consumption of the driving stages of the switches. The threshold mismatch compensation is performed digitally after the baseband offset compensation. Traditionally the baseband effect has been overlooked in the $IM_2$ generation process either because of the application of an active mixer (which isolates the RF and baseband section), or by injecting a DC signal at the baseband to compensate the $IM_2$ signal of both the RF and baseband stages together, which will not be stable.

Buffers driving the mixers should be carefully sized to provide equal-width pulses to the switches since the mismatch in the width is considered a pulse with twice the frequency that would result in the $IM_2$ generation in the baseband. Also there is a trade-off in the crossing point of the consecutive waveforms. By lowering the crossing points, the sensitivity to the mismatches in the mixer is reduced and the sensitivity to the RF node parasitic capacitance will increase simultaneously, since in this way, overlapped time would be reduced and the fluctuation on the RF node will increase.

### C. Calibration Procedures

The proposed self-calibration schemes perform two calibrations mechanisms: base-band offset compensation and mixer threshold mismatch calibration. The detail calibration procedure is introduced as follows.

The calibration engine first detaches the RF section by a switch at the input of a TIA as shown in Fig. 3, which shorts the mixers' output and performs offset compensation of the baseband amplifier through a state machine that ramps the DC voltage across the small resistor (20 Ω) in the source of the input transistors of the amplifier. When the zero crossing occurs, the comparator sends the stop command, the voltage is saved and the switch after the mixer opens which ends the base-band offset compensation.This will be performed for both I and Q channels separately.

Fig. 5. Block diagram of threshold mismatch calibration logic flow.

The mixer threshold mismatch digital calibration is performed next in two steps. The first calibration procedure is to compensate transistors threshold mismatch in the positive RF side, Q1 and Q2 as shown in Fig. 4 (a). During this procedure, transistors in the negative RF side, Q3 and Q4 are turned off. When the mismatch compensation is completed, Q1 and Q2 will be turned off, and then the second calibration procedure starts for Q3 and Q4.

The threshold mismatch compensation is done by a DAC, and the simplified schematic of Mixer threshold DAC is shown in Fig. 4 (b). By injecting a DC current from the RF side (positive or negative sides) and stepping the voltage at gates of the transistors, one can compensate the threshold mismatch between transistor Q1 and Q2, or Q3 and Q4 through detecting the zero crossing at the output of the compensated TIA and a comparator connected to its output.

The resistors' value (Fig. 4 (b)) is kept small (22 Ω) to minimize the thermal noise contribution to the receiver and as well as the sweeping resolution. The threshold voltage-sweeping step is kept in the range of 100 μV-200 μV, which depends on the DC current and can be controlled by the gate voltage VC1. The simplified logic flow of the mixer threshold mismatch digital calibration for each procedure is shown in Fig. 5. When baseband calibration is finished; the enable signal (Cal_EN) is sent out to both sections by a D-flip flop.

Performing calibration in these steps assures independent cancellations of different mechanisms responsible for $IM_2$ generation and hence the result is expected to be stable over different blocker profiles. This is key metrics since once the filter is removed, one cannot simply perform calibration for each individual blockers to squeeze as much of the linearity as possible out of the receiver.

## IV. MEASUREMENT RESULTS

The prototype circuit is fabricated in a 90 nm CMOS technology, and the chip has been packaged using 32-pin MLF package and mounted on a FR4 board as shown in Fig. 6 (a). The die microphotograph of the chip is shown in Fig. 6 (b), and the total chip area is $1.9 \times 1.95$ mm$^2$. Fig. 7 shows the PCB measurement setup. The 5 GHz LO clock signal feeding the divide-by-two circuitry and the low frequency clock of 2 MHz for calibration engine is brought from off-chip.

978-1-4673-6145-3/13 $31.00 © 2013 IEEE

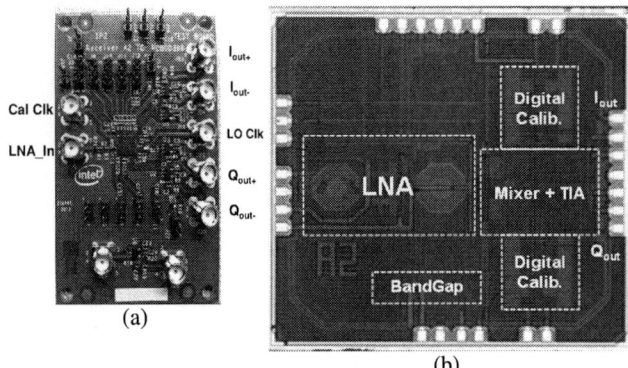

Fig. 6. (a) PCB board, an (b) chip die microphotograph.

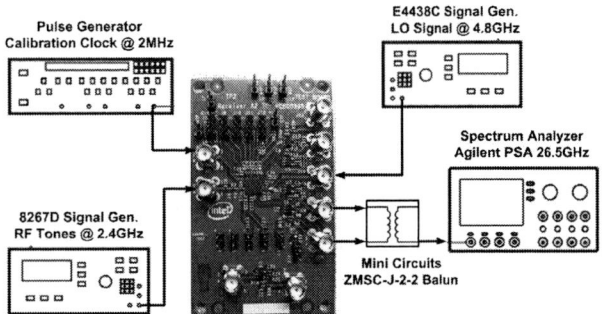

Fig. 7. Measurement setup.

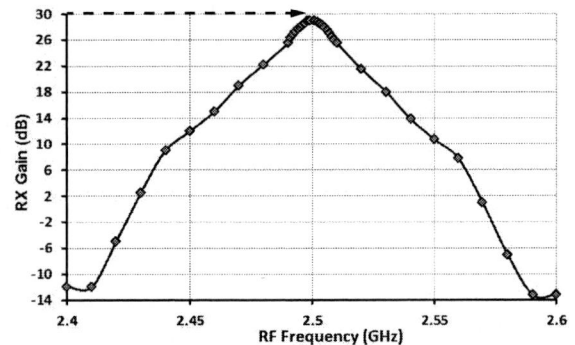

Fig. 8. Measured RX gain vs. blocker freq. with LO = 2.5 GHz.

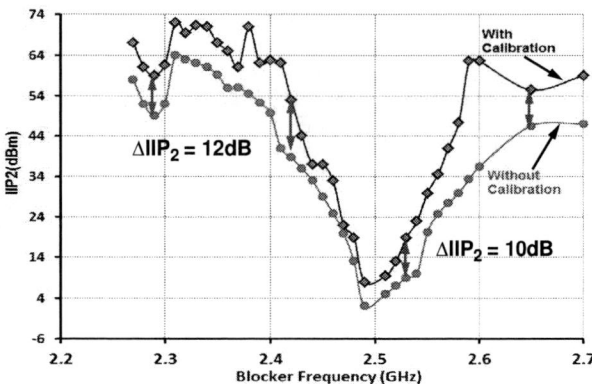

Fig. 9. Measured IIP$_2$ w/wo calibration vs. the blocker frequency.

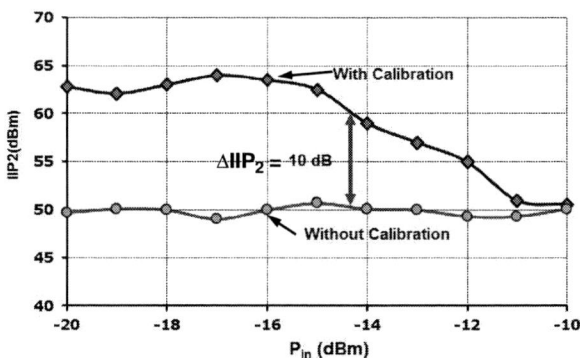

Fig. 10. Measured IIP$_2$ w/wo calibration vs. different power level.

The gain response of the receiver system is shown in Fig. 8, and the measured 30 dB gain shows a good agreement with the extracted simulation. The IP$_2$ measurements have been performed first by sweeping the blocker frequency from 2.27 GHz to 2.7 GHz with different tone separations. The measured IIP$_2$ results with and without IP$_2$ calibration at 1 MHz spacing is shown in Fig. 9. The measured data supports that an IIP$_2$ of more than 60 dBm can be achieved with RF tone of 100 MHz away from LO frequency and an average of 10 dB improvement on IP$_2$ can be achieved with calibration. Additional measurment on the IIP$_2$ at 5 MHz and 10 MHz spacing with the same settings of 1 MHz separation have also been measured, and it shows an average of 10 dB IIP$_2$ improvement with proposed calibration scheme.

Fig. 10 shows the IIP$_2$ performance with and without the proposed calibration scheme by sweeping the blocker power from -20 dBm up to -10 dBm with RF tones at 2.4 GHz and 1 MHz spacing, LO at 2.5 GHz. The measurements confirm that the proposed scheme achieves a larger than 10 dB IIP$_2$ improvement and a better than 60 dBm IIP$_2$ up to a block power level of -14 dBm.

## V. CONCLUSION

A fully integrated highly linear receiver with automatic IP$_2$ calibration scheme for multi-standard applications is proposed in this paper. A direct down-conversion receiver is implemented in 90 nm CMOS. The measurement results demonstrate that a 60 dBm stable IP$_2$ and at least 10 dB improvement on the receiver IP$_2$ are achieved with the blocker power of up to -14 dBm.

## REFERENCES

[1] D. Manstretta et al., "Second-order intermodulation mechanisms in CMOS downconverters", IEEE J. Solid-State Circuits, Vol. 38, No. 3, pp. 394-406, Mar., 2003
[2] H. Darabi et al., "An IP2 Improvement Technique for Zero-IF Down-Converters," ISSCC Dig. Tech. Papers, pp. 1860-1869,Feb., 2006.
[3] M. Chen et al., "Active 2nd-order intermodulation calibration for direct-conversion receivers ," ISSCC Dig. Tech. Papers, pp. 1830-1839,Feb., 2006.
[4] N. Poobuapheun, W. H. Cheng, Z. Boos and A. Niknejad, "A 1.5-V 0.7–2.5-GHz CMOS quadrature demodulator for multiband direct-conversion receivers," Solid-state Circuits, IEEE Journal, pp. 1669-1677, 2007.
[5] T. H. Lee, "The design of CMOS radio-frequency integrated circuits," Cambridge university press, 2003.

978-1-4673-6145-3/13 $31.00 © 2013 IEEE

# 40V MESFETs Fabricated on 32nm SOI CMOS

William Lepkowski[1,2], Seth J. Wilk[1,2], J. Kam[1], and Trevor J. Thornton[1,2]

[1]Electrical, Computer and Energy Engineering, Arizona State University, Tempe, AZ 85287

[2]SJT Micropower Inc., Fountain Hills, AZ 85268

*Abstract-* **This article describes 40V N-channel MESFETs fabricated at a commercial 32nm SOI CMOS foundry without changing any of the process flow or including additional mask steps. The 32nm technology node is the most advanced technology node to date for MESFET fabrication and builds upon previous work completed at other process nodes. High voltage MESFETs were measured with current drives of 110mA/mm. The devices are suitable for RF development and have peak cut-off frequency, $f_T$, of 30.5GHz and maximum oscillation frequency, $f_{max}$, of 34.5GHz.**

## I. INTRODUCTION

The feature size and operating voltage of CMOS devices continue to scale with each shrinking process node. On the 32nm process used here, the standard and "high voltage" MOSFETs had ratings of less then 1V and 2V respectively. This places stringent constraints for on-chip analog and RF circuits. For example, a power amplifier (PA) has particular challenges because the low supply voltage then requires high current to be used to maintain output power resulting in very small output impedances. Issues including the number of bondwires needed to handle the current drive and lower efficiency in the output match must then be resolved [1]. Nevertheless, the industry continues to demand higher integration, smaller and faster digital, and for more of the circuit to be designed in silicon to reduce costs and external component count. Solutions to increasing the maximum voltage such as modifying the CMOS process [2] or including an LDMOS [3] have been used and are viable, but they require additional masks steps and increase cost. Stacking of transistors to distribute a high supply over multiple transistors is also an option, but it requires a complex gate biasing scheme [4].

In this paper, Si-MESFETs are shown to be a low cost, yet easy and effective solution to allow the incorporation of high voltage applications directly into silicon. As was the case with previous demonstrations at different technology nodes [5-9], the native process was not altered nor were additional mask layers used. Device operation is shown in excess of 40V on extended drain MESFET geometries while the most compact MESFET aims at higher speed applications with a 7V soft breakdown and peak $f_T$ and $f_{max}$ of 30.5 and 34.5GHz respectively. The DC and RF results of MESFET structures with different geometries on the 32nm process are also presented. Finally, these results are compared to MESFETs at the 45 and 150nm nodes to see how the performance improves from one node to the next.

## II. MESFET TECHNOLOGY DESCRIPTION

A cross-section of the MESFET is shown in Fig. 1. A Schottky gate was created by incorporating the metal silicide step over a lightly doped n-well. The other critical process step was in the patterning of the spacer regions ($L_{aS}$ and $L_{aD}$). The separation between the spacers defines the gate length, $L_G$, and its minimum distance is determined by the design rules of the technology used. On the 32nm SOI process, $L_G$ was limited to 200nm without breaking the design rules. Since that was about 6X bigger than minimum feature size of the process, short channel effects (SCE) in the MESFET were not a concern. MESFETs at other nodes have shown SCE becoming apparent when $L_G$ is less than 1.5 to 2X of the feature size [5].

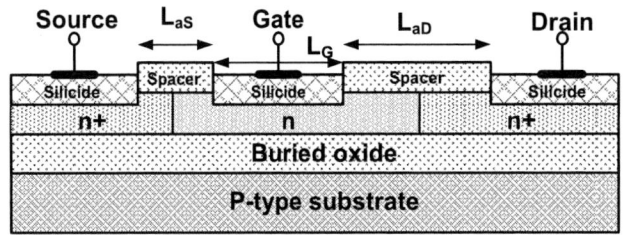

Fig. 1. Cross section of a partially depleted MESFET structure.

The spacers are also needed to separate and prevent shorting between the Schottky gate and the silicide used to form low resistance contacts at the source and drain terminals. The minimum sized access length possible of $L_{aS}$ (distance between gate and source, Fig. 1) and $L_{aD}$ (distance between gate and drain, Fig. 1) was also 200nm. Therefore, the minimum sized MESFET on the 32nm process had $L_G = L_{aS} = L_{aD} = 200$nm. As it will be shown in Section III and IV, the spacers can be optimally sized to fit different applications based on $f_T$ verse voltage breakdown requirements.

## III. DC MEASUREMENTS

The MESFETs high voltage characteristic is credited to its Schottky gate which can tolerate high current flow and its non-self-aligned structure. Without a thin gate oxide, MESFETs avoid electric field gate oxide breakdown and snapback which are common causes of device failure in MOSFETs. Instead, it is thought that breakdown in the MESFET is caused mostly by avalanche ionization and tunneling mechanisms. When the MESFET approaches soft breakdown, the surface electric field becomes large enough to lower the barrier height at the gate and allow electrons to tunnel into the channel from the gate metal. The device begins to see an exponential increase in drain-to-gate current [5, 9].

978-1-4673-6145-3/13 $31.00 © 2013 IEEE

The soft breakdown event does not damage the MESFET and is completely reversible; although a high enough drain voltage will eventually lead to a hard breakdown.

By increasing the length of $L_{aS}$ and $L_{aD}$ the electric field at the gate-source and gate-drain junctions can be reduced, which in turn increases the breakdown voltage capability. The key junction though is the gate-drain since the drain voltage, $V_D$, can be much higher than the gate voltage. The tradeoff of increasing $L_{aS}$ and $L_{aD}$ is that it will increase the parasitic resistance and consequently lower the current drive and degrade the RF performance [9]. There is also a die area penalty.

Figure 2 shows the family of curves (FOC) measurements for four MESFETs with different $L_{aS}$ and $L_{aD}$ and highlights the high voltage capability of the device. The minimum sized device, $L_G = L_{aS} = L_{aD} = 200$nm, in Fig. 2a, had a current drive around 110mA/mm when biased at a gate-to-source voltage, $V_{GS}$, of +0.5V. The device shows good drain curve characteristics to about 6V before it begins to roll up and starts to approach the soft breakdown region. The current drive in Fig. 2b is slightly reduced when $L_{aD}$ is increased to 2000nm, but the device now shows good output characteristics to about 33V. The output is further enhanced to about 36V in Fig. 2c with $L_{aS} = 500$nm and $L_{aD} = 5000$nm and finally to 40V in 2d with $L_{aS} = 1000$nm and $L_{aD} = 10,000$nm. Lastly, the negative slope seen in the drain current of Fig. 2b-d for the two uppermost curves biased at $V_{GS}$ = +0.25 and +0.5V is most likely due to a heating effect and could be eliminated by pulsing the gate voltage.

## IV. RF MEASUREMENTS

Scattering parameter measurements of the MESFETs were completed out to 15GHz using an Agilent 8510C network analyzer on devices fabricated in Ground-Signal-Ground (GSG) pad configurations and the parasitics due to the pads were de-embedded. Both $f_T$ and $f_{max}$ were determined by fitting H21 and maximum available gain (MAG) with a 20dB per decade response and extrapolating to 0dB.

The RF performances is shown in Figs. 3-4 for the MESFET with $L_G = L_{aS} = L_{aD} = 200$nm (Fig. 2a) and a MESFET with larger drain access length, $L_{aD} = 2000$nm (Fig. 2b). The minimum sized MESFET comprises the least parasitics and thus has the highest peak $f_T$ and $f_{max}$, (30.5 and 34.5GHz) of any of the devices that were fabricated on the 32nm process. For the device with longer $L_{aD}$ the $f_T$ and $f_{max}$ decreased to 22 and 26GHz, but it is a very attractive device for higher power PA applications due to its > 30V operation.

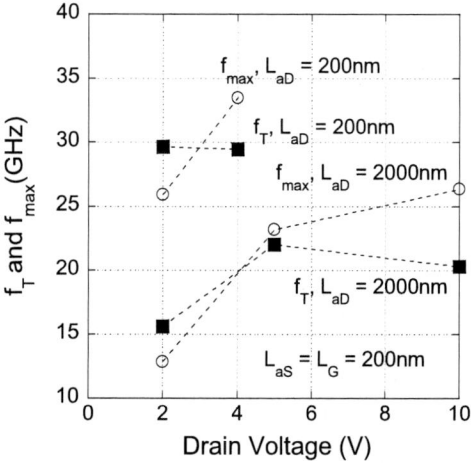

Fig. 3. Compares measured $f_T$ and $f_{max}$ for $L_{aD}$ of 200 and 2000nm with $L_G = L_{aS} = 200$nm when different drain biases are applied. For each measurement $V_{GS}$ = +0.1V.

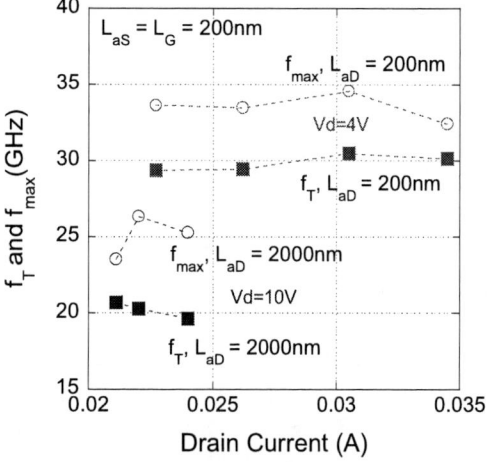

Fig. 4. Compares measured $f_T$ and $f_{max}$ when different gate biases are applied for two devices with $L_G = L_{aS} = 200$nm. For the $L_{aD} = 200$nm, $V_D = 4$V while $V_D$ = 10V for $L_{aS} = 2000$nm.

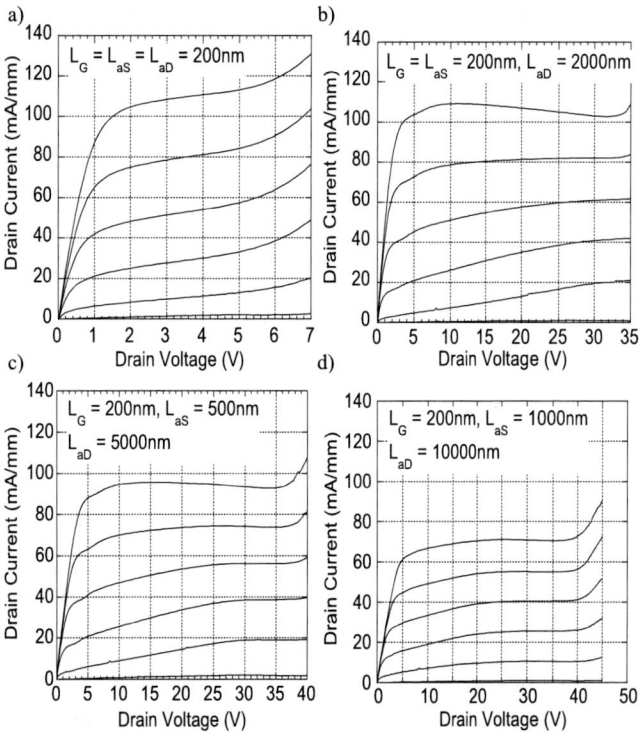

Fig. 2. Compares FOC plots of MESFETs with $L_G = 200$nm and different $L_{aS}$ and $L_{aD}$ sized from 200 to 10,000nm. In each graph the $V_{GS}$ is stepped from +0.5V (uppermost curve) to -0.75V in 0.25V steps.

## V. SCALING MESFETS

One question that often arises is how well do the MESFETs scale with each new technology node. Comparing nodes provides valuable insight into future MESFET performance. In [5], the MESFETs were compared at the 150 and 350nm nodes. Here a similar comparison is made between the same 150nm process with the next most scaled processes; 32 and 45nm. Both the 32 and 45nm devices were fabricated at the same foundry, but not the same one as the 150nm MESFETs. The minimum $L_G$, $L_{aS}$, $L_{aD}$ on the 45nm processes was also 200nm [6]. The design rules allowed the 150nm process to have $L_G$ as small as 150nm, but $L_{aS}$ and $L_{aD}$ were limited to 300nm. In Figs. 5-6, Gummel and FOC measurements are compared for the minimum sized devices on the 32 and 45nm process ($L_G = L_{aS} = L_{aD} = 200$nm) and a device with an $L_G = 200$nm and $L_{aS} = L_{aD} = 300$nm on the 150nm process. As was the case in [5], the overall performance continues to improve even if the $L_G$ remains the same. The MESFET from the 150nm process has a higher current drive than the 45nm device (Fig. 6), but it is because the 150nm has a much more negative threshold voltage, $V_t$, due to SCE which allowed it to have a larger gate overdrive. Also, SCE effects are manifested as a low output resistance, $R_O$, for the drain curves (Fig. 7).

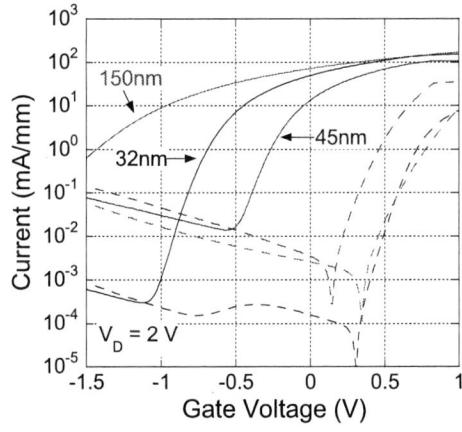

Fig. 5. Compares the Gummel plots of MESFETs with $L_G = 200$nm fabricated on a (a) 32nm, (b) 45nm, and (c) 150nm process. The solid curves represent drain current while the dashed represent the gate leakage current.

From the 45 to the 32nm process the same sized MESFET from the same foundry makes significant improvement with respect to drain current, output resistance and gate leakage. Interestingly, that while the gate leakage is lower on the 32nm process the $V_t$ is more negative. One might think that a lower leakage should make the MESFET easier to pinch off and thus be less depletion mode, but $V_t$ is also dependent on other parameters such as Si-channel thickness, doping and the silicide step. Exact details of these are unknown because the process details are propriety to the foundry which makes it difficult to pinpoint the exact reason for the different $V_t$ between processes.

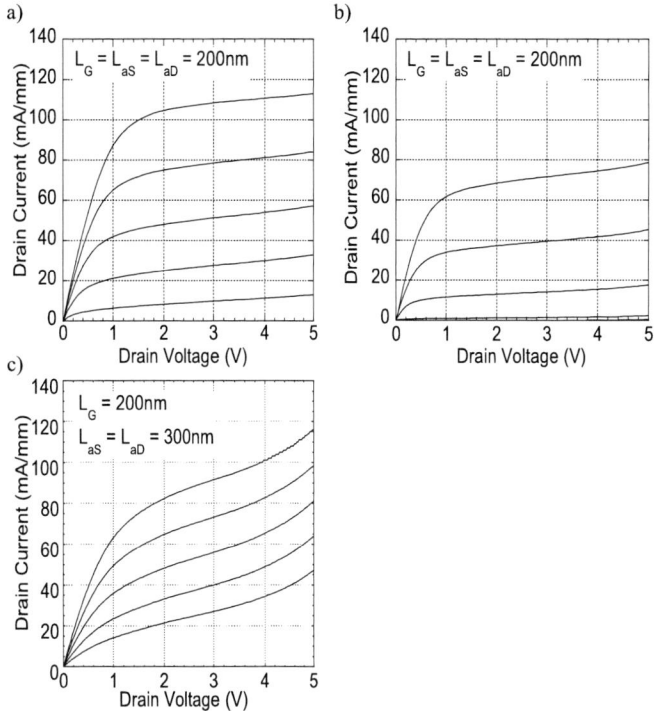

Fig. 6. Compares FOC plots of MESFETs with $L_G = 200$nm fabricated on a (a) 32nm, (b) 45nm, and (c) 150nm process. In each graph the $V_{GS}$ is stepped from +0.5V (uppermost curve) to -0.5V in 0.25V steps.

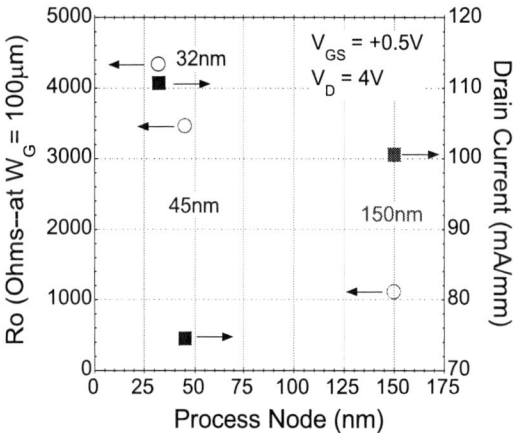

Fig. 7. Extracted $R_O$ and drain current at $V_{GS} = +0.5$V and $V_D = 4$V from the FOC in Fig. 6.

Figure 8 shows the $f_{max}$ of the 3 MESFETs in Figs. 5-7. There is a small but systematic increase in peak $f_{max}$ as the technology scales with values of 32.5, 33, and 34.5GHz on the 150, 45, 32nm processes respectively. The improvement with each scaled node is most likely the result of progressively smaller contact sizes which allowed more contacts on each gate finger and thus lowered parasitic gate resistance. While it is not shown, the peak $f_T$ is roughly the same on the 32 and 150nm [5] processes at about 30.5GHz. That is slightly higher than the 27.5GHz [6] from the 45nm process. The closeness in values is not too surprising since $f_T$ is most dependent on $L_G$ which is the same on all 3 devices.

978-1-4673-6145-3/13 $31.00 © 2013 IEEE

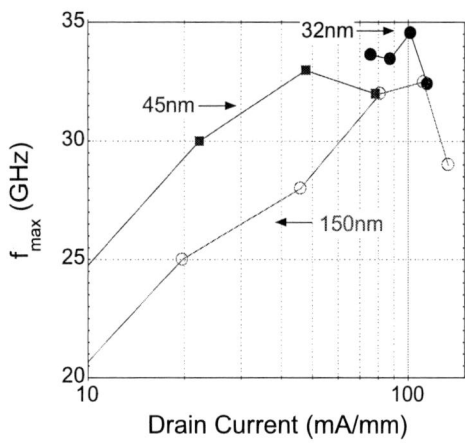

Fig. 8. Compares measured $f_{max}$ from the 3 MESFETs in Figs. 5-7.

More substantial improvement to the drain current, $f_T$ and $f_{max}$ could be achieved for MESFETs on both the 32 and 45nm processes if the design rules allowed $L_G$ to be less than 200nm. Since these devices were taped out as part of a multi-wafer process run the design rules were strictly followed. At other foundries, special permission has allowed MESFETs to break key design rules that limit $L_G$ as long as they do not interfere with the CMOS fabrication. On a dedicated process run reducing $L_G$ to 100nm on each of these processes may be feasible. Again to avoid SCE the $L_G$ would mostly likely need to be at least 2X of the feature size. Nevertheless, the improvement seen from the 45nm to the 32nm node was similar to the results when scaling from 350nm to 150nm [5]. While the sizing of $L_G$, $L_{aD}$, and $L_{aS}$ and the $V_t$ of the device play a major role in its RF and DC performance so do the parasitics associated with each technology.

## VI. CONCLUSIONS

This paper shows the first demonstration of MESFETs fabricated on a scaled 32nm SOI CMOS process with high breakdown devices operating up to 40V. The devices offer many exciting opportunities to integrate analog and RF circuitry alongside the leading edge digital CMOS. Of specific interest are interface I/O circuitries, power management [10] and the wireless power amplifiers targeting Watts of power [1] and GHz operation. Current research also includes statistical analysis of the devices on multiple fabrication runs.

## ACKNOWLEDGMENTS

This work was supported in part by DARPA and NASA contracts. The authors would also like to thank the Trusted Access Programs Office (TAPO) for access to the 45 and 32nm processes.

## REFERENCES

[1] S.J. Wilk, *et al*, "32 dBm Power Amplifier on 45 nm SOI CMOS," *IEEE Microwave and Wireless Components Letters*, vol.23, no.3, pp.161-163, 2013.

[2] M. Apostolidou *et al.*, "A 65 nm CMOS 30 dBm class-E RF power amplifier with 60% PAE and 40% PAE at 16 dB back-off," *IEEE J. S. S. Circuits*, vol. 44, no. 5, pp. 1372–1379, May 2009.

[3] A. Tombak, *et al.*, "High-efficiency cellular power amplifiers based on a modified LDMOS process on bulk silicon and silicon-on-insulator substrates with integrated power management circuitry," *IEEE Trans. Microw. Theory Tech.*, vol. 60, no. 6, pp. 1862–1869, Jun. 2012.

[4] S. Pornpromlikit *et al.*, "A watt-level stacked-FET linear power amplifier in silicon-on-insulator CMOS," *IEEE Trans. Microw. Theory Tech.*, vol. 58, no. 1, pp. 57–64, Jan. 2010.

[5] W. Lepkowski, *et al*, "Scaling SOI MESFETs to 150 nm CMOS Technologies," *IEEE Trans. on Electron Devices*, vol. 58, no. 6, pp. 1628-1634, June 2011.

[6] W. Lepkowski, *et al*, "High Voltage SOI MESFETs at the 45nm Technology Node", *IEEE Int. SOI Conf.*, 2012.

[7] J. Yang, *et al*, "Silicon-Based Integrated MOSFETs and MESFETs: A New Paradigm for Low Power, Mixed Signal, Monolithic Systems using Commercially Available SOI," *Int. J. of High Speed Electronics and Systems*, vol. 16, pp. 723 - 732, 2006.

[8] W. Lepkowski, et al, "SOI MESFETs Fabricated Using Fully Depleted CMOS Technologies," *IEEE Electron Device Letters*, vol. 30, pp. 678-680, 2009.

[9] J. Ervin, *et al*, "CMOS-Compatible SOI MESFETs With High Breakdown Voltage," *IEEE Trans. Electron Devices*, vol. 53, pp. 3129-3135, 2006.

[10] W. Lepkowski, et al, "An integrated MESFET voltage follower LDO for high power and PSR RF and analog applications," *IEEE Custom Integrated Circuits Conference*, pp. 1-4, Sept. 2012.

978-1-4673-6145-3/13 $31.00 © 2013 IEEE

# Recent Advances in GaN Power Electronics

## Karim Boutros, Rongming Chu, Brian Hughes
### HRL Laboratories, LLC
### 3011 Malibu Canyon Road, Malibu, CA, USA, 90265

*Abstract* - **Gallium Nitride power devices are poised to replace silicon-based MOSFETs in power switching applications having weight and volume constraints, while simultaneously needing a high overall efficiency. With its projected 100x performance advantage over silicon, GaN is a game changing technology for energy-efficient power electronics. This paper reviews the advantages of GaN material and devices, the performance of these devices in power circuits, and the potential applications for this technology.**

**Index Terms - Gallium Nitride; GaN; GaN-on-Si; hetero-structure field-effect-transistor; HFET; power switch; energy-efficient electronics; power electronics.**

## I. INTRODUCTION

Power management applications, such as point of load (PoL) DC-to-DC converters and voltage regulators require low-loss power semiconductor switches. Presently, these circuits rely almost exclusively on silicon-based power devices. However, silicon power devices are rapidly approaching maturity. That means further enhancements are incremental, while the demand for advancement in efficiency, miniaturization, and increased functionality continues to grow at a high rate. Particularly demanding are circuit applications where size/weight/power and efficiency trade-offs are limited, such as server power management, battery chargers, and micro-inverters for solar farms. These applications demand unprecedented power density (>500W/in$^3$), specific power (10kW/lb) and total PoL power (>1000W), at efficiencies > 95%. New electronic materials and device structures enabling game changing power management solutions with significantly higher efficiency and differentiating system-level advantages are therefore needed.

Gallium Nitride (GaN) switches are projected to have a 100x performance advantage over silicon-based devices and 10x over SiC, owing to their excellent material properties such as high electron mobility, high breakdown field, and high electron velocity. Recent progress in GaN epitaxial growth on silicon substrates allowed the manufacturing of GaN-on-Si epi-wafers with high quality and low cost. Recent progress in GaN power switch design resulted in normally-off operation, low conduction and switching losses, and nano-second switching speeds. These advances enable superior performance and affordable manufacturing. System-level advantages, and reduced system cost can be realized with energy-efficient GaN power electronics, making this technology a game changer for many applications.

## II. KEY ATTRIBUTES OF GaN FOR ENERGY EFFICIENCY

### A. Normally-off operation at high junction temperature

High energy efficiency applications require low stand-by current, fail-safe devices with high current/voltage capability, and minimum cooling. HRL has developed the technology for insulating-gate normally-off GaN-on-Si heterostructure-field-effect-transistors (HFETs) with a threshold voltage ($V_T$) of ~1V, which is maintained up to a junction temperature of 200°C, a maximum current ($I_{max}$) of ~200mA/mm, and high breakdown voltage ($V_B$) >600V with low off-state leakage ($I_{dss}$) of <5µA/mm [1]. Figures 1 and 2 show the transfer and off-state characteristics of the HRL devices at 200°C, respectively.

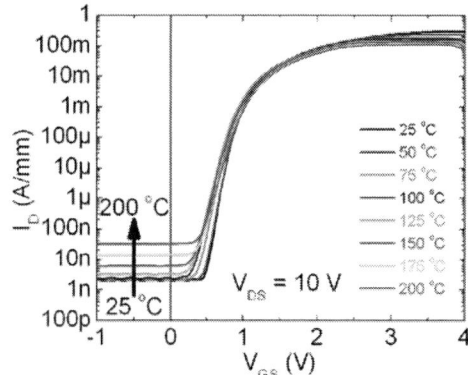

Figure 1. Transfer characteristics of normally-off insulating gate GaN HFET vs. temperature, showing the device maintains normally-off operation up to 200°C.

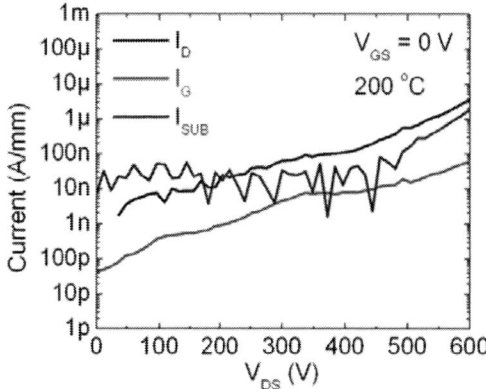

Figure 2. Off state I-V characteristics at 200°C and Vgs=0V showing that the device maintains a low leakage < 5µA/mm up to 600V.

### B. High $V_B^2/R_{on,specific}$ ratio

978-1-4673-6145-3/13 $31.00 © 2013 IEEE        287

A key material figure of merit (FOM) used for power devices represents the trade-off between breakdown voltage ($V_B$) and specific on-resistance ($R_{on,sp} = R_{on}*$Active area). The intrinsic properties of GaN enable it to have the highest theoretical value of $V_B^2/R_{on,sp}$=5000 MW/cm$^2$ compared to 40 for Si super-junction MOSFETs and 640 for SiC. The performance advantage of normally-off power GaN-on-Si devices has been demonstrated for the first time by HRL Laboratories in 2009 with a 5A/1100V GaN-on-Si device [2]. Fig. 3 shows the trend line for GaN power transistors (both normally-on and normally-off) from the published literature. The inset of the figure shows the dependence of $V_B$ on the gate-to-drain spacing of devices fabricated on the same wafer. HRL's normally-off devices have a $V_B^2/R_{on,sp}$ ratio of >260MW/cm$^2$ and a breakdown field of 90V/µm, which are among the highest in the literature, respectively.

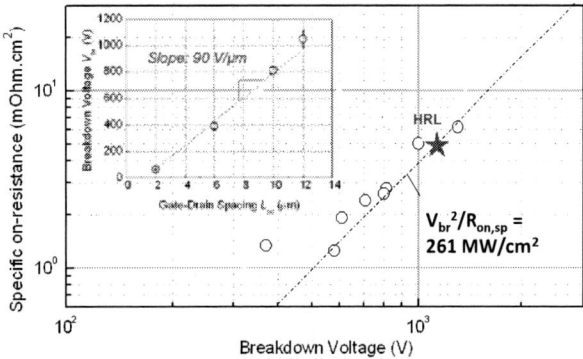

Figure 3. Relationship between specific resistance and breakdown voltage showing GaN power switch performance. Inset: Breakdown voltage dependence on gate-to-drain spacing.

### C. Low dynamic on-resistance

Hysteresis is frequently observed in III-V insulating gate FETs where the gate-insulator/semiconductor interface has a high trap density. A small hysteresis is very important for stable operation of the HFETs. Furthermore, GaN HFETs usually show a degradation of the on-resistance ($R_{on}$) when the devices are switched from the off-state to the on-state, especially at high-voltage operation. The dynamic $R_{on}$ degradation is attributed to field-assisted electron trapping in the drift region between the gate and the drain, and in the semiconductor bulk [3]. The dynamic Ron ($R_{on,dynamic}$) degradation has been a major obstacle for GaN FETs to achieve high-efficiency power switching at high voltages. HRL devices exhibit low hysteresis in the I-V characteristics and low $R_{on,dynamic}$. In the HRL device, we employed a field-shaping structure, which enabled a dynamic_Ron/Static_Ron ratio of <1.2 at 350 V switching. Fig. 4 compares this ratio for HRL devices with commercial GaN and Si MOSFET parts [4].

### D. Nano-second switching

GaN HFET properties, such as high mobility and high electron velocity allow it to operate at more than 10x the switching frequency of Si devices with the same switching loss. A 10x higher frequency proportionately reduces the size

of the passive components, resulting in smaller, lighter and cheaper systems. HRL has achieved switching times of <5 ns with 6A/600V devices switching at 300V in a boost converter [5]. Switching time as small as 1.2ns is seen in Fig. 5 for switching at 400V. This corresponds to a switching speed of 325V/ns, which is the highest reported for 600V power switches [6].

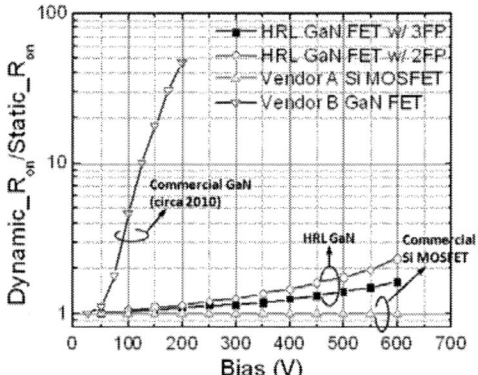

Figure 4. Ratio between dynamic on-resistance and static on-resistance at varied switching voltages for different devices.

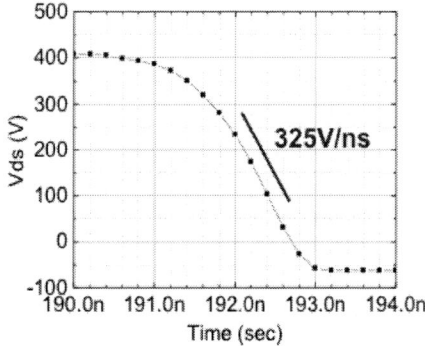

Figure 5. Ultra-fast turn-on with HRL GaN-on-Si switches showing a τon of 1.2ns and a slew rate of 325V/ns for switching at 400V.

### E. Ability for Integration

Due to the lateral device layout of GaN-on-Si HFETs, integration of multiple devices with different functions can be achieved on the same chip. This can be accomplished with the same semiconductor fabrication techniques used in CMOS IC fabrication. It is worth noting that, due to its high critical electric field (10x higher than silicon), high voltage lateral GaN devices can be fabricated with a relatively small drift region (6-10µm for 600V blocking). This approach is not practical with silicon power devices because the area required would be 10x larger as a results of the much lower critical field.

By eliminating the package inductance of discrete components, and minimizing the inductance in both the gate and power switching loops, a monolithically integrated phase leg (or ½ bridge) will realize the full potential of GaN switches and achieve high efficiency circuits with ultra-high slew rates and low voltage overshoot. Furthermore, an integrated phase leg with the gate driver is also attractive

978-1-4673-6145-3/13 $31.00 © 2013 IEEE

from the thermal point of view. Thermal balancing, as well as compatibility between the devices is accomplished with monolithic integration of devices made with the same GaN material.

Fig. 6 shows the concept circuit schematic of the GaN phase leg integrating the upper and lower sides of the half bridge with the gate driver transistor. As discussed later in the paper, no freewheeling diode is used with the GaN topology. For comparison, a conventional phase leg with discrete silicon IGBTs is shown with the required freewheeling diode.

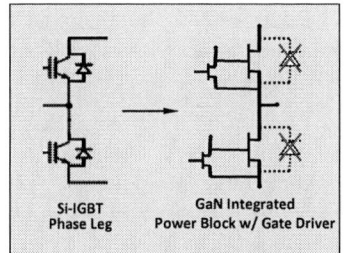

Figure 6. Schematic of a conventional discrete Si IGBT phase leg with connected freewheeling diodes vs. the monolithic integrated GaN phase leg with integrated gate driver and no freewheeling diodes.

### F. Elimination of the free-wheeling diode

High power motor drive inverter modules, such as those used in electric vehicles, require a freewheeling diode in parallel with the silicon IGBT switch. To improve performance, low-loss SiC diodes have replaced slower, less efficient silicon PiN diodes in such inverter modules. The GaN switch does not require the use of a freewheeling diode, therefore eliminating the loss and the added cost of the diode used in silicon-based inverter modules. In the third quadrant of the I-V operation (negative I and V), normally-off GaN switches act like a diode with a turn-on voltage equal to ($V_{GS} - V_T$) as shown in Fig. 7. The GaN channel turns on for $V_{DS}$ more negative than $V_{GS}-V_T$. For $V_{GS}$ greater than $V_T$, the GaN switch acts as a resistor in both the first and third quadrants of the I-V operation. This feature eliminates the need for freewheeling diodes in inverters and in synchronous converters. To test this principle, HRL has demonstrated the operation of both inverters and synchronous converters without freewheeling diodes [5].

### III. HIGH EFFICIENCY CIRCUITS

Demonstration of the full potential of GaN requires a paradigm shift in circuit design and implementation. Significant reduction of parasitics in the power loop as well as between the gate driver and the power stage is required to achieve nano-second speed switching with acceptable voltage overshoot and switching losses. Conventional power module designs will limit the performance of GaN-based power electronics due to their relatively large interconnect parasitics, and low level integration. A shift to monolithic integration of

the GaN phase leg, and possibly also of the gate driver may be necessary to achieve high efficiency/high frequency power management as described in the previous section. New circuit topologies and implementation schemes will emerge to take advantage of the high performance of the GaN devices. For example, new PoL power supply designs – enabled with higher frequency switching and higher bandwidth – could open the door for higher functionality systems with faster response and reduced losses.

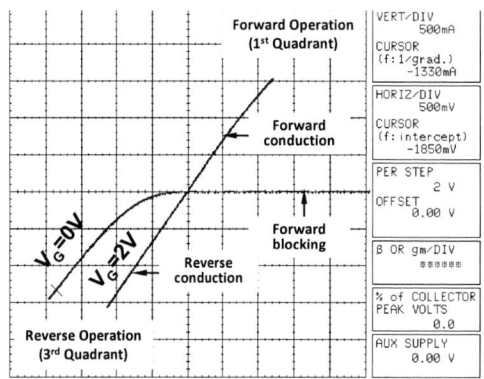

Figure 7. I-V operation of normally-off GaN switches showing behavior in the first and third quadrant. In the first quadrant the switch operates in a conducting or a blocking mode, and in the third quadrant the switch operates as a diode.

Boost converters were constructed at HRL using surface mount devices and high speed gate drivers. Packaging was done on a ceramic substrate, and special care was taken in the circuit construction to minimize the distance between components, and reduce interconnect parasitics to reduce parasitic losses. The first boost converter was operated at 1MHz. The measured boost converter efficiency versus output power for a range of output voltages is show in Fig. 8. A maximum efficiency of 95% was achieved with 425W of output power and an output voltage of 360V [7]. The converter demonstrated a power density of ~ 180W/in$^3$.

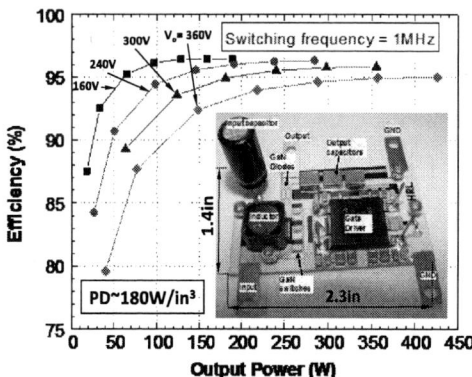

Figure 8. Efficiency vs. output power for a GaN-based 1MHz boost converter at various switching voltages. Efficiency greater than 95% is maintained up to 430W of output power. The inset shows the converter module with dimensions equivalent to a power density of 180W/in$^3$.

A larger power boost-converter was constructed using similar integration techniques. The hard-switched boost converter was operated at 500 kHz. The switching waveform of the converter and the switching energy versus current are shown in Fig. 9 and 10, respectively. The measured boost converter efficiency was 96% at 2kW (400V/10A) and a 50% duty cycle. The switching energy ($E_{on}+E_{off}$) was 27µJ at 300V/10A switching. This efficiency, power, and switching frequency combination is among the highest reported in the literature for kW-scale DC-DC converters, demonstrating the capability of GaN power devices for high speed switching.

## IV. CONCLUSION

GaN power electronics promise significantly lower conduction and switching loss compared to silicon devices. To realize the benefits of fast switching GaN devices, a paradigm shift in circuit design and implementation is required. Significant reduction of parasitics is needed to achieve the nano-second switching speeds promised with GaN. Several challenges still face the technology, both from the performance and reliability points of view. Acceptance in the application will ultimately be determined by the ability to manufacture GaN devices in large volume, at a competitive price compared to silicon, while maintaining high performance and reliability.

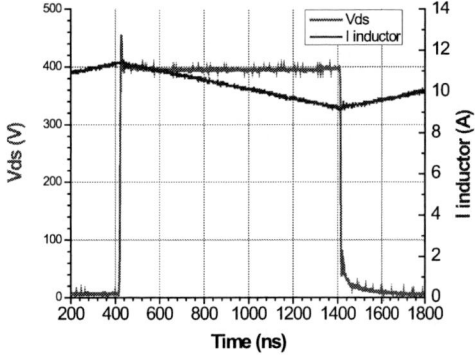

Figure 9. 2kW (10A/400V) 2:1 GaN boost converter waveform showing nano-second switching and small overshoot. At 50% duty cycle and switching frequency of 500kHz the converter had an efficiency of 96%.

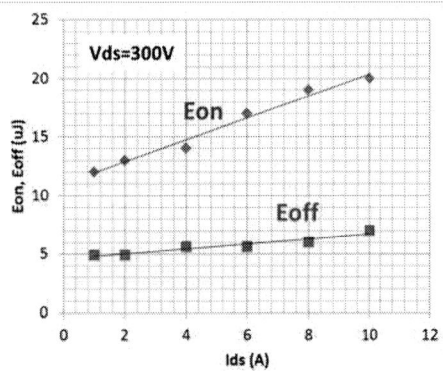

Figure 10. Switching energy of GaN boost converter vs. current at 300V switching

## V. ACKNOWLEDGEMENTS

This work was partially funded by the ADEPT program of Advanced Research Projects Agency – Energy (ARPA-E), U.S. Department of Energy, program managers Rajeev Ram and Timothy Heidel. The authors would also like to acknowledge General Motors and the Boeing Company for their continued support.

## VI. REFERENCES

[1] "Normally-off GaN-on-Si MISFET with 600-V blocking capability at 200 °C." Rongming Chu ; D. Brown, D. Zehnder, Xu Chen, A. Williams, R. Li, M. Chen, S. Newell, K. Boutros. 24th International Symposium on Power Semiconductor Devices and ICs (ISPSD), 2012, Page(s): 237- 240.

[2] "Normally-off 5A/1100V GaN-on-silicon device for high voltage applications." K.S. Boutros, S. Burnham, D. Wong, K. Shinohara, B. Hughes, D. Zehnder, C. McGuire. IEEE International Electron Devices Meeting (IEDM), 2009. Page(s): 1- 3.

[3] "High power AlGaN/GaN HFET with a high breakdown voltage of over 1.8 kV on 4 inch Si substrates and the suppression of current collapse." N. Ikeda, S. Kaya, Li Jiang, Y. Sato, S. Kato, S. Yoshida. 20th International Symposium on Power Semiconductor Devices and IC's, 2008. Page(s): 287- 290.

[4] "1200-V Normally Off GaN-on-Si Field-Effect Transistors With Low Dynamic on–Resistance." Rongming Chu, A. Corrion, M. Chen, R. Li, D. Wong, D. Zehnder, B. Hughes, K. Boutros. Electron Device Letters, IEEE. Volume: 32, Issue: 5, 2011. Page(s): 632- 634.

[5] "GaN HFET switching characteristics at 350V/20A and synchronous boost converter performance at 1MHz." B. Hughes, J. Lazar, S. Hulsey, D. Zehnder, D. Matic, K. Boutros. Twenty-Seventh Annual IEEE Applied Power Electronics Conference and Exposition (APEC), 2012. Page(s): 2506 - 2508.

[6] "Normally-Off GaN-on-Si Transistor Enabling Nanosecond Power Switching at One Kilowatt." Rongming Chu, Brian Hughes, Mary Chen, David Brown, Ray Li, Sameh Khalil, Steve Chen, Adam Williams, Austin Garrido, Marcel Musni, and Karim Boutros. Submitted for publication at the 2013 IEEE Device Research Conference.

[7] "A 95% efficient normally-off GaN-on-Si HEMT Hybrid-IC boost-converter with 425W output power at 1MHz," B. Hughes, Y.Y. Yoon, D.M. Zehnder, K.S. Boutros. Digest of the 2011 IEEE Compound Semiconductor Integrated Circuit Symposium, Waikolowa, HI, pp. 154-6, 2011.

# Prospective for Nanowire Transistors

## Jean-Pierre COLINGE[1] and Sang H. DHONG[2]

<table>
<tr><td>1: TSMC</td><td>2: TSMC</td></tr>
<tr><td>168, Park Ave. 2, Hsinchu Science Park</td><td>8, Li-Hsin Rd. 6, Hsinchu Science Park</td></tr>
<tr><td>Hsinchu, Taiwan 300-75, R.O.C.</td><td>Hsinchu, Taiwan 300-78, R.O.C.</td></tr>
<tr><td>jcolinge@tsmc.com</td><td>sang_dhong@tsmc.com</td></tr>
</table>

*Abstract* - **The multigate nanowire FET architecture allows for ultimate short-channel control and push Moore's law down to sub-5nm gate lengths. This paper reviews nanowire transistor device physics as well as circuit prospects in the fields of CMOS logic, memory, analog, RF and integrated sensor applications.**

## I. INTRODUCTION

The multigate (double-, triple- and quadruple- gate) architecture allows one to improve the control of carriers in the channel by the gate and to reduce short-channel effects. The first demonstration of a reduction of short-channel effects dates back to a 1984 paper predicting lower $V_{TH}$ roll-off in double-gate SOI transistors than in bulk MOSFETs.[1] Three years later, "volume inversion", where full inversion of the bulk of a thin silicon film is achieved, was predicted.[2] Volume inversion was experimentally observed in gate-all-around (GAA) transistors in 1990.[3]

The first triple-gate silicon nanowire "quantum wire" MOSFET dates back to 1995, reporting low-temperature conductance oscillations caused by quantum confinement.[4] Improved triple gate "3+" gate devices include the Π-gate MOSFET and the Ω-gate MOSFET.[5-6]

GAA Nanowire (GAANW) transistors can be fabricated either horizontally or vertically. [7,8,9,10] and can be made using III-V materials.[11,12,13] One decisive advantage of the nanowire architecture is that it enables the fabrication of junctionless transistors using either SOI or bulk-like approaches.[14,15,16] Figure 1 presents a overview of the multigate FET pantheon.

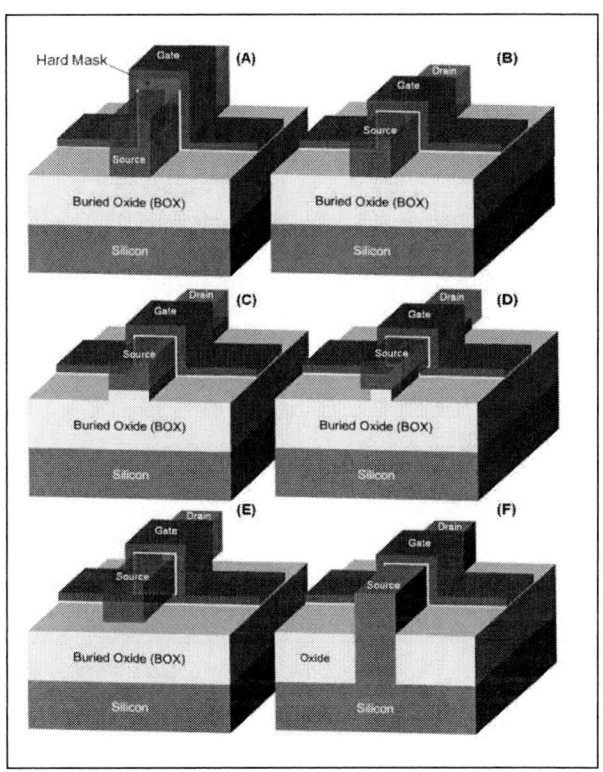

FIGURE 1. Schematics of multigate FET configurations: (A) Double-gate FinFET. (B) Triple-gate (or tri-gate) MOSFET. (C) Π-gate MOSFET. (D) Ω-gate MOSFET. (E) GAA nanowire MOSFET. (F) Bulk tri-gate MOSFET.[14]

## II. MULTIGATE FET DEVICE PHYSICS

Short-channel effects can be minimized by improving the control of the gate over the channel region. This control can be derived from Poisson's equation. A MOSFET is free of short-channel effects if the gate length is at least 4 to 6 times larger than the "natural length" of the device.[17] The natural length, $\lambda_n$, represents the penetration of field lines from the source and drain in the channel region. Assuming a square device cross-section (Figure 1) for simplicity, the natural length is given by the following expression:

$$\lambda_n = \sqrt{\frac{\varepsilon_{si}}{n\,\varepsilon_{ox}} t_{ox} t_{si}}$$

where $n$ is the equivalent number of gates ($n$=2 for a double gate device, $n$=3 for a trigate device, $n$=4 for a GAANW transistor).[18,19]

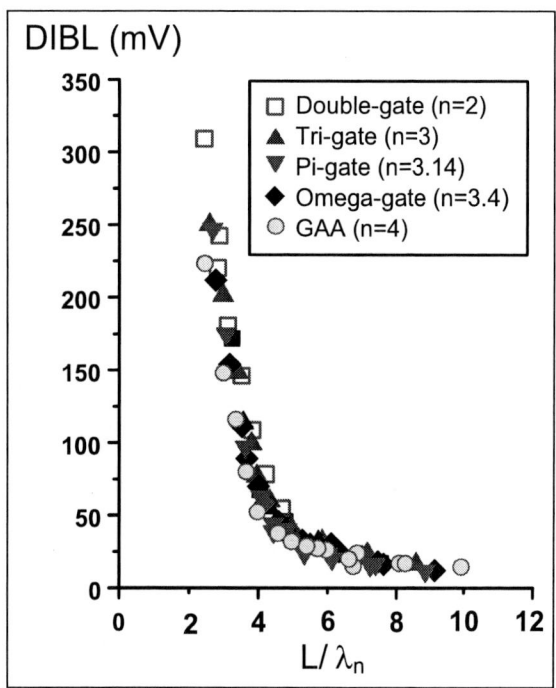

FIGURE 2. DIBL vs. gate length, $L$, normalized to $\lambda_n$ for different gate architectures.[18]

The universality of the natural length concept is demonstrated in Figure 2 for drain-induced barrier lowering (DIBL). Similar curves can be obtained for the subthreshold slope variation vs. $L/\lambda_n$, [20]. The natural length concept of has been verified experimentally in a variety of multigate transistor structures.[21,22] The best electrostatic control is obtained in GAA devices with a cylindrical channel, which is the most widely accepted structure for vertical transistor technology. The reduction of DIBL and subthreshold slope (SS) are extremely important for low-supply voltage circuits, as it improves $I_{Dsat}/I_{OFF}$ and $I_{EFF}/I_{ON}$.

## III. THE GAA NANOWIRE MOSFET

GAANW FETs deliver performance that is compatible with ITRS requirements. Key processing challenges are the reduction of S&D resistance and fabrication of smooth wires with good diameter control. The performance of unstrained silicon GAANW FETs are shown in Table 1. One can note, in particular, the good SS and DIBL values achieved. Strain can be added for enhanced performance, and mobilities of 1200 and 2000 cm$^2$/Vs are predicted for 2GPa strain levels for electrons and holes, respectively.[23] It is worth noting, though, that in small-diameter and short-channel devices, drain current may be limited by the density of states (number of subbands) and carrier ballisticity rather than by mobility.[24,65] In terms of AVt variability, nanowire transistors rank better than bulk FETs, FinFETs and FDSOI.[25]

|  | Ref. 26 | Ref.27 | Ref. 27 | Ref. 22 |
|---|---|---|---|---|
| Polarity | nMOS | nMOS | pMOS | nMOS* |
| NW diameter (nm) | 10 | 7 | 6 | 8 |
| Gate length (nm) | 8 | 50 | 50 | 20 |
| EOT (nm) | 4 | 3 | 3 | 1.2 |
| VDD (V) | 1.2 | 1 | -1 | 0.9 |
| $I_{ON}$ (mA/mm) @ $I_{OFF}$ = 100nA/mm | 3500 | 2800 | 3200 | - |
| $I_{ON}$ (mA/mm) @ $I_{OFF}$ = 10nA/mm | 3000 | 2100 | 2800 | - |
| $I_{ON}$ (mA/mm) @ $I_{OFF}$ = 1nA/mm | 2500 | 1500 | 2500 | 1200 |
| SS (mV/dec) | 75 | 67 | 64 | 80 |
| DIBL (mV/V) | 22 | 6 | 6 | 50 |

TABLE 1: Performance of silicon GAANW transistors (* $\Omega$-gate).

GAANW transistors can be either horizontal or vertical. Horizontal devices have a layout that is compatible with Fin/trigate FETs and can be stacked in parallel arrangements that increase current drive per real estate area.[28,29,30,31]. The vertical architecture, which is already used with polysilicon channels for NAND flash memory, is more challenging for logic since it requires re-design. It offers, on the other hand, promises in terms of co-integration of different

978-1-4673-6145-3/13 $31.00 © 2013 IEEE       292

semiconductor materials, high gate density and 3D integration (Figure 3).[32,33]

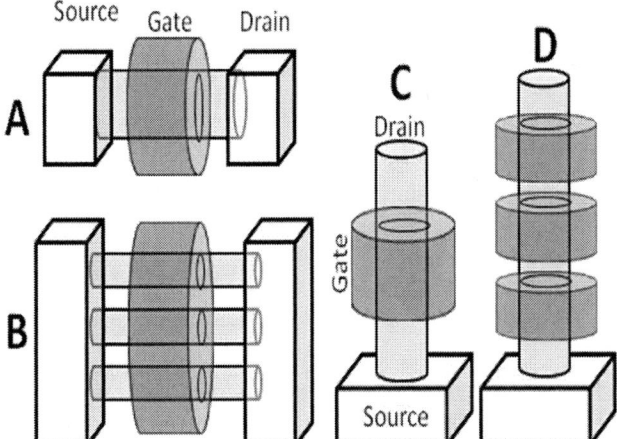

FIGURE 3. GAANW transistors. A: Horizontal, B: Multiple nanowires in parallel, C:Vertical, D: Multiple gates in series.

## IV. CIRCUITS: LOGIC, SRAM, ANALOG AND RF

As in the case of conventional FETs, GAANW design can be optimized for performance and power consumption. Several studies suggest that these devices can meet ITRS projections for 16nm gate length and below.[34]

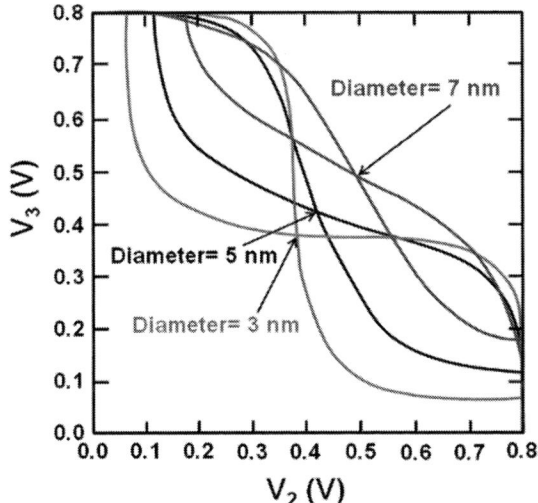

FIGURE 4. SRAM butterfly curves for a junctionless GAANW SRAM cell with L=10nm and different NW diameters.[35]

Nanowires are particularly suited to CMOS SRAM cells. Because of their small DIBL, SS and body effect, GAANW SRAM cells can achieve large SNM values. Decreasing the diameter of the nanowire reduces short-channel effects and, as a result, improves the SNM (Figures 4-5)

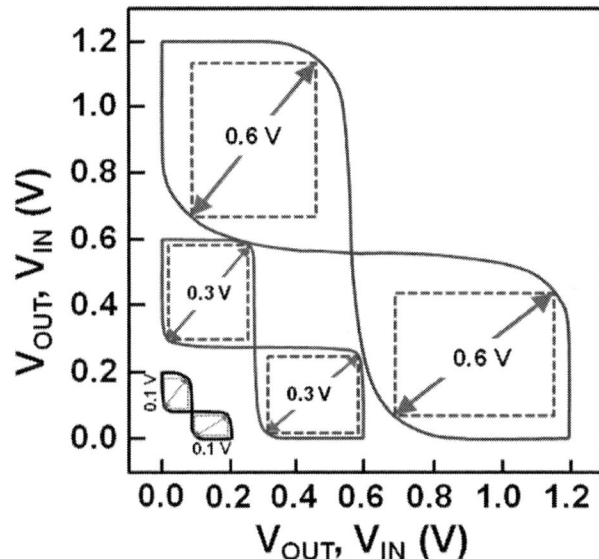

FIGURE 5. SRAM butterfly curves for a GAANW SRAM cell with $L_G$=350 nm, a NW diameter of 7nm and EOT=5 nm,, for supply voltages of 1.2V, 0.6V and 0.2V.[36]

From an analog device standpoint the excellent control of the channel by the GAA/multigate architecture yields a low output conductance, a large Early voltage, and a large $g_m/I_D$ ratio, all leading to excellent performance, especially in low-power operation mode.[37,38,39]
BSIMSOI modeling of low-power nMOS and pMOS vertical GAANW FETs with 10 nm channel length biased with $V_{DD}$ = 0.5 V predict 36 THz and 25 THz unity-current-gain cutoff frequencies, and 120 THz and 100 THz maximum frequency of oscillations, respectively. Using these FETs a single-stage CMOS GAANW amplifier dissipating 5µW power and providing 5THz bandwidth with a voltage gain of 16 has been designed. The amplifier has a linear output voltage swing of 0.5V, and a distortion better than 3% from a 1.8V power supply and a 20aF capacitive load. The 2nd and 3rd order harmonic distortions of

the amplifier are -40dBm and -52dBm, respectively, and the 3rd order intermodulation is -24dBm for a two-tone input signal with 10mV amplitude and 10GHz frequency spacing. [40,41]. The linearity of GAANW FETs can be further improved using a junctionless architecture. [42]

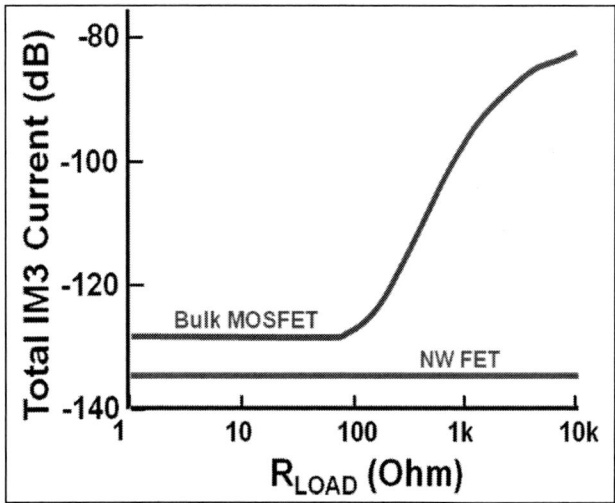

FIGURE 6. Third-order intermodulation distortion (IM3) current in a junctionless GAA nanowire transistor and a bulk MOSFET.[42]

Further improvement in linearity can be obtained in ultrascaled GAANW FETs operating in ballistic transport mode.[43] A high degree of linearity is maintained in RF operation. [44,45,46,47]

## V. FLASH MEMORY

Using a simple modification of the gate insulator, GAA NW FETs can readily be transformed into high-density SONOS NAND flash memory arrays. These have been demonstrated using horizontal nanowires [48,49], vertical nanowires with one or two gates per nanowire [32,33], and vertical nanowires with a single gate but storing 2 bits per gate.[50]

## VI. NANOWIRE TUNNEL FETS (TFETS)

The reduction of power supply, and therefore, of power consumption in CMOS circuits is limited by the unability to scale the SS below 60 mV/dec at room temperature in a MOSFET. Band-to-band tunneling (BTBT) at the source junction of a lubistor (gated P-i-N diode) can be used to achieve SS below 60 mV/dec. Since extremely sharp control of the electric field and carrier concentrations are required for achieving low SS values, the GAA nanowires has become the preferred architecture for the fabrication of tunnel FETs with sub-60 mV/dec subthreshold slopes [51,52,53,54]. Improvement of SS with reduction of nanowire diameter can be clearly seen in Figure 7.

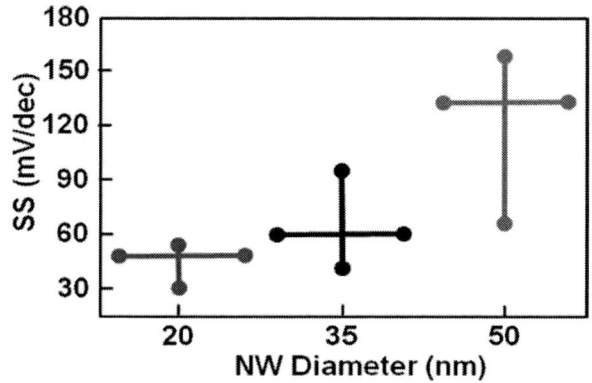

FIGURE 7. Subthreshold slope in silicon GAA nanowire TFETs as a function of nanowire diameter.[54]

Impact-ionization (II) MOSFETs (IMOS) can also be used to obtain a SS below 60 mV/dec. In classical SOI MOS devices, a drain voltage of at least 3-4 V is usually needed to generate enough impact ionization to observe this phenomenon. It has, however, been recently observed that a drain voltage of only 1.75V (i.e. approximately 1.5x the bandgap energy) can give rise to a sub-60 mV/dec in junctionless nanowire transistors.[55,56,57]. Unlike BTBT-based TFETs, IMOS devices deliver at least as much ON current than MOSFETs. Other TFET devices based on resonant barrier tunneling rather than BTBT or II and utilizing the excellent gate control provided by the GAANW structure, have been proposed. Such devices can, in theory, operate at low drain bias and deliver a drain current comparable to that of a MOSFET.[58]

## VII. Nanowire sensing devices

For the sake of completion one should mention that the large surface-area/volume ratio of nanowires makes them excellent candidates for chemical sensing and biosensor applications. Such devices can be combined with nanowire logic and mixed-mode devices for medical "more-than-Moore" applications.[59,60,61, 62,63]

## VIII. Conclusions

The multigate/GAA silicon nanowire FET architecture has the potential of keeping Moore's law "rolling" down to sub-5nm gate lengths. Well behaved devices with a gate length of 3nm have been demonstrated by ab-initio simulation.[64] The GAA architecture offers optimal channel control by the gate, resulting in low short-channel effects (excellent DIBL, SS, $gm/I_D$, $g_D$) which are critical for LVLP operation of logic, memory, analog and RF circuits. GAANW FETs can also be used for flash memory and sensing applications.

## References

[1] T. Sekigawa and Y. Hayashi, "Calculated threshold-voltage characteristics of an XMOS transistor having an additional bottom gate", Solid-state Electronics, **27**, pp. 827-828, 1984

[2] F. Balestra, S. Cristoloveanu, M. Benachir, J. Brini, T. Elewa, "Double-gate silicon-on-insulator transistor with volume inversion: A new device with greatly enhanced performance", IEEE Transactions On Electron Devices, **8**, pp. 410-412, 1987

[3] J.P. Colinge, H.H. Gao, A. Romano-Rodriguez, H.E. Maes, C. Claeys, "Silicon-on-insulator gate-all-around device", *Technical Digest of International Electron Device Meeting (IEDM)*, pp. 595-598, 1990

[4] X. Baie, J.P. Colinge, V. Bayot, E. Grivei, "Quantum-wire effects in thin and narrow SOI MOSFETs", *Proc. IEEE International SOI Conference*, 1995, pp. 66-67

[5] J.T. Park, J.P. Colinge and C. H. Diaz, "Pi-gate SOI MOSFET", IEEE Electron Device Letters, **22**, pp. 405-406, 2001

[6] F.-L. Yang, D.-H. Lee, H.-Y. Chen, C.-Y. Chang, S.-D. Liu, C.-C. Huang, T.-X. Chung, H.-W. Chen, C.-C. Huang, Y.-H. Liu, C.-C. Wu, C.-C. Chen, S.-C. Chen, Y.-T. Chen, Y.-H. Chen, C.-J. Chen, B.-W. Chan, P.-F. Hsu, J.-H. Shieh, H.-J. Tao, Y.-C. Yeo, Y. Li, J.-W. Lee, P. Chen, M.-S. Liang, C. Hu, "5nm-Gate Nanowire FinFET", *Digest of Technical Papers, Symposium on VLSI Technology*, pp.96-197, 2004

[7] N. Singh, A. Agarwal, K.L. Bera, T.Y. Liow, R. Yang, S.C. Rustagi, C.H. Tung, R. Kumar, G.Q. Lo, N. Balasubramanian, D.L. Kwong, "High-performance fully depleted silicon nanowire (diameter < 5 nm) gate-all-around CMOS devices", IEEE Electron Device Letters, **27**, p. 383-386, 2006

[8] Dong-Il Moon, Sung-Jin Choi, J.P. Duarte, Yang-Kyu Choi, "Investigation of Silicon Nanowire Gate-All-Around Junctionless Transistors Built on a Bulk Substrate", IEEE Transactions on Electron Devices, **60-4**, pp. 1355-1360, 2013

[9] V. Schmidt, H. Riel, S. Senz, S. Karg, W. Riess, U. Gösele, Realization of a Silicon Nanowire Vertical Surround-Gate Field-Effect Transistor, Small, **2**, pp. 85-88, 2006

[10] G. Larrieu, X.-L. Han, "Vertical nanowire array-based field effect transistors for ultimate scaling", Nanoscale, **5**, pp. 2437-2441, 2013

[11] K. Tomioka, M. Yoshimura, T. Fukui, A III-V nanowire channel on silicon for high-performance vertical transistors, Nature, **488**, pp. 189-192, 2012

[12] J. J. Gu, X. W. Wang, H. Wu, J. Shao, A.T. Neal, M.J. Manfra, R.G. Gordon, P.D. Ye, "20-80nm Channel Length InGaAs Gate-all-around Nanowire MOSFETs with EOT=1.2nm and Lowest SS=63mV/dec", *Technical Digest of International Electron Device Meeting (IEDM)*, pp. 634-637, 2012

[13] K. S. Im, Y. W. Jo, K. W. Kim, D. S. Kim, H. S. Kang, C. H. Won, R. H. Kim, S. M. Jeon, D. H. Son, Y. M. Kwon, J. H. Lee, J. H. Lee, S. Cristoloveanu, "Comparison of Junctionless GaN FinFET and AlGaN/GaN FinFET", *Proceedings EUROSOI Conference*, 2013

[14] J.P. Colinge, Chi-Woo Lee, A. Afzalian, N. Dehdashti Akhavan, Ran Yan, I. Ferain, P. Razavi, B. O'Neill, A. Blake, M. White, A.M. Kelleher, B. McCarthy, R. Murphy, Nanowire transistors without junctions", Nature Nanotechnology, **5-3**, pp. 225-229, 2010

[15] Ming-Hung Han, Chun-Yen Chang, Hung-Bin Chen, Jia-Jiun Wu, Ya-Chi Cheng, Yung-Chun Wu , "Performance Comparison Between Bulk and SOI Junctionless Transistors", IEEE Electron Device Letters, 34-2, pp. 169-171, 2013

[16] H.X. Guo, X. Zhang, Z. Zhu,E.Y.J. Kong, Y.-C. Yeo, "Junctionless Π-gate transistor with indium gallium arsenide channel", Electronics Letters, 49-6 , pp. 402-404, 2013

[17] R.H. Yan, A. Ourmazd, K.F. Lee, "Scaling the Si MOSFET: from bulk to SOI to bulk", IEEE Transactions on Electron Devices, 39, pp. 1704–1710, 1992

[18] I. Ferain, C.A. Colinge, J.P. Colinge, "Multigate transistors as the future of classical metal–oxide–semiconductor field-effect transistors", Nature, 479, pp. 310–316, 2011

[19] K.J. Kuhn, "Considerations for Ultimate CMOS Scaling", IEEE Transactions on Electron Devices, 59, pp. 1813-1828, 2012

[20 Chi-Woo Lee, Se-Re-Na Yun, Chong-Gun Yu, Jong-Tae Park, J.P. Colinge, "Device design guidelines for nano-scale MuGFETs", Solid-State Electronics, 51, pp. 505-510, 2007

[21] S. Bangsaruntip, G. M. Cohen, A. Majumdar, Y. Zhang, S. U. Engelmann, N. C. M. Fuller, L. M. Gignac, S. Mittal, J. S. Newbury, M. Guillorn, T. Barwicz, L. Sekaric, M. M. Frank, J. W. Sleight, "High Performance and Highly Uniform Gate-All-Around Silicon Nanowire MOSFETs with Wire Size Dependent Scaling", *Technical Digest of International Electron Device Meeting (IEDM)*, pp. 297-300, 2009

[22] S. Barraud, R. Coquand, M. Cassé, M. Koyama, J.-M. Hartmann, V. Maffini-Alvaro, C. Comboroure, C. Vizioz, F. Aussenac, O. Faynot, T. Poiroux, "Performance of Omega-Shaped-Gate Silicon Nanowire MOSFET With Diameter Down to 8 nm", IEEE Electron Device Letters, 33-11, pp. 1526-1528, 2012

[ 23 ] Y.M. Niquet, C. Delerue, C. Krzeminski, "Effects of Strain on the Carrier Mobility in Silicon Nanowires", Nano Letters, 12, pp. 3545-3550, 2012

[24] V.H.Nguyen, F. Triozon, F.D.R. Bonnet, Y.M. Niquet, "Performances of Strained Nanowire Devices: Ballistic Versus Scattering-Limited Currents", IEEE Transactions on Electron Devices, 2013, DOI 10.1109/TED.2013.2248734

[25] K.J. Kuhn, M.D. Giles, D. Becher, P. Kolar, A. Kornfeld, R. Kotlyar, S.T. Ma, A. Maheshwari, S. Mudanai, "Process Technology Variation", IEEE Transactions on Electron Devices, 58-8, pp. 2197-2208, 2011

[26 ] Y. Jiang, T.Y. Liow, N. Singh, L.H. Tan, G.Q. Lo, D.S.H. Chan, D.L. Kwong, "Performance Breakthrough in 8 nm Gate Length Gate-All-Around Nanowire Transistors using Metallic Nanowire Contacts", *Symposium on VLSI Technology Digest of Technical Papers*, pp. 34-35, 2008

[27] Yi Song, Qiuxia Xu,Jun Luo, Huajie Zhou, Jiebin Niu, Qingqing Liang, Chao Zhao, "Performance Breakthrough in Gate-All-Around Nanowire n- and p-Type MOSFETs Fabricated on Bulk Silicon Substrate", IEEE Transactions on Electron Devices, 59-7, pp. 1885-1890, 2012

[28] T. Ernst, C. Dupré, C. Isheden, E. Bernard, R. Ritzenthaler, V. Maffini-Alvaro, J.-C. Barbé, F. De Crecy, A.Toffoli, C. Vizioz, S. Borel, F. Andrieu, V. Delaye, D. Lafond, G. Rabillé, J.-M. Hartmann, M. Rivoire, B. Guillaumot, A. Suhm,P. Rivallin, "Novel 3D integration process for highly scalable Nano-Beam stacked-channels GAA (NBG) FinFETs with HfO2/TiN gate stack", *Technical Digest of International Electron Device Meeting (IEDM)*, 2006

[29] T. Ernst, Member, IEEE, E. Bernard, C. Dupré, A. Hubert, S. Bécu, B. Guillaumot, O. Rozeau, O. Thomas, P. Coronel, J.-M. Hartmann, C. Vizioz, N. Vulliet, O. Faynot, T. Skotnicki, S. Deleonibus, "3D multichannels and stacked nanowires technologies for new design opportunities in nanoelectronics", *IEEE International Conference on Integrated Circuit Design and Technology and Tutorial (ICICDT)*, pp. 265-268, 2008.

[30] L.K. Bera, H.S. Nguyen, N. Singh, T.Y. Liow, D.X. Huang, K.M. Hoe, C.H. Tung, W.W. Fang, S.C. Rustagi, Y.Jiang, G.Q. Lo, N. Balasubramanian, D.L. Kwong, "Three Dimensionally Stacked SiGe Nanowire Array and Gate-All-Around p-MOSFETs", *Technical Digest of International Electron Device Meeting (IEDM)*, pp. 339-342, 2006

[ 31 ] E. Dornel, T. Ernst, J.C. Barbé, J.M. Hartmann,V. Delaye, F. Aussenac, C. Vizioz,

S. Borel, V. Maffini-Alvaro, C. Isheden, J. Foucher, "Hydrogen annealing of arrays of planar and vertically stacked Si nanowires", Applied Physics Letters, 91, 233502 2007

[32] Y. Sun, H.Y. Yu, N. Singh, K.C. Leong, E. Quek, G.Q. Lo, D.L. Kwong, "Demonstration of Memory String with Stacked Junction-Less SONOS Realized on Vertical Silicon Nanowire" , *Technical Digest of International Electron Device Meeting (IEDM)*, pp. 223-226, 2011

[33] Y. Sun, H.Y. Yu, N. Singh, T.T. Le, E. Gnani, G. Baccarani, K.C. Leong, G.Q. Lo, D.L. Kwong, "Junction-Less Stackable SONOS Memory Realized on Vertical-Si-Nanowire for 3-D Application", *International Symposium on VLSI Technology, Systems and applications (VLSI-TSA), Proceedings of Technical Program*, pp.154-155, 2011

[34] Yuchao Liu, Ru Huang, Runsheng Wang, Jing Zhuge, Qiumin Xu, Yangyuan Wang, "Design Optimization for Digital Circuits Built With Gate-All-Around Silicon Nanowire Transistors", IEEE Trans. On Electron Devices, vol. 59, 2012, pp. 1844-1850

[35]   Yi-Bo Liao, Meng-Hsueh Chiang, Keunwoo Kim, Wei-Chou Hsu, "Assessment of structure variation in silicon nanowire FETs and impact on SRAM", Microelectronics Journal, **43**, pp. 300–304, 2012

[36]   N. Singh, K.D. Buddharaju, S.K. Manhas, A. Agarwal, S.C. Rustagi, G.Q. Lo, N. Balasubramanian, Dim-Lee Kwong, "Si, SiGe Nanowire Devices by Top–Down Technology and Their Applications", IEEE Transactions on Electron Devices, **55-11**, pp. 3107- 3118, 2008

[37] A. Vandooren, J. P. Colinge, D. Flandre, "Gate-All-Around OTA's for Rad-Hard and High-Temperature Analog Applications", IEEE Transactions on Nuclear Science, **46-4**, pp. 1242-1249, 1999

[38] C. Davanzzo Gomes dos Santos, M.A. Pavanello, J.A. Martino, "Analysis of silicon thickness reduction on analog parameters of GC GAA SOI transistors operating up to 300°C", ECS Transactions, **4-1**, pp. 283-291, 2006

[39] R. Trevisoli Doria, M.A. Pavanello, R. Doria Trevisoli, M. de Souza, Chi-Woo Lee, I. Ferain, N. Dehdashti Akhavan, Ran Yan, P. Razavi, Ran Yu, A.

Kranti, J.P. Colinge, "Junctionless Multiple-Gate Transistors for Analog Applications", IEEE Transactions On Electron Devices, **58-8**, pp. 2511-25187, 2011

[40] S. Hamedi-Hagh, Sooseok Oh, A. Bindal, Dae-Hee Park, "Design of Next Generation Amplifiers Using Nanowire FETs", Journal of Electrical Engineering & Technology, **3-4**, pp. 566-570, 2008

[41] S. Hamedi-Hagh, A. Bindal, "Spice Modeling of Silicon Nanowire Field-Effect Transistors for High-Speed Analog Integrated Circuits", IEEE Transactions on Nanotechnology, **7-6**, pp. 766-775,2008

[42] Tao Wang, Liang Lou, Chengkuo Lee, "A Junctionless Gate-All-Around Silicon Nanowire FET of High Linearity and Its Potential Applications", IEEE Electron Device Letters, **34-4**, pp.478, 480, 2013

[43] A. Razavieh, N. Singh, A. Paul, G. Klimeck, D. Janes, J. Appenzeller, "A New Method to Achieve RF Linearity in SOI Nanowire MOSFETs", *IEEE Radio Frequency Integrated Circuits Symposium (RFIC)*, pp. 1-4, 2011

[44] D. Ghosh, M.S. Parihar, A. Kranti, "Optimizing nanoscale MOSFET architecture for low power analog/RF applications", IEEE 5th International Nanoelectronics Conference (INEC), pp. 22-3, 2013

[45] Sunhae Shin, In Man Kang, Kyung Rok Kim, "Extraction Method for Substrate-Related Components of Vertical Junctionless Silicon Nanowire Field-Effect Transistors and Its Verification on Radio Frequency Characteristics", Japanese Journal of Appl. Phys., **51**, 06FE20, 2012

[46] Yi Song, Jun Luo, Xiuling Li, "Vertically stacked individually tunable nanowire field effect transistors for low power operation with ultrahigh radio frequency linearity", Appl. Phys. Lett. **101**, 093509, 2012

[47] Yuchao Liu, Ru Huang, Runsheng Wang, Jiaojiao Ou, Yangyuan Wang, "Improving Analog/RF Performance of Multi-gate Devices through Multi-dimensional Design Optimization with Awareness of Variations and Parasitics", *Technical Digest of International Electron Device Meeting (IEDM)*, pp. 339-342, 2012

[48] Sung-Jin Choi, Dong-Il Moon, J. P. Duarte, Sungho Kim, Yang-Kyu Choi, "A Novel Junctionless

All-Around-Gate SONOS Device with a Quantum Nanowire on a Bulk Substrate for 3D Stack NAND Flash Memory", *Symposium on VLSI Technology Digest of Technical Papers*, pp. 74-75, 2011

[49] Kyoung Hwan Yeo, Keun Hwi Cho, Ming Li, Sung Dae Suk, Yun-young Yeoh, Min-Sang Kim, Hyunjun Bae, Ji-Myoung Lee, Suk-Kang Sung, Jun Seo, Bokkyoung Park, Dong-Won Kim, Donggun Park, Won-Seoung Lee, "Gate-all-around Single Silicon Nanowire MOSFET with 7 nm width for SONOS NAND Flash Memory", *Symposium on VLSI Technology Digest of Technical Papers*, pp. 138-139, 2008

[50] Y. Sun, H.Y. Yu, N. Singh, K.C. Leong, G.Q. Lo, and D.L. Kwong, "Junctionless Vertical-Si-Nanowire-Channel-Based SONOS Memory With 2-Bit Storage per Cell", IEEE Electron Device Letters, **32-6**, pp. 725-727 , 2011

[51] A.M. Ionescu, H. Riel, "Tunnel field-effect transistors as energy-efficient electronic switches", Nature, **479**, pp. 329-337, 2011

[52] A. Vandooren, D. Leonelli, R. Rooyackers, K. Arstila, G. Groeseneken, C. Huyghebaert, Impact of process and geometrical parameters on the electrical characteristics of vertical nanowire silicon n-TFETs, Solid-State Electronics, vol. 72, pp.82-87, 2012

[53] G. Ramanathan, Z. X. Chen, N. Singh, K. Banerjee, S. J. Lee, "Vertical Si Nanowire n-type Tunneling FETs with Low Subthreshold Swing ($\leq 50$ mV/decade) at Room Temperature", IEEE Electron Device Letters, **32-4**, pp. 437-439, 2011

[54] G. Ramanathan, Z. X. Chen, N. Singh, K. Banerjee, S. J. Lee, "CMOS Compatible Vertical Silicon Nanowire Gate-All-Around p-type Tunneling FETs with $\leq$ 50 mV/decade Subthreshold Swing", IEEE Electron Device Letters, **32-11**, pp. 1504-1506, 2011)

[55] M. Singh Parihar, D. Ghosh, G.A. Armstrong, Ran Yu, P. Razavi, A. Kranti, "Bipolar effects in unipolar junctionless transistors", Applied Physics Letters, **101**, 093507, 2012

[56] M. Singh Parihar, D. Ghosh, G.A. Armstrong, A. Kranti , "Bipolar snapback in junctionless transistors for capacitorless dynamic random access memory", Appl. Physics Letters, **101**, 263503, 2012

[57] S. M. Lee, H. J. Jang, J. T. Park, "Analysis of subthreshold slope with substrate bias in Junctionless

multiple gate MOSFETs", *Proceedings EUROSOI Conference*, 2013

[58] O.Moldovan, F. Lime, B. Nae, B. Iñiguez, "A simple compact model for the junctionless Variable Barrier Transistor (VBT)", *Spanish Conference on Electron Devices (CDE)*, pp. 67-70, 2013

[59] R. Yu, Y. M. Georgiev, O. Lotty, B. McCarthy, N. Petkov, D. O'Connell, A. Nightingale, S. Das, J. D. Holmes, "Si Junctionless Transistor for Sensing Application: Subthreshold Region Sensor" , *Proceedings EUROSOI Conference*, 2013

[60] Kuan-I Chena, Bor-Ran Li, Yit-Tsong Chen, "Silicon nanowire field-effect transistor-based biosensors for biomedical diagnosis and cellular recording investigation", Nano Today, **6**, pp. 131-154, 2011

[61] P.R. Nair, M.A. Alam, "Design Considerations of Silicon Nanowire Biosensors", IEEE Transactions on Electron Devices, **54-12**, pp. 3400-3408, 2007

[62] R. Gautam, M. Saxena, R.S. Gupta, M. Gupta, "Numerical Model of Gate-All-Around MOSFET with Vacuum Gate Dielectric for Biomolecule Detection", IEEE Electron Device Letters, **33-12**, pp. 1756-1758, 2012

[63] Jieun Lee, Jin-Moo Lee, Jung Han Lee, Won Hee Lee, Mihee Uhm, Byung-Gook Park, Dong Myong Kim, Yong-Joo Jeong, Dae Hwan Kim,"Complementary Silicon Nanowire Hydrogen Ion Sensor With High Sensitivity and Voltage Output", IEEE Electron Device Letters, **33-12**, pp. 1768-1770, 2012

[64] L. Ansari, B. Feldman, G. Fagas, C. Martinez Lacambra, M.G. Haverty, K. J. Kuhn, S. Shankar, J.C. Greer, "First Principle-based Analysis of Single-Walled Carbon Nanotube and Silicon Nanowire Junctionless Transistors", to be published in IEEE Transactions on Nanotechnology (TNANO) , on the Web at:

http://arxiv.org/ftp/arxiv/papers/1303/1303.3755.pdf

[65] Yeonghun Lee, K. Kakushima, K. Shiraishi, K. Natori, H. Iwai," Trade-off between density of states and gate capacitance in size-dependent injection velocity of ballistic n-channel silicon nanowire transistors" , Applied Physics Letters, **97**, 032101, 2010

# Interconnect and Package Design of a Heterogeneous Stacked-Silicon FPGA

Ephrem Wu, Khaldoon Abugharbieh, Bahareh Banijamali, Suresh Ramalingam, Paul Wu, Chris Wyland
Xilinx, Inc.
San Jose, USA

*Abstract*—This paper reviews the interconnect and package design of a heterogeneous stacked-silicon FPGA. Up to five dice from two die types are mounted on a passive silicon interposer. A hardware- and software-scalable FPGA family can be created by mixing different combinations of these two die types. The FPGA, inside a low-temperature co-fired ceramic (LTCC) package, consists of two silicon die types—up to three FPGA ICs having a total of seventy-two 13.1-Gb/s transceivers (943.2 Gb/s full-duplex) and up to two GTZ ICs having up to sixteen 28.05-Gb/s transceivers (448.8 Gb/s full-duplex). Two types of interconnects are discussed: those joining the ICs through wires in the silicon interposer, and those connecting the 28-Gb/s transceivers through TSVs in the interposer to the package balls. An end-to-end 28.05-Gb/s channel simulation is discussed in the context of silicon interposer resistivity as well as package material and stack-up. In addition, this paper reviews 3D thermal-mechanical analysis confirming the reliability of heterogeneous stacked silicon.

## I. INTRODUCTION

Compact and flexible solutions for emerging multi-100-Gb/s and 400-Gb/s optical line interfaces call for not only hardware-programmability but also high-speed electrical transceivers. This paper describes an FPGA design approach that addresses bandwidth and logic scalability, and in particular reviews interconnect and packaging design.

Figure 1 illustrates a 2x100-Gb/s optical line interface typical of router line card. Here, the FPGA has eight 28.05-Gb/s electrical transceivers interfacing two 100-Gb/s CFP2 or CFP4 optical modules over short PCB traces. The insertion loss is of these traces is specified by the CEI-28G-VSR standard. The 13.1-Gb/s transceivers connect with network processors. Figure 2 illustrates an FPGA with sixteen 28.05-Gb/s transceivers (GTZs) and up to 72 13.1-Gb/s transceivers (GTHs) for a 4x100-Gb/s line interface. For a 1x400Gb/s line interface, the sixteen 28.05-Gb/s transceivers can connect to a single 400-Gb/s optical module.

These FPGAs share only two unique IC designs, each implemented with a silicon process optimized for power and performance. In particular, the first IC is monolithic FPGA design with twenty-four 13.1-Gb/s transceivers and 290K logic cells. The second IC is an ASIC with eight 28.05-Gb/s transceivers without any FPGA elements. Figure 3 shows up to five of these ICs stacked in a co-planar fashion on a silicon interposer to addresses a range of applications that require different amounts of transceiver bandwidth and logic cells.

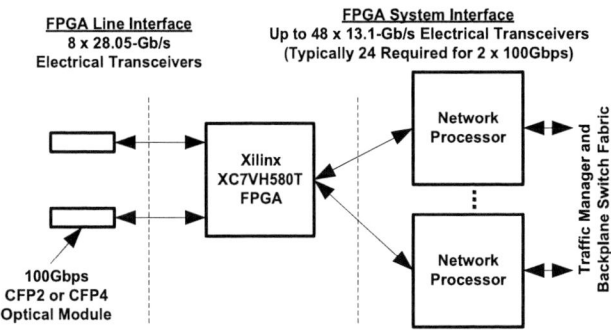

Figure 1    A 2x100-Gb/s Optical Line Interface with Eight 28.05-Gb/s CEI-28G-Very-Short-Reach (VSR) Electrical Transceivers

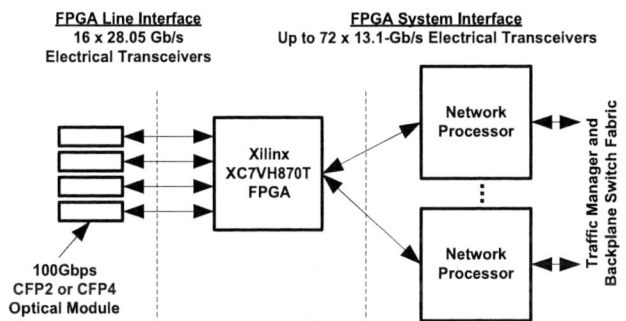

Figure 2    A 4x100-Gb/s Line Interface

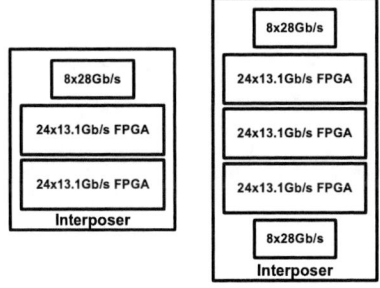

Figure 3    Up to Five ICs from Two Unique Designs Make Multiple FPGA Products Possible

## II. INTER-IC SIGNAL INTEGRITY

The ICs inside the FPGA package communicate amongst each other via fine-pitch metal wires inside a silicon interposer, and care must be taken to analyze the behavior of these wires to ensure inter-IC signal integrity. Figure 4 illustrates the cross-section of a heterogeneous stacked-silicon FPGA device. This FPGA device consists of three active-silicon ICs: one 8x28-

978-1-4673-6145-3/13 $31.00 © 2013 IEEE          299

Gb/s IC and two 24x13.1-Gb/s FPGA ICs. These three ICs communicate with each other with metal wires in the passive silicon interposer. There are approximately 4000 wires between the 8x28-Gb/s IC and the neighboring FPGA IC, and roughly 10,000 wires between the two FPGA ICs.

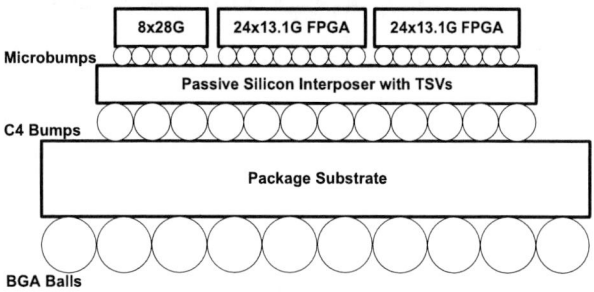

Figure 4    Cross-section of a Heterogeneous Stacked-Silicon FPGA

Inter-IC wires behave somewhere between on-chip silicon wires, which often can be analyzed with RC-only models, and printed-circuit board (PCB) wires, which are modeled as transmission lines. Figure 5a shows the cross-section of a parallel set of wires sandwiched between reference planes whereas there are no metal reference planes in Figure 5b. With an RLC wire model, Figure 6a and Figure 6b illustrate the waveform of a victim signal corresponding to models in Figure 5a and Figure 5b.

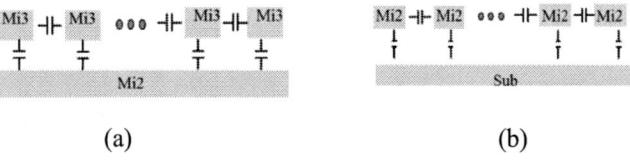

(a)                            (b)

Figure 5    Interposer Wire Modeling (a) With Ground Reference Planes (b) Without Ground Reference Planes

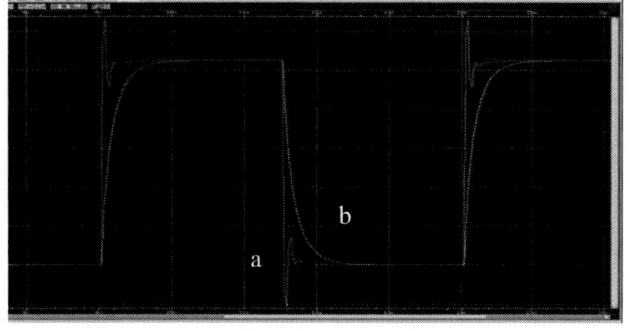

Figure 6    (a) Waveform without Ground Reference (b) Waveform with Ground Reference

Static timing analysis (STA) tools did not yet support RLC interconnect models during the development of this. Figure 7 shows the wire length distribution of over 4,000 wires between the 8x28-Gb/s IC and the FPGA IC. Clearly, an automated method such as STA was necessary for completing timing analysis productively. As a result, inter-IC wires were laid out to be tools-friendly. In addition to a ground reference plane, capacitive side shields were implemented due to the dense inter-IC routing pitch. The inter-IC router layers in the silicon interposer were assigned as follows, from top to bottom:

redistribution, signal, ground reference, and signal. Capacitive side shields were interleaved between inter-IC signals. Figure 8 illustrates signal waveforms with and without these side shields. Inter-IC timing calibration results between SPICE simulations using 3D RLC interconnect models and STA using RC-only interconnect models confirmed that the signaling was not in transmission-line mode and STA with RC-based interconnects would suffice.

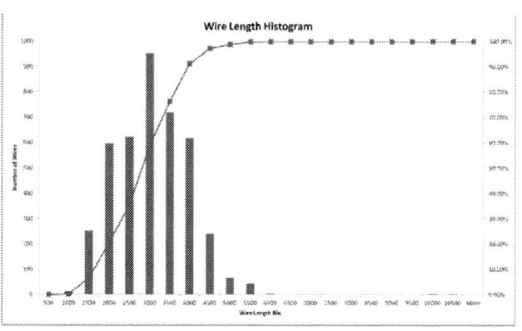

Figure 7    Wire-Length Distribution Between 8x28-Gb/s ICand FPGA IC

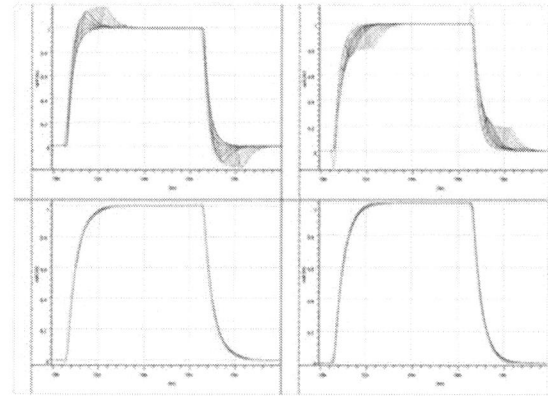

Figure 8    Top: Waveforms Showing Coupling without Side Shields. Bottom: Waveforms with Side Shields

III.    CROSS-SECTIONAL BANDWIDTH DENSITY

Cross-sectional bandwidth density (CBD) is the sum of ingress and egress bandwidth divided by the width of a cross-sectional area through which traffic enters and exits. In other words,

$$CBD = \frac{B_1 + B_2}{W}$$

where $B_1$ is the ingress bandwidth, $B_2$ is the egress bandwidth, and $W$ is the width of the cross-sectional area perpendicular to traffic flow. Figure 9 illustrates cross-sectional bandwidth density with a three-dimensional drawing. The block in Figure 9 in the context of the previous section represents a silicon interposer.

For a silicon interposer cross-section between two active top dice, the CBD is the sum of the bandwidth—which for a synchronous interface is typically the clock frequency multiplied by the number of wires in the interface—divided by the width of the cross-section. The CBD between the 8x28-Gb/s IC and the FPGA IC is 121 Gb/s/mm.

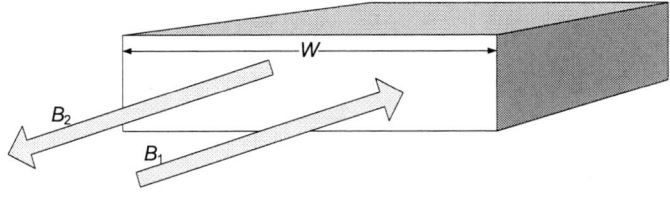

Figure 9    Cross-sectional Bandwidth Density

While the CBD of 121 Gb/s/mm is sufficient for Virtex 580HT and 870HT, future generations of inter-die interfaces will undoubtedly achieve much higher CBD. TABLE I summarizes the CBD of optical and electrical interfaces relevant to this product. The first four rows refer to 100-Gb/s optical interfaces. In particular, a standard 370-mm face plate can accommodate four CFP modules, eight CFP2 modules, 16 CFP4 modules, or 32 CFP4 modules (in a belly-to-belly configuration) [4].

TABLE I    CROSS-SECTIONAL BANDWIDTH DENSITY EXAMPLES

| Interface | Cross-sectional Bandwidth Density (Gb/s/mm) |
|---|---|
| CFP Optical Module | 2.2 |
| CFP2 Optical Module | 4.3 |
| CFP4 Optical Module | 8.6 |
| Belly-to-belly CFP4 Optical Module | 17.3 |
| Package Balls (1-mm pitch, 2 signals deep) | 15.3 |
| Package Balls (1-mm pitch, 4 signals deep) | 30.6 |
| C4 Bumps | 125 |
| Inter-die Silicon Interposer Interface | 121 |

While reviewing CBD values for just one side of an interface such as those in TABLE I is interesting, a more useful exercise is to compare the CBD values of both sides of an interface. When CBD values match, routing across an interface is easy since the wires can run straight across the interface. On the contrary, when the CBD values differ significantly, jogs must be introduced so that wires can converge onto the side of the interface with a lower CBD value. These jogs widen the channel between the two communicating devices across the interface. For instance, consider the two silicon interposers in Figure 10. In Figure 10(a), IC1 has a higher CBD value than IC 2. As a result, routing is congested as wires go from IC 2 to IC 1. In Figure 10(b), the CBD values match and the routes are straight. Clearly, the interposer in Figure 10(a) is taller than that in Figure 10(b).

Mismatch in CBD values across a passive silicon interposer interface does not just affect the size of the interposer. Because wires between two top dice on an interposer cannot be buffered as there are no active devices on a passive silicon interposer, simply widening the channel between two ICs with different CBD values may create wires so long that a reliable communication channel can no longer be established. For an IC that is expected to communicate with many different devices, its *maximum achievable* CBD should be tuned as high as possible. When two such identical ICs are put on a silicon interposer, this maximum CBD can be utilized 100%. The utilization of this maximum achievable CBD can be reduced for this IC to communicate with other ICs, for instance, memory, with a lower CBD value.

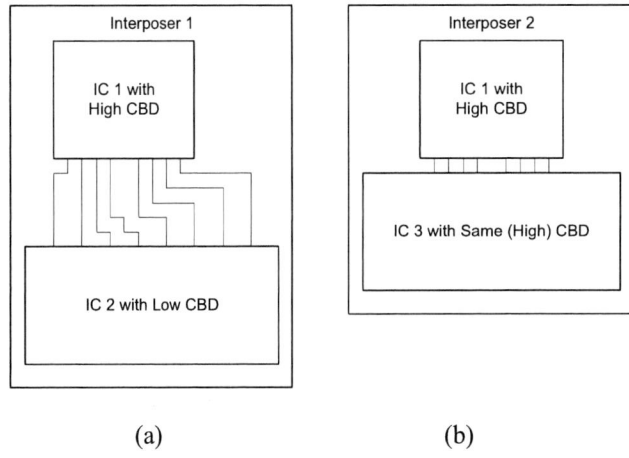

(a)                                (b)

Figure 10  (a) CBD mismatch causes routing congestion due to wire jogs. A large channel between the two ICs is needed. The resulting interposer is tall. (b) The CBD values values of both sides of the interface match, resulting in straight routes and a smaller interposer.

IV.    SILICON INTERPOSER OPTIMIZATION

A. *Interposer-to-Package Connection Modeling*

To determine the process recipe for the silicon interposer capable of 28-Gb/s signal transmission, a lumped RLC model was developed. As opposed to a broadband S-parameter model, the lumped model is more intuitive in representing physical structures in the silicon interposer [1].

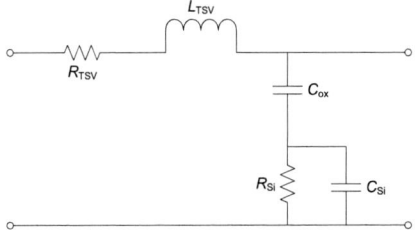

Figure 11  Lumped RLC Model of the Passive Silicon Interposer

The lumped RLC model in Figure 11 consists of two parts: the silicon RC model and a TSV RLC model. The RC model of the silicon ($R_{Si}$ and $C_{Si}$) represents the lossy silicon substrate [5]. The cylindrical copper TSV with copper core radius $r$,

978-1-4673-6145-3/13 $31.00 © 2013 IEEE        301

length $l$, with oxide liner thickness $t_{ox}$ is modeled with resistance $R_{TSV} = \rho l / \pi r^2$, inductance $L_{TSV} = \frac{\mu_0}{2\pi} \left( l \ln \left( \frac{1}{r} \left( l + \sqrt{l^2 + r^2} \right) \right) - \sqrt{l^2 + r^2} + r \right)$ [2], and oxide liner capacitance $C_{ox} = 2\pi\varepsilon_{ox} l / \ln((r + t_{ox})/r)$ [3].

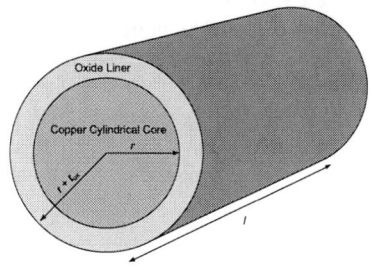

Figure 12  TSV Dimensions

To model high-speed signals from a top die connected to the package via the silicon interposer, the following elements are modeled:

- C4 bumps connecting between the package balls to the TSVs

- TSVs,

- four metal layers in the interposer, and

- micro-bumps for connecting the top die to the silicon interposer.

A broadband S-parameter model up to 50GHz was generated for the above elements with a 3D field solver. In addition, silicon interposer test chips were characterized to validate the model.

## B. Silicon Substrate Resistivity

To support signaling at 28 Gb/s, optimizing the effective capacitance and conductance is necessary. Two silicon substrates were evaluated: one with 10 ohm-cm, typically available from foundries, and the other with the higher 20 ohm-cm resistivity. The 20 ohm-cm version turned out to be beneficial to high-speed signaling [5].

Figure 13 compares silicon measurement (dotted lines) and simulation results (solid lines). Figure 13a illustrates the effective capacitance and Figure 13b shows the effective conductance. The green and the magenta lines are for the 10-ohm-cm silicon substrate whereas the blue and the red lines are for the 20-ohm-cm silicon substrate. The 10-ohm-cm silicon substrate shows about 205.2fF effective capacitance at low frequency and 19.4fF at high frequency, compared to the lower 115.8fF at low frequency and 12.4fF at high frequency for the 20-ohm-cm silicon substrate. Measured and simulated results generally agree.

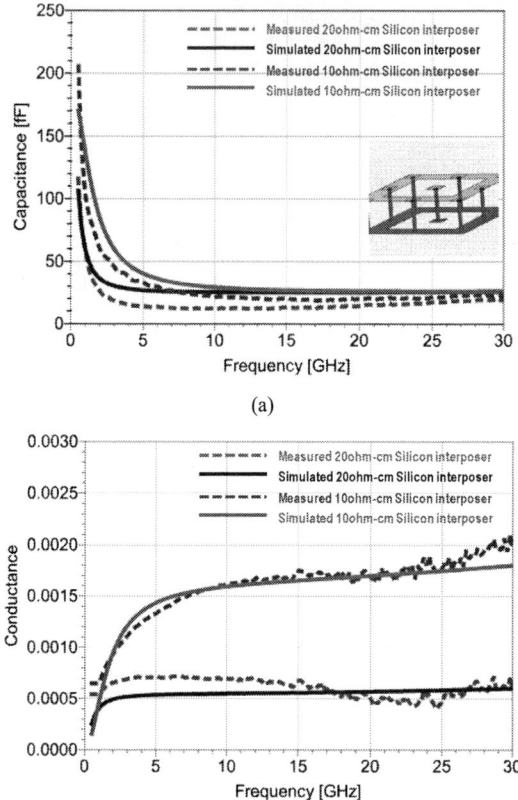

(a)

(b)

Figure 13  (a) Effective Capacitance          (b) Effective Conductance

Aside from effective capacitance and conductance evaluation as a function of silicon resistivity, insertion loss and return loss were also measured for high-speed signals serially connected to TSVs. In Figure 14a, the S21 insertion loss for the 20-ohm-cm silicon substrate is -0.414dB at 14GHz (the Nyquist frequency for 28-Gb/s signaling) and -0.822dB for the 10-ohm-cm silicon substrate. The 20-ohm-cm silicon substrate generally shows less insertion loss over the range of frequency of interest, which benefits high-speed signaling. As shown in Figure 14b, the return loss values above 10GHz are similar and the 20-ohm-cm silicon substrate exhibits more return loss below 10GHz, which again is beneficial for high-speed signaling.

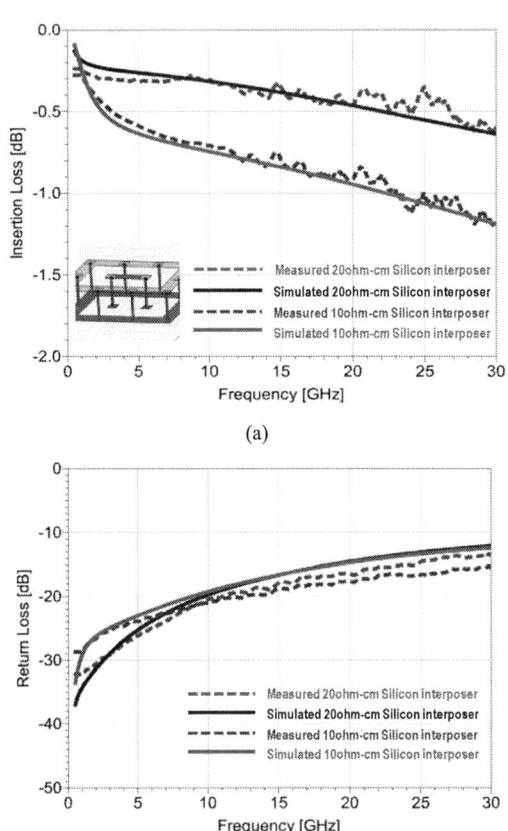

(a)

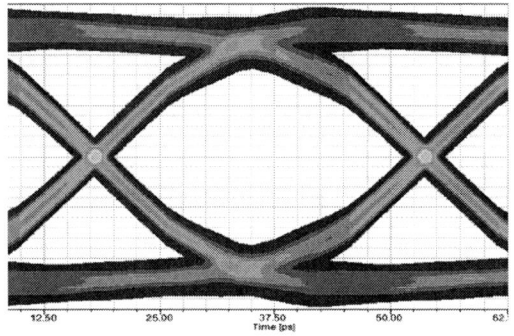

(a)

(b)

Figure 14 (a) Insertion Loss (b) Return Loss

### C. 28-Gb/s Signaling

Once the interposer and TSVs were well understood, the most important aspect of channel design was the minimization of attenuation through the package and printed circuit board. This involved the selection of low loss materials, like copper for traces, LTCC for package substrates and Megtron6 laminate for PCBs. Minimization of impedance discontinuities between substrates and vias were also intensively designed to minimize losses. This meant impedance matching for capacitive vias, solder balls and solder lands. These package and PCB interconnects were modeled in ANSYS HFSS.

To optimize the performance, an end-to-end channel model was simulated and compared to measurement. This channel simulation included a behavioral TX silicon model, lumped element interposer model, S-parameter package and printed circuit board models and a behavioral receiver model. Each section of the channel was individually modeled and optimized to achieve the desired performance. The channel was simulated in Agilent ADS.

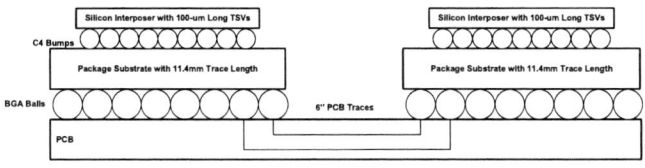

Figure 15 Simulation Test Bench

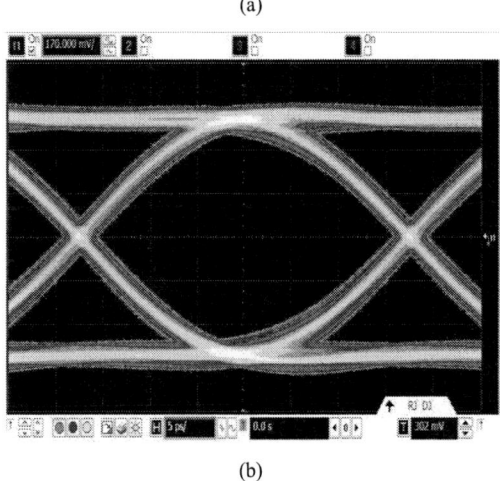

(b)

Figure 16 (a) Simulated 28-Gb/s Eye (b) Measured 28-Gb/s Eye

## V. PACKAGE DESIGN AND ASSEMBLY

Figure 17 Assembled Stacked-Silicon
Two FPGA ICs and One 8x28.05-Gb/s Transceiver IC on an Silicon Interposer Mounted on Package Substrate (Photo courtesy TSMC)

The Virtex-7 580HT and 870HT products were assembled using the CoWoS™ manufacturing flow offered by TSMC.

978-1-4673-6145-3/13 $31.00 © 2013 IEEE      303

Figure 18 is a schematic illustration of TSMC's CoWoS™ process flow.

Through-Silicon-vias (TSV) are first formed on a silicon interposer wafer using a DRIE etch process followed by dielectric liner deposition and copper plating. The four interconnect layers on top of the TSV follow typical 65-nm BEOL design rules and processes.

Interposer wafer backside thinning, TSV reveal, passivation and bump formation are then completed. The FPGA slices and the ASIC die with micro-bumps are then attached to the interposer wafer using a chip-on-wafer assembly process.

The chip-on-wafer is then singulated into individual chip-stacks using a dicing process and these chip-stacks are then flip-chip attached to the HiTCE substrate.

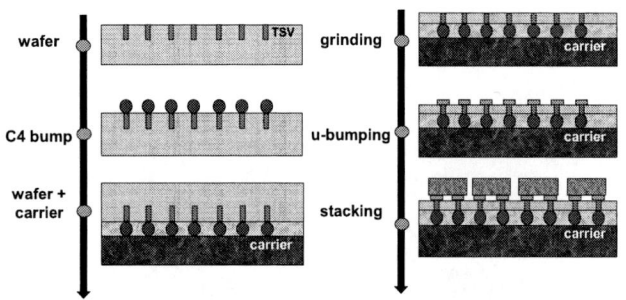

Figure 18   Schematic of CoWoS™ flow (courtesy Ting Lo et al. from TSMC preesented at IEDM 2012)

## VI. THERMO-MECHANICAL PROPERTIES

3D models were constructed using the commercial software ABAQUS to perform package thermo-mechanical simulation and analysis. Owing to the symmetry of the package, only one quarter was modeled to increase computational efficiency. The package was assumed to be stress-free at an underfill cure temperature of 150°C. Furthermore, sub-modeling techniques were used and detailed local models were solved at selected locations on the global model.

Different substrate sizes and designs, lid designs (forged-lid vs. stamped-lid), lid materials (copper vs. AlSiC) and underfill materials were investigated in order to optimize warpage, low-k stress and micro-bump and C4 bump reliability. In addition, LTCC ceramic package reduces fatigue in C4 bumps while increasing the risk for BGA balls to fail in thermal stressing. BGA ball reliability was also studied later on in this paper. Simulations showed using an AlSiC lid improved the reliability of C4 bumps, BGA balls, low-k and micro-bumps. In addition, warpage was below the JEDEC spec for 45-mm package sizes. Figure 19 shows the warpage simulation results.

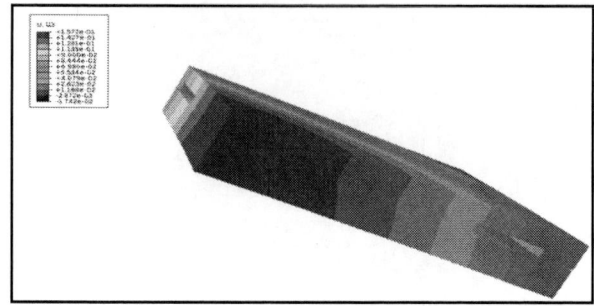

Figure 19   Warpage  at 25°C

The results shown in Figure 20 and TABLE II indicate that for the given package, TSV interposer attributes, and metallization, the micro-bumps and C4 bumps undergo acceptable reliable solder inelastic strain and fatigue.

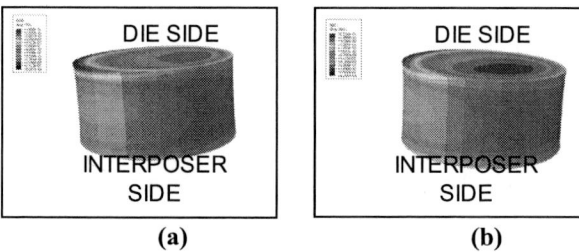

(a)                                (b)

Figure 20   Micro-bump inelastic strains after x TC cycles at 125°C (a) Logic slices (b) SerDes die

TABLE II        SUMMARY OF SOLDER FATIGUE STUDY

|  | After x cycles | |
| --- | --- | --- |
|  | @ 125°C | @ -55°C |
| Micro-bump solder inelastic strains | 0.13 | 0.39 |
| Micro-bump solder inelastic energy | 3.4 | 13.3 |
| C4 solder inelastic strains | 0.09 | 0.2 |

In order to study fatigue for BGA balls, inelastic strains and energy were obtained after  temperature cycles of 0°C/+100°C.  Figure 21 shows simulation results for corner BGA balls.

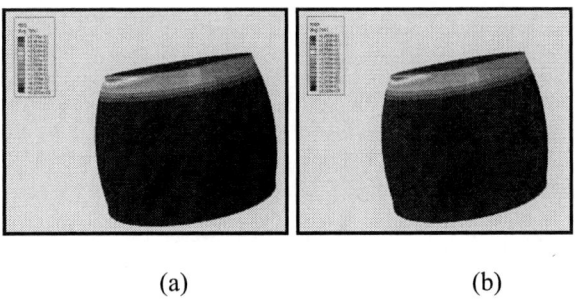

| (a) | (b) |

Figure 21  BGA balls inelastic strains and energy after x TC cycles at 100°C
(a) inelastic strains (b) inelastic energy

Furthermore to increase BGA reliability, different C4 underfill materials were considered. Figure 22 illustrates that a higher Tg underfill material improves BGA reliability as well as reducing inelastic energy in C4 bumps.

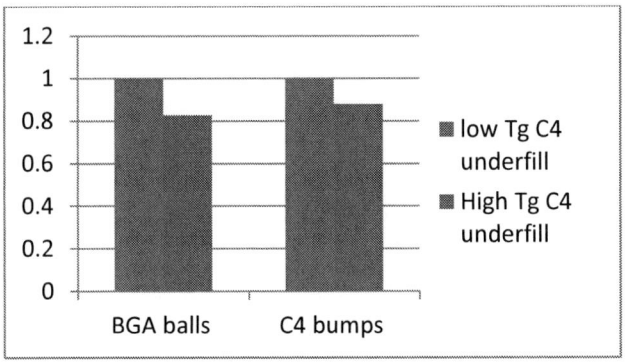

Figure 22  Inelastic energy  after three TC cycles at 100°C

## VII.   3D THERMAL ANALYSIS

3D non-linear models were constructed using the commercial software Flotherm to perform the package thermal simulation and analysis.

An accurate measurement or estimation of power was required to drive the simulation. The Xilinx Power Estimator (XPE) tool was used to estimate the FPGA power [5]. The tool is typically used in the pre-design and pre-implementation phases of a project. XPE assisted with architecture evaluation and device selection and helped select the appropriate power supply and thermal management components which might be required for an application. XPE considered the design's resource usage, toggle rates, I/O loading, and many other factors which it combined with the device models to calculate the estimated power distribution. The device models were extracted from measurements, simulation, and/or extrapolation.

The package was assumed to be initially at ambient temperatures of 55°C, 85°C and 100°C. A JEDEC 2s2p board was considered here.

The material properties of package compounds, conductivity and specific heat, are temperature-dependent. At room temperature, conductivity is shown in TABLE III.

TABLE III        MATERIAL PROPERTIES OF PACKAGE COMPOUNDS AT ROOM TEMPERATURE

| Materials | Conductivity W/mK) |
|---|---|
| LTCC ceramic | 2.0 |
| Prepeg | 0.35 |
| Si | 117.5 |
| Copper | 389 |
| Solder mask | 0.24 |
| TIM1 | 0.92 |

In order to study heat transfer to a high level of accuracy and without the need to model every micro-bump, C4 bump and BGA ball, a representative conductivity and specific heat of micro-bumps/micro-bump underfill, C4 bumps/C4 underfill and BGA level were calculated using composite material property equations. In addition, considering TSV pitch, attributes and material properties, equivalent TSV silicon interposer conductivity and specific heat were estimated. The 3D non-linear model was solved for applied boundary conditions and corresponding power. The package was assumed to be initially at ambient temperature, and temperatures, temperature gradients and thermal resistances were obtained after the system reached a steady-state condition.

TABLE IV demonstrates thermal simulation results for the optimized design for 3 different ambient conditions of 55°C, 85°C and 100°C when every logic chip consumes the same amount of power. Simulation results confirm that a heatsink is required to cool down the package and for the selected passive heatsink, the FPGA package is thermally reliable and meets thermal specifications.

TABLE IV    THERMAL SIMULATION RESULTS

| Virtex7 DEVICE | Power (W) | Ta (C) | JB (C /W) | JC (C /W) | Eff. JA (C/W) | Max Junction temp. ( C ) |
|---|---|---|---|---|---|---|
| No HS | 36 | 55 | 4.2 | 0.07 | 7.6 | 297.2 |
| No HS | 37 | 85 | | | 7.4 | 338.4 |
| No HS | 44 | 100 | | | 6.9 | 355.2 |
| w *HS @ 250 LFM | 36 | 55 | | | 0.6 | 81.2 |
| w *HS @ 250 LFM | 37 | 85 | | | 0.6 | 107.1 |
| w *HS @ 250 LFM | 44 | 100 | | | 0.6 | 130.3 |
| *HS : Xilinx reference heatsink | | | | | | |

## VIII. SUMMARY

This paper outlined the design of inter-die interconnects on a silicon interposer, interposer-to-substrate channel modeling, silicon interposer resistivity optimization, end-to-end channel simulations of 28 Gb/s off-chip interconnects over TSVs, stacked-silicon packaging and assembly flow, as well as package thermal and mechanical properties.

The focus of the inter-die interconnects was signal integrity and cross-sectional bandwidth density. Careful wire layer assignment to avoid inductive effects and crosstalk was discussed. An RC silicon interposer wire model was calibrated against an RLC model. After matching the RLC model, the RC model was used for inter-die timing closure because the static timing analyzer at the time did not support RLC wire models.

The cross-sectional bandwidth density metric was introduced to predict interposer routing congestion. The authors advocate matching cross-sectional bandwidth density values on both sides of a silicon interposer interface to reduce wire length and interposer size.

The focus of the 28-Gb/s interconnects was signal integrity and the agreement between channel simulations and measurements. In terms of work done for the silicon interposer,

TSV modeling and silicon interposer resistivity were critical in achieving high signal integrity.

Currently Xilinx uses the TSMC CoWoS$^{TM}$ manufacturing flow to assemble stacked-silicon products. In particular, the Virtex-7HT interposer and top dice are housed in an LTCC ceramic package for optimal 28-Gb/s signal integrity.

Thermal and mechanical analyses were carried out to ensure package reliability and estimate maximum junction temperatures under different system conditions. Various substrate and lid designs, lid materials, and underfill materials were investigated to optimize warpage, low-k stress, as well as micro-bump and C4 bump reliability. Simulations showed using an AlSiC lid improved the reliability of C4 bumps, BGA balls, low-k stress and micro-bumps, while keeping warpage in check. Last but not least, 3D non-linear models were created to confirm system conditions under which maximum junction temperatures stay within reliability limits.

### ACKNOWLEDGMENT

The authors would like to thank the Virtex-7HT team for many technical contributions to creating the world's first heterogeneous stacked-silicon FPGA product.

### REFERENCES

[1] N. Kim, Z. D. Wu, A. Rahman, D. Kim, and P. Wu, "Through Silicon Via (TSV) Design Considering Technology Challenges for Very High-Speed Signal Transmission," *DesignCon 2011*, Santa Clara, Jan. 31–Feb. 3, 2011.

[2] F.B.J. Leferink, "Inductance Calculations; Methods and Equations," *Proc. IEEE Int. Electromagn Compat. Symp.*, Aug. 1995, pp. 16–22.

[3] T. Bandyopadhyay, R. Chatterjee, D. Chung, M. Swaminathan, and R. Tummala, "Electrical modeling of through silicon and package via," in *IEEE Int. Conf. 3D System Integration (3DIC)*, San Francisco, CA, 2009.

[4] CFP MSA, "Next-Gen PMD CFP MSA Baseline Specifications," http://www.cfp-msa.org/Documents/CFP_MSA_baseline_specifications.pdf, Apr. 2011.

[5] N. Kim, D. Wu, D. Kim, A. Rahman, P. Wu, "Interposer Design Optimization for High Frequency Signal Transmission in Passive and Active Interposer using Through Silicon Via (TSV)," IEEE/ECTC, Orlando, Florida, June 2011, pp. 1160–1167.

# A 5 Gb/s 3.2 mW/Gb/s 28 dB Loss-compensating Pulse-Width Modulated Voltage-Mode Transmitter

Saurabh Saxena, Romesh Kumar Nandwana, and Pavan Kumar Hanumolu

School of EECS, Oregon State University, Corvallis, Oregon, USA

Email: ssaurabh0204@gmail.com

*Abstract*—A voltage mode transmitter employs pulse width modulation (PWM) based equalization of NRZ input data at 5 Gb/s and compensates 28 dB channel loss at 2.5 GHz. Fabricated in a 90 nm CMOS process, the proposed transmitter achieves a horizontal eye opening of 0.3 UI with BER< $10^{-12}$ and consumes only 16 mW power of which 2.5 mW is consumed by the digital PLL.

## I. INTRODUCTION

High speed communication across backplanes is difficult because of signal attenuation and distortion caused by lossy channels. Such channel limitations on the transmitted signal are countered by applying the inverse channel response in the signal path using filter blocks such as feed forward equalizer (FFE), continuous time linear equalizer (CTLE), and decision feedback equalizer (DFE). For higher channel loss compensation, a combination of these equalization schemes are used in the same transceiver [1] [2]. For high channel losses, decision feedback equalizer implemented on the receiver (RX) side has been shown to be very effective. However, this complicates the design of the clock and data recovery (CDR) circuit. On the other hand, transmitter (TX) side equalization eases the design of the CDR but implementing equalization in a power efficient voltage-mode (VM) driver is complex. Equalization with current mode (CM) output drivers is a simpler but it is a power hungry option.

When equalization is implemented within the VM output driver, the complexity in implementing equalizer taps while maintaining fixed output impedance increases. As a channel termination requirement, the VM driver should maintain its output impedance irrespective of the output voltage levels. These voltage levels depend on the equalizer taps and maximum output swing of the TX. Also, the driver should have a fine control of the equalizer taps for optimized channel loss compensation. These problems are tackled by using a variety of source-series-terminated (SST) drivers with defined output impedance. Multiple instances of a basic output driver cell as shown in Fig. 1(a) implement the equalizer taps and compensate for PVT variations [3] [4]. For different output swings, the driver output impedance can be regulated by using voltage regulator with drivers and pre-drivers both as depicted in Fig. 1(b) [5]. Nevertheless, the resolution of the equalizer coefficient is coupled with impedance control in the output drivers. And, it increases the complexity and area to implement fine equalizer taps with same output impedance under different TX output swings. This artifact can be avoided

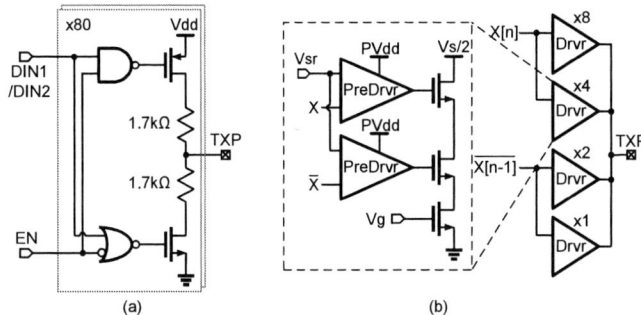

Fig. 1. Conventional voltage mode drivers with (a) multi-cell based equalizer, and (b) supply regulated impedance control for output drivers.

if the equalization is independent of the output impedance control.

The proposed transmitter decouples the equalization and impedance control in VM drivers by employing pulse width modulation (PWM) based de-emphasis. Earlier, PWM based de-emphasis introduced with current mode drivers [6] has a poor power efficiency for the transmitter. This inhibits its use to compensate large channel loss compared to other power efficient equalization schemes. Section II elaborates on the PWM based equalization scheme. Employing PWM equalization with VM drivers achieves a power efficient 28 dB channel loss compensation as demonstrated in the proposed transmitter architecture in Section III. Measured results from the prototype transmitter are shown in Section IV followed by conclusions in Section V.

## II. PWM BASED EQUALIZATION

The PWM performs equalization by varying the duty cycle of the transmit pulse of time period T. It equalizes channel response similar to 2-tap FIR filter where the former employs time as an equalizer variable and the latter uses voltage variable. In contrast to this FIR filter based de-emphasis, PWM provides selective frequency amplification beyond the Nyquist frequency (1/2T) as shown by impulse response of the two filters in Fig. 2. It is to be noted that a high pass frequency response of an equalization filter over larger frequency range helps in compensating channel loss beyond 1/2T and thereby reducing distortion and ISI. In addition, duty-cycling exploits the fine time resolution in lower technology nodes while always transmitting the peak power available unlike voltage based equalization.

978-1-4673-6145-3/13 $31.00 © 2013 IEEE

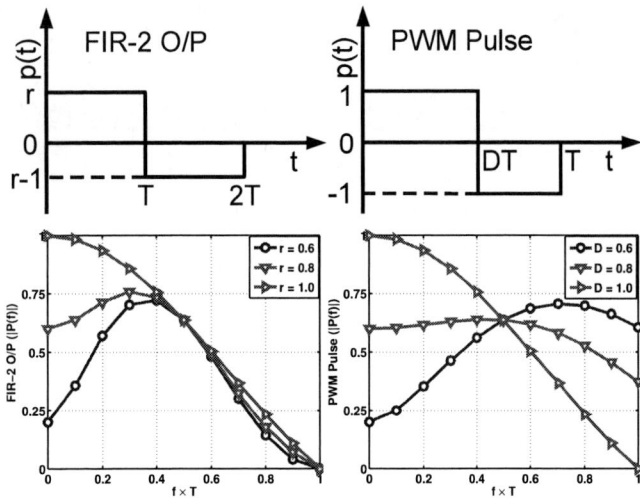

Fig. 2. Impulse response of a FIR-2 filter and a pulse width modulation based filter.

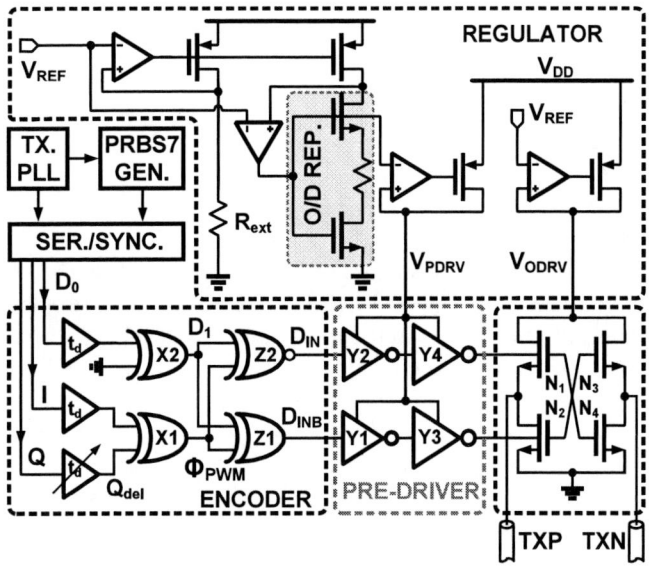

Fig. 3. Proposed transmitter block diagram.

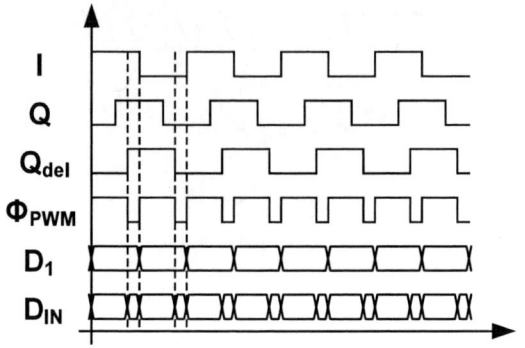

Fig. 4. Signal waveforms for PWM of input data in the encoder.

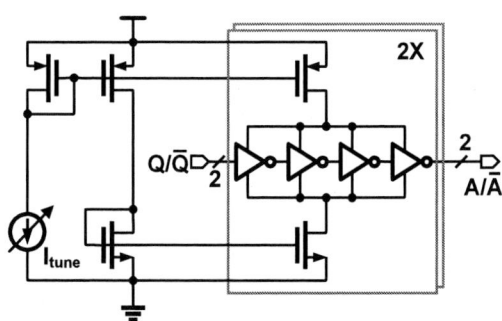

Fig. 5. Current controlled delay cell.

### III. PROPOSED VOLTAGE MODE TRANSMITTER

Fig. 3 shows the block diagram of the proposed pseudo-differential voltage mode transmitter. Quadrature clock signals (I/Q) generated by a digital TX PLL (DPLL) are used to serialize half-rate PRBS data and generate 5 Gb/s non-return-to-zero (NRZ) data, $D_0$. Equalization is implemented by modulating synchronized 5 Gb/s NRZ data $D_1$, with 5 GHz-gating signal $\Phi_{PWM}$ as depicted by waveforms in the encoder path in Fig. 4. XOR gate X1 generates $\Phi_{PWM}$ from 2.5 GHz I/Q phases and obviates the need for an explicit 5 GHz clock signal. XOR gate X2 is used to match delays in the clock and data paths. Variable duty cycle is obtained by first adding delay $\Delta t_d$ between I and Q phases and later combining with XOR gates to get $\Phi_{PWM}$. Current controlled delay cell shown in Fig. 5 varies $t_d$ controlling the amount of equalization. It

consists of even number of current starved inverters to limit the duty-cycle variation across $I/\bar{I}$ ($Q/\bar{Q}$) going through two independent delay lines coupled by cross-coupled inverters.

The data modulation is done using XOR gates Z1 (Z2). Implemented in a differential symmetric topology (Fig. 6) these gates are critical to the design as they modulate the data with inverse channel response. Compared to a current mode logic (CML) based XOR implementation, this architecture offers superior power efficiency but its design must be optimized to minimize data dependent transmit jitter and duty cycle distortion caused by the limited bandwidth and asymmetric rise and fall transitions. To this end, parasitics were minimized by careful transistor sizing and layout design.

The encoded data $D_{IN}(D_{INB})$ drives the pseudo differential voltage mode output drivers through inverter based pre-drivers. Voltage-mode output driver is implemented using only NMOS transistors ($N_1 - N_4$) and the output swing is set to a reference voltage, $V_{REF}$, by using an on-chip regulator. The pre-driver composed of inverters ($Y1 - Y4$) operating with a regulated supply, $V_{PDRV}$, drives the output stage. Because the input control of the pre-driver is a rail-to-rail signal ([$0, V_{DD}$]) and its supply is regulated to $V_{PDRV}$, the pre-driver introduces large duty cycle distortion on the input data. To mitigate this, inverters sizes are optimized. $V_{PDRV}$ is generated by a bias circuit consisting of a replica of the output driver to force its output impedance to be equal to the termination impedance ($\approx 50\,\Omega$).

The impact of clock jitter on equalization is minimized by generating low jitter clocks using an on-chip digital PLL

978-1-4673-6145-3/13 $31.00 © 2013 IEEE

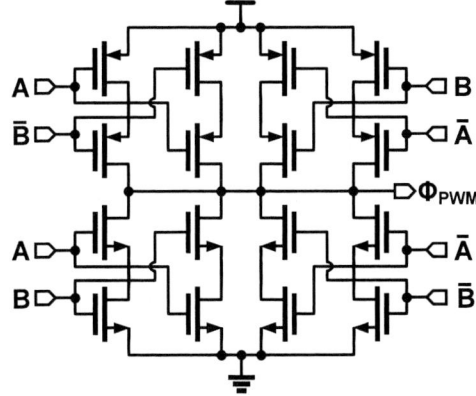

Fig. 6. Differentially symmetric XOR gate.

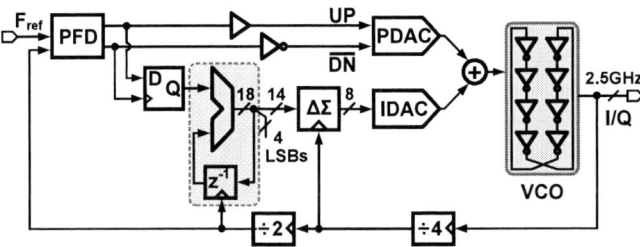

Fig. 7. Digital phase locked loop (DPLL).

(DPLL) shown in Fig. 7. The DPLL operates with a 312.5 MHz reference clock and generates 2.5 GHz quadrature clocks. The proportional path is implemented using a 3-state phase frequency detector (PFD) and directly driving the oscillator with a 3-level current mode DAC. D flip-flop acts as a bang-bang phase detector on the PFD outputs and generates the sign of the phase error which is integrated by an 18-bit accumulator. Lower LSBs of the accumulator output are ignored to suppress jitter caused by limit cycles and the rest of 14 MSBs are truncated to 8-bits using digital $\Delta\Sigma$ modulator. A thermometer-coded DAC (IDAC) converts these 8 bits into equivalent current and drives the oscillator.

## IV. MEASURED RESULTS

The proposed transmitter was implemented in a 90 nm TSMC CMOS process and occupies an active area of 0.13 mm$^2$ (Fig. 8). The measured long-term absolute jitter of the 2.5 GHz PLL output is 2.4 ps r.m.s. and 19.3 ps pk-pk (Fig. 9). The channel used to characterize the transmitter includes 2 mm bondwire, QFN package parasitics, 1.5" on-board microstrip line, SMA connectors, and a differential FR4 stripline. Fig. 10 shows the measured channel loss with 50 Ω source and sink termination. The channel is driven with 0.66 V peak-to-peak differential swing 7-bit PRBS data at the TX output. The duty cycle of the modulating clock is tuned by externally varying current in the delay cell to maximize the received eye opening. The measured output eye diagrams at the end of three different channels with losses 16 dB, 24 dB, and 28 dB at Nyquist frequency of 2.5 GHz, are shown in Fig. 11. Bathtub plots measured using 80SJNB BER analysis software with

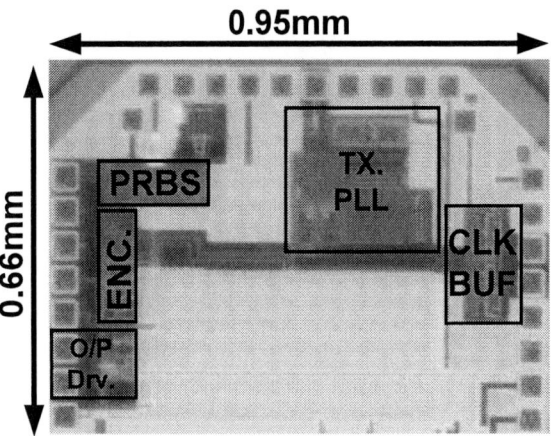

Fig. 8. Die photo of the proposed transmitter.

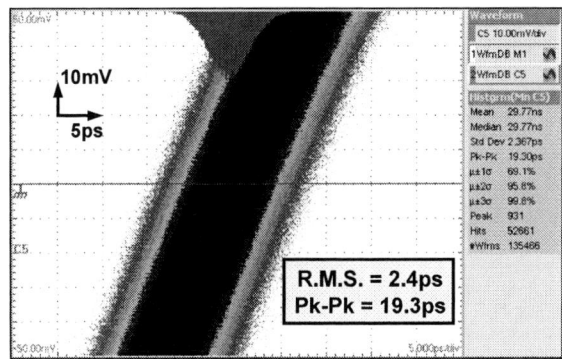

Fig. 9. Jitter histogram of the DPLL output.

Tektronix DSA8200 sampling oscilloscope and depicted in Fig. 12 show that the transmitter achieves a BER $< 10^{-12}$ with 0.6-0.3 UI time margin for channel loss ranging from 16-to-28 dB, respectively. Variation of horizontal eye opening with different transmit amplitudes for BER $= 10^{-12}$ is shown in Fig. 13. The variation is roughly constant for low channel loss where the received signal is higher than the minimum sensitivity of the BER tester. The total power dissipation is 16 mW of which the PLL consumes 2.5 mW. It translates to a power efficiency of 3.2 mW/Gb/s. The prototype performance is compared with other state-of-the-art transmitters compensating similar amount of channel loss in Table. I. This work achieves 6.9 times better power efficiency compared to [6] employing PWM scheme with current mode drivers. Taking channel loss in account, the figure-of-merit defined as FOM $=$ DR(Gb/s) $\times 10^{(\text{loss}/10)}$/Power(mW) with value 197 is the best for the proposed transmitter.

## V. CONCLUSION

A power efficient and high channel loss compensation for backplane channels is demonstrated by employing PWM based de-emphasis in voltage mode transmitter. It overcomes the voltage sensitivity of equalization filter and exploits the fast switching transistors in lower technology nodes. The implemented voltage mode transmitter achieves power efficiency

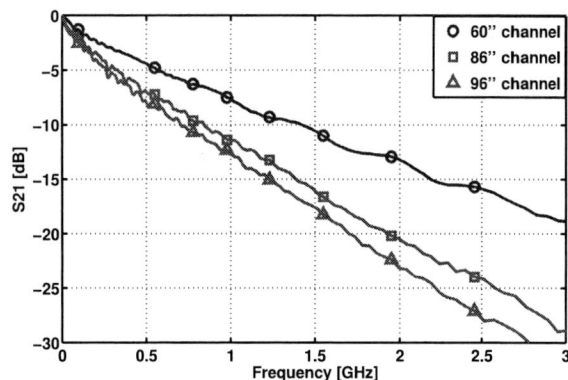

Fig. 10. Measured channel loss.

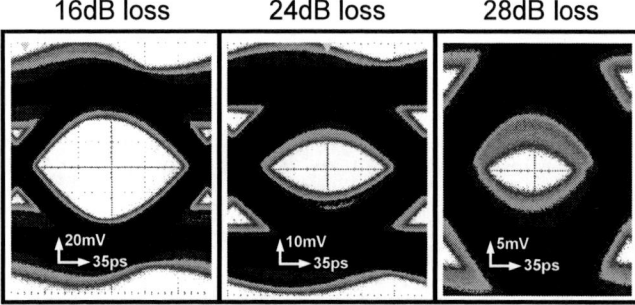

Fig. 11. Eye diagram at the output of three different channels.

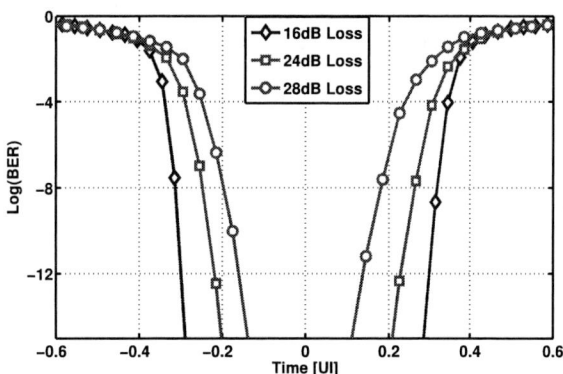

Fig. 12. Measured BER bathtub plots.

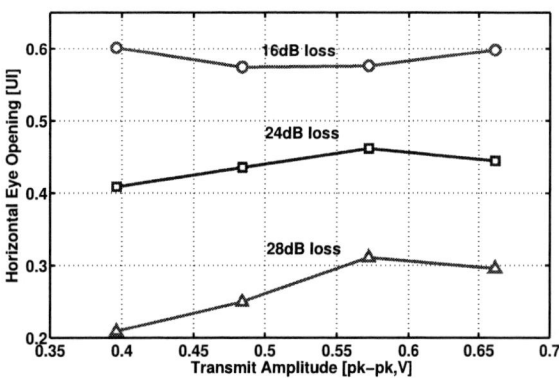

Fig. 13. Horizontal eye opening vs peak-to-peak transmit amplitude for BER = $10^{-12}$.

TABLE I
PERFORMANCE COMPARISON.

| | [6] | [7] | [1] | This work | | |
|---|---|---|---|---|---|---|
| **Technology (nm)** | 90 | 130 | 90 | 90 | 90 | 90 |
| **Supply voltage (V)** | 1.2 | 1.8/3 | 1.2 | 1.25 | 1.25 | 1.25 |
| **Data rate (Gb/s)** | 5 | 16 | 8 | 5 | 5 | 5 |
| **Channel loss (dB)** | 33 | 30 | 37 | 16 | 24 | 28 |
| **Vertical eye (mV)** | 15 | 10 | - | 78 | 18 | 8 |
| **Horizontal Eye (UI)** | 0.75 | 0.25 | 0.11 | 0.60 | 0.45 | 0.30 |
| **Power (mW)** | 110 | 1578 | 232 | 15.5 | 16 | 16 |
| **FOM (mW/Gb/s)** | 22 | 97 | 29 | 3.1 | 3.2 | 3.2 |
| **FOM2\*** | 91 | 11 | 165 | 13 | 78 | 197 |

$$^{*}\text{FOM2} = \text{DR(Gb/s)} \times 10^{(\text{loss}/10)}/\text{Power(mW)}$$

of 3.2 mW/Gb/s while compensating 28 dB channel loss. The equalization scheme can be used to compensate for different channel losses and data rates using a single control variable (duty cycle).

ACKNOWLEDGEMENT

This research was partly supported by Kawasaki Microelectronics America Inc., and NSF under CARRER Award EECS-0954969. We thank Berkeley Design Automation for providing Analog Fast Spice (AFS) simulator.

REFERENCES

[1] K. Fukuda, H. Yamashita, F. Yuki, M. Yagyu, R. Nemoto *et al.*, "An 8Gb/s Transceiver with 3x-Oversampling 2-Threshold Eye-Tracking CDR Circuit for -36.8dB-loss Backplane," in *IEEE Int. Solid-State Circuits Conf. (ISSCC) Dig. Tech. Papers*, 2008, pp. 98–598.
[2] A. Agrawal, J. Bulzacchelli, T. Dickson, Y. Liu, J. Tierno *et al.*, "A 19Gb/s serial link receiver with both 4-tap FFE and 5-tap DFE functions in 45nm SOI CMOS," in *IEEE Int. Solid-State Circuits Conf. (ISSCC) Dig. Tech. Papers*, 2012, pp. 134–136.
[3] M. Kossel, C. Menolfi, J. Weiss, P. Buchmann, G. von Bueren *et al.*, "A T-Coil-Enhanced 8.5Gb/s High-Swing source-Series-Terminated Transmitter in 65nm Bulk CMOS," in *IEEE Int. Solid-State Circuits Conf. (ISSCC) Dig. Tech. Papers*, 2008, pp. 110–599.
[4] S. Fallahi, D. Cui, D. Pi, R. Zhu, G. Unruh *et al.*, "A 19 mW/lane Serdes transceiver for SFI-5.1 application," in *Proc. IEEE Custom Integrated Circuits Conf.*, 2011, pp. 1–4.
[5] J. Poulton, R. Palmer, A. Fuller, T. Greer, J. Eyles *et al.*, "A 14-mW 6.25-Gb/s Transceiver in 90-nm CMOS," *IEEE J. Solid-State Circuits*, vol. 42, no. 12, pp. 2745–2757, 2007.
[6] J.-R. Schrader, E. A. M. Klumperink, J. Visschers, and B. Nauta, "Pulse-width modulation pre-emphasis applied in a wireline transmitter, achieving 33 dB loss compensation at 5-Gb/s in 0.13-$\mu$m CMOS," *IEEE J. Solid-State Circuits*, vol. 41, no. 4, pp. 990–999, 2006.
[7] A. Carusone, H. Cheng, and F. Musa, "A 32/16-Gb/s Dual-Mode Pulsewidth Modulation Pre-Emphasis (PWM-PE) Transmitter With 30-dB Loss Compensation Using a High-Speed CML Design Methodology," *IEEE Trans. Circuits Syst. I*, vol. 56, no. 8, pp. 1794–1806, 2009.

# Current-Steering Pre-Emphasis Transmitter with Continuously Tuned Line Terminations for Optimum Impedance Match and Maximum Signal Drive Range

Gerrit W. den Besten[1], Harold G. Hanson[2], Ranjeet K. Gupta[3]

[1] NXP Semiconductors, Eindhoven, The Netherlands, gerrit.den.besten@nxp.com
[2] NXP Semiconductors, Tempe AZ, USA
[3] NXP Semiconductors, Bangalore, India

*Abstract* — **A 24-segment current-steering transmitter with 3-tap pre-emphasis and linear auto-tuned active line terminations is presented. A linearized resistor-MOSFET termination structure for continuous impedance tuning is proposed to maintain an accurate output voltage level ($\sigma$=1%) and good impedance matching ($\sigma$=2%). This enables larger signal drive strengths thanks to more efficient supply utilization. The circuit is implemented in 0.16 μm CMOS technology for data rates up to 6 Gbps and consumes 5-43 mA from a 1.8V supply for 50-1200 mVppd output signal.**

## I. INTRODUCTION

For a long time, current-steering drivers have been widely used for serial interfaces thanks to their superior differential signal quality, low electro-magnetic emission (EME), good source termination, and high achievable speed for a given process technology [1]. The main disadvantage is their relatively high power consumption, especially for double terminated links. Recently, voltage-mode drivers have become more popular, primarily to reduce power consumption, and enabled by advanced CMOS nodes which provide good low-impedance switches [2]. Disadvantages are limited output swing due to the combination of source-series termination and low supply voltage, the difficulty to achieve good source termination in particular during transitions, and more common-mode noise, making EME worse. These drawbacks motivated the design of hybrid solutions [3]. Nevertheless, when process technology speed is a bottleneck, current steering enables the highest speed [1]. For termination often fixed integrated resistors are used, but variations in drive current and termination resistance may cause large output voltage inaccuracy, which compromises the maximum drive strength in order to keep the driver pairs in saturation. Previous solutions for impedance tuning have speed or linearity limitations [4,5] and digitally calibrated on-die termination (ODT) additionally suffers from ripple or drift [6].

The proposed current-steering transmitter with continuously tuned terminations combines high accuracy output voltage and impedance with maximum speed in a process technology.

In section II the transmitter architecture is presented. Section III provides a detailed description of the proposed line termination structure. Measurement results are presented in Section IV. Section V provides the conclusions.

## II. TRANSMITTER ARCHITECTURE

Figure 1 shows the architecture of the proposed segmented driver consisting of 24 cells, with 1mA output drive current each, sub-divided in 3 banks. Every cell contains a buffer and a 2:1 multiplexer to select the input signal. This configuration allows all cells to be driven by input signal A, while 12 cells can be driven by B, and 6 by C, with 6 more inputs (T) remaining for test purposes. With signals A, B and C being time-shifted copies of a serial bit stream, this provides a fully flexible 3-tap pre-emphasis programming with 1mA resolution. Each cell additionally contains measures to control slew rate and to reduce power consumption for lower data rates. The summed output currents flow through active termination resistors which are continuously tuned to 50 Ohm.

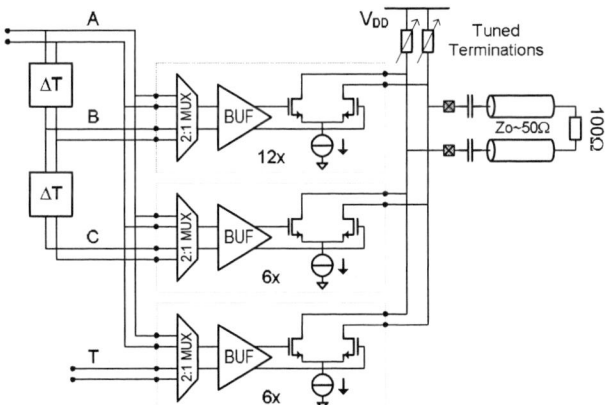

Fig. 1. Architecture of the proposed segmented current-steering transmitter with tuned terminations

Circuit details of the transmitter cells are shown in Figure 2. A solution for both CML (black) or CMOS (grey) input signals is provided given different data source types; a continuous CML bitstream with analog delays, or clocked CMOS logic with register delays. In the prototype both were implemented and multiplexed inside the buffer. Each individual cell can be disabled with a logic signal CD, driving the nodes bp and bn low, thereby implicitly turning off that output stage without speed penalty. Signal MD disables MUX and buffer for the CML path (and MDX for CMOS) This enables flexible output signal level setting without residual power in disabled cells.

978-1-4673-6145-3/13 $31.00 © 2013 IEEE

The required power in the buffer stage increases for higher data rates, while for lower data rates it is preferable to reduce power and slow down transitions. Therefore a scheme is proposed with adaptive buffer bias current and load resistors. This also provides output slew-rate control by the adaptable RC time constants at nodes bp and bn. Furthermore an adaptive extra common mode resistor is added to optimally bias the output stages for a maximum signal drive range.

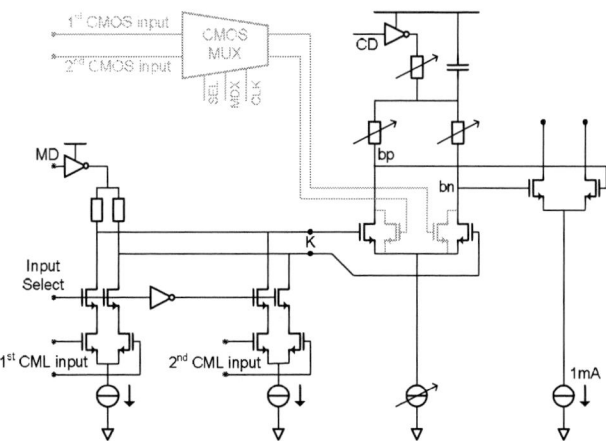

Fig. 2. Circuit details of transmitter cell

## III. Line Termination Topology

The circuit topology of the proposed tuned line impedance consists of a poly-resistor $R_{PS}$ in series with a PMOS transistor in triode region, a capacitor, and a bias resistor, as shown in Figure 3. The operation principle of this structure can be best understood using simple square-law MOS characteristics for triode region operation. If the gate would have a fixed voltage, the output impedance linearity becomes poor for large signal swings due to MOSFET output current saturation, especially for lower gate overdrive levels. However, if half of the drain signal voltage is added to the gate node, the MOSFET impedance becomes roughly linear up to its saturation point, as shown by the equation:

$$I_{out} \cong KV_{DS}\left(V_{GT} - \tfrac{1}{2}V_{DS}\right)$$
$$= KV_{DS}\left(V_{GT0} + \tfrac{1}{2}v_{DS} - \frac{V_{DS0}+v_{DS}}{2}\right)$$
$$= KV_{DS}\left(V_{GT0} - \tfrac{1}{2}V_{DS0}\right) = \frac{V_{DS}}{R_{PMOS}}.$$

The superposition is accomplished with coupling capacitance and high-impedance resistive gate bias. The implicitly resulting cut-off frequency is positioned well below the signal spectrum. As the PMOS operates in triode region, most of the required capacitance is intrinsically present as $C_{DG}$ of the PMOS, so only a fraction needs to be explicitly added. Connecting this capacitor to the output instead of the drain further reduces the required capacitance value and relaxes the parasitic pole at the drain. The effective capacitive loading of the termination structure on the output is negligible compared to the impact of the driver cells.

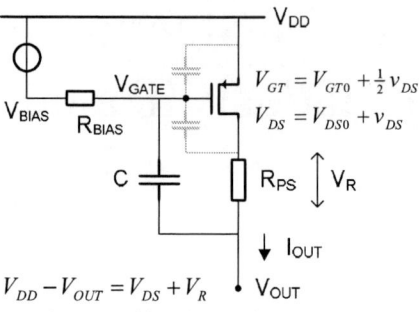

$$V_{DD} - V_{OUT} = V_{DS} + V_R$$

Fig. 3. Tunable active termination topology

Although real MOSFETs do not show ideal square-law behavior of course, they are still non-linear, so the same concept can be utilized to significantly reduce non-linearity by proper choice of capacitor value. The total termination impedance is $R_{PMOS}+R_{PS}$, providing a resistive divider, which multiplies the linear impedance signal range by $1+R_{PS}/R_{PMOS}$.

Figure 4 illustrates the signals at different nodes of the termination circuit when a DC plus sine wave current flow through it.

The driver common-mode (CM) currents through the terminations cause a voltage $V_{CM}$. The I-V curve of the linearized termination structure certainly encompasses the CM bias point and the origin, so the resulting AC impedance approaches $V_{CM}/I_{CM}$. Because $V_{CM}$ is controlled by $V_{BIAS}$, and $I_{CM}$ equals half the driver current, both swing and impedance can be set accurately.

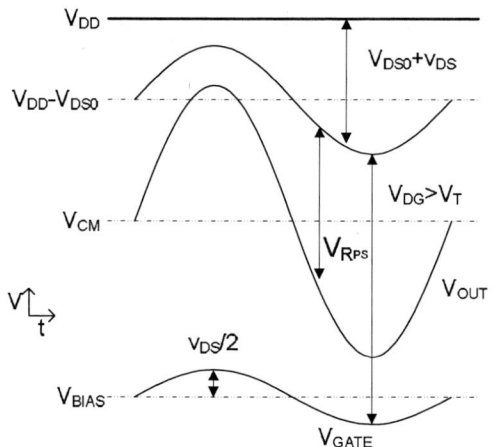

Fig. 4. Signals in termination structure for sinusoidal excitation

Figure 5 shows the control loop that generates the gate bias by making the voltage across a scaled termination replica equal to the desired CM level. Note that the output CM level, and therefore the required $V_{BIAS}$, depends on the number of enabled driver cells. The appropriate CM reference level is selected from a bandgap-referenced resistor ladder $R_{CMS}$ in 25 mV steps (50Ω×1mA/2) and the replica scaling is adapted for the programmed drive current. The control loop continuously adapts $V_{BIAS}$ to the correct value over all PVT conditions. Driver cell currents are derived from the bandgap reference too using an integrated digitally-calibrated reference resistor.

978-1-4673-6145-3/13 $31.00 © 2013 IEEE        312

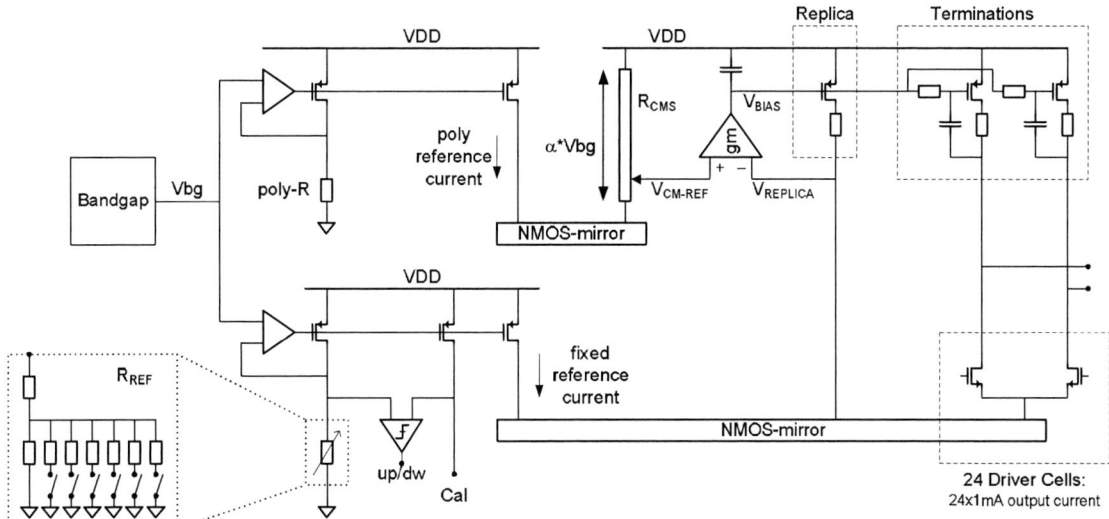

Fig. 5. The termination bias control loop with reference and calibration circuitry

This reference resistor can be easily calibrated at test time with an integrated comparator and an external reference current or precision resistor connected to the Cal node.

An additional advantage of this termination topology is that it can be easily disabled by connecting $V_{BIAS}$ to $V_{DD,}$ thereby utilizing the same PMOS also as disabling switch.

Typically, 60% of the impedance is placed in the resistor and 40% in the MOSFET. This allows sufficient tuning range to cover all bias and PVT conditions, for a wide range of drive strengths, while maintaining good linearity. Note that for 20% spread in poly resistance, the MOSFET may contribute between 28-52% of the impedance. The MOSFET is allowed to contribute such a substantial fraction of the impedance because its impedance is linearized. Therefore its size can be kept small, which beneficially reduces parasitic load.

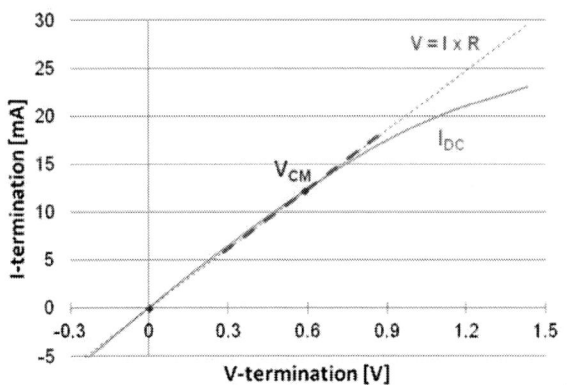

Fig. 6. Measured I-V characteristics of the termination

## IV. MEASUREMENT RESULTS

The proposed transmitter topology with tuned terminations was implemented for 1-6 Gbps applications in 0.16 μm CMOS with 1.8V supply. The design occupies less than 0.05 mm² of silicon, excluding pads. At maximum drive strength, the σ accuracy of the output voltage is 1% and for the termination impedance 2%. These numbers slightly degrade for lower swings (i.e. lower number of enabled driver cells).

Figure 6 shows the measured DC response of the termination with all cells active. The linearized impedance AC work line has been added which crosses the DC curve at the CM bias point. Note the DC current limiting characteristic for higher DC voltages as the PMOS moves towards and beyond its saturation point. This behavior is advantageous as it aids in protection against short circuit conditions.

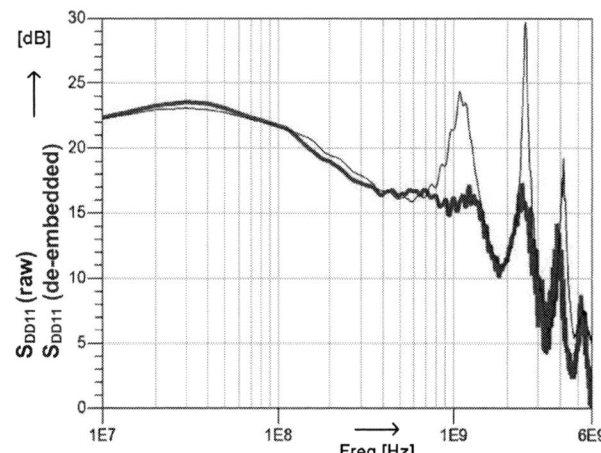

Fig. 7. Output return loss of the packaged transmitter measured on a test board (blue: raw data, red: with first-order de-embedding of test board)

Figure 7 shows the measured output return loss, which is 23 dB for low signal frequencies, more than 15 dB below 1 GHz and 10 dB up to about 3 GHz. The low frequency return loss is good but lower than expected from termination accuracy. This is caused by the limited output impedance of the differential pairs. For this test the output differential pair gates were driven to $V_{DD}$ (MD=1), which reduces their output impedance compared to normal bias conditions, so it is expected that this has significantly less impact during normal operation. The slight degradation of return loss at very low frequency results from the AC coupling in the terminations.

Several speeds, drive strengths, and pre-emphasis combinations were tested to evaluate signal quality. Some examples of measured eye diagrams with PRBS7 data at 5Gbps at maximum drive strength both with and without pre-emphasis are shown in Figure 8. Note the accurate voltage levels (tens of mVs) and low ISI (8 ps).

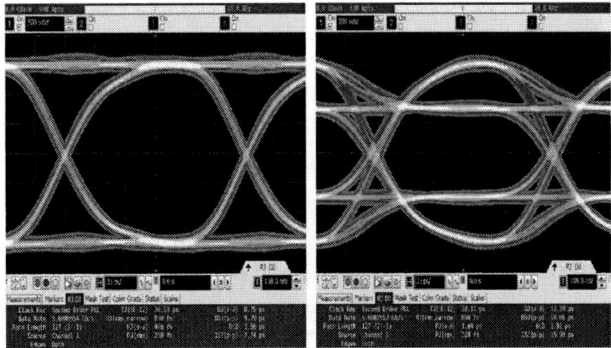

Fig. 8. Measured eye diagrams at 5Gbps with 1200mVppd swing (24 cells enabled) for 0dB and 6dB pre-emphasis. Some cross-talk from the neighboring channel is visible. The subtle knee in the settling can be attributed to the measurement set-up.

For maximum data rate, the driver consumes 1.75 mA per cell (so 5-43mA for 2-24 cells). The tuned active terminations only consume bias current for the control loop, which was about 1mA in this implementation. This can be reduced, as it only requires DC control. The significant cell power overhead ($I_{DD}/I_{OUT}-1$) of 75% is caused by tapering factor limitations imposed by the technology in order to achieve 6 Gbps over all temperature and process corners at high signal integrity levels. For a transmitter redesign in this process which is suitable up to 5Gpbs the power overhead was reduced to 40%. Applying the same transmitter and termination topology in more advanced CMOS nodes will enable higher speeds and/or further reduction of pre-driver overhead power.

A photo of the transmitter on silicon is shown in Figure 9.

## V. CONCLUSIONS

In this paper a linear active line termination concept has been presented for current steering line drivers. Good linearity, low parasitic capacitance, good impedance accuracy and high drive level accuracy are achieved simultaneously.

Driver cells, suitable for both CML and CMOS input signals were proposed, including adaptive buffer bias to optimize the trade-off between speed, power, and slew-rate in the application.

Although the achievable transmitter data rate of the prototype is limited to 6Gbps for worst cost PVT conditions due to the 0.16μm process technology, signal integrity is not degraded by the terminations thanks to the proposed topology.

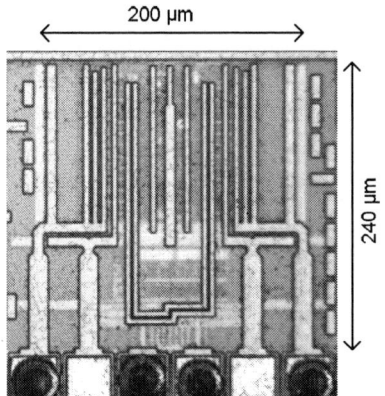

Fig. 9. Die photo of the transmitter

## ACKNOWLEDGMENTS

The authors want to thank Cor Speelman, Stefan Kwaaitaal, and Hans Schmitz for measurement support, Bob Frost for layout work, and Johan Klootwijk for making a die photo.

## REFERENCES

[1] G. Ono et al., "A 10:4 MUX and 4:10 DEMUX Gearbox LSI for 100-Gigabit Ethernet Link," *IEEE Journal of Solid-State Circuits*, Vol. 46, No. 12, pp 3101-3112, December 2011.

[2] C. Menolfi et al., "A 28Gb/s Source-Series Terminated TX in 32nm CMOS SOI," *ISSCC Digest of Technical Papers*, pp. 334-335, February 2012.

[3] A. K. Joy et al., "Analog-DFE-Based 16Gb/s SerDes in 40nm CMOS That Operates Across 34dB Loss Channels at Nyquist with a Baud Rate CDR and 1.2Vpp Voltage-Mode Driver," *ISSCC Digest of Technical Papers*, pp. 350-351, February 2011.

[4] H. Conrad, "2.4 Gbits/s CML I/Os with integrated line termination resistors realized in 0.5-μm BICMOS technology," *Proceedings of Bipolar/BiCMOS Circuits and Technology Meeting*, pp. 120–122, September 1997.

[5] Y. Fan and J.E. Smith, "On-Die Termination Resistors With Analog Impedance Control for Standard CMOS Technology," *IEEE Journal of Solid-State Circuits*, Vol. 38, No. 2, pp 361-364, February 2003.

[6] H.Y. Song et al., "A 1.2Gb/s/pin Double Data Rate SDRAM with On-Die-Termination," *ISSCC Digest of Technical Papers*, pp. 314 - 496, February 2003.

978-1-4673-6145-3/13 $31.00 © 2013 IEEE

# Design Techniques for CMOS Backplane Transceivers Approaching 30-Gb/s Data Rates
## (*Invited Paper*)

### John F. Bulzacchelli

### IBM Research Division, T. J. Watson Research Center, Yorktown Heights, NY 10598 USA

*Abstract*—Serial link transceivers with sophisticated equalization are needed for data transmission over high-loss electrical channels such as backplanes. This paper highlights design techniques for extending the data rates of such circuits by describing a 28-Gb/s transceiver implemented in 32-nm SOI CMOS technology. Equalization is provided by a 4-tap feed-forward equalizer (FFE) in the transmitter and a two-stage peaking amplifier with active feedback topology and 15-tap decision-feedback equalizer (DFE) in the receiver. The transmitter employs a source-series terminated (SST) driver topology with double the speed of previous designs. The use of capacitive level-shifters allows a single current-integrating summer to drive the parallel paths used for speculating the first two DFE taps. Error-free signaling at 28 Gb/s is demonstrated over a 35-dB loss channel with a power consumption of 693 mW/lane.

## I. Introduction

Both technology trends and new applications are fueling a rapidly increasing demand for communication bandwidth at all levels of the I/O hierarchy, including intra-chip, chip-to-chip across the same circuit board and across backplanes, rack-to-rack, and system-to-system (the latter two of which are increasingly optical). As CMOS technologies are scaled to finer dimensions, and the density of digital computing cores rises, the aggregate I/O system bandwidth must be increased to harness all of the computing power available. The growing popularity of advanced network services such as multimedia-on-demand and the fast adoption of cloud computing are powerful drivers in expanding data traffic.

To address the I/O needs of future computing and network systems, serial link data rates are now being pushed up to 25-28 Gb/s, as exemplified by standards such as OIF CEI-25G-LR and CEI-28G-SR [1], 32GFC [2], IEEE 802.3bj (100GbE over backplane and copper cable) [3], and InfiniBand EDR [4]. These standards address both short-reach (SR) and long-reach (LR) serial links. For short-reach links (with roughly 15 dB or less of channel loss), reliable signaling can be achieved with relatively simple and power-efficient transceivers such as that described in [5]. For long-reach links such as backplanes, however, the channel losses are much higher, so more complex transceivers with sophisticated equalization are needed. This latter class of transceivers is the main focus of this paper, which describes design techniques for extending their data rates to 28 Gb/s (and beyond).

The paper is organized as follows. Section II begins with a review of the equalization requirements for an electrical backplane operating at these data rates and then presents the architecture of a 28-Gb/s serial link transceiver implemented in 32-nm silicon-on-insulator (SOI) CMOS technology [6]. Sections III and IV detail key circuits of the transmitter and receiver, respectively. The measurement results of the transceiver are discussed in Section V, and Section VI concludes the paper.

## II. System-Level Considerations

### A. Channel Equalization

The physical structure of a backplane channel is shown in Fig. 1(a). The transmission lines of the passive backplane are used to transfer data from a chip on one line card to a chip on another line card. The bandwidth of the electrical channel is limited by several physical effects, including skin effect, dielectric loss, and reflections due to various impedance discontinuities. Even with advanced backplanes employing low-loss materials, counterbored vias (for shorter stubs), and best-in-class connectors, the loss at 14 GHz (half-baud frequency at 28 Gb/s) is usually at least 30 dB for a 1-meter-long channel [7]. The frequency response of a test channel [8] being studied for the IEEE 802.3bj (25 Gb/s) standard is plotted in Fig. 1(b). The 30.1-dB loss shown in the plot does not include degradations due to the transmitter and receiver packages. With typical package losses, the total channel loss is easily 35 dB. As illustrated in Fig. 1(c), the impulse response of such a lossy channel extends for many unit intervals (UIs), so each transmitted data bit is broadened and overlaps with several of its neighbors. The large amount of intersymbol

978-1-4673-6145-3/13 $31.00 © 2013 IEEE

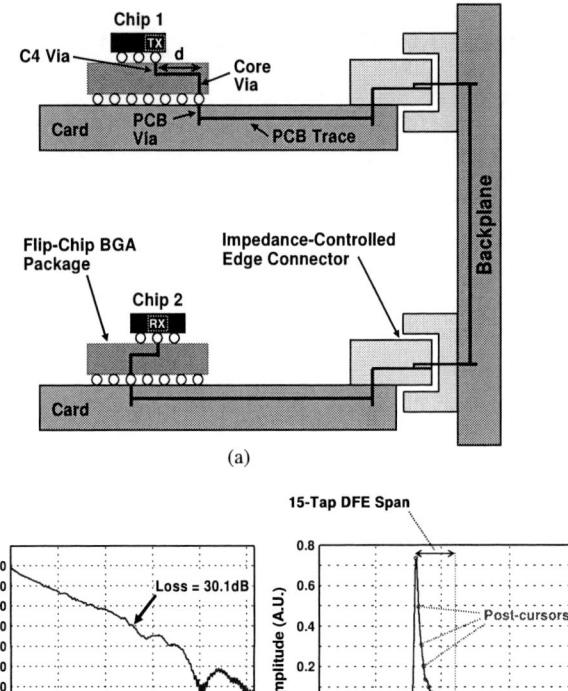

Fig. 1. Backplane channel. (a) Physical structure. (b) Channel frequency response. (c) Channel impulse response.

interference (ISI) completely shuts down the received data eye, so equalization is required to recover the data bits.

An effective method for equalizing such high-loss channels is to implement a decision-feedback equalizer (DFE) in the receiver. A fundamental advantage of a DFE over a linear equalizer such as a peaking amplifier is that it is able to compensate for ISI without amplifying noise or crosstalk [9]. A DFE uses the history of recent data decisions to cancel post-cursor ISI (interference terms from previous bits in the sequence) in the received input signal. As an example, a 15-tap DFE would be helpful in compensating a channel with the impulse response of Fig. 1(c), as most of the significant post-cursors occur within 15 UIs (at 25 Gb/s) of the peak (main cursor). A DFE of such complexity is also effective in dealing with package escape reflections. Let $d$ be the package trace length (as labeled in Fig. 1(a)). Due to the impedance discontinuities introduced by the core via of the package, the solder ball, and the printed circuit board (PCB) via, a reflection is generated which is not fully absorbed by the transmitter and appears at the receiver with an extra delay equal to $2d/c$, where $c$ is the wave velocity inside the package. Assum-

ing typical package materials and a maximum trace length $d$ of 25 mm, this extra delay may be as large as 450 ps, which fits within the time span of a 15-tap DFE operating at data rates close to 30 Gb/s.

A DFE alone, however, is not usually sufficient for equalizing a high-loss channel as it has some inherent limitations. It cannot compensate pre-cursor ISI (interference terms from later bits in the sequence), as well as post-cursor ISI outside its time span. Therefore, its operation is usually complemented by some linear equalization in the form of a feed-forward equalizer (FFE) in the transmitter or a continuous-time linear equalizer (CTLE) such as a peaking amplifier in the receiver. To deal with higher channel losses, recent backplane transceivers [10]-[12] have employed the configuration of Fig. 2, in which a DFE is combined with both a transmit-side FFE and a receive-side CTLE. Such a configuration has also been adopted here for the 28-Gb/s transceiver, whose architecture is described next.

### B. Transceiver Architecture

Fig. 3 presents the top-level architecture of the transceiver, which is configured as a four-port I/O core. Two phase-locked loops (PLLs) with 2:1 dividers generate the half-rate (C2) clocks which are distributed to the four transmitters and four receivers, which all employ half-rate architectures. Each PLL includes two different LC voltage-controlled oscillators (VCOs) so that the oscillator frequency can be varied over a range of 14-28.05 GHz.

The data path of the transmitter includes a 32:4 serializer and a shift register that produces time-delayed quarter-rate tap data streams for a baud-spaced 4-tap FFE. The tap data streams are then distributed to a set of weighted source-series terminated (SST) driver segments, which perform the final serialization to the data output. Asymmetric T-coils are used to compensate for driver output capacitance and parasitics of the electrostatic discharge (ESD) device (low-capacitance silicon-controlled rectifier) and to provide wideband impedance matching [13].

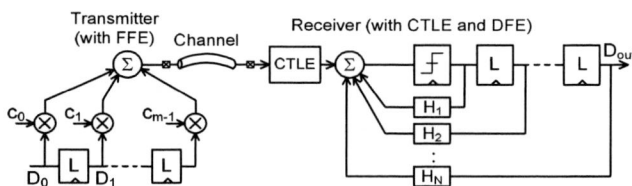

Fig. 2. Serial link equalization with FFE in transmitter and CTLE and DFE in receiver.

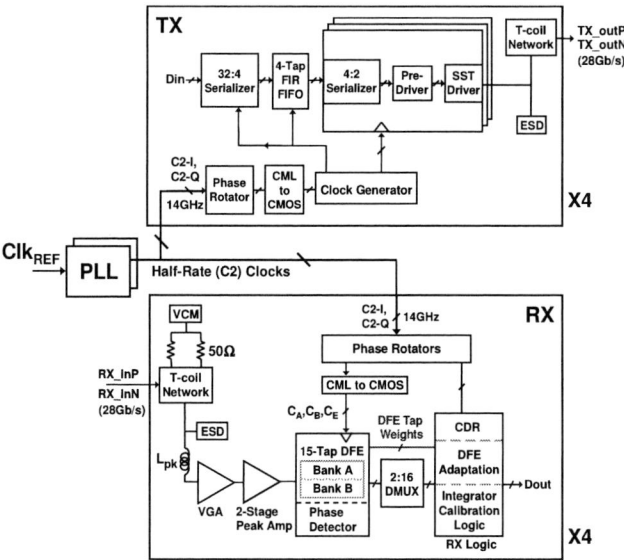

Fig. 3.    Top-level transceiver architecture.

The clock generator produces the sub-rate clocks needed in the serializer stages and provides a mechanism for adjusting the duty cycles of the performance-critical half-rate clocks. For this prototype, the duty cycle control bits are set manually (after measuring the transmitter outputs with an oscilloscope), but adding closed-loop duty cycle correction (DCC) such as that described in [12] would be straightforward. The power supplies for the data (AVDDt) and clock (CVDD) paths have the same nominal value (1.05 V) but are kept separate up to the board level in order to minimize supply noise-induced jitter.

The major functional blocks of the receiver are similar to those in [12], but their underlying circuits are extensively modified to support higher data rates. Inductive peaking (both shunt and series) is used heavily to extend the bandwidths of the variable gain amplifier (VGA) and peaking amplifier. Another inductor $L_{pk}$ (actually, pair of inductors since the signals are differential) is placed in series with the VGA input to provide some fixed passive peaking (about 3-4 dB boost at 12.5-14 GHz). To save area, the 12 peaking inductors of the receiver data path are realized as stacked spirals. The two-stage peaking amplifier (CTLE) provides up to 12 dB boost at half-baud frequency.

The 15-tap DFE employs two redundant banks (A and B), each of which is realized as a half-rate structure. As in the design of [12], the two banks can be swapped between the functions of data detection and adaptation/calibration. CML-based phase rotators generate the half-rate clocks for the two DFE banks and the phase detector that provides edge samples for a digital clock and data recovery (CDR) loop. In contrast with the transmitter, the receiver features closed-loop DCC of clocks $C_A$, $C_B$, and $C_E$, based on the circuits presented in [12]. The analog receiver circuits are powered from a single supply (AVDDr), with a nominal value of 1.05 V. The synthesized logic used for CDR, DFE adaptation, and analog circuit calibrations operates from the main digital supply (VDD), nominally 0.85 V.

## III. TRANSMITTER CIRCUITS

### A. SST Driver

Many recent serial link transmitters [5], [11], [13], [14] have employed SST (voltage-mode) drivers, which offer reduced power consumption (compared with CML output stages) and multiple termination options. Fig. 4 depicts the optimization steps which have been applied in doubling the speed of existing half-rate SST drivers [14]. Fig. 4(a) shows the original structure along with the associated 28-Gb/s eye, which suffers from limited slew rate, incomplete settling, and data-dependent jitter. The root cause of this degradation is parasitic capacitors within the multiplexing stack which may become undriven and store data-dependent charges. As an example, during a pull-up operation, the parasitic capacitor highlighted in gray is charged upwards relatively slowly through the pull-down resistor. This particular source of slow settling can be eliminated by converting the separate pull-up and pull-down resistors to a single shared resistor, as shown in Fig. 4(b). The corresponding eye is substantially improved but still exhibits some data-dependent components, which are due to other parasitic capacitors (gray in Fig. 4(b)) becoming undriven when a clock transistor is turned off. In the final step of the optimization (Fig. 4(c)), the clock transistors are relocated between the even/odd branches and the single shared resistor. There are no undriven circuit nodes in this very simple structure. The clean data eye confirms the superior performance of the proposed circuit, which has been adopted here for the transmitter driver segments.

978-1-4673-6145-3/13 $31.00 © 2013 IEEE        317

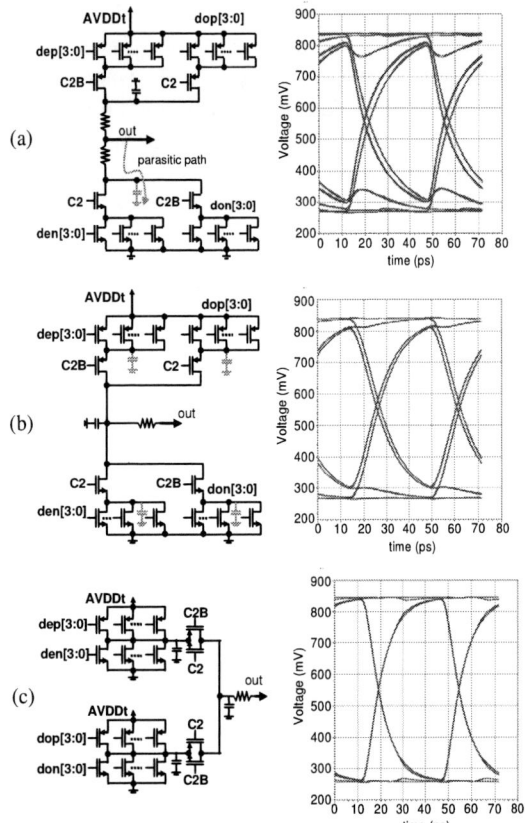

Fig. 4. Evolution of the SST driver speed optimization. (a) Original stacked structure. (b) Stacked structure with single shared linearization resistor. (c) Final structure.

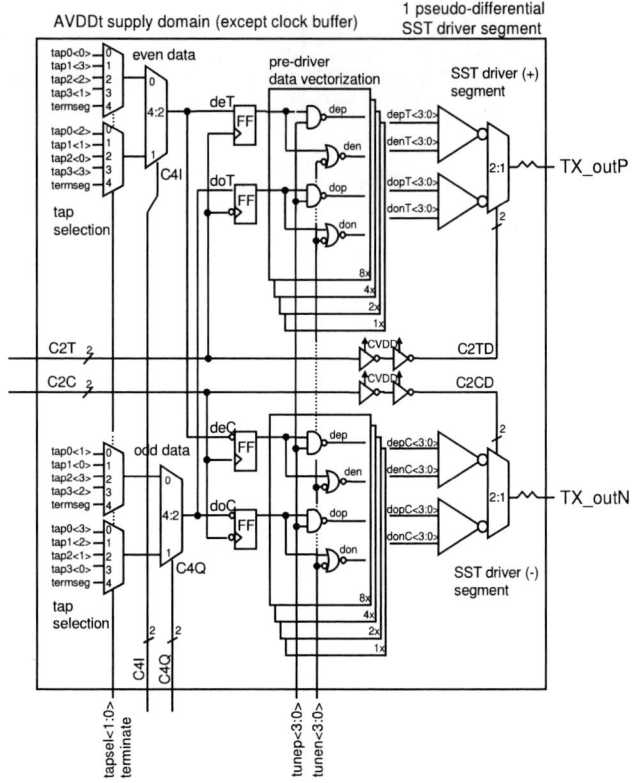

Fig. 5. Tap selection and pre-driver circuitry for one SST driver segment.

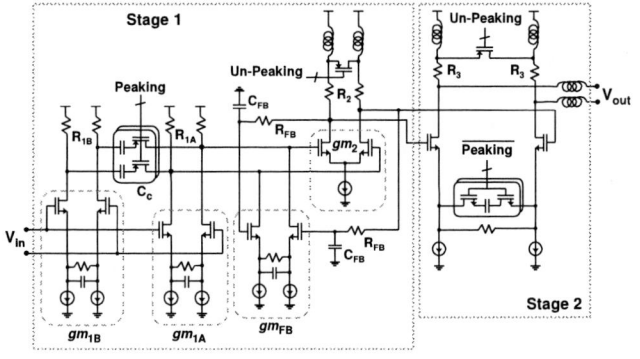

Fig. 6. Schematic of two-stage peaking amplifier.

## B. Tap Selection and Pre-Driver Circuitry

The tap selection and pre-driver circuitry for one SST driver segment is shown in Fig. 5. A static tap selection MUX configures each driver segment as one of the four FFE taps or as a terminating static high or low segment. After being converted to half-rate by 4:2 MUXes and re-timed, the data bits are multiplied by pull-down (tunen<3:0>) and pull-up (tunep<3:0>) impedance tuning vectors in the pre-driver and delivered to the SST driver circuit. The complete driver is composed of 24 weighted SST driver segments. A driver segment weighting of 8x8, 4x4, 4x2 and 8x1 segments has been chosen, which results in an SST driver with 96 equivalent segments.

## IV. RECEIVER CIRCUITS

### A. Peaking Amplifier

Fig. 6 shows the schematic of the two-stage peaking amplifier. The second stage employs a conventional zero-peaked topology with switched capacitive degeneration. A

fundamental limitation of this topology is that its high-frequency gain cannot exceed the dc gain of a non-degenerated CML stage, and even that gain (~6 dB) cannot be obtained given bandwidth limitations. Better peaking is achieved in the first stage by adopting a structure with capacitively-coupled parallel input stages and active feed-back, whose operation is now explained.

Let each differential stage of the peaking amplifier be identified by its transconductance. At low frequencies, input stage $gm_{1B}$ is isolated from the rest of the circuit by capacitor $C_c$, and the active feedback structure operates like the broadband amplifier described in [15]. Negative feedback from stage $gm_{FB}$ reduces the dc gain of stage 1; the ratio of $gm_{FB}$ to $gm_{1A}$ is chosen so that this dc gain is at least 0 dB across all process, voltage, and temperature (PVT) corners.

At high frequencies, capacitor $C_c$ couples together the outputs of the parallel input stages. Because stage $gm_{1B}$ is four times stronger (lower impedance) than stage $gm_{1A}$, with strong capacitive coupling, it overwhelms the feedback from the much weaker stage $gm_{FB}$, and the high-frequency gain of stage 1 approaches the dc gain of *two* cascaded CML stages: $(gm_{1A} + gm_{1B})(R_{1A} \| R_{1B})gm_2 R_2$. The $R_{FB}C_{FB}$ filters reduce the feedback factor at high frequencies, further enhancing the maximum peaking. As depicted in Fig. 7, the peaking is adjusted by switching the value of capacitor $C_c$. In addition to the Peaking control bits, there are Un-Peaking control bits for reducing the peaking; when asserted, these bits connect differential resistances across the shunt inductors, thereby de-Qing them. In total, 17 levels of peaking are available.

### B. DFE

A critical circuit affecting the performance and power efficiency of a DFE is its summing amplifier (summer), which adds the DFE feedback signals to the received input signal. Earlier DFEs [9] employed the resistively loaded CML summer shown in Fig. 8(a), but the need to maintain a small RC time constant for fast settling leads to high power dissipation. Previous works [16], [17] have shown that power consumption can be reduced with the use of current-integrating summers, in which the load resistors are replaced with PMOS reset switches (Fig. 8(b)). However, integrating the analog input signal for 1 UI introduces a 3.9-dB loss penalty at half-baud frequency, which is undesirable in a receiver intended for equalizing high-loss channels. Reference [18] showed that this 3.9-dB loss penalty could be avoided by placing a passgate sample-and-hold (S/H) in front of the integrator, as shown in Fig. 8(c). Including a S/H has a couple of drawbacks, though. The kT/C noise of a low-capacitance sampler may degrade SNR, and kickback from the sampling switch disturbs the previous stage, which may have difficulty recovering by the next sampling interval (especially at these data rates). In this work, the S/H and its associated difficulties are eliminated by adopting the summer of Fig. 8(d), in which the input stage is peaked with an RC degeneration network, whose values are tuned to provide about 3.9 dB of peaking at half-baud frequency.

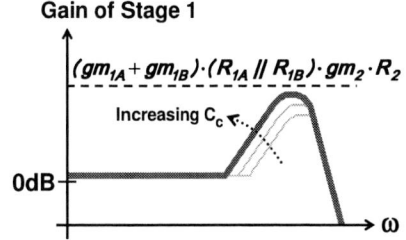

Fig. 7. Frequency response of stage 1 of peaking amplifier.

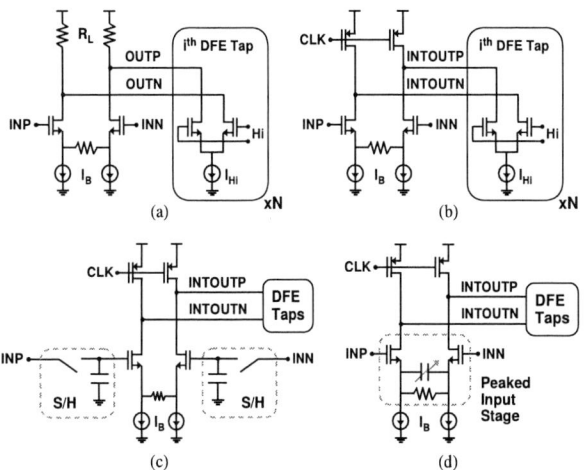

Fig. 8. Evolution of DFE summers. (a) Resistively loaded CML summer. (b) Current-integrating summer. (c) Sampled current-integrating summer. (d) Peaked current-integrating summer.

To relax DFE feedback timing requirements, the first two taps (H1 and H2) are realized speculatively (loop unrolled) in this design. In a previous current-integrating DFE with one speculative tap [16], a separate summer was employed for each speculative path. This overhead quickly becomes excessive as more taps are speculated. In principle, the dc offsets representing the H1 and H2 compensation can be added into the decision-making latches themselves, but inserting extra devices into a latch increases its internal parasitics, which is undesirable at these data rates. As shown in Fig. 9, which presents the block diagram of a DFE half (of one bank), the dc offsets in this design are stored across series capacitors placed between the integrator output and the CML buffers driving the latches. Using a single current-integrating summer to drive all four parallel paths eliminates potential mismatches between summers and saves area. Sense amplifiers are used as the decision-making latches, and the DFE feedback logic is implemented in domino and static CMOS circuitry.

Fig. 10 shows a detailed schematic of the DFE summer along with its timing diagram. The H3-H15 tap circuits employ a return-to-zero (RZ) structure, which minimizes

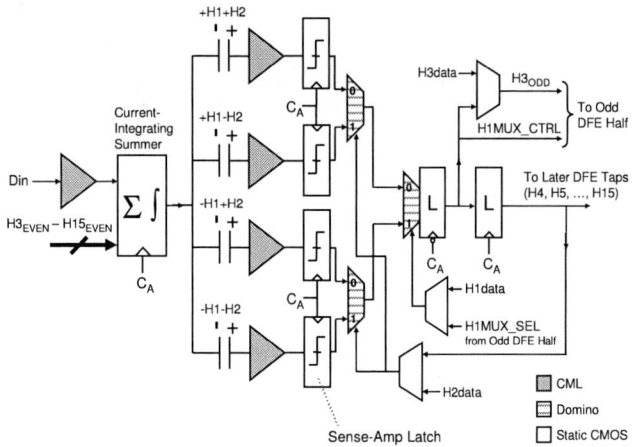

Fig. 9. Block diagram of even DFE half.

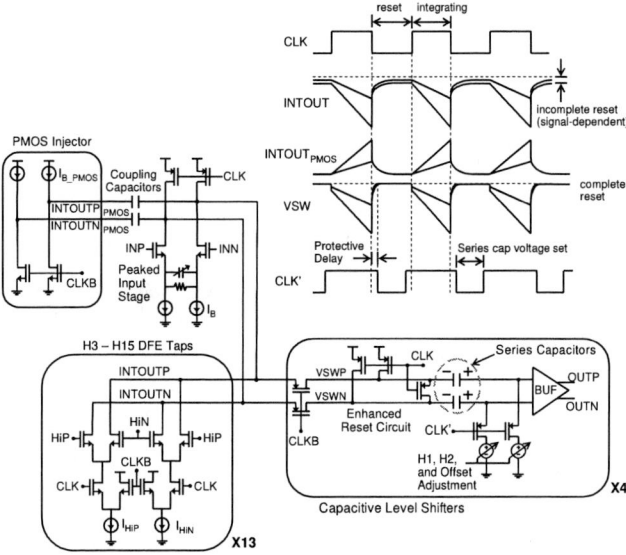

Fig. 10. Detailed schematic and timing diagram of DFE summer.

setup time requirements on the DFE feedback signals [12]. At high data rates, integration times are short, and integrator gain is reduced. Boosting integration currents can help restore gain but causes excessive common-mode drop on the summer output, degrading linearity. This limitation is overcome by introducing a PMOS injector circuit that drives currents (through coupling capacitors) into the summer output nodes, which restores their common-mode. The switches inside the box labeled Capacitive Level Shifters are used to charge the series capacitors. During integrator reset, the left sides of the capacitors are pulled up to the supply; an Enhanced Reset Circuit ensures that signal VSW is accurately reset even when INTOUT is reset incompletely. After VSW is reset, the right sides of

the capacitors are connected to bias voltages representing the desired H1, H2, and offset compensation. Intentional skew between the falling edges of clock signals CLK and CLK' provides a protective delay against making these connections too early.

The integrator bias currents must be calibrated to maintain the gain and linearity of the DFE summer over PVT variations and different data rates. The calibration circuit of [16] included a replica integrator whose output common-mode (at the end of integration) was compared to a reference voltage with a clocked comparator. This technique must be modified for an integrator with a PMOS injector, as there is an infinite number of NMOS and PMOS bias current combinations that yield a given output common-mode. To solve this problem, the integrator calibration loop of Fig. 11 calibrates the NMOS and PMOS bias currents separately in a two-step process. In the first step (with NMOSONLY=1), the PMOS injector bias currents ($I_{B\_PMOS}$) are shut off in the replica integrating summer, and only 2/3 of the NMOS tail devices are active. A 6-bit NCALDAC value is then adjusted so that the bias currents of the active NMOS tail devices yield the desired common-mode drop. In the second step (with NMOSONLY=0 and NCALDAC fixed at its calibrated value), all of the NMOS tail devices are activated, and the 50% increase in total NMOS current drops the output common-mode below the reference level. The PMOS injector bias currents are also turned on, and a 6-bit PCALDAC value is then adjusted so that the PMOS injector produces the desired level of common-mode restoration.

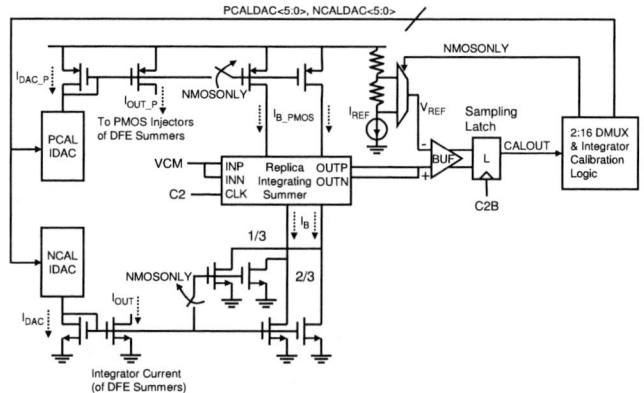

Fig. 11. Integrator calibration loop for setting NMOS and PMOS bias currents of DFE summer.

978-1-4673-6145-3/13 $31.00 © 2013 IEEE          320

## V. EXPERIMENTAL RESULTS

Fig. 12 shows a micrograph of the four-port I/O core, which was fabricated in a 32-nm SOI CMOS process. With the PLL overhead amortized over four lanes, the area of a single transmitter/receiver pair is 0.81 mm². The test chip was packaged in a flip-chip plastic BGA module mounted on a socketed evaluation board. Fig. 13 shows a measured 28-Gb/s differential output eye diagram generated by the transmitter. The measured random jitter (RJ) is 450 fs rms, about twice that predicted in circuit simulations. Such jitter is not a fundamental limitation of the LC-VCO-based PLLs in this technology, as significantly lower RJ (~250 fs rms) has been recently achieved with an updated version of the PLL (including layout refinements).

Fig. 14 shows the measured integrator calibration codes as a function of data rate. As expected, the calibrated values of both NCALDAC and PCALDAC scale roughly linearly with data rate. CLOAD bits control switches (not shown in Fig. 10) which connect extra load capacitors to the integrator output nodes. Increasing the load capacitance raises the calibrated values of both NCALDAC and PCALDAC; this is helpful at lower data rates (where small integration currents may be affected by leakage). At high data rates (> 21 Gb/s), though, the load capacitance should be set to minimum value (CLOAD=0).

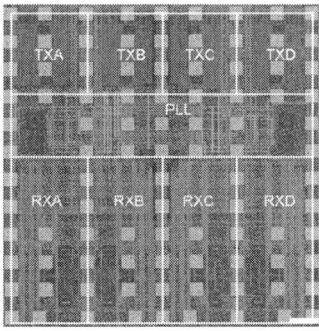

Fig. 12.    Micrograph of four-port I/O core.

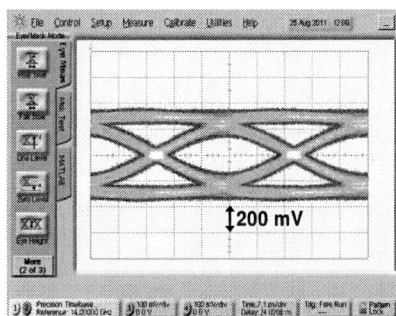

Fig. 13.    Measured 28-Gb/s transmit eye diagram with PRBS31 pattern.

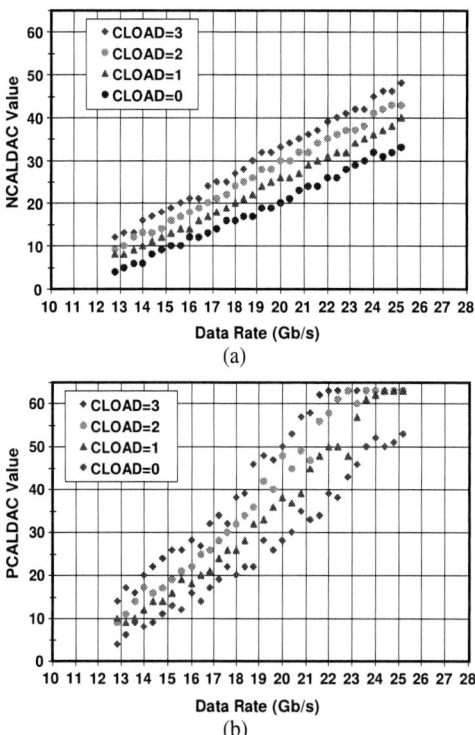

Fig. 14.    Measured integrator calibration codes as function of data rate. (a) NCALDAC codes for NMOS bias currents. (b) PCALDAC codes for PMOS injectors.

Over a short channel, the measured input sensitivity of the receiver at 28 Gb/s is 15 mVppd at a bit error rate (BER) of $10^{-9}$. The equalization performance of the full transceiver is demonstrated at 28 Gb/s over a test channel comprising a 15-inch PCB trace, interconnect cables, and an evaluation card. S-parameter measurements (Fig. 15(a)) of this channel show a loss of 29 dB at 14 GHz; the losses of the transmitter and receiver packages bring the total to 35 dB. Fig. 15(b) shows the equalized bathtub curve with a 28-Gb/s PRBS31 data pattern. The horizontal eye opening is 35.6% at a BER of $10^{-9}$, and operation is error-free (BER < $10^{-13}$) at eye center. The measured power consumption is 693 mW per lane (211 mW for transmitter, 392 mW for receiver, and 90 mW for amortized PLL). The use of known power management schemes (such as shutting down one of the DFE banks when it is not needed) could reduce this power but was not exercised for this prototype.

## VI. CONCLUSION

Data transmission over high-loss electrical channels such as backplanes requires serial link transceivers with advanced equalization. This paper has illustrated a num-

978-1-4673-6145-3/13 $31.00 © 2013 IEEE        321

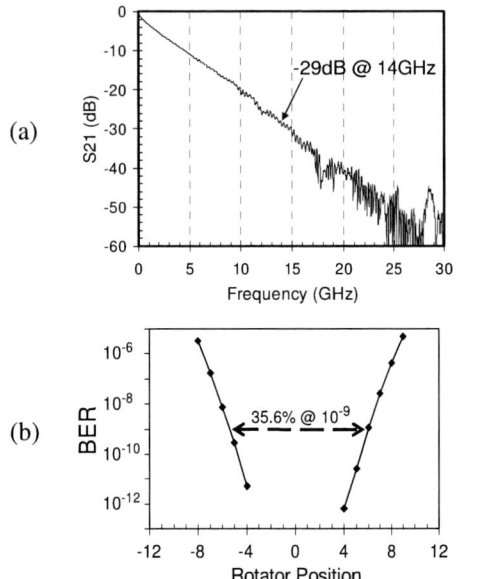

Fig. 15. (a) Frequency response of 15-inch PCB trace, interconnect cables, and evaluation card. (b) Equalized bathtub curve with 28-Gb/s PRBS31 data pattern.

ber of design techniques for extending the data rates of such circuits by describing a 28-Gb/s 4-tap FFE/15-tap DFE transceiver implemented in 32-nm SOI CMOS technology. The proposed SST driver topology eliminates the main speed bottlenecks of previous half-rate designs. The peaking amplifier based on an active feedback structure provides greater high-frequency gain than a conventional zero-peaked differential amplifier. The use of capacitive level-shifters facilitates efficient implementation of DFE architectures with multiple speculative taps. The introduction of a PMOS injector enhances the gain of a current-integrating summer at high data rates. The equalization performance of the transceiver at 28 Gb/s has been demonstrated with error-free operation over a channel with 35-dB loss.

### ACKNOWLEDGMENT

The design of the 28-Gb/s serial link transceiver was a large team effort, so the author would like to thank all the co-authors of [6], as well as other contributing members from IBM Research, IBM STG, and Miromico.

### REFERENCES

[1] Common electrical I/O (CEI) – Electrical and jitter interoperability agreements for 6G+ bps, 11G+ bps and 25G+ bps I/O. Optical In-

ternetworking Forum, Sep. 2011. [Online] Available: http://www.oiforum.com/public/documents/OIF_CEI_03.0.pdf

[2] Fibre Channel Solutions Guide Book 2010. Fibre Channel Industry Association (FCIA) [Online]. Available: http://www.fibrechannel.org/documents

[3] J. D'Ambrosia, IEEE 802.3WG Closing Plenary Report, IEEE P802.3bj 100 Gb/s Backplane and Copper Cable Task Force, Mar. 2012 [Online]. Available: http://www.ieee802.org/3/minutes/mar12/0312_bj_close_report.pdf

[4] InfiniBand Roadmap. InfiniBand Trade Association (IBTA) [Online]. Available: http://www.infinibandta.org/content/pages.php?pg=technology_overview

[5] M. Harwood et al., "A 225mW 28Gb/s SerDes in 40nm CMOS with 13dB of analog equalization for 100GBASE-LR4 and optical transport lane 4.4 applications," IEEE ISSCC Dig. Tech. Papers, pp. 326-327, Feb. 2012.

[6] J. F. Bulzacchelli et al., "A 28-Gb/s 4-tap FFE/15-tap DFE serial link transceiver in 32-nm SOI CMOS technology," IEEE J. Solid-State Circuits, vol. 47, no. 12, pp. 3232-3248, Dec. 2012.

[7] A. Healey and C. Morgan, "A comparison of 25 Gbps NRZ & PAM-4 modulation used in legacy & premium backplane channels," Proc. IEC DesignCon, Jan. 2012.

[8] P. Patel and B. Barnett, Experimental Test Fixture S-Parameters 100 Gb/s Backplane Study Group, May 2011 [Online]. Available: http://www.ieee802.org/3/100GCU/public/ChannelData/IBM_11_0518

[9] T. Beukema et al., "A 6.4-Gb/s CMOS SerDes core with feed-forward and decision-feedback equalization," IEEE J. Solid-State Circuits, vol. 40, no. 12, pp. 2633-2645, Dec. 2005.

[10] A. K. Joy et al., "Analog-DFE-based 16Gb/s SerDes in 40nm CMOS that operates across 34dB loss channels at Nyquist with a baud rate CDR and 1.2Vpp voltage-mode driver," IEEE ISSCC Dig. Tech. Papers, pp. 350-351, Feb. 2011.

[11] F. Zhong et al., "A 1.0625 ~ 14.025 Gb/s multi-media transceiver with full-rate source-series-terminated transmit driver and floating-tap decision-feedback equalizer in 40 nm CMOS," IEEE J. Solid-State Circuits, vol. 46, no. 12, pp. 3126-3139, Dec. 2011.

[12] G. R. Gangasani et al., "A 16-Gb/s backplane transceiver with 12-tap current integrating DFE and dynamic adaptation of voltage off-set and timing drifts in 45-nm SOI CMOS technology," IEEE J. Solid-State Circuits, vol. 47, no. 8, pp. 1828-1841, Aug. 2012.

[13] M. Kossel et al., "A T-coil-enhanced 8.5 Gb/s high-swing SST transmitter in 65 nm bulk CMOS with < -16 dB return loss over 10 GHz bandwidth," IEEE J. Solid-State Circuits, vol. 43, no. 12, pp. 2905-2920, Dec. 2008.

[14] C. Menolfi et al., "A 16Gb/s source-series terminated transmitter in 65nm CMOS SOI," IEEE ISSCC Dig. Tech. Papers, pp. 446-447, Feb. 2007.

[15] S. Galal and B. Razavi, "10-Gb/s limiting amplifier and laser/modulator driver in 0.18-μm CMOS technology," IEEE J. Solid-State Circuits, vol. 38, no. 12, pp. 2138-2146, Dec. 2003.

[16] M. Park, J. Bulzacchelli, M. Beakes, and D. Friedman, "A 7Gb/s 9.3mW 2-tap current-integrating DFE receiver," IEEE ISSCC Dig. Tech. Papers, pp. 230-231, Feb. 2007.

[17] L. Chen, X. Zhang, and F. Spagna, "A scalable 3.6-to-5.2mW 5-to-10Gb/s 4-tap DFE in 32nm CMOS," IEEE ISSCC Dig. Tech. Papers, pp. 180-181, Feb. 2009.

[18] T. O. Dickson, J. F. Bulzacchelli, and D. J. Friedman, "A 12-Gb/s 11-mW half-rate sampled 5-tap decision feedback equalizer with current-integrating summers in 45-nm SOI CMOS technology," IEEE J. Solid-State Circuits, vol. 44, pp. 1298-1305, Apr. 2009.

# Design Considerations for Low-Power Receiver Front-End in High-Speed Data Links

S. Shekhar, J. E. Jaussi, F. O'Mahony, M. Mansuri and B. Casper

Intel Corporation
JF2-04, 2111 NE 25th Ave, Hillsboro, OR, 97124 USA

**Abstract-This paper presents different design considerations for the receiver front-end (RXFE) in low-power, high-speed data links. Specifications for the RXFE are defined and explained in detail, including their impact on the overall link performance. Based on these specifications, low-power RXFE topologies are then analyzed to illustrate the design and performance tradeoffs. Techniques to properly characterize and measure the RXFE specifications are also provided, supplemented with measurement results from three different low-power links operating at 10Gb/s, 16Gb/s and 20Gb/s.**

## I. INTRODUCTION

As the aggregate data rate for high-speed I/O links starts to exceed Tb/s for microprocessor systems, designing transceiver circuits to consume low active power has become critical. Based on the link requirements, a co-design of channel interconnects and circuits yields the best power/performance tradeoff [1]. For a specific interconnect, it is power efficient to design the link to operate at a moderate per-pin data rate that is enabled with simple equalization schemes and achieve high aggregate bandwidth (BW) using wider interfaces. These simpler low-order equalization schemes, like transmit pre-emphasis (TXPE) and/or continuous-time linear equalization (CTLE) at the receiver (RX), can compensate for 10-18dB of loss at the Nyquist frequency. For channels that have higher loss and discontinuities, multiple taps of RX decision-feedback equalization (DFE) are typically used. Multiple DFE taps degrade the power efficiency of the link, more so if the taps are "loop-unrolled" to overcome circuit BW limitations because speculative samplers must be used.

Power efficiency can be improved by using low-power clocking techniques like injection-locking [2] or CMOS clocking [3]. A large portion of link power comes from the transmitter (TX) driver, owing to the low impedance drive requirement as well as high signal swing at its output. Significant power savings can be obtained if the TX is designed with a low-swing driver, in conjunction with a highly sensitive RX. Fig. 1 illustrates the conceptual power consumption trade-off between TX swing and RX sensitivity. If the TX swing is reduced, the RX sensitivity must improve. The net result is a reduction in TX power and an increase in RX power. There generally exists an optimal operating point for the link power consumption with respect to the TX swing and RX sensitivity. This in turn, necessitates a good understanding of RX design, specifications and tradeoffs to optimize the power/performance of the link.

Section II of this paper describes the broad criteria behind

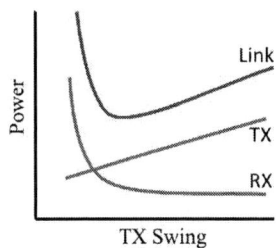

Fig. 1. Power consumption of the TX, RX and entire link as a function of TX swing

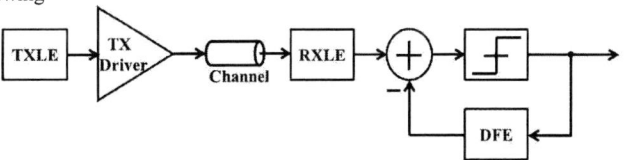

Fig. 2. A high-speed link showing different linear and non-linear equalization schemes

different receiver front-end (RXFE) architectures selection. Section III defines different specification metrics for an RXFE and explains their impact on the performance of the link. A few examples of different low-power RX architectures are then presented in Section IV, and design tradeoffs are analyzed. Section V presents techniques to characterize and measure various specifications.

## II. RXFE ARCHITECTURE SELECTION

### A. Equalization

Fig. 2 shows a high speed link with different possible equalization schemes at the TX and RX. Both TXLE and RXLE are examples of linear equalization (LE), whereas DFE is an example of non-linear equalization. TXLE is popularly implemented in discrete-time as TXPE, while RXLE can be implemented using a CTLE or feed-forward equalizer (FFE). The single-bit pulse response of a channel is shown in Fig. 3, with the cursor at $N$th unit-interval (UI) and non-zero magnitude of ISI over other UIs due to dispersion. A TXPE mitigates ISI to improve the overall pulse response. In Fig. 3, a 3-tap equalizer cancels the immediate pre-cursor and post-cursor ISI. However, for a fixed total transmit power, TXPE attenuates the cursor magnitude. In the frequency domain, it attenuates the low-frequency content of the signal. An active CTLE in the RXFE, on the other hand, provides gain peaking at higher frequency [4]. In the time domain, this is reflected as an increase in the main cursor and a reduction in the pre- and post-cursors.

978-1-4673-6145-3/13 $31.00 © 2013 IEEE

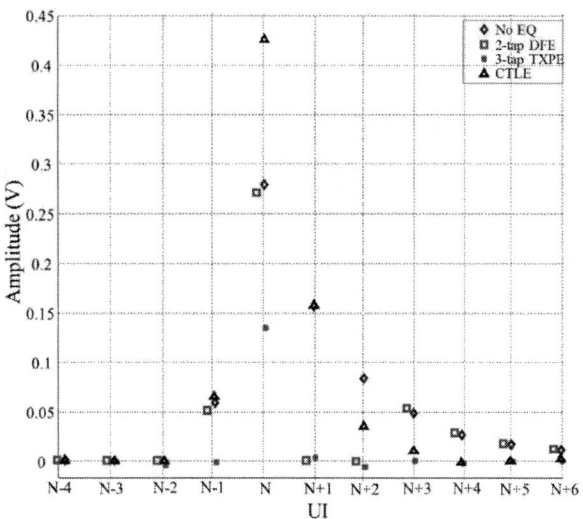

Fig. 3. Single-bit pulse response of a channel without equalization, and with 2-Tap DFE, 3-tap TXPE, and CTLE, respectively

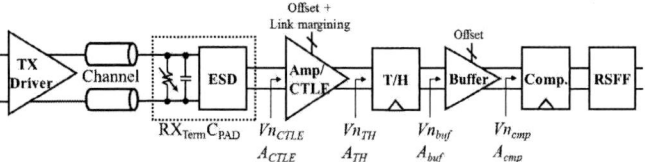

Fig. 4. A low-power, high-speed link showing the TX driver and RXFE

A DFE, being a feedback mechanism, is only able to cancel the post-cursor ISI, but has the advantage of equalizing without any enhancement of noise or crosstalk. It is also easier to independently set and adapt each DFE tap. A DFE is suited to equalize channels that have significant discontinuities.

### B.    Sub-rate Implementation

An RXFE can be implemented at the full rate RX clock frequency ($f_{clock} = f_{data-rate}$), or at a sub-rate clock frequency ($f_{clock} = f_{data-rate}/N$) where $N$ comparator-slices are time-interleaved using $N$ phases of the clock. Popular implementations include half-rate [1] and quarter-rate architectures [3], [4]. An advantage of using a $1/N$-rate implementation is that the decision time for the slices increases by $N$, relaxing their BW and settling time requirements. Note that interleaving slices does not relax the feedback delay in a DFE. For a mixed-mode design, such as a comparator slice in a given CMOS process, the rate of increase in power consumption significantly increases beyond a certain operating frequency. By operating each of the multiple slices below this threshold frequency, an overall low-power design can be achieved at a high aggregate data-rate. However, a power penalty is paid in generating and distributing multiple clock phases, correcting duty-cycle and phase errors [5], driving an increased load seen by the front-end CTLE, etc.

### III.    RXFE Specifications

Fig. 4 shows a typical implementation of a low-power high-speed link with a TX and RXFE. The signal from the transmit driver is attenuated and dispersed as it passes through the channel. The RXFE may consist of RX termination, ESD, an optional CTLE, a track-and-hold (T/H) switch, a buffer amplifier, regenerative comparator(s) and precharge-removing flip-flop(s). The RX termination provides termination for the channel to minimize reflections, and the ESD circuits provide discharge paths for charged-device model (CDM) and human body model (HBM) events. The CTLE provides gain-peaking at higher frequencies to compensate for the low-pass response of the channel. The T/H switch sets up the timing aperture [6] to reduce the impact of jitter; the buffer amplifier provides additional gain and attenuates the kickback from the comparator back to the T/H switch. The comparator provides further gain and resolves the analog signal to full rail. The pre-charge removal circuit converts the signal into NRZ signal.

Next, we discuss different RXFE specifications in detail. The specifications described below apply to a low-power RXFE with DFE as well, where the amplifier/CTLE block in Fig. 4 can model an automatic gain control (AGC) and/or summer. The summer is typically implemented using a differential amplifier with linear or integrating load [7].

### A.    Input Impedance

The characteristic impedance of the channel is 45-50Ω (single-ended), ideally up to the Nyquist frequency of transmission. The RX termination ($RX_{TERM}$) is matched to this characteristic impedance to minimize reflection, and creates an input pole with the total capacitance at this node. The total capacitance ($C_{PAD}$) is the sum of pad capacitance, ESD capacitance and the parasitic circuit capacitance looking into the CTLE. For large values of $C_{PAD}$, $RX_{TERM}$ may be designed to be slightly lower (30~45Ω) to increase the -3dB BW at the RX input. However, there are two tradeoffs associated with lower $RX_{TERM}$ – it can degrade the voltage margin because of reduced input swing to the RX with a voltage-mode TX driver, and the impedance mismatch can cause reflections leading to ISI. The latter may not be severe if the return path from the nearest discontinuity has high loss.

$RX_{TERM}$ can be implemented with passive resistances, or with active devices operating in linear region. Two concerns regarding the implementation are linearity and parasitic capacitance. Passive resistances have the benefit of high linearity. However, if the signal swing at the $RX_{TERM}$ is low, active transistors may be chosen for termination. Based on the layout rules and the CMOS technology used, the latter may be implemented with lower parasitic capacitance overhead. The termination value can be easily reconfigured to compensate for PVT variations by implementing the resistances in unit array with series-switches.

### B.    ESD

The size of the ESD device depends upon the specification. For example, JEDEC Class II standard specifies a 200–500V of CDM (Charged device model) ESD protection. This necessitates ESD diodes and clamps that come at the expense of large parasitic capacitance. Fig. 5 shows the insertion loss of a channel comprising seven inches of FR4 with sockets and package vs. frequency, as the total pad capacitance is scaled up

978-1-4673-6145-3/13 $31.00 © 2013 IEEE

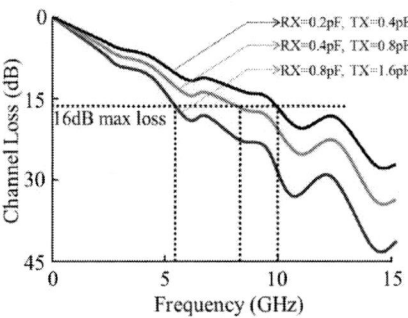

Fig. 5. Effect of pad-capacitance on channel loss

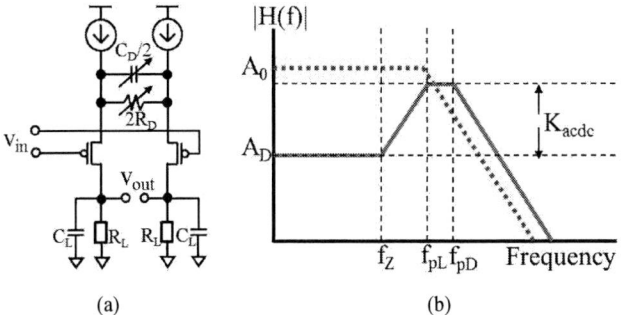

(a)                                        (b)

Fig. 6. (a) A differential amplifier based CTLE and (b) its frequency response with and without source degeneration.

as 0.6pF, 1.2pF and 2.4pF. On the assumption that the link can equalize up to 16dB of loss at the Nyquist frequency with reasonable power consumption, the maximum achievable data rate (MADR) degrades as 20Gb/s, 16Gb/s and 12Gb/s, respectively. Clearly, large pad capacitance severely degrades the loss at the Nyquist frequency, and limits the MADR.

Several layout techniques are usually employed to decrease both the device and routing capacitances of the ESD devices [8]. On-die transmission lines are used to mitigate the effect of lumped capacitances by distributing the routing capacitance and matching the characteristic impedance of the on-die trace to the package [1]. Additional improvement can be obtained by tuning out the pad capacitance with inductive-peaking techniques [9], [10]. The peaking can be designed to also provide some amount of equalization.

*C.    Gain and Bandwidth*

The voltage gain of an RXFE can be divided into linear and regenerative regions. The linear gain is calculated from the input of the CTLE amplifier, through the buffer amplifier and the linear gain stage of comparator. When the voltage difference between the output nodes of the strong-arm latch reach a threshold, the positive feedback initiates the regenerative action, thereby, providing further gain [11].

For a given data rate, the -3dB BW of the amplifier in the CTLE as well as the buffer amplifier is usually designed to be about ~0.7X the sampling frequency. This is to ensure that the loss near the Nyquist frequency is minimal ($\approx$ 0dB). The BW of the T/H switch is designed to be 2~3X the sampling frequency for fast settling and low residual error. Sizing up the sampling switch has a tradeoff of lower ON resistance to higher clock feed-through and power consumption.

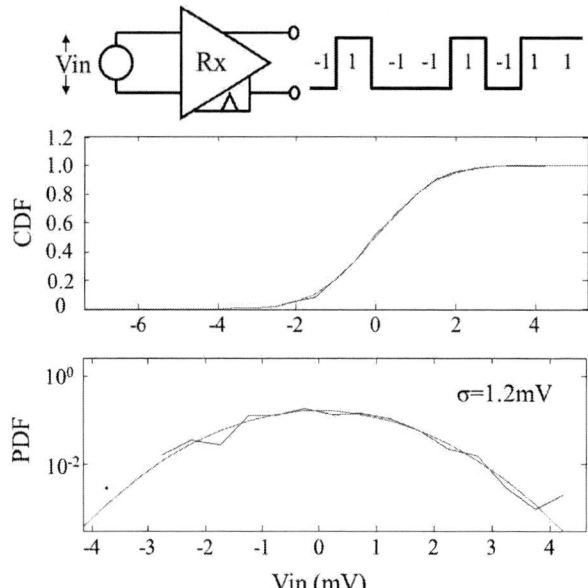

Fig. 7. Simulation methodology for calculating the input-referred thermal noise of the RXFE

Fig. 6(a) shows a popular topology for CTLE where $R_D C_D$ degeneration is used in a differential amplifier to provide a high frequency boost [12]. The differential amplifier without the degeneration has a single pole response given by the transfer function:

$$H(s) = \frac{g_m R_L}{1 + s R_L C_L} \quad (1)$$

The low frequency gain is given by $A_0 = g_m R_L$, and the -3dB bandwidth is given the pole frequency, $f_{pL} = 1/(2\pi R_L C_L)$. Adding the $R_D C_D$ degeneration introduces another pole ($f_{pD}$) and zero ($f_z$) in the transfer function:

$$H(s) = \frac{g_m R_L (1 + s R_D C_D)}{(1 + s R_L C_L)(1 + g_m R_D + s R_D C_D)} \quad (2)$$

The low frequency gain is reduced to $A_D = g_m R_L/(1 + g_m R_D)$. By positioning the zero, $f_z = 1/(2\pi R_D C_D)$, to be smaller than the pole, $f_{pL}$, gain boosting at high frequency can be achieved. The ratio of maximum high frequency gain to low frequency gain is defined as $K_{acdc}$, and is a measure of equalization provided by the CTLE. $K_{acdc}$ is controlled by the separation of $f_z$ from $f_{pL}$. Fig. 6(b) shows the frequency response of the amplifier without and with $R_D C_D$ degeneration.

*D.    Input-Referred Thermal Noise*

Thermal noise in the RX degrades the voltage margin at the input of the RX and degrades the RX sensitivity. For the overall RX noise ($Vn_T$), the thermal noise of each block is integrated over frequency, and referred to the input of the RX, according to the Friis Equation:

$$Vn_T = \sqrt{Vn_{CTLE}^2 + \left(\frac{Vn_{TH}}{A_{CTLE}}\right)^2 + \left(\frac{Vn_{buf}}{A_{CTLE}A_{TH}}\right)^2 + \left(\frac{Vn_{cmp}}{A_{CTLE}A_{TH}A_{buf}}\right)^2} \quad (3)$$

Here, $Vn_{CTLE}$ ($A_{CTLE}$), $Vn_{TH}$ ($A_{TH}$), $Vn_{buf}$ ($A_{buf}$) and $Vn_{cmp}$ denote the total input-referred integrated RMS noise (low frequency voltage gain) of the CTLE, T/H, buffer and comparator, respectively. From the equation, it is apparent that the voltage gain of the CTLE and the buffer amplifier helps in decreasing the input referred RX noise. Moreover, the input referred noise of the CTLE should be minimized in the RXFE to decrease $Vn_T$. The RMS noise of the T/H can be shown to be $\sqrt{KT/C}$ where $C$ is the total hold capacitance, including the parasitics and $A_{TH} \approx 1$ [13]. The noise of stages following the regenerative comparator is negligible owing to its high gain.

The thermal noise of the regenerative comparator can be quantified using a time-domain simulation by including a model for transient thermal noise. Fig. 7 shows the simulation methodology. DC voltage sources, $Vcm \pm Vin/2$, are applied to the differential inputs of the regenerative comparator where Vcm is the nominal common-mode voltage and $Vin$ is the differential offset. The output of the comparator is then sampled and all the 1's and 0's are counted over time. $Vin$ is then swept across a range of voltages to generate a noise CDF and PDF. A Gaussian fit of this PDF yields the input referred noise for the comparator.

### E. Input-Referred Residual Offset

Random and systematic device variations cause mismatch in the two halves of a differential circuit, resulting in an input offset voltage error. In an RXFE, this voltage offset is present in each stage of the amplification – CTLE, T/H, buffer amplifier and comparator. This offset voltage can be cancelled by introducing a deliberate offset in the opposite direction using voltage- or current-based DACs, or device-mismatch [14], [15], limited by the LSB of offset cancellation. The overall input-referred offset voltage is given by

$$Vos_T = Vos_{CTLE} + \frac{Vos_{TH}}{A_{CTLE}} + \frac{Vos_{buf}}{A_{CTLE}A_{TH}} + \frac{Vos_{cmp}}{A_{CTLE}A_{TH}A_{buf}} \quad (4)$$

where $Vos_{CTLE}$, $Vos_{TH}$, $Vos_{buf}$ and $Vos_{cmp}$ are the residual input referred offset for the CTLE, T/H, buffer and comparator after its offset cancellation, respectively. Similar to input-referred noise, $Vos_T$ can be reduced by increasing the gain of the CTLE. Any residual $Vos_T$ directly degrades the sensitivity of RXFE.

There is a tradeoff between the range and resolution of offset voltage cancellation. A small LSB results in low residual voltage, while a large range helps in covering $\pm 3\sigma$ of expected offset-voltage mismatch. Beyond the range needed for offset cancellation, additional voltage offset range, typically $\pm 50\text{-}100\text{mV}$, is needed for link margining (Section V).

### F. Power Supply Noise and Common-Mode Noise

Because of the impact of RX sensitivity to link margin, its susceptibility to power supply and common-mode variations must be taken into account. Power supply noise can be self-induced or coupled from other circuits sharing the same supply. Determined by package inductance and on-chip capacitance, the on-chip resonant supply noise is typically in the 50-300MHz frequency range. Such a long period of supply noise, along with large amplitudes, adversely affect the chip performance by altering sampling apertures [15], [16] and adding timing and voltage uncertainty to the link eye. Power supply noise can be reduced by increasing the bypass capacitor to track Vcc-Vss noise and/or adding supply regulators.

Power-supply sensitivity ratio (PSSR) is defined as the ratio of the change in input-referred offset to the change in the power supply voltage.

$$PSSR = \frac{\Delta V_{offset}}{\Delta V_{supply}} \quad (6)$$

Similar to power-supply noise, any common-mode noise present in the RX degrades performance. Common-mode noise can originate from varying output voltage of the regulator in a voltage-mode TX due to data-dependent current flow, common-mode coupling of noise from an RF power-amplifier in an SOC, cross-talk in the interconnects, etc. Common-mode sensitivity ratio (CMSR) is defined as the ratio of the change in input referred offset to the change in the input common-mode voltage.

$$CMSR = \frac{\Delta V_{offset}}{\Delta V_{cm}} \quad (7)$$

Both PSSR and CMSR can be simulated by sweeping the induced voltage-offset across the entire range and tracking the change in induced offset at a given step at varying levels of Vcc and Vcm. The resulting simulated curves are similar in nature to the measured plots shown later in Section V (Fig. 16).

For an ideal differential receiver, the impact of these noise sources on the differential output should be zero because of the circuit symmetry. However, PSSR and CMSR deteriorate significantly when effects of device variations are taken into account. Typical offset calibration algorithms cancel offset by inducing a deliberate offset in the opposite direction. However, this method does not optimally compensate for the incremental impact of power supply or common-mode noise due to circuit imbalances caused by variation. Therefore, an optimal DC cancellation of input referred offset does not result in zero or even optimal PSSR or CMSR [17]. PSSR (CMSR) can be quantified in simulation by introducing random device variations in the circuit, cancelling the input referred DC offset and then performing the PSSR (CMSR) simulations as mentioned above.

### G. Comparator Metastability

When a regenerative comparator such as a strong-arm latch (SAL) (Fig. 8) is operating in the positive feedback mode, driven by the two inverters, the voltage difference between its output nodes change exponentially. In this time period ($T_{latch} < UI/2$), this voltage difference must reach sufficiently close to $Vcc$ ($\sim\Delta V_{logic}$) to be resolved properly by the succeeding RS F/F, in order to avoid any metastability. A minimum required signal at the input of the latch is necessary to prevent this metastability, given by [18]

$$\Delta V_{met} = \Delta V_{logic} e^{-\frac{G_m}{C_L}T_{latch}} \quad (8).$$

where $G_m$ is the transconductance of each inverter, and $C_L$ is the load capacitance. The metastability voltage can be reduced by increasing the SAL current, reducing the load capacitance or increasing the gain in the preceding stages. The input-referred metastability voltage is given by

$$\Delta V_{met,i} = \frac{\Delta V_{met}}{A_{CTLE}A_{T-H}A_{buf}} \quad (9).$$

978-1-4673-6145-3/13 $31.00 © 2013 IEEE

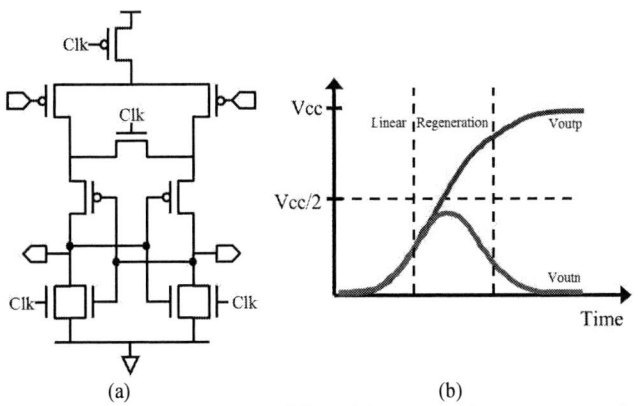

Fig. 8. (a) Schematic and (b) differential output voltage response of a strong-arm latch

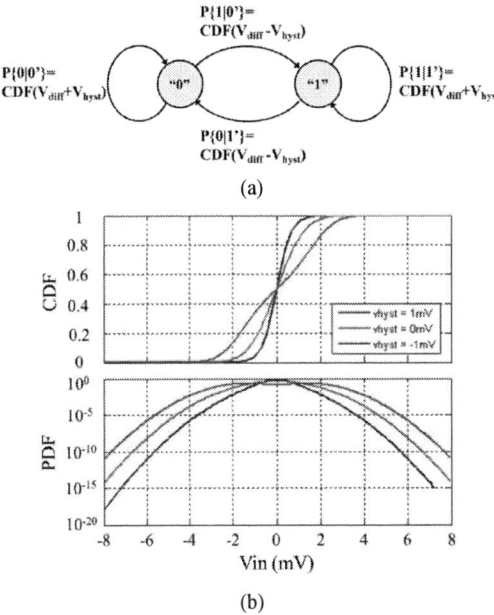

Fig. 9. (a) State-transition diagram showing the effect of hysteresis, and (b) its PDF and CDF profile

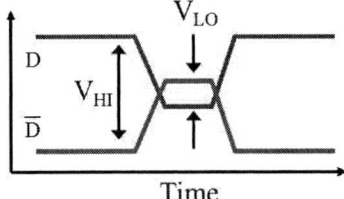

Fig. 10. Lone pulse differential input for simulating RXFE hysteresis and ISI.

### H. Hysteresis

Hysteresis in the RX arises from any bias in the clocked comparator based on the previously resolved state. Hysteresis can be either positive – a bias to the previous state, or negative – a bias towards the complement of the previous state. This memory effect arises from phenomena such as inadequate BW, charge feed-through and kickback, insufficient comparator precharge, non-linearity, etc.

Hysteresis and random noise both modify the RX uncertainty PDF but they can be separated because of their different signatures. Fig. 9(a) shows a state transition diagram showing the effect of hysteresis. Combining hysteresis with the thermal noise in a comparator results in a distorted normal distribution, as shown in Fig. 9(b). Note that the hysteresis distorts the center of the PDF, but not the tails. Residual ISI of the RXFE, arising due to circuit gain-BW limitations, is affected by the hysteresis. One method to characterize hysteresis [19] and residual ISI (due to circuit gain-BW limitations) in the RXFE is to use a lone pulse transient response. Fig. 10 shows the differential lone pulse to the input of the RXFE. With the expected $Vcm$ and $V_{HI}$ at the input of the RXFE, and an optimal sampling point determined, $V_{LO}$ is swept such that the ratio $V_{LO}/V_{HI}$ is made smaller. The smallest (worst) $V_{LO}$ that the RXFE can resolve gives the hysteresis and residual ISI voltage ($V_{hyst}$).

### I. Overall RX Sensitivity

Sensitivity of the RX is defined as the minimum signal required at the input of the RX to attain a specified bit-error ratio (BER). Worst-case RX sensitivity can be shown to be

$$Vsen_{pp} = 2Q|Vn_T| + |Vos_T| + |\Delta V_{met,i}| + |V_{hyst}| + |PSSR \cdot V_{psn}| + |CMSR \cdot V_{cmn}| \quad (10)$$

where the $Q$ parameter is about 7 for BER=$10^{-12}$, and a DC balanced NRZ input signal with equiprobable ones and zeros and Gaussian-distributed thermal noise is assumed. A *high-sensitive* RX actually implies a low $Vsen_{pp}$.

Apart from the factors discussed earlier that are accounted for in (10), there are other link parameters that impact RX sensitivity as well, such as jitter, non-linearity, input common-mode range (ICMR), etc. Any uncorrelated jitter between the RX clock and data, or jitter beyond the jitter tracking-BW of the clock recovery circuitry [5] can be translated into voltage noise. Linearity of the CTLE becomes important when the signal swing at its output is large, as the non-linearity degrades eye-margin, especially eye-height. Linearity becomes more critical in RXFEs where the CTLE utilizes a multi-stage cascaded topology and/or active inductors, and in RXFEs employing DFE. Even though DFE itself is a non-linear equalization scheme, it requires AGC and summers to be linear. Common-mode voltage at the input of the RXFE determines the operating point of front end amplifier. Beyond a certain ICMR, the performance of the amplifier degrades, directly affecting all the above link parameters. For links operating at low supply voltages for power considerations, the ICMR can be severely restricted for amplifiers or SALs.

## IV. RXFE EXAMPLES

Next we discuss examples of low-power RXFE and relate the above design considerations to the architecture selection. Fig. 11(a) shows a quarter-rate RXFE architecture operating at 20Gb/s in 90nm CMOS [4]. The link uses 4 tap TXPE and a full rate CTLE to equalize 16dB of channel loss at the Nyquist frequency of 10GHz. A primary design intent of this RX was to show a high-speed operation up to 20Gb/s. Current-based DACs (I-DACs) are used for per-slice voltage offset cancellation in the buffer amplifier, and for offset cancellation

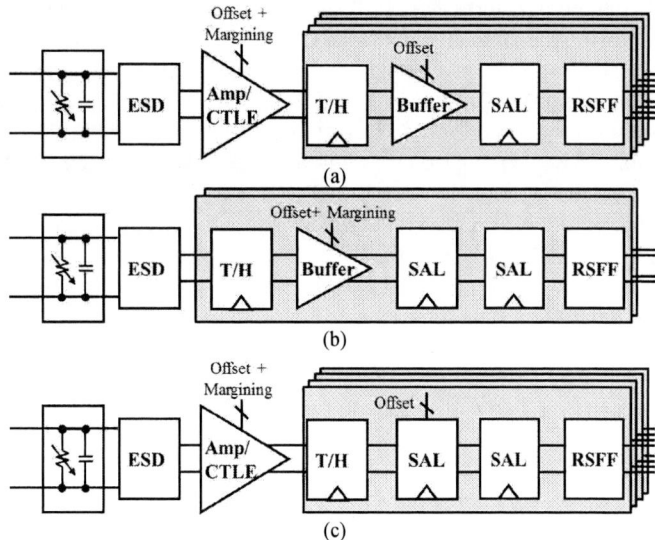

(a)

(b)

(c)

Fig. 11. Examples of RXFE Architecture: (a) Quarter-rate 20Gb/s in 90nm CMOS [4], (b) half-rate 10Gb/s in 45nm CMOS [1], and (c) Quarter-rate 16Gb/s in 32nm LP CMOS [3]

TABLE I
COMPARISON BETWEEN DIFFERENT RXFES OF FIG. 11

| Reference | [4] | [1] | [3] |
|---|---|---|---|
| CMOS Process | 90nm | 45nm | 32nm LP |
| Supply [V] | 1.2 | 0.8 | 1.08 |
| RX Pad Cap [fF] | <100 | 480 | 350 |
| TX Pad Cap [fF] | <400 | 520 | 380 |
| Data Rate [Gb/s] | 20 | 10 | 16 |
| Loss @ Nyq [dB] | 16 | 8-12 | 13/18* |
| TX Swing [mV$_{ppdiff}$] | 900 | 150 | 360 |
| CTLE Equalization | Yes | No | Yes |
| $Vn_T$ [mV rms] | 1 | 1.3 | 1.2 |
| Voltage Offset Range [mV] | ±200 | ±100 | ±100 |
| Resolution | 8b | 6b | 6b |
| Power [mW] | 24 | 2.42 | 12.64 |
| Normalized Power [pJ/b] | 1.2 | 0.24 | 0.79 |

*Including RXFE BW limitations

and link margining in the CTLE, respectively. Quarter rate operation was necessary to achieve 20Gb/s in 90nm CMOS process.

Fig. 11(b) shows a half-rate RXFE architecture operating at 10Gb/s in 45nm CMOS [1]. The design intent of this RX was to reduce the overall link power consumption, while supporting 8-12 dB of loss at the Nyquist frequency of 5GHz. Signaling link analysis showed that 2-tap of TXPE was sufficient for the link, and an RX-side CTLE was not required. Having no CTLE in the RXFE eliminated the power overhead of a full-rate amplifier and eased the sizing of the T/H and input of buffer amplifiers in the slices. It, however, resulted in non-unified link margining. Therefore, careful efforts were needed in designing the buffer amplifiers to still maintain high RX sensitivity (low $Vsen_{pp}$) without the CTLE.

Fig. 11(c) shows a quarter-rate RXFE architecture operating at 4-16Gb/s in 32nm low-power (LP) CMOS [3]. The design intent of this RX was to show a scalable data-rate operation (4 - 16Gb/s) with power supply (0.6 – 1.08V), at a low power dissipation. The buffers in each of the slices are removed for

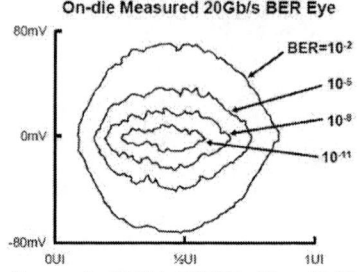

Fig. 12. BER eye-diagram for 20Gb/s RXFE in 90nm CMOS [4]

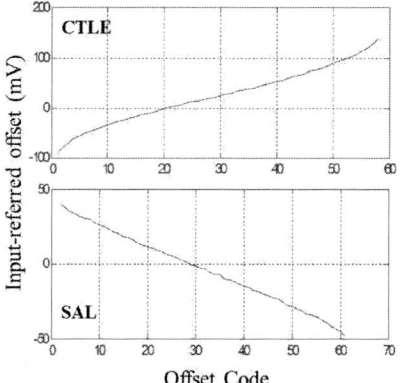

Fig. 13. Measured input-referred offset voltage vs. the offset code for CTLE and SAL for 16Gb/s RXFE in 32nm LP CMOS

two reasons: (a) they are the highest power consuming block in each slice and (b) they do not scale well at lower power supply (< 0.7V). The full-rate CTLE is included to provide equalization up to 12Gb/s and improve the sensitivity. The CTLE together with 3-tap TXPE provides the necessary equalization to account for 13dB (18dB with RXFE BW limitations) loss at 8GHz (16Gb/s). Per-slice offset cancellation is done in the SAL using I-DACs. Lack of buffer between the T/H and SAL results in charge kickback. Therefore, careful layout was needed to reduce its effects. Table I compares the three different RXFE.

## V. RXFE CHARACTERIZATION AND MEASUREMENTS

In this section, we discuss techniques to characterize the RXFE for different specifications and include measurement results from three test chips [1], [3], [4].

### A. Voltage-Offset Calibration and Link-Margining

Consider the RXFE shown in Fig. 11(b) which has the capability to do offset cancellation on a per-slice basis in the buffer amplifier. There are multiple ways to cancel the voltage offset of these slices. (1) With the TX transmitting a "1" DC pattern at nominal swing, the offset code of the buffer amplifier is swept until the slice output is tripped to the opposite state. This offset code is recorded. Next, the offset code with the TX transmitting a "0" DC pattern is calculated. An average of the two offset codes gives the required slice offset setting [20]. This procedure is then repeated for both the slices. (2) If possible, the two differential outputs (inputs) of the TX (RX) are shorted and set to the nominal Vcm. Or, (3) The TX is turned off and the input of the RX is floated. With the RX

978-1-4673-6145-3/13 $31.00 © 2013 IEEE    328

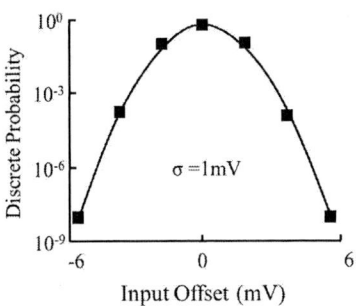

Fig. 14. Measured input-referred noise for 20Gb/s RXFE in 90nm CMOS

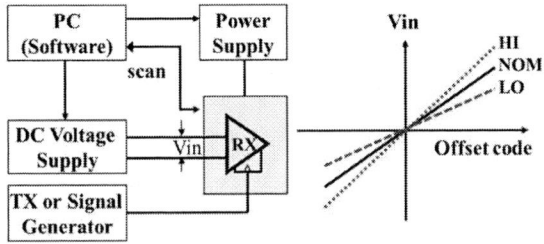

Fig. 15. Setup for PSSR or CMSR measurements.

error-checker set to expect all "1"s or "0"s, the buffer offset code is swept, counting the number of bits that resolve to "1" or "0" to generate a CDF and PDF. The code corresponding to the mean value of the related PDF gives the offset code for the slice. This process is repeated for all the slices. Note that the sigma of the PDF will signify the input referred noise (Fig. 7).

A two-step method of offset cancellation is required for RXFEs shown in Fig. 11(a) or (c). First, the offset code of the CTLE is set at the center point. Step (2) as described earlier is carried out for each of the slices. Next, the CTLE offset code is incremented and (2) is repeated to minimize the mean of the offset codes of the four slices. The corresponding offset codes are taken to be the nominal codes for the CTLE and the four slices.

In order to do link-margining, the CTLE and the four slices are set to their offset-calibrated codes. Next, the offset of the CTLE is swept and the BER at the corresponding voltage margin is calculated. In order to generate the BER eye diagram, the sampled clock is swept in phase over the entire UI as well. Fig. 12 shows the BER eye generated for the RXFE [4] of Fig. 11(a), captured using an on-die oscilloscope [14].

Fig. 13 shows the input-referred offset as the offset code is swept across the entire range of the CTLE and one of the comparator-slices for the 16Gb/s RXFE [3] (Fig. 11(c)). For a given code, the input-referred offset is the amplitude of input differential voltage to the RXFE that results in a trip-point at the output.

### B. Input-referred Noise

The total input-referred thermal noise for an RXFE can be calculated by plotting the PDF of receiving a "1" as explained before in Section II-D and Section V-A. The standard deviation of the PDF gives the RMS voltage of the noise. Alternatively, the slope of the Gaussian PDF tails can be calculated by converting the PDF plot to $Q$-scale, which linearizes the tails. Fig. 14 shows the input-referred noise measured at 20Gb/s for the RXFE [4] (Fig. 11(a)).

### C. PSSR and CMSR

Fig. 15 shows the measurement setup for PSSR (CMSR) in which Vcc (Vcm) is swept from a nominal to HI or LO value, for e.g., +/- 5%. First, the offset calibration codes are calculated for each of the slices and the self-offset is cancelled. A DC voltage supply is then used to determine the transfer function between the CTLE offset code and input differential voltage for each of the slices. Next, Vcc (Vcm) is changed to LO or HI value and the process is repeated. The PSSR (CMSR) is calculated using (6) ((7)). Fig. 16 (a) shows the PSSR measurement for each of the four slices of the 20Gb/s RXFE (Fig. 11(a)) with offset-voltage calibration. Here, Vcc is changed ±100mV from the nominal 1.2V supply. Clearly, optimal offset calibration does not provide optimal PSSR. The PSSR is degraded by up to 40dB relative to its optimal point. Fig. 16(b) shows the CMSR measurement result, where Vcm is changed from ±100mV from nominal 0.3V. Up to 50dB degradation is seen in CMSR relative to its optimal point.

Fig. 17 shows another measurement setup for PSSR where a pulse generator is used to inject sinusoidal noise at a specific frequency into the RX. The TX acts as a voltage DAC and transmits a static waveform to the RX, whose differential voltage is swept to generate PDF and CDF curves. A static waveform into the RX minimizes the impact of clock jitter due to power supply noise. As the CTLE offset is swept, the PDF

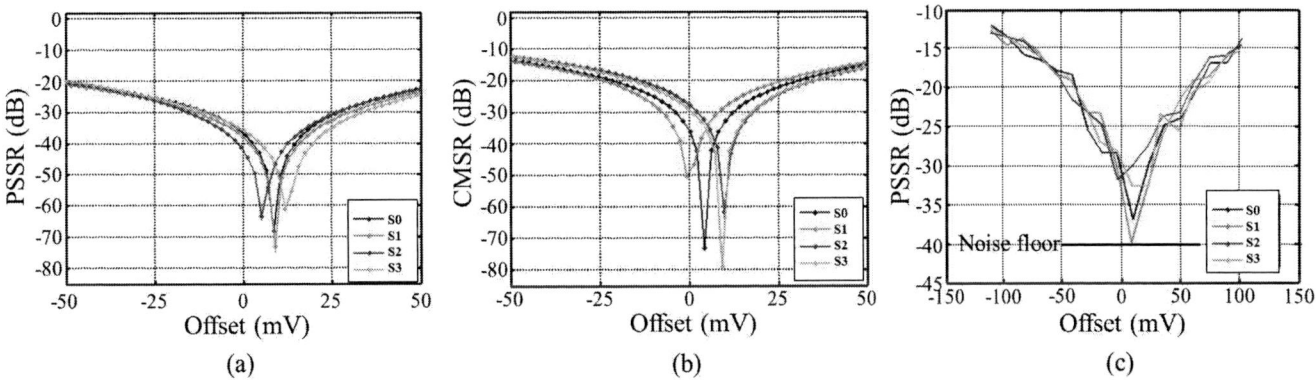

Fig. 16. (a) PSSR, (b) CMSR and (c) *dynamic* PSSR measurements with offset-voltage calibration for 20Gb/s RXFE in 90nm CMOS

978-1-4673-6145-3/13 $31.00 © 2013 IEEE

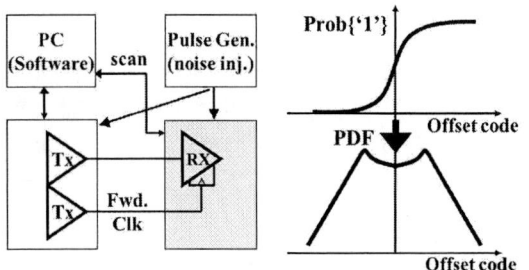

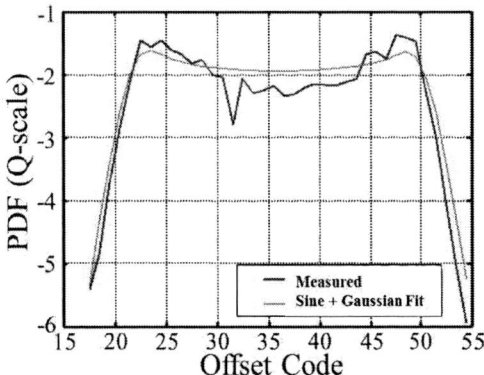

Fig. 17. Setup for PSSR measurement with power supply noise injected at a specific frequency

Fig. 18. Measured noise PDF for slice-0 at 200MHz power supply noise with CTLE offset code for 20Gb/s RXFE in 90nm CMOS

curves that are generated are fitted to a convolution of Gaussian distribution with a sinusoid. Fig. 18 shows the PDF generated for this dynamic PSSR measurement with power supply noise of $138mV_{pp}$ at 200MHz for one slice of the RXFE [4]. Fig 17(c) shows the PSSR measurement for all four slices with offset-voltage calibration.

## V. CONCLUSION

Low-power link design warrants a co-optimization of circuits and interconnects to minimize circuit complexity and remove excess link margins for a given BER. For a specific interconnect, low-order equalization schemes operating at moderate per-pin data-rate keep the power consumption low. Considerable power savings can be achieved by reducing the TX swing and improving the sensitivity of the RXFE. Apart from the gain and BW, various other factors like thermal noise, power-supply and common-mode noise, variations, hysteresis, metastability etc., affect the sensitivity of the RXFE. Future link specifications must benefit from these optimizations to address the needs of low-power I/O operating at Tb/s and above.

## ACKNOWLEDGMENT

The authors would like to thank G. Balamurugan, T-C. Hsueh, S. Hyvonen, R. Inti, J. Kennedy, G. Keskin, R. Mooney, T. Musah, C. Roberts. S. Sen and C. Thakkar for collaboration and helpful discussions.

## REFERENCES

[1] F. O'Mahony, et al., "A 47x10 Gb/s 1.4 mW/Gb/s parallel interface in 45nm CMOS," *IEEE J. Solid-State Circuits*, vol. 45, pp. 2828-2837, Dec. 2010.

[2] S. Shekhar, et al., "Strong injection locking in low-Q LC oscillators: modeling and application in a forwarded-clock I/O receiver," *IEEE Trans. Circuits and Systems I*, vol. 56, pp. 1818-1829, Aug. 2009.

[3] M. Mansuri, et al., "A scalable 0.128-to-1Tb/s 0.8-to-2.6pJ/b 64-lane parallel I/O in 32nm CMOS," *ISSCC Dig. Tech. papers*, pp. 402-403, Feb. 2013.

[4] B. Casper, et al., "A 20Gb/s forwarded clock transceiver in 90nm CMOS," *ISSCC Dig. Tech. papers*, pp. 18-20, Feb. 2006.

[5] B. Casper and F. O'Mahony, "Clocking analysis, implementation and measurement techniques for high-speed data links – a tutorial," *IEEE Trans. Circuits and Systems-I*, vol. 56, pp. 17-39, Jan. 2009.

[6] H. Johansson and C. Svensson, "Time resolution of NMOS sampling switches used on low-swing signals," *IEEE J. Solid-State Circuits*, vol. 33, pp. 237-245, Feb. 1998.

[7] M. Park, J. Bulzacchelli, M. Beakes and D. Friedman, "A 7Gb/s 9.3mW 2-tap current-integrating DFE receiver," *ISSCC Dig. Tech. papers*, pp. 230-231, Feb. 2007.

[8] S. Hyvonen and E. Rosenbaum, "Diode-based tuned ESD protection for 5.25-GHz CMOS LNAs," *Proc. EOS/ESD Symp.*, pp. 9-17, Sep. 2005.

[9] S. Galal and B. Razavi, "Broadband ESD protection circuits in CMOS technology," *IEEE J. Solid-State Circuits*, vol. 38, pp. 2334-2340, Dec. 2003.

[10] S. Shekhar, J. S. Walling and D. J. Allstot, "Bandwidth extension techniques for CMOS amplifiers," *IEEE J. Solid-State Circuits*, vol. 41, pp. 2424-2439, Nov. 2006.

[11] J. Kim, B. S. Leibowitz, J. Ren and C. J. Madden, "Simulation and analysis of random decision errors in clocked comparators," *IEEE Trans. Circuits and Systems I*, vol. 56, pp. 1844-1857, Aug. 2009.

[12] R. Farjad-Rad, et al., "0.622-8.0Gbps 150mW serial macrocell with fully flexible preemphasis and equalization," *IEEE VLSI Circuits Symposium Dig. Tech. papers*, pp. 63-66, Jun. 2003.

[13] K. Kundert, "Simulating switched-capacitor filters with SpectreRF," v. 6c, July 2006, http://www.designers-guide.org/analysis/sc-filters.pdf

[14] B. Casper, A. Martin, J. E. Jaussi, J. Kennedy and R. Mooney, "An 8-Gb/s simultaneous bidirectional link with on-die waveform capture," *IEEE J. Solid-State Circuits*, vol. 38, pp. 2111-2120, Dec. 2003.

[15] T. Toifl, et al., "A 22-Gb/s PAM-4 receiver in 90-nm CMOS SOI technology," *IEEE J. Solid-State Circuits*, vol. 41, pp. 954-965, Apr. 2006.

[16] M. Jeeradit, et al., "Characterizing sampling aperture of clocked comparators," *IEEE VLSI Circuits Symposium Dig. Tech. papers*, pp. 68-69, Jun. 2008.

[17] K-L. J. Wong and C-K. J. Yang, "Offset compensation in comparators with minimum input-referred supply noise," *IEEE J. Solid-State Circuits*, vol. 39, pp. 837-840, May 2004.

[18] D. A. Johns and K. Martin, *Analog Integrated Circuit Design*. Wiley, 1996.

[19] M-J. E. Lee, W. J. Dally and P. Chiang, "Low-power area-efficient high-speed I/O circuit techniques," *ISSCC J. Solid-State Circuits*, pp. 1591-1599, Nov. 2000.

[20] J. E. Jaussi, et al., "8-Gb/s source-synchronous I/O link with adaptive receiver equalization, offset cancellation, and clock de-skew," *IEEE J. Solid-State Circuits*, vol. 40, pp. 80-88, Jan. 2005.

978-1-4673-6145-3/13 $31.00 © 2013 IEEE

# An 8mW Frequency Detector for 10Gb/s Half-Rate CDR using Clock Phase Selection

Mohammad Sadegh Jalali[1], Ravi Shivnaraine[1], Ali Sheikholeslami[1], Masaya Kibune[2], Hirotaka Tamura[2]

[1]Department of Electrical and Computer Engineering, University of Toronto, Canada, [2]Fujitsu Laboratories Limited, Japan

*Abstract*—**A half-rate single-loop CDR with a new frequency detection scheme is introduced. The proposed frequency detector selects between the clock phases (I and Q) to reduce cycle slipping, hence improving lock time and capture range. This frequency detector, implemented within a 10Gb/s CDR in Fujitsu 65nm CMOS, consumes only 8mW, but improves the capture range by up to 3.6×. The measured capture range with the FD is from 8.675Gb/s to 11Gb/s.**

## I. INTRODUCTION

Most clock and data recovery (CDR) circuits include two locking mechanisms: one for frequency and one for phase. Conventional frequency detectors (in reference-less CDRs) are based on either the analog [1-2] or the digital implementation of the quadricorrelator architecture [3-7]. Digital quadricorrelator-based frequency detectors (DQFD) for half-rate CDRs typically sample four phases of the clock ($0°$, $45°$, $90°$ and $135°$) at each data edge to uniquely identify phase rotation, as illustrated in Fig. 1. Accordingly, as the data edge crosses the quadrant boundaries of the clock (CK), the FD asserts pulses, directly contributing to the charge-pump (CP) current, slowing down phase rotation, and pushing the VCO towards lock [3-4]. When frequency error is close to zero, the FD becomes inactive, and the PD takes full control of the VCO. Due to concurrent operation, the two loops can interfere with each other [5, 6] and delay phase locking. In [6], the FD and PD loops are uncoupled, where the FD changes the VCO frequency by switching capacitors in and out of the tank. However, this complicates the VCO design.

In contrast, we propose an embedded FD, shown in Fig. 1, that uses one loop and two phases of the clock (I and Q). The proposed embedded FD affects the CP through the PD, instead of directly controlling the CP current, enabling the PD to deal with frequency offset. Prior to lock, the clock phases ($CK_I$ and $CK_Q$) into the PD are continually swapped (in a manner described in Section II) to reduce frequency error. We will show in this paper that this clock phase selection (CPS) scheme has a much lower power consumption and complexity than the DQFD. We implement the proposed FD within a half-rate CDR in 65nm CMOS, and demonstrate that the proposed FD, using only 8mW, on average increases the capture range by 3.6× to 8.675Gb/s - 11Gb/s.

The rest of this paper is organized as follows. Section II introduces the basic idea of CPS-based frequency detectors. In section III, the FD implementation, along with the circuits used in the CDR are shown. Simulation and experimental results

from a test-chip fabricated in a 65-nm CMOS process are included in Section IV. Finally, section V concludes the paper.

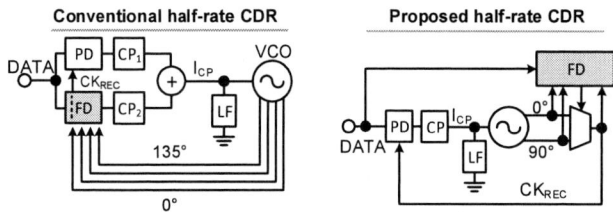

Fig. 1. Basic architecture of conventional and proposed half-rate FD

## II. PROPOSED FREQUENCY DETECTION TECHNIQUE

Fig. 2 compares the operation principle of a conventional half-rate DQFD [3-4] with that of the proposed FD in the presence of a large frequency offset. Assuming a positive frequency offset ($f_{DATA} > f_{CK}$), the data phase in terms of the clock phase rotates clockwise, as shown in Fig. 2(a). In a conventional CDR, in regions 3-4 and 7-8, phase error is positive while in the other regions, it is negative. As a result, the PD output increases the VCO frequency in half of the regions and decreases it in the other half. This makes the net PD output near zero and causes cycle slipping. The DQFD, however, is able to detect the direction of this phase rotation by sampling all clock phases on the data edge and tracking the change in polarity of these samples. The FD then asserts pulses to oppose phase rotation. The sum of the PD and the FD outputs reduces frequency error [3-4].

Fig. 2(b) shows the CP current ($\propto \Phi_{err}$) if $CK_I$ or $CK_Q$ is selected as the PD clock ($CK_{REC}$ in Fig. 1). In this figure, the net charge into the loop filter due to $CK_I$ and $CK_Q$ are shown in blue (light) and red (dark), respectively. Here, similar to the conventional case, cycle slipping occurs regardless of whether $CK_I$ or $CK_Q$ is used. However, we make a critical observation: since $CK_I$ and $CK_Q$ are 90° apart, at any given time, phase error with respect to either $CK_I$ or $CK_Q$ will move the VCO frequency in the correct direction. Therefore, if $CK_{REC}$ could switch between $CK_I$ and $CK_Q$ at appropriate times, it is possible to reduce cycle slipping (ideally avoid it altogether), and the need for a secondary FD loop is obviated. This is done by comparing $CK_I$ and $CK_Q$ on the rising edge of the data in regions 1-2 and 5-6 and choosing the one closest to the eye center. We disable switching in regions 3-4 and 7-8. This is conceptually demonstrated in Fig. 2(b). We observe that the PD output in regions 1-2 and 5-6 averages to zero, while in regions 3-4 and 7-8, the PD output averages to a positive

978-1-4673-6145-3/13 $31.00 © 2013 IEEE

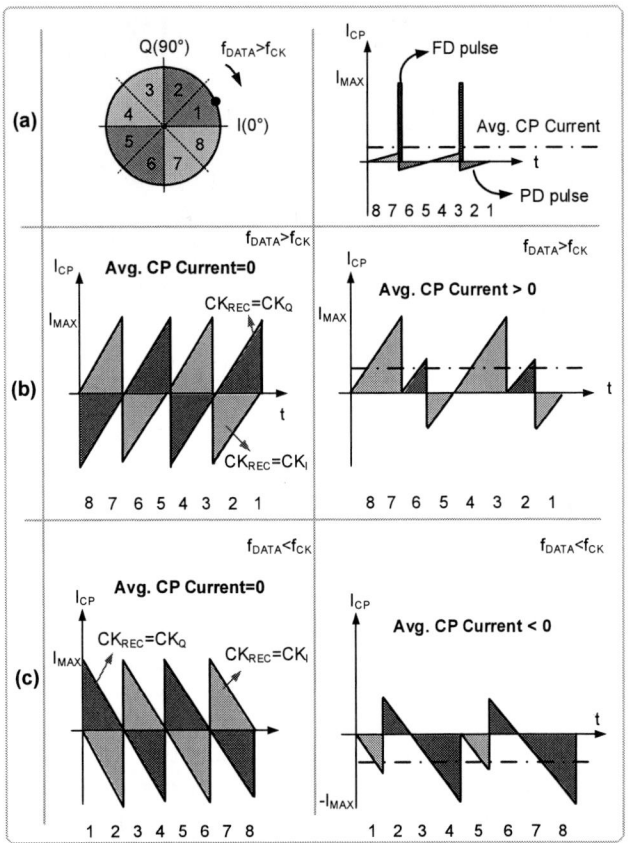

Fig. 2. Operation principle of conventional (a) and proposed half-rate FD (b) and (c)

value, therefore, overall, the VCO frequency moves in the direction of reducing the frequency offset over every cycle slip period. Similarly, as shown in Fig. 2(c), when $f_{DATA} < f_{CK}$, switching between $CK_I$ and $CK_Q$ results in a net negative CP current, again moving the VCO frequency in the direction of reducing the frequency offset.

As we will see later in this paper, compared to half-rate DQFDs, the proposed FD is simpler to implement, offers reduced lock time and increased capture range while consuming less power.

## III. CDR ARCHITECTURE AND IMPLEMENTATION

Fig. 3 shows a block level implementation of the proposed half-rate FD embedded in the PD loop. We use a linear half-rate PD [8], along with a differential ring VCO. After a data rising edge, one of the two clock phases $CK_I$ and $CK_Q$ are chosen by the proposed FD, and is fed back to the PD. If $CK_{REC}$ is positive at the rising edge of the data, the higher (in amplitude) of $CK_I$ and $CK_Q$ is chosen, while if it is negative, the lower of the two is chosen. Fig. 3 also shows an example where the edge falls in region 6, and the PD clock is swapped from $CK_I$ to $CK_Q$. This simple FD logic removes the need for a secondary FD loop, uses a total of only 45 transistors, and consumes $4.75\times$ less power in simulations than the half-rate FDs in [3] and [4]. After the CDR locks, the FD will

become inactive since the data edge will occur at the same phase and the swapping stops.

While in a conventional FD, after phase locking, phase can deviate from its locked position by $1UI_{pp}$ without activating the FD, in a CPS-based FD, this zone is reduced to $0.5UI_{pp}$. This is because the CPS-based FD uses the instantaneous phase information to feed the desirable clock phase to the phase detector. To solve this problem, the $FDlock$ signal shown in Fig. 3 can be used to turn the CPS-based FD off.

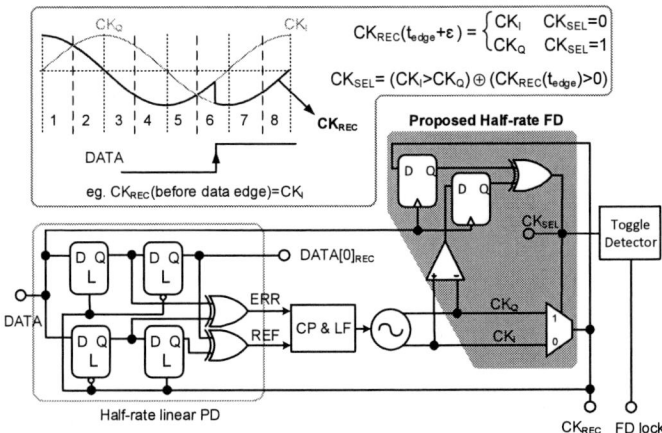

Fig. 3. Half-rate linear CDR with the proposed half-rate FD

Fig. 4 shows the circuit implementation of the VCO. The VCO delay cell is based on a differential pair with a cross-coupled stage. The delay of each stage is controlled by $V_{TUNE}$, which adjusts the trans-conductance of the negative-gm stage, varying delay. $V_{TUNE}$ is used in a differential fashion to maintain a constant common-mode at the VCO output. The single-ended to differential converter circuit converts the single-ended $V_{CNT}$ to the differential $V_{TUNE}$.

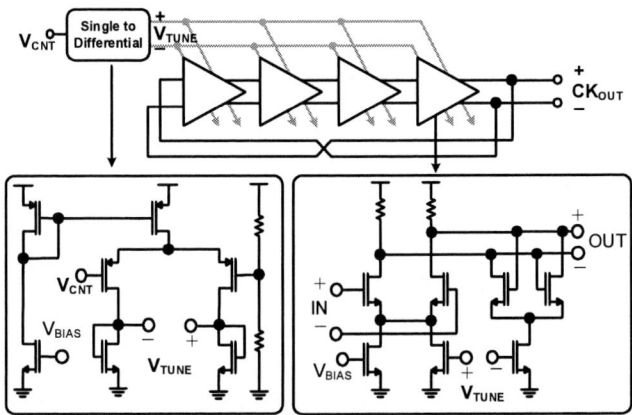

Fig. 4. Half-rate linear CDR with the proposed half-rate FD

Fig. 5 shows the circuit implementation of the charge-pump. The current associated with the error signal is twice as large as the current associated with the reference signal, due to the half-rate nature of the PD [8]. An on-chip DAC is used to adjust the CP current during capture range measurements.

978-1-4673-6145-3/13 $31.00 © 2013 IEEE

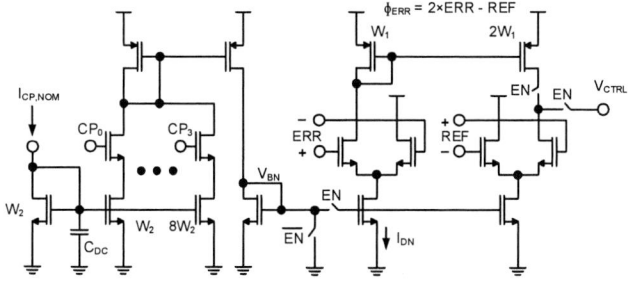

Fig. 5. Circuit diagram of the charge-pump

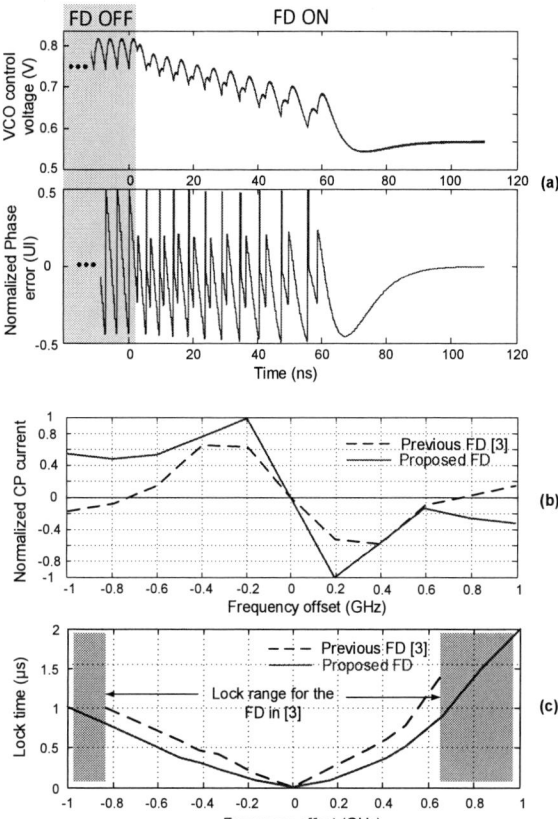

## IV. SIMULATION AND EXPERIMENTAL RESULTS

Fig. 6(a) shows the behavioral simulation results of the system with a 3% frequency offset. The figure shows that the CDR does not acquire lock until the FD is turned on. Fig. 6(b) characterizes the response of the proposed FD versus frequency offset with a PRBS7 10Gb/s input and compares it to the response of the previous FD [3] with the same inputs, CP currents, and loop filter values. The average gain of the proposed FD is 1.9 times that of the conventional FD. Fig. 6(c) shows the lock time of the CDR with a PRBS7 10Gb/s input (defined as the time it takes for the CDR to start producing error free data) with a conventional half-rate FD [3] and a CPS-based FD where both systems are simulated under the same conditions. On average, the CPS-based FD locks 1.8 times faster than the conventional frequency detector. Also, the CDR with the conventional FD fails to lock for very large frequency offsets (indicated by the blue region) which is predicted by Fig. 6(b) results. Also, our behavioral simulations show that the proposed FD locks 2.7 times faster than the FD in [4].

Fig. 7 shows the measured tuning range of the ring VCO which is from 3.94GHz to 6.25GHz.

Fig. 8(a) shows the measured recovered demultiplexed eye of the CDR with and without the FD. In both cases, we initialize the VCO frequency to 5GHz. With the FD off, the CDR does not lock to a PRBS7 input at 8.675Gb/s, while locking is acquired once the frequency detector is turned on. Fig. 8(b) shows the spectrum of the recovered CK before and after lock where the incoming data rate is 9.7Gb/s. In one case, CDR loop is opened and the VCO frequency is held at 10Gb/s (5GHz). This forced frequency error causes constant swapping of $CK_I$ and $CK_Q$, creating spurs in the clock spectrum. Closing the loop locks the CDR; here, clock spectrum is clean even though the FD is on. This is because the freq. error is zero and clock swapping no longer occurs.

Fig. 9 shows the measured capture range of the CDR (defined as $(f_{max}$-$f_{min})$/10Gbps) with and without the FD and its jitter tolerance (JT). The VCO frequency is initialized to 5GHz, and PRBS7 data frequency is swept. As expected, increasing CP current improves the capture range. The maximum locking range of the CDR without the frequency detector is from 9.5Gb/s to 10.15Gb/s, while with the FD this range is increased to 8.675Gb/s to 11Gb/s. Due to the bandwidth limitation of the test fixture, the maximum reliable data rate

Fig. 6. Behavioral simulation results for the proposed FD (a) the system locks only after the FD is turned on, (b) normalized charge-pump current versus frequency error with a PRBS7 pattern (c) lock time of the CDR with PRBS7 pattern

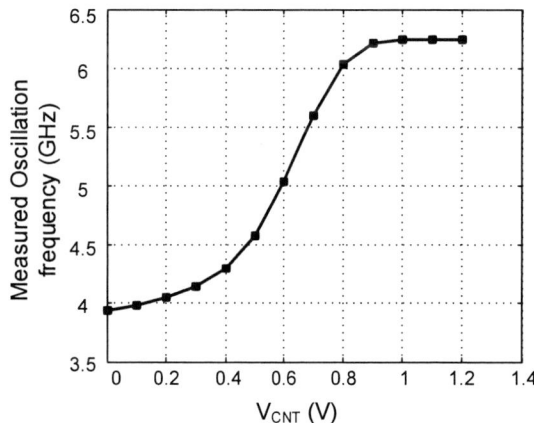

Fig. 7. Measured VCO tuning range

for measurements was found to be 11Gb/s. Limited CP swing, caused by charge-pump current sources entering triode, results in the CDR capture range being less than the VCO tuning range. The use of a linear phase detector further limits the CP swing. A bang-bang PD can be used if an even higher capture range is needed. The measured JT of the CDR for a BER less that $10^{-12}$ at 10Gb/s is $0.2UI_{pp}$ at high frequency.

The chip is fabricated in Fujitsu's 65nm CMOS process. The die photo is shown in Fig. 10. The CDR area is $350 \times 400 \mu m^2$,

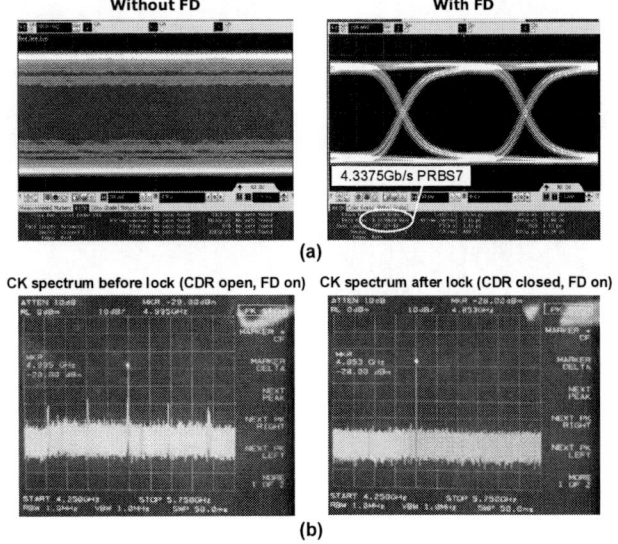

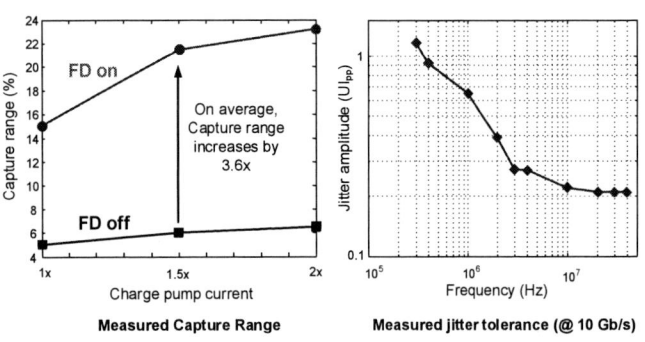

Fig. 8. (a) Demuxed eye for the CDR with and without the FD with an 8.675Gb/s PRBS7 input, (b) clock spectrum before and after lock

Fig. 9. Capture range and JT measurement results

of which $100 \times 65 \mu m^2$ is occupied by the proposed FD. At a 1.2V supply and operating at 10Gb/s, the chip consumes a total power of 37.2mW when the FD is on and 28.8mW when the frequency detector is off. Hence the CPS-based FD consumes only 8.4mW.

Finally, Table I summarizes the results and compares this paper against previous work. Also, note that simulating all frequency detectors in 65nm CMOS at 10Gb/s (with the same gates) result in a power consumption of 6mW for the proposed FD and 29.5mW and 28.6mW for the FDs in [3] and [4], respectively. Since the details of the design in [6] and [7] are not available, they cannot be simulated. Also exact transistor count cannot be obtained.

## V. CONCLUSION

In this paper, a novel clock phase selection based frequency detector for half-rate CDRs is introduced. It was shown that by changing the PD clock (switching between $CK_I$ and $CK_Q$) at the right time, the phase detector can be capable of dealing with frequency offset itself, removing the need for a secondary FD loop. Based on this idea, a chip was fabricated in Fujitsu 65nm CMOS process. It was shown that the FD increases the

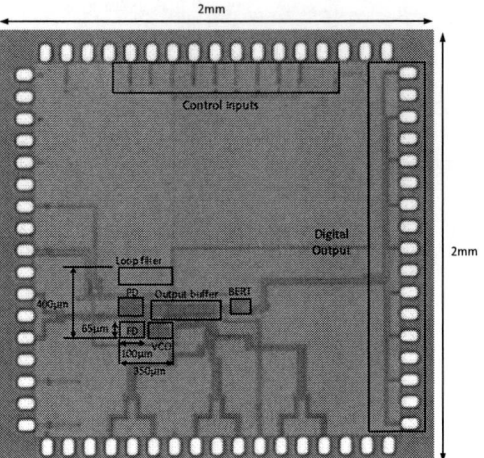

Fig. 10. Die photo

TABLE I
COMPARISON OF CDR RESULTS

| FD | Type | Tech. (nm) | Lock Rate (Gb/s) | Lock range (%) | No. of trans- istors | FD power (mW) |
|---|---|---|---|---|---|---|
| [3] | Half-rate | 180 | 3.125 | 11.52 | 156 | 30.6 |
| [4] | Half-rate | 180 | 10 | 14.3 | 147 | 42.2 |
| [6] | Full-rate | 65 | 10 | 30 | >73 | NA |
| [7] | Full-rate | 180 | 3.125 | 16 | >76 | 15.5 |
| **This work** | **Half-rate** | **65** | **10** | **23.25** | **45** | **8.4** |

capture range to 23.25% while consuming only 8mW. Our simulation and measurement results show that this inherent FD consumes much less power and area than its conventional equivalents.

## ACKNOWLEDGMENT

The authors would like the acknowledge CMC Microsystems for the provision of test equipment and CAD tools.

## REFERENCES

[1] D. Richman, "Color-Carrier Reference Phase Synchronization Accuracy in NTSC Color Television," *Proceedings of the IRE*, vol. 42, pp. 106-133, Jan. 1954.

[2] B. Razavi, "A 2.5-Gb/sec 15-mW Clock Recovery Circuit," *IEEE Journal of Solid-State Circuits*, vol. 31, pp. 472-480, April 1996.

[3] R. Yang, S. Chen, and S. Liu, "A 3.125-Gb/s Clock and Data Recovery Circuit for the 10-Gbase-LX4 Ethernet," *IEEE Journal of Solid-State Circuits*, vol. 39, pp. 1356-1360, Aug. 2004.

[4] J. Savoj and B. Razavi, "A 10-Gb/s CMOS Clock and Data Recovery Circuit with a Half-Rate Binary Phase/Frequency Detector," *IEEE Journal of Solid-State Circuits*, vol. 38, pp. 13-21, Jan. 2003.

[5] D. Dalton, S. Fallahi, M. Kargar, M. Khanpour, and A. Momtaz, "A 12.5-Mb/s to 2.7-Gb/s Continuous-Rate CDR With Automatic Frequency Acquisition and Data-Rate Readback," *IEEE Journal of Solid-State Circuits*, vol. 40, pp. 2713-2725, Dec. 2005.

[6] N. Kocaman, K. Chai, E. Evans, M. Ferriss, D. Hitchcox, P. Murray, S. Selvanayagam, P. Shepherd, and L. DeVito, "An 8.511.5Gbps SONET Transceiver with Reference less Frequency Acquisition," *IEEE Custom Integrated Circuits Conference*, pp. 1-4, Sep. 2012.

[7] M. Lee, and T. Lee, "A clock and data recovery circuit with wide linear range freq. detector," *IEEE International Symposium on VLSI design, automation and test*, pp. 121-124, April 2008.

[8] J. Savoj, and B. Razavi, "A 10-Gb/s CMOS Clock and Data Recovery Circuit with a Half-Rate Linear Phase Detector," *IEEE Journal of Solid-State Circuits*, vol. 36, pp. 761-768, May 2001.

# A 60 GHz Linear Wideband Power Amplifier using Cascode Neutralization in 28 nm CMOS

Siva V Thyagarajan[*], Ali M Niknejad[*], Christopher D Hull[†]

[*]Berkeley Wireless Research Center, Dept. of EECS, University of California Berkeley

[†]Intel Mobile Communications, Intel Corporation

Email: {sivavth, niknejad}@eecs.berkeley.edu, christopher.d.hull@intel.com

*Abstract*—**The rapid scaling of CMOS technology in the last decade has enabled the design of high speed and efficient digital CMOS circuits. However, the design of RF and mm-wave systems has become more challenging due to inaccuracies in modeling and increased losses in the active and passive devices. This paper presents the design of a 60 GHz linear wideband power amplifier (PA) in deeply scaled 28 nm CMOS technology. The PA utilizes cascode drain-source neutralization to improve stability and low-k transformer techniques to achieve high bandwidth. Using transmission line power combining, the PA delivers a saturated output power of 16.5 dBm with a peak power added efficiency (PAE) of 12.6%. The three stage PA achieves an overall bandwidth of 11 GHz with a peak gain of 24.4 dB.**

## I. INTRODUCTION

The 7 GHz of unlicensed spectra in the 60 GHz band is a potential solution for high data-rate communication systems. The low cost and scaling of CMOS technology has enabled the design of several 60 GHz transceivers that achieve Gbps of data rates with reasonable efficiency numbers [1], [2]. One of the critical blocks in a mm-wave transceiver is the design of an efficient wideband power amplifier (PA) [3]-[6]. In CMOS technologies, the rapid supply scaling and low breakdown voltages of the transistors have severely affected the performance of the PAs. Since wireless systems target low cost, efficient and fully integrable solutions, the understanding of the performance of PAs at these scaled technology nodes become important. In order to cover this entire band and also account for process variations, the PA needs to be designed as a wideband system with high efficiency and gain. The scaling of technology provides a partial benefit due to improved transition frequencies of the devices. However, there are other design challenges which need to be addressed in order to design a high performance PA.

In this paper, we discuss the design of a linear PA implemented in 28 nm bulk CMOS technology. We discuss a cascode neutralization technique which is critical for maintaining the stability of the PA due to increased parasitic coupling between the drain-source nodes. The design also utilizes low-k transformer techniques to achieve a wideband PA with 11 GHz bandwidth. Finally, to achieve high output power levels, the PA uses on-chip power combining using transmission lines. To the authors' knowledge, this is the first mm-wave PA designed in 28 nm CMOS technology. Section II describes the impact of technology scaling on the design. Section III describes the challenges faced and provides the circuit details of the PA. The measurement results are shown in Section IV and concluding remarks are provided in Section V.

## II. TECHNOLOGY SCALING : ACTIVES AND PASSIVES

The scaling of technology to the 28 nm node offers benefits with respect to the $f_t$ of the devices and these typically range around 250 GHz. This is especially important in mm-wave applications where the required bandwidth is high (high data rates) and also there is no convenient way to compensate for temperature and process variations. However, the $f_{max}$ of the 28 nm technology node is still comparable with 65 nm. Hence, the maximum achievable gain $G_{MAX}$ of the device is around ∼11-12 dB per amplification stage if one utilizes a common-source structure. The scaled power supply and low breakdown voltages offered by CMOS technology also severely limit the achievable output power levels and efficiency of PAs. In order to achieve high output power, one has to resort to on-chip power combining techniques where power from several unit PAs are combined using on-chip passive networks. Fortunately, this technology node offers thick metal layers whose current carrying capacity is more or less comparable to that of the 65 nm node. However, the electromigration rules for the lower metal layers are a factor of 2X worse. Also, their sheet resistance is a factor of 2 to 3X higher when compared to the 65 nm technology node. This requires strapping of the lower metal layers and thus results in higher layout parasitics. The shrink in the metal stack especially for the lower layers also increases the loss contribution from the conductive silicon subsrate. This technology node also has stringent requirements with regard to the metal density. Hence, passive devices such as inductors/transformers must include dummy metal layers from Metal 1 to the highest metal layer. Even though the density of this fill is kept to its minimum, this still adds an approximate loss of 0.3 − 0.4 dB per matching stage. Complicated design rules along with the above mentioned issues make mm-wave design in this technology node challenging.

## III. POWER AMPLIFIER DESIGN

Fig. 1 shows the complete circuit diagram of the 60 GHz power amplifier (PA). The design comprises of three stages that are cascaded together using low coupling coefficient (low-k) transformer-based matching networks. A transmission line based power combiner combines the power from the individual PA units (output stage + interstage) to generate the required

978-1-4673-6145-3/13 $31.00 © 2013 IEEE

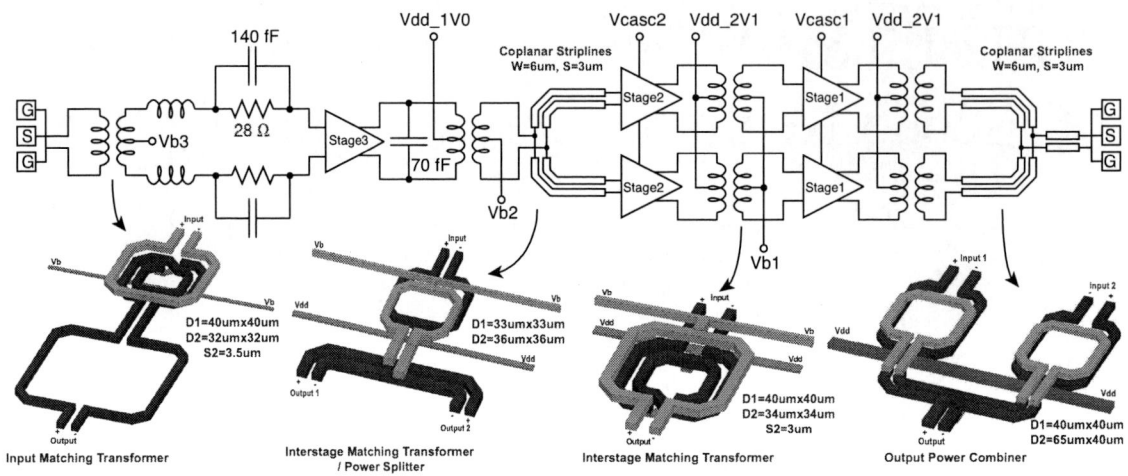

Fig. 1. Circuit diagram of the overall power amplifier with the matching network structures

output power. The output PA and the interstage drivers utilize drain-source neutralized cascode stages that achieve better stability and allow high output voltage swings. The design uses a finger width of $0.65\,\mu m$ for all the transistors. This value gives the best tradeoff for $f_{max}$ with respect to the gate resistance and the increased layout parasitics as one scales to lower finger widths. A single pre-driver stage operating out of a 1 V supply voltage drives both the interstage drivers. The predriver comprises of a simple differential pair ($NF = 36$). A parallel RC network ($28\,\Omega||140\,\text{fF}$) is added at the input of the predriver for stability at lower frequencies. The PA is stabilized for common mode oscillations using resistors at the center taps (Vb1, Vb2, Vb3) of the transformers.

### A. Drain-Source Neutralized Cascode Stages

In order to achieve high gain and output power, the design utilizes a cascode topology for the output and interstage PA units. This allows one to increase the supply voltage to upto twice the nominal value, thereby providing higher output voltage swings. In this case, we used a supply voltage of 2.1 V. To avoid gate-bulk breakdown issues, a triple-well structure (with the source tied to the bulk) has been used for all the devices. This also provides isolation between the transistors. The implementation of a cascode topology however, has implications in terms of stability. The classically known instability issue is the oscillation through the gate node of the cascode device. At mm-wave frequencies, the cascode device is degenerated at the source by a capacitive impedance and this results in a negative resistance as seen from the gate. Hence, a small parasitic inductance at the gate of the cascode device is sufficient to cause common mode oscillations. This is usually mitigated by reducing the lead inductance and by adding a small capacitance very close to the transistor gate node. This causes the frequency of oscillation (if any) to fall outside the $f_{max}$ of the device, thereby preventing oscillations.

The implementation of a cascode topology in this technology leads to an additional stability concern on the output side due to significant drain-source ($C_{ds}$) capacitance of the device.

The fringe capacitance in the Metal 1 layer of the device is the major contributor to this capacitance due to the reduced metal pitch in this technology. In order to understand the effect of a finite $C_{ds}$, let us consider the cascode device and the small signal circuit shown in Fig. 2. The output admittance $y_{in}$ of the network is calculated to be

$$y_{in} = sC_{neut} \tag{1}$$
$$+ \frac{1}{2}\left[\frac{(g_{ds1} + s(C_1 + 2C_{neut}))(g_{ds2} + s(C_2 - C_{neut}))}{g_m + g_{ds1} + g_{ds2} + s(C_1 + C_2 + C_{neut})}\right]$$

where $g_m$ is the transconductance of the cascode device, $g_{ds1}$ and $C_1$ the net conductance and capacitance looking into the drain of the differential pair, $g_{ds2}$ and $C_2$ the output conductance and drain-source capacitance of the cascode device and $C_{neut}$ the neutralization capacitance. Let us now consider the case when $C_{neut} = 0$. The $g_m/g_{ds}$ ratio for this technology node is almost twice when compared to the 65 nm case. Hence, from (1), this causes the real part of the output impedance of the device to be very high. Also, at reasonably high frequencies where $sC_1$ and $sC_2$ are comparable to $g_{ds1}$ and $g_{ds2}$, the real part of $y_{in}$ becomes negative. This causes the output reflection coefficient $S_{22}$ to become greater than 0 dB potentially causing stability issues. In order to account for the mismatch in the antenna impedance, the $S_{22}$ should be at least better than $-4$ to $-5$ dB (40 % absorption of the reflected wave) to avoid standing waves in the antenna. Part of this can be obtained from the insertion loss of the output matching network. In order to have a realiable design, we propose a drain-source neutralized output stage as shown in Fig. 2. The drain-source capacitance of the cascode device is neutralized using capacitors $C_{neut}$ implemented as MOM capacitors. From (1), when $C_{neut} = C_2$, the real part is positive but it is still fairly high and tends to make the $S_{22}$ close to unity. Hence, in this design, we have chosen $C_{neut} > C_2$ to obtain a lower positive value for the real part of the output impedance. This implementation is more efficient compared to directly adding a resistor at the output of the network. The simulated value of $S_{22}$ for various matching conditions

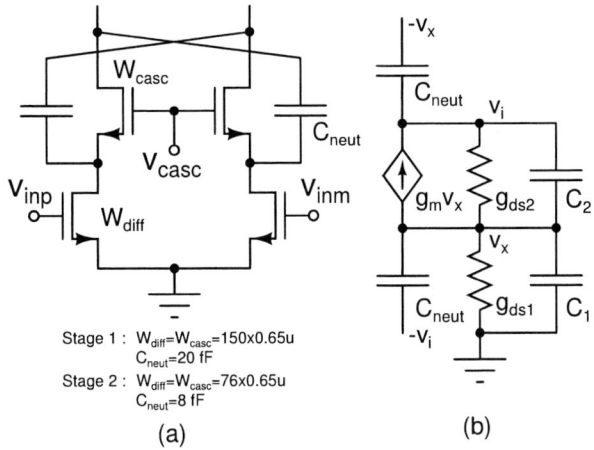

Stage 1 : $W_{diff}=W_{casc}=150\times0.65u$
$C_{neut}=20$ fF
Stage 2 : $W_{diff}=W_{casc}=76\times0.65u$
$C_{neut}=8$ fF

(a)

(b)

Fig. 2. (a) Circuit diagram of the output and interstage networks (b) Small signal equivalent circuit

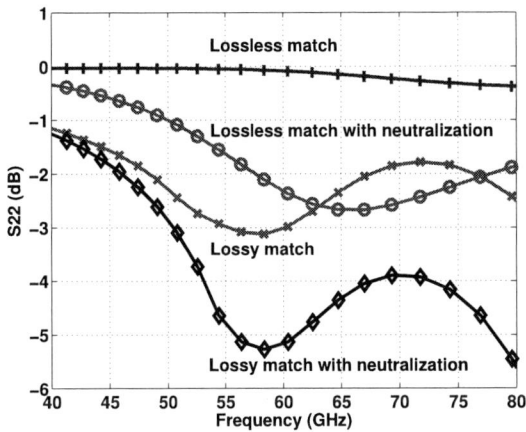

Fig. 3. Simulated output reflection coefficent $S22$

is shown in Fig. 3. For a lossless match, we observe that the $S_{22}$ is pretty close to unity thereby having potential stability issues. With neutralization, the $S_{22}$ improves to around $-2$ dB. With only a lossy transformer matching network, the value is close to $-3$ dB. In this design, the value of the neutralization capacitance is adjusted taking into consideration the output matching network loss, to obtain a net $S_{22}$ of around $-5$ dB as shown.

### B. Low Coupling Coefficient Transformer Networks

In order to obtain high bandwidth, this design utilizes loosely coupled (low k) transformers for matching the successive PA stages. A transformer network loaded with capacitors on the primary and secondary can be treated as a four pole network (two conjugate pairs). Under a high coupling coefficient case ($k \sim 0.7 - 0.8$), the second conjugate pole pair occurs at a frequency much higher than the resonant frequency of the system. This results in a response similiar to that of a second order system. When the coupling coefficient is reduced ($k \sim 0.2 - 0.3$), the second pole pair comes in-band

and an appropriate design choice could allow the synthesis of various filter networks. In this design, a maximally flat Butterworth response has been chosen. To implement this transfer function, the transformer is designed using two spiral inductors of comparable quality factors $Q_1$ and $Q_2$. In some cases, the quality factor needs to be adjusted by adding an external capacitor. The coupling factor $k$ of the transformer is then chosen such that $k \approx 1/Q_1 \approx 1/Q_2$. In this design, the output stage and the interstage match transformer/power splitter have been implemented using low-k transformers. The use of these low-k matching network along with the $f_t$ benefit of the technology helps the PA achieve a bandwidth of 11 GHz.

### C. Power Combining/Splitter

In order to achieve high output power levels, the design utilizes transmission lines to perform on-chip parallel power combining. In parallel power combining, each unit PA is presented with a load impedance that scales up in proportional to the number of combining stages. As the optimal load pull impedance of each unit PA element also scales up (due to reduction in device size), this mode of power combining is preferred over series power combining. In this design, a two-way power combiner has been implemented using coplanar striplines (CPS). As the output load is capacitive (due to pad capacitance), the length of the line plays a critical role in the output matching network design. The length of the line is chosen considering physcial constraints in the layout and also its impact on the output transformer network design. The CPS lines have a width of $6\,\mu m$ and spacing of $3\,\mu m$ with a characteristic impedance of $Z_0 = 28\,\Omega$. The simulated loss of the CPS lines is 1.1 dB/mm. A similar approach is employed in the pre-driver stage of the PA. Here, the output from the pre-driver stage is split into two branches using CPS lines. The power splitting is performed at the pre-driver output as opposed to the interstage driver to avoid efficiency degradation due to matching network loss. The interstage driver has a higher impact on the overall efficiency as compared to the pre-driver stage.

### IV. MEASUREMENT RESULTS

The PA has been fabricated in 28 nm bulk CMOS process. Fig. 4 shows the die photo of the chip. The chip occupies a total area of $0.64\,\mathrm{mm}^2$ and is pad limited. The core area of the PA is $0.122\,\mathrm{mm}^2$.

The measured S-parameters of the PA is shown in Fig. 5. The PA achieves a peak gain of 24.4 dB with a 3 dB bandwidth of 11 GHz extending from 56 GHz to 67 GHz. This is mainly due to the increased $f_t$ of the process and the application of low-k transformer techniques for the matching networks. The reverse isolation is better than $-40$ dB for the indicated range of frequencies. Due to the neutralization technique discussed above, the output reflection coefficient of the PA is maintained to be less than $-5$ dB for the inband frequencies. The measured stability factor of the PA is also greater than unity for all frequencies thereby providing the necessary condition for the overall stability of the amplifier.

978-1-4673-6145-3/13 $31.00 © 2013 IEEE

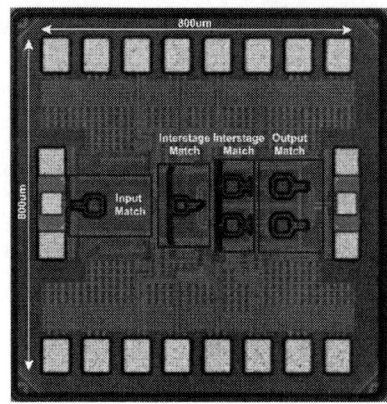

Fig. 4.   Chip Microphotograph

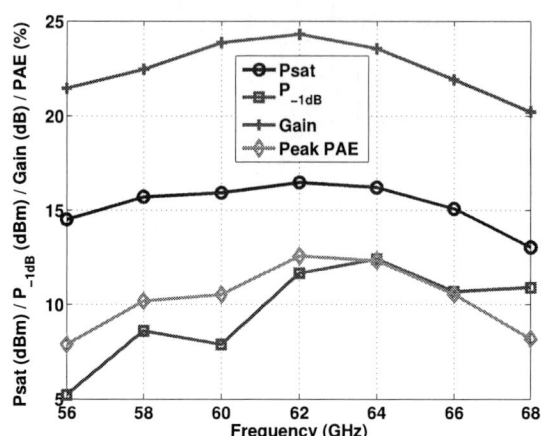

Fig. 7.   Measured Small signal gain, Psat, $P_{-1dB}$ and Power-Added Efficiency as a function of frequency

TABLE I
COMPARISON TABLE OF 60 GHz CMOS POWER AMPLIFIERS

|  | **This Work** | [3] | [4] | [5] | [6] |
|---|---|---|---|---|---|
| Process | **28** nm | 65 nm | 65 nm | 65 nm | 90 nm |
| Gain (dB) | **24.4** | 14.3 | 19.2 | 20.3 | 20.6 |
| 3 dB BW (GHz) | **11** | 15 | - | 9 | 8 |
| Psat (dBm) | **16.5** | 16.6 | 17.7 | 18.6 | 19.9 |
| $P_{-1dB}$ (dBm) | **11.7** | 11 | 15.1 | 15 | 18.2 |
| Peak PAE (%) | **12.6** | 4.9 | 11.1 | 15.1 | 14.2 |

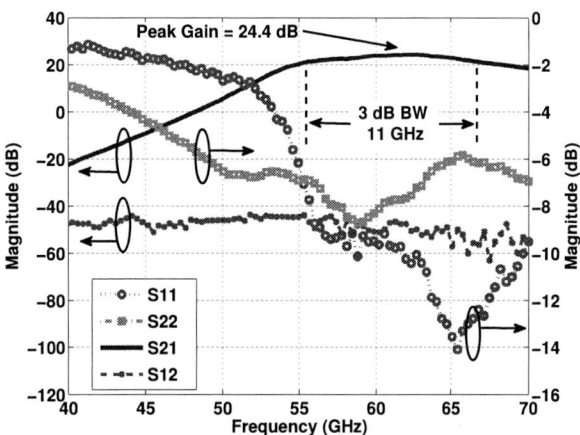

Fig. 5.   Measured S-parameters

$P_{-1dB}$ and PAE as a function of frequency. The average saturated output power of the PA is around 15.5 dBm within the band of interest. The average PAE is around 10.5 %. Table I shows a comparison table of the state-of-art 60 GHz CMOS PAs published in literature.

## V. CONCLUSION

A V-band power amplifier is demonstrated in 28 nm bulk CMOS technology. The PA achieves a peak gain of 24.4 dB with a bandwidth of 11 GHz due to the low-k transformer matching networks and the benefit gained from the $f_t$ of the technology. A drain-source neutralization technique was also demonstrated and is essential in maintaining the stability of the PA. The PA achieves a saturated output power of 16.5 dBm with a peak PAE of 12.6%.

## REFERENCES

[1] M. Tabesh, J. Chen, C. Marcu, Lingkai Kong, Shinwon Kang, A.M. Niknejad, E. Alon, "A 65 nm CMOS 4-element Sub-34mW/element 60 GHz phased-array transceiver," *IEEE Journal of Solid-State Circuits*, vol. 46, no. 12, pp. 3018-3032, Dec 2011.

[2] F. Vecchi et. al., "A wideband mm-Wave CMOS receiver for Gb/s communications employing interstage coupled resonators," *ISSCC*, pp. 220-221, Feb 2010.

[3] B. Martineau, V. Knopik, A. Siligaris, F. Gianesello, D. Belot, "A 53-to-68 GHz 18 dBm power amplifier with an 8-way combiner in standard 65nm CMOS," *ISSCC*, pp. 428-429, Feb 2010.

[4] J.-W Lai, A. Valdes-Garcia, "A 1V 17.9 dBm 60 GHz power amplifier in standard 65nm CMOS," *ISSCC*, pp. 424-425, Feb 2010.

[5] J. Chen, A. M. Niknejad, "A compact 1V 18.6 dBm 60GHz power amplifier in 65nm CMOS," *ISSCC*, pp. 432-433, Feb 2011.

[6] C.Y. Law, A.-V. Pham, "A high-gain 60 GHz power amplifier with 20 dBm output power in 90nm CMOS," *ISSCC*, pp. 426-427, Feb 2010.

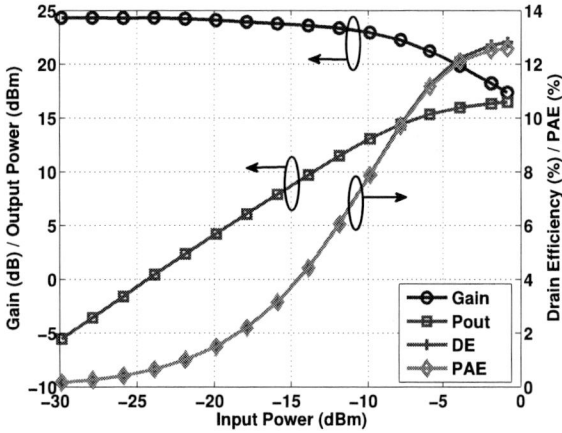

Fig. 6.   Measured Gain, Output Power, Drain Efficiency and Power-Added Efficiency as a function of the Input Power at 62 GHz

Fig. 6 shows the measured gain, output power, drain efficiency and power-added efficiency (PAE) of the PA at 62 GHz. The PA achieves a saturated output power of 16.5 dBm with a PAE of 12.6 %. The measured $P_{-1dB}$ at 62 GHz is 11.7 dBm. Fig. 6 shows the measured gain, saturated output power,

978-1-4673-6145-3/13 $31.00 © 2013 IEEE

# Compact High-Power 60 GHz Power Amplifier in 65 nm CMOS

Payam M. Farahabadi, *Member, IEEE*, Kambiz Moez, *Senior Member, IEEE*

iCAS Laboratory, University of Alberta, CANADA

*Abstract*— This paper presents a compact 60 GHz power amplifier utilizing a novel 4-way multi-conductor power combiner and splitter. The proposed topology provides the capability of combining the output power of four individual power amplifier cores in a compact die area of 0.025 mm$^2$ with the advantage of lower insertion loss and higher efficiency compared to the conventional distributed active transformer topology. Each power amplifier core consists of a three-stage common-source amplifier with transformer-coupled impedance matching networks. Fabricated in 65 nm CMOS process, the measured gain of the 0.19 mm$^2$ power amplifier is 18.8 dB at 60 GHz with 3dB band width of 4 GHz while consuming 424 mW from a 1.4V supply. A maximum saturated output power of 18.3dBm is measured with the 15.9% peak power added efficiency at 60 GHz.

## I. INTRODUCTION

The ever-increasing demand for high data rate short-range wireless communication devices motivates the designers to develop new wireless devices capable of fulfilling these consumer demands. Driven by the need for higher data-rates, the continuous 7 GHz bandwidth around 60 GHz is a promising contender because it is unlicensed and well suited for high speed indoor wireless personal area network (WPAN) applications [1].One particular feature of the 60 GHz band is that the oxygen molecules absorb electromagnetic energy at these specific frequency at higher rate compared to low GHz frequencies [1]. The signal attenuation requires millimeter wave (mmW) transmitter to transmit higher output powers than their low GHz counterparts in order for the receiver to successfully detect the transmitted signal. The need for high output power along with the operation of transistors near cut-off frequencies makes the design of 60 GHz power amplifiers (PA) very challenging.

Continuous scaling of CMOS technology has resulted in significant improvement of $f_{max}$ enabling the low cost integration of 60 GHz front-ends on a single chip. Nevertheless, design of efficient CMOS PAs remains challenging because of low gain of transistors at these frequencies, low breakdown voltage of CMOS transistors and losses of on-chip passive power combiners in deep sub-micron CMOS process [1]. To date, there have been several CMOS PAs targeting high gain and output power at 60 GHz [2]-[11]. Considering the fact that the gain of the power MOSFETs

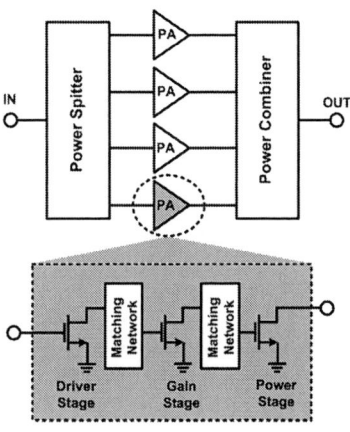

Fig. 1. 60 GHz PA building blocks.

with large channel width is relatively low at mmW frequencies, a single-stage PA cannot deliver a high gain and high power simultaneously [3]. Therefore, in all reported 60 GHz CMOS PAs, different power combining techniques such as using transformers, Wilkinson power combiners and distributed active transformers (DAT) are proposed to increase the output power of CMOS PAs [2]-[11]. However, the power combiners themselves cause additional losses limiting the maximum achievable output power and PAEs of CMOS PAs.

This paper presents a 60 GHz PA topology with a novel multi-conductor transformer-based power combiner. The proposed topology provides a power combining solution in a compact area with minimum insertion loss. In addition to perform the power combining, the network matches the 50-Ω load to the optimum load impedance of the PA to maximize the power/gain performance as well eliminating the need of additional impedance matching networks.

## II. DESIGN OF 60 GHz PA

The methodology for the design and implementation of 60 GHz PA based on the multi-conductor combiner is presented in this section. The block diagram of the proposed PA is illustrated in Fig. 1. A distributed 4-way parallel power combiner is proposed based on multi-conductor cross-coupled transformer architecture to combine the output current of the four individual PAs to achieve higher output power compared to the reported designers. The output power, gain, and the efficiency are the parameters which must be simultaneously

---

978-1-4673-6145-3/13 $31.00 © 2013 IEEE      339

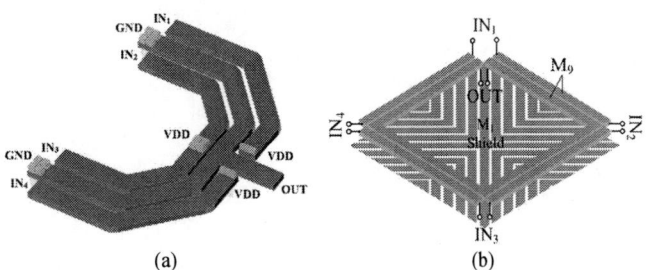

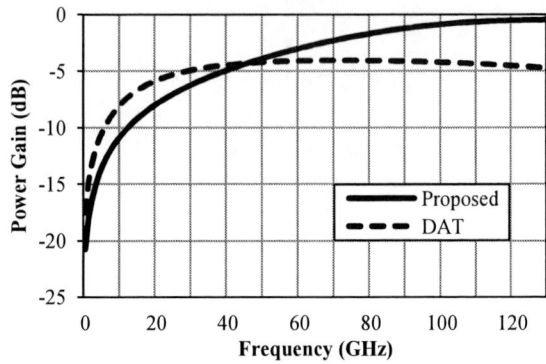

Fig. 2. Power combiners. a) Proposed Topology b) Conventional DAT

optimized while designing a PA. The output power of the four PA cores is combined by a 4-way transformer-based power combiner. Each PA cores is constructed by cascading three stages of common-source amplifiers. The output power is determined by the current handling capability of the last stage of the PA cores. Driver stages and gain stages are necessary to increase the total voltage gain of the amplifier. Also, the power splitters are providing the necessary input power for the input stages. By the use of power combiner/splitter, a PA can be derived by high input power which is resulting in a higher output power than a single stage. The smaller transistors are needed because of the low input power of each PA core, so that the transistors can be designed for achieving high $f_{max}$.

### A. Multi-conductor mmW power combiner

In order to combine the output power of the four separated PA cores, a 4-way multi-conductor transformer is proposed as shown in Fig. 2.a. The most challenging aspect of designing the transformer was to choose a topology which can satisfy both the impedance transformation and the bias requirements. The proposed topology can simultaneously provide impedance transformation, power combination, and biasing for the power transistors at output. Utilizing the multi-conductor topology, more electromagnetic waves can be prevented to penetrate into the conductive substrate which results in higher power transfer efficiency compare to the conventional DAT structure which is basically two coupled planar inductors.

Two thick metal layers ($M_9$ as ultra thick and $M_8$ as thick) are available for fabricating transmission lines and on-chip inductors in TSMC 1P9M CMOS technology. The thickness and width of the metal layers are the parameters which determine the quality factor of the inductors while the gap between two layers determines the coupling factor between two metal layers. The ultra-thick metal layer ($M_9$) with thickness of 3.4 μm is preferred to maximize the quality factor of the windings. The minimum distance (2 μm) between two metal lines on the same layer and the dielectric gap between two different metal layers (0.85 μm between $M_8$ and $M_9$) are the process restrictions which limit the performance of mmW transformers. Primary and secondary metal widths, (8 μm and 6 μm respectively), and the gap between two windings are adjusted to avoid the additional lossy tuning capacitors. By eliminating the tuning capacitors, the proposed transformers are guaranteed to have a higher efficiency compared to the typical stacked transformer-based network. Another major

Fig. 3. Insertion loss of the proposed topology vs. conventional DAT

advantage of the proposed transformer is the adjustable gap between to windings which can provide different coupling factors. Moreover, the proposed topology has the advantage of being very compact.

3D EM simulations are performed to extract the S-parameters and evaluate the proposed topology. Fig. 2.a illustrates the proposed topology. A conventional DAT structure like Fig. 2.b is designed and simulated to be compared in terms of efficiency and loss. In comparison with the same size conventional DAT topology, higher power gain and lower insertion loss is achieved for the proposed power combiner at frequencies higher than 60 GHz as illustrated in Fig. 3. The other advantage of the proposed topology is the capability of using a same topology as input power splitter which is a solution to avoid the use of lossy Wilkinson power splitters and matching components.

### B. Power stage transistors

In addition to lossy passive power combiners, a major challenge in the design of a fully integrated mmW PA is dealing with the low power gain of the large transistors which are used at such high frequencies [1]. The most important parameters that characterize the frequency dependant gain performances of a MOSFET are the cut-off frequency ($f_T$) and the maximum frequency of operation ($f_{max}$) [1]. In addition to this, a high level of output current is required for power amplifier application which poses the challenge of using transistors with large channel width. For mmW MOSFETs, the $f_{max}$ is primarily limited by the series gate resistance and the losses in the drain and source connections [1][3]. To provide a high level of output power, a transistor with a large channel width is required. To achieve the wider channel width and higher current, the multi-finger parallel configuration must be used as shown in Fig. 4.a. Using multiple fingers is a common way in order to build high power MOSFETs. However, this has the negative effect that increasing the total number of fingers with a fixed finger width leads to a large layout. In this case, although the poly-silicon gate resistance per finger remains fixed, large inter-connections are needed to connect the gates and drains of multiple fingers, which cause additional resistive losses and leads to a lower $f_{max}$ and gain. In addition, the I/O impedances drop when the number of fingers

978-1-4673-6145-3/13 $31.00 © 2013 IEEE

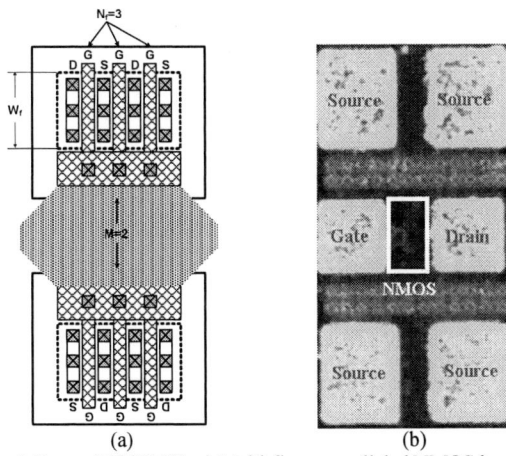

(a)                    (b)

Fig. 4. Power MOSFETS. a) Multi-finger paralleled NMOS layout
b) Micrograph of the fabricated NMOS.

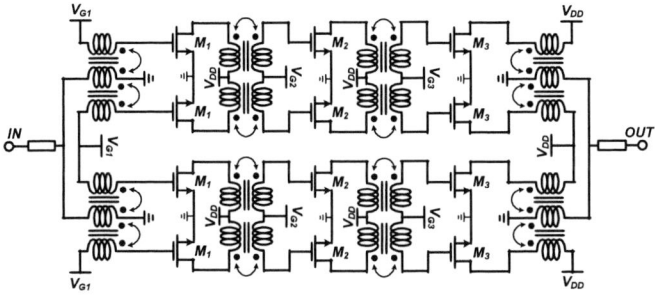

Fig. 6. Schematic of the 60 GHz power amplifier.

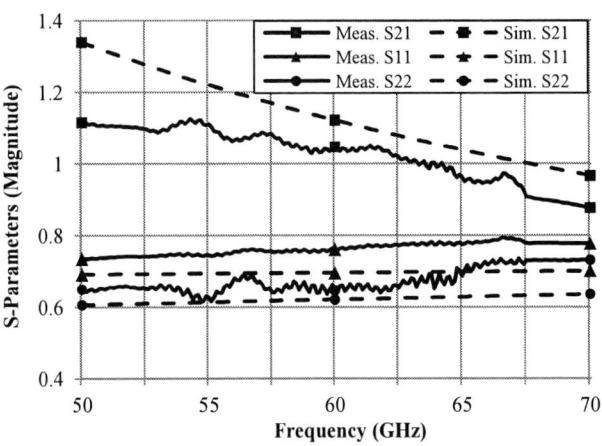

Fig. 5. Simulated and measured S- parameter of the power MOSFET.

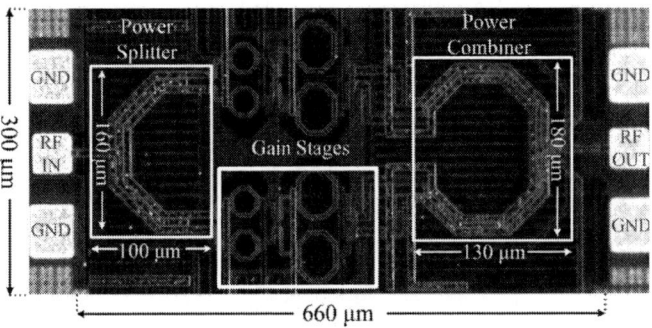

Fig. 7. Chip micrograph of the 60 GHz PA.

is increased which leads to higher matching network losses because the higher transformation ratio is needed. A power MOSFET with 120 um channel width is designed and fabricated in 65 nm CMOS technology shown in Fig. 4.b. The output current of 63 mA is measured using a 1.4V supply and the gate voltage of 0.8 V. Simulation and measured results of the fabricated MOSFET are illustrated in Fig. 5. According to Fig. 5, using a total gate width of 120 μm keeps the $f_{max}$ above 60 GHz.

## C. Gain stages and Inter-stage matching

The schematic of the 60 GHz CMOS power amplifier is presented in Fig. 6. The transistors were laid out in two parallel 32 fingers with 1.8μm wide fingers for the last stage and with 32 fingers, 1.2 μm and 1.6 μm for driver and gain stages respectively. To ensure a reasonable voltage gain, the driver stage and the gain stage were cascaded before the main power stage. Bias conditions were set to achieve the maximum current density at frequencies near $f_T$. The gate bias ($V_{GG}$) voltage is set to be 0.8V while the supply voltage ($V_{DD}$) is 1.4 V for all the stages.

To ensure a reasonable voltage gain, the driver stage and the gain stage were cascaded before the main power stage. For inter-stage matching circuits, stacked transformers were modeled in Advanced Design System (ADS) and fabricated on the $M_9$ and $M_8$ metal layers with radii of 15 μm and 20 μm and conductor width of 6 μm. Utilizing the $M_8$ and $M_9$ metal layers, a coupling factor of 0.75 and maximum quality factor

of 15 is achieved based on EM simulations. Simulated gain and I/O reflections of the 60 GHz PA are illustrated in Fig. 8.

## III. MEASURED RESULTS

The 60 GHz PA is fabricated in TSMC 1P9M 65nm CMOS process. The chip micrograph of the PA is shown in Fig. 7. The PA only occupies a core area of 0.19 mm² by utilizing a compact power combiner and splitter. The S-parameter measurements are performed using Agilent N5251A vector network analyzer (VNA) solution which uses an E8361 power network analyzer (PNA), millimeter wave test controller and broadband frequency extenders. The measured S-parameters are compared with the simulation results in Fig. 8. Applying 1.4V supply, a power gain of 18.8 dB and a 3dB bandwidth of 4 GHz (58 to 62 GHz) are measured. With the stability factor of greater than unity, the amplifier is unconditionally stable over the frequency range of operation.

The 60 GHz power measurements have been performed using Rohde-Schwarz NRP-Z power sensors and the R&S ZVA67 VNA. The measured results are shown in Fig. 9. With 1.4V supply voltage, the measured 1dB compressed output power ($P_{1dB}$) is 16.9 dBm and the saturated output power ($P_{sat}$) is 18.3 dBm. The measured peak PAE is 15.9%. Compared to the reported 60 GHz PAs, the narrow-band PA is smaller in size for comparable higher output power while improving the linearity based on the 16.9 dBm measured 1dB compression point. Table. I summarizes a comparison with the state-of-the-art 60 GHz PAs in 65nm CMOS technology.

978-1-4673-6145-3/13 $31.00 © 2013 IEEE

## TABLE. I
### COMPARISON WITH PUBLISHED 60 GHZ PAS IN 65NM CMOS.

| | B.W. [GHz] | Stages | Architecture | $P_{sat}$ [dBm] | $P_{1dB}$ [dBm] | $G_{max}$ [dB] | $PAE_{max}$ [%] | $P_{DC}$ [mW] | Size [mm²] | Power/Area [mW/mm²] | Ref. |
|---|---|---|---|---|---|---|---|---|---|---|---|
| 60 GHz PAs in 65nm CMOS Technology | 57-65 | 2 | Single-ended/C.S | 13 | 8.9 | 8 | 11 | 64.8 | 0.29 | 45 | [2] |
| | 58-65 | 3 | Differential/C.S | 11.5 | 2.5 | 15.8 | 11 | 43.5 | 0.05 | 230 | [3] |
| | 55-65 | 2 | Differential/C.S/4x | 17.9 | 15.4 | 18.6 | 11.7 | 460 | 0.83 | 22 | [4] |
| | 53-68 | 2 | Single/Cascode/8x | 18.1 | 11.5 | 15.5 | 3.6 | 1504 | 0.46 | 140 | [5] |
| | 55-65 | 1 | Single-ended/C.S | 9 | 6 | 4.5 | 9 | 27.6 | 0.27 | 29 | [6] |
| | - | 4 | Single-ended/C.S | 13.8 | 12.2 | 13.4 | 7.6 | 300 | 1.28 | 19 | [7] |
| | 59-67 | 2 | Single-ended/C.S | 10.6 | 9.2 | 13.2 | 8.9 | 80 | 0.29 | 40 | [8] |
| | 56-62 | 3 | Differential/C.S | 14.6 | 10 | 23.2 | 16.3 | 135 | 0.60 | 48 | [9] |
| | 58-64 | 3 | Single-ended/C.S/2x | 17.8 | 13.8 | 11 | 12.6 | - | 0.28 | 214 | [10] |
| | 58-62 | 3 | Single-ended/C.S/4x | 18.3 | 16.9 | 18.8 | 15.9 | 424 | 0.19 | 360 | This work |

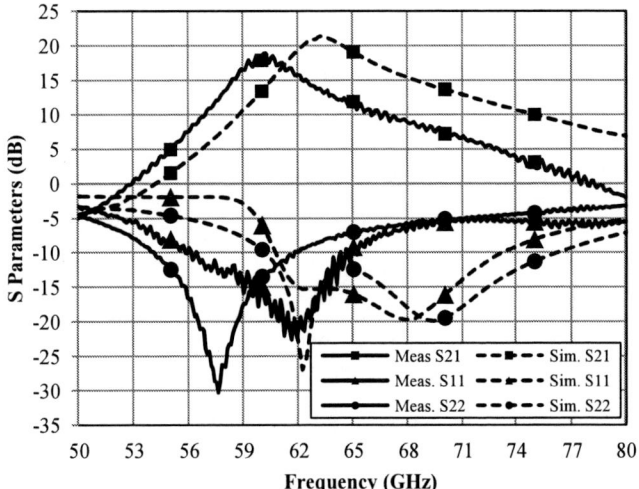

Fig. 8. Measured and simulated S-parameters

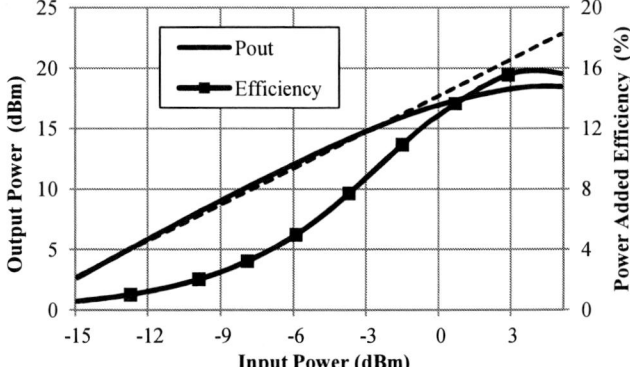

Fig. 9. Measured output power and efficiency

## IV. CONCLUSION

A novel multi-conductor power combiner for an integrated 60 GHz CMOS PA has been proposed. The proposed topology provides a compact 4-way power combining capability to combine the output power of four individual PA cores. The PA core consists of cascaded common source stages with inter-stage matching. The peak power gain of 18.8 dB and 18.3dBm saturated output power are measured over the 3dB bandwidth of 4 GHz. The 0.19 mm² die area consumes 424 mA from a 1.4V supply and presents 15.9% PAE at saturation.

## ACKNOWLEDGMENT

The chips are fabricated by TSMC Semiconductors Corp. through the Canadian Microelectronic Center (CMC). The authors acknowledge Rohde & Schwarz Canada Inc. for providing the power measurements test setup.

## REFERENCES

[1] Yikun Yu, Peter G. M. Baltus, Anton de Graauw, Edwin van der Heijden, Cicero S. Vaucher, and Arthur H. M. van Roermund, "A 60 GHz Phase Shifter Integrated With LNA and PA in 65 nm CMOS for Phased Array Systems," *IEEE Journal of Solid-State Circuits*, vol. 45, no. 9, pp. 1697–1709, Sep. 2010.

[2] Aloui, E. Kerhervé, D. Belot, and R. Plana, "A 60GHz, 13dBm Fully Integrated 65nm RF-CMOS Power Amplifier," in *NEWCAS*, Montreal, Canda, 2008, pp. *93- 96.*

[3] Wei L. Chan, John R. Long, "A 58–65 GHz Neutralized CMOS Power Amplifier With PAE Above 10% at 1-V Supply," *IEEE Journal of Solid-State Circuits*, vol. 45, no. 3, pp. 554–564, Mar. 2010.

[4] J. W. Lai and A. Valdes-Garcia, "A 1V 17.9dBm 60GHz power amplifier in standard 65nm CMOS," in *ISSCC, Dig. Of Technical Papers*, San Fransisco, CA, 2010, pp. *424- 425.*

[5] B. Martineau, V. Knopik, A. Siligaris, F. Gianasello, and D. Belot, "A 53-to-68GHz 18dBm power amplifier with an 8-way combiner in standard 65nm CMOS," in *ISSCC, Dig. Of Technical Papers*, San Fransisco, CA, 2010, pp. *428- 429.*

[6] A. Valdes-Garcia, S. Reynolds, and J. O. Plouchart, "60 GHz Transmitter Circuits in 65nm CMOS," in *RFIC Symposium*, Atlanta, GA, 2008, pp. *641- 644.*

[7] T. Quémerais, L. Moquillon, S. Pruvost, J.M. Fournier, P. Benech, "A CMOS Class-A 65nm Power Amplifier for 60 GHz Applications," in *SiRF*, New Orlean, LA, 2010, pp. *120- 123.*

[8] Sofiane Aloui, Eric Kerherve, Robert Plana, Didier Belot, "A 59GHz to 67GHz 65nm CMOS High Efficiency Power Amplifier," in *New Circuit and Systems*, Bordeaux, France, 2011, pp. *225- 228.*

[9] Hiroki Asada, Kota Matsushita, Keigo Bunsen, Kenichi Okada, and Akira Matsuzawa, "A 60 GHz CMOS Power Amplifier Using Capacitive Cross-Coupling Neutralization with 16% PAE," in *EuMC*, Manchester, 2011, pp. *1115- 1118.*

[10] Jiashu Chen, Ali M Niknejad, "A Compact 1V 18.6dBm 60GHz Power Amplifier in 65nm CMOS," in *ISSCC, Dig. Of Technical Papers*, Melbourne, VIC, 2011, pp. *432- 433.*

[11] Chi Y Law, Anh-Vu Pham, "A High-Gain 60GHz Power Amplifier with 20dBm Output Power in 90nm CMOS," in *ISSCC, Dig. Of Technical Papers*, San Fransisco, CA, 2010, pp. *426- 427.*

# CMOS Low-Power Transceivers for 60GHz Multi Gbit/s Communications

## Invited

Vojkan Vidojkovic[1], Viki Szortyka[1,2], Khaled Khalaf[1,2], Giovanni Mangraviti[1,2], Bertrand Parvais[1], Kristof Vaesen[1], Steven Brebels[1], Annachiara Spagnolo[1,4], Michael Libois[1], John Long[3], Kuba Raczkowski[1], Praveen Raghavan[1], André Bourdoux[1], Min Li[1], Charlotte Soens[1], Vito Giannini[1], Piet Wambacq[1,2]

[1] Imec, Leuven, Belgium

[2] Vrije Universiteit Brussel, Brussels, Belgium

[3] Delft University of Technology, Delft, The Netherlands

[4] University of Salento, Italy

*Abstract*—**The availability of 9GHz bandwidth around 60GHz in combination with simple modulations schemes, low-cost radio ICs and small antenna size, allows for multi Gbit/s wireless communications. In this article the potential of 60GHz wireless communications is evaluated from system, application and user point of view. Further, design challenges for 60GHz CMOS transceivers are identified. State-of-the-art designs show that short-range high-datarate radio links based on CMOS ICs can be made, potentially helped with beamforming.**

## I. INTRODUCTION

In the last two decades the datarates in wireless communications increased exponentially, as shown Fig. 1. This improvement in the speed of wireless communications happened thanks to joint improvements in CMOS technology, memories, analog and digital design including new circuit topologies, methodologies and tools. As a consequence smartphones, tablets and cloud computing became reality. Considering crossing the border of 1Gbit/s communication speed, the IEEE802.11ac and IEE802.11ad standards are of particular interest [2],[3]. They can provide datarates up to 7Gbit/s.

The IEEE802.11ac standard operates in the band around 5GHz. High data rates are obtained by combining high order modulation schemes (QAM 64 and QAM256), wide channel bandwidths (80MHz and 160MHz), MIMO spatial division multiplexing (up to 8 spatial streams) and short guard interval of 400ns. A signal spatial stream can provide a datarate of 867Mbit/s using a QAM256 modulation and channel bandwidth of 160MHz. It is important to note that occupying a bandwidth of 160MHz means using 8 WLAN channels of 20MHz.

The IEEE802.11ad standard provides high datarates thanks to availability of four wide channels in the 60GHz band with a spacing of 2.16GHz. It has both single-carrier (SC) and OFDM modes with a sample rate of 1.76GHz and 2.64GHz, respectively. IEEE802.11ad also supports fast session transfer to other IEEE802.11 standards operating in the 2.4GHz and 5GHz bands. Considering operation at 60GHz the main idea is to exploit the very wide channel bandwidth and to use simple modulation schemes as BPSK, QPSK and QAM16 in order to alleviate circuit design.

Concerning applications, both IEEE802.11ac and IEEE802.11ad standards aim to provide multi Gbit/s wireless connectivity among displays, PCs and handheld portable devices. Several applications

are labeled with mass market and large volume potential, e.g. kiosk allowing fast download of multi-Gbyte content to smartphones and tablets, wireless docking stations, wireless USB and HDMI.

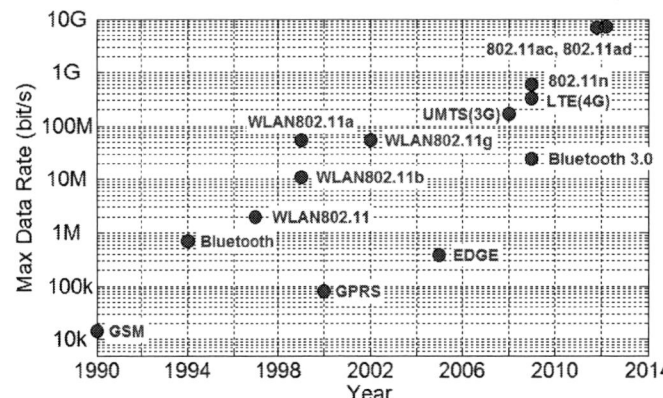

Fig. 1: Evolution of data rates in wireless communications versus time [1].

It is interesting to compare these two standards in terms of hardware challenges and from the user perspective. First, considering the operating frequency, IEEE802.11ac is easier. Compared to 5GHz, 60GHz designs are more sensitive to parasitics and have less design margin due to the operation of the transistors close to $f_T$ and the larger bandwidth. On the other hand, IEEE802.11ac uses more complex modulation schemes up to QAM256, which impose tight specifications on the radio [2]. Further, 5GHz WLAN transceivers achieve a long range. In order to reach a high datarate they occupy up to 8 standard WLAN channels. Therefore, multiple users have to share the same bandwidth reducing the effective throughput. On the other hand, 60GHz communication is short-range due to a large free-space path loss. Hence, users can reuse the same channels even in the same room without throughput degradation. Finally, the size of 5GHz antennas is substantially larger compared to 60GHz antennas. Therefore it would be difficult to use MIMO transceivers in applications that need a small form factor (e.g. smartphones). To summarize, the two standards are complementary. Actually, it is reasonable to expect multi-mode chips compliant to 802.11ac, 802.11ad and maybe even other 802.11 standards as 11n and 11g, operating around 2.4GHz, 5GHz and 60GHz.

978-1-4673-6145-3/13 $31.00 © 2013 IEEE

Another aspect in comparing communication in the low-GHz range to communication around 60GHz is the communication efficiency expressed in energy per bit. An IEEE802.11ad compliant transceiver targeting portable applications achieving datarates up to 1.5Gbit/s on a distance of 1m is presented in [5]. Its communication efficiency is compared to various IEEE802.11n transceivers in Table 1.The IEEE802.11ad transceiver consumes the lowest energy per bit.

Table 1

COMPARISON BETWEEN 802.11ad AND 802.11n TRANSCEIVERS

| Reference | Panasonic 802.11ad [5] | Intel 802.11n [8] | Broadcom 802.11n [6] | Atheros 802.11n [7] |
|---|---|---|---|---|
| Energy/bit (J/Gbit) | 1.2 | 6.2 | 3 | 2.3 |

To increase the range of a mm-wave link to distances beyond 1m [5], beamforming can be used. In this article beamforming approaches are reviewed. The challenge here is to combine this functionality with the classical radio functionality without heavily increasing the power consumption such that the energy consumption per bit remains attractive.

The organization of the paper is the following. Section 2 provides system considerations including an estimation of smartphone power consumption when using a 60GHz transceiver. In section 3 different beamforming strategies and transceiver architectures are discussed. Next, sections 4, 5, 6 , 7 and 8 provide information about the RX front-end, TX front-end, LO synthesis, beamforming and high-speed analog to digital converters. Some experimental results are discussed in section 9. Conclusions are drawn in section 10.

## II. SYSTEM CONSIDERATIONS

Nowadays smartphones and tablets employ complex architectures that combine wireless transceivers, baseband, embedded memory, an application and graphic processor, storage and display. The goal is to match and synchronize performance of all these blocks such that when operating together they yield optimal performance. Therefore, system considerations are started by evaluating performance of 60GHz transceiver in a smartphone.

A possible architecture of a future smartphone is shown in Fig. 2. It consists of a "processor", a display and storage. The processor is a system in package incorporating different dies (radio, baseband, memory, ...). Compared to the situation today [9], the 60GHz radio and baseband are added. Considering storage, flash memory is used due to its high density and non-volatile nature. The flash memory is not integrated together with the processor because it scales differently from conventional CMOS and it requires relatively high programming voltages [10].

Next we estimate the power consumption of the smartphone when it uses 60GHz technology to reach multi Gbit/s datarates. The 1.2J/Gbit consumption of [5] (see Table 1) can be split into 0.4J/Gbit for the radio and 0.8J/Gbit for the digital baseband part. For the power consumption of the storage we take Intel's 60Gbytes 500Mbit/s 520 series solid-state drive [11] as an example. It consumes approximately 800mW, which corresponds to 1.6J/Gbit. This brings the total energy consumption of the transceiver and memory to 2.8J/Gbit. Taking into account a datarate of 1.5Gbit/s [5], a power consumption of 4.2W is calculated. After adding 0.5W

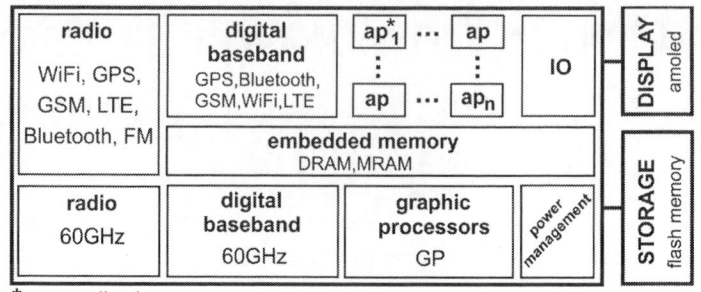

*ap - application processor

Fig. 2: Future smartphone architecture.

which is consumed by other smartphone circuitry [12], the estimated smartphone power consumption when using 60GHz technology is 5W. Assuming a battery capacity of 8.6Wh [4], it is possible to transfer 1.1Tbytes of data without recharging, which is equivalent to 100 hours of HD movie [15]. Consequently, transferring a 1 hour HD movie (11Gbytes) takes 1 minute, using only 1% of the battery capacity [4]. When using an IEEE802.11n chip [7] this takes 5 minutes while consuming 108mWh. In [13] the power consumption of a Galaxy S3 smartphone is estimated for different applications. The most power hungry are camera recording and gaming applications consuming up to 1.5W, which is more than three times lower than the consumption for the 60GHz mode. Therefore further power consumption reduction is required in all three domains of a 60GHz system: radio, digital baseband and storage.

Next, we consider the 60GHz radio link budget. Table 2 shows the result of our link budget analysis for SC QPSK and QAM16. To reach 4m and 0.7m with QPSK and QAM16, respectively, 4 antenna paths with beamforming are used at both TX and RX sides, TX $P_{1dB}$ per antenna path has to be 10dBm and the RX NF per antenna path must be 8dB. Beamforming in both RX and TX improves the link budget with $30*\log(N)$, where N is the number of antenna paths. Without this 4-way beamforming, TX $P_{1dB}$ should be around 28dBm to reach the same distance, which is very difficult to realize.

Table 2

LINK BUDGET FOR SC QPSK AND QAM16 MODULATED SIGNALS

| QPSK modulation | | QAM16 modulation | |
|---|---|---|---|
| Required SNR=11dB Required TX EVM=-15dB | | Required SNR=17dB Required TX EVM=-21dB | |
| TX specs | RX specs | TX specs | RX specs |
| $P_{1dB}$/ant. path 10dBm | RX sensitivity -61dBm | $P_{1dB}$/ant.path 10dBm | RX sensitivity -45.7dBm |
| # ant. paths 4 | # ant. paths 4 | # ant. paths 4 | # ant. paths 4 |
| Gain/ant.elem. 2dB | Gain/ant.elem. 2dB | Gain/ant.elem. 2dB | Gain/ant.elem. 2dB |
| PA back off 4dB | NF/ant. path 8dB | PA back off 5dB | NF/ant. path 8dB |
| Insertion loss 2dB | Insertion loss 2dB | Insertion loss 2dB | Insertion loss 2dB |
| Distance=4m | | Distance=0.7m | |

## III. ARCHITECTURES AND DESIGN APPROACH

To save power consumption and to limit the duplication of the radio functionality over the different antenna paths, beamforming is best implemented in the analog domain, rather than in the digital domain. Analog beamforming can be realized in the RF

[17][19][20], LO [18][21] or baseband (BB) part [16][22] of a transceiver (see Fig. 3 [24]).

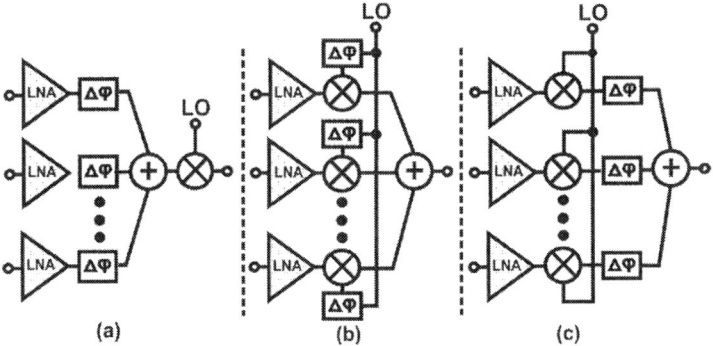

Fig. 3: Beamforming strategies: (a) RF signal path; (b) LO path; (c) at analog BB.

RF beamforming has the smallest hardware footprint. It requires one single signal path after the beamforming. Practically, only the LNA and PA are duplicated as these are between the beamforming circuitry and the antenna. However, the summation of signals has to be performed at 60GHz. Further, RF phase shifters introduce a loss which depends on the phase shift, requiring a variable-gain RF amplifier that must be calibrated. RF phase shifters often use transmission lines [24], leading to a bulky layout.

Transceiver performance is less sensitive to amplitude variations in the LO path compared to the signal path as long as the LO amplitude is sufficiently large. This makes LO phase shifting an interesting beamforming approach [33]. Compared to RF beamforming, the required bandwidth is lower than the bandwidth of the signal path [16]. The hardware footprint is moderate. The LNA/PA-mixer combination has to be duplicated. Nevertheless, the LO distribution becomes more complex and power hungry while the layout gets more bulky. Considering the baseband beamforming the hardware footprint is the same as in the LO beamforming while the LO distribution is simplified. The main advantage is high robustness because the beamforming operations (phase shifting and signal combination) are performed at low frequencies.

The three main beamforming strategies discussed above can be combined. For example, in the combination of LO and baseband phase shifting of [27], coarse phase shift steps of 90° are realized in the baseband while the fine phase shift between 0° and 90° is realized in the LO path.

Considering the IF selection, two main approaches are found in literature: sliding IF [17][19][20][42] and zero IF architectures [16][18][22][23][41]. With the first approach the design of the PLL and the LO distribution is relaxed but the signal path is longer, which constrains the dynamic range in the signal path.

Following the strategy to minimize signal processing at 60GHz, the architecture of [22] combines zero-IF with baseband beamforming (see Fig. 4). Thanks to the use of one PLL per two antenna paths with one central reference frequency the LO buffering is relaxed. In addition, the phase noise performance improves up to 3dB per doubling of the number of PLLs because the uncorrelated phase noise from two different PLLs is averaged. Further, with baseband beamforming, signal operations at 60GHz are kept to a minimum and sensitivity to small layout parasitics is lower than at 60GHz. Further, the inevitable parasitic interconnect capacitances arising from bringing together the antenna paths are absorbed in the baseband filter capacitors. In this way, buffering stages can be

avoided, which preserves the dynamic range and leads to a lower power consumption. The fourth-order baseband filtering is made with a biquadratic section combined with two passive poles, realized in VGA$_1$ in Fig. 4 and the combiner.

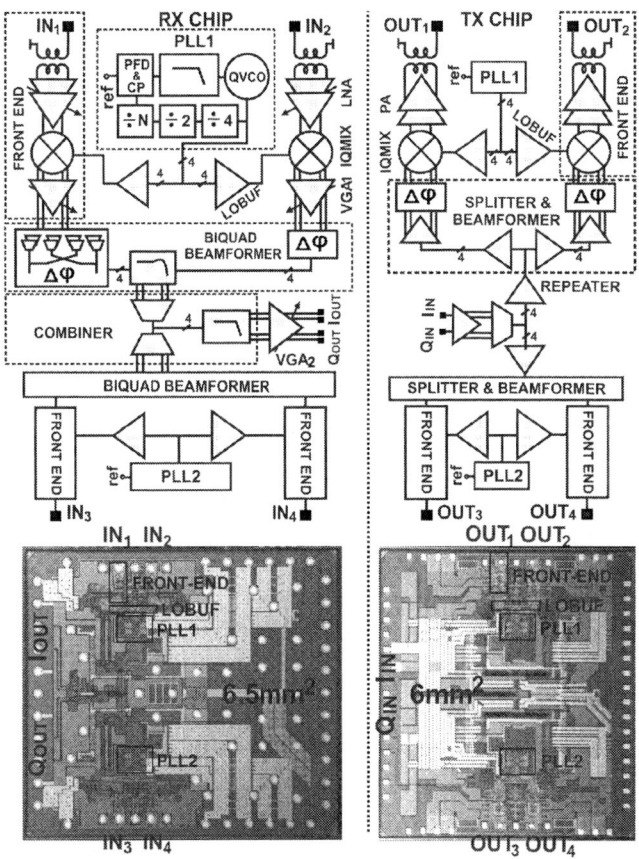

Fig. 4: Four antenna-path RX and TX beamforming architectures and die photos.

The smaller wavelength at 60GHz (about 2mm for on-chip transmission lines) and the smaller distance between the operating frequency and $f_T$ yield some modifications in the design style compared to low-GHz range design. First, parasitic capacitance is often tuned out with inductors. This can be done unpunished as the footprint of inductors is only a few tens of microns in diameter. Further, the lower gain per stage at 60GHz necessitates the use of multiple stages which are usually AC coupled. As shown in Fig. 5 this coupling can be capacitive [26] or via a transformer [28][45]. The latter approach, which is too bulky for the low-GHz range, results in a very compact layout, particularly for differential circuits.

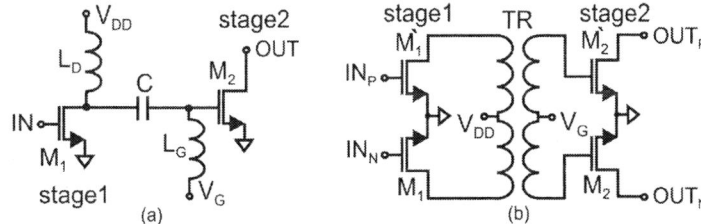

Fig. 5: Coupling in mm-wave multi-stage circuits.

Further, interconnects longer than 10μm are best considered as transmission lines (TL), which can be simulated accurately using pseudo- or full-3D simulators. In Fig. 6 the input impedance of an on-chip transmission line (L and C per unit length are 50pH and 8fF

978-1-4673-6145-3/13 $31.00 © 2013 IEEE        345

per 100μm, respectively), loaded with capacitors of 20fF and 50fF, and a 200Ω resistor, is plotted versus length . As a comparison, the line is also modeled as a lumped capacitance to ground, as obtained from a classical layout parasitic extraction. Clearly, the latter model starts to become inaccurate for lines longer than 10μm.

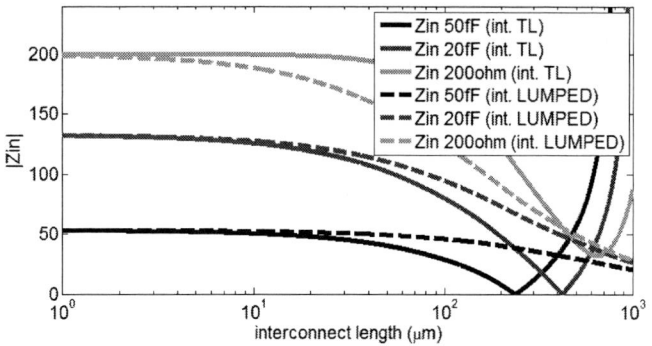

Fig. 6: Influence of the interconnect length to input impedance.

## IV. RECEIVER FRONT-END

At 60GHz two LNA topologies are mostly used (see Fig. 7): common source [16][17][41][42] and inductively-degenerated common-source [29]. Cascode transistor $M_2$ is often added to improve stability. However, at mm-wave frequencies, its noise contribution is no longer negligible as its noise source undergoes less degeneration from the relatively low output impedance of $M_1$. To compensate the high parasitic capacitance at the source of $M_2$ and the drain of $M_1$, an inductor in between these nodes can be added [54] or in parallel [29] as shown in Fig. 7(b) at the expense of layout complication.

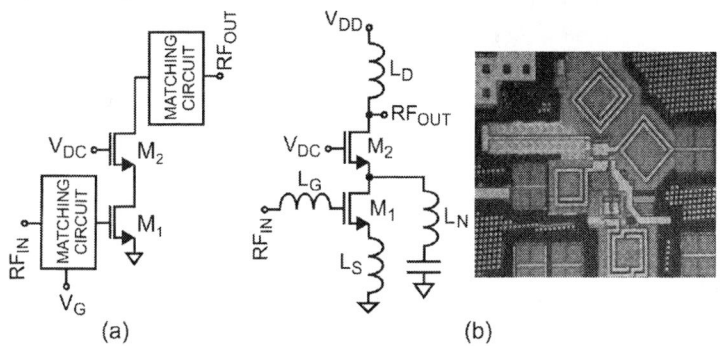

Fig. 7: LNA topologies: (a) common-source (b) inductively-degenerated common-source (schematic and layout) [29].

The matching network in a common-source LNA is implemented either using transmission lines [41][42] or using coils and capacitors [16], both resulting in a fairly bulky layout. In addition, ESD protection circuitry needs to be added. An example is shown in Fig. 8 [42]. The area of this LNA is 0.35mm².

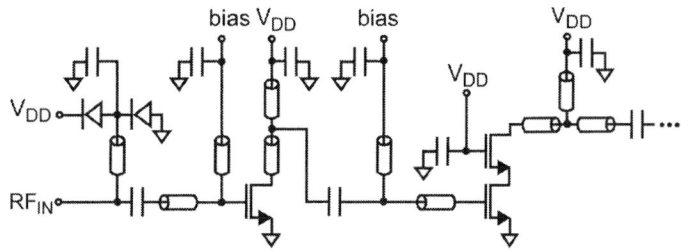

Fig. 8: Input stage of the LNA of [42], containing ESD protection diodes.

As an alternative to a single-ended topology, a differential LNA can be used with an input transformer acting as a balun and featuring ESD protection. A block diagram of a receiver front-end containing such LNA and a passive mixer is shown in Fig. 9 [23]. The differential nature of the LNA improves robustness to common-mode disturbances and reduces the LO feedthrough to the antenna. In addition the layout is very compact, occupying only 0.04mm². The input transformer has a 1:3 turn ratio and a coupling factor of 0.5. It provides wideband input impedance matching and contributes to the overall wideband response of the LNA. A similar approach is used in [30]. There the wideband response is obtained thanks to the use of capacitively coupled resonators. It results in a larger layout compared to inductively coupled resonators used in [23]. The front-end of Fig. 9 achieves a NF of 5.5dB while front-ends in [29][42] have a NF larger than 7dB.

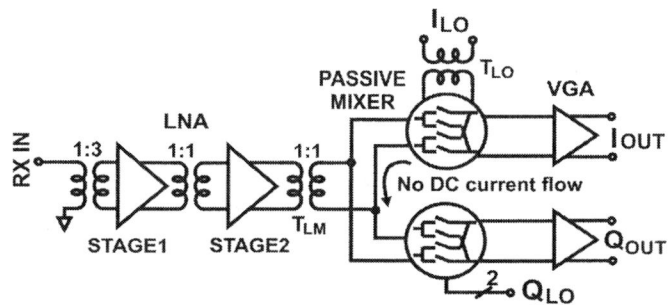

Fig. 9: Block diagram of the zero-IF RX front-end of [22].

The different stages of the LNA of Fig. 9 use pseudo-differential amplifiers [28] with cross-coupled capacitors $C_C$ that neutralize $C_{GD}$, yielding higher gain while maintaining stability (see Fig. 10).

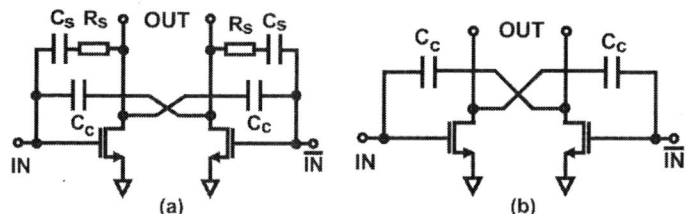

Fig. 10: Schematics of the LNA (a) first stage (b) second stage.

In addition they contribute to simultaneous impedance and noise matching. The condition for simultaneous input impedance matching and noise matching can be expressed as:

$$Z_{NM} = Z_{IN}^* \qquad (1)$$

$Z_{NM}$ is the driving impedance for the first LNA stage yielding minimum NF and $Z_{IN}^*$ is the complex-conjugate input impedance of the LNA. $Z_{IN}$ can be approximated as:

$$Z_{IN} = R_g + \frac{1}{j\omega(C_{GS}+G_{ST1}C_{GD})} \qquad (2)$$

Here $R_g$ is the gate resistance of the transistor $M_1$, $C_{GS}$ and $C_{GD}$ are gate-source and gate-drain parasitic capacitances, respectively and $G_{ST1}$ is the voltage gain of the first stage. $Z_{NM}$ can be calculated as:

$$Z_{NM} = 2\sqrt{\frac{R_g}{\gamma g_m}} + \frac{j}{\omega(C_{GS}+C_{GD})} \qquad (3)$$

Based on equation (1), the real and imaginary parts of $Z_{IN}$ and $Z_{NM}$ have to be equal. The imaginary parts are equal when $C_{GD}=0$ and

978-1-4673-6145-3/13 $31.00 © 2013 IEEE

this is exactly what neutralization does when choosing $C_C = C_{GD}$. In addition to neutralization, shunt feedback is added to reduce the difference between the real parts of $Z_{NM}$ and $Z_{IN}$.

In Fig. 9 a passive IQ mixer downconverts the RF signals to voltages that are sensed by differential source-degenerated amplifiers with a differential load that can be programmed for a variable gain. The center tap terminal of the transformer $T_{LM}$ is left open. Therefore, there is no DC current through the mixer switches, yielding less flicker noise. The mixer LO port is coupled to the LO buffer via the 1:1 transformer $T_{LO}$. The DC voltage at the gates of the mixer switching transistors can be adjusted via the center tap of $T_{LO}$.

## V. TRANSMITTER FRONT-END

The transmitter front-end consists of a PA and upconversion mixers (see e.g. Fig. 11). The PA is the most power hungry block in a transceiver. Therefore, a high PA efficiency is important for low-power consumption. Most 60GHz PAs are differential transformer-coupled multistage amplifiers [28][42][45].

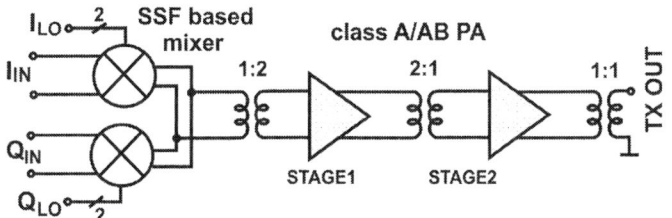

Fig. 11: Block diagram of the zero-IF TX front-end of [22].

The active part of the PA can again be based on the neutralized pseudo-differential amplifier of Fig. 12 (a) [28][45]. Stability is obtained thanks to neutralization. In [42] stability is obtained thanks to layout parasitics while in [31] an RC network is added in series with the gate as shown in Fig. 12 (b). Alternatively, single-ended PAs are sometimes used, such as in [41], although they intrinsically yield a lower output power. In this PA the matching networks are based on transmission lines.

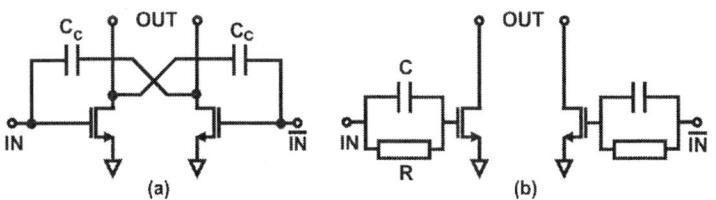

Fig. 12: Techniques for stability improvement in mm-wave PAs.

Most 60GHz PAs operate in class A. The class A PA of [45] achieves a peak power added efficiency (PAE) of 18%. Unfortunately, the PAE at 5dB back-off, which is needed for QAM16 modulation, is only 2.5%. A straightforward way to improve PAE at back-off is to operate the PA in class AB mode (see Fig. 13 (a)) as in [22], where a PAE of 7.4% at 5dB back-off is obtained with an output 1dB compression of 8dBm. However, gain is lower in class AB than in class A (see Fig. 13 (a)).

A high PAE requires a careful layout of the passive components. For example, the layout of the 2:1 octagonal transformer of Fig. 14, used in [22] has a quality factor of the primary and secondary windings at 60GHz of 15 and 13, respectively. The insertion loss at 60GHz is only 2.3dB in a digital CMOS technology with 7 metal layers. With this high Q, sufficient gain per PA stage can be realized which can operate both in class A and class AB (see Fig. 13 (a)). The load is optimized for operation in class A and class AB with little performance penalty. The wanted mode of operation can be set by a DAC on the gate bias voltage.

The TX output 1-dB compression point can be expressed as:

$$\frac{1}{P_{1dB,out,TX}} = \frac{1}{P_{1dB,out,PA}} + \frac{1}{P_{1dB,out,MIX} G^2{}_{PA}} + \frac{1}{P_{1dB,out,BB} G^2_{PA} G^2_{MIX}} \quad (4)$$

This formula shows that the overall TX output 1-dB compression point might be deteriorated if the PA gain is not high. This problem can be alleviated by using high-gain high-linearity upconversion mixer such as in [23] where a current-switching mixer uses a transconductor built around super-source follower.

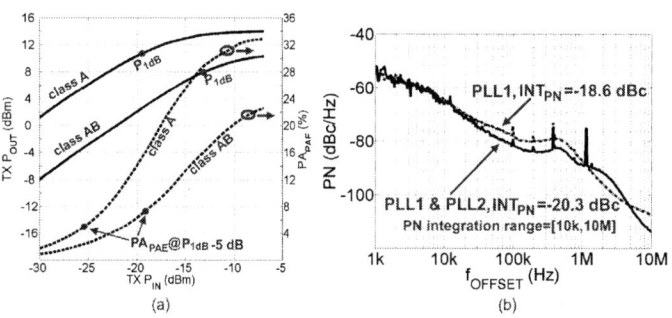

Fig. 13: Measured (a) transmitter large signal transfer function and PA power added efficiency of [22];(b) phase noise of PLL1 and PLL1&PLL2 in channel 1.

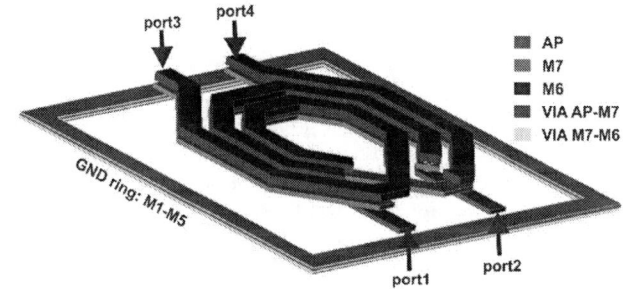

Fig. 14: 2:1 transformer used in the PA. Its size is 85µm x 85µm.

## VI. LO SYNTHESIS AND DISTRIBUTION

A zero-IF radio architecture requires a higher LO frequency than superheterodyne and relies on an LO signal in quadrature format, which is more challenging for the design of the LO synthesis and distribution circuitry. For this reason we limit the discussion here to the zero-IF case.

For the IEEE802.11ad standard, four LO frequencies need to be synthesized, namely 64.8GHz – n∗2.16GHz (n = 0, ..., 3). These frequencies can be generated with an integer N PLL. With a fractional PLL [25] the designer has extra freedom in the choice of the reference frequency but fractional spurs and phase noise at high offset frequencies require careful attention.

Considering phase noise specification, in first order, the integrated phase noise should be better than the opposite of the transmit EVM (see Table 2), namely -21dB for QAM16 and at least 3dB better than the required SNR at the RX output (17dB for QAM16), as the RX sees the phase noise impact of both the TX and the RX PLL.

978-1-4673-6145-3/13 $31.00 © 2013 IEEE

For the LO synthesis there are several possibilities (see Fig. 15). The first one (see Fig. 15 (a)) is similar to the approach that is classically used in the low-GHz range which is built around a PLL operating at the double of the required frequency. However, this option is not preferred because the VCO and the divider have to operate at 120GHz consuming a lot of power and having poor performance.

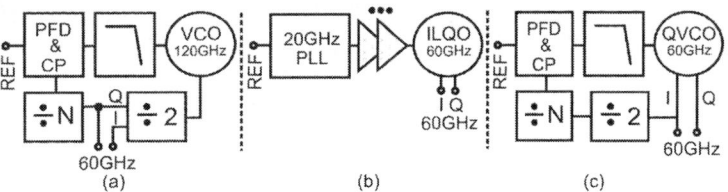

(a)  (b)  (c)

Fig. 15: Different approaches for 60GHz LO generation.

The second possibility (see Fig. 15 (b)) is to use a 60GHz quadrature injection-locked oscillator (QILO) [23][34][41]. In [34] and [41] a signal at one third of the target LO frequency is used to lock the QILO. The approach achieves a low phase noise at 60GHz of -96dBc/Hz at 1MHz offset, consuming 77.5mW [41]. An issue with this approach is the locking range, which should be wide enough to guarantee reliable operation without complicated calibration. In [23] a QILO is reported with an off-chip generated injecting signal around 20GHz. A very wideband locking range of more than 8GHz is obtained with a phase noise better than -96dBc/Hz at 1MHz offset in all four 60GHz channels. The QILO and 60GHz buffers together consume 77mW.

The third possibility is to use an integer N 60GHz PLL [16][32][36] built around a VCO or QVCO. In [16] the PLL is built around a VCO and I/Q generation is performed by using a hybrid in the LO distribution. A low phase noise of -82dBc/Hz at 1MHz offset is achieved with a power consumption of only 29mW including the LO distribution. The 60GHz PLL of [22] is built around a QVCO (Fig. 15 (c)). Similar to the solution in [36], two QVCOs with overlapping tuning range are used to cover the complete 60GHz band. The divider consists of an injection-locked divide by-4 (prescaler), a static divide-by-2, and a programmable divide-by-N with division ratios between 64 and 1023. This PLL achieves a phase noise of -87dBc/Hz at 1MHz offset and consumes 72mW. The integrated phase noise is better than -18.6dBc. When two such PLLs are used in a beamforming architecture, then the effective phase noise is reduced, theoretically with 3dB as already pointed out in Section III. Here the effective phase noise has decreased to -20.3dBc (see Fig. 13 (b)).

Next to the (Q)VCO, the mm-wave prescaler is an another challenging block in the 60GHz PLL. Most 60GHz dividers are injection locked [38][39]. The main advantage is low power consumption. The divider presented in [39] consumes only 4.8mW while achieving a locking range of more than 20GHz. Static dividers based on a master-slave connection of CML latches can also be realized at 60GHz. They are more robust but power consumption is larger [37]. To reach the high operating frequency, the load impedance of the CML stages, which is usually resistive, is adapted with inductors and/or capacitors to give a peak around 60GHz [37]. A more compact, inductorless divider can be made by a ring oscillator into which the 60GHz LO signal is injected. In [40] a three-stage divide-by-four prescaler in 90nm is demonstrated, consuming 25mW. The third harmonic in the ring mixes with the injected signal, to produce an output at one fourth of the injected

frequency. However, a 4-stage ring oscillator (see Fig. 16) is more compatible with differential injection, yielding more robust operation with a sufficient locking range. It is similar to the 65nm divider of [38], which consumes 6.5mW. The differential 60GHz signal is injected via the tail current source. The oscillation frequency is adjusted by changing a differential pMOS load. The circuit consumes 10mW in 40nm LP CMOS.

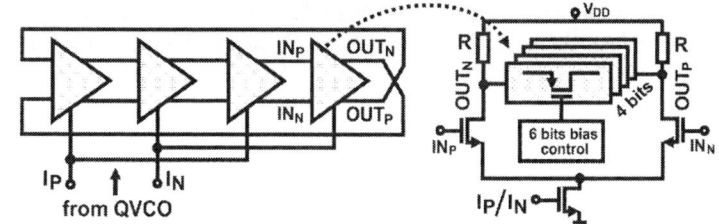

Fig. 16: Block diagram of the prescaler and schematic of the active stage

## VII. ANALOG BASEBAND SECTION

The analog baseband section has to provide filtering, both in the TX and the RX, as well as variable-gain amplification in the RX. If, in addition, beamforming is also implemented at analog baseband as in [22], then all these functions should be combined while keeping a bandwidth of at least 1.76/2 GHz for the SC mode of IEEE802.11ad. In [47] all these functions are entangled (see Fig. 4 and Fig. 17). This is the main difference with respect to the analog baseband beamformer of [16], which uses the same concept of baseband beamforming but without filtering. The biquad uses a gm-C topology which offers good high-speed performance [42][43]. The phase shifting functionality is added in the input transconductor of the biquad.

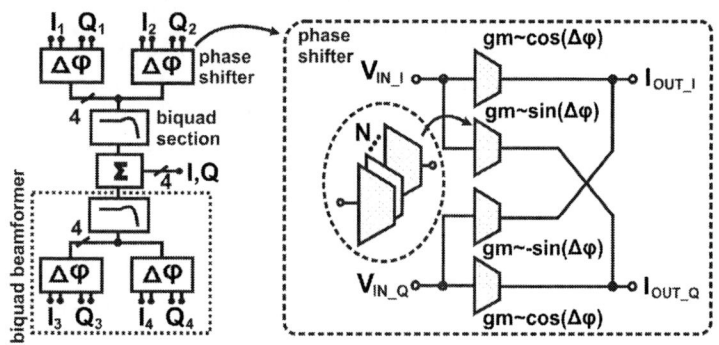

Fig. 17: Concept of analog baseband beamforming used in the receiver.

Phase shifting is applied to an RX baseband signal using the following operations [44] (see Fig. 17):

$$I_{OUT\_I} = V_{IN\_I} \cos(\Delta\varphi) - V_{IN_Q} \sin(\Delta\varphi) \qquad (5)$$

$$I_{OUT\_Q} = V_{IN\_I} \sin(\Delta\varphi) + V_{IN\_Q} \cos(\Delta\varphi) \qquad (6)$$

The gains proportional to $\cos(\Delta\varphi)$ and $\sin(\Delta\varphi)$ are implemented using programmable transconductors [16][46]. The other possibility is to use current-mode VGAs [44]. This approach is applied in the TX beamformer of [48]. Returning to the transconductors of Fig. 17, these are implemented as a parallel connection of multiple small transconductor cells. The implementation with multiples of identical cells instead of one programmable transconductor avoids calibration and is robust to PVT variations. The choice of the number of gm-cells is based on the required phase resolution and amplitude variation. In this particular case 20 unit gm-cells are used, yielding a phase resolution of 3.1° and an amplitude variation of 0.4dB [47].

978-1-4673-6145-3/13 $31.00 © 2013 IEEE          348

The beamformer of [16] has a phase resolution of $5°$ and amplitude variations of 0.5dB.

The cutoff frequency of the fourth-order filtering transfer function implemented in Fig. 17 is 1GHz, while the gain is programmable between 10.6dB and 30dB with an output IP3 above 10dBm and an SFDR higher than 31dB. Power consumption is less than 42mW in 40nm LP CMOS. The 90nm CMOS baseband section for 60GHz of [49] implements sixth-order filtering and a 0.1-19.6dB variable gain but no beamforming and it has a power consumption of 21.6mW.

## VIII. ANALOG-TO-DIGITAL CONVERTERS

The high sample rates of the different modes provided in IEEE802.11ad are challenging for the data converters, especially for the analog-to-digital converters (ADCs). In single-carrier mode, an oversampling ratio of two is common [5] to ease synchronization, yielding a sample rate of 3.52GHz. The effective number of bits (ENOB) at Nyquist rate depends on the complexity of the constellation scheme. For QAM16 five effective bits are acceptable.

High-speed data converters are often realized with flash architectures [5], [53], often in combination with time interleaving. For example, the 6 bit 40nm design of [53] uses two channels and consumes 12mW. To obtain a high speed combined with a low power consumption, the input capacitance of the comparators should be minimized, which conflicts with low offset requirements. This can be resolved using offset calibration [50][53]. To further lower power consumption, alternative architectures are in use such as a combination of interleaving, folding, flash and pipelining as in the 6bit 40nm LP CMOS ADC of [50] yielding a power consumption of 2.6mW at a 2.2GHz sample rate. The asynchronous SAR 8 bit ADC of [51] in 32nm SOI only consumes 3.1mW without interleaving at a sample rate of 1.2GHz.

## IX. MESURED PERFORMANCE OF 60GHZ TRANSCEIVERS

Measured performance of different state of the art 60GHz RX and TX chips is shown in Fig. 18 and Fig. 19, respectively. The first observation is that CMOS implementations do not lag behind SiGe implementations when a TX output power up to 10 dBm is required. Next, in terms of RX and TX efficiency for a large number of antennas, the solution of [17] achieves the highest values. For a small number of antennas (up to 4) the TRX chip set of [22] reaches the highest efficiency. It is the first complete TX/RX radio chipset with direct conversion architecture and baseband beamforming, which in combination with an array of patch antennas implemented on PCB, demonstrates a 60GHz wireless link. The solution of [16] achieves good performance at low power consumption but does not include channel filtering and it does not demonstrate a wireless link.

Considering integration, in [22] the chips are mounted on a Nelco N4000-13 PCB with a 4x2 array of broadside radiating patch antennas on the back. The maximum gain of the complete antenna array is 7.5dBi. The antenna patterns and scanning pattern are shown in Fig. 20. Clearly, the beam becomes narrower with more antenna paths while at the same time it has extra antenna gain. The 3dB scan angle range is around $120°$. In [53] a fixed endfire antenna element is used. It is built on a multilayer organic substrate and placed next to the transceiver in the same package. The maximum gain of the antenna is 10dBi.

In [5], [22] and [53] a 60GHz link is demonstrated using TX and RX modules. The 802.11ad transceiver of [5] achieves a datarate of 1.8Gbit/s at a distance of 40cm and consumes 621mW. The link of

[22] reaches datarates of 2.31Gbit/s (QPSK) and 4.62Gbit/s (QAM16) at a distance of 3.6m and 70cm, respectively. An EVM better than −13.1dB is obtained in the 3 channels that can be reached. Total power consumption is 980mW. The transceiver in [53] is able to achieve a datarate of 3.1 Gbit/s at a distance of 1.8m, while consuming 541mW.

| RX | UC Berkeley [16] | SiBeam [17] | IBM [20] | Imec [22] |
|---|---|---|---|---|
| Tech. | 65nm CMOS | 65nm CMOS | 0.12um SiGe | 40nm LP CMOS |
| # antenna paths | 4 | 32 | 16 | 4 |
| Gain / ant. path (dB) | 24 | NA | 70 | 45 |
| NF / ant. path (dB) | 6.8 | <10 | 7.4 | 7.9 - 8.7 (SSB) |
| RF BW (GHz) | 6.5 | NA | 6 | >8.3 |
| LNA | SE | NA | SE | DIFF |
| BB filter + VGA | NO | YES | NO | YES |
| $P_{DC}$ / ant. path (mW) | 34.2 | 39 (32 ant.) 177 (4 ant.) | 112.5 | 99.25 |
| Area / ant. path (mm$^2$) | 2.2/2 | 2.4 | 2.3 | 1.63 |

Fig. 18: RX performance comparison to state of the art.

| TX | UC Berkeley [16] | SiBeam [17] | IBM [19] | Imec [22] |
|---|---|---|---|---|
| Tech. | 65nm CMOS | 65nm CMOS | 0.12um SiGe | 40nm LP CMOS |
| # antenna paths | 4 | 32 | 16 | 4 |
| Gain / ant. path (dB) | NA | NA | 35 | 31 (PA class A) 22.5 (PA class AB) |
| $P_{1dB}$ / ant. path (dBm) | -1.5 (Psat) | 6.4 | 9 | 10.8 (PA class A) 8 (PA class AB) |
| PA max.PAE (%) | 20 | 22 | 9 | 32.7 (class A) 22.5 (class AB) |
| PA PAE @$P_{1dB}$ - 5dB (%) | NA | NA | NA | 4.9 (class A) 7.4 (class AB) |
| $P_{DC}$ per antenna path (mW) | 34.2 | 56 (32 ant.) 112 (8 ant.) | 237 | 181 (PA class A) 146 (PA class AB) |
| TX efficiency $P_{1dB}/P_{DC}$ (%) | 2.58 | 7.7 (32 ant.) 3.9 (8 ant.) | 3.3 | 6.6 (PA class A) 4.3 (PA class AB) |
| Area / ant. path (mm$^2$) | 2.2/2 | 2.2 | 2.7 | 1.5 |

Fig. 19: TX performance comparison to state of the art.

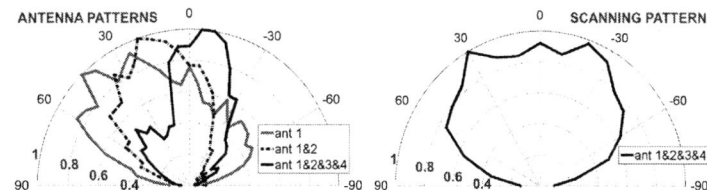

Fig. 20: Antenna patters and scanning pattern

## X. CONCLUSIONS

In this paper the potential of 60GHz communications is evaluated from system, application and user point of view. The conclusion is that 60 GHz communications can be used for short-range high-datarate links. It is reasonable to expect soon multi-mode chips compliant to 802.11ac, 802.11ad and possibly other 802.11 standards such as 11n and 11g, operating in the 2.4GHz, 5GHz and 60GHz bands. Considering efficiency, state of the art 60GHz CMOS transceivers, which are reviewed in this paper, show good performance. Helped by beamforming, the trend of reducing the power consumption will continue in the future making a massive deployment of multi Gbit/s wireless connectivity more realistic.

## REFERENCES

[1] "Comparison of wireless data standards", [Online] Available: http://en.wikipedia.org/wiki/Comparison_of_wireless_data_standards

[2] IEEE draft standard P802.11ac™/D5.0, "Draft STANDARD for Information Technology - Telecommunications and information exchange between systems - Local and metropolitan area networks - Specific requirements. Part 11: Wireless LAN Medium Access Control (MAC) and Physical Layer (PHY) specifications. Amendment 4: Enhancements for Very High Throughput for Operation in Bands below 6 GHz", January 2013

[3] IEEE Std 802.11ad™-2012, "Draft STANDARD for Information Technology - Telecommunications and information exchange between systems - Local and metropolitan area networks - Specific requirements. Part 11: Wireless LAN Medium Access Control (MAC) and Physical Layer (PHY) specifications. Amendment 3: Enhancements for Very High Throughput in the 60 GHz Band", December 2012

[4] "galaxys3 specifications",[Online] Available: http://www.samsung.com/global/galaxys3/specifications.html

[5] T. Tsukizawa et. al., "A Fully Integrated 60GHz CMOS Transceiver Chipset Based on WiGig/IEEE802.11ad with Built-In Self Calibration for Mobile Applications," ISSCC Dig. Tech. Papers, pp. 230–231, February 2013.

[6] A. Behzad et. al., "A Fully Integrated MIMO Multi-Band Direct-Conversion CMOS Transceiver for WLAN Applications (802.11n)," ISSCC Dig. Tech. Papers, pp. 560–561, February 2007.

[7] L. Nathawad et. al., "A Dual-Band CMOS MIMO Radio SoC for IEEE 802.11n Wireless LAN," ISSCC Dig. Tech. Papers, pp. 358–359, February 2008.

[8] S. Gross et. al., "Dual-Band CMOS Transceiver with Highly Integrated Front-End for 450Mb/s 802.11n systems," RFIC Proceedings, pp. 351-354, 2010.

[9] " Snapdragon", [Online] Available: http://www.qualcomm.com/snapdragon/processors/s4/specs

[10] M. She, "Semiconductor Flash Memory Scaling," PhD thesis, University of California, Berkeley, 2003.

[11] "Intel Solid-State Drive 520 Series," Product Specifications, Intel, Feb. 2012.

[12] A. Carol, "An Analysis of Power Consumption in a Smartphone," USENIX Annual Technical Conference, 2010.

[13] X. Chen et Al., "How is Energy Consumed in Smartphone Display Applications,"ACM HotMobile Conference, 2013.

[14] "Samsung Galaxy S4 review", [Online] Available: http://reviews.cnet.com/samsung-galaxy-s4/

[15] "A Digital Video Primer: Understanding and Using High-Definition Video", [Online] Available: http://ifsstech.files.wordpress.com/2008/06/hdprimer_0306.pdf

[16] M. Tabesh et al., "A 65 nm CMOS 4-Element Sub-34 mW/Element 60 GHz Phased-Array Transceiver," vol. 46, no. 12, pp. 2988–3004, 2011.

[17] S. Emami et al., "A 60GHz CMOS Phased-Array Transceiver Pair for Multi-Gb/s Wireless Communications", ISSCC Dig. Tech. Papers, pp. 164-165, 2011.

[18] W. Chen et al., "A 60GHz-Band 2×2 Phased-Array Transmitter in 65nm CMOS", ISSCC Dig. Tech. Papers, pp. 42-43, 2010.

[19] A. Valdes-Garcia et al., "A SiGe BiCMOS 16-element phased-array transmitter for 60GHz communications", ISSCC Dig. Tech. Papers, pp. 218-219, 2010.

[20] A. Natarajan et al., "A Fully-Integrated 16-Element Phased-Array Receiver in SiGe BiCMOS for 60-GHz Communications", IEEE JSSC, vol. 46, no. 5, pp. 1059 –1075, 2011.

[21] A. Natarajan et al., "A 77-GHz Phased-Array Transceiver With On-Chip Antennas in Silicon: Transmitter and Local LO-Path Phase Shifting", IEEE JSSC, vol. 41, no. 12, pp. 2807 –2819, 2006.

[22] V. Vidojkovic et al., "A low-power radio chipset in 40nm LP CMOS with beamforming for 60GHz high-datarate wireless communication," ISSCC Dig. Tech. Papers, pp. 236–237, 2013.

[23] V. Vidojkovic et al., "A low-power 57-to-66GHz transceiver in 40nm LP CMOS with −17dB EVM at 7Gb/s," ISSCC Dig. Tech. Papers, pp. 268–269, 2012.

[24] A. Niknejad, "mm-Wave Phased-Array Receivers", Short Course, ISSCC, 2013.

[25] W. Wu et al., " A 56.4-to-63.4GHz Spurious-Free All-Digital Fractional-N PLL in 65nm CMOS," ISSCC Dig. Tech. Papers, pp. 352–353, 2012.

[26] K. Raczkowski et al., " 50-to-67GHz ESD-protected power amplifiers in digital 45nm LP CMOS," ISSCC Dig. Tech. Papers, pp. 382–383, 2009.

[27] K. Raczkowski et al.," A four-path 60GHz phased-array receiver with injection-locked LO, hybrid beamforming and analog baseband section in 90nm CMOS," RFIC Symposium Proceedings, pp. 431-434, 2012.

[28] W. Chan et al., "A 60GHz band, 1V 11.5 dBm Power Amplifier with 11% PAE in 65nm CMOS," ISSCC Dig. Tech. Papers, pp. 380-381, 2009.

[29] J. Borremans et al., "A Digitally Controlled Compact 57-to-66GHz Front-End in 45nm Digital CMOS," ISSCC Dig. Tech. Papers, pp.492-493, 2009.

[30] F. Vecchi et al., "A Wideband Receiver for Multi-Gbit/s Communications in 65 nm CMOS", IEEE JSSC, vol. 46, no. 3, pp. 551 –561, 2011.

[31] D. Chowdhury et al., "Design Considerations for 60 GHz Transformer-Coupled CMOS Power Amplifiers," IEEE JSSC, vol. 44, no. 10, pp. 2733–2744, 2009.

[32] K. Scheir et al., "A 57-to-66GHz Quadrature PLL in 45nm Digital CMOS," ISSCC Dig. Tech. Papers, pp. 493-495, 2009.

[33] K. Scheir et al., " A 52GHz Phased-Array Receiver Front-End in 90nm Digital CMOS," ISSCC Dig. Tech. Papers, pp. 184-185, 2008.

[34] A. Musa et al., "A 58-63.6GHz Quadrature PLL Frequency Synthesizer in 65nm CMOS," IEEE ASSCC Proceedings, 2010.

[35] B. Parvais et al.," A 40 nm LP CMOS PLL for high-speed mm-wave communication," ESSCIRC Proceedings, pp. 254-257, 2010.

[36] Y. Zhang et al.,"A Low-Phase-Noise LC QVCO with Bottom-Series Coupling and Capacitor Tapping," IEEE ISCAS, 2008, pp. 1000-1003.

[37] L. Li et al.," A 60GHz 15.7mW Static Frequency Divider in 90nm CMOS," ESSCIRC Proceedings, pp. 246-249, 2010.

[38] A. Ghilioni et al., "A 6.5mW Inductorless CMOS Frequency Divider-by-4 Operating up to 70GHz," ISSCC Dig. Tech. Papers, pp. 282-283, 2011.

[39] A. Ghilioni et al., "A 4.8mW Inductorless CMOS Frequency Divider-by-4 with more than 60% Fractional Bandwidth up to 70GHz," CICC Proceedings 2012.

[40] H. Hoshino et al, "A 60GHz phased-locked loop with inductorless prescaler in 90nm CMOS", ESSCIRC Proceedings, pp. 427-475, September 2007.

[41] K. Okada et al., "A 60GHz 16QAM/8PSK/QPSK/BPSK Direct-Conversion Transceiver for IEEE 802.15.3c," ISSCC Dig. Tech. Papers, pp. 160-161, 2011.

[42] A.Siligaris et al., "A 65nm CMOS fully integrated transceiver module for 60GHz wireless HD applications," ISSCC Dig. Tech. Papers, pp.160–161, 2011.

[43] M. Hosoya et al., "A 900-MHz bandwidth analog baseband circuit with 1-dB step and 30-dB gain dynamic range," ESSCIRC Proceedings, 2010.

[44] K. Raczkowski et al., "A wideband beamformer for a phased-array 60GHz receiver in 40nm digital CMOS," ISSCC Dig. Tech. Papers, pp. 40–41, 2010.

[45] M. Boers, "A 60GHz transformer coupled amplifier in 65nm digital CMOS," RFIC Symposium Proceedings, pp. 343–346, May 2010.

[46] C. Marcu et al., "A 90 nm CMOS Low-Power 60 GHz Transceiver With Integrated Baseband Circuitry," IEEE JSSC, vol. 44, no. 12, pp. 3434 –3447, Dec. 2009.

[47] V. Szortyka et al.,"A 42mW Wideband Baseband Receiver Section with Beamforming Functionality for 60GHz Applications in 40nm Low-Power CMOS," RFIC Symposium Proceedings, pp. 261-264, 2012.

[48] V. Szortyka et al., "Analog baseband beamformer for use in a phased-array 60 GHz transmitter," SiRF Proceedings, pp. 167-170,2012.

[49] S. D'Amico et al., "A 9.5mW analog baseband RX section for 60GHz communications in 90nm CMOS," RFIC Proceedings, pp.1-4, 2011.

[50] B. Verbruggen et al., "A 2.6mW 6b 2.2GS/s 4 times interleaved fully dynamic pipelined ADC in 40nm digital CMOS," ISSCC Dig. Tech. Papers, pp.296-297, 2010.

[51] L. Kull et al., "A 3.1mW 8b 1.2GS/s Single-Channel Asynchronous SAR ADC with Alternate Comparators for Enhanced Speed in 32nm Digital SOI CMOS," ISSCC Dig. Tech. Papers, pp. 468-469, 2013.

[52] A. Valdes-Garcia, et al., "A SiGe BiCMOS 16-element phased-array transmitter for 60GHz communications", ISSCC Dig. Tech. Papers, pp. 218-219, 2010.

[53] K. Okada et al., "A Full 4-Channel 6.3Gb/s 60GHz Direct-Conversion Transceiver with Low-Power Analog and Digital Baseband Circuitry," ISSCC Dig. Tech. Papers, pp. 218-219, 2012.

[54] T. Yao et al., "60-GHz PA and LNA in 90-nm RF-CMOS," RFIC Symposium Proceedings, 2006.

# A CMOS 21-48GHz Fractional-N Synthesizer Employing Ultra-Wideband Injection-Locked Frequency Multipliers

Alvin Li, Shiyuan Zheng, Jun Yin, Howard C. Luong
Department of Electronics and Computer Engineering
The Hong Kong University of Science and Technology
Hong Kong

Xun Luo
Microwave Branch, 2012 lab
Huawei Technologies Co. Ltd.
Shenzhen, China, 518129

*Abstract* — **Higher-order LC tanks with proper design parameters are proposed to widen the phase response to enhance the frequency locking range of mm-Wave injection-locked frequency multipliers (ILFMs). Employing a chain of such ILFMs at the output, a complete ultra-wideband fractional-N frequency synthesizer is demonstrated. Fabricated using a 65nm CMOS process, the synthesizer prototype measures a continuous output frequency tuning range of 80.2% from 20.6GHz to 48.2GHz when locked to a 4.5GHz to 6.1GHz fractional-N PLL with excellent phase noise < -107dBc/Hz at 1MHz offset while consuming 148 mW.**

## I. Introduction

Injection-locked frequency multipliers (ILFM) are one of the key building blocks that enable sub-harmonically-injected phased-locked loops (SHI-PLL) to achieve wider output frequency range and superior phase noise performance at mm-wave (mmW) frequencies compared to direct synthesis methods [1]. In direct synthesis methods, the output phase noise is dominantly contributed by that of a mmW oscillator, which scales worse than the theoretical $20\log(\omega_o/\omega_{in})$dB with frequency due to the diminishing capacitor quality factor Q at mmW frequencies [2]. In addition, the increasing dominance of layout parasitics and the need to drive high-speed dividers further limit the achievable tuning range of mm-wave PLLs.

In SHI-PLL architectures, superior phase noise performance is achieved by injection-locking the frequency-multiplier chain to a low-phase-noise signal generated from a PLL at RF frequencies where the tank quality factor is reasonably high. By doing so, the frequency multiplied output signal tracks the input phase noise by up to an offset defined by the locking range $\omega_L$ regardless of the multiplier's free-running phase noise performance [1]. However, due to the narrow locking range of conventional frequency multipliers, PVT variations and modeling inaccuracies at mmW frequencies, frequency calibration to align the ILFM's operating frequency to the injected signal's harmonic is necessary to ensure frequency locking [3]. This frequency alignment is typically implemented using an additional PLL for each multiplier used and is very costly in terms of extra chip area, increased routing complexity and parasitic capacitance, and additional loading when implemented at high frequencies.

To address this problem, a novel technique is proposed to enhance the locking range of ILFMs such that the frequency

alignment circuitry in SHI-PLLs can be eliminated at mmW frequencies.

Employing this proposed technique, a synthesizer prototype demonstrates both ultra-wide frequency tuning range of 80.2% from 20.6GHz to 48.2GHz and low phase noise of < -107dBc/Hz at 1MHz offset simultaneously, which is suitable for point-to-point (P2P) backhaul communications as defined in [4] and shown in Figure 1.

| Band | Frequency (GHz) |
|------|-----------------|
| 1 | 24.5 - 27.5 |
| 2 | 27.5 - 29.5 |
| 3 | 31.0 - 31.3 |
| 4 | 31.5 - 33.4 |
| 5 | 37.0 - 39.5 |
| 6 | 40.5 - 43.5 |

Figure 1: Allocated fixed wireless access frequency bands for P2P systems [4]

## II. Injection-Locked Frequency Multipliers

Motivated by the need for robust operation under PVT frequency variations and to cover multiband or wideband standards, recent works have focused on widening the frequency locking range. The schematic of a 2nd-order LC-type injection-locked frequency tripler is shown in Figure 2. The cross-coupled pair, $M_3$ and $M_4$, generates a negative feedback loop to ensure start-up and sustain self-oscillation of the frequency multiplier at $\omega_o$ when there is no input signal.

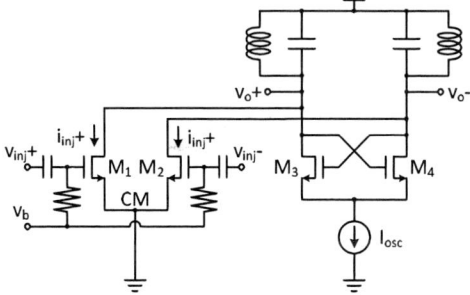

Figure 2: Conventional injection-locked frequency tripler

Given sinusoidal input signals, injection currents with fundamental and harmonic components (1) are generated from the nonlinearities of $M_1$ and $M_2$ and injected into the resonant tank.

$$i_{inj} = k_0 + k_1 \cdot v_{inj} + k_2 \cdot v_{inj}^2 + k_3 \cdot v_{inj}^3 + \cdots \quad (1)$$

978-1-4673-6145-3/13 $31.00 © 2013 IEEE

Since the 2nd-order LC-based resonant tank is centered around $\omega_0 \approx \omega_{in}$, the third harmonic current component generated from $k_3$ is used to lock the tripler. Based on the resonant tank's magnitude and phase response in conjunction with Barkhausen's criterion for stable oscillation:

$$|G_m R| = 1$$
$$\angle G_m R = 2\pi n, n \in 0,1,2 \dots \quad (2)$$

the locking range of the ILFM is derived to be approximately:

$$\omega_L \approx \frac{\omega_0}{2Q} \cdot \frac{i_{inj}}{i_{osc}} \cdot \frac{1}{\sqrt{1 - \frac{i_{inj}^2}{i_{osc}^2}}} \quad (3)$$

To achieve a wide locking range, it is desirable to maximize the ($i_{inj}/i_{osc}$) ratio. Although (3) seems to indicate minimizing the self-oscillation current $i_{osc}$ would lead to a wider locking range, this is slightly misleading as it would also reduce the output amplitude until the lower limit to ensure start up is reached determined by the tank Q.

Nevertheless, both [2] and [5] attempt to widen the locking range by increasing the ($i_{inj}/i_{osc}$) ratio while also reducing the LC tank's quality factor and maintaining acceptable output amplitude. In [2], this is achieved by biasing the input devices $M_1$ and $M_2$ at the boundary of sub-threshold region for class-B operation which increases the 3rd order nonlinearity, $k_3$, to maximize the $3\omega_{in}$ injection current $i_{inj}$. In [5], the non-linearity based input stage is replaced with a more efficient self-mixing input stage resulting in a large $3\omega_{in}$ injection current generated from mixing with the input signal with it's common mode 2nd harmonic.

An alternative approach to obtaining a wide locking range is to flatten the resonant tank's phase response around 0° over a wide frequency range using higher order LC tanks. Although [6] attempts to achieve a wideband locking range using a 0° plateau in the phase response, this stipulation is too restrictive compared to a flattened or rippled phase response around 0° which limits the achievable locking range. Furthermore, their presented phase noise spectrum indicates a locking range of less than 100kHz. The enhanced locking range can be understood through the ILFM model and corresponding phasor diagram shown in Figure 3a. Furthermore, the magnitude and phase plots for a 2nd, 4th and 6th order LC tank are shown in Figure 3b. For locking to be possible at an output frequency of $\omega_{out} = N \cdot \omega_{in}$, the gain and phase conditions for steady state oscillation (2) must first be met. From the gain condition, the resonant tank's amplitude response at $\omega_{out}$ must be large enough to ensure start-up given a self-oscillation current $i_{osc}$ through the -gm cell. To satisfy the phase condition, the phase $\phi$ between $i_{total}$ and $i_{osc}$ must be large enough to compensate for the phase shift $\theta$ introduced by the resonant tank at $\omega_{out}$ such that the total phase around the loop is $2\pi$. The maximum phase shift $\phi_{max}$ that $i_{total}$ can compensate for can be found by representing $i_{total}$ in a phasor diagram as the vector sum of the self-oscillation current $i_{osc}$ and the injected current $i_{inj}$ at $\omega_{inj} = N \cdot \omega_{in}$ and occurs when $i_{total}$ is 90° with $i_{inj}$.

The phase responses of a 2nd, 4th, and 6th-order resonant tanks are shown in Figure 3b along $|\phi_{max}|$. The frequency

range over which $|\theta| \le |\phi_{max}|$ can be identified as the range where the phase condition can be satisfied. In order to achieve a wide locking range, it is therefore desirable to both maximize the phase angle between $i_{total}$ with $v_{out}$ by increasing the $i_{inj}/i_{osc}$ ratio and to minimize the phase shift across the resonant tank over the desired frequency range by using higher-order LC tanks. In the case where multiple frequencies satisfy the gain and phase conditions, the steady-state output frequency tends to favor the frequency with a larger loop gain. To avoid a strong self-oscillation at an undesirable frequency and to ensure that the ILFM tracks the injection frequency when higher-order resonators are used, it is recommended to design the resonant tank such that the amplitude variation across the locking range is minimized and to increase the injection current so that it is large enough to compensate for the loop gain difference at $\omega_{inj}$ and the other possible resonant frequencies.

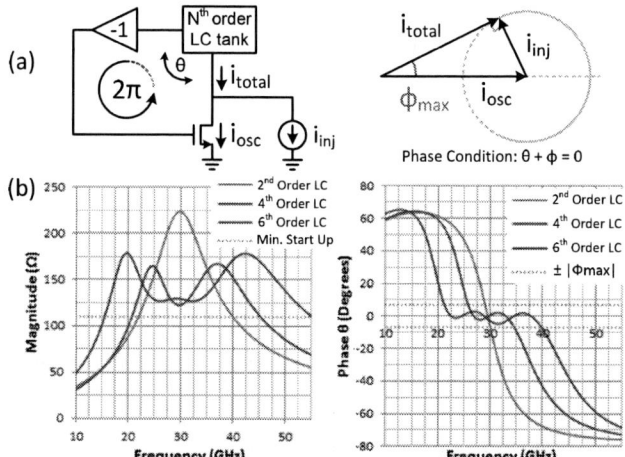

Figure 3: (a) ILFM model and corresponding phasor, (b) Magnitude and phase response of 2nd, 4th, and 6th order LC tank.

## III. FRACTIONAL-N PLL AND WIDEBAND ILFM CHAIN

The proposed frequency synthesizer is shown in Figure 4. The low frequency fractional-N PLL with 5kHz loop bandwidth generates a low phase noise signal using a 2.5V supply class-C VCO from 4.5 to 6.1GHz with a frequency resolution of 24.4kHz. The PLL's output tuning range is further widened from 30.2% to 57.5% by utilizing a multi-modulus regenerative divider capable of ÷3 and ÷4 division ratios with quadrature output before injection locking the ILFM chain. The multiplier chain consists of four ILFMs arranged such that ILFM3 and ILFM4 are in parallel to further widen the output frequency range to cover all the P2P frequency bands defined in Figure 1.

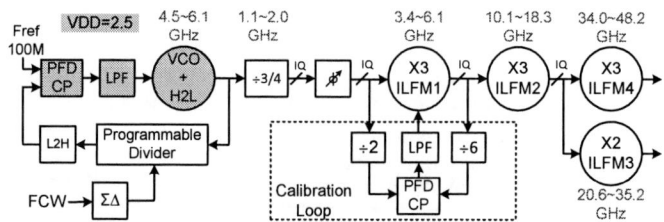

Figure 4: Proposed wideband ILFM Chain with low frequency PLL

The first frequency tripler, ILFM1, utilizes a narrowband I/Q topology with large SCA and varactor tuning to achieve a wide locking range from 3.4-6.1GHz. This is viable since the output phase noise depends solely on the input signal when locked. At this low operating frequency, the PLL based frequency calibration loop (FCL) for frequency alignment does not occupy much chip area as no inductor is required for the TSPC dividers, nor does the additional capacitive loading significantly reduce the tuning range or output amplitude. A 10MHz loop bandwidth is chosen to enable fast calibration time and a small loop filter capacitance. At higher frequencies, however, this topology becomes increasing inefficient as the diminishing MIM capacitor and varactor Q limits the achievable tuning range. Additionally, the FCLs required for frequency alignment would require increasingly larger chip area, complex routing and additional capacitive loading due to the need for wideband high frequency injection locked dividers. Therefore, the subsequent frequency multipliers in the chain utilize higher-order LC tanks and efficient injection stages to achieve wideband locking ranges without the need for FCLs.

The second frequency tripler, ILFM2, utilizes a quadrature-in-quadrature-out topology with $6^{th}$ order LC tank implemented using a transformer. The in-phase schematic is shown in Figure 5. The $6^{th}$ order LC tank creates a rippled phase response around $0°$, similar to the one shown in Figure 3, resulting in an output locking range from 10.1-18.3 GHz. To enhance the injection current at $3\omega_{in}$, a push-push mixing input stage is proposed as shown in Fig. 6a. A large $2\omega_{in}$ voltage is created at the common mode node CM using a push-push pair, $M_1$ and $M_2$, biased at the boundary of sub-threshold region and driven by the in-phase input signal at $\omega_{in}$. The upper differential pair, $M_3$ and $M_4$, mixes the $2\omega_{in}$ tone with the quadrature input signal at $\omega_{in}$ to generate a differential injection current at $3\omega_{in}$ which is injected into the resonant tank. To generate the quadrature injection currents, the in-phase input signals are swapped with the quadrature-phase input signals.

Compared to the conventional case where $3\omega_{in}$ is generated from the differential pair's nonlinearity, simulations show that the enhanced input stage improves the $3\omega_{in}$ current from 0.14mA to 1.24mA for the same power with a 400mV input swing at $3\omega_{in}$=15GHz. The simulated differential injection current is plotted in Figure 6b for various phase differences between the "$I_{inj}$" and "$Q_{inj}$" input signals. The plot shows that the push-push and mixing input signals should be $90°$ apart to maximize the conversion efficiency and is quite robust even with phase delays of $±20°$. Since ILFM2 has a very wide locking range, a strong $2\omega_{in}$ current may cause false locking when trying to lock $3\omega_{in}$ to the top portion of the locking range. To address this, an additional $2\omega_{in}$ current steering path is utilized to cancel the $2\omega_{in}$ tone before it reaches the resonant tank. A cascade device is also placed in the injection stage current path to increase the impedance seen by the resonant tank.

The quadrature output of ILFM2 is used to drive both the frequency doubler ILFM3 and the frequency tripler ILFM4. Although this increases the loading to ILFM2, it further widens the achievable locked output frequency range to 80.23%. The schematic of ILFM3, shown in Figure 7, uses a 6th order LC tank and a quadrature-to-differential push-push input stage to obtain a locking range from 20.6 to 35.2 GHz. Meanwhile, the schematic of ILFM4, shown in Figure 8, utilizes a 4th order LC tank with the previously discussed injection stage (Figure 6).

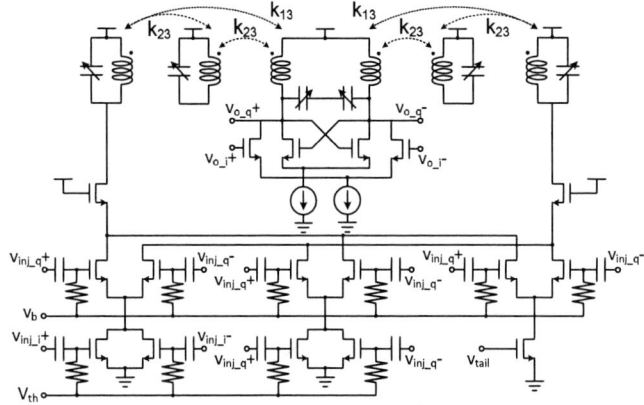

Figure 5: Schematic of ILFM2 with $6^{th}$ order LC tank.

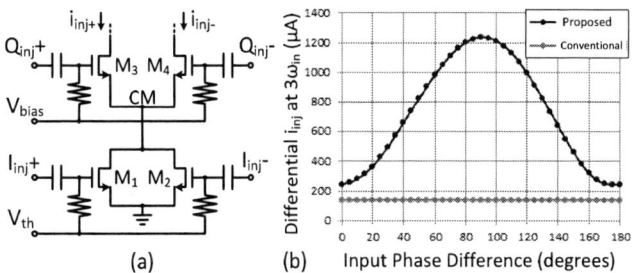

Figure 6: (a) Proposed $3\omega_{in}$ input stage, (b) $i_{inj}$ vs. I/Q phase accuracy.

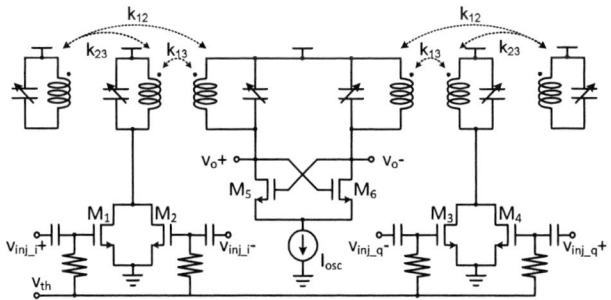

Figure 7: Schematic of ILFM3 with $6^{th}$ order LC tank

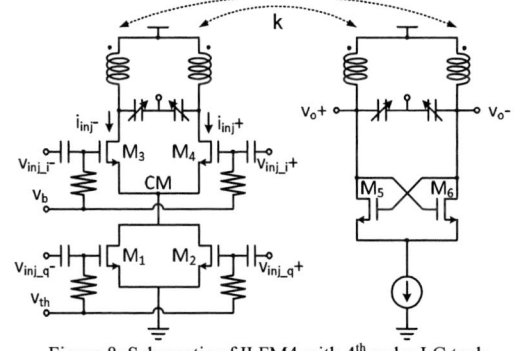

Figure 8: Schematic of ILFM4 with $4^{th}$ order LC tank

## IV. EXPERIMENTAL RESULTS

The wide-band ILFM chain is fabricated along with the fractional-N PLL in a 65nm GP CMOS process with 1 poly and 9 metal layers. Figure 9 shows the chip micrograph, which occupies a core area of 1850 x 1130 $\mu m^2$.

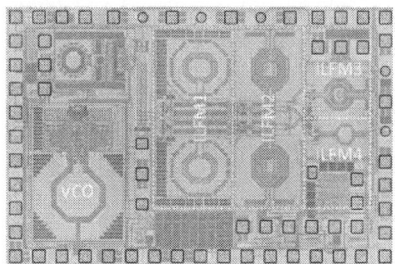

Figure 9: Die micrograph of ILFM chain and PLL

To verify the locking range and phase noise tracking properties of injection locking, the ILFM chain is measured using the low frequency PLL as an injection signal source. An overall continuous output locking range of 80.23% from 20.6 to 48.2GHz is achieved with a 1.6GHz overlap between the ILFM3 and ILFM4 output frequency bands. The corresponding output frequency spectrums are shown in Figure 10.

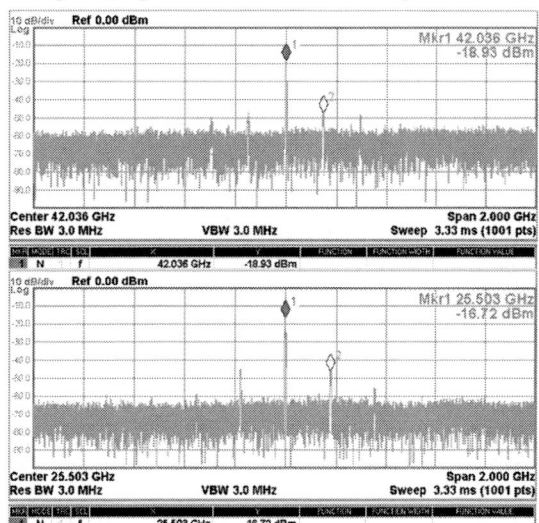

Figure 10: Output spectrum of ILFM3 at 25.5 GHz and ILFM4 at 42.0 GHz

The measured phase noise plots for output frequencies of 25.5GHz and 42GHz are shown in Figure 11. Each plot includes the four curves taken at different points along the ILFM chain. Starting from the bottom curve and moving up, they correspond to the output of DIV3/4, ILFM1, ILFM2 and either ILFM3 or ILFM4 respectively. The frequency synthesizer measures output phase noise of -83.99, -88.78, -83.80, -80.11, -84.82 dBc/Hz at 100kHz offset and -112.36, -113.89, -113.10, -106.96, -108.13 dBc/Hz at 1MHz offset for output frequencies of 25.5, 28.5, 32.0, 38.0 and 42.0 GHz respectively. It is important to note that the 10MHz offset is dominated by the noise floor and that the actual phase noise performance at this offset can be interpolated from the 1MHz data assuming -20dB/decade. It can be seen that the phase noise tracking between subsequent stages closely follows $20\log(\omega_{out}/\omega_{inj})$ as expected from wideband locking until the noise floor is reached. This is a key indication that frequency locking has indeed been achieved.

The frequency synthesizer consumes 148.3mW of power and measures a RMS jitter of 1.056ps integrated from 10kHz to 10MHz at 42GHz. Figure 12 and 13 summarize and compare the performance of the frequency multipliers and of the synthesizer, respectively.

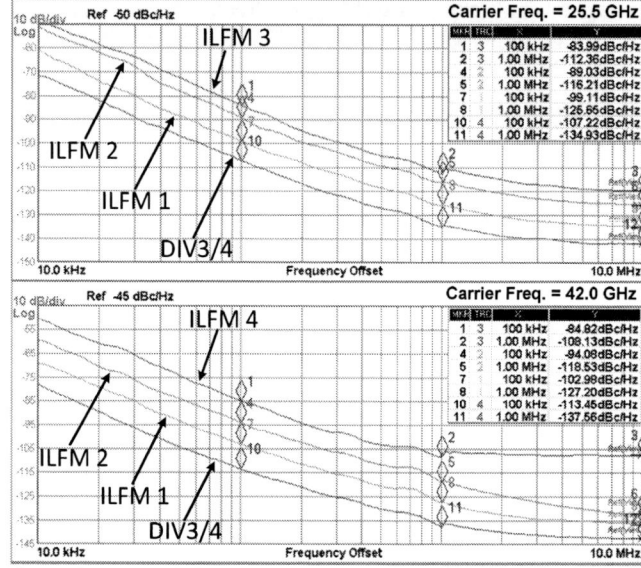

Figure 11: Measured phase noise of ILFM3 at 25.5GHz and ILFM4 at 42GHz

| ILFM | [4] | M.Chen MTT 2008 | [5] | [6] | **ILFM2** | **ILFM3** | **ILFM4** |
|---|---|---|---|---|---|---|---|
| Fout (GHz) | 60 | 60 | 46.1 | 60 | 14.2 | 28.1 | 41.1 |
| Multiply Ratio | 3 | 3 | 3 | 3 | 3 | 2 | 3 |
| Power (mW) | 9.6 | 1.86 | - | 37 | 23.3 | 21.8 | 16.8 |
| Locking Range (GHz) | 56.5-64.5 (13.2%) | 1.42 (2.4%) | 42.75-49.5 (14.6%) | 52-66 (15%) | 10.1-18.3 (57.7%) | 20.6-35.5 (53.1%) | 33.9-48.2 (34.8%) |
| Input power | 0 dBm | 6 dBm | 0 dBm | n/a | n/a | n/a | n/a |
| Tech. | 90nm CMOS | 0.13um CMOS | 65nm | 40nm CMOS | 65nm CMOS | 65nm CMOS | 65nm CMOS |

Figure 12: ILFM performance summary and comparison

| | **This Work** | D. Murphy JSSC 7/2011 | O. Richard ISSCC 2010 | A. Musa ASSC 11/2011 | S. Pellerano ISSCC 2008 |
|---|---|---|---|---|---|
| Frequency (GHz) | **42.0** | 43.20 | 20.88 | 60.48 | 41.247 |
| Output Frequency Range (GHz) | **20.6- 48.2** | 42.1 - 53 | 17.5 - 21 / 35 - 41.9 | 58 - 63 | 39.1 - 41.6 |
| fref (MHz) | **100** | 54 | 36 | 36 | 50 |
| Out-band phase noise (dBc/Hz) | **-108.13 @ 1MHz** | -85.67 @ 1MHz | -100 @ 1MHz | -96 @ 1MHz | -90 @ 1MHz |
| Integrated Jitter (s) | **1.056 ps** | n/a | n/a | n/a | n/a |
| Power (mW) | **148.3** | 72 | 80 | 77.5 | 64 |
| Process | **65nm CMOS** | 65nm CMOS | 65nm CMOS | 65nm CMOS | 90nm CMOS |
| Architecture | **Fractional-N (VCO @ 4.67GHz)** | Integer-N (VCO @ 50.11GHz) | Integer-N (QVCO @ 20.88GHz) | Integer-N (VCO @ 20GHz) | Fractional-N (VCO @ 41.247GHz) |

Figure 13: Synthesizer performance summary and comparison

## REFERENCES

[1] J. Lee, H. Wang, "*Study of Subharmonically Injection-Locked PLLs,*" IEEE J. Solid-State Circuits, vol. 44, no. 5, pp. 1539-1553, May 2009.

[2] W. L. Chan, J. R. Long, "*A 56-65 GHz Injection-Locked Frequency Tripler with Quadrature Outputs in 90-nm CMOS,*" IEEE J. Solid-State Circuits, vol. 43, no. 12, pp. 2739-2746, Dec. 2008.

[3] B. Razavi, "*A study of injection locking and pulling in oscillators,*" IEEE J. Solid-State Circuits, vol. 39, no. 9, pp. 1415-1424, Sep. 2004.

[4] ECC&CEPT Communication Standard: REC0104, REC1101, REC0202, and TR1201E.

[5] W. Liang, A. Li, H. Luong, "*A 4-path 42.8-to-49.5GHz LO generation with automatic phase tuning for 60GHz phased-array receivers,*" ISSCC Dig. Tech. Papers, pp.270-272, Feb. 2012.

[6] G. Mangraviti, et al., "*A 52-66GHz Subharmonically Injection locked quadrature oscillator with 10GHz locking range in 40nm LP CMOS,*" IEEE RFIC, pp.309-312, June 2012.

978-1-4673-6145-3/13 $31.00 © 2013 IEEE

# A 75.7GHz to 102GHz Rotary-traveling-wave VCO by Tunable Composite Right /Left Hand T-line

Shunli Ma[1], Wei Fei[2], Hao Yu[2*], and Junyan Ren[1]

[1]State Key Laboratory of ASIC and System, Fudan University, Shanghai, P.R China, 200433.
[2]School of Electrical and Electronic Engineering, Nanyang Technological University, 50 Nanyang Ave, Singapore 639798

*Abstract*-With the use of tunable composite-right/left-hand (CRLH) transmission line (T-line), this paper provides a wide frequency-tuning-range (FTR) mechanism for Mobius-ring rotary-traveling-wave (RTW) VCO in millimeter-wave region. CRLH T-line is implemented in RTW-VCO with inductor-loaded transformer to realize sub-band selection over a wide FTR. Each sub-band is further covered by a varactor for fine-tuning. The chip was fabricated in GF 65nm RF-CMOS process with area of 0.08mm². The measured results show a current consumption of 14mA under supply voltage of 1V, a tuning range of 29.5% with center frequency at 89.3GHz, and a phase noise from -100.08dBc/Hz to -98.7dBc/Hz with 10MHz offset. A state-of-art figure-of-merit FOM$_T$ of -177.78dBc/Hz is demonstrated.

## I. INTRODUCTION

Many big-data communication and imaging applications have been recently demonstrated by CMOS based millimeter-wave integrated circuits (MMICs) [1-6]. Multi-phase and quadrature oscillators are essential building blocks in these applications, which are normally realized by traveling wave to generate multi-phase clock outputs with good phase noise performance [5-8]. Mobius-ring rotary-traveling-wave (RTW) VCO topology is commonly adopted due to its advantages such as easy placement of cross-coupled transistors, good matching of differential blocks and compact area [8].

Traditionally, RTW-VCO is implemented with conventional right-handed (RH) transmission line (T-line), with a phase delay directly proportional to the T-line physical length [7-9]. Since a total phase delay of 360-degree is required to for oscillation, a large area is induced. Recently, left-handed (LH) T-line has shown to provide a superior performance at high frequency [10], and also unique features such as nonlinear dispersion curve [11]. Due to large parasitic capacitors from cross-coupled transistors that are RH in nature, the actual implemented T-line is a composite right/left hand (CRLH) T-line. By merging the phase shifts from LH and RH components together, CRLH T-line provides a phase delay independent of its physical size, and thus can be designed to be much more compact than conventional RH T-line for VCO.

As big-data communication or imaging applications require a wideband to ensure high data rate and also to cover process variation by CMOS MMIC at advanced technology, tuning ability of RTW-VCO has not been thoroughly studied and achieved as far. Conventional RTW-VCO is mostly tuned by varactor and capacitor bank due to the RH topology [7-9]. Due to constraint tuning ability of varactor and capacitor bank

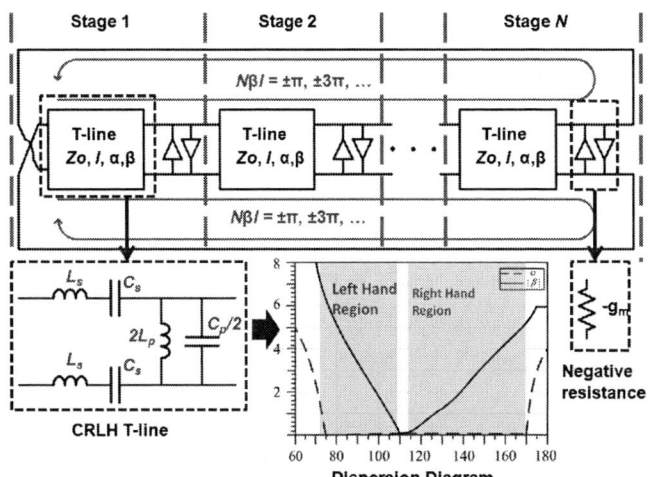

Fig. 1. CRLH T-line based Mobius-ring RTW-VCO.

in millimeter-wave region, the achieved FTR is quite limited [7-9]. CRLH T-line, on the other hand, provides more choices of tunable elements to achieve a wide FTR in RTW-VCO design, but is not well explored at millimeter-wave region [10]. In this work, a tunable CRLH T-line is studied for RTW-VCO to achieve wide FTR.

The paper is organized as follows. Section II presents the design and analysis of the tunable CRLH T-line based RTW-VCO. A VCO prototype is designed with measurements results in Section III and conclusions in Section IV.

## II. TUNABLE CRLH T-LINE BASED RTW-VCO

### A. CRLH T-line based RTW-VCO

The topology for Mobius-ring RTW-VCO is shown in Fig. 1. A Mobius-ring is evenly divided into $N$ stages, with each stage loaded with a cross-coupled transistor pair. As wave travels along the Mobius-ring, certain phase delay must be fulfilled to create a positive feedback for VCO oscillation. At the same time, cross-coupled transistors should generate enough power to compensate the loss from the T-line. In summary, the start-up condition of Mobius-ring RTW-VCO is

$$g_m > \frac{2\exp(\alpha \cdot l)}{z_o}; \beta l = \frac{M\pi}{N}. \tag{1}$$

where $g_m$ is the transconductance of the cross-coupled pair, $z_o$, $l$, $\alpha$, $\beta$ are T-line characteristic impedance, physical length, attenuation constant, and phase constant, respectively. $N$ is the stage number, and $M=\pm1, \pm3,...$ is an odd integer number.

---

This work was sponsored by Singapore MOE Tier-1 fund RG 26/10 and China National NSFC grant 61176028. * haoyu@ntu.edu.sg, +65-67904509.

978-1-4673-6145-3/13 $31.00 © 2013 IEEE

In this work, CRLH T-line is deployed in the Mobius-ring RTW-VCO for compact size. CRLH T-line is one kind of metamaterial that is composed of LH and RH components [12-13] with equivalent circuit of one unit-cell shown in Fig.1. Serial capacitors ($C_s$) and parallel inductors ($L_p$) form its LH portion with a negative phase constant ($\beta_L$), while parallel capacitors ($C_p$) and serial inductors ($L_s$) form its RH portion with a positive phase constant ($\beta_R$). With a balanced design ($C_s \cdot L_s = C_p \cdot L_p$), phase constant for CRLH T-line can be simplified as

$$\beta = \beta_R + \beta_L = \omega\sqrt{L_s C_p} - \frac{1}{\omega\sqrt{L_p C_s}}. \quad (2)$$

Note here all components ($L_s$, $L_p$, $C_s$, $C_p$) are normalized with respect to the unit-cell length.

With (1) and (2), the oscillation frequency for an $N$-stage Mobius-ring RTW-VCO by CRLH T-line can be obtained

$$\omega_{CRLH} = \frac{\pi}{2Nl\sqrt{L_s C_p}} \times \left( \sqrt{1 + \frac{4N^2 l^2}{\pi^2}\sqrt{\frac{L_s C_p}{L_p C_s}}} \pm 1 \right). \quad (3)$$

Here only the fundamental resonant condition $M=\pm1$ is considered for simplicity of illustration. The plus and minus signs in (3) correspond to CRLH T-line working in the RH region and LH region, respectively.

What is more, phase noise is an important specification for VCO design. Generally, for $N$-stage RTW-VCO, the phase variation $<\Phi^2(t)>$ is proportional to $1/N$ [7, 14-16], which is reduced by $1/N$ when compared to single stage.

In this work, the LH operation is selected for compact size and superior performance when implemented in multiple stages [10]. However, there is no study on how to tune the CRLH T-line based RTW-VCO, which will be addressed in the next part.

### B. Wideband tuning for CRLH T-line based RTW-VCO

Note that in (3), there are 4 components that may be used for tuning: $L_s$, $L_p$, $C_s$, $C_p$. For easy analysis, we represent the product of the LH components ($L_p C_s$) as $P_L$; and represent the product of the RH components ($L_s C_p$) as $P_R$. Then, the oscillation frequency in the LH region becomes

$$\omega_{CRLH\_LH} = \frac{\pi}{2Nl\sqrt{P_R}} \times \left( \sqrt{1 + \frac{4N^2 l^2}{\pi^2}\sqrt{\frac{P_R}{P_L}}} - 1 \right). \quad (4)$$

Conventionally, $P_R$ is used to realize FTR by varactor as part of $C_p$ [10]. Unfortunately, with the omitted $L_s$ component and thus small $P_R$ value in [10], the tuning ability by $P_R$ is very limited, not to mention the already constraint tuning ability as well as the limited quality factor of varactor at high frequency. In fact, for a small $P_R / P_L$ value, $\omega_{CRLH\_LH}$ approaches the operation frequency of a pure LH T-line based RTW-VCO

$$\omega_{CRLH\_LH}\big|_{\frac{P_R}{P_L} \to 0} = \omega_{LH} = \frac{Nl}{\pi\sqrt{P_L}} \quad (5)$$

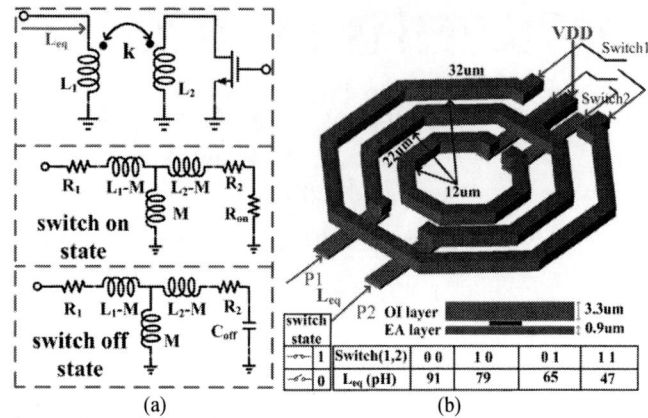

Fig. 2. (a) Equivalent circuits of an inductor-loaded transformer with the loaded switches. (b) Layout implementation for inductor-loaded transformer where tuned inductance is determined by states of two switches

which is independent of $P_R$ with poor tuning ability.

Intuitively, a wider FTR should be obtained by tuning $P_L$ since the LH-components dominate in the LH region. Since $\frac{\delta\omega_{CRLH}}{\delta P_L}$ stays positive for all $P_L$ values, the FTR can be calculated

$$\Delta\omega_{CRLH_{LH}} =$$
$$\frac{\pi}{2Nl\sqrt{P_R}} \times \left( \sqrt{1 + \frac{4N^2 l^2}{\pi^2}\sqrt{\frac{P_R}{P_{L\_min}}}} - \sqrt{1 + \frac{4N^2 l^2}{\pi^2}\sqrt{\frac{P_R}{P_{L\_max}}}} \right). \quad (6)$$

The extreme condition forms for a pure LH T-line with

$$FTR_{LH} = \frac{\frac{1}{\sqrt{P_{L\_min}}} - \frac{1}{\sqrt{P_{L\_max}}}}{\frac{1}{\sqrt{P_{L\_min}}} + \frac{1}{\sqrt{P_{L\_max}}}} \times 2 \approx \frac{\alpha_{P_L}}{2} \quad (7)$$

where $\alpha_{P_L} = \frac{\Delta P_L}{P_L}$ measures the tunability of components in $P_L$. As (7) shows, $FTR_{LH}$ is directly proportional to $\alpha_{P_L}$.

However, since the loss in $C_s$ adds directly into the signal path, it is not feasible to tune $C_s$. On the other hand, one can realize a wide FTR by tuning $L_p$ with a loaded transformer structure [4]. More specifically, inductive-loaded transformer can achieve a large $\alpha_{P_L}$, which is adopted in this work.

The mechanism for inductive-loaded transformer can be explained in Fig. 2(a), where a transformer is loaded with switch on its secondary coil. By turning the switch on or off, the equivalent inductance ($L_{eq}$) looking into transformer primary coil can be effectively changed as:

$$\begin{cases} L_{eq\_on} = L_1 - \frac{\omega^2 M^2 L_2}{(R_2 + R_{on})^2 + \omega^2 L_2^2} \approx (1 - k^2)L_1 \\ L_{eq\_off} = L_1 + \frac{\omega M^2(\frac{1}{\omega C_{off}} - \omega L_2)}{R_2^2 + (\frac{1}{\omega C_{off}} - \omega L_2)^2} > L_1 \end{cases} \quad (8)$$

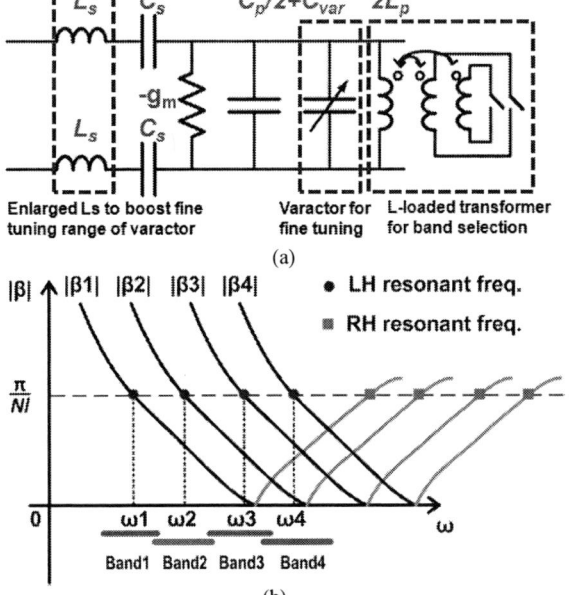

(a)

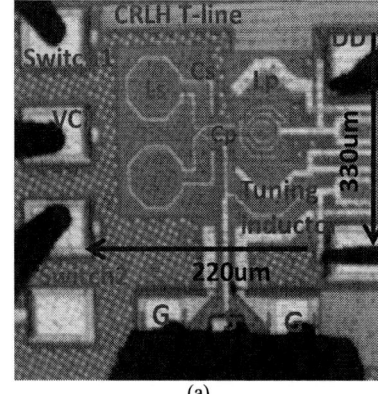

Fig. 3. Tuning mechanism for the proposed tunable CRLH T-line based Mobius-ring RTW-VCO: (a) Equivalent circuit, (b) Dispersion diagram.

where $L_{eq\_on}$ and $L_{eq\_off}$ represent different $L_{eq}$ values when the switch is turned on or off, respectively. $L_1$, $L_2$, $M$, and $k$ are transformer primary, secondary, mutual inductances, and coupling factor. The loss of transformer is represented by two serial resistances $R_1$ and $R_2$. Equivalent circuits for the switch during its on and off states are modeled by a resistor $R_{on}$ and a capacitor $C_{off}$, respectively. As (4) indicates, a large $\alpha_{P_L}$ can be easily obtained by implementing a large coupling factor $k$ for the transformer. Moreover, since the tuning element is not directly included in the signal path, the phase noise degradation is low. What is more, multiple inductors can be switched on and off to further increase $\alpha_{P_L}$ with a wide FTR achieved by creating multiple sub-bands.

The designed switched coupled-inductor for inductive tuning is shown in Fig. 2(b). Inductors are realized by the top Cu layer to guarantee a high quality factor. Two transformers loaded with two switches are used to realize 4 sub-bands. As summarized in the tables shown in Fig. 2(b), the resulted $L_{eq}$ can be varied over a large range from 47pH to 91pH. As such, wide FTR can be realized with 4 sub-bands: (75.67-83.11GHz), (79.65-87.78GHz), (86.18-94GHz) and (93.89-102.01GHz). To realize a continuous tuning, fine-tuning by varactor is used in each sub-band. To increase the tuning ability of varactor as (5) indicates so as to fully cover each sub-band, a relatively large $L_s$ value is adopted in this design.

The resulted tuning mechanism for the proposed CRLH T-line based RTW-VCO can be explained in Fig. 3. Inductive-loaded transformer creates multiple sub-bands by shifting the dispersion curve to different resonant frequency points. Each sub-band is then covered with fine-tuning by a varactor.

## III. VCO IMPLEMENTATION AND MEASUREMENT

Fig. 4(a) shows the on-chip implementation for the proposed tunable CRLH T-line based Mobius-ring RTW-

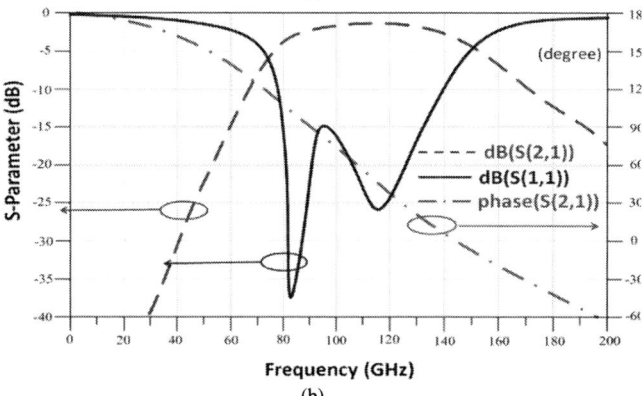

(a)

(b)

Fig. 4. (a) Die micrograph of the proposed VCO. (b) The EM simulation for one CRLH T-line unit-cell. Note that $L_p$ is shared by two unit-cells and its value is purposely designed doubled.

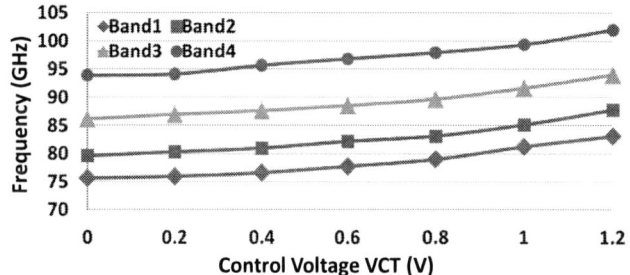

Fig. 5. Measured frequency tuning range by 4 sub-bands.

VCO. To push the cut-off frequencies away from operation frequency region, each stage is implemented with 2 distributed CRLH T-line unit-cells. As a result, 180-degree phase shift is required due to the Mobius-ring connection, which leads to a 90-degree phase shift in each unit-cell. The EM simulation results for the designed unit-cell are shown in Fig. 4(b). At the frequency of interest 100GHz, one unit-cell can provide 90-degree phase-shift with loss at -1.86dB, which is compensated by the negative resistors realized by a cross-coupled pair. Note the unit-cell is biased to operating in the LH region, with the resonant mode in the RH region (-90-degree phase shift) highly suppressed.

The proposed VCO is fabricated 65nm CMOS Global Foundries 1P8M RF CMOS process. The VCO core area is about $0.0812\text{mm}^2$. The output spectrum is measured by E4408B spectrum analyzer through one 11970W harmonic

978-1-4673-6145-3/13 $31.00 © 2013 IEEE    357

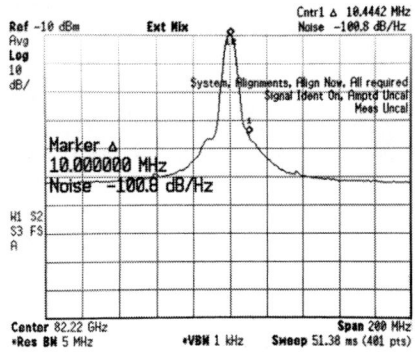

Fig. 6. Measured phase noise at 82.22GHz center frequency with 10 MHz offset frequency

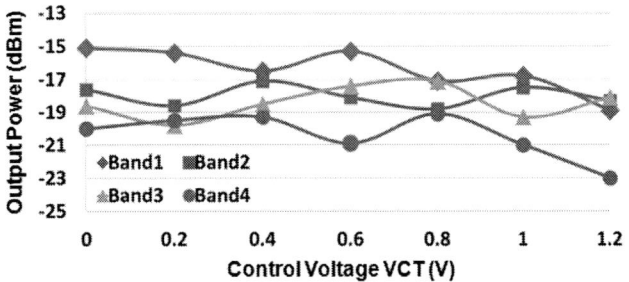

Fig. 7. Measured output power across the entire frequency tuning range.

TABLE I.  PERFORMANCE SUMMARY AND COMPARISON

| Parameters | [1] | [2] | [3] | This Work | Unit |
|---|---|---|---|---|---|
| $f_{osc}$ | 95.7 | 101 | 70.4 | 89.3 | GHz |
| $VDD_{core}$ | 1.5 | 0.8 | 1.2 | 1 | V |
| $P_{DC}$ | 9 | 11.9 | 5.4 | 14 | mW |
| Phase Noise ($PN$) @10MHz | -106 | -104.5 | -106.1 | -100.8 | dBc/Hz |
| FTR (%) | 3.6 | 11.2 | 9 | **29.6** | % |
| $FOM_T$ | -167.1 | -176.5 | -175.4 | **-177.8** | dBc/Hz |
| Tech. | CMOS 65 | CMOS 65 | SOI CMOS 65 | CMOS 65 | nm |

$$FOM_T = PN - 20\log\left(\frac{f_{osc}}{\Delta f} \times \frac{FTR}{10}\right) + 10\log(P_{DC}/1mW)$$

mixer. The supply voltage for buffer is 1.2V and for VCO is 1V. The measured current for the core VCO is 14mA. As shown in Fig. 5, by using the proposed tunable CRLH-T-line, a wide FTR of 29.6% is achieved from 76.59GHz to 102.01GHz, with a center frequency at 89.33GHz. The full FTR is formed from the four sub-bands controlled by two switches: (75.67-83.11GHz), (79.65-87.78GHz), (86.18-94GHz) and (93.89-102.01GHz). With a tuning voltage for varactor from 0V to 1.2V, each sub-band is fully covered. The measured phase noise varies from -100.08dBc/Hz to -98.7dBc/Hz with a sample plot shown in Fig. 6. The measured output power is from -23dBm to -15dBm as shown in Fig. 7. The output power variation is about 8dBm.

As summarized in Table I, the performance of the proposed VCO is further compared with other published millimeter-wave VCOs in 65nm CMOS technology. According to the Table I, the phase noise is comparable with others, and the widest FTR and the best $FOM_T$ are achieved by the proposed VCO.

## IV. CONCLUSION

A tunable CRLH T-line based millimeter-wave Mobius-ring RTW-VCO has been demonstrated in this paper with a wide FTR of 75.7-102GHz. Inductor-loaded transformer is implemented in CRLH T-line to realize 4 sub-bands. Each sub-band is covered by a varactor with fine-tuning. The proposed tunable CRLH T-line based RTW-VCO is fabricated in 65nm GF RF-CMOS process with area of 0.0821mm². The measured results show a current consumption of 14mA under supply voltage of 1V, a FTR of 29.5% with center frequency at 89.3GHz, and a phase noise of -100.8dBc/Hz with 10MHz offset at 82.2GHz center frequency. The state-of-art figure-of-merit $FOM_T$ is demonstrated at -177.78dBc/Hz.

## ACKNOWLEDGMENTS

The authors would like to thank Mr. Lim Wei Meng for the support of testing work. The authors also thank Integrand Software for providing the EM simulation tool EMX.

## REFERENCES

[1] N. Zhang and K. K. O, "94 GHz Voltage Controlled Oscillator With 5.8% Tuning Range in Bulk CMOS," *IEEE Microwave and Wireless Components Letters*, vol. 18, pp. 548-550, Aug 2008.

[2] X. Yi, C. C. Boon, J. F. Lin, and W. M. Lim, "A 100 GHz transformer-based varactor-less VCO with 11.2% tuning range in 65nm CMOS technology," *IEEE ESSCIRC*, pp. 293-296, Sept. 2012.

[3] D. D. Kim, J. Kim, J. O. Plouchart, C. Cho, W. Li, D. Lim, R. Trzcinski, M. Kumar, C. Norris, and D. Ahlgren, "A 70GHz Manufacturable Complementary LC-VCO with 6.14GHz Tuning Range in 65nm SOI CMOS," *ISSCC Dig. Tech. Papers*, pp. 540-620, Feb. 2007.

[4] J. Yin and H. C. Luong, "A 57.5-to-90.1GHz magnetically-tuned multi-mode CMOS VCO," *IEEE CICC*, pp. 1-4, Sept. 2012.

[5] J. Borremans, M. Dehan, K. Scheir, M. Kuijk, and P. Wambacq, "VCO design for 60 GHz applications using differential shielded inductors in 0.13 um CMOS," *RFIC Symposium*, pp. 135-138, 2008.

[6] A. Mazzanti, E. Monaco, M. Pozzoni, and F. Svelto, "A 13.1% tuning range 115GHz frequency generator based on an injection-locked frequency doubler in 65nm CMOS," *ISSCC Dig. Tech. Papers*, pp. 422-423, Feb. 2010.

[7] F. Ben Abdeljelil, W. Tatinian, L. Carpineto, and G. Jacquemod, "Design of a CMOS 12 GHz Rotary Travelling Wave Oscillator with switched capacitor tuning," *IEEE RFIC Symposium*, pp. 579-582, June 2009.

[8] N. Nouri and J. F. Buckwalter, "A 45-GHz Rotary-Wave Voltage-Controlled Oscillator," *IEEE Trans. on Microwave Theory and Techniques*, vol. 59, pp. 383-392, Feb. 2011.

[9] A. Moroni, R. Genesi, and D. Manstretta, "A distributed "hybrid" wave oscillator array for millimeter-wave phased-arrays," *IEEE CICC*, pp. 1-4, Sept. 2012.

[10] G. Li and E. Afshari, "A Low-Phase-Noise Multi-Phase Oscillator Based on Left-Handed LC-Ring," *IEEE JSSC*, vol. 45, pp. 1822-1833, Sept. 2010.

[11] S.-W. Tam, H.-T. Yu, Y. Kim, E. Socher, M. C. F. Chang, and T. Itoh, "A dual band mm-wave CMOS oscillator with left-handed resonator," *IEEE RFIC Symposium*, pp. 477-480, June 2009.

[12] T. J. Cui, D. R. Smith, and R. Liu, *Metamaterials: Theory, Design, and Applications*: Springer, 2009.

[13] W. Fei, H. Yu, K. S. Yeo, X. Liu, and W. M. Lim, "A 44-to-60GHz, 9.7dBm P1dB, 7.1% PAE Power Amplifier with 2D Distributed Power Combining by Metamaterial-based Zero-Phase-Shifter in 65nm CMOS", *IEEE IMS*, June 2012.

[14] A. Hajimiri and T. H. Lee, "A general theory of phase noise in electrical oscillators," *IEEE JSSC*, vol. 33, pp. 179–194, Feb. 1998.

[15] D. Ham and A. Hajimiri, "Virtual damping and einstein relation in oscillators," *IEEE JSSC*, vol. 38, no. 3, pp. 407-418, Mar.2003.

[16] R. Kubo and D. Ter Haar, "Fluctuation, relaxation and resonance in magnetic systems," *Oliver and Boyd, Edinburgh*, p. 23, 1962.

# Transformer-Based Dual-Band VCO and ILFD for Wide-Band mm-Wave LO Generation

Yue Chao, Howard C. Luong

The Hong Kong University of Science and Technology, Clear Water Bay, Hong Kong

*Abstract*- Switched-transformer and transformer-distribution design techniques are proposed for wideband mm-Wave dual-band VCOs and ILFDs. Fabricated in a 65nm CMOS process, a dual-band VCO prototype measures a tuning range of 22.3% from 62.1GHz to 78.3GHz with a phase noise of -112.0dBc/Hz at 10MHz offset while consuming 7.7mW, corresponding to FoM and FoM$_T$ of -180.4dBc/Hz and -187.4dBc/Hz, respectively. A proposed ILFD prototype achieves locking range from 58GHz to 77.8GHz with 1.44mW power and FoM of 13.75GHz/mW. A cascade of the proposed VCO and ILFD is also demonstrated with an overall locking range of 15.7% from 62.1GHz to 73.3GHz.

## I. INTRODUCTION

Aggressive scaling of CMOS technologies has made it more feasible to implement CMOS phased-lock loops (PLLs) at mm-Wave (mmW) frequencies, such as 57−66GHz for WPAN and 76−77GHz for automotive radars. However, there remain great challenges in designing high-performance VCOs and ILFDs for these PLLs. Specifically, due to process variation and inaccurate device modeling at mmW frequencies, the operation frequencies of VCOs and ILFDs tend to be misaligned, which results in very small overlapping frequency.

To cover required communication bandwidth with sufficient margins, wide-band mmW VCOs and ILFDs are desired. However, due to poor quality factor Q of varactors at mmW frequencies, the tuning range of typical single-band VCOs is limited to about 10% [1][2]. Dual-band VCO topology can achieve wide tuning range [3], but expensive mmW MUX is required to select one of the two VCO outputs to drive the next ILFD stage. Moreover, two cross-coupled pairs are needed in [3], which would contribute more parasitic capacitance. For the existing mmW ILFDs, several techniques have been proposed to enhance the locking range [4-7], but they all require relatively large power, which results in FoM much smaller than 10GHz/mW. Finally, multiple inductors are adopted in the existing ILFDs [4-6], which cost large chip area and are difficult for integration with VCO.

This paper presents two novel techniques used in mmW VCO and ILFD. Switched-transformer (ST) technique is proposed for a dual-band VCO (DB-VCO). Fabricated in 65nm CMOS, the proposed DB-VCO has only a single output and achieves wide tuning range (62.1GHz-78.3GHz) and good phase noise (-112dBc/Hz at 10MHz offset). In addition, transformer-distribution (TD) technique is also proposed to enhance the locking range of ILFD with low power consumption. A TD-based ILFD is demonstrated with locking range from 58GHz to 77.8GHz consuming only 1.44mW,

corresponding to FoM of 13.75GHz/mW. Besides, a single transformer is used for the proposed TD-ILFD, which occupies small chip area and is easy for integration with VCO. Finally, a cascade of the proposed DB-VCO and TD-ILFD is demonstrated with a combined overlapping frequency range from 62.1GHz to 73.3GHz.

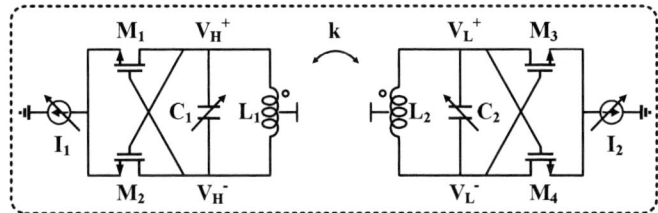

Fig. 1. Schematic of conventional DB-VCO

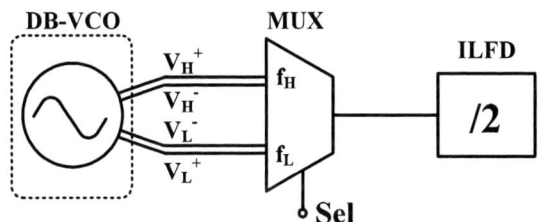

Fig. 2. Conventional DB-VCO driving ILFD with a MUX

## II. PROPOSED SWITCHED-TRANSFORMER DB-VCO

Fig. 1 shows the schematic of a conventional DB-VCO used in [3] to enhance the tuning range, which has two major problems when used at mmW frequency. Firstly, there are two separate outputs for the low band and the high band. When $I_1$ is turned on and $I_2$ is turned off, the VCO works in the high band, and $V_H$ is chosen as the output. When $I_1$ is turned off and $I_2$ is turned on, VCO works in the low band, and $V_L$ is selected as the output. With two separate outputs $V_H$ and $V_L$, a mmW MUX is needed to select one of the two outputs to drive the next ILFD stage as shown in Fig. 2. However, such a mmW MUX is quite expensive in terms of power and chip area. Secondly, two cross-coupled pairs are needed for the high and low bands. Although only one of them needs to be enabled at any time, the disabled cross-coupled pair still contributes parasitic capacitance and inevitably degrades the overall VCO tuning range and phase noise.

Fig. 3 shows the schematic of the proposed switched-transformer DB-VCO to solve these problems. For this DB-VCO, all the switching current from the cross-coupled pair flows into $L_1$. So for both the high and low bands, they share

978-1-4673-6145-3/13 $31.00 © 2013 IEEE          359

the same output $V_O$, and no mmW MUX is needed to drive ILFD. Besides, there is only one cross-coupled pair ($M_1$ and $M_2$), which helps reduce parasitic capacitance. To select the operation band, two switches $S_1$ and $S_2$ are added across the primary coil $L_1$ and secondary coil $L_2$. Stability circuit $M_3$ and $M_4$ are also added to improve the stability of DB-VCO when it operates in high band. 2-bit AMOS varactor banks ($C_1$ and $C_2$) are used to tune the oscillation frequency.

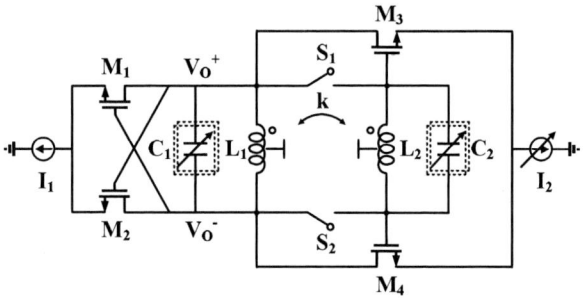

Fig. 3. Schematic of the proposed switched-transformer DB-VCO

When $S_1$ and $S_2$ are turned off, the DB-VCO operates in high-band mode as shown in Fig. 4(a). The peak frequency of the transformer tank can be calculated as follows assuming that $L_1=L_2=L$ and $C_1=C_2=C$ for simplicity:

$$f_L = \frac{1}{2\pi\sqrt{LC(1+k)}} \ , \ f_H = \frac{1}{2\pi\sqrt{LC(1-k)}} \tag{1}$$

To ensure that the DB-VCO operates at $f_H$, the transformer tank is designed with the input impedance $|Z_{in}(f_H)|>|Z_{in}(f_L)|$ as shown in Fig. 4(a) (solid line). In addition, stability circuit $M_3$ and $M_4$ are also activated in the high band to avoid instability problem due to the competition of the two oscillation modes. $M_3$ and $M_4$ together with the transformer tank form a two-port oscillator providing negative $G_{m2}$. When $I_2$ is turned on, $|Z_{in}(f_H)|$ increases and $|Z_{in}(f_L)|$ decreases as shown in Fig. 4(a) (dashed line), which helps improve the stability of DB-VCO and further ensures the proposed DB-VCO operating in the high frequency band.

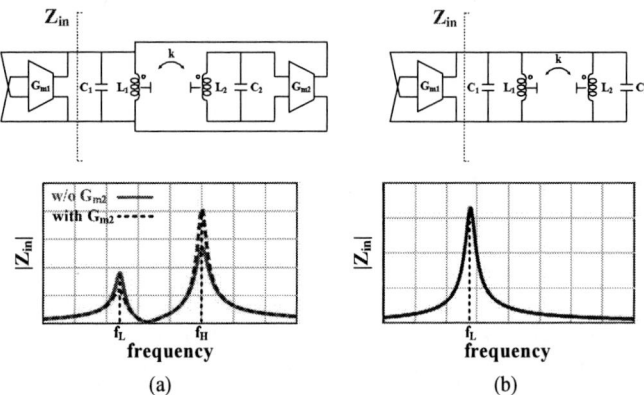

Fig. 4. Proposed DB-VCO operating in: (a) high band and (b) low band

When $S_1$ and $S_2$ are turned on and $I_2$ is turned off, the proposed DB-VCO operates in the low band as shown in Fig. 4(b). The turned-on switch shorts the two coils of transformer

and the two peaks of $|Z_{in}|$ will be merged into one in the low band as calculated below and shown in Fig. 4(b):

$$f_L = f_H = \frac{1}{2\pi\sqrt{LC(1+k)}} \tag{2}$$

As a result, the DB-VCO will operate at $f_L$. As there is only one peak of $|Z_{in}|$ when switches are turned on, DB-VCO has no instability problem, and no stability circuit is needed in the low band.

## III. Proposed Transformer-Distribution ILFD

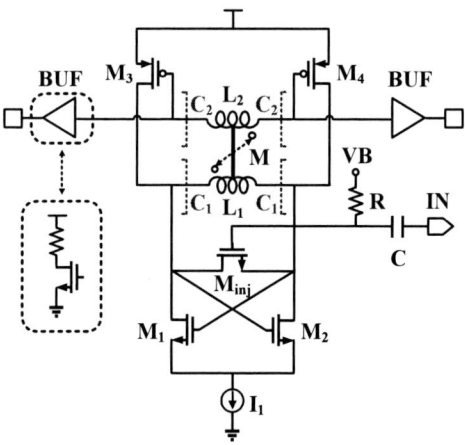

Fig. 5. Schematic of the proposed TD-ILFD

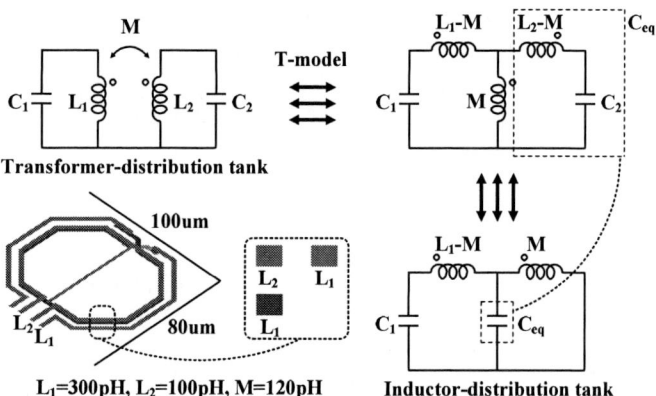

Fig. 6. Analysis of the proposed TD technique

Fig. 5 shows the proposed TD-ILFD, in which a transformer is used to distribute the parasitic capacitance from the primary coil $L_1$ to the secondary coil $L_2$ for less capacitive loading to enhance the locking range. In addition to parasitic capacitance of the output buffer, to distribute more capacitance to the secondary coil for even less capacitive loading, $M_3$ and $M_4$ are split from the cross-coupled pair ($M_1$ and $M_2$) and connected to the secondary coil. $M_3$ and $M_4$ form a two-port oscillator with the transformer tank to provide negative $g_m$ to compensate for the tank loss. To save power consumption, $M_3$ and $M_4$ are designed as PMOS transistors which reuse the DC current of $M_1$ and $M_2$.

The detailed analysis of proposed TD technique is shown in Fig. 6. $C_1$ is the parasitic capacitance presented at the primary coil $L_1$ while $C_2$ is the distributed parasitic

978-1-4673-6145-3/13 $31.00 © 2013 IEEE     360

capacitance at the secondary coil $L_2$ (including parasitic capacitance of buffer and split cross-coupled pair $M_3$ and $M_4$). T-model can be used to replace the transformer to obtain the equivalent circuit similar to inductor-distribution tank [4] except that an inductor $L_2$-M is added in series with the distributed capacitance $C_2$, which can be simply viewed as an equivalent capacitance $C_{eq}$ as follows:

$$C_{eq} = \frac{C_2}{1 - \omega^2(L_2 - M)C_2} \qquad (3)$$

Interestingly, when $L_2$-M is negative, $C_{eq}$ is smaller than $C_2$, which helps to further reduce the capacitive loading and enhance the locking range. To increase the mutual coupling M for negative $L_2$-M and also for larger loop gain of the distributed two-port oscillator, stacked transformer is designed as shown in Fig. 6. $L_1$ has two turns while $L_2$ has one turn. The second turn of $L_1$ is designed right blow $L_2$ for more mutual coupling. Compared with inductor-distribution technique in [4], the proposed TD technique has wider locking range and lower power consumption due to cross-coupled pair splitting, negative inductance $L_2$-M, and current reusing. Besides, instead of using multiple inductors in [4], only a transformer is used for TD technique, which would help decrease chip area and relax the difficulty of integration with VCO.

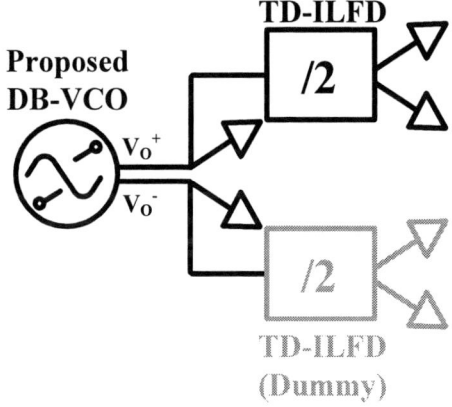

Fig. 7. Cascade of proposed DB-VCO and TD-ILFDs

In addition to be demonstrated as stand-alone blocks, the proposed TD-ILFD is also connected in cascade with the proposed DB-VCO as shown in Fig. 7. As the output of DB-VCO is differential, a dummy TD-ILFD is added so that the differential outputs of VCO have the same loading. Open-drain buffers are used at the outputs of DB-VCO and the TD-ILFD for testing purposes.

## IV. MEASUREMENT RESULTS

The proposed DB-VCO and TD-ILFD are fabricated in a 1P9M 65nm CMOS process. Fig. 8 (a) shows the chip photo of the stand-alone TD-ILFD while Fig. 8 (b) shows the chip photo of the DB-VCO+TD-ILFD cascade. The ILFD, the DB-VCO, and the DB-VCO + TD-ILFDs cascade occupy core chip areas of 130um x 100um, 130um x 90um, and 200um x 280um, respectively. The layout of proposed DB-VCO and TD-ILFD are quite compact due to the use of transformer.

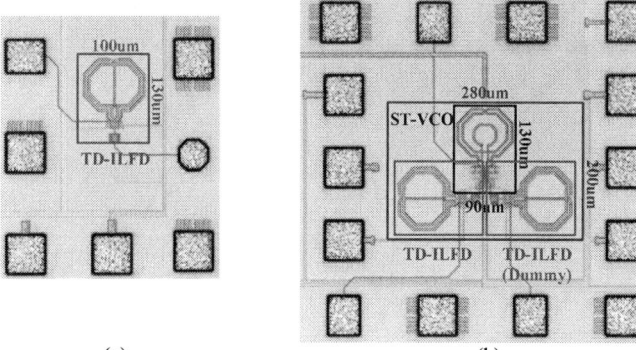

(a)                              (b)

Fig. 8. Chip photos of: (a) TD-IFLD, and (b) DB-VCO + TD-ILFD cascade

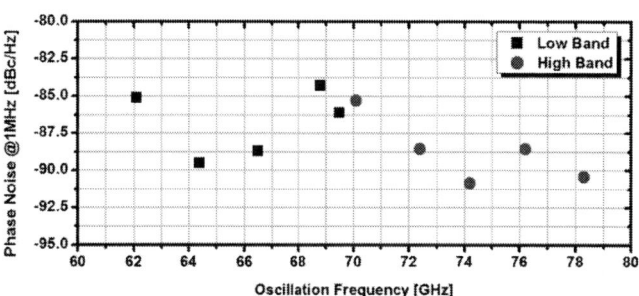

(a)

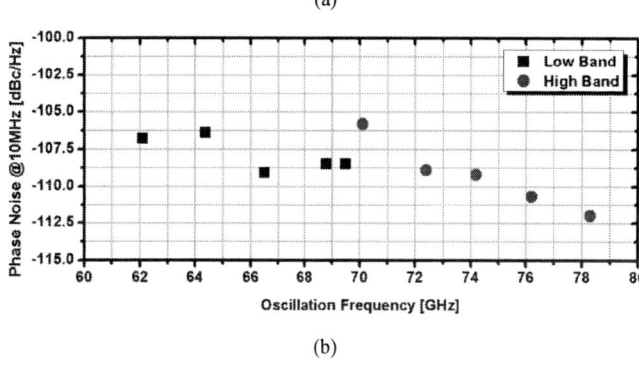

(b)

Fig. 9. Measured phase noise of proposed DB-VCO: (a) at1MHz offset, and (b) at 10MHz offset.

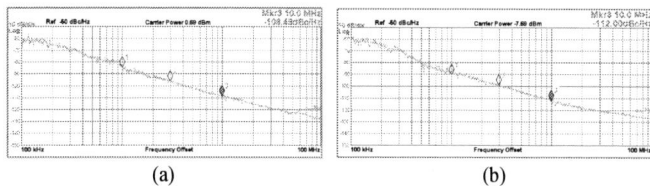

(a)                              (b)

Fig. 10. Measured phase noise plot: (a) in the low band (68.8GHz) and (b) in the high band (78.3GHz).

The proposed DB-VCO measures a tuning range from 62.1GHz to 69.5GHz in the low band and from 70.1GHz to 78.3GHz in the high band. Fig. 9 shows the measured phase noise of the DB-VCO, which is from -84.3dBc/Hz to -90.9dBc/Hz at 1MHz offset and from -105.8dBc/Hz to -112.0dBc/Hz at 10MHz offset. The proposed DB-VCO draws a total current of 7-8mA from a 1.1V supply, corresponding to FoM from -173.8dBc/Hz to -180.4dBc/Hz and FoM$_T$ from -180.8dBc/Hz to -187.4dBc/Hz using the phase noise measured at 10MHz offset. Fig. 10 shows the phase noise plot measured in the low band (68.8GHz) and in the high band (78.3GHz). The noise corner is around 1MHz. Table I summarizes and compares the measured performance of the proposed DB-VCO.

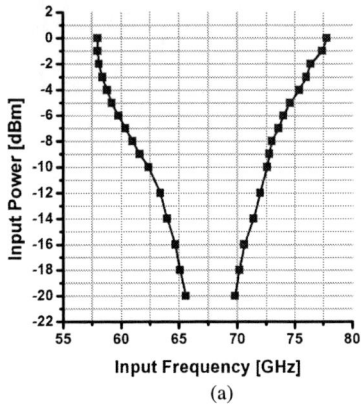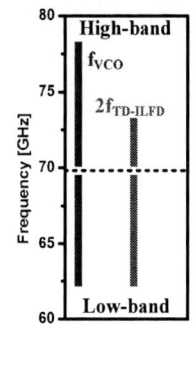

(a)                                    (b)

Fig. 11.  (a) Measured input sensitivity curve of the TD-ILFD, and (b) measured locking range of the DB-VCO+TD-ILFD cascade

Fig. 11(a) shows the measured input sensitivity curve of proposed TD-ILFD, which measures a locking range from 58GHz to 77.8GHz with 0dBm input power. The TD-ILFD can self-oscillate at 33.9GHz. Consuming only 1.44mW from a 1.2V power supply, the proposed TD-ILFD achieves much better performance in terms of locking range and FoM as compared with other existing solutions as shown in Table II.

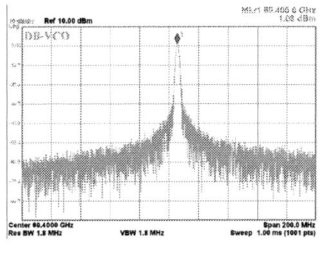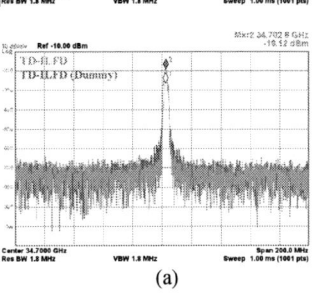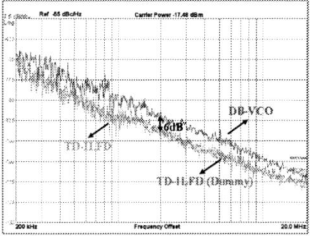

(a)                                    (b)

Fig. 12.  Measurement of DB-VCO in cascade with TD-ILFD locked at 69.4GHz: (a) spectra, and (b) phase noise plot.

The cascade of proposed DB-VCO and TD-ILFD is also measured. Fig. 11(b) shows the measured total locking range of DB-VCO + TD-ILFD. The measured locking range for the DB-VCO + TD-ILFD is from 62.1GHz to 73.3GHz. Fig. 12(a) shows the measured output spectra of DB-VCO and TD-ILFD (including dummy TD-ILFD) locked at 69.4GHz. Fig. 12(b) shows the measured phase noise of DB-VCO and TD-ILFD locked at 69.4GHz. As expected, a 6dBc difference can be observed between the two outputs.

## TABLE I. VCO PERFORMANCE SUMMARY AND COMPARISON

| VCOs | [1] | [2] | [3] | This Work |
|---|---|---|---|---|
| Output | NO DIV | NO DIV | NO DIV | **DIV** |
| Frequency [GHz] | 50.9-57.1 | 55.7-61.1 | 57.5-90.1 | **62.1-78.3**[(1)] |
| TR [%] | 11.5 | 9.3 | 44.2 | **22.3** |
| Phase Noise [dBc/Hz] | -118 @10MHz | -91 @1MHz | -104.6 ~ -112.2 @10MHz | **-105.8 ~ -112 @10MHz** |
| FoM [dBc/Hz][(2)] | -184 | -177.2 | -172 ~ -180 | **-173.8 ~ -180.4** |
| FoM$_T$ [dBc/Hz][(2)] | -185.2 | -176.6 | -184.2 ~ -192.2 | **-180.8 ~ -187.4** |
| P$_{diss}$ [mW] | 7.2 | 8.1 | 8.4 ~ 10.8 | **7.7 ~ 8.8** |
| V$_{DD}$ [V] | 1.2 | 0.7 | 1.2 | **1.1** |
| Area [mm²] | 0.23 | 0.01 | 0.03 | **0.012** |
| Process [CMOS] | 65nm | 90nm | 65nm | **65nm** |

(1) Low band: 62.1GHz~69.5GHz, High band: 70.1GHz~78.3GHz.

(2) $\text{FoM}=\text{PN}-20\log(\frac{f_0}{\Delta f})+10\log(\frac{P_{diss}}{1mW})$,  $\text{FoM}_T=\text{PN}-20\log(\frac{f_0}{\Delta f}\cdot\frac{TR}{10\%})+10\log(\frac{P_{diss}}{1mW})$

## TABLE II. ILFD PERFORMANCE SUMMARY AND COMPARISON

| ILFDs | [4] | [5] | [6] | [7] | This Work |
|---|---|---|---|---|---|
| Frequency [GHz] | 90.7 | 55.7 | 66.4 | 62.8 | **67.9** |
| LR [GHz, %] | 10.5 (11.6%) | 14.4 (25.9%) | 26 (39.2%) | 18.3 (29%) | **19.8 (29.2%)** |
| P$_{IN}$ [dBm] | 0 | 0 | 0 | 0 | **0** |
| P$_{diss}$ [mW] | 3.50 | 1.60 | 2.90 | 1.90 | **1.44** |
| V$_{DD}$ [V] | 1.2 | 1 | 0.8 | 0.8 | **1.2** |
| FoM [GHz/mW][(1)] | 3.14 | 8.73 | 8.97 | 9.53 | **13.75** |
| Area [mm²] | 0.34 | 0.48 | 0.126 | 0.023 | **0.013** |
| Process [CMOS] | 90nm | 65nm | 65nm | 65nm | **65nm** |

(1) FoM=Locking Range/P$_{diss}$

## V. CONCLUSION

In this work, ST and TD techniques are proposed to enhance the tuning range and locking range of mmW VCOs and ILFD. With ST technique, the proposed DB-VCO measures 22.3% tuning range with FoM and FoM$_T$ of -180.4dBc/Hz and -187.4dBc/Hz respectively. A prototype of ILFD based on TD technique is implemented, which measures ~30% locking range with only 1.44mW. A cascade of the proposed DB-VCO and TD-ILFDs is also demonstrated with an over locking range of 15.7% from 62.1GHz to 73.3GHz.

### REFERENCES

[1] S. Bozzola, et al, "An 11.5% frequency tuning, -184 dBc/Hz noise FOM 54 GHz VCO," *Proc. IEEE RFIC Symp. Dig.,* 2008, pp. 657–660.

[2] L. Li, et al., "Design and Analysis of a 90 nm mm-Wave Oscillator Using Inductive-Division LC Tank", *IEEE J. Solid-State Circuits,* vol. 44, pp. 1950-1958, Jul. 2009.

[3] J. Yin, et al., "A 57.5-to-90.1GHz Magnetically-Tuned Multi-Mode CMOS VCO", Proc. CICC, pp. 1-4, Sep. 2012.

[4] K.-H. Tsai, L.-C. Cho, J.-H. Wu, and S.-I. Liu, "3.5mW W-Band Frequency Divider with Wide Locking Range in 90nm CMOS Technology," *ISSCC Dig. Tech. Papers,* Feb. 2008, pp. 466-467.

[5] K. Takatsu, et al, "A 60-GHz 1.65mW 25.9 % locking range multi-order LC oscillator based injection locked frequency divider in 65nm CMOS," *Proc. CICC,* pp. 1-4, June 2011.

[6] Y. Chao, et al, "A 2.9mW 53.4-79.4GHz Frequency-Tracking Injection-Locked Frequency Divider with 39.2% Locking Range in 65nm CMOS", Proc. *IEEE RFIC Symp. Dig.,* 2012, pp. 337–340.

[7] J. Yin, et al, "A 0.8 V 1.9 mW 53.7-to-72.0 GHz self-frequency-tracking injection-locked frequency divider", *Proc. IEEE RFIC Symp. Dig.,* 2012, pp. 305–308.

# Concurrent Design of ESD Protection and ICs for Optimization and Prediction

*A Tutorial at IEEE CICC 2013, San Jose, CA*

Prof. Albert Wang
Dept. of Electrical Engineering
University of California
Riverside, CA 92521 USA          Email: aw@ee.ucr.edu
Tel: 1-951-827-2555               URL: http://lics.ee.ucr.edu

Copyright © 2013 by Albert Wang. All Rights Reserved

Albert Wang, Univ. of California, IEEE CICC 09242013

---

# OUTLINE

- ❏ Introduction to ESD protection

- ❏ Mixed-mode ESD design by CAD

- ❏ Whole-chip ESD protection design

- ❏ ESD+IC co-design method

- ❏ Emerging ESD protection concepts

- ❏ Summary

Albert Wang, Univ. of California, IEEE CICC 09242013

# ESD: Fun or Risk?

*Albert Wang, Univ. of California, IEEE CICC 09242013*

# ESD Failure: A Billion-$ Problem!

- ❑ ESD = Electrostatic Discharge
- ❑ ESD phenomena: HUGE current & voltage transients (~2kA & 70kV!)
- ❑ ESD failures: anywhere, anytime, any device & unavoidable!
  - ✓ Hard failure: instant malfunction,
  - ✓ Soft failure: degradation & lifetime.
- ❑ Multi-billion-$ annual losses to IC industry
  - ✓ EOS/ESD failures → up to 50% IC failures
  - ✗ ESD losses to electronic system products?

☞ ESD & other surge protection required!

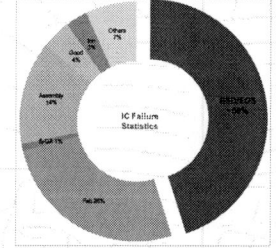

*Albert Wang, Univ. of California, IEEE CICC 09242013*

# ESD Protection Solutions

❑ IC level: on-chip ESD protection:
- ✓ Integrated ESD protection into ICs

❑ System level: TVS/EMI/Lightning protection:
- ✓ TVS = system level ESD/transient protection,
- ✓ EMI = electromagnetic interference: EMI + ESD integration,
- ✓ Other surge protection: lightning, burst, etc.

❑ Sub-system module ESD protection:
- ✓ System-in-package (SiP),
- ✓ Multi-chip module (MCM),

➢ Whole-chip/module/system surge protection
- ✓ On-Chip + In-Module + On-Board ESD protection

Albert Wang, Univ. of California, IEEE CICC 09242013

---

# ESD Testing Models

❑ HBM – human body induced ESD failures,
❑ MM – machine related ESD failures,
❑ CDM – device self-charge/discharge ESD failure,
❑ IEC – system level ESD failures,
- ✓ ESD, IEC61000-4-2
- ✓ Lightning, IEC61000-4-5
- ✓ Fast Transient/Burst, IEC61000-4-4

❑ TLP – transmission line pulsing testing
❑ Emerging ESD testing models

Albert Wang, Univ. of California, IEEE CICC 09242013

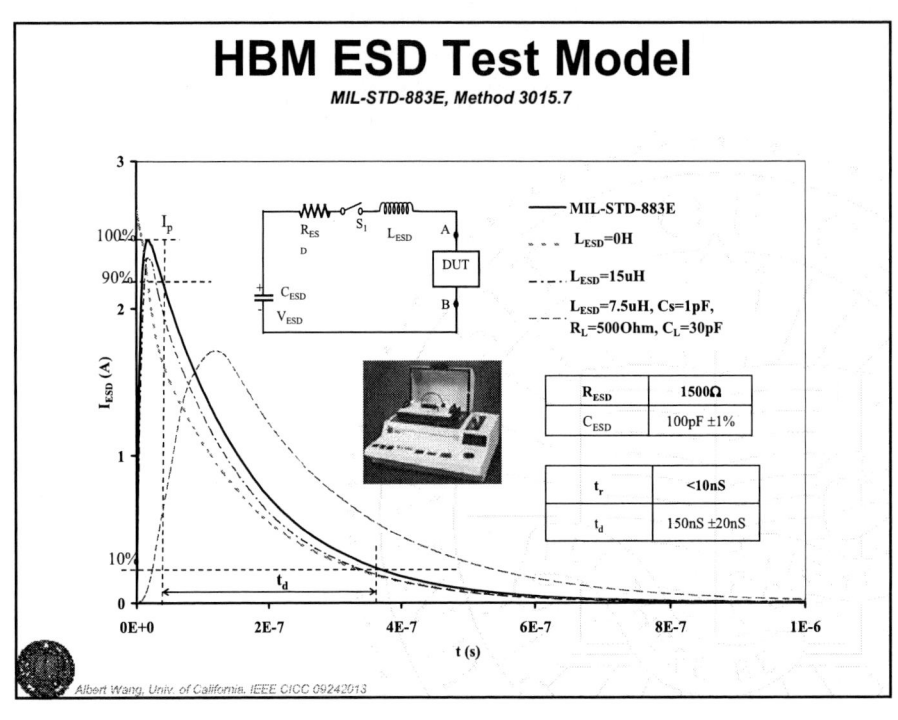

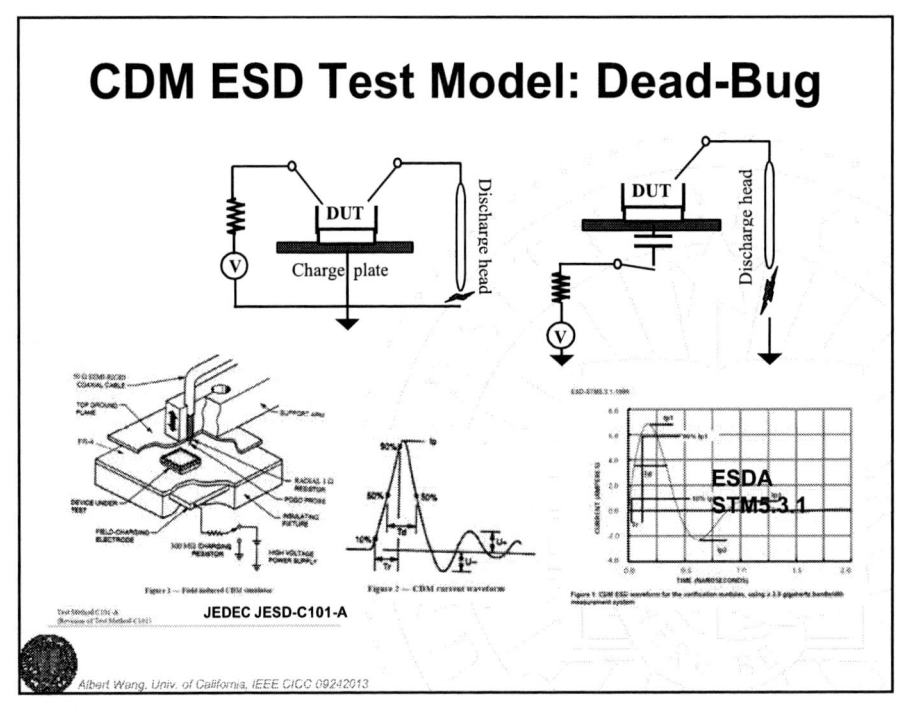

# CDM ESD Model Challenges: Repeatability & Comparability

| Standards | CDM Voltage (V) | Peak-I $I_{p1}$ (A) | $I_{p2}$ | $t_r$ | $t_d$ |
|---|---|---|---|---|---|
| JESD22-C101-A | 500 | 5.75+/-15% | <50% of $I_{p1}$ | <400ps | 1.0+/-0.5ns |
| ESDA STM5.3.1 | 500 | 7.5 +/-20% (4pF module) | <50% of $I_{p1}$ | <200ps | <400ps |
| ESDA SP5.3.2 (32pin DIP test fixture) | 500 | 4.8 | - | 700ps | 4.0ns (1st positive half cycle) |
| ESDA SP5.3.2 25x25 PGA test fixture) | 500 | 3.6 | - | 1.0ns | 10.8ns (1st positive half cycle) |

| ESD class | ESD Voltage (v) | | |
|---|---|---|---|
| | JESD22-C101-A | ESDA STM5.3.1 | ESDA SP5.3.2 |
| 1 | <200 | 125 | 250 |
| 2 | 200-500 | 250 | 500 |
| 3 | 500-1000 | 500 | 750 |
| 4 | >1000 | 1000 | 1000 |
| 5 | | 1500 | 1250 |
| 6 | | 2000 | |

- ❑ parasitic C/L will severely affect tester waveforms
- ❑ CDM test is not repeatable and not comparable!
- ❑ Does CDM test really make any sense?!

  ➢ If customers want it, you have to do it!

*Albert Wang, Univ. of California, IEEE CICC 09242013*

---

# IEC ESD Test Model: System to Chip
## *IEC61000-4-2, Ed1.2, 2001-04*

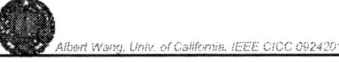

| $V_{ESD}$ (V) | Peak current $I_p$ (A) ±10% | Current at 30ns (A) | Duration (ns) | $t_r$ (ns) |
|---|---|---|---|---|
| 1000 | 7.5 | 4 | ~ 80 | 0.7-1.0 |
| 4000 | 30 | 16 | ~ 80 | 0.7-1.0 |

IEC model waveform parameters

| Level | Test voltage | Peak current | $t_r$ (ns) | I(60ns) |
|---|---|---|---|---|
| 1 | 2kV | 7.5A | 0.7-1 | 2A |
| 2 | 4kV | 15A | 0.7-1 | 4A |
| 3 | 6kV | 22.5A | 0.7-1 | 6A |
| 4 | 8kV | 30A | 0.7-1 | 8A |

IEC test classification

| Contact discharge | | Air discharge | |
|---|---|---|---|
| Level | Test voltage | Level | Test voltage |
| 1 | 2kV | 1 | 2kV |
| 2 | 4kV | 2 | 4kV |
| 3 | 6kV | 3 | 8kV |
| 4 | 8kV | 4 | 15kV |
| x | open | x | open |

*Albert Wang, Univ. of California, IEEE CICC 09242013*

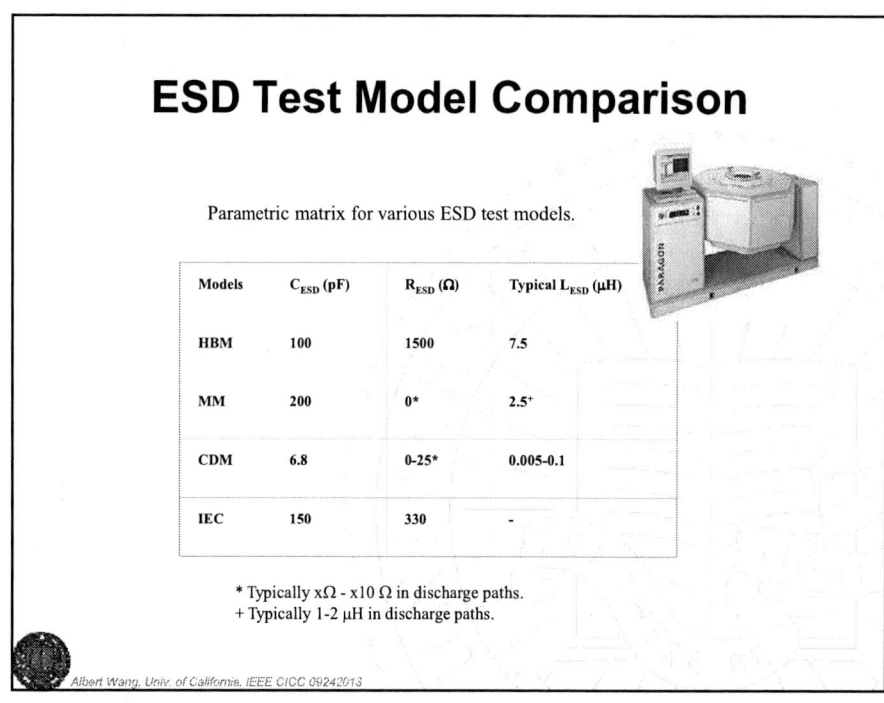

# ESD Test Model Comparison

Parametric matrix for various ESD test models.

| Models | $C_{ESD}$ (pF) | $R_{ESD}$ ($\Omega$) | Typical $L_{ESD}$ ($\mu$H) |
|---|---|---|---|
| HBM | 100 | 1500 | 7.5 |
| MM | 200 | 0* | 2.5+ |
| CDM | 6.8 | 0-25* | 0.005-0.1 |
| IEC | 150 | 330 | - |

\* Typically x$\Omega$ - x10 $\Omega$ in discharge paths.
+ Typically 1-2 $\mu$H in discharge paths.

---

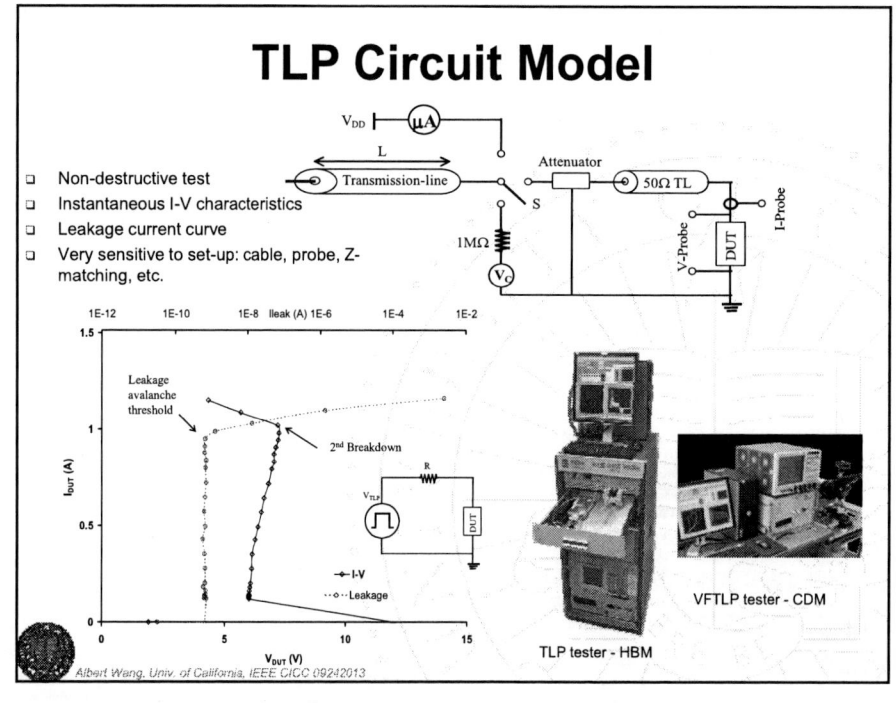

# TLP Circuit Model

- Non-destructive test
- Instantaneous I-V characteristics
- Leakage current curve
- Very sensitive to set-up: cable, probe, Z-matching, etc.

VFTLP tester - CDM

TLP tester - HBM

# ESD Failure & Protection

❏ **Two types of ESD damages:**
- ✓ Thermal damage: heating in Si/metal ← high ESD current pulse
- ✓ Dielectric rupture ← high electric field ← high ESD voltage pulse

❏ **Two ESD protection principles:**
- ✓ To discharge hi-current safely without generating too much HEAT,
- ✓ To clamp pad voltage to a sufficiently low level, <BV.

---

# On-Chip ESD Protection Principles

❏ Simple turn-on I-V,
❏ Snapback I-V.
❏ Protect EVERY I/O pad on chip!

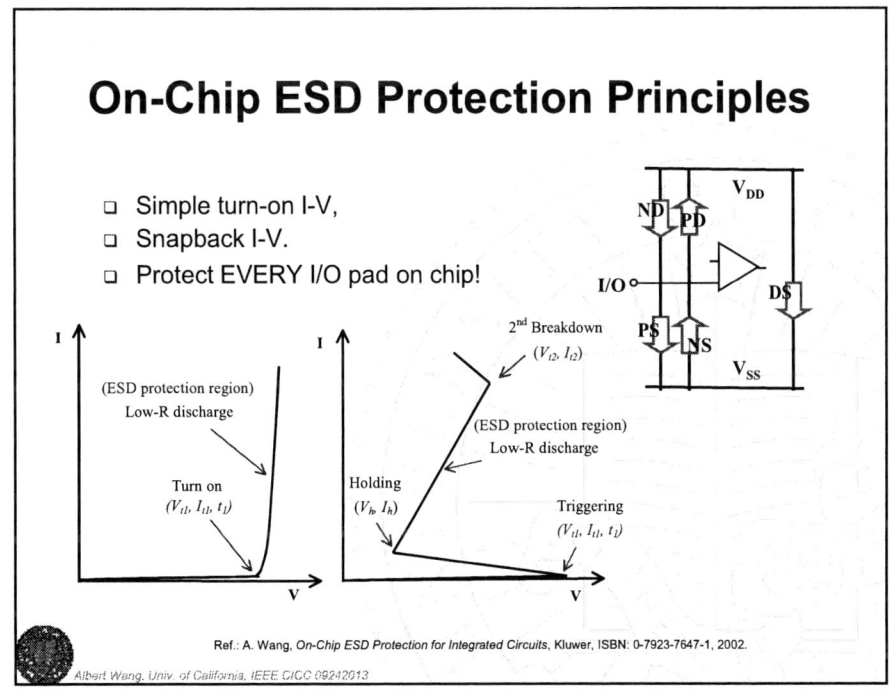

Ref.: A. Wang, *On-Chip ESD Protection for Integrated Circuits*, Kluwer, ISBN: 0-7923-7647-1, 2002.

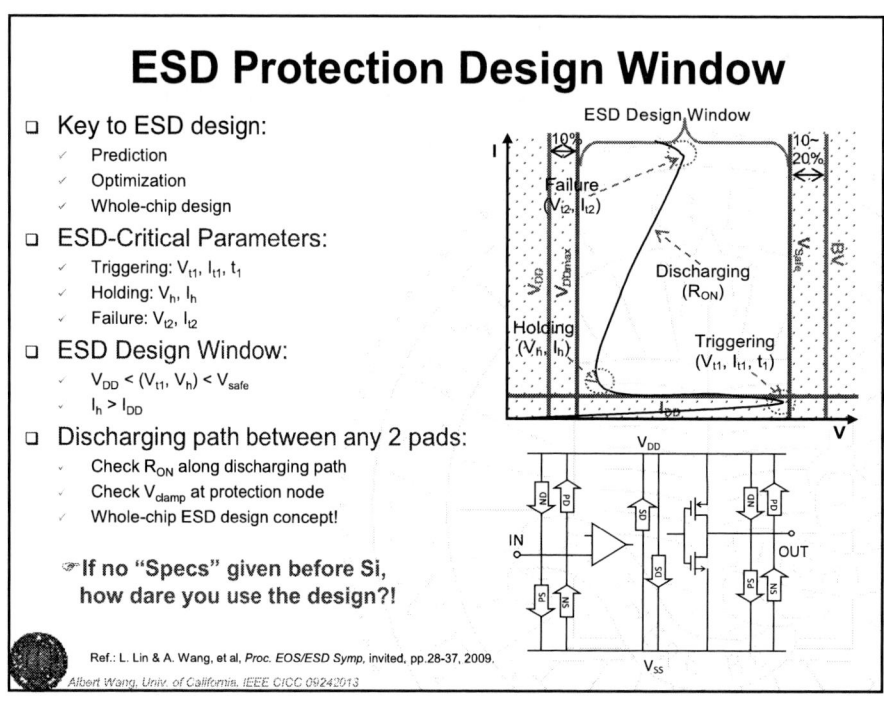

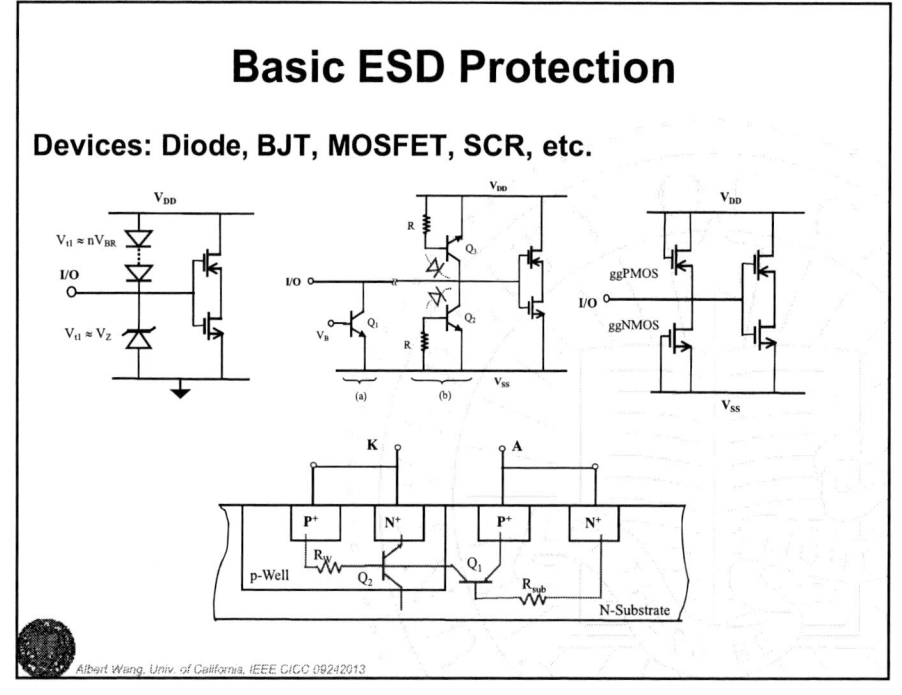

# ESD Protection: Simple or Complex?!

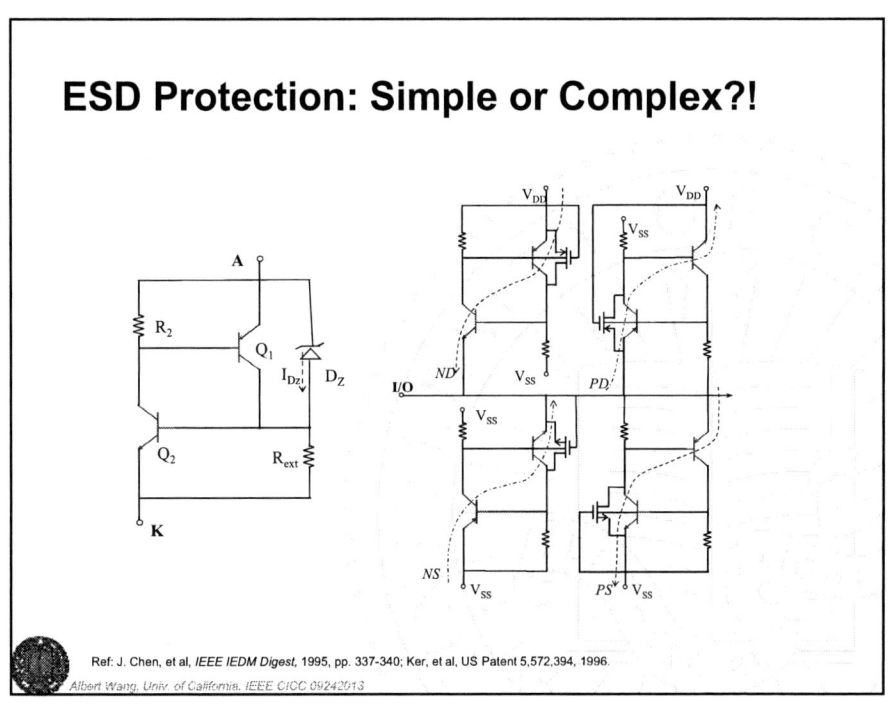

Ref: J. Chen, et al, *IEEE IEDM Digest*, 1995, pp. 337-340; Ker, et al, US Patent 5,572,394, 1996.

Albert Wang, Univ. of California, IEEE CICC 09242013

# ESD Design Roadmap and Challenges

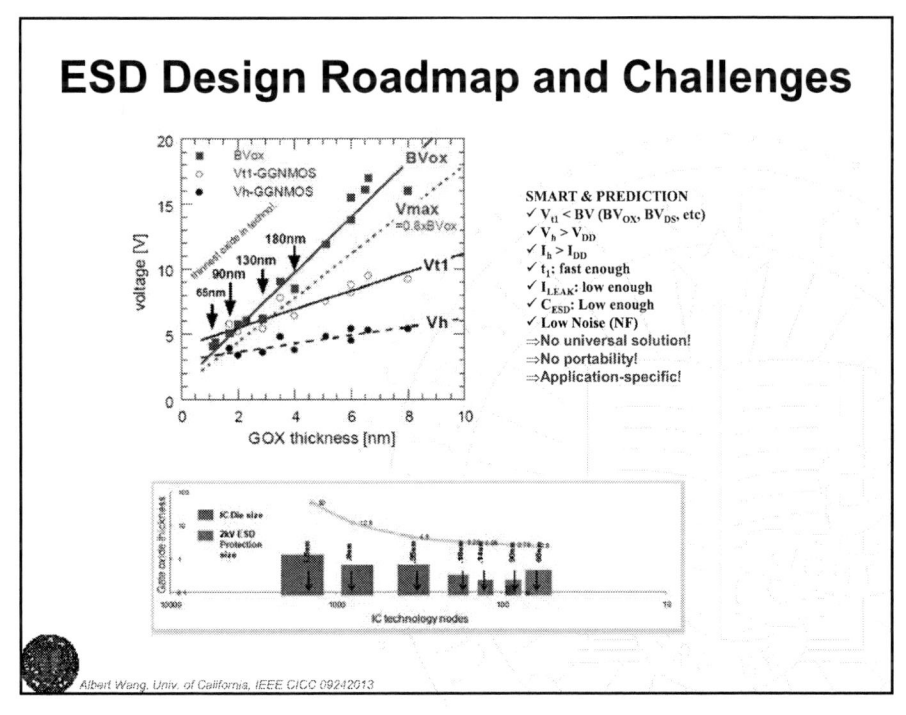

Albert Wang, Univ. of California, IEEE CICC 09242013

# ESD Design Challenges

- Design prediction by simulation

- Design optimization by simulation

- 3D ESD protection device modeling

- Whole-chip ESD design methodology

- New CAD for ESD synthesis and verification

- ESD protection for RF/AMS/HV/Broadband/Gbps ICs

- RFIC+ESD co-design

- ESD protection for nano technologies

- Field-Programmable ESD

- Above-IC ESD

Albert Wang, Univ. of California, IEEE CICC 09242013

---

# Mixed-Mode ESD Protection Design

2D/3D Mixed-Mode ESD Simulation-Design Methodology:
- Electro-thermal-process-device-circuit-layout coupling effects
- Static & transient ESD simulation
- Whole-chip ESD design

⇒ ESD design optimization, no trial-and-error!
   no over/under-design!

⇒ ESD design prediction, not backward analysis!

⇒ Compact ESD protection designs

⇒ Minimize ESD-induced parasitic effects

⇒ Explore novel ESD structures

Ref: A. Wang, et al, IEEE Trans Elec. Dev., v52, n7, p1304, 2005.
H. Feng, et al, IEEE JSSC, v38, n6, p995, 2003.
H. Xie, MS Thesis, IIT, 2004.

Albert Wang, Univ. of California, IEEE CICC 09242013

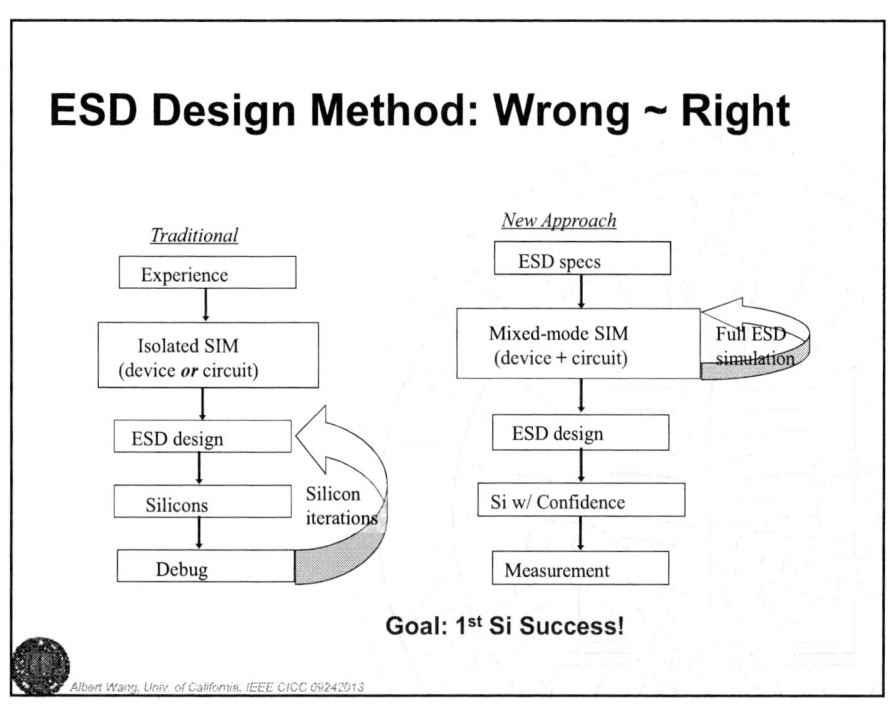

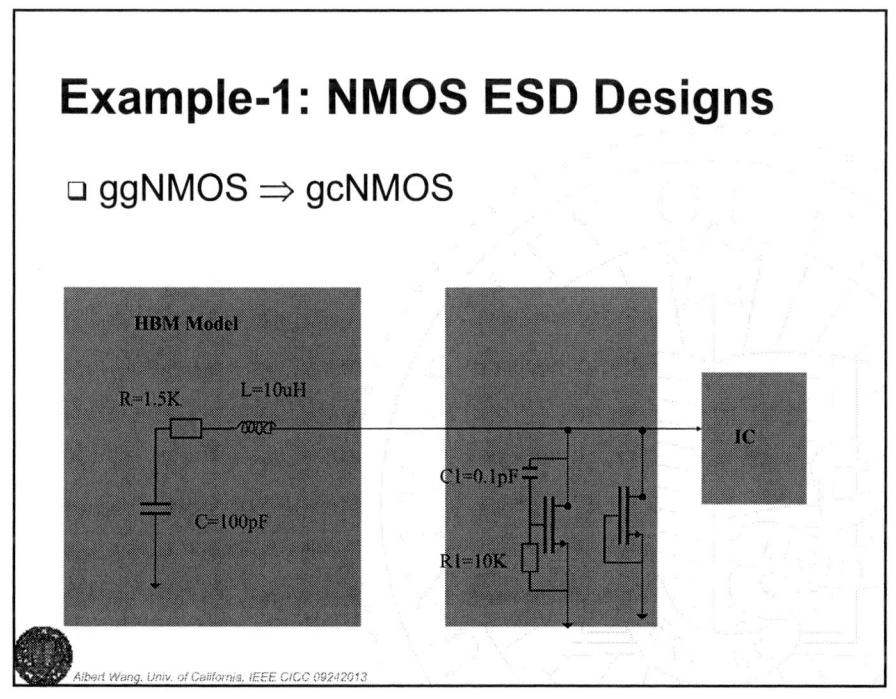

# Example-1: ggNMOS ~ gcNMOS ESD

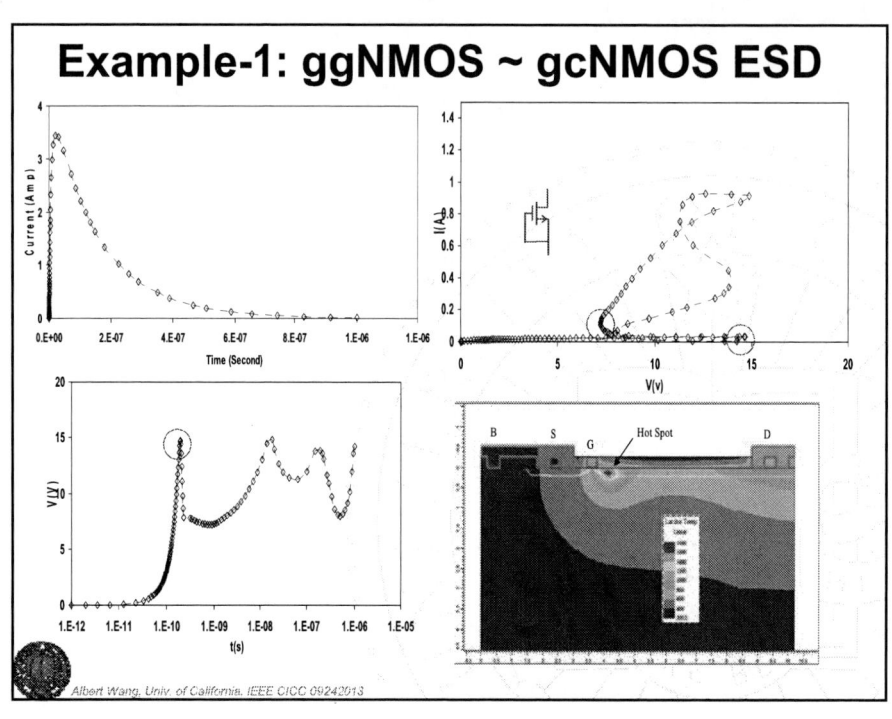

Albert Wang, Univ. of California, IEEE CICC 09242013

# Example-1: ggNMOS ~ gcNMOS ESD

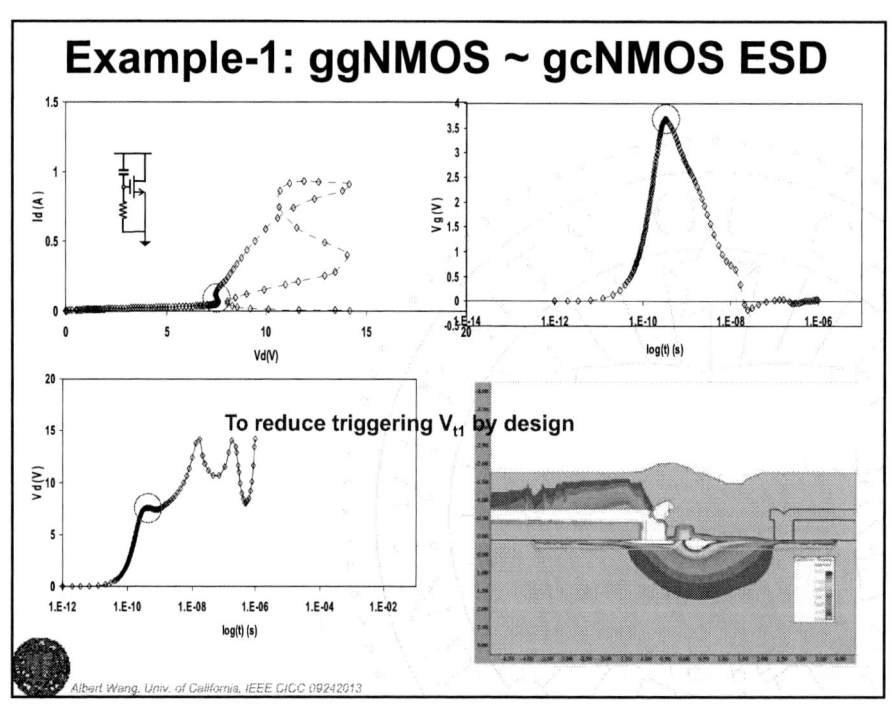

To reduce triggering $V_{t1}$ by design

Albert Wang, Univ. of California, IEEE CICC 09242013

# Example-1: ggNMOS ~ gcNMOS ESD

|  | GGNMOS | | GCNMOS | |
|---|---|---|---|---|
|  | SIM. | TEST | SIM. | TEST |
| $V_{t1}(V)$ | 14.68 | 12.56 | 7.54 | 6.66 |
| $t_1(ns)$ | 0.2 | - | 0.42 | - |
| $V_h(V)$ | 6.92 | 6.48 | 7.41 | 6.08 |
| $V_G(V)$ | - | - | 3.67 | - |
| $t_G (ns)$ | - | - | 0.32 | - |

❑ $V_{t1}$ reduction by simulation,

❑ Good design prediction,

❑ 1st Si success.

---

# Example 2: ESD + RF

❑ RF output buffer block,

❑ Differential buffer with open collector,

❑ 5kV SCR ESD protection

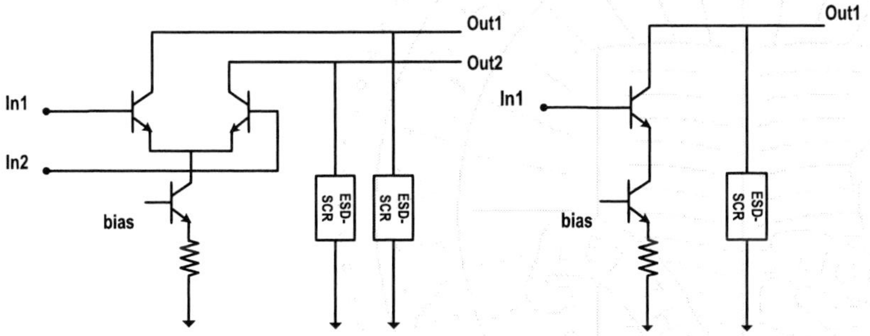

Ref: H. Feng, et al, IEEE JSSC, V38, N6, p995, 2003.

# Example-2: RF+ESD Design Verification

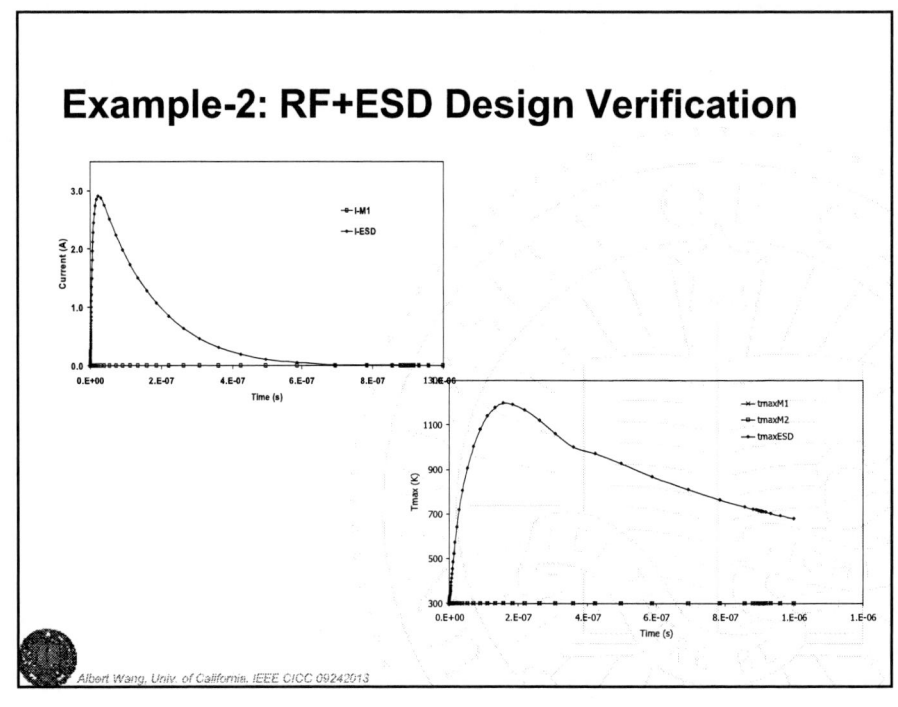

# RF & Broadband ESD Design

❑ What's Unique for RF ESD protection?!
  ✓ RF IC is sensitive to ESD-induced parasitics
  ✓ Need accurate RF ESD characterization
  ✓ Low-parasitic compact RF ESD protection design
  ✓ Whole-chip ESD protection circuit design concept

❑ New & Critical: ESD-Circuit Interactions
  ✓ ESD-to-Circuit Influences
  ✓ Circuit-to-ESD Influences

❑ RFIC+ESD co-design

Ref: A. Wang, et al, invited, IEEE Proc. CICC, 2002, pp411-418.

# ESD Parasitic: $C_{ESD}$

❑ **Circuit performance may be affected by ESD circuitry:**
  - ✓ ESD-induced parasitic $C_{ESD}$ (up to ~pF) & $R_{ESD}$,
  - ✓ $C_{ESD}$ $R_{ESD}$ delay $\Rightarrow$ signal integrity, clock corruption, …
  - ✓ $C_{ESD}$ $\Rightarrow$ loading effect, Z-matching, power efficiency, BW, …
  - ✓ $\Delta C_{ESD}$, $\Delta R_{ESD}$ ~ frequency, biasing, temperature, …

☞ **Unique Challenge:**

Accurate $C_{ESD}$ estimation,
Including $C_{ESD}$ in RF IC design,
Reduce $\Delta C_{ESD}$ over $\Delta f_{RF}$

Albert Wang, Univ. of California, IEEE CICC 09242013

---

# ESD Parasitic: Noises

❑ **Substrate noise coupling effect due to $C_{ESD}$:**
  - ➢ Incident noises at I/O coupled into substrate,
  - ➢ Substrate noises $\Rightarrow$ I/O $\Rightarrow$ signal path

❑ **ESD self-generated noises:**
  - ➢ Thermal noises,
  - ➢ Flicker noises,
  - ➢ Shot noises, etc.

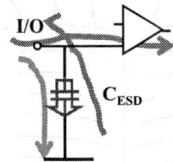

☞ **Unique Challenge:**

ESD noises into RF ICs.

Albert Wang, Univ. of California, IEEE CICC 09242013

# Example-3: ESD + RFIC Co-Design

ESD affects RF and broadband IC substantially:

- 5GHz LNA for dual-band WLAN transceiver
  - ✓ CE-CB cascode topology
  - ✓ High/low gain switching
  - ✓ Unique double shutdown function
- 0.18μm SiGe BiCMOS
- 2KV ESD protection

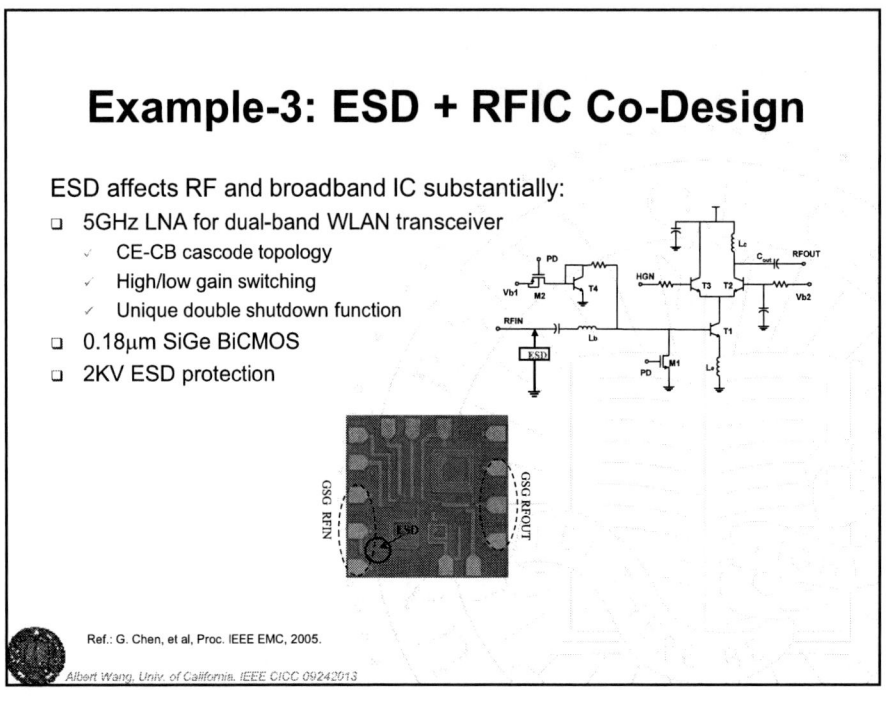

Ref.: G. Chen, et al, Proc. IEEE EMC, 2005.

---

# Example-3: LNA Noise ~ ESD

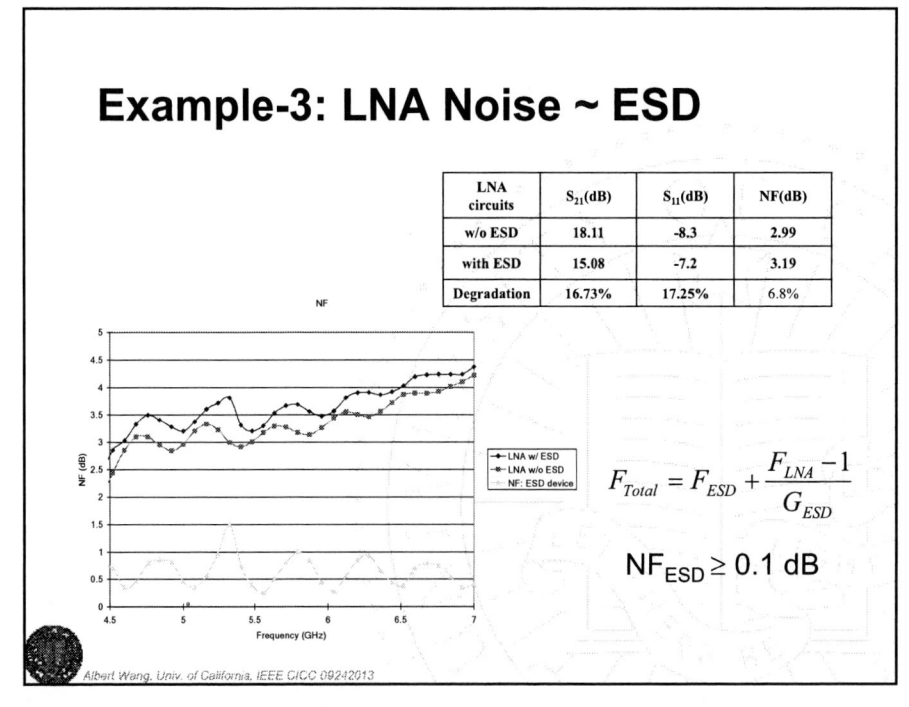

| LNA circuits | $S_{21}$(dB) | $S_{11}$(dB) | NF(dB) |
|---|---|---|---|
| w/o ESD | 18.11 | -8.3 | 2.99 |
| with ESD | 15.08 | -7.2 | 3.19 |
| Degradation | 16.73% | 17.25% | 6.8% |

$$F_{Total} = F_{ESD} + \frac{F_{LNA} - 1}{G_{ESD}}$$

$$NF_{ESD} \geq 0.1 \text{ dB}$$

# Reference-1: 2.4GHz LNA+ESD

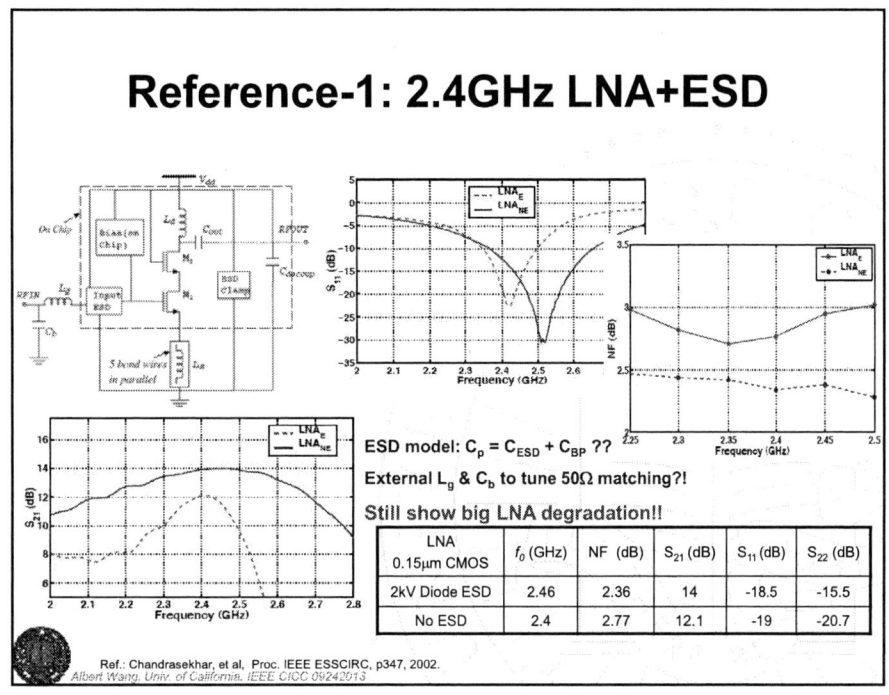

ESD model: $C_p = C_{ESD} + C_{BP}$ ??

External $L_g$ & $C_b$ to tune 50Ω matching?!

Still show big LNA degradation!!

| LNA 0.15μm CMOS | $f_0$ (GHz) | NF (dB) | $S_{21}$ (dB) | $S_{11}$ (dB) | $S_{22}$ (dB) |
|---|---|---|---|---|---|
| 2kV Diode ESD | 2.46 | 2.36 | 14 | -18.5 | -15.5 |
| No ESD | 2.4 | 2.77 | 12.1 | -19 | -20.7 |

Ref.: Chandrasekhar, et al, Proc. IEEE ESSCIRC, p347, 2002.

Albert Wang, Univ. of California, IEEE CICC 09242013

---

# RF ESD Design Characterization

□ Comprehensive & accurate RF ESD characterization:

- ✓ S-parameter measurement,
- ✓ Noise measurement,
- ✓ 1-port *vs.* 2-port,
- ✓ Q-factor?  $Q = \dfrac{1}{2\pi f C_{ESD} R_{ESD}}$
- ✓ Critically affect I/O z-matching of RF ICs!
- ✓ Never trust foundry ESD models yet!

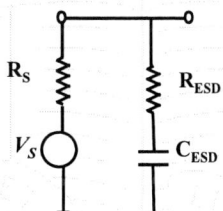

Ref.: G. Chen, et al, Proc. IEEE RFIC Symp., pp347, 2003. Wang, et al, IEEE Trans ED, V52, N7, p1304, 2005.

Albert Wang, Univ. of California, IEEE CICC 09242013

---

978-1-4673-6145-3/13 $31.00 © 2013 IEEE

# Example-4: RF ESD Characterization

❏ Most commonly ESD protection structures

  ✓ ggMOS
  ✓ SCR
  ✓ dSCR
  ✓ Diode string: Dx1, Dx2, Dx3, Dx4, Dx5, … Dxn

❏ Designed and fabricated in 0.35μm BiCMOS

❏ 2kV/5kV ESD protection

❏ Design optimization by mixed-mode ESD simulation

❏ Simulation matches measurement well

*Albert Wang, Univ. of California, IEEE CICC 09242013*

# Example-4: 2kV $C_{ESD}$ by Simulation & Test

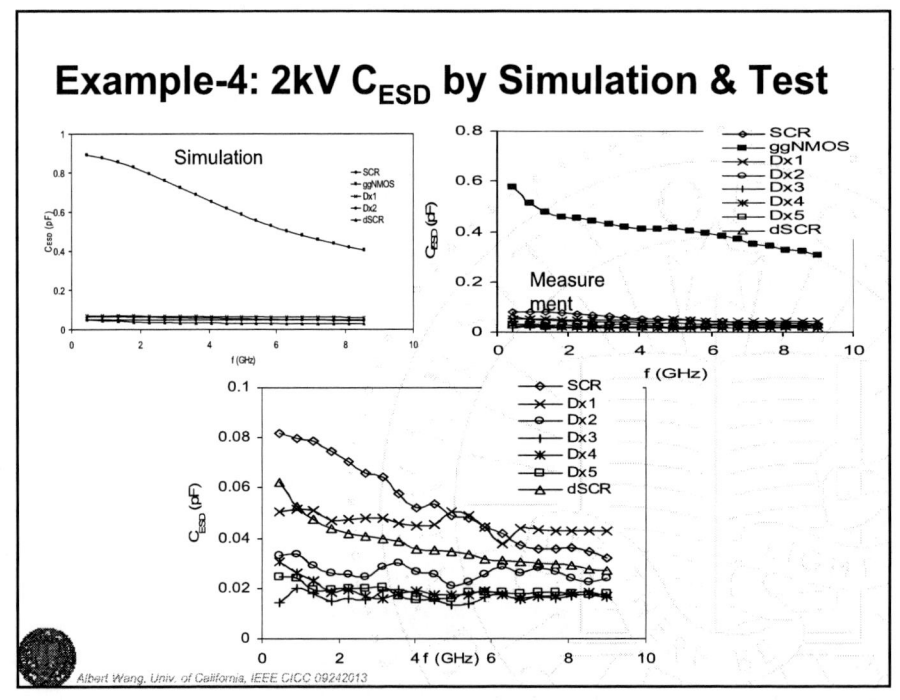

*Albert Wang, Univ. of California, IEEE CICC 09242013*

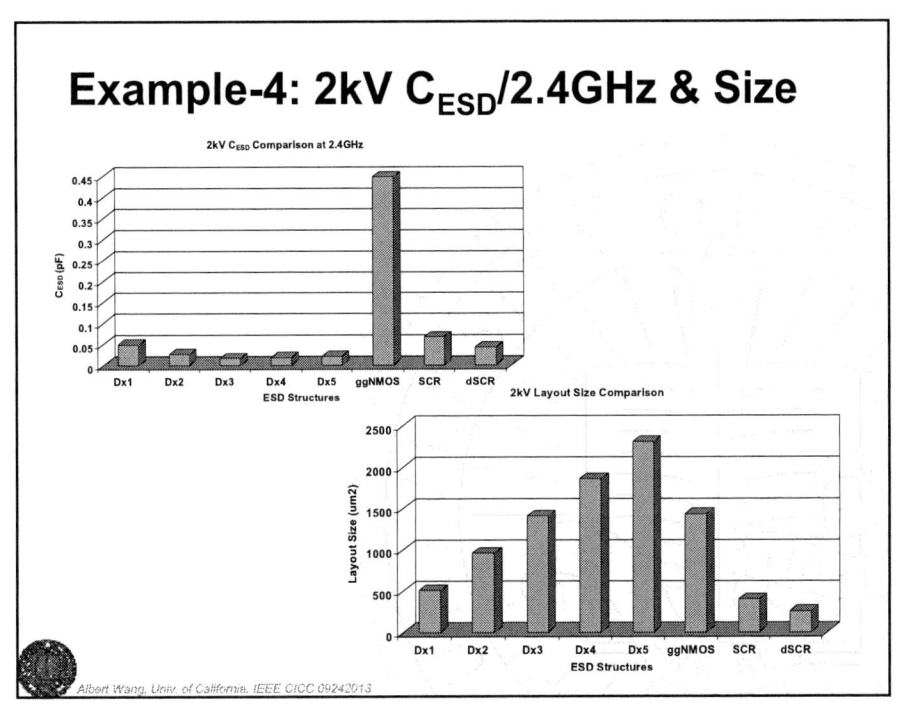

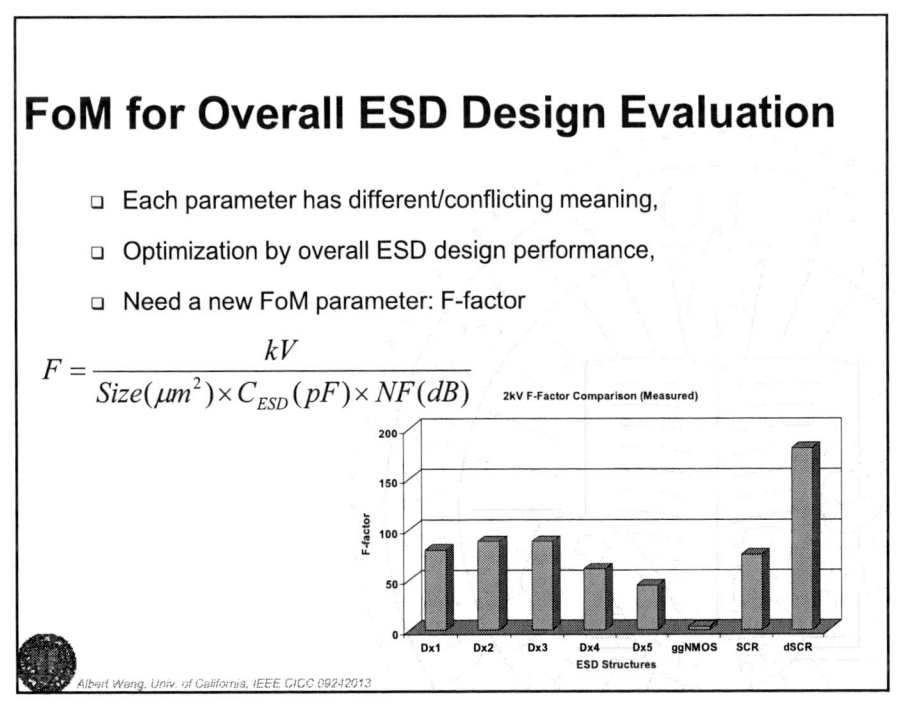

# FoM for Overall ESD Design Evaluation

- Each parameter has different/conflicting meaning,
- Optimization by overall ESD design performance,
- Need a new FoM parameter: F-factor

$$F = \frac{kV}{Size(\mu m^2) \times C_{ESD}(pF) \times NF(dB)}$$

# Example-4: Number Talks

❑ Extracted testing data at 2.4GHz,
❑ Optimized ESD design by simulation for min-size, min-parasitics,
❑ Some ESD designs just not good for RF ICs,
❑ Trial-and-error designs and "rich" experiences do not work here!
☞ You still need ESD-IC co-designs!!

| ESD | Dx1 | Dx2 | Dx3 | Dx4 | Dx5 | ggNMOS | SCR | dSCR |
|---|---|---|---|---|---|---|---|---|
| $C_{ESD}$ (fF) | 48.7 | 26.3 | 16.3 | 18.7 | 21.0 | 448.3 | 66.6 | 42.4 |
| Size ($\mu m^2$) | 506 | 956 | 1405 | 1855 | 2305 | 1433 | 396 | 257 |
| F (V/$\mu m^2$ pF) | 79.4 | 88.1 | 88.1 | 60.6 | 44.1 | 3.1 | 75.0 | 180.4 |

---

# Real-World ESD Design: Funny?

# RF and Broadband ESD SOLUTIONS

❑ No universal RF ESD design!

❑ Any ESD protection $\Rightarrow$ RF ESD given that

ESD-Circuit interactions $\downarrow\downarrow\downarrow$

➢**Goal for RF ESD $\Rightarrow$ ANY NOVEL structures:**
- ✓ **Ultra-fast ESD switching,**
- ✓ **Novel triggering mechanisms,**
- ✓ **Hi-ESDV/Si ratio,**
- ✓ **Small size,**
- ✓ **Low-parasitic,**
- ✓ **Multiple-mode ESD protection,**

Albert Wang, Univ. of California, IEEE CICC 09242013

---

# Novel ESD Protection Design Helps:
**From 1-directional to multi-directional**

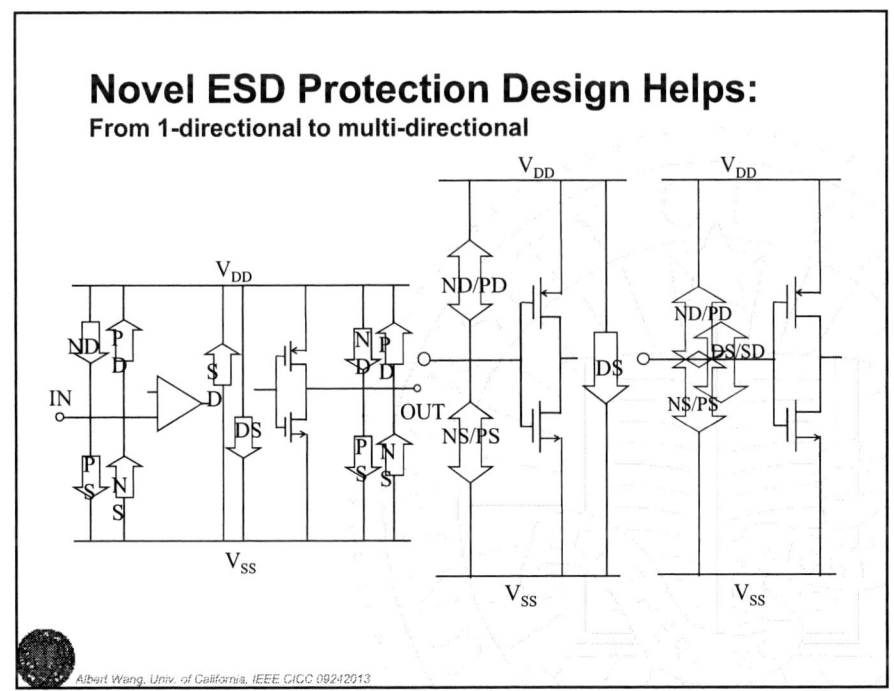

Albert Wang, Univ. of California, IEEE CICC 09242013

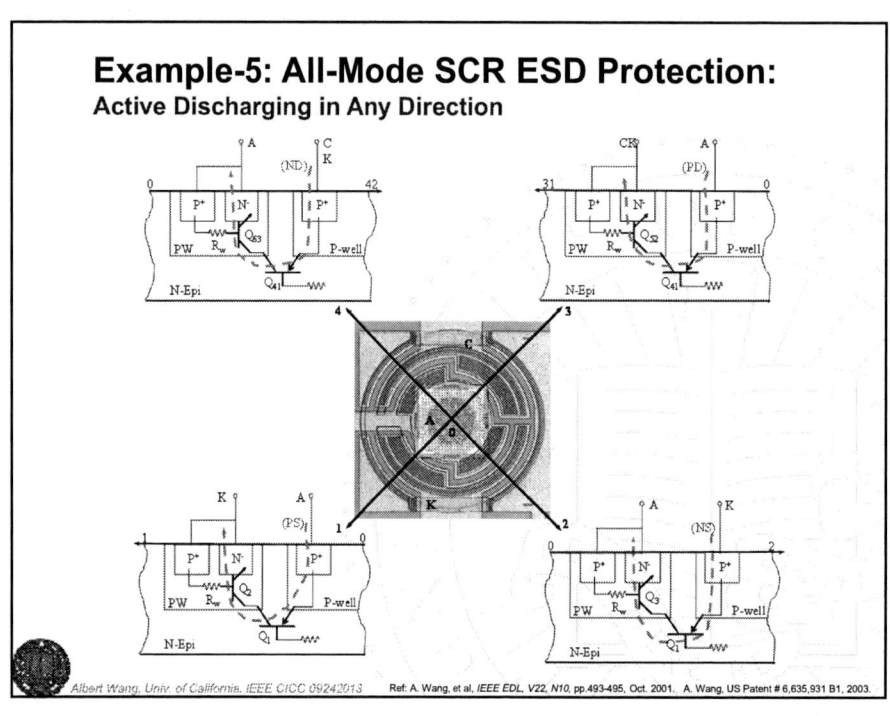

# Example-5: All-Mode SCR ESD Protection:
## Active Discharging in Any Direction

Albert Wang, Univ. of California, IEEE CICC 09242013    Ref.: A. Wang, et al, *IEEE EDL, V22, N10*, pp.493-495, Oct. 2001.   A. Wang, US Patent # 6,635,931 B1, 2003.

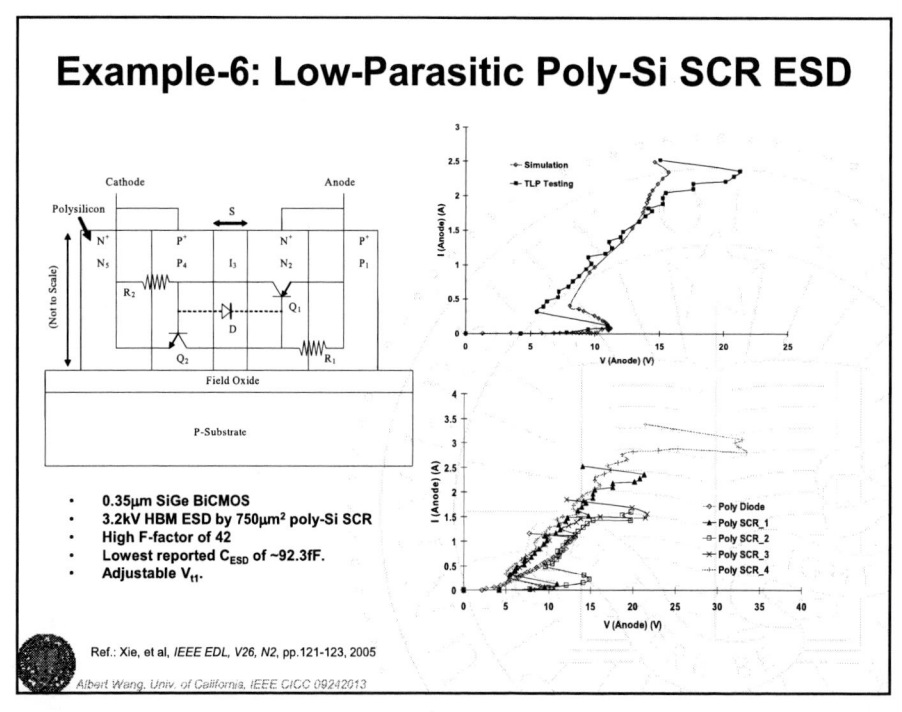

# Example-6: Low-Parasitic Poly-Si SCR ESD

- **0.35µm SiGe BiCMOS**
- **3.2kV HBM ESD by 750µm² poly-Si SCR**
- **High F-factor of 42**
- **Lowest reported $C_{ESD}$ of ~92.3fF.**
- **Adjustable $V_{t1}$.**

Ref.: Xie, et al, *IEEE EDL, V26, N2*, pp.121-123, 2005

Albert Wang, Univ. of California, IEEE CICC 09242013

# Whole-Chip ESD Protection Scheme-1

❑ Whole-chip ESD protection principles:
- Low-R discharging path between ANY two pads,
- Estimate $R_{on}$ in the longest path – worst case,

❑ Solution 1:
- Pad + clamp scheme
- 1-direction ESD

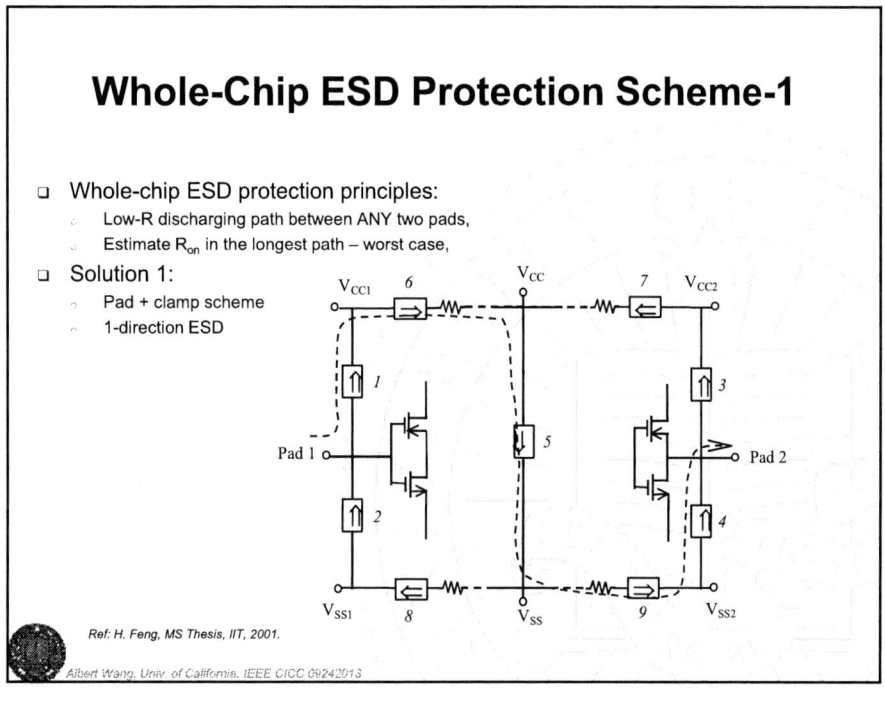

Ref: H. Feng, MS Thesis, IIT, 2001.

# Whole-Chip ESD Protection Scheme-2

❑ Using a common ESD discharging path,
❑ Dual-directional ESD

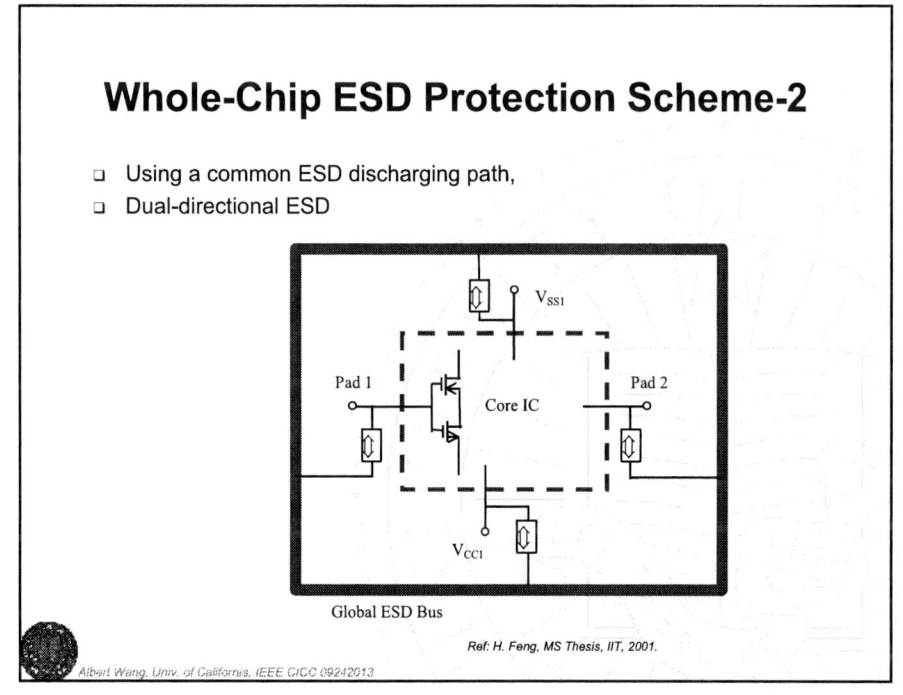

Ref: H. Feng, MS Thesis, IIT, 2001.

# ESD Protection for MS/HV ICs

- No universal ESD solutions!
- No one $V_{t1}$ fits the whole chip!
- Multi-$V_{DD}$/$V_{SS}$ $\Rightarrow$ locally-optimized $V_{t1}$ for different I/Os,
- Need a safety margin for $V_{t1}$:
  - $V_{t1}$ of 5V fits $V_{DD}$=3.3V blocks,
  - $V_{t1}$ of 23V good for $V_{DD}$=15V blocks.

> Challenge: multi-$V_{t1}$ ESD design in RF/AMS ICs
> > $\Rightarrow$ whole-chip ESD design optimization,
> > $\Rightarrow$ on-chip local ESD design optimization

*Albert Wang, Univ. of California, IEEE CICC 09242013*

---

# Keys to HV & Multi-$V_{DD}$ ESD Design

- Must meet the **ESD Design Window**!
- Local ESD design optimization in each $V_{DD}$-domain
- Flexible ESD triggering for HV/LV
- Avoid latch-up in HV

> Understand process Specs (BV=?)
> Know IC circuit Specs ($V_{DD}$ =?, multi-$V_{DD}$?, different $V_{DD}$ ~ I/O?, etc)
> Set ESD Design Window
>   - $V_{DD} < V_{t1} < BV$; $V_{DD} < V_h < BV$; $I_h > I_{DD}$
>   - Specs for $t_1$, $I_{LEAK}$, $C_{ESD}$, NF, etc.
> Full-Circuit ESD protection schemes
> ESD design prediction by CAD
> Never universal ESD solution!

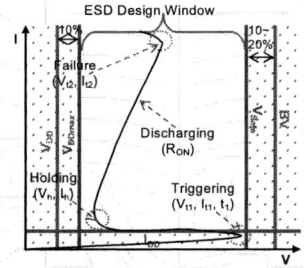

Ref.: A. Wang, et al, *Proc. EOS/ESD Symp*, pp.28-37, 2009

*Albert Wang, Univ. of California, IEEE CICC 09242013*

# ESD+RFIC Co-Design

① ESD protection design optimization by mixed-mode simulation:
  ✓ Minimum sizes,
  ✓ Minimum parasitics,
  ✓ Rational & accurate design

② Accurate RF ESD design characterization:
  ✓ S-parameter using co-planar and de-embedding technique
  ✓ Extract $C_{ESD}$, $R_{ESD} \sim f$
  ✓ Get NF $\sim f$ by noise analyzer

*Albert Wang, Univ. of California, IEEE CICC 09242013*

---

# ESD+RFIC Co-Design

③ ESD + RFIC co-design for RF circuit optimization:
  ✓ Design I/O Z-matching for RF ICs (noises, power or efficiency, etc)
  ✓ Include ESD-parasitics in RF circuit simulation
  ✓ Evaluate I/O mismatching corruption by ESD
  ✓ RF I/O *Re-matching* technique with ESD parasitics
  ✓ Post-simulation verification
  ✓ New *S-parameter ESD+RF co-simulation* technique

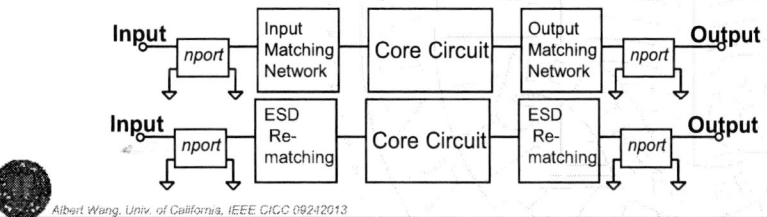

*Albert Wang, Univ. of California, IEEE CICC 09242013*

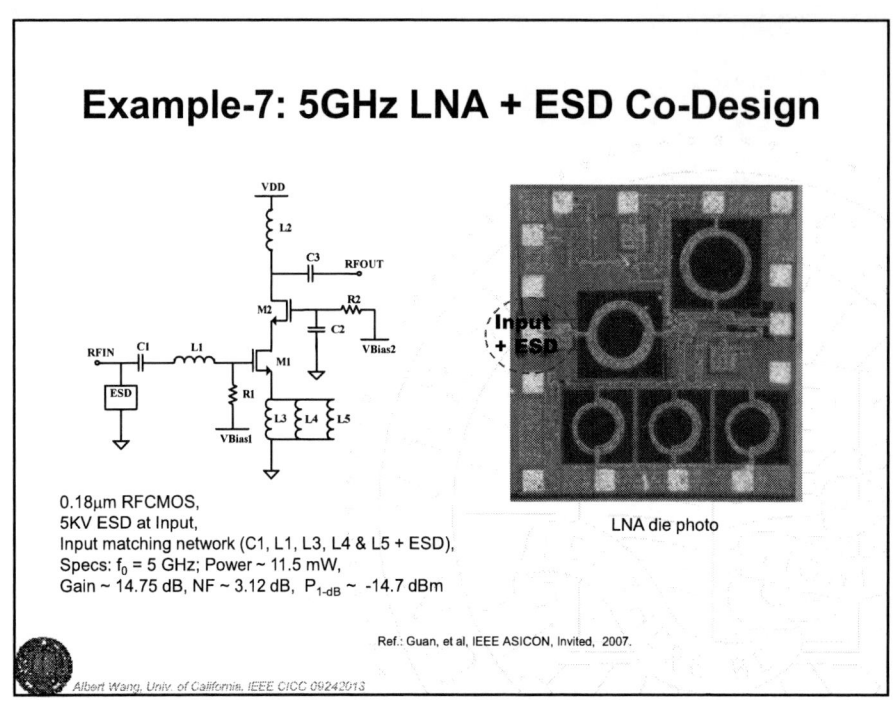

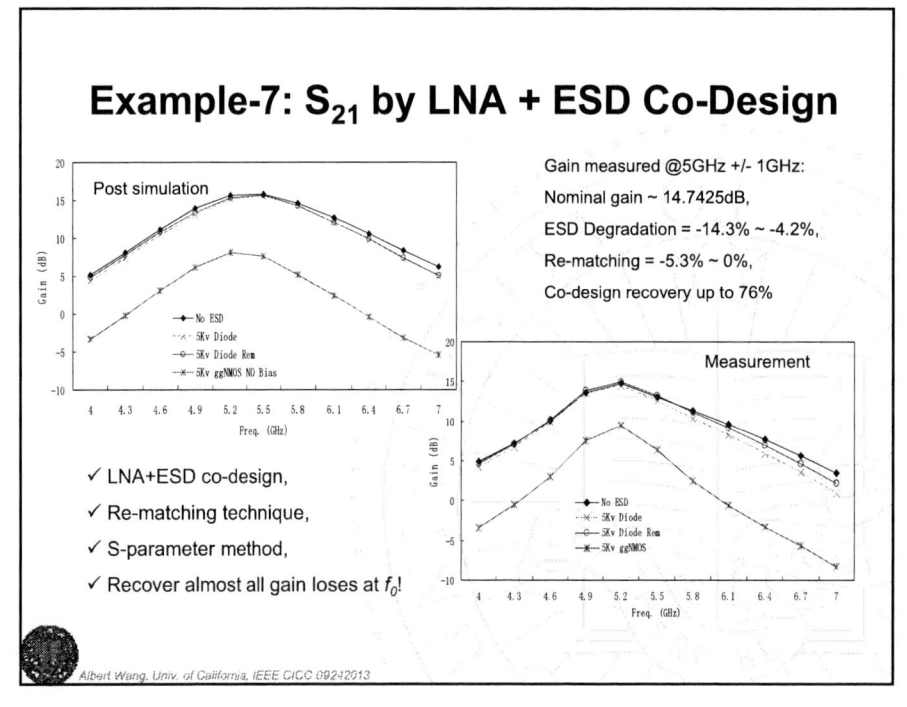

# Example-7: NF by LNA + ESD Co-Design

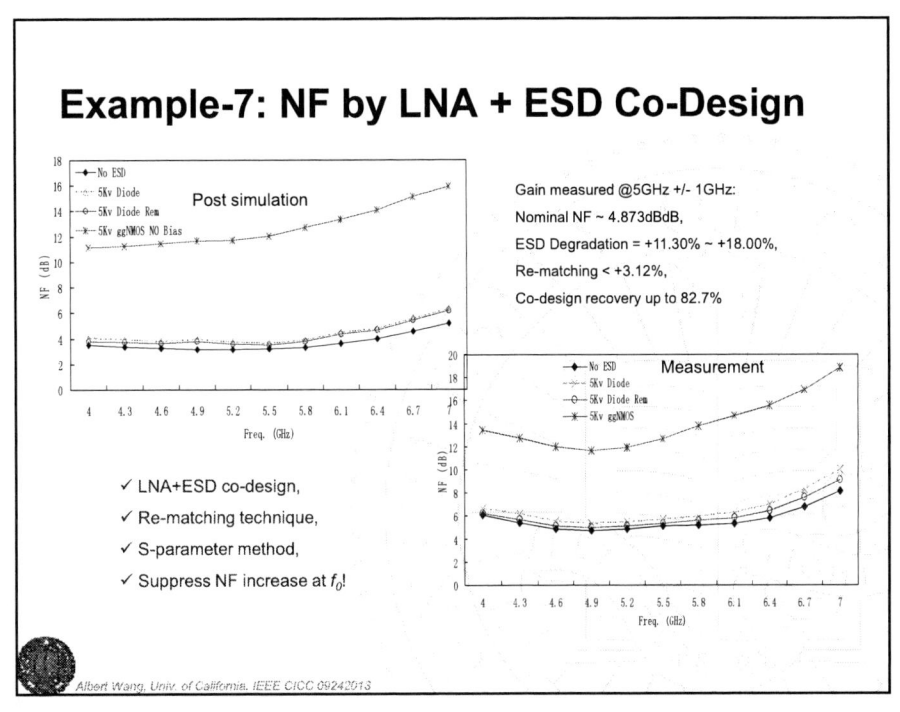

Gain measured @5GHz +/- 1GHz:

Nominal NF ~ 4.873dBdB,

ESD Degradation = +11.30% ~ +18.00%,

Re-matching < +3.12%,

Co-design recovery up to 82.7%

✓ LNA+ESD co-design,

✓ Re-matching technique,

✓ S-parameter method,

✓ Suppress NF increase at $f_0$!

---

# Example-7: $S_{11}$ by LNA + ESD Co-Design

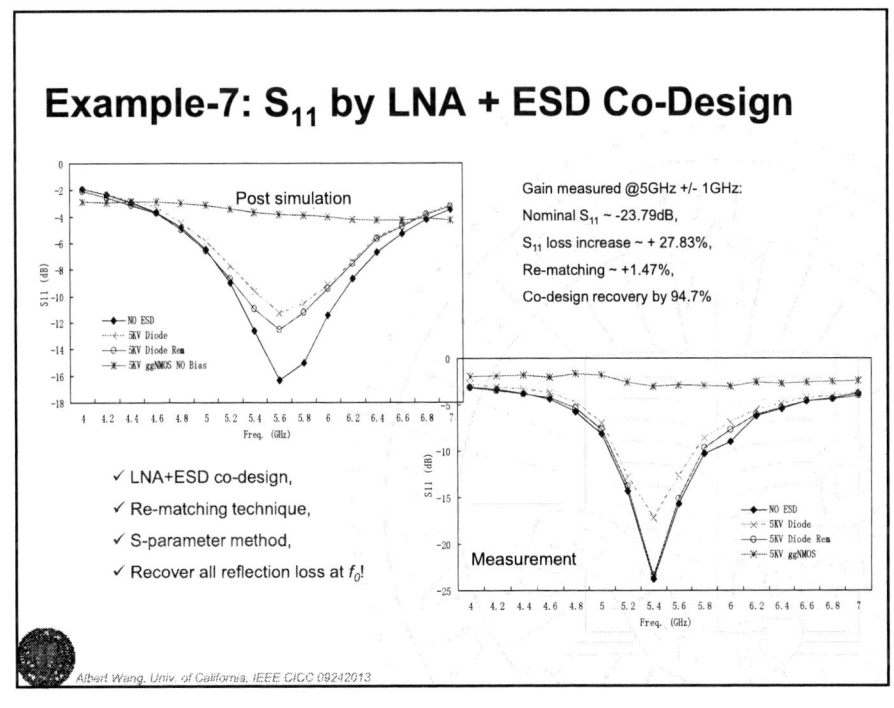

Gain measured @5GHz +/- 1GHz:

Nominal $S_{11}$ ~ -23.79dB,

$S_{11}$ loss increase ~ + 27.83%,

Re-matching ~ +1.47%,

Co-design recovery by 94.7%

✓ LNA+ESD co-design,

✓ Re-matching technique,

✓ S-parameter method,

✓ Recover all reflection loss at $f_0$!

# Broadband IR-UWB SoC + ESD Protection

❑ Where does UWB need ESD protection?

- ✓ 3.1-10.6GHz bandwidth,
- ✓ Input pins,
- ✓ Output pins,
- ✓ Power rails,
- ✓ Control ports.

# Broadband UWB Tx + ESD Co-Design

❑5th-order Gaussian PG
- ✓ pulse combination & single delay line

❑Digital control

❑Integrated BPSK modulation

❑Programmable and very short pulse width for Gbps

❑PSD complies with FCC mask

❑CMOS-friendly topology

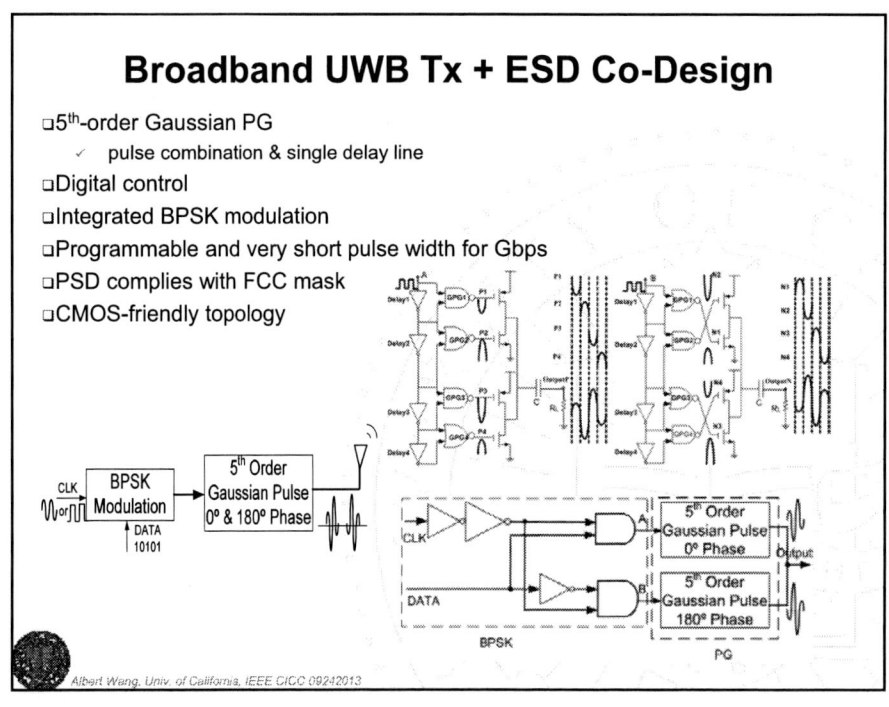

# Example-8: UWB Tx + ESD in Si

❑ESD at output port
❑Full 5kV ESD protection
　　　Output↔Vdd, Output↔GND, Vdd↔GND
❑0.18µm MS/RF CMOS, 0.25mm².
❑Full measurements

# Example-8: UWB Tx Measurement

❑Pulse width: ~400ps; Pulse amplitude: 506mV; Power: 0.14pJ/p-mV @ 100MHz PRF.

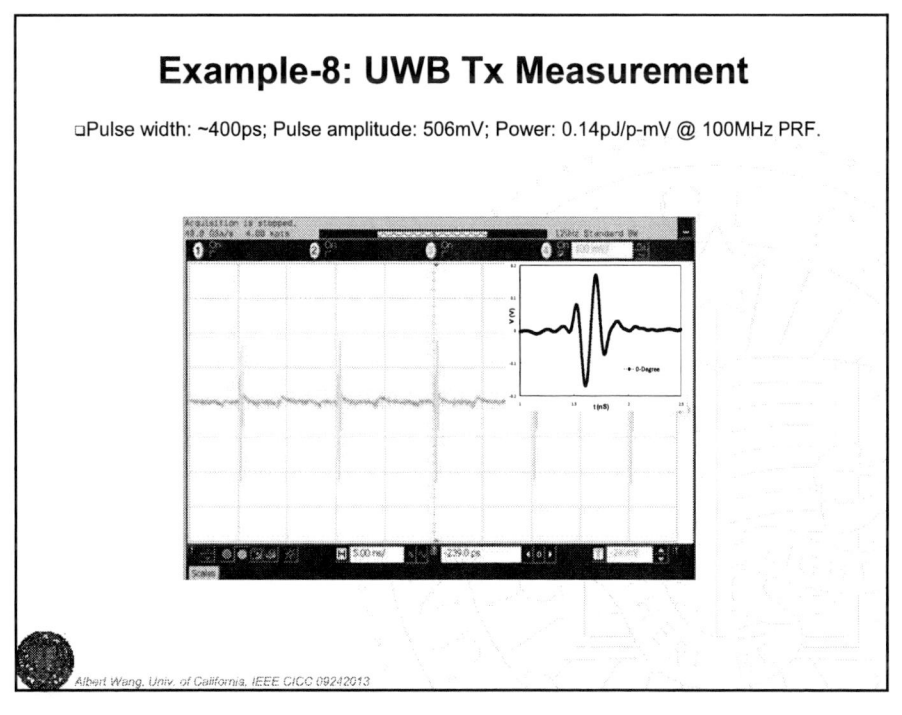

# Example-8: UWB Tx+ESD Co-Design by Post SIM

❏ No-ESD, with ESD, ESD co-design

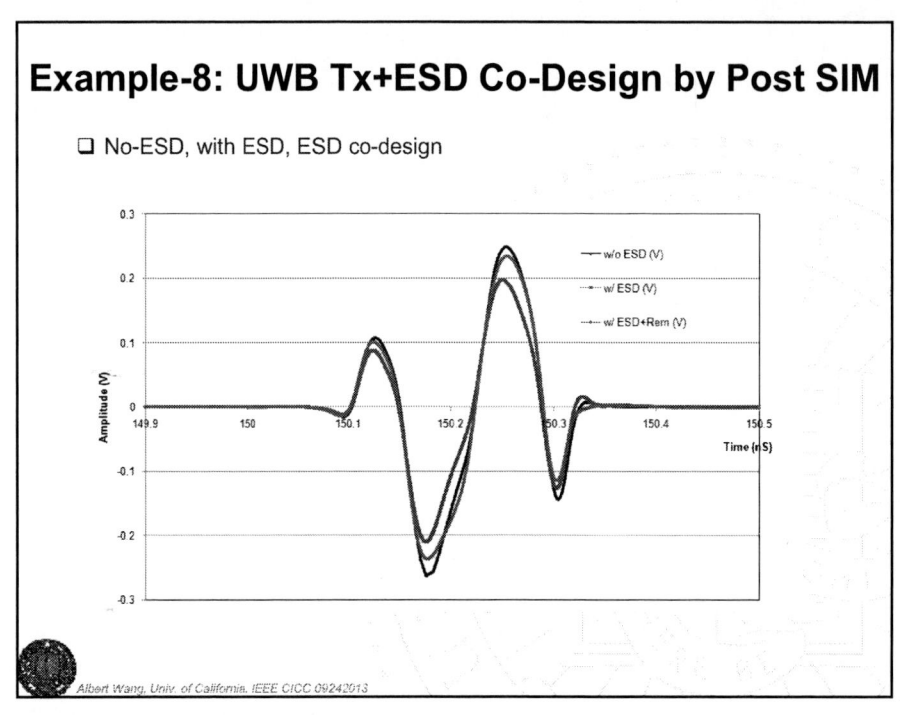

# Example-8: UWB Tx+ESD Co-Design by Testing

❏ Ideal Gaussian, ESD and no ESD

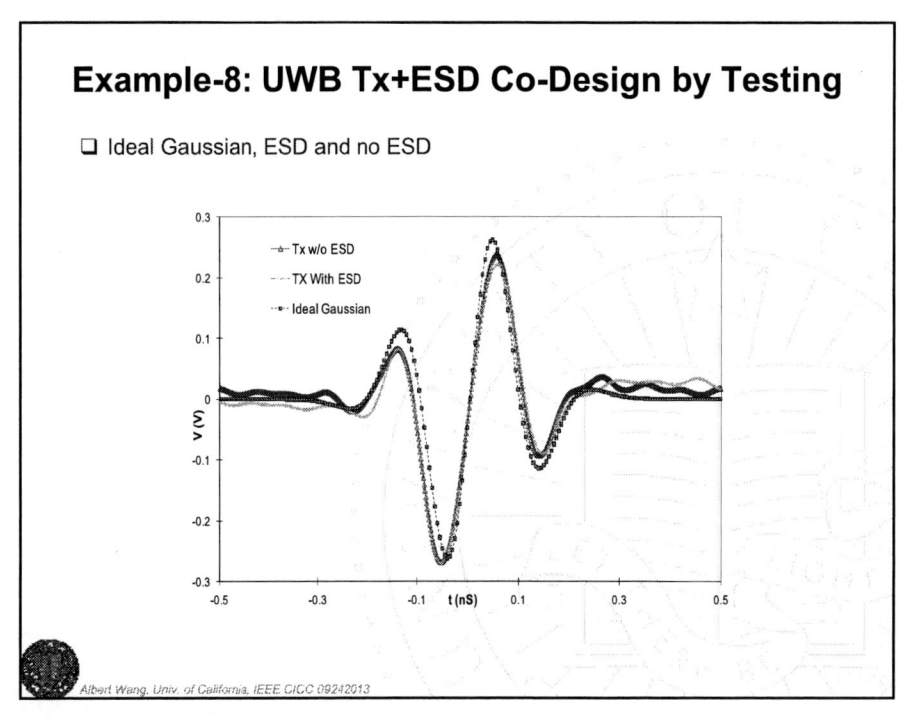

# Example-8: UWB Tx ESD Testing

❑Complete full-chip ESD testing using TLP
- ✓ I/O-GND, I/O-$V_{DD}$ and $V_{DD}$-GND
- ✓ Positive & negative

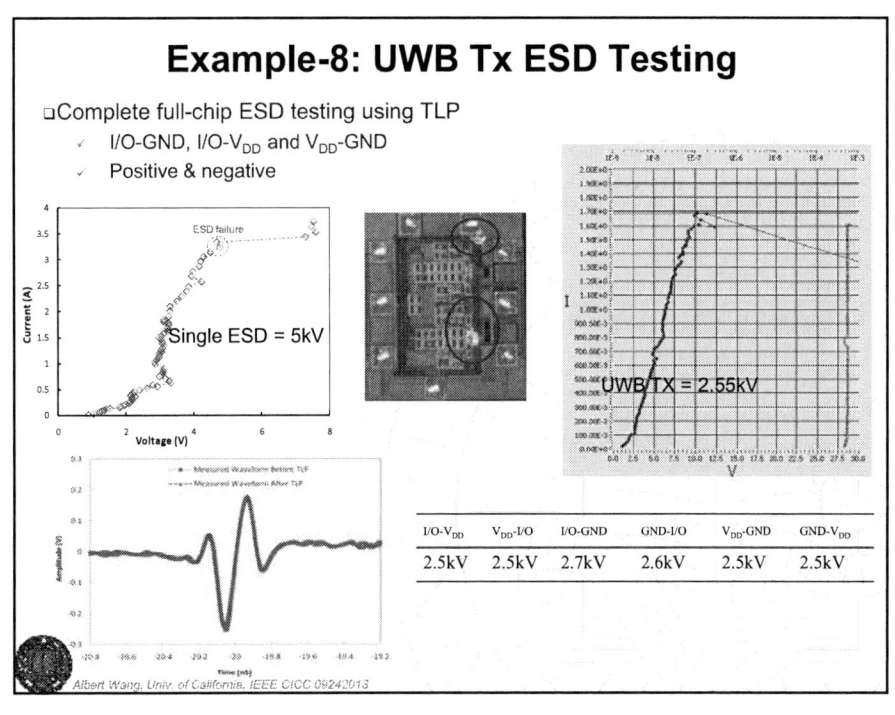

| I/O-$V_{DD}$ | $V_{DD}$-I/O | I/O-GND | GND-I/O | $V_{DD}$-GND | GND-$V_{DD}$ |
|---|---|---|---|---|---|
| 2.5kV | 2.5kV | 2.7kV | 2.6kV | 2.5kV | 2.5kV |

# Example-9: UWB LNA + ESD Co-Design

❑A single-stage, cascode, shunt-series FB

❑Cascode M2 for isolation

❑Shunt-series FB covers 3.1-10.6GHz

❑Shunt peaking inductor Ld compensate gain attenuation

❑Full 8kV ESD protection

❑0.18µm RF CMOS, 0.8mm²

❑Design splits:
- ✓ LNA w/o ESD;
- ✓ LNA w/ diode ESD;
- ✓ LNA w/ diode string;
- ✓ LNA w/ ggNMOS.

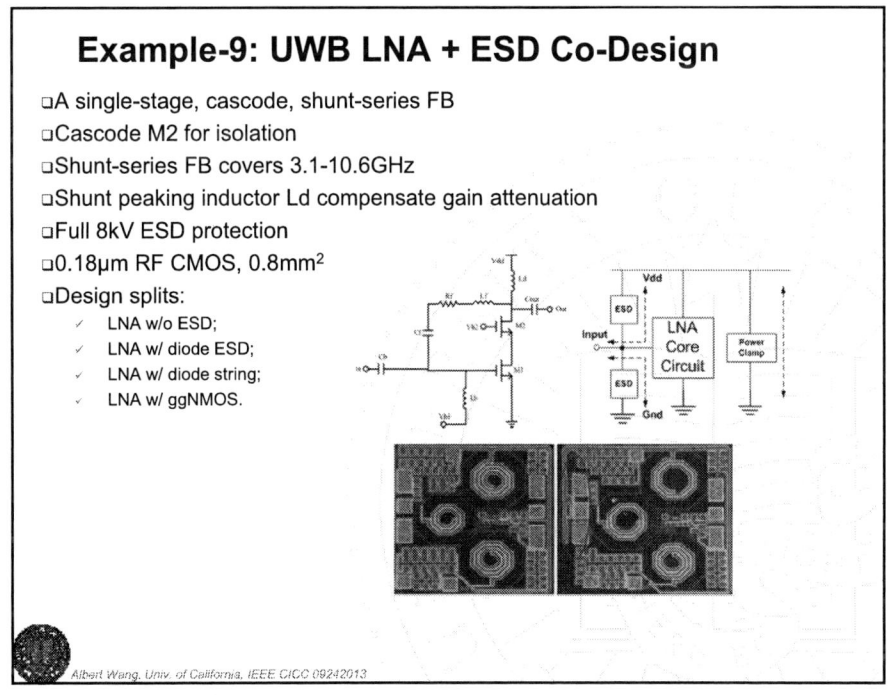

978-1-4673-6145-3/13 $31.00 © 2013 IEEE

# Example-9: UWB LNA + ESD Measurement

❑$S_{21}$~10.6dB, good flatness of ~3.64%/GHz across 3.1-10.6GHz
❑ggNMOS ESD causes gain drop of 45.9%

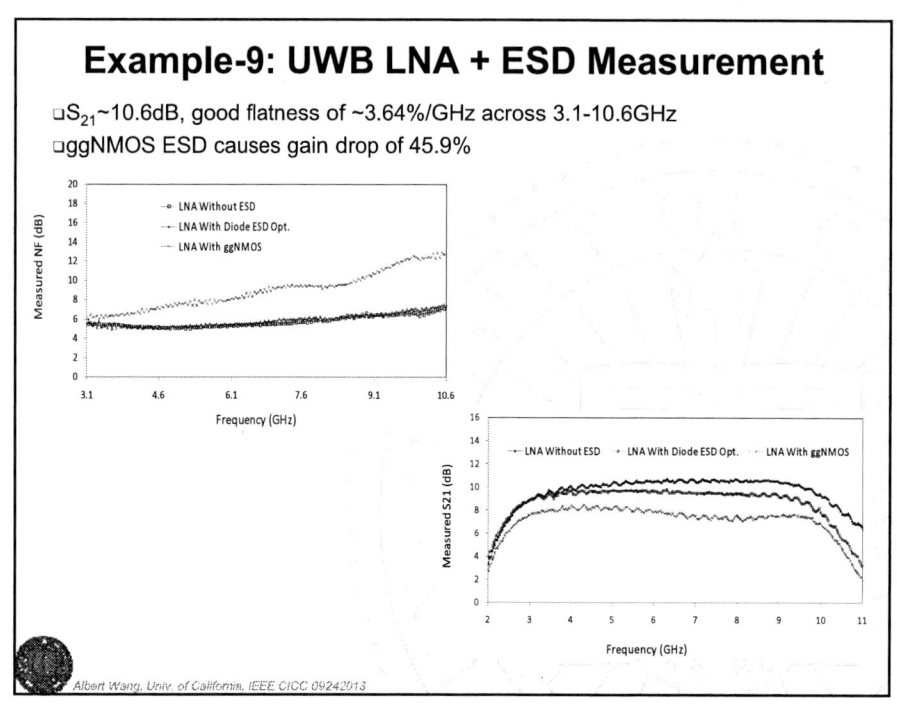

# Example-9: UWB LNA ESD Testing

❑ESD testing by TLP

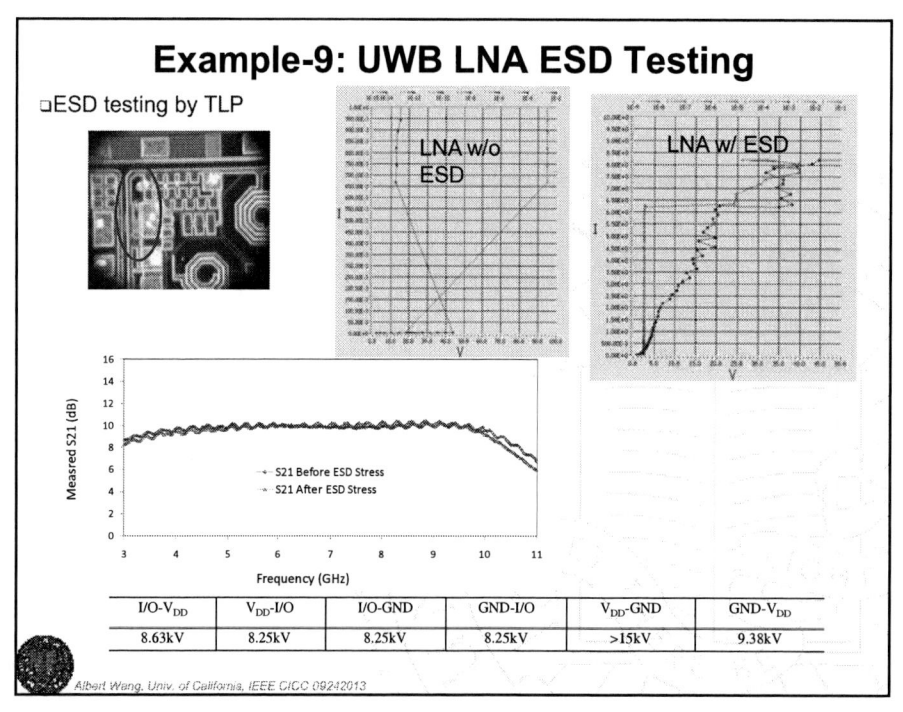

| I/O-$V_{DD}$ | $V_{DD}$-I/O | I/O-GND | GND-I/O | $V_{DD}$-GND | GND-$V_{DD}$ |
|---|---|---|---|---|---|
| 8.63kV | 8.25kV | 8.25kV | 8.25kV | >15kV | 9.38kV |

# Field-Programmable ESD Design & Example 10

❑ Traditional ESD: everything is fixed after Si done

❑ Need for programmable ESD:
- ✓ Fine-tune in Si to address PVT variation
- ✓ Fine-tune ESD specs for complex system level ESD protection

❑ Novel Field-Programmable ESD Mechanisms
- ➢ SONOS-based ESD structures
- ➢ Nano Crystal Dots based ESD structures
  - ✓ Implemented in CMOS flows

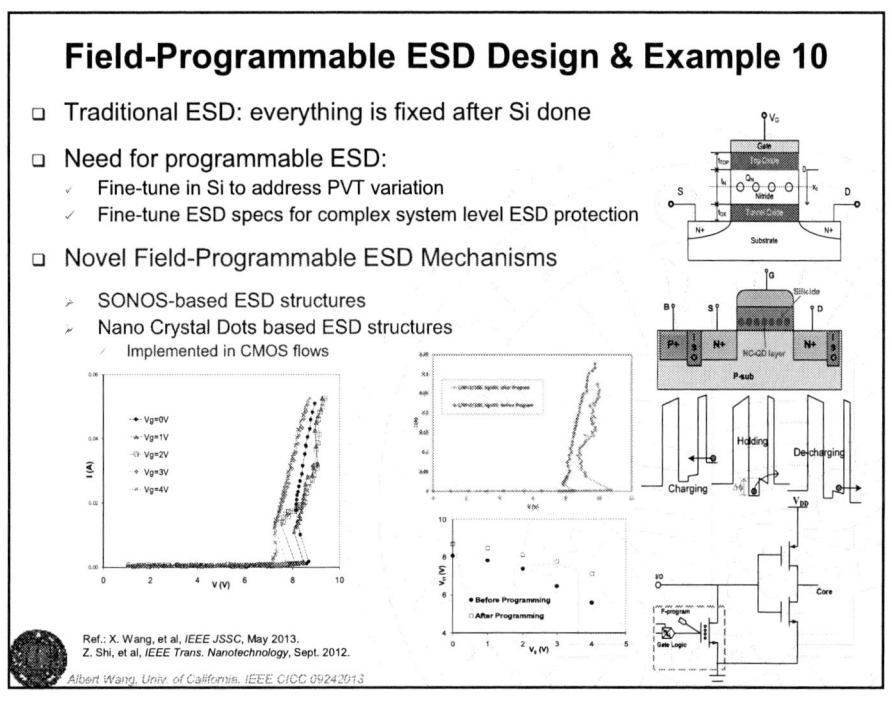

Ref.: X. Wang, et al, *IEEE JSSC*, May 2013.
Z. Shi, et al, *IEEE Trans. Nanotechnology*, Sept. 2012.

Albert Wang, Univ. of California, IEEE CICC 09242013

# Above-IC ESD Protection Design & Example 11

❑ Traditional ESD has embedded PN junctions

❑ ESD-induced parasitic effects:
- ✓ Leakage, capacitance, mis-triggering, etc.
- ✓ ESD size and layout issues

❑ Above-IC ESD: non-traditional and revolutionary!
- ➢ Nano Crossbar ESD Array
  - ✓ Nano phase switching mechanism
  - ✓ Dispered local tunneling model
  - ✓ Fabricated in IC backend
  - ✓ $I_{leak}$ <2pA, $I_{t2}$ >8A for 5x5 array by TLP
  - ✓ Dual-directional ESD discharging
  - ✓ Behavior model for SPICE simulation

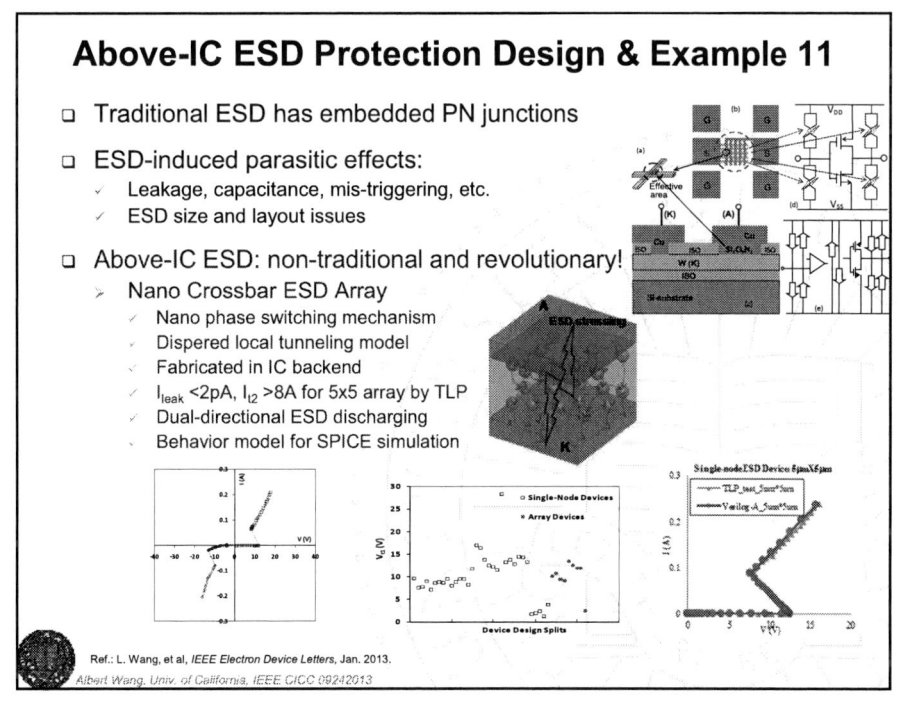

Ref.: L. Wang, et al, *IEEE Electron Device Letters*, Jan. 2013.

Albert Wang, Univ. of California, IEEE CICC 09242013

# SUMMARY

❑ ESD failure is a killing factor to ICs

❑ On-chip ESD protection required for ICs

❑ Mixed-mode ESD design method for prediction

❑ ESD-IC co-design is important & feasible!

❑ Novel ESD design concept is the future

# REFERENCES

- X. Wang, et al, "Post-Si Programmable ESD Protection Circuit Design: Mechanisms and Analysis", *IEEE JSSC*, pp.1237, May 2013.
- L. Wang, et al, "Dual-Directional Nano Crossbar Array ESD Protection Structures", *IEEE Electron Device Letters*, pp.111, Jan. 2013.
- Z. Shi, et al, "Programmable on-Chip ESD Protection Using Nano Crystal Dots Mechanism and Structures", *IEEE Trans. Nanotechnology*, pp.884-889, September 2012.
- J. Liu, et al, "Field Programmable SONOS ESD Protection Design", *Proc. IEEE CICC*, pp. 1-4, 2012.
- X. Wang, et al, "Recent Advances in ESD Protection Design for Ultra Wideband High Data Rate ICs", *Proc. IEEE ISCAS*, 2012.
- J. Liu, et al, "Design and Analysis of Low-Voltage Low-Parasitic ESD Protection for RF ICs in CMOS", *IEEE JSSC*, pp.1100, May 2011.
- L. Lin, A. Wang, et al, "Novel Nanophase-Switching ESD Protection", *IEEE EDL, V32, N3*, pp.378-380, March 2011.
- X. Wang, A. Wang, et al, "FCC-EIRP-Aware UWB Pulse Generator Design Approach", *Proc. IEEE ICUWB*, pp.592-596, 2009.
- X. Wang, A. Wang, et al, "ESD-Protected Power Amplifier Design in CMOS for Highly Reliable RF ICs", *IEEE Trans. Industrial Electronics, V58, N7*, pp2736-2743, July 2011.
- H. Xie, A. Wang, et al, "A 52mW 3.1-10.6GHz Fully Integrated Correlator for IR-UWB Transceivers in 0.18μm CMOS", *IEEE Trans. Industrial Electronics,V57, N5*, pp. 1546-1554, May 2010.
- L. Lin, A. Wang, et al, "Whole-Chip ESD Protection Design Verification by CAD", *Proc. EOS/ESD Symp*, pp.28-37, 2009
- X. Guan, et al, "ESD-RFIC Co-Design Methodology", Invited, *Proc. IEEE RFIC*, pp467-470, 2008.
- A. Wang, et al, "A Review on RF ESD Protection Design", *IEEE Trans. Electron Devices, V2, N7*, p.1304, July 2005.
- H. Xie, et al, "A New Low-Parasitic Polysilicon SCR ESD Protection Structure for RF ICs", *IEEE EDL*, pp.121, February 2005.
- R. Zhan, et al, "ESDInspector: A New Layout-level ESD Protection Circuitry Design Verification Tool Using A Smart-Parametric Checking Mechanism", *IEEE Trans on CAD of Integrated Circuits and Systems, V23, N10*, p.1421, Oct. 2004.
- G. Chen, et al, "Characterizing Diodes For RF ESD Protection", *IEEE Electron Device Letters, V25, N5*, p.323, May 2004.
- A. Wang, *On-Chip ESD Protection For Integrated Circuits*, Kluwer Academic Publishers, Boston, ISBN: 0-7923-7647-1, 2002.
- A. Wang, et al, "ESD Protection Design for RF Integrated Circuits: New Challenges", *Invited, IEEE CICC*, p.411, 2002.
- A. Wang, "A Study of Parasitic Effects of ESD Protection on RF ICs", *IEEE Trans. Microwave Theory & Tech.*, p.393, Jan. 2002.
- H. Feng, et al, "A Mixed-Mode ESD Protection Circuit Simulation-Design Methodology ", *IEEE J. Solid-State Circuits*, p.995, June 2003.
- R. Zhan, et al "ESDExtractor: A New Technology-Independent CAD Tool For Arbitrary ESD Protection Device Extraction", *IEEE Trans on CAD of Integrated Circuits and Systems, V22, N10*, p.1362, October 2003.
- A. Wang, et al, " An on-Chip ESD Protection Circuit with Low Trigger-Voltage in BiCMOS Technology", *IEEE J. Solid-State Circuits, V36, N1*, p.40, January 2001.

978-1-4673-6145-3/13 $31.00 © 2013 IEEE      396

*Educational Session 2. CICC 2013, San Jose, CA, September 24th 2013*

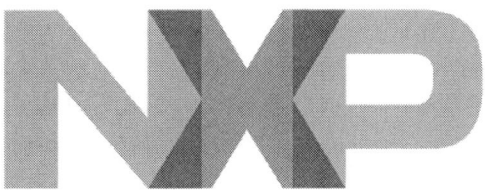

## characterization of matching, variability
## and low-frequency noise
## for mixed-signal technologies

## Hans Tuinhout

NXP Semiconductors    Operation / Design Platforms
High Tech Campus 37, Eindhoven, the Netherlands
email: hans.tuinhout@nxp.com

---

## acknowledgement

this lecture was made with help of many colleagues at
**Philips** &  Semiconductors

**in particular:**

- Nicole Wils
- Maarten Vertregt
- Marcel ("the") Pelgrom
- Maurice Meijer
- Adrie Zegers
- Andries Scholten

and special thanks to Albert Uderzo

**High Tech Campus Eindhoven, the Netherlands**

## hans' perspective (disclaimer)

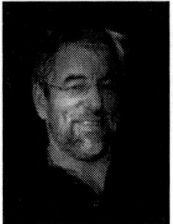

*this lecture is based on more than 20 years
of device matching studies and analog characterization
of Philips' and NXP's mixed-signal IC-technologies*

the focal point of this lecture
is on results coming from many different **characterization studies**
resulting in **an overview of observations** on variability effects
from **many different technology nodes.**

Magnitudes of such effects vary over technology nodes (& fabs)
so most shown numbers (although never in A.U.'s)
should be treated (used) with care.
Verification with your own process owner is always essential !

## theme & objectives

variability of semiconductor device parameters is
one of the toughest barriers to beat on the "more Moore"
shrink path for advanced CMOS technologies

taste
this magic
variability soup

**the term variability
is used for many different effects
and is often a bit mystified...**

## objectives of this lecture:

- clarify definitions
- provide orders of magnitude & rules of thumb
- enhance understanding of variability
- provide realistic quantitative perspectives
- hints on "how to cope with" variability

## outline

- introduction

- random parametric fluctuations ("matching")

- systematic & environmental variability effects
    - distance
    - WPE & STI
    - metal related stress

- low frequency noise variability

- wrap-up

---

## the analog signal playing field...

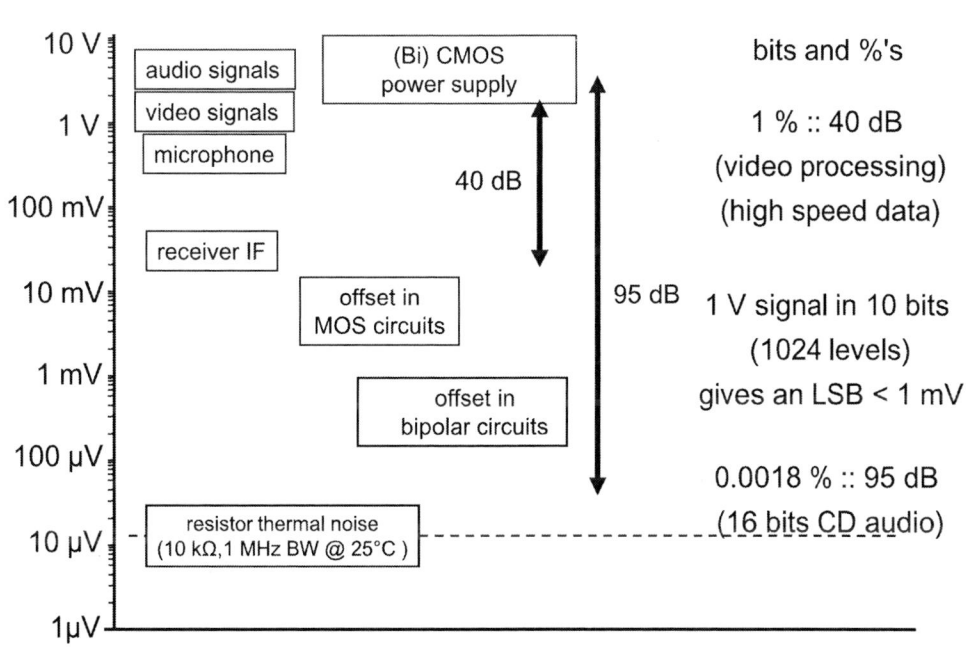

## knock, knock, digital variability comes in...

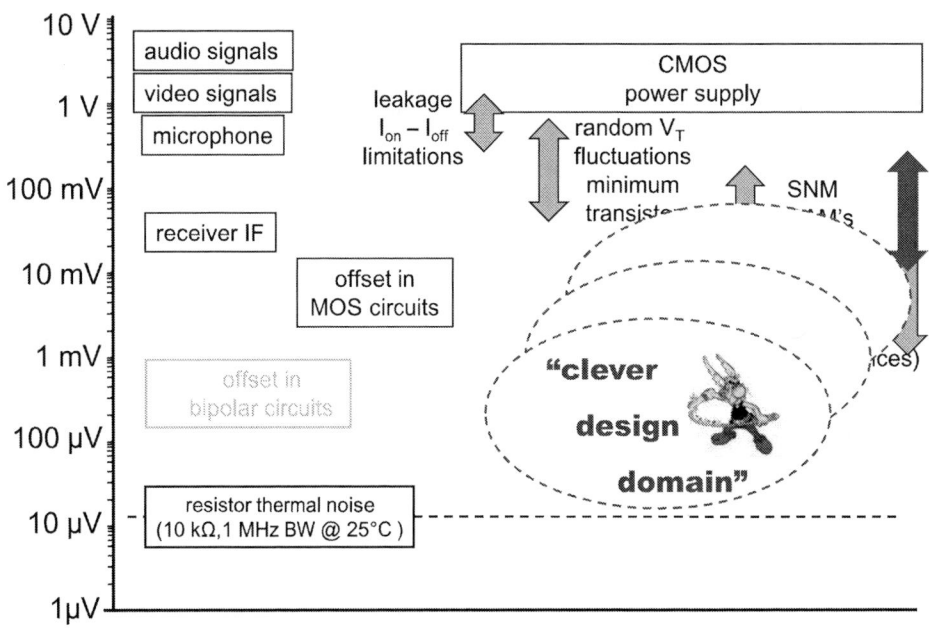

## but also the floor is rising !

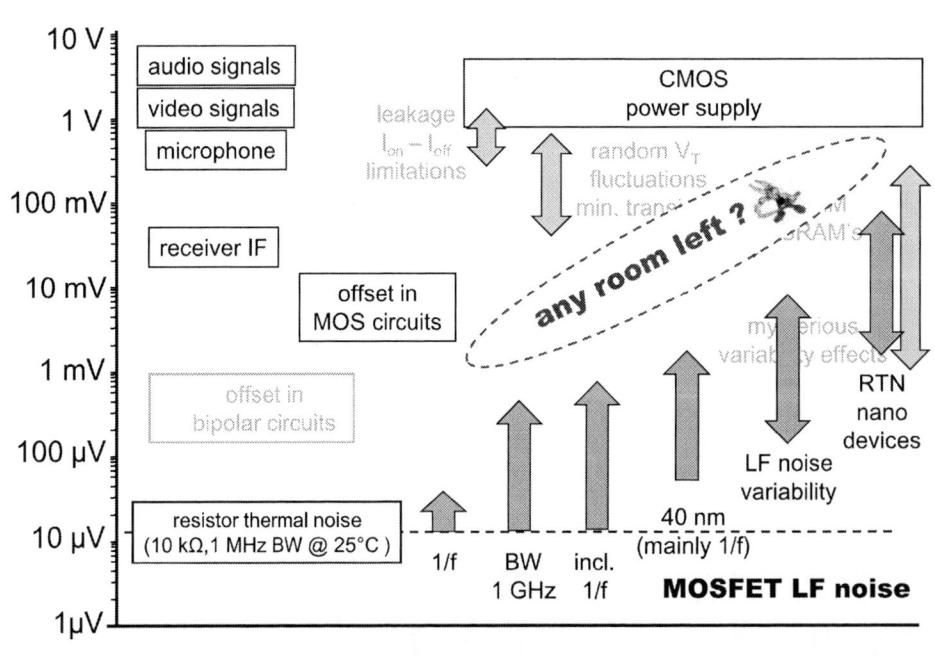

let's begin with
the terminology of
variability ingredients

---

**spread / systematic effects / matching**

## 1. wafer spread & process spread

- fabrication equipment settings (recipes)
- fabrication equipment drift & control within fab
- fab-to-fab differences (different equipment, different culture)
- economics
  - trade-off product performance vs. yield
  - trade-off wafer costs vs. line yield
  - 2nd sourcing fab requirements
- parametric spec ranges; n-sigma corners

## 2. intra-chip systematic effects

- performance differences between not exactly equal devices
- not a-priori recognized by circuit design tools
- incomplete modeling of novel device degradation effects
- "performance prediction inaccuracies"

## 3. matching

- random microscopic device architecture fluctuations

## wafer spread

performance variation of IC components across wafer

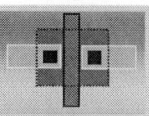

e.g. MOSFET:

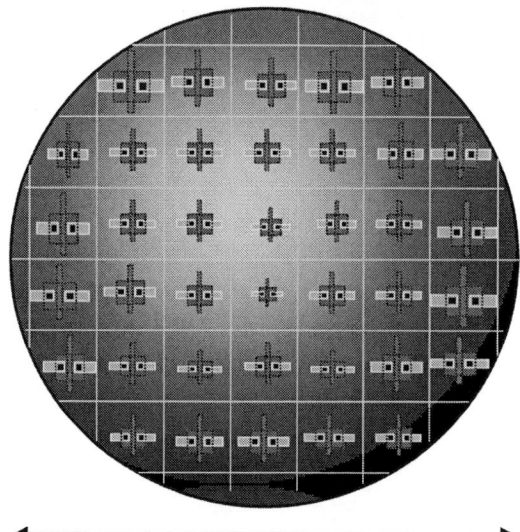

physical variations:

- Critical Dimensions    $\Delta CD$
- layer thickness    $\Delta t$
- furnace Temperatures    $\Delta T$
- uniformity of chemicals    $\Delta C$

**"deterministic parametric gradients"**

*strictly speaking these variations are not random*

100 - 150 - 200 - 300 mm

## process spread

performance variation of IC components
across wafer / wafer-to-wafer / lot-to-lot

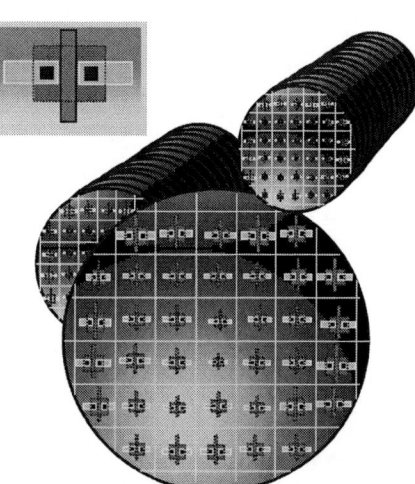

**"global variations"**

physical variations:

- Critical Dimensions    $\Delta CD$
- layer thickness    $\Delta t$
- furnace Temperatures    $\Delta T$
- uniformity of chemicals    $\Delta C$

total electrical variations
incl. wafer-to-wafer & lot-to-lot

currents: ~ 5 - 30 %

voltages: 10 - 100 mV

*these variations are (also) not random*  *but for circuit designers they are...*

## spread / systematic effects / matching

### 1. wafer spread & process spread
- fabrication equipment settings (recipes)
- fabrication equipment drift & control within fab
- fab-to-fab differences (different equipment, different culture)
- economics (politics…)
  - trade-off product performance vs. yield
  - trade-off wafer costs vs. line yield
  - 2nd sourcing fab requirements
- parametric spec ranges; n-sigma corners

### 2. intra-chip systematic effects
- performance differences between not exactly equal devices
- not a-priori recognized by circuit design tools
- incomplete modeling of novel device degradation effects
- "performance prediction inaccuracies"

### 3. matching
- random microscopic device architecture fluctuations

---

## intra-chip device variations

parametric variations across die (reticle) & "ACLV" (Across Chip Line width Variation)

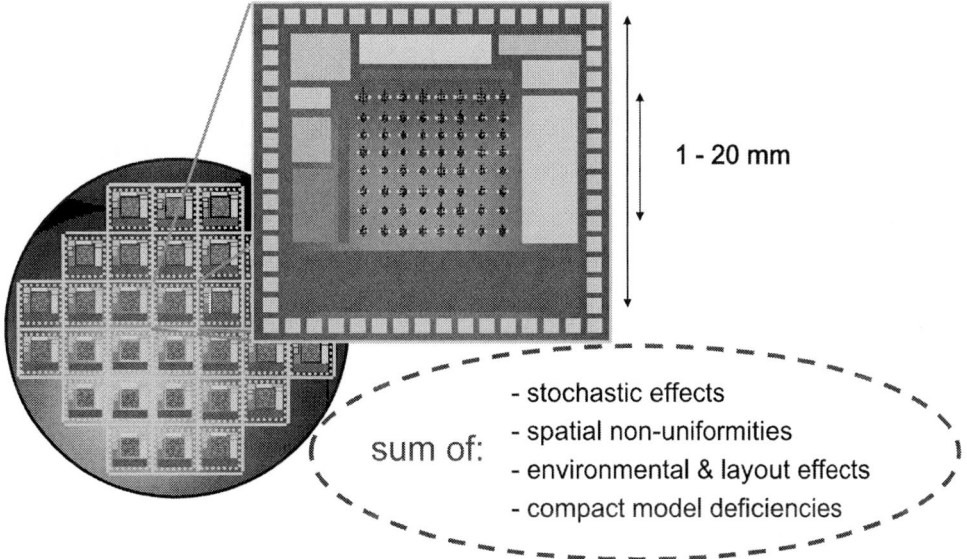

1 - 20 mm

sum of:
- stochastic effects
- spatial non-uniformities
- environmental & layout effects
- compact model deficiencies

**this combinations is often (ab)used as "VARIABILITY"**

## variability = much more than 'spread'

All un~~expected~~ parameter variation between supposedly identical transistors or components or circuit blocks in one circuit intra-die variations

### confusion: "variability" is also (ab)used for other effects

- lithographical printing deviations (feature corner rounding etc.)
- OPC mask manipulations
- on-chip electronic signal integrity
- mechanical stress (by wiring or STI)
- temperature gradients across working chip
- substrate noise
- clock jitter
- NBTI & hot carrier reliability degradation
- not properly (or incompletely) modeled effects

---

## matching problems

# 80 % of the alleged "matching problems" are design- or layout- asymmetry errors

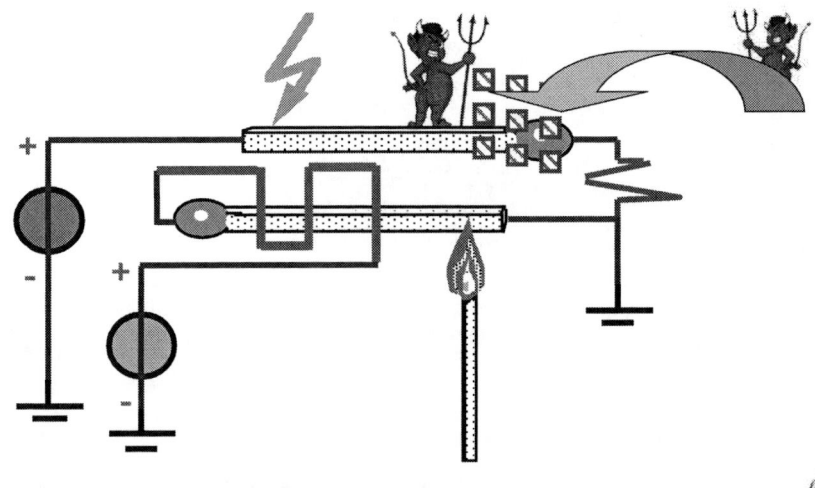

## spread / systematic effects / matching

### 1. wafer spread & process spread
- fabrication equipment settings (recipes)
- fabrication equipment drift & control within fab
- fab-to-fab differences (different equipment, different culture)
- economics
    - trade-off product performance vs. yield
    - trade-off wafer costs vs. line yield
    - 2nd sourcing fab requirements
- parametric spec ranges; n-sigma corners

### 2. intra-chip systematic effects
- performance differences between not exactly equal devices
- not a-priori recognized by circuit design tools
- incomplete modeling of novel device degradation effects
- "performance prediction inaccuracies"

### 3. matching
- random microscopic device architecture fluctuations

---

## spread / deterministic effects / matching

### 1. wafer spread & process spread
- fabrication equipment settings (recipes)
- fabrication equipment drift & control within fab
- fab-to-fab differences (different equipment, different culture)
- economics
    - trade-off product performance vs. yield
    - trade-off wafer costs vs. line yield
    - 2nd sourcing fab requirements
- parametric spec ranges; n-sigma corners

### 2. intra-chip deterministic effects
- performance differences between not exactly equal devices
- not a-priori recognized by circuit design tools
- incomplete modeling of novel device degradation effects
- "performance prediction inaccuracies"

### 3. matching
- random microscopic device architecture fluctuations

"global" variations

"local" variations

**outline**

- introduction

- random parametric fluctuations ("matching")

- systematic & environmental variability effects
  - distance
  - WPE & STI
  - metal related stress

- low frequency noise variability

- wrap-up

---

**relevance**

many analogue and mixed-signal applications are based
on availability of pairs or multiples of 'supposedly identical' circuit elements

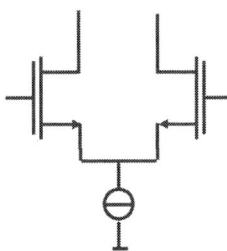

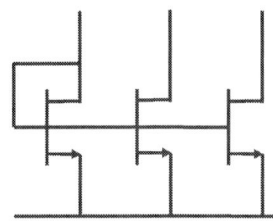

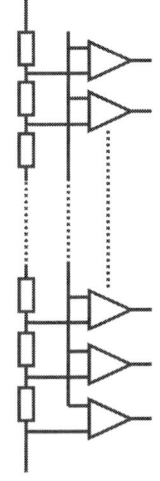

current mirrors
amplifiers,
opamps,
comparators
differential pairs

D/A & A/D converters, PLLs
band gap voltage references
switched capacitors / filtering

## mismatch in A/D converters: yield

yield of an N-bit converter as a function of the comparator mismatch

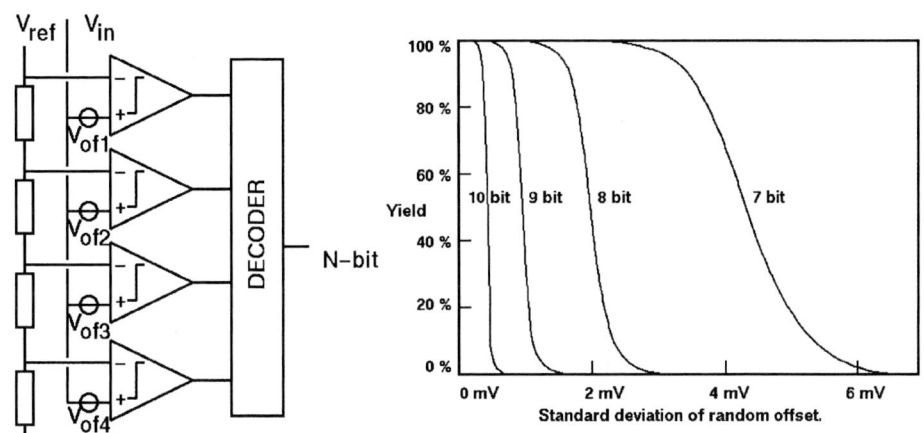

...analog circuit designers

go for the ppm's or micro volts

---

## "...memories are the biggest analogue circuits"

45 nm CMOS node
threshold voltage fluctuation

L ≈ 40 nm (effective)
W ≈ 80 nm (effective)

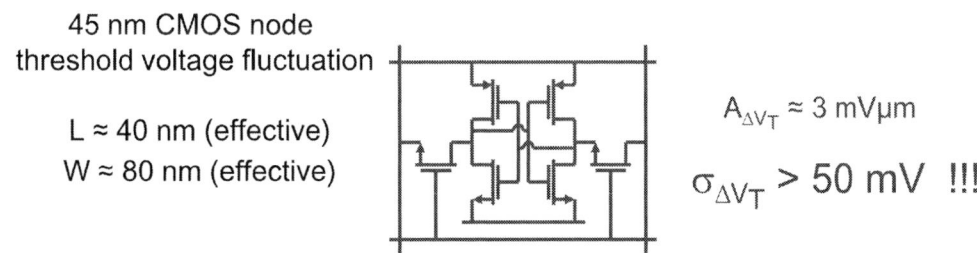

$A_{\Delta V_T} \approx 3$ mVμm

$\sigma_{\Delta V_T} > 50$ mV  !!!

in a multi Mb (embedded) SRAM
a $V_T$ mismatch fluctuation standard deviation
of over 50 mV
means that there will be cells
for which the threshold voltage difference
between two IDENTICAL transistors in ONE CELL
is more than ~~300 mV~~

!!! > 500 mV (+/- 5 sigma)  !!!

## an SRAM yield issue!

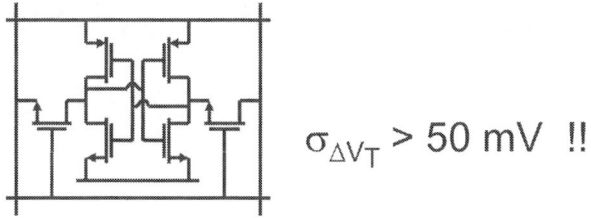

$\sigma_{\Delta V_T} > 50$ mV !!!

- $V_{DD}$ power supply = 1.1 V

- typical $V_T$ = 0.2 to 0.5 V (leakage in the off state vs. speed & stability)

- > 500 mV $V_T$ difference between 2 transistor : noise margin problem

- Static Noise Margin (SNM) is a major design issue for SRAMs

- **random phenomenon: yield issue (looks like defects ! )**

yield of SRAMs is limited by random parametric fluctuations

## but also a digital timing issue

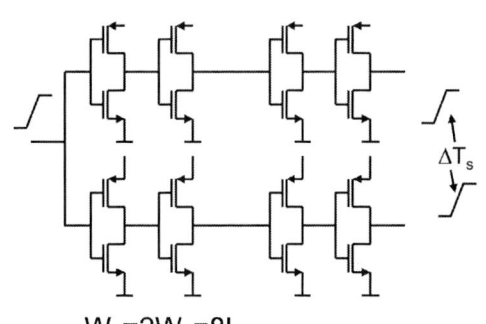

$\Delta T_s$

jitter / clock period
goes up
from 0.16 % to 18 % (6.4 %)
(1 sigma!)

$W_p = 2W_n = 8L_{min}$

| circuit simulation | 0.25 μm | 0.18 μm | 0.13 μm | 90 nm | 65 nm |
|---|---|---|---|---|---|
| $\sigma_{\Delta Ts}$    $C_{load}$ = 50 fF | 16 ps | 21 ps | 38 ps | 68 ps | 88 ps |
| $\sigma_{\Delta Ts}$    $C_{load}$ = | 16 ps   50 fF | 16 ps   35 fF | 22 ps   25 fF | 33 ps   20 fF | 32 ps   15 fF |
| clock period | 10 ns | 5 ns | 2 ns | 1 ns | 0.5 ns |

## Monte-Carlo circuit simulation: CMOS 90 nm W/L=10/5

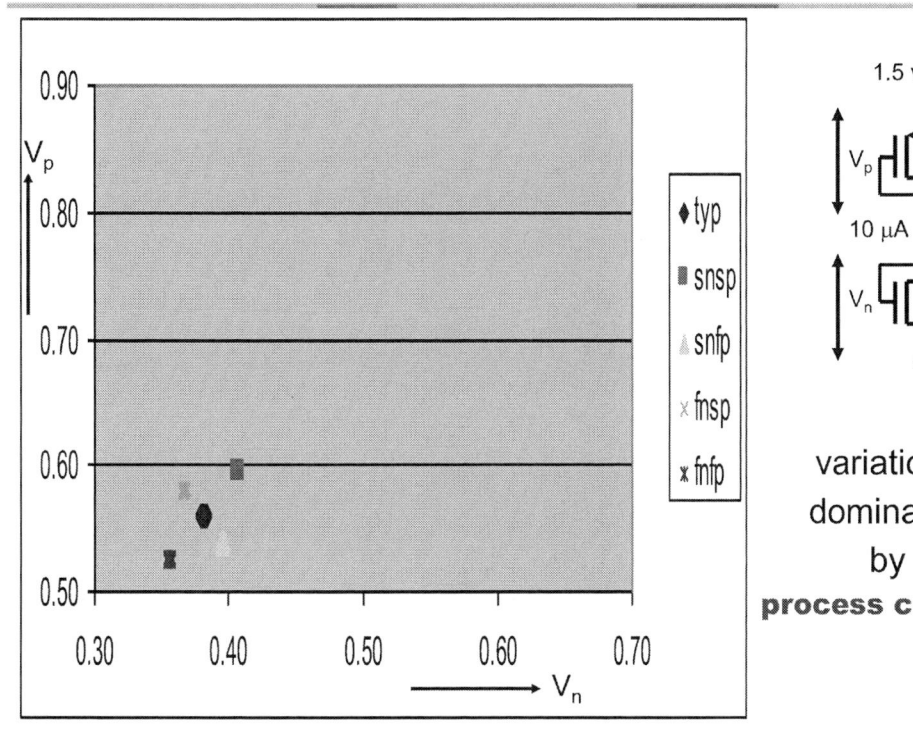

variations
dominated
by
**process corners**

## Monte-Carlo circuit simulation: CMOS 90 nm W/L=1/0.5

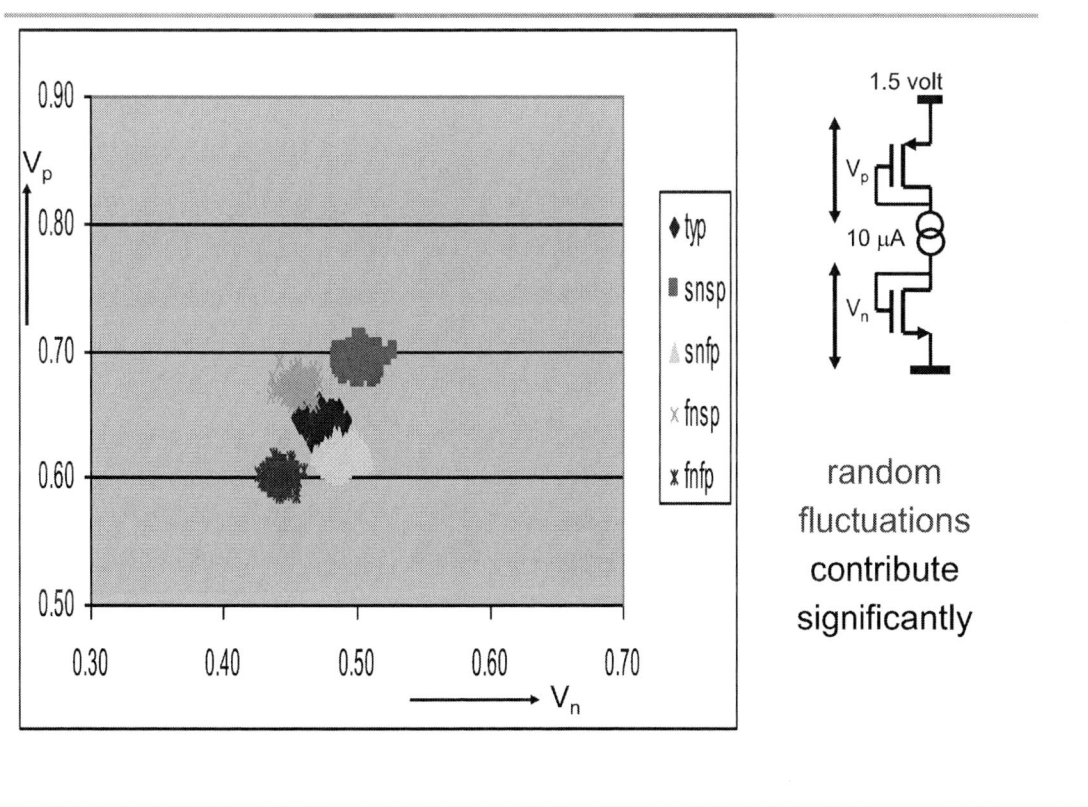

random
fluctuations
contribute
significantly

## Monte Carlo circuit simulation: CMOS 90 nm W/L=0.2/0.1

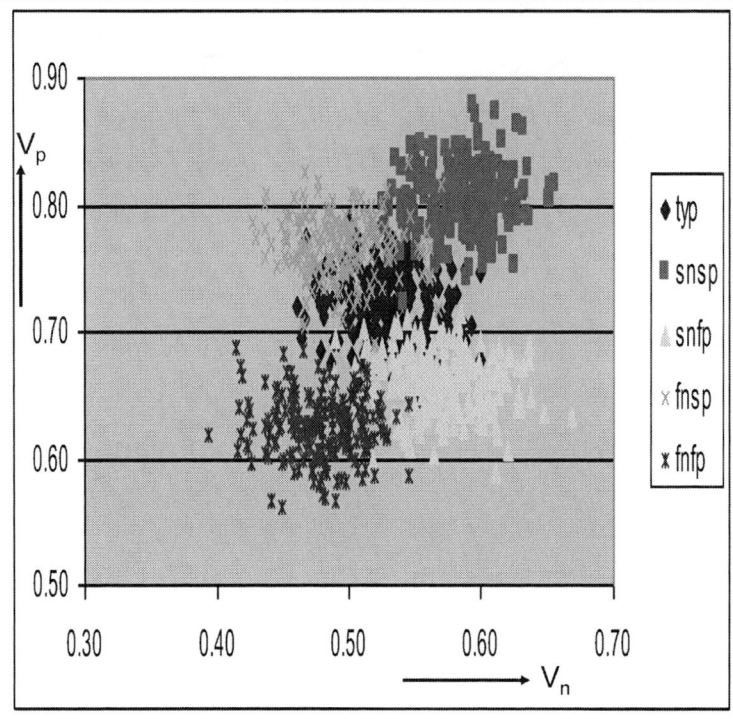

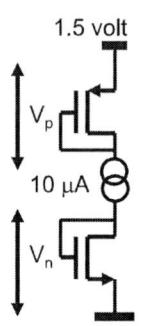

1.5 volt

$V_p$

10 µA

$V_n$

random
fluctuations
**> 100 mV**
major portion of
total
variations window

## Monte Carlo circuit simulation: CMOS 90 nm W/L=0.2/0.1

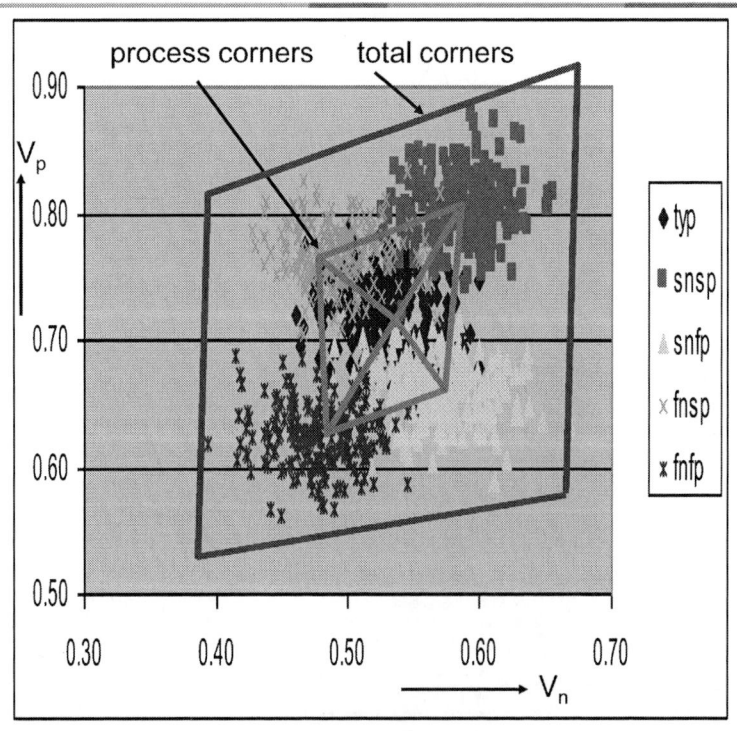

process corners    total corners

process
corners
vs.
total
corners

random fluctuations
dominate over
process corners
for sub 100nm
technology nodes

**matching**

*the basis of this soup is made from random local device variations caused by stochastic microscopic device architecture fluctuations*

---

**matching**

matched pair

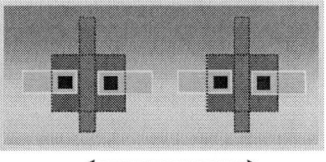

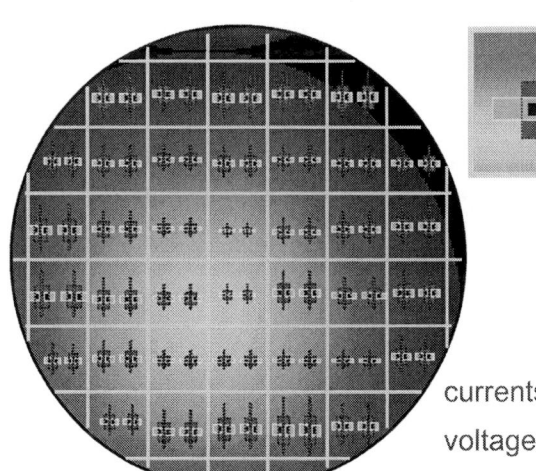

10 - 100 µm   e.g. memory cell,
band gap, opamp, etc.

electrical differences:

currents: ~ 0.1 - 2 %

voltages:  200 µV - 2 mV for large devices

(up to 100 mV for small devices)

spacing small enough to mitigate spatial parametric variation

differences caused by STOCHASTIC (random) effects
related to microscopic device architecture fluctuations

**atomistic effects: counting atoms**

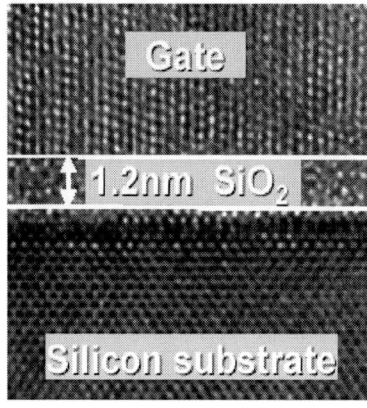

Source: Intel website

we can not only count the silicon atoms but the number of
device performance defining construction elements
(dopant atoms, fixed charges, interface states)
is becoming countable and subject to random variations

**random dopant fluctuations (RDF)**

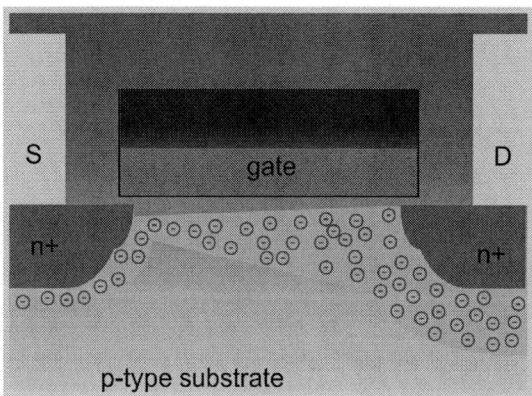

MOSFET threshold voltage ($V_T$) is determined by

- gate oxide thickness
- dopant concentration of depleted region
- (oxide charges, interface states, work function of gate)

## number of active dopants in a transistor

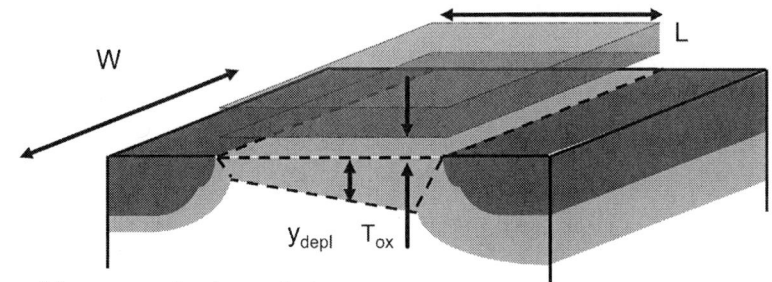

### in a 40 nm node transistor

L = 0.04 μm ; W = 0.08 μm

$y_{depl} :: \dfrac{1}{\sqrt{N_a}} \approx 0.02$ μm

$N_a = 1.5 \times 10^{18}$ cm$^{-3}$ (not uniform)

$N_{act,depl} \approx 100$ dopant atoms

( + $3.2 \times 10^6$ Si-atoms)

RDF (Poisson statistics) : $1/\sqrt{N_{act,depl}} = 10$ *(= 10 %, 1 sigma)*

similar calculations can be made with charges, interface states, grain boundaries, work function fluctuations, etc.

## MOSFET RDF variability model

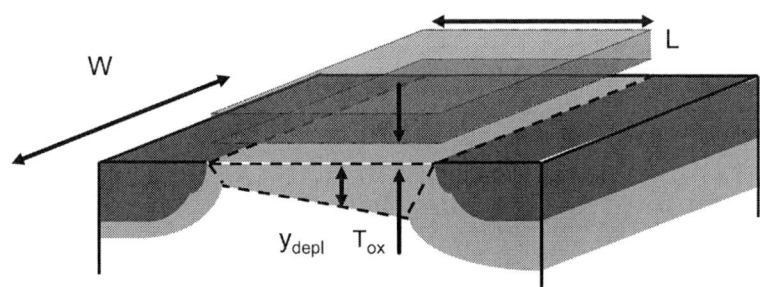

$$V_T - V_{FB} = \frac{Q_{bulk}}{C_{gate}} = \frac{qN_a WL y_{depl}}{C_{ox} WL}, \quad q = 1.6 \times 10^{-19} C$$

$$y_{depl} = \sqrt{2\varepsilon_o \varepsilon_{si} \phi_b / qN_a}, \quad C_{ox} = \varepsilon_o \varepsilon_{ox} / T_{ox}$$

$$\sigma_{VT} \propto \frac{T_{ox} \sqrt[4]{N_a}}{\sqrt{W \times L}} \quad \Rightarrow \quad \sigma_{\Delta VT} = \frac{A_{\Delta VT}}{\sqrt{W \times L}}$$

matched pairs
⇒ factor $\sqrt{2}$

practical rule of thumb:
$A_{\Delta VT} \approx$
1 to 2 mVμm/nm $T_{ox}$

## area scaling: theory of 2 μm CMOS

**(Pelgrom et al. IEEE J-SSC, 1989)**

still holds for 65 nm

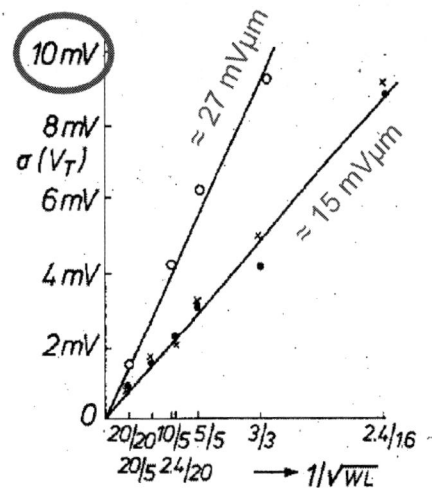

(Pelgrom meant $\sigma_{\Delta VT}$ in this graph...)

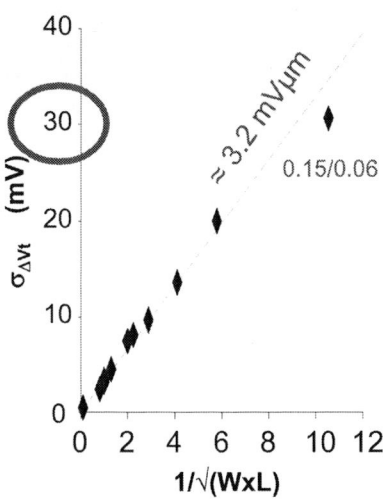

- \> 500x smaller devices
- 20x thinner $T_{ox}$

---

## CLT & parameter fluctuations

generalized application of the Central Limit Theorem:

1. random parametric variations are caused by single events of device architecture fluctuations

2. the effect of a single event on a parameter is so small that contributions of many single events can be summed

3. the effects have a correlation distance that is much smaller than the active area of the devices

when random parametric fluctuations are caused by independent random fluctuations of microscopic device construction elements (e.g. dopant atoms)

$$\sigma_{\Delta P} = \frac{A_{\Delta P}}{\sqrt{(W \times L)}}$$

$A_{\Delta P}$ is characteristic for microscopic device architecture

## first rule of matching: large devices match better

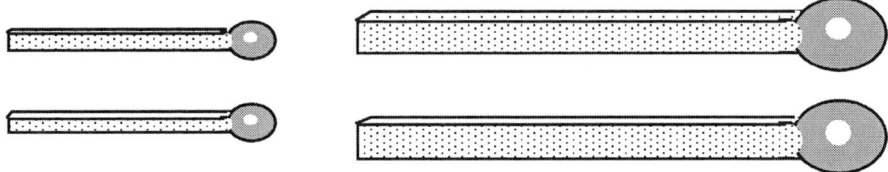

true stochastic mismatch is caused by random fluctuations
of device properties (doping, charges, grains, etc.)

these 'average out'
when the active areas of the matched components are enlarged.

universal area scaling law: $\sigma_{\Delta P} = \dfrac{A_{\Delta P}}{\sqrt{(W \times L)}}$

## Pelgrom plot

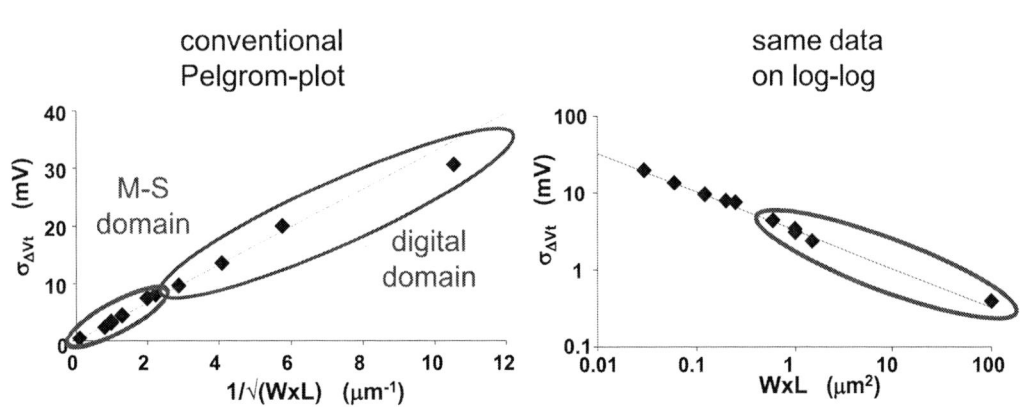

theory also holds for BJTs, MOSFETs, Rs, Cs
(when the CLT assumptions are met)

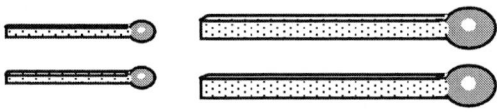

*** and this is what to expect in contemporary (40 nm) M-S technologies

---

**an example for 40 nm technology node**

**our characterization approach:**

measure & characterize parametric variations
of mismatch, spread & variability effects
**with matched pairs** across full 300 mm wafers

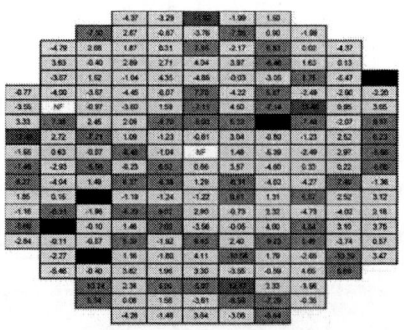

## the ideal MOSFET matched pair test structure

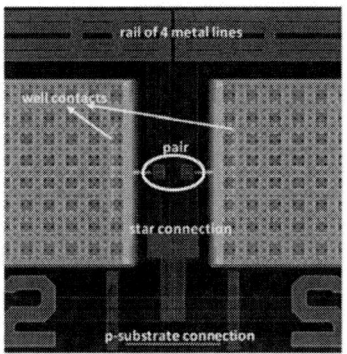

## example Pelgrom plots 40 nm CMOS

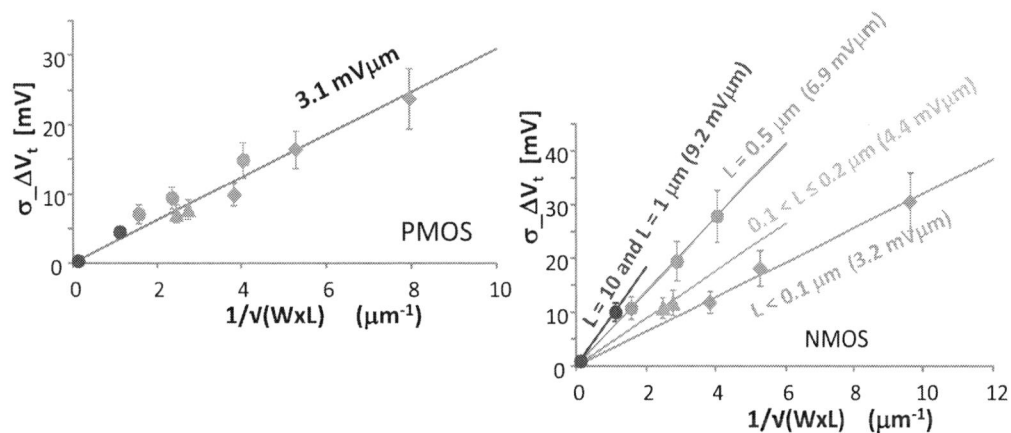

- clear **length effect** for NMOS (minor effect for PMOS)
- A-factor of long NMOS devices much higher

series connection of short transistors matches better than long device !

***

> hah, this magic will make them think it's the end of Pelgrom's good old theory...

CLT requirements are not met in this case!
heavy pocket (halo) implants (also used for $V_T$ adjust)
- non-uniform doping profile along the channel
- highest non-uniformity for long transistors
- potential barriers @ S&D
- not all disturbances have equal impact on mismatch

## modeling: length dependent A-factor

same (NMOS) data presented in another way

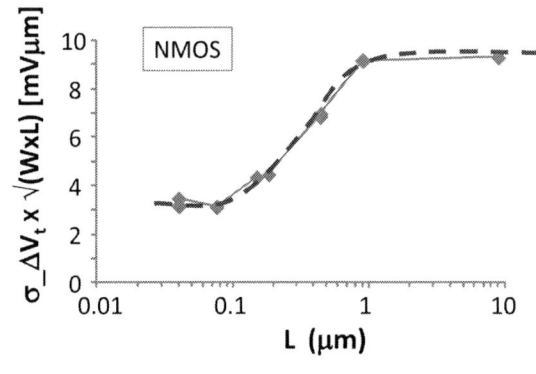
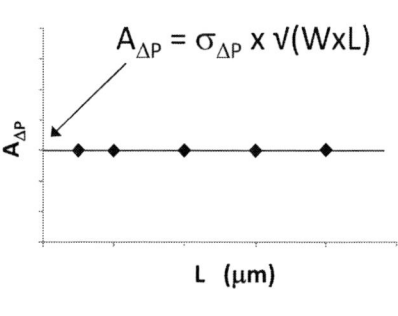

$$A_{\Delta P} = \sigma_{\Delta P} \times \sqrt{(W \times L)}$$

- the $A_{\Delta VT}$ increases for L > 0.1 µm and levels off for L > 1 µm

  - in PDK (compact model) described by a heuristic A=f(L)

  - the A-factor ≈ 3 mVµm for short (digital devices)

    (remember the SRAM example?)

## $I_{on}$ mismatch fluctuations

matching: $I_{on}$ mismatch area scaling (+ comparison with other foundry)

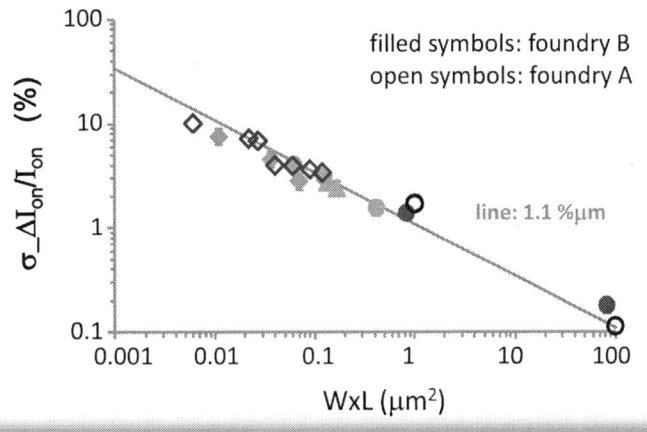

filled symbols: foundry B
open symbols: foundry A

line: 1.1 %µm

**≈ 1 %µm
is again
a useful number
to remember!**

- only minor L- effect for $I_{on}$ mismatch fluctuations

    (drain profile engineering is focused on optimal $I_{on}$ performance)

- remarkably good correspondence between two foundries

- this is good news for IP-library builders (and the digital designers)

...

compact models
probably don't describe
mismatch fluctuations
very well...

## intermezzo: mismatch analysis

N x left transistor:

N x right transistor:

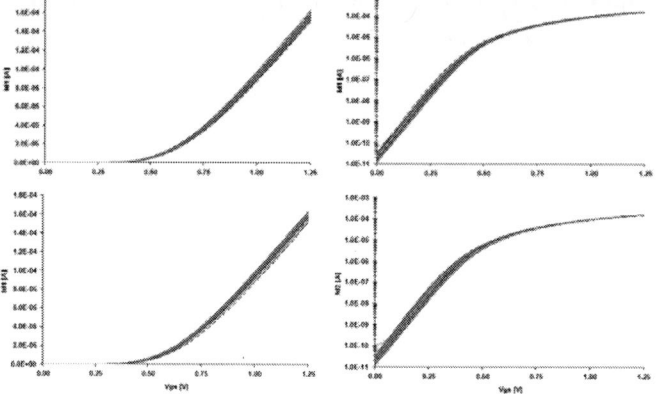

> ➤ parameter extractions
>> ➤ $V_{tlin}$, $V_{tsat}$, $V_{tmaxgm}$, fixed current $V_T$, $V_{T3pt}$, K' (β), $I_{on}$, $I_{off}$, SS, DIBL, $g_{DS}$
>> ➤ all use the exactly the same I-V data points (& interpolations) !

> ➤ fluctuation sweeps (mismatch signatures)

---

## intermezzo: mismatch sweep

• complete V-sweep: drain current of $T_1$ and $T_2$ measured ($V_{ds}$=50 mV)

• calculate drain current mismatch

for each bias point: $\dfrac{\Delta Id}{Id} = \dfrac{(I_d(T_1) - I_d(T_2))}{(I_d(T_1) + I_d(T_2))}$ X 200%

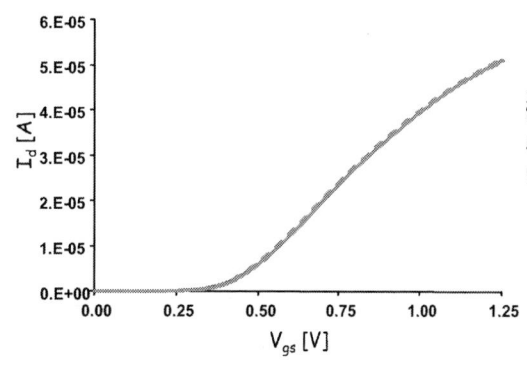

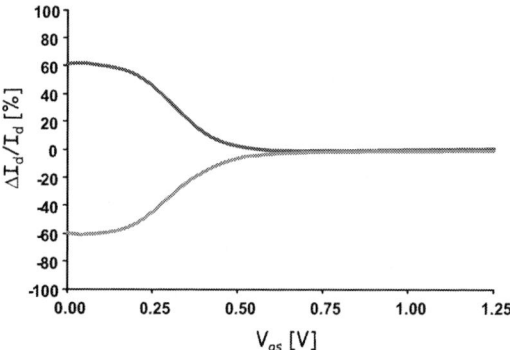

• position 1
• position 2

## intermezzo: mismatch fluctuation sweep

➡ N mismatch sweeps (N = population size = 119)

➡ mismatch median (μ) and mismatch fluctuation (σ) sweeps

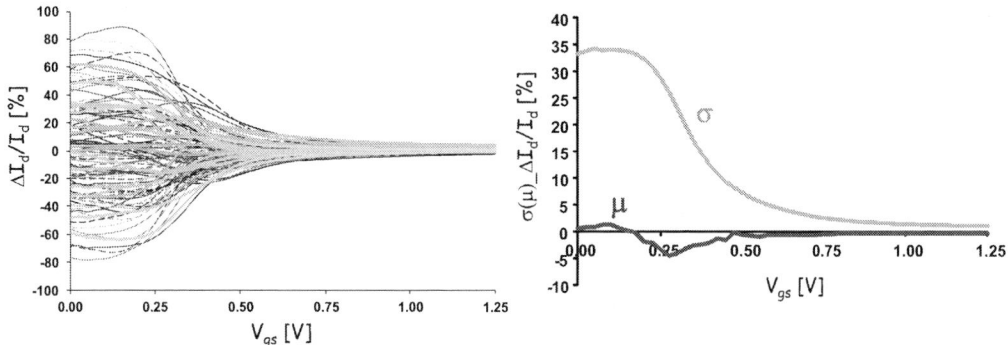

**note** : same $I_d$-$V_{gs}$ data are used to calculate (mismatch) parameters
such as all sorts of $V_T$'s, β's, $I_{on}$, $I_{off}$, subthreshold slope, DIBL, etc.

## measured vs. simulated currents & fluctuations

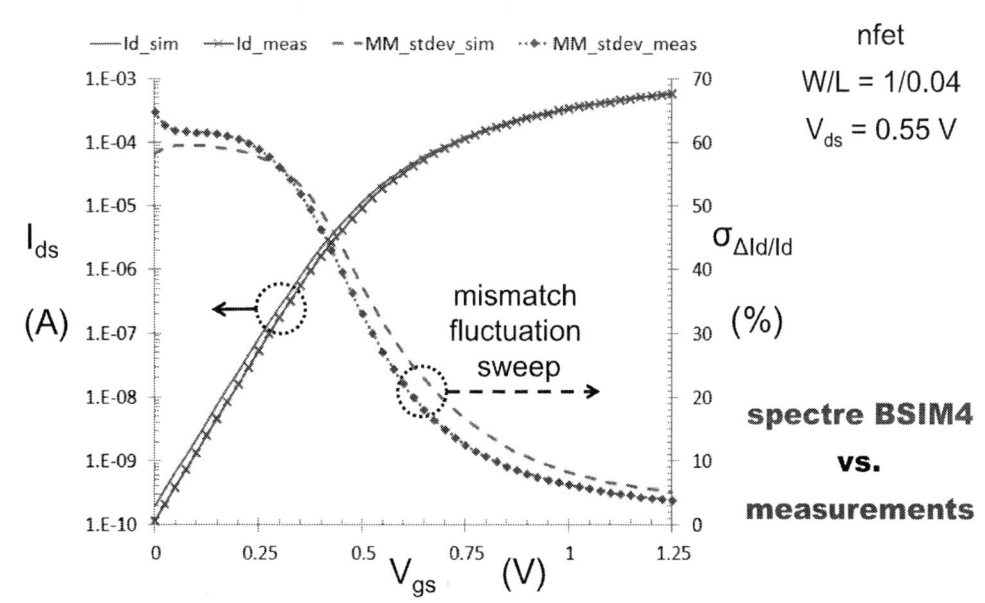

## measured vs. simulated currents & fluctuations

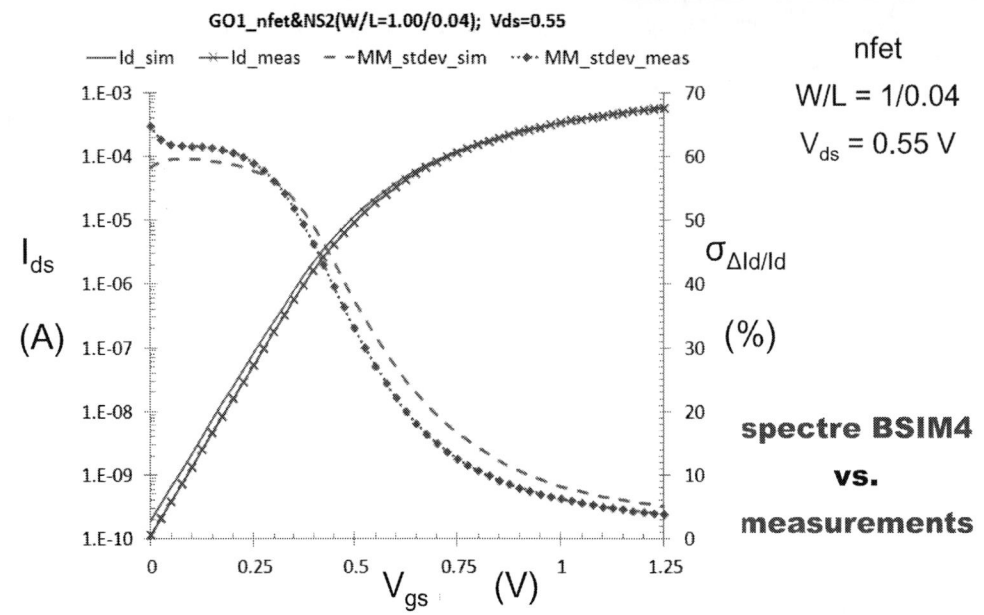

**nfet**

W/L = 1/0.04

$V_{ds}$ = 0.55 V

**spectre BSIM4**

**vs.**

**measurements**

- **excellent agreement** : well modeled (**& processed right on target** !)
- excellent prediction of $I_{on}$ & $I_{off}$ fluctuations (upturn due to GIDL fluctuations)

## analog designers: be careful !

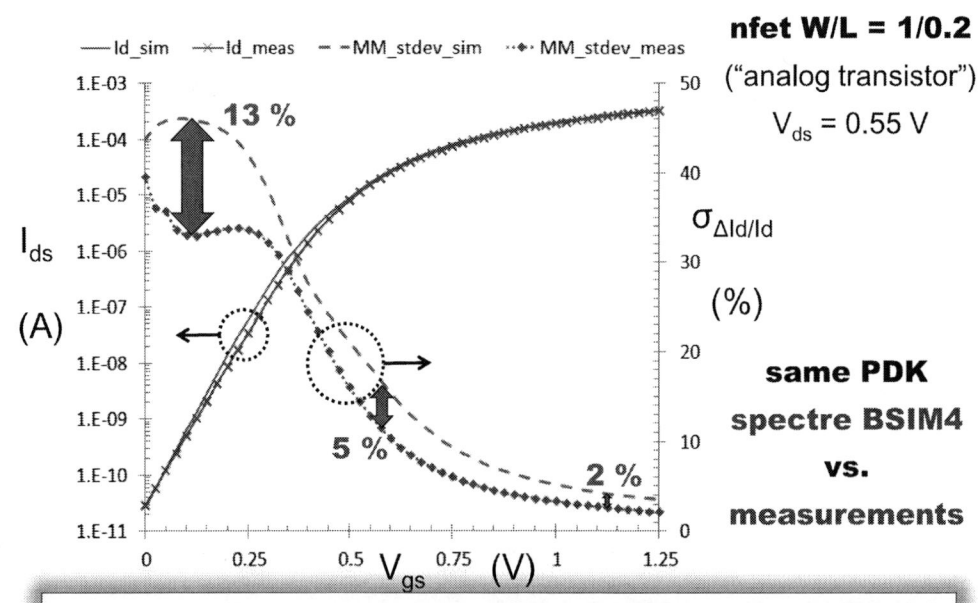

**nfet W/L = 1/0.2**

("analog transistor")

$V_{ds}$ = 0.55 V

**same PDK**

**spectre BSIM4**

**vs.**

**measurements**

- substantial differences between the mismatch fluctuation sweeps (DC=OK)
- low power designs: just above and well below $V_T$ up to 2x deviations

> what's brewing for
> FinFETs, TriGates
> & UTB-SOI

---

## hans' perspective...

- CMOS technology is heading towards 10 nm dimensions
- $V_{dd}$ will reduce to below 0.5 V & low power will be the key
- the industry will go for systems with over $10^9$ transistors

- drastic reductions of variability causes possible ?
  - "with dope no hope" : nano wires, UTB-SOI, hi-K & metal gates
  - better dielectrics "alchemy"
  - S/D & contacts "wizardry"

- W/L = 0.03/0.01 transistor has a "variability multiplier" ($1/\sqrt{A}$) of $\approx 60$
- $A_{\Delta VT}$ not reduced much below 1 mV$\mu$m & $A_{\Delta Idsat/Idsat}$ remains $\approx 1$ %$\mu$m
- this results in **one sigma** values $\sigma_{\Delta VT}$ & $\sigma_{\Delta Id/Id}$ of 60 mV and 60 %

- circuit design: variability resilient architectures inevitable
  - less aggressive scaling of M-S functions
  - trimming ; D.E.M. ; auto-zero, etc.
  - redundancy (codes, multiple signal paths, etc.)

**outline**

- introduction

- random parametric fluctuations ("matching")

- systematic & environmental variability effects
    - distance
    - WPE & STI
    - metal related stress

- low frequency noise variability

- wrap-up

**parametric gradients**

performance variation of IC components across wafer

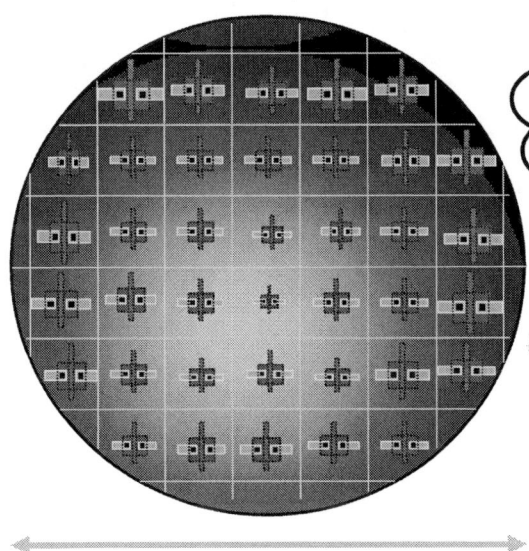

100 - 150 - 200 - 300 mm

**bold statement:**
**parametric gradients**
**are negligible**
**for (most) circuits**

**rule of thumb:**

parametric gradients
across wafer (& within die)

< 1% / cm  (= 1 ppm/μm)

< 1-10 mV / cm (= 0.1 - 1 μV/μm)

## "parametric gradients are negligible for circuits"

### (data spread out over a full 300 mm wafer)

- $I_{on}$ variation (with respect to median value): -3.5 % to + 1.7 % (extremes)
- transistor dimensions 10 x 10 μm²; equivalent $T_{ox} \approx 2$ nm

wafer map ($I_{on}$ spread (%) ) 10/10 NMOS

**< 1 %/cm !!!**
**(100 μm :== 0.01 %)**
**(usually negligible)**

5.2 % of 2 nm ($\approx$ 0.1 nm ) corresponds to about 1/5th of an atomic layer !

## wafer maps **left** transistors (NMOS)

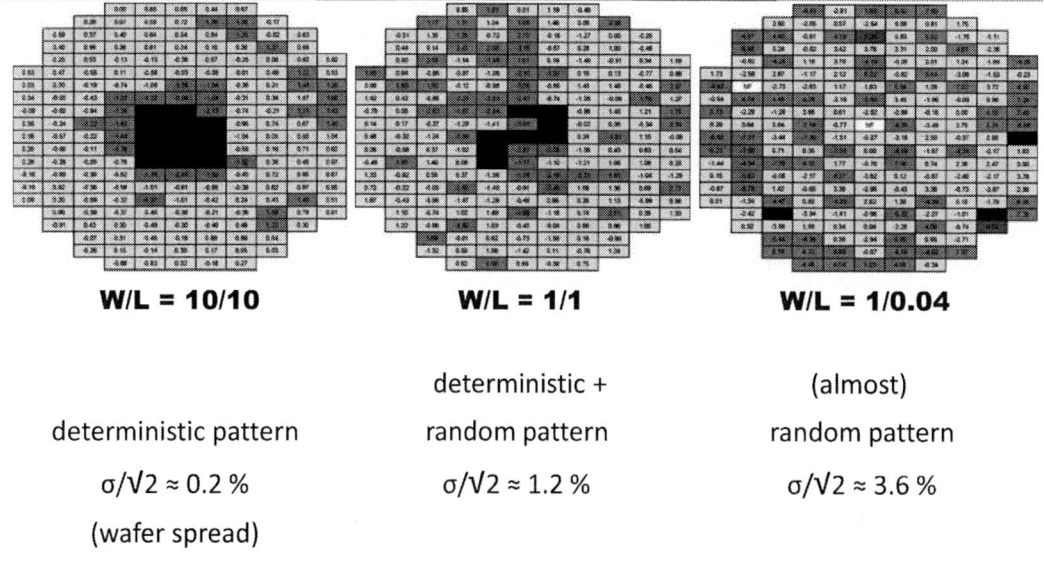

| W/L = 10/10 | W/L = 1/1 | W/L = 1/0.04 |
|:---:|:---:|:---:|
| | deterministic + | (almost) |
| deterministic pattern | random pattern | random pattern |
| σ/√2 ≈ 0.2 % | σ/√2 ≈ 1.2 % | σ/√2 ≈ 3.6 % |
| (wafer spread) | | |

## wafer maps matched pairs (NMOS)

### all patterns random!

| **W/L = 10/10** | **W/L = 1/1** | **W/L = 1/0.04** |
|:---:|:---:|:---:|
| random pattern | random pattern | random pattern |
| $\sigma/\sqrt{2} \approx 0.2\ \%$ | $\sigma/\sqrt{2} \approx 1.2\ \%$ | $\sigma/\sqrt{2} \approx 3.6\ \%$ |

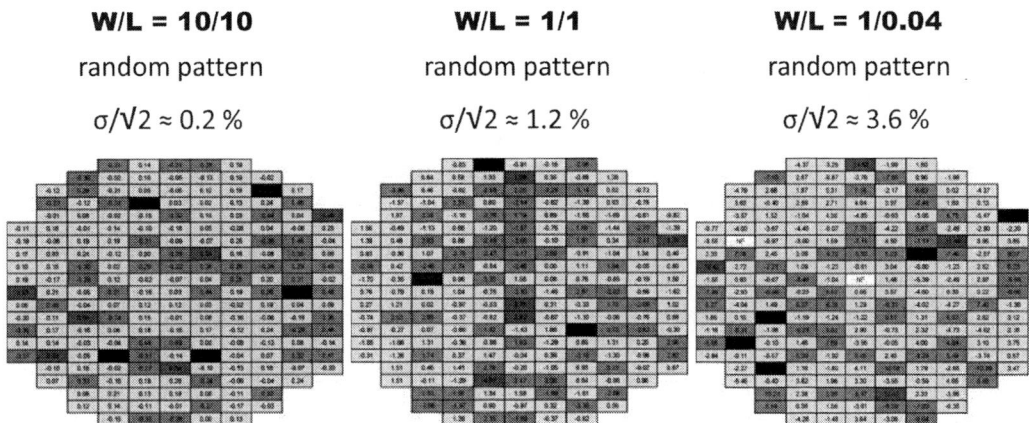

## variability maps left transistor vs. matched pairs

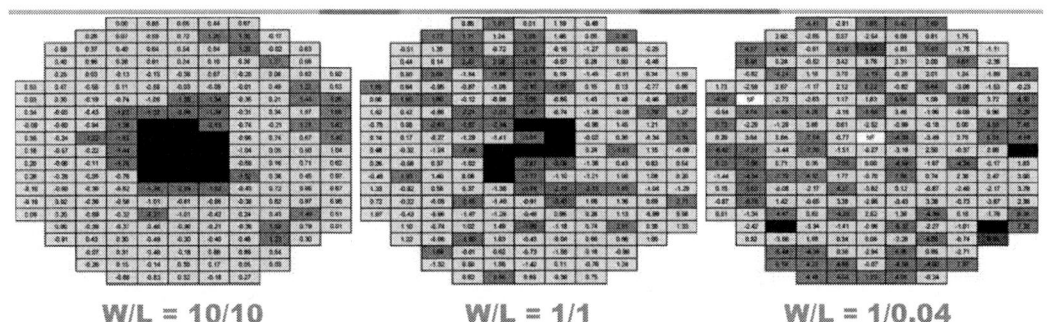

| W/L = 10/10 | W/L = 1/1 | W/L = 1/0.04 |
|:---:|:---:|:---:|

- need really big device to assess deterministic wafer spread
- 1 µm² represents total active area of an IP cell of 80 transistors

## multiple placements of identical pairs (40 nm tech.)

### study "distance effects"

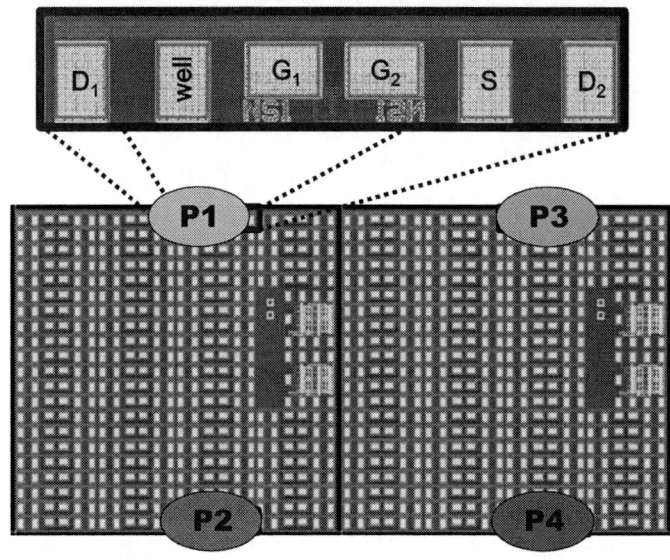

## mismatch fluctuation sweeps for PMOS pairs

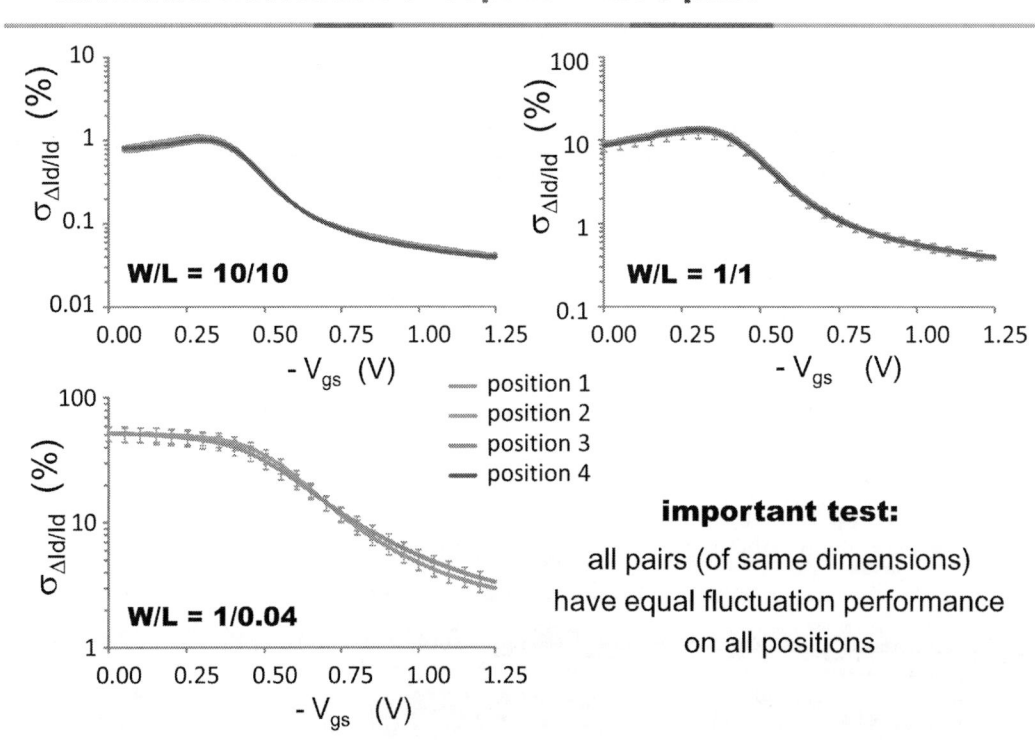

**important test:**

all pairs (of same dimensions)
have equal fluctuation performance
on all positions

**multiple placements of identical pairs (40 nm tech.)**

study "distance effects"

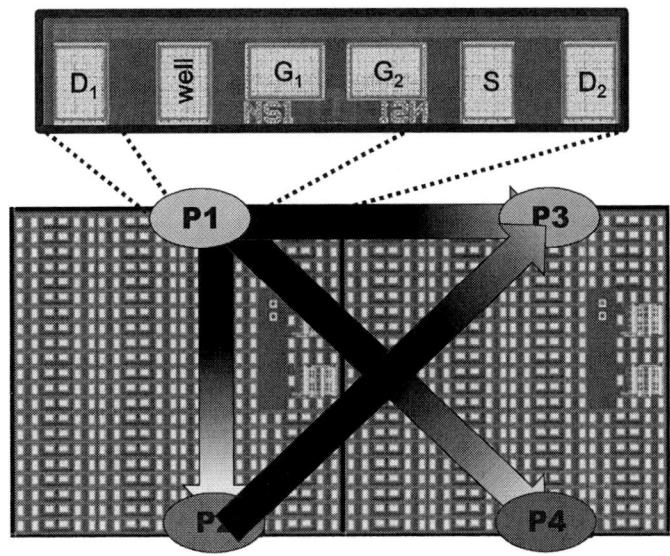

**"distance pair"**

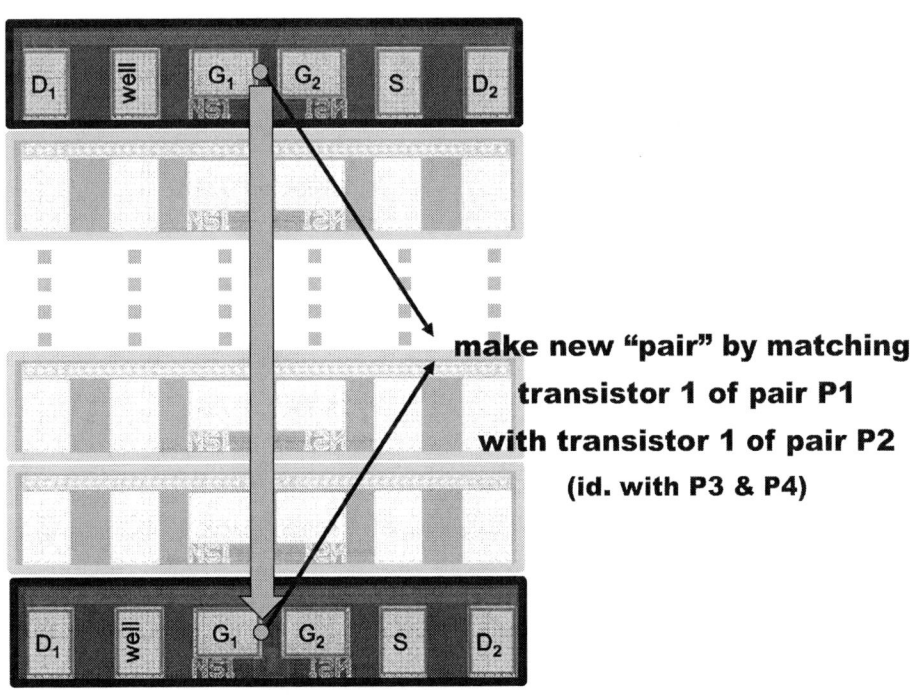

make new "pair" by matching
transistor 1 of pair P1
with transistor 1 of pair P2
(id. with P3 & P4)

## mismatch fluctuation sweeps for **combined pairs**

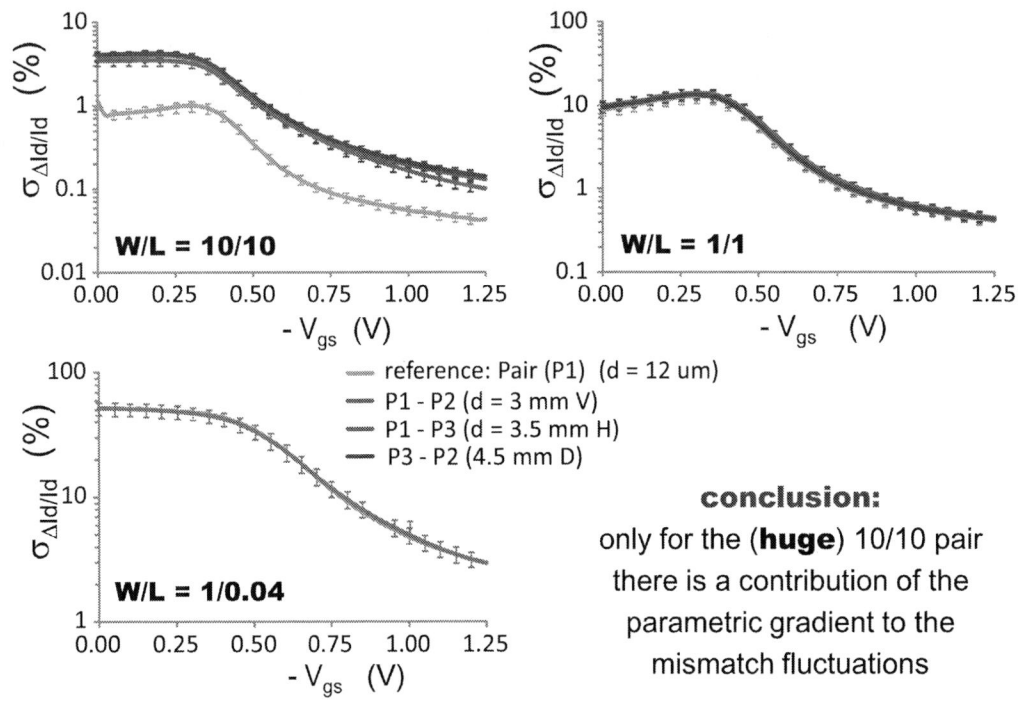

reference: Pair (P1) (d = 12 um)
P1 - P2 (d = 3 mm V)
P1 - P3 (d = 3.5 mm H)
P3 - P2 (4.5 mm D)

**conclusion:**
only for the (**huge**) 10/10 pair there is a contribution of the parametric gradient to the mismatch fluctuations

## take-away points

- parametric gradients are generally well below 1 %/cm (1ppm/µm)

- most M-S (& digital) blocks are spread out over less than 2 to 5 mm

- most IP-blocks are smaller than 100 x 100 µm² & most digital IP-blocks have < 1µm² total active area (= 80 transistors): variability is dominated by random fluctuations

- only use common centroid layouts when very big components
$\qquad$ (> 1000 um²) are required

- hans' "wisdom" (experience): **distance does not matter**
except :
- local heating / temperature gradients
- layout environment differences

### when matching (device equality) matters
### make both devices equally bad...

## where's lithography in this story?

- our experience (with relatively mature Mixed-Signal technologies):
  never encountered substantial reticle (mask) or wafer stepper induced
  variability (ACLV; across chip line width variation) nor LER (line edge
  roughness)

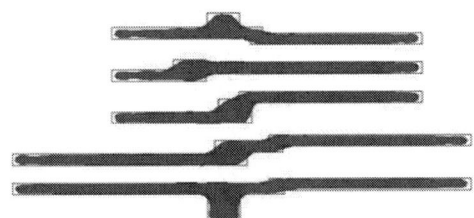

Litho simulation
Solid line: gds2 lay-out
Red: expected pattern on die

- use dummies to eliminate differences between dense and solitary lines

- hans' "wisdom" (experience): when litho limits product yield
  - litho engineers work harder
  - better quality masks are ordered

  "product yield improvement by litho is (legal form of) printing money"

## outline

- introduction

- random parametric fluctuations ("matching")

- systematic & environmental variability effects
  - ~~distance~~
  - WPE & STI
  - metal related stress

- low frequency noise variability

- wrap-up

## WPE: well proximity effect

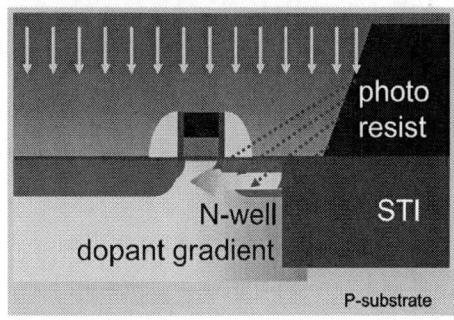

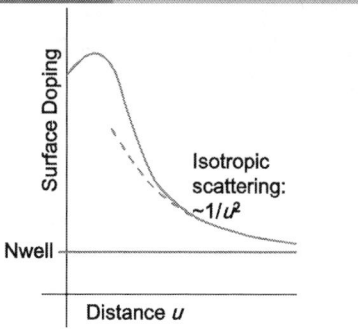

see: T.B. Hook et al.,
IEEE Trans. on Electron Devices,
Vol. 50, No. 9, pp. 1946-1951, 2003.

▸ related to:

  – distance of well-edge to active-region

  – well implant dose density and energy

  – oxide thickness & channel doping profile of the devices

  – slightly W and L dependent

▸ WPE impacts:

  – $V_T$ (up to 30mV per well edge), and thereby $I_{on}$ & $I_{off}$

  – narrow / long (P)MOSFETs are most severely affected

▸ Well-to-active-region distance >1μm generally sufficient to avoid problems

## study WPE with asymmetric "matched" pairs

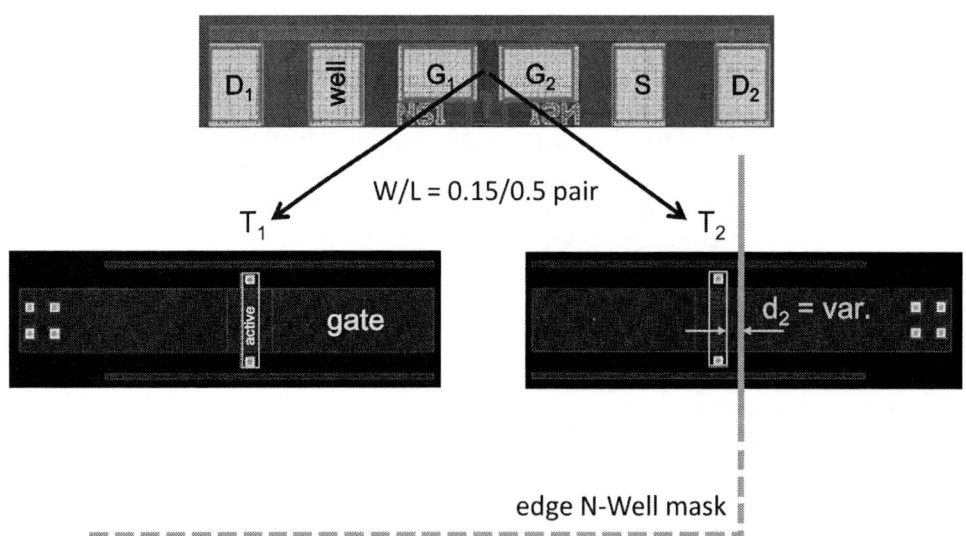

$d_2$ = 0.12, 0.48 and >> 2 μm (=reference)

## WPE-results: mm fluctuation sweeps

0.15/0.5 PMOS (a-symmetrical)

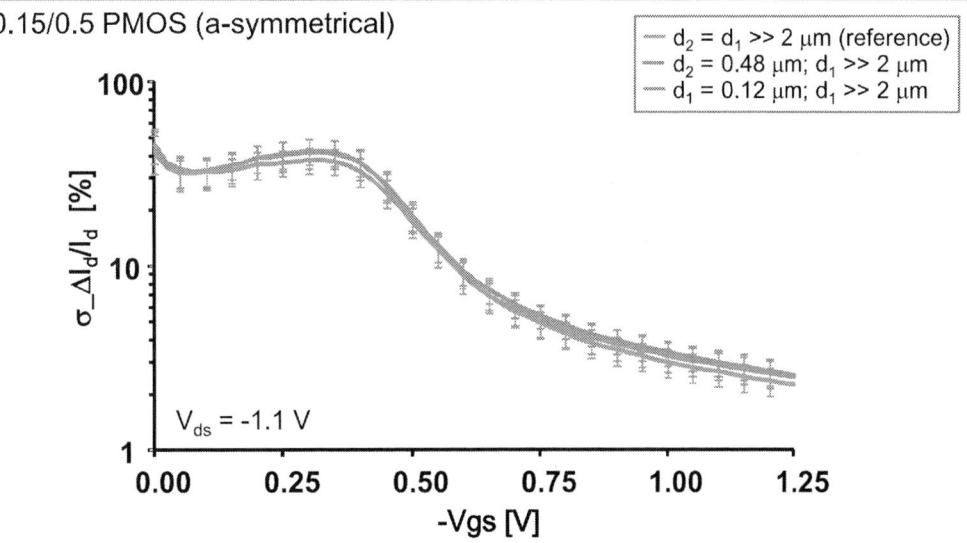

Legend:
- $d_2 = d_1 \gg 2$ µm (reference)
- $d_2 = 0.48$ µm; $d_1 \gg 2$ µm
- $d_1 = 0.12$ µm; $d_1 \gg 2$ µm

$V_{ds} = -1.1$ V

- no significant impact of the WPE effect **on mismatch fluctuations !**
- relatively small increase of channel dopant concentration ($\sigma_{\Delta VT}$ goes with $N^{-1/4}$)

## intermezzo: the NSCP plot (Q-Q plot, NQ plot)

### normal scaled cumulative probability plot

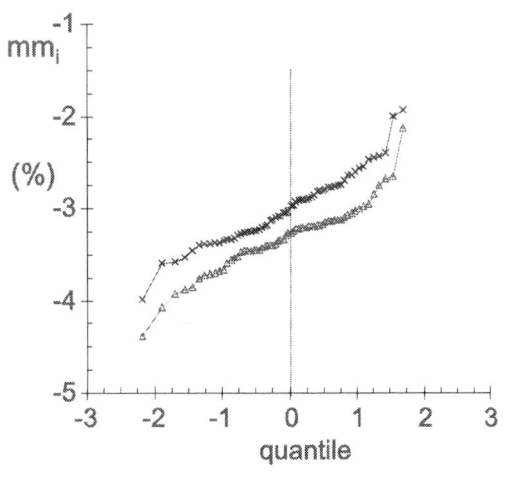

- N samples
- sorted lo – hi
- scatter plot
    $mm_i$ vs. quantile

use
*normsinv(i/(N+1))*
for the quantile axis

➢ slope = standard deviation (unconventional axes choice)
➢ straight line: normal (Gaussian) distribution
➢ medians: 'quantile=0'

## WPE-results: $V_T$ mismatch 0.15/0.5 PMOS (a-symmetrical)

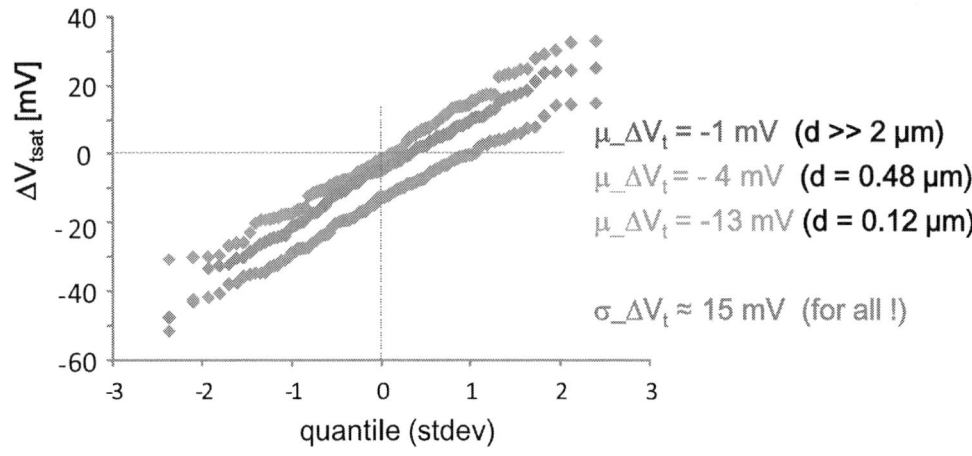

$\mu\_\Delta V_t = -1$ mV  (d >> 2 µm)

$\mu\_\Delta V_t = -4$ mV  (d = 0.48 µm)

$\mu\_\Delta V_t = -13$ mV  (d = 0.12 µm)

$\sigma\_\Delta V_t \approx 15$ mV  (for all !)

- 13 mV can be devastating for many analog circuits; digital designers don't care!
- note that all are standard deviations (slopes) are equal!

**(WPE is constant across the wafer)**

## WPE-results: $I_{on}$ mismatch 0.15/0.5 PMOS (a-symmetrical)

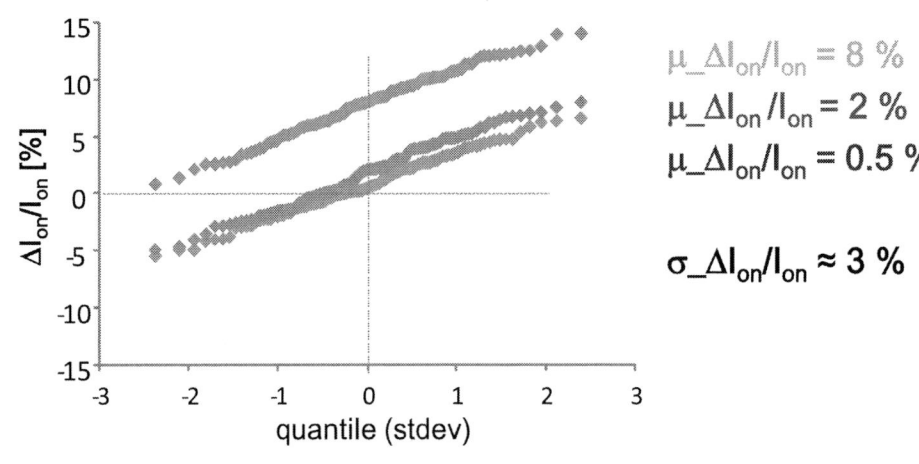

$\mu\_\Delta I_{on}/I_{on} = 8$ %

$\mu\_\Delta I_{on}/I_{on} = 2$ %

$\mu\_\Delta I_{on}/I_{on} = 0.5$ %

$\sigma\_\Delta I_{on}/I_{on} \approx 3$ %

➢ a few % can be devastating for many analog circuits

➢ standard deviations equal

➢ digital designers hardly care

**(compact models capture WPE effects !)**

## STI stress

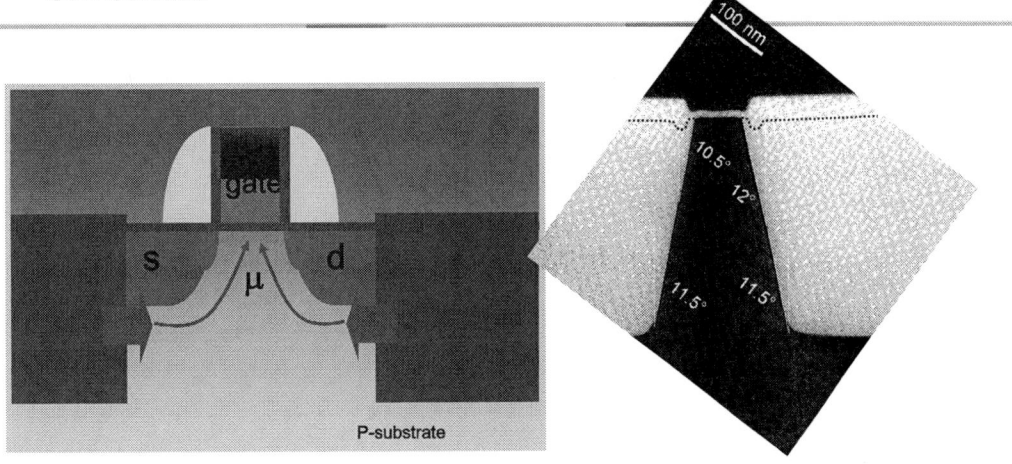

STI field isolation creates compressive mechanical stress
- increases hole $\mu$, reduces electron $\mu$
- stress effect proportional to inverse of distance
- makes device current dependent on layout environment

## study STI-effects with asymmetric "matched" pairs

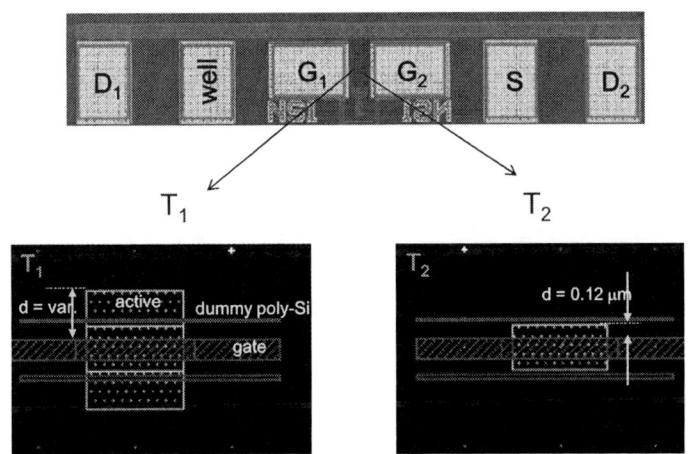

W/L = 1/0.2 pair (asymmetrical)

d = 0.12 (= reference), 0.14, 0.27 and 0.5 $\mu$m

## STI-result: $I_{on}$ mismatch 1/0.2 NMOS (a-symmetrical)

$I_{dsat}$ distributions (cumulative probability plots) of $\Delta ion/Ion$ with different distance to STI edge

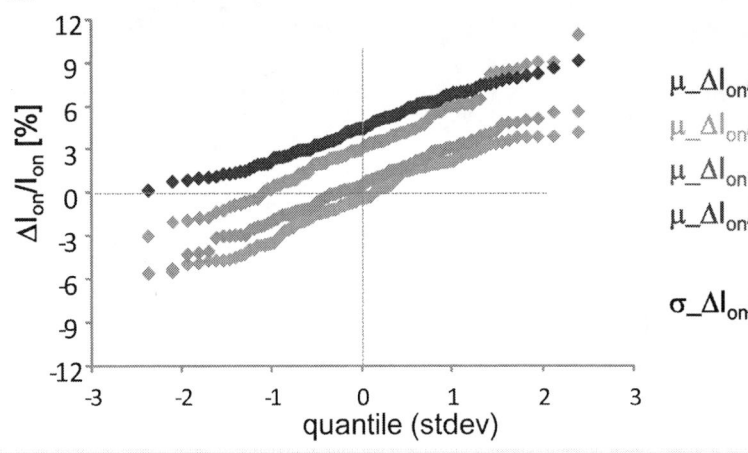

$\mu\_\Delta I_{on}/I_{on} = 4.5\ \%$

$\mu\_\Delta I_{on}/I_{on} = 3.2\ \%$

$\mu\_\Delta I_{on}/I_{on} = 0.5\ \%$

$\mu\_\Delta I_{on}/I_{on} = -0.3\ \%$

$\sigma\_\Delta I_{on}/I_{on} \approx 2.5\ \%$

> **a few % can be devastating for many analog circuits**
>                          (current mirrors in one active region)
> all standard deviations equal (**STI effects is constant across the wafer**)
> digital designers hardly care (**& compact models include STI effects !**)

---

## talking about stress

metal coverage is a fantastic ingredient for variability

## impact of aluminum routing on systematic mismatch

a useful set of experiment with current mirrors and distant stress sources

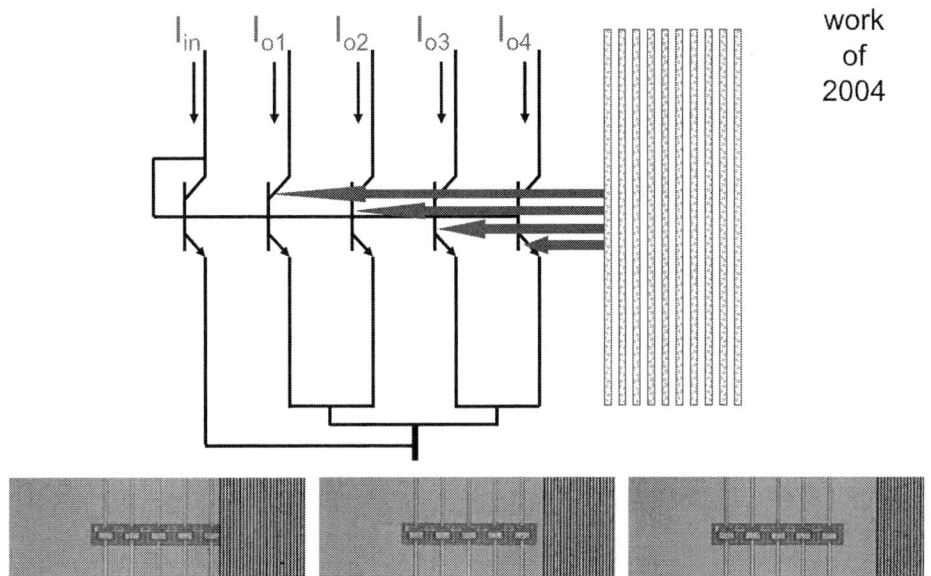

work of 2004

distance between $T_4$ and asymmetry is varied from minimum to 36 microns

## current mirror offset vs. distance

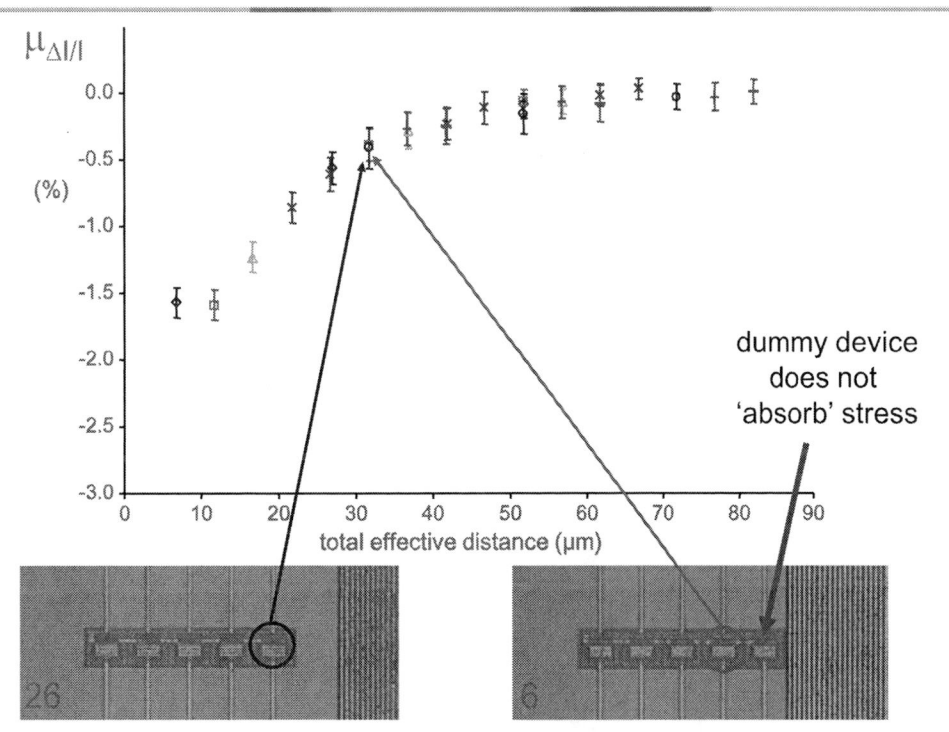

dummy device does not 'absorb' stress

## span of mechanical stress

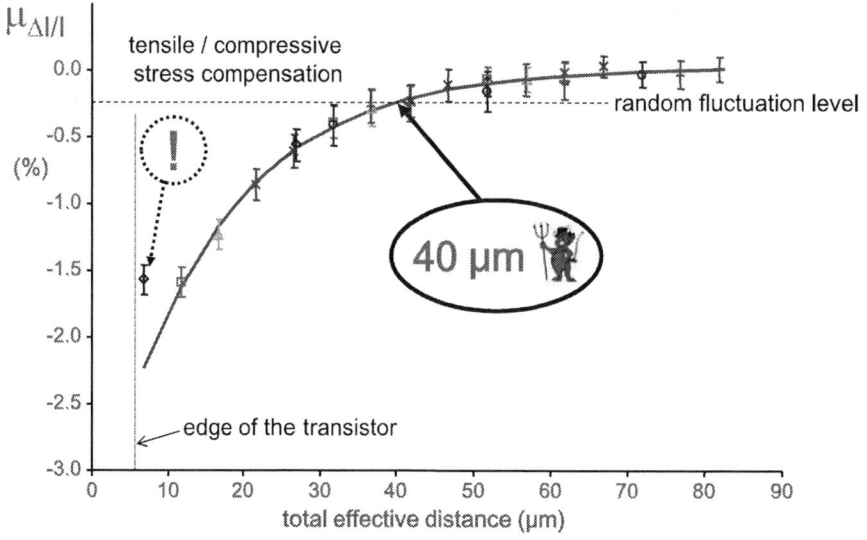

systematic mismatch decreases a factor **2** (1 bit)
for each additional spacing of about **10 µm**

## variability question

## study metallization stress with asymmetric "matched" pairs

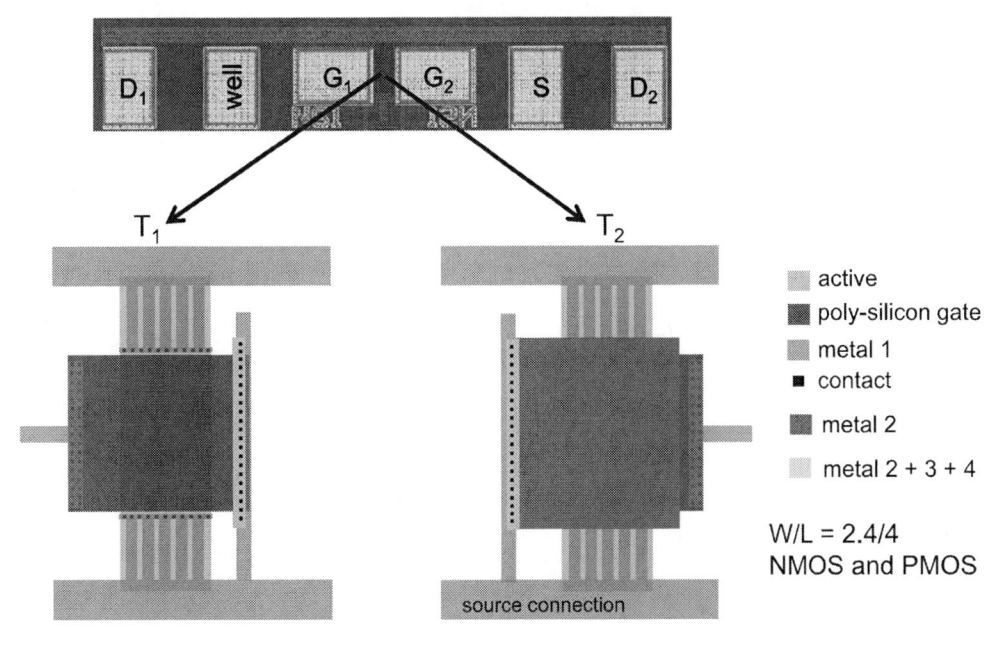

## mismatch fluctuation sweeps with M1 & M3 routing

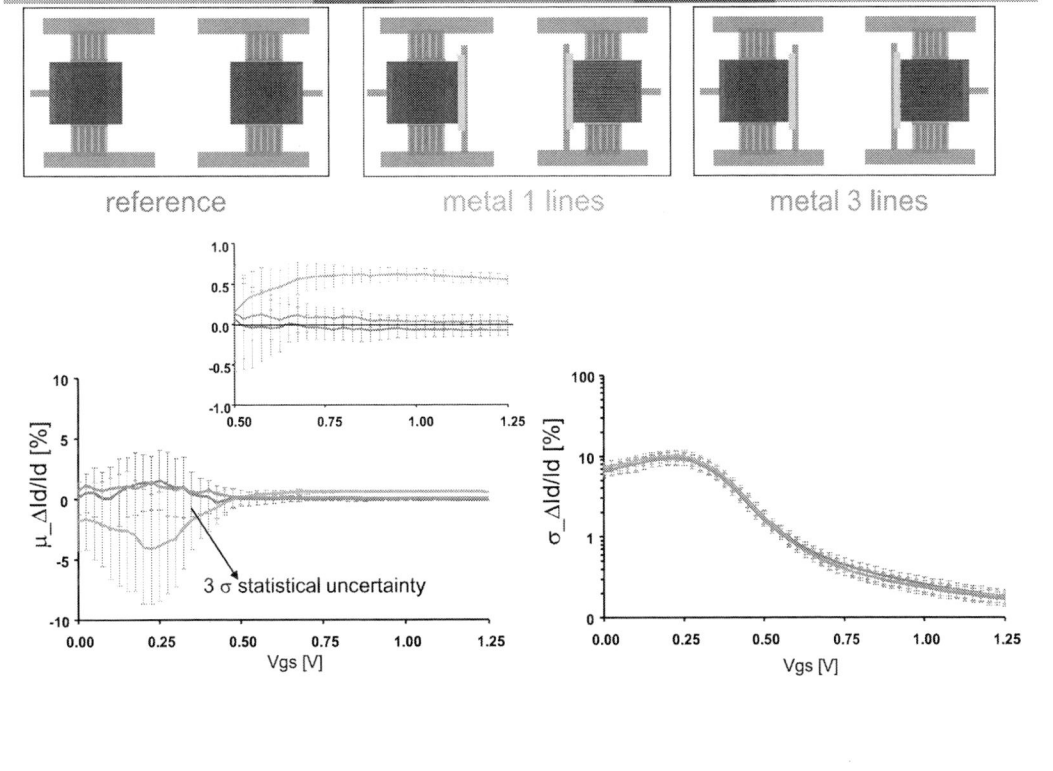

## take-away points

- metal device connection and routing lines give mechanical stress
- typical impact is between 1 and 3 % (depends on device & stack M1 … M3)
- need large devices (>> 1 µm² !) to see this (mitigate random fluctuations)
- stress impact reduces with factor 2 per 10 µm extra distance
- dummy device does not absorb stress
- possible to compensate compressive through tensile stress (other line)
- dense metal patterns (CMP tiles !) hence average-out the stress
- remaining (mismatch) effect due to metal on top of transistor typical 1 %
- also true for contemporary Cu damascene metallization

hans' perspective:

- digital designs are not seriously hampered by this 1 %
  (no worries about routing and CMP dummy tile placement)
- an unexpected 1 % offset can be devastating for analog circuit
- to my knowledge, such adjacent layout effects are usually not
  captured in compact models

## outline

- introduction

- random parametric fluctuations ("matching")

- systematic & environmental variability effects
  - distance
  - WPE & STI
  - metal related stress

- low frequency noise variability

- wrap-up

## LF noise: new kid on the variability battle field

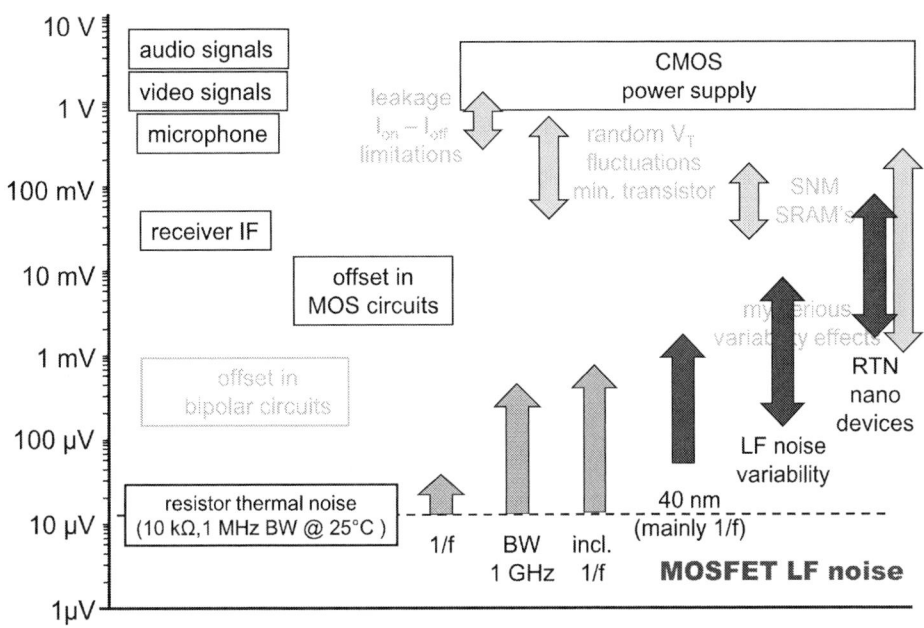

## low frequency noise (LFN)

LFN is a limitation for many high performance mixed-signal circuits
- audio: LFN limits usable amplification of small signals (S/N ratio)
- RF: LFN can determine phase noise in GHz oscillators (VCO's)

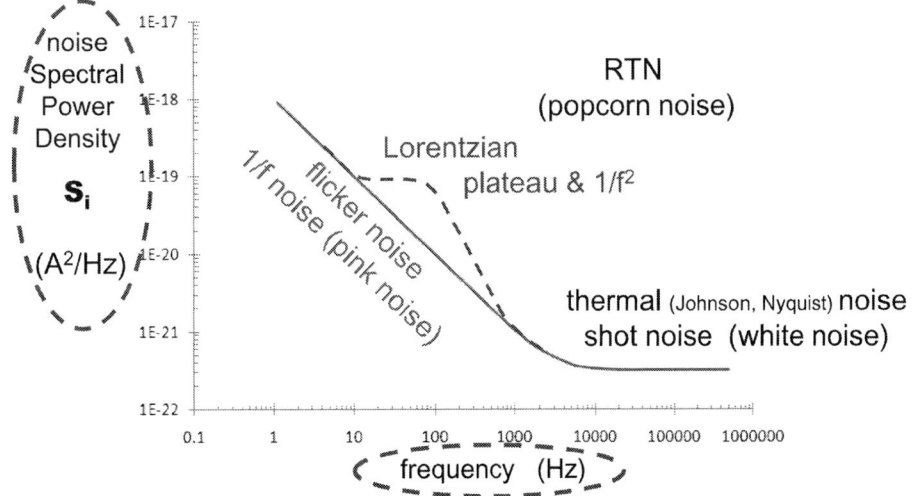

**variability of LFN (and its modeling) is becoming a design issue**

## Figure of Merit for LF noise

noise spectral power density $S_{id}$ scales with $I_D^2$ and 1/(device area)

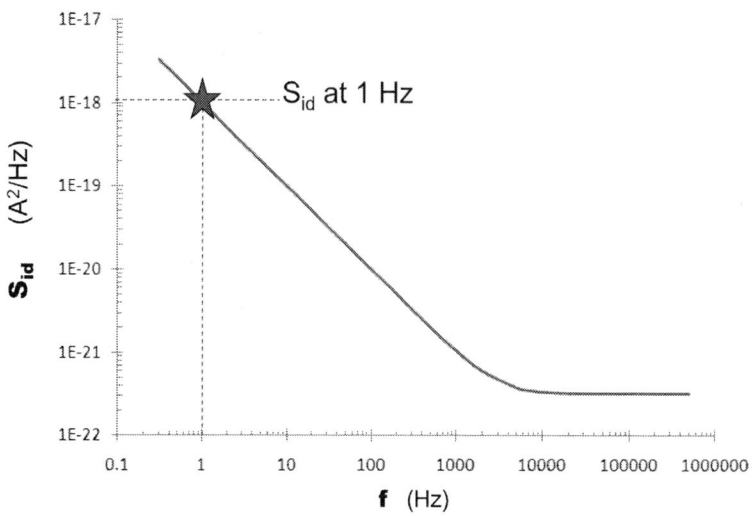

used FoM :  Area $\times$ $S_{id}$(@ 1 Hz) / $I_D^2$   ($\mu m^2$/Hz)

## talking about variability...

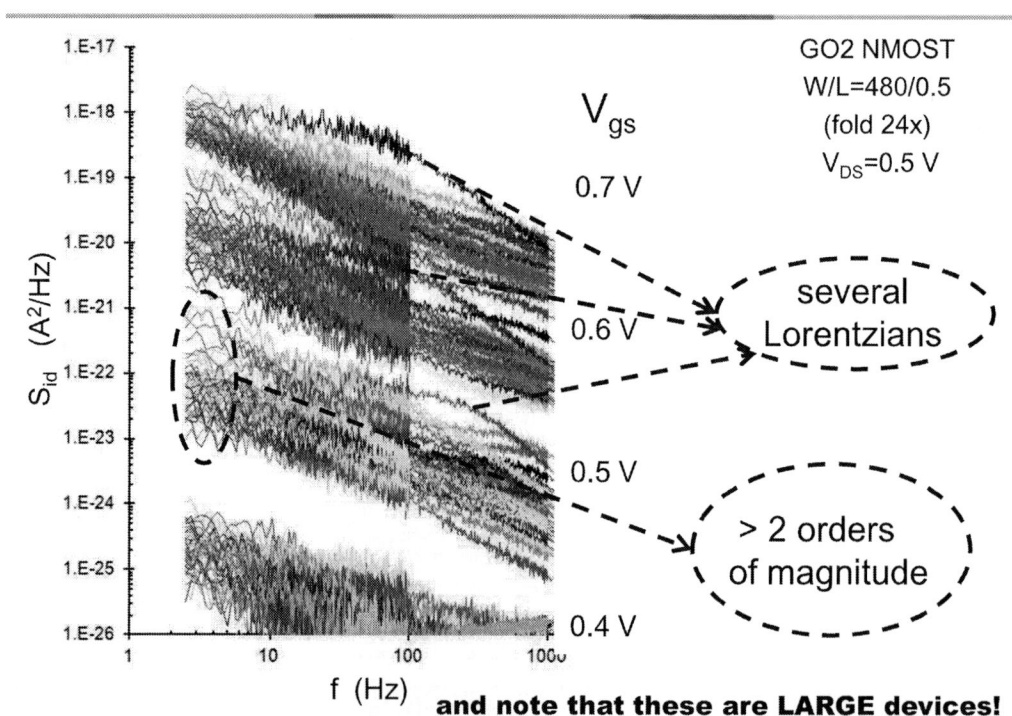

and note that these are **LARGE** devices!

## (large) device operation in subthreshold

- product study for an audio product: simulated minimum LF noise
  for very wide transistor (W ≈ 500 μm) in subthreshold region ($I_D$= few μA's)

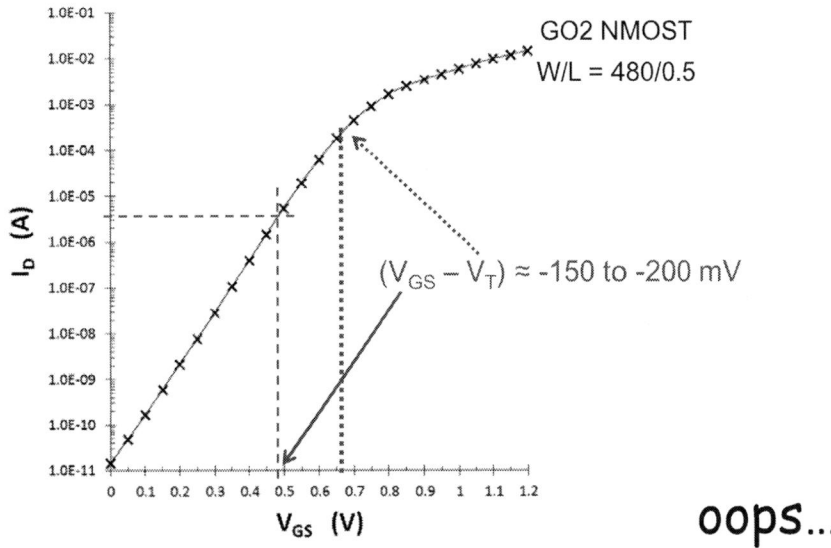

oops...

## oops...

- LFN models for large devices are based on standard area scaling rules

- device modeling usually done for devices up to a few μm (max. 10 μm)

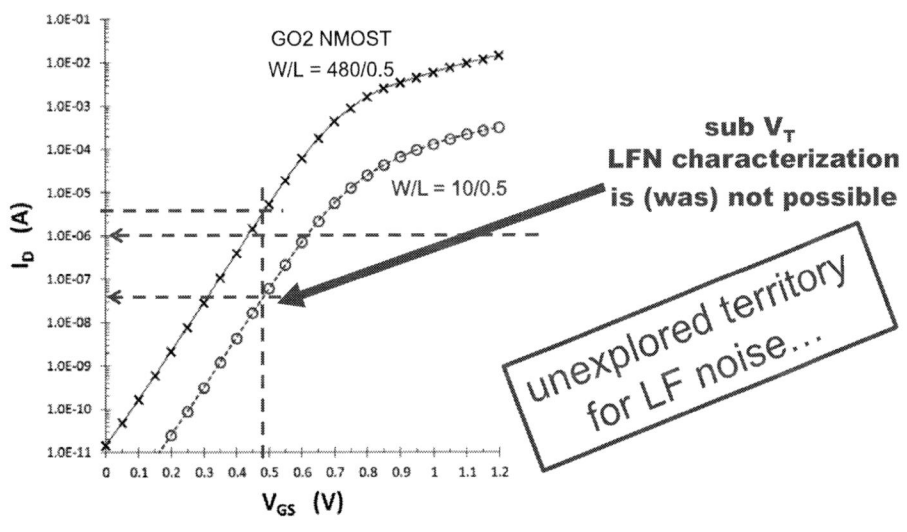

## 1/f noise FoM typical compact model prediction

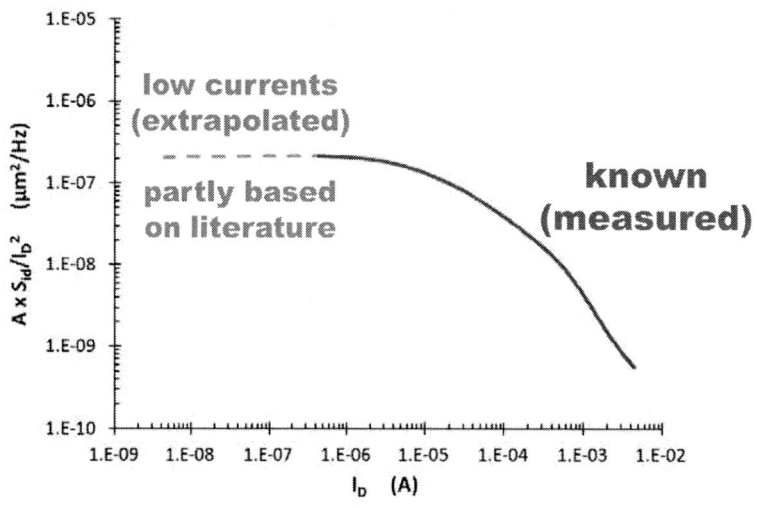

## noise variability upturn ?

variability increases in subthreshold, but how much?

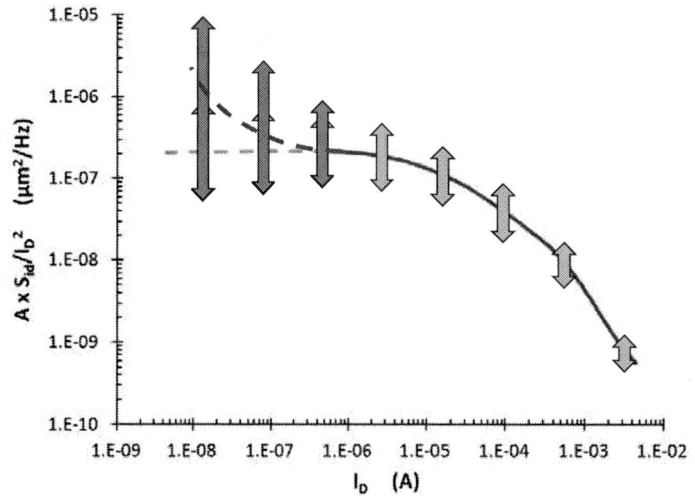

## example FoM vs. $I_D$

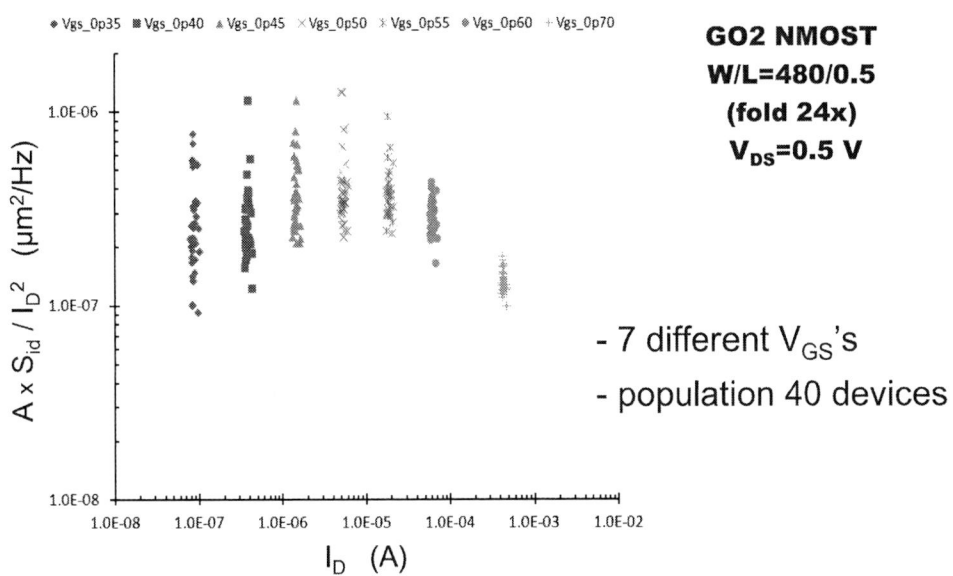

**GO2 NMOST**
**W/L=480/0.5**
**(fold 24x)**
**$V_{DS}$=0.5 V**

- 7 different $V_{GS}$'s
- population 40 devices

**this is indeed roughly the anticipated behavior...**

## observations

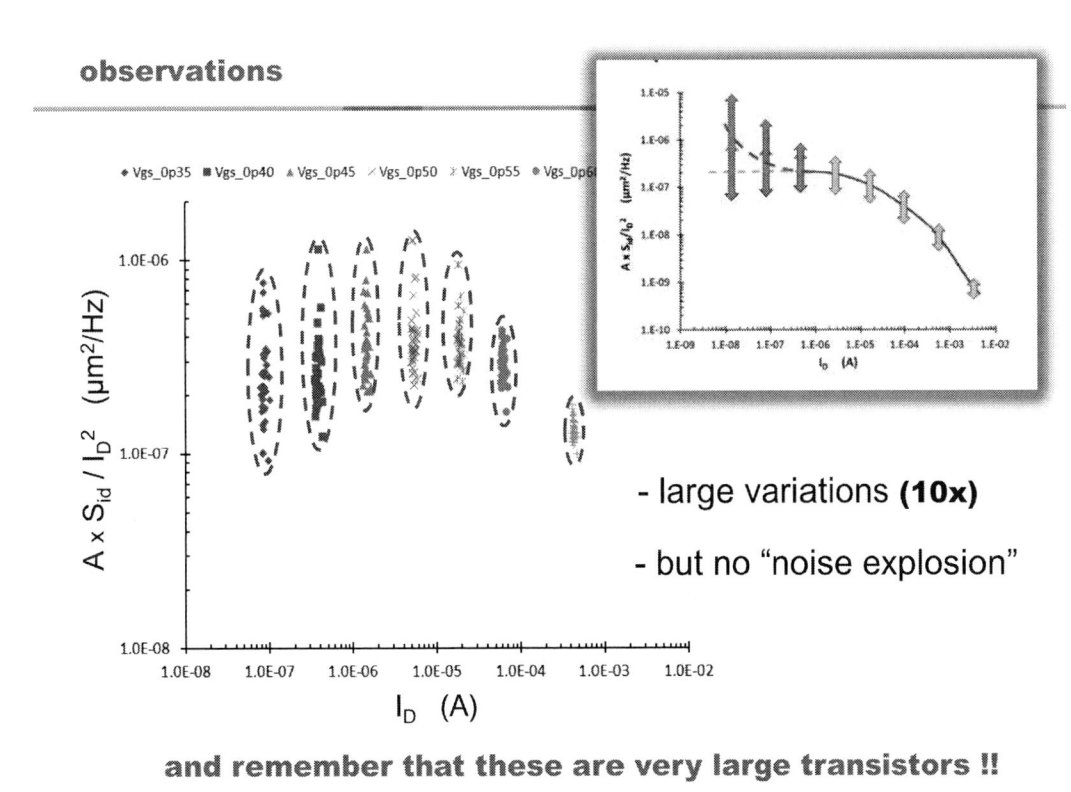

- large variations **(10x)**

- but no "noise explosion"

**and remember that these are very large transistors !!**

...

> and this is what to expect in contemporary (40 nm) M-S technologies

---

### the challenge: chaos

- one wafer, 54 positions, 4 biases, small nfet (W/L=1/0.04)

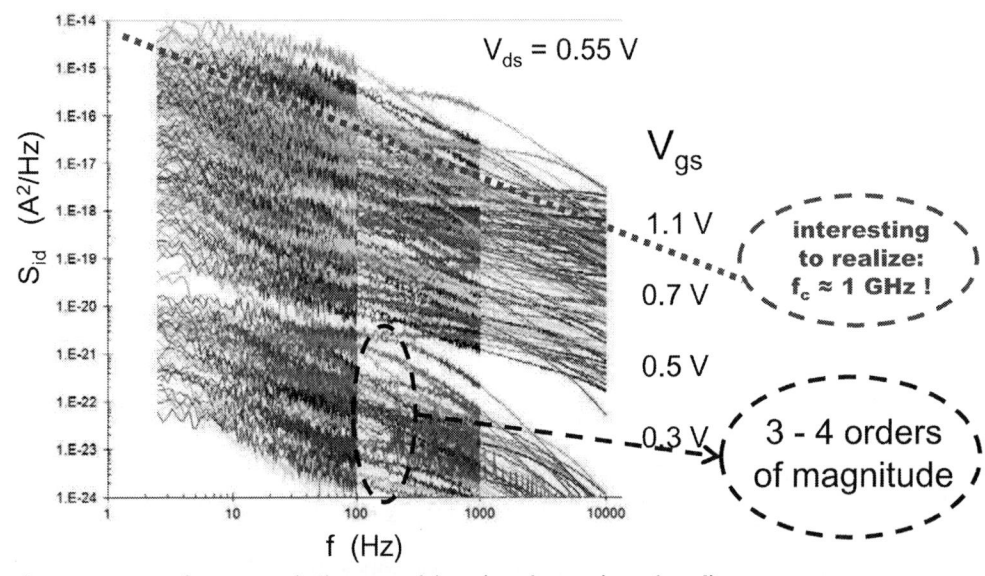

- large range; large variation per bias (and overlapping !);

- different effects: RTN Lorentzians (& measurement system BW limitations)

## translated into FoM plot for 9 V$_{gs}$'s

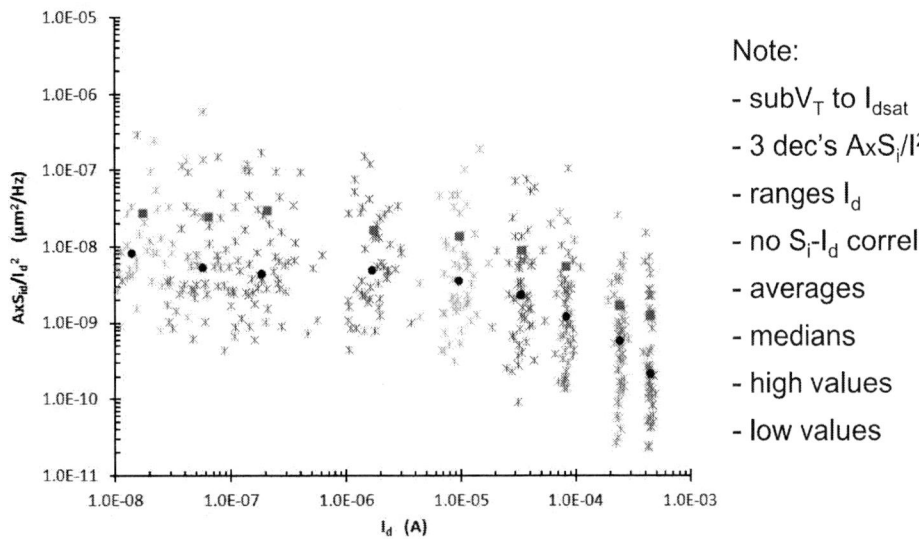

Note:
- subV$_T$ to I$_{dsat}$
- 3 dec's A$_x$S$_i$/I$^2$
- ranges I$_d$
- no S$_i$-I$_d$ correl.
- averages
- medians
- high values
- low values

## same data: FoM connected per device

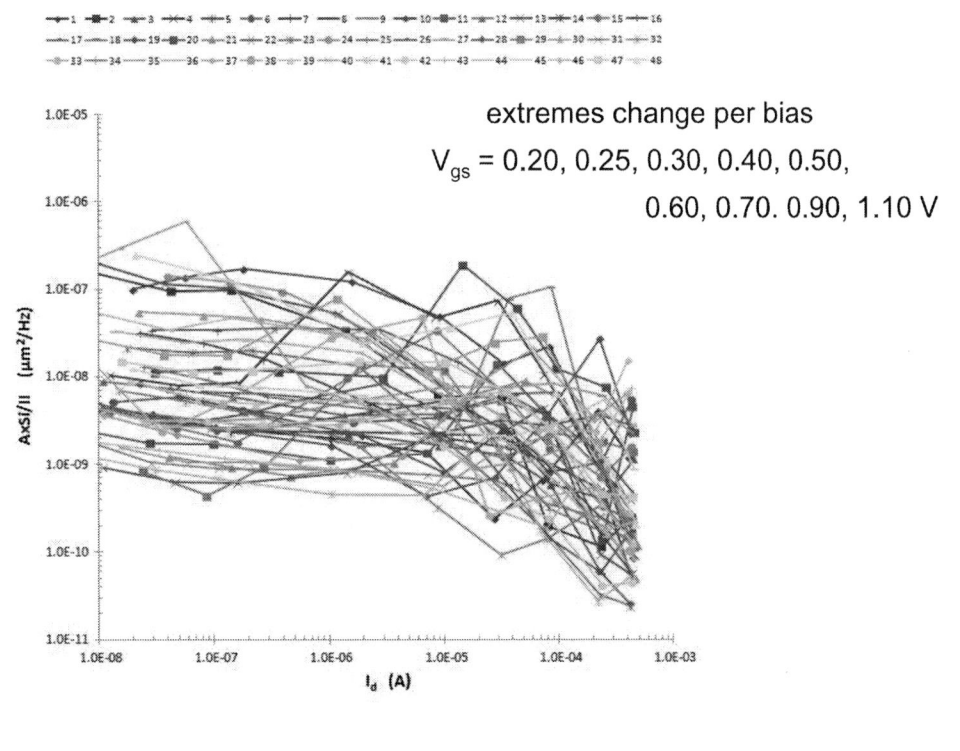

extremes change per bias

V$_{gs}$ = 0.20, 0.25, 0.30, 0.40, 0.50,
0.60, 0.70. 0.90, 1.10 V

## example: (dev 20)

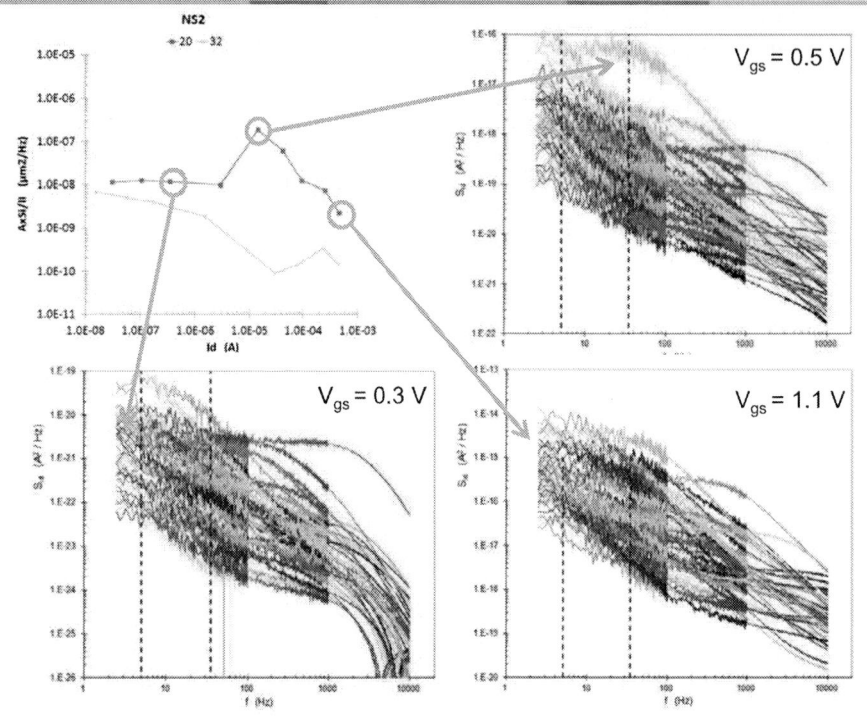

## the good news is: the compact model captures this!

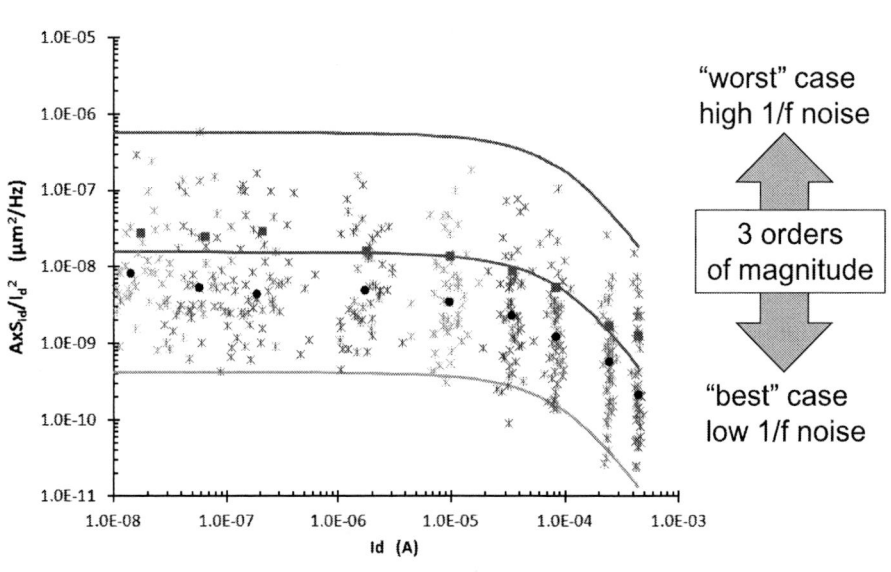

"worst" case
high 1/f noise

3 orders
of magnitude

"best" case
low 1/f noise

well done by the foundry!

### what's ahead?

- ... CMOS technology is heading towards 10 nm dimensions

- RTN observations 20 nm technology (Tega et al. VLSI-2009):

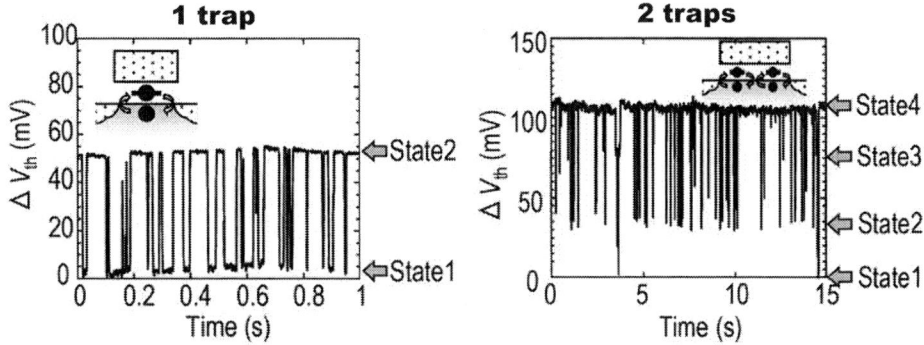

- equivalent $V_T$ steps of up to 100 mV!

- time scales: RTN trapping times range from nano seconds to
  kilo seconds !!

- affects yield of SRAMs (SNM)

---

### outline

- introduction

- random parametric fluctuations ("matching")

- systematic & environmental variability effects
  - distance
  - WPE & STI
  - metal related stress

- low frequency noise variability

- wrap-up

## from high precision analog design to digital ULSI

parametric mismatch fluctuations and systematic mismatches
have always been crucial performance- and/or yield- limiters
for **high precision analog** electronic circuits

## but...

with the introduction of

- reduced supply voltages
- reduced device dimensions
- massive numbers of components per chip
- large embedded memories
- mixed-signal integration demands

random parametric fluctuations

have become limitations for **all ULSI SoC's**

---

## device **equality** categories

- **10 %** ($\approx\approx$ 1 - 10 mV; circuit level: 20 dB, 3.3 b); minimum device area's;
  **(compact models like PSP & BSIM4 cover main effects)** ;
  can only go wrong through design errors or process disasters

- **1 %** ($\approx\approx$ 0.1 - 1 mV; 40 dB, 7 b); 1 to 10 $\mu m^2$ devices;
  **(compact models at the edge; beware sub $V_T$)** ; decent layout required;
  symmetry 10 $\mu m$

- **0.1 %** ($\approx\approx$ 10 - 100 $\mu V$; 60 dB; 10 b); 10 to 100 $\mu m^2$ d
  **(don't rely on compact models)**; very careful la
  sym          $\mu m$

- **0.01 %** ($\approx\approx$ 1 - 10 $\mu V$; 80          ry very large area devices
  (100 - 10000 $\mu m^2$          eme layout symmetry (> 100 $\mu m$)
  and circuit tri          . data weighted averaging, A.Z. etc)

- **0.001%** ($\approx\approx$ 0.1 - 1 $\mu V$; 100 dB; 17 b) **forget it...** (use          resolution)

*only the paranoid can hope to succeed*

*even for the paranoid*

## summary / conclusions / take-away points

- a "device characterization view" on parametric and noise variability

- summarized definitions, orders of magnitude and some rules of thumb

- microscopic device architecture fluctuations ("matching") are
    the dominant parametric spread contribution **(<100 nm nodes)**
- many other "systematic" variability effects affect devices by "a few percent"

- blindly adding all possible effects explodes design margins,
    but with a bit of common sense a lot of the panic can be avoided

- a lot of the effects are captured (largely) in compact models
    but beware of the near- and sub-threshold regions

- noise variability and RTN are the new trending topics…
- the variability buzz is certainly not over (the worst is yet to come)

## the end

# Testability Improvement for 12.8 GB/s Wide IO DRAM Controller by Small Area Pre-bonding TSV Tests and a 1 GHz Sampled Fully Digital Noise Monitor

Takao Nomura, Ryo Mori, †Munehiro Ito, Koji Takayanagi, Toshihiko Ochiai, Kazuki Fukuoka, Kazuo Otsuga, Koji Nii, Sadayuki Morita, Tomoaki Hashimoto, Tsuyoshi Kida, Junichi Yamada and Hideki Tanaka

Renesas Electronics Corporation, 5-20-1 Josuihon-cho Kodaira, Tokyo 187-8588, Japan
†Renesas Micro Systems Corporation, 3-1 Kinkou-choYokohama Kanagawa 221-0056, Japan

*Abstract*—**We developed a Wide IO DRAM controller chip with Through Silicon Via (TSV) technology. Test circuitry is embedded in the micro-IOs placed between the fine pitch TSVs which can reject TSV connectivity failures prior to stacking process. In order to reduce Vmin degradation induced by 512 DQs simultaneously switching noise, we introduce a package-board impedance optimization method utilizing a full digital noise monitor. We achieved 12.8 GB/s operation, while IO power was reduced by 89 % compared to LPDDR3.**

*Keywords- TSV, Wide IO DRAM, pre-bonding test, fully digital noise monitor, simultaneous switching noise, impedance optimization*

## I. INTRODUCTION

The needs of high bandwidth DRAMs are rapidly rising in the mobile market due to higher definition displays for smartphones and tablets. Conventional DRAMs can achieve sufficiently high bandwidth, though it is difficult for it to be adopted in mobile application because of the constraints in thermal design and battery life time due to high power. One solution to this is Wide IO DRAM [1]. By using fine pitch TSV arrays, the interface bus width can be expanded to up to 512 Data signals (DQ), which makes 12.8 GB/s achievable by a 200 MHz SDR operation. In addition, TSV has low connection capacitance compared to conventional stacking such as Package on Package (PoP) stacking, and the active IO power can be reduced significantly.

However in employing Wide IO DRAM, there are two major challenges. One is the testability of the TSV connectivity. Issues for TSV probing during assembly are shown in Fig. 1. Before back grinding (BG), probing can be done only from the front side because the TSVs are not exposed until BG process. TSVs are exposed after BG, though probing at this point require techniques for probing 50 μm thin wafers, and probing on both sides of the wafer at once, which is not a realistic solution. TSV connectivity tests after stacking is defined in the Wide IO DRAM specification [1], which checks TSV connectivity indirectly by testing data transfer between DRAM and the logic chip. Rejecting all of the TSV defects at final test leads to bad assembly yield. In order to improve assembly yield, methods for testing TSV connectivity which can be done before BG, with no other cooperating

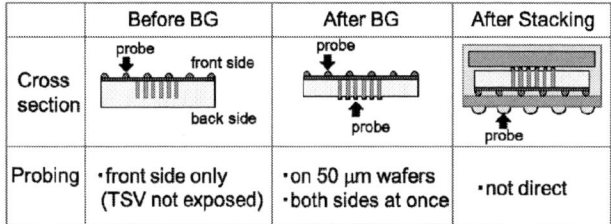

Fig. 1 Issues for TSV probing during assembly

device is needed [2]. Another challenge is noise immunity, where 512 DQs simultaneously switching noise degrades max operation frequency of the DRAM. In order to evaluate switching noise, an effective method is to apply a voltage sensor [3, 4]. The voltage sensor implemented in the prior work has a high resolution, though requires analog reference voltage or has an analog output, which is difficult to fit into tests in mass production. In order to address these challenges, we have designed a TSV stacked DRAM controller chip with dedicated test circuitry, which gives solutions to these issues.

## II. TSV STACKED CONTROLLER CHIP

Fig. 2 depicts the cross section view of the 3D stacked chips. The controller die is stacked face-down, and connected to the DRAM die by back-to-face using micro-bumps. The microphotograph of the stacked chips and the specification of the designed chip are shown in Fig. 3 and Fig. 4 respectively. The DRAM die is a 4 Gbit 512 DQs monolithic Wide IO DRAM. We adopted via middle TSV process, and the TSVs are formed in a fine pitch array (40 μm x 50 μm) in the 50 μm thin controller die. Fig. 5 shows the block diagram of the controller chip. It consists of a controller logic block and an array of TSVs associated with micro-IO cells. Micro-IO cells including dedicated test circuitry, needs to be embedded into the fine pitch TSV array to minimize the area overheads. The whole micro-IO cell has 793 μm², which is small enough to fit into the fine pitch TSV array as in Fig. 6. The test results are captured and collected by the boundary scan chain. A noise monitor circuitry is proposed for monitoring internal IO power lines, VDDQ and VSSQ, and is controlled by a sensor controller [5].

978-1-4673-6145-3/13 $31.00 © 2013 IEEE

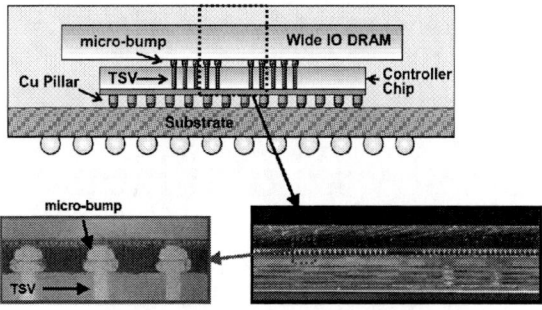

Fig. 2 Cross section view of the stacked chips

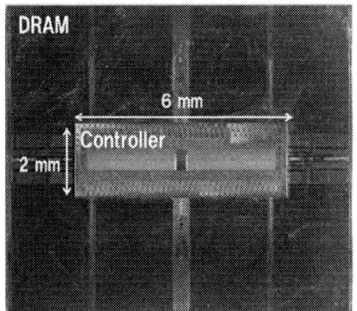

Fig. 3 Microphotograph of the stacked chips

| Supply Voltage | Core: 1.0 V<br>Internal IO:1.2 V<br>External IO:1.8 V<br>For DRAM:<br>VDD1=1.8 V<br>VDD2=VDDQ=1.2 V |
|---|---|
| Process | 28 nm technology |
| Metallization | 9Cu-1Al (Logic) |
| TSV process | Via-Middle |
| TSV pitch | 40 µm (vertical)<br>50 µm (horizontal) |
| Wafer Thickness | 50 µm (Controller) |
| Chip area | 2 mm x 6 mm (logic) |
| Stacked DRAM | 4 Gb Wide IO DRAM<br>(Manufacured by Elpida<br>Memory Inc. , SK hynix Inc.) |
| Package | BGA |

Fig. 4 Specification of the designed chip

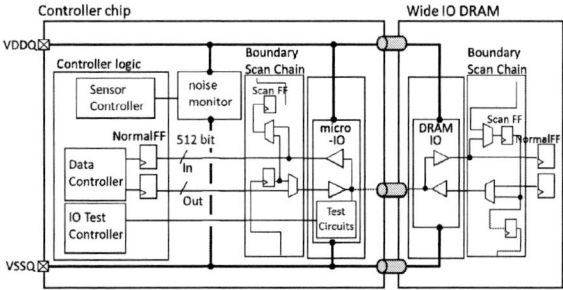

Fig. 5 Block diagram of the designed chip

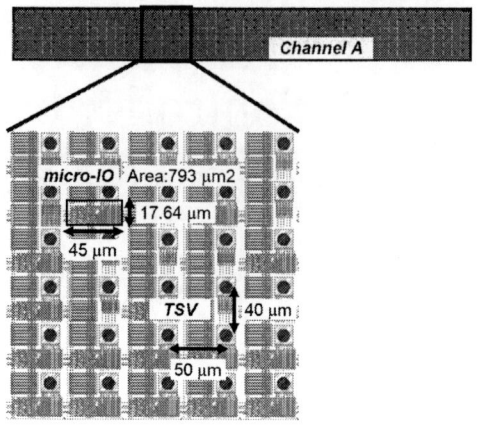

Fig. 6 Layout of a Wide IO channel

## III. TESTABILITY OF TSV CONNECTIVITY

We have developed two dedicated circuitries for testing TSV connectivity before stacking. The first test circuitry is for detecting Cu popup [6], which is one failure mode due to thermal stress, and induces metal-to-TSV connection opens. The schematic diagram is shown in Fig. 7. Pull up and pull down test drivers are connected through the TSV node, and the driver strength is imbalanced. By pulling up strongly and pulling down weakly, a high output should be obtained for a good TSV connection. If Cu popup occurs and the connection metal become open, the pull up driver become absent, and a low output will be obtained. The results are collected by a scan chain and the defect positions can be identified. Another test mode we have applied is a TSV charge/discharge test, and the schematic diagram for the charge test case is shown in Fig. 8. The TSVs are charged/discharged by weak drivers, and the charge/discharge time is evaluated. The product of driver MOS on resistance ($R_{on}$), and TSV connection capacitance determines the delay. For the before stacking case, the connection capacitance equals to the TSV-to-substrate capacitance ($C_{TSV}$), although for the after stacking case, the capacitance includes $C_{TSV}$, the micro-bump capacitance and the input capacitances of the DRAM IOs. In case of open defects, we see smaller delay which corresponds to smaller capacitance. In case of weak short, a larger delay will be observed in a charging test. We show a histogram of a charging test for a sample with open defects in Fig. 9. 14 signals had 80 ns delay when all other had delays around 150 ns. We have confirmed that the 14 signals with 80 ns delay had a defect in connectivity. Fig. 10 shows the cross section of the connections corresponding to the 14 signals, where the connections have open micro-bumps. It should be noted that the tests were done after bonding by setting DRAM IOs to Hi-z, though the capacitance deltas were well detected, which shows that inefficient Cu fillings and capacitance changing voids shall be detected simultaneously in pre-bonding tests.

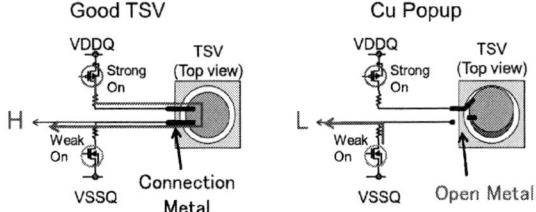

Fig. 7 Proposed TSV Cu popup test circuit

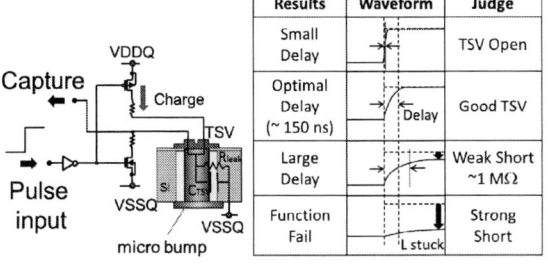

Fig. 8 A TSV charging test

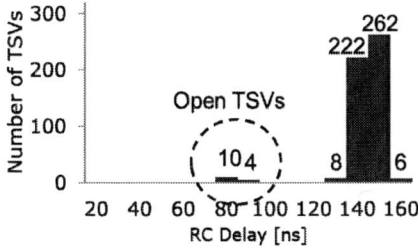

Fig. 9 RC delay histogram for a TSV defect sample

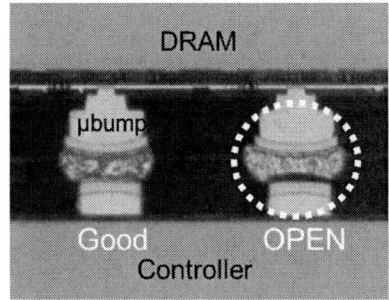

Fig. 10 Cross section of an open connection

## IV. IMPEDANCE OPTIMIZATION BY A NOISE MONITOR

512 DQs switching rapidly during 200 MHz DRAM operation should be a serious noise source. In order to confirm degradation, we have evaluated in a noisy test environment under write and read operation. Fig. 12 shows the cumulative IO fail counts for various toggle rates with a 200 MHz access. The data shows that Vmin improve as the toggle rate decrease, with a 30 mV difference for a 4 ch (=512 bit) 100 % toggle case and a 1 ch (=128 bit) 100 % toggle case. We propose an impedance optimization scheme utilizing a TDC-based voltage sensor as a noise monitor [5].

It can be easily calibrated, and simply implemented due to the full digital structure, and thus is suitable for mass production. The resolution is less than 10 mV, and the sampling rate is 1 GHz, which is sufficient for monitoring 200 MHz DRAM operation. We show a block diagram of the noise monitor in Fig. 13. The power lines of the micro-IO array are attached to the sensor, and the sensor converts the power voltage to a thermal code, and contains the encoded values as digital codes. The stored data can be scanned out in chronological order, and a digital waveform can be obtained. Fig. 14 is the measured digital waveform for the power noise of the micro-IOs during the worst noise case. There are 2 types of fluctuation are seen in the waveform excluding static drops, where one is the 200 MHz (short term) ripple seen in both of the waveforms caused by impedance inside the package, and the other is a ~10 MHz (long term) ripple caused by board impedance. For a not optimized case, we see a 25 mV drop caused by the ~10 MHz ripple, and we see an extra 20 mV drop from 200 MHz ripples and have a total 75 mV drop including the 30 mV static drop. We have proposed an impedance optimization scheme shown in Fig. 15, where we select the most effective methods according to the noise monitor results. After impedance optimization, ~10 MHz ripples are suppressed and 200 MHz ripples become dominant. In this case, we have added 47 μF onboard capacitors for reducing the ~10 MHz ripples. We confirmed Vmin improvement by reducing ~10 MHz ripples, though the 200 MHz ripples still remain where further suppression require adding on chip decoupling capacitors. The Shmoo plot with improved test environment is shown in Fig. 16. We have confirmed 12.8 GB/s operation with Vmin 1.07 V which has enough margins to the specification. As a result, by utilizing the proposed noise monitor, we have successfully extracted voltage drops for each frequency range, which gives us knowledge to select effective countermeasures.

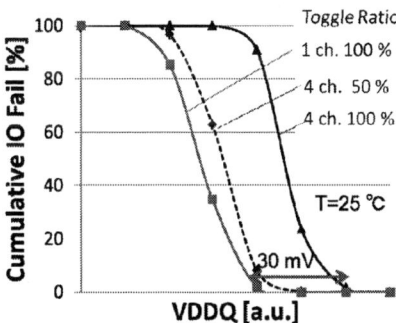

Fig. 12 Fail dependence on VDDQ

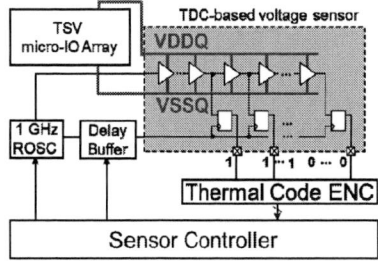

Fig. 13 Block Diagram of the TDC-based voltage sensor

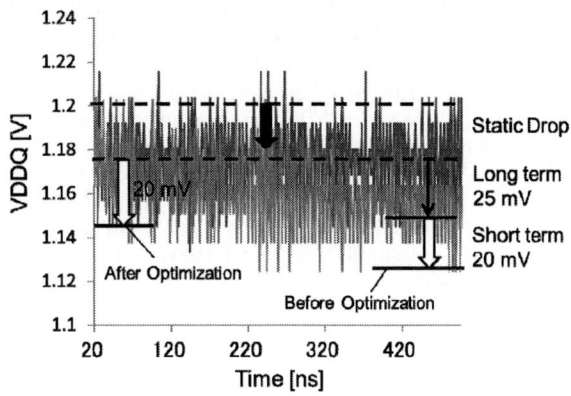

| | Drops | |
|---|---|---|
| Optimization (Color Online) | Before (in blue) | After (in red) |
| Static ⬇ | 30 mV | 30 mV |
| Long term (~10 MHz) ⬇ | 25 mV | 0 mV (-25 mV) |
| Short term (~200 MHz) ⬇ | 20 mV | 20 mV |
| Total | 75 mV | 50 mV (-25 mV) |

Fig. 14 Measured noise during burst write operation

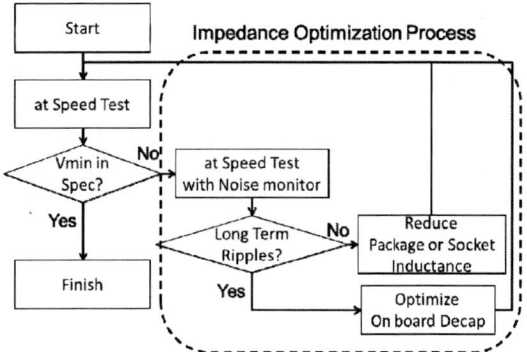

Fig. 15 Impedance optimization scheme

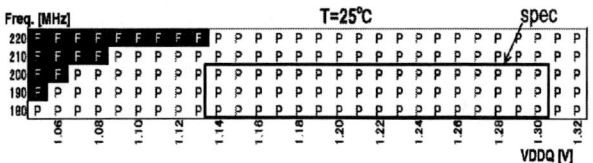

Fig. 16 Shmoo plot obtained after optimization

## V. POWER CONSUMPTION

Power consumption was evaluated by measuring IDD4R, which is a 12.8 GB/s burst read current with 50 % Data toggle. The power per bandwidth was 0.41 mW/Gbps in average. We have confirmed 89 % reduction of IO power compared to 6.4 GB/s LPDDR3 [7] shown in Fig. 17.

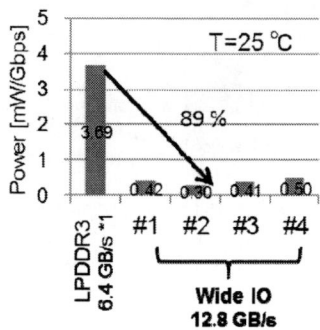

Fig. 17 Measured IO power consumption

## VI. CONCLUSION

Wide IO DRAM was successfully implemented with an 89 % IO power reduction under a 12.8 GB/s operation. The proposed test circuitry can be used to extract TSV connectivity defects prior to stacking process. The proposed noise monitor circuitry can extract voltage drops for each frequency range, and by performing the proposed impedance optimization flow, we have confirmed 30 mV Vmin improvement.

## ACKNOWLEDGEMENT

We thank Elpida Memory Inc. and SK hynix Inc. for the manufacturing and the providing of the Wide IO DRAM.

## REFERENCES

[1] JEDEC Standard Wide IO SDR specification. Dec. 2011.

[2] Y. Lou, Z. Yanm F. Zhang, P. D. Franzon, "Comparing Throug-Silicon-Via (TSV) void/pinhole defect self-test methods," *Journal of Electronic Testing*, vol. 28 No. 1, pp. 27-38, 2011.

[3] I. Savidis, S. Kose, E. G. Friedman, "Power Noise in TSV-Based 3-D Integrated Circuits," *proceedings of ISSCC*, vol. 48, No. 2, pp. 587-597, 2013.

[4] S. Takaya, M. Nagata, A. Sakai, T. Kariya, S. Uchiyama, H. Kobayashi, H. Ikeda, "A 100GB/s Wide IO with 4096b TSVs Through an Active Silicon Interposer with In-Place Waveform Capturing" *ISSCC Dig. Tech. Papers*, pp. 434-436, Feb. 2013.

[5] K. Otsuga, M. Onouchi, Y. Igarashi, T. Ikeya, S. Morita, K. Ishibashi, K. Yanagisawa,"An On-Chip 250 mA 40 nm CMOS Digital LDO Using Dynamic Sampling Clock Frequency Scaling with Offset-Free TDC-Based Voltage Sensor" *SOCC conf.* pp. 11-14, 2012.

[6] S. Kang, S. Cho, K. Yun, S. Ji, K. Bae, W. Lee, E. Kim, J. Kim, J. Cho, H. Mun Y-L. Park, "TSV Optimization for BEOL Interconnection in Logic Process" *proceedings of 3DIC*, Jan. 2012.

[7] Y. Bae, J-Y Park, S. Rhee, S. Ko, Y. Jeong, K-S. Noh, Y. Son,J. Youn, Y. Chu, H. Cho, M. Kim, D. Yim, H-C. Kim, S-H. Jung, H-I. Choi, S. Yim, J-B. Lee, J. Choi, K. Oh, "A 1.2 V 30 nm 1.6 Gb/s/pin 4Gb LPDDR3 SDRAM with Input Skew Calibration and Enhanced Control Scheme" *ISSCC Dig. Tech Papers*, pp. 44-45, Feb. 2012.

# A 12.8GS/s Time-Interleaved SAR ADC with 25GHz 3dB ERBW and 4.6b ENOB

## Yida Duan and Elad Alon

### Department of Electrical Engineering and Computer Sciences, University of California, Berkeley

*Abstract* - **This paper presents a 12.8GS/s 32-way hierarchically time-interleaved SAR ADC with 4.6-bit ENOB in 65nm CMOS. The prototype utilizes multi-stage sampling and a cascode sampler circuit to enable greater than 25GHz 3dB effective resolution bandwidth (ERBW). We further employ a pseudo-differential SAR ADC to save power and area. The core circuit occupies only 0.23 mm² and consumes a total of 162mW from dual 1.2V/1.1V supplies.**

## I. INTRODUCTION

Advanced backplane receivers with DSP-based equalization rely on ADCs with sampling rates of >10GS/s [1], and emerging coherent optical receivers require even higher sampling speeds (~40GS/s) [2, 3]. Most state-of-art solutions with a high degree of interleaving require either relatively high power consumption [2] or achieve relatively low 3dB ERBW [4]. This tradeoff is due to two issues with traditional time-interleaved ADC architectures. Specifically, such ADCs typically consist of a broadband input buffer that directly drives all of the parallel sampling switches. To meet the stringent jitter requirements required to achieve high ERBW, an excessive amount of power must be spent to distribute the many clock signals that turn on/off all of these switches. Furthermore, the input buffer must charge each sampling capacitor through the series resistance of the sampling switch during track mode, thus severely limiting the bandwidth and the sampling rate of the converter. To overcome these issues, we leverage a hierarchical sampling architecture [5] and propose a cascode sampler circuit.

The overall proposed ADC architecture and hierarchical sampling approach are described in Section II. In section III, the circuit level implementations of the cascode samplers and the sub-ADCs are discussed. Measurement results are then finally presented in Section IV.

## II. ADC ARCHITECTURE

The overall ADC architecture is shown in Fig. 1. This design adopts a hierarchical sampling approach in order to reduce the number of low-jitter clocks. Specifically, although the 12.8GS/s clock of the font-end sample and hold circuit (stage-1) must have low jitter to avoid high-frequency SNDR degradation, the following sampler stages effectively see constant input voltages at the instant of sampling. Therefore, jitter on the clocks of the succeeding stages does not further degrade SNDR [5]. The main function of the later samplers (stage-2 and stage-3) is to simply de-multiplex the input analog samples down to the rate of the sub-ADCs. A 1-to-4 de-multiplexer is used in stage-2, and four 1-to-8 de-multiplexors are used in stage-3. Finally, thirty-two sub-ADCs convert the samples into digital codes at 400MS/s. Figure 2 details the relative timing of the sampling clocks of each level. All clock signals are generated on-chip from

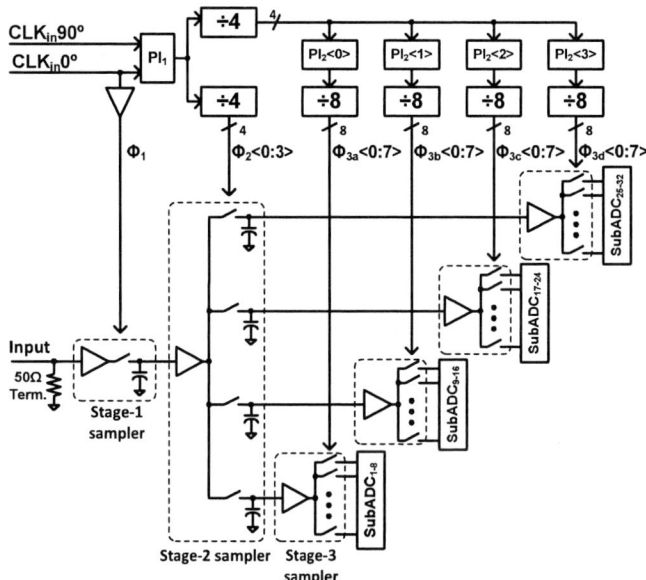

Fig. 1. ADC architecture

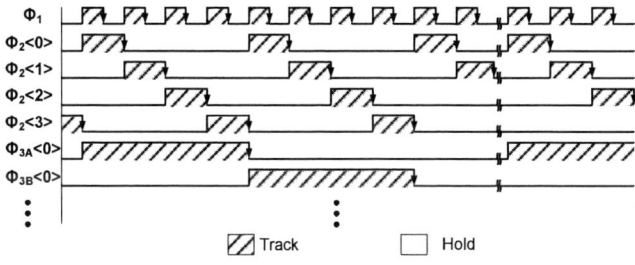

Fig. 2. Timing diagram

quadrature 12.8GHz input clocks. The sampling clock for Stage-1 ($\Phi_1$) is simply a buffered version of one of the 12.8GHz clock inputs; the Stage-2 samplers use 25% duty-cycle 3.2GHz quadrature clocks with non-overlapping track phases so that each sampler drives only one load capacitor at time. In order to maximize the tracking time available for the Stage-2 sampler, the falling edges of $\Phi_2<0:3>$ should occur right before the rising edges of $\Phi_1$. This phase alignment is accomplished using a 7-bit phase interpolator ($PI_1$). Sampling clocks for the four Stage-3 samplers are generated in a similar manner. A total of six frequency dividers and five phase interpolators are used to generate all necessary clock signals. As mentioned earlier, although thirty-seven clock signals must be distributed to all the samplers, only the sampling clock for the stage-1 sampler is jitter-critical.

## III. CIRCUIT IMPLEMENTATION

In this section, we will discuss the implementation of a proposed cascode sampler circuit and explore its advantages over traditional sampler designs. We will then further discuss circuit design considerations of the sub-ADC.

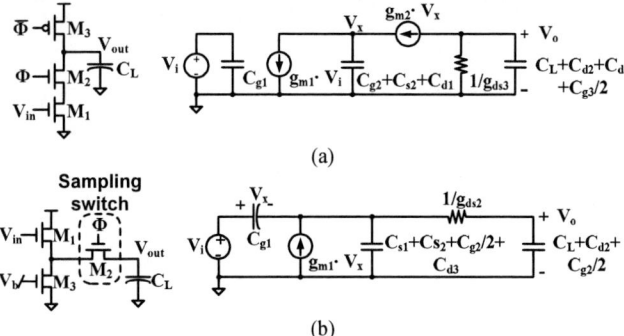

(a)

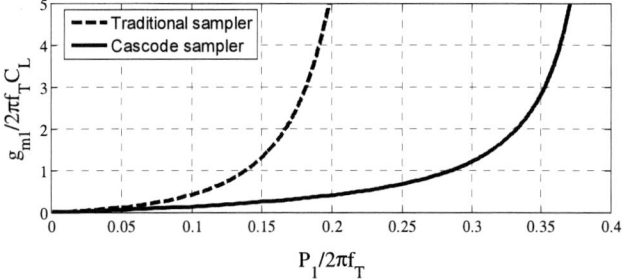

(b)

Fig. 3. Simplified schematic and small signal model of (a) proposed cascode sampler (b) traditional sampling circuit

Fig. 4 Trade-off between normalized trans-conductance, $g_{m1}/2\pi f_T C_L$, and normalized dominant pole, $P_1/2\pi f_T$, for traditional and cascode sampler.

## A. Casocde sampler

As shown in Fig. 3(b), typical sampling circuit designs consist of a source follower and a sampling switch. The final load capacitance $C_L$ is driven by the sum of the output resistance of the source follower and the switch resistance. For sampling periods that approach the several fanout-of-four inverter delays (typically > ~10GS/s), constant-$V_{GS}$ sampling techniques [6] do not perform well because of the increased rise time of the switch control signal. The circuit's settling time must therefore be maintained even under the worst-case (signal-dependent) switch resistance, leading to substantially increased power consumption.

To improve the tradeoff between sampler speed and power consumption, we propose the cascode sampler circuit shown in Fig. 3(a). When $\Phi$ is high, $M_{1,2,3}$ form a cascoded common source amplifier, with $M_3$ acting as a triode load resistor. $M_1$ and $M_3$ are sized to provide a DC gain of ~1. When $\Phi$ is low, both $M_2$ and $M_3$ are off and the output voltage is held on $C_L$. The key advantage of this architecture is that as long as the cascode device operates in saturation and has sufficiently high $f_T$ relative to the operating rate, the dominant pole of the amplifier is set only by the output node resistance and capacitance. In other words, in contrast to the traditional sampling circuit, the addition of the sampling switch does not directly affect the settling time.

To more rigorously highlight the benefits of cascode sampling for high-speed designs, we will next analyze the input device $g_m$ required to achieve a given dominant pole location for the two designs. This analysis will be performed utilizing the small signal models shown in Fig. 3. The dominant pole of the cascode sampling circuit is approximately:

$$P_1 = -\frac{g_{ds3}}{C_L + C_{d2} + C_{d3} + C_{g3}/2} \qquad (1)$$

To provide some numerical comparisons, we will assume that the $f_T$ of the PMOS transistors is half that of the NMOS transistors, and that the ratio between $C_{d,s}$ and $C_g$ is 1 for all transistors. We will further assume (as is the case in our particular technology) that the maximum triode $g_{ds}$ of a transistor is roughly twice the maximum saturation $g_m$. With all of these assumptions combined, if $f_T$ is the unity current-gain frequency of the NMOS transistors, then $g_{ds3}/C_{g3} = 2\pi f_T$. Finally, $g_{ds3}$ is equal to $g_{m1}$ for unity DC gain. With these assumptions, (1) can be rewritten as:

$$P_1 = \frac{2\pi f_T}{\frac{2\pi f_T C_L}{g_{m1}} + \frac{5}{2}} \qquad (2)$$

We next examine the dominant pole of the traditional sampler, which can be approximated as:

$$P_1 \cong \frac{1}{\frac{C_{g1}+C_{s1}+C_{d3}+C_{d2}+C_{s2}+C_{g2}+C_L}{g_{m1}} + \frac{C_L+C_{d2}+C_{g2}/2}{g_{ds2}}} \qquad (3)$$

Utilizing the same assumptions as stated earlier for the cascode sampler, (3) becomes:

$$P_1 = \frac{2\pi f_T}{\frac{15}{4} + 3\beta + \left(\frac{1}{\beta}+1\right)\frac{2\pi f_T C_L}{g_{m1}}} \qquad (4)$$

Where $\beta = W_2/W_1$ is the ratio of the widths of $M_2$ and $M_1$. The dominant pole achieves its maximum value when $\beta = \sqrt{3g_{m1}/2\pi f_T C_L}$, and this optimal $P_1$ is:

$$P_1 = \frac{2\pi f_T}{\frac{15}{4} + \frac{2\pi f_T C_L}{g_{m1}} + 2\sqrt{3 \cdot \frac{2\pi f_T C_L}{g_{m1}}}} \qquad (5)$$

With these expressions in hand, Fig. 4 compares the dominant poles of the two designs as a function of the $g_m$ of their input stages. The advantage of the cascode sampler is most apparent as the circuit bandwidth approaches $f_T$. Specifically, for $P_1=\frac{1}{8}\cdot 2\pi f_T$ – which is the target for the stage-1 sampler in our ADC design – the proposed cascode sampler requires over four times less $g_m$ (and hence power) than the traditional sampler. It is important to point out that the cascode sampler will suffer from distortion due to the nonlinear $g_m$ of the input transistor, $M_1$. However, with ~300mV differential input swing, this issue does not significantly degrade the SNDR of the ADC at the 5-bit target ENOB.

Figure 5(a) shows the complete schematic of the stage-1 sampler utilizing the proposed cascode sampling technique. In order to maintain a constant input capacitance, the tail current of $M_{1,2}$ is steered away to un-used branches $M_{7-10}$ when $\Phi_1$ is low. Note that if the sampling rate is doubled to 26.4GS/s, $M_{7-10}$ would be utilized to act as a sampler operating on $\overline{\Phi_1}$. Note that although the clocks connected to the gates of $M_{3,4}$ and $M_{7,8}$ are level-shifted to achieve peak voltages above $V_{dd}$ through AC-coupling capacitors (in order to keep those transistors in saturation), the $|V_{gs}|$ and $|V_{ds}|$ of those devices is kept below $V_{dd}$ at all times for reliable operation.

As shown in Fig. 5(b), the stage-2 demultiplexer is

978-1-4673-6145-3/13 $31.00 © 2013 IEEE

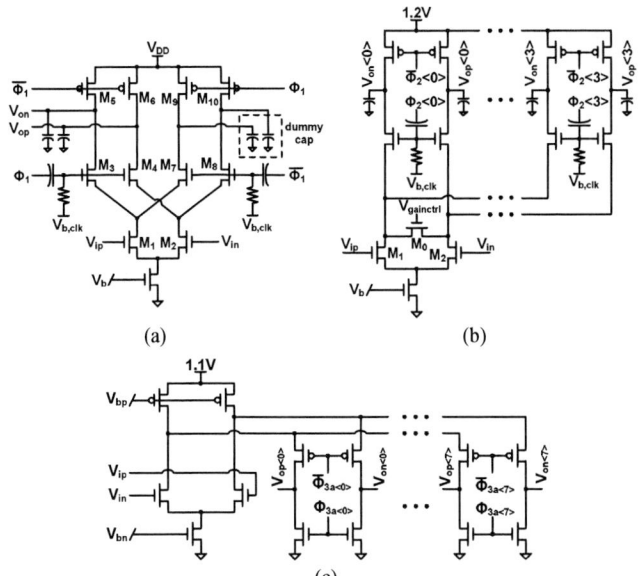

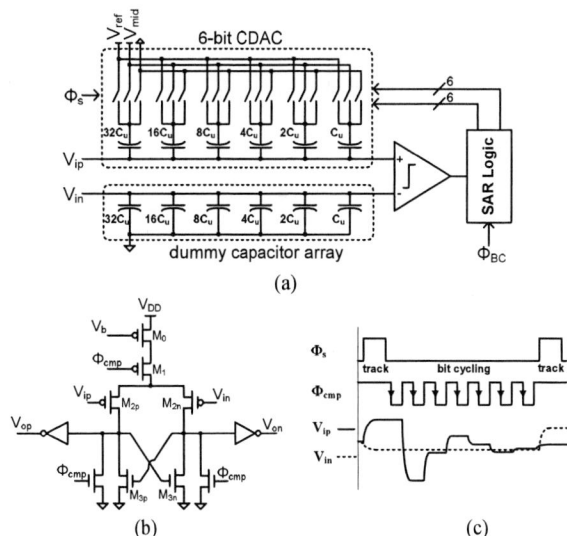

Fig. 5. Schematics of (a) stage-1 sampler, (b) stage-2 sampler, and (c) stage-3 sampler.

implemented in a similar manner. The four cascode branches are turned on successively in order to steer signal currents into one of the de-multiplexed outputs ($V_{op,on}$<0:3>). $M_0$ is added to enable DC gain tuning and maintain roughly unity gain for the entire sampler chain across process variation. The folded-cascode structure shown in Fig. 5(c) is used for stage-3 to lower the common-mode output voltage to a level that is compatible with the sub-ADCs.

### B. Sub-ADC

In ADC designs with a high degree of interleaving, both the power and the area consumed by each sub-ADC must be carefully considered. The importance of sub-ADC power is perhaps self-evident, but sub-ADC area can be equally important since a large sub-ADC implies longer wiring to route the inputs and clocks. These long wires can lead to significant parasitic loading and hence substantially increased sampler/clock distribution power. Thus, due to its energy-efficiency and relative compactness, for this design we have chosen an SAR-based sub-ADC.

Figure 6(a) highlights the 7-bit 400MS/s SAR sub-ADC. Extra bits beyond the target ENOB were included to enable digital calibration of cross-channel gain and offset. Since our target resolution is moderate, the capacitor matching constraints are relatively relaxed, enabling small unit capacitors (1fF) and reduced loading (64fF) on the stage-3 samplers.

To further save power and area, we also employed a single-ended DAC switching technique. During the tracking phase, the positive input ($V_{ip}$) is sampled on a 6-bit capacitor DAC while the negative input ($V_{in}$) is sampled onto a matched dummy capacitor array. As shown in Fig. 6(c), during bit cycling, $V_{ip}$ is forced to converge to $V_{in}$ in a binary fashion. Compared to conventional differential SAR ADCs, this technique saves half of the DAC switches and reduces the routing and fan-out of the SAR logic.

One drawback of single-ended switching is that the input

Fig. 6. (a) Sub-ADC block diagram (b) Sub-ADC comparator schematic. (c) Waveform at the comparator's inputs.

common-mode of the comparator does not remain constant during bit cycling. The comparator must therefore be designed so that its offset has minimal dependence on the input common-mode [7]. This is achieved using the circuit shown in Fig. 6(b); as in [7], $M_0$ is added as a current source to make the drain currents of $M_{1,2}$ relatively independent of input common-mode. This comparator design does not provide full swing outputs, and thus skewed inverters are added to recover to full digital levels. As a whole, each sub-ADC consumes only 1.14mW at 400MS/s and occupies 22μm×81μm of die area.

### III. MEASUREMENT RESULT

The proposed ADC was implemented in a 65nm GP CMOS technology. With the proposed cascode sampler design, the input capacitance of the ADC itself is <25fF. The 12.8GHz clock inputs are provided by an external signal generator and the ADC inputs are provided through a 0-40GHz differential RF-probe. The differential full-scale voltage of ADC before clipping is 335mV. Single tone tests were performed with 300mV differential peak-to-peak sinusoids with cable losses as well as cable length mismatches calibrated out at each input frequency. To generate a 4096-pt FFT and calculate SNDR, the input frequency was set to $N/4096 \cdot f_s$, where $N$ is an odd number. The ADC outputs were then subsampled at ~39kHz and read out through a shift register.

Foreground cross-channel gain and offset calibration are performed off-chip using a pilot input tone of ~3.1GHz; no non-linear correction was used. Fig. 7 shows the DNL and INL of the 7-bit raw output codes for each sub-ADC. Third order nonlinearity of the samplers is responsible for the shape of the INL curve. Fig. 8 shows the spectrum of the ADC for an input frequency of ~3.1GHz before and after gain/offset calibration. Calibration removes most of the inter-modulation tones caused by mismatch, increasing the SFDR to 32.4dB. This post-calibration SFDR is limited by third order distortion, and the remaining tones are caused by second order distortion and intermodulation between the input signal and the 3.2GHz Stage-2 demultiplexer clocks. As shown in Fig. 9, the prototype achieves 29.5dB SNDR at low input frequencies and

978-1-4673-6145-3/13 $31.00 © 2013 IEEE 459

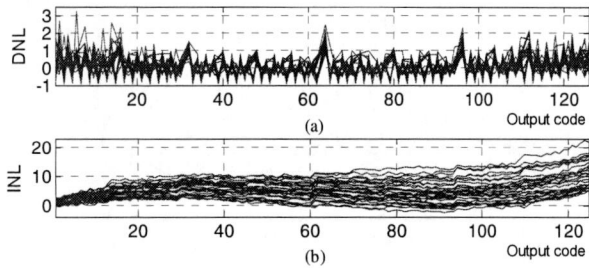

Fig. 7. (a) DNL and (b) INL vs. 7-bit raw output codes for each sub-ADC.

26.4dB at 25GHz. The SNDR remains relatively flat and above 26dB over the entire 25GHz bandwidth. The total ADC power is 162mW excluding digital I/O and initial clock buffers necessary only to interface with external instruments. The samplers consume 32mW, the sub-ADCs consume 36mW, and the clock distribution network consumes 83mW. The FOM is 0.74pJ/conv-step, which is comparable to [8] and [9] while achieving the highest -3dB ERBW published to date.

## IV. CONCLUSION

This work shows that the combination of hierarchical sampling and cascode sampler circuits can enable time-interleaved ADCs to achieve extremely broad ERBW and high sampling rates. The prototype demonstrates a 4.6-bit 12.8GS/s design with 25GHz ERBW. This architecture can be extended to support 25.6GS/s by exploiting the unused branch ($M_{7-10}$) in the stage-1 sampler along with replicating the rest of the ADC, or up to 51.2GS/s by replicating the entire 25.6GS/s structure with quadrature samples. Based on measurements of this prototype and simulated per-component power, we predict that the 25.6GS/s / 51.2GS/s designs would consume ~285mW / 570mW while retaining 25GHz ERBW.

## ACKNOWLEDGEMENTS

The authors would like to thank BWRC sponsors for support and the TSMC University Shuttle Program for chip fabrication.

## REFERENCES

[1] M. Harwood, et al., "A 12.5Gb/s SerDes in 65nm CMOS Using a Baud-Rate ADC with Digital Receiver Equalization and Clock Recovery", IEEE ISSCC. 2007.

[2] Y. Greshishchev, et al., "A 40GS/s 6b ADC in 65nm CMOS", IEEE ISSCC 2010.

[3] D. Crivelli, et al., "A 40nm CMOS Single-Chip 50Gb/s DP-QPSK/BPSK Transceiver with Electronic Dispersion Compensation for Coherent Optical Channels", IEEE ISSCC 2012.

[4] C.C. Huang, et al., "A CMOS 6-Bit 16-GS/s Time-Interleaved ADC with Digital Background Calibration", IEEE VLSI Circ. 2010.

[5] S. Gupta, et al., "A 1GS/s 11b Time-Interleaved ADC in 0.13um CMOS", IEEE ISSCC 2006.

[6] A. Abo, et al., "A 1.5V, 10-it, 14.3-MS/s CMOS Pipeline Analog-to-Digital Converter", IEEE JSSC, Vol. 34, No. 5, May 1999.

[7] C-C. Liu, et al., "A 10-bit 50-MS/s SAR ADC With a Monotonic Capacitor Switching Procedure", IEEE JSSC, Vol. 45, No. 4, April 2010.

[8] M. El-Chammas, et al., "A 12-GS/s 81-mW 5-Bit Time-Interleaved Flash ADC with Background Timing Skew Calibration", IEEE VLSI Circ. 2010.

[9] S. Verma, et al., "A 10.3GS/s 6b Flash ADC for 10G Ethernet Applications", IEEE ISSCC 2013.

[10] W. Cheng, et al., "A 3b 40GS/s ADC-DAC in 0.12um SiGe", IEEE ISSCC 2004.

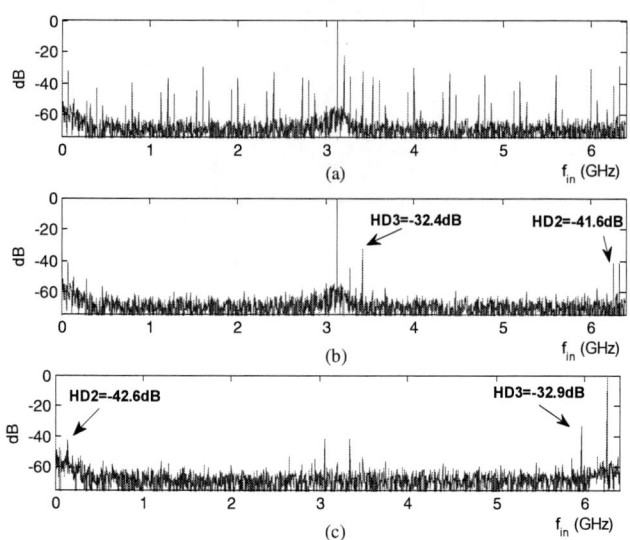

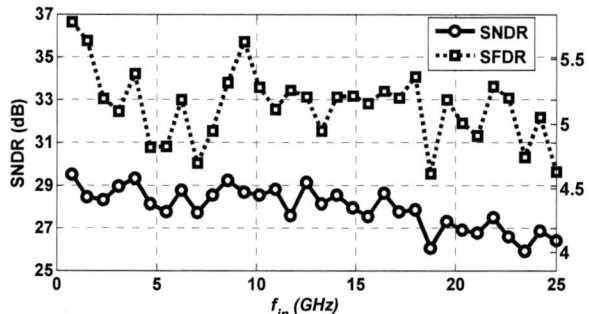

Fig. 8. ADC output spectrum with $f_{in} \approx 3.1$GHz (a) before calibration, (b) after calibration, and (c) $f_{in} \approx 6.25$GHz after calibration

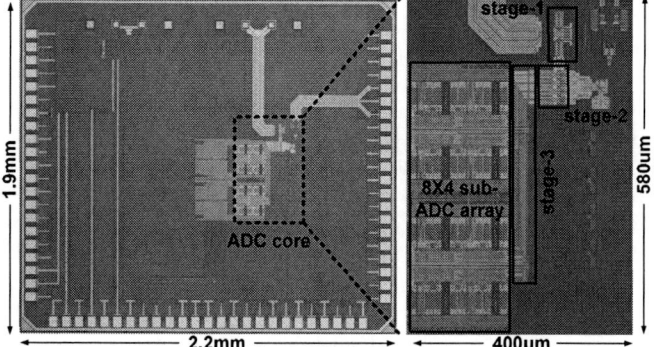

Fig. 9. Input frequency vs. SNDR and SFDR.

Fig 10. Die photo.

## TABLE I
### PERFORMANCE COMPARISON

|  | $f_s$ (GS/s) | BW (GHz) | SNDR @ BW (dB) | Power (mW) | FOM (pJ/c-s) |
|---|---|---|---|---|---|
| *This work* | 12.8 | 25 | 26.4 | 162 | 0.74 |
| [2] | 40 | 18 | 25.1 | 1500 | 2.5 |
| [3] | 25 | 9 | 25.8 | 500 | 1.25 |
| [4] | 16 | 3 | 28 | 435 | 0.96 |
| [8] | 12 | 6 | 25.1 | 81 | 0.46 |
| [9] | 10.3 | 5 | 33 | 240 | 0.7 |
| [10] | 40 | 20 | 18.6 | 3800 | 13.6 |

# A 10GS/s 6b Time-Interleaved ADC with Partially Active Flash sub-ADCs

Xiaochen Yang[1], Robert Payne[2], and Jin Liu[1]

[1]University of Texas at Dallas, Richardson, TX, 75080
[2]Texas Instruments, Inc., Dallas, TX 75243

*Abstract*-A 10GS/s 6b time-interleaved ADC in 65nm CMOS is presented in this paper. A partially-active flash sub-ADC structure is proposed to improve the ADC power efficiency and a source-follower based boot-strap T&H circuit is proposed to reduce input kickback and improve the ADC bandwidth. The four-phase 2.5GHz clocks for the sub-ADCs are derived from a 5GHz Nyquist frequency input clock. This leads to accurate timing skew calibration based on duty-cycle calibration, improving the ADC effective resolution at high input frequencies. Measured SNDR is 34.3dB at low input frequencies and 32.0dB at the Nyquist input frequency. The ADC including the input clock buffer consumes 83mW with a FOM of 197fJ/conv-step.

## I. INTRODUCTION

In high-speed wire-line communication systems, ADC based receivers can allow more sophisticated equalization and complex timing recovery approaches in the following digital signal processors to enhance the spectral efficiency of transmission media. The key design challenge for these ADCs is to ensure sufficient effective resolution across the input signal bandwidth such that critical information of the received signal is preserved, while maintaining low power consumption. This calls for ADCs with high power-efficiency. To achieve larger than 10GS/s sampling rate in current CMOS technologies, the ADCs need to adopt time-interleaving architectures and require multi-phase clocks for the time-interleaved channels. Timing skew among the multi-phase clocks can degrades ADC effective resolution at high input frequencies, leading to degraded power efficiency. The state of the art >10GS/s ADC designs in CMOS have employed pipeline [1], SAR [2] and flash [3][4] sub-ADC architectures. In this paper, a new sub-ADC architecture, partially active flash (PA flash), is presented to achieve a better power efficiency. Also presented are a new source-follower based boot-strap track-and-hold (T&H) circuit to reduce the input kickback and interference among the interleaved channels and a multi-phase clock generation scheme with accurate timing skew calibration based on duty-cycle calibration. Table I compares the performance of this work with the state of the art ADCs with larger than 10GS/s sampling rate. The design in [4] had the best power efficiency with a FOM of 350fJ/conv-step by using background timing skew calibration for the sub-ADC multi-phase clocks. This work achieved a better power efficiency with a FOM of 197fJ/conv-step in the same technology, while achieving better resolution.

Table I. Comparison with the state of the art >10GS/s ADCs

|  | [1] | [2] | [3] | [4] | **This work** |
|---|---|---|---|---|---|
| CMOS (nm) | 90 | 65 | 65 | 65 | **65** |
| Sub-ADC | Pipeline | SAR | Flash | Flash | **PA-Flash** |
| Fs (GS/s) | 10 | 40 | 10 | 12 | **10** |
| ENOB@DC | 5.8 | 5.5 | 4.9 | 4.3 | **5.4** |
| ENOB@Fs/2 | 5.1 | ~3.9 | 4.5 | 3.9 | **5.0** |
| Power (mW) | 1600 | 1500 | 330 | 81 | **83** |
| FOM$^{\dagger}$ (fJ/cs) | 3200$^{\ddagger}$ | 3000$^{\ddagger}$ | 1105 | 348 | **197** |

$^{\dagger}$ FOM=power/($2^{\text{ENOB@DC}}$*min(Fs, 2*ERBW))
$^{\ddagger}$ ERBW estimated based on ENOB plot in the paper

## II. ADC ARCHITECTURE

Fig. 1 shows the block diagram of the 10GS/s 6b four-way time-interleaved ADC. Each of the four channels runs at 2.5GS/s and is clocked by one of the 2.5GHz four-phase clocks $\Phi 1$ to $\Phi 4$. The detailed clock generation and skew calibration schemes will be discussed in section IV. Each channel consists of the proposed source-follower based boot-strap (SF-BS) T&H circuit and 6b partially active flash sub-ADC. In this design, the sampling rate in each channel is 2.5GS/s which is power efficient at 65nm CMOS process. Capacitive DACs [5] are used for comparator offset calibration for the sub-ADCs. For higher data rate, wide-bandwidth calibration schemes such as that in [6][7] can be used. The digital outputs of the four channels are decimated and multiplexed to generate serialized outputs for measurement.

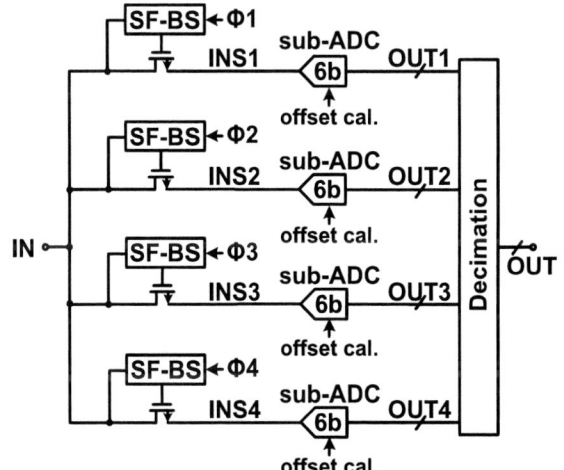

Fig. 1. Block diagram of the 10GS/s four-way time-interleaved ADC.

## III. PARTIALLY ACTIVE FLASH SUB-ADC

Fig. 2 shows the proposed power-efficient partially active flash sub-ADC design. A 6b flash ADC is divided into four slices each with 4b resolution. For one input sample, only one slice is activated for comparison based on the decision of a 2b coarse flash ADC, while the rest slices are off to save power. The 2b and 4b decisions are combined to achieve a total of 6 bits. Upon the availability of comparison result of the coarse 2b ADC, one of the four clock signals, CLK<1:4>, will be active and trigger the corresponding 4b slice inside the 6b partially-active flash ADC. This is different from a folding or a two-step ADC which requires analog residue to be generated for the second step and long settling time to meet the accuracy requirement.

Fig. 3 presents the timing diagram for the PA flash sub-ADC. The clock for each sub-ADC is designed to have a quarter of the cycle for tracking and three quarters for holding. A cycle starts with 100ps tracking time. At 150ps, the 2b ADC starts to regenerate. The decision of the 2b coarse ADC asynchronously activates one of the CLK<1:4> signals to choose a corresponding 4b slice by turning on its preamplifiers which are designed to have a fast response time. The regeneration of the 4b slice starts at about 350ps into a cycle and lasts 200ps. During this 200ps, the preamplifiers in the 4b slice are in reset, so that comparator regeneration is possible though the track-and-hold circuit is back to the tracking mode. The choice of 25%-75% duty cycle for the track-and-hold clocks, $\Phi i$, is important. It allows the input signal to see only one sub-ADC at a time to decrease the input loading and reduce interference among interleaved channels. In addition, it allows two comparisons in one clock period, each with reasonable regeneration time. For applications requiring very low meta-stability rate, a meta-stability detector can be used. When meta-stable state is detected for a particular comparator in the coarse ADC, the input level must be very close to the reference level of that comparator and we can directly set the ADC output to the code corresponding to that reference level. For the pre-amplifiers, since they conduct current continuously, half of their tail current is switched off to save power when they are in reset and not in use for amplification. Half the current is still on to shorten the wake-up time from reset to amplification.

## IV. CLOCK GENERATION AND SKEW CALIBRATION

Fig. 4 shows the clock generation and skew calibration schemes for the sub-ADCs. A differential-to-single-ended circuit (DTS) converts the half-rate 5GHz differential input clocks to complementary single-ended ones, C and CB. Dividing C and CB by 2 generates 2.5GHz clocks C1 and C2. Passing C and CB through transmission gates clocked by C1, C2 and their complementary phases generates the T&H clocks, $\Phi 1$ to $\Phi 4$. The critical sampling-point timing for $\Phi 1$ to $\Phi 4$ is at the transition from the track mode to the hold mode indicated by rising arrows in Fig. 4. This timing is derived directly from edge 1 for $\Phi 1$, $\Phi 3$ and derived from edge 2 for

$\Phi 2$, $\Phi 4$, through transmission gates clocked by C1, C2 or their complementary phases. Since $\Phi 1$ and $\Phi 3$ come from the same clock edge, edge 1, they should have minimal timing skew between them. The only possible timing skew between them is introduced when they pass through the transmission gates. In this design, the transmission gates are large enough such that skews between $\Phi 1$, $\Phi 3$ and between $\Phi 2$, $\Phi 4$ are controlled within 100fs. C1 and C2 are just control signals providing passing windows and its timing variation will not affect the critical timing of $\Phi 1$ to $\Phi 4$. The remaining source of timing error is the duty cycle distortion on C and CB caused by the DTS and buffer circuits, as illustrated by the timing variation of edge 1 in the dotted line. This duty cycle distortion can cause timing skew between the group $\Phi 1/\Phi 3$ and the group $\Phi 2/\Phi 4$. Noted that the cross-coupled inverters used in DTS keep C and CB closely complimentary despite local mismatch effect. Then, with simple digital duty cycle calibration, C and CB can simultaneously obtain a duty cycle very close to 50%, thereby making the timing distance between edge 1 and 2 very close to the sampling distance of 100ps and meeting the required timing accuracy. Therefore, accurate four-phase timing skew calibration can be accomplished by the simple duty cycle calibration.

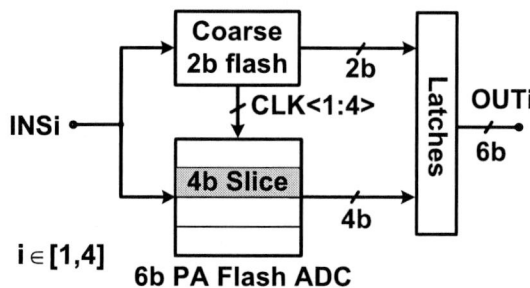

Fig. 2. Proposed partially-active flash sub-ADC structure.

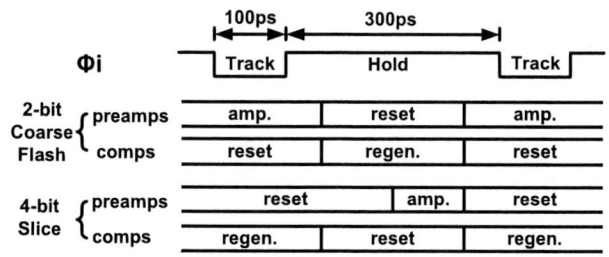

Fig. 3. Timing diagram for the PA flash sub-ADC.

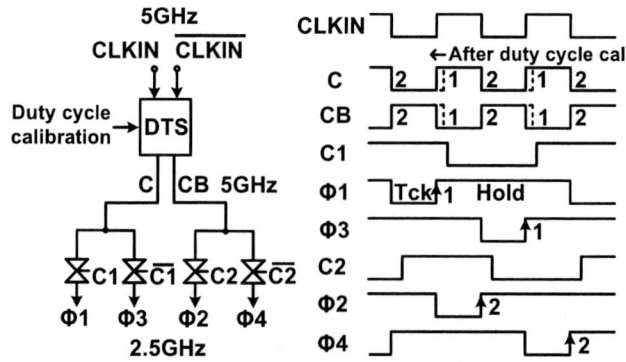

Fig. 4. Clock generation and skew calibration schemes.

Fig. 5 shows the detailed circuits for input clock buffer and the clock generation scheme. The input clock buffer adopts that in [8]. To limit duty cycle distortion, the incoming differential 5GHz clocks first pass through a capacitor-degenerated peaking amplifier and AC coupled inverters with resistor feedback. An inverter buffer chain is followed with cross-coupled inverters to ensure the single-ended clock outputs, C and CB, to be complementary. The interleaved T&H sampling clocks, $\Phi1$ to $\Phi4$, are derived directly from C and CB through transmission gates. As shown in the figure, the gates pass through every other pulse from the 5GHz, C or CB, to generate the 2.5GHz sampling clocks. Compared with conventional clock generation methods, such as those using dividers, transmission gates have less timing delay, leading to less timing skew among the interleaved channels. As discuss earlier, the residual duty cycle distortion of C and CB needed to be corrected. We adopt a floating current source design [8] to digitally calibrate the clock duty cycle distortion to be within the required timing accuracy.

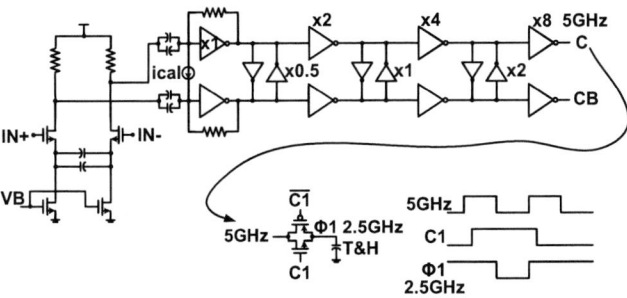

Fig. 5. Detailed circuits for the input clock buffer and the proposed four-phase clock generation scheme.

## IV. SOURCE-FOLLOWER BASED BOOT-STRAP T&H CIRCUIT

For input Nyquist frequency as high as 5GHz, boot-strap sampling technique is needed. The schematic of the proposed source-follower based boot-strap T&H circuit is shown in Fig. 6. In a conventional boot-strap T&H circuit, many internal nodes draw charges from the input node, causing kickback noise and affecting other interleaved channels, especially at high input frequencies. In the proposed T&H circuit, we use a PMOS source-follower, M1, to accomplish the boot-strap sampling. Besides the sampling switch M0, the input signal sees only a PMOS source follower M1. This eliminates the need of charging the boot-strap internal nodes and therefore reduces input kickback. With M2 on and M3 off, the PMOS source follower level-shifts the input voltage by approximately 1.0V to accomplish a constant Vgs across the sampling switch for boot-strap sampling. With M3 on and M2 off, the gate of M0 is grounded and the input is held at the output node INSi. The proposed circuit requires a higher voltage of 2.3V. A switch-cap voltage booster doubles the clock voltage to turn on and off M2. Cascode transistors are placed to ensure safe voltage levels for all devices. Although the input voltage level-shifting is not strictly 1.0V for different input voltages, the proposed source-follower based T&H

circuit can achieve the required linearity, as demonstrated by design simulation and later measurement results. Fig. 7 presents the simulated SFDR of the complete ADC at 10GS/s. It demonstrates that the proposed source-follower based boot-strap T&H is able to provide needed accuracy for a 6b converter.

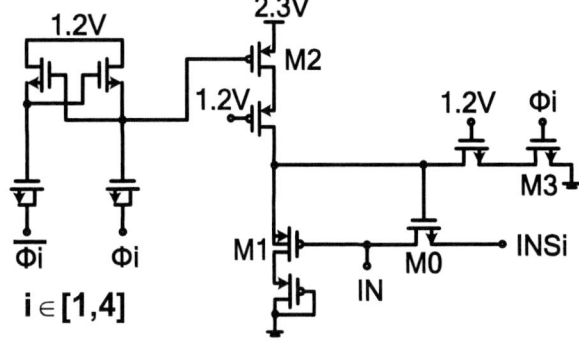

Fig. 6. Proposed source-follower based boot-strap T&H circuit.

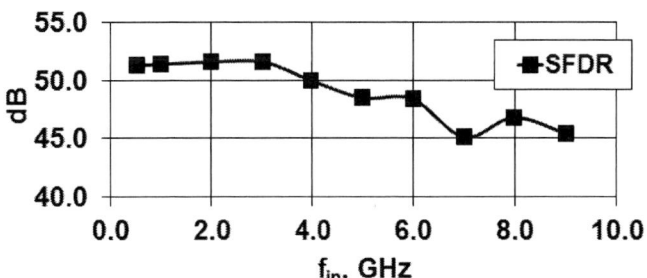

Fig. 7. Simulated SFDR of the time-interleaved ADC at 10GS/s.

## V. MEASUREMENT RESULTS

The prototype chip was fabricated in a standard 65nm CMOS process. The chip micrograph is shown in Fig. 8 and the ADC occupies an active die area of $0.2\text{mm}^2$. Digital offset calibration is performed at the ADC startup with the input signals set at the common mode. The offset calibration can reduce DNL from -1LSB/2.5LSB to -0.8LSB/0.8LSB and reduce INL from -2.0LSB/1.5LSB to -0.7LSB/0.5LSB.

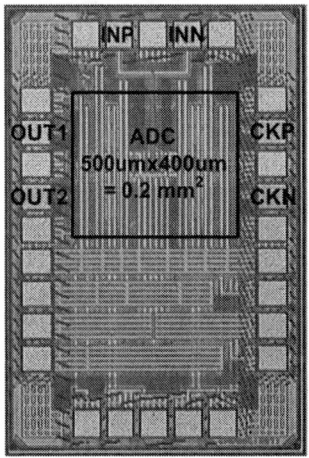

Fig. 8. Chip micrograph.

Fig. 9 shows the measured spectrum of the decimated ADC output at 10GS/s with a 5GHz Nyquist input signal. Before the duty cycle calibration, the duty cycle distortion of the input clocks results in timing skew among interleaved channels causing a large spur at Fs/2-Fin as shown in Fig. 9 (a), where Fs is the sampling frequency and Fin is the input frequency. After the duty cycle calibration, the spur is reduced from -26.8dB to -48.4dB, corresponding to a timing skew reduction from 2ps to 0.25ps, as shown in Fig. 9 (b). The resulting SNDR improvement is from 25.5dB to 32.0dB while the SFDR is improved by 18dB from 26.8dB to 44.7dB. The figure also shows that there is no significant spur at Fin-Fs/4 and 3/4Fs-Fin locations before or after the duty cycle calibration, indicating that the skew between Φ1, Φ3 and the skew between Φ2, Φ4 introduced by the transmission gates in the clock generation scheme are indeed minimal.

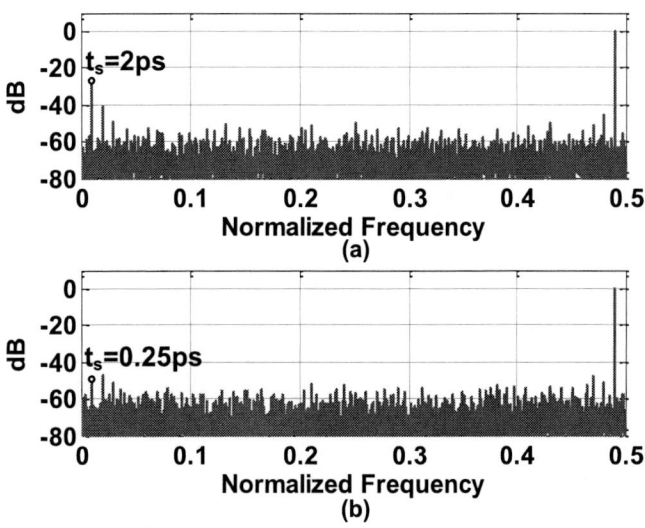

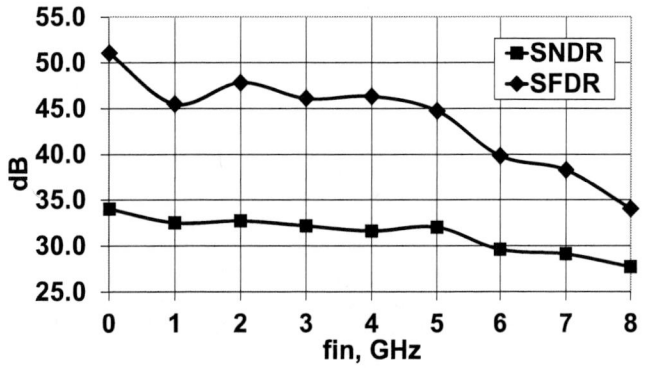

Fig. 9. Measured spectrum of the decimated ADC output at 10GS/s with 5GHz input (a) before and (b) after duty cycle calibration.

Fig. 10 shows the measured SNDR and SFDR versus input frequency at 10GS/s sampling rate. At low input frequencies, the measured SNDR is 34.3dB (5.4-bit ENOB) and the measured SFDR is 51.0dB. At Nyquist input frequency of 5GHz, the measured SNDR is 32.0dB (5.0-bit ENOB) and the measured SFDR is 44.7dB.

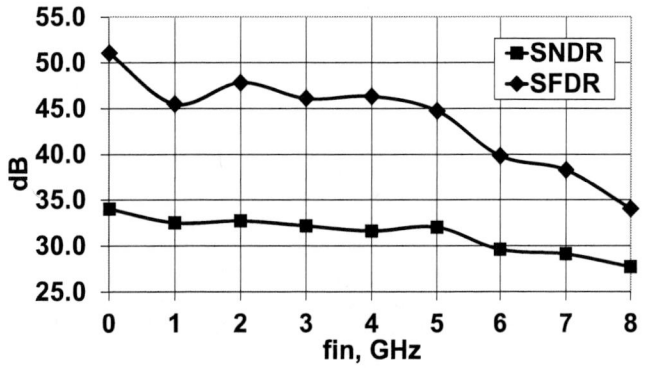

Fig. 10. Measured SNDR and SFDR vs. input frequency at 10GS/s.

Fig. 11 is the ADC power breakdown at 10GS/s. The total ADC power consumption is 83mW, with 25mW for the input clock buffer and multi-phase clock generation circuits, 9mW for each 6b PA flash ADC, 3mW for each 2b coarse ADC, and 2.5mW for each SF-BS T&H circuit. The resulted ADC FOM is 197fJ/conv-step at 10GS/s.

## IV. CONCLUSION

This paper presents a 10GS/s 6b time-interleaved ADC with a proposed partially active flash sub-ADC structure to improve the ADC power efficiency. To achieve better effective resolution at high input frequencies, a new source-follower based boot-strap T&H is presented. The four-phase 2.5GHz clocks for sub-ADCs are generated from a half-rate 5GHz clock and accurate skew calibration is accomplished by simple duty-cycle calibration, resulting minimal skew among the multi-phase clocks to improve the ADC effective resolution across the wide input frequency bandwidth.

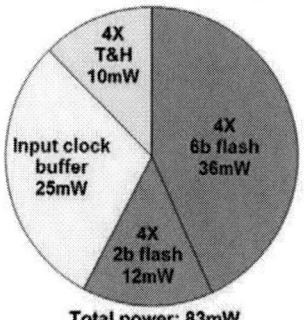

Fig. 11. Power breakdown.

## ACKNOWLEDGEMENT

This work was supported by Semiconductor Research Corporation (SRC).

## REFERENCES

[1] A. Nazemi, et. al., "A 10.3GS/s 6bit time-interleaved/pipelined ADC using open-loop amplifiers and digital calibration in 90nm CMOS," in *IEEE VLSIC Dig. Tech. Papers*, 2008, pp. 18-19.

[2] Y. M. Greshishchev, et. al., "A 40GS/s 6b ADC in 65nm CMOS," in *IEEE ISSCC Dig. Tech. Papers*, Feb, 2010, pp. 390-391.

[3] J. Cao, et. al., "A 500mW ADC-based CMOS AFE with digital calibration for 10Gb/s serial links over KR-backplane and multimode fiber," *IEEE J. Solid-State Circuits*, vol. 45, pp. 1172-1185, Jun. 2010.

[4] M. El-Chammas and B. Murmann, "A 12-GS/s 81-mW 5-bit time-interleaved flash ADC with background timing skew calibration," in *IEEE VLSIC Dig. Tech. Papers*, 2010, pp. 157-158.

[5] G. Van der Plas, et. al., "A 0.16pJ/conversion-step 2.5mW 1.25GS/s 4b ADC in a 90nm digital CMOS process," in *IEEE ISSCC Dig. Tech. Papers*, Feb. 2006.

[6] J. Yao, J. Liu and H. Lee, "Bulk voltage trimming offset calibration for high-speed flash ADCs," *IEEE Trans. Circuits Syst. II*, vol. 57, pp. 110-114, Feb. 2010.

[7] J. Yao and J. Liu, "A 5-GS/s 4-Bit flash ADC with triode-load bias voltage trimming offset calibration in 65-nm CMOS," in *Proc. IEEE CICC*, Sept. 2011.

[8] C. Menolfi, et. al., "A 28Gb/s source-series terminated TX in 32nm CMOS SOI," in *IEEE ISSCC Dig. Tech. Papers*, Feb. 2012.

# An 8-Bit 4-GS/s 120-mW CMOS ADC

Hegong Wei, Peng Zhang[1], Bibhu Datta Sahoo[2], and Behzad Razavi
University of California, Los Angeles, CA, USA
[1]Tsinghua University, Beijing, China
[2]Amrita University, Amritapuri, Kerala-India

## Abstract

**A four-channel time-interleaved pipelined ADC employs a new timing calibration technique to suppress mismatch-induced spurs and achieve a Nyquist-rate SNDR of 44.4 dB. Designed in 65-nm CMOS technology, the ADC draws 120 mW, providing an FOM of 219 fJ per conversion step.**

## I. Introduction

The design of gigahertz ADCs greatly benefits from time-interleaving if the interchannel mismatches are corrected efficiently and reproducibly. While offset and gain mismatches can be readily removed by long-term averaging and correction in the digital domain, timing mismatches pose other difficulties. For example, the background calibration proposed in [1] requires digital multipliers and hence a high complexity for more than two channels.

This paper describes a four-channel time-interleaved CMOS ADC that maintains an SNDR of 44 dB up to the Nyquist rate while drawing 120 mW. This performance is achieved through the use of a new background timing mismatch calibration technique requiring no multipliers.

Section II presents the proposed timing mismatch detection technique and its underlying mathematical principles. Section III describes the ADC implementation, including the architecture and the timing mismatch correction circuitry. Section IV summarizes the experimental results for a prototype designed in 65-nm CMOS technology.

## II. Timing Mismatch Detection

It is possible to measure the timing mismatch by means of only digital adders. Consider the scenario shown in Fig. 1(a), where sample $y_2$ is offset from its ideal point in time by $T$. Noting that the time difference between $y_1$ and $y_2$ is $2\,T$ seconds greater than that between $y_2$ and $y_3$, let us construct $y_{21}=|y_2 - y_1|$ and $y_{32}=|y_3 - y_2|$. We expect that, with no timing mismatch, $y_{21}$ and $y_{32}$ exhibit equal averages and surmise that the average value of $y_{21} - y_{32}$ is proportional to $T$. It is difficult to prove this conjecture directly, but if we approximate the absolute value operation by a squaring function, then the proof is as follows. We wish to prove that the average difference between $(y_2 - y_1)^2$ and $(y_3 - y_2)^2$ is proportional to $T$. To this end, we write the expectation of

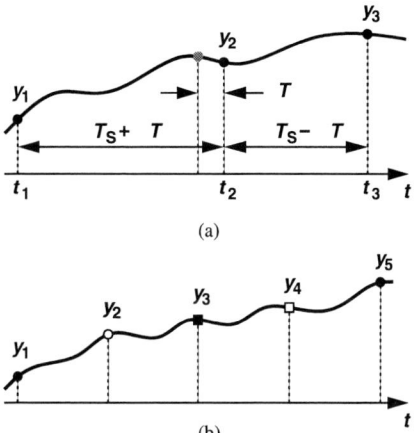

Fig. 1. Waveforms showing effect of timing error in a (a) two-channel or (b) four-channel interleaved ADC system.

$(y_2 - y_1)^2$ as

$$
\begin{aligned}
E[(y_2 - y_1)^2] &= E[y_2^2] + E[y_1^2] - 2E[y_2 y_1] \quad (1) \\
&= \sigma_{y2}^2 + \sigma_{y1}^2 - 2E[y(t_1 + T_S + T)y(t_1)],
\end{aligned}
$$

where $\sigma^2$ denotes the average power. We recognize the expectation on the right-hand side as the autocorrelation of $y(t)$, $R(\tau)$, evaluated at $T_S + T$. That is,

$$
E[(y_2 - y_1)^2] = 2\sigma_y^2 - 2R(T_S + T). \quad (2)
$$

Similarly, the average value of $(y_3 - y_2)^2$ is given by

$$
E[(y_3 - y_2)^2] = 2\sigma_y^2 - 2R(T_S - T). \quad (3)
$$

For a small $T$, the two terms on the right-hand side can be approximated by their Taylor series around $T_S$, yielding

$$
E[(y_2 - y_1)^2] - E[(y_3 - y_2)^2] \approx -4\,T\frac{dR}{d\tau}. \quad (4)
$$

Thus, the average difference between the squares or between $|y_2 - y_1|$ and $|y_3 - y_2|$ can serve as a measure of the mismatch if $dR/d\tau$ does not vanish around $\tau = T_S$.

The above computation of $T$ can be generalized to more than two interleaved channels. Fig. 1(b) shows the case for four. Here, we first consider $|y_3 - y_1|$ and $|y_5 - y_1|$ and detect the timing mismatch between the first and third channels. Once this mismatch is corrected, the third channel can

978-1-4673-6145-3/13 $31.00 © 2013 IEEE    465

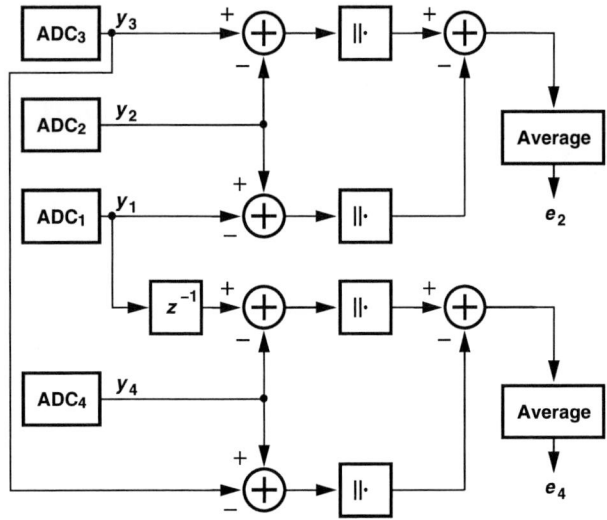

Fig. 2. Timing mismatch detection in a 4-channel ADC system.

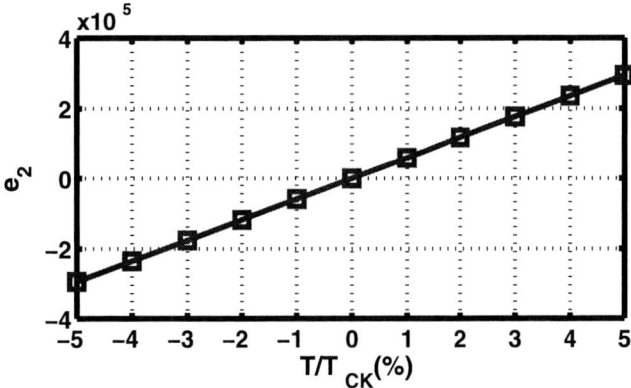

Fig. 3. Simulated error as a function of timing mismatch (arbitrary vertical scale).

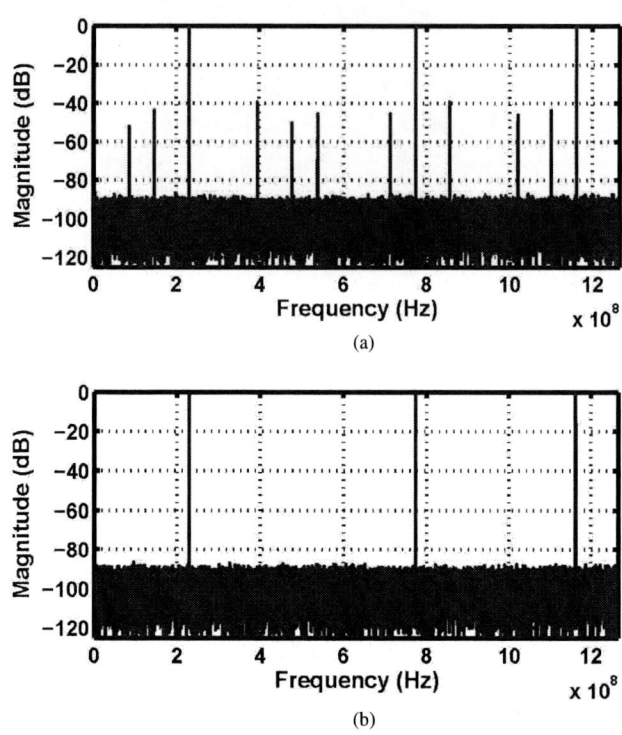

Fig. 4. Simulated ADC output spectrum (a) before, and (b) after calibration for a multi-tone input.

## III. ADC IMPLEMENTATION

### A. Architecture

Shown in Fig. 5, the ADC design incorporates four pipelined channels, a phase generator, and a phase correction circuit. Based on the design in [2], each channel consists of a 4-bit first stage, seven 1.5-bit stages, and a 2-bit last stage. The downsampled outputs of the channels are carried off-chip for per-channel gain error calibration, interchannel offset and gain

be considered ideal. Now, as depicted in Fig. 2, we compute the timing mismatch in $y_2$ by forming the difference between $|y_2 - y_1|$ and $|y_3 - y_2|$ and that in $y_4$ by forming the difference between $|y_4 - y_3|$ and $|y_5 - y_4|$. Fig. 3 plots, as a example, $e_2$ versus the displacement of $y_2$ from its ideal position in time for a random band-limited analog input signal. The error displays a monotonic dependence on $T$. In another behavioral multi-tone test, Fig. 4 shows the output spectrum for a four-channel ADC before and after calibration with infinite mismatch calibration resolution. We return to the choice of this resolution in Section III.

The digital functions in Fig. 2 require only delay elements and adders. (The absolute value is realized by changing the sign bit). Moreover, to reduce the power consumption, these functions can be enabled for a short time once every few milliseconds to perform the error measurement and kept dormant for the rest of the time [1].

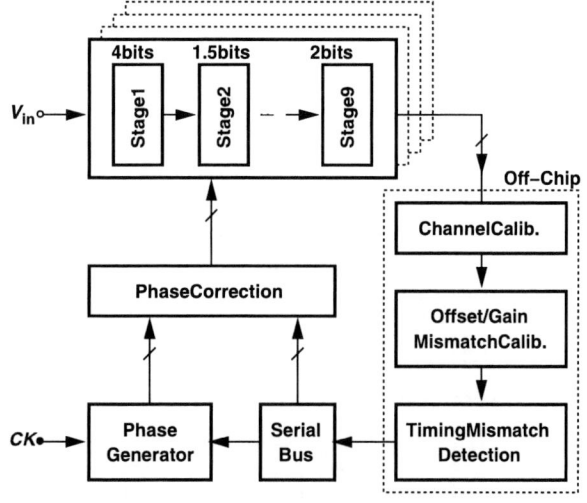

Fig. 5. ADC architecture.

978-1-4673-6145-3/13 $31.00 © 2013 IEEE

mismatch correction, and timing mismatch detection according to the scheme in Fig. 2. The result of this detection travels back to the chip on a serial bus and adjusts the phase correction circuit.

### B. Phase Generation and Correction

The ADC generates four 1-GHz clock phases with a duty cycle of about 25%[1] (Fig. 6). These phases are then adjusted according to the results of timing mismatch detection.

The phase correction circuit must provide (1) a wide enough tuning range to accommodate the maximum anticipated mismatch, $T_{max}$, and (2) a sufficiently fine step size, $T_{min}$, to minimize the SNDR penalty due to the residual timing mismatch. It is important to note that the mismatch arises in both the clock paths and the analog signal paths leading to the channels. In this work, we select $T_{max}$=3.5 ps and $T_{min}$=30 fs.

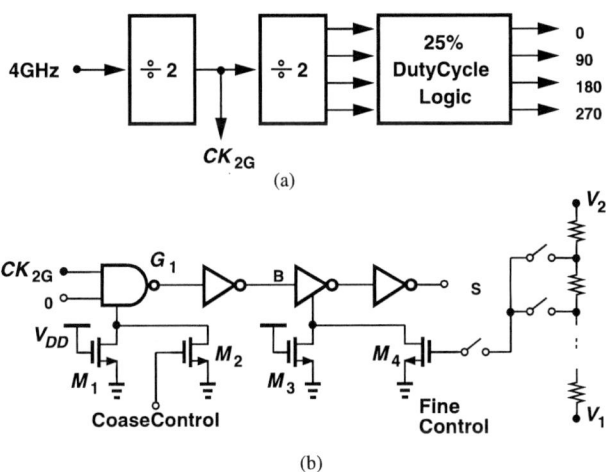

(a)

(b)

Fig. 6. (a) Phase generation, and (b) phase adjustment circuits.

In order to achieve the above values for $T_{max}$ and $T_{min}$ with minimal phase noise, power, and complexity, we employ, in the clock path of each channel, a short variable delay line (VDL) with a 1-bit coarse control and a 6-bit fine control. Illustrated in Fig. 6(b), the VDL incorporates transistors $M_1$ and $M_2$ to adjust the strength of the NAND gate $G_1$. The relative widths of $M_1$ and $M_2$ determine the absolute delay step that the latter can create. In this design, $W_2$=1.6$W_1$ and $L_1$=$L_2$.

The fine delay adjustment is realized by $M_4$, whose gate voltage can be varied from $V_1$ ($\approx V_{TH}$) to $V_2$ in 64 steps. With $W_3$=2.25$W_4$ and $L_3$=$L_4$, this scheme provides a $T_{min}$ of 30 fs. Note that $CK_{2G}$ gates the phase entering this circuit so as to remove the skew and phase noise contributed by the second $\div 2$ circuit and the logic in Fig. 6(a). The ADC employs four VDLs to adjust $\phi_0$-$\phi_{270}$. The phase generation and correction circuits consume a total of 17.7 mW at full rate.

---

[1]Each channel allocates 25% of its clock cycle to sampling and 75% to conversion.

## IV. EXPERIMENTAL RESULTS

The four-channel ADC has been fabricated in 65-nm digital CMOS technology. Fig. 7 shows the die photograph of the prototype, which occupies an active area of 900 $\mu$m $\times$1500 $\mu$m. All of the measurement results are reported for a sampling rate of 4 GHz.

Fig. 8 plots the measured DNL and INL after per-channel gain error calibration, indicating a maximum of $-0.75$ LSB for the former and $-1.5$ LSB for the latter after calibration.

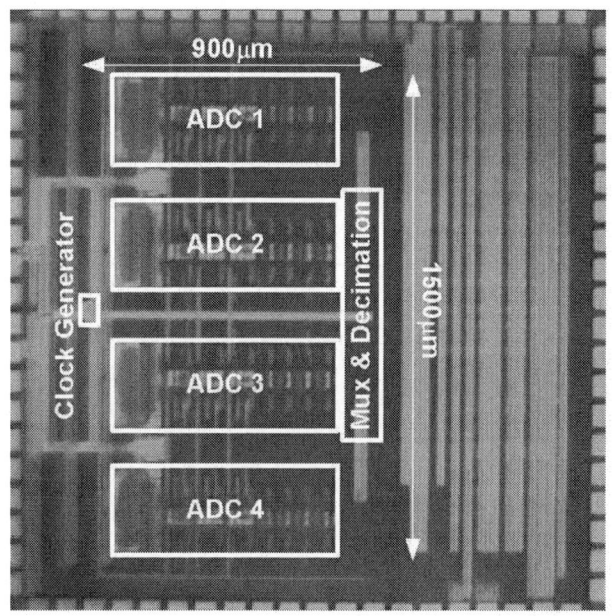

Fig. 7. Prototype ADC micrograph.

Fig. 9 shows the convergence of the phase adjustment codes for channel 2 and 4 in response to a 1.9-GHz sinusoidal input after channel 3 is calibrated as explained in Section II and Fig. 10 depicts the corresponding improvement in the SNDR. Plotted in Fig. 11 is the subsampled output spectrum before and after timing mismatch calibration, demonstrating that the mismatch-induced spurs can be suppressed to about $-60$ dB.

Fig. 12 plots the SNDR as a function of the analog input frequency. The ADC consumes 120 mW: 57 mW in the analog section, 46 mW in the digital section, and 16 mW in the reference ladders. Table 1 compares our prototype's performance to that of recent gigahertz ADCs with an SNDR range of 44 to 49 dB.

## Acknowledgments

Research supported by the DARPA HEALICS Program, Realtek Semiconductor, and Pullman Lane Productions. We gratefully acknowledge the TSMC University Shuttle Program for chip fabrication.

## REFERENCES

[1] B. Razavi, "Problem of Timing Mismatch in Interleaved ADCs," *Proc. CICC*, pp. 1-8, Sept 2012.

[2] B. Sahoo and B. Razavi, "A 10-Bit 1-GHz 33-mW CMOS ADC," *Symposium on VLSI Circuits,* vol. 38, pp. 2138-2146, December 2003.

[3] C. Chen, et al., "A 12-Bit 3 GS/s Pipeline ADC With 0.4mm2 and 500 mW in 40 nm Digital CMOS," *IEEE J. Solid-State Circuits,* vol. 47, no. 4, pp. 1013-1021, Apr 2012.

[4] E. Janssen, et al., "An 11b 3.6GS/s Time-Interleaved SAR ADC in 65nm CMOS," *ISSCC Dig. Tech. Papers,* pp. 464-465, Feb 2013.

[5] K. Doris, et al., "A 480mW 2.6GS/s 10b 65nm CMOS Time-Interleaved ADC with 48.5dB SNDR up to Nyquist," *ISSCC Dig. Tech. Papers,* pp. 180-181, 2011.

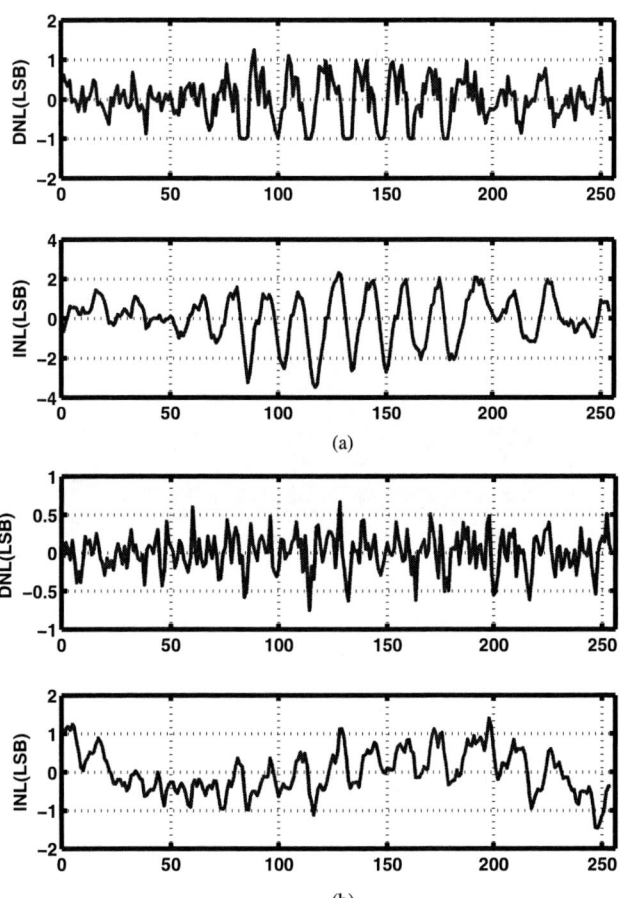

(a)

(b)

Fig. 8. Measured DNL and INL at a clock rate of 4 GS/s, (a) before , and (b) after per-channel calibration.

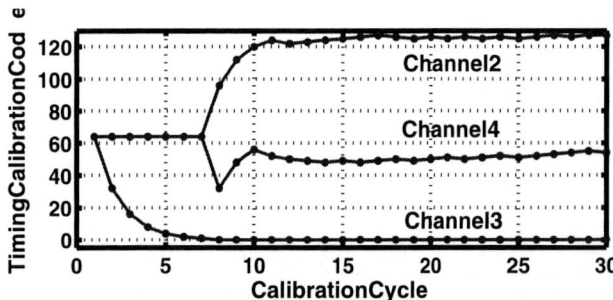

Fig. 9. Measured timing calibration code convergence.

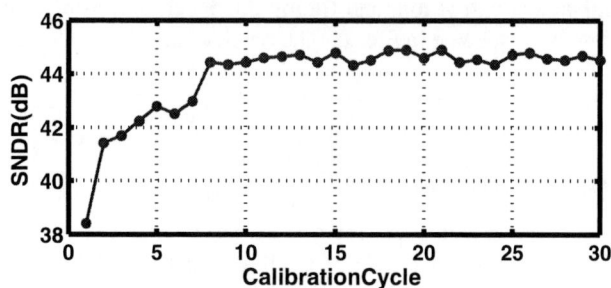

Fig. 10. Measured SNDR during convergence of timing calibration.

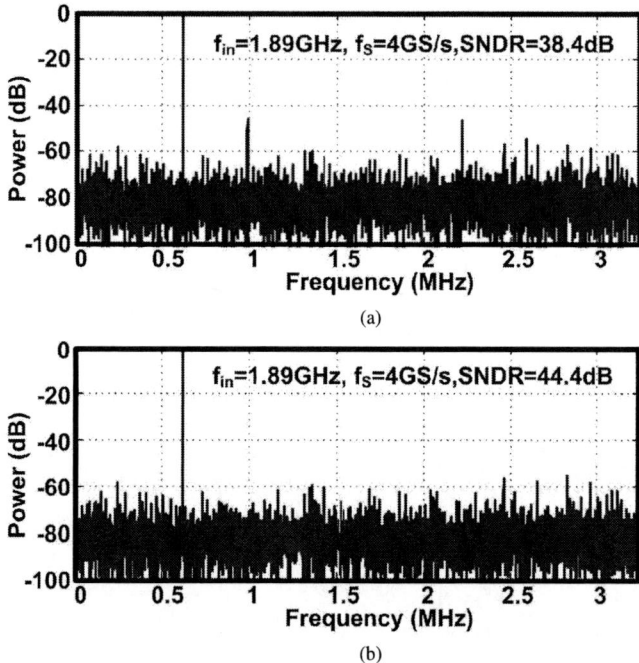

(a)

(b)

Fig. 11. Measured output spectrum (a) before, and (b) after timing mismatch calibration (decimated by a factor of 625).

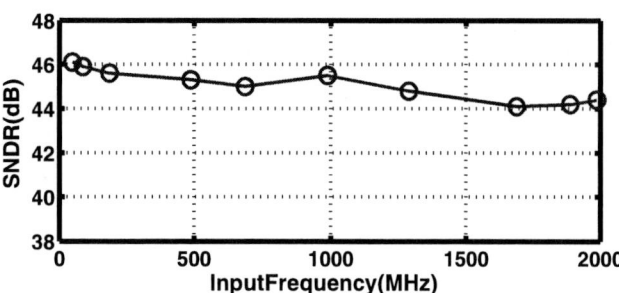

Fig. 12. Measured SNDR as a function of $f_{in}$ at 4 GS/s.

|  | This Work | JSSC'12 [4] | ISSCC'13 [5] | ISSCC'11 [6] |
|---|---|---|---|---|
| f(GS/s) | 4 | 3 | 3.6 | 2.6 |
| SNDR @Nyq.(dB) | 44.4 | 49 | 47.5 | 48.5 |
| Supply(V) | 1.2/1.4 | 2.5 | 1.2/2.5 | 1.2/1.3/1.6 |
| Power(mW) | 120 | 500 | 795 | 480 |
| FOM(fJ) | 219 | 724 | 1140 | 1000 |
| Tech.(nm) | 65 | 40 | 65 | 65 |
| Area(mm)$^2$ | 1.35 | 0.4 | 7.4 | 5.12 |

Table 1. Comparison with state-of-the-art designs.

# A 7.1-mW 1-GS/s ADC with 48-dB SNDR at Nyquist Rate

Sedigheh Hashemi and Behzad Razavi
Electrical Engineering Department
University of California, Los Angeles

## Abstract

A two-stage pipelined ADC employs a double-sampling residue amplifier, two interleaved precharged DACs, and a new calibration scheme to correct for residue gain error, offset, and nonlinearity. Realized in 65-nm CMOS technology and sampling at 1 GHz, the prototype exhibits an FOM of 25 fJ/conversion-step while drawing 7.1 mW from a 1-V supply.

## I. INTRODUCTION

Pipelined analog-to-digital converters (ADCs) have reached gigahertz sampling rates at resolutions around 10 bits while consuming 30 to 40 mW [1, 2] and exhibiting a figure of merit (FOM) of 70 to 90 fJ/conversion-step. Most pipelined architectures avoid the use of a multi-bit *second* stage as its high input capacitance would slow down the residue generation in the first stage. However, a two-stage pipelined architecture remains attractive because it would eliminate the residue amplifiers necessary in the subsequent stages, potentially saving power.

This paper presents a two-stage pipelined ADC that employs 33 comparators and one differential amplifier in 65-nm CMOS technology to achieve a resolution of 9 bits at 1 GHz with an FOM of 25 fJ/conversion-step. A new calibration technique corrects the gain error and nonlinearity of the open-loop residue amplifier as well as the offsets of the comparators, leading to a low-power design.

Section II introduces the proposed ADC architecture and Section III describes the precision techniques employed in this work. Section IV presents the experimental results.

## II. ADC ARCHITECTURE

### A. High-Level View

Shown in Fig. 1 (a) is a functional single-ended diagram of the system. The converter consists of a 5-bit coarse flash sub-ADC ($ADC_1$), two resistor-ladder DACs (sharing the same ladder), a residue amplifier, and a 5-bit fine flash sub-ADC ($ADC_2$). At a sampling rate of 1 GHz with realistic clock transitions and non-overlap times, the overall system must acquire and convert in about 950 ps. In order to relax the trade-off of the residue amplifier, the ADC employs three techniques: (1) a double-sampling network composed of $C_{S1}$ and $C_{S2}$ in the front end, (2) two interleaved resistor-ladder DACs, and

(3) pipelining after the amplifier. As a result, the amplifier has one full clock period to settle. To interleave the two DACs, each coarse comparator is followed by two time-interleaved paths whose outputs drive two switches tied to one tap of the resistor ladder. One bit of redundancy accommodates comparator offsets and timing mismatches in the front end.

The timing diagram in Fig. 1 (b) illustrates the operation at a high level. In the first half cycle (from 0 to $t_1$), $C_{S1}$ and the coarse sub-ADC sample the input while $DAC_1$ is precharged [2]. The next half cycle is entirely allocated to $ADC_1$'s conversion, after which one ladder tap voltage is selected. . The following complete cycle (from $t_2$ to $t_4$) is dedicated to the settling of $V_{DAC1}$ and $V_{res}$. At $t = t_4$, switch $S_F$ turns off so as to freeze the residue, and $ADC_2$ begins to convert. Additionally, at $t = t_2$, $C_{S2}$ and $ADC_1$ start sampling the input while $DAC_2$ is precharged. $DAC_2$ and $C_{S2}$ then produce and hold the residue at the input of the amplifier for one clock cycle.[1]

### B. Detailed Architecture

Figure 2 shows the ADC architecture in greater detail, highlighting the power-saving methods applied to this design. The 5-bit coarse stage utilizes a comparator as a polarity detector, allowing a twofold reduction in the number of comparators and hence their kick-back noise and power consumption. Once the sampling of the analog input is completed, this comparator is clocked and its decision slides the reference of the 15 comparators to either the top or the bottom half of the full scale, i.e., $[V_{REF1}^+ \ V_{mid}]$ or $[V_{REF1}^- \ V_{mid}]$, respectively. After 100 ps, these 15 comparators are clocked and their collective decision is used to update $V_{DAC1}$ or $V_{DAC2}$ (which have been precharged to $V_{in}$ during the sampling mode). In addition to reducing the power and hardware, the above reference sliding scheme also presents only half of the analog input swing to the comparators. As a result, the overall ADC can accommodate rail-to-rail input swings. These benefits accrue at the cost of 170 ps in the conversion time of the coarse ADC.

By virtue of pipelining, the fine ADC has a half cycle for conversion, and hence, first detects the polarity of the amplified residue by means of three comparators.[2] Based on this decision, either the top or the bottom bank of comparators is clocked [3]. To save power, the fine ADC shares the inter-

---

[1]To remove dynamic errors, a small portion of the cycle is assigned to resetting the input nodes of the residue amplifier.

[2]The use of three comparators rather than one relaxes the offset requirement.

978-1-4673-6145-3/13 $31.00 © 2013 IEEE

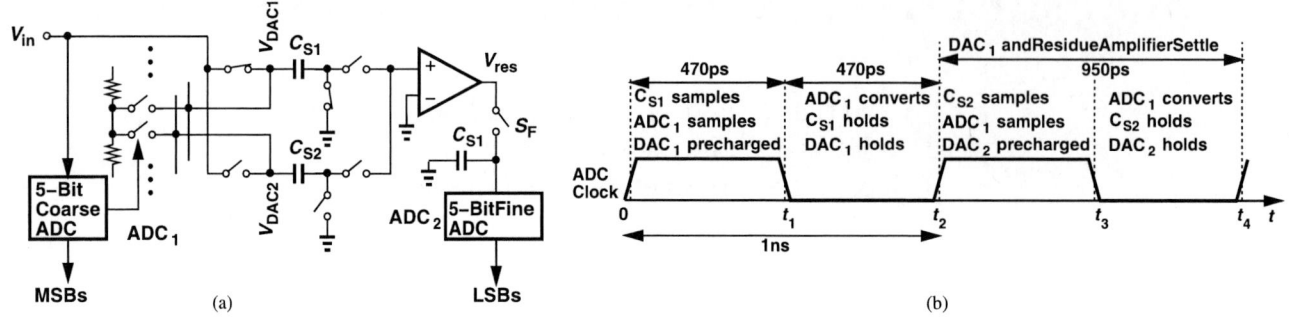

Fig. 1. ADC (a) high-level view, and (b) timing diagram.

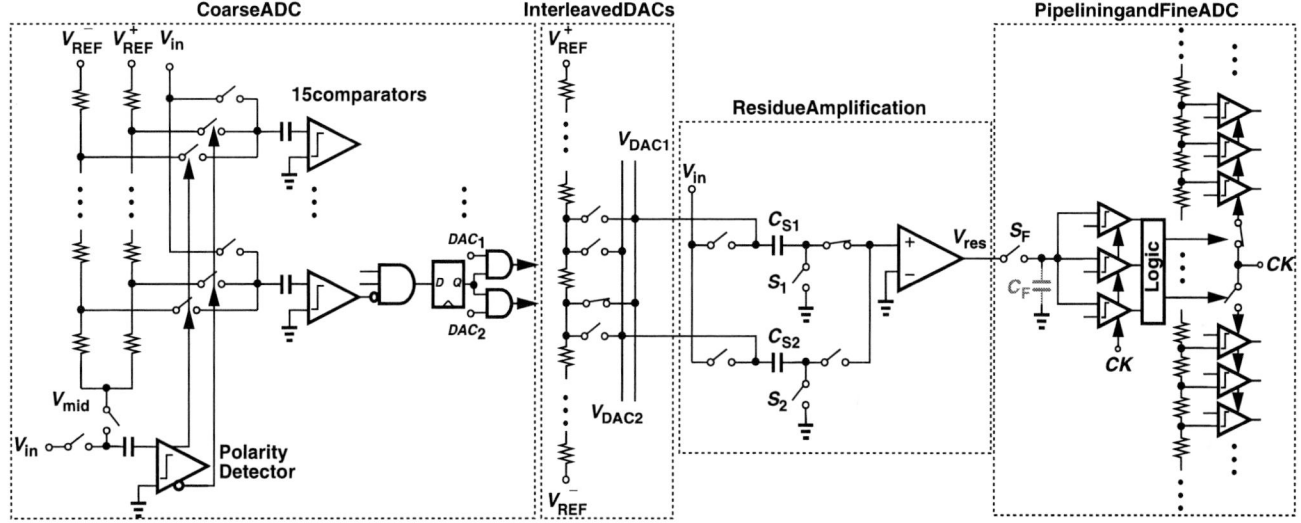

Fig. 2. Detailed ADC architecture.

leaved DACs' resistor ladder even though Fig. 2 shows two separate ones for clarity.

In order to reduce the effect of the timing mismatch between the samples taken by $C_{S1}$ and $C_{S2}$, this design employs an additional clock phase within the switch bootstrapping circuits, thus reducing the mismatch to that of two switches.

## III. PRECISION TECHNIQUES

A fast, power-efficient design naturally incorporates small, poorly-controlled devices, thus necessitating additional means to improve the precision. A key issue is the hardware and power overhead associated with such means. This section describes two techniques that ultimately lead to an SNDR of 51 dB at low input frequencies and 48 dB at the Nyquist rate with a conversion rate of 1 GHz.

### A. Offset Cancellation

At a clock frequency of 1 GHz and with a response time of about 150 ps, 33 comparators can consume considerable power - unless their design is linearly scaled down to the point where their electronic noise limits the performance. However, with nearly-minimum size transistors, the offset grows quite large, demanding a proportionally wide correction range and

prohibitive overhead. Additionally, the offset correction devices themselves may contribute a large noise and/or offset and slow down the circuit.

For the offset cancellation technique to negligibly degrade the speed of a comparator, it is desirable to avoid tying appendages (e.g., programmable capacitors [3, 4]) to the internal nodes. In this work, the coarse and fine ADCs employ a cancellation technique that resides entirely outside the comparator. Illustrated in Fig. 3(a), our approach calibrates comparator number $j$ for a decision threshold of $V_j$ as follows: (1) connect one input to $V_j$, (2) change the other input by means of $DAC_j$ until the comparator output changes, and (3) freeze the $DAC_j$ content. To minimize the effect of comparator noise on the calibration, this procedure is repeated 10 times and the average value is chosen.

The dedicated DACs in Fig. 3 (a) may appear to add great complexity. However, we recognize that $DAC_j$ need only generate a moderate range around $V_j$. The DACs can therefore be embedded within the main resistor ladder as depicted in Fig. 3(b), provided that the ladder yields sufficient resolution for offset cancellation. With 64 taps and differential implementation, the offset of coarse comparators is reduced to 4 LSB. The fine ADC offsets are corrected in a similar manner.

978-1-4673-6145-3/13 $31.00 © 2013 IEEE

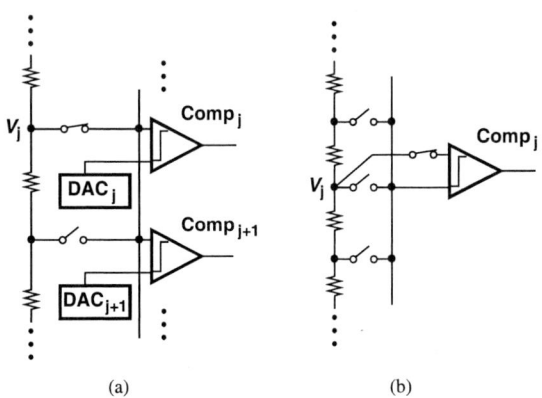

Fig. 3. (a) Conceptual offset calibration technique, and (b) actual implementation.

### B. Proposed Calibration Technique

The open-loop residue amplifier in Fig. 2 introduces gain error, offset, and nonlinearity, the most critical impact of which is a residue range that does not match the full scale of $ADC_2$. We propose a low-complexity calibration technique that "programs" the fine stage so as to match the residue range. Depicted in Fig. 4(a), the idea is to adjust the decision thresholds of the fine ADC so as to remove comparator offsets as well as reach a full scale that is aligned with the minimum and maximum residue values.

Calibration proceeds as follows. In the sampling mode, one of the precision tap voltages produced by the resistor ladder DAC, $V_k$, is impressed on the top plate of $C_{S1}$ in Fig. 2. Next, the top plate switches to zero, and the resulting voltage is amplified and applied to the fine ADC. This voltage experiences the gain error, nonlinearity, and offset of the residue amplifier [Fig. 4(b)], but, for correct operation, it must trip a known comparator, number k. Thus, the reference voltage of comparator k is iteratively adjusted by $DAC_k$ until it trips for a sampled voltage of $V_K$. This procedure is repeated for all of the fine comparators, covering a total range of $\pm 36$ LSB around the nominal voltage tap. As with the coarse ADC, $DAC_k$ is in fact embedded within the resistor ladder [Fig. 4(c)].

## IV. EXPERIMENTAL RESULTS

The prototype ADC has been designed and fabricated in standard 65-nm CMOS technology in an active area of $350\,\mu$m $\times\,280\,\mu$m. Figure 5 shows the die photograph. The foreground offset cancellation and calibration are performed off-chip, with the results returned to on-chip registers through a serial bus. The ADC has been tested in a chip-on-board assembly. All of the measurement results reported below are for a sampling rate of 1 GHz with a 1-V supply.

Figure 6 plots the differential nonlinearity (DNL) and integral nonlinearity (INL), before and after calibration, indicating maximum values of $-0.87$ LSB and $+1.8$ LSB, respectively.

Figure 8 shows the output spectrum for an analog input frequency of 490 MHz, demonstrating an SNDR of 48 dB and an SFDR of 58 dB at Nyquist rate. Figure 7 plots the SNDR as a function of the input frequency, revealing a low-frequency value of 51 dB. Table I summarizes the measured performance

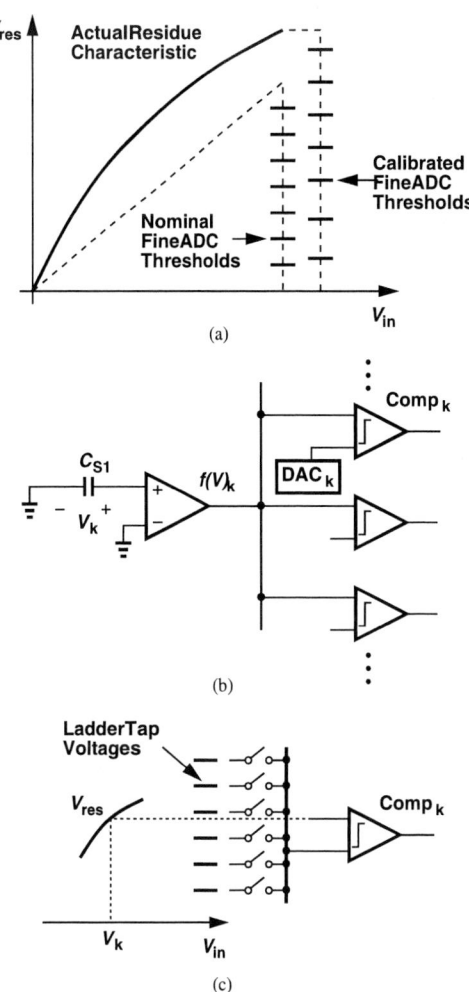

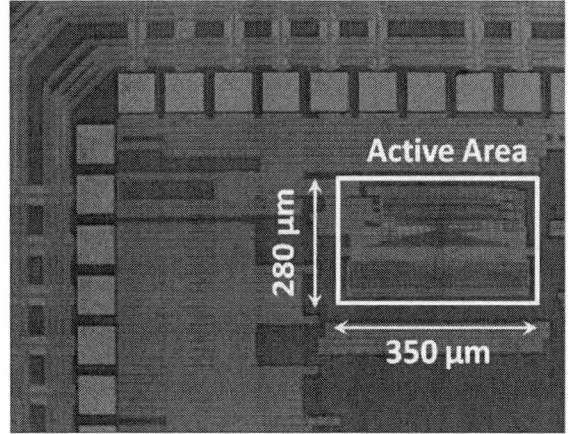

Fig. 4. (a) Calibration of fine ADC decision thresholds to correct for residue errors, (b) conceptual implementation, and (c) use of main ladder to realize voltage taps.

Fig. 5. ADC die photograph.

of the prototype and compares it with that of recent prior art in the range of 8 to 10 bits.

## V. CONCLUSION

The use of a double-sampling front end, interleaved precharged DACs, and a new comparator-based calibration scheme allows two-stage pipelined ADCs to operate at high speeds with good dynamic performance. A 60-nm 1-GHz prototype exhibits an SNDR of 48 dB at Nyquist-rate while consuming 7.1 mW.

## Acknowledgment

This research was supported by the DARPA HEALICs program and Realtek Semiconductor. The authors gratefully acknowledge the TSMC University Shuttle Program for chip fabrication.

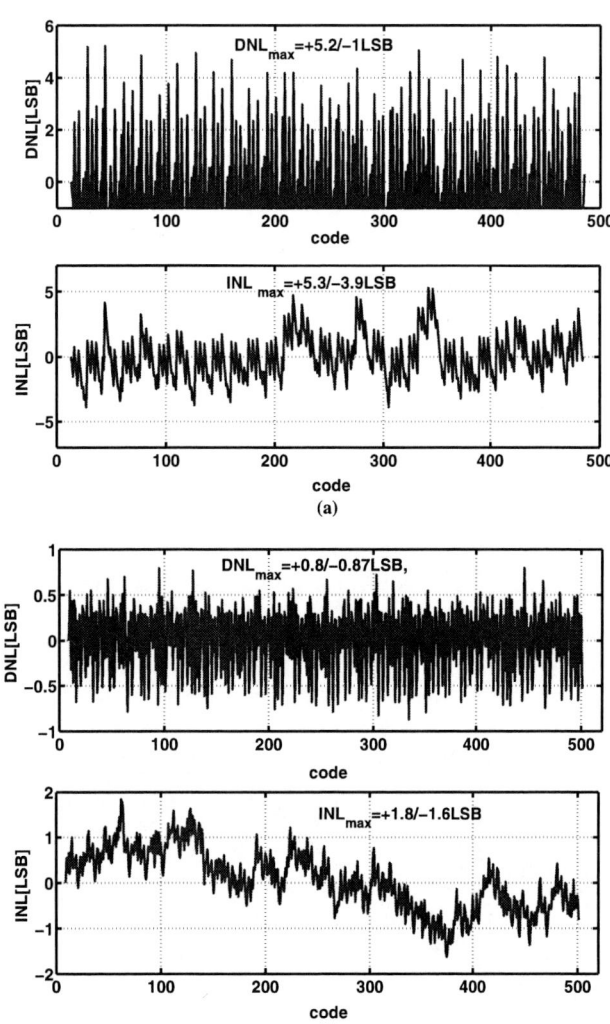

Fig. 6. Measured DNL and INL (a) before, and (b) after calibration.

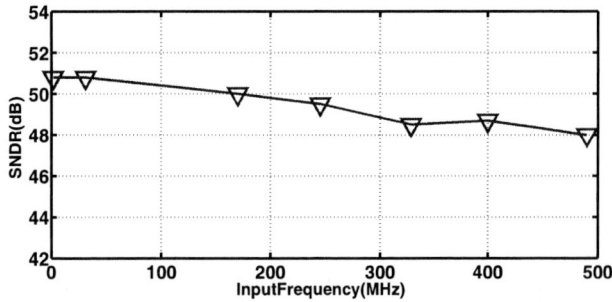

Fig. 7. Measured SNDR versus input frequency at a sampling rate of 1 GHz.

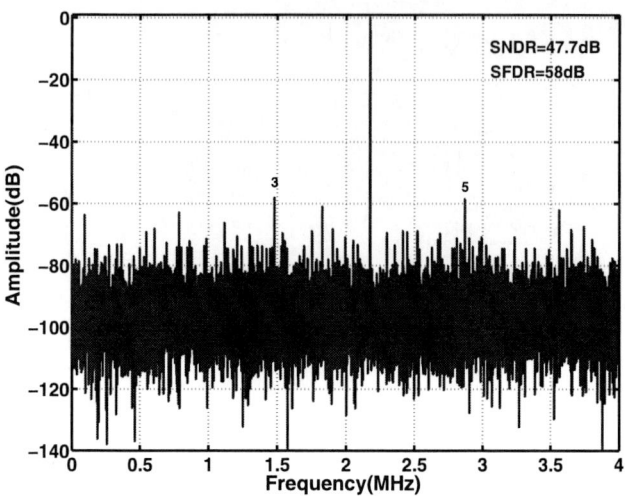

Fig. 8. ADC output spectrum at a sampling rate of 1 GHz and an input frequency of 490 MHz (downsampled by a factor of 125).

Table I. ADC PERFORMANCE SUMMARY AND COMPARISON WITH PRIOR ART

|  | Sahoo VLSI2012 | Hashemi CICC2012 | Hong ISSCC2013 | Lien VLSI2012 | This Work |
|---|---|---|---|---|---|
| Resolution(bit) | 10 | 10 | 9 | 8 | 9 |
| InputCap.(pF) | NA | 0.7 | 0.7 | NA | 0.45 |
| SNDR(dB) LowFreq./Nyquist | 56/52 | 57/53 | 54/51 | 45/43 | 51/48 |
| $f_s$(MHz) | 1000 | 1000 | 900 | 750 | 1000 |
| Power(mW) | 33 | 36 | 10.8 | 4.5 | 7.1 |
| FoM(fJ/CS) @LowFreq. | 75 | 70 | 30 | 41 | 25 |
| FoM(fJ/CS) @Nyquist. | 97 | 100 | 40 | 50 | 34 |
| Technology(nm) | 65 | 65 | 45 | 28 | 65 |
| Supply(V) | 1.2 | 1.2 | 1.2 | 1.0 | 1.0 |
| Activearea(mm)$^2$ | 0.225 | 0.175 | 0.038 | 0.004 | 0.1 |

## References

[1] B. D. Sahoo and B. Razavi, "A 10-Bit 1-GS/s 33-mW ADC," *IEEE Symp. on VLSI Circuits*, pp. 30–31, June 2012.

[2] S. Hashemi and B. Razavi, "A 10-Bit 1-GS/s CMOS ADC with FOM = 70 fJ/Conversion," *IEEE Custom Integrated Circuits Conference*, pp. 296–297, Sept, 2012.

[3] B. Verbruggen et. al., "A 2.6-mW 6b 2.2GS/s 4-times Interleaved Fully Dynamic Pipelined ADC in 40nm Digital CMOS," *IEEE Int. Solid-State Circuits Conference*, pp. 1–4, Feb. 2010.

[4] G.. Van der Plas and B. Verbruggen, "A 150 MS/s 133 $\mu$W 7 bit ADC in 90 nm Digital CMOS," *IEEE J. Solid-State Circuits*, vol.43, No.12, pp. 2631–2640, Dec. 2008.

# A 0.55 V 7-bit 160 MS/s Interpolated Pipeline ADC Using Dynamic Amplifiers

James Lin, Daehwa Paik, Seungjong Lee, Masaya Miyahara, and Akira Matsuzawa

Department of Physical Electronics, Tokyo Institute of Technology

2-12-1-S3-27, Ookayama, Meguro-ku, Tokyo 152-8552, Japan

E-mail: james@ssc.pe.titech.ac.jp

*Abstract*— This paper presents a 0.55 V, 7-bit, 160 MS/s pipeline ADC using dynamic amplifiers. In this ADC, dynamic amplifiers with a common-mode detection technique are used as residual amplifiers to increase its robustness against supply voltage lowering. These amplifiers also remove the unnecessary static power consumption achieving clock-scalability in power performance. The 7-bit prototype ADC fabricated in 90 nm CMOS demonstrates an ENOB of 6.0 bits at a conversion rate of 160 MS/s with an input close to the Nyquist frequency. At this conversion rate, it consumes 2.43 mW from a 0.55 V supply. The resulting FoM is 240 fJ/c.-s.

## I. INTRODUCTION

System-on-a-chip (SoC) designs have significantly reduced the cost and the form factor of modern electronics by combining the analog interface circuits with digital computing and signal processing circuits on the same die. As the digital circuits often occupy the majority of the area in a SoC, the technology selection and system design choices are mainly driven by the digital circuits' requirements.

As the feature sizes of advanced nanoscale CMOS technologies continue to reduce, the maximum supply voltage is expected to reduce to as low as 0.55 V in the next decade or so [1] for reliability reasons. Such supply voltage scaling is extremely beneficial to digital circuits and memory in reducing the heating issues as well as increasing the energy efficiency at the cost of slower operation speed. To overcome the reduced speed, parallelism is an effective method for digital circuits [2].

For analog circuits, there are many challenges in addition to the speed reduction such as the smaller voltage headroom, the reduced SNR, and the increased effects of transistor variation at low voltages. To address these, several techniques have been reported such as sub-threshold operation [3], body driven circuits [4], and SAR-based operation [5], [6]. These techniques were all very successful and have achieved very good performance at ultra-low voltage (ULV); however, they all share a common drawback: slow operation speed.

Unfortunately for analog circuits, the excessive use of parallelism have some disadvantages such as an increase of area, a reduction of PVT margin, a degradation of performance, and an increase of input drive difficulty [7]. Therefore, in order to realize high-speed SoCs using advanced technologies, ULV high-speed analog design techniques are necessary. In this paper, we propose an ULV clock-scalable high-speed pipeline ADC using dynamic amplifiers.

## II. CIRCUIT DESIGN

### A. Interpolated Pipeline Architecture

The pipeline architecture is suitable for high speed and moderate resolution conversion. However, the insufficient OpAmp gain of the scaled CMOS technologies makes designing high performance pipeline ADCs challenging. To address this issue, a pipeline ADC with a capacitive interpolation has been developed in [8] to shift the gain requirement from absolute accuracy to relative accuracy between open-loop amplifiers. By harnessing this relative gain accuracy property, dynamic amplifiers that are suitable for ULV operation can be used as residual amplifiers to realize an ULV high-speed pipeline ADC.

Fig. 1 shows the block diagram of the proposed ADC with dynamic amplifiers as its residual amplifiers. A fully differential scheme is implemented; however, a single-ended scheme is used in the paper for simplicity. In the first stage, an input signal, $V_{in}$, and a reference voltage, $V_{ref}$, are sampled by a pair of capacitor arrays. Then a 3-bit sub-ADC (CMP1) generates the first set of conversion data that are used to control the switches of the capacitor arrays. With the conversion data, these capacitor arrays behave like a pair of capacitor DACs (CDACs) generating the required residual voltages for the next pipeline stage.

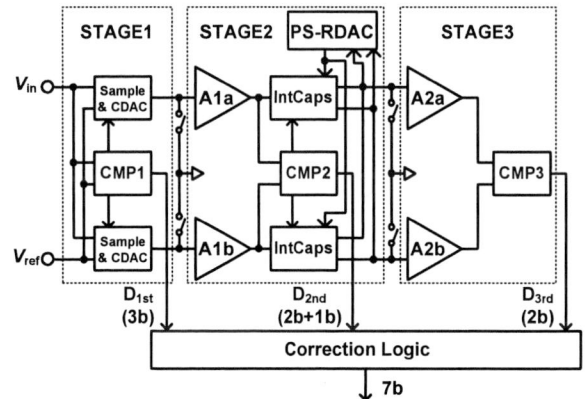

Fig. 1. Block diagram of the proposed ADC using dynamic amplifiers as residual amplifiers.

978-1-4673-6145-3/13 $31.00 © 2013 IEEE

At the second stage, the residual voltages are first amplified by the dynamic amplifiers, A1a and A1b. The output signals of the amplifiers are stored on the interpolation capacitor arrays (IntCaps) and compared by another 3-bit sub-ADC (CMP2) using gate-width-weighted interpolation [9]. Again, the second set of conversion data controls the switches of the interpolation capacitors providing the required residual voltages for the final pipeline stage. The final stage consists of two more dynamic amplifiers, A2a, and A2b, with a 3-bit sub-ADC (CMP3) providing the final set of conversion data of the pipeline ADC.

Fig. 2 shows the capacitor arrays of the interpolation circuits. The input signals, $V_{ia}$ and $V_{ib}$, are amplified by the dynamic amplifiers resulting in internal output voltages of $V_{xa}$ and $V_{xb}$, respectively. These voltages are sampled on the interpolation capacitor arrays. During the interpolation phase, each capacitor is controlled by the sub-ADC to either connect to the input of the next pipeline stage or remain connected to the reference voltage provided by the pseudo-static RDAC (PS-RDAC) thus providing the interpolated values.

### B. Dynamic Amplifier with Common-Mode Detection

In this ADC, dynamic amplifiers are used as residual amplifiers to eliminate the unnecessary static current. Furthermore, dynamic operation allows the power consumption to be clock-scalable. In this work, a dynamic amplifier with a common-mode detector from [10] is modified for high-speed operation. The capacitive common-mode detector (CMD) is replaced by the inverter-based CMD, as shown in Fig. 3, to reduce the capacitive load on the amplifiers. The two inverters with the shorted outputs approximate the output common-mode voltage while the last inverter's threshold voltage determines when the triggering signal is activated. The gain of this amplifier is designed to approximately 3 times, which is sufficient for the interpolated pipeline architecture.

Fig. 4 shows the operation of the dynamic amplifiers with the interpolation capacitors directly acting as the load. When the clock is low, the output nodes are reset and pre-charged by the PS-RDAC and the internal output nodes, $V_{xp}$ and $V_{xn}$, are reset to $V_{DD}$. When the clock turns high, $V_{xp}$ and $V_{xn}$ are discharged proportionally to the input voltages, $V_{inp}$ and $V_{inn}$, until the internal output common-mode voltage ($V_{xc}$) crosses a pre-determined threshold voltage. Upon crossing this threshold voltage, the CMD is activated terminating the discharging providing stable output voltages for the interpolation capacitor arrays.

### C. Calibration

To realize the proposed design, the increased mismatch between transistors at ULV and the amplifier's sensitivity to the input common-mode voltage ($V_{ic}$) all require compensation. This ADC uses the double-tail latched comparators from [11] with the timing calibration [7] to suppress the offset voltages of the comparators. To address the amplifier's sensitivity to $V_{ic}$, a 5-bit PS-RDAC is used to modify the output common-mode voltage ($V_{oc}$) of the interpolation capacitor arrays. The schematic of the PS-RDAC and its operation diagram are shown in Fig. 5. This PS-RDAC draws much less static current than the required instantaneous current from the interpolation capacitors. As a result, a large capacitor is placed at its output node to act as a tank. When the amplifiers are activated, the interpolation capacitors draw current from the tank lowering the tank voltage, $V_{RDAC}$. When the amplifiers reset, the tank is slowly restored by PS-RDAC's weak static current. The final $V_{oc}$ of the interpolation capacitors is automatically tuned to the ADC's common-mode voltage, $V_{com}$, by adjusting the PS-RDAC using comparators and logic during the startup calibration.

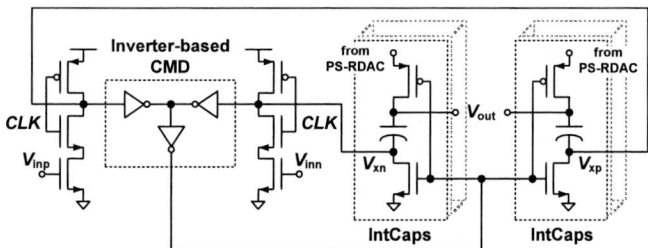

Fig. 3 Schematic of the dynamic amplifier with an inverter-based common-mode detector

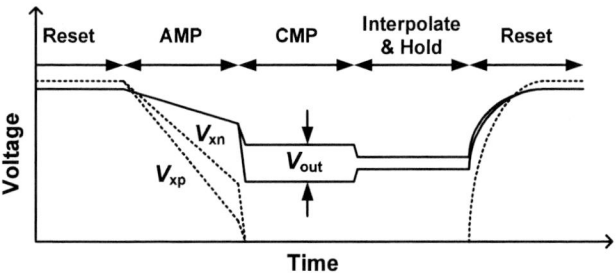

Fig. 4. Operation waveform of the modified dynamic amplifier with the interpolation capacitors as its load.

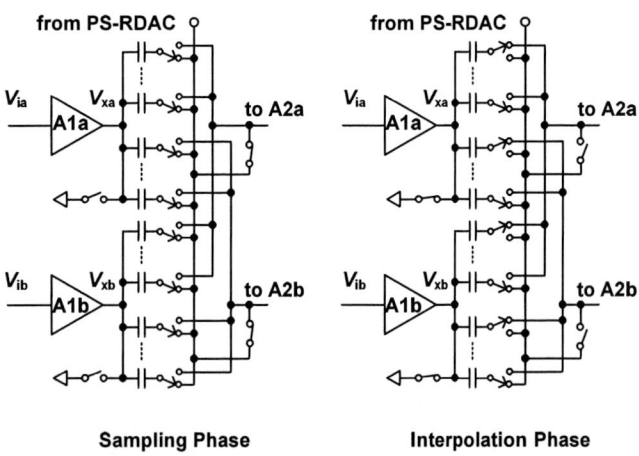

Fig. 2. Capacitor arrays to perform the interpolation.

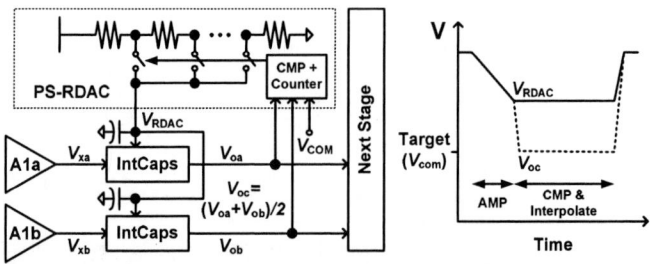

Fig. 5. Schematic of the proposed pseudo-static RDAC for the common-mode voltage control and its operation waveform.

## D. Self-Clocking Scheme

Asynchronous operation allows the circuit to allocate just the right amount of time to function correctly. This maximizes the overall speed of the circuit [12]. Fig. 6 shows the simulated timing diagram of the self-clocking scheme with some key steps. In step 1, the dynamic amplifier uses its triggering signal to initiate the following sub-ADC and to reset the following pipeline stage. Likewise, the sub-ADC is designed to notify the following stage's amplifiers upon its completion as shown in step 2. This asynchronous triggering mechanism continues in a pipeline fashion as illustrated by step 3. To utilize the conversion time effectively, a 75% duty cycle clock is generated on-chip from a $2 \times f_s$ off-chip signal source. The ON cycle is self-allocated among the three operations: amplify, compare, and interpolate and hold. Upon completing each operation, it will initiate the subsequent operation resulting in a fully self-timed behavior maximizing the speed of the proposed ADC.

## III. MEASUREMENT RESULTS

The prototype ADC is fabricated in 90 nm CMOS technology with the low threshold voltage and the deep N-well options. The supply voltages are 0.55 V for the analog part and 0.5 V for the digital part. The power consumption is 2.43 mW at a conversion rate of 160 MS/s. The breakdown of the consumed power is as follows: 1.03 mW for the analog part, 1.33 mW for the digital part including the clock generator and the clock buffers, and 0.07 mW for the references and the peripheral circuits such as the PS-RDAC. Fig. 7 shows the measured results. DNL and INL are +0.63/-0.41 LSB and +0.42/-0.42 LSB, respectively. SNDR of at least 38 dB is measured up to 160 MS/s. The effective resolution bandwidth (ERBW) is up to 80 MHz. The resulting FoM is 240 fJ/c.-s. The occupied area is 0.25 mm$^2$ as illustrated by the chip photo in Fig. 8. From the measurement results, the proposed design achieves the highest speed among other state-of-the-art ULV ADCs as shown in the comparisons in Fig. 9 and Table I.

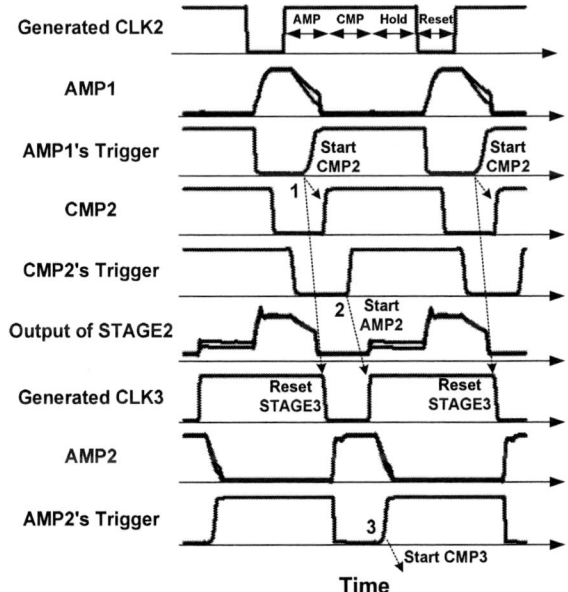

Fig. 6. Simulated timing diagram of the self-clocking scheme.

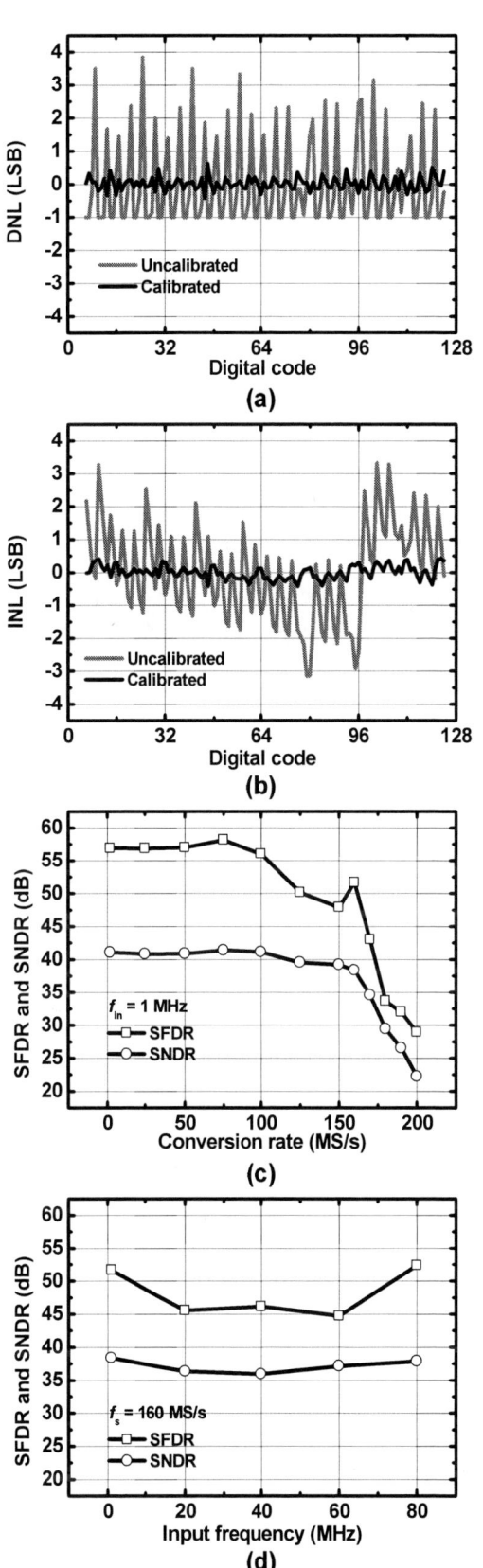

Fig. 7. Measured (a) DNL, (b) INL, (c) SFDR and SNDR vs. conversion rate, and (d) SFDR and SNDR vs. input frequency.

TABLE I: PERFORMANCE COMPARISON WITH OTHER STATE-OF-THE-ART ULTRA-LOW-VOLTAGE HIGH-SPEED ADCs

| | [13] | [14] | [15] | [16] | This work |
|---|---|---|---|---|---|
| Architecture | Flash | Pipeline | Pipeline | Pipeline | Pipeline |
| Resolution (bit) | 5 | 8 | 10 | 12 | 7 |
| Supply voltage (V) | 0.6 | 0.5 | 0.5 | 0.6 | 0.55/0.5* |
| $f_s$ (MS/s) | 60 | 10 | 10 | 10 | **160** |
| Power consumption (mW) | 1.3 | 2.4 | 3.0 | 0.56 | 2.43 |
| ENOB (bit) | 4.01 | 7.7 | 8.5 | 10.8 | 6.0 |
| FoM (fJ/c.-s.) | 1060 | 1150 | 825 | 30.9 | 240 |
| Technology (nm) | 90 | 90 | 130 | 65 | 90 |
| Active area (mm$^2$) | 0.11 | 1.44 | 0.98 | 0.36 | 0.25 |

* 0.55 V for analog and 0.5 V for digital.

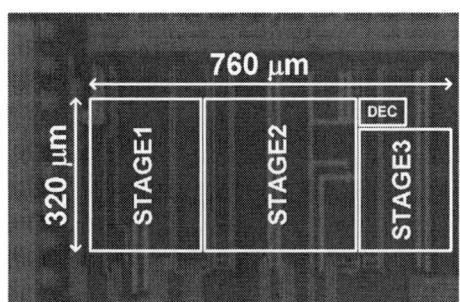

Fig. 8. Chip photo of the prototype ADC.

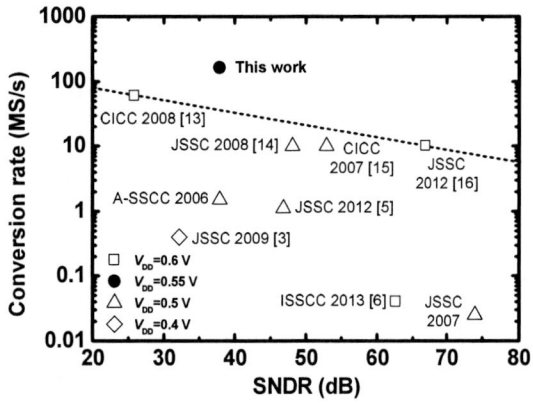

Fig. 9. Performance comparison chart showing the state-of-the-art ULV ADCs.

## IV. CONCLUSION

In conclusion, this paper presents an ULV, clock-scalable, high-speed interpolated pipeline ADC that operates up to 160 MS/s. The proposed dynamic amplifier enables both clock scalability and high-speed operation at 0.55 V. The proposed ADC demonstrates the feasibility of ultra-low voltage high-speed analog circuit design.

## ACKNOWLEDGMENT

This work was partially supported by NEDO, MIC, CREST in JST, STARC, Huawei, Berkeley Design Automation for the use of the Analog FastSPICE(AFS) Platform, and VDEC in collaboration with Cadence Design Systems, Inc.

## REFERENCE

[1] International Technology Roadmap for Semiconductors. 2012. [Online]. Available: http://www.itrs.net/

[2] M. E. Sinangil, M. Yip, M. Qazi, R. Rithe, J. Kwong, and A. P. Chandrakasan, "Design of low-voltage digital building blocks and ADCs for energy-efficient systems," *IEEE Trans. Circuits and Systems II: Express Briefs*, vol. 59, pp. 533-537, 2012.

[3] D. C. Daly and A. P. Chandrakasan, "A 6-bit, 0.2 V to 0.9 V highly digital flash ADC with comparator redundancy," *IEEE J. of Solid-State Circuits*, vol. 44, pp. 3030-3038, 2009.

[4] S. Chatterjee, Y. Tsividis, and P. Kinget, "0.5-V analog circuit techniques and their application in OTA and filter design," *IEEE J. of Solid-State Circuits*, vol. 40, pp. 2373-2387, 2005.

[5] A. Shikata, R. Sekimoto, T. Kuroda, and H. Ishikuro, "A 0.5 V 1.1 MS/sec 6.3 fJ/Conversion-Step SAR-ADC With Tri-Level Comparator in 40 nm CMOS," *IEEE J. of Solid-State Circuits, IEEE Journal of,* vol. 47, pp. 1022-1030, 2012.

[6] P. Harpe, E. Cantatore, and A. van Roermund, "A 2.2/2.7fJ/conversion-step 10/12b 40kS/s SAR ADC with data-driven noise reduction," in *IEEE ISSCC*, pp. 270-271, Feb. 2013.

[7] M. Miyahara, J. Lin, K. Yoshihara, and A. Matsuzawa, "A 0.5 V, 1.2 mW, 160 fJ, 600 MS/s 5 bit flash ADC," in *IEEE A-SSCC*, pp. 177-180, Nov. 2010.

[8] M. Miyahara, H. Lee, D. Paik, and A. Matsuzawa, "A 10b 320 MS/s 40 mW open-loop interpolated pipeline ADC," in *IEEE Symp. on VLSI Circuits*, pp. 126-127, Jun. 2011.

[9] Y. Asada, K. Yoshihara, T. Urano, M. Miyahara, and A. Matsuzawa, "A 6bit, 7mW, 250fJ, 700MS/s subranging ADC," in *IEEE A-SSCC*, pp. 141-144, Nov. 2009.

[10] J. Lin, M. Miyahara, and A. Matsuzawa, "A 15.5 dB, wide signal swing, dynamic amplifier using a common-mode voltage detection technique," in *IEEE ISCAS*, pp. 21-24, May 2011.

[11] M. Miyahara, Y. Asada, D. Paik, and A. Matsuzawa, "A low-noise self-calibrating dynamic comparator for high-speed ADCs," in *IEEE A-SSCC*, pp. 269-272, Nov. 2008.

[12] R. Kapusta, J. Shen, S. Decker, H. Li, and E. Ibaragi, "A 14b 80MS/s SAR ADC with 73.6dB SNDR in 65nm CMOS," in *IEEE ISSCC*, pp. 472-473, Feb. 2013.

[13] J. E. Proesel and L. T. Pileggi, "A 0.6-to-1V inverter-based 5-bit flash ADC in 90nm digital CMOS," in *IEEE CICC*, pp. 153-156, Sep. 2008.

[14] J. Shen and P. R. Kinget, "A 0.5-V 8-bit 10-Ms/s pipelined ADC in 90-nm CMOS," *IEEE J. of Solid-State Circuits*, vol. 43, pp. 787-795, 2008.

[15] Y. J. Kim, H. C. Choi, S. W. Yoo, S. H. Lee, D. Y. Chung, K. H. Moon, et al., "A re-configurable 0.5V to 1.2V, 10MS/s to 100MS/s, low-power 10b 0.13um CMOS pipeline ADC," in *IEEE CICC*, pp. 185-188, Sep. 2007.

[16] S. Lee, A. P. Chandrakasan, and H. S. Lee, "A 12 b 5-to-50 MS/s 0.5-to-1 V voltage scalable zero-crossing based pipelined ADC," *IEEE J. of Solid-State Circuits*, vol. 47, pp. 1603-1614, 2012.

978-1-4673-6145-3/13 $31.00 © 2013 IEEE

# A 95-MS/s 11-bit 1.36-mW Asynchronous SAR ADC with Embedded Passive Gain in 65nm CMOS

Jae-Won Nam, David Chiong, Mike Shuo-Wei Chen

University of Southern California, Los Angeles, CA, USA

Email: jaewon.nam@usc.edu, swchen@usc.edu

*Abstract*— **An 11b asynchronous successive approximation register analog to digital converter with embedded passive gain architecture is proposed and prototyped in 65nm CMOS. The proposed passive gain technique is integrated in the sampling capacitor network as part of the SAR conversion, and provides a signal gain of 2x prior to the comparator without consuming static current. It thus reduces the comparator noise impact as well as enhancing the overall ADC conversion speed and power efficiency. The ADC prototype demonstrates a peak SNDR of 63.1dB and SFDR of 75.2dB when sampling at 95MS/s. Both measured differential and integral nonlinearities of the prototype are less than 0.84 LSB. It occupies an active area of 0.073mm² and dissipates 1.36mW from 1.1V supply. Keywords: analog-to-digital, SAR, sub-radix.**

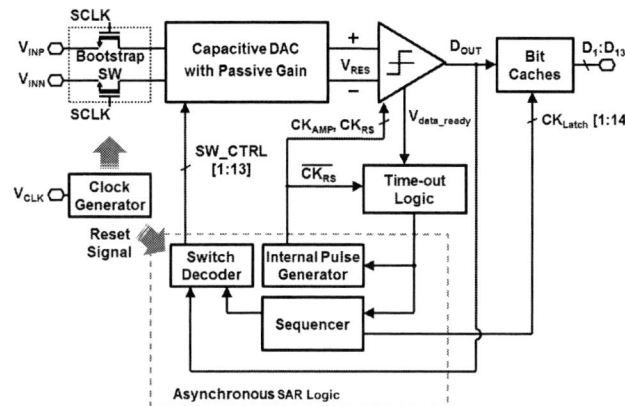

Fig. 1. Top-level block diagram of the proposed ADC

## I. INTRODUCTION

The successive approximation register (SAR) ADC has been proven a power efficient and scaling friendly architecture as it contains mostly digital implementation and only a single comparator. The conversion speed constraint is further mitigated by asynchronous SAR architecture, which was proposed in [1] and demonstrated for high-speed and medium resolution. This work aims to further advance the asynchronous SAR architecture into high resolution, high sampling rate regime, where sampling noise (KT/C) and comparator noise contributions become dominant factors that result in significant speed and power overhead. To alleviate this bottleneck, an asynchronous SAR ADC with embedded passive gain architecture is proposed to not only speed up the SAR conversion but also relax the comparator noise specification. Additionally, a comparator time-out mechanism is incorporated in the asynchronous SAR logic to enhance the robustness over the comparator metastability issue.

The key idea of the proposed passive gain technique is to provide a signal gain of 2x prior to the comparator without power penalty to a) relax the comparator noise requirement, b) speed up the comparator due to enlarged signal size, and c) reduce the size of charge redistribution capacitor array for faster DAC settling time. The SPICE simulation shows that the comparator consumes approximately more than twice the power dissipation without the proposed passive gain techniques given the target noise/speed specification. The passive gain stage is fully integrated in the capacitor sampling network via split capacitor arrays and charge pumps. Furthermore, the capacitor array is non-integer ratioed to implement the non-binary radix for accommodateing incomplete DAC settling issues.

This paper is organized as follows: Section II describes about the proposed ADC architecture with relevent architectural advantages. Details of the circtuit implementation of the prototype ADC are discussed in Section III. Measurement results together with the comparison with the state of arts are presented in Section IV, and Section V draws the conclusion.

## II. PROPOSED ADC ARCHITECTURE

The block diagram of the proposed ADC is shown in Fig. 1. The ADC input sampling is performed by a bootstrapped switch for better linearity. The non-binary capacitor array with embedded passive gain is followed by the two-stage comparator that also generates the data ready signal. To eliminate metastable conditions, a time-out logic is employed. The output signal then drives the internal pulse generator block to create reset pulses for the comparator and switch decoder to change the connection of capacitors properly. The sequencer block generates the multi-phase clocks that strobe an array of bit caches to store the comparator results.

Figure 2 shows the concept of the proposed passive gain technique. Firstly, the analog input signal is sampled on both capacitors via switch $M_{S1}$ and $M_{S2}$. When the two capacitors ($C_{S1}$ and $C_{S2}$) are stacked up, the differential signal is doubled without active amplifiers. Unlike capacitive charge pumps for residue signal amplification in a pipelined ADC [2], nonlinear parasitic ($C_{P1}$ and $C_{P2}$) effects are mitigated by the fact that non-binary SAR operations ultimately force comparator input, i.e. voltage across $C_{P2}$, within just one LSB around a fixed common mode voltage. Together with the minimized switch size via clock doublers, the non-linear parasitic capacitance induced gain error achieves ~12bit accuracy according to SPICE simulations. Note that an additional level shifting func-

978-1-4673-6145-3/13 $31.00 © 2013 IEEE

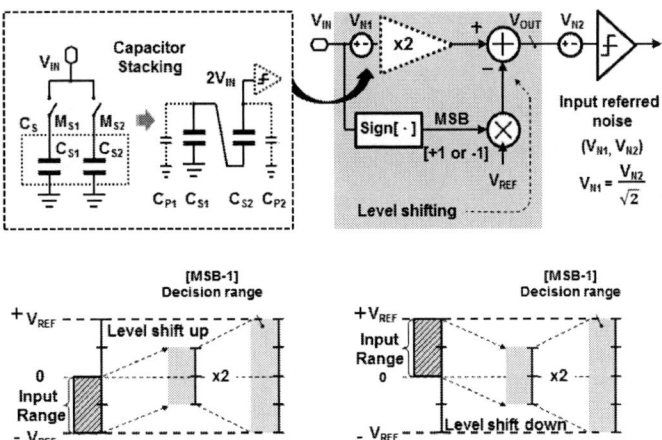

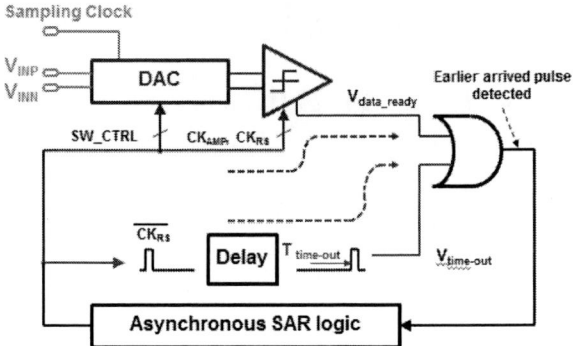

Fig. 4. Block diagram of the time-out logic

Fig. 2. The concept of the proposed passive gain technique

tion is implemented together with the passive gain network for the proper SAR algorithm operation.

From the noise perspective, the split sampling capacitors ($C_{S1}$ and $C_{S2}$) are chosen precisely half of the intended sampling capacitance ($C_S$) to maintain the same SNR after passive amplification. Since the level shifting capacitor ($C_{BAT}$) will introduce additional thermal noise, it should be designed sufficiently large so that the SNR degradation is negligible. Thanks to the passive gain occurring prior to the comparator, it effectively reduces the noise contributions from the comparator and hence relaxes its noise specification.

III. CIRCUIT IMPLEMENTATION

The circuit implementation of the sampling network with embedded passive gain stage is illustrated in Fig. 3. During the sampling phase, the capacitor, $C_{BAT}$, is pre-charged with the reference voltage, $V_{BAT}$, and the input signal is sampled on the bottom plates of $C_{S1}$ and $C_{S2}$. To minimize voltage droop and noise contribution from $C_{BAT}$, capacitor size of 4pF was chosen

to ensure the associated error less than 0.1-LSB over the entire conversion period. For the MSB ($D_1$) comparison, the comparator detects the signal polarity based on the sampled voltage on $C_{S1}$. Once $D_1$ is obtained, $C_{BAT}$ is connected in series with $C_{S1}$ and $C_{S2}$ with the polarity depending on $D_1$. The sampled value is then shifted and amplified, and its polarity is compared to generate MSB-1 ($D_2$). The clock boosting circuit is further applied to double the voltage swing presented at the gate of the series switches associated with $C_{BAT}$. It effectively reduces the turn-on resistance to speed up the charge redistribution time without increasing the switch size and hence the parasitic capacitances. From the third MSB comparison ($D_3$) to the very last LSB, the SAR algorithm is performed via conventional charge redistribution sequence on capacitor array $C_{S2}$.

In this ADC, non-binary radix is adopted by converting two extra redundant comparisons [3]. First of all, it allows incomplete DAC settling and hence shortens the overall conversion time. Secondly, since the passive gain will not occur until the MSB-1 comparison, the MSB comparison can still suffer from the larger comparator noise. That is, the decision result can be erroneous when the input signal is within the comparator input referred noise range. The conversion redundancy then corrects for this comparison error using the following bits. The non-

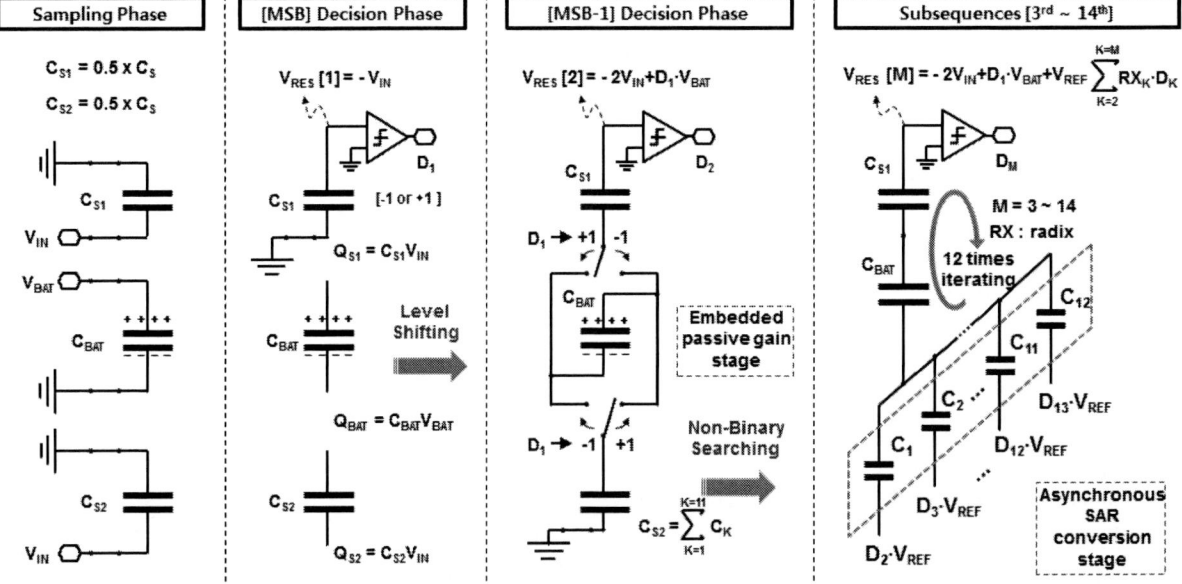

Fig. 3. Proposed implementation and conversion sequence of the sampling capacitor network with embedded passive gain

978-1-4673-6145-3/13 $31.00 © 2013 IEEE        478

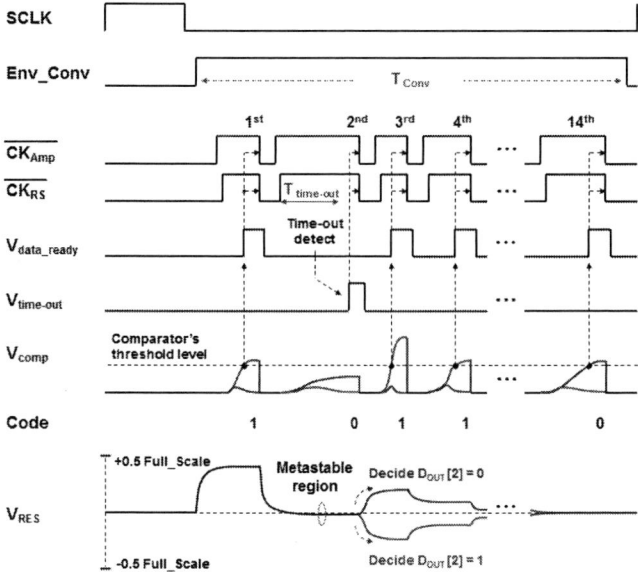

Fig. 5. Timing diagram of SAR ADC with the occurrence of comparator metastability

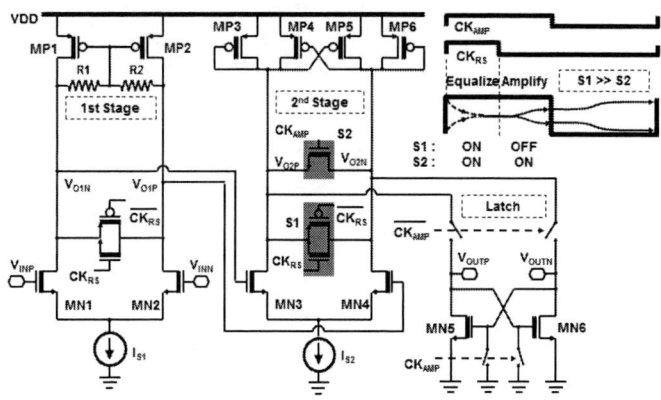

Fig. 6. Circuit implementation of the comparator

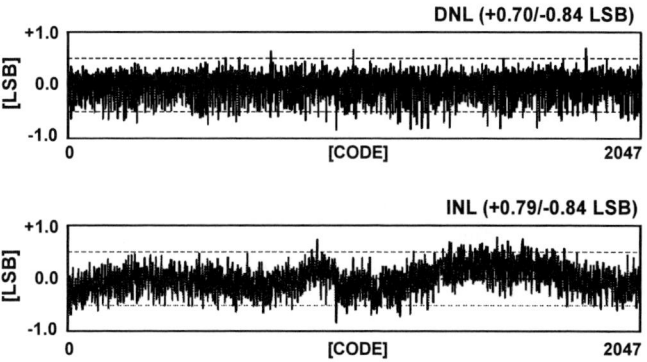

Fig. 7. Measured DNL and INL at 95MS/s

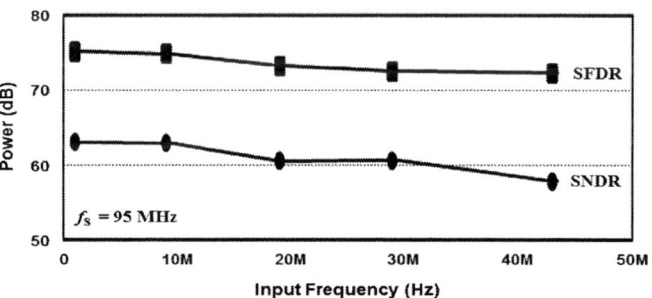

Fig. 8. Measured dynamic performance

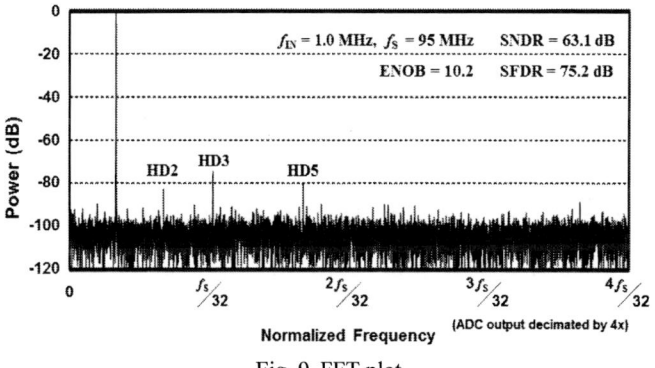

Fig. 9. FFT plot

binary radix is implemented by properly sized metal-oxide-metal capacitor array with non-integer ratio. The actual non-binary radix and linear gain error due to dissimilar parasitic of $C_{BAT}$ are calibrated via post processing, similar to [1]. Furthermore, the precision of the pre-charged voltage ($V_{BAT}$) stored on $C_{BAT}$ is tolerant up to 24% from the nominal value owing to this calibration scheme. Note that the reference voltages of this non-binary capacitor array are simply the power supply and ground of the ADC to avoid the extra reference buffer and reduce the switch turn-on resistance by adding extra capacitances.

To enhance the robustness over comparator metastability, the time-out circuit is implemented to force the advancement of asynchronous conversion whenever the comparator falls into the metastable state. It is accomplished by strobing the SAR logic with the earlier rising edge between the data ready signal ($V_{data-ready}$) from the comparator and predetermined delay pulse ($V_{time-out}$), as shown in Fig. 4. The delay time of $V_{time-out}$, is determined by the tolerable bit-error-rate (BER) of the comparator. Since the metastability occurs only when the differential input of the comparator is smaller than one LSB by design, the comparator decision result does not cause conversion error. Figure 5 shows an example of asynchronous SAR operation with the occurrence of metastability. During the second conversion cycle, the output of the comparator ($V_{comp}$) would have taken excessive time to resolve and trigger the following digital logic, which will lead to incomplete SAR conversion. Instead, the time-out circuit generates $V_{time-out}$ pulse to terminate the current comparison cycle and proceed forward such that a total of 14 comparison cycles will be complete within the allocated conversion time.

For the comparator design, a pulse generator is used to generate two clock pulses ($CK_{Amp}$ and $CK_{RS}$, as shown in Fig. 5) of various durations that drive large and small reset switches (S1 and S2) individually to mitigate the comparator memory

978-1-4673-6145-3/13 $31.00 © 2013 IEEE

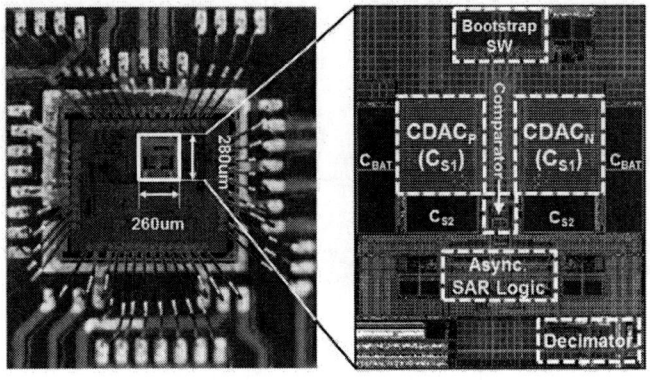

Fig. 10. Die micrograph.

from the signal source are illustrated in Fig. 8-9. In the dynamic performance versus the input frequency plot, the SNDR at 1.0MHz input frequency achieves 63.1dB and maintains 57.8dB up to 43MHz. The proposed ADC's FoM, defined as

$$FoM = \frac{Power}{2^{ENOB} \times f_{Sampling}},$$

It achieves 12fJ/conversion-step at 1.0MHz input, and 22fJ/conversion-step at Nyquist frequency with 95MS/s. Figure 10 shows the chip micrograph of the ADC prototype. The measurement summary and performance comparison with the state-of-the-art published ADCs are demonstrated in Table I.

effect while reducing the required reset time, as shown in Fig. 6. The switch, S1, is designed with low resistance to provide voltage equalization, while S2 provides the proper load resistance for the pre-amplification to speed up comparison time.

## IV. MEASUREMENT RESULTS

The ADC prototype is fabricated in 65nm 1P6M CMOS process and occupies an active die area of 0.073mm². It consumes 1.36mW through a 1.1V supply, while the analog and digital sections dissipate 0.55mW and 0.81mW, individually. As shown in Fig. 7, the measured DNL and INL are within +0.70/-0.84 LSB and +0.79/-0.84 LSB respectively. The measured FFT plots and dynamic performance of the ADC with an external input band-pass filter to minimize the noise

## V. CONCLUSIONS

This paper presents an 11-bit asynchronous SAR ADC with the embedded passive gain stage, which is fully integrated in the proposed sampling capacitor network. It not only relaxes the comparator noise constraint but also enhances the conversion speed with low power consumption. The proposed time-out mechanism further enhances the ADC robustness over the comparator metastability issue. As a result, it achieves the lowest FoM among the recently published ADCs with similar specification (>10 ENOB, >10MS/s) [10].

## ACKNOWLEDGEMENTS

The authors would like to thank Dr. Jerry Shiao and National Nano Device Laboratories (NDL), Taiwan for chip fabrication and assistance, and ONR for funding support.

TABLE I. ADC PERFORMANCE COMPARISON
(> 10 ENOB & 10MS/s)

| | [4] | [5] | [6] | [7] | [8] | [9] | This work |
|---|---|---|---|---|---|---|---|
| ADC Topology | Pipelined | Synch. SAR | Pipelined SAR | Asynch. SAR | Over-sampled SAR | Pipelined | Asynch. SAR |
| $f_{Sampling}$ (MS/s) | 50 | 50 | 50 | 80 | 90 | 100 | 95 |
| ENOB | 10.0 | 11.2 | 10.67 | 11.55 | 10.0 | 10.2 | 10.2 |
| Signal Bandwidth (MHz) | 25 | 25 | 25 | 40 | 11 | 50 | 47.5 |
| Supply (V) | 1.2 | 1.2 | 1.3 | 1.2 | 1.2 | 1.2 | 1.1 |
| SFDR (dB) | 68 | 84.0 | 78 | 80.3 | 72 | 74 | 75.2 |
| SNDR (dB) | 62 | 69.1 | 66 | 71.3 | 62 | 63.2 | 63.1 |
| Power (mW) | 4.5 | 4.2 | 3.5 | 3.11 | 0.8 | 6.2 | 1.36 |
| Area (mm²) | 0.3 | 0.097 | 0.16 | 0.55 | 0.03 | 0.32 | 0.073 |
| Process (nm) | 90 | 90 | 65 | 65 | 65 | 90 | 65 |
| FoM (fJ/conv) | 87.5 | 32 | 51.8 | 129.5 | 35.8 | 52.7 | 22 |

## REFERENCES

[1] M. S. W. Chen and R. W. Broderson, "A 6-b 600-MS/s 5.3-mW Asynchronous ADC in 0.13-μm CMOS," *IEEE J. Solid-State Circuits*, vol. 45, pp. 2669-2680, Dec. 2006.

[2] I. Ahmed, J. Mulder, and D. A. Johns, "A 50MS/s 9.9mW pipelined ADC with 58dB SNDR in 0.18μm CMOS using capacitive charge-pumps", *ISSCC Dig. Tech. Papers*, pp. 164-165, Feb. 2009.

[3] F. Kuttner, "A 1.2 V 10b 20MSample/s nonbinary successive approximation ADC in 0.13μm CMOS", *ISSCC Dig. Tech. Papers*, pp. 176-177, Feb. 2002.

[4] L. Brooks, and H.-S. Lee, "A 12b 50MS/s fully differential zero-crossing-based ADC without CMFB", *ISSCC Dig. Tech. Papers*, pp. 166-167, Feb. 2009.

[5] T. Morie, T. Miki, K. Matsukawa, Y. Bando, T. Okumoto, K. Obata, S. Sakiyama, and S. Dosho, "A 71dB-SNDR 50MS/s 4.2mW CMOS SAR ADC by SNR enhancement techniques utilizing noise", *ISSCC Dig. Tech. Papers*, pp.272-273, Feb. 2013.

[6] C. C. Lee, and M. P. Flynn, "A SAR-assisted two-stage pipeline ADC," *IEEE J. Solid-State Circuits*, vol. 46, pp. 859-869, Apr. 2011.

[7] R. Kapusta, J. Shen, S. Decker, H. Li, and E. Ibaragi, "A 14b 80MS/s SAR ADC with 73.6dB SNDR in 65nm CMOS", *ISSCC Dig. Tech. Papers*, pp. 472-473, Feb. 2013.

[8] J. Fredenburg, and M. P. Flynn, "A 90-MS/s 11-MHz bandwidth 62-dB SNDR noise-shaping SAR ADC," *IEEE J. Solid-State Circuits*, vol. 47, pp. 2898-2904, Dec. 2012.

[9] J. Chu, L. Brooks, and H.-S. Lee, "A zero-crossing based 12b 100MS/s pipelined ADC with decision boundary gap estimation calibration," in *IEEE VLSI Circuits Symposium*, pp. 237–238, Jun. 2010.

[10] B. Murmann, "ADC Performance Survey 1997-2012," [Online]. Available:http://www.stanford.edu/~murmann/adcsurvey.html.

978-1-4673-6145-3/13 $31.00 © 2013 IEEE

# AUTHOR INDEX

Abdelfattah, K. .................................... 204
Abdelhalem, S. ....................................... 49
Abdul-Latif, M. ...................................... 37
Abugharbieh, K. .................................... 299
Agrawal, V. .......................................... 236
Ahmadi, M. .......................................... 485
Ahmed, K. ............................................ 113
Aitken, R. ............................................. 61
Akré, J. ............................................... 712
Alam, S. .............................................. 799
Ali, T. ................................................. 37
Alioto, M. ............................................ 263
Alldred, D. .......................................... 837
Alon, E. .............................................. 457
Amourah, M. ......................................... 885
Anceau, F. ........................................... 712
Andre, T. ............................................. 799
Aouini, S. ............................................ 636
Arora, S. ............................................. 101
Arsovski, I. .......................................... 825
Asada, K. .............................. 188, 212, 660
Atalla, E. ............................................ 624
Bakhshiani, M. ......................................... 1
Bakishev, T. ......................................... 236
Balasubramanian, S. ................................ 684
Balsara, P. .......................................... 624
Banijamali, B. ....................................... 299
Bawa, G. ............................................. 620
Becker, B. ........................................... 640
Bellaouar, A. ........................................ 624
Belostotski, L. ...................................... 580
Ben-Hamida, N. ..................................... 636
Besten, G. ........................................... 311
Bhagavatula, S. ..................................... 117
Billoint, O. ......................................... 712
Bingert, R. .......................................... 152
Blaauw, D. ............................. 13, 252, 263
Bo, Y. ................................................ 716
Boling, E. ........................................... 244
Borna, A. ............................................ 279
Borremans, J. ....................................... 156
Bourdoux, A. ........................................ 343
Bousquet, J. ......................................... 636
Boutros, K. .......................................... 287
Bowers, J. ........................................... 889
Braswell, B. ......................................... 676
Brebels, S. .......................................... 343
Brooks, D. ..................................... 109, 708
Brooks, T. ........................................... 204
Buckwalter, J. ...................................... 576
Bull, D. .............................................. 692
Bulzacchelli, J. ..................................... 315
Burbach, G. .......................................... 684
Burg, A. ............................................. 256
Cai, S. ............................................... 704

Caldwell, T. ......................................... 837
Cao, P. .............................................. 700
Casper, B. ........................................... 323
Catli, B. .............................................. 37
Cha, S. .............................................. 156
Chan, M. ............................................. 105
Chandra, V. .......................................... 61
Chao, Y. ............................................. 359
Chen, A. ............................................. 204
Chen, B. ............................................. 616
Chen, C. ............................................. 664
Chen, D. ....................................... 192, 664
Chen, F. ............................................. 267
Chen, G. ............................................. 224
Chen, J. ............................................. 271
Chen, M. ........................... 176, 477, 584
Chen, S. ....................................... 148, 656
Chen, W. ......................................... 29, 200
Chen, Y. ......................................... 29, 263
Chen, Z. .............................. 168, 208, 248
Chi, B. .............................................. 168
Chi, T. .............................................. 652
Chiang, P. ..................................... 224, 656
Chiong, D. .......................................... 477
Chippa, V. .......................................... 696
Chiu, Y. ............................................. 133
Cho, S. .............................................. 873
Choi, J. ............................................. 640
Chou, M. ............................................ 200
Chu, R. .............................................. 287
Chun, K. ............................................. 821
Chung, Y. ............................................ 481
Ciftcioglu, B. ....................................... 592
Cimaz, L. ............................................. 81
Clark, L. ............................................ 236
Cline, B. .............................................. 61
Cogal, Ö. ............................................ 256
Cojbasic, R. ......................................... 256
Colinet, E. .......................................... 712
Colinge, J. .......................................... 291
Craninckx, J. ....................................... 156
Darabi, H. ........................................... 160
Das, S. .............................................. 692
Dasika, G. ........................................... 692
De Bock, M. ......................................... 881
Deng, C. ....................................... 700, 704
Dhong, S. ...................................... 291, 817
Ding, H. ............................................. 734
Do, K. ............................................... 640
Dong, Z. ......................................... 53, 144
Driesen, J. .......................................... 248
Duan, J. ............................................. 192
Duan, Y. ............................................. 457
Dudek, P. ............................................ 632
Elad, D. ............................................. 275

# AUTHOR INDEX

Ema, T. .................................................. 236
Emami-Neyestanak, A. .................................. 5
Esumi, A. ............................................... 260
Fallahi, S. .............................................. 37
Farahabadi, P. .......................................... 339
Fei, W. ................................................. 355
Feng, P. ................................................ 271
Ferriss, M. ............................................. 129
Fick, D. ................................................ 252
Fischer, P. ............................................. 628
Fojtik, M. .............................................. 252
Friedman, D. ...................................... 129, 275
Fu, Y. .................................................. 176
Fujiwara, H. ............................................ 813
Fukuoka, K. ............................................. 453
Furtner, W. ............................................. 81
Galal, S. ............................................... 204
Galayko, D. ............................................. 712
Gande, M. .......................................... 720, 730
Gao, L. ................................................. 208
Garrity, D. ............................................. 676
Garverick, S. ........................................... 596
Garzia, F. .............................................. 248
Gerber, D. .............................................. 93
Ghosh, A. ............................................... 833
Ghosh, S. ............................................... 680
Giannini, V. ............................................ 343
Giridhar, B. ............................................ 252
Gogl, D. ................................................ 799
Goldbach, M. ............................................ 684
Goss, J. ................................................ 825
Groeseneken, G. ......................................... 148
Grzymkowski, P. ......................................... 825
Gu, C. ............................................. 845, 869
Gudem, P. ............................................... 49
Guerber, J. ............................................. 730
Guo, Q. ................................................. 17
Gupta, P. ............................................... 746
Gupta, R. ............................................... 311
Hamashita, K. ........................................... 720
Han, J. ................................................. 716
Hanson, H. .............................................. 311
Hanumolu, P. ............................................ 307
Hashemi, S. ............................................. 469
Hashimoto, T. ........................................... 453
Haslett, J. ............................................. 580
Hebig, T. ............................................... 825
Hellings, G. ............................................ 148
Hernes, B. .............................................. 196
Hershberg, B. ........................................... 720
Hill, B. ................................................ 608
Hiseh, Y. ............................................... 176
Holdø, C. ............................................... 196
Homayoun, A. ............................................ 172
Hong, Z. ................................................ 224

Hori, M. ................................................ 236
Horowitz, M. ............................................ 857
Houssameddine, D. ....................................... 799
Hsiao, C. ............................................... 137
Hsiao, S. ............................................... 176
Hsu, W. ................................................. 176
Hu, J. .................................................. 592
Hu, W. .................................................. 656
Huang, A. ............................................... 620
Huang, J. ............................................... 841
Huang, Z. ............................................... 208
Hughes, B. .............................................. 287
Hull, C. ........................................... 279, 335
Hung, C. ................................................ 176
Iizuka, T. ......................................... 188, 660
Ishizone, Y. ............................................ 660
Ismail, Y. .............................................. 220
Ito, M. ............................................ 260, 453
Jalali, M. .............................................. 331
Janesky, J. ............................................. 799
Jang, J. ........................................... 184, 853
Jariwala, D. ............................................ 628
Jaussi, J. .............................................. 323
Javidan, M. ............................................. 712
Jayakumar, H. ........................................... 696
Jeng, M. ................................................ 137
Jeong, S. ............................................... 13
Jiang, X. ............................................... 204
Joseph, A. .............................................. 53
Joshi, V. ............................................... 684
Jou, S. ................................................. 29
Juillard, J. ............................................ 712
Jung, B. ................................................ 117
Jung, D. ................................................ 572
Jung, M. ................................................ 628
Jung, S. ........................................... 121, 572
Kaald, R. ............................................... 196
Kam, J. ................................................. 283
Kapusta, R. ............................................. 724
Karmazin, R. ............................................ 608
Kawaguchi, H. ........................................... 612
Kepler, N. .............................................. 236
Khalaf, K. .............................................. 343
Kibune, M. .............................................. 331
Kida, T. ................................................ 453
Kidd, D. ................................................ 236
Kim, B. ................................................. 588
Kim, C. ........................................ 220, 588, 821
Kim, D. ................................................. 267
Kim, H. ................................................. 640
Kim, J. ......................................... 33, 37, 121, 267, 853
Kim, K. ......................................... 121, 640, 873
Kim, T. ................................................. 121
Kim, W. ................................................. 109
Kim, Y. ................................................. 640

# AUTHOR INDEX

Kinget, P. .................................................93
Kline, M. .................................................93
Komatsu, S. ............................................212
Koo, J. ...................................................648
Korniienko, A. .........................................712
Kotani, K. ...............................................240
Kousai, S. ...............................................644
Krishnan, G. ...........................................236
Krishnapura, N. .......................................672
Krishnegowda, S. .....................................885
Kumamoto, T. ..........................................232
Kuo, A. ..................................................244
Kuo, M. ..................................................817
Lam, H. ..................................................604
Larson, L. ................................................49
Le, C. .....................................................93
Leblebici, Y. .....................................228, 256
Lee, C. ............................................29, 600
Lee, E. .....................................................9
Lee, H. ........................................89, 97, 730
Lee, J. ...................................................121
Lee, K. ...................................................829
Lee, M. ..................................................176
Lee, R. ...................................................664
Lee, S. ....................................152, 244, 473, 877
Lee, W. ..................................................572
Lee, Y. .........................................33, 200, 263, 628
Lei, P. ....................................................664
Leinonen, P. .............................................81
Lepkowski, W. ..........................................283
Leshner, S. .............................................236
Levantino, S. .......................................41, 524
Li, A. .....................................................351
Li, H. .......................................152, 481, 616, 656
Li, J. .....................................................576
Li, K. .....................................................260
Li, M. ...............................................343, 861
Li, W. ....................................................208
Li, X. ...............................................129, 869
Li, Y. ...............................................267, 716
Li, Z. .....................................................837
Liao, S. ..................................................857
Libois, M. ...............................................343
Liempd, B. ..............................................156
Lim, I. ...................................................676
Lim, S. ...................................................628
Lin, C. ..............................................137, 817
Lin, D. ...................................................267
Lin, H. ...................................................799
Lin, J. ...................................................473
Linten, D. ...............................................148
Littow, M. ................................................81
Liu, C. ...................................................604
Liu, J. ...................................................461
Liu, L. ..............................................700, 704

Liu, R. ...................................................224
Liu, W. ..................................................271
Liu, Y. .........................................37, 200, 275
Liu, Z. ....................................................97
Loh, M. .....................................................5
Løkken, I. ...............................................196
Long, J. ..................................................343
Lopich, A. ...............................................632
Lu, F. ...............................................53, 144
Lu, M. ...................................................604
Lu, N. ....................................................152
Luo, X. ..................................................351
Luong, H. ..........................................351, 359
Lyden, C. ................................................724
Ma, K. ...................................................176
Ma, R. ...............................................53, 144
Ma, S. ...................................................355
Macpherson, A. .........................................580
Maeder, T. ..............................................256
Majerus, S. .............................................596
Majidzadeh, V. .........................................228
Mangraviti, G. .........................................343
Manohar, R. ............................................608
Mansuri, M. .............................................323
Martens, E. .............................................156
Mastrangelo, C. .........................................17
Matsuo, M. .............................................576
Matsuzawa, A. .....................................473, 877
McAndrew, C. ..........................................676
Meadows, W. ...........................................799
Mehr, I. ..................................................204
Meinerzhagen, P. ......................................256
Mikhemar, M. ..........................................160
Mirzaei, A. .............................................160
Mitani, J. ...............................................236
Mitra, S. ................................................640
Miura, S. ................................................660
Miura, Y. ...............................................232
Miyahara, M. .....................................473, 877
Miyano, S. ..............................................612
Moez, K. ................................................339
Mohapatra, D. .....................................628, 696
Mohseni, P. ...............................................1
Mok, P. ...........................................105, 489
Moon, U. ...........................................720, 730
Mori, R. .................................................453
Morita, S. ...............................................453
Moriwaki, S. ...........................................236
Morrow, P. ..............................................628
Muhammad, K. ..........................................176
Mukai, H. ...............................................576
Mukhopadhyay, S. ......................................113
Murakami, Y. ...........................................660
Murmann, B. .......................................688, 934
Murphy, D. ..............................................160

# AUTHOR INDEX

Nakura, T. .................................................. 188
Nam, J. ....................................................... 477
Namgoong, W. ............................................ 485
Nandwana, R. ............................................. 307
Natarajan, A. .............................................. 129
Nazemi, A. .................................................... 37
Nii, K. ................................................. 453, 813
Nomura, T. ................................................. 453
Ochiai, T. ................................................... 453
Oh, S. ......................................................... 180
Oh, T. ......................................................... 720
O'Mahony, F. ............................................. 323
Otero, C. .................................................... 608
Otis, B. ............................................... 648, 772
Otsuga, K. .................................................. 453
Paik, D. ...................................................... 473
Pamarti, S. ................................................. 833
Pan, L. ....................................................... 152
Pao, C. ....................................................... 616
Papadopoulos, N. ........................................ 600
Park, D. ........................................................ 89
Park, J. ....................................... 572, 644, 652
Parker, B. ........................................... 129, 275
Parlak, M. .................................................. 576
Parvais, B. .................................................. 343
Patch, J. ..................................................... 825
Pavan, S. ...................................................... 57
Payne, R. .................................................... 461
Piazza, G. ................................................... 648
Pietromonaco, D. .......................................... 61
Pileggi, L. .................................................. 129
Pitchumani, V. ............................................ 628
Pivonka, D. ................................................. 184
Plouchart, J. ............................................... 129
Poon, A. ..................................................... 184
Putnam, C. ................................................. 152
Qi, N. ......................................................... 168
Qin, Y. ........................................................ 224
Raczkowski, K. ........................................... 343
Raghavan, P. ............................................... 343
Raghunathan, A. ......................................... 696
Rajesh, N. ..................................................... 57
Ramalingam, S. .......................................... 299
Ranade, P. .................................................. 236
Randall, M. ................................................. 152
Rascoe, J. ................................................... 734
Razavi, B. ........................... 172, 465, 469, 738
Ren, J. ........................................................ 355
Reynolds, S. ............................................... 275
Rhee, W. ..................................................... 267
Rizzo, N. .................................................... 799
Roberts, N. ................................................. 180
Rogenmoser, R. ................................... 236, 244
Rombouts, P. ............................................... 881

Roy, K. ....................................................... 696
Roy, R. ....................................................... 236
Ryu, K. ....................................................... 572
Ryu, S. ......................................................... 33
S, R. ........................................................... 672
Sachdev, M. ................................................ 600
Sadhu, B. ................................................... 129
Sahoo, B. .................................................... 465
Sakakibara, K. ............................................ 232
Samori, C. .................................................... 41
Sanders, S. ................................................... 93
Sanduleanu, M. ................................... 129, 275
Sauer, M. .................................................... 640
Saxena, S. .................................................. 307
Schächer, S. .................................................. 81
Schenker, R. ................................................. 77
Schmid, A. .................................................. 228
Scorletti, G. ................................................ 712
Segovai-Fernandez, J. .................................. 648
Senning, C. ................................................. 256
Seomun, J. .................................................. 640
Shahidi, G. ................................................... 69
Shan, C. ..................................................... 712
Sharma, P. .................................................. 584
Sheikholeslami, A. ................................. 21, 331
Sheinman, B. ............................................... 275
Shekhar, S. ................................................. 323
Shi, Z. ........................................................ 144
Shimanouchi, M. .......................................... 861
Shinohara, H. .............................................. 612
Shivashankar, K. ......................................... 692
Shivnaraine, R. ........................................... 331
Shlafman, S. ............................................... 275
Sholz, M. .................................................... 148
Sim, J. .................................................. 13, 263
Singh, V. ...................................................... 77
Sinha, S. ...................................................... 61
Slater, C. .................................................... 256
Slaughter, J. ............................................... 799
Sobue, K. .................................................... 720
Soens, C. .................................................... 343
Someya, T. .................................................. 188
Son, S. ......................................................... 33
Song, D. ..................................................... 121
Song, S. ...................................................... 821
Song, T. ...................................................... 628
Song, Z. ...................................................... 168
Spagnolo, A. ............................................... 343
Springer, S. ................................................ 152
Su, D. ......................................................... 101
Su, K. ......................................................... 137
Su, M. .......................................................... 29
Subramanian, C. .......................................... 799
Sun, L. ......................................................... 53
Sun, N. ....................................................... 829

# AUTHOR INDEX

Sun, S. .....................................129, 869
Surapaneni, R. .....................................17
Suster, M. .....................................1, 17
Suys, H. .....................................156
Sylvester, D. .....................................13, 252, 263
Szortyka, V. .....................................343
Takahashi, O. .....................................817
Takamiya, M. .....................................612
Takayanagi, K. .....................................453
Takeuchi, K. .....................................807
Tamura, H. .....................................21, 331
Tan, C. .....................................604
Tanaka, H. .....................................453
Tanaka, M. .....................................813
Tanaka, S. .....................................813
Tandon, J. .....................................212
Tanimoto, S. .....................................232
Taylor, G. .....................................628
Tazzoli, A. .....................................648
Tekin, A. .....................................204
Telstø, F. .....................................196
Thornton, T. .....................................283
Thyagarajan, S. .....................................335
Tian, Y. .....................................604
Tierno, J. .....................................129
Tong, T. .....................................109, 708
Townsend, K. .....................................192
Tripathi, V. .....................................688
Tsai, C. .....................................616
Tsai, T. .....................................200
Tse, J. .....................................608
Tsukamoto, Y. .....................................813
Tsuruta, T. .....................................236
Tuinhout, H. .....................................397
Vaesen, K. .....................................343
Valdes-Garcia, A. .....................................129, 275, 734
VanBentum, R. .....................................684
Venkatram, H. .....................................720, 730
Venugopalan, S. .....................................684
Vidojkovic, V. .....................................343
Wachnik, R. .....................................152
Wakayama, S. .....................................236
Wambacq, P. .....................................343
Wan, D. .....................................664
Wan, L. .....................................224
Wan, Y. .....................................628
Wang, A. .....................................53, 144, 363
Wang, D. .....................................53, 700
Wang, F. .....................................129
Wang, G. .....................................133
Wang, H. .....................................271, 279, 628, 644, 652
Wang, K. .....................................176
Wang, L. .....................................53, 144
Wang, M. .....................................817
Wang, P. .....................................817

Wang, S. .....................................716
Wang, X. .....................................53, 53, 144
Wang, Y. .....................................248, 279
Wang, Z. .....................................168, 267
Webb, C. .....................................628
Wei, G. .....................................109, 708
Wei, H. .....................................465
Wei, S. .....................................700, 704
Wentzloff, D. .....................................180
Whately, M. .....................................885
Wilk, S. .....................................283
Wojko, M. .....................................236
Wong, W. .....................................600
Woods, W. .....................................734
Wooley, B. .....................................101
Wu, E. .....................................299
Wu, H. .....................................592, 861
Wu, M. .....................................481
Wu, N. .....................................271
Wu, P. .....................................29, 299
Wyland, C. .....................................299
Xie, J. .....................................168
Xu, J. .....................................216, 592
Xu, W. .....................................588
Xu, Y. .....................................168, 168
Xu, Z. .....................................877
Yabuuchi, M. .....................................813
Yakovlev, A. .....................................184
Yaldiz, S. .....................................129
Yamada, J. .....................................453
Yamada, T. .....................................236
Yamaguchi, T. .....................................212
Yang, C. .....................................220
Yang, H. .....................................208
Yang, L. .....................................656
Yang, M. .....................................600
Yang, P. .....................................817
Yang, S. .....................................841, 853
Yang, X. .....................................461
Yang, Z. .....................................216, 656
Yeo, H. .....................................33
Yeric, G. .....................................61
Yin, J. .....................................351
Yin, S. .....................................700, 704
Yoon, Y. .....................................829
Yoshimoto, M. .....................................612
Yoshimoto, S. .....................................612
Youn, S. .....................................121
Young, D. .....................................17
Yu, H. .....................................355
Yu, W. .....................................873
Yuan, J. .....................................841
Yuan, M. .....................................200
Yue, C. .....................................53
Zamudio, L. .....................................684

# AUTHOR INDEX

Zeng, T. ...................................................................192
Zeng, X. ...................................................................716
Zhang, C. ..................................................................144
Zhang, D. ..................................................................208
Zhang, F. ..................................................................624
Zhang, J. ..................................................................604
Zhang, P. ..................................................................465
Zhang, W. ..................................................................704
Zhang, X. ..................................................109, 708, 799
Zhao, D. ...................................................................236
Zhao, H. ...................................................................144
Zheng, S. ..................................................................351
Zhou, C. ...................................................................604
Zhu, H. ....................................................................724
Zhu, M. ....................................................................700
Zhu, W. ....................................................................208
Zhuo, H. ...................................................................267
Zianbetov, E. .............................................................712

9781467361453